JO ANNE STOLAROFF COTSEN PRIZE

IMPRINT

The Excavation of the Prehistoric Burial Tumulus at Lofkënd, Albania
Volume 1: Text

UCLA Cotsen Institute of Archaeology Press
MONUMENTA ARCHAEOLOGICA

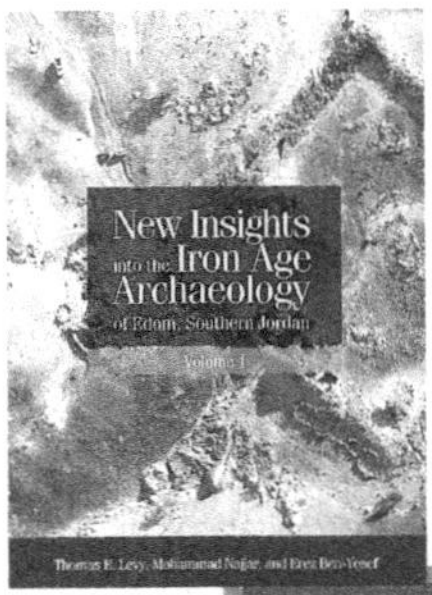

VOLUME 35 *New Insights into the Iron Age Archaeology of Edom, Southern Jordan*
By Thomas E. Levy, Mohammad Najjar, and Erez Ben-Yosef

VOLUME 33 ***Formative Lifeways in Central Tlaxcala Volume 1: Excavations, Ceramics, and Chronology***
Edited by Richard G. Lesure

VOLUME 32 ***Integrating Çatalhöyük: Themes from the 2000–2008 Seasons, Volume 10***
Edited by Ian Hodder

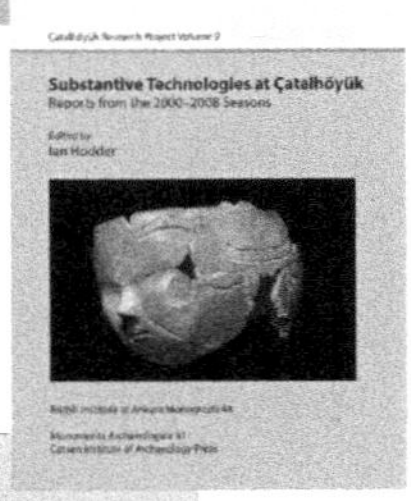

VOLUME 31 ***Substantive Technologies at Çatalhöyük: Reports from the 2000–2008 Seasons, Volume 9***
Edited by Ian Hodder

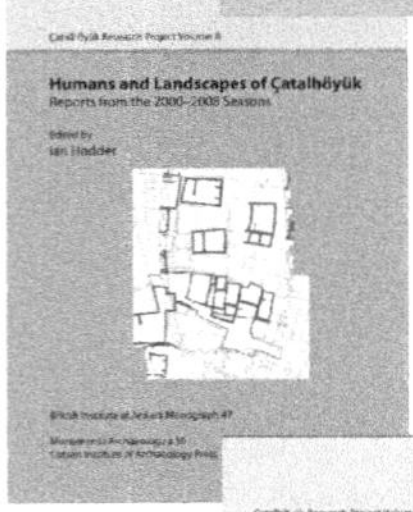

VOLUME 30 ***Humans and Landscapes of Çatalhöyük: Reports from the 2000–2008 Seasons, Volume 8***
Edited by Ian Hodder

VOLUME 29 ***Çatalhöyük Excavations: The 2000–2008 Seasons, Volume 7***
Edited by Ian Hodder

VOLUME 28 ***Light and Shadow: Isolation and Interaction in the Shala Valley of Northern Albania***
Edited by Michael L. Galaty, Ols Lafe, Wayne E. Lee, and Zamir Tafilica

The Excavation of the Prehistoric Burial Tumulus at Lofkënd, Albania
Volume 1: Text

John K. Papadopoulos, Sarah P. Morris,

Lorenc Bejko, and Lynne A. Schepartz

With contributions by

Esmeralda Agolli, Jamie D. Aprile, Brian N. Damiata, J.E. Foss, Shpresa Gjongecaj, Christopher Johanson, Rovena Kurti, Richard MacDonald, John M. Marston, Samantha Martin-McAuliffe, Laura Menez, Vanessa Muros, Yannis Mylonas, Seth Pevnick, David A. Scott, Sarah C. Sherwood, John Southon, Lyssa C. Stapleton, George Theodorou, Evi G.Vardala-Theodorou, Evangelos Velitzelos

Survey, site plans, and drawings by Max Farrar and Samantha Martin-McAuliffe

Object drawings by Ilir Zaloshnja

Photography by Richard MacDonald (2004–2007),
Anna MacDonald (2006–2007),
and Ian Coyle (2008)

Monumenta Archaeologica 34
Cotsen Institute of Archaeology Press
University of California, Los Angeles
2014

The Cotsen Institute of Archaeology Press is the publishing unit of the Cotsen Institute of Archaeology at UCLA. The Cotsen Institute is a premier research organization dedicated to the creation, dissemination, and conservation of archaeological knowledge and heritage. It is home to both the Interdepartmental Archaeology Graduate Program and the UCLA/Getty Master's Program in the Conservation of Archaeological and Ethnographic Materials. The Cotsen Institute provides a forum for innovative faculty research, graduate education, and public programs at UCLA in an effort to impact positively the academic, local and global communities. Established in 1973, the Cotsen Institute is at the forefront of archaeological research, education, conservation, and publication and is an active contributor to interdisciplinary research at UCLA.

The Cotsen Institute Press specializes in producing high-quality academic volumes in several different series, including *Monographs, World Heritage and Monuments, Cotsen Advanced Seminars,* and *Ideas, Debates, and Perspectives.* The Press is committed to making the fruits of archaeological research accessible to professionals, scholars, students, and the general public. We are able to do this through the generosity of Lloyd E. Cotsen, longtime Institute volunteer and benefactor, who has provided an endowment that allows us to subsidize our publishing program and produce superb volumes at an affordable price. Publishing in nine different series, our award-winning archaeological publications receive critical acclaim in both the academic and popular communities.

Edited by Barbara Kohl
Designed by Carol Leyba
Cover design by Randi Danforth
Index by Matthew White

This publication was undertaken with the assistance of the Institute for Aegean Prehistory

Library of Congress Cataloging-in-Publication Data

Papadopoulos, John K., 1958-
The excavation of the prehistoric burial tumulus at Lofkënd, Albania / by John K. Papadopoulos, Sarah P. Morris, Lorenc Bejko, and Lynne A. Schepartz ; with contributions by Esmeralda Agolli, Jamie D. Aprile, Brian N. Damiata, J.E. Foss, Shpresa Gjongecaj, Christopher Johanson, Rovena Kurti, Richard MacDonald, John M. Marston, Samantha Martin-McAuliffe, Laura Menez, Vanessa Muros, Yannis Mylonas, Seth Pevnick, David A. Scott, Sarah C. Sherwood, John Southon, Lyssa C. Stapleton, George Theodorou, Evi G.Vardala-Theodorou, Evangelos Velitzelos ; survey, site plans, and drawings by Max Farrar and Samantha Martin-McAuliffe ; object drawings by Ilir Zaloshnja ; photography by Richard MacDonald (2004-2007), Anna MacDonald (2006-2007), and Ian Coyle (2008).
pages cm — (Monumenta archaeologica ; 34)
Includes bibliographical references.
ISBN 978-1-938770-00-5 (hardback)
1. Illyria—Antiquities. 2. Illyria—Social life and customs. 3. Illyria—Environmental conditions. 4. Lofkënd (Albania)—Antiquities. 5. Excavations (Archaeology)—Albania—Lofkënd. 6. Mounds—Albania—Lofkënd. 7. Bronze age—Albania—Lofkënd. 8. Iron age—Albania—Lofkënd. 9. Social archaeology—Albania—Lofkënd. 10. Antiquities, Prehistoric—Albania—Lofkënd. I. Morris, Sarah P., 1954- II. Bejko, Lorenc. III. Schepartz, Lynne Alison. IV. Agolli, Esmeralda. V. Title.

DR39.5.P36 2014
939'.865—dc23

2014015415

For

Muzafer Korkuti

Charlie Steinmetz

and

Lloyd Cotsen

(for making it happen)

Sketch of the Lofkënd tumulus from southeast by Anna MacDonald

There is no heap so eloquent as tumulus,
for no farewell is quite as definite
as that of the grass-growing earth
Freya Stark (1949:126)

Contents

Part II The Population of the Tumulus

Part III Analyses of Materials from the Tumulus

Part IV The Tumulus in Its Context

LIST OF TABLES
(IN VOLUME 1: TEXT)

List of Figures

(in Volume 2: Illustrations)

Unless otherwise noted, all photographs are by Rich MacDonald and all drawings are by Ilir Zaloshnja.

Abbreviations

AIR	air (= USGS standard for reporting isotopic values)
AMS	accelerator mass spectrometry
ASL	above sea level (elevation, in meters)
C	carbon (and isotopes thereof: C3, C4, δ13, δ15 etc.)
cal	calibrated (radiocarbon date)
cm	centimeters
Cu	copper
D	diameter
den	dendrochronology (dendrochronological)
DF	dark fabric
DISH	diffuse idiopathic skeletal hyperostosis
DJD	degenerative joint disease
DNA	deoxyribonucleicacid
E	east
EDC	energy dispersive spectroscopy
EIA	Early Iron Age
FD	fine dark (ceramic fabric)
FL	fine light (ceramic fabric)
ext	exterior
fr	fragment
frr	fragments
g	grams
GIS	Geographic Information Systems
GPS	global positioning system
H	height
ha	hectar(e)
HM [ceramics]	handmade
ICMPS	inductively coupled mass spectrometry
int	interior
kg	kilograms
km	kilometers
kV	kilo-volts
L	length
LBA	Late Bronze Age
LF	light fabric
Lt	left
m	meter
max	maximum
mm	millimeters
N	nitrogen (Chapter 7)
N	north (Chapter 3)
P	pottery (inventory number)
PDB	Pee Dee Belemnite (for delta values of carbon isotopes)
PH	preserved height
pH	pH factor (measure of acidity/alkalinity)
PL	preserved length
PW	preserved width
pXRF	portable X-ray fluorescence
RH	relative humidity
Rt	right
S	south
SA	subadult [appears in tables in Chapter 6, Figures]
SC	semi-coarse (ceramic fabric)
SEM	scanning electron microscope
SF	small find (inventory number)
SU	stratigraphic unit
Th	thickness
W	width (in catalogues: Chapters 9–10)
W	west (in grave descriptions: Chapter 3)
WM	wheelmade
Wt	weight

Preface and Acknowledgments

Sarah P. Morris, John K. Papadopoulos, Lorenc Bejko, and Lynne Schepartz

Our exploration of the pre- and proto-historic tumulus at Lofkënd owes its genesis to Professor Muzafer Korkuti, who brought the three directors to the site in July 2003 (Morris, Papadopoulos, and Bejko), and suggested it as an excellent project for an international collaboration. As then Director of the Institute of Archaeology, Academy of Sciences, he not only gave the project his blessing but also supported our annual applications to the Institute for permission to excavate, survey, and study material, over the next five years. Since that first visit, he has also served as official consultant to the project, spending part of each season with our team, taking a keen interest in our discoveries from tumulus and landscape, guiding our analysis of the lithics in particular, and in general advising us on all matters, archaeological and administrative, related to our research, its results, and our interpretation. It is therefore with deep affection and gratitude that we dedicate this volume to him, for initiating the project and guiding it through to its completion.

In recognition of her many contributions to Lofkënd, as one of the few team members to have been on the project from beginning to end, and for serving as the bioarchaeologist of a project whose human remains are crucial, Lynne Schepartz became a project co-director and co-author. The four project directors acknowledge the many and varied contributions of the scholars who penned the chapters that make up this volume. It has been a privilege and pleasure to work with so many talented contributors.

The Institute of Archaeology, Tirana, not only sponsored our project but generously contributed services, equipment, specialists, and logistics over five summers, including accommodation at Apollonia. Hospitality at these facilities granted us not only museum, storage, and study space for five annual campaigns, but also housing for our team and visitors in a unique ancient and modern setting. We thank not only the successive Directors of the Institute (Muzafer Korkuti and Shpresa Gjongecaj) for making this possible, but the Director of the Museum at Apollonia, Vangjel Dimo, and the Director of the Archaeological Park at Apollonia, Marin Haxhimihali, for their gracious support and help. We are also grateful to Institute members Ilir Gjipali, Skënder Muçaj, and Lami Koço for manifold assistance and advice on site and elsewhere in Albania, over the duration of our project. Our thanks are also due to the two different owners of the land—Baki Ymeri and Rrapi Malaj—on which our tumulus lies (how there came to be two landowners is explained in the Epilogue).

For inspiring our initial visit to Albania, hosting us at Apollonia, and encouraging us to begin a new archaeological project in the area, John Papadopoulos and Sarah Morris are particularly grateful to Jack Davis and Shari Stocker of the University of Cincinnati. They set a high standard for scientific, collaborative research with the Mallakastra Regional Archaeological Project (co-directed with Michael Galaty, Skënder Muçaj, Lorenc Bejko and Muzafer Korkuti), which provided an important foundation for further fieldwork in the area, and guided us towards a closer understanding of the landscape, as well as through local logistics. Over the years, Mike Galaty has generously provided help and advice to our team and its members; his continued fieldwork in Albania has helped make ancient Illyria an important arena for international archaeological research. Richard Hodges and his colleagues at the archaeological site of Butrint, in particular Oliver Gilkes, welcomed annual visits by our team and guided us through their ongoing excavations and finds. Skender Aliu hosted several visits by team members to the Korçë region, initiated us into the archaeology of southeastern Albania, and provided critical publications along with valuable counsel on the interpretation of finds in the final study season. Curtis Runnels kindly advised on prehistoric stone tools and their implications. We are also grateful to Raiko Krauss for visiting the site and his discussions of the material with us. We would never have solved the mystery of our modern gun cartridges and bullets (Appendix 3) without the expertise and generosity of time graciously provided by Yannis Mylonas of the War Museum of Greece in Athens, who stepped in past his retirement to identify each and every one. We are also grateful to the members of the Experiential Technologies Center (ETC) at UCLA, not least Christopher Johanson and Itay Zaharovitz, for spearheading the construction of the digital model of Lofkënd. Among other members of the ETC, Stanislav Parfenov provided our maps and Steven Hicks and Marie Saldaña contributed their digital expertise. Many scholars and friends, both in Albania and around the world, contributed in various ways by providing us access to material in their care and for discussion on various topics. They are more fully thanked in the various contributions that make up this volume. We are especially grateful to Bettina Arnold and the anonymous reviewers of this volume for reading the entire volume and for their many suggestions. For all errors of fact and judgment that remain we alone are responsible.

In the United States, the support of the Cotsen Institute of Archaeology, both spiritual and in substance (through Ahmanson Field Research Grants), was critical to our project, and we are grateful to its faculty, students, and research associates who served as team members on site and in labs, and as audiences for our reports. We would like to thank Charles Stanish and a succession of staff for their time and professional contributions to our research, as well as their willingness to bring this project to fruition in the form of a final publication. The Friends of Archaeology and its Director's Council, together with the Steinmetz Family Foundation, made it possible for our graduate students to participate in this project, by providing travel grants each summer to bring them to the field. In turning UCLA's Institute of Archaeology into a world-class research institution, Lloyd Cotsen made this and future projects possible.

Other funding agencies in the United States include the Kress Foundation, which supported travel and stipends for conservators, and the UCLA-Getty MA Program in the Conservation of Archaeological and Ethnographic Materials, which provided us with summer interns and the invaluable time and expertise of Vanessa Muros. The Mellon Foundation and the Detroit Institute of Arts made it possible for Naoko Fukumaru to join us as a conservation intern for the first summer of excavation.

Without the existence of the Steinmetz Family Foundation, and its wisdom and generosity in bestowing on UCLA an endowed chair in archaeology and material culture, this project would never have taken place. Beyond the generous funding of fieldwork, student travel, and follow-up research expenses, the personal involvement of Charlie Steinmetz and Laura Steinmetz as team members in 2005, and their continued enthusiasm and interest in archaeology in Albania, made all the difference. We hope that the final result makes them as proud as we are of the project that they inspired, and are glad that Charlie's name joins those of Muzafer Korkuti and Lloyd Cotsen on our dedication page.

In June 2011, Crawford Hallock Greenewalt, Jr. ("Greenie" to his friends) quoted out loud Freya Stark's memorable sentence on the tumulus to one of the project directors, less than a year before his untimely death. The relevant passage appears on the page following the dedications. We are grateful to Greenie for citing Stark and for his humanity.

Finally, the enthusiasm, collegiality, and friendship of the many local communities in Albania—in Pojan, Fier, and the villages of Ngrançija and Gjinoqara in the vicinity of the tumulus—gave us all a set of experiences and memories we will never forget. A list of all official participants can only be a partial tribute to their many contributions.

LIST OF TEAM MEMBERS
(in addition to authors of this volume)

Alison Adams (2008)
Esmeralda Agolli (2005, 2006, 2007, 2008)
Frosejda Angjellari (2007)
Jamie Aprile (2004, 2005, 2007, 2008)
Alma Bardho (conservation, 2005, 2006, 2007, 2008)
Nermin Bartulli (object photography, 2005, 2006, 2007)
Lorenc Bejko (co-director, 2003, 2004, 2005, 2006, 2007, 2008)

Michelle Bonogofsky (bioarchaeology, 2004)
Sarah Cleary (conservation, 2004)
Ian Coyle (photography, 2008)
Brian Damiata (^{14}C dating and sampling, 2007)
Andon Dimo (logistics, 2005, 2006, 2007, 2008)
Arjan Dimo (2004, 2005, 2006)
Mimoza Dimo (cook, housekeeper, 2005, 2006, 2007, 2008)
Max Farrar (surveyor, architect, draughtsman: plans, 2004, 2005, 2006, 2007)
John Foss (soil scientist, 2005, 2008)
Naoko Fukumaru (conservation, 2004)
Rexhep Halili (2004)
Ergys Hasa (2008)
Rahman Hasa (2005, 2006)
Sami Hasa (manager, driver, 2004, 2005, 2006, 2007, 2008)
Artan Hoxha (2004)
Lavdosh Iaupaj (2008)
Alket Islami (aerial photography, 2005)
Chris Johanson (digital archaeology, 2005)
Ilir Koço (2004)
Klodiana Kontos (2006)
Muzafer Korkuti (consultant, 2003, 2004, 2005, 2006, 2007, 2008)
Kushtrim Kuçi (2004)
Shauna Kullmann (2008)
Rovena Kurti (2004, 2005, 2006, 2007)
Jessica Langenbucher Brooks (2004)
Surja Lela (2004, 2005)
Allison Lewis (conservation, 2006)
Anna MacDonald (photography, 2005, 2006)
Richard MacDonald (photography, 2004, 2005, 2006, 2007)
Samantha Martin-McAuliffe (draughtsman: plans and landscape, 2007, 2008)
John McCampbell Marston (2004, 2005)
Laura Menez (DNA analysis, 2004)
Albana Meta (2004, 2005, 2006)
Sonila Meta (2004, 2005, 2006)
Rita Mitre (housekeeper, cleaner, 2005, 2006, 2007, 2008)
Sarah Morris (co-director, 2003, 2004, 2005, 2006, 2007, 2008)
Vanessa Muros (conservation, 2006, 2007, 2008)
Caitlin O'Grady (conservation, 2005)
John Papadopoulos (co-director, 2003, 2004, 2005, 2006, 2007, 2008)
Genti Përzhita (2006)
Seth Pevnick (2004, 2006, 2007, 2008)
Joanna Potenza (2006)
Rudenc Ruka (2004, 2005, 2006, 2007)
Eugen Ruzi (2008)
Lynne Schepartz (co-director, 2004, 2005, 2006, 2007, 2008)
Lyssa Stapleton (2005, 2006, 2008)
Charlie Steinmetz (2005)
Laura Steinmetz (2005)
Sarah Stoutamire (bioarchaeology, 2005)
George Theodorou (geologist, paleontologist, 2008)
Evi Vardali-Theodorou (malacologist, 2008)
Serena Vartarian (2008)
Donika Xherro (housekeeper, 2004)
Luajana Xherro (cleaner, 2004)
Milika Xherro (cook, 2004)
Baki Ymeri (foreman and guard, 2004, 2005, 2006, 2007)
Itay Zaharovitz (digital archaeology, 2005)
Ilir Zaloshnja (draughtsman: objects, 2004, 2005, 2006, 2007, 2008)

Workers

Enriko Ajupaj (2006, 2007), Shëlqim (Ndriçim) Beqiri (2005, 2006, 2007), Artur Dervishi (2005, 2006), Asgeri Dervishi (2007), Mentor Dervishi (2006), Sazan Dervishi (2005), Barjam Kapllani (2005, 2006, 2007), Adriatik Malaj (2004, 2005, 2006), Arben Malaj (2004, 2005, 2006), Denis Malaj (2005), Marjol Malaj (2007), Mynyr Malaj (2006, 2007), Përparim Mesuti (2007), Zaim Mesuti (2007), Albana Ymeri (2005, 2006), Edlir Ymeri (2004, 2005), Elton Ymeri (2007), Sulu Ymeri (2007)

Contributors

ESMERALDA AGOLLI received her Ph. D. in Archaeology from the University of California, Los Angeles, and specializes in ceramic technology and theories of cultural evolution and transmission in prehistory.

JAMIE APRILE is a Lecturer in the Department of Classics at the University of Texas, Austin, with research interests in political economy, lithic analysis, and prehistoric Greece.

LORENC BEJKO is Associate Professor and head of the Department of Archaeology and Cultural Heritage Management at the University of Tirana, Albania, and past Director of the Institute of Cultural Monuments, Albania, and the International Centre for Albanian Archaeology. He has co-directed and published many field projects in Albania.

BRIAN DAMIATA is a geophysicist and affiliate of the Cotsen Institute of Archaeology at UCLA who conducts research in near-surface geophysical imaging of archaeological sites, paleo-diet reconstruction based on stable isotope analyses, and paleoclimatic studies as applied to sea-level change and coastal reconstruction.

MAX FARRAR is a freelance archaeological surveyor, 3D model builder and illustrator based in Oxford, England.

JOHN E. FOSS is Professor Emeritus in the Department of Plant and Soil Science at the University of Tennessee, Knoxville. A soil scientist specializing in soil mineral weathering and soil profile development, he helped integrate the study of soil history with archaeology, and continues to research soil-landscape relationships.

SHPRESA GJONGECAJ is Director Emerita of the Institute of Archaeology, Academy of Science, Tirana (Albania) and a specialist in numismatics of the classical world.

CHRISTOPHER JOHANSON is Assistant Professor in Classics and Digital Humanities, Co-Director of the Experiential Technologies Center at UCLA, and Director of RomeLab, a multi-disciplinary research group that studies the interrelationship between historical phenomena and their spatial contexts.

ROVENA KURTI specializes in costume and identity in pre- and proto-historic Europe as a researcher in the Prehistory Department of the Institute of Archaeology, Albania, and is a Ph. D. candidate at the École Pratique des Hautes Études, Paris.

SAMANTHA MARTIN-McAULIFFE is Lecturer in the School of Architecture at University College Dublin, where her teaching and research focus on the built environment of classical antiquity and the phenomenology of landscapes.

JOHN McCAMPBELL MARSTON is Assistant Professor of Archaeology and Anthropology at Boston University, where he directs the Environmental Archaeology Lab and conducts research in paleoethnobotany, agricultural systems, and climate change adaptation, primarily in the Mediterranean and Near East.

ANNA and RICHARD MacDONALD are practicing artists and instructors in studio arts and photography at Stony Brook School, Stony Brook (Long Island), New York.

SARAH MORRIS is Steinmetz Professor of Classical Archaeology and Material Culture in the Department of Classics and the Cotsen Institute of Archaeology at UCLA, where she specializes in Greek archaeology and literature.

VANESSA MUROS is Conservation Specialist and Lecturer in the UCLA/Getty Program in the Conservation of Archaeological and Ethnographic Materials.

YANNIS MYLONAS is curator emeritus at the War Museum in Athens, Greece, where he is a specialist in the history of modern European weapons and uniforms.

JOHN PAPADOPOULOS is Professor in the Department of Classics and the Cotsen Institute of Archaeology at UCLA. He has excavated and published widely on material culture and burial customs of Bronze and Iron Age Greece, Italy, and Albania.

SETH PEVNICK is Chief Curator as well as Richard E. Perry Curator of Greek and Roman Art at the Tampa Museum of Art, and a specialist in Greek vase painting.

LYNNE SCHEPARTZ is Associate Professor and Head of Biological Anthropology (School of Anatomical Sciences) in the Faculty of Health Sciences, University of Witwatersrand, Johannesburg, South Africa, and a Consulting Scholar with the University of Pennsylvania Museum of Archaeology and Anthropology. She is a specialist in human skeletal biology and paleoanthropology.

DAVID SCOTT is Professor in Art History and Archaeology at UCLA, and teaches in the MA Program in the Conservation of Archaeological and Ethnographic Materials at the Getty Conservation Institute and the Cotsen Institute of Archaeology, UCLA.

JOHN SOUTHON is a research scientist in the Department of Earth System Science (School of Physical Sciences) at the University of California, Irvine, where he is co-founder and director of the W.M. Keck Carbon Cycle AMS Laboratory and conducts research on biogeochemical cycles.

LYSSA STAPLETON is a doctoral candidate in Archaeology at UCLA, where she completed her MA with a mortuary analysis of the burials at Lofkënd, and is writing a doctoral thesis on cultural heritage and the market in illicit Chinese antiquities.

GEORGE THEODOROU is Professor of Paleontology and Stratigraphy in the Department of Historical Geology and Paleontology, and Director of the Museum of Paleontology and Geology, at the National Kapodistrian University of Athens.

EVI VARDALA-THEODOROU is Curator Emerita at the Goulandris Museum of Natural History and a malacologist specializing in mollusca of the Mediterranean.

Part I
The Excavation of the Tumulus

Chapter 1
Introduction

John K. Papadopoulos, Sarah P. Morris, Lorenc Bejko, and Lynne A. Schepartz
With contributions by Seth Pevnick and Esmeralda Agolli

Chapter 1.1
Lofkënd: The Site and Archaeological Objective

John K. Papadopoulos, Sarah P. Morris, Lorenc Bejko, and Lynne A. Schepartz

The region of Mallakastër, and southwestern Albania in general, is home to a number of burial tumuli, spanning the Bronze and Iron Ages of later prehistory (see Cabanes et al. 2008:34, 46, fig. 1; see also Chapter 17) (**Figs. 1.1–1.2, 17.1**). Some were robbed long ago, others reused for modern burials, and a few excavated under scientific conditions, most notably at Patos some 14 km away (Korkuti 1981). Indeed, modern understanding of the pre- and proto-history of Illyria has largely been shaped by the contents of such burial mounds, as explored later in this volume (Chapter 17), while the Illyrian past still displays its most evocative images today in these "mounds of memory" (Papadopoulos 2006; see further Chapter 20). Against this history, the mound that bears the name of a nearby village, Lofkënd, signals both a distinctive profile within its landscape (**Figs. 1.3–1.4**) and an undisturbed opportunity for archaeology. The tumulus also lies within a region well known for sites ranging from Neolithic (Cakran) to Bronze Age (Margëlliç), "proto-urban" (Gurëzezë, Mashkjezë), Classical/Hellenistic (Apollonia, Klos-Nikaia) and Roman through Byzantine (Byllis, Ballsh) periods of antiquity (**Fig. 1.2**). Thus the site offered us not only a chance to excavate, with multiple modern scientific methods not fully available to earlier projects, an intact ancient burial ground in full, but also to revisit the significance of this tumulus and its fellows for the emergence of urbanism and complexity in ancient Illyria. Given its manifest surface indications of occupation in the Bronze and Iron Ages (see Chapter 1.2), this tumulus invited a modern archaeological project that promised to yield results in periods familiar to the directors, yet in dire need of new, controlled case studies.

The Site and Its Environmental and Historical Setting

The western coastal region of Albania has been shaped by the flow and deposits of major rivers, themselves fed by smaller waterways that carved a path through sandstone, limestone, and shale (**Fig. 1.1**; Chapter 16.4). The Gjanicë River is one of these, as it flows from southeast to northwest to join the Seman River near Apollonia and reach the sea, and its valley thus belongs to the hinterland of the coastal regions that attracted Greek colonies (**Fig 1.2**). Targeted explorations of the ancient sites in this region now not only locate Greek cities like Apollonia, but have also revealed long-term settlements such as Margëlliç (**Fig. 1.5a–c**; see also **Fig. 20.6**), whose ancient name remains unknown. One of Albania's first systematic surface surveys explored the environs of Apollonia, but was named for the region of Mallakastër where the tumulus lies, and has extended the reach of local history from the Paleolithic through modern times. (For the Mallakastër

Regional Archaeological Project [MRAP], see *http://river.blg.uc.edu/mrap/MRAP_en.html*; and Davis et al. 2003–2004. See also the pioneering work of Praschniker 1922–1924.)

To a modern traveler on the north–south highway from modern Fier to Ballsh, the area displays cultivated fields along the river, higher slopes and terraces both planted and for pasture, and scattered villages, punctuated by the rise and fall of oil pumps and the scent of tar (**Fig. 1.6**; see also Chapter 15 on local hydrocarbon resources). In prehistoric times, the area must have been heavily forested (Fouache 2007:9) to support large game, according to the copious finds of stone tools in both ancient deposits and the present surface (Chapters 13 and 18). From the Bronze Age to the Byzantine era and beyond, it supported settlements, cemeteries, and sanctuaries, some named in antiquity or only known by their modern place names, explored in modern times, some informally/illegally, yet one remained untouched. More generally, this area is associated by Classical authors with one or more of the tribes belonging to southern Illyria and Epirus.

Hekataios of Miletos, writing in the later sixth or early fifth century BC, provides the names of a number of tribes in Epirus and southern Illyria. According to Nicholas Hammond (1967a:458), the following are listed with the references to the fragments of Hekataios: "Dexari, Chaones, Encheleae (F 103), Taulantii (F 99 and F 101), Sesarethii (F 100), Orestae (F 107), Perraebi (F 137), Molossi (F 108), Dodonaei (F 108), Athamanes (F 133), Eordi (F 372), Amphilochi (F 102c), and also, if we are correct in ascribing some passages of Strabo to Hecataeus, Thesproti, Cassopaei, Tymphaei from Tymphe (Strabo 7.7.6), Pororaei from Pararaea (Strabo 7.7.6), and Talares and Aethices (Strabo 9.5.12)." Theopompos (apud Strabo 7.7.4–5 [C323–324]) counted 14 Epirote tribes, and Strabo himself (7.7.4.–12 [C323– 329]), some 300 years later, records the names of 11 of these: Molossians (Μολοττοί), Chaonians (Χάονες), Thesprotians (Θεσπρωτοί), Athamanes or Athamanians (Ἀθαμᾶνες), Aithikes (Αἴθικες), Tymphaioi (Τυμφαῖοι), Orestai (Ὀρέσται), Paroraioi or Parauaeans (Παρωραῖοι), Atintanes (Ἀτιντᾶνες), Amphilochoi (Ἀμφίλοχοι), and Kassopaioi (Κασσωπαῖοι) (see further Cross 1932:5, n. 1; Lepore 1962: 16–33; Malkin 1998:135; Klotzsch 1911:10, n. 1). The tribal borders have been discussed by a number of scholars (not least Cross 1932; Dakaris 1964:11, fig. 1; Evangelides 1947:8–9; see also Winnifrith 2002:43, and Ceka 2009–2010, especially for the Atintanes), but defining these borders—if we can call them borders—with any clarity or consensus remains difficult. An attempt to place these Illyrian and Epirote tribes on the modern map of Albania and northwest Greece is rendered on **Figure 1.7**.

As for where precisely in this topography of tribes of southern Illyria and Epirus Lofkënd lies, there is no clear information, except that the region is north of the Epirote tribes and thus in southern Illyria. The tumulus is located somewhat to the north and east of ancient Chaonia—usually located between the Thyamis River in Greek Epirus (the modern Kalamas) and the Akrokeraunian headland, the modern Karaburun peninsula that defines the western edge of the bay of Vlorë, the ancient Aulon, in Albania (for Chaonia, see Malkin 1998:138; and for its prehistory, Prendi 1993)—and to the north and west of Atintania (see **Fig. 1.7**). More importantly, in his description of the Illyrian tribes near the southern part of the mountainous country and those above the Ionian Gulf, Strabo (7.7.8 [C326]) specifically names the Bylliones (Βυλλίονες), Taulantii (Ταυλάντιοι), Parthini (Παρθῖνοι), and the Brygi (Βρῦγοι). Similarly, in locating ancient Nymphaion (Nymphaeum)—famous for its bitumen (see Chapter 15)—which is in the broader territory of Apollonia (for Strabo and the Illyrians, see Šašel Kos 2011), the Roman natural historian, Pliny the Elder (3.23.145; see further Hammond 1967a:232–233) has it in the area inhabited by the native tribes of the Amantes and Bylliones (Buliones in Latin). The proximity of Lofkënd to both Byllis and Nymphaion establishes that the tumulus was in the territory of the Bylliones. The site of Byllis itself lies at the northern edge of the territory of the Atintanes (cf. Cabanes 2011). But there is no further information on this culture or on any settlements that it may have embraced, prior to Greek colonization.

Clear contacts with the Aegean survive from the Late Bronze Age at Margëlliç by way of Mycenaean pottery, perhaps as early as Late Helladic IIIA (Ceka 1990b:especially 141, fig. 4; with further discussion in Bejko 1993:103, 110, 120, fig. 2c, d; Bejko 2007; for the Bronze Age in Illyria, see also Prendi 1998; for Mycenaean pottery and influences in FYROM [Former Yugoslav Republic of Macedonia], see Videski

2007), and eventually colonization by Corinth placed a Greek city at the (then) coastline and river delta at Apollonia. According to Strabo (8.316) and Pseudo-Scymnus (439), Apollonia was a joint foundation of Corinth and Kerkyra; Thucydides (1.26.2) describes it as a Corinthian *apoikia* (cf. Stephanos Byzantinos, sv. "Apollonia," who only mentions 200 Corinthians under the oikist Gylax), whereas Pausanius (5.22.4) refers to Kerkyra as the sole *metropolis* (Graham 1964:130). Indeed, in all of Albania, Apollonia and Epidamnos were the only colonies in the true Greek sense, that is, formal *apoikiai* of a sponsoring mother-city or *metropolis* (Stocker 2009 on the *ktisis* of Apollonia). Other sites on the coast of Illyria are only linked with legendary Greek settlers returning from Troy. Thus Bouthrotos (Buthrotum, Butrint) was founded by Helenos, one of the sons of Priam (Stephanos Byzantinos quoting Teuker of Kyzikos; see Hammond 1967a:385, 474); Orikon (Oricum) was founded by Lokrians and Euboians, whose ships were driven there by storms on their way home from Troy (Pseudo-Scymnus 442–443; for Orikon, see further Bereti et al. 2011); Thronion/Thronium is also linked with Lokrians and Euboians returning from Troy (Pausanias 5.22). Between these points in time and space—the Mycenaean era, on the one hand, and the establishment of the Greek colonies, around 600 BC, on the other—the hinterland still offered more questions.

Some 25 km east and slightly southeast of Apollonia and about 357 m above sea level (ASL), a prominent sandstone ridge running from northeast to southwest forms a narrow terrace sloping off sharply to the southeast and southwest, more gently to the northwest, terminating at its south end in a natural point that attracted an ancient community to bury their dead, beginning sometime in the Late Bronze Age (**Figs. 1.3–1.4**, and especially **Fig. 2.1**). More precisely, the Lofkënd tumulus is located in UTM Zone 34T, Easting 391927, Northing 4500348. Some six centuries of use as a cemetery (plus 15 early modern internments inserted at the flatter northeast end of the mound) turned the peak into a prominent mound now visible from a distance of several kilometers (**Fig. 1.3**). The tumulus is most conspicuous when seen from the lower stretches of the surrounding valleys to the east, west, and south, and particularly from the valley of the Gjanicë River a few kilometers to the south, along the banks of which lies the road that still connects Albania and Greece. Presumably the compact (clay-heavy) mass of the mound discouraged removal (see Chapters 16.4 and 22), and its prominence limited cultivation around it to the west and south, and kept it intact over centuries. In 2003 (**Fig. 1.8**), this site showed stark erosion on its steep, southern side that exposed human bone, and its surface held scattered sherds of the Bronze and Early Iron Ages, along with stone tools of much earlier eras, but no signs of recent or earlier looting. The prominence of the tumulus was not just the result of the human-made process of continuously heaping earth above the deceased. As subsequent excavation was to show (Chapters 2 and 22), the spot chosen for the tumulus was a natural sandstone outcrop (**Fig. 1.9**).

As one of the oldest villages in the region, modern Lofkënd, which lies on a prominent ridge almost 3 km to the east, gave its name to the ancient tumulus. Several more recent villages now ring the site: Ngrançija, a short distance to the southeast, Gjinoqara to the northwest, together with their various offshoots (Ngrançija Sinaj and Ngrançija Arizaj, Gjinoqara Laçaj, Gjinoqara Berhamaj, and Gjinoqara Mashkullorë). Tumuli of the same or similar periods are known in this area, especially at Mashkullorë, the latter today overlain by Ottoman burials, marked by plain tombstones (**Fig. 18.4**), and, as at Lofkënd, Mashkullorë also had a modern Muslim cemetery immediately below the tumulus (**Figs. 20.13–20.14**). The prototype, however, for both tumulus and project at Lofkënd was offered by that at Patos (**Fig. 1.10**), rescued in the 1970s (Korkuti 1981), and whose objects in the Fier Museum offered us our first sample of likely artifacts from a newly excavated necropolis of the same era (**Fig. 1.11**).

The region in which Lofkënd lies thus belongs to one of the richest archaeological areas of Albania, now more clearly understood thanks to the work of the MRAP team (although the tumulus lies outside the area covered by the MRAP survey). The latter focused on the broader *chora* of Classical Apollonia, while Mallakastër, together with its modern regional center at Ballsh, is located well to the east of the MRAP survey area. More importantly, the Lofkënd tumulus is ringed by some of the most significant fortified proto-urban and Classical/Hellenistic centers, including Margëlliç (**Fig. 1.2**) to the west-northwest (Ceka 1977–1978, 1983a, 1985a, 1986a, 1987,

1990a:138–141), Gurëzezë (Anamali 1949; Ceka 1983a, 1985a, 1990b; Praschniker 1922–1924:64– 68) and Mashkjezë (Ceka 1977–1978, 1983a, 1983b) to the south and west, Byllis and Klos-Nikaia to the south (at Byllis there is little that clearly predates the fourth-century BC and Hellenistic period [see Gilkes 2013:130–142]; for the identification of Klos with the ancient Nikaia, see Robert 1928:433; and for work at the site, see Ceka 1977–1978, 1985a, 1990a; Papajani 1974, 1976), and Dimal (Kalaja e Krotinës) to the northeast (Dautaj 1965, 1971, 1973, 1974, 1976; Praschniker 1922–1924; see also Myrto 1998:73–75). Given the evident date of the Lofkënd tumulus in the Late Bronze and Early Iron Ages, its relationship to local fortified hilltops (explored in Lafe 2003) was a particular concern. Meanwhile, the proximity of Lofkënd to the Greek colonial foundation at Apollonia brings the site into the hinterland of one of the largest Greek *poleis* and Roman centers in the Mediterranean (Amore 2003–2004, 2004, 2011; Amore et al. 1995; Cabanes and Lamboley 2004; Cabanes et al. 1994, 1997, 1999, 2000; Davis et al. 2006; Delouis et al. 2007; Dimo, Lenhardt, and Quantin 2007; Gilkes 2013:39–55; Gjongecaj and Picard 2003–2004; Lamboley and Vrekaj 2003–2004; Mano 1971, 1974, 2003–2004; Rey 1932). Important prehistoric sites in this region include the Bronze and Early Iron Age tumulus at Patos (14 km to the west-northwest) already referred to—which, like Lofkënd, overlooks the Gjanicë valley (Korkuti 1981)—the open-air Bronze Age site and tumulus at Drenovë (Ceka 1976:366, 1985a, 1990b), and the Neolithic settlement at Cakran (Korkuti and Andrea 1974).

Thus the modest rise on the horizon of the Gjanicë valley marked a place rich in the long-term history of its surroundings, and one poised to deliver fresh news of the past.

Objectives of the Project

What drew the authors to this site, and inspired its systematic exploration, was more than mere opportunity or the promise of an unplundered necropolis (see *http://www.sscnet.ucla.edu/ioa/staff/papadopoulos/lofkend/aims.html*). Recent research in this part of the southern Balkans has moved archaeological investigation of pre- and proto-history beyond pursuing the ethnogenesis of "Illyrians" and Albanians (Bejko 1998; Martin 2006; Tsonos 2009; Veseli 2006; Chapter 17), or the origins of Hellenic cities and colonies, to consider the evolution of parallel ways of life, whether pastoral or mixed in economy, settled or partly mobile, against larger and more distant developments of "Classical" history. The Lofkënd site thus offered itself as a further link in recent explorations of this period in Illyria and Epirus. In southeastern Albania, the tumulus of Kamenicë was partially excavated (Bejko 2004; Bejko, Amore, and Aliu 2006) and succeeded by a regional survey of the Korçë River basin (see *http://www.icaa.org.al/Anglisht/kobas_2005.html*, for preliminary reports on the KOBAS [Korça Basin Archaeological Survey] Project). In northwest Greece across the border, the Liatovouni cemetery, rescued in salvage operations of the 1990s, spans similar centuries from the Late Bronze to Early Iron Ages, and in fact through the Classical era (**Fig. 1.7**; Douzougli and Papadopoulos 2010), which adds considerable time depth to the cemetery and associated settlement (*komē*) excavated by Ioulia Vokotopoulou at Vitsa Zagoriou (Vokotopoulou 1986) (the Vitsa cemetery begins in the ninth century BC). More than time links these locales, for Liatovouni lies near the convergence of three Greek rivers—the Aoös, the Voidomatis, and the Sarantaporos—that flow across the Greek border to form the Vjosë that leads to the sea near Apollonia, along an ancient cultural and historical highway from the Pindos Mountains to the Ionian Sea.

The power of these rivers as natural routes of communication, especially between the mountainous interior of northwest Greece into southern Illyria and on to the Adriatic, cannot be underestimated. More particularly, the valley of the Aoös/Vjosë was a route that must have been exploited for millennia, as it still is today (the modern border between Greece and Albania at Melissopetra/Tre Urat lies some 7.5 km north of the confluence of the Aoös/Vjosë and Voidomatis rivers; see Sivignon 2007). Hammond (1931–1932) imagined the Dorian invasion along these paths, and Lévêque revived them in his study of the campaigns of Pyrrhos (see below). Beaumont (1952) discusses the important land route mentioned by Thucydides (1.26.2) that enabled the Corinthians in 435 BC to get troops to Epidamnos, via Apollonia, without having to sail around Kerkyra. Beaumont traces the route from Ambrakia up the Louros valley toward the plain of Hellopia—that is, the plain of Ioannina (Zachos 1997)—which was controlled by the Molossians, and from there northward to the Drinos valley, through Chaonian territory, to the Aoös/Vjosë valley, up the river to Apollonia (Beaumont 1952:64–65). This is the route followed by the modern road from Ioannina to Gjirokastër/Argyrokastro and Tepelenë; it was also the path used by Lord Byron in 1809. By using this artery, some of the most mountainous terrain of southeastern Europe was easily bridged by a large body of people moving along river valleys. In later history, the same landscape, particularly the Aoös/Vjosë Gorge, featured prominently in the Roman campaign against Philip V in the Second Macedonian War of the early second century BC (Hammond 1966). A century earlier, King Pyrrhos of Epirus (319–272 BC) controlled a network of urban centers from Ambrakia in the south to Antigoneia in the north by means of the river valleys of northwest Greece and southern Albania (for Pyrrhos, see Franke 1955:87–88; Garoufalias 1979; Klotzsch 1911:153–218; Nenci 1953; Nilsson 1909:69–77; von Hassell 1947; Lévêque 1957:92, pl. I). A millennium and a half later, Ali Pasha (1741–1822)—of Tepelenë and of Ioannina—expanded his territory to include most of Albania, western Greece, and much of the Peloponnese, by exploiting the same river valleys (for Ali Pasha, see, most recently, Fleming 1999; on his forts, see Smiris 2001). In recent wars, both in the early twentieth century (**Fig. 1.12**) and again during World War II, Greek troops were moved by boat up these rivers to meet the enemy, and as we learned from tumulus finds (Appendix 3, following Chapter 10), some may even have reached Löfkend in the early 1940s.

In prehistory, leading questions link these areas through their related materials, particularly in the form of handmade painted pottery and metal artifacts; thus a new set of grave contexts promised fresh answers to the questions of cultural contact and exchange. More broadly, we hoped to recover some of the tribal dimensions of these local cultures (Galaty 2002, and various other papers in Parkinson 2002; cf. Galaty et al. 2011, 2013), not for their proper names in Classical sources, but in order to begin to understand what kind of lifestyle supported them, whether specialized pastoralism or a mixed economy based on farming, and what kind of exchange processes characterized their interrelations. We shall return to these questions in the Epilogue that concludes this volume.

The life of the Lofkënd tumulus, according to its surface finds, spanned some of the most elusive yet important centuries of this process, during the Late Bronze and Early Iron Ages. Those phases were conspicuously scarce in the regional survey of the Mallakastër region (Papadopoulos, Bejko, and Morris 2007:110; *http://classics.uc.edu/mrap/mrap_en.html*), while framing the critical phase after the demise of Bronze Age citadels like Margëlliç (**Fig. 1.5a–c**) and prior to the rise of fortified citadels such as those at Gajtan (Shkodër) (for Gajtan, see Cabanes et al. 2008: 48; Islami and Ceka 1965), and closer to Lofkënd at Gurëzezë (**Figs. 1.1–1.2**). It was our hope to fill in this gap with finely textured and time-calibrated materials including information on the health and kinship of human populations, environmental proxy data, burial practices, and material technologies for this era, and ultimately to situate a complex of detailed evidence against patterns in social evolution from the Bronze to the Iron Age. Particularly pressing was the need for a scientifically established data set providing an absolute chronology for artifacts, hitherto largely calibrated via stylistic parallels to, and the relative chronology of, a handful of stratified sites and a number of burial complexes with untested sequences. Our quest for signs of an accompanying settlement of the period, ever puzzling in the historical record, eventually led us to add an intensive field survey in the area, with unexpected results for the prehistoric period, spiraling back to the Paleolithic era (Chapter 18). A major priority guiding this agenda was to consider the rise of more permanent settlements transcending the size and complexity of extended-family villages, and whether these emerged as an indigenous development rather than an imitation of Greek cities.

The excavation of a tumulus such as Lofkënd allowed us to investigate a number of interrelated phenomena. First, there was the issue of linking the burial mound to a particular group, or groups, of people, as well as to rethink the nature of settlement in south-central Albania against the backdrop of historical pastoralism and agriculture. As one of us outlines in the Epilogue, the archaeological patterns for the Late Bronze and Early Iron Ages that have recently emerged from Epirote sites (e.g., Vitsa Zagoriou, Pogoni, and Liatovouni), when taken together with the material coming to light in Albania, suggests that from the Neolithic period through the years of the fifth and earlier fourth centuries BC, an economy based on sedentary mixed farming and associated localized herding remained the primary subsistence strategy. The people burying their dead at Lofkënd were no exception.

One of the original goals of the project was to explore the relationship between the people buried at Lofkënd and the colonists of Apollonia, and more particularly the contrast between the Greek *polis* and its citizens, on the one hand, and the tribal, clan-based political and social organization based primarily on kinship, on the other (see Galaty 2002; Hammond 2000:345 for the use of the term *ethne*). As we shall see, however, our tumulus may have ceased to function as a burial ground some 200 years before the foundation of Apollonia (see Chapter 4 for the relative and absolute chronology), but not without some tantalizing connections (as in Appendix 2, following Chapter 9).

In addition, one of the most critical material developments in the broader region of Lofkënd was the establishment of a new type of site in the Illyrian hinterland, especially during the developed Iron Age, what has come to be known as the "proto-urban" center (Ceka 2011; Lafe 2003). It is not yet fully understood whether sites such as Margëlliç, Gurëzezë, and Mashkjezë represent true towns, hilltop refuges, or regional trading and meeting places. At most, if not all of these sites, evidence for permanent year-round habitation has yet to be furnished, but this is not a phenomenon limited to southern Illyria. At the site of Emporio on the Aegean island of Chios, a dry-stone fortification wall—probably built in the eighth century BC—running around the top of the hill enclosed an area of approximately 2.4 ha, and the dearth of habitation finds inside the wall, as well as the area immediately outside, suggests that the fortification was a refuge (Boardman 1967:4–5; Frederiksen 2011:137–138, fig. 31).

Also unclear are the processes by which these Illyrian proto-urban sites developed, as well as the relationship of these centers with one another and with the coastal colonies. A particularly vexing problem with all of these proto-urban sites is the issue of their chronology, and conclusive evidence for the initial stages of their period of use is usually lacking. At Mashkjezë, the earliest cultural layers are dated to the Archaic period on the basis of imported Greek pottery, and a similar date is suggested for Margëlliç, although the latter has furnished some evidence of use in the Mycenaean period. The exact chronology of Gurëzezë is far from clear, while at Klos-Nikaia there is little that clearly predates the fifth century BC (for the development of occupied hilltops to proto-urban centers, see Ceka 1985a, 1985b; Harding 1992:22–27; Korkuti 1982b). Despite recent research (both survey and excavation) of fortified sites in Chaonia (Butrint, Kalivo, and Kepi i Stillos), few yielded more than mixed deposits of prehistoric ceramics, without absolute dates (Gjipali 2011; Lima 2013). Meanwhile, for the first time, a proto-urban citadel (in northern Albania) has been firmly dated by AMS radiocarbon dates to an initial fortification in the Early Iron Age (ca. 800 BC, calibrated), and occupied thereafter through the Early to Middle Iron Ages (Galaty 2007; Galaty et al. 2011, 2013).

The issue of the development of cities in Illyria and northwest Greece needs to be seen in a broader context, and here the precept of the Italian peninsula offers a useful comparison. Among others, Anthony Snodgrass (1971:91) remarked long ago that in Sicily and southern Italy permanent settlement only begins in the eighth century BC. The assumption that urbanism came to parts of Italy as a result of Greek colonization during the historic era has been remarkably long-lived, though it may well be overstated. Some scholars have suggested that Mycenaean trade was in part responsible for the urban development of Apulia (e.g., Whitehouse 1973). Other scholars, not least David Ridgway, in dealing with the Late Bronze Age in Italy, point to a basic distinction between primary (coastal) and secondary (inland) reception points (e.g., Ridgway 1992:7). A similar pattern is seen in the historic period, when there is a difference between the coastal Greek *poleis* and the indigenous hinterland on both the Ionian and Tyrrhenian coasts of Italy (see various papers in Descoeudres 1990; see also Papadopoulos 2001 and Malkin 2011). The same is true for Illyria, where in the Archaic period, the *poleis* of Apollonia and Epidamnos differ significantly from the indigenous hinterland. The question of how prominent burial places like Lofkënd functioned in such a contested landscape takes on added significance when seen in the light of contemporary settlements.

But one problematic issue remained: What do habitation sites of the Late Bronze and Early Iron Ages in this part of Illyria look like? The dearth of sites of this period has been one of the features of the archaeology of Albania. As far as we know, the only purported habitation site of the Late Bronze and Early Iron Age in this broader area is located at Kraps not far from the tumulus at Patos, originally reported by Ceka (1990b:142–143, pl. 4), and where a test excavation was carried out by the MRAP team (*http://classics.uc.edu/mrap/mrap01.html*, Site 038; Stocker 2009: 584–589). While the site at Kraps looks nothing like the fortified hilltop "proto-urban" centers, and may not have been permanently occupied, it lies on a hill forming part of a narrow ridge above the Gjanicë River, and separating the valley of the Gjanicë from the floodplain of the Seman River. Although material of the later Bronze Age is clearly represented at the site, evidence of the Early Iron Age, contemporary with both Patos and Lofkënd, appears to be more circumscribed (information supplied by Bejko, one of the excavators of the site at Kraps). The MRAP survey was not the only one that brought to light a dearth of Late Bronze and Early Iron Age settlements. Our own more modest, intensive survey, which was limited to the area immediately around the Lofkënd tumulus, failed to yield any clear evidence of habitation of the period (see Chapter 18).

Consequently, in the absence of clearly visible settlement sites of the Late Bronze and Early Iron Ages in the area, the Lofkënd tumulus came to represent the proverbial elephant in the room. So we decided to explore, through systematic excavation, what was the most apparent feature of the period in the landscape. And here, our tumulus, like all tumuli, emerged as a monument in every sense of the word: a structure deliberately designed and built over an extended period of time that was part of a landscape socialized through human action. It was—and in many ways still is—a focus of memory and identity (Papadopoulos 2006; see further Alcock 2002:28; Bender 1993:11; and especially Martin-McAuliffe, Chapter 20). The location of any burial ground is a conscious and carefully conceived activity on the part of the living (Parker Pearson 1999:124). This is especially true for tumuli, which, as monuments of some permanence, dominate the landscape of later generations, imposing themselves on the consciousness of later generations (cf. Bradley 1985:9). It is this visibility of the Lofkënd tumulus (see **Fig. 1.3**) that made it such an obvious target of investigation: to understand this landscape, one had to understand this tumulus.

The implications for the nature of settlement in this region and beyond may be more far-reaching and may contribute to a clearer understanding both of the transition from unwalled villages—*komai* in our Classical sources (e.g., Thucydides 1.5, 1.10, 3.94; Herodotos 5.98; Aristotle, *Poetica* 1448a.36; *Politica* 1261a.28), which are characteristic of an *ethnos*, denoting a tribal state based on kinship (Galaty 2002:109; Hammond 2000:345)—to fortified "cities," and to what extent these cities relied on, or differed from, the Hellenic model (for a recent discussion of this development in the Ioannina Basin, see Pliakou 2007, 2011). In that liminal space between the Greek city-state and the Illyrian tribes—Epirus—two Molossian cemeteries have been fully excavated together with their settlement, at Vitsa Zagoriou (Vokotopoulou 1986) and Liatovouni (Douzougli and Papadopoulos 2010; for the later Molossian cities, see Dausse 2007), while the neighboring Korçë-Kolonjë region of southeastern Albania saw new settlements created on naturally defended hilltops in the later Bronze and Early Iron Ages (Bejko, Amore, and Aliu 2006; Karaiskaj 1976; for tumuli in this area, see Aliu 1984, 1987, 1995, 1996, 2004; Andrea 1985; Bejko 2004). At both Vitsa and Liatovouni, the chronological conclusions are clear enough: both cemeteries extended through various stages of the Classical period, and it was only at the end of the fifth or during the early fourth century that both cemeteries went out of use and their settlements were abandoned.

Just on the Albanian side of the Greek-Albanian border, not far from Kakavia, the site of Tekke Melan—excavated because it was originally thought to be medieval—has yielded important Late Bronze and Early Iron Age material (Muçaj and Hobdari 2005). Only a small sondage was excavated (for stratigraphy, see Muçaj and Hobdari 2005:33, fig. 3), but the material recovered can be dated, on the basis of Aegean comparanda, from ca. 1350–1200 BC down to about 850–800 BC, dates remarkably in keeping with those of Lofkënd (for the pottery, see Muçaj and Hobdari 2005:39–57). The settlement itself is small, resembling a *kome*, not unlike Vitsa and Liatovouni, but closer in date to the latter. Despite the small scale of the site and the circumscribed area of it explored, it represents one the best-preserved insights we have from Albania of the critical period of transition between the Late Bronze and Early Iron

Age (for the Early Iron Age in Thesprotia, see Forsén 2009; Tzortzatou and Fatsiou 2009).

Our questions include the following: What brought about these changes? How did tumuli or other burial grounds function in relation to a settlement, or group of settlements, and what was their contemporary significance in the region? What were the patterns of settlement and/or partial mobility in the era before cities? (For "urbanization" in the Mediterranean between the ninth and sixth centuries BC, see Andersen et al. 1997.) And what can we learn about the city outside the historical contours of the Classical *polis*? These questions are not limited to Illyria. What is the nature of Early Iron Age settlement in many parts of Greece during the period before the Classical era, especially in Macedonia, Thrace, and even parts of Thessaly? We should not lose sight of the fact that despite the numerous burials in the cemetery of the tumuli at Vergina, for instance, little, if anything, is known of the contemporary settlement (Andronikos 1969; Bräuning and Kilian-Dirlmeier 2013; Petsas 1961–1962; Radt 1974).

In a similar vein, the recent discovery in Tumulus 10 at Apollonia of prehistoric graves yielding Bronze Age ceramics, bronze and bone pins, a bronze sword and knife, as well as a terracotta violin-shaped figurine of the Early Bronze Age, together with tombs of the Classical period, has started a new chapter in the history of the city and territory of Apollonia (Amore 2010, 2011; Damiata et al. 2007–2008). As the excavators of Apollonia Tumulus 10 have noted, the "new challenge is now to locate the settlement of this early population" (Amore, Bejko, and Dimo 2006). As in many parts of neighboring Greece in the later Bronze and especially during the Early Iron Age, cemeteries are much more conspicuous than settlements.

Supporting this interconnected set of intellectual goals was a commitment to base our investigation on a shared enterprise in terms of human resources, skills, and responsibilities, by assembling new partners from the United States and Albania to prepare future professionals in the manifold disciplines we had adopted. Thus we agreed to bring together students and specialists from a number of institutions in both countries, pair them as teams for each sector or trench in the field and on common lab enterprises, and share equipment, supplies, facilities, and experience. In particular, we sought to initiate practice and training in disciplines relatively new to Albania—though already practiced at some projects, as at Butrint and Apollonia—such as conservation, by bringing to the field and the lab not only trained conservators but student trainees from the United States and Albania. This happened successfully each summer, and in our final study season, a University of California, Los Angeles (UCLA) field school brought together a fresh set of American and Albanian students for training in surface survey, identifying and cataloguing of finds, analysis of shell and bone, and flotation and sorting of plant remains from soil samples. For the final publications, primary materials and topics such as burial customs, pottery, small finds, lithics, organic remains, and the survey itself were assigned to students from Albania and the United States, in some cases working as co-authors (Chapters 8, 9, 10, 13, 16.1, 18), thus fulfilling our goals of training the next generation of archaeologists in both countries.

METHODOLOGIES

The Lofkënd project pledged itself to full recovery of tumulus materials and analysis of them according to the highest standards possible. To this end, all material was dry-sieved, soil samples saved for flotation from every unit and grave, and archaeological contents counted, weighed, and recorded for each unit on a daily basis (Chapter 2.1). Our field and object recording system was based on that developed by the Albanian Rescue Archaeology Unit (ARAU) and adapted for use by MRAP as well as the excavation of the Apollonia necropolis, designed for data entry in FileMaker Pro (see Amore 2010:857–864, for sample forms). For a more comprehensive understanding of the origins and formation of the mound, we enlisted a soil scientist to investigate not only the matrix of the tumulus, its composition, and method of mounding, but the surrounding landscape for its soils and geomorphology, thanks to two seasons with John Foss (Chapter 16.4). Moreover, we developed a close interest in non-artifact contents of tumulus fill, particularly the many fragments of fired clay which derived from wattle-and-daub structures of unknown ancient date (Chapter 14), our best—indeed our only—evidence for local prehistoric indigenous architecture. Documentary standards aimed high: architect Max Farrar served as site surveyor, recorded and downloaded data daily from a Total Station, and drew every grave on site for scanning and digitizing into final form (whose drawings were finalized by Samantha Martin-McAuliffe; see Chapter 3). We were fortunate to have as photographers two talented artists, Rich and Anna Macdonald, who captured in color both site and objects for four seasons (2004–2007). Their images live on the Internet (*http://www.sscnet.ucla.edu/ioa/staff/papadopoulos/lofkend/photos.html*), and Richard's panorama of the site on July 4, 2006, is included in this volume (Chapter 2.6). For our final season in 2008, Ian Coyle joined our team for a study season of truly heroic work. Finally, the Lofkënd project is proud to have inaugurated low-level aerial photography in Albania in 2005, thanks to the high-flying artwork of Alket Islami and his paraglider (Chapter 2.5; Papadopoulos, Bejko, and Morris 2007:130–131, figs. 24–25; cf. Faustmann and Palmer 2005).

Bioarchaeology was a high priority, and once we enlisted Professor Lynne Schepartz as co-investigator of the Lofkënd tumulus, our project gained a fourth co-author, thanks to her constant presence on the project, interventions on site for complex identifications and extractions, and thorough analysis of each skeleton following cleaning and sorting of the human bone. Previous work by Todd Fenton on the human populations of the Kamenicë tumulus (partially excavated: Bejko, Fenton, and Foran 2006) and by Schepartz on the cemeteries of Apollonia (partially studied: Amore 2010) had already activated this discipline in Albania, but at Lofkënd we had the opportunity to recover and analyze the skeletal total of an entire tumulus (Chapter 6). Two additional initiatives sought to bring novel methods to this material, inspired by previous ventures in Albania (Bejko, Fenton, and Foran 2006). We invited the Wellcome Laboratory at Oxford University to conduct a DNA analysis of human bone in our first season (Laura Menez in Papadopoulos et al. 2007; Appendix 1, following Chapter 6). Later, Brian Damiata joined our team to sample human bone, teeth, and charcoal for AMS (accelerator mass spectrometer) analysis at the Keck Carbon Cycle AMS Laboratory, University of California, Irvine, a co-investigation with its director, John Southon (Damiata et al. 2007–2008; Chapter 4). The latter enterprise proved more successful than the former in extracting and identifying adequate samples, and produced the first absolute dates not just for graves from Lofkënd but for several from Apollonia as well. The same human bone and teeth samples also provided some of the earliest strontium isotope figures for human and animal populations of ancient Albania (Chapter 7), following those initiated at Apollonia (Stallo et al. 2010).

Organic remains were targeted from the onset, although a cemetery of largely inhumed bodies (only two cremations were found) with no traces of fire did not promise high preservation of seeds, plants, or wood. Nevertheless, over 250 kg of soil, representing just under one-fifth of the volume of the tumulus fill, was collected and floated, yielding primarily modern plant material (Chapter 16.1). Charcoal samples were tiny and often disintegrated, but allowed for the recognition of two primary species, oak and maple, in antiquity (Chapter 16.1), and 18 samples from graves provided absolute radiocarbon dates (Chapter 4.2). Animal bones were not found in great numbers, beyond single skeletal remains from common domestic species, largely from tumulus fill and possibly (in the case of rodents) intrusive, except for three animal (sheep) burials evidently deposited as part of a ritual in connection with the modern graves (Chapters 3.2 and 16.1). Shell remains were also collected systematically in every context, largely those of modern land gastropods attracted to dark, cool cavities after antiquity (Chapter 16.2). Although results of these collection and analysis strategies gave minimal evidence for the kind of nuanced environmental and human history possible at waterlogged sites such as Sovjan (Allen 2002, 2005), for instance, or at settlements with multiple fire-affected phases and well-preserved hearths, floors, and rubbish pits, this was a new element in the formation of a burial mound not previously quantified. In short, our team was pleased to establish all of these as excavation protocols, and introduce several specialists from Greece and the United States to the material and project in novel case studies.

An unexpected bonus of our conservation team was a series of materials analysis studies initiated by Vanessa Muros, and co-conducted with faculty and students of the Cotsen Institute of Archaeology at UCLA. In the first instance, objects such as glass beads received her special attention and analysis (Chapter 10), but the metal finds attracted some of the most comprehensive and advanced analytical techniques (Chapter 11), analyses of which will continue in the future. A portable X-ray fluorescence device allowed us to identify properly certain materials for the first time (most notably, the "gold" earrings, **10/11–12**, turned out to be made of electrum, but we also identified elements unsuspected in other objects), which initiated further material studies. A total of 162 metal artifacts were analyzed, beginning with optical microscopy on site, enhanced by pXRF and X-ray diffraction analysis, and 23 copper alloy objects were examined for their metallography. Finally, the detection of trace elements via a portable X-ray fluorescence device (pXRF) led to a new minimally destructive method of sampling metals for ICPMS (inductively coupled plasma mass spectrometry) lead isotope analysis, by soaking objects in EDTA (ethylenediaminetetraacetic) acid and extracting samples from solution (Muros, Lehner, and Bardho 2009). Perhaps most valuable of all, some 35 textile pseudomorphs left in corrosion products primarily on iron jewelry were detected, photographed, and analyzed for likely weaves of lost fabrics that once enveloped the bodies of the dead (Chapter 12).

From our first season, we longed to understand and visualize more of the long-term history of the area around the tumulus, in particular the Gjanicë River valley below to the west, its slopes below the tumulus, and the surrounding hills. Several places (Mashkullorë, Gjinoqare, and Mezhdat e Kuqe; see Chapter 18) were long known as ancient sites, and as is common, neighbors, visitors, and team members would report discoveries or spot artifacts in the environs. Thus we were pleased to expand our project with a systematic survey of several tracts on the slopes and in the valley, and to invite Jamie Aprile to conduct it (Chapter 18), as well as to analyze the lithics from the tumulus (Chapter 13) related to those from the survey.

RECONSTRUCTION AND AFTERLIFE: WHOSE MOUND, WHOSE MEMORY?

From the onset of the project, if not from our first visit to the site, the commanding profile of the tumulus, visible from many points in the landscape, demanded our respect and deserved a commitment to its original shape. To that end, we not only remounded it every season by backfilling it up to original ground level (Chapter 2.1), but undertook a final comprehensive, and rewarding, exercise in rebuilding the tumulus out of its own materials. Mobilizing the same workmen who had dismantled its layers, now enlisted as mound-builders, we organized and supervised the manufacture of some 2,000 mud bricks out of the clay matrix removed from the mound, mixed with water and straw by human hand and foot (Papadopoulos, Bejko, and Morris 2008, 2009; Chapter 22). This allowed us to complete a full restoration of an ancient mound, within its landscape and from its own materials, and return to present and future communities a major symbol of their past and their environs. Its status as a monument called for more, however, and we were fortunate to persuade Samantha Martin-McAuliffe to participate in our final two seasons as field archaeologist and draftsman, and most of all, to give this mound the keen and imaginative eye of an architectural historian (Martin-McAuliffe forthcoming; Chapter 20).

A different type of reconstruction is offered by one of the co-directors in Chapter 21. In reviewing all the burials from the tumulus, two stood out not only as non-normative, but totally unique at Lofkënd and, as far as we know, at any tumulus in Albania. Ironically, one of these was among the earliest, and probably *the* earliest burial (Tomb I) in the mound, and one was among the latest (Tomb LXXIV), although not the final burial at the site. The idiosyncratic circumstances of both tombs permitted the reconstruction of a formative event that led to the establishment of the mound, as well as a reconstruction of an event that led to its final demise. Whereas the circumstances of Tomb I provide a glimpse of the beginning of the Lofkënd tumulus, and thus of a community seeking to assert its own identity within a landscape, the case of Tomb LXXIV tells a very different story, one of likely violence, and the simultaneous loss of three male members of the community, whose bodies were treated in a manner so enigmatic that it was initially impossible to date the tomb as ancient or modern.

Finally, last but not least, modern techniques allow novel ways of reimagining, and reviewing, an ancient site: thus three-dimensional imaging was initiated in 2005 by the Experiential Technologies Center at UCLA (C. Johanson and I. Zaharovitz in Papadopoulos, Bejko, and Morris 2007:135–138; Chapter 19; *http://www.sscnet.ucla.edu/ioa/staff/papadopoulos/*

lofkend/mapping.html). While the publication of our results in this volume is fixed in time, the three-dimensional model is a living entity, one that can be amended and added to by and for future generations of scholars and students.

Any archaeological project active in the twenty-first century recognizes postmodern priorities beyond modern scientific responsibilities. In Albania, cultural heritage display for this type of site was inaugurated in an exemplary way at the tumulus of Kamenicë, in a public opening attended by members of the Lofkënd team that inspired the future of other projects (*http://www.kamenicatumulus.org/*). But how can and does a modern team practice the kind of "reflexive archaeology" (Hodder 2000) called for in this millennium?

At Lofkënd, our engagement with the mound did not end with its reburial, as we sought to integrate its history into the region via a cultural biography of sorts (Epilogue). This called for synchronizing its indices of cultural activity with Corinthian colonization (Appendix 2, following Chapter 9), early modern reuse for minority burials (Chapter 3.2), and more recent visitation by twentieth-century soldiers (Appendix 3, following Chapter 10). Such a biography would be incomplete without the active life of the mound in the human imagination, a particularly rich dimension of archaeology in the Balkans. For the local villagers, the mound held the remains of those who were always just beyond the grasp of memory, the fallen in a war in the distant past: the First World War, or one of the Balkan Wars, for whose recovery modern fiction provides the most vivid text (Kadare 1970; Papadopoulos 2006: 49). A more historicizing eye might consider its competing identities for the multicultural Balkans (as in Andric 1945; see Chapter 3.2), while a longer-term view puts it more simply: "Albania was and is the homeland of heroes" (Hammond 1974:132; cf. Hammond 1976; Stocker and Davis 2006). But every hero needs an antihero, or a victim: to archaeologists attuned to cultural history and ethnography, certain aspects of burials in the Lofkënd tumulus called for comparison to the Albanian *Kanun*, or feudal code of honor and revenge, and a consideration, however speculative, of its prehistory (Chapter 21). For each of those buried in the tumulus at Lofkënd, followed by those who visited their graves or found their resting place a quiet haven for their own dead, and finally those who brought them to modern light, contributed to a continuing history of place (see Epilogue).

Chapter 1.2
Surface Collection of Material on the Lofkënd Tumulus

John K. Papadopoulos, Seth Pevnick, and Esmeralda Agolli

Prior to the excavation of the tumulus, two rounds of surface collection were conducted on and immediately around the tumulus. The first of these was conducted on July 10, 2003, by Bejko, Korkuti, Morris, and Papadopoulos. All of the material collected in 2003 was ancient and included ceramics, lithics, and daub; clearly modern material (plastic, glass, etc.) was intentionally not collected, and the primary aim was to gauge some idea of the chronological range of the tumulus in the premodern past. A good deal of human skeletal fragments, much of it sun-bleached, was especially noted off the exposed south scarp of the tumulus, and this was the primary area of erosion of the tumulus (**Fig. 1.8**).

With the initiation of excavations in 2004, a full surface collection of all materials, ancient and modern, was undertaken on June 21–22. In addition to the ancient pottery, lithics, and daub, a variety of fragments of modern glass, iron, plastic, foam, spent gun cartridges, and even cloth, was collected. This material provided a classic case study of intensive site surface collection prior to the full-scale excavation of the site. The material is presented below in a series of separate tables, one dealing with the ancient pottery and daub, together with modern roof tile, one with the lithics, and another with the modern material. The quantity and range of the material was interesting, given the fact that the tumulus was located some distance from the nearest villages, although it is only a short distance from the nearby modern cemetery. Much of the modern surface material was likely generated by the family of Baki Ymeri (from the nearby village of Gjinoqara), who, during the communist and postcommunist eras, farmed the land immediately to the east, northeast, and southeast of the tumulus. The prominence of the tumulus in the landscape, the splendid views it commands of the neighboring countryside, and the fact that the mound itself was difficult to plow or plant, made it an attractive place to rest during the farming seasons.

In the precommunist period, the land on which the tumulus was located was near a local Tekke, identified in the intensive survey as Site S008 (Bektashi Teqe and Tyrbe), immediately to the southwest of our mound (Chapter 18, **Fig. 18.1**). Part of the modern material on the surface of the tumulus may thus have derived from another local landowner, whose property lies immediately to the east of the tumulus, Rrapi Malaj—the self-styled local representative of the now defunct Tekke—who had planted an especially resilient pear tree on the south side of the tumulus sometime before the excavations were initiated, and who attempted to plant several more trees on the eastern edge of the mound in 2004–2005 (one of which damaged Tomb LVII: Chapter 3). In modern Albania, both families could lay claim to the farming land around the tumulus, though in reality, one family largely farmed only to the west of the mound, the other to the east.

Much of the collected ceramic material was ancient—primarily Late Bronze and Early Iron Age—including the following six inventoried pieces (all pottery is handmade unless otherwise noted) (**Fig. 1.13**): one base fragment (P001), one rim fragment (P004), and two handle fragments of light fabric (P005, P006); one coarseware rim (P003); and one wheelmade base fragment, perhaps an ancient Greek import (P002) (fuller descriptions of the pottery and the ceramic typology are provided in Chapter 9). The noninventoried pieces also included a considerable amount of prehistoric material, much of which can now be easily classified into the four prehistoric fabric groups that have been established by the closer study of the excavated tumulus pottery, including the complete vessels from tombs and the more fragmentary, but considerably more numerous sherd material from the tumulus fill (**Figs. 1.14a–b**). In addition, many pieces of daub, also familiar from the tumulus and some grave fills, were found, together with a vitrified sherd, a modern tile fragment, and two pieces of light fabric not demonstrably ancient. All of these pieces are presented below in tabular form (**Table 1.1**), both quantitatively and by weight, in order to establish the relatively large amount of prehistoric material visible on the surface of an unexcavated burial mound. Diagnostic fragments are noted in the final (right) column, as are joining fragments; all other pieces are body fragments.

The lithic material collected in 2003, amounting to five inventoried pieces, included examples that could be assigned plausibly to the Paleolithic and Mesolithic periods, as well as the Neolithic or Bronze Age. The presence of such early lithics on a burial mound that yielded material largely of the Late Bronze and Early Iron Ages initially came as something of a surprise, until it was subsequently considered in the light of the Middle and Early Upper Paleolithic and Mesolithic finds from the region of Apollonia and Fier, particularly at the Kryegjatë B open-air site (see Papadopoulos, Bejko, and Morris 2007:127–128, 134, fig. 29; for Kryegjatë, see Runnels et al. 2004; for earlier overviews of the Paleolithic and Mesolithic in Albania, see Korkuti and Petruso 1993; Korkuti 1995b, 1998:21–23, 40–41), and elsewhere in the MRAP survey area. The quantity and chronological range of these tools cannot be easily accounted for, although the subsequent survey of the Lofkënd region by Aprile (Chapter 18) brought to light a substantial Paleolithic open-air hunting station in the vicinity of Ngrançija, with isolated nodes of activity occurring near the tumulus, on the lower ridges, and near the north bank of the Gjaniçë River. The lithics from the tumulus surface are presented in **Table 1.2** and in **Figure 1.15** (see also Papadopoulos, Bejko, and Morris 2007:127–128, 134, fig. 29; all pieces are chert, unless otherwise noted), and a fuller analysis is provided in Chapter 13.

All of the modern material was collected on the first two days of fieldwork in the 2004 season and includes an array of iron, glass, plastic, cloth, and other materials (**Table 1.3**, **Figs. 1.16a–b**). Of these, only the spent gun cartridges and fragments of iron sheet were also encountered in the topsoil units, and were analyzed along with other small finds in Chapter 10, with the spent gun cartridges and bullets in Appendix 3, following Chapter 10.

Table 1.1 Ancient materials collected on the Lofkënd tumulus (pottery, daub, and roof tiles)

Collection Date	Fabric	Number of fragments	Weight (g)	Notes
July 10, 2003	Vitrified	1	17.2 g	cf. P139
	Wheelmade	1	4.5 g	
	Fine dark	3	3.7 g	incl. 2 joining from rim
	Fine light	10	51.2 g	incl. 1 neck/shoulder, 4 handles, 1 rim
	Semi-coarse	12	94.6 g	incl. 1 handle
	Coarse	9	39.3 g	incl. 4 joining
	Daub	6	31.2 g	
June 21–22, 2004	Modern tile	1	82.5 g	
	Wheelmade	2	5.5 g	
	Modern light(?)	2	13.6 g	
	Fine dark	3	16.3 g	incl. 1 rim
	Fine light	10	38.8 g	incl. 1 handle, 1 rim, 1 base
	Semi-coarse	15	85.5 g	incl. 2 joining, 1 base
	Coarse	8	46.9 g	incl. 4 joining
	Daub	11	18.6 g	

Table 1.2 Ancient lithics collected on the Lofkënd tumulus (2003)

Collection Date	Inventory Number	Dimension	Weight (g)	Notes (by Korkuti and Runnels)
July 10, 2003	**14/3**	L: 0.025 W: 0.014	1.6 g	Neolithic or Early Bronze Age
	14/37	L: 0.033 W: 0.024	7.4 g	Cannot be precisely dated
	14/89	L: 0.020 W: 0.047	7.8 g	Middle Paleolithic, broken Levallois flake
	14/82	L: 0.023 W: 0.020	3.6 g	Possibly Upper Paleolithic
	14/59	L: 0.025 W: 0.027	12.8 g	Perhaps Mesolithic

Table 1.3 Modern material (other than ceramic) collected on the Lofkënd tumulus prior to excavation, June 21–22, 2004

Collection Date	Fabric	Number of fragments	Weight (g)	Notes
June 21, 2004	Iron	2	23.7 g	1 fragment heavy (object unidentified) 1 fragment from iron attachment
	Gun cartridges (spent)	3	29.5 g	Cf. catalogued examples from topsoil units
	Glass bottle fragments	3	6.7 g	Green bottle glass, wall fragments
	Glass jar fragment	1	16.9 g	Clear glass rim fragment
	Plastic tubing	2	20.7 g	Both fragments green
	Plastic bottle cap	1	1.5 g	White
	Mattress foam	1	5.0 g	Yellow
	Cloth	1	9.4 g	White, with floral pattern
June 22, 2004	Gun cartridges (spent)	2	23.1 g	Cf. catalogued examples from topsoil units
	Glass bottle fragments	21	136.1 g	Green bottle glass, including 1 complete rim and neck and 2 base fragments
	Glass jar fragments	3	16.5 g	Clear glass, including 1 base fragment
	Plastic tubing	1	1.5 g	Off-white
	Cloth	2	7.8 g	Two strips, both red

Chapter 2

The Excavation of the Tumulus

John K. Papadopoulos, Lorenc Bejko, and Sarah P. Morris
With a contribution by Richard MacDonald

Chapter 2.1
The Excavation and Stratigraphy of the Tumulus

John K. Papadopoulos, Lorenc Bejko, and Sarah P. Morris

The Excavation of the Tumulus

Excavations were initiated at Lofkënd on June 20, 2004. The first task was to prepare a contour map of the tumulus and its immediate surrounds. Although Albanian Army maps of the area were available, labeled in both Albanian and Russian, their scale permitted limited resolution of the less clear topographical features, and locating the tumulus on these military maps was not straightforward even for those who were familiar with the landscape. A classic case in point is the 1942 Albanian Army map labeled in Russian (**Fig. 2.1**) at 1:50,000. It shows the Vjosë River running diagonally through the central portion of the map, with the Gjanicë River running more or less parallel to the north. The immediate region around Lofkënd is shown in the upper right corner. Although the village of Lofkënd to the east of tumulus that bears its name is clearly indicated, as are the villages of Ngrançija and Gjinoqara, with their various offshoots, the largest village in the immediate region at the time was Visokë. A more detailed Albanian Army map of 1942 (updated in 1972) at 1:25,000 provides rather more detail. The 1942 maps show a landscape crossed by roadways no longer in use, and a limited number of villages, significantly fewer and smaller than today. Although the course of rivers and streams did not change in the second half of the twentieth century, some other bodies of water, like the small dams now dotting the landscape, had not yet been constructed or were of rather different form. In the end, a 1:10,000 Albanian Army map of 1986 proved the most useful, especially for the "big picture" of the region around our site. It formed the base map for John Foss's study of regional soils (**Fig. 16.18**), as well as Jamie Aprile's extensive survey of the area around the tumulus (**Fig. 18.1**), and Samantha Martin-McAuliffe's study of Lofkënd as a cultivated place (**Fig. 20.1**). However useful these maps proved to be, they simply did not provide the desired resolution for the purposes of excavating the burial mound.

On the first day of the first season of fieldwork in 2004, a series of concrete fixed points were established by the project surveyor, Max Farrar, as follows:

ST01:	E1000.00, N1000.00, Z100.00
ST02:	E1000.00, N1058.90, Z107.74
ST03:	E930.28, N1023.39, Z96.90 (lost)
ST04:	E669.81, N1155.15, Z88.99
ST05:	Vacant
ST06:	Vacant
ST07a:	E957.08, N975.03, ZZ89.64 (lost)
ST07b:	E953.79, N1002.84, Z99.26

Given the fact that the fields surrounding the tumulus to the north, south, and west were plowed and harvested, from 2004 onward, with heavy agricultural machinery (see **Fig. 20.7a**), the decision was taken to establish the datum point (ST01) at a place on the landscape where it was unlikely to be damaged. Consequently, ST01 was established immediately to the south and slightly east of the tumulus, across the small road leading up to the site and villages to the north, in a fenced off area with a small water hole. The ST02, on the same longitude and 58.9 m to the north, proved to be the most used point for daily measurements during the excavations. At this locale, the project Total Station was set up daily for all topographical measurements and for plotting all small finds. Two points were established to the south and west of the mound (ST03 and ST07a), with an additional point (ST07b) farther south, on a lower terrace; of these, ST03 and ST07b were lost due to heavy machinery accessing the fields to the west at harvest time in mid-summer. Just over 330 m to the west of the tumulus, and on a small spur unlikely to be plowed, we established ST04; two additional points (ST05 and ST06) were set aside and labeled but never established. The surviving points allowed us to triangulate all of the lost points at any time, and re-establish, or relocate, them as necessary. From these various points, Max Farrar was able to construct a 2D and 3D contour plan of the immediate area surrounding the tumulus, over which a 100×100 m grid was laid (**Fig. 2.2**). By zooming in on the same digital map, one had access to the more detailed plan of the tumulus, which was itself overlain by a 1-m^2 grid (**Fig. 2.3**). The UCLA Experiential Technologies Center team acquired GPS coordinates at each marker, and these points were used to transform the digital model from the arbitrary coordinate system to a projected geographical coordinate system. The Lofkënd tumulus is located in UTM Zone 34T, Easting 391927, Northing 4500348.

Once the mapping points were established, and following the preliminary mapping of the area, as well as the collection of surface materials on the tumulus (Chapter 1.2), the tumulus was divided into four sectors, or "trenches," labeled 1 to 4, separated by baulks of 0.5 m. The four trenches/sectors were numbered in a clockwise direction, beginning with Sector 1 at the northeast, 2 in the southeast, 3 in the southwest, and 4 in the northwest. At the beginning of the 2006 season, Sectors 2 and 3 were combined into one, labeled Sector 2/3. The excavation of each sector was co-supervised by American and Albanian graduate students. All field notes were recorded in English, and we used the ARAU (Albanian Rescue Archaeology Unit) database, which was a specially formatted FileMaker Pro database labeled in Albanian and used for many projects in the country. The field data and feature sheets were in both Albanian and English (as were the bulk find sheets where we recorded all the non-inventoried pottery and other small finds).

In 2004 we employed four workmen to assist with the excavations, and by 2007 we employed 11 (all participants are named in the Foreword). A variety of picks, trowels, and scrapers were used in the excavation of the tumulus fill, which was often dry and very hard-packed, especially when exposed to sunlight. All burials, however, were excavated almost exclusively with wooden tools to minimize damage to the human remains (**Fig. 2.4**). Brushes, thin "souvlaki" sticks (wooden skewers), and a range of dental tools were used for more detailed cleaning. All material was dry-sieved using diverse screen-size meshes, and samples of each soil unit were water-sieved. Samples collected for flotation and subsequent paleoethnobotanical analysis were approximately 8–10 liters in volume, except for smaller discrete samples, such as the contents of particular pots deposited in graves (Chapter 16.1). Samples were also taken from each grave and all other identified features in the mound, as well as from each stratigraphic layer. All observed burned patches or lenses, concentrations of ash, and areas with visible charcoal flecks were sampled. More substantial pieces of charcoal or burned matter were gathered in the field for ^{14}C dating following the protocols laid out in Chapter 4 (see also Chapter 16.1 for identifications of the wood charcoal remains). As anticipated, burned plant remains were negligible in a cemetery where inhumation predominated. Of the samples water-sieved in the first season, only a solitary carbonized seed was recovered. Subsequent water-sieving of all the samples from 2005 to 2008 did not add to the quantity of flora remains (Chapter 16.1). All small finds from the tumulus fill, including scraps retrieved from dry-sieving, were examined, recorded, and stored. Metal finds, pottery fragments preserving painted or plastic decoration, or those in need of mending or conservation, were treated directly in the conservation laboratory (Chapter 5). All other objects of clay or stone were washed, sorted, and inventoried either as individual objects or as bulk finds, with all non-inventoried material counted and weighed for each excavation unit. The material retrieved was entered in the project database, with copies residing at UCLA and at the Institute of Archaeology at Tirana.

Human bone was lifted in the field and packed in absorbent paper first and then aluminum foil to be subsequently cleaned and sorted by the physical anthropologists (Chapter 6); many of the tombs were lifted with the assistance of physical anthropologists, with each bone identified and labeled—as far as it was possible, given their preservation—in the field. In addition, skeletal material was examined in the field and laboratory for possible signs of postmortem treatment (in particular, crania). Human skeletal remains from 24 burials excavated during the 2004 season were viewed under natural light by Michelle Bonogofsky using an 8×-handheld lens to examine the human bone for evidence of cutting, defleshing, and sanding, as well as postmortem applications of plastic modeling substances such as those found on skulls in the Neolithic Near East (Bonogofsky 2001a, 2001b, 2003, 2005). This material was, in many cases, highly fragmented, with soil adhering to all exposed surfaces, and extensive root action in the latest burials of the tumulus that were at the highest levels of the mound had severely eroded the cortical bone. No clear evidence of postmortem modeling on human skeletal remains was found, and this was also the case for all the skeletons recovered from the tumulus by the end of the 2007 season. In 2004 a pilot project tested for possible consanguinity or relationship among individuals buried in the tumulus using mitochondrial and, if possible, nuclear DNA, the limited results of which are described in Appendix 1 (following Chapter 6).

In addition to the human remains, some animal bone was recovered, but this was relatively infrequent (fully analyzed in Chapter 16.1). In contrast, shells of land snails were ubiquitous throughout virtually every level of the tumulus and often in the graves. The surviving shell represents several common land molluscs often attracted to human remains and/or cool, damp underground environments (Chapter 16.2).

A total of 28 burials were uncovered during the first season in the uppermost meter or so of the tumulus. A view of the tumulus at the very beginning of excavations and at the end of the 2004 season is shown in **Figure 2.5a–b**. With the continuation of excavations in 2005, the number of cleared graves increased to 62, and an additional four burials, some of which extended into one or other of the baulks separating the sectors/trenches, were uncovered but not excavated by the conclusion of the second season. The characteristic oval shape of the tumulus was clear, especially in the upper levels, throughout the 2004 and 2005 seasons (see **Fig. 2.3**), although this was to change by the end of the 2005 season and in the course of the two final seasons of excavation in 2006 and 2007 (see below). More burials were excavated in 2006, and by the conclusion of the 2007 season—at which time the tumulus was completely cleared—a total of 100 graves had been excavated, many of them multiple burials containing two, three, or sometimes more individuals. The human remains are presented in Chapter 6 and the results of stable isotope analyses of the human bone in Chapter 7, while the prehistoric burial customs are fully discussed in Chapter 8 (the burial customs of the modern burials are treated in Chapter 3.2).

The excavated burials belong to two primary phases. A modern phase yielded ^{14}C dates all clustering around 1800, although with a high range of ± 100 years at the 2-σ spread (Chapter 4). This modern phase was largely confined to the northeastern portion of the mound (**Fig. 3.287**; see also Chapter 19) and was characterized by inhumations of infants and several adults, all oriented east to west, with the crania to the west, facing east, as well as a few animal inhumations (Chapter 3.2). The existence of the modern burials was not anticipated because of the presence of a modern Muslim cemetery located on a prominent bluff immediately below the tumulus and associated with the nearby village of Ngrançija (**Fig. 2.6**). This later Muslim cemetery was established in the twentieth century, whereas the modern burials in the tumulus are earlier, representing a Christian community that lived in the area. The modern burials numbered 15 in all, including the two animal burials. By far the largest number of tombs (85 in all) was prehistoric. Thanks to a robust program of radiocarbon dates based largely on human bone collagen, it was possible to date these from ca. 1400 to 800 BC (Chapter 4). As is fully discussed in Chapter 4, parts 1 through 3, this chronology is about 200 to 300 years earlier than that based on the conventional chronology of Albanian archaeology.

In the course of the four years of the excavation of the tumulus (2004–2007), individual tombs were numbered, as they were found, with Arabic numerals. Once we began the study of the material at the conclusion of excavations, at the end of the 2007 season and throughout the course of the 2008 campaign, it was necessary to renumber all the graves chronologically, beginning with the earliest. In Chapter 3 the tombs are indeed presented chronologically, beginning with the earliest. As all the records and the database were kept according to the original Arabic numbering, the process of renumbering all the tombs with Arabic numerals would have created havoc in the entire system. Consequently, we decided to renumber all the tombs with Roman numerals, but keeping the original grave number in brackets for ease of cross-reference. Hence, the earliest burial was Tomb I (64) and the latest prehistoric burial was Tomb LXXXV (10); the earliest of the modern burials was LXXXVI (22) and the latest C (48). In hindsight, numbering the graves as they were encountered in the ground with Roman numerals, and renumbering them at the conclusion of excavations with Arabic numerals would have been far more convenient. In some of the schematic plans of the burials, such as **Figures 4.1** and **4.2**, we retained the original Arabic numerals, as the more cumbersome Roman numerals would have been more difficult to align and read. For easier cross-reference, a concordance of all tomb numbers follows (see Chapter 2.2).

The baulks separating the excavation sectors, which had been kept largely intact to provide sections through the mound (see below), stayed in place until the very beginning of the final season of excavation in 2007. Several parts of the baulks had been explored during earlier excavation campaigns in order to expose more fully individual burials or features, but these had been photographed, with sections drawn, prior to their excavation. **Figure 2.7** shows part of the baulk separating Sectors 1 and 4 at the end of the 2006 season. The various incursions into this baulk include the cutting immediately above Arben Malaj's head for the excavation of Tomb LXXXIV (2); the large central tunnel was necessary to clear Tomb XLIV (65), while the smaller tunnel to the left (south) filled with stones exposed Tomb LXIII (35); another small cutting to the right (north) of Malaj was for Tomb XVII (72), at an even lower level. Due to a variety of factors—including the location of numerous graves, particularly the central grave (Tomb I [64])—and to clarify a number of stratigraphic features, it was necessary to remove the baulks altogether. The fact that a 0.50-m baulk did not collapse with the excavators standing on top of it is testimony to the compact nature of the fill, which is discussed more fully below. Once the baulks were removed, only the bedrock base of the tumulus was visible (see **Figs. 22.3–22.4**). We had originally decided to leave intact a 0.50-m square pillar of unexcavated earth at the highest point of the tumulus and near its center as a remnant of one section of the stratigraphic sequence, but the location of the central grave and fear of collapse of the resulting pillar made us abandon this idea. In order to be able to record the position of all finds encountered in the baulks, it was decided to give each baulk a sector number: the one separating Sectors 1 and 4 was labeled Sector/Trench 5, that between 1 and 2 Sector/Trench 6, between 2 and 3 Sector/Trench 7, and between 3 and 4 Sector/Trench 8.

The final shape of the tumulus in plan is indicated in **Figure 2.8**. By the end of the 2004 season the oval shape was clear, measuring approximately 13.5 m in length north-south and 8 m wide east-west. By the end of the 2005 campaign, the tumulus edge was wider in plan, about 11.5 m east-west, but instead of the more characteristic oval form, it had assumed a more tongue-like plan, with the northern end of the mound imperceptibly merging with the flatter ground to the north, a feature already visible during prior excavations (see **Fig. 1.4**). By the end of the 2007 season, the bedrock platform on which the earthen tumulus was heaped lay fully exposed (**Fig. 2.9**; see also **Figs. 1.9, 22.3a–b, 22.4**), at which point it resembled more of a promontory or prominent tongue of land, with the ground sloping away sharply to the south, east, and for much of the western side, but with the bedrock merging with the ground to the north. This is best seen in **Figure 2.10** (see also **Fig. 22.4**). The exposed bedrock platform had a width of almost 20 m east-west and an approximate length of about 25 m.

In the course of the excavations, the bedrock was quickly reached in the northern portion of the tumulus, as well as down the slope to the northeast and northwest. Although the scarp of the mound rose sharply to the southwest, here too bedrock was expeditiously reached, and the earliest tombs were exposed at a relatively high level. The exception was the southeast sector, and here both tumulus fill and tombs continued to a considerable depth, and in an area well off what was to be the later mound. In order to ensure that no tombs were overlooked, we decided to do two things. First of all, we did what many excavators of Late Bronze and Early Iron Age cemeteries do in the Aegean (e.g., Popham, Sackett, and Themelis 1979–1980:102, note 2; Brouskari 1980:19; Papadopoulos and Smithson 2002:157, fig. 5; Papadopoulos 2005:21): we dug well into the bedrock, often to the excavators' frustration and dismay, in order to ensure that no tombs were overlooked. Happily, the bedrock was easily dug in most cases, although, as John Foss expands (in Chapter 16.4), there were some strongly cemented lithic beds that occurred in the bedrock on which the mound was placed, generally less than 0.5 m in thickness, and with many thinner bands (>1 cm) of calcium carbonate, which provided a straightforward method of determining when the excavators encountered undisturbed bedrock. According to the geological map (a large hand-colored map produced by the Albanian geological service), the site is located in Miocene sediments that are characterized as clays, aleurolites (silt stones), and sandstones. Within this larger matrix, the bedrock of the tumulus occurred in sandstone that was loosely cemented by carbonates (for further detail, see Chapter 16.4). The excavation of the bedrock was straightforward enough, and the final form of the bedrock platform, the very top of which we, in places, shaved flat, was very much the product of our digging into the rock. In some places we excavated almost a meter, and in others less, but we did not stop until the soft stone was completely sterile. The idiosyncratic surface of the bedrock platform seen in **Figure 2.9** (cf. also **Fig. 2.10**) includes a lip of bedrock to the right (southwest) that shows more or less the original level of the bedrock, although even here we dug into the rock to be certain that there were no hidden burials.

The second thing we did was to lay out and excavate, to bedrock, a 10×4-m trench to the southeast of the tumulus, in the area of the greatest concentration of early tombs, on flat ground, to ensure that there was no significant extension of graves in that direction; this was designated Trench 9 (**Fig. 2.11a–b**). We began by excavating the northernmost 5 m of this trench and quickly reached bedrock and determined that there was no extension of tombs. To be absolutely certain, we excavated the remainder of the trench to bedrock and into bedrock, and it was clear that this area lay well beyond that used for burials (the scarp seen in **Fig. 2.11b** and on the left on **Fig. 2.11a** represents bedrock that was dug into). Only two stratigraphic units were cleared in Trench 9: topsoil (here SU 0587) and bedrock (SU 0589).

Apart from the burials, the only significant feature we encountered was a line of stones, intentionally placed on tumulus fill (**Fig. 2.12**; the wall is also just visible in **Fig. 2.8** and in several of the aerial photographs [see below]). As seen in Chapter 17, many Illyrian tumuli had a ring of stones defining the outer limit of the mound (e.g., Barç and Çinamak), often with the burial—or burials—lined and covered with stone (see **Figs. 17.2–17.3**). In the case of the larger tumulus at Kamenicë, a large stone ring, together with multiple smaller stone rings, and more substantial constructions of stone, were encountered (**Fig. 17.6**; cf. some of the stone tumuli of Shtoj: Koka 2012:23–52). In other parts of the Balkans, burial mounds with similar stone constructions were common, such as the tumuli of Pogoni in Epirus (e.g., Andreou 1981, 1982a, 1982b, 1983, 1985, 1997; Andreou and Andreou 1987) or even the mounds largely constructed of dry stone, like those at the base of Biokovo Mountain in Croatia (e.g., Olujić 2011; for stone and earth tumuli in northern Italy, see Borga and Càssola Guida 2007). In this respect, the Lofkënd tumulus more closely resembled the Patos mound excavated by Korkuti, in that it was largely made of earth.

Wall 1 at Lofkënd was not substantial. Comprising some 25 stones, the largest section of the wall ran diagonally from west-southwest to east-northeast in Sector/Trench 4, continuing into Sector/Trench 1 (see **Figs. 2.8** and **2.12a**); this stretch of Wall 1 breaks off in Sector/Trench 1, with a few stones farther to the east, closer to the tumulus edge, defining what appeared to be—or may have been—the continuation of the same feature. The stones at the point where Wall 1 breaks off in the central portion of Sector/Trench 1 begins to define a slight curve, and if the stones farther to the east belong to the same feature—as they appear to do, though this was not absolutely certain—then the "wall" was clearly curved. In order to draw the wall in section, Farrar established two string lines, one at 355.0 m above sea level (ASL), the other at 345.5 m ASL; the highest point of the wall was encountered at 355.25 m ASL, the lowest at 354.4 m ASL. The stones were, for the most part, medium to small in size; the largest stone, in the central stretch of the wall encountered in Sector/Trench 4, measured slightly more than a meter in length, a width of 0.45 m, and a depth of 0.28 m.

What, precisely, Wall 1 represents is difficult to glean. It was not a foundational event. Far from it, Wall 1 was constructed at a late stage of the tumulus. One of the burials that lay directly above the central portion of the wall in the baulk separating Sector/Trench 4 and 1 was Tomb LXXI (28), which had a calibrated AMS ^{14}C date of 852 ± 44 BC and belonged to Phase Va (see Chapter 4), the penultimate of the prehistoric phases. Other tombs that were stratigraphically related with the wall were even later and include Tombs LXXXII (9) and LXXXIV (2), both of the final phase of the prehistoric burials, Phase Vb; the wall was overlain by at least one modern burial, Tomb XC (14). None of the prehistoric burials were located clearly below Wall 1. What was, however, originally considered as the likely robbed-out section of the wall in Sector/Trench 1 yielded Tomb XXXVIII (79)—the double burial of an inhumed adult male and a cremation, probably adult female—of Phase III (eleventh–tenth centuries BC). The highest point of this tomb was first encountered at 354.39 m ASL, 1 cm below the lowest point of the wall encountered elsewhere, while the lowest point of the grave was at 354.23 m ASL. Had the portion of wall above Tomb XXXVIII been robbed out, it surely would have caused some damage to the earlier tomb, which was clearly not the case. It seems most likely, therefore, that the wall did not continue into the central portion of Sector/Trench 1, as corroborated by the stratigraphy immediately above Tomb XXXVIII (79).

One of the most interesting features of Wall 1 was that it was constructed only to a single course in height, although at one or two points a smaller stone overlaid one of the larger stones (**Fig. 2.12**). The fact that it was located in one of the highest and most stable parts of the tumulus made it an unlikely retention wall; such a retention wall would have been much more useful at the south, east, or west scarps of the mound, not in the central northern section. The fact that prehistoric (and modern) tombs were located to the north and south of the wall prompted us to look more closely at all differences between the tombs on either side of the feature. No clear patterning was apparent, either chronologically or in terms of other aspects of mortuary customs (for which, see further, Chapter 8). Even if one disregards Tombs XXVII (82) and XXVIII (77) of Phase II, together with Tomb LI (Phase III), which were located north of Wall 1, but also to the west of it, there were numerous tombs north of Wall 1 of all chronological phases of the tumulus, except for those of Phase I (the following tombs were located north of the wall: XVII [72], XVIII [73], XXVI [74] [Phase II]; XXXIX [66], XLVIII [52], XLIX [51] [Phase III]; LV [53], LVI [43], LVII [58], LXVI [31] [Phase IV]; LXXIV [29], LXXV [33] [Phase Va]; LXXVII [18], LXXXIII [7], LXXXV [10] [Phase Vb]). The fact that all tombs of Phase I are located to the south of the wall is of little significance, as all of these tombs were located at depths well below the base of Wall 1. Of the tombs dating to the final prehistoric phase of the tumulus, three were located to the north of the wall (Tombs LXXVII [18], LXXXIII [7], LXXXV [10]), five to the south (Tombs LXXVI [6], LXXVIII [5], LXXIX [6], LXXX [4], LXXXI [1]), while two were located immediately above the wall (Tombs LXXXII [9], LXXXIV [2]). In the same way that there was no chronological differentiation between burials located to the north or south of Wall 1, there is no clear patterning in terms of age or sex or number of individuals interred or any clear patterning with regard to the deposition of grave goods. Wall 1, therefore, remains something of an enigma. That it was intentionally constructed is clear enough, but its function continues to elude us.

The stones of Wall 1 were not the only stones within the tumulus. A number of stones, ranging from large to small, were directly associated with individual prehistoric graves, sometimes placed immediately over the cranium or upper torso of the deceased (e.g., Tombs XXVI [74], LXXIV [29]), or else over the lower body of the dead (such as Tomb XIX [54]), occasionally over much of the body of the deceased (e.g., Tomb IV [98]; see also the stones covering Tomb VIII [100]); all such cases are enumerated and discussed in Chapter 8. Far more substantial stone covers and linings were noted in the modern burials, especially those of adults, although they were also encountered in the case of some infants (e.g., Tombs LXXXVI [22], XCII [23], XCIX [45]; cf. Tomb XCVII [39]; see also the infant inhumation, Tomb XCVIII [36]) (**Figs. 3.288–3.289, 3.290a, 3.305–3.305, 3.306a–b, 3.320a, 3.321a–b, 3.322a, 3.323–3.324, 3.326a, 3.327a–b**).

In other parts of the tumulus, individual stones were noted throughout the fill and were often kept in situ on a small pedestal to see whether there was a tomb or some other feature below them. A classic case in point is the large stone illustrated on **Figure 2.13**, dubbed "the menhir," in the field, and located in the baulk, Sector/Trench 8, in the upper levels of the tumulus. Despite its size and the fact that the stone had the appearance of being intentionally placed upright, it was clear that it was not associated with any particular tomb or other feature. Elsewhere in the tumulus, a localized dump of stones was encountered in the western sector of the mound, near Tomb XXVIII (77) but not clearly associated with it (see **Fig. 3.89a**). A more substantial area of concentrated stone was noted in the southeast sector of the tumulus, in Sector/Trench 2; it was originally referred to as a "stone feature," and was even labeled "Wall" 2, although it was never a clearly constructed wall (see **Fig. 3.14**). This stone feature was in the general area of many of the earliest (Phase I) tombs of the tumulus, but was not clearly associated with any particular tomb. Several of the stones in this sector of the tumulus were either *naturally occurring* or represent outcrops of bedrock, such as those at the southeast of the mound, off the tumulus proper, visible in **Figure 2.11a–b**.

The number of stones encountered in the fill of the tumulus, together with those of Wall 1 and the stone feature labeled "Wall" 2, can be gleaned from the photograph showing all the stone from the tumulus used in its backfill (**Fig. 2.14**). Virtually all stones encountered in the tumulus fill, including those from the more substantial stone-covered and -lined modern burials, were reburied when the mound was reconstructed (see Chapter 22). **Figure 2.14** shows that the stones comprise only a relatively small portion of the total fill of the tumulus.

Stones serving as possible tomb markers in situ were exceedingly rare, and only a few instances were noted, as follows (and these are discussed more fully in Chapter 3):

Tomb XV (80) (**Figs. 3.45a, 3.46b**)
Tomb XXVI (74) (**Figs. 3.82–3.83**)
Tomb XLV (60) (**Figs. 3.146–3.147**)

The only other substantial feature encountered in the tumulus was a channel, whether natural or intentionally cut into bedrock, referred to as the "gulley" (**Figs. 2.12b, 2.15**). Roughly oriented east to west, the gulley is visible both in the plan of the tumulus (**Fig. 2.8**) as well as the north-south sections through the tumulus (**Figs. 2.16–2.17**; see also the photograph, **Fig. 3.4**). It has a length of about 10.6 m and an average width of approximately 0.5 m, usually 0.6 m toward the top, narrowing to 0.35–0.4 m at the bottom, though in places, especially toward the east (**Fig. 2.15**), it was somewhat wider, approaching 0.75 m. The average depth of the gulley was about 0.3–0.45 m. Although located close to Wall 1, the gulley was located slightly to the north of Wall 1, and it was straighter, lacking the curved plan of the wall. Moreover, it was located 0.75–1.2 m below the base of Wall 1. Although Wall 1 and the gulley were in close proximity to one another, they were unrelated. The fill of the gulley was essentially a somewhat softer version of the surrounding bedrock, and it was clearly sterile. This, combined with the thin bands of calcium carbonate visible in **Figure 2.15**, both in the area of the gulley and the surrounding bedrock, suggests that this was a natural feature of the bedrock platform of the tumulus. The final form of the gulley, as illustrated in **Figure 2.15** may partly be the result of too zealous an excavation of the feature.

The Stratigraphy and Formation of the Tumulus

As we have seen, the Lofkënd tumulus, like the Patos tumulus, was essentially an earthen mound with a palimpsest of tombs, many of them stratigraphically related to one another (see Chapter 4). The tumulus was formed over a period of some 600 years (ca. 1400–800 BC), and this is reflected in its complex stratigraphy. The fact that the mound was opened and reopened on numerous occasions to accommodate individual burials added to its distinctive stratigraphy. Indeed, two of the most pressing issues addressed by the Lofkënd project were the character of the tumulus fill and the issue of its formation. The stratigraphy of the tumulus is indicated on **Figures 2.16–2.18**. As both the north-south sections (**Figs. 2.16–2.17**) and the east-west section (**Fig. 2.18**) show, the mound was formed over a period of time by a relatively limited number of distinct soil units (i.e., limited in terms of their composition; see Chapter 16.4), which tended to be characteristic over large areas of the tumulus.

By the end of the final excavation season, the tumulus was excavated using a total of 595 stratigraphic units (listed in Chapter 2.3). A simplified Harris matrix focusing on the burials and the primary deposits of tumulus fill is presented in **Figure 2.19a–c** according to the three sectors of the tumulus as excavated: Sectors/Trenches 1, 2/3, and 4.

Although the process of mounding appears to have begun at a relatively early stage in the creation of the tumulus, this was clearly not the case in the earliest period of burials (Phase I), where all tombs, with the exception of what is considered the earliest grave (Tomb I [64]), extended to the southeast, in an area well down-slope from the later tumulus (see **Figs. 2.18**, **4.2**, and **19.2–19.3**). Mounding proper appears to begin with Phase II (see Chapters 4 and 19). Indeed, the contours of the surface prior to excavation were to an extent followed stratigraphically as we proceeded through the mound, from top to bottom, revealing that a smaller tumulus existed during the earlier stages of its history, except for the tombs of Phase I.

It is also fairly clear that in some areas of the mound, in certain time periods, the ground had been leveled to receive a burial and earth was subsequently piled on top of the tomb, or series of tombs. In many cases, a clear tomb pit was not encountered, and in some cases a pit was found only partially lining the grave of the deceased that it accommodated (see Chapter 2.3 and Chapter 3). Consequently, many tombs were encapsulated or enveloped in a particular level. In many cases, earlier tombs were reopened to accommodate additional deceased, often with the skeletal remains of the earlier burial pushed to one side (for further details, see Chapters 3 and 8). At certain points, clear examples of tomb pits cutting through or into an existing level or levels were encountered (a phenomenon also seen with the modern burials). The pits for prehistoric tombs with noticeable cuttings were limited, however, and in some cases the actual cutting itself could only be partially followed in the ground, and rarely were the tomb cuttings substantial. To be sure, there are several exceptions, but comparatively deep prehistoric tomb pits were the exception, not the rule. A classic case in point was the pit for Tomb LXIV (61), which cut into part of the cranium and much of the torso of the body in earlier Tomb XLVII (41) (see **Figs. 3.151–3.153**). Although several of the prehistoric burials in the northeast sector of the tumulus were cut, disturbed, and occasionally substantially damaged by the modern burials, this was one of the rare cases of a prehistoric burial being significantly cut by another prehistoric tomb.

The other feature of the tumulus that can be partly ascertained from the sections, but was also clear in the process of excavations, was that in certain areas throughout the tumulus, there were distinct localized dumps of earth, which are perhaps more visible in the north-south sections (**Figs. 2.16–2.17**). The process of following these variations—sometimes slight, sometimes prominent—in the color and texture of the earth proved at times frustrating and, in some cases, elusive. Occasionally, a clearly distinct type of fill could be followed for some depth, whereas elsewhere it would disappear, only to reappear within what proved to be a larger matrix of fill. What seemed to be a most confusing stratigraphy was, to some degree, clarified inadvertently by our own annual backfilling. Because the tumulus was backfilled at the conclusion of each season and reopened at the beginning of the following season, certain parts of the baulks that needed to be cleaned because of the location of the grave were left backfilled, while the trenches themselves were cleared. Looking at the backfilled sections, one could observe a stratigraphy not unlike—and in some cases very similar to—that of the tumulus itself, and in these cases the process of formation was clear enough, as it consisted of variations in the soil that were the result of separate loads of earth piled on top of one another. In re-excavating our own backfill, these variations in soil became even more tangible, and—whether as wheelbarrow or bucket loads—their similarity to the actual stratigraphy of the tumulus was striking. Although it was natural to see these different loads as distinct strata, they were nevertheless part of a single event. Something very similar was true in the case of the Early Bronze Age mound at Skelhøj in Denmark, where various stages of construction can be seen in the stratigraphy of a "one-phase" tumulus (see Randsborg and Christensen 2006:6–7, fig. 1). Consequently, the broader outlines of the stratigraphy were at times more telling than the minor variations. At Lofkënd, the continuous process of digging new tombs, reopening old ones, filling, and piling earth on top of burials, repeated over time, gave the tumulus its distinctive stratigraphic profile (cf. Karkanas et al. 2012).

An important realization was that the mound existed as a tumulus at least as early as the end of deposition for the Phase II tombs, growing with time as more burials were added or cut into the fill. The formation of a burial tumulus differs significantly from a normal cemetery, which consists of tombs dug through relatively flat earth. In some respects it more closely resembles the processes in the formation of a settlement mound—a *toumba* or *magoula* as they are known in Greece, or *tell* in the Near East—though with significant differences, not only in scale. Like our burial tumulus, *toumbes* and *magoules* only take on the appearance of a mound with time, on account of successive phases of construction and leveling.

As for the fill of the mound, a good deal of human-made material was recovered from the Lofkënd tumulus. Some of this material must represent objects that were displaced from tombs on account of the continual process of digging for new tombs, or to reopen older burials. For example, a number of bronze and iron finds of clearly prehistoric types were noted in the various contexts of the tumulus fill, and sometimes in topsoil. These are tabulated in **Table 2.1**.

Of the prehistoric metal objects presented in **Table 2.1**, several stand out as objects that must have been disturbed from funerary contexts. The small bronze spectacle fibula (**10/18**) or the bronze head of a bimetallic pin (**10/51**) are likely displaced and damaged funerary objects, as are the cluster of bronze objects comprising four rings (**10/75–10/78**) and a perforated bronze boss (**10/61**), all found in close proximity to one another in Sector/Trench 2, Unit 0202, and conceivably from the same object. Bronze and iron jewelry, together with iron weapons and tools, such as spearheads and knives, were found in the tumulus fill of Tumulus I at Shuec in the Korçë basin (Andrea 2009–2010:279, pl. X). In addition to metal objects, there were quite a number of terracotta objects that were found in contexts outside of tombs at Lofkënd, usually in tumulus fill but occasionally in topsoil. Indeed, 7 of the 10 terracotta objects presented in Chapter 10 derive from non-funerary contexts. These are tabulated in **Table 2.2**.

The situation with the terracotta objects is somewhat less certain than the case of prehistoric metal finds. The terracotta spindlewhorls, beads, or buttons may well have been displaced from tombs, but the fact that the majority of them were encountered not in tombs but in the tumulus fill is noteworthy, and it is not impossible that some of these objects may represent material brought in with the earth from elsewhere, together with the fragmentary pottery noted throughout the tumulus fill. It should be noted, however, that at least three terracotta spindlewhorls, beads, or buttons were found in tombs, two in Tomb XXX (86) and one in Tomb LXVII (12). If the metal

TABLE **2.1** Bronze, iron, and bimetallic objects and fragments of objects of prehistoric type found in various contexts of tumulus fill and topsoil

Excavation unit	Catalogue number	Object type (Chapter 10)
1.0047 (tumulus fill)	**10/125**	Bronze pin fragment
1.0067 (tumulus fill)	**10/93**	Iron tubular bead
1.0070 (tumulus fill)	**10/126**	Fragment bronze sheet
1.0279 (tumulus fill)	**10/32**	Bronze pin shaft
2.0002 (topsoil)	**10/40**	Iron dress pin
2.0003 (topsoil)	**10/88**	Copper wire (modern?)
2.0003 (topsoil)	**10/128**	Fragment iron sheet
2.0040 (tumulus fill)	**10/39**	Iron dress pin
2.0040 (tumulus fill)	**10/51**	Bronze head of bimetallic pin
2.0066 (tumulus fill)	**10/49**	Iron pin fragments (modern?)
2.0117 (tumulus fill)	**10/30**	Bronze pin shaft
2.0118 (tumulus fill)	**10/131**	Unidentified fragment of bronze
2.0202 (tumulus fill)	**10/61**	Perforated bronze boss
2.0202 (tumulus fill)	**10/75**	Bronze ring
2.0202 (tumulus fill)	**10/76**	Bronze ring
2.0202 (tumulus fill)	**10/77**	Bronze ring
2.0202 (tumulus fill)	**10/78**	Bronze ring
2.0274 (topsoil)	**10/123**	Iron knife fragment
2.0474 (tumulus fill)	**10/130**	Unidentified fragment of bronze
4.0004 (topsoil)	**10/18**	Small bronze spectacle fibula
4.0004 (topsoil)	**10/99**	Iron tubular bead
4.0004 (topsoil)	**10/110**	Glass bead
4.0035 (tumulus fill)	**10/129**	Unidentified fragment of bronze
4.0201 (tumulus fill)	**10/111**	Glass bead
4.0204 (tumulus fill)	**10/31**	Bronze pin shaft
4.0204 (tumulus fill)	**10/127**	Bronze fragment (resembling sprue?)
4.0204 (tumulus fill)	**10/134**	Fragment of iron pin
4.0286 (tumulus fill)	**10/124**	Fragment of worked bone

TABLE 2.2 Terracotta objects of prehistoric type found in various contexts of tumulus fill and topsoil

Excavation unit	Catalogue number	Object type (Chapter 10)
1.0007 (tumulus fill)	**10/5**	Terracotta spindlewhorl, bead, or button
1.0039 (tumulus fill)	**10/4**	Terracotta spindlewhorl, bead, or button
1.0039 (tumulus fill)	**10/9**	Spindlewhorl made from pot sherd?
1.0067 (tumulus fill)	**10/8**	Terracotta spindlewhorl, bead, or button
1.0201 (topsoil)	**10/7**	Terracotta spindlewhorl, bead, or button
2.0239 (tumulus fill)	**10/3**	Terracotta spindlewhorl, bead, or button
3 (baulk cleaning)	**10/10**	Possible loomweight?

and terracotta objects listed above were not displaced from tombs, then the critical question is: where do they come from? To begin to answer this question, we must also account for the material in the fill and topsoil levels of the tumulus that was most plentiful, including the fragmentary pottery, lithic tools (together with debitage), and daub.

In Chapter 20, Samantha Martin-McAuliffe makes the important point that the term "fill"—as in "tumulus fill"—reinforces the idea that a mound has both an inside and an outside. The fill deep within the tumulus is unproblematic, but as one moves closer to the surface, it becomes increasingly difficult to distinguish between fill proper and the outer layers of the mound; in a sense, interior and exterior coexist. This is particularly the case at the interface between the stratification units that have been referred to collectively as tumulus fill and those as topsoil. As anticipated, a good deal of prehistoric material was encountered in topsoil and on the surface of the tumulus (Chapter 1.2). But given the sloping contour of the mound, the distinction between topsoil and tumulus fill could be challenging at the edges of the tumulus. Occasionally, a modern object was encountered within a context otherwise clearly part of the prehistoric tumulus fill. Particularly effective in working their way down into prehistoric tumulus fill were the spent gun cartridges and bullets of World War II catalogued and discussed in Appendix 3 following Chapter 10. Three such objects were encountered in contexts that were otherwise prehistoric: **A3/8** (Trench 1, Unit 399), **A3/17** (Trench 4, Unit 11), and **A3/6** (Trench 4, Unit 286). All such cases, however, were usually located at the very edges of the tumulus at the interface between topsoil and tumulus fill.

One of the largest components of material found throughout the tumulus fill was fragmentary pottery of the Late Bronze and Early Iron Ages, contemporary with the material deposited in tombs. All the catalogued pottery is presented in Chapter 9, and all pottery fragments given a Special Find (SF) number are presented below (Chapter 2.4). A total of 433 pottery fragments were inventoried and given "P" numbers, and of these 341 are catalogued in Chapter 9. A much larger quantity of pottery was recorded in the bulk finds, quantified and weighed, but not inventoried or catalogued (see below).

Some soil units contained slightly more pottery than others, but fragmentary pottery was fairly standard throughout the tumulus fill and topsoil; the larger quantities of pottery noted in some units (e.g., *topsoil*—Trench 2, Unit 2; Trench 4, Unit 201; *tumulus fill*—Trench 1, Units 39, 70 [with 279]; Trench 2, Unit 202; Trench 4, Units 204, 286) were the result of larger excavated units in terms of the volume of material excavated. Discrete clusters of pottery were designated "ceramic units" in the field, but these were not very common. Some of the "ceramic units" comprised small fragments clustered together; others were more sizable, such as the large coarse vessel (P283; **9/259**, **15/12**; see **Fig. 3.214**, and especially **Fig. 15.4**)—fragmentary, but largely complete and preserving an entire profile—found a few centimeters north of Tomb LXIII (35) and more or less at the same level as the lower level of the burial (at ca. 355.08 m ASL). The vessel was found on its side, its base toward the west and mouth stopped by a stone. It evidently contained nothing except for liquid bitumen that had dried (identified as bitumen using the Raspail test described in Odegaard, Carroll, and Zimmt 2005:158–159; see also Chapter 5). As no formal grave cutting for Tomb LXIII (35) was found, and none for the coarse vessel, it was difficult to associate the vessel with the tomb, but its location so close to the deceased suggests that the vessel may have been intended for this tomb. This bitumen pot was, however, an exception, not the rule, especially for its size and the manner of its deposition in the tumulus fill.

Since the Lofkënd tumulus is one of the few burial mounds in Albania to have been systematically excavated, it afforded us the opportunity to quantify the weight and total number of pottery fragments from a completely excavated tumulus. In addition to the inventoried pottery already noted, all of the non-inventoried material was quantified according to the following categories (beginning with the heaviest component, proceeding to the lightest component, and ending with the non-diagnostic pottery):

- Handmade coarse
- Handmade semi-coarse
- Handmade fine
- Wheelmade undecorated
- Wheelmade decorated
- Non-diagnostic (mostly scraps of pottery)

In the case of the non-diagnostic pottery, a straight count of the pieces, most of them minuscule scraps that would often break up or disintegrate, served little purpose, so for the non-diagnostic category only weight is tabulated in **Table 2.3**.

The wheelmade material was paltry and largely confined to the topsoil and surface units. Of the non-inventoried wheelmade undecorated pottery, the three fragments from the surface consisted of two body fragments from a modern vessel or vessels, and a handle fragment of a modern water jar; similarly, all the material from topsoil was modern, except for one fragment that was possibly, but not certainly, Hellenistic. Unfortunately, very little more could be said of the latter. The non-inventoried wheelmade decorated pottery was even less, amounting to five fragments, weighing about that many grams. Three of the sherds are clearly Corinthian, and the other two probably are. All five fragments were recorded in the uppermost level of the tumulus fill but very close to the interface with the topsoil or at the edge of the tumulus.

In terms of the number of fragments, the quantity of handmade coarse sherds (2,436) was almost the same as that of the handmade semi-coarse (2,535), but the former was, as one might anticipate, more than twice as heavy as the latter (22.281 kg as opposed to 9.661 kg). Hence the importance of counting and weighing all the material, although, as already noted, counting the minuscule scraps and crumbs of the non-diagnostic, non-inventoried pottery was virtually impossible as many of the pieces would break up or disintegrate to the touch, or through friction with other fragments in their bag.

TABLE 2.3 Summary quantification in terms of total weight and number of fragments, non-inventoried pottery from Lofkënd tumulus

Stratigraphic unit	Weight	Number of fragments
	Handmade coarse	
Surface (2004)	120 g	10
Topsoil	3587 g	417
Grave fills (26 tombs)	578 g	75
Tumulus fill	17,996 g	1,934
Total	22.281 kg	2,436
	Handmade semi-coarse	
Surface (2004)	133 g	24
Topsoil	2,080 g	530
Grave fills (46 tombs)	282 g	169
Tumulus fill	7,166 g	1,812
Total	9.661 kg	2,535
	Handmade fine	
Surface (2004)	15 g	2
Topsoil	226 g	66
Grave fills (11 tombs)	24 g	20
Tumulus fill	464 g	217
Total	0.729 kg	305
	Wheelmade undecorated	
Surface (2004)	112 g	3
Topsoil	37 g	23
Total	149 g	26
	Wheelmade decorated	
Tumulus fill	6 g	5
Total	6 g	5
	Non-diagnostic	
Surface (2004)	18 g	
Topsoil	1,129 g	
Grave fills (22 tombs)	81 g	
Tumulus fill	4,594 g	
Total	5,822 g	
Grand total (all pottery)	38.642 kg	5,307 fragments + many scraps

The total weight of all the non-inventoried pottery encountered in the tumulus fill, individual grave fills, topsoil, and on the surface was just under 40 kg (38.641 kg). The total count of all pottery fragments that could be counted was 5,307, and the figure would be greatly augmented if the non-diagnostic scraps could be meaningfully counted.

In the context of the stratigraphy and processes of tumulus formation, the pieces of daub (see Chapter 14) from the tumulus were, in many ways, as surprising, perhaps more so, than the quantity of Paleolithic, Neolithic, and Bronze Age chipped-stone tools from the mound (for which see Chapter 13). A selection of daub pieces are catalogued and discussed in Chapter 14. Entered onto the project database as "fired clay not pottery," many of the pieces and lumps of this material were amorphous, but many preserved reed, rod, or stake impressions, indicating that the clay had been used as an integral component of wattle-and-daub architecture. In addition to the few catalogued pieces of daub, a staggering 15,875 lumps of the material, weighing some 40 kg, were recovered from the topsoil and tumulus fill, as well as from some of the grave fills. In terms of a straight count of pieces, the daub outnumbered the pottery, while in terms of weight it was about the same as that for the pottery, though slightly more.

Light and durable, the daub actually encountered in the Lofkënd tumulus may represent the tip of the iceberg, as what remains unknown is the quantity of unfired—and thus not preserved or invisible—daub that may have made its way into the tumulus. Occasionally, a lump of what must be daub has been found in other burial tumuli in the Balkans, such as the fragment from Planje Tumulus 1, Grave 3, in Bosnia Herzegovina (Benac and Čović 1956:59, pl. XXXV, no. 11), but such material has not been generally noted for the numerous tumuli excavated in Albania. Having observed the daub from our tumulus, Korkuti later confirmed that similar pieces were found throughout the fill of the Patos tumulus but were never published (personal communication). Although daub is plentiful in various prehistoric settlement sites in Albania, the fact is that it has simply not been recorded from other burial tumuli. Whether this is an accurate reflection of the presence/absence of daub in burial mounds, or whether it has simply been overlooked, we may never know.

Although not recovered in the same quantities as the daub, the number of chipped-stone tools from the tumulus was significant. What was perhaps most striking about the lithic artifacts was their chronological range, dating from the Middle Paleolithic and Mesolithic periods through the Neolithic and Bronze Age. Although the presence of such early lithics need not come as a surprise, especially since Middle and early Upper Paleolithic and Mesolithic finds are well represented in the region of Apollonia and Fier (see Runnels et al. 2004; cf. Runnels 2009; for earlier overviews, see Korkuti and Petruso 1993; Korkuti 1983b, 1995b, 1998:21–23, 40–41; Metallinou 2008:30–35 [Paleolithic], 36–39 [Mesolithic]; for projects elsewhere in Albania focusing on the Paleolithic and Mesolithic, see, among others, Fistani 1989; Galaty 2006; Gjipali 2006), the quantity of the lithic artifacts within and on a burial tumulus demanded explanation. Indeed, the sheer quantity of the lithics and daub was one of the reasons that led to the survey of the area around the Lofkënd tumulus in 2007 and 2008. Unlike daub, which has largely gone unnoticed, the quantity of lithic artifacts recovered from the fill of other Illyrian tumuli is noteworthy. They were first noted by Frano Prendi in Tumulus 1 at Vajzë, near Vlorë (Prendi 1957: 91–92, fig. 17 [11 illustrated examples, dated to the Neolithic period]), and since then, lithics have been reported by a number of scholars from several other tumuli (e.g., Amore 2010:621–626, nos. 11.14–11.58 [Apollonia, Tumuli 9, 10, 11, and the tumulus appendices]; Bodinaku 1981:90, pl. I, nos. 1–23 [Pazhok], and 2001–2002:79, pl. III, no. 8; Jubani 1995:89, pl. XIV, nos. 1–10 [Shkrel]). Be that as it may, the Lofkënd tumulus yielded no fewer than 589 lithic artifacts, of which 94 are identified and catalogued tools, including cores, and 495 pieces of debitage (fully described and discussed in Chapter 13). The only Balkan burial mound that we know of to have yielded a number of lithic artifacts close to the quantity of those recovered at Lofkënd is the Illyrian Graveyard site in western Serbia, where more than 450 flint tools of various types, functions, and raw materials were collected in all sections of the mound (Filipovic and Starovic 2005).

As for organic material recovered from the tumulus, the situation was interesting. Although not great, there was a steady quantity of animal remains from the tumulus: 358 pieces in all. Although the two animal skeletons interred with the modern burials (see Chapter 16.1) and much of the fragmentary animal bones were in the upper parts of the tumulus (see **Table 16.5**), animal bone was essentially encountered throughout the tumulus, including in the fill of Tomb I, which is considered the earliest burial. In stark contrast, floral material was negligible, so much so that its absence meant that it was never part of the funerary ritual, unless, of course, seeds were used and deposited in an unburned state. The only other significant organic component of the tumulus—including tumulus fill, grave fills, and the topsoil levels—was the mollusca, which was almost exclusively land shells, belonging to 13 different species (Chapter 16.2). Apart from a solitary brackish-water mollusc, their occurrence is attributed to natural processes that do not involve human agency. Carefully recorded and quantified, the mollusca from the Lofkënd tumulus had a total weight of 7.835 kg, of which 2.308 derive from the topsoil or surface levels, 3.552 from the tumulus fill, and the remainder from the fill of some 70 tombs (see **Table 16.5**, and Chapter 16.2 for further details of quantification and identification). As far as we know, this is the first time that all shells have been collected, recorded, and studied from an Illyrian, or indeed Balkan, tumulus.

The three-way combination of fragmentary pottery, chipped-stone tools, and remnants of wattle-and-daub architecture, together with the smaller quantity of animal bone remains, suggests that such material was either plentiful enough in the nearby landscape to be used as tumulus and grave fill, or it raises the intriguing possibility that those burying the dead intentionally brought material from sites that were not in the immediate vicinity of the tumulus, as well as sites that were already ancient (cf. Papadopoulos 2006). If this was the case, however, the material was not consistent with what might be expected from "normal" prehistoric habitation sites. For example, it is clear that the fill of the tumulus does not contain any quantity of carbonized floral remains or significant quantities of animal bone (see Chapter 16.1). Moreover, confirming or disproving that the earth brought to the tumulus was ever subject to curation, meaning that the earth was intentionally chosen or even screened for the presence/absence of certain materials, is a challenging prospect (the process of the construction, repair, and maintenance of mounds is a feature noted in several mound sites along the Mississippi River in the United States; see, most recently, Saunders et al. 2006, who deal with the site of Watson Brake).

When focusing on individual components of the fill of the tumulus, it is possible to suggest the process by which such material made its way into the fill of a burial mound. For instance, Jamie Aprile, in her study of the lithic artifacts from the tumulus (Chapter 13), together with her intensive survey of the landscape surrounding the tumulus (Chapter 18), suggests that lithic tools of the Neolithic and Bronze Age periods arrived in the mound through interment activities or settlement soil packages, while those dating to the Paleolithic and Mesolithic periods were collected with local clay-rich soils used to stabilize the tumulus (for which see Chapter 16.4). As we have already noted, the excavation of individual tombs in the Lofkënd tumulus was exacerbated by the dense, hard, cement-like quality of the soil, which John Foss determined derived from the clayey fine-textured sediment from soils weathered from shale, located immediately around the site (especially in the area approximately 30 m north of the tumulus) (Chapter 16.4). The basic parent material for soils occurring at Lofkënd is weakly cemented Pliocene sandstone. The fine and very fine sands that weathered from the calcareous sandstone are quite erosive and would pose difficulties for stabilizing the earth matrix during individual burials, as well as for the completed tumulus. It was to this parent material, in addition to soil brought in from elsewhere in the landscape, that the clayey sediment deriving from shale was added throughout the tumulus to control erosion (for detailed descriptions of the soils examined at the site and in the general vicinity of the tumulus, see Foss and Timpson 2007: 140–144, tables 1–3; Foss, Chapter 16.4). Indeed, the interplay, as it were, between the fine-textured shale sediment obtained off-site that had formed in Miocene sediments (Foss's Unit 1), with the weakly cemented Pliocene sandstone that was the basic parent material of Lofkënd (Unit 2), is noted in Chapter 2.3, where the two types of soil continuously recur throughout the tumulus.

As for the presence of daub, this is discussed in further detail by Martin-McAuliffe in Chapter 20, who alludes to the stationary, yet more ephemeral function of the material. She raises the possibility that the Lofkënd daub did have architectural origins and was possibly curated, perhaps symbolic of a lost, impermanent form of dwelling that was preserved in an "eternal dwelling," a house of the dead. She also notes the curious attribute of the daub, that it was hardened through a firing process, and she brings to the discussion the case of a Neolithic settlement in Calabria in southern Italy, where hardened clay material was likely fired on purpose by its creators. As Martin-McAuliffe notes (Chapter 20), as a recycled material, "fired daub would not only lend increased durability to new structures, but it would also reduce the amount of new, fresh daub that was required for building" and later, "the daub may have been recycled yet again and permanently preserved as tumulus fill."

The combination, and sheer quantity, of fragmentary pottery, lithic artifacts, and daub from the tumulus is matched by the dearth of such materials—with the exception of the Paleolithic and Mesolithic stone tools—in the landscape immediately surrounding the tumulus, and it suggests that people of the Gjanicë River valley who buried their dead at Lofkënd between approximately 1400 and 800 BC brought material with them from their current, or even older, settlement(s) in the area to accompany the deceased. Such a phenomenon only bolsters the commemorative aspects of the tumulus, as it suggests that survivors of the dead deliberately and selectively incorporated components of their lived environment into a growing memorial for their dead.

Chapter 2.2
Concordance of Grave and Tomb Numbers

John K. Papadopoulos

Grave 1	Tomb LXXXI
Grave 2	Tomb LXXXIV
Grave 3	Tomb LXXXVIII
Grave 4	Tomb LXXX
Grave 5	Tomb LXXVIII
Grave 6	Tomb LXXIX
Grave 7	Tomb LXXXIII
Grave 8	Tomb LXXXVII
Grave 9	Tomb LXXXII
Grave 10	Tomb LXXXV
Grave 11	Tomb LXXXIX
Grave 12	Tomb LXVII
Grave 13	Tomb LXVIII
Grave 14	Tomb XC
Grave 15	Tomb XCI
Grave 16	Tomb LXXVI
Grave 17	Tomb LXX
Grave 18	Tomb LXXVII
Grave 19	Tomb XCIII
Grave 20	Tomb XCVI
Grave 21	Tomb XCV
Grave 22	Tomb LXXXVI
Grave 23	Tomb XCII
Grave 24	Tomb LXXII
Grave 25	Tomb XCIV
Grave 26	Tomb LXXIII
Grave 27	Tomb LXIX
Grave 28	Tomb LXXI
Grave 29	Tomb LXXIV
Grave 30	Tomb LXV
Grave 31	Tomb LXVI
Grave 32	Tomb LXII
Grave 33	Tomb LXXV
Grave 34	Tomb LXI

Grave 35	Tomb LXIII
Grave 36	Tomb XCVIII
Grave 37	Tomb LVIII
Grave 38	Tomb LIX
Grave 39	Tomb XCVII
Grave 40	Tomb LIV
Grave 41	Tomb XLVII
Grave 42	Tomb XLVI
Grave 43	Tomb LVI
Grave 44	Tomb LX
Grave 45	Tomb XCIX
Grave 46	Tomb L
Grave 47	Tomb XXII
Grave 48	Tomb C
Grave 49	Tomb XIII
Grave 50	Tomb XX
Grave 51	Tomb XLIX
Grave 52	Tomb XLVIII
Grave 53	Tomb LV
Grave 54	Tomb XIX
Grave 55	Tomb XXI
Grave 56	Tomb XXIII
Grave 57	Tomb XLI
Grave 58	Tomb LVII
Grave 59	Tomb XLII
Grave 60	Tomb XLV
Grave 61	Tomb LXIV
Grave 62	Tomb XLIII
Grave 63	Tomb LIII
Grave 64	Tomb I
Grave 65	Tomb XLIV
Grave 66	Tomb XXXIX
Grave 67	Tomb XL

Grave 68	Tomb XVI
Grave 69	Tomb LII
Grave 70	Tomb XXX
Grave 71	Tomb XIV
Grave 72	Tomb XVII
Grave 73	Tomb XVIII
Grave 74	Tomb XXVI
Grave 75	Tomb XXXVI
Grave 76	Tomb XXXVII
Grave 77	Tomb XXVIII
Grave 78	Tomb LI
Grave 79	Tomb XXXVIII
Grave 80	Tomb XV
Grave 81	Tomb III
Grave 82	Tomb XXVII
Grave 83	Tomb XXIX
Grave 84	Tomb XXXV
Grave 85	Tomb XXIV
Grave 86	Tomb XXXI
Grave 87	Tomb XXXIV
Grave 88	Tomb XII
Grave 89	Tomb XXXII
Grave 90	Tomb XXV
Grave 91	Tomb II
Grave 92	Tomb XXXIII
Grave 93	Tomb XI
Grave 94	Tomb IX
Grave 95	Tomb X
Grave 96	Tomb V
Grave 97	Tomb VI
Grave 98	Tomb IV
Grave 99	Tomb VII
Grave 100	Tomb VIII

Chapter 2.3
List of Lofkënd Stratification Units

Unit	Trench/ sector	Description	Date
0001	1	Topsoil	23/6/2004
0002	2	Topsoil	23/6/2004
0003	3	Topsoil	24/6/2004
0004	4	Topsoil	24/6/2004
0005	2	Hard-packed, off-white, with white specks	24/6/2004
0006	2	Loose brown soil	24/6/2004
0007	1	Loose olive-brown soil	24/6/2004
0008	3	Loose sandy soil (topsoil)	24/6/2004
0009	1	Topsoil	25/6/2004
0010	2	Loose sandy soil	25/6/2004
0011	4	Compact clay deposit, below topsoil	25/6/2004
0012	3	Tomb LXXXI (1) (grave)	25/6/2004
0013	3	Tomb LXXXI (1) (fill)	25/6/2004
0014	3	Tomb LXXXI (1) (skeleton)	25/6/2004
0015	4	Lighter, sandier soil than SU 0011	25/6/2004
0016	1	Tomb LXXXIV (2) (grave)	26/6/2004
0017	1	Tomb LXXXIV (2) (fill)	26/6/2004
0018	1	Tomb LXXXIV (2) (first skeleton)	26/6/2004

Unit	Trench/ sector	Description	Date
0019	1	Tomb LXXXVIII (3) (grave)	26/6/2004
0020	1	Tomb LXXXVIII (3) (fill)	26/6/2004
0021	1	Tomb LXXXVIII (3) (skeleton)	26/6/2004
0022	8	Topsoil (baulk separating Trenches 3–4)	28/6/2004
0023	2	Yellowish clay	28/6/2004
0024	2	Light sandy soil (pit?)	28/6/2004
0025	5	Topsoil	28/6/2004
0026	5	Loose olive-brown soil	28/6/2004
0027	7	Topsoil (baulk separating Trenches 2–3)	28/6/2004
0028	8	Tomb LXXXI (1) (fill, continued)	29/6/2004
0029	7	Tomb LXXXI (1) (fill, continued)	29/6/2004
0030	4	Tomb LXXX (4) (grave)	29/6/2004
0031	4	Tomb LXXX (4) (fill)	29/6/2004
0032	4	Tomb LXXX (4) (skeleton)	29/6/2004
0033	2	Darker, sandy earth	30/6/2004
0034	4	Ceramic deposit, with flecks of carbon	30/6/2004
0035	4	Sandy, loose deposit	30/6/2004
0036	2	Lighter, granular deposit	30/6/2004
0037	2	Sandy earth	30/6/2004
0038	1	Tomb LXXXIV (2) (second skeleton)	1/7/2004
0039	1	Darker soil, more brown than SU 7	1/7/2004
0040	2	Deposit in SE corner	1/7/2004
0041	2	White, rectilinear (cut?)	1/7/2004
0042	2	Dark deposit in north of trench	1/7/2004
0043	4	Sandy olive deposit in Tomb LXXX (4) fill	2/7/2004
0044	4	Deposit with decomposed fire-affected clay in Tomb LXXX (4) fill	2/7/2004
0045	3	Sandy earth below Tomb LXXXI (1)	2/7/2004
0046	3	Clayey soil in east of trench	2/7/2004
0047	1	Loose brown soil	2/7/2004
0048	4	Deposit with decomposed ceramic/ fire-affected clay	3/7/2004
0049	2	Tomb LXXVIII (5) (grave)	3/7/2004
0050	2	Tomb LXXVIII (5) (fill)	3/7/2004
0051	2	Tomb LXXVIII (5) (skeleton)	3/7/2004
0052	1	Tomb LXXIX (6) (grave)	5/7/2004
0053	1	Tomb LXXIX (6) (fill)	5/7/2004
0054	1	Tomb LXXIX (6) (skeleton)	5/7/2004
0055	3	Clayey deposit, similar to SU 0046	5/7/2004
0056	3	Deposit of white clay and sand in NE part of trench	5/7/2004
0057	4	Sandy, loose deposit (as SU 0035)	6/7/2004
0058	4	Tomb LXXXIII (7) (grave)	6/7/2004
0059	4	Tomb LXXXIII (7) fill	6/7/2004
0060	4	Tomb LXXXIII (7) skeleton	6/7/2004
0061	3	Dark clayey deposit in south of trench	6/7/2004
0062	5	Topsoil	6/7/2004
0063	*Vacat*	*Vacat*	*Vacat*
0064	3	Topsoil on west and south edges	7/7/2004
0065	2	Tomb LXXVIII (5) (cut)	7/7/2004
0066	2	Loose, dark yellow clay	7/7/2004
0067	1	Clayey brown soil	7/7/2004
0068	3	Clayey brown deposit in west of trench	7/7/2004
0069	3	Sandy deposit in east of trench	7/7/2004
0070	1	Yellow sandy deposit	7/7/2004

Unit	Trench/ sector	Description	Date
0071	2	Yellowish deposit with white calcareous material	8/7/2004
0072	2	Dark deposit under SU 0040	8/7/2004
0073	2	Dark deposit in southwest of trench	8/7/2004
0074	2	Dark square feature	8/7/2004
0075	2	Whitish calcareous deposit	8/7/2004
0076	3	White strip in east within SU 0068	9/7/2004
0077	3	Brown deposit south of SU 0061	9/7/2004
0078	2	Sandy deposit	9/7/2004
0079	4	Yellowish brown deposit	9/7/2004
0080	1	Animal Tomb LXXXVII (8) (grave)	9/7/2004
0081	1	Animal Tomb LXXXVII (8) (fill)	9/7/2004
0082	1	Animal Tomb LXXXVII (8) (skeleton)	9/7/2004
0083	4	Tomb LXXXII (9) (grave)	10/7/2004
0084	4	Tomb LXXXII (9) (fill)	10/7/2004
0085	4	Tomb LXXXII (9) (skeleton)	10/7/2004
0086	4	Deposit below SU 0011	12/7/2004
0087	4	Dark deposit (perhaps fill of Tomb LXXXII to the east)	12/7/2004
0088	1	Tomb LXXXV (10) (grave)	12/7/2004
0089	1	Tomb LXXXV (10) (fill)	12/7/2004
0090	1	Tomb LXXXV (10) (skeleton)	12/7/2004
0091	1	Small pot fragments	12/7/2004
0092	3	Sandy deposit in middle of trench	12/7/2004
0093	4	Topsoil around skull in Tomb LXXXII (same as SU0004)	12/7/2004
0094	1	Tomb LXXXIX (11) (grave)	12/7/2004
0095	1	Tomb LXXXIX (11) (fill)	12/7/2004
0096	1	Tomb LXXXIX (11) (skeleton)	12/7/2004
0097	2	Deposit of lighter clay	13/7/2004
0098	3	Tomb LXVII (12) (grave)	13/7/2004
0099	3	Tomb LXVII (12) (fill)	13/7/2004
0100	3	Tomb LXVII (12) (skeleton)	13/7/2004
0101	2	Tomb LXVIII (13) (grave)	13/7/2004
0102	2	Tomb LXVIII (13) (fill)	13/7/2004
0103	2	Tomb LXVIII (13) (skeleton)	13/7/2004
0104	4	Deposit below SU 0048	13/7/2004
0105	4	Tomb LXXXII (9) (second skeleton)	13/7/2004
0106	4	Tomb LXXXII (9) (third skeleton)	13/7/2004
0107	1	Tomb XC (14) (grave)	13/7/2004
0108	1	Tomb XC (14) (fill)	13/7/2004
0109	1	Tomb XC (14) (south skeleton)	13/7/2004
0110	1	Tomb XC (14) (north skeleton)	13/7/2004
0111	1	Human bone (no grave)	13/7/2004
0112	1	Animal bone (no grave)	13/7/2004
0113	7	West baulk	14/7/2004
0114	8	South baulk	14/7/2004
0115	4	Deposit south of stone and cup	14/7/2004
0116	8	Deposit north of stone in baulk	14/7/2004
0117	2	Variegated deposit	14/7/2004
0118	2	Gray deposit with white specks	14/7/2004
0119	2	Inner grave cut of Tomb LXVIII (13)	16/7/2004
0120	1	Tomb XCI (15) (grave)	16/7/2004
0121	1	Tomb XCI (15) (fill)	16/7/2004
0122	1	Tomb XCI (15) (skeleton; complete, articulated)	16/7/2004
0123	4	Tomb LXXVI (16) (grave)	16/7/2004
0124	4	Tomb LXXVI (16) (fill)	16/7/2004

Unit	Trench/ sector	Description	Date
0125	4	Tomb LXXVI (16) (skeleton)	16/7/2004
0126	3	Tomb LXX (17) (northern skeleton)	16/7/2004
0127	1	Compact clay feature, olive brown	16/7/2004
0128	1	Tomb XCI (15) (second skeleton; incomplete)	16/7/2004
0129	4	Deposit, southern end	16/7/2004
0130	4	Tumulus fill	16/7/2004
0131	4	Tumulus fill	16/7/2004
0132	4	Deposit of heavy clay	16/7/2004
0133	3	Tomb LXX (17) (grave)	17/7/2004
0134	3	Tomb LXX (17) (fill)	17/7/2004
0135	*Vacat*	*Vacat*	*Vacat*
0136	2	Tomb LXVIII (13) (second skeleton, lower)	17/7/2004
0137	4	Tomb LXXVII (18) (grave)	19/7/2004
0138	4	Tomb LXXVII (18) (fill)	19/7/2004
0139	4	Tomb LXXVII (18) (skeleton)	19/7/2004
0140	5	Topsoil (same as SU 0004 in baulk, Trench 5)	19/7/2004
0141	1	Topsoil below surface at top of tumulus	19/7/2004
0142	1	Animal Tomb XCIII (19) (grave)	19/7/2004
0143	1	Animal Tomb XCIII (19) (fill)	19/7/2004
0144	1	Animal Tomb XCIII (19) (skeleton)	19/7/2004
0145	1	Tomb LXXXV (10) (second skeleton)	19/7/2004
0146	1	Tomb CXVI (20) (grave)	19/7/2004
0147	1	Tomb CXVI (20) (fill)	19/7/2004
0148	1	Tomb CXVI (20) (skeleton)	19/7/2004
0149	2	Tomb LXVIII (13) (third skeleton)	21/7/2004
0150	2	Tomb LXVIII (13) (skeleton fragments; third skeleton?)	21/7/2004
0151	1	Tomb CXV (21) (grave)	21/7/2004
0152	1	Tomb CXV (21) (fill)	21/7/2004
0153	1	Tomb CXV (21) (skeleton)	21/7/2004
0154	1	Tomb LXXXVI (22) (grave)	21/7/2004
0155	1	Tomb LXXXVI (22) (fill)	21/7/2004
0156	1	Tomb LXXXVI (22) (skeleton)	21/7/2004
0157	1	Tomb XCII (23) (grave)	21/7/2004
0158	1	Tomb XCII (23) (fill)	21/7/2004
0159	1	Tomb XCII (23) (skeleton)	21/7/2004
0160	1	Tomb LXXII (24) (grave)	21/7/2004
0161	1	Tomb LXXII (24) (fill)	21/7/2004
0162	1	Tomb LXXII (24) (skeleton)	21/7/2004
0163	1	Tomb CXIV (25) (grave)	22/7/2004
0164	1	Tomb CXIV (25) (fill)	22/7/2004
0165	1	Tomb CXIV (25) (skeleton)	22/7/2004
0166	2	Outer grave cut of Tomb LXVIII (13)	22/7/2004
0167	1	Tomb LXXIII (26) (grave)	22/7/2004
0168	1	Tomb LXXIII (26) (fill)	22/7/2004
0169	1	Tomb LXXIII (26) (skeleton)	22/7/2004
0170	2	Dark brown soil in northwest of trench	23/7/2004
0171	3	Tomb LXVII (12) (second skeleton)	23/7/2004
0172	1	Tomb LXIX (27) (grave)	24/7/2004
0173	1	Tomb LXIX (27) (fill)	24/7/2004
0174	1	Tomb LXIX (27) (skeleton)	24/7/2004
0175	1	Tomb LXXI (28) (grave)	24/7/2004
0176	1	Tomb LXXI (28) (fill)	24/7/2004
0177	1	Tomb LXXI (28) (skeleton)	24/7/2004
0178	5	Loose brown soil	24/7/2004
0179	5	Clayey brown soil	24/7/2004

Unit	Trench/ sector	Description	Date
0180	6	Topsoil	24/7/2004
0181	6	Same as SU 0070 (extending into Trench 6)	24/7/2004
0182	2	Animal bone	24/7/2004
0183	1	Tomb LXXXVI (22) (cut)	26/7/2004
0184	3	Fill beneath Tomb LXVII (12)	26/7/2004
0185	1	Tomb XCII (23) (cut)	26/7/2004
0186	7	Topsoil	27/7/2004
0187	7	Deposit beneath and south of SU 0186 (probably topsoil)	27/7/2004
0188	7	Clayey deposit with calcareous specks	27/7/2004
0189	7	Clayey deposit south of SU 0187 and 0188	27/7/2004
0190	7	Deposit beneath SU 0188 and 0189; abundant calcareous chunks	27/7/2004
0191	7	Deposit south of SU 190, beneath SU 0189, yellowish brown, with calcareous chunks	28/7/2004
0192	7	Deposit beneath SU 0190: brown, clayey	28/7/2004
0193	7	Very pale brown deposit over entire trench	28/7/2004
0194	7	Brown deposit beneath SU 0193; same as SU 0061, 0071, 0116	28/7/2004
0195	7	Yellowish brown, sandy deposit; possible lining for grave cut SU 0133 (Tomb LXX [17])	28/7/2004
0196	7	Tomb LXX (17) (fill continuation to east; same as SU 0134)	28/7/2004
0197	7	Tomb LXX (17) (second skeleton)	28/7/2004
0198	2	Deposit atop "bone platform"	21/6/2005
0199	2	Yellow clayey deposit (= SU 0036)	21/6/2005
0200	2	Darker sandy deposit (= SU 0037)	21/6/2005
0201	4	Topsoil (continuation of SU 0004)	22/6/2005
0202	2	Generic brown deposit (classic tumulus fill)	22/6/2005
0203	4	Yellow-brown deposit	22/6/2005
0204	4	Light yellow-brown deposits (continuation of SU 0086)	23/6/2005
0205	4	Light red/brown sandy deposit (perhaps = SU 130)	23/6/2005
0206	4	Ceramic deposit	23/6/2005
0207	4	Continuation of SU 0129	23/6/2005
0208	4	Tomb LXXIV (29) (grave)	24/6/2005
0209	4	Tomb LXXIV (29) (fill)	24/6/2005
0210	4	Tomb LXXIV (29) (skeleton)	24/6/2005
0211	1	Tomb LXV (30) (grave)	24/6/2005
0212	1	Tomb LXV (30) (fill)	24/6/2005
0213	1	Tomb LXV (30) (skeleton)	24/6/2005
0214	4	Tomb LXVI (31) (grave)	24/6/2005
0215	4	Tomb LXVI (31) (fill)	24/6/2005
0216	4	Tomb LXVI (31) (skeleton)	24/6/2005
0217	2	Dark soil in southwest of Trench 2	25/6/2005
0218	2	Dark soil = Tomb LXII (32) (grave fill)	25/6/2005
0219	2	Tomb LXII (32) (grave)	25/6/2005
0220	2	Tomb LXII (32) (skeleton)	25/6/2005
0221	1	Tomb LXXV (33) (grave)	25/6/2005
0222	1	Tomb LXXV (33) (fill)	25/6/2005
0223	1	Tomb LXXV (33) (skeleton)	25/6/2005

Unit	Trench/ sector	Description	Date
0224	2	Tomb LXI (34) (grave)	27/6/2005
0225	2	Tomb LXI (34) (skeleton)	27/6/2005
0226	2	Deposit of whitish soil	27/6/2005
0227	2	Ceramic deposit	28/6/2005
0228	2	Dark soil around SU 0227	28/6/2005
0229	2	Yellowish soil	28/6/2005
0230	4	Tomb LXXIV (29) (central skeleton)	28/6/2005
0231	4	Tomb LXXIV (29) (southernmost skeleton)	28/6/2005
0232	4	Tomb LXIII (35) (grave)	28/6/2005
0233	4	Tomb LXIII (35) (fill)	28/6/2005
0234	4	Tomb LXIII (35) (skeleton)	28/6/2005
0235	2	Orange soil	29/6/2005
0236	1	Tomb XCVIII (36) (grave)	29/6/2005
0237	1	Tomb XCVIII (36) (fill)	29/6/2005
0238	1	Tomb XCVIII (36) (skeleton)	29/6/2005
0239	2	Dark brown soil	29/6/2005
0240	1	Yellowish brown soil (10 YR 5/6)	30/6/2005
0241	1	Tomb XCVIII (36) (cut)	30/6/2005
0242	2	Tomb LVIII (37) (grave)	30/6/2005
0243	2	Tomb LVIII (37) (skeleton)	30/6/2005
0244	2	Tomb LVIII (37) (fill)	1/7/2005
0245	4	Tomb LIX (38) (grave)	1/7/2005
0246	4	Tomb LIX (38) (fill)	1/7/2005
0247	4	Tomb LIX (38) (skeleton)	1/7/2005
0248	1	Tomb XCVII (39) (grave)	1/7/2005
0249	1	Tomb XCVII (39) (fill)	1/7/2005
0250	1	Tomb XCVII (39) (skeleton)	1/7/2005
0251	2	Tomb LIV (40) (grave)	4/7/2005
0252	2	Tomb LIV (40) (skeleton)	4/7/2005
0253	2	Brown clayey soil	4/7/2005
0254	2	Tomb LIV (40) (fill)	4/7/2005
0255	2	Tomb XLVII (41) (grave)	4/7/2005
0256	2	Tomb XLVII (41) (skeleton)	4/7/2005
0257	2	Tomb XLVII (41) (fill)	4/7/2005
0258	8	Baulk cleaning	5/7/2005
0259	2	Dark soil by Tomb XLVII (41)	5/7/2005
0260	2	Rectilinear cut by SU 0259 (=Tomb LXIV [61] [cut])	5/7/2005
0261	7	Baulk trimming	5/7/2005
0262	1	Tomb XCVII (39) (cut)	6/7/2005
0263	2	Skeleton by Tomb XLVII (41) = Tomb XLVI (42) (skeleton)	6/7/2005
0264	2	White/pink plaster-like layer below Tomb LIV (40)	6/7/2005
0265	2	Tomb XLVI (42) (grave)	6/7/2005
0266	2	Tomb XLVI (42) (fill)	6/7/2005
0267	4	Tomb LVI (43) (grave)	6/7/2005
0268	4	Tomb LVI (43) (fill)	6/7/2005
0269	4	Tomb LVI (43) (skeleton)	6/7/2005
0270	2	Cut around Tomb XLVI (42)	6/7/2005
0271	2	Fill of cut around Tomb XLVI (42)	6/7/2005
0272	2	Tomb LX (44) (grave)	7/7/2005
0273	2	Tomb LX (44) (skeleton)	7/7/2005
0274	2	Tomb LX (44) (fill)	7/7/2005
0275	1	Tomb XCIX (45) (grave)	7/7/2005
0276	1	Tomb XCIX (45) (fill)	7/7/2005
0277	1	Tomb XCIX (45) (skeleton)	7/7/2005
0278	1	Topsoil	7/7/2005
0279	1	Same as SU 0070 (1.5 m below surface of tumulus)	7/7/2005
0280	4	Ceramic unit north of Tomb LXIII (35)	7/7/2005
0281	1	Same as SU 0039 (1.5 m below surface of tumulus)	7/7/2005
0282	4	Tomb L (46) (grave)	8/7/2005
0283	4	Tomb L (46) (fill)	8/7/2005
0284	4	Tomb L (46) (skeleton)	8/7/2005
0285	1	Tomb XCIX (45) (cut)	8/7/2005
0286	4	Deposit below SU 0205	12/7/2005
0287	2	Hard calcareous soil, Tomb XLII (59) (fill)	12/7/2005
0288	2	Variegated dark soil	12/7/2005
0289	4	Carbonized area (resembling post-hole?) north of Wall 1	12/7/2005
0290	4	Cut of carbonized area SU 0289	12/7/2005
0291	1	Ceramic deposit 11.3–11.5 m north of southwest corner	12/7/2005
0292	2	Dark soil = Tomb XIX (54) (fill)	12/7/2005
0293	2	Tomb XXII (47) (grave)	13/7/2005
0294	2	Tomb XXII (47) (skeleton)	13/7/2005
0295	2	Tomb XXII (47) (fill)	13/7/2005
0296	4	Tomb LVI (43) (lower skeleton)	13/7/2005
0297	2	Dark soil	13/7/2005
0298	2	Cut surrounding SU 0292 = Tomb XIX (54) (cut)	13/7/2005
0299	1	Tomb C (48) (grave)	13/7/2005
0300	1	Tomb C (48) (cut)	13/7/2005
0301	1	Tomb C (48) (fill)	13/7/2005
0302	1	Tomb C (48) (skeleton)	13/7/2005
0303	5	Same as SU 0070 in Trench 5	13/7/2005
0304	2	Tomb XIII (49) (grave)	13/7/2005
0305	2	Tomb XIII (49) (skeleton)	13/7/2005
0306	2	Tomb XIII (49) (fill)	13/7/2005
0307	2	Tomb XX (50) (grave)	13/7/2005
0308	2	Tomb XX (50) (skeleton)	13/7/2005
0309	2	Tomb XX (50) (fill)	13/7/2005
0310	4	Deposit at north end = SU 0203	14/7/2005
0311	4	Burned area within SU 0204 = Tomb XLIX (51) (upper fill)	14/7/2005
0312	2	Tomb XX (50) (cut)	16/7/2005
0313	4	Tomb XLIX (51) (grave)	16/7/2005
0314	4	Tomb XLIX (51) (lower fill)	16/7/2005
0315	4	Tomb XLIX (51) (skeleton)	16/7/2005
0316	1	Tomb XLVIII (52) (grave)	18/7/2005
0317	1	Tomb XLVIII (52) (fill)	18/7/2005
0318	1	Tomb XLVIII (52) (skeleton)	18/7/2005
0319	1	Tomb LV (53) (grave)	18/7/2005
0320	1	Tomb LV (53) (fill)	18/7/2005
0321	1	Tomb LV (53) (skeleton)	18/7/2005
0322	2	Tomb XIX (54) (grave)	18/7/2005
0323	2	Tomb XIX (54) (skeleton)	18/7/2005
0324	4	Tomb XXI (55) (grave)	18/7/2005
0325	4	Tomb XXI (55) (fill)	18/7/2005
0326	4	Tomb XXI (55) (skeleton, adult)	18/7/2005
0327	4	Tomb XXIII (56) (grave)	18/7/2005
0328	4	Tomb XXIII (56) (fill)	18/7/2005
0329	4	Tomb XXIII (56) (skeleton)	18/7/2005
0330	2	Tomb XLI (57) (grave)	19/7/2005

Unit	Trench/ sector	Description	Date
0331	2	Tomb XLI (57) (skeleton)	19/7/2005
0332	2	Tomb XLI (57) (fill)	19/7/2005
0333	2	Calcareous soil in Tomb XIX (54) (fill)	19/7/2005
0334	1	Tomb LVII (58) (grave)	20/7/2005
0335	1	Tomb LVII (58) (fill)	20/7/2005
0336	1	Tomb LVII (58) (skeleton)	20/7/2005
0337	2	Tomb XLII (59) (grave)	20/7/2005
0338	2	Tomb XLII (59) (skeleton 1, adult)	20/7/2005
0339	2	Tomb XLII (59) (skeleton 2, child)	20/7/2005
0340	2	Tomb XLII (59) (skeleton 3, child?)	20/7/2005
0341	4	Tomb XXI (55) (skeleton 2, younger individual)	20/7/2005
0342	1	Cut into Tomb LVII (58) from modern tree planting	21/7/2005
0343	1	Tomb XLV (60) (grave)	21/7/2005
0344	1	Tomb XLV (60) (fill)	21/7/2005
0345	1	Tomb XLV (60) (skeleton on top, in situ)	21/7/2005
0346	2	Tomb LXIV (61) (grave)	22/7/2005
0347	2	Tomb LXIV (61) (skeleton?)	22/7/2005
0348	2	Tomb XLIII (62) (grave)	22/7/2005
0349	2	Tomb XLIII (62) (skeleton)	22/7/2005
0350	2	Tomb XLIII (62) (fill)	22/7/2005
0351	1	Tomb XLV (60) (skeleton on bottom, in situ)	23/7/2005
0352	1	Tomb XLV (60) (skeleton pushed to one side)	23/7/2005
0353	1	Tomb XLV (60) (cut)	23/7/2005
0354	2	Tomb XLII (59) (cut)	23/7/2005
0355	2	Skeleton under Tomb XLVI (42)	23/7/2005
0356	2	Tomb XLIII (62) (second skeleton)	23/7/2005
0357	1	Tomb LIII (63) (grave)	23/7/2005
0358	1	Tomb LIII (63) (fill)	23/7/2005
0359	1	Tomb LIII (63) (skeleton)	23/7/2005
0360	1	Tomb I (64) (grave)	23/7/2005
0361	1	Tomb I (64) (fill)	23/7/2005
0362	1	Tomb I (64) (skeleton)	23/7/2005
0363	4	Tomb XLIV (65) (grave)	23/7/2005
0364	4	Tomb XLIV (65) (fill)	23/7/2005
0365	4	Tomb XLIV (65) (skeleton)	23/7/2005
0366	1	Tomb XLVIII (52) (second skeleton)	23/7/2005
0367	1	Tomb XXXIX (66) (grave)	25/7/2005
0368	1	Tomb XXXIX (66) (fill)	25/7/2005
0369	1	Tomb XXXIX (66) (skeleton)	25/7/2005
0370	2	Dark soil by Tomb XLIII (62) = Tomb XVI (68) (fill)	25/7/2005
0371	2	Tomb XLIII (62) (third skeleton)	25/7/2005
0372	1	Tomb XLVIII (52) (third skeleton)	26/7/2005
0373	1	Tomb XLVIII (52) (cut, continued)	26/7/2005
0374	4	Wall 1 (at middle of Trench 4)	26/7/2005
0375	1	Topsoil/backfill (at start of 2006 season)	22/6/2006
0376	1	Bedrock at north end of trench	22/6/2006
0377	1	Tumulus fill at south end of trench	22/6/2006
0378	1	Dark soil at approximately center of trench	22/6/2006
0379	2	Tumulus fill (majority of trench at start of 2006)	22/6/2006
0380	2	Dark soil in southeast area	22/6/2006

Unit	Trench/ sector	Description	Date
0381	2	Tomb XL (67) (grave)	23/6/2006
0382	2	Tomb XL (67) (fill)	23/6/2006
0383	2	Tomb XL (67) (skeleton)	23/6/2006
0384	4	Ceramic deposits (in western part of SU 0204)	23/6/2006
0385	2	Tomb XVI (68) (fill)	23/6/2006
0386	2	Tomb XVI (68) (skeleton)	23/6/2006
0387	1	Ceramic deposit	24/6/2006
0388	1	Tomb LII (69) (grave)	24/6/2006
0389	1	Tomb LII (69) (fill)	24/6/2006
0390	1	Tomb LII (69) (skeleton)	24/6/2006
0391	2	Tomb XVI (68) (cut)	24/6/2006
0392	1	Dark soil at center of trench	26/6/2006
0393	1	Tomb I (64) (cut, continued)	26/6/2006
0394	4	Tomb XLIV (65) (disarticulated skeleton)	27/6/2006
0395	2	Dark clayish soil south of stone feature in SU 0380	27/6/2006
0396	2	Tomb XXX (70) (grave)	28/6/2006
0397	2	Tomb XXX (70) (fill)	28/6/2006
0398	2	Tomb XXX (70) (skeleton)	28/6/2006
0399	1	Deposit west of topsoil	28/6/2006
0400	4	Tomb XIV (71) (fill) east side of trench, at baulk, below Tomb XXX (70)	28/6/2006
0401	4	Tomb XIV (71) (cut)	28/6/2006
0402	1	Tomb XXXIX (66) (cut)	28/6/2006
0403	2	Dark soil in northeast of trench = Tomb XV (80) (fill)	28/6/2006
0404	4	Tomb XIV (71) (cut)	28/6/2006
0405	4	Tomb XIV (71) (skeleton)	29/6/2006
0406	1	Tomb XVII (72) (grave)	29/6/2006
0407	1	Tomb XVII (72) (fill)	29/6/2006
0408	1	Tomb XVII (72) (skeleton)	29/6/2006
0409	4	Tomb XVIII (73) (cut)	30/6/2006
0410	4	Tomb XVIII (73) (fill)	30/6/2006
0411	4	Tomb XVIII (73) (grave)	30/6/2006
0412	4	Tomb XVIII (73) (skeleton)	30/6/2006
0413	4	Sandy, red soil (tumulus fill)	30/6/2006
0414	1	Tomb XXVI (74) (grave)	30/6/2006
0415	1	Tomb XXVI (74) (fill)	30/6/2006
0416	1	Tomb XXVI (74) (skeleton)	30/6/2006
0417	1	Tomb XXVI (74) (cut)	30/6/2006
0418	2	Tomb XXXVI (75) (grave)	1/7/2006
0419	2	Tomb XXXVI (75) (fill)	1/7/2006
0420	2	Tomb XXXVI (75) (skeleton)	1/7/2006
0421	2	Tomb XXXVI (75) (second skeleton)	1/7/2006
0422	2	Tomb XXXVII (76) (grave)	3/7/2006
0423	2	Tomb XXXVII (76) (fill)	3/7/2006
0424	2	Tomb XXXVII (76) (skeleton)	3/7/2006
0425	4	Tomb XXVIII (77) (fill = yellowish brown soft soil in SU 0286)	3/7/2006
0426	4	Tomb XXVIII (77) (cut)	3/7/2006
0427	4	Tomb XXVIII (77) (fill, harder, grayish soil, with heavy clay content)	4/7/2006
0428	4	Tomb XXVIII (77) (grave)	4/7/2006
0429	4	Tomb XXVIII (77) (skeleton)	4/7/2006
0430	4	Stones (possible tomb cover?)	4/7/2006
0431	2	Dark soil in northeast corner	4/7/2006

Unit	Trench/ sector	Description	Date
0432	4	Tomb LI (78) (grave)	5/7/2006
0433	4	Tomb LI (78) (fill)	5/7/2006
0434	4	Tomb LI (78) (skeleton)	5/7/2006
0435	4	Tomb LI (78) (cut)	5/7/2006
0436	1	Tomb LII (69) (cut)	5/7/2006
0437	1	Tomb XXXVIII (grave)	5/7/2006
0438	1	Tomb XXXVIII (fill)	5/7/2006
0439	1	Tomb XXXVIII (skeleton)	5/7/2006
0440	1	Dark soil near north baulk	5/7/2006
0441	4	Ceramic deposit (western part of SU 0286)	6/7/2006
0442	2	Dark, sandy soil, bordering stone feature	6/7/2006
0443	2	Tomb XV (80) (grave)	6/7/2006
0444	2	Tomb XV (80) (skeleton)	6/7/2006
0445	2	Tomb XV (80) (cut)	6/7/2006
0446	2	Tomb XV (80) (fill, continued)	6/7/2006
0447	1	Tomb XXXVIII (79) (cremated skeleton)	6/7/2006
0448	4	Skeletal unit (animal jaw) in SU 0286	6/7/2006
0449	4	Compact yellow soil associated with SU 0450	7/7/2006
0450	4	Skeletal unit (animal jaw and teeth)	7/7/2006
0451	4	Cut associated with SU 0449	7/7/2006
0452	2	Tomb III (81) (grave)	7/7/2006
0453	2	Tomb III (81) (fill)	7/7/2006
0454	2	Tomb III (81) (skeleton)	7/7/2006
0455	4	Tomb XXVII (82) (grave)	7/7/2006
0456	4	Tomb XXVII (82) (fill)	7/7/2006
0457	4	Tomb XXVII (82) (skeleton to southeast)	7/7/2006
0457a	4	Tomb XXVII (82) (additional teeth)	
0458	2	Fill above, and west of Tomb III (decomposed bedrock?)	7/7/2006
0459	1	Tomb XVII (72) (cut)	8/7/2006
0460	1	Darker soil west of topsoil	8/7/2006
0461	4	Tomb XXVII (82) (skeleton to west)	8/7/2006
0462	1	Dark unit against east-west baulk	8/7/2006
0463	4	Tomb XXVII (82) (cut)	10/7/2006
0464	2	Tomb III (81) (cut)	10/7/2006
0465	1	Bedrock in southern half of trench	10/7/2006
0466	1	Dark unit north of Tomb LII (=Tomb XXIX [83] fill)	10/7/2006
0467	2	Stone feature in SU 0380 (designated Wall 2)	10/7/2006
0468	1	Tomb XXIX (83) (grave)	10/7/2006
0469	1	Tomb XXIX (83) (skeleton)	10/7/2006
0470	1	Tomb XXXV (84) (grave)	10/7/2006
0471	1	Tomb XXXV (84) (fill)	10/7/2006
0472	1	Tomb XXXV (84) (skeleton)	10/7/2006
0473	2	Cut into bedrock, with dark soil, beneath Wall 2	10/7/2006
0474	2	Dark fill in cut SU 0473	10/7/2006
0475	2	Dark soil north of skeleton SU 0454 (of Tomb III), within cut SU 0479 (= upper fill of Tomb XXIV [85])	2/07/2006
0476	1	Tomb XXIX (83) (cut)	12/072006
0477	2	Tomb XXIV (85) (grave)	13/7/2006
0478	2	Tomb XXIV (85) (skeleton)	13/7/2006
0479	2	Tomb XXIV (85) (cut)	13/7/2006
0480	2	Sandy soil on western side, beneath cut SU 0479	13/7/2006
0481	2	Dark soil at eastern edge, beneath SU 0479 (= Tomb XXXI [86] fill)	13/7/2006
0482	1	Soil unit east of SU 0378	13/7/2006
0483	1	Cut associated with SU 0482	13/7/2006
0484	1	Tomb XXXV (84) (cut)	13/7/2006
0485	2	Lower fill of cut SU 0473	13/7/2006
0486	1	Cut associated with SU 0378	13/7/2006
0487	2	Transitional soil at interface of topsoil and SU 0431	13/7/2006
0488	2	Tomb XXXI (86) (grave)	13/7/2006
0489	2	Tomb XXXI (86) (skeleton)	13/7/2006
0490	2	Rectangular cut for Tomb XII (88)	17/7/2006
0491	2	Dark fill in cut SU 490 = Tomb XII (88) (fill)	17/7/2006
0492	1	Cut associated with SU 0460	17/7/2006
0493	2	Ceramic deposit within SU 0487	17/7/2006
0494	2	Tomb XII (88) (grave)	17/7/2006
0495	2	Tomb XII (88) (skeleton 1, head to north)	17/7/2006
0496	1	Tomb XXXIV (87) (grave)	17/7/2006
0497	1	Tomb XXXIV (87) (fill)	17/7/2006
0498	1	Tomb XXXIV (87) (skeleton)	17/7/2006
0499	2	Tomb XII (88) (skeleton 2, head to south)	17/7/2006
0500	1	Unit northeast of Tomb XXXIV (87), heavy clay content	18/7/2006
0501	2	Tomb XII (88) (skeleton 3, head on chest of SU 0499)	18/7/2006
0502	2	Tomb XXXI (86) (cut)	18/7/2006
0503	1	Clay area south of Tomb XXXIV (87)	19/7/2006
0504	1	Tomb XXXIV (87) (skeleton of second individual)	19/7/2006
0505	2	Tomb XII (88) (northern fill)	20/7/2006
0506	2	Tomb XXXII (89) (grave)	20/7/2006
0507	2	Tomb XXXII (89) (fill)	20/7/2006
0508	2	Tomb XXXII (89) (south skeleton)	20/7/2006
0509	1	Dark ashy-colored unit west of topsoil, 3 m north of Trench 6	20/7/2006
0510	2	Patch of dark soil at south end = Tomb XXV (90) (fill)	21/7/2006
0511	2	Tomb XXV (90) (grave)	21/7/2006
0512	2	Tomb XXV (90) (skeleton)	21/7/2006
0513	2	Tomb XXV (90) (cut)	22/7/2006
0514	2	Tomb II (91) (grave)	22/7/2006
0515	2	Tomb II (91) (fill)	22/7/2006
0516	2	Tomb II (91) (skeleton)	22/7/2006
0517	1	Isolated(?) cranial bones in SU 0051 (at south end)	22/7/2006
0518	1	Soil unit running northeast in east-west baulk	22/7/2006
0519	1	Tomb II (91) (cut)	22/7/2006
0520	2	Dark brown soil at west half	24/7/2006
0521	1	Tomb XXXIII (92) (grave)	24/7/2006
0522	1	Tomb XXXIII (92) (fill)	24/7/2006
0523	1	Tomb XXXIII (92) (skeleton)	24/7/2006
0524	2	Tomb XXXII (89) (north skeleton)	24/7/2006

Unit	Trench/ sector	Description	Date
0525	2	Sandy soil north of Tomb XXV (90), above northern fill of Tomb II (91)	24/7/2006
0526	1	Cut associated with SU 0446	24/7/2006
0527	2	Sandy soil beneath fill of Tomb XII (88)	24/7/2006
0528	2	Tomb XXXII (89) (cut)	25/7/2006
0529	7	Deposit below topsoil in Trench 7	25/7/2006
0530	7	Dark orange soil from baulk cleaning	25/7/2006
0531	7	Gray soil from baulk cleaning	25/7/2006
0532	2	Deposit in area of SU 0380	29/6/2007
0533	2	Deposit in area of SU 0487	29/6/2007
0534	5	Clayey calcareous unit	29/6/2007
0535	5	Loose yellow sandy soil	29/6/2007
0536	5	Dark calcareous soil, with clumps at south end	29/6/2007
0537	5	Tomb LXXX (4) (fill, continued)	29/6/2007
0538	2	Dark soil with bone fragments =Tomb XI (93) (fill)	29/6/2007
0539	2	Tomb XI (93) (grave)	29/6/2007
0540	2	Tomb XI (93) (skeleton)	29/6/2007
0541	6	Soil unit below topsoil, west end	30/6/2007
0542	6	Soil unit below SU 0541, west end	30/6/2007
0543	2	Patch of dark soil with bone fragments (pit within cut SU 0563)	30/6/2007
0544	2	Clayey soil at northeast, below SU 0532 and topsoil	30/6/2007
0545	2	Tomb IX (94) (grave)	2/7/2007
0546	2	Tomb IX (94) (fill)	2/7/2007
0547	2	Tomb IX (94) (skeleton)	2/7/2007
0548	5	Dark soil underlying SU 0535	2/7/2007
0549	5	Tomb LXIX (27) (fill, continued)	2/7/2007
0550	6	Soil below topsoil, 2.85 m west and beyond	2/7/2007
0551	5	Tomb LXIX (27) (second skeleton)	2/7/2007
0552	2	Tomb X (95) (fill): soil north of Tomb IX (94), bordering topsoil; very sandy	3/7/2007
0553	2	Tomb X (95) (grave)	3/7/2007
0554	2	Tomb X (95) (skeleton)	3/7/2007
0555	6	Soil below SU 0550, 2.85–3.75 m west of center	3/7/2007
0556	6	Dark soil below SU 0555	3/7/2007
0557	6	Soil below SU 0556	3/7/2007
0558	5	Ceramic unit	3/7/2007
0559	6	Soil east of SU 0557	3/7/2007
0560	6	Soil below and east of SU 0559	3/7/2007
0561	5	Same deposit as Trench 6 SU 0542	4/7/2007
0562	6	Soil unit just above bedrock	4/7/2007
0563	2	Cut for pit of SU 0543	5/7/2007
0564	2	Triangular shelf of soil, north of Pit SU 0473–0474, east of bedrock	5/7/2007
0565	2	Loose sandy soil southeast of Pit SU 0473–0474, east of bedrock	5/7/2007
0566	2	Sandy soil, close to bedrock, in east part of trench	5/7/2007
0567	2	Tomb V (96) (grave)	5/7/2007
0568	2	Tomb V (96) (fill)	5/7/2007
0569	2	Tomb V (96) (skeleton)	5/7/2007
0570	2	Soil at southeast corner, below topsoil	5/7/2007
0571	2	Rectangular patch of dark soil = Tomb VI (97) (fill)	6/7/2007
0572	2	Tomb VI (97) (grave)	6/7/2007
0573	2	Tomb VI (97) (skeleton)	6/7/2007
0574	2	Tomb VI (97) (cut)	6/7/2007
0575	2	Tomb V (96) (cut)	6/7/2007
0576	2	Soil below SU 0543 and cut SU 0563	6/7/2007
0577	2	Soil beneath SU 0563 continuing from Trench 1	7/7/2007
0578	2	Ovoid area of dark soil at east slope, within SU 0570 = Tomb VII (99) (fill)	7/7/2007
0579	2	Tomb IV (98) (grave)	9/7/2007
0580	2	Tomb IV (98) (fill)	9/7/2007
0581	2	Tomb IV (98) (skeleton)	9/7/2007
0582	2	Tomb VII (99) (grave)	9/7/2007
0583	2	Tomb VII (99) (skeleton)	9/7/2007
0584	2	Tomb VII (99) (cut)	9/7/2007
0585	2	Loose sandy soil beneath Tomb V (96) and northwest of large stone	9/7/2007
0586	2	Dark soil, pit south of Tomb VII (99), at northwest end of rock = Tomb VIII (100) (fill)	10/7/2007
0587	9	Topsoil	10/7/2007
0588	2	Tomb VIII (100) (cut)	10/7/2007
0589	9	Bedrock	10/7/2007
0590	2	Tomb VIII (100) (grave)	10/7/2007
0591	2	Tomb VIII (100) (skeleton)	10/7/2007
0592	2	Fill beneath Tomb IV (98)	13/7/2007
0593	2	Skeletal unit assigned to loose bone from SU 0040	26/7/2007
0594	2	Skeletal unit assigned to loose bone from SU 0202	26/7/2007
0595	2	Skeletal unit assigned to loose bone from SU 0202	26/7/2007

CHAPTER 2.4
INVENTORIED FINDS BY CONTEXT

Sarah P. Morris and John K. Papadopoulos

The following list includes only those excavation units that were topsoil or tumulus fill. Material from the fill of individual tombs is presented separately in Chapter 3. Individual items composed of fragments from different units are listed under both units, as are joining fragments from the same unit. Units that yielded no material are not listed here.

Surface		
2003	Pottery	**9/11, 9/28, 9/29, 9/88, 9/279, 9/330**
	Lithics	**13/3, 13/37, 13/59,** SF007, SF008
2005	Modern tile	**3/1**
Ngrançija	Modern tiles	**3/2**

Topsoil		
Trench 1		
Unit 1	Pottery	**9/44, 9/65, 9/78, 9/251, 9/296, 9/305,** P015
	Daub	SF011, SF012
	Lithics	**13/27, 13/30, 13/44, 13/83,** SF031, SF033, SF207
	Iron (modern)	**10/137**, SF001, SF002
Unit 7	Pottery	**9/99, 9/102, 9/120, 9/131, 9/178, 9/260, 9/262, 9/293**
	Daub	SF017
	Lithics	SF094, SF148
Unit 9	Pottery	**9/48, 9/234**
	Lithics	**13/38, 13/49, 13/61**, SF088, SF096, SF097
	Gun cartridge/bullet	**A3/3, A3/4**
Unit 141	Pottery	**9/26, 9/79, 9/97, 9/132, 9/216, 9/229**, P101, P102, P385
	Daub	**14/17**
	Lithics	**13/32**, SF185
Unit 278	Pottery	**9/41, 9/59, 9/239, 9/333, 9/341**, P263, P264, P274
	Lithics	**13/69, 13/71**, SF233, SF239
	Modern tile	SF285 (see Chapter 3.2)
	Modern iron	**10/138**
Unit 375	Pottery	**9/130, 9/336**, P294, P312
	Lithics	SF319
	Gun cartridge/bullet	**A3/9, A3/13, A3/18, A3/23, A3/28**
	Modern glass	SF401
Trench 2		
Unit 2	Pottery	**9/9, 9/37, 9/56, 9/139, 9/185, 9/200, 9/207, 9/249, 9/253, 9/267, 9/270, 9/306, 9/314, 9/331, 9/338**, P162, P407, P415, P425
	Lithics	**13/9, 13/18, 13/23, 13/28, 13/53, 13/72, 13/78, 13/90**, SF35, SF187, SF235, SF322, SF323, SF328, SF361, SF368, SF375, SF379, SF397, SF399, SF441
	Ancient metal	**10/40**
	Modern iron and glass	**10/137, 10/141, 10/142, 10/144**, SF004, SF088, SF211, SF374
	Gun cartridge/bullet	**A3/1, A3/5, A3/7, A3/12, A3/15, A3/16, A3/20, A3/22, A3/30, A3/32**, SF183, SF184, SF186, SF192, SF254, SF397, SF399

Topsoil		
Trench 2 continued		
Unit 3	Pottery	**9/211, 9/231**, P374
	Lithics	**13/24, 13/64**, SF237, SF369
	Ancient metal (?)	**10/88, 10/128**
	Modern metal	**10/143**, SF392
	Gun cartridge/bullet	**A3/24, A3/27**
Unit 274	Modern metal (?)	**10/123** found near Tomb LX
Backfill	Lithics	SF327, SF329
Trench 3		
Unit 3	Pottery	**9/308**, P009
	Daub	SF011
Unit 8	Lithics	**13/4**, SF027
Unit 45	Lithics	SF053
Unit 64	Pottery	**9/239**
	Modern iron	SF060
	Spindlewhorl	**10/10**
Trench 4		
Unit 4	Pottery	**9/62, 9/222**
	Lithics	**13/1**
	Ancient metal	**10/99**
	Glass bead	**10/110**
	Gun cartridge/bullet	**A3/2, A3/11**, SF087
	Modern iron	SF090
Unit 201 (=Unit 4)	Pottery	**9/42, 9/126, 9/129, 9/165, 9/192, 9/210, 9/218, 9/244, 9/247, 9/275, 15/19**, P152, P265, P286, P348
	Bronze fibula	**10/18**
	Spindlewhorl	**10/7**
	Lithics	**13/21, 13/64, 13/75**, SF178, SF185, SF275
	Gun cartridge/ bullet	**A3/21**
Trench 5		
Unit 1	Pottery	**9/50**, P396
Trench 6		
Unit 2	Pottery	**9/19, 9/314**
	Lithics	SF432
	Gun cartridge/bullet	**A3/31**
	Modern iron	SF432, SF435
Unit 141	Gun cartridge/bullet	**A3/32**

Topsoil		
Trench 7		
Unit 2	Lithics	SF400
Unit 27	Lithics	**13/60**
Units 186–187	Pottery	**9/223**
Trench 8		
Unit 22	Pottery	**9/182, 9/241**
	Petrified wood	SF048
Unit 201	Pottery	**9/57**
	Modern glass	SF407
Tumulus fill		
Trench 1		
Unit 35	Lithics	SF100
Unit 39	Pottery	**9/39, 9/45, 9/61, 9/69, 9/71, 9/108, 9/125, 9/133, 9/190, 9/193, 9/266**, P129, P130, P148, P183, P184
	Spindlewhorl, disk	**10/4, 10/9**
	Lithics	**13/19, 13/36**, SF034, SF068, SF102, SF143
	Daub	**14/6, 14/7**
	Iron fragment	SF109
Unit 47	Pottery	**9/146, 9/183, 9/225, 9/264**
	Bronze pin fragment	**9/125**
	Lithics	SF047
	Daub	**14/4**
Unit 67	Pottery	**9/127, 9/151, 9/184, 9/199, 9/219, 9/312**
	Spindlewhorl	**10/8**
	Iron bead	**10/93**
	Lithics	**13/6, 13/31**, SF062, SF063, SF198
	Daub	14/13, SF166
Unit 70	Pottery	**9/12, 9/31, 9/34, 9/36, 9/54, 9/55, 9/72, 9/75, 9/76, 9/79, 9/83, 9/86, 9/100, 9/109, 9/143, 9/148, 9/170, 9/180, 9/181, 9/194, 9/197, 9/208, 9/213, 9/214, 9/235, 9/237, 9/249, 9/252, 9/285, 19/288, 9/300, 9/301, 9/304, 9/320, 9/321, 9/327, 9/332, 9/340**, P138, P177, P187, P188, P200
	Bronze fragments	**10/126**
	Lithics	**13/7, 13/20, 13/51, 13/68**, SF075, SF082, SF120, SF 200
	Daub	**14/10**

Tumulus fill		
Trench 1, continued		
Unit 91 (ceramic unit within 39)	Pottery	**9/6**
Unit 112	Iron object	**10/136**
Unit 240	Pottery	**9/27, 9/66, 9/70, 9/74, 9/196, 9/246, 9/248, 9/258, 9/333**, P270
	Lithics	SF279
Unit 279 (= Unit 70, below 1 m)	Pottery	**9/10, 9/27, 9/47, 9/52, 9/55, 9/83, 9/119, 9/142, 9/247, 9/257, 9/290, 9/310, 9/322, 9/333**
	Bronze pin fragment	**10/32**
Unit 281 (= Unit 39, below 1 m)	Pottery	**9/89, 9/161**
Unit 291 (ceramic unit)	Pottery	**15/11, 9/161**
Unit 377	Pottery	**9/67, 9/82, 9/123, 9/176, 9/187, 9/292, 9/302, 9/319**, P308
	Lithics	**13/52**
	Daub	SF355
Unit 378	Lithics	**13/40**, SF383
	Daub	SF355
Unit 378	Lithics	**13/40**, SF383
Unit 387	Pottery	**9/307**
Unit 399 (west of topsoil; probably topsoil)	Pottery	**9/232, 9/311**
	Lithics	SF398
	Gun cartridge/bullet	**A3/8**
Unit 440	Pottery	**9/58, 9/79, 9/116, 9/195, 9/329**, P355, P358
	Lithics	SF358
Unit 509	Pottery	**9/335**
Trench 2		
Unit 6	Pottery	**15/2, 9/152** (2 frr), **9/163**
Unit 10	Pottery	**9/149**
	Bronze fragment	SF417
Unit 23	Pottery	**9/261**
	Daub	**14/11**

Tumulus fill		
Trench 2, continued		
Unit 33	Pottery	**9/30**
Unit 37	Lithics	SF058
Unit 40	Ancient metal	**10/39, 10/51,** SF418
Unit 41	Pottery	**9/111, 9/140**
Unit 66	Pottery	**9/236**
	Ancient metal	**10/49**
	Lithics	SF099
Unit 71	Pottery	**9/134**
	Daub	**14/5**
Unit 78	Pottery	**9/55, 9/101, 9/188, 9/212**
	Bronze fragments	SF419, SF420, SF421
	Lithics	**13/41**, SF140
Unit 97	Lithics	SF131
Unit 117	Ancient metal	**10/30**
	Lithics	**13/56, 13/80**
Unit 118	Ancient metal	**10/131**
	Lithics	**13/92**
Unit 202	Pottery	**9/2, 9/113, 9/128, 9/161, 9/166, 9/168, 9/169, 9/171, 9/256, 9/269, 9/280, 9/289, 9/291, 9/299, 9/317, 15/10,** P144, P225, P433
	Ancient metal	**10/61**
	Lithics	**13/16, 13/46, 13/58, 13/77,** SF174, SF175, SF176, SF177, SF191, SF196, SF201, SF202, SF203, SF204, SF205, SF 216, SF222, SF223, SF224
	Daub	**14/12**, SF219
Unit 217	Pottery	**9/13**
Unit 226	Lithics	**13/42**
Unit 227	Pottery	**9/174, 9/263**
Unit 228	Pottery	**9/323**
	Lithics	**13/2**
Unit 229	Lithics	**13/42**
Unit 235 (=Interface Topsoil and Tumulus Fill)	Lithics	SF271
	Modern tile	SF268 (see Chapter 3.2)
Unit 239	Pottery	**9/103, 9/147**
	Terracotta	SF269
	Lithics	**13/43**
Unit 253	Pottery	P190
	Lithics	SF179
Unit 288	Pottery	**9/106, P241**

Tumulus fill		
Trench 2 continued		
	Lithics	**13/10**
Unit 297	Pottery	**9/104**
	Lithics	**13/11**, SF248, SF270, SF277
Unit 380	Pottery	**9/175, 9/276**, P297
	Daub	SF309
	Lithics	**13/54, 13/73**
Unit 395	Pottery	**9/112, 9/217**
	Lithics	SF345, SF363
Unit 403	Pottery	**9/292, 9/337**
Unit 431	Pottery	**9/273, 9/313, 9/334**, P329, P330, P331, P359
	Lithics	**13/45, 13/67, 13/84**, SF365, SF367
Unit 474 (fill of pit)	Pottery	**9/122**
	Bronze fragment	**10/130**
	Lithics	**13/15, 13/65, 13/81**, SF376
Unit 487 (transitional from topsoil)	Pottery	**9/326**
	Lithics	SF403
Unit 493	Pottery	**9/92**
Unit 520	Pottery	**9/233**
Unit 532	Pottery	**9/135**
	Lithics	**13/94**
	Shell (mother-of-pearl)	SF433 (see Chapter 16.2)
Unit 543	Lithics	SF427
Trench 3		
Unit 45	Pottery	**9/118, 9/209**
Unit 68	Pottery	**9/100**
	Lithics	SF95
Unit 69	Lithics	SF061
Trench 4		
Unit 11 (Interface Topsoil/ Tumulus Fill)	Pottery	**9/32, 9/114, 9/117, 9/201, 9/238, 9/294**, P025
	Lithics	**13/5, 13/62**, SF021, SF030, SF038, SF055, SF071
	Gun cartridge/ bullet	**A3/17**
Unit 34 (ceramic deposit, cf. Tr. 5, Unit 558)	Pottery	**9/245**

Tumulus fill		
Trench 4, continued		
Unit 35	Pottery	**9/84**
	Lithics	**13/33**
Unit 48	Pottery	**9/221**
	Lithics	**13/50**, SF101
Unit 57	Lithics	SF076
Unit 79	Pottery;	**9/17, 9/22**
	Lithics	13/74
Unit 86 (below Unit 11)	Pottery	**9/60, 9/167, 9/203**
	Lithics	**13/26**, SF122, SF142
Unit 104 (below Unit 48)	Pottery	**9/33, 9/162**
	Lithics	**13/63**
Unit 129	Pottery	**9/110, 9/295**
	Lithics	SF112, SF115, SF146
Unit 130	Pottery	**9/189, 9/250**
Unit 203	Pottery	**9/77, 9/80, 9/141, 9/150**
Unit 204	Pottery	**9/20, 9/38, 9/43, 9/51, 9/68, 9/105, 9/107, 9/137, 9/153, 9/220, 9/242, 9/286, 9/287, 9/324**, P164, P298
	Ancient metal	**10/31, 10/134**
	Lithics	**13/8, 13/48, 13/70**, SF188, SF189, SF221, SF281
	Daub	SF286
Unit 205	Pottery	**9/18, 9/138**
Unit 206 (ceramic unit)	Pottery	**9/265, 9/274**
Unit 207 (= Unit 129)	Lithics	SF208
Unit 280 (ceramic unit)	Pottery	Bitumen pot **9/259, 15/12**
Unit 286 (below Unit 205)	Pottery	**9/20, 9/87** (2 frr)**, 9/121, 9/144, 9/173, 9/198, 9/215, 9/255, 9/303, 9/328, 9/339**, P266, P292, P300, P311, P315, P319, P341, P352
	Worked bone	**10/124**
	Lithics	**13/17**, SF243, SF273, SF334, SF347
	Daub	**14/15**, SF330, SF331, SF332, SF333
	Gun cartridge/ bullet	**A3/6**
Unit 441 (ceramic unit in 286)	Pottery	**9/186**
	Daub	**14/16**

Tumulus fill		
Trench 5		
Unit 240	Pottery	P379
Unit 278	Pottery	P378
Unit 279	Pottery	9/278
Unit 535	Pottery	**9/115, 9/145, 9/205**
Unit 536	Pottery	**9/53, 9/172, 9/271, 9/272, 9/282**
	Bronze fragment	SF426
Unit 548	Pottery	**9/191, 9/230, 9/297**
	Daub	**14/9**
(Units 534, 535, 536, 548)	Pottery	P390, P393, P402, P409, P410, P427
Unit 558 (cf. ceramic Unit 34)	Pottery	**9/254**
Unit 570	Daub	**14/8**
Trench 6		
Unit 181 (= Unit 70)	Pottery	**9/55**
Unit 240	Pottery	**9/160**
Unit 541	Pottery	**9/277**, P403
Unit 542	Pottery	**9/72**
	Lithics	**13/66**, SF429
Unit 550	Pottery	**9/49, 9/86**
Unit 556	Pottery	**9/290**, P419
Unit 559	Pottery	**9/63, 9/204**
Unit 562 (above bedrock)	Pottery	P412, P420
Trench 7		
Unit 192	Pottery	**9/87**
Unit 202	Pottery	**9/161, 15/10**
Trench 8		
Unit 129	Lithics	**13/55**
Unit 204	Pottery	P381, P382
	Bronze fragment	**10/127**
Unit 207	Lithics	SF406

CHAPTER 2.5
AERIAL PHOTOGRAPHY

John K. Papadopoulos and Lorenc Bejko
(Photographs by Alket Islami)

In the course of the 2005 campaign it became clear that low-level aerial photography would greatly enhance the process of recording the tumulus and some of its exposed features, such as individual tombs, as well as other exposed remains, like Wall 1. In neighboring Greece, one could employ a variety of different techniques to achieve high-quality aerial photography, from balloon photography to photographs taken from airplanes or helicopters. The simplicity and relatively inexpensive way of making low vertical photographs showing impressive detail from a tethered balloon, used to great effect in Crete and other parts of the Greek world (e.g., Myers, Myers, and Cadogan 1992:5–14; see also Myers 1978; Myers and Myers 1980, 1990), proved difficult in the context of Albania in 2005. First of all, there was the issue of coaxing a reputable balloon photographer to the wilds of the region between Ballsh and Fier, and then there was the problem of procuring necessary supplies, not least helium, which had to be imported. Moreover, although balloon photography provided high-quality vertical images, the possibility of also having non-vertical aerial photography that captured the tumulus within its landscape from different angles, was greatly minimized. The other expedient—renting a light airplane or helicopter—was also not straightforward. At the time, the only non-military helicopter belonged to the Prime Minister of Albania, which later unexpectedly crashed into the Adriatic on a medical emergency flight to Italy, killing all on board. And renting a light aircraft in nearby Greece and flying it over Albanian airspace did not seem like the wisest, or safest, of options. It is important to add that our excavations were conducted in the era before the straightforward availability of small drones for archaeological photography.

In researching the options of aerial photography in 2005, Bejko came across Alket Islami, who was, at the time, preparing a volume of 101 photographs of Albania from the air, under the auspices of the Albanian Ministry of Tourism (Islami 2006). At the very conclusion of the 2005 season, the day before the last day of the season, Islami arrived with his engine-powered paraglider (often referred to as a paramotor) and began taking low-level aerial photographs of the site (**Fig. 2.20**). A two-person paramotor—as opposed to our one-person version—was used to great effect in a much more comprehensive aerial survey of Armenia (Faustmann and Palmer 2005). As it turned out, this was the first time that Islami took aerial photographs of an archaeological project during excavations, and the result was such a success that he was, and still is, in great demand by both Albanian and foreign archaeological projects in the country; so much so, that we were unable to secure his services in 2006 and 2007 to photograph later phases of exposure of the Lofkënd tumulus, due to his growing commitments.

At the time Islami arrived at the site to take the aerial photographs in 2005, the only exposed features were Wall 1 and Tomb XLVIII (52). Both are clearly visible in the aerial view from the north-northeast (**Fig. 2.21**), as well as in the more vertical view from above west (**Fig. 2.22**). In well under an hour, Islami took some 60 photographs, using his digital camera in one hand, steering the paraglider with the other hand. Two additional aerial photographs well capture the tumulus in its greater landscape. In **Figure 2.23** the mound is at the center of the frame, with the modern Muslim cemetery on the natural terrace below, with the ground sloping away to the south and southeast; Margelliç is visible on the skyline to the left. In contrast, **Figure 2.24** shows the tumulus from due north, with the access road to the east (left), the surrounding fields, recently harvested, and the mountains in the distance to the south.

CHAPTER 2.6
TWO MINUTES IN THE LIFE OF LOFKËND: A PANORAMA OF THE SITE BY RICHARD MACDONALD, JULY 4, 2006

Richard MacDonald and John K. Papadopoulos

Throughout the course of the four years in which Richard (Rich) MacDonald was an integral member of the Lofkënd team, he was continuously experimenting using diverse techniques in photographically recording the excavations. Of the numbered document photos of individual tombs and other features, there was a great deal of experimentation with different lighting effects. The baulks of the tumulus, particularly once the excavations reached a certain depth, presented challenging shadows that required some sort of even lighting, such as that achieved by shading the item framed with clear plastic or white sheeting, to prevent affecting the true color of the deposits excavated. Direct sunlight, whenever possible, was the preferred solution, particularly as shadows picked out details of individual bones. But burials either close to a baulk, or those exposed by means of a tunnel through a baulk (e.g., Tomb XLIV [65], **Fig. 3.143a–b**), did not permit direct and even sunlight. The process of experimenting with various lighting effects can be seen in four views of a single tomb taken on the same morning, some only minutes apart, but employing different shading (**Fig. 3.168a–d**, Tomb LII [69]). In addition to the many numbered and unnumbered photographs that MacDonald took of the site and of the people who excavated and visited the place (see *http://www.sscnet.ucla.edu/ioa/staff/papadopoulos/lofkend/slides/slideshow.html*), he took detailed video footage of the excavation, as well as of the making of the mud bricks for the reconstruction of the tumulus (Chapter 22).

Among other things, he began stitching together photographs that presented a panoramic view, and these were among his most successful experiments. Among the many overlapping exposures he took, we decided upon one image, **Figure 2.25**, which documented, in a single panorama, no more than two minutes in the life of the tumulus. The panorama was shot on July 4, 2006, an auspicious day for the American members of the team. MacDonald placed his mounted camera directly above the highest point of the tumulus, and by positioning the camera at the place where the baulks intersected, an equal view into all four sectors or trenches of the tumulus was achieved. Composed from around a dozen photographs taken over a two-minute time span, the panorama caught many of the everyday excavation tasks. At the far left, MacDonald's shadow, together with that of his camera, is appropriately captured; not even the photographer can escape documentation! From left to right, two workmen, Enriko Jaupaj and Ndriçim Beqiri, are cleaning the ground immediately to the west of the tumulus in preparation for photography. One of the student excavators, Joanna Potenza, in Sector/Trench 4, is holding the prism, while Esmeralda Agolli and Mandi Hasa, in the far distance, near the excavation shed, are taking readings from the Total Station; and between the two stands the project architect/surveyor, Max Farrar, recording the readings. Immediately to the east (right), standing right at the baulk, Papadopoulos is writing in his notebook. In the center of the panorama, looking into Sector/Trench 1 with its many umbrellas, the frame is dominated by Lorenc Bejko reclining while excavating a tomb, assisted by Rovena Kurti, who is partially obscuring him. In the distance, obscured by the umbrellas, another workman (Barjam Kapllani) is walking from north to south. To the right (east of Bejko and Kurti), Arben Malaj and Lyssa Stapleton, the latter obscured by the baulk, are excavating another tomb. To the right, looking into Sector/Trench 2, Rudenc Ruka is excavating the upper levels of yet another tomb, assisted by Artur Dervishi, while in the foreground another workman, Adriatik Malaj, is digging into the tumulus fill. To the far right, Seth Pevnick stands alone, near the baulk, taking notes (or a Munsell reading).

The video footage taken by MacDonald, together with other panoramas of the site, are made available on the project website through the Cotsen Institute of Archaeology at UCLA (*http://www.sscnet.ucla.edu/ioa/staff/papadopoulos/lofkend/index.html*).

CHAPTER 3

CATALOGUE OF TOMBS AND THEIR CONTENTS

John K. Papadopoulos, Lorenc Bejko, and Sarah P. Morris
With a contribution by Shrepsa Gjongecaj

CHAPTER 3.1

CATALOGUE OF THE PREHISTORIC TOMBS AND THEIR CONTENTS

John K. Papadopoulos, Lorenc Bejko, and Sarah P. Morris

TOMB DEPOSIT I

Pit tomb, multiple secondary inhumation (three adult males)

Phase I (AMS Charcoal ^{14}C: Cal 1373 ± 57 BC)

Figs. 3.1–3.8

0.0–2.0E, 1.0N–2.0N (high: 107.64 m = 354.64 m ASL; low: 106.62 m = 353.62 m ASL)

Sector 1; SU 360 (grave); 361 (fill); 362 (skeletal remains); 393 (cut)

Excavated as Grave 64 (June 28, 2006–July 19, 2007)

The circumstances of this tomb were unique. The grave pit was first encountered near the beginning of the 2006 season, and by the end of that season the deposit continued to greater depth; it was not until a relatively advanced stage of the 2007 campaign that bedrock was reached and the human remains cleared. In the end, some nine arbitrary levels of bone were recorded and cleared; these are illustrated in both plans (**Figs. 3.1a–f, 3.2**) and photographs (**Figs. 3.3a–h, 3.4**). At the uppermost level a clearly defined pit was discerned only on the north, south, and east sides; toward the west the grave cut was less clear. The pit fill, however, was darker than the surrounding deposits and easily distinguishable. At this upper level, the pit was irregular in shape, and it was only with depth that the pit assumed its more or less rectangular shape, with rounded corners. Within the pit, at the uppermost level, the bones recovered were in almost total disarray, which was to continue throughout all levels. Clearly distinguishable pieces from northeast-southwest included fragments of humerus, scattered rib bone fragments, one complete lower rib, several hand and foot bones, six vertebrae and, perhaps, a fragmentary lower arm bone. At a slightly lower level, a portion of a pelvis was noted, in addition to several more vertebrae and fragments of hand. Throughout all of the levels, the bone was concentrated in the north, east, south, and central portions of the tomb; the western portion of the tomb pit was devoid of human remains, and bedrock was reached in the western quarter of the pit at a higher level than in the rest of the pit (see plan, **Fig. 3.1e**, Level VIII). The first mandible was encountered in Level III (**Fig. 3.1b**), and another in Level IV (**Fig. 3.1b**); the first mandible was almost 0.35 m below the highest bone found. It was also at Level III that the pit assumed its shape as a rectangle with rounded corners. At its greatest extent, the pit measured 2.20 m long, 1.20 m wide, and 1.02 m deep; the actual cut in the bedrock was less, 0.69 m deep. This was considerably larger than any of the other tomb pits in the tumulus.

The total weight of human bone recovered from the pit was 2.55 kg (**Fig. 3.8**). The hand elements consisted of 12 carpals, 4 first metacarpals, 18 other, and 34 phalanges. Foot elements comprise four tarsals, two first metatarsals, two other metatarsals, and two phalanges (there were more hand than feet bones). Very durable bone, such as femora, tibiae, and crania, was not represented. The three individuals, first identified on the basis of mandibles, were all probably male: SU 362 was an adult aged 25–30 years, SU 362a was a younger adult male aged 18–23 years, and SU 362b was a mature adult, 45 years or older (cf. the three individuals in Tomb LXXIV [Grave 29]). In addition to the human remains, the deposit yielded a comparatively large number of animal bones, particularly given the relatively low number of human remains. On the basis of Marston's preliminary analysis of the animal bone remains, the grave yielded several pieces of pig, including molar and podial fragments, in addition to a scapula blade fragment, several fragments of sheep/goat, and the long bone of an unidentified mammal. Such a quantity of faunal material was extremely rare in other tombs at Lofkënd. The two only "grave goods," if they can be referred to as grave goods, **TI-1** and **TI-2** (**Figs. 3.5–3.6**), both fragmentary bone pins, were not recognized as artifacts in the field but considered to be animal bone fragments. This was exacerbated by their fragmentary state and the fact that neither pin was decorated in any way (the better-preserved **TI-1** terminated in a plain head). In addition to the fragments of bone pins, the deposit yielded a small number of pottery fragments (inventoried fragments: P 316, P 354, P 366), a few of which are illustrated (**Fig. 3.7a–c**), a piece of daub preserving at least two rod or stake impressions (SF 397, **Fig. 3.7d**), and a chert blade (SF 396, **Fig. 3.7e**), which is Paleolithic. To what extent the two bone pins were directly associated with the remains of any of the deceased in the pit, or represent residual material, like the daub, pottery, and lithic pieces that made their way into the fill, is difficult to establish given the nature of the human remains.

The unique elements of this tomb deposit are noteworthy: the pit itself was both larger and deeper than all other grave pits in the tumulus. Moreover, the pit was centrally located, particularly in terms of the bedrock ridge that underlay the tumulus. The pit was among the earliest, if not *the* earliest feature established at the site. All the human remains were disarticulated and clearly represented secondary inhumation. Furthermore, although remains from at least three individuals were identified, there was nothing even approaching a complete skeleton, and it almost appears as if certain bones, such as femora, tibiae, and crania, may have been deliberately overlooked by those who collected them in antiquity (cf. Liston 2007). In addition, the pit contained human *and* animal bone remains, with the faunal pieces intermingled with the human remains. Another interesting aspect of the pit was that the disarticulated bone was recovered in some nine arbitrary layers; there appeared to be no system in the manner in which the bone was deposited. These numerous unique characteristics of Tomb I are such that they suggest this may well have been a foundational event: the conceptual beginning of the tumulus. This, and other aspects of the tomb, including what it may tell us of the burying population, is considered more fully elsewhere in this volume (Chapter 22 and Epilogue).

As for chronology, two elements combine to indicate that Tomb I (64) is among the very earliest, if not *the* earliest tomb of the tumulus. First of all, the grave pit was overlain by no fewer than four tombs (Tombs LXXIX, LXIX, LXIII, XLV [Graves 6, 27, 35, 60]; the feet of Tomb LXXII [Grave 24] were located right at the edge of the pit for Tomb I) and may have also been stratigraphically interrelated. One of the burials which clearly overlay Tomb I, Tomb XLV [Grave 60], itself yielded a tenth-century BC calibrated AMS ^{14}C date. No other burial in the mound was overlain by so many other prehistoric burials. Second, a ^{14}C date deriving from a charcoal sample from the grave pit gave a calibrated date of 1373 ± 57 BC. We hasten to add that this sample does not derive from human bone, but from a small piece of charcoal of oak found in the fill of the tomb pit. The charcoal may well be earlier (note that the tomb also yielded one lithic tool that is Paleolithic, **Fig. 3.7e**) or, conceivably, even later than the human bone, having made its way into the fill well after the human remains had been recovered from their original resting place and reburied in Tomb I.

The two following pins were found intermingled with the human and animal bone recovered from the grave and may well represent material that made its way into the fill (like the lithic tool, fragment of daub, and pottery sherds, **Fig. 3.7a–e**) rather than grave goods directly associated with one or other of the deceased. Both pins were recovered as part of the skeleton, SU 362, rather than the grave fill (SU 361).

TI-1 (SF 411) **Fig. 3.5**

Fragmentary bone pin, Type IV.1 (**10/52**).
PL: 0.155.

TI-2 (SF 412) **Fig. 3.6**

Fragmentary shaft, bone pin (**10/56**).
PL: 0.059.

Grave fill: inventoried: SU 361

P 316 Body fr PH: 0.051; PW: 0.065 (**9/156**).

P 354 Base fr PH: 0.043; PW: 0.040 (**9/318**).

P 366 Rim fr, open vessel. PH: 0.042; PW: 0.050 (**9/227**).

SF 387 Daub fr. PH × PW × PTh: 0.055 × 0.051 × 0.029 (**14/14**).

SF 396 Lithic blade (**13/14**).

Grave fill: not inventoried: HM ceramics (fine, 1 fr = 1 g; semi-coarse, 24 frr = 100 g; coarse, 11 frr = 108 g); daub (312 frr = 863 g); lithics (1 nucleus, 1 chert chip; 5 debitage); shell (2 bags).

Tomb II

Pit tomb, single inhumation (mature adult male)

Phase I (AMS ^{14}C: 1374 ± 58 cal BC)

Figs. 3.9–3.11

4.0E–5.0E, 7.0S–8.0S (high: 106.65 m = 353.65 m ASL; low: 106.50 m = 353.5 m ASL)

Sector 2; SU 514 (grave); 515 (fill); 516 (skeleton)

Excavated as Grave 91 (July 21–25, 2006)

Tomb II was stratigraphically located below Tomb XXV (Grave 90). There was no clearly discernible pit and the fill around the skeleton was almost pure sand, not clearly distinguishable from the sandy variety of bedrock encountered in the southeast sector of the tumulus. Indeed, so sandy was this deposit and so like bedrock that the cranium and torso were encountered *under* what was originally considered to be bedrock! There was a small stone to the northeast of the cranium and another, at a higher level (about 0.40 m), above the knees of the deceased. The body of the deceased (SU 516), identified as a mature adult male aged 45+ years at death, was laid out in a fetal position, lying on the right side, the cranium facing west. The arms were flexed, the elbows pointing southeast, the hands under the chin, with the left arm above the right; the legs were flexed, the knees pointing west, with the right femur directly under the left. The deceased was oriented northwest-southeast (340°), head to the northwest. The bone was well preserved, with most parts in situ, except for the greater part of both feet, which had eroded off the tumulus edge. The spinal column was curved. As preserved, the bone in situ measured 1.50 m long, at least 0.40 m wide, and 0.15 m deep. There were no grave goods. An AMS ^{14}C date from a sample taken from the right tibia of the deceased gave a date squarely in the 14th century: 1374 ± 58 cal BC.

Grave fill: SU 515: 1 scrap daub.

Tomb III

Pit tomb, single inhumation (adult male)

Phase I (AMS Charcoal ^{14}C: 1363 ± 50 cal BC)

Figs. 3.12–3.14

3.0E–5.0E, 5.0S–7.0S (high: 107.27 m = 354.27 m ASL; low: 107.14 m = 354.14 m ASL)

Sector 2; SU 452 (grave); 453 (fill); 454 (skeleton); 464 (cut)

Excavated as Grave 81 (July 7–12, 2006)

Tomb III consisted of a large and relatively clearly defined pit much larger than the deceased placed within it; the cut was visible on all sides except the southern half, where the tomb was largely uncovered with the removal of topsoil on account of the proximity of the tumulus edge, and where it was clear that the cut was damaged by erosion. The grave cut on the north side, both to the northeast and northwest, was bounded by stones that were outside the tomb pit (**Figs. 3.12, 3.13b**). At this lower level of Sector 2, there were a number of stones, many of them outcrops of the bedrock, and to what extent these stones were directly associated with the tomb was difficult to determine. Farther to the northwest a larger concentration of stones was encountered that had clearly nothing to do with this or any other burial and represents an outcrop of bedrock (**Fig. 3.14**). The tomb pit fill consisted of a darker soil clearly distinguishable from the surrounding bedrock and tumulus deposits. To the east, the relationship of the tomb pit to that of Tomb XXIV (Grave 85) was difficult to establish, particularly as the southeast extent of the tomb pit of Tomb III was unclear at this point, but if anything, it appears that the pit for

Tomb III was either cut by the pit for Tomb XXIV or, alternatively, was lost due to erosion off the tumulus edge. With the subsequent excavation of Tomb XXIV, it was determined that Tomb XXIV cut into Tomb III and that Tomb XXIV was therefore later.

Within the pit, the body of the deceased (SU 454), identified as an adult male aged 35+ years, was laid with the torso and cranium lying on the right side, head facing east-northeast, the arms folded in front of the body. The legs were flexed, the knees pointing east-northeast. Most of the bone survives, except for the hands, feet, and ribs; the cranium, which is comparatively well preserved, was found more or less crushed flat. The deceased was oriented southeast-northwest (about 150°), cranium to the southeast. As preserved, the tomb pit measured approximately 1.90 m long, though the southeast edge of the tomb was far from clear, about 1.50 m wide and 0.13–0.19 m deep. There were no grave goods.

A ^{14}C date based on a charcoal sample from the fill of this tomb provided an early date, 1363 ± 50 cal BC, but as with the ^{14}C sample from Tomb 64, this was based on charcoal, not bone, and the burial itself may be later or conceivably even earlier than the sample analyzed. Nevertheless, a date in the Late Bronze Age for this tomb would not be unusual, given its position in the tumulus (cf. Tomb XIII [Grave 49]).

Grave fill: SU 453: HM ceramics: inventoried: P 346 (body fr, incised decoration, PH: 0.031; PW: 0.039), not inventoried (fine, 2 frr = 3 g; semi-coarse, 14 frr = 27 g; coarse, 1 fr = 20 g); daub (123 frr = 189 g); lithics (2 debitage); shell.

TOMB IV

Pit tomb, single inhumation (adult male)

Phase I

Figs. 3.15–3.18

9.0E–10.0E, 9.0S–11.0S (high: 104.63 m = 351.63 m ASL; low: 104.45 m = 315.45 m ASL)

Sector 2; SU 579 (grave); 580 (fill); 581 (skeleton)

Excavated as Grave 98 (July 9–12, 2007)

Tomb IV was the southernmost and easternmost of all graves at Lofkënd, and was also one of the most substantially covered with stone. The tomb was covered by a number of small to medium-sized stones, all of the local sandstone, one approaching large, located above the pelvis of the deceased (**Fig. 3.17a–b**). There was a much larger outcrop of bedrock to the north of the tomb (**Fig. 3.16**). Once the stones were removed, the disposition of the skeletal remains, although comparatively well preserved, seemed widely spaced, perhaps on account of the weight of the overlying stones and deposits (**Figs. 3.15a–b, 3.17c**). A clear cut for the tomb was not discerned, and it almost appeared as if the tomb floor was prepared by leveling the bedrock. The grave fill was harder than the surrounding sandy bedrock, especially to the west. The body of the deceased (SU 581), identified as a younger adult male aged 20–25 years, was placed on the level ground. The torso and pelvis were supine; the cranium was lying on the right side, mouth wide open, facing east and southeast. The left arm was folded across the lower torso, the flexed right arm had the elbow extended to the east; the hands appeared to be clasped, or alternatively the bones of the hands were intermingled, the result of one hand placed above the other. The legs were flexed, the knees pointing east. The deceased was oriented south-southeast to north-northwest (170°), head to the south. As preserved, the tomb measured about 1.85 m (north to south), 1.05 m (east to west), and 0.18 m deep. The only grave good was the bone pin, **TIV-1**, found broken between the ribs and right scapula of the deceased, on the upper torso, immediately below the cranium.

TIV-1 (SF 443) **Fig. 3.18**

Bone pin, Type IV.3, with beaded head (**10/54**). PL: 0.068.

Grave fill: SU 580: HM ceramics (semi-coarse, 1 rim fr = 2 g); daub (9 frr = 10 g).

TOMB V

Pit tomb, single inhumation (adult female)

Phase I

Figs. 3.19–3.21

7.0E–8.0E, 8.0S–9.0S (high: 105.44 m = 352.44 m ASL; low: 105.24 m = 352.24 m ASL)

Sector 2; SU 567 (grave); 568 (fill); 460 (skeleton); 575 (cut)

Excavated as Grave 96 (July 5–9, 2007)

Tomb V was located in the far southeast sector, in an area that was originally thought to have been beyond the tumulus proper; only Tomb IV (Grave 98) was located even farther to the south and east. The tomb consisted of a clearly defined, almost straight, cut along the west side of the grave and along part of the north side. The tumulus edge was immediately to the south, and a small portion of the uppermost cranium located right at the tumulus edge was inadvertently broken in the course of excavation. Three stones along the east side, closer to the north edge of the cut, possibly extended beyond the grave, but this could not be determined with certainty, as there was no clear preserved cut on the east side. The grave fill consisted of a darker soil than the surrounding fill, but it was still loose, soft, and sandy. As preserved, the tomb measured 1.55 m long, 0.60 m wide, and 0.20 m deep.

Within the pit, the body of a young adult female aged 20–25 years appears to have been laid out on her left side, rolling or falling slightly to the west in a partly prone position, the left arm under the torso, the left hand under the skull; both arms and hands folded up toward the skull; the cranium faced west. The legs were flexed, the knees pointing west and west-southwest. The state of preservation of the skeleton (SU 460) was among the very best of the prehistoric burials, with such bones as the phalanges and both patellae preserved. On the basis of the spinal column, which was slightly curved, the deceased was oriented south-southwest to north-northeast (approximately 185°), the head to the south-southwest.

The only grave good was a bone pin, **TV-1**, found together with the smashed fragments of the cranium at the southern edge of the tomb. Consequently, the pin was most likely worn in the hair; alternatively, it may have been held in the left hand, but this seems most unlikely as most of the hand was well preserved underneath the cranium of the deceased. It is clear that the pin was not associated with the upper torso, and thus the likelihood it served as a dress pin is remote.

TV-1 (SF 437) **Fig. 3.21**

Bone pin with incised decoration, Type IV.2, perhaps hairpin (**10/53**).

PL: 0.157.

Grave fill: SU 568: HM ceramics, inventoried: P 418 (handle fr, matt-painted, PL: 0.023; PW: 0.031); daub (6 frr = 11 g).

TOMB VI

Pit tomb, single inhumation (infant)

Phase I

Figs. 3.22–3.24

5.0E–7.0E, 3.0S–5.0S (high: 106.43 m = 353.43 m ASL; low: 106.32 m = 353.32 m ASL)

Sector 2; SU 572 (grave); 571 (fill); 573 (skeleton); 574 (cut)

Excavated as Grave 97 (July 6, 2007)

This tomb consisted of a clearly defined pit cut into loose sandy earth with calcareous streaks; the latter, together with a deposit of hard shaley earth at the northwest end of the pit indicated bedrock or proximity to bedrock. The pit measured 1.37 m long (southeast-northwest), 0.80 m wide (southwest-northeast), and 0.11 m deep. The grave fill was sandy, but darker than the surrounding fill and bedrock and clearly distinguishable. All that survived of the skeletal remains was what appeared as a small scatter of teeth, fragments of cranium, and at least one small long bone at the very center of the pit. The deceased (SU 573) was an infant aged 3 (± 1) years at death, the bones of which had almost completely disintegrated. Given the state of preservation of the bone, it was impossible to determine the orientation and disposition of the deceased; the pit itself was oriented 135° southeast-northwest. A bronze dress pin, **TVI-1**, found together with the bone, was oriented east to west (90°), head to the east. The tomb was stratigraphically located directly below the cremation, Tomb XXX (Grave 70).

TVI-1 (SF 438) **Fig. 3.24**

Bronze dress pin, Type I.1, with small rounded head and incised decoration (**10/27**).

L: 0.127.

Grave fill: SU 571: HM ceramics: inventoried: P 426 (shoulder fr, PH: 0.031; PW: 0.056), not inventoried (semi-coarse, 4 frr = 5 g; coarse, 2 frr = 12 g); daub (4 frr = 7 g).

TOMB VII

Pit tomb, single inhumation (child)

Phase I

Figs. 3.25–3.27

8.0E–9.0E, 4.0S–5.0S (high: 106.13 m = 353.13 m ASL; low: 105.41 m = 352.41 m ASL)

Sector 2; SU 582 (grave); 589 (fill); 583 (skeleton); 584 (cut)

Excavated as Grave 99 (July 9–11, 2007)

Tomb VII was one of several early graves in large and deep pits containing the remains of a child or infant that did not fill the tomb pit entirely. The pit measured 1.53–1.68 m long, 1.33 m wide, and 0.15–0.74 m deep. It was first encountered as a clearly distinct pit, ovoid to rectangular in shape, with a dark, medium loose fill, clearly distinguishable from the surrounding bedrock and tumulus fill, with patches of decomposed shale throughout. The pit was considerably deeper at the west, with the cut at least 0.60 m deep, shallower to the east, where it ranged from 0.15 to 0.22 m. Within the pit, the body of a child (SU 583) aged 7 (± 2) years, was laid out with the cranium lying on its left side facing west. The legs were flexed, with the knees, which are not preserved, pointing to the west-southwest. It is likely that the deceased was lying on the left side (cf. Tomb VIII [Grave 100]), rather than with the torso in a supine position. The deceased was oriented south-southeast to north-northwest (160°), head to the south-southeast. The floor of the tomb was leveled flat, which accounts for the deeper cut on the west side. The only grave good was the bronze dress pin, **TVII-1**, beside and immediately to the northeast of the cranium, evidently worn on the right shoulder. The pin was oriented southeast-northwest (about 115°), head to the southeast. There was no cultural material in the tomb fill (SU 598).

TVII-1 (SF 440) **Fig. 3.27**

Bronze dress pin, Type I.2, with concave disk head (**10/28**).
L: 0.175.

TOMB VIII

Pit tomb, single inhumation (infant)

Phase I

Figs. 3.28–3.30

8.0E–10.0E, 6.0S–8.0S (high: 105.79 m = 352.79 m ASL; low: 105.25 m = 325.25 m ASL)

Sector 2; SU 590 (grave); 586 (fill); 591 (skeleton); 588 (cut)

Excavated as Grave 100 (July 11–13, 2007)

As with Tomb VII, Tomb VIII consisted of a comparatively large and deep pit, cut into bedrock, substantially deeper on the west side, where the bedrock of the ridge on which the tumulus was constructed sloped sharply down from west to east. The tomb cut was clear on all sides except the south. The tomb measured 1.33 m (north to south), 0.88 m wide (east to west), and 0.54 m deep. The fill was a sandy, medium to loose, and darker-colored soil to that of the surrounding bedrock and tumulus fill, and easily distinguishable. Within the pit, the body of an infant (SU 591), aged 3 (± 1) years at death, was laid out in a fetal position, torso and cranium lying on the left side, the cranium facing east-southeast. The right arm was folded across the upper torso, with the right hand, which was not preserved, in front of the cranium; the left arm extended straight by the side of the body. The legs were flexed, the knees pointing east-southeast. The deceased was oriented northeast-southwest (40°), head to the northeast. The state of preservation of the bone was remarkably good for an individual of this age; only the feet and hands were largely not preserved. The entire tomb was irregularly covered by a large number of small to medium-size stones, with several stones lining the tomb cut along the east and west sides of the tomb. The only grave good was the minuscule bronze spiral bead, **TVIII-1**, found near or on the pelvis, perhaps originally connected with the left hand, which is largely not preserved, though this could not be established with certainty.

TVIII-1 (SF 444) **Fig. 3.30**

Minuscule bronze spiral coil bead (**10/65**).
L: 0.005.

Grave fill: SU 586: HM ceramics (semi-coarse, 1 fr = 2 g; coarse 1 fr = 15 g); daub (5 frr = 25 g).

TOMB IX

Pit tomb, single inhumation (adult female)

Phase I

Figs. 3.31–3.32

5.0E–6.0E, 8.0S–9.0S (high: 106.04 m = 353.04 m ASL; low: 105.95 m = 352.95 m ASL)

Sector 2; SU 545 (grave); 546 (fill); 547 (skeleton)

Excavated as Grave 94 (July 2–4, 2007)

Tombs IX and X were found close to one another, laid out on what appeared to be a ledge of bedrock at the southeast side of the tumulus. A prominent shelf of bedrock was located immediately to the north of the feet of Tomb IX and the cranium of Tomb X (**Figs. 3.31–3.32a**). Despite their close proximity to one another, the two skeletons were clearly part of two separate tombs, as they were oriented in very different directions. Although what survives of the bone in Tomb IX is well preserved, nothing of the cranium and upper body of the skeleton was preserved, these having fallen victim to erosion off the south slope of the tumulus (**Fig. 3.32a** shows the proximity of the tumulus edge). Both femora, tibiae, and fibulae were preserved, together with parts of the feet to the north, as well as a very small portion of the pelvis. A clear tomb cut was not discerned, and it appears almost as if the ground was simply leveled for the interment of both Tombs IX and X, which must be more or less contemporary. The darker soil encountered immediately to the south of Tomb IX was topsoil, and the rest of the tomb fill was in no way distinguishable from the surrounding sandy deposits. A small band of calcareous material below the skeletal remains indicated that bedrock was very near. As preserved, the tomb measured 1.20 m north to south (i.e., from the ledge of bedrock to the north to the tumulus edge to the south), about 1.00 m east to west, and approximately 0.10 m deep. The deceased (SU 547), identified as an adult female, was laid out with the legs tightly flexed, lying on the right side, the knees pointing east. On the basis of the surviving leg bones, the skeleton was probably oriented south-southeast to north-northwest (approximately 170–175°), the head, which is missing, to the south-southeast. There were no grave goods.

Grave fill: SU 546: HM ceramics (semi-coarse, 1 handle fr = 6 g); daub (5 frr = 7 g); shell.

TOMB X

Pit tomb, single inhumation (adult female)

Phase I

Figs. 3.33–3.34

6.0E–7.0E, 8.0S (high: 106.20 m = 353.20 m ASL; low: 106.02 m = 353.02 m ASL)

Sector 2; SU 553 (grave); 552 (fill); 554 (skeleton)

Excavated as Grave 95 (July 3–4, 2007)

Tomb X was uncovered on what appeared to be the same ledge of bedrock on which Tomb IX was laid out, and at more or less the same level, though fractionally higher. As with Tomb IX, only the northern portion of the grave was preserved, the remainder lost to erosion off the south-southeast scarp of the tumulus, located a short distance to the south; the tomb was immediately to the south of the same shelf of bedrock that was also located to the north of Tomb IX. No tomb cut survived and the fill was not distinguishable from the surrounding sandy deposits. As preserved, the bone defined an area of 0.50 × 0.50 m, and about 0.18 m deep. Of the skeletal remains of the deceased (SU 554), identified as an adult female aged 30–40 years at death, only the cranium, the upper torso, including both clavicles, the upper vertebrae, and much of the right arm and ribs were preserved, as were portions of the left arm; small portions of the pelvis also were found. The legs and feet were completely lost. The cranium of the deceased was found lying on the right side, facing west, but determining the original position of the upper torso in situ was difficult. The excavator noted the possibility that the torso was laid out supine, the right arm tightly flexed over the right shoulder. Alternatively, the position of the arm and the curved vertebrae might suggest that the body was laid out in a fetal position very similar to that of the deceased in Tomb II (Grave 91), the arms in front of the body and the hands under the chin (cf. also Tombs V and IV [Graves 96 and 98]). Tomb X was more or less oriented in the opposite direction to Tomb IX: the orientation of the deceased in Tomb X, based on the vertebrae, was north-northwest to south-southeast (340°), head to the north-northwest. There were no grave goods.

Grave fill: SU 552: daub (3 frr = 10 g); shell.

TOMB XI

Pit tomb, single inhumation (infant)

Phase I

Figs. 3.35–3.36

5.0E, 3.0S (high: 106.66 m = 353.66 m ASL; low: 106.54 m = 353.54 m ASL)

Sector 2; SU 539 (grave); 538 (fill); 540 (skeleton)

Excavated as Grave 93 (June 29–30, 2007)

Although one of the earliest burials, Tomb XI is in a very poor state of preservation and its circumstances are a little unusual. All that survives is a portion of the cranium of an infant (SU 540) aged 9 (± 3) months at death, and the position of this is itself unclear. The eye sockets were not visible in situ and the mandible was to the north-northeast, which appeared to be disturbed. Nothing whatsoever of the postcranial bone survives and it was initially thought that disturbance to the tomb may have been the result of the cutting for Tomb XII (Grave 88) to the north-northwest. Although close, Tomb XII was to the north-northwest and, since no cut for Tomb XI was discerned, such disturbance could not be established, and this was exacerbated by the fact that the fill immediately around the cranium was not distinguishable in any way from the surrounding tumulus fill. Consequently, as there is nothing of the body preserved, both the orientation of the tomb and the position of the body cannot be determined. Even the measurements of the tomb are a matter of speculation, since only the cranium survives. As preserved, it covers an area 0.20 × 0.20 m, and was 0.12 m deep. There were no grave goods.

Grave fill: SU 538: HM ceramics (fine, 2 frr = 6 g); daub (6 frr = 10 g); shell.

TOMB XII

Pit tomb, multiple inhumation (adult female, child, and infant)

Phase I

Figs. 3.37–3.39

5.0E–6.0E, 1.0S–3.0S (high: 106.84 m = 353.84 m ASL; low: 106.63 m = 353.63 m ASL)

Sector 2; SU 494 (grave); 490 (cut); 491, 505 (fill); 499 (skeleton, adult female); 495 (skeleton, child); 501 (skeleton, infant)

Excavated as Grave 88 (July 17–22, 2006)

Tomb XII was among the best preserved and most interesting of the multiple inhumations at Lofkënd. The tomb consisted of a large, roughly rectangular pit, slightly wider at the northwest end, measuring 1.40 m long, 0.52–0.65 m wide, and 0.21 m deep. The clearly distinguishable tomb fill was characterized by a soil that was darker and looser than the tumulus fill. At the wider northwest end, the fill extended slightly beyond the area of the cut. Within the pit, the skeletal remains of three individuals were encountered: SU 499 was identified as a younger adult female aged 20–25 years; SU 495 was a child aged 8–9 years; and SU 501 was an infant aged 12–18 months. The primary burial, that is, the one interred first, was the female, SU 499. She was laid out with her torso supine, the cranium lying on its right side facing north. The left arm was folded across the lower torso, the right arm was doubled back toward the shoulder; the legs were flexed, the knees pointing northeast. The skeleton was oriented southeast-northwest (150°), cranium to the southeast. The skeletal remains were very well preserved, and all of the bone was represented except for parts of the hands and feet. The second skeleton, SU 495, that of the child, was oriented in the opposite direction to the female, northwest-southeast (336°), head to the northwest. The skeleton was found with the cranium lying on its right side and partially on its crown, the mandible slightly to the southeast; the arms, some ribs, and the upper vertebrae were unclear, but it appears as if the arms were flexed in front of the body, suggesting that the individual may have been laid out in a fetal position, lying on the right side, though this was not absolutely clear; the legs were flexed, the knees pointing southwest. The bones of SU 495 were also fairly well preserved, with all of the bones represented except for parts of the torso and perhaps the upper arm. The third skeleton, that of the infant (SU 501), was placed above the right upper torso and shoulder of the female. Clearly visible in situ were only the pieces of the cranium. It was not clear whether SU 501 or SU 495 was first placed over the female, SU 499, as there was minimal overlap of the skeletal remains of the child and infant; if anything, it is more likely that SU 501 was interred before SU 495. The fact that the interment of SU 495 and SU 501 caused little, if any, disturbance to the primary skeleton of the female suggests that the time between the first and third interment was not great, and it is even possible that all three were interred at the same time. The position of the female, SU 499, in relation to the infant, SU 501, the latter placed above her right shoulder over her arm, with her head pointing down toward the infant as if nuzzling it, seemed to provide a poignant image of mother and child.

The only grave good was the bone pin, **TXII-1**, found on the upper central torso of the female (SU 499), below the neck and immediately beneath one of the cranial fragments of the infant (SU 501). The pin was oriented roughly northeast-southwest, head to the northeast. Its position was such that it was unclear whether the pin was associated with the infant or female, but its position on the torso of the female (cf. the bone pin, **TIV-1**, associated with upper right torso of the young adult male in Tomb IV [Grave 98]) may suggest that it belonged with the female rather than the infant, though either is possible. Tomb XII was stratigraphically located below Tomb LX (Grave 44).

TXII-1 (SF 404) **Fig. 3.39**

Bone pin, Type IV.4 (**10/55**).
L: 0.084; head: 0.012 × 0.013.

Grave fill: SU 491: HM ceramics (semi-coarse, 7 frr = 18 g; coarse, 1 fr = 16 g); daub (31 frr = 56 g); lithics (1 debitage, 3 pebbles); shell.

SU 505: daub (2 frr = 5 g).

Tomb XIII

Pit tomb, single inhumation (adult female)

Phase I (AMS ^{14}C: 1299 ± 87 cal BC)

Figs. 3.40–3.41

1.0W–0.0, 4.0S–5.0S (high: 107.77 m = 354.77 m ASL; low: 107.52 m = 354.52 m ASL)

Sector 2; SU 304 (grave); 305 (skeleton); 306 (fill)

Excavated as Grave 49 (July 14–19, 2005)

This tomb was stratigraphically located below Tomb XLVII (Grave 41) and, therefore, among the earliest burials on the tumulus. This was later confirmed by AMS ^{14}C dating, taken from a sample of bone from the right hand of the deceased, which gave an absolute date of 1299 ± 87 cal BC, placing the tomb squarely in the closing stages of the Late Bronze Age. Although some of the bone was slightly displaced, including both of the lower arms, except for the left humerus, and with the bones of the left hand scattered over various parts of the tomb, the skeletal remains, as with several other burials in this same area, were remarkably well preserved. The deceased (SU 305), identified as an adult female aged 25–35 years, appears to have been laid out in a fetal position, lying on the left side; the cranium had fallen slightly forward, toward the west. The legs were flexed, the knees pointing south. The displacement of the arm and hand bones was such that it was difficult to reconstruct the original position of the arms. The deceased was oriented northeast-southwest (60°), head to the northeast. A clearly defined tomb cut was not discerned. The skeletal remains as found measured 1.20 m long, 0.55 m wide, and some 0.25 m deep. There were no grave goods.

Grave fill: SU 306: HM ceramics (1 fr, coarseware = 3 g; 2 non-diagnostic scraps = 1 g); shell.

Tomb XIV

Pit tomb, single inhumation (infant)

Phase II

Figs. 3.42–3.44

1.0W, 1.0N–2.0N (high: 107.36 m = 354.36 m ASL; low: 107.01 m = 354.01 m ASL)

Sector 4; SU 404 (grave); 400 (fill); 405 (skeleton), 404 (cut)

Excavated as Grave 71 (June 28–30, 2006)

This tomb was stratigraphically located below the western portion of Tomb XLIV (Grave 65), which was itself located below Tomb LXXX (Grave 4); it was also located below the bitumen pot (SU 280) immediately to the north of Tomb LXIII (Grave 35). The tomb consists of a very clearly defined pit, roughly rectangular in plan, but slightly curved, especially on the west side, that was relatively deep. The pit measured 1.26 m long, 0.59 m wide, and 0.26–0.35 m deep. The fill was a rich, loose-textured, dark-colored soil, which, according to Foss, shows evidence of once being topsoil. Indeed, the color of this fill was darker and richer than all the other soils noted in the tumulus. Within the pit, the poorly preserved skeleton of an infant (SU 405) aged 6 (± 3) months was uncovered. The largely crumbled cranium was more or less in the center of the pit, with fragments of two leg bones to the north. The tomb pit was considerably larger than the deceased (cf. Tomb XVI [Grave 68]). The skeleton was so poorly preserved that it was not possible to discern the original position of the body, though the excavator noted the possibility that the infant was laid out on its back and in a supine position. The pit, together with the poorly preserved remnants of the skeleton, was oriented south to north (180°), the head to the south. The only grave good was a small, crude, one-handled vessel, **TXIV-1**, on its side, immediately to the east of the cranium, rim facing east-northeast. The grave fill yielded a couple of pieces of Paleolithic stone tools.

TXIV-1 (P 303) **Fig. 3.44a–b**

Crude one-handled vessel, light fabric, SC Type 1 (**9/159**).

H (to rim): 0.077; H (max): 0.107; D (rim): 0.076.

Grave fill: SU 400: HM ceramics (semi-coarse, 20 frr = 58 g; coarse, 1 fr = 29 g); daub (49 frr = 116 g); lithics (SF 312, SF 314, plus 1 nucleus); shell.

Tomb XV

Pit tomb, single inhumation (infant)

Phase II

Figs. 3.45–3.47

2.0E–3.0E, 1.0S–1.0N (high: 107.46 m = 354.46 m ASL; low: 107.16 m = 354.16 m ASL)

Sectors 2, 6, 1; SU 443 (grave); 403 and 446 (fill); 444 (skeleton), 445 (cut)

Excavated as Grave 80 (July 6–18, 2006)

This burial was very similar to Tombs XIV and XVI (Graves 71 and 68), and all three should be considered roughly contemporary. The tomb was first encountered in Sector 2 to the south, extending for a short distance into Sector 1 (**Fig. 3.46a**) and necessitated a significant cut into the baulk, Sector 6, in order for it to be fully exposed. The tomb consisted of a small pit, very well defined, cut into bedrock to a depth of 0.30 m. In plan, the roughly rectangular pit measured 1.02 m long and 0.60 m wide. The clearly distinguishable fill was a darker color and looser texture than the tumulus fill. Unlike Tomb XIV (cf. Tomb XVI), the pit for Tomb XV was not significantly larger than the body placed within it. The deceased (SU 444), identified as an infant aged 6 (± 3) months at death, appeared to be oriented south-southwest to north-northeast (200°), head to the south-southwest. Parts of the cranium, one scapula, and some ribs were preserved, but very few long bones; all the bone was extremely fragile. Too little survived to determine the original position of the skeleton, but the excavator noted the possibility that the deceased was laid out in a supine position, not unlike the skeleton of Tomb XXIX (Grave 83). At a higher level, a large stone partially covered the grave pit (**Fig. 3.46b**). Between the stone and the level of the human remains, a small one-handled feeder, **TXV-1**, was found which should belong to this burial, but was found at a higher level than the human bone. The tomb was stratigraphically located below Tombs LXVIII, LXXII, and XLV (Graves 13, 24, and 60).

TXV-1 (P 362) **Fig. 3.47**

Spouted vessel (spout not preserved) or "baby feeder," SC Type 2 (**9/160**).

H (to rim): 0.052–0.057; H (max): 0.072; D (rim): 0.034–0.045.

Grave fill: SU 446: daub (3 frr = 3 g); shell.

Tomb XVI

Pit tomb, single inhumation (child)

Phase II

Figs. 3.48–3.49

1.0E–2.0E, 1.0S–2.0S (high: 107.42 m = 354.42 m ASL; low: 107.30 m = 354.30 m ASL)

Sector 2; SU 385 (grave); 370 (fill); 386 (skeleton), 391 (cut)

Excavated as Grave 68 (June 23–26, 2006)

Relatively deep, neatly cut, and roughly rectangular, the pit measured 1.25–1.40 m long, about 0.80 m wide, and 0.12 m deep. Like the nearby Tomb XIX (Grave 54), this was one of the best preserved and clearest grave pits. Within the center of the pit, the body of a child (SU 386), aged approximately 6 (± 2) years, was laid out on its right side, the cranium and body facing north, the arms and legs sharply flexed. The pit was significantly larger than the body of the child. The skeleton was oriented east to west (105°), head to the east. The grave fill was sandy, but darker and of a looser texture to the surrounding tumulus fill, becoming even sandier toward the bottom and edges. A medium-sized stone (0.24 × 0.14 m) was placed immediately to the north of the cranium. The floor of the tomb was hard-packed and calcareous. Most of the cranium was well preserved, including some teeth; postcranial bones were considerably more fragile. In addition to the cranium, most of the arms, ribs, and legs were preserved, including bits of the pelvis, but no hands and feet. There were no grave goods. The tomb was stratigraphically located below Tomb LXXVIII (Grave 5) and should be roughly contemporary with the nearby Tombs XIX and XX (Graves 54 and 50).

Grave fill: SU 370: HM ceramics (semi-coarse, 5 frr = 6 g; coarse, 4 frr = 18 g); daub (56 frr = 148 g); lithics, SF 325 (chert flake, 2 debitage); shell.

TOMB XVII

Pit tomb, double inhumation (two children)

Phase II

Figs. 3.50–3.54

1.0E–0.0, 5.0N (high: 107.40 m = 354.40 m ASL; low: 107.32 m = 354.32 m ASL)

Sectors 1 and 5; SU 406 (grave); 407 (fill); 408 (skeleton); 408a (maxilla); 459 (cut)

Excavated as Grave 72 (June 29–July 8, 2006; 5 July, 2007)

Papadopoulos 2010a:36–37, figs. 2–3

This tomb consists of a clearly defined pit, visible on the north and part of the south sides; the grave cut clearly peters out on the east side. To the west, the pit was less clear, as the western extent of the tomb was in the baulk, Sector 5. On the north side, the pit cuts into bedrock, whereas on the south it cuts into tumulus fill (SU 377). The pit fill was a loose, sandy, gray-colored soil, clearly distinguishable from the tumulus fill. Within the pit the remains of a child (SU 408) aged 7 (± 2) years were oriented east to west (93°), head to the east. The bone was very poorly preserved, though most of the cranium survives, as do parts of the vertebrae and legs, small portions of the arms, pelvis, and probably also a clavicle; hands, feet, and most of the ribs are not preserved. From what survives, the torso and probably also the cranium were laid out in a supine position; the arms may have been folded across the lower torso, immediately above the pelvis, but little is preserved. The legs were flexed, knees pointing south-southwest. The remains of a second child (SU 408a), aged 4 years (± 12 months) at death, represented only by dentition, including the maxilla and some scattered cranial fragments, were encountered immediately to the northeast of the cranium of SU 408. As preserved, the tomb averaged a width of 0.60–0.65 m and a depth of about 0.08 m. The length of the grave cut was difficult to determine because it was so unclear to the east and west. As exposed in 2006, the tomb measured approximately 1.40 m long, the bone extending about 1.05 m. The original length of the tomb was probably closer to 1.65 m.

All grave goods found in 2006 were clustered around the cranium of the more complete skeleton, SU 408. A fineware, handmade, matt-painted kantharos, **TXVII-1**, was placed upright to the south-southwest of the cranium. A bronze headband (often referred to as a "diadem"), **TXVII-2**, was worn around the cranium (**Fig. 3.51a–c**). Two bronze earrings, **TXVII-3** and **TXVII-4**, together with two small bronze spiral coils, conceivably beads, **TXVII-5** and **TXVII-6**, were found overlying the headband, and were originally thought to be part of it, but these were probably worn by the deceased separate from the headband. Although the pot and the headband clearly belonged with the more complete skeleton, SU 408, it is not impossible that some of the other jewelry was originally displaced from the more fragmentary individual, SU 408a, represented by the maxilla and scattered cranial fragments, though this could not be established with any conviction. It should be noted that the earrings and one of the small spiral coils were found about 0.05 m higher than the headband itself.

In 2007, the baulk, Sector 5, separating Sectors 1 and 4 was cleared, and in the process an intact dark-fabric, one-handled pot, **TXVII-7**, was found 0.17 m to the west of the lower leg bones of the deceased, SU 408. The base of the pot was precisely at the level of the tomb (**Fig. 3.51d**), the rim almost exactly at the level of the pot placed by the cranium (**TXVII-1**). This vessel could only belong to Tomb XVII, though its position, by the feet of the deceased, was somewhat unusual (the majority of complete vessels were placed near the cranium of the deceased). Careful excavation of the baulk failed to yield any clear traces of the western end of the grave cut, so it is possible that this vessel may have been an offering outside the tomb, which is also most unusual for burial customs at Lofkënd. The possibility suggested itself that the more incomplete skeleton, SU 408a, may have been the original occupant of the tomb and that the grave was reopened, its better-preserved remains, including the maxilla, pushed toward the east in order to accommodate the interment of SU 408. If this was the case, then it is possible that **TXVII-7**, and some of the bronze jewelry near the headband, were displaced from SU 408a. The tomb was stratigraphically located below Tomb LXXXV (Grave 10), and the pot, **TXVII-7**, was located more than 1.50 m below the level of Tomb LXXXIII (Grave 7).

TXVII-1 (P 350) **Fig. 3.52a**

Fine, light-fabric, matt-painted kantharos, FL Type 1 (**9/1**).
H (to rim): 0.074; H (max): 0.094; D (rim): 0.055–0.070.

TXVII-2 (SF 349) **Fig. 3.53**

Bronze headband (**10/84**).
L (all fragments lined up): 0.495; L (main fragment at front = D of headband): 0.210.
Papadopoulos 2010a:38, figs. 4a–b.

TXVII-3 (SF 313a) **Fig. 10.30**

Bronze earring (**10/80**).
D (max): 0.029.
Papadopoulos 2010a:39, fig. 5.

TXVII-4 (SF 313c) **Fig. 3.54**

Bronze earring (**10/81**).
D (max): 0.025.
Papadopoulos 2010a:39, fig. 5.

TXVII-5 (SF 313b = 348) **Fig. 10.26**

Small bronze spiral coil (**10/66**).
PL: 0.019; D: 0.004–0.005.
Papadopoulos 2010a:39, fig. 5

TXVII-6 (SF 313d) **Fig. 3.54**

Small bronze spiral coil (**10/67**).
L: 0.018–0.019; D: 0.004.
Papadopoulos 2010a:39, fig. 5.

TXVII-7 (P 422) **Fig. 3.52b**

Dark-fabric, one-handled vessel, DF Type 2 (**9/90**).
H (to rim): 0.088; H (max): 0.111; D (base): 0.047–0.051; D (rim): 0.070–0.076.

Grave fill: SU 407: HM ceramics (semi-coarse, 5 frr = 43 g); daub (12 frr = 60 g); shell.

Tomb XVIII

Pit tomb, double inhumation (at least two children)

Phase II

Figs. 3.55–3.58

1.0E–2.0E, 4.0–5.0N (high: 107.32 m = 354.32 m ASL; low: 107.17 m = 354.17 m ASL)

Sector 4; SU 411 (grave); 409 (cut); 410 (fill); 412 (north skeleton); 421 (south skeleton); 421a (remains of possible third skeleton)

Excavated as Grave 73 (June 30–July 2, 2006)

Papadopoulos 2010a:39, fig. 6

Tomb XVIII was stratigraphically located below Tombs LXXXIII, LXVI, and XLIX (Graves 7, 31, and 51). It consisted of a small but clearly defined pit measuring 1.23 m long, 0.63 m wide, and 0.15–0.17 m deep. In plan, the pit was almost rectangular, more or less straight on the east, west, and south sides, but with the north side tapering slightly toward the west. The pit fill was characterized by a dark, clayey, and comparatively compact soil, clearly distinguishable from the tumulus fill. The borders of the pit were marked by clear lines of black organic matter; although this material resembled charcoal, it was clearly not charcoal. Within the pit, and largely concentrated to the east and southeast, were the disintegrated bones of at least two children. One of these, designated SU 412, to the north, was clear on account of the bronze headband, **TXVIII-1**, worn around the head of the deceased (**Fig. 3.56a–b**); this individual was identified as a child aged 3 (± 1) years at death. Immediately to the south, another cluster of very poorly preserved cranial bones were clearly part of a second individual, designated SU 421, and identified as a child of similar age (3 ± 1 years). The teeth of another possible individual, SU 412a (an infant/child aged 2 years ± 8 months) were noted in the lab; given the surviving remains, these teeth probably represent a full-fledged third inhumation. As for the postcranial bones, these were very poorly preserved, and little could be said about the original positions of the deceased on the basis of the little that survived. The tomb pit, together with the preserved crania as encountered, established that the tomb was oriented roughly east to west (90°), the crania to the east. In addition to the headband worn by SU 412, a bronze disk, **TXVIII-2**, was found at the center of the east end of the grave; the disk could not be directly associated with either of the skeletons.

TXVIII-1 (SF 317) **Fig. 3.57**

Bronze headband (**10/85**).

L (all fragments lined up together): 0.444; L (longest preserved fragment = close to D of headband): 0.140.

Papadopoulos 2010a:40, fig. 7.

TXVIII-2 (SF 318) **Fig. 3.58**

Bronze disk, with repoussé decoration (**10/60**). D: 0.047–0.050.

Papadopoulos 2010a:40, fig. 8.

Grave fill: SU 410: HM ceramics (semi-coarse, 1 fr = 3 g); daub (13 frr = 20 g); shell.

TOMB XIX

Pit tomb, single inhumation (adult female)

Phase II

Figs. 3.59–3.60

2.0W–1.0E, 2.0S–3.0S (high: 107.51 m = 354.51 m ASL; low: 107.34 m = 354.34 m ASL)

Sectors 3, 7, and 2; SU 322 (grave); 292 (fill); 333 (southeast fill); 323 (skeleton); 298 (cut)

Excavated as Grave 54 (July 18–22, 2005)

Papadopoulos, Bejko, and Morris 2007: 123, 127, fig. 21

With the exception of Tomb I, the pit for this grave was among the most substantial of any of the prehistoric burials and significantly larger than the body placed within it. The pit in plan was roughly rectangular, with rounded corners, and measured 2.20 m long, 1.08 m wide, and just under 0.20 m deep, though the depth from the top of the stone cover to the floor of the pit was 0.41 m deep; the grave fill was clearly distinguishable from the tumulus fill, and generally of a darker color, though at an upper level streaks and patches of white calcareous material were noted (**Fig. 3.60a**). Within the pit, the deceased (SU 323), identified as a mid-age adult female, was laid out southwest-northeast (245°), head to the southwest. The torso and cranium were supine, with the cranium lying slightly on the left side, facing northeast; the right arm was folded across the lower torso, the left arm by the side of the body, the lower arm and hand below the pelvis. The legs were flexed, the knees pointing north (**Fig. 3.60d**). The bone was well preserved. In the process of excavation, a dark stripe was clearly distinguished running in a straight line along the south side of the cranium and torso (**Fig. 3.60b**), which should represent the decomposed remnants of a shroud or garment worn by the deceased.

The tomb was partially stone-covered: four large stones at the northeast end covered the lower legs and feet of the deceased, with one of these stones, that to the west, covering the skeleton directly (**Fig. 3.60c**), the other three at a higher level but clearly within the grave cut (**Fig. 3.60a**). Three additional stones bordered the grave cut but lay outside it: two at the southwest corner, and one to the northeast not far from the feet of the deceased. The tomb was stratigraphically located below Tombs LXX and LXVII (Graves 17 and 12). There were no grave goods, but the deep fill yielded a number of sherds, a few of which are indicated on the plan and visible on the photograph of the skeleton fully exposed (**Figs. 3.59a–b, 3.60c–d**).

Grave fill: SU 292: HM ceramics: inventoried: P 271 (semi-coarse rim fr: PH: 0.021; PW: 0.029); P 273 (body fr, matt-painted: 0.014 × 0.017); P 238 (coarse strap handle: PH: 0.060); P 281 (coarseware base: H: 0.028; D [base]: 0.090–.100). HM ceramics: not inventoried: (12 frr, semi-coarse, including 1 handle and 1 base = 70.5 g; 9 frr, coarse, including 1 handle = 64 g; non-diagnostic, 16 scraps = 6 g); daub (21 frr = 38 g); shell.

SU 333: HM ceramics (2 frr, semi-coarse, including 1 rim = 16 g; 2 frr, coarse = 35 g; non-diagnostics 1 scrap = 0.5 g); daub (2 frr = 1 g); shell (19/7/2005).

TOMB XX

Pit tomb, single inhumation (adult female)

Phase II

Figs. 3.61–3.62

1.0E–3.0E, 2.0S–3.0S (high: 107.72 m = 354.72 m ASL; low: 107.54 m = 354.54 m ASL)

Sector 2; SU 307 (grave); 308 (skeleton); 309 (fill); 312 (cut)

Excavated as Grave 50 (July 14–21, 2005)

Morris 2006:102, fig. 6; Papadopoulos, Bejko, and Morris 2007:123–24, 129, fig. 23

This tomb bordered an area in the southeast portion of the tumulus that was relatively free of graves. The deceased (SU 308), a younger adult female aged

20–25 years, was placed in a pit that was clearly visible from a higher level. The pit was partially stone-lined (**Fig. 3.61b**), particularly on the north and west sides of the cranium and lower torso. At a higher level a number of stones within the cutting of the pit served as a partial covering for the tomb (**Fig. 3.61a**). Two additional stones lined the pit on the north and south sides of the legs. At its greatest extent, the tomb pit measured 2.0 m long and 1.0 m wide; the tomb cut was about 0.18 m deep. The skeleton was very well preserved, with no disturbed bone except for the hands. The torso and cranium were laid out in a supine position, the cranium lying on the left side facing southwest; the arms folded over the lower torso. The legs were only slightly flexed, the knees pointing southwest. The deceased was oriented east to west (100°), head to the east. There were no grave goods, although there was a "layer" of bitumen (SF 252).

The left arm, lower ribs, and part of the pelvis were covered by an amorphous strip and patch of bitumen (**Figs. 3.61b, 3.62b**), which was unique among all of the tombs in the tumulus. What at first appeared to be a lump or patches of bitumen were encountered near the left side of the pelvis, extending, in the form of a thin strip, over the lower arms and onto the ribcage; a small isolated patch of bitumen was also noted on the right pelvis; the bitumen was designated SF 252. What this bitumen represents was difficult to ascertain, both in the field and later in the lab. When first discovered near the left pelvis, the possibility was raised that this was an object, a grave good, made of bitumen. Further excavation, however, rendered this possibility unlikely, and given that small patches of bitumen were found over the ribs and lower arms, it is possible that the material represents the remnants of a partial bitumen lining of a shroud or some other item of clothing or ornament worn by, or placed with, the deceased. Alternatively, the strips of bitumen were perhaps used as an adhesive in some kind of organic jewelry or item of personal ornament, such as leather or bark, otherwise not preserved. This, and other aspects of bitumen encountered in the mound, are more fully discussed in Chapter 15.

Grave fill: SU 309: HM ceramics (non-diagnostic scrap = 0.5 g); daub (1 fr = 1 g); 1 fr bitumen; shell.

Tomb XXI

Pit tomb, double inhumation (adolescent female; adult male)

Phase II

Figs. 3.63–3.73

2.0W–3.0W, 1.0N–2.0N (high: 107.57 m = 354.57 m ASL; low: 107.49 m = 354.49 m ASL)

Sector 4; SU 324 (grave); 325 (fill); 326 (first skeleton); 341 (second skeleton)

Excavated as Grave 55 (July 18–21, 2005)

Papadopoulos, Bejko, and Morris 2007:121, 125, fig. 18; Papadopoulos 2010a:41, fig. 9

Tomb XXI was uncovered in the western sector of the tumulus, south of Wall 1 (**Fig. 3.64b**), and was stratigraphically located below Tombs LXXVI and LIX (Graves 16 and 38). The remains of two individuals were placed within a clearly defined grave pit measuring 1.48 m long, 0.65 m wide, and about 0.10 m deep. The fill was characterized by a soil of a similar color to that of the surrounding tumulus fill, but of a looser texture. The better-preserved and more fully articulated individual (SU 326) was identified as a robust younger adult male, aged 20–25 years at death. The skeleton was oriented northeast-southwest (50°), head to the northeast. The poorly preserved torso and the cranium, the latter largely crushed, appear to have been laid out in a supine position, with the arms probably folded across the lower torso. As for the legs, although the femora were clearly articulated with the pelvis, the tibiae and fibulae were not only underneath the femora, disarticulated, but at 90° to them, and it was therefore clear that the lower legs were not in original position or were slightly disturbed. The very poorly preserved remains of the second skeleton were encountered in the northeast corner of the grave, to the east-southeast of SU 326. Designated SU 341, this individual was identified as an adolescent female aged 15 (± 3) years. All that survived of this individual were parts of the cranium, including the maxillary mandibular teeth. The cranium appeared to share the same orientation as that of the male, but nothing could be said about the position of the deceased.

Grave goods were plentiful. A dark-fabric kantharos, **TXXI-1**, was found to the southeast and beside the left arm of the male (SU 326), but the vessel was also just to the southwest of the cranium of the female (SU 341); consequently, **TXXI-1** was difficult to assign to either the male or female on the basis of its location in situ. Probably associated with the male was the bronze fibula of Cassibile type, **TXXI-2**, found directly over the upper torso of SU 326, as well as the poorly preserved and much corroded fragments of an unidentified object of iron, **TXXI-3**, a short distance to the southeast and over the mid torso. Associated with the female were the headband, **TXXI-4**, a bronze button or small boss, **TXXI-5**, perhaps associated with the headband, as well as two tubular iron beads, **TXXI-6** and **TXXI-7**, a glass paste bead, **TXXI-8**, and part of an iron coil or spiral, **TXXI-9**. In the process of cleaning the skeletal remains in the lab, the fragmentary bronze ring, **TXXI-10**, was found associated with the male (SU 326).

TXXI-1 (P 253)**Fig. 3.65**

Kantharos of dark fabric, FD Type 4 (**9/91**).
H (to rim): 0.099; H (max): 0.119; D (base): 0.054–0.060; D (rim): 0.079.

TXXI-2 (SF 251) **Fig. 3.66**

Bronze fibula, Type I.2b, Cassibile type (**10/21**).
L: 0.089.

TXXI-3 (SF 256) **Fig. 3.67**

Fragments of unidentified iron sheet object (**10/132**).
PL × PW (largest fragment): 0.017 × 0.018.

TXXI-4 (SF 255) **Fig. 3.68**

Bronze headband (**10/86**).
L (all fragments lined up together): 0.495; L (longest preserved group of joining fragments): 0.146.
Papadopoulos 2010a:42, fig. 10.

TXXI-5 (SF 259) **Fig. 3.69**

Bronze boss, probably associated with the headband (**TXXI-4**) (**10/63**).
D (est.): 0.032–0.035.
Papadopoulos 2010a:42, fig. 11.

TXXI-6 (SF 257a) **Fig. 3.70a**

Tubular iron bead (**10/92**).
L: 0.017; D: 0.013.

TXXI-7 (SF 257b) **Fig. 3.70b**

Tubular iron bead with attached iron coil (**10/91**).
L (without attached iron coil): 0.018; D: 0.015.

TXXI-8 (SF 258) **Fig. 3.71**

Spherical fluted opaque white glass bead (**10/114**).
H: 11.22 mm; D: 11.56 mm.

TXXI-9 (SF 264) **Fig. 3.72**

Fragmentary iron coil/spiral (**10/72**).
PL: 0.024.

TXXI-10 (SF 292a) **Fig. 3.73**

Fragments of bronze ring (**10/79**).
PL: 0.020; D (est.): 0.024.

Grave fill: SU 325: HM ceramics (3 frr, coarse, including handle = 13 g; 1 fr, semi-coarse = 0.5 g); daub (10 frr = 5.5 g); shell.

Tomb XXII

Pit tomb, single inhumation (adult male)

Phase II

Figs. 3.74–3.75 (cf. Fig. 3.152)

1.0E–0.0, 5.0S (high: 107.73 m = 354.73 m ASL; low: 107.59 m = 354.59 m ASL)

Sectors 3 and 7; SU 293 (grave); 294 (skeleton); 295 (fill)

Excavated as Grave 47 (July 13–14, 2005)

The very top of the tomb was first noted with the final photography of Tombs XLVI and XLVII (Graves 42 and 41) (**Fig. 3.152**), immediately to the south of Tomb XLVI (Grave 42) and between it and the south scarp of the tumulus. With further excavation, it was established that a portion of this tomb was stratigraphically located below Tomb XLVI (Grave 42) and that the tomb was, therefore, relatively early among the prehistoric burials in the tumulus. As with the nearby Tombs XLVI and XLVII, what survives of the skeleton is remarkably well preserved, and includes part of the spinal column, a clavicle, all of the right arm and hand, and a portion of the left forearm and hand. The entire southern portion of the tomb had probably eroded off the steep south scarp of the mound. The spine of the deceased was oriented east to west (80°), with the head, nothing of which survives, to the east. The preserved human remains were identified as a mid-aged adult male (SU 294). The original position of the body was difficult to determine from the surviving remains. The excavator stated that the deceased was probably laid out in a fetal position on the right side, the arms flexed and the skeleton facing north, but the possibility that the torso was supine, with the left arm lying on top of the right, cannot be dismissed, although it seems less likely. A distinct grave cut was not noted, and this is not surprising given the proximity of the tumulus edge. As preserved, the tomb measured 0.50 × 0.40 m, and about 0.14 m deep. There were no grave goods.

Grave fill: SU 293 and 295: HM ceramics (1 fr, coarseware = 5 g; 9 non-diagnostic scraps = 1 g); lithic debitage = 0.8 g.

TOMB XXIII

Pit tomb, single inhumation (child)

Phase II (AMS Charcoal ^{14}C: 1070 ± 59 cal BC)

Figs. 3.76–3.78

1.0W, 3.0N (high: 107.72 m = 354.72 m ASL; low: 107.60 m = 354.60 m ASL)

Sector 4; SU 327 (grave); 328 (fill); 329 (skeleton)

Excavated as Grave 56 (July 20–21, 2005)

The tomb was located only 0.11–0.20 m below Tomb L (Grave 46), parallel to Wall 1, with the deceased on the same east to west orientation (80°), head to the east. As with Tomb L, the skeletal remains of Tomb XXIII were very poorly preserved, with the bone fragmented and splintering. Portions of the cranium and teeth, upper parts of both arms, as well as smaller portions of the lower left arm, together with parts of the legs were noted in situ, but very little of the actual torso. The deceased (SU 329) was a child aged 9 (± 1) years, and although difficult to determine because of the preservation of the bone, the cranium and upper body appeared to have been laid out supine, and the left arm was folded across the abdomen; the legs were probably flexed, the knees evidently pointing north-northwest. The tomb measured 1.10 m long, 0.70 m wide, and about 0.12 m deep. An iron dress pin, **TXXIII-1**, was found oriented north to south on the chest of the deceased immediately to the west of the chin (it was not possible to determine in the field whether the pin head was to the north or south).

A clear continuous grave cut from Tomb L to Tomb XXIII was not encountered, so it remains difficult to associate the two burials directly, but the fact that they were directly one above the other, that both were children laid out precisely on the same orientation, with similar grave goods, suggests that the two were related. Although not clearly part of Tomb XXIII, two stones on the northern side of the deceased (SU 329), another two on the south side, and one on the west seem to surround the tomb. Something of a grave cut was visible along the north and south sides of the grave (**Fig. 3.77**). A charcoal sample from the tomb yielded a calibrated ^{14}C date of 1070 ± 59 BC.

TXXIII-1 (SF 260) **Fig. 3.78**

Iron dress pin, Type II.2, with disk finial (**10/42**). PL: 0.142.

Grave fill: SU 328: HM ceramics (semi-coarse, 1 fr = 0.3 g; non-diagnostic, 7 scraps = 1 g); daub (18 frr = 15.5 g); shell.

TOMB XXIV

Pit tomb, single inhumation (infant)

Phase II

Figs. 3.79–3.80

5.0E, 5.0S–6.0S (high: 106.86 m = 353.86 m ASL; low: 106.66 m = 353.66 m ASL)

Sector 2; SU 477 (grave); 475 (upper fill); 480 (lower fill); 478 (skeleton); 479 (cut)

Excavated as Grave 85 (July 13, 2006)

The grave pit for Tomb XXIV was clearly defined on the north and west sides, where it was cut into bedrock. As preserved, the pit measured 1.40–1.50 m long (north to south) × 1.00 m wide (east to west), and was 0.32 m deep. The eastern edge was very far from clear and was subsequently shown to have been cut into by the pit for Tomb XXXI (86). The west edge of the pit for Tomb XXIV was so clearly defined that it had cut into the east side of the cut for Tomb III (Grave 81). Although these three tombs—Tombs III, XXIV, and XXXI (Graves 81, 85, and 86)—all appeared to be early, it seemed that Tomb XXXI cut into Tomb XXIV, which cut into Tomb III. Within the center of this large and irregularly shaped pit, fragments of the cranium, primarily dentition, of an infant aged 3 years (± 12 months) were found, designated SU 478; too little survived to venture a guess on the original position or orientation of the deceased. The upper fill (SU 475) was only slightly darker than the surrounding tumulus fill; the lower fill (SU 480) was sandy immediately around the bone, with a patch of decomposed shale at the northwest corner. Both the upper and lower fills proved to be sterile. There were no grave goods.

Tomb XXV

Pit tomb, single inhumation (child)

Phase II

Figs. 3.9, 3.81a–c

5.0E, 8.0S (high: 106.66 m = 353.66 m ASL; low: 106.57 m = 353.57 m ASL)

Sector 2; SU 511 (grave); 510 (fill); 512 (skeleton); 519 (cut)

Excavated as Grave 90 (July 21–25, 2006)

Tomb XXV, together with Tomb II, was the last of the graves to be encountered and excavated in the 2006 season. It consisted of a clearly defined pit at the southern edge of the tumulus, marked by a cut into sandy earth and characterized by a dark-colored fill, clearly distinguishable from tumulus fill (**Fig. 3.81a**). There was a medium-sized stone to the west and a smaller crescent-shaped stone at the northern edge of the cut (which overlay the pelvis of Tomb II). Tomb XXV was stratigraphically located above Tomb II (**Fig. 3.81b–c**). As preserved, the pit measured 1.10 (north to south) × 1.10 m (east to west), and 0.11 m deep. All that survives of the skeleton are parts of the leg bones, which are clearly flexed, the knees evidently pointing east-northeast, and although little bone survives, what is preserved is in fairly good condition. What survives of the deceased was identified as belonging to a child aged 6–10 years. The remainder of the tomb fell victim to proximity to the tumulus edge. Little else can be said about the original position and orientation of the deceased. There were no grave goods.

Grave fill: SU 510: HM ceramics (semi-coarse, 1 fr = 5 g); fired clay (7 frr = 14 g); shell.

Tomb XXVI

Pit tomb, single inhumation (mature adult female)

Phase II

Figs. 3.82–3.84

3.0E–5.0E, 4.0–5.0N (high: 107.56 m = 354.56 m ASL; low: 107.24 m = 354.24 m ASL)

Sector 1; SU 414 (grave); 415 (fill); 416 (skeleton); 417 (cut)

Excavated as Grave 74 (June 30–July 5, 2006)

Stratigraphically located below Tomb XLVIII (Grave 52), as well as below two of the modern graves, Tombs XCII and XCIII (Graves 23 and 19), Tomb XXVI consists of a neatly defined, roughly rectangular pit with rounded corners that was clear on all sides except for the west-northwest. The pit measured 1.90 m long, with a maximum width of about 0.80 m; from the top of the stone to the bottom of the tomb pit, the depth was 0.32m, the depth of the actual cut about 0.10 m. The fill was characterized by a loose-textured, dark-colored soil clearly distinguishable from the surrounding tumulus fill. Within the pit, the body of the deceased (SU 416), identified as a mature adult female aged 45+ years, was placed with the torso and cranium supine, the cranium facing up; the arms were folded across the lower abdomen. The legs were flexed, the knees pointing southwest. Most of the skeleton was preserved, except for the feet and most of the left ribs, but the bone was noted even in the field as being light and porous. The deceased was oriented southeast-northwest (125°), head to the southeast. A small piece of black material—thought to be either bitumen or daub in the field, and which proved to be daub—was found next to the outermost of the right ribs, and a small piece of daub was also found above the left knee. A large stone (0.42 m long) was placed at the southeast end of the grave, overlying about one-quarter of the cranium; the stone may well have served as a grave marker. The only grave good was the one-handled matt-painted vessel with distinctive "biforata" handle, **TXXVI-1**, placed upright beside the left shoulder of the deceased, about 0.10 m to the southwest of the cranium.

Grave fill: SU 415: HM ceramics (coarse, 2 frr = 13 g); daub (5 frr = 7 g); lithics (1 piece); shell.

TXXVI-1 (P 326) **Fig. 3.84**

One-handled, matt-painted, light fabric vessel, FL Type 2A (**9/5**).

H (max): 0.158; D (base): 0.050–0.052; D (rim): 0.103.

TOMB XXVII

Pit tomb, multiple inhumation (at least two adults plus additional teeth)

Phase II

Figs. 3.85–3.87

5.0W–6.0W, 3.0N–4.0N (high: 106.55 m = 353.55 m ASL; low: 106.40 m = 353.40 m ASL)

Sector 4; SU 455 (grave); 456 (fill); 457 (skeleton to southeast); 461 (skeleton to west); 463 (cut); 457a (additional teeth)

Excavated as Grave 82 (July 7–10, 2006)

Tomb XXVII was the westernmost of all burials on the tumulus and the last of a cluster of three tombs—together with Tombs XXVIII and LI (Graves 77 and 78)—to be excavated, located in an area that was originally thought to lie beyond the actual tumulus. Tomb XXVII was stratigraphically below Tomb LI (Grave 78) (the latter had cut into the southwest portion of Tomb XXVIII). Tomb XXVII consists of a pit cut into bedrock, with the cut visible on the northeast and northwest sides; what appeared at first as a clear cut on the southeast side perhaps followed the original contour of the bedrock of the ridge on which the tumulus was constructed. To the west and southwest, the cut was truncated due to erosion off the tumulus edge. The tomb was first encountered with the removal of topsoil, and the pit fill was mixed, patchy, and rather compact. As preserved, the tomb measured at least 1.70 m long, and was probably originally closer to 1.85 m, with a maximum width of 1.16 m; the depth of the tomb deposit was 0.15 m.

Within the pit, the skeletal remains of two individuals were discerned in the field. The southeast skeleton (SU 457), identified as an adult aged 35+ years, and of indeterminate sex, was laid out with the cranium and torso supine. The cranium was lying on the left side, facing toward the south; the right arm was folded across the lower torso; the left arm appears to have been tightly flexed, but the lower arm bones may be out of their original position. The legs were flexed, the knees pointing southeast. The deceased was oriented northeast-southwest (37°), head to the northeast. The cranium and upper arm of the west or northwest skeleton (SU 461), identified as an adult female, 35+ years at death, was in situ and shared a similar orientation to SU 457; the remainder of the bone of this individual was largely disarticulated, particularly toward the southwest. The relationship of the two inhumations led the excavator to conclude that SU 461 was the earlier of the two, that the tomb was opened at a later time to accommodate SU 457, and that the leg bones and pelvis of SU 461 were pushed toward the southwest, where they were found piled up; the cranium and right arm remained, however, articulated and in situ. In the process of cleaning the skeletal remains in the lab, Dr. Schepartz identified the teeth only of a third individual, a young adult of indeterminate sex aged 18–25 years. The dentition of this individual was not noted in situ, but the teeth may be those marked on the drawing immediately to the southwest of the cranium of SU 457.

The only grave good was the bronze dress pin, **TXXVII-1**, found on the torso of SU 457, oriented northeast-southwest, head to the northeast, and clearly associated with the individual with which it was found. Although located on the western periphery of the tumulus and at a greater depth than many other tombs (and over 0.50 m deeper than Tomb XXVIII), this is not an indicator of an early date, and it appears likely that Tombs XXVIII and XXVII are not separated by any great length of time (the two tombs are considered to belong to the same chronological phase).

TXXVII-1 (SF 354) **Fig. 3.87a–b**

Bronze dress pin, Type I.3, with rolled head (**10/29**).
L: 0.128.

Grave fill: SU 456: HM ceramics: inventoried: P 349 (base fr: PH: 0.019; PW: 0.039), not inventoried (semi-coarse, 3 frr = 15 g); daub (27 frr = 34 g); shell.

TOMB XXVIII

Pit tomb, single inhumation (infant/child)

Phase II

Figs. 3.88–3.96

4.0W–5.0W, 3.0N–4.0N (high: 107.11 m = 354.11 m ASL; low: 107.00 m = 354 m ASL)

Sector 4; SU 428 (grave); 425 (NE fill); 426 (cut); 427 (SW fill); 429 (skeleton)

Excavated as Grave 77 (July 4, 2006)

This was the first of a cluster of three tombs that were located in the far west of Sector 4, in an area that was, at first, thought to lie outside the actual tumulus. Tomb XXVIII was stratigraphically interrelated with Tombs LI and XXVII (Graves 78 and 82). The tomb consisted of a well-defined and comparatively deep pit. All the edges of the grave cut were clear with the exception of the southwest side, which was encountered directly under topsoil. The edges on the northwest and southeast sides were marked by bands of black organic material at the upper level of the tomb (**Figs. 3.88, 3.89b**). From what was clear of the grave cut, the pit was at least 0.90 m long, and the original pit may have had a length closer to 1.05–1.10 m; the width of the pit ranged between 0.27 and 0.43 m, and its maximum depth was 0.16 m. To the northeast of the tomb pit, a pile of stones was uncovered (**Fig. 3.89a**) that originally was thought may have been connected with the tomb, but with further excavation it was clear that the stones were not directly associated. The fill within the grave varied considerably. To the northeast (SU 425), the fill was of a looser texture and lighter color to the surrounding tumulus fill; the fill to the southwest was darker, somewhat harder, and more clayey. The poorly preserved cranium of an infant/child (SU 429), aged 3 years (± 24 months), was encountered not far from the center of the pit and closer to the northeast end. All of the small finds, including bronze ornaments and a variety of glass and stone beads, were clustered together on or beneath the remnants of the cranium. There was no trace of any postcranial bone and, as such, it was impossible to determine the original position of the deceased. The grave cut and the individual within it were oriented northeast-southwest (50°), but the lack of postcranial bone led to some uncertainty as to whether the cranium was to the northeast or southwest, though it is more likely that the cranium was to the northeast. With the subsequent excavation of the nearby Tomb LI (Grave 78), it became clear that Tomb LI cut into and destroyed the southwest end of the pit of Tomb XXVIII; the southwest fill of Tomb XXVIII (SU 427) was the same as the fill of Tomb LI (also designated SU 433). In the process of studying the skeletal remains in the lab, a solitary tooth of a young adult of indeterminate sex was found and labeled SU 429a; this tooth is almost certainly displaced from the skeleton of Tomb LI (Grave 78) and after closer study was shown to belong to that individual (SU 434).

Among the many ornaments associated with the cranium of the deceased, the bronze spectacle pendant, **TXXVIII-1**, was found on the northeast side of the cranium, the bronze wheel pendant, **TXXVIII-2**, was right next to it, to the southwest, and the bronze fibula, **TXXVIII-3**, was immediately to the southwest of the wheel pendant; the three were found clearly in line with one another. The small blue glass bead, **TXXVIII-4**, was found to the northeast of the spectacle ornament; the core-formed glass bead, **TXXVIII-5** and the stone bead, **TXXVIII-6**, were found next to each other and partially underlying the wheel pendant; two more glass beads, **TXXVIII-7** and **TXXVIII-8** were located next to these, immediately to the southeast, their positions clearly marked on the plan (**Fig. 3.88**). Two more glass beads, **TXXVIII-9** and **TXXVIII-10**, were found in many fragments in the lab in the process of cleaning the block-lifted cranium. Also found in the lab was the bronze tube (designed SF 343), which was part of the wheel ornament, **TXXVIII-2**. In the process of cleaning the cranium and its ornaments, a dark organic material was noted associated with the wheel pendant, **TXXVIII-2** (**Figs. 3.90–3.91**), which could be either a wooden backing for the pendant, or modern roots. Similar organic material noted with two of the bronze headbands analyzed in the labs of the Getty Conservation Institute was clearly not leather or anything protein-based.

TXXVIII-1 (SF 335) **Fig. 3.92**

Bronze spectacle ornament (**10/58**).
L: 0.035.

TXXVIII-2 (SF 336 + 343) **Fig. 3.93**

Bronze wheel pendant (perhaps with backing of organic material) (**10/59**).
D (outer circle): 0.052–0.055.

TXXVIII-3 (SF 337) **Fig. 3.94**

Bronze fibula, "Cassibile" Type (I.2a) (**10/20**).
L: 0.094.

TXXVIII-4 (SF 338) **Fig. 3.95c**

Small blue glass bead (**10/112**).
H: 5.41 mm; D: 7.16 mm.

TXXVIII-5 (SF 339) **Fig. 3.95a–b**

Rod-formed, dark green glass bead with trails of opaque white glass (**10/115**).

H: 16.52 mm; D (max): 14.56 mm.

TXXVIII-6 (SF 340) **Fig. 3.95d**

Bead of crypto-crystalline quartz (carnelian or sardonyx) (**10/102**).
D: 0.018.

TXXVIII-7 (SF 341) **Fig. 3.96a**

Opaque orange-brown glass bead (**10/108**).
H: 10.20 mm; D: 10.51 mm.

TXXVIII-8 (SF 342) **Fig. 3.96b**

Opaque yellow-brown glass bead (**10/106**).
D: 0.030.

TXXVIII-9 (SF 344) **Fig. 3.96c**

Fragments of opaque golden-yellow glass bead (**10/116**).
Original dimensions NR; PL (largest fragment): 0.008.

TXXVIII-10 (SF 351) **Fig. 3.96d**

Fragments of white opaque glass bead (**10/117**).
Dimensions NR.

Grave fill: SU 425: HM ceramics (semi-coarse, 3 frr = 5 g); daub (9 frr = 10 g); shell. SU 427: 1 scrap non-diagnostic ceramics; shell.

TOMB XXIX

Pit tomb, single inhumation (infant)

Phase II

Figs. 3.97–3.98

5.0E, 1.0N–2.0N (high: 107.30 m = 354.30 m ASL; low: 107.14 m = 354.14 m ASL)

Sector 1; SU 468 (grave); 466 (fill); 469 (skeleton); 476 (cut)

Excavated as Grave 83 (July 11–13, 2006)

This tomb consisted of a well-defined pit, roughly oval in plan, measuring 1.02–1.10 m long, 0.45 m wide, and 0.16 m deep, cut into bedrock. The clearly distinguishable fill was a dense clayey, compact, and calcareous soil. Within the center of the pit, the body of an infant aged 6 (± 3) months was laid out in a fully extended supine position, cranium facing up, arms by the side of the body, legs bent at the knees. This was among the most fully articulated and best preserved of the prehistoric infant burials, with most of the bones surviving except for the hands and feet. The deceased was oriented southeast-northwest (140°), head to the southeast. The tomb was stratigraphically located beneath Tomb LIII (Grave 63). There were no grave goods.

Grave fill: SU 466: no material except for shell.

TOMB XXX

Pit tomb, secondary cremation (infant)

Phase II

Figs. 3.99–3.102

6.0E, 4.0S (high: 107.18 m = 354.18 m ASL; low: 107.08 m = 354.08 m ASL)

Sector 2; SU 396 (grave); 397 (fill); 398 (skeleton, infant), 398a (skeleton fragments, adult)

Excavated as Grave 70 (June 28–29, 2006)

Tomb XXX was one of only two cremations in the entire tumulus and the first cremation to be excavated. The southeast portion of the grave was first encountered with the clearance of topsoil and was located only some 0.40 m from the eastern edge of the mound. The cremated remains defined a roughly oval area measuring 0.45 m long southeast-northwest × 0.30 m southwest-northeast, and 0.10–0.15 m deep, and were identified as belonging to an infant aged 9 (± 3) months. It was noted in the field that some of the bones were decalcinated (white), others were blackened, while some were differentially burned more on one side than the other. Fragments of an olive pip (**Fig. 3.102**) were also found with the cremated remains. A single stone was found 0.14 m to the south of the bone. The entire circumscribed area of cremated bone was pedestalled and block-lifted for careful cleaning and analysis of the cremated remains in the lab. In the process of cleaning in the lab, Dr. Schepartz noted three fragments of long bone of an adult of indeterminate sex that were not cremated intermingled with the cremated remains of the infant; these fragments were designated SU 398a and are clearly intrusive.

The subsequent discovery of the better-preserved cremation in Tomb XXXVIII (Grave 79) indicated that both cremations were secondary, that is, that the process of cremation was performed elsewhere, beyond the tumulus. The cremated bone was then collected and must have been bundled in some sort of organic material (cloth, leather, or skin), which does not survive, and this was placed in a pit dug in the tumulus. There were no grave goods. Tomb XXX was stratigraphically located immediately above Tomb VI (Grave 97), which was the inhumation of an infant.

Grave fill: SU 397: HM ceramics (semi-coarse, 4 frr = 9 g; coarse, 3 frr = 29 g); daub (25 frr = 35 g); lithics (10 debitage, 1 nucleus); shell.

TOMB XXXI

Pit tomb, single inhumation (adult, indeterminate sex)

Phase II

Figs. 3.103–3.105

5.0E–7.0E, 5.0S–7.0S (high: 106.70 m = 353.70 m ASL; low: 106.55 m = 353.55 m ASL)

Sector 2; SU 488 (grave); 481 (fill); 489 (skeleton); 502 (cut)

Excavated as Grave 86 (July 14–18, 2006)

Evident only on the west side, especially to the southwest, the pit cut into sandy soil resembling bedrock. It was this western edge of the tomb that cut into the earlier Tomb XXIV (Grave 85). There was no clear cut on any of the other sides. The pit fill was a dark, loose soil, clearly distinguishable to the west and southwest, merging with topsoil to the east and southeast as the tumulus edge was approached. As preserved, the tomb was 2.00–2.10 m long (north to south) × 1.70–1.80 m wide (east to west); the maximum depth of the cut was 0.26 m, the depth of skeletal remains 0.15 m. Within the pit, the poorly preserved remains of the deceased (SU 489), identified as an adult of indeterminate sex, aged 40+ years at death, were encountered. Only a small portion of the cranium and torso were clearly distinguishable as found, as well as portions of one arm bone; parts of the legs and pelvis were somewhat better preserved. A line of three small to medium stones extended from northwest to southeast from the area of the pelvis; a large stone, probably an outcrop of bedrock, was under the left femur, and what appears to be another outcrop of bedrock was noted to the east of the legs. Given the poor state of the bone, it was difficult to establish the precise original position of the deceased, but it appears that the cranium was probably lying on its left side, facing west; the legs were flexed, the knees pointing west. It was the excavator's impression that the body was probably laid out in a fetal position on the left side, but this was unclear given the poor preservation of the torso. The deceased was oriented south to north (180°), head to the south.

Two terracotta spindlewhorls, beads, or buttons, **TXXXI-1** and **TXXXI-2**, were found next to one another against the northeast face of what appeared to be the left femur (that to the southwest).

TXXXI-1 (SF 388) **Fig. 3.105a**

Biconical terracotta spindlewhorl, bead, or button (**10/2**).

H: 0.025; D: 0.023.

TXXXI-2 (SF 389) **Fig. 3.105b**

Biconical terracotta spindlewhorl, bead, or button (**10/1**).

H: 0.027; D: 0.023–0.024.

Grave fill: SU 481: HM ceramics (fine, 1 rim fr = 3 g; semi-coarse, 2 frr = 10 g; coarse, 3 frr = 5 g); daub (29 frr = 45 g); lithics (1 debitage, 5 pebbles); shell.

TOMB XXXII

Pit tomb, double inhumation (two adult males)

Phase II

Figs. 3.106–3.108

7.0E–8.0E, 2.0S–3.0S (high: 106.63 m = 353.63 m ASL; low: 106.51 m = 353.51 m ASL)

Sector 2; SU 506 (grave); 507 (fill); 508 (southern skeleton); 524 (northern skeleton); 528 (cut)

Excavated as Grave 89 (July 20–26, 2006)

This tomb consisted of a clearly defined cut to the west and for part of the north side of the grave; to the northeast, east, and south the cut was largely obliterated on account of the proximity to the tumulus edge (i.e., erosion). The pit fill was a dark-colored soil difficult to distinguish in places from the topsoil. As preserved, the tomb measured about 2.00 m long (northwest-southeast) and 1.50 m wide (east to west); the cut was about 0.35 m deep, the depth of the bone about 0.12 m. A medium-sized stone was found at the western edge of the tomb, partly overlying portions of both skeletons, and there was another smaller stone to the east. The highest level of the western stone was 106.91 m. As the top of this stone was almost 0.30 m higher than the level of the bone, it may have served as something of a grave marker. Within the pit, the skeletal remains of two individuals were encountered. The fully articulated skeleton (SU 508) to the south was identified as a relatively young adult male aged 20–30 years. The cranium and torso, poorly preserved on account of proximity to topsoil and to the tumulus edge, were laid out in a supine position, with the arms clearly folded across the lower torso, the right arm above the left, but not stacked; too little survived of the cranium to determine in which direction it faced. The legs were flexed, the knees pointing northeast. The deceased was oriented southeast-northwest (150°), head to the southeast. The second individual, SU 524, identified as a younger adult male aged 18–25 years, was largely disarticulated, long bones in fragments in the western and northern part of the tomb, around the legs of SU 508. The respective positions of the two skeletons suggested that SU 524 was the earlier of the two interments, and that the remains of the deceased were pushed to the west and north in order to accommodate SU 508.

The only finds associated with the grave were two iron projectile points, **TXXXII-1** and **TXXXII-2**, both found on the torso of the more fully articulated later individual (SU 508), **TXXXII-1** to the east, **TXXXII-2** to the west, the latter in the area of the abdomen, the former on mid-right torso. These were the only two arrowheads in the entire tumulus, and their position on the torso of the deceased were such that they could either be interpreted as two arrows placed on—or even held by—the deceased, or else as projectiles that led to the demise of the individual through inter-personal violence. The bone was too fragmentary to preserve any clear evidence for death by arrowheads.

TXXXII-1 (SF 410) **Fig. 3.108a**

Iron arrowhead (**10/119**).
PL: 0.042.

TXXXII-2 (SF (409) **Fig. 3.108b**

Fragmentary iron arrowhead (**10/120**).
PL: 0.031.

Grave fill: SU 507: HM ceramics: inventoried: P 373 (matt-painted handle fr, PH: 0.027; W [max]: 0.033), not inventoried (semi-coarse, 3 frr = 7 g; coarse, 5 frr = 20 g); daub (42 frr = 70 g; lithics (1 debitage, 3 pebbles). Small fr of modern glass recovered from topsoil in the immediate vicinity.

TOMB XXXIII

Pit tomb, single inhumation (adolescent, female?)

Phase II

Figs. 3.109–3.110

8.0E–9.0E, 1.0S–1.0N (high: 106.45 m = 353.45 m ASL; low: 106.31 m = 353.31 m ASL)

Sectors 1, 2, and 6; SU 521 (grave); 522 (fill); 523 (skeleton)

Excavated as Grave 92 (July 5–6, 2007)

The lower leg bones of the deceased in this tomb were first encountered in Sector 1 at the end of the 2006 season, but as they clearly continued into the baulk, Sector 6, and into Sector 2, the decision was taken to excavate the tomb in 2007. The tomb was found in a poor state of preservation, lying as it did near the interface of topsoil and the tumulus edge to the east and the tumulus fill to the west. Proximity to the tumulus edge and topsoil obliterated any signs of the grave cut. The skeleton was very poorly preserved; in addition to the cranium, which was the best preserved of the skeletal remains, some of the long bones of the arms and legs survived, as did a small portion of the pelvis. The torso was largely not preserved, and there was nothing of the hands and feet. The deceased (SU 523), identified as an adolescent, probably female, aged 16 ± 2 years, was laid out with the torso in supine position, and although virtually nothing of the torso was noted in situ, the fact that the arms were folded across the lower torso indicated the position of the upper body. The excavator noted that the cranium was lying on the right side, facing northeast. The legs were tightly flexed, the knees pointing east-northeast. The deceased was oriented southeast-northwest (150°), head to the southeast. As preserved, the tomb measured 1.20 m long, 0.80 m wide, and 0.14 m deep. There were no grave goods.

Grave fill: SU 522: HM ceramics (coarse, 4 frr = 40 g).

TOMB XXXIV

Pit tomb, single or double inhumation (one or two adult males)

Phase III

Figs. 3.111–3.113

6.0E–7.0E, 1.0N–3.0N (high: 106.89 m = 353.89 m ASL; low: 106.73 m = 353.73 m ASL)

Sector 1; SU 496 (grave); 497 (fill); 498 (skeleton); 484 (cut); 504 (second individual)

Excavated as Grave 87 (July 14–20, 2006)

The relationship of this tomb with Tomb XXXV (Grave 84) is discussed further in the description of Tomb XXXV. The pit for Tomb XXXIV was the same cut as that for Tomb XXXV, visible on all sides except the southeast. The dimensions of the cut were the same: 1.20 m long and 0.62 m wide, with the overall maximum depth of the cut being 0.63 m, though the depth of the deposit of Tomb XXXIV was only about 0.16 m. The fill was a loose, sandy, dark-colored soil, clearly distinguishable from the surrounding deposits. The surrounding deposits included, to the northwest, a large patch of soft clayey, orangey soil, designated SU 500, which was roughly ovoid and had a maximum length of 0.75 m; a smaller, but similar patch (SU 503) was encountered to the south of Tomb XXXIV.

Within the tomb pit, the cranium and torso of the deceased (SU 498), identified as an adult male aged 45–55 years, were laid out in a supine position, the cranium lying on the right side, facing northeast. The arms were folded across the lower torso and the legs were flexed, the knees pointing northeast. The skeleton was oriented southeast-northwest (132°), head to the southeast. The position and orientation of the deceased closely mirrored those of the individual immediately above, in Tomb XXXV. As was the case with Tomb XXXV, the skeletal remains of Tomb XXXIV were comparatively well preserved, with most of the bone represented except for the right ribs and feet. A small scatter of bone, designated SU 504, was found to the southwest of the left shoulder and cranium of SU 498. The only grave good was the iron dress pin, **TXXXIV-1**, found lying across the upper torso oriented northeast-southwest, head to the northeast.

The critical aspect of this tomb was its relationship to Tomb XXXV, which essentially followed the cut for Tomb XXXIV, even though much of the bone of the deceased in Tomb XXXV, particularly the torso and feet, were at a significantly higher level. In the process of excavation, it was not immediately clear whether Tombs XXXV and XXXIV were one multiple burial or two separate graves that happened to be stratigraphically related. Two scenarios were suggested: the first is that the two tombs were unrelated; the deceased of Tomb XXXV happened to be buried on top of Tomb XXXIV, following the same orientation. At some later time, the pelvis and legs of Tomb XXXV collapsed into the pit of Tomb XXXIV, whereas the torso and feet remained at a higher level. The alternative, and more likely, scenario was that the two tombs were intentionally laid out at the same time, which accounts for the lack of disturbance to the earlier burial. Although the torso of the deceased in Tomb XXXV was at a higher level, sloping sharply from southeast to northwest, no part of it extended beyond the visible cut for Tomb XXXIV; and the same was true for the leg bones. The intentional layout of these two skeletons was perhaps the most marked feature of this grave, the position of the bodies mirroring one another. There was no material recorded from the grave fill.

TXXXIV-1 (SF 390) **Fig. 3.113**

Iron dress pin, Type II.1, with rolled head (**10/35**).

PL: 0.122.

Tomb XXXV

Pit tomb, single inhumation (younger adult male)

Phase III

Figs. 3.114–3.116

6.0E–7.0E, 1.0N–3.0N (high: 107.23 m = 354.23 m ASL; low: 106.82 m = 353.82 m ASL)

Sector 1; SU 470 (grave); 471 (fill); 472 (skeleton); 484 (cut)

Excavated as Grave 84 (July 11–17, 2006)

Tomb XXXV consisted of a clearly defined pit partially cut into bedrock and partially into tumulus fill at the eastern edge of the mound, long and narrow in plan, with cuts visible on all sides except the southeast. The fill was easily distinguishable except to the southeast, where the tomb was uncovered with the removal of the topsoil. As preserved, the tomb measured at least 1.48 m long (the preserved tomb cut measuring 1.20 m in length), 0.62 m wide, with the cut, as discerned, at least 0.16 m deep, though the depth from the highest point of the preserved bone to the lowest was 0.41 m. Within the pit, the body of the deceased (SU 472), identified as a younger adult male, aged 23–27 years at death, was laid out with the torso supine, the arms folded across the lower torso; the legs were flexed, the knees pointing northeast. The skeleton was oriented southeast-northwest (141°), the head, which was missing, to the southeast. The bone was comparatively well preserved, with most parts of the body represented except for the cranium, which fell victim to the proximity of the tumulus edge (only a few teeth were recovered). What was most unusual was that the torso was propped up at the southeastern end of the tomb and was found at a significantly higher level than the legs of the deceased. The feet, as opposed to the legs, were also encountered at a higher level, near the northwest corner of the tomb, and it was soon clear that the pit was smaller than the individual interred, resulting in the position of the feet and the propped-up cranium and upper torso.

At the northwestern end of the tomb, the leg bones of another skeleton were encountered, just below the level of the leg bones of SU 472 but somewhat lower than the feet. Since only these leg bones were first encountered, the remains were allocated a separated grave number—Tomb XXXIV—but it was soon evident once the skeletal remains of SU 472 were lifted that the individual of Tomb XXXIV, rather than representing a separate grave, was interred in the same pit. Consequently, although Tomb XXXV was located immediately above Tomb XXXIV, the two individuals are better regarded as part of the same grave, which was a multiple burial of two adult males within the same tomb pit, in much the same manner as the two adult males in Tomb XLV (Grave 60). For the purposes of this report, and to distinguish between the two, we keep them here as separate graves, though they are best regarded as one. The fact that the individual of Tomb XXXV was propped up at both the southeast and northwest now seemed to make sense, since the deceased was interred into a pit originally made for another individual. It is interesting that the position of the individual in Tomb XXXV mirrors precisely that of the individual in Tomb XXXIV, and this, together with the minimal disturbance of the lower burial, suggests that the interment of the two was probably contemporary.

The only material object encountered with the deceased were the fragments of iron of an unidentified object, **TXXXV-1**, perhaps part of a blade, found on the upper torso resting on the vertebrae, immediately below the clavicle. The proximity of topsoil at this uppermost part of the tomb was such that the object may well have been intrusive, though it had the appearance of being in situ. There was nothing in the grave fill directly associated with Tomb XXXV that was clearly distinguishable from the topsoil.

TXXXV-1 (SF 384) **Fig. 3.116**

Fragments of iron (3) of unidentified object (**10/133**).
PL (largest fragment): 0.018.

TOMB XXXVI

Pit tomb, single inhumation (adult, indeterminate sex)

Phase III

Figs. 3.117–3.119

3.0W, 4.0S (high: 107.26 m = 354.26 m ASL; low: 107.16 m = 354.16 m ASL)

Sector 3; SU 418 (grave); 419 (fill); 420 (skeleton)

Excavated as Grave 75 (July 1, 2006)

This was one of the most, if not the most poorly preserved of all graves, the bones mostly encountered in topsoil, with the tomb largely lost due to erosion off the steep southwest scarp of the tumulus. No clear cut or distinguishable grave fill could be discerned. All that survived were fragments of two long bones running roughly east to west, with fragments of a few more bones to the north-northwest running roughly northwest-southeast. Nothing could be said about the position and orientation of the deceased, and it was not absolutely clear that the bone was even in situ. The extent of the bone was roughly 0.50 × 0.40 m, and about 0.10 m deep. The human remains (SU 420) were identified as an adult of indeterminate sex. There were no grave goods. A tiny land snail shell fragment (SF 346, **Fig. 3.119a–b**) was found in the fill of the grave and thought to have been possibly pierced and therefore an item of personal ornament, but was shown by Dr. Vardala-Theodorou to be the operculum of a land gastropod with a small hole at center that was not intentional.

Grave fill: SU 419: inventoried: SF 346 (mollusc with hole [natural], **Fig. 3.119a–b**); HM ceramics (semi-coarse, 1 fr = 5 g; 1 non-diagnostic scrap); daub (11 frr = 10 g); shell.

TOMB XXXVII

Pit tomb, single inhumation (adult, indeterminate sex)

Phase III

Figs. 3.120–3.121

4.0W–5.0W, 2.0S (high: 107.16 m = 354.16 m ASL; low: 106.92 m = 353.92 m ASL)

Sector 3; SU 422 (grave); 423 (fill); 424 (skeleton)

Excavated as Grave 76 (July 3–6, 2006)

This was less of a tomb and more of a scatter of bone fragments with no clear articulation. The burial, if it can be called a tomb, was encountered within topsoil at the steep western edge of the tumulus; as was the case with Tomb XXXVI, it was not absolutely clear that the bone was in situ, though the human remains (SU 424) all appeared to belong to an adult of indeterminate sex. Given the state of the bone, which was eroding down the side of the mound, nothing could be said about the position or orientation of the deceased. As it was finally exposed, the bone scatter covered an area approximately 0.60 m (east to west) × 0.70 m (north to south); given the active erosion off the west scarp of the mound, the bone occurred over a depth of about 0.24 m. There were no grave goods.

Grave fill: SU 423: HM ceramics (fine, 1 fr, matt-painted; semi-coarse, 7 frr = 12 g; non-diagnostic, 7 frr = 4 g); daub (9 frr = 25 g); lithics: SF 362 (flake) + 4 debitage.

TOMB XXXVIII

Pit tomb, double burial, inhumation and cremation (two adults: inhumed male and cremated female?)

Phase III

Figs. 3.122–3.126

2.0E–4.0E, 3.0N–4.0N (high: 107.39 m = 354.39 m ASL; low: 107.23 m = 354.23 m ASL)

Sector 1; SU 437 (grave); 438 (fill); 439 (inhumed skeleton); 447 (cremation)

Excavated as Grave 79 (July 4–10, 2006)

Papadopoulos 2010b:237, fig. 3

This was one of the most interesting of all tombs, comprising a double burial of an inhumed adult male (SU 439) aged 30–40 years, and a cremated individual, possibly female (SU 447). The remarkably well-preserved body of the male, described as large and robust, was laid out with the torso supine, the cranium lying on the right side, facing south. The left arm was folded across the chest, the right arm sharply folded over the right shoulder. The legs were very tightly flexed, artificially doubled back on themselves, the lower left leg under the right femur, the knees pointing almost due east (east-southeast). The position of the legs was intentional, that is, the legs were not drawn up and subsequently fell to one side, and the degree of flexing was such that the body may well have been bundled, a possibility that seemed more likely as the tomb was studied. The deceased was oriented west-northwest to east-southeast (290°), cranium to the west-northwest. Although no real grave cut was discerned, the inhumation was clearly defined by lines of black organic material, very clear on the south side, variable on the north side, but extending to a slightly higher level and somewhat more irregular on west; there was a much less clearly visible line on the east (**Fig. 3.124a**). These thin black lines, almost certainly the remnants of a textile or skin, perhaps a shroud, seemed to encase the inhumation, and perhaps also the cremation. The area thus defined measured 1.45–1.50 m long, 0.46–0.50 m wide, and 0.16 m deep.

The cremation itself was secondary—that is, the body of the deceased was cremated elsewhere, the burned remains collected and deposited in this tomb—and was placed to the southeast of the inhumation in a clearly defined oval area, itself demarcated by lines of organic material suggesting that the cremated remains were wrapped separately in textile or leather; a smaller scatter of cremated bone was found directly overlying the right pelvis, which had been disturbed from the main concentration of cremated remains to the southeast. The fill immediately surrounding the inhumation and cremation was patchy, but essentially little different from the surrounding tumulus fill.

In the process of lifting the skeletal remains of the inhumation, a fragment of an iron knife blade, **TXXXVIII-1**—the actual tip of the knife—was found immediately under the thoracic vertebrae

(**Fig. 3.125**). The fragmentary state of the blade, coupled with its position in situ, suggested that the adult male may well have died by being stabbed in the abdomen, side, or back. This combination of features—a tightly flexed and seemingly bundled inhumation within what appeared to be a substantial textile or skin, a secondary cremation interred together with the inhumation, and the possibility of a violent death—suggests that both individuals may have died at some distance from the tumulus and were transported, bundled in a textile or skin, to their final resting place.

Tomb XXXVIII was located more or less where the continuation of Wall 1 may have been had it continued into this portion of Sector 1, and it is even possible that the interment necessitated the removal of some of the stones of the wall. The tomb was stratigraphically located below one of the modern inhumations, Tomb XCI (Grave 15).

TXXXVIII-1 (SF 370) **Fig. 3.126a–b**

Fragment preserving the tip only of an iron knife (**10/122**).
PL: 0.039.

Grave fill: SU 438: HM ceramics (semi-coarse, 2 frr = 6 g); daub (3 frr = 8 g).

TOMB XXXIX

Pit tomb, single inhumation (adult female)

Phase III

Figs. 3.127–3.131

5.0E–6.0E, 3.0N–5.0N (high: 107.41 m = 354.41 m ASL; low: 107.30 m = 354.30 m ASL)

Sector 1; SU 367 (grave); 368 (fill); 369 (skeleton); 402 (cut)

Excavated as Grave 66 (June 19–July 1, 2006)

This tomb, located at the eastern edge of the tumulus, was first encountered at the end of the 2005 season and excavated early in the 2006 campaign. It consisted of a neatly defined pit, clearly distinguishable on the southwest and northwest sides, and for a portion of the northeast side; the remainder of the cut to the southeast was obliterated by topsoil and proximity to the tumulus edge and modern surface. The fill of the grave was generally slightly darker and looser than the surrounding tumulus fill, containing pebbles and shells. Within the pit, the comparatively well-preserved skeleton of an adult female (SU 369) aged 25–35 years, was oriented southeast-northwest (140°), head to the southeast. The cranium and torso of the deceased were laid out in a supine position, the head facing up; the right arm was folded over the lower torso, the left arm sharply folded vertically over the left shoulder. The legs were tightly flexed, the knees lying on the left side, pointing west. Most of the bones were preserved and identified in situ, except for the ribs, particularly on the left side. As preserved, the tomb, including the grave cut, measured 1.75 m long; the cut was at least 0.95 m wide, but was originally wider to the southeast; the bones of the deceased as laid out were 0.75 m wide, and the tomb was 0.11 m deep.

Grave goods included a one-handled matt-painted tankard, **TXXXIX-1**, placed upright to the southwest of the cranium and at the same level. A bronze spectacle fibula, **TXXXIX-2**, was found in situ on the left shoulder, with traces of the disintegrated organic remains of the garment encountered as darker soil in the immediate vicinity, and must have pinned a textile at the left shoulder. The fibula was overlaid by the finger bones of the left hand, as if the deceased was holding the bronze. A large matt-painted sherd, P 324, was noted in the fill of the grave to the west of the left arm (**Fig. 3.131**).

TXXXIX-1 (P 304) **Fig. 3.129**

Light fabric, matt-painted, one-handled vessel, Type 2A (**9/3**).
H (to rim): 0.115–0.118; H (max): 0.143; D (base): 0.052; D (rim): 0.105–0.108.

TXXXIX-2 (SF 315) **Fig. 3.130**

Bronze spectacle fibula, Type I.1c (**10/16**).
L (not including pin): 0.091.

Grave fill: SU 368: HM ceramics: inventoried: P 324 (**9/75**) (**Fig. 3.131**: wall fr, lower body, matt-painted vessel, PH: 0.055; PW: 0.065), not inventoried (semi-coarse, 3 frr = 5 g; coarse, 2 frr = 10 g); daub (7 frr = 14 g); lithics (3 pieces of nucleus); shell.

TOMB XL

Pit tomb, single inhumation (adult, probably male)

Phase III

Figs. 3.132–3.133

0.0W–0.0E, 6.0S (high: 107.47 m = 354.47 m ASL; low: 107.40 m = 354.40 m ASL)

Sector 7; SU 381 (grave); 382 (fill); 383 (skeleton)

Excavated as Grave 67 (June 23–24, 2006)

This tomb, as preserved, was entirely located in the baulk, Sector 7, at the southern edge of the tumulus, with most of the burial destroyed due to erosion off the steep south slope of the tumulus. Topsoil immediately to the south of the tomb contained some loose bone that may derive from this burial. What little survives of the bone was difficult to distinguish in situ, although a femur, fragments of ribs, and a long bone from an arm appeared to be in situ, together with a scapula and clavicle. The skeletal remains, designated SU 383, were identified as an adult, probably male, aged 30+ years at death. Although too little survived to determine the position of the deceased, the excavator noted that the skeleton was probably oriented west to east (about 285°), head, which did not survive, to the west. As preserved, the tomb measured 0.54 m long, 0.38 m wide, and 0.07 m deep. The grave fill was loose and sandy, with most of the bone resting on calcareous materials, suggesting proximity to bedrock. There were no grave goods.

Grave fill: SU 382: HM ceramics (1 fr, semi-coarse = 1g); daub (1 fr = 1 g); lithics (1 debitage).

TOMB XLI

Pit tomb, single inhumation (adult, indeterminate sex)

Phase III

Figs. 3.134–3.135

3.0E, 6.0S (high: 107.68 m = 354.68 m ASL; low: 107.56 m = 354.56 m ASL)

Sector 2; SU 330 (grave); 332 (fill); 331 (skeleton)

Excavated as Grave 57 (July 19–20, 2005)

Only the cranium, portion of the upper torso, and the greater part of both arms and hands of the deceased survive. The remainder of the skeleton was lost due to erosion on the steep south side of the tumulus, and some of the exposed postcranial bone found off the south side of the mound prior to excavation may have derived from this tomb. What survives of the bone is remarkably well preserved, in keeping with other nearby tombs. No grave cut or fill clearly distinguishable from the tumulus fill was discerned. The deceased (SU 331), identified as an adult of indeterminate sex aged 20–30 years, was oriented northwest-southeast (335°), head to the northwest. The torso and cranium were laid out so the deceased was lying on the right side, the arms flexed, in front of the cranium, the latter facing toward the west-southwest. As preserved, the tomb measured 0.70 m long, 0.60 m wide, and 0.12 m deep. There were no grave goods, nor was there any material recovered from the grave fill.

TOMB XLII

Pit tomb, multiple inhumation (adult female and two infants)

Phase III

Figs. 3.136–3.139

2.0W–4.0W, 1.0S–2.0S (high: 107.65 m = 354.65 m ASL; low: 107.49 m = 354.49 m ASL)

Sector 3; SU 337 (grave); 287 (fill); 354 (cut); 338 (adult skeleton); 339 (south skeleton); 340 (central skeleton)

Excavated as Grave 59 (July 12, 20–25, 2005)

This tomb consisted of a clearly defined pit much larger than the bodies placed within it, measuring 2.20 m long, 1.15 m wide, and 0.16–0.25 m deep, though the original depth of the pit, as with Tomb 61, may have been deeper. Within the pit, the skeletal remains of three individuals were found: to the north, the somewhat disturbed remains of an adult female (SU 338) aged 20–25 years at death; the southernmost skeleton (SU 339) was an infant/child aged 4 (± 1) years, and the central skeleton (SU 340) was an infant aged 2 years (± 8 months). It seems clear that the grave was originally for the adult female. At some later time the tomb was reopened, the bones of the adult moved to the north, but following the orientation of the tomb and the bodies of two infants placed in the central portion of the tomb pit. Both infants were found lying on their left sides against each other and partly on top of one another, their heads facing north; their legs were flexed, and although the leg bones of the central skeleton (SU 340) were disturbed, it appears that originally the knees of both infants were pointing to the north-northwest. The upper right arms of both infants were visible, and the lower right arm of the central infant (SU 340) was folded over its lower torso. Although the adult was disturbed, it appeared that all three skeletons shared the same southwest-northeast (240–245°) orientation, heads to the southwest. Two stones at the western edge of the tomb may be a partial lining; the larger stone at the east extends beyond the cut of the pit.

The only grave goods were two bronze spiral ornaments, **TXLII-1** and **TXLII-2**, thought to be earrings, found immediately to the west and northwest of the southern skeleton (SU 339); **TXLII-2** was found on a vertebra of the adult female. Neither object could be directly associated with any one of the three individuals in the tomb, but it is possible that both were in their original position, that is, that the ornaments were worn by the adult, assuming that she was originally in the central portion of the grave, and remained where they were once the cranium and the remainder of her skeleton was pushed to the north to accommodate the two infants. The possibility that they were associated with one or both of the infants cannot be ruled out categorically but seems less likely. Consequently, both objects are best associated with the adult female.

TXLII-1 (SF 267) **Fig. 3.138**

Fragmentary bronze spiral ornament (earring?), found north of SU 339 (**10/68**).
D: 0.010.

TXLII-2 (SF 265) **Fig. 3.139**

Fragmentary bronze spiral ornament or five small rings (earring?), found west of SU 339 (**10/69**).
PL (largest fragment): 0.007; D (est.): 0.010.

Grave fill: SU 287: inventoried: SF 415 (tiny frr copper alloy, probably parts of **TXLII-1** and **TXLII-2**).

Lithics: SF 240 (flake with retouch); SF 280 (chert flake); debitage = 3.5 g.

HM ceramics: inventoried: P 259 (body and neck fr, matt-painted vessel: PH: 0.041; PW: 0.045); not inventoried (9 frr, semi-coarse = 43 g; coarse, 9 frr, including 2 joining frr, 1 handle; non-diagnostic, 28 scraps = 18 g); daub: 51 frr = 170 g.

TOMB XLIII

Pit tomb, multiple inhumation (probably three infants)

Phase III

Figs. 3.140–3.141

1.0E–2.0E, 1.0S–0.0 (high: 107.63 m = 354.63 m ASL; low: 107.56 m = 354.56 m ASL)

Sector 2; SU 348 (grave); 350 (fill); 349 (east skeleton); 356 (west skeleton); 371 (other skeletal remains)

Excavated as Grave 62 (July 22–25, 2005)

Although a clear grave cut was not discerned, a well-defined grave fill was distinguished, characterized by a darker soil different from the surrounding tumulus fill. A single stone standing upright at the southern edge may have served as a grave marker, but this is not certain. Of the skeletons little survives, and some of the bone may not be in its original place. On the basis of the preserved remains, there appear to be parts of three crania: the eastern skeleton (SU 349), which included the cranium and some postcranial bones, was identified as an infant aged 18 (± 6) months; the western skeleton (SU 356), which consisted of the larger part of the cranium and teeth, was identified as an infant of similar age (18 ± 6 months); a smaller group of teeth and cranial bones (SU 371) were found together with SU 349 in the eastern part of the grave and were identified as belonging to an infant aged 2 years (± 8 months) at death. The possibility that the skeletal remains of SU 356 and SU 371 were originally part of one individual, somewhat displaced, cannot be categorically dismissed but is highly unlikely. It should be noted that the western skeleton (SU 356) is some 0.20 m to the west-northwest of SU 349 and SU 371 and appeared to be slightly separate from the rest; this suggested the possibility that there were originally two graves, but the fact that all of the bone was within the same well-defined grave fill indicates a single grave with multiple individuals. As preserved, the eastern skeleton (SU 349), and probably also SU 371, was oriented east-southeast to west-northwest (125–130°), crania to the east-southeast. The western skeleton may have been oriented north to south, head to the north, but too little survived to establish this with certainty. The state of preservation of all three infants was such that little could be said about the position of any of them. As preserved, the tomb measured 0.90 m long, 0.50 m wide, and 0.07 m deep. There were no grave goods.

Grave fill: No finds encountered except shell.

TOMB XLIV

Pit tomb, double inhumation (adult male and adult female)

Phase III

Figs. 3.142–3.145

1.0W–1.0E, 2.0N–3.0N (high: 107.64 m = 354.64 m ASL; low: 107.40 m = 354.40 m ASL)

Sectors 4, 5, and 1; SU 363 (grave); 364 (fill); 365 (articulated skeleton); 394 (disarticulated skeleton)

Excavated as Grave 65 (June 23–28, 2006)

As was the case with Tomb LIII (Grave 63), Tomb XLIV was first noted at the very end of the 2005 season and was excavated in the course of 2006. The tomb was stratigraphically located below Tomb LXXX (Grave 4) and above Tomb XIV (Grave 71), and, given its location straddling the baulk between Sectors 4 and 1, it necessitated a tunnel through the baulk in order to expose the tomb fully, in the same way that Tomb LXIII (Grave 35) was excavated. A clearly defined grave cut was not discerned, but the tomb fill was characterized by patches of darker soil immediately around the bone; on the north side of the grave, a clay line was noted near the pelvis, legs, and lower torso of SU 365, which seemed to demarcate the tomb on this side.

The burial contained the remains of two individuals: SU 365 was a mid-age adult male, aged 35–45 years; SU 394 was a younger adult female aged 18– 25 years. The bones of SU 365 were reasonably well preserved, particularly the cranium, clavicle, legs, and feet. This individual was laid out in a supine position, the cranium facing up, the arms folded across the lower torso; the legs were flexed very tightly, the knees pointing north-northwest. The skeleton was roughly oriented east to west (75°), head to the east. The skeletal remains of the female (SU 394) were less well preserved, especially the cranium, and disarticulated, with the leg bones of the deceased encountered both on the northeast and west sides of the tomb; the cranium was located immediately to the east-southeast of the cranium of the male (SU 365), with fragments of the pelvis and ribs to the north and east. It was, therefore, clear that the original burial was SU 394; at a later time, the bones of the female were moved to the northeast and west in order to accommodate the interment of the male (SU 365), the latter laid out fully articulated. As preserved, the tomb measured 1.67 m long, 0.70 m wide, and 0.24–0.27 m deep.

Grave goods included the fragments of an iron pin, **TXLIV-1**, found in three pieces immediately to the south of the cranium and left shoulder of the male (SU 365) and clearly associated with him. Another iron pin, **TXLIV-2**, was found oriented northeast-southwest among the disarticulated ribs, femur, and pelvis to the north of the cranium of SU 365 and more likely associated with the female, SU 394.

TXLIV-1 (SF 304a–c) **Fig. 3.144**

Fragments of iron dress pin (**10/46**).
PL (fragment a–b): 0.073; PL (fragment c): 0.036.

TXLIV-2 (SF 305) **Fig. 3.145**

Iron dress pin, Type II.2 (**10/43**).
PL: 0.143.

Grave fill: SU 364: HM ceramics: inventoried: P 299 (rim fr, fine, PH: 0.021; PW: 0.022); not inventoried (semi-coarse, 7 frr = 50 g; coarse, 3 frr = 25 g); daub (91 frr = 164 g); lithics (see bulk finds 23/6/06); shell.

TOMB XLV

Pit tomb, multiple inhumation (adult female, young adult male, old adult male)

Phase III (AMS ^{14}C: 953 ± 53 cal BC)

Figs. 3.146–3.148

2.0E–3.0E, 1.0N–2.0N (high: 107.69 m = 354.69 m ASL; low: 107.61 m = 354.61m ASL)

Sector 1; SU 343 (grave); 344 (fill); 353 (cut); 345 (top skeleton); 351 (bottom skeleton); 352 (disarticulated skeleton)

Excavated as Grave 60 (July 21–26, 2005)

This tomb consists of a neatly cut rectangular pit measuring 1.50 × 0.70 m, about the right size for a single inhumation. Within the pit, the remains of three individuals were deposited. The uppermost skeleton (SU 345) was identified as an older adult male aged 55+ years at death and among the oldest of the individuals buried in the Lofkënd tumulus. The torso and cranium of the deceased were laid out in a supine position, the cranium lying slightly on the left side, facing southwest; the left arm was folded across the lower torso, the right arm only gently bent, with the right hand over the pelvis. The legs were flexed, the knees pointing northwest. The deceased was oriented east-southeast to west-northwest (about 105°), head to the east-southeast. At the east-southeast end of the tomb, a single stone was found upright at the edge of the grave cut, the bottom of the stone level with the top of the grave. This was one of the few possible instances in the entire tumulus of a stone serving as a grave marker. The only grave good in the tomb, an iron spearhead, **TXLV-1**, and one of the very few weapons in the entire tumulus, was found beside the cranium of SU 345, on the right side and to the north; although the spearhead was directly above the cranium of the lower skeleton (SU 351), it seems more likely to be associated with the older male (SU 345). Moreover, the position of SU 345, particularly his right arm and hand, was such that it suggests he may have been holding the now missing wooden shaft of the spear in his right hand.

Immediately below SU 345, and sharing a similar orientation and position, was a younger adult male (SU 351) aged 18–25 years. The lower arms of this individual were both folded across his lower torso; his legs were flexed, the knees pointing north. At the southeast corner of the grave cut were the disarticulated skeletal remains of an adult female (SU 352) age 30–40 years at death. In the course of excavation, this individual was considered to be the earliest of the burials and the original occupant of the grave pit, her remains later moved to the southeast when the tomb was reopened for the interment of the two males. It is interesting to note that her cranium lines up with the other crania, as if intentionally placed, with the postcranial bones, including the legs and pelvis, neatly piled up in the southeast corner of the tomb. An alternative possibility is that the remains of the female were bundled and interred as a secondary burial at the same time that the two males were buried; such a scenario does not require the reopening of the tomb pit, and the neat cut of the pit makes such a possibility attractive. The fill of the grave, although not significantly different from that of the surrounding tumulus fill, was clearly distinguishable, particularly at the eastern end of the grave, composed of a fine, sandy soil. The tomb cut was only about 0.08 m deep, but the depth from the top of the stone that may have served as a marker to the bottom of the tomb pit was about 0.30 m.

Tomb XLV was stratigraphically located below Tomb LXXII (Grave 24) and above the central grave, Tomb I. A ^{14}C radiocarbon date based on collagen from the left tibia of the adult female (SU 352) gave an absolute date of 953 ± 53 cal BC.

TXLV-1 (SF 263) **Fig. 3.148a–b**

Iron spearhead, with remnants of wooden shaft inside socket (**10/118**).
PL: 0.182.

Grave fill: SU 344: HM ceramics (coarse, 1 handle fr = 9 g;, semi-coarse, 5 frr including 1 base); daub (3 frr = 2 g).

TOMB XLVI

Pit tomb, single inhumation (younger adult female)

Phase III

Figs. 3.149–3.150, 3.152

2.0E–0.0, 5.0S (high: 107.88 m = 354.88 m ASL; low: 107.74 m = 354.74 m ASL)

Sectors 2 and 7; SU 265 (grave); 266 (fill); 263 (skeleton); 271 (dark soil surrounding tomb pit)

Excavated as Grave 42 (July 6–12, 2005)

Papadopoulos, Bejko, and Morris 2007: 123, 128, fig. 22; Papadopoulos, Bejko, and Morris 2008:690, fig. 5

Tomb XLVI was only a short distance to the south-southeast of Tomb XLVII (Grave 41) (**Fig. 3.152**) and also of a younger adult female (SU 263) aged 20–25 years at death. It consisted of a clearly defined pit, the edges of which were lined on the north, east, and south sides by a thin band of dark soil, which not only neatly defined the grave but effectively separated the burial from Tomb XLVII only 0.15 m to the north. Within the pit, the deceased was laid out in a fetal position (like Tomb XLVII), on the left side, legs flexed, with the knees pointing south; the arms were tightly flexed, the hands under the chin. The upper torso on the left side had fallen toward the south, effectively covering the only grave good, a handmade kantharos, **TXLVI-1**, which had been placed below the chin of the deceased. The skeleton was oriented east-southeast to west-northwest (120° head to feet; 140° spine), head to the east-southeast. The tomb, including the grave cut, measured 1.50 m long, 0.70 m wide (max), and 0.14 m deep. The skeleton was remarkably well preserved and, together with the deceased of Tomb XLVII, among the best preserved of all the prehistoric graves. The grave fill was characterized by patches of totally sterile sand, both to the north and south of the cranium. The tomb was stratigraphically located above Tomb XXII (Grave 47).

TXLVI-1 (P 223) **Fig. 3.150**

Kantharos, FD Type 6 (**9/93**)
H (to rim): 0.082; H (max): 0.099; D (rim): 0.084.

Grave fill: SU 266: HM ceramics (1 coarse fr = 4 g; 1 semi-coarse = 5 g); daub (2 frr = 2 g); shell.

TOMB XLVII

Pit tomb, single inhumation (younger adult female)

Phase III

Figs. 3.151–3.153

1.0E–0.0, 4.0S (high: 107.96 m = 354.96 m ASL; low: 107.78 m = 354.78 m ASL)

Sectors 2 and 7; SU 255 (grave); 257 (fill); 256 (skeleton)

Excavated as Grave 41 (July 4–11, 2005)

Papadopoulos, Bejko, and Morris 2007:123, 128, fig. 22; Papadopoulos, Bejko, and Morris 2008:690, fig. 5

This tomb was only a short distance to the north of Tomb XLVI and more or less at the same level (**Fig. 3.152**), though the bones of Tomb XLVII were at a slightly higher level than those of Tomb XLVI. The cranium and much of the torso were cut by the roughly rectangular pit for Tomb LXIV (Grave 61), which was bordered by stones to the east and west sides and was marked by a distinctly darker fill, designated SU 259 (the cut for the pit of Tomb LXIV was SU 260). Although several of the prehistoric burials in the northeast sector of the tumulus were cut, disturbed, and sometimes substantially damaged by the modern burials, this was one of the rare cases of a prehistoric burial being cut by another tomb of the same broad period.

As with Tomb XLVI, the skeleton of Tomb XLVII (SU 256), a younger adult female aged 18–21 years, was very well preserved. The deceased was laid out in a fetal position, lying on the left side, the legs tightly flexed, with the knees pointing to the south. The arms were sharply folded in front of the body, with the hands under the chin. The skeleton was oriented east to west (100°), head to the east. The tomb measured 1.25 m long, 0.65 m wide, and 0.11–0.18 m deep. In addition to having been cut by Tomb LXIV (Grave 61), Tomb XLVII was stratigraphically located above Tomb XIII (Grave 49), the latter being one of the earliest burials in the tumulus. There were no grave goods.

Grave fill: SU 257: chipped stone debitage (1 flake = 0.9 g); shell.

TOMB XLVIII

Pit tomb, multiple inhumation (adolescent female, adult male, and one or two more adults)

Phase III

Figs. 3.154–3.159

2.0E–4.0E, 5.0N–6.0N (high: 107.73 m = 354.73 m ASL; low: 107.52 m = 354.52 m ASL)

Sector 1; SU 316 (grave); 317 (fill); 373 (cut); 318 (top skeleton); 366 (bottom skeleton); 366a (cranial remains at top); 372 (disarticulated bone)

Excavated as Grave 52 (July 18–27, 2005)

Papadopoulos, Bejko, and Morris 2007: 119–123, fig. 15

This tomb was stratigraphically located below Tomb LXXXV (Grave 10) (and the modern Tombs XCII and XCIII [Graves 23 and 19]) and above Tomb XXVI (Grave 74). It consisted of a clearly cut pit along the east-southeast, south-southwest, and west-northwest sides, as well as the eastern portion of the north-northeast side; the pit was filled with a dark, compact soil clearly distinguishable from the surrounding tumulus fill. The tomb pit measured 1.63 m long, 0.63–0.70 m wide, about 0.21 m deep, though at its deepest point it was about 0.32 m. There was a noticeable downward slope of the grave from west-northwest to east-southeast, with the crania of the upper two skeletons lying at a lower level than the feet. The skeletal remains of at least three individuals were noted in the field. The little that survived of the earliest of the three skeletons, considered to be the original burial in the tomb (SU 372), was moved to the western side when the grave was reopened for the interment of the other two individuals. Consequently, SU 372 was disarticulated and fragments of its cranium were found together with some piled-up postcranial bones; this individual was identified as a mid-age adult of indeterminate sex. The top two skeletons (SU 318 and 366) appear to have been laid out more or less contemporaneously, since both appeared to be fully articulated. Both shared the same east-southeast to west-northwest orientation (110°), heads to the east-southeast. The uppermost skeleton (SU 318) was identified as an adolescent female aged 15–17 years at death; her torso and cranium were laid out in a supine position, the head lying on the left side, pointing southwest. The arms were folded across the lower torso and the legs were flexed, knees pointing north, though the legs were not exactly parallel. The position of the lower skeleton (SU 366) was obscured, but it was evidently very similar to the position of the top skeleton; this individual was identified as an adult male aged 30– 35 years. According to the excavator, the male may have been slightly disturbed by the burial of the adolescent female, who was clearly interred last, but any disturbance was minor. There were, in addition to these two fully articulated skeletons and the bones of SU 372, the very poorly preserved remains of what may be a fourth skeleton (SU 366a), consisting of dentition only (10 teeth); these were distinguished in the lab. Since only teeth were noted, they probably belong to the same individual as SU 372, found at the western side of the tomb, rather than to a fourth individual. Laboratory analysis of the remains as they had been lifted indicated that they were isolated at the northeast corner of the tomb. Consequently, the tomb was almost certainly the grave of three individuals.

Grave goods included the one-handled matt-painted tankard, **TXLVIII-1**, found right next to the right side of the cranium of the top skeleton (SU 318) on the north side of the grave. Another complete vessel, the dark-fabric kantharos, **TXLVIII-2**, was found only 0.06 m to the south of the cranium of the top skeleton but right up against the left shoulder of the bottom skeleton (SU 366). Although **TXLVIII-1** clearly belonged with the adolescent female (SU 318); the kantharos, **TXLVIII-2**, may have been associated with either SU 318 or SU 366. The gold/electrum foil ornaments, **TXLVIII-3** and **TXLVIII-4**, were found in situ, one on the right, the other on left, of the cranium at the position of the ears. Their position indicated that they may have been earrings, but how they were attached remains an issue. Alternatively, they may have served as ornaments of some other item of personal decoration, conceivably even a headband, made of some organic material otherwise lost. The only grave good clearly associated with the adult male (SU 366) was the tubular iron bead, **TXLVIII-5**, found intermingled with the fragmentary cranium of the skeleton and almost certainly worn around the neck. As with several other graves (Tombs LXX, XXI [Graves 17, 55]), the individual most richly adorned in this tomb was the adolescent female (SU 318).

TXLVIII-1 (P 277) **Fig. 3.157a**

One-handled matt-painted vessel, FL Type 2B (**9/8**).
H (to rim): 0.095; H (max): 0.115; D (base): 0.048; D (rim): 0.095.

TXLVIII-2 (P 276) **Fig. 3.157b**

Kantharos, dark fabric, FD Type 3 (**9/96**).
H (to rim): 0.089–0.092; H (max): 0.114; D (base): 0.040; D (rim): 0.068–0.073.

TXLVIII-3 (SF 290) **Fig. 3.158**

Gold/electrum foil ear ornament (right side of cranium) (**10/11**).
D: 0.038; Wt: 1.0 g.
Papadopoulos, Bejko, and Morris 2007:124, fig. 16.

TXLVIII-4 (SF 291) **Fig. 3.158**

Gold/electrum foil ear ornament (left side of cranium) (**10/12**).
D: 0.038; Wt: 0.9 g.
Papadopoulos, Bejko, and Morris 2007:124, fig. 16.

TXLVIII-5 (SF 288) **Fig. 3.159**

Iron tubular bead (**10/94**).
L: 0.025; D: 0.016.

Grave fill: SU 317: Inventoried: P 279 (matt-painted handle: PH: 0.041; PW: 0.025); P 280 (matt-painted rim: PH: 0.027; PW: 0.035).

Not inventoried ceramics (2 frr coarse = 24 g; 1 fr, semi-coarse = 1 g; 8 non-diagnostic scraps); daub (2 frr = 6 g); shell (317, 18/7, 27/7); faunal (SU 318) found in lab.

TOMB XLIX

Pit tomb, single inhumation (child)

Phase III

Figs. 3.160–3.161

1.0W–2.0W, 5.0N–6.0N (high: 107.80 m = 354.80 m ASL; low: 107.66 m = 354.66 m ASL)

Sector 4; SU 311 (upper fill); 313 (grave); 314 (lower fill); 315 (skeleton)

Excavated as Grave 51 (July 15–21, 2005)

This tomb was stratigraphically located below Tombs LXXXIII and LXVI (Graves 7 and 31) and above Tomb XVIII (Grave 73). The circumstances and general appearance of this grave were not unlike those of Tombs XXIII and L (Graves 56 and 46). Of the skeleton, all that survives are nine teeth and poorly preserved fragments of the cranium, all at the east end of the pit; there was no clearly preserved postcranial bone encountered in situ. Thanks to the teeth, the deceased (SU 315) was identified as a child aged 4 (± 1) years at death, and although virtually nothing could be determined about the precise position of the deceased, the skeletal remains were oriented east to west (75°), head to the east. At its lower level the tomb pit was roughly rectangular, extending slightly into the baulk, Sector 5 (the baulk was not cleared in order to get a section through the east end of the grave); at the higher level, the tomb was more triangular in appearance, but this was largely defined by the fill. At its greatest extent, the pit measured 1.40 m long, 0.75 m wide, and 0.14–0.17 m deep. A significant part of the grave was covered by a blackened material; a substantial stripe of this material along the southern end appeared to contain white calcareous material, whereas toward the north it was more consistently black. This material clearly overlies what survives of the bone, creating a concave surface from north to south on the elevation section (**Fig. 3.160**). Although samples were taken for analysis, the material remains unidentified (what is clear is that there was little if any charcoal in the material). There were no grave goods.

Grave fill: SU 311 (upper fill): pottery (3 fr, semi-coarse, including 1 rim = 12 g); daub (12 frr = 18 g). SU 314 (lower fill): daub (3 frr = 0.5 g); shell.

TOMB L

Pit tomb, single inhumation (child)

Phase III

Figs. 3.162–3.164

1.0W–2.0W, 3.0N (high: 107.95 m = 354.95 m ASL; low: 107.83 m = 354.83 m ASL)

Sector 4; SU 282 (grave); 283 (fill); 284 (skeleton)

Excavated as Grave 46 (July 7–9, 2005)

The tomb was encountered south of Wall 1 and parallel to it and located directly above Tomb XXIII (Grave 56), which was also the inhumation of a child. The location of these two tombs, one directly on top of the other, and the fact that they were separated by only 0.11–0.20 m may indicate that the two were contemporary or near contemporary. Tomb L was clearly defined by more or less regular lines of dark soil, thought to represent carbonized material, along the north, south, and east sides (**Fig. 3.163a**). A thin layer of the same dark material was present also in the central portion of the grave, above the bone. Although the samples taken from Tomb L did not yield sufficient material for ^{14}C dating, a charcoal sample from Tomb XXIII immediately below yielded a date in the twelfth or eleventh century BC (1070 ± 59 cal BC).

The deceased (SU 284), a child aged 4 (± 1) years at death, was oriented east to west (80°), head to the east. Given the extremely poor state of preservation of the bone (only the mandible with teeth and some unidentified long bones were noted in situ), it was not possible to establish the original position of the body with any conviction (**Fig. 3.163b**). As preserved, the tomb measured 1.20 m long, 0.55 m wide, and 0.12 m deep. A fragmentary iron pin, **TL-1**, oriented northwest-southeast, was found in the area of the upper body, near the mandible.

TL-1 (SF 225) **Fig. 3.164**

Fragmentary iron dress pin, undetermined type (**10/48**).

PL: 0.076.

Grave fill: SU 283: HM ceramics (1 fr, semi-coarse = 2 g; 1 scrap non-diagnostic = 0.5 g); daub (4 frr = 28 g); shell.

Tomb LI

Pit tomb, single inhumation (young adult, probably male)

Phase III

Figs. 3.165–3.166

4.0W–5.0W, 3.0N–4.0N (high: 107.07 m = 354.07 m ASL; low: 106.94 m = 353.94 m ASL)

Sector 4; SU 432 (grave); 433 (fill); 434 (skeleton); 435 (cut)

Excavated as Grave 78 (July 5–6, 2006)

This tomb was first encountered while digging topsoil, and it soon became evident that the northeast end of the skeleton was associated with a tomb fill identical to the southwest fill of Tomb XXVIII (Grave 77). Further excavation made it clear that the pit for Tomb LI had cut across the southwest portion of the pit for Tomb XXVIII, which accounts nicely for the two different fills noted in Tomb XXVIII (SU 427 of Tomb XXVIII was the same as SU 433 of Tomb LI). The only bone encountered was the poorly preserved cranium and parts of the uppermost torso and arms of an individual (SU 434) identified as a young adult, probably male. It appears that the cranium was found lying on the left side, facing south, but little else could be said about the position of the body, since the entire west end of the grave fell victim to erosion, being located at the western edge of the tumulus. The orientation of the tomb was difficult to ascertain, but the excavator noted that it was approximately east to west (about 100°), the cranium to the east. As preserved, the tomb measured 0.56 m long southeast-northwest × 0.4–50 m wide northeast-southwest, and 0.13 m deep. In addition to cutting across the pit for Tomb XXVIII, Tomb LI was stratigraphically located above Tomb XXVII (Grave 82) and was therefore later than both Tombs XXVII and XXVIII. There were no grave goods.

Grave fill: SU 433: HM ceramics (semi-coarse, 2 frr = 7 g; 3 non-diagnostic scraps); daub (1 fr = 1 g).

Tomb LII

Pit tomb, single inhumation (adolescent male)

Phase III

Figs. 3.167–3.168

4.0E–5.0E, 2.0N–0.0 (high: 107.52 m = 354.52 m ASL; low: 107.40 m = 354.40 m ASL)

Sectors 1 and 6; SU 388 (grave); 389 (fill); 390 (skeleton), 436 (cut)

Excavated as Grave 69 (July 1–7, 2006)

This tomb was first encountered not far from the tumulus edge, extending slightly into the baulk, Sector 6, necessitating a small extension into the baulk separating Sectors 1 and 2. The tomb consisted of a clearly defined pit, with the cut fully preserved along the northeast and southeast sides, but with no real cut discerned on the southwest and northwest sides, though a darker fill may represent the southwest side of the grave. The grave fill was sandy, looser, and darker than the tumulus fill and clearly distinguishable. Within the pit, the body of a large and robust individual (SU 390), identified as an adolescent male aged 15 (± 3) years at death, was oriented southeast-northwest (125°), the head to the southeast. The deceased was laid out with the cranium and torso supine, the head facing up; the left arm was folded across the chest, the right arm tightly flexed, with the right hand on the shoulder. The legs were flexed, the knees pointing northeast. The skeletal remains were comparatively well preserved, the only bones missing being the feet; and the maxilla was slightly damaged during the early stages of the excavation of the tomb. The right pelvis was found resting on a stone. The tomb measured 1.40 m long, 0.58 m wide, and 0.12–0.14 m deep. There were no grave goods.

Grave fill: SU 389: HM ceramics (semi-coarse, 1 fr = 7 g); lithics (4 frr, including 1 nucleus); shell.

Tomb LIII

Pit tomb, multiple inhumation (adult, indeterminate sex, and child)

Phase III

Figs. 3.169–3.177

5.0E, 1.0N–2.0N (high: 107.59 m = 354.59 m ASL; low: 107.45 m = 354.45 m ASL)

Sector 1; SU 357 (grave); 358 (fill); 359 (adult skeleton); 359a (child)

Excavated as Grave 63 (June 19–23, 2006)

This tomb was one of several graves first uncovered in the 2005 season and excavated in the course of

2006. The tomb was stratigraphically located above Tomb XXIX (Grave 83). Although a distinct grave cut was not discerned, the grave fill consisted of a darker and looser-textured soil, different from the surrounding tumulus fill, immediately over the body of the deceased and particularly in the central portion of the grave. As preserved, the tomb measured 1.20 m long, with a maximum width of approximately 0.50 m, and about 0.14 m deep. The skeletal remains were very poorly preserved, splintering and friable to touch. The cranium was crushed, with teeth encountered both to the east and west sides of the cranium. Although poorly preserved, parts of the torso, arms, pelvis, and legs, in addition to the cranium, were clearly distinguishable in situ. The deceased (SU 359), identified as an adult of indeterminate sex, aged 25–35 years at death, appeared to have been laid out lying on the right side, rather than supine, facing northeast; the arms and legs were flexed, the knees probably pointing northeast. The skeleton was oriented southeast-northwest (145°), head to the southeast. In addition to the skeleton of the adult (SU 359), teeth and cranial fragments of a child aged 4 years (± 16 months) were identified in the lab; designated SU 359a, these skeletal remains were not discerned in situ.

Grave goods were plentiful: a dark-fabric one-handled vessel with distinctive so-called "biforata" handle, **TLIII-1**, was flush against the cranium on the northeast side, placed upright. Bronzes included a discoid button or small boss, **TLIII-2**, found on the upper torso, immediately to the northwest of the cranium; an earring, **TLIII-3**, and a spectacle pendant ornament, **TLIII-4**, were found under the upper torso, in the area of the sternum; a small bronze spiral coil, **TLIII-5**, was found in the lab during the cleaning of the cranium, which had been block-lifted. Three spherical beads were found in line in situ on the lower torso, immediately above the pelvis: the bead to the west, **TLIII-6**, was of glass; that in the center, **TLIII-7**, was of stone, carnelian or sardonyx, and that to the east, **TLIII-8**, was of glass paste. About 0.12 m to the southeast of these beads, a fourth bead, **TLIII-9**, was found between the bones of the lower arm, made of faience. This bead was more elongated, almost melon-shaped.

TLIII-1 (P 322) **Fig. 3.171**

One-handled vessel of dark fabric, FD Type 1 (**9/94**).

H (to rim): 0.068; H (max): 0.075; D (base): 0.042–0.045; D (rim): 0.098–0.100.

TLIII-2 (SF 297) **Fig. 3.172**

Bronze disk, perforated in the center (button or small boss) (**10/62**).

D: 0.033.

TLIII-3 (SF 299) **Fig. 3.173**

Bronze earring (**10/82**).

H: 0.031; D: 0.031.

TLIII-4 (SF 300) **Fig. 3.174**

Bronze spectacle ornament; pendant (**10/57**).

L (max): 0.028; H: 0.020.

TLIII-5 (SF 307) **Fig. 3.175**

Small bronze spiral coil (**10/70**).

D: 0.010.

TLIII-6 (SF 294) **Fig. 3.176a**

Spherical opaque reddish brown glass bead (**10/109**).

D: 0.014.

TLIII-7 (SF 296) **Fig. 3.176b**

Spherical bead of crypto-crystalline quartz (carnelian or sardonyx) (**10/103**).

H: 0.010; D: 0.016.

TLIII-8 (SF 295) **Fig. 3.176c**

Fragmentary spherical opaque golden yellow glass paste bead (**10/107**).

PL (largest piece): 0.011.

TLIII-9 (SF 298) **Fig. 3.177**

Cylindrical whitish yellow with light blue-green glaze faience bead, almost melon-shaped, corrugated (**10/105**).

H: 0.017; D: 0.009.

Grave fill: SU 358: HM ceramics (fine 1 fr = 1 g; semi-coarse, 6 frr, including 1 rim, and 1 handle = 14 g; coarse, 2 frr = 48 g); daub (7 frr = 16 g); lithics (1 debitage); shell.

Tomb LIV

Pit tomb, single inhumation (adult, indeterminate sex)

Phase III

Figs. 3.178–3.179

2.0W–3.0W, 2.0S (high: 107.92 m = 354.92 m ASL; low: 107.81 m = 354.81 m ASL)

Sector 2; SU 251 (grave); 254 (fill); 252 (skeleton)

Excavated as Grave 40 (July 4–6, 2005)

Perhaps the most interesting aspect of this tomb was the floor of the burial, composed of four parallel stripes of calcareous material, the best preserved of which was about 0.75 m long, with a width ranging between 0.05 and 0.09 m. Following the analysis of our soil scientist, John Foss, it was clear that these were natural stripes indicating that the bedrock was being approached. The surviving bone, identified as a mid-age adult of indeterminate sex, was comparatively well preserved, but the skull and a large portion of the upper torso were lost due to erosion; the proximity of the tumulus edge is clearly visible on **Figure 3.178**. The right femur also was missing, along with part of the right arm and much of the lower torso. The deceased was laid out southwest-northeast (240–245°), with the head, which is not preserved, to the southwest. The legs were clearly flexed, with the knees originally pointing to the southeast. The possibility was noted in the field that the deceased may have been laid out in a fetal position, lying on its left side, with the left arm tightly folded under the torso, but given what little survives of the torso, and the fact that parts of the pelvis appear to have been slightly disturbed, the original position is difficult to determine. The tomb measured 1.10 m long, 0.60 m wide, and at least 0.11 m deep. Although the edges of the grave cut were difficult to discern, a distinct grave fill was encountered along the northeast portion of the grave, composed of a light yellow clay; this petered out toward the southwest as the tumulus edge was approached. The tomb may have been located below the original extent of Tomb LVIII (Grave 37). There were no grave goods.

Grave fill: SU 254: HM ceramics inventoried: P 169 (rim fr, small open vessel, PH: 0.055; PW: 0.067); daub; flint debitage (2 flakes); faunal specimen; shell.

Tomb LV

Pit tomb, single inhumation (child)

Phase IV

Figs. 3.180–3.186

4.0E–5.0E, 7.0N (high: 107.76 m = 354.76 m ASL; low: 107.63 m = 354.63 m ASL)

Sector 1; SU 319 (grave); 320 (fill); 321 (skeleton)

Excavated as Grave 53 (July 18–23, 2005)

Papadopoulos 2010b:238–239

This was the northeasternmost of all prehistoric burials in the tumulus and roughly contemporary with Tomb LVI (Grave 43). The skeleton was found in a very poor state of preservation, the bones fragmented and splintering, the poor state of preservation partially due to proximity to both the tumulus edge and the modern surface. All that survived were parts of the cranium, arms, and legs belonging to a child (SU 321) aged 8 (± 1) years at death. The deceased was oriented southeast-northwest (120–130°), head to the southeast. Determining the original position was difficult, as virtually nothing of the torso or lower arms was preserved. The cranium appeared to have been laid out supine, as the mandible was facing up; the left humerus was by the side of the body and the right arm was probably bent across the torso; the legs were flexed, the knees evidently facing southwest. There was no clear grave cut or fill discerned during excavation. The tomb, as preserved, measured 0.90 m long, 0.24–0.40 m wide, and 0.13 m deep.

Despite the poor state of preservation of the bone, this proved to be one of the richest burials in the tumulus in terms of the quantity of material deposited with the deceased. A small one-handled vessel, **TLV-1**, was found to the north-northeast of the cranium, more or less upright. The distinctive iron fibula of Lofkënd type (Papadopoulos 2010b), **TLV-2**, was found in situ beside the pot, roughly oriented east-west, and was clearly worn over the right shoulder of the deceased, and a bimetallic figure-of-eight fibula, **TLV-3**, was found over the left shoulder, beside the mandible. Two iron tubular beads, **TLV-4** and **TLV-5**, were found, one on either side of the mandible, and another two iron beads, **TLV-6** and **TLV-7**, together with a glass bead, **TLV-8**, were found with the cranium in the conservation lab, as the cranium had been block-lifted.

TLV-1 (P 256) **Fig. 3.182**

One-handled matt-painted vessel, FL Type 3 (**9/14**).

H (to rim): 0.054; H (max): 0.089; D (base): 0.040; D (rim): 0.098.

TLV-2 (SF 261) **Fig. 3.183**

Iron fibula, Type II.2 (**10/23**).

L: 0.126.
Papadopoulos 2010b:239, fig. 4.

TLV-3 (SF 262) **Fig. 3.184a–b**

Bimetallic (iron and bronze) fibula, Type III.1 (**10/25**).
L: 0.084.

TLV-4 (SF 266a) **Fig. 3.185a**

Tubular iron bead (**10/98**).
PL: 0.016; D: 0.015.

TLV-5 (SF 266b) **Fig. 3.185b**

Tubular iron bead (**10/97**).
L: 0.019; D: 0.013–0.015.

TLV-6 (SF 287a) **Fig. 3.185c**

Tubular iron bead (**10/95**).
L: 0.019–0.020; D: 0.013.

TLV-7 (SF 287b) **Fig. 3.185d**

Tubular iron bead (**10/96**).
L: 0.018; D: 0.014.

TLV-8 (SF 283) **Fig. 3.186**

Translucent dark green glass bead (**10/113**).
H: 0.003; D: 0.005–0.006.

Grave fill: SU 320: HM ceramics (5 frr, coarseware = 20 g; 6 non-diagnostic scraps = 2 g); shell rare.

TOMB LVI

Pit tomb, double inhumation (younger adult female and male)

Phase IV

Figs. 3.187–3.190

1.0W–1.0E, 7.0N (high: 108.02 m = 355.02 m ASL; low: 107.81 m = 354.81 m ASL)

Sectors 4, 5, and 1; SU 267 (grave); 268 (fill); 269 (upper skeleton); 296 (lower skeleton)

Excavated as Grave 43 (July 6–15, 2005)

This was the northernmost of all the prehistoric burials, and was stratigraphically located below prehistoric Tomb LXXVII (Grave 18), and perhaps also Tomb LXXIV (Grave 29); it was also located below the modern Tomb XCVII (Grave 39). The tomb measured 1.45 m long, 0.65 m wide, and 0.21 m deep. A partial grave cut was noted along the north side of the burial at the eastern end. Associated with this cut was a distinct fill characterized by a dark clayey soil, with a high content of calcareous material. On the north side, the cut and associated fill was first encountered at a much higher level, dipping down to the level of the skeleton. One large flat stone (0.57 × 0.38 m) was found immediately to the north of the legs of the lower skeleton at the northwest corner of the tomb. Within the pit, two individuals were interred, both oriented east to west (80°), crania to the east. The upper of the two skeletons (SU 269) was an adult female aged 20–25 years; the lower skeleton (SU 296) was of an adult male, also aged 20–25 years. The comparatively poor state of preservation of both skeletons, coupled with the fact that some of the bones were somewhat displaced, made it difficult to determine the precise positions of the two individuals. The cranium and torso of the upper skeleton were supine, head facing up, arms folded across the lower chest; the legs were probably flexed, but disturbed. The position and articulation of the top skeleton establish that it is the later of the two, but it remains difficult to determine the time between the two interments. The cranium of the lower skeleton (SU 296) was immediately to the southeast of the cranium of the upper skeleton, but its torso appeared to have been displaced by the interment of SU 269. Although the legs of both skeletons had the appearance of being intermingled, those of the female (SU 269) were above those of the male (SU 296).

Grave goods included a kantharos of dark fabric, **TLVI-1**, located over the left shoulder and immediately to the south of the cranium of the female (SU 269) (**Fig. 3.188b**). Although the vessel was also right next to the cranium of the male (SU 296), the disturbed state of the latter made it almost certain that the pot had been placed as an offering associated with SU 269. A coarse pot fragment (not catalogued) was found over the upper torso of the female just to the west of the mandible (the sherd is visible in **Figs. 3.187** and **3.188a**). A bimetallic figure-of-eight fibula, **TLVI-2**, was found directly under the right humerus of the upper skeleton (SU 269) and was probably worn near the right shoulder of the individual (**Figs. 3.187, 3.188c**).

TLVI-1 (P 228) **Fig. 3.189**

Kantharos, FD Type 3 (**9/95**).
H (to rim): 0.086–0.088; H (max): 0.108; D (base): 0.037; D (rim): 0.065.

Papadopoulos, Bejko, and Morris 2007:124, fig. 17.

TLVI-2 (SF 231) **Fig. 3.190a–b**

Bimetallic (bronze and iron) figure-of-eight fibula, Type III.1 (**10/26**).
L (est.): 0.068.

Grave fill: SU 268: HM ceramics (2 coarse frr = 11 g; 4 semi-coarse frr = 9 g); daub (25 frr = 21 g); lithics (SF 213: scraper, Paleolithic); abundant shell. Found with the upper skeleton (SU 269) were the following: HM ceramics (1 fr, semi-coarse = 3 g); faunal specimen; shell.

TOMB LVII

Pit tomb, single inhumation (adult, indeterminate sex)

Phase IV

Figs. 3.191–3.192

5.0E–6.0E, 6.0N (high: 107.62 m = 354.62 m ASL; low: 107.45 m = 354.45 m ASL)

Sector 1; SU 334 (grave); 335 (fill); 336 (skeleton)

Excavated as Grave 58 (July 20–22, 2005)

This tomb, located at the northeast edge of the tumulus, had been largely destroyed by a small pine tree planted in the winter of 2004–2005. The digging of the pit for the tree went straight through the torso of the deceased, scattering postcranial bone throughout the vicinity. The cranium, despite its many fragments preserved in situ, was destroyed by the tree pit, and any traces of the tomb cutting or fill were obliterated. As for the postcranial bones, these were virtually impossible to identify in the field, and it is unclear whether the legs survived; it is possible they were already disturbed and mostly destroyed by proximity to the east scarp of the tumulus and to the modern surface. On the basis of the surviving remains, the tomb appeared to be roughly oriented east to west, with the head to the west, though the exact orientation and any details of the original position of the deceased were impossible to reconstruct on account of the damage caused by the tree. The deceased (SU 336) was identified as a mid-age adult of indeterminate sex. The preserved bone as excavated covered an area 0.80 m long, 0.45 m wide, and 0.17 m deep. There were no grave goods.

Grave fill: SU 332: HM ceramics (2 frr, coarse = 3 g); daub (1 fr = 0.5 g).

TOMB LVIII

Pit tomb, single inhumation (adolescent)

Phase IV

Figs. 3.193–3.201

3.0W, 1.0S–2.0S (high: 108.01 m = 355.01 m ASL; low: 107.92 m = 354.92 m ASL)

Sector 2; SU 242 (grave); 244 (fill); 243 (skeleton); 227 (ceramic deposit over bone)

Excavated as Grave 37 (June 30–July 4, 2005)

The human remains in this tomb were found in an extremely poor state of preservation, with the bone tending to disintegrate upon touch. Fragments of the right arm and possible ribcage could be determined in the field, in addition to fragments, much weathered and fragile, of the skull. No distinct grave cut or fill was noted, and the poor state of preservation was in part due to the proximity of the tumulus edge (**Figs. 3.193**, **3.194b**), the entire southwestern portion of the tomb having eroded off the side of the mound. The deceased, as preserved, appeared to have been laid out northeast-southwest (20°), head to the northeast; the right arm may have been slightly bent or folded over the pelvis, but little else could be said about the position of the body. The human remains were those of an adolescent of indeterminate sex aged 15 (± 3) years at death. The tomb as preserved measured 0.80 m long, 0.45 m wide, and only about 0.010 m deep. The original tomb may have overlapped slightly Tomb 40, which was located immediately to the southeast at a slightly lower level.

A bronze spectacle fibula, **TLVIII-1**, was found in situ on the left shoulder, and an iron pin, **TLVIII-2**, oriented north-northeast to south-southwest, head to the north-northeast, was worn on the right shoulder. Given the state of the cranium and upper torso, these were block-lifted, and another iron pin, **TLVIII-3**, was found together with the fragmentary iron chain or coil, **TLVIII-4**, in the conservation lab.

A concentration of non-joining pottery fragments, designated SU 227, consisting primarily of coarseware sherds (**Fig. 3.200**) but including the handle fragment P 205 (**Fig. 3.199**), was found overlying the cranium and uppermost torso of the deceased, between the fibula (**TLVIII-1**) and the dress pin (**TLVIII-2**). These fragments were associated with a layer of what resembled fire-affected clay, which was quite different from the much more common daub. Despite care in block-lifting this clay, it disintegrated into a mass of fragments and chips, and further handling only resulted in even more fragments. Consequently, counting the numerous fragments served little purpose, but collectively they weighed 482.2 g, and **Figure 3.201**, taken of all the pieces together, provides an idea of what the material looks like. The numerous fragments and chips ranged in color from black through various shades of brown and reddish brown, closest to reddish brown and yellowish red (5 YR 4/4–4/6) on the Munsell scale. Interpreting this concentration of fire-affected clay is not clear-cut, but it does appear as if a layer of clay was deliberately and carefully placed over the cranium and upper torso of the deceased (**Fig. 3.194a**) which somehow was set on fire. The burial was stratigraphically located above Tomb LIV (Grave 40).

TLVIII-1 (SF 170) **Fig. 3.195**

Bronze spectacle fibula, Type I.1b (**10/14**).
L: 0.080.

TLVIII-2 (SF 171) **Fig. 3.196**

Iron dress pin, Type II.1, with rolled head (**10/33**).
PL: 0.101.

TLVIII-3 (SF 226) **Fig. 3.197**

Iron dress pin, Type II.4, with bent back head (**10/45**).
L: 0.147.

TLVIII-4 (SF 172) **Fig. 3.198**

Fragment of iron coil resembling linked chain (**10/71**).
PL: 0.085.
Ceramic deposit SU 227:
Fragment of horned handle, P 205 (**9/174**) (**Fig. 3.199**: PH: 0.029; W: 0.039), plus 26 non-joining fragments of coarseware pottery (**Fig. 3.200**).

TOMB LIX

Pit tomb, single inhumation (adult male)

Phase IV

Figs. 3.202–3.203

3.0W, 1.0N (high: 108.03 m = 355.03 m ASL; low: 107.82 m = 354.82 n ASL)

Sector 4; SU 245 (grave); 246 (fill); 247 (skeleton)

Excavated as Grave 38 (July 1–5, 2005)

With the exception of the cranium, the skeleton was very poorly preserved, with most of the bones splintered and disintegrated, in part due to proximity to the tumulus edge and to the modern surface. No distinct grave cut or grave fill significantly different from the surrounding tomb fill was noted. The cranium and upper torso of the deceased, identified as an adult male aged 35–45 years, appears to have been laid out in a supine position, the head facing up; the upper arms were by the side of the body, the left arm at a lower level than the right. Below the torso, there was a jumble of leg and other bones to the southwest, so little can be said about the position of the lower body and legs. On the basis of the surviving bone, the tomb was oriented northeast-southwest (45–50°), head to the northeast. The tomb measured 0.88 m long, 0.40 m wide, and 0.11–0.21 m deep (the greater depth due to the left humerus, which was at a lower level). The tomb was stratigraphically located above Tomb XXI (Grave 55). There were no grave goods.

Grave fill: SU 246: HM ceramics: inventoried: P226 (**9/325**) (fr of indeterminate shape, PH: 0.051; PW: 0.040); P227 (**9/284**) (body fr, PH: 0.032; PW: 0.030); not inventoried (coarse, 1 fr = 10 g; semi-coarse, 5 frr = 14 g); daub (21 frr = 49 g).

TOMB LX

Pit tomb, single inhumation (young adult, indeterminate age and sex)

Phase IV

Figs. 3.204–3.206

5.0E, 2.0S–3.0S (high: 107.82 m = 354.82 m ASL; low: 107.70 m = 354.70 m ASL)

Sector 2; SU 272 (grave); 274 (topsoil and grave fill); 273 (skeleton)

Excavated as Grave 44 (July 13–15, 2005)

This was among the most poorly preserved of all burials. Noted in the field was a tibia, some cranium fragments, including several teeth, and little more than splinters of postcranial bone. The tomb was encountered during the excavation of the topsoil right at the tumulus edge, and the damage caused to it was the combined result of proximity to the edge of the mound to the east and the modern surface immediately above. A grave fill distinct from topsoil was not noted, though topsoil in the immediate vicinity of the tomb was given a separate unit number (SU 274). An iron blade, SF 230 (**Fig. 3.206**), was found in the topsoil and was considered by the excavator to be a possible grave good displaced from this burial, but this could not be determined with certainty, and it is equally likely that the blade is post–Early Iron Age. The bone as preserved measured 0.55 × 0.45 m, and was about 0.12 m deep. The original position or orientation of the deceased cannot be determined on the basis of the surviving bone (the tibia in situ was oriented 60° northeast-southwest). The preserved remains were of a young adult of indeterminate age or sex. The tomb was stratigraphically located above Tomb XII (Grave 88).

Grave fill: SU 273: daub (8 frr = 8.5 g). Topsoil SU 274: iron blade (**10/123**: **Fig. 3.206**); shell.

Tomb LXI

Pit tomb, single inhumation (adult)

Phase IV

Figs. 3.207–3.208

5.0E, 4.0S (high: 108.09 m = 355.09 m ASL; low: 108.01 m = 355.01 m ASL)

Sector 2; SU 224 (grave); 225 (skeleton)

Excavated as Grave 34 (June 27–28, 2005)

The circumstances of this tomb were not unlike those of Tomb LXII (Grave 32), which was located just over 2.0 m to the southwest, and the two are probably more or less contemporary. The tomb was located right at the scarp of the tumulus and had largely eroded down the steep south side of the mound. A discernible grave cut or fill were not encountered, and the poorly preserved bones were revealed within topsoil (SU 202). As preserved, the tomb measured only 0.40 m long, 0.30 m, and 0.08 m deep. All that can be said about the position of the body is that the legs were flexed, the knees pointing west-southwest. The original orientation of the tomb was probably southeast-northwest, with the head, nothing of which survives, to the southeast. The human remains (SU 225) were those of an adult of indeterminate age or sex. There were no grave goods and no clear grave fill.

Tomb LXII

Pit tomb, single inhumation (adult, indeterminate age and sex)

Phase IV

Figs. 3.209–3.210

2.0E–3.0E, 5.0S (high: 108.15 m = 355.15 m ASL; low: 108.10 m = 355.10 m ASL)

Sector 2; SU 219 (grave); 218 (fill); 220 (skeleton)

Excavated as Grave 32 (June 25–27, 2005)

This grave was very poorly preserved, located at the very southern edge of the scarp of the tumulus (**Fig. 3.210**), with most of the burial having eroded down the side (much of the scattered and eroded bone noted on the south side of the tumulus prior to excavation may derive from this tomb and others like it). Very little can be said about the orientation and position of the body; the long bones as preserved (including tibia, femur, patella, and one ramus) were oriented 90° east to west, but the original orientation and position of the remainder of the body is difficult to ascertain. The remains were those of an adult of indeterminate age and sex. The grave fill (SU 218) extended in a rather irregular manner to the north, but was much damaged at this point by a particularly tenacious pear tree. The combination of the tree and tumulus edge obliterated anything that may have survived of the grave cut. The tomb as preserved, including the clearly discernible fill, measured 0.85 m long and 1.10 m wide. There were no grave goods.

Grave fill: SU 218: HM ceramics (coarse, 1 fr = 15 g; non-diagnostic, 1 fr = 2 g); shell.

Tomb LXIII

Pit tomb, single inhumation (mature adult, probably female)

Phase IV

Figs. 3.211–3.217

1.0W–1.0E, 1.0N–0.0 (high: 108.16 m = 355.16 m ASL; low: 108.08 m = 355.08 m ASL)

Sectors 4, 5, 1, and 8; SU 232 (grave); 233 (fill); 234 (skeleton)

Excavated as Grave 35 (June 28–July 20, 2005)

This tomb was first uncovered in the southeast corner of Sector 4 and clearly extended in the baulks both to the east (Sector 5) and south (Sector 8). This necessitated a small extension in the south baulk, Sector 8, but in order to reveal the tomb in its entirety, the central and deepest portion of the baulk, Sector 5, had to be either removed or tunneled through. Given our knowledge of the soils of the tumulus (see Foss and Timpson in Papadopoulos, Bejko, and Morris 2007:140–44), and the importance of keeping the baulk intact, the decision was taken to excavate a tunnel through Sector 5 into Sector 1 in order to expose the burial. The excavation of the tunnel and the careful cleaning of the grave proved to be laborious and took over three weeks to complete.

The human remains were encountered in an extremely poor state of preservation, with much of the torso and cranium, as well as other parts of the skeleton, surviving in splintered and disintegrated pieces. The deceased, a mature adult aged 45 or more years and probably female, was oriented northeast-southwest (70°), head to the northeast. The torso appears to have been laid out in a supine position, with the arms folded across the pelvis. Judging by the cranium as preserved, the head was probably facing up. Although the legs were perhaps originally flexed, as encountered they were disturbed and not in their original position, so that the original position cannot be determined. As preserved, the tomb measured 1.28 m in length, about 0.50 m wide, and 0.08 m deep. A clearly defined grave cut was not noted. Grave goods included a one-handled matt-painted tankard, **TLXIII-1** (**Fig. 3.212**), placed upright next to the cranium on the south side, and two iron pins, **TLXIII-2** and **TLXIII-3**, both in situ over the left and right shoulders, respectively, and clearly in their original position as worn. The first, **TLXIII-2**, was oriented east to west, head to the east; **TLXIII-3** was oriented north-northeast to south-southwest, though the orientation of the head was not clear in the field, but was probably toward the north-northeast. The eastern end of Tomb LXIII was stratigraphically located above the west end of the tomb pit of Tomb I.

Tomb LXIII was located immediately to the south and at a slightly higher level than the large coarseware amphora (P 283, SU 280), broken in situ, containing bitumen (see Papadopoulos, Bejko, and Morris 2007:121, 126, fig. 20) (**Fig. 3.214**). The vessel was found on its side, its base toward the west and its mouth evidently stopped by a stone. As no formal grave cutting for Tomb LXIII was found, it was difficult to associate the bitumen amphora directly with the tomb, but its location next to the deceased suggests that the vessel may have been intended for this tomb. The significance of the vessel and the bitumen is more fully discussed in Chapter 15 (see also Morris 2006:100–101, fig. 5).

TLXIII-1 (P 166) **Fig. 3.215**

Matt-painted one-handled tankard, FL Type 2B (**9/7**).

H (to handle): 0.160; D (base): 0.057; D (rim): 0.103.

Papadopoulos, Bejko, and Morris 2007:126, fig. 19.

TLXIII-2 (SF 254) **Fig. 3.216**

Iron dress pin with rolled head, Type II.1 (**10/34**).

PL: 0.134.

TLXIII-3 (SF 253) **Fig. 3.217**

Iron dress pin, Type II.3, perhaps with animal head finial (**10/44**).

PL (two non-joining frr): 0.140.

Grave fill: SU 233: HM ceramics (coarse, 4 frr = 34 g; semi-coarse, 15 frr = 28 g); daub (62 frr 121 g); lithics (1 debitage = 0.1 g); shell.

TOMB LXIV

Pit tomb, single inhumation (child)

Phase IV

Figs. 3.218–3.219

1.0E–2.0E, 3.0S–4.0S (high: 107.96 m = 354.96 m ASL; low: 107.46 m = 354.46 m ASL)

Sector 2; SU 346 (grave); 259 (fill); 260 (cut); 347 (skeleton)

Excavated as Grave 61 (July 6, 20–25, 2005)

Pit noted in Papadopoulos, Bejko, and Morris 2007:128, fig. 22

The cut for this grave was first noted at a significantly higher level, at which point several stones, some large, partially lined the pit. Three stones lined the tomb pit on the east side and one on the west; the latter may well have served as something of a grave marker. The southwest corner of the pit, together with the west edge, cut through Tomb XLVII (Grave 41), severing the cranium of that tomb more or less in half. Tomb LXIV is, therefore, later than Tomb XLVII. The fill of the grave was a distinct dark brown to reddish brown calcareous soil, different from the surrounding tumulus fill, with smaller pieces of sandstone encountered within the fill at a lower level, particularly along the southwest and northwest sides. The fill tended to be more homogenous at a higher level and more variable at a lower level. The shape of the pit was somewhat squarer at the top and more rectangular at the bottom. When first encountered (**Figs. 3.152, 3.219a**), the pit measured 0.97 × 0.98 m; its final dimensions, at the bottom, were 1.60 m long and 0.95 m wide. The depth of the pit was 0.50–57 m.

All that survives of the skeleton (SU 347), identified as a child aged 4 (± 1) years at death, are the very poorly preserved pieces of the cranium. The poor state of preservation of the bone is in contrast to the excellent condition of the skeletal remains in Tombs XLVI and XLVII, though this was the skeleton of a child, and graves of children were, on the whole, less well preserved than those of adults. Isolated fragments of the cranium were encountered at a higher level in the central portion of the grave; none of the postcranial bones were encountered in situ. The cranium was uncovered at the west-northwest end of the tomb, and it appears that the body of the deceased, together with the tomb pit, was oriented west-northwest-east-southeast (290°), the head to the west-northwest. Given the state of preservation of the bone, little could be said about the position of the deceased, though the excavator noted the possibility that the cranium may have been supine, facing up. A number of stones lined the tomb pit, particularly along the north and south sides of the tomb, and at the western corner, near the cranium. There were no grave goods.

Grave fill: SU 259: HM ceramics: Inventoried: P 267 (rim fr, large coarse/semi-coarse vessel, PH: 0.035; PW: 0.045), P 268 (rim fr, fine fabric, PH: 0.015; PW: 0.030); not inventoried (semi-coarse, 9 frr, including 1 rim and 2 handles; coarse, 11 frr, including 1 rim = 111 g; non-diagnostic, 2 scraps = 2.5 g); daub (30 frr = 62 g); lithics (1 debitage).

Tomb LXV

Pit tomb, single inhumation (adult female)

Phase IV

Figs. 3.220–3.222

2.0E, 2.0N–3.0N (high: 108.32 m = 355.32 m ASL; low: 108.25 m = 355.25 m ASL)

Sector 1; SU 211 (grave); 212 (fill); 213 (skeleton)

Excavated as Grave 30 (June 24–27, 2005)

Papadopoulos 2010b:240, fig. 5

The cranium, together with the upper cervical vertebrae and upper right torso, including the right arm, of the deceased of Tomb LXV were completely destroyed by the cut for the modern Tomb LXXXVI (Grave 22), though what survived of the human remains was comparatively well preserved. Some of the scattered human bone (including cranial fragments) noted in the area of Tombs LXXXVI and XCII (Graves 22 and 23) may well derive from this tomb. The southwestern portion of the tomb was located below Tomb LXXIX (Grave 6). The skeleton was oriented southeast-northwest (130°), the head, which was not preserved, to the southeast. The torso of the deceased (SU 213), an adult female aged 18–35 years, was laid out in a supine position, the left arm, and probably also the right, folded across the lower chest; the legs were flexed, the knees pointing southwest. The tomb measured 1.12 m long, 0.60 m wide, and 0.07 m deep. The iron fibula, **TLXV-1**, was found over the left shoulder of the deceased, where it would have been worn.

TLXV-1 (SF 162) **Figs. 3.222**

Iron fibula, Type II.2, arched, with triangular catch-plate (**10/24**).

L: 0.150.

Papadopoulos 2010b:241, fig. 6.

Grave fill: SU 212: HM ceramics (semi-coarse, 9 frr = 16 g; non-diagnostic, 9 scraps = 7 g); daub (5 frr = 25 g); shell.

Tomb LXVI

Pit tomb, single inhumation (adolescent, probably female)

Phase IV (AMS ^{14}C: 863 ± 44 cal BC)

Figs. 3.223–3.225

1.0W–2.0W, 4.0–5.0N (high: 108.29 m = 355.29 m ASL; low: 108.12 m = 355.12 m ASL)

Sector 4; SU 214 (grave); 215 (fill); 216 (skeleton)

Excavated as Grave 31 (June 24–29, 2005)

This grave was stratigraphically located below Tomb LXXXIII (Grave 7) and above Tomb XVIII (Grave 73). The deceased (SU 216), an adolescent, perhaps female, aged approximately 16 years at death, was oriented northeast-southwest (70°), head to the northeast. The torso was laid out in a supine position, arms folded across the abdomen, the head lying on the right side, facing north. The legs were flexed, knees pointing south-southwest. The tomb measured 1.20 m long, 0.58 m wide, and 0.150–0.170 m deep. Although a distinct grave cut was not noted during excavation, a slim lens of dark organic material, resembling burned matter, was revealed east of the cranium, extending along the southern side of the skeleton. This material was difficult to sample due to its fugitive state, but rather than carbonized material, it may represent remnants of a textile garment. The tightly flexed position, the slight inclination downward west to east, and the general appearance of the skeleton might support the idea that the deceased was bundled for transportation to, and burial in, the tumulus. Two tubular iron beads, **TLXVI-1** and **TLXVI-2**, were found beneath the right clavicle and mandible. An AMS ^{14}C date from a sample taken from a humerus of the deceased gave a date of 863 ± 44 cal BC.

TLXVI-1 (SF 164a) **Fig. 3.225a**

Tubular iron bead (**10/101**).

L: 0.018; D: 0.016.

Both beads preserve traces of the cord inside as iron corrosion.

TLXVI-2 (SF 164b) **Fig. 3.225b**

Tubular iron bead (**10/89**).

L: 0.020; 0.012.

Grave fill: SU 215: HM ceramics (2 frr, semi-coarse = 5 g; 2 non-diagnostic = 2 g); daub (1 fr = 10 g); faunal specimen; shell.

TOMB LXVII

Pit tomb, double inhumation (adult male and female)

Phase Va

Figs. 3.226–3.228

1.0W–2.0W, 3.0S–4.0S (high: 108.44 m = 355.44 m ASL; low: 108.17 m = 355.17 m ASL)

Sector 3; SU 98 (grave); 99 (fill); 100 (1st skeleton, male); 171 (2nd skeleton, female)

Excavated as Grave 12 (July 13–26, 2004)

When first encountered, the upper portion of this tomb was thought to have been part of the same burial as Tomb LXX (Grave 17), but as excavation proceeded, the two tombs were shown to be clearly distinct (**Fig. 3.227a**). Tomb LXVII was the first tomb to be uncovered in Sector 3 and partially overlapped Tomb XIX (Grave 54), and was therefore later than that tomb. A large portion of the grave, including much of the south and parts of the east sides, were lost due to erosion off the steep southern edge of the tumulus. Two individuals were placed in the tomb, an adult male (SU 100) older than 35 years at death, who was buried first, and a younger adult female (SU 171) aged 23–27 years. The male was oriented east-northeast to west-southwest (75°), head to the east-northeast; whereas the cranium of the female (largely missing, though some cranial fragments were recovered in 2005), was originally to the west-southwest. In the course of excavation, the excavator noted the possibility that the female may have disturbed the earlier male, but given the fact that much of the tomb had subsided, it is possible that the two individuals were interred at the same time, one above the other. The upper body of SU 171 was not preserved, but the legs were clearly flexed, the knees pointing north, the feet mostly above the chest of SU 100. The torso of the male (SU 100) appears to have been laid out in a supine position, the arms crossed over the chest; as found, the cranium was on its side, facing north, though the mandible was found facing up. As encountered, the bone in situ measured 1.12 in length, 0.83 m in width, and had a depth of about 0.27 m.

The only grave good, **TLXVII-1**, a terracotta spindlewhorl, bead, or button, was found to the west of the westernmost femur and was clearly associated with the female (SU 171).

TLXVII-1 (SF 130) **Fig. 3.228**

Terracotta spindlewhorl, bead, or button (**10/6**).

H: 0.022; D (max): 0.027.

Papadopoulos, Bejko, and Morris 2007:132, fig. 26c.

Grave fill: SU 99: HM ceramics (coarse, 7 frr = 60 g; semi-coarse, 3 frr = 10 g; non-diagnostic scraps = 18 g); daub (39 frr = 61 g); 9 bags shell, 1 faunal.

TOMB LXVIII

Pit tomb, multiple inhumation (adult female and three young adults of indeterminate sex)

Phase Va (AMS ^{14}C: 876 ± 46 cal BC)

Figs. 3.229–3.234

3.0E–4.0E, 1.0S–2.0S (high: 108.819 m = 355.819 m ASL; low: 108.304 m = 355.304 m ASL)

Sector 2; SU 101 (grave); 102 (fill); 103 (1st skeleton); 119 (cut); 136 (2nd skeleton); 140 (3rd, eastern skeleton); 150 (eastern fill); 103a (4th skeleton); 166 (outer cut)

Excavated as Grave 13 (July 13–22, 2004)

The circumstances of the discovery of this tomb were interesting. The first thing to be uncovered was the very poorly preserved remains of a skull (including the mandible and at least one tooth) of a mid-adult female (SU 103). Associated with this individual was the one-handled, matt-painted vessel (**TLXVIII-1**) lying more or less upright along the southeast of the cranium, and the small iron conical boss (**TLXVIII-2**), in situ on the north side of the cranium (**Fig. 3.230a**). Lab study of the human remains from this upper level of the tomb revealed the mandible and four molars of another individual (designated SU 103a), being a young adult of indeterminate sex. The orientation of SU 103 and 103a could not be determined due to the lack of postcranial bones. Clearance of these human remains, together with the associated grave goods (**TLXVIII-1** and **TLXVIII-2**), brought to light a larger pit, containing the remains of at least two more individuals: SU 136 was a young adult of indeterminate sex, aged 20–25 years, and SU 149, another young adult of similar age and indeterminate sex. The somewhat better-preserved SU 136 was oriented northwest-southeast (305°), head to the northwest; the cranium of SU 149 was to the southeast. The bronze spectacle fibula, **TLXVIII-3**, was associated with SU 136 and was found immediately to the south of the cranium of that individual (**Fig. 3.230b**); the iron pin, **TLXVIII-4**, was found among the postcranial bones (femora and humerus) of SU 149 and is most likely associated with that individual (the pin head was pointing northeast) (**Fig. 3.230c**). The state of preservation of the bone of both SU 136 and 149 was such that the position of the bodies as placed in the tomb could not be determined with any accuracy. Indeed, so poorly preserved were the skeletal remains of all the individuals in the tomb that the excavator speculated on the possibility of secondary burial, but enough articulation was noted to suggest primary interment, though proximity to both the modern surface and the tumulus edge may have contributed to the state of survival of the bone.

The human remains were encountered within a clearly defined pit, the edges of which appeared to have been lined with a fine clay, particularly on the north, west, and part of the south sides, but not on the east side, where the grave was right at the edge of the tumulus; the grave cut petered out toward the east-southeast as topsoil and the tumulus edge were approached. The extent of the human remains measured 1.40 m long and 0.67 m wide; the grave cut was 0.34 m deep; including the upper skeleton 103 and its associated grave goods, the overall depth of the tomb approached 0.50 m. The northwest portion of the grave cut partially overlapped Tomb XV (Grave 80). An AMS ^{14}C date from a sample taken from a long bone of the deceased gave a date of 876 ± 46 cal BC.

TLXVIII-1 (P 077) **Fig. 3.231**

One-handled vessel, FL Type 2A (**9/4**).
H (max): 0.155; H (to rim): 0.114–0.120; D (base): 0.055; D (rim): 0.113.

TLXVIII-2 (SF 092) **Fig. 3.232**

Small iron conical boss (**10/64**).
H: 0.021; D: 0.047–0.049.

TLXVIII-3 (SF 106) **Fig. 3.233**

Bronze spectacle fibula, Type I.1c (**10/15**).
L: 0.114.
Papadopoulos, Bejko, and Morris 2007:119, fig. 10.

TLXVIII-4 (SF 116) **Fig. 3.234**

Iron dress pin, Type II.1, with rolled head (**10/38**).
PL: 0.121.

Grave fill: SU 102: Inventoried: ceramics: P 081 (coarseware rim, PH: 0.035; PW: 0.046); lithics: SF 121 (Levallois tool); SF 141 (chert flake); HM ceramics (coarse, 3 frr = 37 g; semi-coarse, 11 frr = 22 g; fine, 2 frr = 4 g; non-diagnostic scraps = 32 g); daub (69 frr = 100 g); 7 bags shell. SU 150: daub, 5 frr = 8 g.

TOMB LXIX

Pit tomb, multiple inhumation (adult, indeterminate sex; child; possible second adult)

Phase Va (AMS ^{14}C: 863 ± 44 cal BC)

Figs. 3.235–3.240

1.0E–0.0, 2.0N (high: 108.57 m = 355.57 m ASL; low: 108.54 m [2004] = 355.54 m ASL)

Sectors 1 and 5; SU 172 (grave); 173 (fill); 174 (adult); 551 (child); 549 (fill 2007)

Excavated as Grave 27 (July 24–29, 2004; July 2–3, 2007)

The eastern portion of Tomb LXIX was encountered and excavated late in the 2004 season in Sector 1, while the remainder of the tomb extended into the baulk separating Sectors 1 and 4 (Sector 5). Given the fact that it was difficult to remove this central portion of the baulk in order to reveal the entire tomb, the decision was taken in 2004 to record and lift that part of the burial exposed in Sector 1 and to retrieve the remainder of the burial when the time came to remove the baulk. With the excavation of Sector 4 to a level below that of Tomb LXIX, it was clear that the grave did not extend to the west beyond the confines of the baulk. Consequently, the western portion of the tomb was excavated in 2007.

The condition of the human remains exposed in 2004 and 2007 was extremely fragile, with a tendency to splinter; in parts, the bone had disintegrated almost to powder. Traces of a cranium, including a row of comparatively well-preserved teeth, were exposed to the east, and parts of the pelvis were clearly visible right at the baulk, but nothing was encountered in 2004 of the lower legs. Designated SU 174, this skeleton was of a young adult of indeterminate sex, aged 20–25 years at death. In 2007 the leg bones of SU 174 were revealed, including the feet, and it was clear that this individual was oriented east to west (90°), cranium to the east. Too little of the cranium survived in situ to determine the position of the head, but the torso was laid out in a supine position, with the upper arms, at least, by the side of the body (it is possible, but not clear, that the lower arms were folded across the torso); the legs were flexed, with the knees pointing north. The only grave good clearly associated with this individual in 2004 was the fragmentary iron pin, **TLXIX-4**, which was found in the bone lab to be associated with the torso; the pin was not noted in situ, since the human remains had been block-lifted.

With the excavation of that part of the tomb located in the baulk, Sector 5, in 2007, it was clear that the tomb contained two individuals. Of the second skeleton encountered in 2007 (SU 551), the only clearly preserved bone in situ was the fragmentary cranium, including part of the dentition, representing the remains of a child aged 6 (± 2) years. Cranial fragments, including three teeth, of another individual were isolated in the lab in 2007 and designated SU 551a; these were of an adult of indeterminate sex. The relationship of SU 551a to SU 174 was not immediately clear, but it is unlikely that they belong to the same individual. A bronze fibula, **TLXIX-1**, together with a bronze earring, **TLXIX-3**, were found by the right side of the cranium of the child (SU 551) at the southwest corner of the tomb, and a small spectacle fibula, **TLXIX-2**, was uncovered a little to the east, along the south side of the grave. The overall dimensions of the tomb on the basis of the bone found in 2004 and 2007 were 1.22 m long and 0.43 m wide; there was no clearly discernible grave cut or fill that was significantly different from the surrounding tumulus fill.

Although it was not clear on account of the generally poor state of preservation of both individuals encountered in situ, one scenario was that the child (SU 551) may have been the earlier of the two burials that was pushed aside toward the west for the inhumation of the adult (SU 174). Another alternative, and much more likely, was that the two individuals were interred at the same time and that SU 551 was articulated in situ (the cranium seemed to be largely in place, and what appeared to be poorly preserved ribs extended below the long bones of SU 174). If so, then the child was oriented west to east (270°), head to the west; the head and torso may have been in a supine position, but little of the lower body was discerned in situ. Either way, it seems reasonably clear that **TLXIX-1**, **2**, and **3**, all belong with the child. If in situ, the earring, **TLXIX-3**, would be where it should be, by the side of the cranium; the fibula, **TLXIX-1**, would have fastened a garment at the upper right shoulder of the individual, and the spectacle fibula, **TLXIX-2**, would be over the right torso of the deceased; it is difficult to imagine this fibula, or any of the other bronzes, associated with the adult. In addition, a tiny shell fragment (cf. SF 346 in Tomb XXXVI [Grave 75]) was found with the skeleton (SU 551) in 2007; thought to have been possibly pierced and therefore an item of personal ornament, closer analysis by Dr. Vardala-Theodorou showed it to be the operculum of a land gastropod with a small hole at the center that was not intentional. Tomb LXIX was one of several graves stratigraphically located above the central grave, Tomb I. An AMS ^{14}C date from a sample taken from the left femur of the deceased gave a date of 863 ± 44 cal BC.

TLXIX-1 (SF 431a) **Fig. 3.237**

Bronze fibula of "Cassibile" type (Type I.2a), with incised decoration (**10/19**).
L: 0.157.

TLXIX-2 (SF 434) **Fig. 3.238**

Bronze spectacle fibula, Type I.1c (**10/17**).
L: 0.073.

TLXIX-3 (SF 431b) **Fig. 3.239**

Bronze earring (**10/83**).
D (max): 0.024.

TLXIX-4 (SF 155) **Fig. 3.240**

Fragments of iron dress pin (**10/47**).
PL: 0.067.

Grave fill: SU 173: HM ceramics (semi-coarse, 2 frr, including 1 rim = 5g; non-diagnostic, 2 scraps); small fragment of operculum of land gastropod with small hole at center (not intentional); cf. SF 346.

Tomb LXX

Pit tomb, double inhumation (adolescent female, mature male)

Phase Va

Figs. 3.241–3.250

1.0W–1.0E, 2.0S–3.0S (high: 108.54 m = 355.54 m ASL; low: 108.31 m = 355.31 m ASL)

Sectors 3 and 7; SU 133 (grave); 134, 196 (fill); 126 (1st skeleton, female); 197 (2nd skeleton, male)

Excavated as Grave 17 (July 13–29, 2004)

Papadopoulos, Bejko, and Morris 2007:119–120, figs. 11–12; Papadopoulos 2010a:43, fig. 12

This was the richest of the tombs encountered in 2004 and the first to bring to light the interesting phenomenon that was to repeat itself in other burials at Lofkënd, namely that the wealthiest tombs were those of adolescent females (**Fig. 3.242a**). As with other burials, this was a simple pit, with the grave cut clearly visible in the baulk separating Sectors 2 and 3, and with a fill clearly distinguishable from the surrounding tumulus fill. The tomb was first encountered in Sector 2 but continued into the baulk, Sector 7, necessitating partial removal of the baulk, and extending slightly into Sector 2. Within the pit the remains of two individuals were interred. The first to be uncovered and the latest of the two was the adolescent female (SU 126) aged 16–18 years at death (**Fig. 3.242b**). She was oriented southwest-northeast (250°), head to the southwest; her torso was in a supine position, head to the left facing north, arms crossed over the lower chest; the legs were flexed, knees pointing north. The bones of the second individual (SU 197), a mid-age to mature adult male aged 45+ years, appear to have been disturbed by the interment of the female, as if pushed slightly to the north and east to accommodate the later burial, but there was enough articulation among the surviving remains to indicate that he had been interred first (**Fig. 3.242d**). As preserved, his cranium was at the southeast corner of the tomb, together with some postcranial fragments; the remainder was found at the northeast corner, with long bones extending along the northern flank of the grave. Although clearly disturbed, SU 197 may have been originally laid out in the opposite direction to the female (the excavator speculated that the original orientation may have been 70° northeast-southwest). The tomb measured 1.32 m long and 0.60 m wide, with an average depth of 0.23 m, though the pit itself was a little deeper, 0.31 m.

All of the grave goods associated with this burial belonged with this younger female. The stemmed goblet, **TLXX-1**, was found placed on its side, the rim facing east-northeast, beside the left leg and pelvis; a bronze headband with hook-shaped ornaments and other attachments, **TLXX-2**, including the rings, **TLXX-2a**, **2b**, were found above and below the skull and were clearly worn on the head. The jewelry included a large bronze spectacle fibula with triple connecting loops, **TLXX-3**, found beside the upper left humerus, beneath the mandible, and thus was worn at the upper left torso; an iron fibula, **TLXX-4**, was found over the upper right ribcage, and the large bimetallic pin, **TLXX-5**, was uncovered above the upper right humerus, and was thus worn on the upper right torso, a worthy counterpart to the spectacle fibula on the other side. A bead of semi-precious stone, carnelian or sardonyx, **TLXX-6** was found immediately adjacent to the front of the skull on the north side, and a fragmentary iron pin, **TLXX-7**, was found in the lab during the cleaning of the cranium and was thus associated with the upper body. The combination of fibulae and dress pins indicated that the female was buried wearing an elaborate garment, her head crowned with a bronze headband. Traces of textile pseudomorphs were encountered in the iron corrosion of the pin, and clearance of the upper torso and cranium of the female revealed a dark discoloration (**Fig. 3.242c**), thought to be the decomposed remnants of the textile worn by the deceased.

Tomb LXX was stratigraphically located above Tomb XIX (Grave 54), and this, together with the small finds, all of which are best accommodated in the ninth century BC, indicate that the tomb is best placed in a late phase of the Early Iron Age, though it is not among the very latest burials in the tumulus.

TLXX-1 (P 082) **Fig. 3.243**

Stemmed goblet (one-handled), matt-painted, FL Type 4 (**9/15**).
H (max): 0.155–0.158; D (rim): 0.130; D (base): 0.081.
Papadopoulos, Bejko, and Morris 2007:120, fig. 12a.

TLXX-2 (SF 91(a–e) **Fig. 3.244**

Fragmentary bronze headband (**10/87**).
PL (all main joining and nearly joining fragments): 0.586.
Papadopoulos 2010a:44, fig. 13.

TLXX-2a (SF 91d) **Fig. 3.245a**

Bronze ring associated with headband (**10/73**).
D: 0.021.
Papadopoulos 2010a:46, fig. 14.

TLXX-2b (SF 91e) **Fig. 3.245b**

Bronze ring, resembling a finger ring, associated with headband (**10/74**).
D: 0.024.
Papadopoulos 2010a:46, fig. 14.

TLXX-3 (SF 107) **Fig. 3.246**

Large bronze spectacle fibula, Type I.1a, with triple connecting loops (**10/13**).
L: 0.172; Wt: 200 g.
Papadopoulos, Bejko, and Morris 2007:120, fig. 12b.

TLXX-4 (SF 110) **Fig. 3.247**

Iron fibula, Type II.1 (**10/22**).
L: 0.98.
Papadopoulos, Bejko, and Morris 2007:120, fig. 12d.

TLXX-5 (SF 111) **Fig. 3.248a–b**

Bimetallic (bronze and iron) dress pin, Type III.1 (**10/50**).
PL: 0.278.
Papadopoulos, Bejko, and Morris 2007:120, fig. 12c.

TLXX-6 (SF 113) **Fig. 3.249**

Spherical bead of crypto-crystalline quartz (carnelian or sardonyx) (**10/104**).
D: 0.008.
Papadopoulos, Bejko, and Morris 2007:120, fig. 12e.

TLXX-7 (SF 414) **Fig. 3.250**

Fragmentary iron dress pin (**10/41**).
PL: 0.067.

Grave fill: SU 134: HM ceramics (coarse, 1 fr = 15 g; non-diagnostic scraps = 75 g); daub (9 frr = 25 g); 3 bags shell.

TOMB LXXI

Pit tomb, single inhumation (child)

Phase Va (AMS ^{14}C: 852 ± 44 cal BC)

Figs. 3.251–3.252

1.0E–0.0, 4.0N (high: 108.52 m = 355.52 m ASL; low: 108.50 m = 355.50 m ASL)

Sector 1; SU 175 (grave); 176 (fill); 177 (skeleton)

Excavated as Grave 28 (July 24–26, 2004)

This tomb was located above Wall 1, and a very small portion of the south edge of the grave was stratigraphically located below Tomb LXXXIV (Grave 2). The burial was of a child (SU 177) aged 8 (± 2) years in a simple pit, the edges of which were not clearly discernible, measuring 1.10 m long, 0.53 m wide, and about 0.09 m deep. The skeletal remains were encountered in a very poor state of preservation; the cranium was crushed, though the mandible was somewhat better preserved and appeared to be in situ, facing up. Little of the torso or arms survived, but the legs were clearly flexed, the knees pointing north-northwest. The skeleton was oriented east-northeast to west-northwest (80°), the head to the east-northeast. There were no grave goods. An AMS ^{14}C date from a sample taken from the left femur of the deceased gave a date of 852 ± 44 cal BC.

Grave fill: SU 176: HM ceramics (coarse, 3 frr = 10 g; semi-coarse, 1 rim = 10 g); daub (2 frr = 5 g); shell.

Tomb LXXII

Pit tomb, single or double inhumation (adult, indeterminate sex; possible child)

Phase Va

Figs. 3.253–3.254

2.0E–4.0E, 1.0N–0.0 (high: 108.49 m = 355.49 m ASL; low: 108.42 m = 355.42 m ASL)

Sectors 1 and 6; SU 160 (grave); 161 (fill); 162 (adult skeleton); 162a (juvenile)

Excavated as Grave 24 (July 21–26, 2004)

The first thing encountered of this burial was a small area of human teeth noted in the field as probably belonging to a child (**Fig. 3.254b**). Since the discovery of this tomb coincided with the excavation of the larger adult cist graves of the modern era, the remains were covered, and it was only at the end of the season that the tomb was more fully investigated. The human remains of this tomb were found in an extremely poor state of preservation. The primary skeleton (SU 162) proved to be a young adult of indeterminate sex aged 22–28 years (**Fig. 3.254a**). As the teeth of the child were found directly above the poorly preserved cranial remains of SU 162, they were bagged as part of that skeleton. Study in the lab showed that this dentition, together with a few cranial fragments, belonged to a child aged 9 (± 2) years at death (SU 162a). Nothing else of the child was recovered, and as such it remains unclear whether these remains were the very poorly preserved remnants of a second individual in the tomb or intrusive.

As for the primary, adult, skeleton, it was oriented southeast-northwest (120°), head to the southeast, but little could be said about its position except that the legs were flexed, the knees pointing to the north-northeast, and fragments of the upper arms were parallel to the body. The torso and perhaps the cranium were noted by the excavator as being probably supine. The tomb measured 1.66 m long, 0.52 m wide, with a shallow depth of 0.07 m. Tomb LXXII was stratigraphically located above Tomb XLV (Grave 60). There were no grave goods.

Grave fill: SU 161: HM ceramics (semi-coarse, 2 frr, including 1 painted neck fr; coarse, 2 frr = 29 g; non-diagnostic scraps = 2 g); daub (2 frr = 6 g); shell.

Tomb LXXIII

Pit tomb, likely double inhumation (two adults, one probably male)

Phase Va (AMS ^{14}C: 867 ± 45 cal BC)

Figs. 3.255–3.256

4.0E–5.0E, 4.0N (high: 108.31 m = 355.31 m ASL; low: 108.17 m = 355.17 m ASL)

Sector 1; SU 167 (grave); 168 (fill); 169 (adult, probably male); 169a (2nd skeleton)

Excavated as Grave 26 (July 22–24, 2004)

This tomb was located after removal of topsoil, only a few centimeters from the modern surface and at the edge of the tumulus at this higher level. The poorly preserved state of the tomb is likely the result of proximity to the tumulus edge, with the greater part of the skeletal remains having probably eroded off the east side of the mound, and perhaps also because of the modern burial activity in the immediate area. Tomb LXXIII was clearly a late burial on the tumulus, the preserved remains likely located above the continuation of Wall 1, though above that part of the wall were there were no surviving stones. The poorly preserved bone, together with traces of a grave cut, measured 0.90 m long, 0.85 m wide, and

0.14 m deep. The excavator noted that the entire upper body and cranium of the deceased were missing, and that the remains as encountered included parts of the disarticulated left leg, but that the right leg was clearly flexed (SU 169). This individual was an adult and perhaps male. On the basis of the little that was preserved, the excavator noted the possibility that the body may have been originally oriented southeast-northwest (120°), head to the northwest. Parts of two feet were recorded in the field, but study in the lab suggested that these belonged to a second individual (designated SU 169a), an adult of indeterminate sex. There were no grave goods, nor was there any residual material recovered from the grave fill. An AMS ^{14}C date from a sample taken from the right tibia of the deceased gave a date of 867 ± 45 cal BC.

Tomb LXXIV

Pit tomb, triple inhumation (three adult males)

Phase Va (AMS ^{14}C: 867 ± 45 cal BC; 805 ± 18 BC)

Figs. 3.257–3.258

1.0W–3.0W, 6.0–7.0N (high: 108.35 m = 355.35 m ASL; low: 108.10 m = 355.10 m ASL)

Sector 4; SU 208 (grave); 209 (fill); 210 (north skeleton), 230 (center skeleton), 231 (south skeleton)

Excavated as Grave 29 (June 25–July 4, 2005)

Papadopoulos, Bejko, and Morris 2007:118–119, 123, fig. 14

When first encountered, this proved to be one of the most enigmatic of all graves and, with no associated grave goods, virtually impossible to date. The tomb was at the northwestern edge of the tumulus, and although close to the prehistoric Tombs LXXVIII and LVI (Graves 18 and 43), as well as the modern Tomb XCVII (Grave 39), the lack of a clear stratigraphic relationship with any other tomb only made the dating all the more difficult (although not stratigraphically interrelated, Tomb LXXIV is located some 0.25–0.30 m below the level of Tomb LXXVII [Grave 18]). The southwest corner of Tomb LVI (Grave 43) may have been below Tomb LXXIV. The three individuals were laid out side-by-side on the same northeast-southwest orientation (40–50°), heads to the northeast. All three were laid out in a fully extended supine position, with the legs extended. The north skeleton was an adult male aged 25–30 years; the south skeleton was also a younger adult male aged 20–25 years, whereas the central skeleton was an older adult male aged 35–45 years. The arms of the north skeleton were folded across the lower abdomen of the deceased; the left arm of the central skeleton was similarly folded, but the right arm was extended below the left arm of the north skeleton. The arms of the south skeleton were both extended: the right arm over the left pelvis and upper thigh, the left arm below the left pelvis and upper thigh. The left arm of the central skeleton was placed above the right arm of the south skeleton. The overlapping arms clearly indicate that the three individuals were interred at the same time and that, of the three, the south individual was laid out first, the central individual second, and the north last. The feet of all of the skeletons and most of the right leg of the northern skeleton were disturbed or lost due to the proximity of the tumulus edge. The fact that the tomb was located right at the northwest edge of the tumulus, immediately below topsoil, also obliterated any clear traces of the original grave cut. The cranium of the south skeleton was facing up but was tilted laterally toward the central skeleton; the cranium of the central skeleton was facing up, whereas that of the north skeleton was facing southeast toward the central skeleton. As preserved, the tomb measured 1.96 m long, 1.30 m wide, and, not including the stone covering, 0.15–0.25 m deep.

At least eight stones, several of them quite large, were placed as a partial tomb covering over the crania and torsos of the deceased, with two of the stones extending over the pelvis and legs of the central skeleton, and another stone just to the northeast of the central and north skeletons. The weight of the stones, and the fact that there was virtually no soil between them and the skeletons, had resulted in some damage to the bone, with the crania of all three—and particularly the central and north skeletons—having been crushed.

As for the date of the tomb, its orientation and the fact that three individuals were interred side-by-side seemed to preclude the possibility of a modern date, as these were features not encountered in any of the modern burials. In a similar vein, the comparatively good state of preservation of the bone and the lack of any flexing of the legs are features not encountered in other prehistoric burial at Lofkënd. Moreover, where more than one individual is buried in a prehistoric tomb, the position of the deceased is usually one on top of the other or with the earlier human remains pushed to one side to make room for a later interment. That the individuals of Tomb LXXIV were neatly laid out side-by-side is a feature not found in any other tomb uncovered on the tumulus. The issue of chronology of this burial was only resolved with AMS radiocarbon dating (see Chapter 5), with samples taken and analyzed from all three individuals, indicating a date in the second half of the ninth century BC. A total of four samples gave the following results: a sample taken from the right humerus and left tibia of the north skeleton, respectively, gave dates of 805 ± 18 cal BC and 817 ± 21 cal BC; a sample taken from the left tibia of the central skeleton gave a date of 854 ± 43 cal BC; and a sample from the right humerus of the south skeleton gave a date of 867 ± 45 cal BC. The unusual treatment of the three individuals is considered more fully in Chapters 8 and 21.

Grave fill: SU 209: ceramics (semi-coarse, 7 frr = 10.5 g), daub (39 frr = 58 g); lithic debitage (1 small flake = 0.2 g); shell, plus one faunal specimen.

Tomb LXXV

Pit tomb, single inhumation (adult male)

Phase Va (AMS ^{14}C: 821 ± 44 cal BC)

Figs. 3.259–3.260

3.0E–4.0E, 6.0N–7.0N (high: 108.27 m = 355.27 m ASL; low: 108.15 m = 355.15 m ASL)

Sector 1; SU 221 (grave); 222 (fill); 223 (skeleton)

Excavated as Grave 33 (June 25–30, 2005)

This tomb was largely encountered at the tumulus edge after the removal of topsoil, with part of the upper cranium disturbed by proximity to the tumulus edge, which sloped sharply down from west to east. The deceased (SU 223), an adult male aged 30–40 years, was oriented east to west (90°), head to the east. The torso was laid out in a supine position; the left arm was gently flexed over the pelvis; the right arm appears to have been extended by the right side of the body, although the right hand may have rested on the right side of the pelvis. The legs were flexed, knees pointing north. Although disturbed, the mandible was facing north, and this was probably the original position of the cranium. The fill of the grave was not significantly different from the surrounding tumulus fill. The tomb measured 1.42 m long, 0.60 m wide, and 0.12 m deep. There were stones to the south and southeast of the left shoulder and immediately below the left elbow. There were no grave goods. An AMS ^{14}C date from a sample taken from the left tibia of the deceased gave a date of 821 ± 44 cal BC.

Grave fill: SU 222: inventoried: P 113 (strap handle, matt-painted vessel); P 155 (semi-coarse rim fr); HM ceramics (coarse, 3 frr = 13 g; semi-coarse, 4 frr = 7 g); daub (11 frr = 40 g); 1 lithic debitage, 0.8 g; shell.

Tomb LXXVI

Pit tomb, single inhumation (adult, indeterminate sex)

Phase Vb

Figs. 3.261–3.262

3.0W, 2.0N (high: 108.55 m = 355.55 m ASL; low: 108.49 m = 355.49 m ASL)

Sector 4; SU 123 (grave); 124 (fill); 125 (skeleton)

Excavated as Grave 16 (July 16, 2004)

This was among the poorest of the preserved burials, located right at the edge of the tumulus at this upper level. Only small fragments of postcranial bone were encountered to the south-southwest and fragments of the cranium to the northeast of an adult of indeterminate sex. The bone measured 0.43 m in length and had the appearance of being in situ, though too little survived to determine orientation and position of the body. The grave was stratigraphically located above Tomb XXI (Grave 55). Two minuscule fragments of bronze of an unidentified object (SF 104: PL [max]: 0.019) were recovered in the sieve from the immediate area of this grave but could not be clearly associated with it.

Grave fill: SU 124: non-diagnostic ceramic scraps = 0.6 g; shell.

Tomb LXXVII

Pit tomb, single inhumation (adult male)

Phase Vb

Figs. 3.263–3.264

1.0W–1.0E, 6.0N–7.0N (high: 108.57 m = 355.57 m ASL; low: 108.44 m = 355.44 m ASL)

Sectors 1, 4 and 5; SU 137 (grave); 138 (fill); 139 (skeleton);

Excavated as Grave 18 (July 19–20, 2004)

Papadopoulos, Bejko, and Morris 2007:117, fig. 9

This tomb was stratigraphically located below the modern Tomb XCVII (Grave 39) and above the prehistoric Tomb LVI (Grave 43), the latter being the northernmost of all prehistoric burials on the tumulus. The skeletal remains of a mid-age adult male (SU 139) were found splintered and generally in a poor state of preservation. A discernible grave cut was not encountered during excavation, and the fill of the tomb was little different from the surrounding tumulus fill at this upper level. The cranium and torso of the deceased were laid out in a supine position, the head facing up; the right arm was folded across the lower chest; the left arm was folded over the left shoulder. The legs were flexed, the knees pointing north-northwest. The deceased was oriented east-northeast to west-southwest (80°), head to the east-northeast. The tomb measured 1.32 m long, 0.55 m wide and about 0.13 m deep. There were no grave goods.

Grave fill: SU 138: inventoried: SF 117 (small chert flake); SF 137 (specimen of petrified wood[?], found with cranium); HM ceramics (coarse, 8 frr = 56 g; non-diagnostic scraps = 18 g); daub (32 frr = 67 g); lithics (1 debitage = 0.5 g); 1 bag shell, 1 faunal.

TOMB LXXVIII

Pit tomb, single inhumation, adult female

Phase Vb

Figs. 3.265–3.267

2.0E–0.0, 2.0S–1.0S (high: 108.86 m = 355.86 m ASL; low: 108.81 m = 355.81 m ASL)

Sectors 2 and 7; SU 49 (grave); 50 (fill); 51 (skeleton); 65 (grave cut)

Excavated as Grave 5 (July 3–6, 2004)

Tomb LXXVIII, which was stratigraphically located above Tomb XVI (Grave 68), was a simple pit containing the remains of an adult female aged 20–22 years. The tomb was oriented east-southeast to west-northwest (105°), head to the east-southeast. The deceased appears to have been placed within a much larger pit (almost 2.0 m long east to west, and about 1.50 m north to south), and the immediate area of the skeleton was characterized by a darker soil, easily distinguishable from the tumulus fill. This area of darker soil around the skeleton measured 1.44 m long and 0.53 m wide, with a depth of about 0.05–0.08 m. The torso and head of the deceased were laid out in a supine position, the cranium facing up; the arms were folded across the chest, the left arm over the right; the legs were flexed toward the right side of the body, the knees pointing northwest. An iron pin, **TLXXVIII-1**, was found on the upper chest of the deceased near one of the clavicles, the pin head pointing northwest.

TLXXVIII-1 (SF 046) **Fig. 3.267**

Iron dress pin, Type II.1, with rolled head (**10/37**).

PL: 0.140.

Grave fill: SU 50: HM ceramics (coarse, 2 frr = 22 g; semi-coarse, 1 fr = 3 g; non-diagnostic = 9 g; daub (18 frr = 27 g); shell.

TOMB LXXIX

Pit tomb, single inhumation, adult (sex indeterminate)

Phase Vb

Figs. 3.268–3.269

1.0E–2.0E, 2.0N (high: 108.937 m = 355.937 m ASL; low: 108.930 m = 355.930 m ASL)

Sector 1; SU 52 (grave); 53 (fill); 54 (skeleton)

Excavated as Grave 6 (July 5–7, 2004)

This tomb was encountered in an extremely poor state of preservation, the damage almost certainly the result of the modern burial activity in the upper levels of the tumulus (the modern Tombs LXXXVIII and LXXXVII [Graves 3 and 8] were both located a short distance to the north). Consequently, very little can be said about the position and orientation of the deceased, although from what survives (fragments of tibiae, femora, and parts of the pelvis), it appears that the deceased was originally oriented southeast-northwest (130°), with the missing head to the southeast. The legs were probably flexed, the knees pointing north. Although the cranium was missing, teeth attributed to this grave were recovered in the process of sieving. The bone, including poorly preserved traces of skeletal material, encountered in situ measured 0.95 m in length and 0.65 m wide. The deceased was an adult aged 20–24 years, but too little survived to determine sex. This tomb was one of several stratigraphically located above the central grave, Tomb I. There were no grave goods.

Grave fill: SU 53: HM ceramics (coarse, 5 frr = 35 g; semi-coarse, 6 frr = 22 g; non-diagnostic = 11 g; daub (29 frr = 70 g); lithics: debitage, SF 100 (flake); shell.

Tomb LXXX

Pit tomb, single inhumation, adult male

Phase Vb (AMS ^{14}C: 871 ± 45 cal BC)

Figs. 3.270–3.274

1.0W–0.0, 2.0N–0.0 (high: 109.19 m = 356.19 m ASL; low: 108.93 m = 355.93 m ASL)

Sectors 4 and 5; SU 30 (grave); 31, 43, 537 (fill); 32 (skeleton)

Excavated as Grave 4 (June 29–July 1, 2004; June 30, 2007)

This tomb was encountered in the upper level of the tumulus, immediately below topsoil. It is stratigraphically located above Tombs XLIV and XIV (Graves 65 and 71), and must be among the latest of the prehistoric burials. The burial was first encountered in Sector 4 in 2004, at which time the greater part of the postcranial bones were exposed, recorded, and cleared; with the removal of the baulk, Sector 5, in 2007, the eastern extent of the inhumation, including the cranium and upper torso, were fully exposed. The tomb was a simple pit, and although there was no clearly distinguishable grave cut or distinct fill, the floor of the tomb was lined with a highly friable and decomposing layer of fire-affected clay (different from the pieces of daub encountered in the prehistoric tumulus fill). This bedding of lightly fired clay was noted by the different excavators in 2004 and 2007, though similar fire-affected clay was noted in the surrounding soil units (SU 35 and 536). As preserved, the tomb measured 1.40 m long (east to west), 0.50 m wide (north to south), and at least 0.173 m deep (and more in some places).

The deceased, an adult male aged 20–30 years, was laid out east to west (89°), the head to the east; the cranium and upper torso were supine. It appears that the head was originally facing up, but the skull was crushed and the mandible had shifted slightly to the south-southwest due to the pressure of the earth above. The arms were bent at the elbows, and it is likely that the left arm was originally folded over the left shoulder; the legs were flexed. An iron dress pin with a rolled head, **TLXXX-1**, was found in fragments in situ on the upper torso, and there were two iron beads, one by the upper right radius and ulna (**TLXXX-2**), the other (**TLXXX-3**) on the left humerus (see **Fig. 3.270**). In comparison to other graves, this was one of the richest of the male burials. An AMS ^{14}C date from a sample taken from the right humerus/radius of the deceased gave a date of 871 ± 45 cal BC.

TLXXX-1 (SF 423) **Fig. 3.272**

Iron dress pin, Type II.1, with rolled head (**10/36**). L: 0.129.

TLXXX-2 (SF 013) **Fig. 3.273**

Tubular iron bead (**10/100**). L: 0.015.

TLXXX-3 (SF 424) **Fig. 3.274**

Fragmentary tubular iron bead (**10/90**). PL: 0.018.

Grave fill: SU 31: HM ceramics: P 038 (large handle[?], pierced); P 040 (rim of large jar); coarseware (8 frr = 90 g); semi-coarse (14 frr = 137 g); non-diagnostic scraps (50 g); fired clay (many pieces, weighing 177 g, found under lower skeleton in Trench 4, 2004; 47 frr = 245 g in rest of grave fill); shell.

SU 43: daub (8 frr = 9 g).

SU 537: ceramics, semi-coarse (1 fr = 10 g); fine (3 frr, including 1 rim, = 2 g); fired clay (10 frr = 10 g); shell.

Tomb LXXXI

Pit tomb, single inhumation, adult male

Phase Vb (AMS ^{14}C: 871 ± 45 cal BC)

Figs. 3.275–3.276

1.0W–0.0; 1.0S–0.0 (high: 109.16 m = 356.16 m ASL; low: 109.08 m = 356.08 m ASL)

Sectors 3, 7, 8; SU 12 (grave); 13, 28, 29 (fill); 14 (skeleton)

Excavated as Grave 1 (June 25–30, 2004)

This was the first tomb to be excavated on the tumulus, located near the center of the original mound and at no great depth from the modern surface. On account of its location, the tomb was much disturbed. Although no clear grave cut was discerned and the tomb fill could not be distinguished from the surrounding tumulus fill, the bone had the appearance of being in a pit, but the remains were too disturbed to ascertain any clear orientation. As encountered, the human bone covered an area approximately 0.87 m in length (east-west) × 0.70 m (north-south); cranium fragments were encountered to the north-northwest, while the pelvis was uncovered at the southernmost point and long bones of the legs to the east. Although disturbed, there was enough articulation to indicate that the burial was not secondary. It is, therefore, possible that the body was originally laid out roughly east to west, with the cranium to the northwest. The skeletal remains were of a young adult male aged 18–22 at death. There were no grave goods. An AMS ^{14}C date from a sample taken from the left tibia of the deceased gave a date of 871 ± 45 cal BC.

Grave fill: SU 13: HM ceramics (1 rim, fine; 5 scraps non-diagnostic = 8 g); lithics (1 flake); shell. Unit 29: HM ceramics (semi-coarse, P 031: handle fr, matt-painted fabric; 6 frr incl. 2 handle frr; 1 bitumen sherd: 31 g); daub (1 fr, 7 g).

TOMB LXXXII

Pit tomb, multiple inhumation (three young adults, two possibly males)

Phase Vb

Figs. 3.277–3.278

1.0W–2.0W, 3.0N–4.0N (high: 108.72 m = 355.72 m ASL; low: 108.56 m = 355.56 m ASL)

Sector 4; SU 83 (grave); 84 (fill); 85 (1st skeleton); 105–106 (two young adults)

Excavated as Grave 9 (July 10–14, 2004)

As with Tomb LXXXIV (Grave 2), Tomb LXXXII partly overlay Wall 1 in the northern sector of the tumulus and was located very close to the tumulus edge at this upper level, and was among the latest of the Early Iron Age burials. The tomb was a simple pit grave, with a partial cut discerned along the north side; a darker soil marked the eastern extent of the grave, and the soil along the southern edge was also clearly distinguishable from the surrounding tumulus fill. It appears that the tomb contained the remains of three individuals, but with little of each skeleton preserved. One skeleton (SU 85), consisting primarily of cranial fragments, though with fragments of a pelvis and femora, was of a young adult aged 20–24 years, and probably male; another (SU 106), primarily mandible and some postcranial bone, was also of a probable male, aged 26–32 years; the third individual (SU 105) was represented only by posterior portions of the skull, and was of a young adult of indeterminate sex. There was no clear articulation of any of the skeletons in situ (postcranial fragments were found between the various crania), though what survived of the crania was originally clearly articulated though disturbed. As preserved, the bone was oriented east to west (90°), with two mandibles to the east (SU 85 and 106) and fragments of the skull of one individual to the west (SU 105). The human remains, together with the darker soil marking the tomb, defined an area approximately 1.10 m long, 0.55 m wide, and at least 0.16 m deep. It is reasonably clear that the damage to the tomb was the result of its proximity to the tumulus edge and modern surface. There were no grave goods.

Grave fill: SU 84: inventoried: P 075, neck and rim frr, semi-coarse open vessel (PH: 0.030; PW: 0.049; D [rim] est: 0.150–0.180); HM ceramics: semi-coarse (9 frr = 27 g); coarse (2 joining frr = 15 g); non-diagnostic = 15 g; daub (22 frr = 46 g); shell.

TOMB LXXXIII

Pit tomb, single inhumation, subadult (sex indeterminate)

Phase Vb

Figs. 3.279–3.281

1.0W–0.0, 4.0N–5.0N (high: 109.03 m = 356.03 m ASL; low: 108.92 m = 355.92 m ASL)

Sectors 4 and 5; SU 58 (grave); 59 (fill); 60 (primary skeleton); 60a–c (scattered teeth)

Excavated as Grave 7 (July 5–8, 2004)

This tomb was encountered in a very poor state of preservation, only centimeters from the modern surface of the tumulus. Tomb LXXXIII was stratigraphically located above Tombs LXVI, XLIX, and XVIII (Graves 31, 51, and 73) and was clearly one of the latest of the Early Iron Age burials. Preserved in situ were the friable remains of the cranium and upper torso, including parts of the arms of a subadult aged 16–20 years at death; a few scattered leg bones were found farther to the west, but not clearly in situ. The area of primary bone measured approximately 0.80 m long (east to west), but the greater area of the tomb as exposed by the excavator measured 1.40 × 1.40 m, and about 0.11 m deep, though in parts a little deeper. No clear grave cut could be discerned. On the basis of the surviving bone in situ (SU 60), the skeleton was roughly oriented east to west (close to 80°), with the head to the east. The torso and cranium were laid out in a supine position, head facing up, the right arm bent over the lower chest. The legs were probably flexed, but too little survives to establish any details. In addition to the remains of the primary skeleton, 29 teeth were encountered scattered in the greater area of the tomb. These included: SU 60a (seven teeth, belonging to an adult 18–35 years), SU 60b (four teeth belonging to an older adult 35–45 years), and SU 60c (three teeth of a mature adult, 45+ years); an additional 15 teeth could not be attributed to dental sets. The only grave good, a handmade kantharos, **TLXXXIII-1**, was found beside the left humerus of the deceased (SU 60) in an upright position.

TLXXXIII-1 (SF 047) **Fig. 3.281**

HM Kantharos, FL Type 5A (**9/16**).
H (max): 0.115; D (rim): 0.067–0.077; D (base): 0.037–0.042.

Grave fill: SU 53: HM ceramics: semi-coarse (9 frr = 17 g, including one handle); coarse (4 frr = 43 g); non-diagnostic scraps = 12 g; daub (15 frr = 55 g); lithics: S081; 5 debitage = 21 g; shell.

TOMB LXXXIV

Pit tomb, multiple inhumation (two adult males, one child)

Phase Vb (AMS ^{14}C: 821 ± 24 cal BC)

Figs. 3.282–3.284

1.0W–2.0E, 3.0–4.0N (high: 109.11 m = 356.11 m ASL; low: 108.96 m = 355.96 m ASL)

Sectors 1, 4, 5; SU 16 (grave); 17 (fill); 18 (younger adult male), 38 (adult male), 38a (child)

Excavated as Grave 2 (June 26–July 2, 2004)

This tomb, located directly above Wall 1, was clearly was one of the latest of the prehistoric burials. The skeletons of two adult males were clearly discerned in situ within a simple pit, which had a partial cut on the west and south sides. As exposed, the tomb measured 1.92 m long, 0.70 m wide, with a depth of 0.12–0.15 m. The first of the adult male skeletons to be encountered (SU 18) was oriented east to west (95°), head to the east; the cranium and torso were laid out in a supine position, the arms gently folded across the abdomen, the legs lightly flexed, the knees pointing north. The cranium was placed, or had fallen, to the right, facing north. The individual was 23–26 years at death. Located largely below the legs and feet of this skeleton was an earlier skeleton (SU 38), that of an older adult male aged 35–45 years. The cranium of SU 38 was below the knees of SU 18, with most of the postcranial bones located in the western sector of the tomb. It is clear that SU 38 was the earlier of the two inhumations, the bones moved to the west to accommodate the later inhumation, SU 18. As such, the original orientation of SU 38 could not be determined (as preserved, the cranium was to the north-northeast, the remainder of the bones to the south and west). Two small fragments of what appeared to be a third individual were found above the upper legs of SU 18. Cleaning of all the skeletal remains in the lab showed these fragments, together with several more fragments, including those of the pelvis and limbs (femur, tibia, fibula), to be of a child aged 2–5 years (SU 38a). Since this skeleton was largely identified in the lab, nothing can be said about its orientation or original position, and given the paucity of the fragments actually noted in situ, the remains may well be intrusive. An AMS ^{14}C date from a sample taken from the left humerus of one of the males gave a date of 821 ± 24 cal BC.

The only grave good, an iron knife, **TLXXXIV-1**, was found under the leg bones of the earlier male skeleton (SU 38); its position may plausibly indicate that the knife was once sheathed and at the side of the deceased.

TLXXXIV-1 (SF 019) **Fig. 3.284**

Iron knife (**10/121**).

PL: 0.177; W (max): 0.016.

Grave fill: SU 17: HM ceramics: P 037 (handle fr, found near skeleton SU 38); P 042 (body fr, large coarseware vessel); non-diagnostic scraps, 30 g; daub (17 frr = 80 g); lithics: SF 049 (flake with traces of use), 3 flakes (debitage) = 6.5 g; abundant shell.

TOMB LXXXV

Pit tomb, double inhumation (adult female and male)

Phase Vb (AMS ^{14}C: 811 ± 18 cal BC)

Figs. 3.285–3.286

1.0E–2.0E, 5.0N (high: 108.80 m = 355.80 m ASL; low: 108.46 m = 355.46 m ASL)

Sector 1; SU 88 (grave); 89 (fill); 90 (1st skeleton, female); 145 (2nd skeleton, male)

Excavated as Grave 10 (July 14–21, 2004)

This burial, stratigraphically one of the latest of the prehistoric tombs, was located in the heaviest concentration of modern burials. The eastern extent of the tomb was destroyed by the cut of the modern cist, Tomb XCII (Grave 23), and subsequently by the animal burial, Tomb XCIII (Grave 19); and Tombs XCIV, XCV, and XCVI (all modern infant inhumations) were located a short distance to the north. Despite disturbance to the grave by this later activity, it is remarkable that what survives of the tomb is comparatively well preserved, particularly the western portion. The grave cut was not identified, but this is not surprising, given the disturbance in the area. The tomb contained the remains of two individuals: an adult female (SU 90) and an adult male (SU 145). There were also scattered remains of the bones of an infant/perinate (designated SU 145a), primarily located above one of the capstones of Tomb XCII (visible in **Figs. 3.285, 3.286a–b**), which destroyed much of the eastern extent of Tomb 10, but these are almost certainly intrusive and should belong to one or other of the nearby modern infant burials. The adult female (SU 90) was oriented west to east (270°), head to the west; the adult male east to west (90°), head to the east. The cranium and torso of the adult female were laid out in a supine position, arms crossed over the lower chest, the right radius and ulna overlying the male skeleton (SU 145); her legs were flexed, the knees pointing north. The torso and much of the cranium of the male (SU 145) were destroyed by Tombs XCII and XCIII, but the legs were also flexed. Although disturbed, it appears that the two individuals were buried at the same time. As preserved, the skeletal remains defined an area 1.90 m long (east to west), 0.55 m wide, with a depth ranging between 0.10 and 0.34 m. The tomb was stratigraphically located above Tombs XVII and XLVIII (Graves 72 and 52). There were no grave goods. An AMS ^{14}C date from a sample taken from the left tibia of the deceased gave a date of 811 ± 18 cal BC, the latest of the AMS ^{14}C dates taken from a sample of human bone from the tumulus.

Of particular interest is the sheep burial, Tomb XCIII (the identification was by Marston, Chapter 16.1). The animal remains were neatly stacked above the capstones of the modern Tomb XCII (Grave 23), near the area of the cranium of SU 145 and the feet of SU 90. The nineteenth-century excavators of Tomb XCII would not have failed to have noticed that they had disturbed an earlier burial, and the "sacrifice" of the animal—if it can be called a sacrifice—may have been to appease the spirit of the disturbed individual in Tomb LXXXV rather than a funerary ritual directly associated with Tomb XCII.

Grave fill: SU 89: HM ceramics (coarse, 1 fr = 10 g; semi-coarse, 3 frr [including 1 rim] = 4 g; non-diagnostic scraps = 3.5 g); daub, 7 pieces = 10 g.

CHAPTER 3.2
CATALOGUE OF THE MODERN TOMBS AND THEIR CONTENTS

Sarah P. Morris, John K. Papadopoulos, and Lorenc Bejko

With a contribution on the coins by Shpresa Gjongecaj

THE MODERN BURIALS

After antiquity, the tumulus above the Gjanicë River lay relatively undisturbed for many centuries—in fact, for the better part of three millennia—until the later Ottoman period. At some point approximately 200 years ago, its northeast portion (what was to be excavated as Sector/Trench 1) was chosen for a few select, modest burials of individuals, probably Orthodox Christians, from a local community (**Fig. 3.287**). Fifteen inhumations (two of them animal burials) were placed in the gentle slope rising at the north end of the sandstone ridge from relatively level ground close to modern access, and represent a deliberate selection of infants who may not have survived birth or died not long after birth, or whose parents were not able to raise them, along with five adults, laid nearly parallel in a row. Most of them were carefully arranged, in an extended supine position so unlike their prehistoric predecessors, in pits sometimes lined with stones and often covered with them. Their orientation—head to the west but facing east, with some seasonal variants—marks them as Christian rather than Muslim (in Islam, one's head rather than feet should point to Mecca), as does the presence of three coins in the graves: two adults and one infant were provided with fares for the afterlife, in the form of coins placed in the mouth or near the cranium, an ancient Greek and Roman practice inherited in Christian customs. All three coins are illegible, and one too fragmentary to restore, much less read, but their silver-copper alloy and small, thin size make them modern. Two infants were covered with roof tiles as well as stones, and one adult leaned his head against a section of tile set upright to his left as a prop. Otherwise no offerings survive from these graves, beyond the small brass clasp of an infant's garment that indicates s/he was wrapped in some sort of cloth that fastened on the left side. The remains of three animals (all sheep) were also interred among these graves, perhaps as part of a funeral rite or ritual act related to the human burials (see Marston, Chapter 16.1).

All of these burials appeared in the preliminary report for the tumulus (Papadopoulos, Bejko, and Morris 2007:114–117, figs. 7–8), for they emerged exclusively in the first two seasons of excavation. As noted in that report, their appearance was in part a surprise, given the proximity of a modern (Muslim) cemetery at Ngrançija just below the tumulus, but such later inhumations are common in ancient burial tumuli and have been uncovered, for example, in the tumuli at Barç, Kuç i Zi, Rehovë, Prodan, Pazhok, Piskova, and Luaras (see Andrea 2005:359), as well as at Patos (Korkuti 1981) and Apollonia (Damiata et al. 2007–2008). The majority of these burials are either Late Antique or medieval, but some are more recent. At the medieval cemetery at Rëmbec, dated to the ninth–eleventh centuries AD, all the tombs were simple pit graves oriented east to west, crania to the west, facing east, the same orientation as the Lofkënd modern burials (Bejko, Aliu, and Lela 2004; Bejko et al. 2005).

Based on radiocarbon dates obtained from bone samples from three graves, these inhumations cluster around AD 1800, calibrated to include a range of 100–150 years, and thus can best be described as late Ottoman in setting/timing. Whether in reality they represent victims of a single catastrophe, occasion, or family cannot be determined without greater precision in dating, but this modern use of the Lofkënd tumulus was short-lived.

What remains equally unclear is whether the prominent sandstone ridge, enhanced by the artificial mound of earth accumulated 85 burials some 3,000 years earlier, simply attracted attention as a convenient site for a cemetery, or whether later interventions were deliberately directed at an older burial ground. Several Iron Age graves were disrupted by the placement of modern ones (e.g., Tomb XCIII [19] dislodged part of Tomb LXXXV [10]), so one is tempted to say that coincidence and topography brought ancient and modern mourners together to bury their dead at the same spot.

The occurrence of late burials in prehistoric mounds, not only in burial tumuli such as Lofkënd but in settlement mounds (*toumbes* or *magoules* as they are known in Greece), is a common phenomenon. In his publication of the prehistoric mound at Zygouries in the valley of Kleonai in the Peloponnese, Carl Blegen (1928:39) noted:

> Seven graves were discovered within the settlement of Zygouries at various points on the hill. Three of these appeared to date from Byzantine times or later and require only passing mention here. They were simple earth burials in which, as the well-preserved bones showed, the body had been laid out at full length on its back with the head to the west and the arms folded across the breast. No objects were found in these graves, but in one case two iron heel-plates beneath the feet indicated that the deceased had been buried wearing heavy boots.

Similar late burials, nine in all, were uncovered in the central area of the Neolithic and Bronze Age low settlement mound of Sitagroi in northeast Greece (Renfrew, Gimbutas, and Elster 1986:182–184). As was the case with the three late burials at Zygouries, all of the skeletons at Sitagroi were oriented east to west, heads to the west, facing east; their date is uncertain, but the excavators judiciously noted an "Iron Age" date, probably well after Classical antiquity (Renfrew, Gimbutas, and Elster 1986:182–184, pls. XVII, nos. 1–2). Certainly the few items of personal ornament associated with some of the Sitagroi tombs look Late Byzantine or post-Byzantine (Renfrew, Gimbutas, and Elster 1986:pls. XVIII–XIX). A solitary late burial, described as a "mittelalterlicher Grabfund" (Hänsel 1987:113–120, pls. 38–41), and more fully equipped with grave goods, was uncovered in the northwest portion of the excavations at Kastanas in Macedonia. As was the case at Zygouries and Sitagroi, the skeleton was oriented east to west, head to the west facing east, the same orientation as the late burials at Lofkënd (Hänsel 1987:113). Even the massive Neolithic mound at Çatalhöyük in the Konya plain of central Turkey eventually harbored infant burials of the Byzantine period (Matthews 1996:88, 95, fig. 7.16–7.17).

The villagers of nearby Ngrançija and Gjinoqara had no knowledge of these modern burials at Lofkënd; rather, they thought of the tumulus as the collective grave of foreign soldiers, casualties of World War I or the Balkan Wars, events just beyond the grasp of memory (Papadopoulos 2006). The burial customs of the late tombs at Lofkënd—like those at Zygouries, Sitagroi, Kastanas, and Çatalhöyük—indicate that the burial group was Christian. In the Ottoman *defter* (tax register) for Hicrî 835 (i.e., AD 1431), "Lofkende," according to İnalcık (1954:58, no. 148), is listed as a "*hassa çiftlik*" (a private farm of a cavalry member who held a *timar*), with registered produce including a vineyard, olives, figs, and walnuts, with a total tax of 1,032 *akçe* (silver coins). The name of the village—located on the prominent ridge to the east of the tumulus—is transcribed into modern Turkish as "Likofoni" ("wolf's voice" in modern Greek). It was the *timar* (a "prebend in the form of state taxes in return for regular military service" [Zarinebaf, Bennet, and Davis 2005:xxxi]) of a certain Angelos, apparently a Christian. According to the Austrian census of 1916–1918, Lofkënd (subsuming the villages of Bregasi, Granci [Ngrançija], Merkanji, and Verkanji) consisted of 55 dwellings, with some 390 inhabitants, all of whom were Albanian Muslims, except for one Vlach, who was Eastern Orthodox (for a useful glossary of Ottoman Turkish terms, see Zarinebaf, Bennet, and Davis 2005:xxiii–xxxi). Prior to the Ottoman conquest of Albania, the Mallakastër region, with the center at Byllis/Ballsh, was Christian. In AD 431, Bishop Felix of Apollonia and Byllis took part in the Council of Ephesos, together with the bishops of Dyrrhachion and Shkodër, representing the province of Epirus Nova (Ceka and Muçaj 2005:16). By AD 458, Byllis had its own bishop—Philocarus, Episcopus Ballidi—at which time the city was one of the bishoprics of New Epirus. (For the excavations at Byllis, see, among others, Muçaj 1979–1980, 1986, 1987, 1990; annual preliminary reports from 1981– 1990 in *Iliria*; and Anamali 1987. For a useful summary of the monuments, excavations, and earlier literature of Byllis, see Ceka and Muçaj 2005; see also Muçaj and Hobdari 2004.) In the sixth century, during the reign of the Emperor Justinian (AD 527– 565), the Basilica of Ballsh, its ruins and those of the buildings around it, known locally as the Monastery of St. Mary, was built; the name of Ballsh was derived from Byllis (for the excavations at Ballsh, see Anamali 1977–1978, 1981; for a useful overview, see Ceka and Muçaj 2005:101–104). Throughout the ensuing Byzantine period and the Middle Ages, Byllis/Ballsh was one of the most important Episcopal centers of Albania (for the identification of Ballsh and the Monastery of St. Mary with the medieval center of Glavinica, see Ceka and Muçaj 2005:102–104). It is clear that sometime between the fifteenth and the early twentieth century, the majority of the inhabitants of the region converted from Christianity to the Muslim faith, but some remained Christians. And a few of these Christians, conceivably crypto-Christians, were buried in the northeast sector of the prehistoric tumulus.

The fact that the tumulus prior to our excavations was not considered ancient is of interest, and the final word may belong to folk memories in the Balkans, where epic battles have succeeded each other across several millennia of war-torn history, and fallen heroes compete with each other to claim prominence in the landscape. For the workmen who excavated this tumulus and inhabit its surroundings, the mound, as we have noted, was associated with soldiers interred in a collective grave, foreigners from one of the many Balkan wars or the First World War (Papadopoulos 2006; Papadopoulos, Bejko, and Morris 2007). Once restored as it stands since the conclusion of excavations in 2007 (**Fig. 22.13**), the tumulus of Lofkënd can reclaim those memories. Thus, rather than answering the question as to continuity of practices, we close with beliefs instead, in an example from modern fiction set in Bosnia, where a diversity of heroes co-inhabit the same mound, in the local imagination:

> On the left bank, standing alone, immediately above the road, there was a fairly large earthen barrow, formed of some kind of hard earth, grey and almost like stone. On it nothing grew or blossomed save some short grass, as hard and prickly as barbed wire. . . . That was the spot which at one time was called Radisav's tomb. They used to tell that he was some kind of Serbian hero, a man of power. . . . The Turks, on the other hand, have long told that on that spot a certain dervish, by name Sheik Turhanija, died as a martyr to the faith. He was a great hero and defended on this spot the crossing of the Drina by an infidel army.
>
> Ivo Andric, *The Bridge on the Drina* (1945)

Tomb LXXXVI

Cist tomb, single inhumation (mature adult male)

Modern (AMS ^{14}C: cal AD 1810 ± 145)

Figs. 3.288–3.290

2.0E–4.0E, 2.0N–3.0N (high: 108.577 m = 355.577 m ASL; low: 108.257 m = 355.257 m ASL)

Sector 1; SU 154 (grave); 155 (fill); 156 (skeleton); 183 (cut)

Excavated as Grave 22 (July 21–29, 2004)

Papadopoulos 2010b:240, fig. 5

This tomb was the first of the large modern cist burials to have been uncovered. The tomb consisted of a neat, almost rectangular cut, measuring 1.77 m long, 0.46 m wide, and 0.28 m deep. The cut was particularly clear on the east and west sides, where there was no stone lining. The cut was lined on the north and south sides with medium to large stones, all sandstone, set on edge, mainly to a height of one course, though in places where smaller stones were used there are two courses. Within the cist so defined, the body of the deceased (SU 156), identified as a mature adult male aged over 45 years at death, was laid out in a fully extended supine position, head facing up and slightly to the east, arms and legs more or less parallel, the arms placed over the pelvis and upper thighs, because the width of the tomb was barely the width of the deceased. The skeleton was oriented east to west (275°), head to the west. The state of preservation of the bone was remarkable, with all of the phalanges of the hands and feet, and the patellae well preserved. The tomb was covered with several larger stones, the interstices filled with a series of smaller stones, and with a fewer small flat stones placed over the stones lining the pit to support better the capstones. Additional stones were placed over the cranium and upper torso, and there was a single large stone covering the feet. The fill of the grave was generally of a looser texture than the tumulus fill into which the tomb pit was cut. There were no grave goods. As for chronology, an AMS ^{14}C date was determined from a sample taken and analyzed from the right hand of the deceased, which indicated a date of AD 1810 ± 145 years, calibrated.

Grave fill: SU 155: HM ceramics (semi-coarse, 5 frr, including 1 handle = 7 g; coarse, 6 frr = 20 g; non-diagnostic scraps = 6 g); daub (15 frr = 47 g); shell(?).

Animal Burial, "Tomb" LXXXVII

Pit tomb, single animal inhumation (sheep/goat)

Modern

Figs. 3.291–3.292

3.0E, 2.0N–3.0N (high: 108.88 m = 355.88 m ASL; low: 108.84 m = 355.84 m ASL)

Sector 1; SU 80 (grave); 81 (fill); 82 (skeleton)

Excavated as Grave 8 (July 9–14, 2004)

Mentioned in Papadopoulos, Bejko, and Morris 2007:117

The circumstances of this burial deposit were unusual, but with the subsequent excavation of the other modern animal burial, Tomb XCIII (Grave 19), a pattern emerged that showed a common custom. To the southeast of the infant inhumation, Tomb LXXXIX (Grave 11), the comparatively well preserved skeleton of a neonatal sheep was encountered, partly covered by a large stone. The animal was placed immediately above the capstones of the large modern cist tomb, Tomb LXXXVI (Grave 22), more or less above the central and right torso of the deceased of that tomb, though this only became clear once the grave was lifted. As encountered, the animal skeleton measured 0.35 m long and 0.21 m wide in situ, with a depth of about 0.05 m. The burial shared a similar west to east orientation as the nearby modern human inhumations, the excavator noting a west-southwest to east-northeast orientation (245°). There were no grave goods associated with the animal burial and no material from the fill immediately around the bone, which was not in any way different from the surrounding soil.

With the subsequent excavation of Tomb LXXXVI (Grave 22), and with the similar circumstances of Tombs XCIII and XCII (Graves 19 and 23), it appears that both animal burials, Tombs LXXXVII and XCIII (Graves 8 and 19), were specifically associated with the large cists, Tombs LXXXVI and XCII (Graves 22 and 23). One immediate scenario was that both animals were sacrifices or offerings made at the grave site, with the animal slaughtered but not consumed,and, in the case of Tomb LXXXVII (Grave 8), not clearly butchered (though cf. the case of Tomb XXCIII [Grave 19]). It was clear, however, that both Tombs LXXXVI and XCII (Graves 22 and 23) disturbed earlier, prehistoric burials. Tomb LXXXVI cut across the eastern portion of Tomb LXV (Grave 30), destroying the cranium and upper torso of that burial, and Tomb XCII (Grave 23) disturbed the eastern extent of Tomb LXXXV (Grave 10). In both cases, the earlier burials were clearly violated, with a good deal of human bone displaced, and it is possible, therefore, that the deposition of the interred animals was more to appease the earlier violated burials rather than being directly associated with the later tombs (see further, Marston, Chapter 16). It is even possible that the siting of the other modern adult inhumations at the far northern extent of the tumulus was deliberate, in an attempt to avoid disturbance of earlier tombs, although this could not be determined with certainty, as the relative chronology of the modern inhumations was far from clear-cut, all the AMS ^{14}C dates from the modern skeletons analyzed clustering around ca. AD 1800.

TOMB LXXXVIII

Pit tomb, single inhumation (infant)

Modern

Figs. 3.293–3.294

1.0E, 3.0N (high: 109.17 m = 356.17 m ASL; low: 109.12 m = 356.12 m ASL)

Sector 1; SU 19 (grave); 20 (fill); 21 (skeleton)

Excavated as Grave 3 (June 26–28, 2004)

This tomb consisted of a pit, roughly oval in plan, measuring 0.60 m long, 0.25 m wide, and 0.08 m deep, cut into tumulus fill. The grave fill was not clearly distinguishable from the surrounding tumulus fill. Within the pit, the body of an infant (perinatal) was laid out in a fully extended supine position, cranium facing up, arms by the side of the body, legs bent at the knees. Most of the bones survived, except for parts only of the hands and feet. This was the first of the modern burials to have been uncovered, and the well-preserved, fully articulated state of the inhumation immediately raised concerns about its antiquity. The deceased was oriented west-southwest to east-northeast (240°), head to the west-southwest facing east. There were no grave goods. Approximately 0.07–0.08 m above the grave, a small pile of stones was noted that may have been a covering for this tomb, though the stones extended beyond the area of the tomb.

Grave fill: SU 20: HM ceramics (fine, 3 frr; coarse, 9 joining frr = 22 g; non-diagnostic scraps = 12 g); daub (1 piece = 14 g); 1 bag shell.

TOMB LXXXIX

Pit tomb, single inhumation (infant)

Modern

Figs. 3.295–3.299

2.0E–3.0E, 3.0N–4.0N (high: 108.81 m = 355.91 m ASL; low: 108.75 m = 355.75 m ASL)

Sector 1; SU 94 (grave, including roof tile); 95 (fill); 96 (skeleton)

Excavated as Grave 11 (July 9–14, 2004)

Papadopoulos, Bejko, and Morris 2007: 114–115, figs. 7a–b, d

The tomb pit was oval in plan, measuring 0.48 m long, 0.28 m wide, and 0.10 m deep, cut into tumulus fill. The grave fill was not clearly distinguishable from the surrounding tumulus fill. Within the pit, the body of an infant (perinatal) was laid out in a fully extended supine position, cranium facing up, arms by the side of the body, legs bent at the knees. Most of the bones survived except for parts of the hands and feet. The deceased was oriented southwest-northeast (218°), head to the southwest facing east-northeast. Although there were no grave goods per se, the tomb was covered, first by a roof tile, **TLXXXIX-1**, placed directly over the deceased (**Fig. 3.296b**), and this was in turn covered by a series of small to medium stones (**Fig. 3.296a**). The roof tile (**Figs. 3.298–3.299**) is similar to modern traditional roof tiles from the nearby villages of Ngrançija, Lofkënd, and Gjinoqara.

TLXXXIX-1 (SF 084) **Figs. 3.298–3.299**

Fragmentary roof tile (modern)
PL: 0.324; W (max): 0.217; Th: 0.018–0.026.

Grave fill: SU 95: HM ceramics (fine, 5 frr = 3 g; non-diagnostic scraps = 6 g); daub (2 frr = 6 g); 1 bag shell.

TOMB XC

Pit tomb, double inhumation (two infants)

Modern

Figs. 3.300–3.301

2.0E, 4.0N (high: 108.78 m = 355.78 m ASL; low: 108.75 m = 355.75 m ASL)

Sector 1; SU 107 (grave); 108 (fill); 109, 110 (skeletons)

Excavated as Grave 14 (July 14–15, 2004)

The tomb pit was oval, approaching circular in plan, measuring 0.43 m long, 0.35 m wide, and 0.05 m deep, cut into tumulus fill. The grave fill was a dark brown soil mixed with clay, distinguishable from the surrounding tumulus fill, becoming sandy toward the bottom of the tomb pit. Within the pit, the bodies of two infants (both perinatal) were laid out in fully extended supine positions, crania facing up, arms by the side of the body, legs evidently flexed at the knees; the legs of the southern skeleton (SU 109) were pointing northwest, those of the northern skeleton (SU 110) to the southeast. Most of the bones survived except for parts only of the hands and feet. The deceased were oriented west-southwest to east-northeast (240°), heads to the west-southwest facing east-northeast. The infants, who appear to have been buried holding hands, were interred at the same time, and this contemporaneity suggests that they were siblings, perhaps even twins. No clear stone cover/marker was uncovered, although some stones encountered at a higher level in SU 39 may have partially overlain Tomb CX.

Grave fill: SU 108: Lithics (2 chips, debitage).

TOMB XCI

Pit tomb, double inhumation (two infants)

Modern

Figs. 3.302–3.303

3.0E, 4.0N (high: 108.73 m = 355.73 m ASL; low: 108.70 m = 355.70 m ASL)

Sector 1; SU 120 (grave); 121 (fill); 122, 128 (skeletons)

Excavated as Grave 15 (July 16–17, 2004)

This tomb was stratigraphically located above Tomb XXXVIII (Grave 79). The oval tomb pit measured 0.47 m long, 0.26 m wide, and about 0.10 m deep, and was cut into tumulus fill. The grave fill was not clearly distinguishable from the surrounding tumulus fill. Within the pit, the bodies of two infants were interred: that to the south (SU 122), identified as perinatal, was laid out in a fully extended supine position, cranium probably facing up, arms folded across the lower torso and pelvis, legs more or less parallel to one another. Most of the bones of this skeleton survived except for parts only of the hands and feet. The deceased was oriented west-southwest to east-northeast (240°), head to the west-southwest facing east-northeast. To the north of SU 122, a second individual (SU 128), identified as a preterm infant, was placed next to the left arm of SU 122, one of the arm bones overlying the pelvis of SU 122. This individual was very poorly preserved, consisting of cranial fragments to the west-southwest and vertebrae to the east-northeast on the same orientation as the better preserved SU 122. It was not immediately clear in the field whether this skeleton was disturbed or that its poor state of preservation was due to its being a preterm infant. The excavator was of the opinion that SU 128 was interred after SU 122 and that it was at least partly disturbed. There were no grave goods and no material was recovered from the fill of the tomb pit.

TOMB XCII

Cist tomb, single inhumation (older adult male), with remnant of second individual (infant)

Modern

Figs. 3.304–3.310

2.0E–4.0E, 5.0N (high: 108.57 m = 355.57 m ASL; low: 108.096 m = 355.096 m ASL)

Sector 1; SU 157 (grave); 158 (fill); 159 (skeleton); 159a (infant skeleton); 185 (cut)

Excavated as Grave 23 (July 21–29, 2004)

Papadopoulos, Bejko, and Morris 2007: 114–117, figs. 8a–c

The details of this tomb were similar to those of Tomb LXXXVI (Grave 22). The tomb consisted of a neat, almost rectangular cut, but with the west and east ends rounded, measuring 1.67 m long, 0.37 m wide, and 0.34 m deep. As with Tomb LXXXVI (Grave 22), the cut was particularly clear on the east and west side, where there was no stone lining. The cut was lined on the north and south sides with medium to large stones, all sandstone, set on edge, to a height of one course along the north wall, and mostly two courses along the south wall. Within the cist so defined, the body of the deceased (SU 159), identified as an older adult male aged over 50 years at death, was laid out in a fully extended supine position, head facing up and slightly to the east, but tilted toward the north, arms folded across the lower abdomen, the legs more or less parallel. The cranium was partially propped up by a fragment of a roof tile, **TXCII-2**, on the left (north) side, and there was a small stone immediately to the right (south) of the cranium. The skeleton was oriented west-southwest to east-northeast (255°), head to the west-southwest. As was the case with Tomb 22, the state of preservation of the bone was remarkable, though most of the bones of the left hand were displaced. The tomb was covered with three large sandstone rocks, the interstices filled with a series of smaller and medium-size stones. The greatest concentration of stone, including the two courses lining the south side of the tomb cut, was to the west, encasing the cranium and torso of the deceased. The fill of the grave was generally of a texture looser than the tumulus fill into which the tomb pit was cut. In the process of lifting the skeletal remains, a silver-copper alloy coin, **TXCII-1**, was found directly under the skull, beside the neck of the deceased (**Fig. 3.306d**); copper-alloy staining on the maxilla of the deceased (**Fig. 3.306c**) indicated that the coin was originally placed in the mouth of the deceased, and had slipped into its ultimate resting place below the cranium (the coin was only visible when the cranium was lifted).

In the process of cleaning the skeleton in the laboratory, the remains of a second individual (SU 159a), identified as a prenatal infant, were encountered. These were not noted in situ, and according to Schepartz's notes, the remnants of SU 159a, primarily teeth, were located beneath the cranium and under the thoracic vertebrae of the older adult male (SU 159). A few additional fragments of human bone of another individual or individuals were noted above the capstones of the tomb.

TXCII-1 (SF 153) **Fig. 3.308a–b**

Ottoman coin, silver-copper alloy.
D: 0.013; Wt: 0.2 g.

TXCII-2 (SF 154) **Figs. 3.309a–b, 3.310**

Roof tile fragment (modern)
PL: 0.116; PW: 0.109; Th: 0.014–0.020.

Grave fill: SU 158: Lithics: 1 debitage.

ANIMAL BURIAL, "TOMB" XCIII

Pit tomb, single animal inhumation (sheep/goat)

Modern

Figs. 3.311–3.312

2.0E–3.0E, 5.0N (high: 108.61 m = 355.61 m ASL; low: 108.50 m = 355.50 m ASL)

Sector 1; SU 142 (grave); 143 (fill); 144 (skeleton)

Excavated as Grave 19 (July 21–14, 2004)

Papadopoulos, Bejko, and Morris 2007:114–117, fig. 8d

Together with Tomb LXXXVII (Grave 8), this was one of two graves that contained only an animal. The remains of the animal (SU 144), identified as a sheep (Marston, Chapter 16.1), were placed immediately to the east of the westernmost capstone of the modern adult cist tomb, Tomb XCII (Grave 23). Tomb XCIII

(Grave 19) was located above what was to be the lower left torso of the deceased of Tomb XCII (Grave 23) once the capstones were removed. As encountered, the animal skeleton measured 0.25 m long and 0.12 m wide in situ, with a depth of about 0.11 m. Unlike Tomb LXXXVII (Grave 8), where an articulated skeleton was uncovered, the skeletal remains of Tomb XCIII consisted of the well-preserved and complete cranium of the animal, which was buried first (**Fig. 3.312b**), with the long bones of some of the limbs (lower legs, feet) collected over the skull (**Fig. 3.312a**). It was therefore clear that the animal in Tomb XCIII had been butchered. The cranium of the animal faced east. There were no grave goods associated with the animal burial, and the small quantity of material recovered from the fill immediately around the bone, which was not in any way different from the surrounding soil, was prehistoric.

Tomb XCIII was related to Tomb XCII in a manner similar to the relationship of Tombs LXXXVIII and LXXXVI (Graves 8 and 22). In the case of preparing the tomb pit for the construction of the cist of Tomb XCII, the eastern extent of the prehistoric Tomb LXXXV (Grave 10) had been violated. As was noted in the case of Tombs LXXXVII and LXXXVI (the latter disturbing the prehistoric Tomb LXV [Grave 30]), it is possible that the deposition of the interred animal was more intended to appease the earlier violated burial rather than being directly associated with the later tomb.

Grave fill: SU 143: HM ceramics (semi-coarse, 1 handle fr = 4 g; non-diagnostic scraps = 1 g); daub (2 pieces = 1 g); shell.

Tomb XCIV

Pit tomb, single inhumation (infant)

Modern

Figs. 3.313–3.315

3.0E, 6.0N (high: 108.452 m = 355.452 m ASL; low: 108.40 m = 355.40 m ASL)

Sector 1; SU 163 (grave); 164 (fill); 165 (skeleton)

Excavated as Grave 25 (July 22–23, 2004)

Tomb XCIV (Grave 25) was located immediately to the west of Tomb XCV (Grave 21) and to the north of Tomb XCII (Grave 23); the eastern edge of the tomb was cut by the pit for Tomb XCV (Grave 21), and it is likely that some of the damage to the southern part of the burial was the result of the cut for Tomb XCII (Grave 23). Tomb XCIV is therefore clearly earlier than Tomb XCV (Grave 21) and probably earlier than Tomb XCII (Grave 23). As was the case with the nearby Tomb XCV, the tomb consisted of a pit, oval in plan, measuring 0.56 m long, 0.26 m wide, and about 0.05 m deep, cut into tumulus fill. The grave fill was not clearly distinguishable from the surrounding tumulus fill. Within the pit, the body of an infant aged 6 months at death was laid out in what was probably originally a fully extended supine position, cranium facing up or to the east, but with considerable damage caused by the cut for Tomb XCV (Grave 21) (and perhaps Tomb XCII), with some of the long bones of the arms disarticulated to the right of the cranium, and the leg bones also disarticulated, some in the western portion of the pit for Tomb XCV (Grave 21). The deceased was oriented west-southwest to east-northeast (255°), head to the west-southwest facing east. The only grave good was the silver-copper alloy coin, **TXCIV-1**, found immediately to the southeast of the cranium and originally obscured by the long bones piled up near the cranium. There was no material recovered from the fill immediately around the burial. As encountered, several small stones covered the lower cranium and upper torso of the deceased (**Fig. 3.314a**), and it is likely that additional stones noted in the vicinity were originally displaced from the stones covering the lower body.

TXCIV-1 (SF 133) **Fig. 3.315a–b**

Ottoman coin, silver-copper alloy.
D: 0.014; Wt: 0.2 g.

Tomb XCV

Pit tomb, single inhumation (infant)

Modern

Figs. 3.313, 3.316–3.317

3.0E–4.0E, 6.0N (high: 108.46 m = 355.46 m ASL; low: 108.41 m = 355.41 m ASL)

Sector 1; SU 151 (grave); 152 (fill); 153 (skeleton)

Excavated as Grave 21 (July 21–23, 2004)

Tomb XCV (Grave 21) was located immediately to the east of Tomb XCIV (Grave 25), the northwest edge of the tomb pit cutting into the eastern portion of Tomb XCIV, and only a few centimeters to the north of the cut for Tomb XCII (Grave 23). As the southern edge of Tomb XCIV (Grave 25) appears to have been disturbed by the cut for the tomb pit for Tomb XCII (Grave 23), whereas Tomb XCV, which was closer to Tomb XCII, was unaffected, it seems most likely that of the three burials, Tomb XCIV (Grave 25) was interred first, followed by Tomb XCII (Grave 23), and finally Tomb XCV (Grave 21). The tomb consisted of a pit, roughly oval in plan, measuring 0.51 m long, 0.24 m wide, and about 0.08 m deep, cut into tumulus fill. The grave fill was not clearly distinguishable from the surrounding tumulus fill. Within the pit, the body of an infant (perinatal) was laid out in a fully extended supine position, cranium facing up, arms by the side of the body, legs as preserved parallel to one another, though there was some disturbance to the lower right leg. The deceased was oriented west-southwest to east-northeast (250–255°), head to the west-southwest facing east. There were no grave goods, and no material was recovered from the fill immediately around the burial.

TOMB XCVI

Pit tomb, single inhumation (infant)

Modern

Figs. 3.318–3.319

2.0E–3.0E, 6.0N (high: 108.62 m = 355.62 m ASL; low: 108.57 m = 355.57 m ASL)

Sector 1; SU 146 (grave); 147 (fill); 148 (skeleton)

Excavated as Grave 20 (July 20–21, 2004)

This tomb consisted of a pit, roughly oval in plan, measuring 0.44 m long, 0.26 m wide, and 0.08 m deep, cut into tumulus fill. The grave fill was not clearly distinguishable from the surrounding tumulus fill. Within the pit, the body of an infant (perinatal) was laid out in fully extended supine position, cranium facing up, arms by the side of the body, legs probably bent at the knees, though there was some disturbance of both legs. The deceased was oriented west-southwest to east-northeast (240°), head to the west-southwest facing east. There were no grave goods. The tomb pit was partially stone-lined, with two small stones set on edge, one placed on either side of the torso immediately to the north and south of the arms, which supported a large, round, sandstone block, which covered the burial except for the legs (top of stone = 108.73 m = 355.73 m ASL).

Grave fill: SU 147: daub (3 pieces = 10 g).

TOMB XCVII

Pit tomb, single inhumation (older adult male?), with remnants of second individual

Modern (AMS ^{14}C: cal AD 1810 ± 130)

Figs. 3.320–3.322

1.0E–2.0E, 6.0N–7.0N (high: 108.05 m = 355.05 m ASL; low: 107.90 m = 354.90 m ASL)

Sector 1; SU 248 (grave); 249 (fill); 250 (main skeleton); 250a (remnants of second skeleton); 262 (cut)

Excavated as Grave 39 (July 19, 2004; July 1–7, 2005)

A number of stones, first encountered and recorded in 2004 (**Fig. 3.321a–b**), proved to be part of the stone covering for this grave, overlying the central portion of the deceased. Additional cover stones were encountered when the tomb was fully exposed in 2005 (**Fig. 3.322a**), which covered the right side of the cranium and upper torso as well as the right knee, tibia, and both feet. Although stone-covered, Tomb XCVII was not stone-lined, so it is classified here as a pit tomb rather than a cist. The tomb consisted of a well-defined pit, only slightly larger than the body interred within it, cut into tumulus fill. The pit measured 1.67 m long, 0.57 m wide, and 0.26 m deep (the depth of the skeletal remains being 0.15 m). Within the pit, the deceased (SU 250), identified as an older adult aged 50–60, and probably male, was laid out in a fully extended supine position, the head facing up and slightly to the east, the left arm by the side of the body, the right arm over the right side of the pelvis, and the legs and feet parallel to one another. The skeleton was oriented west to east (about 260°), head to the west. The skeleton was very well preserved, although the bones of the left hand were not in situ, and neither were most of the bones of the right hand. The hyoid bone, deriving from the throat of the deceased, was found by the left foot. The excavator thought that the hand bones were displaced over and around the legs of the deceased. In studying the skeletal remains in the laboratory, however, Schepartz noted teeth and hand and foot bones that may derive from a second individual (designated SU 250a), identified as a mature adult, over 45 years at death, of undetermined sex. Some of these bones must derive from the hands of the SU 250, though there appears to be enough duplication among the bone to suggest the remains of a second individual, almost certainly intrusive. There were no grave goods, and no material was recovered from the fill of the grave, which was clearly distinguishable from the surrounding tumulus fill. Tomb XCVII was stratigraphically located above the prehistoric Tombs LXXVII and LVI (Graves 18 and 43). An AMS ^{14}C date determined from a sample taken and analyzed from one of the phalanges of the hand of the deceased indicated a date of AD 1810 ± 130 years, calibrated.

TOMB XCVIII

Pit tomb, single inhumation (infant)

Modern

Figs. 3.323–3.325

2.0E, 8.0N (high: 108.24 m = 355.24 m ASL; low: 108.13 m = 355.13 m ASL)

Sector 1; SU 236 (grave); 237 (fill); 238 (skeleton); 241 (cut)

Excavated as Grave 36 (June 29–30, 2005)

This was the first of the modern graves to be encountered in the 2005 season. The tomb consisted of an almost rectangular-shaped pit, slightly wider at the west, with the corners rounded, cut into tumulus fill, the grave fill clearly distinguishable from the surrounding tumulus fill. The tomb pit, which measured 0.79 m long, 0.32 m wide, and 0.11 to over 0.20 m deep, was only partially lined by four small to medium-size stones, two on the west and two on the east sides of the grave, on either side of the cranium and legs (**Fig. 3.324b**). These stones helped to support the stone cover, which was a more substantial tomb covering than many of the other infant burials (**Fig. 3.324a**). Within the pit, the body of the deceased (SU 238), identified as an infant aged 3–6 months at death, was laid out in a fully extended supine position, cranium facing up and toward the east, arms gently folded over the lower torso and pelvis, legs parallel to one another. The skeleton was generally very well preserved, except for the feet, of which there was no trace; the right tibia was found displaced near the right elbow. The deceased was oriented west to east (255°), head to the west facing east. The only grave good was the small copper-alloy clasp or hook, **TXCVIII-1**, used to fasten clothing, found in situ on the left shoulder of the deceased. There was no material recovered from the fill of the grave.

TXCVIII-1 (SF 167) **Fig. 3.325a–b**

Small copper-alloy clasp.
H: 0.012; W: 0.009; Wt: 0.3 g.

TOMB XCIX

Pit tomb, single inhumation (adult female)

Modern (AMS ^{14}C: cal AD 1808 ± 141)

Figs. 3.326–3.327

1.0E–2.0E, 8.0N–9.0N (high: 107.86 m = 354.86 m ASL; low: 107.79 m = 354.79 m ASL)

Sector 1; SU 275 (grave); 276 (fill); 277 (skeleton); 285 (cut)

Excavated as Grave 45 (July 7–9, 2005)

The details of this grave were very similar to those of Tomb XCVII (Grave 39). The tomb consisted of a well-defined pit only slightly larger than the body interred within it, cut into what appeared to be tumulus fill. The pit measured 1.56 m long, 0.39 m wide, and just over 0.20 m deep (the depth of the skeletal remains being 0.06 m). Within the pit, the deceased (SU 277), identified as an adult female aged 25–35 years, was laid out in a fully extended supine position, the head facing up and slightly to the east, and tilted slightly toward the southeast, the arms gently folded over the lower abdomen/pelvis, the legs and feet parallel to one another, with the feet turned slightly outward. The skeleton was oriented west to east (about 260°), head to the west. The skeleton was very well preserved except for some of the toe bones that were found scattered over the lower legs. The grave was partially covered by a number of stones as shown, two smaller stones over pelvic region, with a series of larger stones placed over the lower body below the pelvis (**Figs. 3.326a, 3.327a–b**). The rest of the stone covering—if the cranium and torso of the deceased were indeed covered, which remains unknown (cf. Tomb C [Grave 48])—was not preserved. Tomb XCIX, together with Tomb C, was so far north that they may have been affected by the modern plow zone. As was the case with Tomb XCVII (Grave 39), Tomb XCIX, although stone-covered, was not stone-lined. There were no grave goods, and only a solitary prehistoric sherd, together with land snails, was noted in the tomb fill, which was distinctly different from the surrounding earth. An AMS ^{14}C date determined from a sample taken and analyzed from one of the bones of the right hand of the deceased indicated a date of AD 1808 ± 141 years, calibrated.

Grave fill: SU 276: HM ceramics (semi-coarse, 1 fr = 2 g); shell.

Tomb C

Pit tomb, single inhumation (adult female)

Modern (AMS ^{14}C: cal AD 1809 ± 127)

Figs. 3.328–3.330

0.0–2.0E, 9.0N–10.0N (high: 107.78 m = 354.78 m ASL; low: 107.63 m = 354.63 m ASL)

Sectors 1 and 5; SU 299 (grave); 301 (fill); 302 (skeleton); 300 (cut)

Excavated as Grave 48 (July 13–16, 2005)

This was the northernmost of all graves at Lofkënd and at a point off the main tumulus where it may have been affected by the modern plow zone. The tomb consisted of a well-defined pit, almost rectangular in shape but with rounded edges to the west and east, cut into earth rather than tumulus fill. The tomb fill, which was clearly distinguishable from the surrounding earth, consisted of a heavy clayey earth to the west, sandier in the east. The pit measured 1.60 m long, 0.35 m wide, and about 0.20 m deep (the depth of the skeletal remains being 0.15 m). Within the pit, the deceased (SU 302), identified as an adult female aged 39–44 years at death, was laid out in a fully extended supine position, the cranium facing up and slightly to the east, the right arm gently folded across the lower abdomen/pelvis, the left arm by the side of the body, the left hand partially curved over the left femur, the legs and feet parallel to one another. The skeleton was very well preserved except for the disturbed bones of the right foot (probably a natural disturbance, with a mouse hole nearby). The skeleton was oriented west-southwest to east-northeast (250–255°), the head to the west-southwest but facing east. The tomb was partially stone-lined, with two medium stones set on edge, one on either side of the torso, and at least three stones, one of which was large, set on either side of the lower legs, defining a narrow space and forcing the legs of the deceased close together. There were a few additional smaller stones lining the north side of the cut west of the larger stone at the lower legs. These stones were configured as if to carry a stone covering, but a stone covering was not encountered. It is possible that the stones of the tomb cover may have been dislodged by plow action, though this is far from certain, and there was certainly no clear damage indicating a plow. The only grave good was the silver-copper alloy coin, **TC-1**, found in several fragments to the right (south) of the cranium and neck of the deceased (cf. the coin in Tomb XCII [Grave 23]), but in extremely poor condition, the alloy completely mineralized and the shape of the coin and any design no longer legible. There was no material recovered from the fill of the tomb pit. An AMS ^{14}C date determined from a sample taken and analyzed from one of the bones of the right hand of the deceased indicated a date of AD 1809 ± 127 years, calibrated, a date consistent with all of the modern burials.

TC-1 (SF 246) **Fig. 3.330**

Ottoman coin, silver-copper alloy.
D (est.): ca. 0.013–0.014; Wt: 0.2g.

Finds Associated with the Modern Burials

Coins (*Shpresa Gjongecaj*)

TXCIV-1 (SF 133) **Fig. 3.315a–b**

Ottoman coin, silver-copper alloy.
TXCIV-1 (SU 1:0165).
D: 0.014; Wt: 0.2 g.
Originally intact, now reconstructed from three joining fragments. Surfaces much worn, with most of the design illegible.
Border of dots visible on both sides, although little else. One side preserves something of a design, but almost impossible to distinguish with certainty.

TXCII-1 (SF 153) **Fig. 3.308a–b**

Ottoman coin, silver-copper alloy.
TXCII-1 (SU 1.0159).
D: 0.013; Wt: 0.2 g.
Looks intact, but it is clear from photo that it is reconstructed from two joining fragments.

Surfaces much worn, with most of the design illegible.
Possible, but unclear, border of dots on one side. Clear traces of design on both sides, but difficult to discern. On one side the design seems to vaguely resemble a knot?

TC-1 (SF 246) **Figs. 3.329a, 3.330**

Ottoman coin, silver-copper alloy.
TC-1 (SU 1.0302).
D (est.): ca. 0.013–0.014; Wt: 0.2 g.
Originally recovered in three joining fragments, now nine joining fragments, though difficult to reconstruct. Surfaces much worn, with little if anything of the design visible.
Illegible as preserved.

Roof tiles (*John K. Papadopoulos)*

The three following roof tile fragments are all modern, deriving respectively from two of the early nineteenth-century burials in the northeast sector of the tumulus and from the surface of the tumulus. The form and fabric—though not the stamped decoration of SF 161—are standard for the local roof tiles of the modern traditional villages of the area, particularly the nearby Lofkënd, Gjinoqara, and Ngrançija. An intact roof tile (SF 382), together with fragments of a broken roof tile, both from Ngrançija (which date to ca. 1990), were provided by Arben Malaj for comparison; the intact example (SF 382) is catalogued and illustrated below.

The roof tiles are readily distinguishable from tiles of earlier periods (Archaic, Classical, Hellenistic, and Roman, many of which have been noted in the survey; see Chapter 18). The roof tiles presented below are made of mainly light-colored coarse clay with many visible impurities of various sizes. The undersides are crudely finished, while the uppersides bear finger smears and other finishing marks. The tiles appear to be exclusively curved, following the ancient Lakonian tradition; the degree of curvature can vary considerably. Specifically articulated tiles, such as eaves tiles, have not been identified, nor are there any specially made pan tiles. It would appear that the traditional roofing system still used today in many of the older houses and huts of the region had been in use for some time (cf. discussion of the Byzantine and post-Byzantine roof tiles of Torone in Chalkidike; see Cambitoglou and Papadopoulos 2001: 178): a pitched roof supported on wooden beams is covered with large reeds, which are, in turn, covered by a layer of clay/mud, into which the tiles are set. There is no distinction in appearance between pan and cover tiles; the same tile could be used upside-down as a pan tile, or the right-way up as a cover tile.

TLXXXIX-1 (SF 084) **Figs. 3.298–3.299**

Fragmentary roof tile (modern)
Used as a partial cover for Tomb LXXXIX (SU 1.0094).
PL: 0.324; W (max): 0.217; Th: 0.018–0.026.
Reconstructed from three joining fragments preserving substantial portion of roof tile, including portions of both long edges and one complete short edge; condition good.
Coarse tile fabric, with numerous small to very large inclusions of various colors, primarily white and light-colored (0.5–15.0 mm), including light gray and light red, and much silvery mica; numerous blowouts, some quite large. Clay core and surfaces evenly fired close to pink (7.5 YR 7/4).
Cover tile that can also be used as a pan tile upside-down, with both of the long edges and the preserved short edge roughly cut; the tile clearly tapers toward the opposite short edge, which is not preserved. Underside essentially unfinished or only very crudely finished; convex upperside smoothed and crudely self-slipped.
The roof tile is very similar to roof tiles from the traditional nearby villages of Lofkënd, Gjinoqara, and Ngrançija; cf. **3/2** (SF 382).

TXCII-2 (SF 154) **Figs. 3.309a–b, 3.310**

Roof tile fragment (modern).
TXCII-2 (SU 1.0157) part of tomb lining, north side of cranium.
PL: 0.116; PW: 0.109; Th: 0.014–0.020.
Single fragment preserving small portion of roof tile, including one corner of tile; upperside scratched, condition otherwise good.
Coarse tile fabric, with numerous small to large white and light-colored inclusions (0.5–3.0 mm) and much silvery mica; numerous blowouts. Clay core and surfaces fired closest to light reddish brown (5 YR 6/4).
Cover tile that can also be used as a pan tile upside-down, with section of preserved short edge cut perpendicularly, and preserved long

edge obliquely cut. Underside essentially unfinished or only very crudely finished; convex upperside smoothed and self-slipped, though not as well finished as other pieces, with numerous scratches and other blemishes, many of them before firing.

Cf. **TLXXXIX-1** (SF 084) and **3/2** (SF 382).

3/1 (SF 161) **Figs. 3.331, 3.332a–b**

Fragment of stamped roof tile (modern).

Tumulus surface collection.

PL × PW: 0.090 × 0.085; Th: 0.013–0.020.

Single fragment, broken on all sides, preserving small portion of roof tile; condition good.

Coarse tile fabric, with numerous small to large white and light-colored inclusions (0.5–7.0 mm), and much mica, primarily silver, though with occasional golden flakes(?); numerous blowouts, particularly on underside. Clay core and surfaces mostly fired close to light brown (7.5 YR 6/4); closer to light reddish brown (5 YR 6/4) for parts of core.

Cover tile that can also be used as a pan tile upside-down; underside essentially unfinished or only very crudely finished; convex upperside nicely smoothed and self-slipped, and stamped with a line of incuse cross-rosettes (portions of four rosettes preserved), as shown. The cross-rosettes (D: 0.014) are deeply impressed, with a plain X in each; they appear to have been originally in line at the apex of the tile.

The fabric and shape of this fragment are identical with the better-preserved roof tile that formed a partial covering for the modern Tomb LXXXIX. For ancient stamped roof tiles in the Archaic and Classical periods, see, among others, Felsch 1979, 1990.

3/2 (SF 382) **Fig. 3.333a–b**

Roof tile from the village of Ngrançija, ca. 1990.

L: 0.400; W (wider of the short ends): 0.175; W (opposite end): 0.140; Th: 0.010–0.023.

Intact, well preserved, some adhering mortar along both of the long edges.

Coarse tile fabric, with numerous small to very large white and light-colored inclusions (0.5–12.0 mm) and a dusting of fine silvery mica; numerous blowouts. Core, as visible at chips, and surfaces evenly fired close to yellowish red (5 YR 5/6), in places approaching red (2.5 YR 5/6).

Cover tile that can also be used as a pan tile upside-down, with all four edges roughly cut, the shorter edges perpendicularly, the longer edges obliquely cut. The tile tapers toward one of the short edges, which has a prominent, but crude groove on the underside, 0.012 from the edge; there is a broader and much shallower groove, as if finger-impressed, at the opposite short edge on the upperside, right at the edge. Underside essentially unfinished or only very crudely finished, with a few small leaf impressions visible at one point; convex upperside smoothed and crudely self-slipped, with finishing marks running mainly parallel to the long edges.

Other inventoried fragments of roof tiles

SF 268	SU 2.0235 (interface of topsoil and tumulus fill)
SF 285	SU 1.0278 (topsoil)

Other (*John K. Papadopoulos*)

Although clearly modern, the basic form of the small copper-alloy clasp enjoys a distinguished prehistory, which goes back at least to the Italian Late Bronze Age, with related types of bronze clasps found farther north in central Europe. I list a few Bronze Age examples in the catalogue entry below.

TXCVIII-1 (SF 167) **Fig. 3.325a–b**

Small copper-alloy clasp.

TXCVIII-1 (SU 1.0238).

H: 0.012; W: 0.009; Wt: 0.3 g.

Intact; condition excellent.

Small clasp made of continuous copper-alloy wire, circular in section, looped at one end and forming a double hook.

Although this example is clearly modern, the basic type in bronze enjoys a remarkably long prehistory, for which, see, among others, Montelius 1904:pl. 190, no. 17 (Vetulonia); Müller-Karpe 1959:pl. 64, G, no. 6 and p. 68, C, no. 9 (Bologna, San Vitale, Graves 488, 661); pl. 78, V, no. 1; X, no. 1; Z, no. 1 (Bologna, Savena, Graves 45, 54, 130); pl. 183, no. 17, right (Grünwald, Grave 1); cf. also Mason 1996:21, fig. 9, bottom right (Slovenia).

Chapter 4
The Relative and Absolute Chronology of the Tumulus

Brian N. Damiata and John Southon
With contributions by John K. Papadopoulos

Chapter 4.1
The Relative Chronology of the Tumulus

John K. Papadopoulos

Once the final burial of the tumulus had been excavated in the summer of 2007, the 2008 season was largely devoted to the intensive surface survey of the area immediately around the tumulus by Jamie Aprile (Chapter 18), as well as the study and recording of all material from the excavations of the mound. In the course of 2007 and 2008, and before the final accelerator mass spectrometry (AMS) ^{14}C dates were prepared and analyzed by Brian Damiata and John Southon, I was able to arrange the tombs into a relative sequence, as well as into chronological phases, based purely on the stratigraphy of the tumulus (Chapter 2). Although not all tombs were stratigraphically interrelated, many were, and in certain parts of the tumulus as many as four or five tombs overlaid one another. The classic case was that of Tomb I, the grave pit of which lay under no fewer than four tombs, all of them prehistoric (Tombs LXXIX, LXIX, LXIII, XLV [Graves 6, 27, 35, 60]), while the feet of another prehistoric burial, Tomb LXXII (Grave 24), were located right at the edge of the pit for Tomb I and may have also been stratigraphically above it.

In a few cases, the pit of one tomb cut across an earlier burial. For example, the cranium and part of the torso of the individual interred in Tomb XLVII (Grave 41) were cut by the roughly rectangular pit for Tomb LXIV (Grave 61), which was bordered by stones to the east and west sides and was marked by a distinctly darker fill (see **Fig. 3.152**). Although several of the prehistoric burials in the northeastern sector of the tumulus were cut, disturbed, and sometimes substantially damaged by the modern burials, Tomb XLVII was one of the rare cases of a prehistoric burial being cut by another prehistoric tomb. Most of the burials were made either in a very shallow pit or on leveled earth that was subsequently covered by a smaller mound limited to that burial. In a few cases, however, the burial pit was substantial in terms of its depth (e.g., Tombs I [64], XII [88], XIV [71], XV [80], XVI [68], XIX [54], LXIV [61], and cf. Tombs VI [97], VIII [100], XXVIII [77]). Consequently, the relative depth of a burial within the tumulus was, by itself, not an infallible indicator of date.

The stratigraphic interrelation of tombs was predictably greatest in the central portion of the tumulus, whereas those burials toward the edges of the mound—which was, over the course of the centuries of the use of the tumulus, something of a moving boundary—were often in areas where there were no other tombs. This was especially the case for those burials in the southeastern portion of the tumulus. The attempt to place some of the burials located at or near the tumulus edge within a meaningful relative sequence was, in some cases, challenging, especially for those graves that were very poorly preserved, most of which had fallen victim to erosion, particularly along the steeper sides of the mound. In many parts of the tumulus, tombs that were not stratigraphically interrelated were nevertheless found either below, within or in relation to the same stratigraphic unit of the tumulus fill, which thus connected them stratigraphically. All instances of the stratigraphic interrelation of the tombs are recorded and discussed in the individual accounts of each tomb in Chapter 3, and are summarized here as follows (in each case the earlier tomb is mentioned first, followed by those above it; modern burials are underlined):

I (64) XLV (60) LXVIII (35) LXIX (27) LXXIX (6)
II (91) XXV (90)
III (81) XXIV (85) XXXI (86)
VI (97) XXX (70)
XII (88) LX (44)
XIII (49) XLVII (41)
XIV (71) XLIV (65) LXXX (4)
XV (80) XLV (60) LXVIII (13) and LXXII (24)
XVI (68) LXXVIII (5)
XVII (72) LXXXIII (7) and LXXXV (10)
XVIII (72) XLIX (51) LXVI (31) LXXXIII (7)
XIX (54) LXVII (12) LXX (17)
XXI (55) LIX (38) and LXXVI (16)
XXII (47) XLVI (42)
XXIII (56) L (46)
XXIV (85) XXXI (86)
XXVI (74) XLVIII (52) XCII (23) XCIII (19)
XXVII (82) LI (78)
XXVIII (77) LI (78)
XXIX (83) LIII (63)
XXXIV (87) XXXV (84)
XXXVIII (79) XCI (15)
XLIV (65) LXXX (4)
XLV (60) LXXII (24)
XLVIII (52) LXXXV (10) and XCII (23) XCIII (19)
XLIX (51) LXVI (31) LXXXIII (7)
LIV (40) LVIII (37)
LVI (43) LXXVII (18) and LXXIV (29)? XCVII (39)
LXV (30) LXXIX (6)
LXVI (31) LXXXIII (7)
LXXI (28) LXXXIV (2)
LXXVII (18) XCVII (39)
LXXXV (10) XCII (23) XCIII (19)

The incidence of stratigraphically interrelated tombs among the modern burials was limited. The clearest cases of one modern burial above another were the two animal graves (Tombs LXXXVII [8] and XCIII [19]) which were intentionally placed above Tombs LXXXVI (22) and XCII (23), respectively (see Chapter 16.1). The only other instance was the tomb pit of Tomb XCV (Grave 21) cutting slightly into that of Tomb XCIV (Grave 25)—both being infant graves—and this relationship was probably intentional, in an attempt to bury two infants as close to one another as possible. Hence:

LXXXVI (22) LXXXVII (8) = animal burial
XCII (23) XCIII (19) = animal burial
XCIV (25) XCV (21)

On the basis of the stratigraphy of the tumulus (as outlined in Chapter 2), coupled with the evidence of the vertical interrelations of the tombs as presented here, it was possible to arrange the burials into a relative sequence of six broad phases, originally designated Phases I–VI (with I being the earliest). Once the calibrated AMS ^{14}C results were available (Chapter 3.2), these were inserted next to the tombs which were thus dated, and the relative and absolute dates were remarkably in accord with one another. The results are tabulated in **Table 4.1**. The only phases where the relative and the absolute dates were not in clear alignment were the two final prehistoric phases. Ironically, these two last phases of the prehistoric period together yielded the largest number of AMS dates, all of which clustered very close to one another. For this reason, the originally designated Phases V and VI were renumbered Phases Va and Vb, and on account of the evidence of the AMS dating, it is clear that these two phases or sub-phases are both largely confined to the ninth century BC. The cogency of the original relative phases nevertheless remains intact, as it is clear that at least some tombs of the final Phase Vb were stratigraphically located above tombs of Phase Va. The clearest example of such a vertical interrelation of tombs was Tomb LXXXIV (2), which was stratigraphically located above Tomb LXXI (28) of Phase Va. In **Figure 4.1**, all of the tombs are shown graphically according to six prehistoric phases (with Phases Va and Vb shown separately), whereas in **Figure 4.2**, Phase Va and Vb are regarded as one. As can be seen, the overall patterning is clearer in **Figure 4.1**, as Tombs LXXI (28) and LXXXIV (2) are in a more meaningful relative sequence.

It is important to stress that the rendering of each of the tombs in **Figures 4.1** and **4.2** is not only schematic, but that the horizontal extent of each tomb is represented at its minimum. In those tombs where a grave pit was clearly encountered, the horizontal extent of the burial was straightforward. In tombs that were poorly preserved, however, especially those near the eroded edges of the mound, only the extent of human bone that was actually encountered is recorded in **Figures 4.1** and **4.2**. Consequently, the stratigraphical interrelation of tombs was likely much greater than that represented here.

TABLE 4.1 The relative chronology of the Lofkënd tombs by phase

Phase I (Late Bronze Age, 14th–13th centuries BC)		
Tomb I	Grave 64	(Charcoal ^{14}C: 1373 ± 57 cal BC)
Tomb II	Grave 9	(Bone ^{14}C: 1374 ± 58 cal BC)
Tomb III	Grave 81	(Charcoal ^{14}C: 1363 ± 50 cal BC)
Tomb IV	Grave 98	
Tomb V	Grave 96	
Tomb VI	Grave 97	
Tomb VII	Grave 99	
Tomb VIII	Grave 100	
Tomb IX	Grave 94	
Tomb X	Grave 95	
Tomb XI	Grave 93	
Tomb XII	Grave 88	
Tomb XIII	Grave 49	(Bone ^{14}C: 1299 ± 87 cal BC)
Phase II (Late Bronze/Early Iron Age, 12th–11th centuries BC)		
Tomb XIV	Grave 71	
Tomb XV	Grave 80	
Tomb XVI	Grave 68	
Tomb XVII	Grave 72	
Tomb XVIII	Grave 73	
Tomb XIX	Grave 54	
Tomb XX	Grave 50	
Tomb XXI	Grave 55	
Tomb XXII	Grave 47	
Tomb XXIII	Grave 56	(Charcoal ^{14}C: 1070 ± 59 cal BC)
Tomb XXIV	Grave 85	
Tomb XXV	Grave 90	
Tomb XXVI	Grave 74	
Tomb XXVII	Grave 82	
Tomb XXVIII	Grave 77	
Tomb XXIX	Grave 83	
Tomb XXX	Grave 70	
Tomb XXXI	Grave 86	
Tomb XXXII	Grave 89	
Tomb XXXIII	Grave 92	
Phase III (Early Iron Age, 11th–10th centuries BC)		
Tomb XXXIV	Grave 87	
Tomb XXXV	Grave 84	
Tomb XXXVI	Grave 75	
Tomb XXXVII	Grave 76	
Tomb XXXVIII	Grave 79	
Tomb XXXIX	Grave 66	
Tomb XL	Grave 67	
Tomb XLI	Grave 57	
Tomb XLII	Grave 59	
Tomb XLIII	Grave 62	
Tomb XLIV	Grave 65	
Tomb XLV	Grave 60	(Bone ^{14}C: 953 ± 53 cal BC)
Tomb XLVI	Grave 42	
Tomb XLVII	Grave 41	
Tomb XLVIII	Grave 52	
Tomb XLIX	Grave 51	
Tomb L	Grave 46	
Tomb LI	Grave 78	
Tomb LII	Grave 69	
Tomb LIII	Grave 63	
Tomb LIV	Grave 40	
Phase IV (Early Iron Age, late 10th–9th century BC)		
Tomb LV	Grave 53	
Tomb LVI	Grave 43	
Tomb LVII	Grave 58	
Tomb LVIII	Grave 37	
Tomb LIX	Grave 38	
Tomb LX	Grave 44	
Tomb LXI	Grave 34	
Tomb LXII	Grave 32	
Tomb LXIII	Grave 35	
Tomb LXIV	Grave 61	
Tomb LXV	Grave 30	
Tomb LXVI	Grave 31	(Bone ^{14}C: 863 ± 44 cal BC)
Phase Va (Early Iron Age, 9th–earlier 8th century BC)		
Tomb LXVII	Grave 12	
Tomb LXVIII	Grave 13	(Bone ^{14}C: 876 ± 46 cal BC)
Tomb LXIX	Grave 27	(Bone ^{14}C: 863 ± 44 cal BC)
Tomb LXX	Grave 17	
Tomb LXXI	Grave 28	(Bone ^{14}C: 852 ± 44 cal BC)
Tomb LXXII	Grave 24	
Tomb LXXIII	Grave 26	(Bone ^{14}C: 867 ± 45 cal BC)
Tomb LXXIV	Grave 29	(Bone ^{14}C: 805 ± 18 cal BC; 817 ± 21 cal BC; 854 ± 43cal BC; 867 ± 45 cal BC)
Tomb LXXV	Grave 33	(Bone ^{14}C: 845 ± 47 cal BC)
Phase Vb (Early Iron Age, latest phase)		
Tomb LXXVI	Grave 16	
Tomb LXXVII	Grave 18	
Tomb LXXVIII	Grave 5	
Tomb LXXIX	Grave 6	
Tomb LXXX	Grave 4	(Bone ^{14}C: 871 ± 45 cal BC)
Tomb LXXXI	Grave 1	(Bone ^{14}C: 871 ± 45 cal BC)
Tomb LXXXII	Grave 9	
Tomb LXXXIII	Grave 7	
Tomb LXXXIV	Grave 2	(Bone ^{14}C: 845 ± 47 cal BC)
Tomb LXXXV	Grave 10	(Bone ^{14}C: 811 ± 18 cal BC)
Modern burials		
LXXXVI	Grave 22	(Bone ^{14}C: 1810 ± 145 cal AD)
LXXXVII	Grave 8	(Animal)
LXXXVIII	Grave 3	
LXXXIX	Grave 11	
XC	Grave 14	
XCI	Grave 15	
XCII	Grave 23	
XCIII	Grave 19	(Animal)
XCIV	Grave 25	
XCV	Grave 21	
XCVI	Grave 20	
XCVII	Grave 39	(Bone ^{14}C: 1810 ± 130 cal AD)
XCVIII	Grave 36	
XCIX	Grave 45	(Bone ^{14}C: 1808 ± 141 cal AD)
C	Grave 48	(Bone ^{14}C: 1809 ± 127 cal AD)

The stratigraphical interrelation of the prehistoric burials was a key component in the establishment of the relative sequence of tombs, and this was validated by the fact that burials of each successive prehistoric phase were overlain by burials of *all* the subsequent prehistoric phases. Tombs of Phase I were overlain by tombs of Phases II, III, IV, Va, and Vb, and in that order. As the Phase I burials were largely located in the southeastern portion of the tumulus, whereas all of the modern burials were located in the northeast, there was no stratigraphic interrelation between the graves of Phase I and those of the modern period. Tombs of Phase II were overlain by burials of Phases III, IV, Va, Vb, and by some modern burials. Similarly, Tombs of Phase III were overlain by burials of Phases IV, Va, Vb, and by modern graves, and tombs of Phase IV were overlain by burials of Phases Va, Vb, and by those of the modern period. The cogency of this relative phasing was bolstered by the fact that at least one tomb of Phase Va was overlain by a tomb of Phase Vb.

All of the burials described in Chapter 3 have been placed into what I consider to be their relative sequence based on stratigraphy. In the final ordering of the burials, I was not swayed to give priority to AMS dates over the stratigraphic order (nor did I need to be). For example, the final burial of my Phase IV, Tomb LXVI (Grave 31), has an AMS date of 863 ± 44 cal BC, which is precisely the same calibrated date range as what I consider to be the seventh tomb in the relative sequence of Phase Va (Tomb LXXIII [26]). And Tomb LXVIII (13)—which I have placed second in Phase Va—has a date range that may potentially be earlier (of course, it could also be 46 years or so later!). As Brian Damiata and John Southon note in Chapter 4.2, the 2-σ spreads in calibrated dates vary between 36 and 338 years, and are a function of the smoothness and slope of the calibration curve, because, for the most part, the ages do not fall on well-defined portions of the curve.

I am confident with regard to the broad parameters of this periodization and of the six phases, but the placement of one tomb before or after another was in some cases clear-cut and in others much less so. Consequently, if there was a more perfect and precise method of dating each and every burial, I would not be surprised if the order of individual tombs within a particular phase shifted a little, but I doubt that any amendments would be radical. The other area where one might expect a little slippage in the relative ordering of the burials is at the cusps of the phases—especially at the cusp between Phases IV and Va, or that between Va and Vb—but the combination of the stratigraphy of the tumulus, coupled with the robust number of tombs that were interrelated vertically, has greatly minimized any significant reordering.

Before considering the ramifications of this chronology for the internal situation at Lofkënd, and particularly the chronological development of the burial ground as a tumulus, and for the broader implications for the archaeology of Albania, it is important to outline the AMS dating that has provided, together with the AMS results from Sovjan (Lera et al. 2011), a baseline for the absolute chronology of Albania in the later Bronze and Early Iron Ages (see also Damiata et al. 2007–2008).

CHAPTER 4.2

THE ABSOLUTE CHRONOLOGY OF THE TUMULUS: RESULTS OF AMS DATING OF HUMAN BONE AND CHARCOAL SAMPLES FROM THE LOFKËND TUMULUS

Brian N. Damiata and John Southon[1]

INTRODUCTION

An initial inspection of the grave goods and fill material from the prehistoric tumulus at Lofkënd indicated that use of the site potentially spanned from the Late Bronze Age into the Early Iron Age, as well as from medieval to modern times. Given the generally good state of preservation of some of the human skeletal remains (i.e., fully articulated with dense-bone fragments), it was decided that radiocarbon dating would be useful to help define the site-specific chronology, with the broader goal to establish a baseline for the absolute chronology of the region. Toward these aims, a radiocarbon dating project was initiated.

This section summarizes the results of AMS dating of human bone and charcoal samples that were collected from the tumulus. The results of carbon (C) and nitrogen (N) stable-isotope analyses of the bone samples are reported separately in Chapter

7. In total, 37 AMS dates were obtained. The analyses were performed in two batches during January and June 2008, respectively. The results provided the first radiocarbon dates for the project and helped to resolve site-specific chronological questions as well as the absolute chronology of the region (see Chapter 4.1 and 4.3; Damiata et al. 2007–2008).

Methods

Selection of samples

The materials available for analyses were human bone and charcoal. Skeletal remains that had been previously exhumed during the 2004–2007 field seasons and stored at the Apollonia Museum were examined in 2007 to assess their overall state of preservation and suitability for dating. In total, 118 bone samples and 27 teeth were collected from 70 sets of remains. Depending on the state of preservation, multiple samples were collected from some of the remains. When possible, samples were selected from thick portions of dense-bone elements (e.g., tibia and humerus) because these, in general, are less susceptible to contamination and are easier to clean. The samples consisted of approximately 1–2-cm segments that were cut using a thin-blade saw. During the sampling process, the bones were evaluated with respect to their relative hardness when cut and as to whether an odor was emitted, which oftentimes indicates that collagen is present. Relatively hard samples from complete skeletons were given priority for analysis (Batch 1), followed by hard to semi-hard bones from scattered remains (Batch 2). A total of 41 bone samples were selected for treatment.

A total of 16 charcoal samples that were collected during the 2005 and 2006 field seasons were selected also for dating. All pieces were identified by John Marston based on preserved wood anatomy (Chapter 16.1). Each piece of charcoal was weighed, and the visible transverse section was examined with a 7.5–75x zoom stereomicroscope. If no clean section was visible, the wood was broken to reveal a fresh section. Additional diagnostic characters in the radial and tangential sections were examined at 100–400x with an incident-light compound microscope. Wood identifications were then based on comparisons with a collection of experimentally carbonized Mediterranean and European wood and with published wood atlases (Schweingruber 1990; Schoch et al. 2004; Schweingruber et al. 2006). All identifications were made to the genus level, noting that it is not usually possible to distinguish between the woods of different species within a genus, except on phytogeographic grounds (Schweingruber 1990).

Three genera were identified: oak (*Quercus* sp.), maple (*Acer* sp.), and buckthorn (*Rhamnus* sp.). Oak is widely distributed within Albania, and oak woodlands are dominant at elevations below 800–1000 m, often in a mixed forest type that includes maple and buckthorn species (Dida 2003). It is likely that all charcoal was from wood grown locally; the area is still forested today where agriculture is not practiced. The life expectancy of oak and maple trees is on the order of hundreds of years, whereas that of buckthorn is less.

Treatment of samples

All samples were treated at the Keck Carbon Cycle AMS facility, University of California, Irvine. The procedure to convert raw samples into graphite targets for AMS dating involves several steps, depending on the type of material. In general, the treatment protocol for bone samples involves the removal of macroscopic contaminants, chemical cleaning, gelatinization and freeze-drying to isolate collagen, combustion, and graphitization. The protocol for charcoal involves chemical cleaning, combustion, and graphitization. A brief summary of the treatment protocols are given below.

The bone samples were initially cleaned using a Dremel drill and dental cutting tools to scrape off macroscopic contaminants. The extraction of collagen was then performed using the modified Longin method (Longin 1971; Brown et al. 1988). Raw samples (~150 mg) were crushed and initially decalcified by application of 6 cc of 0.5N (normal) HCl for approximately 36 hours (i.e., until the fragments looked translucent). The acid was removed and the samples were neutralized with deionized water. The samples were then gelatinized by using 5 cc of 0.01 N HCl for approximately 12 hours at 60° C, followed by two applications of ultrafiltration (>30 kD)/centrifugation for 20 minutes. To help reduce the chloride content, the filtrate was diluted with deionized water and followed by two more applications of ultrafiltration/centrifugation. The remaining filtrate (approximately 1 cc) was then frozen with liquid nitrogen and allowed to freeze-dry under centrifuge for at least 12 hours. Note that of the 41 bone samples selected for treatment, 20 yielded little or no collagen and were not dated (samples from Tomb V [Grave 96], Tomb XIX [Grave 54], Tomb XX [Grave 50], Tomb XXXIV [Grave 87] [2 samples], Tomb XXXIX [Grave 66], Tomb XL [Grave 67], Tomb XLVI [Grave 42] [2 samples], Tomb LIV [Grave 40], Tomb LVI [Grave 43] [2 samples], Tomb LIX [Grave 38], Tomb LXV [Grave 30], Tomb LXVII [Grave 12], Tomb LXX [Grave 17], Tomb LXXVIII [Grave 5], Tomb LXXXII [Grave 9], Tomb LXXXIV [Grave 2], and Tomb XCII [Grave 23]).

Approximately 2 mg of dried sample were then placed in a quartz tube along with cupric oxide to provide oxygen and silver wire to "getter" any impurities that may adversely impact the graphitization process. The tubes were sealed under vacuum using a gas torch and then combusted at 900° C for 3 hours to generate CO_2 gas. The tubes were then placed on a vacuum line, and the gas was cryogenically moved to a vial containing an iron-powder catalyst. The gaseous sample was converted into graphite via the hydrogen-reduction method by heating to 550° C for 3 hours (Vogel et al. 1984). The graphite was then packed into aluminum sample pellets and analyzed by AMS.

As is outlined further in Chapter 7, gas bench aliquots of the dried collagen were also analyzed for C and N stable-isotopes using a Fisons NA-1500NC elemental analyzer equipped with a Delta-Plus CFIRMS stable-isotope mass spectrometer. A total of 25 stable-isotope measurements were obtained.

The 16 charcoal samples were subjected to an acid-base-acid (ABA) chemical cleaning to remove humic acids (Olsson 1986). The procedure involved application of approximately 6 cc of 1N HCl for 30 minutes, followed by 1N NaOH for 1 hour and then 1N HCl for another 30 minutes—all at temperatures around 70° C. The samples were then neutralized with deionized water. Note that the intermediate application of the base solution was repeated until a clear or slightly tan liquid appeared. Typically, five or six rinses were required for most of the samples, but several of the *Quercus* sp. required 12 or more applications. The dried samples (~2 mg) were then placed in quartz vials with cupric oxide and silver, combusted, and graphitized using the protocol previously described for the bone samples.

Results and Discussion of AMS Dating

Tables 4.2 and **4.3** summarize the results for AMS dating of human bone and charcoal samples, respectively, and the results are also presented graphically in **Figures 4.3** and **4.4**. The samples are identified by their ID number which, for the bone samples only, incorporates the actual grave number, and by the designated UCIAMS lab number. The results are given in terms of uncalibrated ages that were converted to calibrated dates using OxCal 4.1.3 (Bronk Ramsey 2009), which incorporates the IntCal04 atmospheric curve (Reimer et al. 2004). These dates are reported in terms of the 2-σ spread (i.e., range at the 95.4% confidence interval) and the simple mean value for the entire range. For completeness, the results of the calibration are graphically presented in **Figures 4.5** through **4.11**. These results are given in terms of the 2-σ spread in dates, the corresponding percentage of likelihood, and the weighted mean and median values of the probability distribution.

The AMS dating yielded consistent uncalibrated ages as noted by the small spread in ages of 30 and 40 years (i.e., twice the stated standard deviation) for charcoal and bone, respectively. The corresponding 2-σ spreads in calibrated dates, however, vary from 36 to 338 years, and are a function of the smoothness and slope of the calibration curve. For the most part, the ages do not fall on well-defined portions of the curve. Plateaus and reversals, caused by increased atmospheric ^{14}C production, are problematic, as they result in a relatively large spread in a calibrated date even when the AMS measurement is of high precision (e.g., modern samples).

Figures 4.3 and **4.4** graphically present the chronological distribution of calibrated dates for bone and charcoal samples, respectively, with modern dates excluded. Of the 21 bone samples, two date to the Late Bronze Age (LB49A and LB91B), 15 to the Early Iron Age or Late Prehistoric/Protohistoric (LB1E, LB2B, LB4C, LB10A, LB13A, LB26A, LB27A, LB28A, LB29A, LB29B, LB29C, LB29F, LB31A, LB33B, and LB60E) and four are modern (LB22A, LB39A, LB45A, and LB48A). The two samples from the Late Bronze Age have relatively large spreads of 116 and 174 years, as their ages encompass a slight reversal in the calibration curve at ca. 1300 cal BC (see **Fig. 4.8**). Similarly, samples from the earliest phases of the Iron Age lie on a plateau at ca. 900–850

TABLE 4.2 AMS dating results for collagen extracted from human skeletal remains from the tumulus[a]

Sample ID (UCIAMS #)	Field season unearthed	Unit	Element[b]	Tomb	Modern fraction	$\Delta^{14}C$ (‰)	^{14}C age (BP)	Calibrated date[c] (2-σ spread)
LB1E (49539)	2004	14	Lt tibia	LXXXI	0.7121 ± 0.0015	−287.9 ± 1.5	2730 ± 20	916–826 (871 ± 45) cal BC
LB2B (49540)	2004	38	Lt humerus	LXXXIV	0.7176 ± 0.0014	−282.4 ± 1.4	2665 ± 20	892–797 (845 ± 47) cal BC
LB4C (49524)	2004	32	Rt humerus/Radius	LXXX	0.7118 ± 0.0015	−288.2 ± 1.5	2730 ± 20	916–826 (871 ± 45) cal BC
LB10A (49525)	2004	145	Rt femur	LXXXV	0.7200 ± 0.0013	−280.0 ± 1.3	2640 ± 20	829–793 (811 ± 18) cal BC
LB13A (49526)	2004	136	Long bone	LXVIII	0.7111 ± 0.0014	−288.9 ± 1.4	2740 ± 20	922–830 (876 ± 46) cal BC
LB22A (43409)	2004	156	Rt hand	LXXXVI	0.9791 ± 0.0020	−20.9 ± 2.0	170 ± 20	1665–1954 (1810 ± 145) cal AD
LB26A (49527)	2004	169	Rt tibia	LXXIII	0.7124 ± 0.0015	−287.6 ± 1.5	2725 ± 20	911–822 (867 ± 45) cal BC
LB27A(49528)	2004, 2007	174	Lt femur	LXIX	0.7127 ± 0.0014	−287.3 ± 1.4	2720 ± 20	907–819 (863 ± 44) cal BC
LB28A (49529)	2004	?	Lt femur	LXXI	0.7152 ± 0.0014	−284.8 ± 1.4	2690 ± 20	896–807 (852 ± 44) cal BC
LB29A (43410)	2005	210	Rt humerus	LXXIV	0.7214 ± 0.0017	−278.6 ± 1.7	2625 ± 20	823–787 (805 ± 18) cal BC
LB29B (49530)	2005	210	Lt tibia	LXXIV	0.7185 ± 0.0014	−281.5 ± 1.4	2655 ± 20	837–796 (817 ± 21) cal BC
LB29C (43411)	2005	230	Lt tibia	LXXIV	0.7146 ± 0.0016	−285.4 ± 1.6	2700 ± 20	897–811 (854 ± 43) cal BC
LB29F (43412)	2005	231	Rt humerus	LXXIV	0.7124 ± 0.0015	−287.6 ± 1.5	2725 ± 20	911–822 (867 ± 45) cal BC
LB31A (49531)	2005	216	humerus	LXVI	0.7129 ± 0.0015	−287.1 ± 1.5	2720 ± 20	907–819 (863 ± 44) cal BC
LB33B (49532)	2005	223	Lt tibia	LXXV	0.7178 ± 0.0014	−282.2 ± 1.4	2665 ± 20	892–797 (845 ± 47) cal BC
LB39A (43413)	2005	250	Mixed phalanges	XCVII	0.9839 ± 0.0021	−16.1 ± 2.1	130 ± 20	1680–1940 (1810 ± 130) cal AD
LB45A (43414)	2005	277	Rt hand	XCIX	0.9817 ± 0.0024	−18.3 ± 2.4	150 ± 20	1667–1949 (1808 ± 141) cal AD
LB48A (43415)	2005	302	Rt hand	C	0.9851 ± 0.0020	−14.9 ± 2.0	120 ± 20	1682–1936 (1809 ± 127) cal AD
LB49A (43416)	2005	305	Rt hand	XIII	0.6862 ± 0.0015	−313.8 ± 1.5	3025 ± 20	1385–1212 (1299 ± 87) cal BC
LB60E (43417)	2005	352	Lt tibia	XLV	0.7060 ± 0.0015	−294.0 ± 1.5	2795 ± 20	1006–900 BC (953 ± 53) cal BC
LB91B (43418)	2006	516	Rt tibia	II	0.6793 ± 0.0015.	−320.7 ± 1.5	3105 ± 20	1431–1316 (1374 ± 58) cal BC

[a] Radiocarbon concentrations are given as fractions of the modern standard, $\Delta^{14}C$, and conventional radiocarbon age, following the conventions of Stuiver and Polach (1977). All results have been corrected for isotopic fractionation, with $\delta^{13}C$ values measured on prepared graphite using the AMS spectrometer. These values can differ from $\delta^{13}C$ of the original material if fractionation occurred during sample graphitization or the AMS measurement, and are not shown. $\delta^{13}C$ measurements on aliquots of ultrafiltered (>30 kD) collagen are reported in Table 7.1. Sample preparation backgrounds have been subtracted based on measurements of ^{14}C-free whalebone.

[b] Lt = left; Rt = right.

[c] The OxCal 4.1.3 results are given as the 2-σ spread of Bayesian statistical analysis, with the simple mean value for all ranges given in parentheses. Weighted mean and median values of the probability distribution are provided in Figures 4.5–4.8.

TABLE 4.3 AMS dating results for charcoal samples from the tumulus[a]

Sample ID (UCIAMS #)	Date sampled	Trench	Unit	Taxon	Grave[b] (Tomb)	Modern fraction	$\Delta^{14}C$ (‰)	^{14}C age (yr BP)	Calibrated date (2-σ spread)[c]
LC1 (43295)	7/18/2006	2	2	*Acer* sp.	—	0.9810 ± 0.0017	−19.0 ± 1.7	155 ± 15	1667–1945 (1806 ± 139) cal AD
LC2 (43296)	7/17/2006	2	489	*Acer* sp.	[86]	0.9776 ± 0.0016	−22.4 ± 1.6	180 ± 15	1666–1953 (1809 ± 144) cal AD
LC12 (43297)	6/28/2005	2	202	*Quercus* sp.	—	0.6820 ± 0.0010	−318.0 ± 1.0	3075 ± 15	1410–1305 (1358 ± 53) cal BC
LC13 (43298)	7/1/2005	2	244	*Rhamnus* sp.	[37]	0.9743 ± 0.0013	−25.7 ± 1.3	210 ± 15	1651–1955 (1803 ± 152) cal AD
LC14 (43299)	7/12/2005	4	289, 290	?	—	0.6996 ± 0.0013	−300.4 ± 1.3	2870 ± 15	1121–995 (1058 ± 63) cal BC
LC15 (43300)	7/22/2005	4	290	*Quercus* sp.	—	0.6948 ± 0.0010	−305.2 ± 1.0	2925 ± 15	1211–1049 (1130 ± 81) cal BC
LC17 (43301)	6/24/2006	4	286	*Acer*	—	0.6886 ± 0.0010	−311.4 ± 1.0	2995 ± 15	1311–1131 (1221 ± 90) cal BC
LC18 (43302)	6/27/2006	4	286	*Acer*	—	0.6918 ± 0.0010	−308.2 ± 1.0	2960 ± 15	1261–1126 (1194 ± 68) cal BC
LC20 (43303)	7/5/2006	2	395	*Acer*	—	0.9729 ± 0.0014	−27.1 ± 1.4	220 ± 15	1648–1954 (1801 ± 153) cal AD
LC22 (43304)	7/10/2006	2	453	*Quercus* sp.	81 (III)	0.6811 ± 0.0010	−318.9 ± 1.0	3085 ± 15	1413–1313 (1363 ± 50) cal BC
LC23 (43305)	7/12/2006	2	475	?	—	0.6625 ± 0.0010	−337.5 ± 1.0	3305 ± 15	1623–1527 (1575 ± 48) cal BC
LC27 (43306)	7/21/2006	1	361	*Quercus* sp.	64 (I)	0.6794 ± 0.0010	−320.6 ± 1.0	3105 ± 15	1430–1316 (1373 ± 57) cal BC
LC28 (43307)	7/21/2006	5	279	?	—	0.7363 ± 0.0010	−263.7 ± 1.0	2460 ± 15	753–416 (585 ± 169) cal BC
LC29 (43308)	7/21/2006	5	279	*Quercus* sp.	—	0.7327 ± 0.0010	−267.3 ± 1.0	2500 ± 15	768–544 (656 ± 112) cal BC
LC30 (43309)	7/25/2006	7	228, 531	?	—	0.6963 ± 0.0011	−303.7 ± 1.1	2910 ± 15	1192–1021 (1107 ± 86) cal BC
LC34 (43310)	7/22/2005	4	329	*Quercus* sp.	56 (XXIII)	0.6975 ± 0.0010	−302.5 ± 1.0	2895 ± 15	1129–1011 (1070 ± 59) cal BC

[a] Radiocarbon concentrations are given as fractions of modern standard, $\Delta^{14}C$, and conventional radiocarbon age, following the conventions of Stuiver and Polach (1977). All results have been corrected for isotopic fractionation, with $\delta^{13}C$ values measured on prepared graphite using AMS. These values can differ from $\delta^{13}C$ of the original material if fractionation occurred during sample graphitization or the AMS measurement, and are not shown.

[b] Bracketed grave numbers indicate dating of fill, which is modern and does not correlate with interpreted date of grave based on grave goods and stratigraphic position.

[c] The OxCal 4.1.3 results are given as the 2-σ spread of Bayesian statistical analysis, with the simple mean value for all ranges given in parentheses. Weighted mean and median values of the probability distribution are provided in **Figures 4.8–4.11**.

cal BC, resulting in spreads of approximately 90 years. Samples from the later phases of the Iron Age have spreads of approximately half of this value. As expected, the modern samples have the largest spreads.

Of the 16 charcoal samples, three are associated with graves (consistent with grave goods and stratigraphic position) and date to the Late Bronze Age (LC22 and LC27) and the Early Iron Age (LC34). Two other samples (LC2 and LC13), which were collected from fill in close proximity to graves, are modern and not reflective of the date of the graves themselves as deduced from grave goods and stratigraphic position. The chronological implications of the AMS dating are discussed in Chapter 4.3, as well as in Damiata et al. (2007–2008).

Chapter 4.3
Ramifications for the Chronology of Southern Illyria in the Bronze and Early Iron Ages

John K. Papadopoulos

The Internal Situation at Lofkënd

The combination of the relative phases and absolute dates presented in **Table 4.1** permits several observations to be made as to the chronological development of the burial ground. First and perhaps foremost, the vast majority of the AMS dates from the Lofkënd burials are based on human bone collagen. In contrast, AMS dates from charcoal samples are limited to two tombs of Phase I and one tomb of Phase II. Moreover, the close synchronism of the AMS dates from human bone collagen with those from charcoal is noteworthy. This is significant, as it minimizes, if not eliminates, the problem of the "old wood" effect that has recently plagued the absolute chronology of the Aegean Late Bronze and Early Iron Age, not least for the important stratified settlements sites of Assiros Toumba and Kastanas in northern Greece (Newton, Wardle, and Kuniholm 2003:185; Wardle, Newton, and Kuniholm 2007: 495–497; Hänsel 1989; Willkomm 1989; and especially Weninger and Jung 2009: 374–380; see further Jung 2006; Toffolo et al. 2013).

In terms of the chronological development of the tumulus as a place of burial, **Figures 4.1** and **4.2** clearly highlight the earliest tombs of Phase I. With the exception of Tomb I, located in the area that was to become the central portion of the mound, and at the highest point in the bedrock, the earliest burials are all clustered close to one another, but with no overlap, in the southeastern segment of the tumulus. Several of these graves, not least Tombs IV and V (Graves 98 and 96), are located at the lowest portion of the bedrock outcrop that formed the base of the later tumulus. The concentration of these tombs to the southeast may not be fortuitous, for this was, as it remains, the warmest sector of the mound, well located in regard to the earliest morning light and attracting the greatest amount of sunshine and heat throughout the year. We may never know what inspired the early Lofkëndians to place their graves in this particular spot, but they certainly chose a warm place.

With Phase II, there is a significant expansion of graves to the north and west of the original cluster, and it is during this phase that the tumulus begins to assume its characteristic "roundness," though it is never a true circle. And this trend is amplified with the burials of Phases III and IV, with the prehistoric tombs of Phase IV expanding in the area farthest to the north, especially Tombs LV (53), LVI (43), and LVII (58), which I would place as the earliest of the burials of this phase. With Phase Va, there is a consolidation of tombs in the central portion of the mound, although at least two burials, Tombs LXXIV (29) and LXXV (33), are also located toward the northern reaches of the tumulus. Of these, the most remarkable is Tomb LXXIV (29), the burial of three adult males without grave offerings, so unique in the context of the tumulus that the date of the tomb—ancient or modern—could not be determined until AMS dating came to the rescue and established it as one of the latest burials at Lofkënd (see further Chapter 21). The last burials in the tumulus, those of Phase Vb, further consolidated the central portion of the mound, being located as they were to the north and south of Tomb I, the putative "central" grave. Changes in the shape of the mound through time are also illustrated with three-dimensional models of the mound prepared by the Experiential Technology Center at UCLA (Chapter 19).

When the relative sequence of tombs is combined with the AMS dates, one feature stands out: the relative paucity of AMS dates from tombs of Phases II, III, and IV. There is only a single AMS date for the burials of Phase II, Tomb XXIII (56), which is more or less in the middle of the Phase II tombs, and there is a solitary date also for Phase III (Tomb XLV [60]) and Phase IV (Tomb LXVI [31]). In the latter case, it is, within the sequence, the last of the burials of Phase IV. In contrast, the number of AMS dates from Phases I, Va, and Vb, is considerably more robust. Four of the 13 tombs of Phase I have good absolute dates, the earliest in the fourteenth century, the latest in the fourteenth or thirteenth century BC. With both Phases Va and Vb, there are even more dates: six of the nine tombs of Phase Va have AMS dates, and 4 out of 10 tombs of Phase Vb can also be dated. The presence/absence of adequate quantities of human bone collagen was, of course, subject to the vicissitudes not only of tomb preservation—the bone from many of the tombs near the eroded edges of the tumulus often being the least well preserved—but also of bone preservation, as this varied from tomb to tomb.

What is remarkable about the absolute dates of the tombs of Phases Va and Vb is the fact that they all fall into the ninth century BC. One or two tombs may plausibly date to the very end of the tenth century, but only just, while the very latest burials, especially Tombs LXXXIV (2) and LXXXV (10), may fall into the first few years of the eighth century BC but no later. If such dating is not an accident of survival and preservation, then there appears to be something of an increase in the quantity of burials in the ninth century BC. There are at least 19 out of 85 prehistoric burials dating to the last 100 years of the tumulus, and this figure may be even greater if some of the Phase IV burials also date to the early ninth century BC, as, for example, Tomb LXVI (31) appears to do (this is the last in the sequence of the Phase IV tombs, and it has a date of 863 ± 44 cal BC). In contrast, the remainder of the 65–66 prehistoric burials covers a period of some 500 years. Put another way, almost one-quarter of the prehistoric burials, some 23%, fall into the last century of the tumulus. Such a quickening in the pace of death in the final century of the use of the mound is significant, as it may well have contributed to the demise of the population that buried their dead in this mound. The ramifications of this observation are considered more fully in Chapter 21 and in the Epilogue.

It goes without saying that the absolute chronology of the Lofkënd tumulus would be on a much more secure footing with additional dates from tombs of Phases II, III, and IV. Despite the comparative paucity of AMS dates from the central periods of the use of the burial mound, the chronology of the beginning and the end of the tumulus is secure and beyond reasonable doubt. The earliest burials fall into the earlier stages of the fourteenth century BC, contemporary with Mycenaean palatial society in the Aegean and with the closing stages of Phase III d1 and the beginnings of Phase III d2 at Maliq (cf. Sovjan Phases 5 c2 and 5 c1), and it is even possible that the earliest burial belongs to the very end of the fifteenth century BC. At the other end of the time scale, the last of the prehistoric occupants of the Lofkënd tumulus were interred sometime around 800 BC. In the context of the conventional chronology of Albanian archaeology, the early date of the first burials was perhaps an even greater surprise than the relatively early date for the demise of the tumulus.

The Absolute Chronology of the Late Bronze and Early Iron Ages of Albania

In a preliminary report on the excavations of Lofkënd published in 2007, which dealt only with the material from the 2004 and 2005 seasons, we were not in a position to speculate as to the earliest of the burials of the tumulus, since only about two-thirds of the burials in the mound (some 62 burials) had been investigated at the time. The discovery, however, of fragments of Corinthian pottery in the topsoil and upper levels of the fill of the tumulus, now fully published by Sarah Morris (Appendix 2, following Chapter 9), and dating to ca. 600 BC, led us to pen the following statement: "These fragments of Corinthian pottery, almost exclusively from a few kotylai, are among the latest securely dated material from the tumulus prior to the modern reuse, and it seems tempting to suppose that the character of human activity in the landscape of Lofkënd was disrupted by the coming of the Greeks" (Papadopoulos, Bejko, and Morris 2007:138).

Although the fragmentary Corinthian pottery was not encountered in any tombs, it seemed reasonable to suppose that the end of the tumulus may have been connected somehow to the establishment of the colony at Apollonia sometime around 600 BC (for which see Graham 1964:130; Stocker and Davis 2006; Papadopoulos, Morris, and Bejko 2007:138). With regard to the burials excavated in the course of the first two seasons, on the basis of the conventional dating of material in Albanian archaeology, the earliest burials seemed to belong to ca. 1100 BC. The latter date was, in part, based on the earliest appearance of iron in southern Illyria, which is generally assumed to be in the eleventh century BC (Prendi 1982:229; see also Prendi 1975 for an overview of the Early Iron Age in Albania), although it is now clear that the earliest burials at Lofkënd contained no iron grave goods.

In order to broaden out the chronological horizons of the project, we decided to include, in the first phase, 32 samples from human bone, human tooth, and charcoal from Lofkënd and Apollonia (fully published in Damiata et al. 2007–2008; for Apollonia, see further Damiata and Southon 2010). Lofkënd provided samples that were originally thought to date to the Early Iron Age, as well as those of the medieval/modern period, whereas the samples from Apollonia were from tombs dated by their contents to the Early

Bronze Age, the Archaic and Classical periods, as well as medieval/modern burials, the latter considered roughly contemporary with those of Lofkënd. The samples analyzed provided, at the time, an unparalleled chronological span and helped to anchor the absolute chronology of several critical periods, now augmented by the first published ^{14}C dates from the settlement at Sovjan (Lera, Oberweiler, and Touchais 2011:49–51, figs. 24–25; for the earlier concordance between Maliq IIIa and Konispol Va, dated by ^{14}C, see Korkuti et al. 1996:197; Korkuti 1998:43).

The AMS dates from charcoal and bone samples at Lofkënd have fine-tuned the chronology of the tumulus and raised the dates of the earliest burials to the Late Bronze Age. The earliest burials are, at least, some 300 years earlier than previously estimated, based on the few finds they held, together with those difficult to date for their lack of grave offerings. To begin with the AMS samples taken from human bone collagen, one of the most important results is the date of the enigmatic Tomb LXXIV (Grave 29), which yielded no grave goods and which could not be firmly assigned on the basis of stratigraphy or tomb type to either the prehistoric or modern period with any conviction. Bone samples taken from each of the three individuals interred in the tomb all yielded ^{14}C dates in the ninth century BC (the ranges were remarkably consistent: 805 ± 18 cal BC and 817 ± 21 cal BC from one skeleton; 854 ± 43 cal BC; 867 ± 45 cal BC), and an Early Iron Age date for this burial is now assured. Another burial, Tomb XLV (Grave 60), clearly Early Iron Age on the basis of both its stratigraphy and an associated iron spearhead (**11/118**), yielded a tenth-century ^{14}C date (953 ± 53 cal BC). Two tombs were clearly earlier on the basis of AMS dating—Tombs XIII (Grave 49) and II (Grave 91)—neither of which had any associated grave goods. Tomb XIII is now dated to the fourteenth or thirteenth century BC (1299 ± 87 cal BC), and Tomb II dates to the fourteenth and perhaps even the very late fifteenth century BC (1374 ± 58 cal BC). These dates are broadly contemporary with Late Helladic IIIA1 and IIIB in the Aegean, and if the tomb dates to the earlier part of the assigned date range, it may even be contemporary with Late Helladic IIB (a date very close to the Early Mycenaean Vapheio cup from Pazhok; see Cabanes et al. 2008:39, fig. 10, no. 6).

A similar fourteenth- or late fifteenth-century ^{14}C date derives from a charcoal sample from what may be the earliest burial in the tumulus, Tomb I (Grave 64) (1373 ± 57 cal BC). This sample does not derive from human bone, but from a small piece of oak charcoal (*Quercus* sp.) found in the fill of the tomb pit; the burial itself may well be earlier or later. Charcoal samples from the fill of Tomb III (Grave 81) provided a similarly early date (1363 ± 50 cal BC), and a twelfth- or eleventh-century BC ^{14}C date was determined for the charcoal sample from Tomb XXIII (Grave 56) (1070 ± 59 cal BC); there were no grave goods in either burial.

Other charcoal samples from various soil units in the general tumulus fill provided similar dates. Of these, there is one charcoal sample that yielded a ^{14}C date of 1358 ± 53 cal BC, but the earliest (LC23) is dated to the sixteenth or perhaps even the seventeenth century BC (1575 ± 48 cal BC). Given the fact that there was, in the fill of the tumulus, Bronze Age material that stylistically predates much of the tomb material, as well as lithics that are considerably earlier (Paleolithic and Mesolithic), these burned samples may have made their way into the fill of the tumulus as part of the more general process through which such earlier material was brought to the site (this is more fully discussed in Papadopoulos 2006; Papadopoulos, Bejko, and Morris 2007:129–130; Chapters 13, 14, and 20).

The two latest charcoal samples from the fill of the tumulus, although falling into a problematic interval of the radiocarbon calibration curve, provided the following dates: 656 ± 112 cal BC and 585 ± 169 cal BC. These are unfortunately broad date ranges, but the general thrust of this evidence accords very nicely with the date of the conventional chronology assigned to the Corinthian kotylai fragments as dating to the sixth century BC. It must be stressed, however, that the latest ^{14}C date from human bone at Lofkënd is in the ninth century BC, and it is possible that this later material represents activity at the mound shortly after (150–250 years) its period of primary use. For the Archaic and Classical periods, the tumuli of Apollonia have provided some corroborating AMS dating evidence, though it is limited to only two samples from human bone (Damiata et al. 2007–2008:160–162; Damiata and Southon 2010:331, table 7.1).

The remainder of the graves at Lofkënd were prehistoric and, on the basis of small finds associated with individual burials, most, if not all, of the tombs apart from those of Phase I and the earliest part of Phase II could be assigned to various stages of the Early Iron Age on the basis of the conventional chronology. The grave goods included various types of handmade pottery common in southern Illyria and northwest Greece in the Early Iron Age (Chapter 9). Although some of the pottery may be assigned to an earlier or later phase in the period of use of the tumulus, whether on the basis of stratigraphy or comparison to material at sites in Epirus (especially Vokotopoulou 1986; Douzougli and Papadopoulos 2010), Macedonia (Heurtley 1923–1925, 1925, 1926–1927, 1927, 1929, 1939; Heurtley and Hutchinson 1925–1926; Heurtley and Radford 1927–1928, 1929–1930), and western Macedonia (unpublished material from Aiani includes matt-painted pottery found in association with both Mycenaean and Early Iron Age material: for preliminary views, see Karamitrou-Mentessidi 2008), the absolute chronology of this pottery is still far from precise. The dates from Lofkënd help anchor this material, at least that from tombs, to one or other phases outlined in **Table 4.1**. Similarly, the various gold/electrum, bronze, iron, and bimetallic jewelry and other implements, as well as those in other materials, such as faience, glass, semi-precious stone, bone, and terracotta (Chapter 10) found in burials that were previously dated only in very broad terms, can now be assigned a more secure date.

The new AMS dates from Lofkënd, together with those from Sovjan (Lera, Oberweiler, and Touchais 2011), allow us to go a little further. Among other things, the prehistory of pins—whether dress pins or hair pins in bronze and bone—can now be taken back well into the Bronze Age in Albania; they do not first appear in the Early Iron Age, as earlier studies suggested (see detailed discussion in Chapter 10). Moreover, some old "rules of thumb"—such as fibula pin shafts that are rhomboidal in section being generally an indicator of late date—fall by the wayside. And the appearance of iron in southern Illyria is now contemporary with that in Greece, dating as it does to the closing stages of the Bronze Age and the transition to the Early Iron Age.

More than this, the new evidence from Lofkënd and Sovjan helps fill in not only a chronological lacuna, but a regional one as well, and for the first time the absolute chronology of the period ca. 1700–1000 BC in Albania can be compared to that of the Aegean and Italy (see Bartoloni and Delpino 2005; Jung 2006; Harding 2007; Toffolo et al. 2013). These dates, together with future determinations, will provide much needed additional material to compare with that from Rocavecchia and Livorno in Italy (for a useful summary, see Jung 2006; Weninger and Jung 2009), with that from the Aegean (Jung 2006; Toffolo et al. 2013), and with the material from the Swiss and southern German lake-side settlements, which have provided robust dendrochronological dates (Weninger and Jung 2009; Papadopoulos, Damiata, and Marston 2011:195–198; dendrochronological studies of timber in Albania are limited to the period after AD 1200—see Westphal et al. 2010). In this context, the possible synchronization of Level 3 at Hauterive-Champréveyres at Lake Neuchâtel, with dates between 1054 and 1037 den BC and timbers from the first phase dating to 1071 and 1034/35 den BC (Weninger and Jung 2009:390) may finally put the transition from the Late Bronze Age to the Early Iron Age in the southern Balkans and the Italian peninsula on much firmer ground.

The raising of the dates of the earliest tombs at Lofkënd is matched, in a much earlier period, to a raising of the absolute dates for the Early Bronze Age from the very late third or earlier second millennium BC to a more secure date around the middle of the third millennium BC, thanks to AMS dating of the prehistoric tombs in Tumulus 10 at Apollonia (Damiata et al. 2007–2008:155–156). Such a date pushes back the Bronze Age prehistory of this part of Albania to an even more remote past, barely represented thus far in this region. As for the latest burials at Lofkënd, the 800 BC date becomes something of a watershed in the later prehistory of Albania. First of all, it decouples the demise of the tumulus from the establishment of the Greek colony at Apollonia. Secondly, it matches, almost precisely, the initial foundation of Early Iron Age sites such as Grunas, Shala, located in the high mountains of north Albania, near the border with Montenegro, a place of refuge during many periods (Galaty, Lafe, and Tafilica 2011:13). What was going on in Albania ca. 800 BC is a question that will be debated for years to come (for further discussion in this volume, see Appendix 2, following Chapter 9, and the Epilogue).

As for the latest reuse of the tumuli at both Lofkënd and Apollonia, the background to this period has already been noted (Chapter 3.2). It was hoped that AMS radiocarbon dating from these late graves, generically identified as "medieval/modern," would establish a more precise date within the period between the fifteenth and twentieth centuries

AD. To this end, two samples from Apollonia were taken (from Tombs 29 and 30) and four from Lofkënd (Tombs LXXXVI [22], XCVII [39], XCIX [45], and C [48]), all from human long bones. The results were unequivocal and neatly clustered to the period AD 1800–1810 but with a broad date range of between 127 and 155 years. The Apollonia samples yielded the following dates: 1803 ± 152 cal AD (Tomb 29), 1800 ± 155 cal AD (Tomb 30); Lofkënd: 1810 ± 145 cal AD (Tomb LXXXVI), 1810 ± 130 cal AD (Tomb XCVII), 1808 ± 141 cal AD (Tomb XCIX), and 1809 ± 127 cal AD (Tomb C). Two further charcoal samples from Lofkënd provided identical dates: one of these (LC13), identified as buckthorn wood (*Rhamnus* sp.), gave a date of 1803 ± 152 cal AD; a second charcoal sample from the upper fill of the tumulus yielded a date of 1801 ± 153 cal AD. These helped provide an absolute (if broad) frame of reference for the later graves at Lofkënd, and attribute them to the Ottoman period, thus assisting in the confirmation of the suspected dates of the coins we found. Moreover, this time frame also led us to critical Ottoman records, which allow a tentative identification of the last occupants of the Lofkënd tumulus as early modern, minority Christians, who sought a quiet place to bury a few of their community (Chapter 3.2).

Acknowledgments

Travel and lab research for this project were funded by the Steinmetz Chair in Classical Archaeology and Material Culture at UCLA. We thank Xiaomei Xu for performing the stable-isotope measurements and Will Beaumont and Kiyo Beverly for their technical assistance.

Chapter 5

Conservation at the Lofkënd Archaeological Project, 2004–2008

Vanessa Muros

Introduction

The conservation of excavated artifacts and their long-term preservation have always been an essential component of the Lofkënd Archaeological Project. The work of the conservators and their collaboration with the archaeologists and other specialists on the project has played an important role not only in the preservation of material but also in the study of the excavated finds. It was the goal of the conservation team to ensure the stability and safe storage of the archaeological material of Lofkënd to allow for research and publication, and in turn to aid in the interpretation of the site and of greater prehistoric Albania.

This report summarizes the objectives and treatment procedures of the Lofkënd conservation team during the 2004–2008 field seasons. The aim of this report is to provide general information on the conservation approaches undertaken. The conservation of any artifact is specific to that particular object, and this report is not meant to serve as a manual or handbook for the conservation of finds at other sites. The hope is that others reading this report can take away information on the overall approach to various preservation issues encountered in the field and to adapt some of the methods described for the conservation of archaeological material at other excavations. The report deals mainly with the conservation of inventoried small finds associated with the graves and not necessarily the bulk ceramic sherds from fill or modern materials found within or on the tumulus.[1]

Laboratory Facilities and Conservation Materials

The first conservation lab at the Lofkënd Archaeological Project was established in 2004 during the initial season within rooms of the Byzantine monastery and museum located at the Apollonia Archaeological Park. The park is also the location of the excavation house of the project where team members lived and worked.

During the first season, some treatments were undertaken, but the primary goal for the conservators at this early stage was to document and examine objects, undertake some minor treatments, and pack and store the artifacts safely to be treated the following season. As with any project, the conservation treatments and approaches evolved and changed as more material was excavated and the conservators adapted their approaches based on the condition of the objects and the materials available. The work during the first season helped to guide future treatments and establish protocols for examination and storage that were carried through all the subsequent seasons. Despite changes that may have been made to the approaches for conserving artifacts in the lab, the overarching goal of the conservation team was always the same: to aid the archaeological interpretation of the site by examining and treating materials, to ensure the preservation of the excavated material, and to implement the highest standards of treatment possible in a field situation, similar to those practiced in museums and other institutions.

[1] "Bulk finds" refer to all non-inventoried material, largely fragmentary pottery and daub. All of this material was washed, quantified, and weighed, and pieces requiring conservation were stabilized.

Problems with the logistics of the first lab space located within the Apollonia museum, such as easy access to running water and expansion of the project to include other specialists, creating the need for additional lab spaces, caused the conservation lab to be moved to the excavation house in 2005. This house was equipped with a room that had been previously used as a photographic darkroom and provided ample space for various conservation activities. The new lab also had a source of running water that allowed the setup of a deionized water column. The basement of the excavation house became the location of the conservation lab throughout subsequent seasons (**Fig. 5.1**).

From the initial season of the project, the conservation team ensured that all materials used for the treatment of excavated objects at Lofkënd were archival quality and conservation grade—that is, inert, not prone to off-gassing, and having good ageing properties. Materials were brought primarily from the United States, but some were purchased in Greece and Albania. Although the project was located in a relatively remote area of Albania with little or no access to conservation suppliers, careful planning prior to commencement of the field season enabled the conservators to maintain a high standard of treatment and storage protocols in the lab. A list of some of the conservation materials used is supplied at the end of this chapter.

CONSERVATION TRAINING

Beginning in 2005, conservation duties at the excavation expanded to include the training of conservators. Training was extended to students currently enrolled in graduate conservation programs as well as undergraduate students with an interest in the field. These interns provided much needed help in the treatment of objects, and also added a new dimension to the work undertaken in the conservation lab.

EXAMINATION AND MATERIALS IDENTIFICATION

The conservation process at Lofkënd began with the examination of finds and identification of materials and techniques used in their manufacture. Initial identification of materials was performed in situ by the archaeologists, and the conservators confirmed or augmented findings with additional technological information observed in the lab. Examination was primarily carried out using a binocular microscope. In the second season, ultraviolet (UV) fluorescence examination was included in the process.

In addition to examination techniques, microchemical testing (Odegaard, Carroll, and Zimmt 2005) was performed to aid in material characterization. Using chemicals and reagents that could be purchased from suppliers in Tirana, the conservators were able confirm previous material identifications. Microchemical testing was also used to assess the condition of objects and identify the nature of deterioration products observed on the artifacts. Implementation of microchemical testing as part of the conservation treatment protocol not only aided in assessing the condition of objects, but also guided conservation treatment decisions.

TREATMENT OF OBJECTS

All inventoried finds excavated from the tumulus underwent some sort of conservation treatment, whether the methods were interventive (e.g., cleaning and reconstruction) or preventive (packing and storage in microenvironments). Treatments were designed to stabilize the finds and to prevent any further deterioration of objects, while at the same time providing the archaeologists with the maximum amount of information possible for use in analysis of the artifacts and site. In the first season, conservation treatments involved some cleaning and reconstruction, primarily of ceramic finds, and limited cleaning of metal finds. The work focused as well on the safe packing/storage of artifacts and the creation of desiccated microenvironments for the metals. This first season allowed for the conservation team to get a feel for the types of finds to be expected in subsequent campaigns to aid in the planning of future conservation and storage needs. Beginning in 2004 and continuing during each subsequent season, all treatments and approaches were reviewed to assess the outcomes and to determine whether modifications were needed to ensure further the long-term preservation of the finds. In addition to reviewing individual treatments from previous seasons to determine their effectiveness and success, general condition issues and treatment approaches were also evaluated to decide whether standardized treatments could be established for objects with similar condition problems.

Ceramics

All excavated ceramic material, including complete vessels (whether intact or not), decorated sherds, and diagnostic fragments, were inventoried and then brought to the conservation lab for examination and possible treatment. Although each treatment approach taken was determined by the condition of the individual object, in general all sherds were cleaned to allow them to be photographed and drawn. Cleaning approaches varied, again according to the condition of the object. They included both dry methods (mechanical cleaning) and wet methods (swabbing with water or chemicals). Fragmentary vessels were reconstructed using conservation-grade resins, such as Paraloid B-72. This resin was also used as a consolidant for friable ceramic materials. Structural fills were used on missing sections of partial or whole vessels to stabilize the piece when necessary, and fills were toned with acrylic paints to integrate the vessel for publication purposes (**Fig. 5.2**).

After microchemical testing was introduced into the treatment protocol in the 2005 season (O'Grady and Bardho 2005:2), the ceramic finds were often tested for the presence of common salts found on archaeological ceramics, such as chlorides, nitrates, sulfates, and carbonates. The results of the tests helped to answer questions about the condition of the object and the burial environment, in addition to guiding conservation treatment decisions.

Geological studies undertaken of the tumulus fill (Foss, Chapter 16.4; Papadopoulos, Bejko, and Morris 2007:140–144), combined with microchemical and pH tests performed in the conservation laboratory (O'Grady and Bardho 2005:2–3), indicated that the tumulus burial environment was slightly alkaline with calcareous deposits prevalent within the tumulus. The presence of these materials within the soil was evident from the white accretions found on much of the pottery and identified through microchemical testing as calcareous (**Fig. 5.3a**). Although the presence of the insoluble accretions on the surface of the object would not cause further deterioration of the ceramic find, in many instances the carbonate deposit obscured a decorative surface of incised lines or matt-painted patterns. For these objects, where the cleaning of the surface was essential to the interpretation of the piece and the decoration, the carbonate accretions were removed (**Fig. 5.3b**). When the carbonate deposit could be removed using gentle mechanical action, this course of action was taken. In many instances, however, the accretions were extremely hard and compacted onto the surface, making mechanical removal difficult without causing damage to the decoration or surface of the ceramic. In these cases, the accretions were removed through the use of dilute acids.

When acids were employed for the cleaning of ceramic finds, a procedure was established to minimize the risk of damage to the vessel or sherd during treatment. The procedure also ensured that all acid residues were removed from the porous material so as not to encourage future deterioration through the creation of soluble salts. All ceramic finds undergoing this type of cleaning were pre-soaked in deionized water prior to application of the acid. The acid was applied in dilute form and locally applied to the area to be cleaned. Head loupes and binocular microscopes were used during the treatment to monitor the cleaning process and observe changes to the surface. After treatment, the ceramics were soaked in successive baths of deionized water in order to wash out any residual acid from the objects and thus "desalinate" the vessel. The baths were changed until no anions specific to the acid used were detected in the wash water using microchemical tests (Odegaard, Carroll, and Zimmt 2005: 108–109), and conductivity readings taken of the water remained relatively low and consistent (Buys and Oakley 1996:97; Unruh 2001:81).

Due to time limitations and the condition of the vessels/sherds, not all finds that had accretions were cleaned in this manner. Priority was given to objects that had obscured decoration and could withstand the treatment. Non-decorated pieces, where the accretion did not obscure any technological or archaeological information or interfere with the interpretation of the piece, were cleaned mechanically when possible, with the accretions slightly reduced or left in place.

Because of the widespread use of bitumen in antiquity in ancient Albania (Morris 2006), as well as naturally occurring deposits of asphaltum in the area, the presence of this material on some ceramic finds was carefully considered. When one partially complete vessel (**9/259** [**15/12**] [P283]) and several sherds from the site were found to contain dark, resinous-looking deposits on the surface (see Chapter 15), the excavators believed that the black material coating the ceramics was probably bitumen and brought it to the conservation lab for further examination and identification. Initial characterization of the material was made through visual examination using a binocular microscope (7–40x magnification). The material found on the surfaces of the pottery was characterized as being black to dark brown in color, brittle, and often had surface craquelure, characteristics consistent with bitumen (O'Grady and Bardho 2005:3). Other than visual identification, bitumen is difficult to characterize without the use of sophisticated analytical equipment, an option not available in the field. The only method available to the conservators was a microchemical test, the Raspail test, used to identify rosin (pine resin) or pitch made from plant materials (Odegaard, Carroll, and Zimmt 2005:158–159), another commonly used natural resin in antiquity. If the microchemical test was negative for rosin, and the material appeared to be bitumen, the surface coating was identified as bitumen and documented. The Raspail test was also performed on modern bitumen samples and fresh tree resin as standards to aid in the correct interpretation of the microchemical tests.

Ceramic fragments or vessels with bitumen were not cleaned, except for the removal of loose soil, to ensure no loss of the material and prevent contamination for future analysis. If fragments containing bitumen required further cleaning or reconstruction for research purposes, the bitumen was sampled and care was taken to prevent contamination of the remaining material on the surface. The samples of bitumen from the Lofkënd objects are intended for future analysis to confirm the initial identification of the material in the field and potential sourcing of the bitumen found in the tumulus through comparisons with regional asphaltum sources. This is an ongoing project.

Metals

The primary goal in the treatment of the excavated metal finds was to stabilize the objects that were actively corroding and prevent further corrosion and deterioration, both after excavation and during storage of the finds at the end of each season. Active corrosion was visible on many of the excavated finds when examined in the conservation lab. For the copper alloy artifacts, "bronze disease" was suspected when discrete areas of powdery light green corrosion were observed on the surface (Cronyn 1990: 218–219). The corrosion was visually identified as a copper trihydroxychloride, possibly paratacamite ($CuCl_2 \cdot 3Cu(OH)_2$) (Scott 2000:125–126; Cronyn 1990:216). The presence of chlorides in the corrosion was confirmed via microchemical testing (Odegaard, Carroll, and Zimmt 2005:108–109). The iron objects seemed to be suffering from spalling and flaking of the surface. Hollow "shells" or spherical corrosion that remained from the "weeping" or "sweating" of the iron suggested that these objects were also actively corroding due to the presence of moisture and chlorides in the burial environment (Cronyn 1990: 195; Selwyn, Sirois, and Argyropoulos 1999:217). The second goal in the treatment of the metallic finds was to clean and reconstruct them in order to clarify the shape of the objects and their function. These two goals helped to form subsequent approaches undertaken for treating these materials.

In the first season of excavation, the treatment of metallic finds consisted of some light cleaning and reconstruction, followed by storage in sealed, desiccated microenvironments with indicating silica gel. The establishment of this type of storage environment became the standard for all the Lofkënd metals due to the uncontrolled environmental conditions in the area where the objects were stored. When the conservation team returned in 2005, a condition assessment of all the metallic finds from the previous season was conducted to determine the condition of objects after being stored within these desiccated microclimates. Objects requiring retreatment were identified during this assessment and the silica gel reconditioned and replaced if required.

The outcome of this condition assessment resulted in modifications to the way that metal objects were stored in order to sustain the desiccated storage throughout the year. These methods are discussed in further detail here in the section dealing with the packing and storage of objects. The condition assessment of all previously excavated metal finds for retreatment or replacement of the desiccant was incorporated into standard practice at the beginning of each season.

As for treatment, many of the same methods and materials were used for the range of metals excavated at the site. After examination, all metals were lightly cleaned mechanically to remove soil and loose corrosion products. The cleaning helped to reveal the shape of the object and made any decoration or technological features more visible. When needed, conservators temporarily consolidated or supported fragile material during cleaning using cyclododecane. All metals were joined using Paraloid B-48N. In cases where joins needed to be supported or fragile material backed, Kozo Japanese paper, reinforced with Paraloid B-48N, was used. For joins that required further support, as on some of the iron objects, Kevlar fibers were employed, again after impregnating the fibers in resin.

Copper alloy objects

Despite showing some signs of active corrosion, the condition of the excavated copper (Cu) alloy objects was, in general, good. Bronze disease was found only in discrete areas and was not prevalent. The corrosion layers were not very thick or voluminous, and the overall shape of the artifact, its form and function, could easily be identified. For publication purposes and to reveal decorative features, almost all copper alloy objects were lightly mechanically cleaned. This cleaning was primarily performed to remove soil or other burial accretions as well as layers of thicker corrosion if present. The soil, burial accretions, and thicker corrosion were easily removed with a scalpel and would cleave off to reveal a slightly smoother, and often darker, underlying stable corrosion layer or patina. Because at this stage of the treatment the objects were easily interpreted and met the needs of the archaeologists, photographers, illustrators, and conservators, no further cleaning of the corrosion was undertaken (**Fig. 5.4**).

In order to prevent further deterioration of the copper alloy objects and any future outbreaks of "bronze disease" due to storage of the metals in a non–climate-controlled area, all the copper alloy objects were treated with a corrosion inhibitor. Success has been reported in the stabilization of copper alloy objects with benzotriazole (BTA) (Scott 2002: 376–381; Sease 1978), and this was the material chosen for the treatment of these objects. After cleaning, the objects were immersed in BTA under vacuum in a glass desiccator for a period of 24 hours (**Fig. 5.5**). The objects were then coated, in this case with Paraloid B-48N, to protect the BTA layer from mechanical damage due to handling (Scott 2002:380). Comparison of BTA-treated and untreated objects from the first two field seasons showed that the treatment of the copper alloy objects with a corrosion inhibitor followed by the application of a coating was effective in preventing outbreaks of bronze disease in the majority of objects, even when the silica gel in the desiccated microenvironment was exhausted. The BTA/B-48N treatment and coating was continued and incorporated into the standard treatment protocol for all copper alloy finds.

Iron

All the iron artifacts excavated at Lofkënd were heavily mineralized and corroded. The iron finds were covered with voluminous orange and brown corrosion products, the majority of which were visually identified as iron oxyhydroxides. Many of the iron finds suffered from splitting or cracking, and some had fragments that had completely spalled from the surface. Almost every iron object showed signs of active corrosion in the form of hollow shells or spheres that form due to weeping iron and the presence of chlorides and moisture (Selwyn, Sirois, and Argyropoulos 1999:221) (**Fig. 5.6**).

As with the copper alloy objects, the iron objects were mechanically cleaned to remove soil, burial accretions such as carbonates, and some of the softer iron corrosion products. Given the hardness of most iron corrosion products and the difficulty of removing them with a scalpel, these objects were only superficially cleaned. Although the corrosion on these finds was somewhat voluminous, the shapes of the objects were comprehensible in almost all cases, and further cleaning was not required in order to aid in the interpretation of the artifacts.

A challenge presented in the cleaning of the iron artifacts was the presence of textile pseudomorphs or mineralized textile fibers (**Fig. 5.7**). On many of the iron pins and fibulae, and in a few cases the copper alloy objects, impressions of fibers, spun threads, and weave patterns were found. Areas that contained these fiber or textile impressions were not cleaned. Loose soil was removed using soft brushing or very gentle cleaning with a bamboo skewer to allow them to be more visible and identifiable, provided that this could be done safely. All textile pseudomorphs were documented in writing and graphically, and analyzed in this volume (Chapter 12).

Due to the flaking and unstable nature of almost all the iron finds, a decision was made to consolidate and coat these objects in order to stabilize them and prevent further deterioration. Much research has been done on the deterioration of excavated iron and its stability in order to determine the effectiveness of various washing techniques and chemical treatments for the removal of chlorides, as well as consolidation and coating. Despite the results of various experiments and case studies, archaeological iron still poses problems post-excavation whether treated or untreated, and the prevention of further deterioration is very difficult to achieve in non–climate-controlled storage. One option has been coating the objects to create a moisture barrier and prevent further corrosion (Keene and Orton 1985:138; Lemmer 1972: 102–107). In the case of the Lofkënd artifacts, the coating would primarily protect the objects from further damage due to repeated handling during examination, photography, and illustration, but it was also the hope that the coating would protect the objects from environmental deteriorants. A program of consolidating and coating iron finds as a stabilization approach was undertaken in the second season to see if further deterioration of the material could be slowed down or prevented altogether.

The coating of iron artifacts began in 2005 using Paraloid B-48N applied through immersion under partial vacuum to ensure penetration of the resin. In 2006, the iron finds from previous seasons were examined in order to compare the condition of the coated (2005 iron finds) and uncoated (2004 iron finds) objects and assess the effectiveness of the consolidant and coating. Many of the untreated iron finds from the first season were still flaking and spalling despite being stored in a desiccated microenvironment with the relative humidity (RH) level within the storage containers at or below 20–30% at the time of examination. Although this RH may be considered low and suitable for other types of metal, it is thought that the RH should be even lower for archaeological iron to prevent corrosion, at around 12% (Watkinson and Lewis 2005). In the case of these uncoated objects, it was not dry enough in the storage containers to prevent the iron from further deteriorating. Due to the lack of environmental controls in storage, the very low RH could not be maintained within the microenvironments over a period of 10–12 months.

The coated iron objects from 2005, however, did not show the same level of flaking or deterioration as the untreated finds, nor were they as crumbly or powdery when handled. Although the silica gel was exhausted when examined and the RH within the microclimates was 20–30%, the objects did not seem to show the same signs or level of deterioration as the untreated iron. Based on these observations, and the comparison of the conditions of the two groups, the consolidation/coating treatment protocol was adopted for all future iron finds, as well as those previously excavated. After coating, the iron artifacts would be stored in desiccated microenvironments and the finds checked at the start of each season for any changes in condition to monitor the efficacy of the treatment protocol.

Electrum

The only other metallic find from the tumulus consisted of a pair of foil ear or head ornaments (Tomb XLVIII-3, 4; Chapter 10: **10/11**, **10/12**; SF 290, SF 291) initially identified as gold, and later as electrum, or native gold, based on analysis of the objects using portable X-ray fluorescence spectroscopy (pXRF) (see Chapter 11). Being made of electrum, these objects did not require any cleaning. One ornament did require, however, some minor treatment to support some tears in the foil to prevent further damage. Kozo Japanese paper coated with Paraloid B-72 was applied to the back of the damaged area to keep the tear from widening and to help support the two halves of the extremely thin electrum foil. After treatment, the ornaments were stored within a custom-built foam support to protect the fragile and delicate objects from damage during handling and storage.

Bone

All excavated bone consisting of animal or human remains was examined, cleaned, and packed by the physical anthropologists on the project after excavation by the archaeologists. Conservation staff dealt with human remains only when they were part of blocks of soil lifted to remove fragile artifacts. In these cases, treatment consisted of removal of the bones from the soil and light cleaning before handing the material over to the specialists. Another instance where the conservators treated human remains was in the reconstruction of complete skulls. For this treatment, the conservation staff used B-72 as the adhesive, along with B-72 resin bulked with glass microballoons when needed to help fill gaps and provide structural support to the joins.

The bone artifacts that underwent conservation consisted of four bone pins. The pins were in relatively good condition, but were fragmentary and in some cases had thick carbonate accretions. The treatment of these artifacts involved cleaning using both mechanical means (soft brushes, bamboo skewers, and, in a few instances, a scalpel), and chemical methods (use of light swabbing with deionized water and ethanol). Paraloid B-72 was used to consolidate fragile areas, and the objects were reconstructed using the same resin. Kozo Japanese paper was used to reinforce joins that required additional support.

Semi-precious stone

The stone jewelry items from the tumulus consisted of three reddish-brown or reddish-orange beads (SF 340 = Tomb XXVIII-6, SF 296 = Tomb LIII-7, SF 113 = Tomb LXX-6; Chapter 10: **10/102**, **10/103**, **10/104**). These were identified as quartz-based stone and probably a type of chalcedony (carnelian or sardonyx). The beads were in very good condition and required no treatment other than removal of loose soil. The perforations of each bead were examined using the binocular microscope prior to any cleaning to make sure there were no remains of fibers or cordage that could have possibly been used to string the beads together. No fiber remains were detected in any of the perforations.

Shell

Although numerous land snail shells were found in the fill of the tumulus, there were few worked shell finds (see Chapter 16.2). One small disk-shaped piece (SF 346), with what appeared to be a possible perforation, was found, but it was not clear whether the appearance of the object was due to the manner in which the shell was damaged or if it was intentionally shaped as a bead. The shell specialist, Dr. Vardala-Theodorou (Chapter 16.2), determined that the perforation was natural and not human made.

Five fragments of mother-of-pearl (SF 433, also discussed in Chapter 16.2) found in the 2007 season appeared to have worked edges. The fragments, however, did not join, and the object could not be clearly identified. The fragments are relatively flat, and several seem to have smooth or finished edges on one side. The pieces have for now been identified as possible inlays or fragments of an object otherwise not preserved. For the treatment of these pieces, the soil was removed by brushing and the surface cleaned with deionized water.

Glass and vitreous materials

The final group of materials treated consisted of 13 beads made from glass or vitreous materials. Seven were identified as glass and exhibited a completely vitrified and amorphous glassy structure. One bead was identified as faience (SF 298 = Tomb LIII-9; Chapter 10: **10/105**). The remaining beads consisted of an opaque, coarser-structured glass-like material, where the silica grains and other unvitrified materials were still visible (see Muros, contribution to Chapter 10). These were also identified as glass beads.

The majority of the beads were in poor condition upon excavation. Several were very fragmentary. The glass matrix of several of the beads had lost cohesion, making them extremely crumbly and friable. When possible, the vitreous beads were block-lifted to keep the fragments in situ until they could be treated in the conservation lab. Soil was gently removed from the bead fragments using bamboo skewers when possible and consolidated with B-72 if needed. Due to the fragility of the beads, they were only lightly cleaned. More stable and robust beads were cleaned with a solution of deionized water and ethanol (1:1 v/v). Bead fragments were reconstructed using Paraloid B-72.

CONSERVATION TREATMENTS IN SITU

For the majority of the very fragile objects excavated, the archaeologists would block-lift the artifacts themselves and bring the artifacts supported in soil to the conservation lab for treatment. The discovery of several extremely fragmentary and fragile copper alloy headbands, however, required stabilization in the field prior to lifting. It was necessary for the conservators to face and lift the objects to transport them back safely to the conservation lab at Apollonia.

In 2005, a highly corroded, mineralized and fragmentary copper alloy headband (Tomb XXI-4; Chapter 10: **10/86**, SF 255) was discovered on the skull of a young female. The headband had been found toward the end of the workday, and due to concerns over looting or disturbance of the site overnight, it could not be left in the ground and had to be excavated. Conservators could not be brought out to the site due to lack of transportation, and since the headband could not be left in situ, the archaeologists lifted the artifact. Because the headband was sitting on a skull, the block of soil had to include the skull and any grave goods found on the head. The block was wrapped and supported as best as possible with the materials available on site to help hold the fragments in place and prevent the soil from shifting. Once in the lab, the conservators were able to extract the fragments, slowly and carefully, and by facing sections of the headband with resin and tissue to keep the remaining associated fragments together as best they could. In subsequent seasons, the headband was cleaned and reconstructed.

When another headband (Tomb XVIII-1; Chapter 10: **10/85**; SF 317) was found in 2006 in highly fragmentary condition, the conservators were brought out to the site to aid in the stabilization and lifting of the headband to prevent any damage during excavation and transport. Treatment in situ also prevented dis-association of the fragments, which would make reconstruction and later treatment more difficult. The archaeologists excavated and exposed the fragments as much as possible without dislodging them from the surrounding soil. Once the extent of the headband was revealed, the conservators stabilized the fragments. Using strips of Kozo Japanese paper impregnated with Paraloid B-48N, the exposed surface of the headband was covered, or faced, to hold the fragments in place and prevent them from moving (**Fig. 5.8**). The facing would also impart strength to the extremely thin and fragile metal and thus prevent damage. After the tissue facing was applied, the headband was block-lifted and taken to the conservation lab where it could be slowly excavated and then treated and reconstructed. The successful facing and block-lifting of this headband was used as a model for future fragile finds excavated from the tumulus, and the same methods were used when another headband was found later in the season.

Packing and Storage

Ensuring the safe storage of excavated artifacts, to prevent further physical damage and deterioration due to environmental factors, is a major challenge for conservators working in the field. In many cases, excavated finds are kept at the excavation house or in some other area that has been modified and turned into a storage space. More often than not, the field project does not have purpose-built artifact storage with environmental controls, nor can the objects be housed in such pre-existing facilities. This requires the conservators to come up with creative solutions to work around storage conditions that are not always ideal for the long-term preservation of the artifacts.

For the finds of the Lofkënd Archaeological Project, the conservation team faced a similar problem. The artifact storage area allocated to the project was located within one of the rooms of the monastery at Apollonia. As with all the monastery buildings, the storage room for the Lofkënd finds is not climate controlled. The room, made of stone, has two openings on one wall, which serve as windows, but have no glass or coverings and remain open. The door to the room is made of wood but is slightly warped and does not completely seal the doorway, letting in light, moisture, and pests. A major portion of this room was excavated to reveal Late Antique and Byzantine foundations, mosaics, and other remains, and left exposed in a deep trench cut into damp soil. An area in the front corner of this room, near the door, was reserved for storage of the Lofkënd material. This area has been fitted with several metal shelving units that were tied into the wall to store the excavated finds from the tumulus (**Fig. 5.9**). The monastery is patrolled 24 hours a day by armed guards, and therefore security is not of great concern. The lack of environmental controls and the ongoing excavation in the storage room make long-term preservation of the artifacts a challenging task.

In the 2006 and 2007 seasons, monitoring of the relative humidity (RH) within the storage area was undertaken to get a sense of the environmental conditions during the summer months and whether the building itself acted as a buffer against exterior conditions. In 2006, monitoring was undertaken using a humidity indicator strip read with a colorimetric scale. The RH in storage that season was generally 70–80% (± 5%). Because the humidity indicator strip only allowed for approximate readings, in the subsequent season a more accurate method of recording humidity was used to allow monitoring 24 hours a day. A data logger (Onset HOBO H-8 logger) recorded the temperature, relative humidity, and light levels throughout the day within the storage area. This allowed the conservators to get a better sense of the environmental parameters within the space, as well as record any fluctuations. The data logger was set to take readings every 30 minutes for a period of 5 weeks, and the data were examined using proprietary software purchased from the manufacturer.

Using the data from the logger, the conservators were able to determine that during the 2007 field season, the average RH in the storage area was 62% (± 5%) and ranged from 36% to 83% (± 5%). The temperature that summer averaged 76° F (± 1.27° F) (24° C ± -17.1° C) in the room, ranging from 71° F to 82° F (± 1.27° F) (22–28° C ± -17.1° C), despite weeks of extremely warm weather with outside temperatures near or above 100° F (± 1.27° F) (38° C ± -17.1° C). Although there were fluctuations observed in the daily temperature readings, they were quite small.

The RH, on the other hand, seemed to cycle daily, starting off drier in the evening, rising in the morning, and then dropping in the afternoon, following the trends observed in the humidity outdoors. Daily fluctuations could be quite extreme: the RH could drop or rise about 30% over a 12- to 24-hour period. The periods of these larger fluctuations seemed to coincide with days when it was extremely hot and dry outside.

Although light levels were not a concern due to the lack of light-sensitive materials, they were monitored nevertheless to see if the levels rose when the doors to the storage room were open. In most areas in storage, the light levels were below the minimum detection level of the logger (2 lux). The only area to receive any light was the section of the shelving unit opposite the door which was illuminated when the door was opened (**Fig. 5.9**). Since all the objects were stored inside sealable containers, however, damage due to light was not an issue.

It was clear from the data collected that the building was not an effective buffer against changes in relative humidity. In order to offset the lack of these, and other, environmental controls, the conservators focused on the archival storage of the objects and the creation of microclimates for artifacts that required desiccated storage. Archival storage materials were brought from the United States or purchased in Europe in order to pack the objects safely. The primary materials used were polyethylene foam (Ethafoam® and Volara®), sealable polyethylene bags (2 mm), cushioning and lining materials such as acid-free tissue and Tyvek®, and polypropylene and polyethylene storage containers purchased locally. All objects were stored using a combination of these materials. The metal finds, which required desiccated storage, were packed with indicating silica gel in order to prevent further corrosion. The specific methods used for packing and storing each material are described in the following sections.

Once packed, the materials were placed in a larger storage container, in this case a lidded Coroplast® board storage box, and then this box placed in a large stackable plastic crate to be placed on the shelving unit in storage. Large polypropylene labels marked with permanent, waterproof ink were attached to each crate to identify the material housed within it, along with other archaeological information. The height of the shelves within the storage locale accommodated two crates. Because of the potential for seismic activity in the area, securing the crates on the shelf was an issue. A lip on the lower portion of the shelf helped to secure the lowest crate and prevent it from sliding off the shelf. For the second stacked crate, a plan was devised to use strapping, in the form of two elastic cords (bungee cords), positioned across the shelves, to hold the upper crates in place (**Fig. 5.9**).

Ceramics

Whole vessels

Whole vessels, or more complete ceramic finds, were packed in small polypropylene "baskets," or non-lidded circular containers, with acid-free tissue and Ethafoam® as padding (**Fig. 5.10**). The baskets were then placed in large polyethylene bags and sealed. The sealed vessels were then placed in lidded Coroplast® boxes and placed within the larger plastic storage crate.

Diagnostic and decorated fragments

Smaller fragments or sherds that were considered diagnostic or were decorated or painted were placed in polyethylene bags, with a label containing all pertinent information (see section titled "Labeling"). The bags of sherds were then placed in Coroplast® boxes within the larger plastic crate.

Metals

Storage supports and packing

The storage of the metal finds involved two steps. The first was to determine the type of housing or packing the object required: whether the item could be placed as is after treatment in a polyethylene bag, or whether the object required some type of storage support. This decision was usually made based on the condition of the piece and how much handling would be required. In the case of reconstructed fragmentary objects that would be extensively handled, supports designed to minimize handling were created.

The storage supports for the metals were made of polyethylene foam blocks with custom-cut cavities to hold the whole artifact or fragments. Small finger holes were cut at the sides of the cavity to make it easier to remove the objects from the foam and reduce the risk of damage. Once cut, the cavities were lined to prevent the foam from scratching the object or to keep the corroded surface of the object from catching on the foam. In the first season, the cavities were lined with acid-free tissue, the only material available at the time. Beginning in 2005, Tyvek® and Volara® foam were used. The tissue linings of the 2004 object supports were eventually replaced with one of these materials (**Fig. 5.11**).

Despite the cavity packing, some objects needed to be secured within the foam block to prevent them from falling out of the support while being handled or moved to storage. To do this, a series of straps or "seat belts" were made that ran across the cavity and gently held the objects in place. These straps were made from strips of Tyvek® that were secured to the foam on one end with archival hot-melt adhesive and were tucked into a slit in the foam on the other side of the cavity. Instructions for how the object should be removed from the support were written on Tyvek® labels that were adhered to the Ethafoam®. After being set in the foam support, the metal objects were placed in resealable polyethylene bags.

After initial packing, the objects were then put into locally purchased polyethylene or polypropylene containers, storing like metals with each other (i.e., only iron or copper alloy objects within one box). Once sealed, the boxes were labeled on the exterior with a list of the finds contained inside. Some objects could not be stored in the boxes due to their size, so the foam support was simply kept in a polyethylene bag. The boxes or bags were then placed within the lidded Coroplast® box and then into the large plastic storage crate to be deposited in the storage room.

Desiccated storage microenvironments

All copper alloy and iron finds were housed in desiccated microenvironments to prevent further deterioration while in storage. Packets of indicating silica gel were put in the polyethylene bag in which the object was stored. If several small objects or fragments were kept together in a small polyethylene or polypropylene box, the silica gel was placed within that box. Indicating humidity strips were included in the clear storage containers and bags to monitor the humidity within the microenvironment.

As stated previously, in 2004 the initial method used for storing the metallic finds did not prevent moisture from entering the storage containers and causing further corrosion to the objects. In the 2005 season, modifications were made to the storage methods to include "double bagging" of the objects in sealable polyethylene bags if they could not be stored in a box due to their size, or placing the polypropylene or polyethylene boxes within two sealable bags (O'Grady and Bardho 2005:3). Double bagging the objects would create another barrier against moisture and help maintain the desiccated storage environment. Comparison of the condition of the finds in 2006 that were double bagged versus those housed after 2004 showed less evidence of active corrosion, as well as better maintenance of the storage environment. For some of the iron finds that seemed to be in especially poor condition and where it seemed that a desiccated environment was not maintained over the course of the year, a second bag of silica gel was often placed within the outer polyethylene storage bag, to further ensure that the desiccated storage environment was maintained.

Anoxic and desiccated microenvironments

Prior to the start of the 2008 season, discussions began regarding the final housing and storage of all the inventoried finds from Lofkënd. In particular, there was concern over the future of the metallic artifacts because the project was coming to an end; after the last season in 2008, there would be no one either to examine the condition of the metals on a regular basis or to replenish any exhausted silica gel in the desiccated microenvironments. Although the system of creating microenvironments using desiccated silica gel worked well, it would not be an effective long-term storage system without monitoring, especially for the iron finds.

To ensure the long-term preservation of the iron artifacts, and to increase the longevity of the desiccated microenvironments, the conservation team opted to use the Revolutionary Preservation (RP) System®. The RP System® consists of an oxygen scavenger used in conjunction with a barrier film with low gas and moisture permeability to create anoxic microenvironments for the storage of cultural material. The RP-A type packets also include a desiccant to create low RH and oxygen environments. Both RP-A and RP-K, which does not include a desiccant, have been used successfully for the storage of a range of materials, including archaeological iron (J.P. Brown, personal communication 2008; Day 2006; C. del Re, personal communication 2008). The barrier film chosen for the storage of the Lofkënd finds was Escal™ because of its low permeability to moisture and most gases, as well as visibility, since the film is clear (Carrió and Stevenson 2003:37; Day 2006:437).

Iron objects, on their foam supports, were heat sealed in custom-sized Escal™ bags with double the amount of RP-A required for the volume of air, as is suggested to maintain the desiccated anoxic environment (Burke 1996:3; Day 2006:438). For smaller objects that were housed together in polyethylene or polypropylene containers, the RP-A was placed in the container and the entire box sealed in Escal™. Keeping the objects either in their foam supports or in boxes not only provided protection during storage, but would also prevent any potential reduction in bag volume that can occur due to the removal of oxygen from damaging or crushing the objects (Burke 1996: 2; Day 2006:139; Shashoua 1999:883). The bags were also made slightly larger than necessary, taking into account this reduction in air volume. In preparing the bags, each side was double sealed as well as sealed diagonally across the corners to prevent leakage (Shashoua 1999:883). A vacuum heat sealer was used to remove some of the air within the bag prior to sealing (Buccellati 2005:1). Humidity indicator strips were placed inside each bag to monitor humidity levels within the enclosure (**Fig. 5.12**).

An oxygen indicator, known as the Ageless Eye®, is often included in the anoxic environment to monitor oxygen levels within the enclosures (Burke 1996: 2; Carrió and Stevenson 2003:33; Day 2006: 437). These color-indicating tablets, depending on the type, will change color when the oxygen levels drop to 0.1–0.3% and are a good visual method of determining whether a low-oxygen environment has been achieved. However, problems have been reported in using the Eyes® where they at times did not change color in low-oxygen environments and failed (Burke 1996:3; Carrió and Stevenson 2003:38; Day 2006: 440). Some research has also shown that the Eyes® lose their sensitivity to oxygen after 6 months of use (Shashoua 1999:886). The use of the Ageless Eye® and its effectiveness as an oxygen indicator is highly dependent on the environment the material is stored in, and fluctuations in temperature and humidity can affect its reliability (Day 2006:440). Because of these factors, and as the work would be undertaken in the field without follow-up, the Eyes® were not used in the enclosures.

The storage microenvironments created for the copper alloy objects were also altered by storing these objects within Escal™ bags with desiccated silica gel (and no RP-A). The use of an Escal™ enclosure would reduce the amount of moisture that permeated into the storage housings and allow the desiccated conditions to be maintained (J.P. Brown, personal communication 2008).

In addition to doubling the amount of oxygen scavenger used when creating an anoxic storage system (Burke 1996:3), bagging the objects inside two bags has also been recommended (Shashoua 1999: 884). Due to the cost of the Escal™ barrier film, however, this could not be done for the Lofkënd metals. The Escal™ enclosures containing the iron and copper alloy objects were placed in two sealable polyethylene bags (2 mm) to extend the life of the microenvironments and help slow oxygen absorption in case of any leaks. After bagging, the objects were repacked in their Coroplast® boxes and then in plastic crates to be placed in storage.

Other materials

All other small finds were packed using a combination of the methods described above. Depending on the condition of the finds, objects were placed directly into sealable polyethylene bags or were placed in custom foam supports that were then placed in labeled bags. Groups of similar materials were then placed in polyethylene or polypropylene boxes, and the boxes stored in a Coroplast® box. These boxes were then placed in plastic crates.

DOCUMENTATION

At Lofkënd, the conservation team established a system for graphically and textually documenting all finds that were brought to the conservation lab for examination and treatment. This documentation included photography and written reports, which were then incorporated into an electronic database.

Photography

Record shots of all excavated finds brought to the lab were taken before, during, and after treatment.

The photo-documentation was used to record the condition of the object during the various stages of treatment to accompany the written conservation report. All images were taken using a digital camera, and the images were archived on various project-owned computers, both in Albania and at UCLA, and backed up on disks and portable hard drives, which are stored both in Tirana and at UCLA.

In addition to record shots, photomicrographs of objects were taken when there was a need to document a particular condition issue, technological feature, or the presence of textile or other pseudomorphs. These photos were also taken with a digital camera and archived with the other images.

Documentation of the objects was also conducted by other members of the Lofkënd project, specifically the two photographers on staff and the illustrator. The photographers were involved in documenting the objects in situ during excavation and after treatment. Unlike the record shots taken by the conservators, the images taken by the photographers served as higher-resolution and color-corrected archived images of the objects showing their condition pre- and post-treatment. These images were also used for publication purposes. All finds were illustrated after treatment, serving as a form of documentation that would be used for recording purposes and publication.

Written reports

All conservation treatments undertaken in the lab, as well as objects brought in for examination, were documented via a written report. The report used by the conservators was adapted from the report form used by the Albanian Rescue Archaeology Unit. Like most conservation reports, it recorded such information as object number, weight, date of excavation, and materials used to manufacture the objects. The remaining sections were used to record the condition of the object, the proposed treatment, the actual treatment undertaken, and information on the photo-documentation.

As the activities of the lab expanded and microchemical tests or different types of treatments were incorporated into the daily protocol, the report template was changed to allow for these tasks to be recorded. For example, a microchemical test checklist was created to indicate what tests were performed on the object (O'Grady and Bardho 2005:2). The results could then be recorded in the section describing the treatment undertaken. In instances where objects required desalination, a chart was created to record information such as the conductivity readings taken and the length of the treatment. At the end of the season, hard-copy duplicates were made of all the conservation record sheets, with one set remaining in Albania at the Institute of Archaeology in Tirana, and the other set brought back to UCLA.

Conservation database

A conservation database was created for the project to provide a searchable electronic version of the written reports, and which could also include supplementary or additional information such as images. The database was created using the software FileMaker Pro® and contained the same information as the written reports. When needed, modifications and additions could be made to reflect changing conservation activities.

In addition to textual information, the database had the advantage of including photographic documentation with each of the records. The images were used to illustrate what the object looked like before, during, and after treatment, but also included photomicrographs that documented technological or condition features. Additional fields for sampling and analytical information were added in 2007 when technical studies were begun on several of the excavated finds from Lofkënd. As with the written reports, a copy of the database was kept at UCLA and in Albania.

Labeling and inventory of artifacts

The conservation team was also responsible for the labeling of artifacts. Inventory numbers were assigned to the excavated finds either in the field or back at the excavation house by the archaeologists and directors. The three-digit inventory number was preceded with a letter to indicate whether the material was pottery ("P") or a small find of another material ("SF"), such as metal, bone, or glass. All pottery was to have the inventory number applied to the object. Small finds would have the number written on a label or the polyethylene bag the object was stored in.

Initially, pottery was labeled using a method commonly used in museums, one in which a layer of resin, acting as a barrier, is brushed onto an inconspicuous area of the object, onto which the number is then written in ink (Segal 1998:68–69). The resin layer prevents the number from being written directly onto the object, rendering the marking "reversible." Paraloid B-72 was used for the barrier layer when labeling began in the first season. For light-colored ceramics, black permanent ink was used to write directly onto the resin layer. For ceramics with a dark fabric, a layer of white paint was applied onto the barrier layer and the number written in black ink on the paint layer.

In the 2005 season, the decision was made to change the labeling system to one that would be more reversible and allow for the numbering to be quite small and even less conspicuous. Although the numbering of objects using ink can be an effective labeling system, when the number needs to be removed the solubilized ink can cause discoloration and staining of the piece. Instead of numbering the object using ink, the inventory numbers were placed on labels that would then be adhered to the ceramic finds (Segal 1998:71–72). This would allow for easier application of the number, especially on small fragments, and allow for the numbers to be smaller and less visible. It could also be easily removed without possible staining caused by the use of ink pens.

To make the labels, pottery inventory numbers were printed on an inkjet printer. The printed sheets were then coated with Acrysol WS-24 acrylic dispersion. Once dry, the coated labels were cut and attached to the object using a small amount of Paraloid B-72 (**Fig. 5.13**).

For objects given small find numbers, the inventory number was initially written on the polyethylene bag in which the object was stored. In addition to the inventory number, the date of excavation and location where the object was found were also included on the bag. Due to handling of the bag during condition assessments or examination, however, the ink often faded or disappeared, causing the associated contextual information to be lost. In order to prevent this, small Tyvek® labels, with the information handwritten in waterproof, permanent ink, was placed within the bag in which the object would be permanently stored. For objects placed in foam supports, the Tyvek® label was attached to the foam support with archival hot-melt adhesive.

All polyethylene and polypropylene boxes housing objects were labeled on the exterior to indicate the contents of that box. These labels were usually made from self-adhesive labels with the information handwritten in permanent ink. The stackable plastic crates in which the Coroplast® boxes were kept were also labeled. As stated previously, polypropylene labels were attached to the crates with information listing the general material contents and season of excavation written in waterproof, permanent ink. The labels were tied onto the end of the crate with nylon string and/or coated wire ties. When the crates were placed on the shelves in the storage area, the label faced outward so that anyone could see what the contents of the crate were without having to remove the crate itself from the shelf (**Fig. 5.9**).

TECHNICAL STUDIES

Beginning in 2007, the Lofkënd conservators embarked on a series of technical studies on various groups of material from the tumulus. The goal of these projects was to identify the raw materials and techniques used in manufacturing the objects.

In conjunction with faculty and staff of the UCLA/Getty Conservation Program, three studies were undertaken, the results of which are reported in various sections of this volume. The first focused on the copper alloy finds from the tumulus to understand the alloy composition, as well as the method of manufacture of the metals (Muros and Scott, Chapter 11). The second study focused on the stone, glass/frit, and faience beads excavated at the site, in order to characterize the materials and their composition (Muros, in Chapter 10). The first stage of both of these studies involved the analysis of the objects using portable X-ray fluorescence (pXRF) spectroscopy for non-invasive compositional analysis in the field. The final study focused on the examination, identification, and documentation of textile pseudomorphs preserved on the metallic finds (Muros, Chapter 12).

SELECTED CONSERVATION SUPPLIES USED IN LOFKËND CONSERVATION LABORATORY

- *Acid-free, double-sided tape:* The tape used consisted of a polyester film coated with a transparent acrylic-based adhesive on both sides and manufactured by 3M (Type #415). Purchased from the supplier University Products (Holyoke, Massachusetts), it was used in the creation of storage supports for the objects.

- *Acrysol (Primal or Rhoplex) WS-24*: An aqueous acrylic colloidal dispersion made from a polyacrylic acid mixed with acrylic copolymers and manufactured by Rohm and Haas, this material was used primarily to coat object labels that were applied to inventoried pottery, but it was purchased for use on the project primarily for the consolidation of damp bone or other excavated material.
- *Archival hot-melt glue*: An ethylene vinyl acetate (EVA) copolymer for use in glue guns, manufactured by Bostik (Thermogrip #6363). This glue was used to make storage supports for objects and to attach labels. It was purchased from the supplier University Products (Holyoke, Massachusetts).
- *Benzotriazole (BTA)* ($C_6H_6NHN_2$): BTA was used as a corrosion inhibitor for copper alloy objects as a solution in ethanol. It was purchased from the supplier Talas (New York).
- *Cerex®*: A spun-bond (non-woven) nylon fabric manufactured by CEREX Advanced Fabrics (Pensacola, Florida), this product was used to back or support joins in combination with an adhesive.
- *Coroplast® storage boxes*: A corrugated board comprised of a copolymer of polypropylene and polyethylene manufactured by Coroplast, these boxes, which had a lid, were used to store inventoried small finds. They were purchased from the supplier University Products (Holyoke, Massachusetts).
- *Cyclododecane*: A waxy cyclic alkane that sublimes at room temperature, Cyclododecane has been used as a temporary consolidant and barrier material in many conservation applications. It was purchased from Kremer Pigmente (Germany) and used as a temporary consolidant for the treatment of the copper alloy headbands block-lifted from the tumulus.
- *Data logger*: A HOBO H8 series data logger, manufactured by Onset Computer Corporation, was used to monitor environmental conditions within storage. The logger used recorded relative humidity, temperature, and light levels.
- *Escal™*: A transparent barrier film made from vacuum-deposited ceramic with an outer layer of polypropylene and an interior sealing layer made of polyethylene, Escal™ is manufactured by Mitsubishi Chemical (Japan).
- *Ethafoam®*: A closed-cell, low-density polyethylene foam manufactured by Dow Chemical Company, this foam was used to create storage supports for the objects.
- *Glass microballoons*: A fine powder comprised of soda-lime/borosilicate glass microspheres, this material was used to bulk resin to fill gaps in joins. The glass microballoons were manufactured by 3M.
- *Humidity indicator cards*: These cards contain a strip of absorbent paper squares impregnated with the indicator cobalt chloride that changes color as it absorbs or releases moisture. The cards change from blue to pink as they absorb moisture, and the relative humidity is read at the square where the color change, from blue to pink, is visible. The cards, manufactured by Humidial Corporation (Colton, California), were used for monitoring conditions within the storage area, in the conservation lab, and in storage containers.
- *Kevlar® fabric or veil*: Available as a woven (fabric) or unwoven (veil) fibers, this is a para-aramid fiber composed of poly(p-phenylene terephthalamide) manufactured by DuPont. Kevlar fibers were used to back and support joins on some iron objects after impregnating or attaching with resin.
- *Kozo Japanese paper*: Paper made from the mulberry tree, Kozo Kashmir (GSM 10) was purchased from the supplier Talas (New York, NY). Kozo paper was used as a facing to stabilize block-lifted objects before removal and to back and support fragile or fragmentary objects after impregnating or attaching with resin.
- *Paraloid B-72* (Acryloid B-72): An ethyl methacrylate and methyl acrylate copolymer manufactured by Rohm and Haas, this resin was used as both an adhesive and consolidant.
- *Paraloid B-48N* (Acryloid B-48N): A methyl methacrylate and butyl acrylate copolymer

manufactured by Rohm and Haas, this resin was used as both an adhesive and a consolidant/coating material.

- *Polyethylene and polypropylene boxes*: Boxes of these inert plastics were purchased locally in Albania. They were used for the storage of small excavated finds as well as to create desiccated microclimates for the storage of metals.
- *Polypropylene labels*: These plastic labels were purchased in Greece for labeling the plastic crates into which the archival Coroplast boxes were placed. The labels were attached to crates using nylon ties or coated wire ties.
- *Revolutionary (RP) Preservation System™*: This oxygen absorber manufactured by Mitsubishi Chemical (Japan) is used to create anoxic storage environments. RP Type A, which contained a desiccant in addition to the oxygen scavenger, was used to store the Lofkënd iron objects.
- *Self-indicating silica gel*: These granules of silica are made to absorb moisture and reduce humidity, indicated by a change in color of the granules that are impregnated with a moisture-sensitive indicator. The silica gel used on the Lofkënd project contains an inorganic iron compound as the indicator; it was supplied by Conservation Resources (United States and United Kingdom).
- *Tyvek®*: This is a material comprised of spunbonded oleofin made from high-density polyethylene fibers, manufactured by DuPont. The Tyvek® used in the lab to line cavities cut in foam, and for other lining or storage purposes, was Tyvek® Soft Structure 1443R, which behaves like a fabric.
- *Tyvek® labels*: These labels, made of a thicker and stiffer structure of Tyvek than that used for lining storage supports, were purchased from University Products (Holyoke, Massachusetts).
- *Volara®*: A closed-cell polyethylene foam manufactured by Voltek, this foam was used for storage supports and as a cushioning material or lining for cavities cut into denser foam.

ACKNOWLEDGMENTS

The author would like to thank all the conservators who worked at the Lofkënd Archaeological Project whose work contributed to this report: Sarah Cleary and Naoko Fukumaru (2004), Caitlin O'Grady (2005), Allison Lewis (2006), and Alma Bardho (2005–2008). The Lofkënd Archaeological Project is grateful to the Mellon Foundation, the Detroit Institute of Arts, the Samuel H. Kress Foundation, and the Steinmetz Family Foundation for providing support for conservation staff and activities over the course of the project. The author would finally like to thank those friends and colleagues who read over drafts of this paper and kindly offered suggestions and comments.

Part II
The Population of the Tumulus

Chapter 6
Bioarchaeology of the Lofkënd Tumulus

Lynne A. Schepartz

Goals of the Study

The first level of bioarchaeological analysis involves determining the basic distribution of age and sex to evaluate how the cemetery sample deviates from an expected population structure modeled from actual data for small-scale human societies. Any differences from the modeled stable and stationary population structure can then be evaluated in terms of differential bone preservation, catastrophic death events, and the factors often of greatest interest in prehistoric studies: selective burial practices involving age, gender, or other variables reflecting social persona and cultural practices, or differential death and disease.

In addition to the fundamental estimation of age and biological sex, the determination of the following information is also desirable: health status and populational patterns of stress with regard to gender or status, the evidence for habitual behaviors that involve specialized uses of teeth or muscle groups, populational affinities and genetic relatedness based on morphological analyses, DNA analysis for familial and populational relationships, and chemical analyses for dietary composition and place of natal origin.

The ultimate goal of a skeletal biology study is to reconstruct the life experiences of the population. While some data are pertinent to specific individuals and may reflect unique life circumstances, environmental and dietary adaptations may be societal responses that are also detectable archaeologically. These reflections of group behavior are of the greatest interest here, and comparative data are used to illustrate the characteristics of Lofkënd within the general region of south-central Albania. At the same time, there are certain individuals whose skeletons reflect life events that are interesting and unique. Details of these conditions are provided in the catalogue of skeletons.

Research Questions

With regard to the specific circumstances of the Lofkënd population, there are several questions that may be effectively addressed through skeletal analysis. Some form of pastoralism is thought to have been the predominant economic system in the Albanian Bronze and Iron Ages (cf. Allen 2002; Gardeisen, Petit, and Piques 2002; Petruso et al. 1996), but the tumulus fill provides almost no evidence of subsistence (Chapter 16.1). If the Lofkënd population was primarily engaged in a pastoral economy, we would expect them to be relatively healthy, with low levels of malnutrition, caries, or infectious disease as compared to agriculturally dependent groups (Schepartz 1989). In the case of Lofkënd, the modern burials can be compared to the prehistoric ones to see if there are any indications of dietary change.

Temporal trends in nutritional stress and local environmental degradation (whether through natural causes or via human over-exploitation of resources or the accumulation of waste) may be detected from changes in the frequency of skeletal health indicators such as porotic hyperostosis, enamel hypoplasias, and bone infections. These conditions are general, rather than specific, indicators of environmental stressors that disrupt growth or normal bone maintenance. Comparisons of prevalences over time, among age cohorts, between sexes, or among populations can elucidate patterns that may be used to infer cultural practices at Lofkënd.

The populational affinities of the Lofkëndis, and specifically their relationship to the prehistoric population from nearby Apollonia, are also of great interest. Apollonia Tumulus 10 provides the only comparable Bronze and Iron Age data from south-central Albania, as skeletal remains from other tumuli in the region are not available. There are also post-medieval/modern burials at Apollonia that are directly comparable with the Lofkënd modern burials. Population relationships can sometimes be derived from biodistance analyses of cranial and dental data or non-metric trait frequencies. A set of potentially informative non-metric traits for the Lofkënd sample, identified during data collection, are included here in a preliminary biodistance analysis.

METHODS

Preparation and curation

Skeletal materials from Lofkënd were cleaned and prepared directly after excavation to ensure maximal retrieval of information from the excavators and to reduce the destructive effects of sediment consolidation. The more fragile and diagnostic elements, such as crania or pelves, were generally brought to the skeletal laboratory en bloc and prepared using wooden satay sticks; the same was true for entire infant burials (**Fig. 6.1**). When necessary, bones were washed and air-dried, although light brushing was the preferred cleaning technique. The fill from cremations was thoroughly screened and hand-sorted in the laboratory.

Particularly fragile material was placed in museum tissue or cotton fiber and stored in botanical tins. Fragmentary or small elements were curated in double-labeled plastic bags and placed in wooden boxes. These techniques have been successful in controlling for mold and mildew growth in the storage conditions at Apollonia where the collection is housed.

Skeletal analysis procedures

Surface bones, as well as isolated bones from the tumulus fill not associated with specific graves, were inventoried by their trench and unit. These materials were then checked for potential associations with disturbed graves. Due to the fragmentary nature of this non-grave bone, it was not possible to relate most of it with specific burials or to factor it into estimations of the Lofkënd sample parameters.

Data were collected using an array of forms designed in FilemakerPro and adapted from *Standards for Data Collection from Human Skeletal Remains* (Buikstra and Ubelaker 1994). These include sheets for dental inventory/metrics, aging, sexing, cranial metrics, postcranial metrics, non-metrics, and pathology.

The methods of aging and sexing used in this study were based largely on those suggested by Buikstra and Ubelaker (1994), with modifications to reflect specific variations observed in the Lofkënd sample. The entire collection was then reassessed to refine aging and sexing estimates based on a fuller understanding of the observed range of variation in the population. Due to the fragmentary nature of the material, a combination of aging techniques based on pubic symphysis changes, the auricular surface, dental development, tooth wear, and sutural closure were applied, with the greatest weight given to the first four methods, as they yield more precise estimates. Seriation of tooth wear was, however, often the only method applicable to poorly preserved burials. For this reason, tooth wear ages were carefully re-evaluated as the study progressed. The methodology for assessing attrition is derived from the techniques of Miles and Molnar, as discussed in Hillson (1996). Specific age estimates were achieved for the cases where dental development and skeletal maturation made more precise aging possible, but individuals were most frequently assigned to age cohorts of 5- or 10-year spans. There were few older adults who could be aged with any precision at >45 years. The creation of general age cohorts was made after the analysis was complete. As these are arbitrary divisions of the sample, they vary according to the question under investigation. The sex estimation was based on pelvic and cranial morphology (Buikstra and Ubelaker 1994), with additional data (both metric and morphological) from postcranial elements for some specimens.

Special attention was given to the dental remains, as the teeth were often the key elements available for individual identification and aging. For the dental analysis, the shorthand notation presented in **Table**

6.1 was generally used for tables and listings, although the terms *mandibular*, *maxillary*, *deciduous*, and *permanent* typically appear in the following text and narrative descriptions. The focus of the dental analysis was fourfold. All teeth were identified and mesial-distal (MD) and buccal-lingual (BL) measures were recorded. The pathological assessment included recording enamel hypoplasias, caries, and abscesses. Standard morphological variation, such as cusp number, incisor shoveling, and congenital absences, along with unusual morphologies, were also noted.

Teeth were judged as lost antemortem when the alveolar bone exhibited substantial remodeling such that the original forms of the sockets were no longer present. Caries presence was tallied by individual tooth. This count includes crowns or roots with active caries as well as teeth that were clearly lost to carious infection. The latter were teeth adjacent to, or occluding with, carious teeth where the decay process involved several teeth in the dentition. If no other evidence for caries was present, these teeth were only counted as antemortem losses. Hypoplasia, assessed by visual inspection, was denoted as present or absent, and the degree of expression was ranked on a scale of slight, moderate, or severe. In cases of clear linear hypoplasia, measurements were taken from the line to the cementoenamel junction. Agenesis was assessed visually and not radiographically. The total dental sample includes both teeth present and those missing where antemortem status could be assessed from the condition of the alveolus. Units of analysis included individual teeth and dentitions.

TABLE 6.1 Dental terminology

Tooth	Terminology
Deciduous central incisor	di_1, di^1
Deciduous lateral incisor	di_2, di^2
Deciduous canine	dc_1, dc^1
Deciduous first molar	dm_1, dm^1
Deciduous second molar	dm_2, dm^2
Permanent central incisor	I_1, I^1
Permanent lateral incisor	I_2, I^2
Permanent canine	C_1, C^1
Permanent first premolar	P_3, P^3
Permanent second premolar	P_4, P^4
Permanent first molar	M_1, M^1
Permanent second molar	M_2, M^2
Permanent third molar	M_3, M^3

Note: Subscript numbers denote mandibular teeth and superscripts denote maxillary teeth.

DEMOGRAPHY

Sample size

A total of 85 prehistoric human graves were identified at Lofkënd. The majority were single interments (68.2%), although there were 12 double burials, 13 triple burials, and 2 that contained four individuals. Multiple burials occur throughout the temporal sequence (**Fig. 6.2**), yet quadruple burials are only identified in Phases IV and V. The multiple interments contain adult pairs, subadult pairs, subadult trios (Tomb XVIII [Grave 73] in Phase II, Tomb XLIII [Grave 62] in Phase III), adult trios (Tomb XLVIII [Grave 52] in Phase III, Tomb LXXIV [Grave 29] in Phase V), and combinations of adults and subadults. Double burials were not recognized in the earliest (Phase I, Late Bronze Age) and latest (Phase Vb) stages of tumulus use. Phase Va is unique in that the number of multiple burials exceeds that of single interments, and double burials are predominant. These patterns could be indicative of behavioral differences or taphonomic factors. It is possible that multiple individuals are not identifiable in graves where preservation is poor or post-depositional disturbances (in the form of displacement by later interments, erosion of the tumulus peripheries, or subsequent agricultural use of the land) have occurred. Conversely, fragmentary individuals could be "added" to burials if they have been displaced from an earlier burial. Therefore, the patterning of these multiple burials should be regarded as representative of a general tradition of multiple burials at Lofkënd throughout the Late Bronze and Early Iron Ages.

One multiple grave is notable for its unusual contents. Tomb I [Grave 64], the probable central or "founding grave" of the tumulus, is comprised of elements from at least three adult males that appear to have been collected together, along with some faunal material, as a secondary deposit. The human bones vary in their levels of preservation from poor to very good. The skull elements include only three mandibles, a fragmentary maxilla, and a right zygomatic bone. No cranial vault elements are present. The postcranial bones include a hyoid, four cervical vertebrae (including three C2s), clavicles, scapulae, rib fragments, humeri, radii, ulnae, fibulae, one innominate, and hand and foot elements. Interestingly, the thoracic and lumbar vertebrae, along with the very large and durable femora and tibiae, are not present. The element representation is therefore quite unusual for a secondary burial deposit (where skulls or crania [Boyd 1993] or combinations of skulls, crania, and long bones are often the focus of redeposition).

Thirteen modern graves (Tombs LXXXVI, LXXXVIII-XCII, XCIV-C) also were discovered that contained human remains (two additional modern burials, Tombs LXXXVII and XCIII, were of animals [see Chapter 16.1]); of these, eight were single and five were double interments. An additional seven individuals were identified from non-grave contexts in the tumulus fill. These represent the remains of graves disturbed by subsequent burial activities or miscellaneous elements moved by natural processes, including animal burrowing, water movement, and other erosional processes. As these individuals were stratigraphically from the prehistoric levels, they are included in that portion of the sample for analyses. The total MNI (minimum number of individuals) for the prehistoric human subsample is therefore 136, and the modern subsample size MNI is 18 (**Table 6.2**).

TABLE **6.2** Graves and numbers of interments

	Grave count	%	Individual count	%
Prehistoric				
Single	58	68.2	58	45.0
Double	12	14.1	24	18.6
Triple	13	15.3	39	30.2
Quadruple	2	2.4	8	6.2
Total	85		129	
Modern				
Single	8	61.5	8	44.0
Double	5	38.5	10	56.0
Total	13		18	
Non-grave individuals			7	
Grand total	**98**		**154**	

Age distribution

Among the prehistoric burials, 12 (9.0%) could only be identifiable as adult individuals without further refinement of their age estimate, 5 (4.0%) could only be described as young adults, 8 (6.0%) as mid-age adults, and 2 (1.0%) as older adults. The remaining adults and subadults (109 [80%] of the total prehistoric subsample) were ultimately assigned to age cohorts spanning approximately 15 years or less, as illustrated in **Figures 6.3** and **6.4**.

The age distribution shows some important deviations from expected prehistoric population parameters. The expected mortality profile has a predominance of infants and children (over 60%), as these cohorts have the highest risk of morbidity and mortality from disease and malnutrition (Weiss 1974). The other at-risk cohort is older adults, although for prehistoric populations, "older" might be viewed as including individuals living beyond 45 years. Beginning with the distribution of individuals by raw counts, Lofkënd seems to differ greatly from this expected pattern, with a striking predominance of younger adult individuals aged 19–34 years. Biological factors that could produce such an unusual mortality profile would be low levels of population replacement due to high mortality of prime-age adults. This would create reduced infant, child, and older adult cohorts. The slight increase in the 3- to 5-year cohort relative to other children could reflect childhood deaths related to weaning stresses and higher infection rates stemming from the shift to gruels and adult foods. Conversely, the Lofkënd mortality profile could be merely the product of selective burial practices that largely favored the interment of prime-age adults.

A different pattern is seen with the modern individuals (**Fig. 6.3**). While the greater representation of infants and older adults resembles the expected distribution, small sample size limits any further interpretations.

The age profile based on the raw individual counts is informative, but the small subsample sizes may make the figure misleading. If the proportional distribution of the various age cohorts is graphed (**Fig. 6.4**), the differences between the prehistoric and modern subsamples are even more dramatic, and the anomalous distribution of the prehistoric subsample is clear. Over 32% of the prehistoric subsample is comprised of young adults, while the older adults and adolescents are relatively rare. The modern subsample is overwhelmingly represented by infants (>60%) and older adults (27.8%).

Interestingly, the pattern in the prehistoric subsample persists through time (**Fig. 6.5**); the only exception is Phase II, where the 3- to 5-year age cohort is the largest, but only by one individual (**Table 6.3**). Again, it is important to look at the proportional representation of the different ages by phase because of unequal sample sizes (**Table 6.3** and **Fig. 6.6**). From these proportions, it is evident that Phase II is clearly different in its representation of children under age 13 (52% of individuals for that phase) and that it is closer to expected population parameters.

Sex distribution

The number of prehistoric individuals for whom sex could be determined is 67, or 49.0% of the total sample. This group includes 71% of the 95 adults and adolescents in the sample. The proportions of females and males are quite different (**Fig. 6.7**). There are 39 males as compared to only 28 females if the questionable assignments are combined with the more definitive estimates. Fairly equal sex ratios are expected in large biological populations; hence the Lofkënd result of 1.39 males to females requires some explanation. The elevated sex ratio could be due to sampling bias, from the perspective that the limited sample sizes for each temporal phase are not truly representative of the biological population. There may also be potential bias for identifying males, particularly when size and rugosity (muscularity) of the skull and long bones are the chief criteria used for the estimation. This is the case for many individuals where the more definitive sexual dimorphic morphology of the pelvis is not preserved. Pelvic morphology was used in determining only 16 (24%) of these sex estimates. Alternatively, the predominance of males may also be indicative of the differential burial of males at Lofkënd. Any of these factors may contribute to the Lofkënd sex ratio, but it is not possible to determine the specific cause or causes.

The adolescents are a difficult group for sex estimation. There are only eight prehistoric individuals in this age cohort, and six of them are sufficiently complete and skeletally mature to warrant sex estimation. They were first compared as a cohort and then compared metrically to fully adult individuals. Tomb LXX (Grave 17), Individual 126 was aged 16–18 years and sexed as female based on cranial morphology and postcranial size. Tomb XLVIII (Grave 52]), Individual 318, aged 15–17 years, was also sexed as female based on postcranial size. Three other adolescents—Tomb XXI (Grave 55), Individual 341; Tomb XXXIII (Grave 92), Individual 523; and Tomb LXVI (Grave 31), Individual 216—were possibly females based on their postcranial size. The only potential male in this age cohort was Tomb LII (Grave 69), Individual 390, aged 15 ± 3 years, who displayed clearly male cranial morphology, a large dentition, and very large rugose postcranial elements. The predominance of female adolescent burials is very interesting, as members of this age cohort are accompanied by some of the most impressive grave goods at Lofkënd (see Chapters 3 and 8).

If we examine the proportions of the sexes by temporal phases (**Table 6.4**), we find that males represent over 30% of the sample in all phases but Phase II, where they are less frequent than both females (22%) and children aged 3–12 (37%) and Phase IV,

TABLE 6.3 Proportional representation of ages by temporal phases

Phase	Infant 0–2 yr	Child 3–5 yr	Child 6–12 yr	Adolescent 13–18 yr	Adult 19–34 yr	Adult 35–45 yr	Adult >45 yr	Adult	Total
I	2 (12%)	2 (12%)	2 (12%)	0	**6 (35%)**	2 (12%)	2 (12%)	1 (6%)	17
II	4 (15%)	**6 (22%)**	4 (15%)	2 (7%)	5 (19%)	5 (19%)	1 (4%)	0	22
III	2 (6%)	6 (18%)	0	2 (6%)	**11 (33%)**	6 (18%)	2 (6%)	4 (12%)	33
IV	0	1 (8%)	1 (8%)	2 (15%)	**4 (30%)**	2 (15%)	1 (8%)	2 (15%)	13
Va	0	0	4 (20%)	1 (5%)	**8 (40%)**	4 (20%)	1 (5%)	2 (10%)	20
Vb	1 (5%)	1 (5%)	0	1 (5%)	**9 (47%)**	3 (16%)	2 (10%)	2 (10%)	19

Note: The largest cohort per phase is in bold.

TABLE 6.4 Proportional representation of sex by temporal phase

Phase	Infant 0–2 yr	Child 3–12 yr	Female	Male	Indeterminate	Total
I	2 (12%)	4 (24%)	5 (29%)	**6 (35%)**	0	17
II	4 (15%)	**10 (37%)**	6 (22%)	4 (15%)	3 (11%)	27
III	2 (6%)	6 (18%)	8 (24%)	**11 (33%)**	6 (18%)	33
IV	0	2 (15%)	4 (30%)	**2 (15%)**	5 (38%)	13
Va	0	4 (20%)	3 (15%)	**7 (35%)**	6 (30%)	20
Vb	1 (5%)	1 (5%)	2 (10%)	**8 (42%)**	7 (37%)	19

Note: The largest cohort per phase is in bold.

where indeterminate adults (38%) and females (30%) are better represented. The temporal subsamples are quite small, however, and no specific conclusions should be drawn from these distributions—except that the bias toward male burial is not limited to any particular phase at Lofkënd.

The distribution of sexes can also be examined for the various prehistoric age cohorts. A total of 61 individuals have both age and sex estimates. The greater number of males is characteristic of all age cohorts except the adolescents. **Figure 6.8** also illustrates the relative scarcity of adult females in the two oldest age cohorts.

Are these differences in male and female representation, whether by temporal phase or by age cohort, significant? If the two largest temporal subsamples (Phase II with more females and Phase III with more males) are compared, they are not significantly different (Fisher's exact test, two-tailed $p = 0.4497$). Similarly, the differences between the numbers of males and females in the two youngest age cohorts (13–34 years) are not significantly different from the two oldest age cohorts (35–45+) (Fisher's exact test, two-tailed $p = 0.2944$).

GENERAL POPULATIONAL FEATURES

Cranial form

The Lofkënd prehistoric crania are fragmentary, and therefore no maximum lengths or breadths can be measured. In situ observations and field photographs document the general characteristics of cranial shape for the population, which can be described as long relative to the breadth (cf. **Fig. 6.9**). This is in clear contrast with the modern subsample where high, short, rounded vaults are documented (**Fig. 6.10**) (for the three measurable modern crania, the cranial indices are 83, 87.7, and 90.6; these are in the brachycephalic [80–85], hyperbrachycephalic [85–90], and ultrabrachycephalic [90–95] ranges [Garson 1886]). The elongated crania shape that characterizes the prehistoric subsample is the most common form observed in Greek prehistoric populations from the Neolithic. For example, Angel (1974) found the four Neolithic Lerna crania ranged from 69.4 to 77.4; the larger Middle Bronze Age Lerna sample had cranial indices ranging between 66.1 and 85.7. The differences between the prehistoric and modern Lofkënd subsamples can best be viewed as conforming to the general trend toward brachycephaly in Europe that has been documented for other Albanian samples (Dhima 1993; Nemeskéri and Dhima 1988).

Stature

Figure 6.11 provides data for the few complete femora from Lofkënd. A small set of femoral lengths ($n = 12$) are available. These include seven prehistoric individuals and five moderns. For the prehistoric individuals, the female mean is 414.7 mm and the male mean is 454.3 mm. Although comparison is limited by these small sample sizes, a pattern emerges: the prehistoric population is clearly taller than the modern group. In fact, the prehistoric females are basically identical to the modern males for this measure. Using stature regression formulae from Trotter (1970) for Europeans, the prehistoric females are 5′1 (155.4 cm), the males are 5′5″ (167.6 cm); the shorter modern females are 4′7″ (143.2 cm) while the males are 5′2″ (158.5 cm). The Trotter formulae are thought to underestimate the stature of Balkan populations (Ross and Konigsberg 2002), and new regression formulae have been developed for Greeks based on cadavers (Eliakis, Eliakis, and Iordanidis 1966). Using these, the stature estimates are slightly greater: prehistoric females 5′2″ (158.5 cm), prehistoric males 5′7″ (173.7 cm); modern females 4′8″ (146.3 cm), modern males 5′4 (164.5 cm). Comparative data from anthropometric studies provide more information on male stature for Albania and northern Greece: southern Albanian males, $n = 85$, 5′5″ (167.7 cm); northern Albanian males, $n = 77$, 5′6″ (170.6 cm) (Tildesley 1933); and for Greek Macedonian males, $n = 200$, 5′5″ (167.6 cm) (Hasluck and Morant 1929). These data reinforce the impression that the modern Lofkënd subsample is derived from a comparatively short population. Reduced stature is a general indicator of stress during growth and development, and it has been used as a marker for health changes in populations (cf. Steckel and Rose 2002; Steckel 1995).

Sexual dimorphism

Sexual dimorphism in human populations has been linked to subsistence behaviors, gender roles, and the consequent stresses that might differentially affect one gender. The level of stature dimorphism is thought to decrease when women are heavily involved in agricultural food production, but prehistoric data do not indicate a clear pattern linking stature dimorphism and subsistence change (Holden and Mace 1999). In some cases, female stature declines with intensive agriculture when protein levels decrease and males have differentially greater access to the protein sources that are available. The result is increased dimorphism for many variables. This is in contrast to the situation where general under-nutrition affects a population. Then the level of dimorphism decreases as males, who are less biologically buffered and have higher nutrient demands, reduce in size. Aside from nutritional concerns, dimorphism of limb midshaft dimensions may be indicative of gender-specific activities as diaphyses are remodeled and augmented by functional demands.

Although the long-bone length data for Lofkënd are limited, it is still possible to evaluate the degree of sexual dimorphism in the population. Stature is one of the most dimorphic features in modern humans, but other measures are also variable. Limb midshaft dimensions are good estimators of body size and are therefore useful for examining sexual dimorphism. Mandibular corpus height, as a component of facial height, is also typically dimorphic. **Table 6.5** presents data for sexual dimorphism based upon mandibular corpus height at P_3, humerus midshaft, and femoral subtrochanteric, midshaft, and length measurements. The index of sexual dimorphism is calculated as the (male mean/female mean) × 100. Sexual dimorphism at Lofkënd varies from 105% to 113%. These results are within the range of values reported in other studies; for example, Frayer and Wolpoff (1985) compiled data on femur length dimorphism that ranged from 103.3% to 110.7%. It is interesting that the higher values are associated with midshaft limb circumferences. This may reflect differential robusticity levels and greater variation in the physical activity or nutritional status of males, as their ranges and standard deviations are greater than they are for females.

General pathology

Pathological conditions provide information on the life experiences of individuals. Some stresses, diseases, or trauma occur early in life, and their impact may or may not be directly observable. This is due to the fact that some stresses affect the growth process

TABLE 6.5 Levels of sexual dimorphism for the prehistoric subsample (all measurements in mm)

Measure	Male *n*	Male mean	Male range	Female *n*	Female mean	Female range	Dimorphism
Mandibular height at P_3	16	30.2	24.8–35.5	11	28.5	22.8–34.0	106.0
Humerus midshaft circumference	14	67.4	40.4–78.0	11	59.5	52.0–71.0	113.0
Femur subtrochanteric circumference	25	81.0	71.0–93.0	29	77.0	68.0–87.0	105.0
Femur midshaft circumference	17	93.1	80.0–109.0	18	82.6	67.5–95.0	113.0
Femur length	4	454.3	435.0–472.0	3	414.7	405.0–420.0	111.0

and leave permanent markers (such as linear enamel hypoplasia of the teeth) while other markers are ephemeral due to bone remodeling (Harris lines in long bones or healed sites of trauma). Stress incidents that are severe enough to produce immediate mortality typically leave no skeletal signatures—but the survivors, who are the more evolutionarily fit individuals, may therefore bear more skeletal markers of stress and disease. Thus "healthy" or pathology-free skeletons are not necessarily indicative of health in life. They may simply represent individuals who succumbed rapidly to a disease that does not affect the skeleton, or individuals who did not have any resistance and died before skeletal markers formed. This is the appropriately named "osteological paradox" (Wood et al. 1992).

The Lofkënd individuals display an array of pathological conditions, ranging from indicators of childhood stress to adult age-related bone and tooth deterioration. The most commonly observed conditions are linear enamel hypoplasias, caries, antemortem tooth losses, porotic hyperostosis, and degenerative joint diseases and osteoarthritis. The following discussion emphasizes pathological conditions that are indicative of population-wide stresses rather than individual-specific conditions, although a few individual cases are described.

Linear enamel hypoplasia

Linear enamel hypoplasia is a disruption to the dental enamel in response to stress. The enamel formation is interrupted, and a line is formed marking the resumption of enamel development. Other, less frequently observed, hypoplasias are expressed as pits or crown defects. The stressors correlated with hypoplasia are multiple and include various pyrogenic childhood diseases and infections, malaria, nutritional deficiencies, and parasitic infections (cf. Hillson 1996). Thus enamel hypoplasias are non-specific stress indicators that cannot be attributed to a particular etiology.

The frequency of linear enamel hypoplasia at Lofkënd is high (56.6%, or 60 of 106 evaluated dentitions) (**Table 6.6**). Most individuals have only mild hypoplasia that does not affect the entire dentition and is not consistently identifiable on dental antimeres as might be expected with a stressor that impacts growth. Only 3.3% of the affected individuals have moderate to severe levels of hypoplasia. Interestingly, the proportion of affected males and females is nearly identical. Approximately 35% of the unaffected individuals were young children who may have died from an initial stress incident and thus had no formation of growth cessation indicators; 51.5% of all children evaluated had slight hypoplasia, but it was almost exclusively observed on the permanent crowns. These results indicate that the Lofkëndis were routinely subject to low levels of stress during childhood when their dentition was forming. In some instances, these stresses were probably acute episodes that contributed to infant and young child mortality.

Cribra orbitalia and porotic hyperostosis

Bones can reflect stress incidents in the form of increased porosity or hypertrophy of marrow-producing regions such as the diploe of the cranial vault and the medullary cavities of long bones (cf. Stravopodi et al. 2009; Stuart-Macadam 1985, 1992). The greater likelihood of cranial vault bone preservation means that the skull is most frequently evaluated. Porosity and hypertrophy are associated with iron-deficiency anemias and with the physiological responses to situations of nutritional insufficiency, infectious diseases, and parasite loads. Certain congenital anemias, such as sickle-cell or thalassemia, are maintained in populations living in areas of endemic malaria due to the conferred resistance that the genetic alterations of red blood cells provide. Although the relationship between congenital anemias and malaria has long been assumed for much of the Mediterranean region (Angel 1966, 1972), more recent research employing microscopic and aDNA techniques has found that congenital anemias, as well as acquired anemias, are often misdiagnosed in skeletal studies

Table 6.6 Hypoplasia prevalence by individuals

	Present	Absent	N/A	*n* Evaluated	% Present	Total
Subadult	17	16	7	33	51.5	33
Male	23	13	3	36	63.9	36
Female	14	8	4	22	63.6	22
Adult	6	9	9	15	40.0	15
Total	60	46	23	106	56.6	106

(Stravopodi et al. 2009). This is particularly true for evaluations of cranial vault porosity and diploe expansion, but it also affects assessment of cribra orbitalia (Wapler, Crubézy, and Schultz 2004). Vault porosity is often produced by taphonomic processes, and diploe expansion or general vault thickening can be due to age-related bone remodeling (Hatipoglu et al. 2008; Stravopodi et al. 2009).

The two anemic responses that affect the cranium, cribra orbitalia of the orbital roof, and porosity on the cranial vault, were evaluated for Lofkënd. Cribra orbitalia is most commonly observed after the age of weaning, when children's nutritional levels change; it is active for a period of time and then begins a slow process of "healing" or bone remodeling that can still be observed in adults. The adult anemic response is porotic hyperostosis or vault porosity. This is often accompanied by vault thickening. Cranial thicknesses exceeding 8 mm (cf. Musgrave 2005) are sometimes interpreted as indicators of anemia, but cranial vault thickening also occurs with aging (Hatipoglu et al. 2008). Thickness was therefore not included in this analysis unless it was independent of potential age-related effects. Neither cribra orbitalia nor cranial porosity can be used to distinguish genetic from acquired anemias.

Although the frequency of linear enamel hypoplasia at Lofkënd indicates childhood stresses were very common, the bone indicators of stress are rare. Only one of the 50 adult prehistoric crania assessed for porotic hyperostosis is possibly porotic, one child has cribra orbitalia, and four adults (three males and one female) potentially have healing or healed cribra orbitalia. The differences between the linear enamel hypoplasia findings and the porotic hyperostosis prevalence may be largely the result of two different processes affecting bone: normal remodeling, and taphonomic destruction of cranial vault surface and the cranium in general. It is also possible that the stresses leading to hypoplasias are of a lesser magnitude than those resulting in porotic hyperostosis and cribra orbitalia; linear enamel hypoplasia may therefore be a more sensitive indicator of these stresses. Additionally, it should be emphasized that the assessment of these conditions is highly subjective and dependent upon the skill and experience of the analyst. In particular, the identification of slight linear enamel hypoplasia by visual observation is difficult, and the levels of moderate and severe linear enamel hypoplasia are probably more indicative of childhood stresses.

Degenerative joint disease (DJD)

Few of the Lofkënd skeletons had well-preserved joint surfaces, so the quantitative assessment of degenerative joint disease is necessarily an underestimate of the true frequency. The most frequent location of joint degeneration was on the vertebral centra. Of the 45 adult individuals that have preserved vertebrae, 8 (17.8%) had clear evidence for osteophytic growths. As DJD is correlated with age, if adults aged <25 years are excluded from the analysis, the frequency rises to 33.3% (8/24). The most commonly affected vertebrae are the cervicals (5/8 cases). The vertebral frequency for the five modern adults is 100%, with all individuals exhibiting DJD of the thoracic vertebrae. The cervical and lumbar vertebrae of the youngest individual (M45 Unit 277, aged 25–35 years) were not affected. Two individuals had osteophytic growths that were extensive and conjoined to produce fused vertebral centra or arches. For the older male Individual 516 from Tomb II (Grave 91), this involved several thoracic and lumbar vertebrae.

A more systemic case of joint deterioration affected Individual 345 from Tomb XLV (Grave 60) (**Fig. 6.12**), among the oldest of the individuals in the Lofkënd tumulus. In addition to osteophytes on his clavicles, scapulae, hand and foot elements, ribs, cervical vertebrae, thoracic vertebrae, and ilia, there were two fused vertebrae with a large bolus formation. No lumbar or sacral elements were preserved. The deterioration of the articular cartilage of the cervical vertebrae, right scapular glenoid fossa, and the distal radius produced eburnation (bone polishing due to loss of articular cartilage on the joint surfaces). The areas of tendon and muscular insertion are rugose on many of the skeletal elements. This suite of features may be due to general degenerative changes (the age estimate is >55 years), but it is also indicative of diffuse idiopathic skeletal hyperostosis (DISH), and that diagnosis cannot be ruled out. Individual 345 also had a possible healed injury to the left ankle involving the tibia and fibula. The latter pathology may have resulted from impaired stability and mobility, although it is important to note that osseous bridging of these two bones is observed in 10% of DISH cases (Rothschild 1985).

Aside from the vertebrae, other joints showing evidence of degenerative joint disease in the prehistoric subsample include the glenoid fossa of the scapula (five cases) and the acetabulum (three cases). In contrast, joint degeneration was quite extensive in the modern adults, and some individuals had systemic alterations of their joint surfaces.

Trauma

Nine of the prehistoric adults suffered traumatic injuries. One of the most seriously affected individuals was a mid-aged adult male (Individual 038, Tomb LXXXIV [Grave 2]) with a dislocation of the right elbow (**Fig. 6.13**). A secondary joint was formed inferior to the original articulation in the olecranon fossa of the humerus. All elbow articular surfaces were altered, and some areas have eburnation from bone-on-bone wear following destruction of the articular cartilage. These features clearly indicate that the arm was used after the injury and continued to be somewhat functional long afterward. Individual 139 from Tomb LXXVII (Grave 18), another mid-aged male, has a healed cranial trauma as a result of a blow to the forehead above the right eye (**Fig. 6.14**). While the external bone surface is remodeled and retains only a discoid depression, there are alterations to the endocranial surface that suggest invasive trauma and possible infection of the periosteum and protective coverings of the brain such as the dura mater. A third adult male (Individual 197 from Tomb LXX [Grave 17]), aged >45 years, has a healed injury involving the fusion of two ribs on the left side of his back near the rib heads and the articulation with the vertebrae.

Less severe injuries to the legs and feet, involving periostitic reactions, are evident for five adults (Individual 32, Tomb LXXX [Grave 4]; Individual 296, Tomb LVI [Grave 43]; Individual 305, Tomb XIII [Grave 49]; Individual 345, Tomb XLV [Grave 60], as discussed above; and Individual 369, Tomb XXXIX [Grave 66]) (**Fig. 6.15**). All of these were young adults, with the exception of Individual 345 who was >55 years. Only one female, Individual 369 (Tomb XXXIX), had evidence of a leg injury. A 30- to 40-year-old male, Individual 439 from Tomb XXXVIII (Grave 79), had a crushed left hallux from a trauma that affected the joint between the two phalanges (**Fig. 6.16**). There is limited evidence of healing and remodeling of this toe.

Infectious disease

Two adult crania exhibit irregular mastoid processes that were altered by infections. The right temporal bone of an older adult male, Individual 516 from Tomb II (Grave 91), has a withered mastoid (**Fig. 6.17**). The surface appears to have vestiges of drainage tracks, and the base of the auditory meatus is potentially affected by the same infectious process. It is expanded and excavated inferiorly, making the inner ear features very visible. The mastoid processes of a young female, Individual 369 from Tomb XXXIX (Grave 66), are very unusual. The right process consists of two distinct projections; the anterior one is small and relatively narrow, while the posterior is a large, broader and bulbous knob that projects more than the anterior (**Fig. 6.18**). No pathological bone structure is evident from visual inspection, but the adjacent sigmoid sulcus is very deep. Bilateral observations were not possible, but while cleaning the matrix block, the fragmentary left mastoid appeared to be complex as well. This suggests that it is probably a morphological variant of the mastoid rather than a pathological condition, although an infection of the oral cavity tissues may produce bilateral alterations of the mastoid sinuses. Similar mastoid morphology is described as examples of possible mastoiditis for two skulls from Ban Chiang, Thailand (Pietrusewsky and Douglas 2002).

Dental pathologies: caries, antemortem tooth loss, crowding, and rotation

The total number of teeth for the prehistoric subsample at Lofkënd is 2,280. These represent 112 individuals of all ages that have 1–42 teeth or tooth crowns preserved. Only 8 adults have a complete dentition of 32 teeth, but 26 adults and adolescents have 28 or more teeth. Twelve adults and 11 subadults have no teeth preserved.

Caries and antemortem tooth loss rates can provide evidence of dietary differences, particularly where diets are based on staple crops that contribute cariogenic carbohydrates to the diet (cf. Larsen, Shavit, and Griffin 1991). For example, significantly different levels of dental health have been documented for Bronze Age Mycenaeans at the Palace of Nestor, Pylos. Dental health is correlated with sex, status, and access to animal protein in this population, where females have significantly more caries and antemortem losses and higher-status individuals from *tholoi* ("beehive" tombs) have healthier dentitions than individuals from chamber tomb burials (Schepartz, Miller-Antonio, and Murphy 2009).

For Lofkënd, the rates of caries and antemortem losses provide some insight regarding possible dietary differences between genders, and between the prehistoric (where levels of dietary animal protein may have been higher) and the modern individuals. Musgrave (2005) compiled useful data on dental caries prevalence in Greece. It is difficult to compare directly results of different studies, so these figures should be regarded as basic indicators of the possible range of prevalences. The values for caries prevalences range from 2.9% for the Mycenae shaft graves (Angel 1944) to 36.9% for Late Minoan Chania from McGeorge (1992). The combined Neolithic prevalence is 12% and the Early Iron Age combined prevalence is 5.5%, although Angel's figures for "carious plus lost teeth" for these time periods are 16.5% and 19.9%, respectively (Angel 1944). Most interestingly, Iron Age individuals from Hungary had fairly low caries rates of approximately 5%; if losses potentially due to caries are factored in, the prevalence rises to approximately 7% (Ubelaker and Pap 1998).

Table 6.7 presents Lofkënd data on caries prevalence. A total of 110 dentitions comprised of 1,996 teeth from 29 subadults and 81 adults were assessed for caries. Caries can occur in deciduous teeth or the erupted permanent dentition of subadults. For this reason, subadult teeth were evaluated if they were erupted and therefore potentially exposed to cariogenic agents. As the years of exposure increase the risk of caries, the subadult individuals should be evaluated separately. While caries in subadults are relatively rare, two prehistoric subadults have one carious infection each (2/29 children).

TABLE **6.7** Frequencies of caries for prehistoric and modern Lofkënd

	n	# Teeth evaluated	# Carious teeth	%
Subadults	29	362	2	0.6
Adults	81	1634	193	11.8
Total sample	110	1996	195	9.8
Males	35	788	104	13.2
Females	24	578	48	8.3
Moderns	6	166	105	63.3

The total adult prevalence, at 11.8%, is relatively high, and is more similar to the Neolithic rather than the Iron Age comparative data. The patterning of caries is also informative. Lofkënd males have significantly more caries than the females (Fisher's exact test, two-tailed $p = 0.0052$), even though fairly similar proportions of each group have carious dentitions (51.4% of males as compared with 45.8% of females). Males tend to have more caries if they have any at all (**Fig. 6.19**). Of the 14 individuals with five or more caries, only 3 are females. One male (Individual 230, Tomb LXXIV [Grave 29]) has 21 carious teeth! If he is removed from the sample, the male prevalence falls to approximately 11%, but the male and female prevalences are still significantly different.

The differences in the adult prevalences may be due to the differential representation of older males at Lofkënd. If we stratify the data by age cohort, adults younger than 35 years have significantly fewer caries than those over 35 (Fisher exact test, two-tailed $p = 0.0001$). Sixty-eight percent of the females are in the younger group as opposed to only 57.6% of the males. Within the younger and older age cohorts, the prevalences of caries for male and females are not significantly different; therefore it is clear that the differences in overall caries prevalence for male and females at Lofkënd are due to the age distribution.

Does the prehistoric prevalence of caries vary over time (**Table 6.8**)? Temporal phases are grouped into pairs to create larger and more equal sample sizes for comparisons. The early prevalence is 12.6%, followed by 13.1% and 10.1% for the most recent phases. If we examine the data by dentitions rather than single teeth, it appears that there is some reduction in the number of individuals who are acarious in Phases III and IV, but the level rises again for Phases Va and Vb. These differences are not statistically significant. The variation in caries prevalence is best explained by fluctuations in sample size and demographic factors.

TABLE **6.8** Prevalence of caries by temporal phases

Phase	*n*	# Teeth evaluated	# Caries	%	% Individuals with no caries
Vb–Va	21	624	63	10.1	57.1
IV–III	29	650	85	13.1	37.9
II–I	21	358	45	12.6	58.1

The caries prevalence for the Lofkënd modern subsample is an astonishing 63.3%, as 105 of 166 evaluated teeth are carious. This is not completely unexpected, as four of the five individuals are mid-age or older.

Antemortem tooth loss is most often associated with advanced caries infection, although trauma, severe attrition, and severe periodontal disease also result in tooth loss. As indicated by the data in **Table 6.9**, tooth loss during life was a common phenomenon for modern Lofkëndis (42.8% of all assessed teeth), but it was relatively rare for the prehistoric subsample (5.3% of all teeth assessed, and affecting 20% of the adults). This is consistent with the caries data. It is important to recognize that antemortem losses are typically the outcome of lengthy decay or disease processes, and thus the frequency is confounded by age. This analysis therefore only considers adult individuals. The modern adults are primarily older adults, and this best explains the higher frequency for that phase.

Dental crowding

Dental crowding refers to conditions where teeth are rotated or displaced from the normal dental arch form. This typically involves the anterior dentition due to mesial drift of teeth from masticatory stresses, although all teeth can be affected. Crowding of the anterior mandibular teeth is the most common manifestation (**Fig. 6.20**). Basically, dental crowding is a result of tooth size exceeding the available space in the alveolar bone. Crowding also increases the risk of caries and periodontal disease (Kuo 1983).

It was once believed that crowding was only characteristic of recent populations, but it is now documented for many prehistoric samples (cf. Mockers, Aubry, and Mafart 2004). There is some dental literature supporting the idea that crowding, as a component of malocclusion, has a genetic basis (Smith and Bailit 1977; Mossey 1999a, 1999b). However, experimental dietary studies in animals have verified that crowding is a result of soft diets and underdevelopment of the masticatory system (Beecher and Corruccini 1981; Ito, Mitani, and Kim 1988; Pottenger 1942). Therefore the prevailing theory is that there is a genetic component to malocclusion and that dental crowding is also indicative of dietary and technological changes in food preparation. Thus some studies have evaluated crowding in order to assess dietary differences in prehistoric groups (cf. Sakashita et al. 1997).

TABLE 6.9 Antemortem tooth loss frequency for the prehistoric and modern subsamples at Lofkënd

Phase	n	# Teeth assessed	# Lost	%
Prehistoric Lofkënd	95	1634	87	5.3
Modern Lofkënd	5	166	71	42.8

The prevalence of dental crowding at Lofkënd is fairly high. Individuals with secondary rotation due to antemortem tooth losses or injury were excluded from this calculation. Of 69 assessed dentitions, 29 (42%) exhibited some form of crowding or tooth rotation. There were approximately equal numbers of affected male and females. Most notably, there is a no clear increase in crowding or rotation over time at Lofkënd, as might be expected with a change in food processing.

FAMILIAL AND POPULATIONAL RELATIONSHIPS THROUGH NON-METRIC TRAIT ANALYSIS

The study of skeletal remains from large cemeteries and shared graves raises the question of familial relationships. There are a number of non-metric traits that are typically used to assess genetic relationships. These are most informative when rare conditions are found to be shared by some individuals. Other, more prevalent, traits might provide information on broader populational affinities. The following features were noted in the Lofkënd skeletal series.

Maxillary lateral incisor (I^2) trait

Variations of the maxillary lateral or second incisors are relatively common (Bailey-Schmidt 1995), but the diversity of I^2 forms at Lofkënd is particularly unusual. These include slight rounding and folding of the lingual surface, a cleft or deep infolding of the crown often associated with a groove that continues down the root (**Fig. 6.21**) (similar to the crown-root groove variant of Hillson [1996:19] but with torsion of the crown), size reduction, peg or barrel morphology (Hillson 1996:19), and complete absence or agenesis. The expression of the trait appears to follow a continuum where the tooth germ is constricted and differentially affected; the crown is increasingly folded and reduced in size, with the most extreme expression being either pegging or total absence. The trait may appear bilaterally, and when that is the case there is considerable asymmetry in its expression. Further research on this morphology might indicate that separate classifications of the forms are warranted. For the purposes of this study, these varying expressions were termed "I^2 trait" and recorded as present or absent for the individual dentition (**Table 6.10**). The frequency of the trait is very high—it characterizes 50% of the prehistoric Lofkënd individuals and 54.6% of the prehistoric Apollonians. Shoveling, a trait more commonly observed on the central maxillary incisors, was also present at Lofkënd, although at a lower frequency (33.3% or 23 of 68 individuals) than the I^2 trait. The two incisor variants may be correlated.

Foramen caecum molare

The foramen caecum molare (f. c. molare) is a pit that can occur on the buccal surface of deciduous and permanent mandibular molars; in some cases, it appears more as an accentuated depression rather than a clear pit (**Fig. 6.22**). Although there is little published about this trait, it is widely recognized and referred to as a "buccal pit." The etiology is not entirely clear. It is most probably a developmental defect of the enamel, as it forms in the buccal groove between the cusps and is correlated with linear enamel hypoplasia in some populations (Capasso and Di Tota 1996). While it is sometimes mistaken for caries (and it is a site for caries formation), the foramen differs in coloration and morphology from carious decay, as it is not a product of enamel destruction but enamel absence. The frequency of f. c. molare varies among populations, making it a potential non-metric marker for examining populational affinities. In the Lofkënd prehistoric population, the f. c. molare was present on 30 of 90 dentitions that could be evaluated for its prevalence (**Table 6.11**). The frequency for prehistoric Apollonia is 56.4%, and it varies from its greatest frequency in the prehistoric phase to 11.8% in the Classical period. Interestingly, the frequency for the small post-medieval/ modern Apollonia subsample is also high, at 45.5%.

Carabelli's trait

This non-metric feature is defined as an accessory cusp or fissure on the mesial lingual aspect of maxillary molars. Both deciduous and permanent teeth can be affected. The trait usually is most strongly expressed on the first molar and then appears in progressively diminished form on the second and third molars if those teeth are affected. Recent research suggests that it is correlated with crown size; larger teeth are more likely to have this accessory cusp (Harris 2007). Eighty-nine prehistoric dentitions with one or more M^1 or dm^1 present could be evaluated for Carabelli's trait. Nineteen dentitions, or 21.3% of the Lofkëndis, displayed the development of a full cusp or a lesser expression of this trait. Interestingly, the affected individuals were predominantly subadults (88%), and there were three instances where two individuals were from the same grave. The prevalence of Carabelli's trait also varied by phase; two cases were from Phase I, four were from Phase II, and Phases III and Va each had five cases.

Carabelli's trait frequency is very low at Lofkënd and Apollonia (**Table 6.12**) in comparison to other values reported for recent European populations, where it is most commonly observed at 75–85% (Scott 1980).

TABLE 6.10 Maxillary lateral incisor (I^2) trait frequency

Phase	# Individuals assessed	# Traits	%
Lofkënd prehistoric	62	31	50.0
Lofkënd modern	3	0	0
Apollonia prehistoric	33	18	54.6
Apollonia post-medieval/ modern	7	2	28.6

TABLE 6.11 F. c. molare frequency

Phase	# Individuals assessed	# Traits	%
Lofkënd prehistoric	90	30	33.3
Lofkënd modern	1	0	0
Apollonia prehistoric	39	22	56.4
Apollonia classical	17	2	11.8
Apollonia post-medieval/ modern	11	5	45.5

TABLE 6.12 Carabelli's trait frequency by temporal phase

Phase	# Assessed	# Traits	%
Lofkënd prehistoric	89	19	21.3
Lofkënd modern	3	0	0
Apollonia prehistoric	37	3	8.1
Apollonia classical	16	2	12.5
Apollonia post-medieval/ modern	9	3	33.3

TABLE 6.13 Frequencies of non-metric traits for Lofkënd prehistoric burials

Trait	# Assessed	# Traits	%
Supernumerary tooth	63	2	3.2
Congenital tooth absence	53	6	11.3
Enamel pearl	77	1	1.3
Incisor shoveling	69	23	33.3
Enamel extension	77	3	3.9
Deciduous tooth retention	66	3	4.5
Metopic suture	33	2	6.0
Multiple infraorbital foramina	20	3	15
Malar tubercle	28	3	10.7
Asterionic bone	28	2	7.1
Parietal notch bone	27	1	3.7
Inca bone	28	3	10.7
Wormian bones	28	11	39.3
Mastoid suture	39	1	2.6
Infraorbital suture	21	2	9.5

There are a few other dental and cranial variants that were observed at Lofkënd. A selection of non-metric traits and their frequencies are given in **Table 6.13**. The frequencies of these traits vary tremendously. Enamel extensions, a condition where the crown enamel extends below the crown and between the roots, were noted for three individuals. Enamel pearls, or isolated beads of enamel on root surfaces, were observed on only one dentition. Three individuals retained deciduous teeth in their adult dentition; these included one canine (**Fig. 6.23**) and two dm_2s. There were two cases of the infraorbital suture running from the inferior orbital border to the infraorbital foramen (**Fig. 6.24**), and a single case of a mastoid suture running down the length of the mastoid process lateral surface. The most common variants are incisor shoveling (33.3%) and wormian bones (39.3%); these are obviously of limited utility for examining within-population genetic affinities, but they can be used to evaluate degrees of relatedness between or among populations.

The next stage in non-metric trait analysis involves the examination of shared traits among individuals in common burials. As presented in **Table 6.14**, there are 30 multiple burials at Lofkënd of which 4 are modern burials. The 71 prehistoric interments are comprised of various combinations of individuals: pairs or trios of males, male/female pairs, adolescents with adults, children with adult pairs, groups of adults, groups of children, and pairs of infants. In the four cases of an identifiable female–male pair, three were composed of young females with older males. In the remaining burial, Tomb LVI (Grave 43), the individuals were both young adults. There is no evidence for pairs or groups of females buried together, but there are at least three clear interments of multiple males. From these observations, one could argue that male–male alliances were important in Lofkënd society, and males were typically paired with younger females after they had established themselves in the society.

In terms of the possible relationships among individuals buried together, it is most parsimonious to assume (a) that adult females buried with children were the mothers, (b) that children interred together are possibly siblings, (c) that male–female pairings represent "married," unrelated individuals, and (d) that males buried together possibly represent groups of siblings or other relatives such as cousins or multigenerational male kindreds. **Table 6.14** indicates cases of shared characteristics that can be recognized for 11 of the burials. Some of these traits are fairly common in the population, and thus they do not provide strong support for genetic similarity. These would include incisor shoveling and the I^2 trait, as they are present in fairly high frequencies and they are possibly not independent traits (cf. Tomb LVI [Grave 43], Tomb XLV [Grave 60], and Tomb XXXII [Grave 89]). However, if these characters occur in combination with other traits, as in Grave 52, then it is more likely that the individuals are related. The two young adults in Tomb LVI (Grave 43) share a rare variant of enamel extensions;

TABLE 6.14 Multiple burial graves and non-metric trait commonalities

Grave	Composition	Shared traits
I (64)	**Male 25–30, male 18–23**, male >45	F. c. molare, crowding
XII (88)	**Child 9 ± 2**, female 20–25, **infant 1 ± 18 mo**	Carabelli's trait
XVII (72)	Child 7 ± 2, child 4 ± 1	None
XVIII (73)	**Child 4 ± 1**, child 3 ± 1, **child 5 ± 1.5**	F. c. molare, canine defect
XXI (55)	Adolescent, male 20–25	None
XXVII (82)	Adult >35, adult 18–25, female >35	None
XXXII (89)	**Male 20–30, male 18–25**	I^2, shoveling
XXXIV (87)	Male 45–55, adult male	N/A
XXXVIII (79)	Male 30–40, adult female	N/A
XLII (59)	Female 20–25, **child 4 ± 1, child 2 ± 8 mo**	Carabelli's, F. c. molare
XLIII (62)	Infants: 18 mo, 1 ± 4 mo, 2 ± 8 mo	None
XLIV (65)	**Female 18–25, male 35–45**	F. c. molare, developmental anomalies
XLV (60)	**Male >55, *male 18–25,*** *female 30–40*	**Shoveling**, I^2
XLVIII (52)	***Female 15 ± 3*, male 20–25,** *Adult mid-age*	**I^2, shoveling,** *M^3 condition*
LIII (63)	Adult 20–25, child 5 ± 1.5	None
LVI (43)	**Female 20–25, male 20–25**	I^2, enamel extension
LXVII (12)	Male >35, female 23–27	None
LXVIII (13)	Female mid-age, male 20–30, adult 20–25, adult 20–25	None
LXIX (27)	Adult 20–25, **child 6 ± 2, child ≥ 6 yr**	F. c. molare
LXX (17)	Female 17–20, male >45	None
LXXII (24)	Adult 25–35, child 10 ± 2.5	None
LXXIII (26)	Adult male, adult	N/A
LXXIV (29)	**Male 25–35, male 35–45, male 20–25**	I^2, crowding, F.c. molare
LXXXI (09)	Male 26–32, young adult, male 26–32	None
LXXXIII (07)	Adolescent, young adult, mid-age adult, older adult	None, N/A
LXXXIV (02)	Male 23–26, male 35–45, child	None, N/A
LXXXV (10)	Female mid-age, adult male, child <1 yr	None, N/A
XC (M14)	Perinate, perinate	N/A
XCI (M15)	Perinate, perinate	N/A
XCVII (M39)	Male >50, older adult	N/A
C (M48)	Female 45–60, adolescent	N/A

Note: Bolded and/or italicized individuals share traits.

thus they are potentially close relatives. The graves with potentially related children (Tomb LXIX [Grave 27], Tomb XLII [Grave 59], Tomb XVIII [Grave 73], and Tomb XII [Grave 88]) may well represent siblings who succumbed to the same acute infectious disease episode. It is interesting that the adults in Tombs LXIX, XLII, and XXII (Graves 27, 59, and 88) do not have traits in common with the children. Also of interest are the graves with potentially related adult males: Tomb LXXIV (Grave 29) and the important secondary burials in Tomb I (Grave 64). These observations result from the initial investigation of the degree of genetic relatedness at Lofkënd, and until a more detailed probability analysis is completed, they should be regarded as reflecting some potentially non-random patterning in the distribution of non-metric traits.

Comparison with the Apollonia Sample

Demography

The prehistoric subsample from Apollonia Tumulus 10, consisting of 60 individuals, and a post-medieval/modern subsample of 33 individuals from Apollonia Tumuli 9, 10, and 11, and Appendices 1 and 3 provide comparative data for understanding the Lofkënd population on a broader regional basis. A comparison of the basic demography of these prehistoric individuals and those from the Lofkënd Tumulus illustrates their commonalities (**Figs. 6.25– 6.29**). The age profiles are very similar. For both groups, there is a peak of young adults and a scarcity of subadults (**Fig. 6.25**). This pattern is more pronounced at Apollonia, where the relative frequencies of infants and young children are lower (**Fig. 6.26**). They are proportionally 30% of the Lofkënd sample and only 20% of the Apollonia sample.

For both localities, males constitute a greater proportion of the sexable adults, and again this is more notable at Apollonia (**Fig. 6.27**). The ratio of males to females at Lofkënd is 39/28 or 1.39; at Apollonia it is 20/9 or 2.2. It is difficult to discern the exact reason for the greater representation of adult males at both these sites. As discussed above for Lofkënd, it may be an important indicator of gender-selective mortuary practices, or it may merely be an artifact of sampling (the difference in the sex ratios is not statistically significant; Fisher's exact test, two-tailed p value = 0.3674). If more males than females were actually interred in Lofkënd during the prehistoric use of the mound, a similar mortuary behavior was practiced at Apollonia Tumulus 10.

The age/sex distribution for the prehistoric burials in Apollonia Tumulus 10 has some important shared characteristics with the distribution at Lofkënd (**Fig. 6.28**). At both sites, the largest cohort is young males (30% for Lofkënd and 29.6% for Apollonia). The proportion of young females is moderately high relative to the remaining cohorts (21.0% at Lofkënd, yet only 14.8% for Tumulus 10). The proportions of mid-age and older females are very similar for the two localities. Both sites have a notable lack of older females. The other major difference between these sites is in the relative proportions of older males, who are almost twice as frequent at Apollonia. From these comparisons, it is evident that neither Tumulus 10 nor Lofkënd has representative population samples for the prehistoric phases. Older males and younger males predominate at Apollonia, and a similar but less extreme pattern characterizes Lofkënd.

At Lofkënd, there is evidence for differential treatment of adolescents and some young children. The burials contain diadems and other distinctive ornaments; one adolescent female had electrum ear or head ornaments (**10/11**, **10/12**; Papadopoulos, Bejko, and Morris 2007:124, fig. 16). An Archaic-period adolescent female from Tumulus 9, Grave 45, at Apollonia was found with distinctive "Illyrian" bracelets, but none of the prehistoric burials of adolescents is similar to the graves at Lofkënd in this regard.

If these age and sex distributions reflect socioeconomic behaviors, then it is also possible that those behaviors were similar at both localities. An example of such a socioeconomic pattern would be transhumance, where certain portions of the population (such as children and mid-aged or older females) remain at the settlement while other group members move seasonally with the animal herds.

A comparison of the more recent post-medieval/modern individuals from Lofkënd and Apollonia (**Fig. 6.29**) illustrates the relative similarities in their age distributions. Peaks of infants and older adults characterize both localities, although the Apollonia data include slightly more individuals in the intervening age cohorts.

Cranial form

Due to the typically poor preservation of the facial skeletons of the prehistoric individuals from Lofkënd and Apollonia, a metrical or morphological comparison of their cranial characteristics is limited to a basic comparison of their cranial vaults. While there are no complete crania where the maximum length and breadth can be measured, the general form was often observed before cleaning. A striking morphological characteristic of both subsamples is a generally narrow and elongated cranial form. This prehistoric cranial form contrasts with the highly brachycephalic or short, broad form of the more complete Apollonia post-medieval/modern individuals (combined sample $n = 9$, cranial index range = 81.2–93.7; mean = 88.3) and the Lofkënd moderns described above. Some of these individuals also exhibit the flattened occipital form seen in modern Balkan populations (Dhima 1993). There are two other fairly complete crania from the Classical/Hellenistic and Hellenistic phases of Apollonia Tumulus 9. They are longer than the post-medieval/modern crania (confidence interval = 74.2 and 76.5, respectively) and more similar to the form of the prehistoric individuals. Based on these observations, it appears that the prehistoric inhabitants of both Lofkënd and Apollonia clearly differed from the post-medieval/modern peoples in terms of cranial form, and the more recent crania from both localities are very similar.

Comparative health

The coastal regions of Albania may have been prime areas for endemic malaria in prehistory. The disease was documented there in recent historical periods (Velo et al. 2002), and changes to the coastline and river courses must have created swampy and still water pools at various times. Human modification of the landscape, whether through irrigation or cistern systems, might have increased suitable breeding environments for mosquitoes, so it is possible that the inhabitants could have been affected by malaria or parasitic diseases due to those culturally mediated circumstances. In contrast, the location of Lofkënd in an inland hilly region with naturally draining topography would suggest that it was not as favorable a natural environment for malaria or waterborne parasites.

A comparison of the Lofkënd and Apollonia rates of cribra orbitalia, porotic hyperostosis, and linear enamel hypoplasia is appropriate for testing this hypothesized difference in disease and parasite risk. Turning first to the bone indicators, these conditions are relatively rare at Lofkënd, with only 1 out of 50 (2%) adults showing possible porotic hyperostosis and 3 (6%) having remnants of childhood cribra orbitalia. One child out of eight (12.5%) with well-preserved orbital portions had cribra orbitalia; none had porotic hyperostosis. A combined frequency of cribra and porotic hyperostosis for adults and children where both conditions could be evaluated would therefore be 8.3%. In contrast, at Apollonia, 5 of 21 (23.8%) individuals had cribra orbitalia or porosity. This difference is statistically significant (Fisher's exact test, two-tailed $p = 0.0311$). The significantly higher rate of cribra orbitalia/porosity at Apollonia may be related to malaria, although there is no specific skeletal evidence for genetic anemia. Differential malnutrition and weaning stresses, other infectious diseases, and intestinal parasites may also have contributed to the higher frequency at Apollonia. The rates for later time periods at Apollonia are even higher than during the prehistoric phase, suggesting that the development of the colony and concentration of population led to increased stresses (Schepartz 2010).

The prevalences of linear enamel hypoplasia reinforce the picture of greater stresses at Apollonia, where 88.5% (or 100 of 113 observed dentitions, representative of approximately 50% of the total sample) of the individuals have hypoplasia. This frequency is greater than the level for Lofkënd (56.6%). Similar to what is observed at Lofkënd, most Apollonia individuals have only mild hypoplasia. Only 7.1% of the assessed individuals have moderate to severe levels of hypoplasia, but this is over twice the proportion seen at Lofkënd. Approximately 50% of the unaffected individuals at Apollonia were young children who may have died from initial stress incidents; the comparative figure at Lofkënd is 35%. These results indicate that both the Lofkëndis and the Apollonians were routinely subject to low levels of stress during childhood when their dentition was forming, yet the Lofkënd population may have been exposed to fewer stresses and stresses of lower intensity.

Dental health and diet

The frequency of caries at Lofkënd is significantly lower than the prehistoric prevalence of 17% for Apollonia (Fisher's exact test, two-tailed $p = 0.0002$). This is again best explained by the confounding effects of age as there are more individuals aged over 45 years for Apollonia (**Fig. 6.26**). The post-medieval/modern subsamples have the highest caries prevalences, with the small Lofkënd group, primarily represented by older males, as the most seriously affected individuals. The Classical through early Hellenistic Apollonia subsamples have intermediate values (**Table 6.15**). It is tempting to suggest that

TABLE 6.15 Adult caries prevalence at Lofkënd and Apollonia

Sample	# Teeth assessed	# Caries	%
Prehistoric Lofkënd	1,624	191	11.8
Prehistoric Apollonia	951	162	17.0
Classical Apollonia	422	84	19.9
Late Classical, Early Hellenistic Apollonia	130	24	18.5
Post-Medieval/modern Apollonia	275	97	35.3
Modern Lofkënd	166	105	63.3

changes in diet related to greater reliance on grains played an additional role in the rise of caries rates for the moderns, but more extensive isotope research is needed to test this hypothesis.

Non-metric trait comparisons

Comparing the traits listed in **Table 6.13** for prehistoric Lofkënd, the frequencies for prehistoric Apollonia are quite similar. Due to the rarity of most traits, testing for statistical differences between the localities is not possible, but one trait, the f. c. molare, does appear to distinguish the samples. The frequency for Apollonia, 56.4% (22/39), is significantly greater (Fisher's exact test, two-tailed $p = 0.0189$). Overall, these non-metric comparisons illustrate the general similarity of Lofkënd and Apollonia.

Biodistance analysis

Biological distance is the study of the phenotypic variation of bones and teeth that can used to determine the biological affinities of populations. Typically an analysis would include non-metric traits, such as those discussed above, that are thought to be representative of variability in the populations. The assumption underlying biodistance is that the traits have a genetic basis and that they are relatively unaffected by environmental effects. Metrical variability can also be used in biodistance calculations (Stojanowski et al. 2007). The degree of similarity, or distance, is used as a proxy for genetic relatedness. The robusticity of any biodistance analysis is dependent upon a number of factors, including the appropriateness of the traits selected (i.e., are they independent, or are they altered by cultural behaviors or environmental adaptations? [Mielke, Konigsberg, and Relethford 2006]), the size of the samples and the variance for the features, the number of traits compared, and the distance statistic that is calculated (cf. Bedrick, Lapidus, and Powell 2000; Harris and Sjøvold 2004; Irish and Konigsberg 2007; Mahalanobis 1936).

Biodistance analysis can be used to investigate the relationship between the prehistoric subsamples from Lofkënd and Apollonia as well as the relationship between these early groups and the post-medieval/modern burials. As an initial attempt to evaluate biodistance, a suite of 13 non-metric traits were compared. These include f. c. molare, Carabelli's trait, I^2 trait, incisor shoveling, supernumerary teeth, crowding/rotation, congenital absence, metopic suture, Inca bone, mastoid suture, enamel pearl, extraneous root, and enamel extensions. The distance statistic calculated is Smith's MMD (mean measure of divergence) with an Anscombe transformation to stabilize the variance, as recommended by Harris and Sjøvold (2004). The result, a problematic negative value (MMD = –0.0496), is an outcome that can occur when the populations being compared are genetically quite similar; however, it also reflects analysis compromised by small sample sizes and the limited variability between the subsamples. Reducing the number of traits to 11 (eliminating the mastoid suture and enamel pearl) in an attempt to constrain the variability yielded a similar result (MMD = –0.0512). In contrast, the comparison of prehistoric Lofkënd and the combined post-medieval/modern subsamples, using the same 11 variables, yields an MMD value of 0.0754, suggesting that these two subsamples are much more genetically distinct. Similarly, the MMD for the prehistoric Apollonia and combined post-medieval/modern subsamples is 0.0518.

These biodistance results reinforce the observations from the other analyses and comparisons presented here: the two prehistoric groups are very similar, and both are distinct from the combined modern subsamples. The potential for further, more refined and robust biodistance analysis with metric and non-metric variables is promising. One option, quantitative R-matrix analysis, may be possible if critical conditions involving the contemporaneity of the localities and the probability of a shared mate exchange network are assumed (cf. Knudson and Stojanowski 2008).

INDIVIDUAL DESCRIPTIONS

Prehistoric tombs

TOMB I (GRAVE 64), TRENCH 1, UNITS 362, 362a, AND 362b, N = 3

Individual 362. Adult male, aged 25–30 years.
Bone condition: good.
Skeletal elements preserved: mandible.
Dentition: 15 permanent teeth, including the R I1, C1–M3; L I1–M3.
Aging criteria: dental attrition.
Sexing criteria: mandibular morphology.
Pathology and morphological variation: f. c. molare; dental crowding with the L I2 rotated distally, linear enamel hypoplasia.

Individual 362a. Adult male, aged 18–23 years.
Bone condition: fair.
Skeletal elements preserved: mandible.
Dentition: 16 permanent teeth, a full mandibular set.
Aging criteria: tooth eruption and attrition.
Sexing criteria: mandibular morphology.
Pathology and morphological variation: f. c. molare, dental crowding with distal rotation of the mandibular canines, linear enamel hypoplasia.

Individual 362b. Adult male, aged >45 years.
Bone condition: fair.
Skeletal elements preserved: maxilla and mandible.
Dentition: 21 permanent teeth, including R I^1–M^2, L I^1–P^4; R I_1–P_4, L I_1–P_3.
Aging criteria: dental attrition.
Sexing criteria: mandibular morphology.
Pathology and morphological variation: caries, n = 2; antemortem losses, n = 5–7.

Tomb I (Grave 64) is considered to be the central burial of the Lofkënd tumulus, and it is an unusual interment. The skeletal elements were haphazardly arranged in the burial pit, constituting a small commingled sample with a total weight of 2.55 kg. The bones have different levels of preservation, ranging from poor to very good. Faunal remains are also associated with the tomb. The MNI for the tomb is based on three mandibles: 362, 362a, and 362b (**Fig. 3.8**). Except for a fragmentary maxilla and an eroded right zygomatic, no other cranial elements were preserved. Femora and tibiae were also not represented. Most of the elements are robust and are probably associated with the three individuals designated above. Due to the unusual composition of the sample, the inventory is provided:

Cranial elements
- Maxilla: 1
- Zygomatic: 1 right
- Mandible: 3

Postcranial elements
- Hyoid: right greater wing with portion of body
- Vertebrae: C1 and 3 C2s, fragments of thoracic and lumbar elements; probable association between C1 and largest C2. The medium-sized C2 has osteophytic lipping around the dens
- Ribs: many fragments
- Pelvis: right innominate with male morphology including broad, large ischial tuberosity, large acetabulum, lack of iliac flair, narrow greater sciatic notch. Aging: Stage 3 (Suchey-Brooks) 28.7 ± -6.5 years and Phase 2 (Lovejoy) 25–29 years. Possible association between this pelvis and Unit 362a based on similarity in age estimate
- Clavicles: 2 right, 1 left
- Scapulae: 1 right, 1 left; both preserve the glenoid fossa with lipping of the border
- Humeri: 2 right, 1 left
- Radii: 2 right, 1 left
- Ulnae: 3 right, 2 left
- Fibulae: 3 shaft fragments
- Hand: 12 carpals (scaphoid, lunate, hamate, trapezoid, capitate represented)
  - 22 metacarpals including identifiable metacarpal 1, n = 4, metacarpal 2, metacarpal 3
  - 34 phalanges, proximal, intermediate, and distal; some with slight osteophyte development
- Foot: 4 tarsals, including talus, calcaneus, cuboid, fragmentary cuneiform
  - 2 metatarsals, including 2 metatarsal 1
  - 2 phalangeal fragments

TOMB II (GRAVE 91), TRENCH 2, UNIT 516, N = 1

Individual 516. Adult male, aged >45 years.
Bone condition: fair.
Skeletal elements preserved: skull, clavicles, scapulae, humeri, radii, ulnae, hand elements, ribs, vertebrae, pelvis, right femur, tibiae, and fibulae.
Dentition: permanent dentition with 19 teeth, including R I^1–M^2, L I^1–P^4; R M_2–M_3, L P_3–M_3.
Aging criteria: dental attrition, cranial sutures, degenerative skeletal changes.

Sexing criteria: skull morphology, postcranial dimensions.

Pathology and morphological variation: caries, n = 2; abscesses, n = 2; antemortem losses, n = 10 minimum; hyper-eruption of maxillary incisors; dental calculus; possible mastoiditis; wormian bones; several fused thoracic and lumbar vertebrae (DISH). (See **Fig. 6.19** for an illustration of the antemortem losses.).

The right temporal bone of Individual 516 has a pathological condition affecting the mastoid process and the external auditory meatus (**Fig. 6.17**). The mastoid appears withered and the surface seems involuted and may have vestiges of drainage tracks. The base of the auditory meatus is potentially affected by the same infectious process. It is expanded and hollowed out inferiorly, with the inner ear features very visible. The articular eminence is flattened. Postmortem damage to the tympanic plate makes it difficult to fully assess the extent of the infection in this temporal bone.

The vertebrae of this individual are fragmentary, but the remaining portions include several fused elements in the thoracic and lumbar region. The spinous processes are fused for at least four vertebrae, and the centra of two to four vertebrae are also fused. This is a possible case of DISH.

TOMB III (GRAVE 81), TRENCH 1, UNIT 454, N = 1

Individual 454. Adult male, aged >35 years.

Bone condition: fair.

Skeletal elements preserved: skull, clavicles, scapulae, humeri, radii, ulnae, ribs, vertebrae, pelvis, femora, tibiae, fibulae.

Dentition: 22 permanent teeth, including the R I^1–C^1, P^4–M^3; L I^1–M^2; R I_2–P_4, L I_2–P_3, M_2.

Aging criteria: dental attrition.

Sexing criteria: skull morphology, postcranial size, and robusticity.

Pathology and morphological variation: caries, n = 1; antemortem losses, n = 6; I^2 trait; incisor shoveling; R I_2 has a pinched root with a flairing mesial side and related crown distortion; linear enamel hypoplasia; wormian bones.

TOMB IV (GRAVE 98), TRENCH 2, UNIT 581, N = 1

Individual 581. Adult male, aged 20–25 years.

Bone condition: fair.

Skeletal elements preserved: skull, clavicles, scapulae, humeri, radii, ulnae, hand elements, vertebrae, ribs, pelvis, femora, left patella, tibiae, fibulae, foot elements.

Dentition: complete permanent dentition, except for the R and L I_1.

Aging criteria: dental attrition, possible incomplete fusion of sternal epiphysis of the clavicle.

Sexing criteria: skull morphology, postcranial size, and robusticity.

Pathology and morphological variation: I^2 trait?; f. c. molare; linear enamel hypoplasia.

TOMB V (GRAVE 96), TRENCH 2, UNIT 569, N = 1

Individual 569. Adult female, aged 20–25 years.

Bone condition: fair.

Skeletal elements preserved: skull, clavicles, scapulae, humeri, radii, ulnae, hand elements, ribs, vertebrae, pelvis, femora, tibiae, fibulae, foot elements.

Dentition: full adult dentition with impacted maxillary canines and a retained deciduous canine on the right.

Aging criteria: dental features, clavicle fusion.

Sexing criteria: postcranial metrical analysis, cranial morphology.

Pathology and morphological variation: caries, n = 2; incisor shoveling; impacted maxillary canines and retention of the deciduous canine; linear enamel hypoplasia; f. c. molare; wormian bones.

The most notable feature of this dentition is the presence of impacted permanent maxillary canines (**Fig. 6.30**). The right tooth was closer to eruption, but it probably did not project through the soft tissue of the palate. This region is damaged, but it appears from the alveolus that the deciduous second molar may still have been present. The opposing mandibular teeth have wear, suggesting that some functional relationship was maintained. The deciduous canine was still in use on the left side (see **Fig. 6.23**).

TOMB VI (GRAVE 97), TRENCH 2, UNIT 573, N = 1

Individual 573. Subadult, aged 3 ± 1 years.

Bone condition: Poor.

Skeletal elements preserved: cranial fragments, including a petrous portion of the temporal bone.

Dentition: mixed dentition with 13 teeth; 7 deciduous (R dm^1, dm^2; L dc^1–dm^2; L dm_1, dm_2) and 6 permanent (R and L I^1, M^1; R and L M_1).

Aging criteria: maxillary I1 and maxillary and mandibular M1 crown development.
Pathology and morphological variation: none.

TOMB VII (GRAVE 99), TRENCH 2, UNIT 583, N = 1

Individual 583. Subadult, aged 7 years ± 2 years.
Bone condition: poor.
Skeletal elements preserved: skull, vertebrae, non-identifiable long bone fragments.
Dentition: mixed dentition with 29 teeth preserved; 7 deciduous (R dm^1, dm^2 and L dm^1, dm^2; R dm_1 and L dm_1, dm_2) and 22 permanent (R I^1,C^1, P^3, M^1, M^2; L I^1–P^3, M^1, M^2; R I_2–P_3, M_1, M_2; L I_2–M_2).
Aging criteria: dental development of permanent molars, incisors, canine and premolars.
Pathology and morphological variation: Carabelli's cusp.

TOMB VIII (GRAVE 100), TRENCH 2, UNIT 591, N = 1

Individual 591. Subadult, aged 3 years ± 1 year.
Bone condition: good.
Skeletal elements preserved: skull, clavicle, scapula, humeri, radii, ulnae, ribs, vertebrae, pelvis, femora, tibiae, fibulae.
Dentition: mixed dentition with 23 teeth, including a complete deciduous set and 3 permanent teeth (R M^1, L I^1 and M^1).
Aging criteria: crown development of the maxillary and mandibular M1s and the I^1; root development of the dm^2s.
Pathology and morphological variation: none.

TOMB IX (GRAVE 94), TRENCH 2, UNIT 547, N = 1

Individual 547. Adult, possible female; no specific age estimation possible.
Bone condition: poor.
Skeletal elements preserved: pelvis, femora, patellae, tibiae, fibulae, tarsals (right talus and calcaneus identifiable).
Dentition: none.
Aging criteria: bone morphology and size.
Sexing criteria: postcranial metric analysis.
Pathology and morphological variation: femoral subtrochanteric flattening.

TOMB X (GRAVE 95), TRENCH 2, UNIT 554, N = 1

Individual 554. Adult female, aged 30–40 years.
Bone condition: poor.
Skeletal elements preserved: skull, left clavicle and scapula, humeri, radii, ulnae, pelvis?.
Dentition: a total of 23 permanent teeth, including the R I^1–C^1, M^1; L I^1–M^1; R I_1–M_2, L I_1–M_1.
Aging criteria: dental attrition.
Sexing criteria: skull morphology (see **Fig. 6.9**), limb shaft dimensions.
Pathology and morphological variation: caries, n = 1; agenesis of the mandibular M3s and right maxillary M3; hyper-eruption of the L P^4.

The maxillary dentition has an undulating wear pattern and a hyper-erupted left P^4 that occupies the space created by a large caries on the left M_1.

TOMB XI (GRAVE 93), TRENCH 2, UNIT 540, N = 1

Individual 540. Subadult, aged 9 months ± 3 months.
Bone condition: poor.
Skeletal elements preserved: cranial vault fragments.
Dentition: deciduous dentition, 16 teeth, including all the molars but missing the R di_1–dc_1 and the L di_1.
Aging criteria: development of tooth roots.
Pathology and morphological variation: none.

TOMB XII (GRAVE 88), TRENCH 2, UNITS 495, 499, AND 501, N = 3

Grave 88 consists of three individuals: one adult and two subadults.

Individual 495. Subadult, aged 9 years ± 2 years.
Bone condition: fair.
Skeletal elements preserved: skull with many elements preserved, clavicles, scapulae, humeri, radii, ulnae (right arm is more complete), hand elements, ribs, vertebrae, pelvic elements, right femora, tibiae, fibulae, foot elements.
Dentition: mixed dentition of 35 teeth, including 15 deciduous (R and L di^2–dm^2; R dc_1–dm_2; L di_1, dc_1–dm_2) and 20 permanent (R I^1–C^1, M^1, M^2; L I^1, I^2, M^1, M^2; R I_1–P_3, M_1, M_2; L I_1, I_2, M_1, M_2; miscellaneous M3 germ fragment).
Aging criteria: dental development, bone growth and maturation.
Pathology and morphological variation: I^2 trait; incisor shoveling; Carabelli's trait; linear enamel hypoplasia.

Figure 6.31 shows the left orbit of Individual 495 (arrow indicates porosity in region where cribra orbitalia occurs); other areas of porosity in the orbit suggest taphonomic damage to the surface.

Individual 499. Adult female, aged 20–25 years.
Bone condition: good.
Skeletal elements preserved: skull, clavicles, scapulae, humeri, ulnae, radii, hand elements, ribs, vertebrae, pelvis, femora, tibiae, fibulae, foot elements.
Dentition: 29 permanent teeth, lacking 3 M3s; the L M_3 is present.
Aging criteria: dental attrition.
Sexing criteria: pelvic morphology, cranial morphology, postcranial size.
Pathology and morphological variation: dental crowding and rotation; hyper-eruption of L M_3.

Individual 501. Subadult, aged 12–18 months.
Bone condition: poor.
Skeletal elements preserved: skull with frontal, parietals, petrous portions of the temporal, occipital; ribs, femora, tibia, miscellaneous long bone fragments.
Dentition: mixed dentition of 23 teeth, with 18 deciduous (R and L di^1–dm^2; R di_1–dm_2, L dc_1–dm_2) and 5 permanent (R I^1 and 4 M_1 crowns).
Aging criteria: dental formation and wear, metopic suture.
Pathology and morphological variation: Carabelli's cusp on the dm^2s and M^1s, small pit on anterior buccal surface of dm_2s.

TOMB XIII (GRAVE 49), TRENCH 2, UNIT 305, N = 1

Individual 305. Adult female, aged 25–35 years.
Bone condition: good.
Skeletal elements preserved: skull, clavicles, scapulae, humeri, radii, ulnae, hand elements, vertebrae, ribs, pelvis, femora, patellae, tibiae, fibulae, foot elements.
Dentition: 31 permanent teeth, lacking only the L M_1 lost antemortem.
Aging criteria: pubic symphysis, auricular surface, dental attrition.
Sexing criteria: skull and pelvic morphology.
Pathology and morphological variation: caries, n = 5; antemortem losses, n = 1; linear enamel hypoplasia; periodontal disease; dental calculus; incisor shoveling; I^2 trait; hyper-eruption; crowding; possible diploe expansion over the mid-parietal region; wormian bones; rocker jaw with raised menton; clavicle sternal ends have pits and inferior projecting knobs—probably congenital as the surfaces are smooth; left tibia has healed midshaft injury.

TOMB XIV (GRAVE 71), TRENCH 4, UNIT 405, N = 1

Individual 405. Subadult, aged 6 months ± 3 months.
Bone condition: poor.
Skeletal elements preserved: skull, scapula, vertebrae, and femur.
Dentition: 15 deciduous tooth crowns, missing the R dc^1, L di^1, R di_1, L di_2, and L dc_1.
Aging criteria: crown development.
Pathology and morphological variation: Carabelli's cusp.

TOMB XV (GRAVE 80), TRENCH 2, UNIT 444, N = 1

Individual 444. Subadult, aged 6 months ± 3 months.
Bone condition: poor.
Skeletal elements preserved: cranial vault fragments and petrous portions of the temporal bones; clavicle, radius, vertebrae, ribs, and femur.
Dentition: 8 deciduous teeth, including R di^1; L di_1–dm_2; 2 fragments of molar crowns.
Aging criteria: dental development.
Pathology and morphological variation: none.

TOMB XVI (GRAVE 68), TRENCH 2, UNIT 386, N = 1

Individual 386. Subadult, aged 6 years ± 2 years.
Bone condition: fair.
Skeletal elements preserved: cranium, including ear ossicles, frontal, parietals, occipital, temporals, sphenoid, left zygomatic, left mandibular ramus; ribs, pelvis, humeri, radii, ulnae, femora.
Dentition: mixed dentition with 41 teeth, including 15 deciduous (R dc^1–dm^2; L di^1–dm^2; R dc_1–dm_2; L di_1, dc_1–dm_2) and 26 permanent (R I^1, I^2, P^4–M^2; L I^1–M^2; R and L I_1–M_2).
Aging criteria: permanent crown and root formation.
Pathology and morphological variation: linear enamel hypoplasia on M_1s in the form of a "waisting".

Most of the tooth surfaces are too damaged to permit enamel quality observations. However, the four M_1s have a notable "waisting" and a possible hypoplastic band approximately halfway down the crown. If this is indeed representative of stress during development, the time this occurred would have been 18–24 months.

TOMB XVII (GRAVE 72), TRENCH 1, UNITS 408 AND 408A, N = 2

Individual 408. Subadult, aged 7 years ± 2 years.
Bone condition: poor.

Skeletal elements preserved: cranium, clavicle, humerus, hand phalanx, pelvis, femora, tibiae, fibulae.
Dentition: mixed dentition totaling 33 teeth with 10 deciduous (R dc^1, dm^2; L dc^1–dm^2; R dc_1–dm_2, L dm_1, dm_2) and 23 permanent (R I^1–M^2; L I^1, C^1–M^2; R I_1–P_3, M_1; L I_1–P_3, M_1).
Aging criteria: dental development of permanent crowns and roots, deciduous wear.
Pathology and morphological variation: caries, n = 1; incisor shoveling; I^2 trait; linear enamel hypoplasia.

Individual 408a. Subadult, aged 4 years ± 1 year.
Bone condition: poor.
Skeletal elements preserved: fragment of right maxilla with the dm^1, P^3 and C^1 crowns unerupted.
Dentition: mixed dentition with 24 teeth, including 11 deciduous (R di^2–dm^2; L dc^1–dm^2; R dm_2, L dc_1–dm_2) and 13 permanent (R I^1, M^1; L C^1, M^1; R I_1–P_3, M_1; L I_1–C_1, M_1).
Aging criteria: dental development and wear.
Pathology and morphological variation: linear enamel hypoplasia.

Tomb XVIII (Grave 73), Trench 4, Units 412, 412a, and 421, N = 3

Individual 412. Subadult, aged 4 years ± 1 year.
Bone condition: poor.
Skeletal elements preserved: cranial fragments with one petrous portion of the temporal bone, fragments of lower limbs.
Dentition: mixed dentition of 25 teeth, including 16 deciduous (R dc^1–dm^2, L di^2–dm^2; R di_2–dm_2, L di_1–dm_2) and 9 permanent (R and L M^1; R I_1, I_2, M_1; L I_1–C_1, M_1).
Aging criteria: permanent crown and root development.
Pathology and morphological variation: enamel hypoplasia pit on R C^1 disto-lingual surface; f. c. molare.

Individual 412a. Subadult, aged 3 years ± 1 year.
Bone condition: very poor.
Skeletal elements preserved: mandible fragment.
Dentition: mixed dentition of 11 teeth with 4 deciduous (R and L di^1 and R dm_1, dm_2) and 7 permanent (R and L I^1, I^2, M^1; R M_1).
Aging criteria: permanent crown development.
Pathology and morphological variation: none.

Individual 421. Subadult, aged 5 years ± 1.5 years.
Bone condition: poor.
Skeletal elements preserved: cranium with petrous portions of the temporals.
Dentition: mixed dentition with 15 teeth, including 3 deciduous (R and L dm^2; R dm_2) and 12 permanent (R I^1, C^1–M^1; L I^1, C^1, M^1; R I_2–P_3; L P_3).
Aging criteria: permanent crown development.
Pathology and morphological variation: hypoplasia pit on the R C^1; f. c. molare.

Tomb XIX (Grave 54), Trench 2, Unit 323, N = 1

Individual 323. Adult female, mid-age.
Bone condition: good.
Skeletal elements preserved: skull, clavicles, scapulae, humeri, radii, ulnae, hand elements, vertebrae, ribs, pelvis, femora, patellae, tibiae, fibulae, foot elements.
Dentition: 14 permanent teeth, including the R P^4; L C^1, P^4; R I_1–P_3, M_2; L I_1–P_4, M_2.
Aging criteria: dental attrition, cranial sutures.
Sexing criteria: cranial morphology, acetabular dimensions.
Pathology and morphological variation: antemortem losses, n = 3; tooth rotation; hyper-eruption; slight development of osteophytes on vertebrae.

Tomb XX (Grave 50), Trench 2, Unit 308, N = 1

Individual 308. Adult female, aged 20–25 years.
Bone condition: good.
Skeletal elements preserved: skull, clavicles, sternum, scapulae, humeri, radii, ulnae, hand elements, vertebrae, ribs, pelvis, femora, patellae, tibiae, fibulae, foot elements.
Dentition: 31 permanent teeth, lacking only the R M_3.
Aging criteria: skeletal maturation, dental attrition.
Sexing criteria: skull morphology, pelvic morphology.
Pathology and morphological variation: caries, n = 1; linear enamel hypoplasia; dental crowding; I^2 trait; f. c. molare; M_3 agenesis and hyper-eruption of opponent, overbite and maxillary incisor crowding; Carabelli's cusp; mastoid suture; wormian bones; asymmetries in the cervical vertebrae.

Individual 308 has asymmetrical morphology for the first four cervical vertebrae, C1–C4. This affects the shape of the neural canal and the positioning and form of the articulations with adjacent vertebra. The right side seems more affected, making the arch slightly compressed. On C3 the superior and inferior articular facets are different in size and shape, as are the superior facets on C4.

The C4 body has lipping and irregular bone on the right near the pedicle.

TOMB XXI (GRAVE 55), TRENCH 4, UNIT 341, 326, N = 1

Individual 341. Subadult, aged 15 years ± 3 years.
Bone condition: very poor.
Skeletal elements preserved: cranium.
Dentition: 23 teeth, including R I^2–M^3, L P^3–M^3; R P_3–M_3, L C_1–M_3.
Aging criteria: dental attrition.
Sexing criteria: canine morphology and size.
Pathology and morphological variation: none.

Individual 341 consists of a very fragmentary skull with metallic remnants of a diadem. No diagnostic portions are preserved. The age estimate is based on dental attrition: there is very limited wear on the canines, third premolars, and second molars, and only slight dentine exposure on the first molars. The third molars appear to be unworn. These data are consistent with an adolescent of approximately 15 years.

The teeth are healthy with no pathology. The moderate size and morphology of the canines might indicate that this is a female, but that assessment must remain tentative.

Individual 326. Adult male, aged 20–25 years.
Bone condition: poor.
Skeletal elements preserved: skull fragment, clavicle, humeri, radii, ulnae, right hand, vertebrae, pelvis, patella, femora, tibiae, fibulae, left foot.
Dentition: 28 teeth, including R I^2–M^3, L I^1–M^3; R I_2–M_3, L C_1–M_2.
Aging criteria: dental attrition.
Sexing criteria: femoral robusticity.
Pathology and morphological variation: caries, n = 1; I^2 trait; dental calculus; linear enamel hypoplasia; femoral subtrochanteric flattening.

TOMB XXII (GRAVE 47), TRENCH 2, UNIT 294, N = 1

Individual 294. Adult, possibly male, mid-age.
Bone condition: fair.
Skeletal elements preserved: right clavicle, scapulae, right humerus, radii, ulnae, hand elements, vertebrae, ribs, and femur.
Dentition: 2 heavily worn permanent teeth: a mandibular incisor and a premolar.
Aging criteria: dental attrition, postcranial degenerative changes.
Sexing criteria: postcranial size and rugosity.
Pathology and morphological variation: osteophytes on right glenoid fossa and cervical vertebrae.

TOMB XXIII (GRAVE 56), TRENCH 4, UNIT 329, N = 1

Individual 329. Subadult, aged 8 years ± 2 years.
Bone condition: very poor.
Skeletal elements preserved: skull and limbs, non-identifiable bone.
Dentition: mixed dentition of 14 deciduous teeth (R dm^1–dm^2; L dc^1–dm^2; R di_1–dm_2; L di_1, dc_1–dm_2) and 28 permanent teeth (lacking the M3s).
Aging criteria: dental attrition, M2 crown formation.
Pathology and morphological variation: linear enamel hypoplasia; incisor shoveling; lingual tubercles on maxillary canines; Carabelli's trait; f. c. molare. (See **Fig. 6.22** for the f. c. molare.)

TOMB XXIV (GRAVE 85), TRENCH 2, UNIT 478, N = 1

Individual 478. Subadult, aged 3 years ± 1 year.
Bone condition: poor.
Skeletal elements preserved: cranium with petrous portions of the temporal bones, possible femur diaphysis.
Dentition: mixed dentition of 23 teeth, with 11 deciduous (R dm^1, dm^2; L di^1–dm^2; R di_1, dm_1, dm_2; L dm_2) and 12 permanent (R I^1, I^2, M^1; L I^1, C^1, M^1; R and L I_1, I_2, M_1).
Aging criteria: permanent crown formation, deciduous wear.
Pathology and morphological variation: linear enamel hypoplasia; Carabelli's cusp; I^2 trait.

TOMB XXV (GRAVE 90), TRENCH 2, UNIT 512, N = 1

Individual 512. Subadult, aged 6–10 years.
Bone condition: poor.
Skeletal elements preserved: left femur diaphysis (over 160 mm preserved), tibiae with portions of the proximal and distal ends preserved, and fibulae.
Dentition: none.
Aging criteria: postcranial size and development.
Pathology and morphological variation: none.

TOMB XXVI (GRAVE 74), TRENCH 1, UNIT 416, N = 1

Individual 416. Adult female, aged >45 years.
Bone condition: poor.
Skeletal elements preserved: skull, clavicles, scapulae, ribs, vertebrae, humeri, ulnae, radii, hand ele-

ments, pelvis, femora, tibiae, fibulae, right foot elements.
Dentition: 12 teeth, including the L P^3–M^1, M^3; R I_1–C_1, P_4; L I_2–P_4.
Aging criteria: dental attrition.
Sexing criteria: skull morphology.
Pathology and morphological variation: caries, n = 9; antemortem losses, n = 5–7 (if M3s were once present, then 7); periodontal disease; hypercementosis on several roots, most notable on L M^3 that is possibly related to tooth losses and hyper-eruption; malocclusion and possible underbite; linear enamel hypoplasia; rotation of the R P_4 due to loss of P_3; osteophytes on vertebrae; subtrochanteric flattening.

TOMB XXVII (GRAVE 82), TRENCH 4, UNITS 457, 457A, AND 461, N = 3

Individual 457. Adult, aged >35 years, no sex determination.
Bone condition: poor.
Skeletal elements preserved: skull, left clavicle, right scapula, humeri, radii, ulnae, hand elements, ribs, pelvis, femora, tibiae, fibulae, left calcaneus and talus, foot elements.
Dentition: 22 permanent teeth present, including the R I^1–P^4, M^3; L I^1–M^1; R I_1–P_4, M_2; L I_2–P_4.
Aging criteria: dental attrition.
Sexing criteria: none.
Pathology and morphological variation: caries, n = 11 possible; linear enamel hypoplasia; rotated R C_1; crowding and malocclusion in the mandible.

Individual 457a. Adult, aged 18–25 years.
Bone condition: poor.
Skeletal elements preserved: maxillary fragment.
Dentition: 6 teeth, including 4 maxillary teeth (R I^1, L C^1, L P^3 and R M^1) and 2 mandibular molars (R M_1 and M_2).
Aging criteria: dental attrition.
Sexing criteria: none.
Pathology and morphological variation: none.

Individual 461. Adult, possible female, aged >35 years.
Bone condition: poor.
Skeletal elements preserved: skull, scapulae, humeri, radii, left ulna, hand elements, ribs, vertebrae, right innominate, femora, tibiae, fibulae.
Dentition: 7 permanent teeth, including the R M^1, M^3; L I^2–C^1, P^4, M^2; L M_3.
Aging criteria: dental attrition.
Sexing criteria: pelvis, postcranial dimensions, and robusticity.
Pathology and morphological variation: linear enamel hypoplasia; dental calculus; osteophytes on cervical vertebrae, glenoid fossa, and acetabulum.

TOMB XXVIII (GRAVE 77), TRENCH 4, UNIT 429, N = 1

Individual 429. Subadult, aged 3 years ± 1 year.
Bone condition: poor.
Skeletal elements preserved: skull with temporal and mandibular portions preserved.
Dentition: mixed dentition with 19 teeth, including 11 deciduous (R di^1–dm^2; L di^1, dm^1, dm2; R dm_1, dm_2; L dm_2) and 8 permanent (R I^1, I^2, M^1; L I^1, I^2, M^1; R M_1; L M_1).
Aging criteria: crown development of the maxillary permanent incisors and the M1s.
Pathology and morphological variation: incisor shoveling; I^2 trait; Carabelli's cusp.

TOMB XXIX (GRAVE 83), TRENCH 1, UNIT 468, N = 1

Individual 468. Subadult, aged 6 months ± 3 months.
Bone condition: fair.
Skeletal elements preserved: skull with frontal, temporals, mandible; right clavicle, humerus, radius and ulna; left humerus and ulna; left scapula; ribs; femora and tibiae.
Dentition: deciduous dentition with 13 teeth, including R di^2, dc^1; L di^1, dc^1, dm^1; Rdi_1, di_2, dm_1, dm_2; L di_2–dm_2.
Aging criteria: deciduous crown development.
Pathology and morphological variation: none.

TOMB XXX (GRAVE 70), TRENCH 2, UNIT 398, N = 1

Individual 398. Subadult, aged 9 months ± 3 months.
Bone condition: cremation, fair (**Fig. 6.32**).
Skeletal elements preserved: cranial vault and temporal petrous portions, limb shafts and vertebrae.
Dentition: deciduous dentition, 13 teeth, including R di^1, dc^1, dm^1, dm^2; L dc^1, dm^1, dm^2; R and L dc_1–dm_2.
Aging criteria: deciduous crown and root development.
Pathology and morphological variation: none.

TOMB XXXI (GRAVE 86), TRENCH 2, UNIT 489, N = 1

Individual 489. Adult, aged >40 years.
Bone condition: poor.
Skeletal elements preserved: cranial vault fragments, non-diagnostic; right mandibular corpus; right and left femoral diaphyses.
Dentition: maxillary molar, probable M^2; R P_3–P_4 roots.
Aging criteria: endocranial sutural closure.
Sexing criteria: none.
Pathology and morphological variation: antemortem loss of R M_1–M_3.

TOMB XXXII (GRAVE 89), TRENCH 2, UNITS 508 AND 524, N = 2

Individual 508. Adult male, aged 20–30 years.
Bone condition: poor.
Skeletal elements preserved: cranial vault and mandible fragments, scapula, miscellaneous limb shafts, pelvis, distal fibula, hand and foot elements.
Dentition: 29 permanent teeth, lacking the R M_3, L I_1, L P_4.
Aging criteria: dental attrition.
Sexing criteria: postcranial size.
Pathology and morphological variation: L P^4 has anomalous form; incisor shoveling; large lingual cusp on L I^1; I^2 trait; congenital absence of L P_4 and retention of dm_2.

Individual 524. Adult male, aged 18–25 years.
Bone condition: poor.
Skeletal elements preserved: femora.
Dentition: 22 permanent teeth, including the R I^1–M^3, L I^1–M^3; R M_1–M_3, L M_1–M_3.
Aging criteria: dental attrition.
Sexing criteria: postcranial size.
Pathology and morphological variation: I^2 trait; incisor shoveling.

TOMB XXXIII (GRAVE 92), TRENCH 1, 2, AND 6, UNIT 523, N = 1

Individual 523. Subadult, possible female, aged 16 ± 2 years.
Bone condition: poor.
Skeletal elements preserved: cranium, right clavicle, humeri, right radius, ribs, pelvis, femora, tibiae, fibulae.
Dentition: 30 teeth, missing only the L I^2 and L M^3.
Aging criteria: dental development and wear.
Sexing criteria: femur and tibia midshaft dimensions.
Pathology and morphological variation: none.

TOMB XXXIV (GRAVE 87), TRENCH 1, UNITS 498 AND 504, N = 2

Individual 498. Adult male, aged 45–55 years.
Bone condition: fair.
Skeletal elements preserved: skull, clavicles, scapulae, humeri, radii, ulnae, hand elements, ribs, vertebrae, pelvis, femora, tibiae, fibulae, foot elements.
Dentition: 24 permanent teeth, including the R I^1–P^4, M^2; L I^1–P^4, M^3; R I_1–P_3, M_2; L I_1–M_2.
Aging criteria: pubic symphysis, auricular surface, dental attrition, cranial sutures.
Sexing criteria: skull morphology, postcranial size.
Pathology and morphological variation: antemortem losses, n = 5.

Individual 504. Adult male.
Bone condition: poor.
Skeletal elements preserved: right temporal bone, left scapula, right calcaneus.
Dentition: none.
Aging criteria: bone morphology.
Sexing criteria: cranial morphology, postcranial size.
Pathology and morphological variation: none.

TOMB XXXV (GRAVE 84), TRENCH 1, UNIT 472, N = 1

Individual 472. Adult male, aged 23–27 years.
Bone condition: fair.
Skeletal elements preserved: skull, left clavicle, left scapula, left humerus, radii, ulnae, hand elements, ribs, vertebrae, pelvis, femora, tibiae, fibulae, foot elements.
Dentition: 5 permanent teeth, including the R C_1–M_2.
Aging criteria: pubic symphysis and dental attrition.
Sexing criteria: pelvic morphology, postcranial rugosity.
Pathology and morphological variation: none.

TOMB XXXVI (GRAVE 75), TRENCH 2, UNIT 420, N = 1

Individual 420. Adult, no age or sex estimation possible.
Bone condition: extremely poor.
Skeletal elements preserved: femur and tibia.
Dentition: none.
Aging criteria: long bone morphology.

Sexing criteria: none.
Pathology and morphological variation: none.

Tomb XXXVII (Grave 76), Trench 2, Unit 424, N = 1

Individual 424. Adult, no specific age or sex estimates.
Bone condition: extremely poor.
Skeletal elements preserved: cranial fragments, radius, lower limbs (femur, tibia, fibula, left talus, and a few metatarsals and phalanges).
Dentition: L P^3 and an irregular M^3 crown (possibly associated).
Aging criteria: dentition and postcranial morphology.
Sexing criteria: none.
Pathology and morphological variation: none.

Tomb XXXVIII (Grave 79), Trench 1, Units 439 and 447 (cremation), N = 2

Individual 439. Adult male, aged 30–40 years.
Bone condition: good.
Skeletal elements preserved: skull, clavicles, scapulae, sternum, humeri, radii, ulnae, hand elements, ribs, vertebrae, pelvis, femora, patellae, tibiae, fibulae, foot elements.
Dentition: 30 permanent teeth, lacking the R M^1 and L M^2 lost antemortem.
Aging criteria: dental attrition, pubic symphysis and auricular surface.
Sexing criteria: pelvic morphology, skull morphology, postcranial size, and rugosity.
Pathology and morphological variation: caries, $n = 6$; antemortem losses, $n = 2$; abscess on R M^1; linear enamel hypoplasia; dental calculus; periodontal disease, especially clear around the left mandibular molars; crowding and rotation of the anterior mandibular dentition; slight osteophyte development on the cervical vertebrae; trauma: left hallux interphalangeal joint healed but crushed and mushroomed; notable rugosity on the right clavicle sternal end, left metacarpal 5 and the capitate.

The phalanges of the left hallux were injured (see **Fig. 6.16**). The head is crushed and mushroomed on the proximal phalanx, while the base of the distal is crushed and much altered. Obviously the injury occurred at the joint between the two bones. There is limited evidence for remodeling.

Individual 447. Adult, probable female, cremation.
Bone condition: fair, approximately 2 kg of bone recovered.
Skeletal elements preserved: skull, humerus, femur, vertebrae.
Dentition: fragments.
Aging criteria: bone and tooth morphology.
Sexing criteria: postcranial size—femoral head and vertebral bodies.
Pathology and morphological variation: none.

Tomb XXXIX (Grave 66), Trench 1, Unit 369, N = 1

Individual 369. Adult female, aged 25–35 years.
Bone condition: fair.
Skeletal elements preserved: skull, scapulae, humeri, radii, ulnae, left hand elements, ribs, vertebrae, pelvis, femora, patellae, tibiae, fibulae, foot elements.
Dentition: 30 permanent teeth, all present except for L I^2 and P^4.
Aging criteria: dental attrition.
Sexing criteria: pelvic morphology, cranial morphology, postcranial rugosity.
Pathology and morphological variation: lingual tubercles and shoveling on I^1s; I^2 trait; dental calculus; "double humped mastoid"; subtrochanteric flattening; tibia injury.

The right mastoid process is very unusual on this individual (see **Fig. 6.18**). It consists of two bulbous projections. The anterior one is small and relatively narrow; the posterior is a large, blunted and bulbous knob that projects more than the anterior. The overall impression is that of a double mastoid—one male in its morphology and one female. Together, this process is extremely large and broad. No pathological bone structure is present. The adjacent sigmoid sulcus is very deep. Bilateral observations are not possible, but while cleaning the matrix block, the left mastoid appeared to be complex as well.

There is a large area of swelling and bone growth on the right distal tibia (see **Fig. 6.15**). The shaft is deformed at this site, appearing to be thickened on the medial and posterior aspects. The affected region extends to the interosseous crest but not the anterior crest. The bone surface is unfortunately not well-preserved, but a remnant of a drainage channel on the medial surface is visible. Unfortunately, the distal end is not complete. The distal fibula does not seem to be affected.

TOMB XL (GRAVE 67), TRENCH 2, UNIT 383, N = 1

Individual 383. Adult, possible male; possibly aged >30 years.
Bone condition: poor.
Skeletal elements preserved: mandible; left clavicle, scapula, humerus, radius, ribs.
Dentition: none.
Aging criteria: glenoid surface deterioration.
Sexing criteria: postcranial size and robusticity.
Pathology and morphological variation: glenoid surface deterioration.

The glenoid surface of the scapula appears to have some ventral edge build-up and possibly some alterations along the dorsal edge as well. The articular surface is very broad and large. The changes to the glenoid fossa are often associated with rigorous activity over a period of years, although extreme use of the arm for specialized activities could produce these changes in younger individuals.

TOMB XLI (GRAVE 57), TRENCH 2, UNIT 331, N = 1

Individual 331. Adult, aged 20–30 years.
Bone condition: fair.
Skeletal elements preserved: skull, clavicles, sternum, scapulae, humeri, radii, ulnae, hand elements, cervical vertebrae, ribs, right patellae, right tibia, and foot elements.
Dentition: complete permanent dentition.
Aging criteria: dental attrition.
Sexing criteria: none.
Pathology and morphological variation: linear enamel hypoplasia; incisor shoveling; I^2s have slight distal rotation.

TOMB XLII (GRAVE 59), TRENCH 2, UNIT 338, 339, AND 340, N = 3

Grave 59 consists of three individuals: one adult (Unit 338) and two subadults (Units 339 and 340).

Individual 338. Adult female, aged 20–25 years.
Bone condition: fair.
Skeletal elements preserved: skull, left clavicle, sternum, scapulae; left humerus, radius and ulna; right radius, hand elements, vertebrae, ribs, pelvis, femora, right patella, tibiae, fibula, and foot elements.
Dentition: 31 permanent teeth, lacking only the L M_3.
Aging criteria: auricular surface, dental attrition.
Sexing criteria: pelvic morphology, cranial morphology, postcranial size, and rugosity.
Pathology and morphological variation: caries, *n* = 5; linear enamel hypoplasia; dental rotation and crowding of the maxillary canines; dental calculus; incisor shoveling.

Individual 339. Subadult, aged 4 years ± 1 year.
Bone condition: good.
Skeletal elements preserved: largely complete, missing portions of the arms, clavicles, scapulae, and the feet.
Dentition: mixed dentition of 27 teeth, with 19 deciduous (the L di^2 is missing) and 8 permanent teeth visible (R M^1; L I^1, I^2, M^1; R M_1, M_2; L I_1, M_1).
Aging criteria: crown and root development, dental attrition and eruption.
Pathology and morphological variation: hypoplastic pits; Carabelli's cusps; f. c. molare.

Individual 340. Subadult, aged 2 years ± 8 months.
Bone condition: fair.
Skeletal elements preserved: skull with right zygomatic, maxilla, mandible (**Fig. 6.33**).
Dentition: mixed dentition of 23 teeth, with a full deciduous set and 3 permanent teeth (R and L M^1; R M_1).
Aging criteria: tooth eruption and permanent crown formation.
Pathology and morphological variation: Carabelli's cusp; f. c. molare, possible mild cribra orbitalia (**Fig. 6.34**).

TOMB XLIII (GRAVE 62), UNITS 349, 356 AND 371, N = 3

Individual 349. Subadult, aged 18 months ± 6 months.
Bone condition: fair.
Skeletal elements preserved: skull with frontal preserving a portion of the orbital roof and unfused metopic suture, temporals, left maxilla and right and left hemi mandibles.
Dentition: mixed dentition with 13 teeth including 7 deciduous teeth in bone (including the L dc^1–dm^2; R and L dm_1–dm_2) and 6 permanent crowns still in bone (L C^1; R I_1–C_1; L I_2–C_1).
Aging criteria: dental development and eruption.
Pathology and morphological variation: none.

Individual 356. Subadult, aged 1 year ± 4 months.
Bone condition: poor.
Skeletal elements preserved: cranial vault and temporal petrous portion fragments, ribs?

Dentition: 16 deciduous teeth, including dR i^2, dm^1; dLi^1–dm^2; dR i_1–dm_2; dL i_1–dc_1, dm_2.
Aging criteria: crown and root development.
Pathology and morphological variation: f. c. molare; medial cleft in base of lingual surface of Rc^1.

Individual 371. Subadult, aged 2 years ± 8 months.
Bone condition: poor.
Skeletal elements preserved: cranial vault fragments and right maxilla.
Dentition: mixed dentition with 9 teeth, including 6 deciduous (R dc^1–dm^2; L di^2, dc^1; L dm_1) and 3 permanent (R and L M^1; R I_1).
Aging criteria: crown development and tooth eruption.
Pathology and morphological variation: none.

TOMB XLIV (GRAVE 65), TRENCH 4, UNITS 394 AND 365, N = 2

Individual 394. Adult female, aged 18–25 years.
Bone condition: poor.
Skeletal elements preserved: skull, clavicles, right scapula, humeri, radii, ribs, vertebrae, pelvis, and right tibiae.
Dentition: 2 deciduous (R and L dm_2) and 23 permanent teeth (R and L I^1–M^2; R C_1, P_3, M_1–M_3, L C_1, P_3, M_1–M_2).
Aging criteria: sacral fusion and dental attrition.
Sexing criteria: pelvic morphology, postcranial size and morphology.
Pathology and morphological variation: caries, n = 1; linear enamel hypoplasia; f. c. molare; I^2 trait; retention of dm_2 and agenesis of P_4s.

An interesting feature of Individual 394 is that 2 fragmentary dm_2s were found with the dentition; one lacks the crown and the other crown is very worn. As the P_4s are missing, it is possible that there were none and that these deciduous teeth were retained into adulthood.

Individual 365. Adult male, aged 35–45 years.
Bone condition: fair.
Skeletal elements preserved: skull, left scapula, humeri, radii, ulnae, hand elements, ribs, vertebrae, pelvis, femora, tibiae, fibulae, tali, right foot elements.
Dentition: 28 permanent teeth, including R I^1–M^3, L I^1–M^1; R I_1–M_1, L I_1–M_3.
Aging criteria: dental attrition, sutural closure.
Sexing criteria: skull morphology (see **Fig. 6.18**), postcranial robusticity.
Pathology and morphological variation: caries, n = 1; antemortem losses, n = 2; crowding and rotation of anterior maxillary teeth; malocclusion; I^2 trait; supernumerary tooth between R I^2 and C^1 (**Fig. 6.35**); linear enamel hypoplasia; dental calculus; periodontal disease; f. c. molare; cribra orbitalia; possible metopic suture; osteophytes on lumbar vertebrae.

TOMB XLV (GRAVE 60), TRENCH 1, UNITS 345, 351, AND 352, N = 3

Tomb XLV (Grave 60) contained a complicated burial of three individuals. The close proximity of Individuals 345 and 351 to each other, and their overlapping placement, made it impossible to assign all the postcranial elements to individual skeletons. Individual 352 was secondarily deposited in the southeast corner of the grave.

Individual 345. Adult male, aged >55 years.
Bone condition: good.
Skeletal elements preserved: skull, hyoid, clavicles, scapulae, sternum?, humeri, radii, ulnae, hand elements, ribs, vertebrae, pelvis, femora, left tibia, left patella, fibulae, foot elements.
Dentition: complete permanent dentition.
Aging criteria: pubic symphysis and auricular surface, dental attrition, cranial sutural fusion, postcranial osteophyte formation and eburnation.
Sexing criteria: pelvic morphology, skull morphology, postcranial robusticity.
Pathology and morphological variation: caries, n = 12; abscesses for most molars; linear enamel hypoplasia; diastema between maxillary I^1s; incisor shoveling; crowding of mandibular anterior dentition including the premolars; L M_3 maloccluded, intruding onto the M_2; periodontal disease (**Figs. 6.37–6.38**); possible healed cribra orbitalia; osteophytes on clavicle, scapula, hand and foot elements, ribs, cervical vertebrae, thoracic vertebrae (DISH, fusion of 2 centra with a large bolus formation), and ilia; eburnation on right scapula, distal radius; healed injury to left ankle involving the tibia and fibula. (See **Fig. 6.12** for the fused thoracic vertebrae, and **Fig. 6.20** for dental crowding and rotation.)

Individual 351. Adult male, aged 18–25 years.
Bone condition: poor.
Skeletal elements preserved: skull, humeri, radii, ulnae, ribs, vertebrae, pelvis, femora, patellae, tibiae, fibulae, foot elements.

Dentition: complete permanent dentition.
Aging criteria: iliac crest with incomplete epiphysis fusion, pubic symphysis, M3 root formation, dental attrition.
Sexing criteria: pelvic morphology.
Pathology and morphological variation: caries, n = 3; slight incisor shoveling; I^2 trait; dental calculus; mandibular I^1 hyper-eruption; f. c. molare.

Individual 352. Adult female, aged 30–40 years.
Bone condition: poor.
Skeletal elements preserved: skull, right humerus, left radius and ulna, femora, tibiae, fibulae.
Dentition: 29 permanent teeth, including the R I^2–M^1, M^3; L I^2–M^3; R I_1–M_3, L I_1–M_3.
Aging criteria: dental attrition, maxillary sutural fusion.
Sexing criteria: cranial morphology, postcranial robusticity.
Pathology and morphological variation: caries, n = 1; abscesses, n = 2; I^2 trait; dental calculus.

TOMB XLVI (GRAVE 42), TRENCH 2, UNIT 263, N = 1

Individual 263. Adult female, aged 20–25 years.
Bone condition: fair.
Skeletal elements preserved: skull, clavicles, scapulae, humeri, radii, ulnae, hand elements, vertebrae, ribs, pelvis, femora, patellae, tibiae, fibulae, foot elements.
Dentition: 31 permanent teeth, lacking only the R I^1.
Aging criteria: auricular surface, dental attrition.
Sexing criteria: skull and pelvic morphology.
Pathology and morphological variation: caries, n = 3; linear enamel hypoplasia; dental calculus; dental crowding and rotation of the R P_3; f. c. molare; extrasutural bone in the squamosal suture.

TOMB XLVII (GRAVE 41), TRENCH 2, UNIT 256, N = 1

Individual 256. Adult female, aged 18–21 years.
Bone condition: fair.
Skeletal elements preserved: cranium, right clavicle, left scapula, left humerus, radii, ulnae, hand elements, ribs, pelvis, femora, tibiae, fibulae, foot elements.
Dentition: 15 permanent teeth, including the R I^1–P^3, L I^1–M^3; L I_1–I_2, P_3.
Aging criteria: bone maturation.
Sexing criteria: cranial morphology, postcranial morphology.
Pathology and morphological variation: incisor shoveling, dental calculus; cribra orbitalia; right clavicle, acromial end has knob on the inferior surface, projecting approximately 4 mm; unusual soleal lines on tibiae.

The left tibia of Individual 256 has a pathological form of the popliteal line. In lieu of the line there is a smooth ridge and an accompanying medial depression. This is a deep groove that parallels the ridge. The right tibia has the same form, but it is not as strongly expressed.

TOMB XLVIII (GRAVE 52), TRENCH 1, UNITS 318, 366, AND 372, N = 3

Individual 318. Subadult, female, aged 15 years ± 3 years.
Bone condition: fair.
Skeletal elements preserved: skull, vertebrae, ribs, left scapula, humeri, radii, ulnae, hand elements, pelvis, femora, tibiae, fibulae, foot elements.
Dentition: n = 31, complete permanent set except for R M^3.
Aging criteria: dental development and skeletal maturation.
Sexing criteria: facial form, mental eminence; postcranial size, and rugosity.
Pathology and morphological variation: caries, n = 3; linear enamel hypoplasia; I^2 trait; incisor shoveling; Carabelli's cusp; agenesis of R M^3; diastema; rotation of mandibular incisors; canine root complexity; dental calculus; multiple mental foramina on mandible.

The mandible is moderately small with a pointed chin, slight notching of the inferior border, and an angled ascending ramus. There are interesting non-metric features preserved: two mental foramina and a clear diastema between the right C_1 and the P_3. The mandibular central incisors are slightly rotated and wing outward distally.

Individual 318 has one of the most interesting dentitions in the Lofkënd population. It is complete with the exception of the right M^3 that was congenitally absent. In addition to this, and the diastema mentioned above, there is a full litany of dental variants (**Fig. 6.21b**). Of particular interest is the morphology of the maxillary incisors. Other Lofkënd individuals have unusual incisor morphology, but Individual 318 constitutes a "textbook" example of irregular morphology and the I^2 trait. The right I^1

has slight shoveling and a narrow cleft or groove in the enamel beginning halfway down its distal border and running down to the root. The left has slight shoveling and a deeper, more prominent cleft in the mesial border running down into the root. These antimeres are similar in crown size, but the adjacent I^2s are very asymmetrical. The right is larger and broader with a trace of shoveling combined with a sizable lingual tubercle. A cleft in the enamel bisects the tubercle and continues down to the root. The left I^2 is reduced mesiodistally and has, in the lingual view, a slight "waist" about midway down the crown. The edges of this tooth appeared rolled, and this creates a shovel shape with a central groove where the edges meet.

Another tooth variant is seen on the right C_1. There is a small secondary lingual rootlet that is lacking on the antimere.

Individual 366. Adult, possible male; aged 20–25 years.
Bone condition: poor.
Skeletal elements preserved: cranium, clavicle, vertebrae, scapula, radius, ulna, tibia, femora, and fibulae.
Dentition: complete permanent dentition.
Aging criteria: dental attrition and root development.
Sexing criteria: mastoid region and dental size.
Pathology and morphological variation: caries, $n = 1$; incisor shoveling; I^2 trait; supernumerary peg tooth; linear enamel hypoplasia; dental calculus; f. c. molare.

The dentition of Individual 366 includes several notable features. Overall, this is a very large set of teeth. The maxillary central incisors have medium shovel shaping. This is more marked on the right. On the fragmentary right lateral incisor, this development is expressed as a crease on the mesial surface, rolling of the lateral crown margins to create a median groove of the lingual surface, and asymmetry of the lingual crown base. Unfortunately the left lateral incisor is even more fragmentary, but the superior portion of a mesial surface crease is clearly visible. Individual 366 also had a supernumerary tooth. This has a peg morphology with a complex tip that has either an irregularity of the enamel or it was coated with dental calculus (**Fig. 6.38**). The tooth even has linear enamel hypoplasia. It is not possible to determine where it was situated in the mouth. Other features of this dentition include a deep crevice or crenulation on the mesial buccal cusp of the maxillary first and second molars, and f. c. molare pits on the M_2s.

Individual 372. Adult, mid-age.
Bone condition: very poor.
Skeletal elements preserved: Cranial fragments, vertebral fragments, humeri, radii, ulnae, carpal (triquetral), femora, tibiae.
Dentition: 16 permanent teeth (R I^1, C^1, P^4, M^1, M^3; L I^1, C^1, M^1, M^2; R I_1–P_4; L P_3, P_4).
Aging criteria: dental attrition.
Sexing criteria: none.
Pathology and morphological variation: possible M^3 impaction; linear enamel hypoplasia.

TOMB XLIX (GRAVE 51), TRENCH 4, UNIT 315, N = 1

Individual 315. Subadult, aged 3 years ± 1 year.
Bone condition: very poor.
Skeletal elements preserved: skull only.
Dentition: mixed dentition of 22 teeth, including R dc^1–dm^2; L dc^1–dm^2; R dm_1–dm_2; L dc_1, dm_1–dm_2; R I^1, I^2, M^1; L I^1, I^2, M^1; R I_2, C_1, M_1; L C_1, M_1.
Aging criteria: dental development of permanent incisors and canines.
Pathology and morphological variation: I^2 trait.

Individual 315 has reduced lateral maxillary incisors (I^2s) that are 68% as wide as the central incisors. As the crowns are only half-completed, it is not possible to determine if they had a cleft or other features associated with incisor reduction in other Lofkënd individuals.

TOMB L (GRAVE 46), TRENCH 4, UNIT 284, N = 1

Individual 284. Subadult, aged 4 years ± 1 year.
Bone condition: very poor.
Skeletal elements preserved: fragments of non-identifiable bone.
Dentition: mixed dentition with fragments of deciduous molars and 11 permanent crowns (L M^1; R and L I_1–P_3, M_1).
Aging criteria: deciduous dental attrition, permanent crown formation.
Pathology and morphological variation: f. c. molare on the M_1s, Carabelli's pit on the M^1, moderately strong linear enamel hypoplastic band on all the M1s.

TOMB LI (GRAVE 78), TRENCH 4, UNIT 434, N = 1

Individual 434. Young adult, possible male.
Bone condition: poor.
Skeletal elements preserved: cranium only, humeri and radius, metacarpal.
Dentition: 10 maxillary teeth, including the right and left C^1–M^2.
Aging criteria: dental attrition.
Sexing criteria: cranial morphology.
Pathology and morphological variation: linear enamel hypoplasia.

TOMB LII (GRAVE 69), TRENCH 1, UNIT 390, N = 1

Individual 390. Subadult, aged 15 ± 3 years, probable male.
Bone condition: good.
Skeletal elements preserved: most skull and post-cranial elements represented, with the exception of the feet (left talus only).
Dentition: a set of 30 teeth, missing the I^2s.
Aging criteria: development of M3s, epiphyseal union.
Sexing criteria: cranial morphology, limb size, and robusticity.
Pathology and morphological variation: caries, *n* = 3; trauma to face and palate with impacted canine; linear enamel hypoplasia; Carabelli's trait (groove); clavicle robusticity.

The anterior palate is irregular and shows signs of trauma (**Fig. 6.39a–b**). On the right, there are clear sockets for the central and lateral incisors and canine. The canine has a slight distal rotation, but otherwise the right maxilla appears normal. The left side has a broken region medial to the central incisor, an irregular, roughened alveolus on the other side of the tooth, and then portions of a socket for a short, single rooted tooth. It is unclear if this is for the lateral incisor, as this tooth is not preserved. In the palate, located behind the central incisor, is an impacted canine. The long root runs diagonally through the palate, and its tip is visible in the labial view just superior and lateral to the incisor. In addition, the intermaxillary suture and incisive canal are displaced by the canine crown. Although it is difficult to say exactly what occurred during life to Individual 390, it appears that there was trauma to the left maxilla while the anterior teeth were still forming. An impact possibly removed the permanent lateral incisor, displaced the developing permanent canine, and led to the retention of a deciduous tooth (the canine?). As several teeth are affected, this seems to be a more likely explanation than merely an impacted canine. The mandible is not affected.

The clavicles are asymmetrical and differ in their size and morphology. The better-preserved left bone is flattened superiorly-inferiorly (measuring 8.46 mm at the midshaft) and quite straight, with little anteriorly-posterior bowing. It has a very stocky appearance (midshaft breadth is 13.2) and a large extended area for the deltoid attachment. The right clavicle is slightly more bowed, and the shaft is rounder, measuring 10.84 mm superior-inferiorly and with a breadth of only 11.8 mm at the midshaft. This clavicular morphology may reflect a habitual activity, such as archery, that places differential stresses on the shoulder musculature. Unfortunately, neither the scapulae nor the arms are sufficiently preserved to evaluate if there was asymmetry in muscular development. Only the ulnae are reasonably complete, and they are both quite large, slightly curved, and rugose.

TOMB LIII (GRAVE 63), TRENCH 1, UNITS 359 AND 359a, N = 2

Individual 359. Adult, aged 25–30 years.
Bone condition: poor.
Skeletal elements preserved: skull, rib head, humerus, radius, ulna, femora, tibia, fibula, calcaneus, and talus.
Dentition: 20 permanent teeth, including R P^3–M^2; L C^1–M^2; R I_2, P_3–M_3; L C_1–M_2.
Aging criteria: dental attrition, rib head.
Sexing criteria: none.
Pathology and morphological variation: caries, *n* = 6; linear enamel hypoplasia; f. c. molare.

Individual 359a. Subadult, aged 5 years ± 1.5 years.
Bone condition: poor.
Skeletal elements preserved: skull, primarily parietals.
Dentition: mixed dentition of 18 teeth, with 2 deciduous (R and L dm^2) and 16 permanent (R I^2–M^1; L M^1, M^2; R I_2–P_3, M_1, M_2; L I_1–C_1, M_1).
Aging criteria: permanent crown development.
Pathology and morphological variation: Carabelli's cusps; incisor shoveling.

TOMB LIV (GRAVE 40), TRENCH 2, UNIT 252, N = 1

Individual 252. Adult, possibly mid-age.
Bone condition: poor to fair.

Skeletal elements preserved: left humerus, radius and ulnae; right radius, hand elements, ribs, lumbar vertebrae, pelvis, left femur, tibiae, fibulae, foot elements.
Dentition: none.
Aging criteria: postcranial morphology and vertebral degenerative changes.
Sexing criteria: none.
Pathology and morphological variation: osteophytes on lumbar vertebra.

The most notable feature of this skeleton is the pathological condition of a lumbar vertebra (probably the L 5). There is significant lipping on both the superior and inferior margins of the centrum. The superior alterations are more extensive. The sacral promontory is somewhat damaged, but it does not appear to be affected.

TOMB LV (GRAVE 53), TRENCH 1, UNIT 321, N = 1

Individual 321. Subadult, aged 7 years ± 2 years.
Bone condition: poor.
Skeletal elements preserved: skull, portions of both arms and legs.
Dentition: mixed dentition with 14 deciduous teeth (R di^1, dc^1–dm^2; L di^1–dm^2; R dc_1–dm_2; L dm_1, dm_2); and 14 permanent teeth (R M^1, M^2; L M^1; R I_1, I_2, P_3, M_1, M_2; L I_1, I_2, P_3–M_2).
Aging criteria: dental development of premolar and M2 crowns.
Pathology and morphological variation: caries, n = 1; f. c. molare.

TOMB LVI (GRAVE 43), TRENCHES 1 AND 4, UNITS 269 AND 296, N = 2

Individual 269. Adult female, aged 20–25 years.
Bone condition: fair.
Skeletal elements preserved: skull, right clavicle, cervical vertebrae, right hand phalanges, pelvis, femora, patellae, tibiae, fibulae, foot elements.
Dentition: complete permanent dentition.
Aging criteria: dental attrition.
Sexing criteria: skull morphology.
Pathology and morphological variation: linear enamel hypoplasia and enamel pitting; rotation and crowding of the anterior mandibular dentition; dental calculus; enamel extensions; I^2 trait; f. c. molare.

Individual 296. Adult male, aged 20–25 years.
Bone condition: fair to good.
Skeletal elements preserved: skull, clavicles, right scapula, humeri, radii, ulnae, hand elements, vertebrae, ribs, pelvis, femora, tibiae, fibulae, foot elements.
Dentition: 27 permanent teeth, including the R I^1–M^1, M^3; L I^1–M^3; R I_2–P_4, M_3; L I_2–M_3.
Aging criteria: dental attrition.
Sexing criteria: cranial morphology, pelvic morphology.
Pathology and morphological variation: caries, $n = 6$; linear enamel hypoplasia and pitting; I^2 trait; enamel extensions; multiple wormian bones including small Inca bone; femoral subtrochanteric flattening; rugosity of right tibia on the interosseous crest at the midshaft extending 76.1 mm.

The 296 cranium is one of the more rugose observed at Lofkënd. This is expressed in the brow form, the temporal markings, and the nuchal region. The brow, of which only the right side is preserved, has a distinct brow ridge delineated by a sulcus and mid-central brow thickening. The orbital border is not thick or particularly blunted, however. The temporal lines are moderately strong, which is not common in the population. The parietals are full and high but not bossed. All sutures are open; the lambdoidal is very complex with many wormian bones and a small triangular one that, as a small Inca bone, completes the occipital at lambda. The occipital is characterized by marked nuchal rugosity and inion forming a torus. There is a vague suprainiac depression. The right temporal is fairly complete; there is a deep, large mandibular fossa, a moderately strong supramastoid crest, and a large, broad, projecting mastoid. There appears to have been a paramastoid process as well.

TOMB LVII (GRAVE 58), TRENCH 1, UNIT 336, N = 1

Individual 336. Adult, mid-age; indeterminate sex.
Bone condition: very poor.
Skeletal elements preserved: cranium, left humerus, left and right ulna, radius, right femur, right patella, tibia, foot phalanges.
Dentition: 19 permanent teeth, including R I^1–M^1, L I^2–P^4, M^2; R I_1–P_3, L I_1, C_1, P_3, M_2.
Aging criteria: dental attrition.
Sexing criteria: none.
Pathology and morphological variation: caries, n = 3; eburnation on foot phalanges.

Tomb LVIII (Grave 37), Trench 2, Unit 243, N = 1

Individual 243. Subadult, aged 15 years ± 3 years.
Bone condition: very poor.
Skeletal elements preserved: skull, upper skeleton, including right arm and hand elements.
Dentition: 21 teeth, including the R I^1–M^3, L P^4–M^3; R C_1–M_2, L C_1, P_4–M_2.
Aging criteria: molar root formation and wear.
Sexing criteria: none.
Pathology and morphological variation: possible I^2 trait; linear enamel hypoplasia.

The individual in Grave 37 was found with a large amount of burned material—clay, ceramics, and possibly wood—and an underlying calcareous sediment. These materials have undoubtedly had an effect on the bone preservation. The bones of the thorax are splintered and flake apart; the cranium and cervical vertebrae are eroded, friable, and spongy. In some fragments where the outer bone cortex is preserved, the bone appears blackened or gray. The right humerus fragments have bronze staining.

Tomb LIX (Grave 38), Trench 4, Unit 247, N = 1

Individual 247. Adult male, aged 35–45 years.
Bone condition: poor.
Skeletal elements preserved: skull, right clavicle; right humerus, radius, and ulna; left radius and ulna; hand elements, vertebrae, and pelvis.
Dentition: 29 permanent teeth, lacking the L M^2 and L I_2, M_1.
Aging criteria: dental attrition, cranial sutures.
Sexing criteria: skull morphology.
Pathology and morphological variation: antemortem losses, $n = 2$; dental calculus, dental crowding and rotation, malocclusion, hyper-eruption, f. c. molare.

Tomb LX (Grave 44), Trench 2, Unit 273, N = 1

Individual 273. Young adult, no further identification possible.
Bone condition: poor.
Skeletal elements preserved: skull, unidentifiable limb fragments, and a few foot elements.
Dentition: 5 permanent teeth, including the L I^1, I^2, P^4, M^1, and the R M_1.
Aging criteria: dental attrition.
Sexing criteria: none.
Pathology and morphological variation: none.

Tomb LXI (Grave 34), Trench 2, Unit 225, N = 1

Individual 225. Adult, no specific age or sex estimation.
Bone condition: very poor.
Skeletal elements preserved: left and right femora, tibiae, and fibulae.
Dentition: none.
Aging criteria: bone morphology.
Sexing criteria: none.
Pathology and morphological variation: none.

Tomb LXII (Grave 32), Trench 2, Unit 220, N = 1

Individual 220. Adult, no age or sex determinable.
Bone condition: very poor.
Skeletal elements preserved: skull, femur, tibia, and left patella.
Dentition: none.
Aging criteria: bone morphology.
Sexing criteria: none.
Pathology and morphological variation: none.

Tomb LXIII (Grave 35), Trenches 4 and 5, Unit 234, N = 1

Individual 234. Adult, probable female, aged >45 years.
Bone condition: very poor.
Skeletal elements preserved: skull, clavicles, scapulae, humeri, radii, ulnae, vertebrae, pelvis, femora, tibiae, fibulae.
Dentition: 21 permanent teeth, including the R I^1–P^3, M^3; L I^1, C^1–P^4; R I_1–I_2, P_4–M_2; L I_1, C_1–M_3.
Aging criteria: dental attrition.
Sexing criteria: skull morphology.
Pathology and morphological variation: caries, $n = 2$; linear enamel hypoplasia; dental calculus; L P_4 rotated distally; f. c. molare.

Tomb LXIV (Grave 61), Trench 2, Unit 347, N = 1

Individual 347. Subadult, aged 4 years ± 1 year.
Bone condition: poor.
Skeletal elements preserved: cranium with petrous portions of the temporals.
Dentition: mixed dentition of 25 teeth, with 14 deciduous (R dm^1, dm^2; L di^2–dm^2; R di_1, dc_1–dm_2; L di_2–dm_2) and 11 permanent (R I^2, M^1, M^2; L I^1, M^1; R and L I_1, I_2, M_1).
Aging criteria: M2 crown development.
Pathology and morphological variation: none.

Tomb LXV (Grave 30), Trench 1, Unit 213, N = 1

Individual 213. Adult female, aged 18–23 years.
Bone condition: poor–fair.
Skeletal elements preserved: left clavicle, left humerus, radius and ulna; hand elements, vertebrae, ribs, pelvis, femora, left patella, tibiae, fibulae, foot elements.
Dentition: none.
Aging criteria: pubic symphysis, bone maturation.
Sexing criteria: postcranial measurements.
Pathology and morphological variation: evidence for habitual hip flexion; left proximal femur diaphysis has extremely dense cortical bone.

The femora are small and gracile, but the gluteal region is rugose with a strong lateral insertion for gluteus maximus and a groove medial to the insertion. This is noted on other specimens and obviously reflects a habitual posture or activity involving hip flexion. The linea aspera is also strongly developed. On the left femur, there is a postmortem break that reveals very dense, thick cortical bone at the upper portion of the diaphysis; inferiorly the interior has much cancellous bone that appears somewhat disorganized.

Tomb LXVI (Grave 31), Trench 4, Unit 216, N = 1

Individual 216. Subadult, aged 15 years ± 2 years.
Bone condition: poor.
Skeletal elements preserved: skull with occipital and mandibular portions, clavicles, humeri, radii, ulnae, hand elements, pelvis, lumbar vertebrae, femora, tibiae, fibulae.
Dentition: 31 teeth, missing the R I^1.
Aging criteria: dental attrition and root development of M3s, epiphyseal fusions.
Sexing criteria: postcranial morphology and size.
Pathology and morphological variation: I^2 trait, moderate to strong linear enamel hypoplasia; femoral subtrochanteric flattening and bowing of the diaphysis, along with evidence of habitual hip flexion.

These femora are characterized by subtrochanteric flattening. They appear to be of moderate size and rugosity, especially when compared to the very robust subadult Individual 390 from Grave 69. However, the area of gluteal muscle attachment seems laterally displaced and rugose, creating a groove or depressed area medial and parallel to the attachment. This morphology is coupled with anterior-posterior diaphyseal bowing in this individual. A similar morphology of the gluteal insertions characterizes the young female Individual 213 from Grave 30, but her leg does not seem as bowed. The linea aspera is strong for the relatively small size of the bone, especially in contrast with the slender humeri.

Tomb LXVII (Grave 12), Trench 3, Units 100 and 171, N = 2

Individual 100. Adult male, mid-age >35 years.
Bone condition: fair.
Skeletal elements preserved: skull, clavicles, scapulae, vertebrae.
Dentition: 21 teeth, including the R C^1, M^1–M^3; L C^1–M^1; R I_1–M_2, L I_1–P_4, M_2.
Aging criteria: dental attrition and cranial sutures.
Sexing criteria: skull morphology and postcranial morphology.
Pathology and morphological variation: caries, n = 6–7 possible; antemortem losses = 3 (L I^2, R P^3, L M_1); several abscesses and severe alveolar loss (**Fig. 6.40**): R I_1 has a clear abscess, 7 other possible abscesses; M_3s agenesis or impacted; dental crowding involving the mandibular I_2s–P_3s; multiple infraorbital foramina; osteophytes on cervical vertebrae.

Individual 171. Adult female, aged 23–27 years.
Bone condition: poor.
Skeletal elements preserved: mandible, left humerus, radii, ulnae, hand elements, femora, tibiae, fibulae, foot elements.
Dentition: 9 permanent mandibular teeth, including the R I_2, P_3; L I_2–M_3.
Aging criteria: dental attrition.
Sexing criteria: mandibular and postcranial morphology; femoral dimensions.
Pathology and morphological variation: linear enamel hypoplasia, periodontal disease; infolding of hallux metatarsal base articular surfaces and corresponding alteration to the articular facet on the first cuniform of the left foot.

The feet have a number of variant morphological features (**Fig. 6.41**): both hallux metatarsal bases show the same feature—a vertical infolding of the articular surface, running from the top to about the mid-section. This creates a somewhat partitioned articular surface that is mirrored on the medial cuniform that is preserved for the left foot. The left talus has an extension of the posterior, inferior articular facet with the calcaneous.

TOMB LXVIII (GRAVE 13), TRENCH 2, UNITS 103, 103a, 136, AND 149, N = 4

This burial has two levels—the higher one containing the fragmentary remains of two individuals (103, 103a), and the lower level with two individuals (136 and 149) with a complete Iron Age vessel and an iron pin.

Individual 103. Adult female, mid-age.
Bone condition: poor.
Skeletal elements preserved: cranial fragments including temporal petrous portion.
Dentition: 2 permanent teeth, including the L I_2 and P_4.
Aging criteria: dental attrition.
Sexing criteria: mandibular morphology and dimensions.
Pathology and morphological variation: caries, $n = 9$; antemortem losses, $n = 8$ (**Fig. 6.42**); linear enamel hypoplasia.

Individual 103a. Adult male, aged 20–30 years.
Bone condition: poor.
Skeletal elements preserved: mandible.
Dentition: 16 permanent teeth, including R M^1 and M^2, L M^1 and M^2; R I_1–P_3, L I_1–M_3.
Aging criteria: dental attrition.
Sexing criteria: mental eminence.
Pathology and morphological variation criteria: caries, $n = 10$ possible on 9 teeth, linear enamel hypoplasia, f. c. molare.

Individual 136. Adult, 20–25 years.
Bone condition: poor.
Skeletal elements preserved: skull, ribs, possibly humerus, femur, tibia, pelvis, foot elements.
Dentition: 19 permanent teeth, including the R I^1–M^3, L I^1–M^3; R M_1–M_3.
Aging criteria: dental attrition.
Sexing criteria: none.
Pathology and morphological variation: caries, $n = 12$ possible on 8 teeth; linear enamel hypoplasia, dental calculus.

Individual 149. Adult, 20–25 years.
Bone condition: poor.
Skeletal elements preserved: skull, humerus, femur.
Dentition: 23 permanent teeth, including the R I^1–M^2, L I^1–M^2; R C_1–P_3, M_1–M_2; L C_1–M_2.
Aging criteria: dental attrition.
Sexing criteria: none.
Pathology and morphological variation: caries, $n = 2$, hypercementosis possible on R C_1, uneven wear of canines suggests possible non-eruption or impaction of L C^1, incisor shoveling; possible expansion of diploe on frontal, parietals.

TOMB LXIX (GRAVE 27), TRENCHES 1 AND 5, UNITS 174, 551, AND 551a, N = 3

Individual 174. Adult, 20–25 years.
Bone condition: very poor.
Skeletal elements preserved: cranium, humeri, ulnae, radii, ribs, femora, tibiae, fibulae, foot elements.
Dentition: 19 permanent teeth, including the R C^1, M^3; L I^2, P^3–M^3; R I_1, C_1–P_4, M_2; L I_1, C_1–M_2.
Aging criteria: dental attrition.
Sexing criteria: none.
Pathology and morphological variation: caries, $n = 1$.

Individual 551. Subadult, aged 6 years ± 2 years.
Bone condition: poor.
Skeletal elements preserved: cranial vault fragments—non-diagnostic.
Dentition: mixed dentition of 33 teeth, including 8 deciduous (R dc^1, dm^2; L dm^1, dm^2; R and L dc_1–dm_2) and 25 permanent (R I^1–M^2, L I^1–P^3, M^1, M^2; R I_1–P_3, M_1, M_2; L I_1, I_2, P_3–M2).
Aging criteria: crown and root formation, wear.
Pathology and morphological variation: I^2 trait; incisor shoveling; f. c. molare on M_1s; Carabelli's trait on dm^2s, M^1s, and M^2; linear enamel hypoplasia.

Individual 551a. Subadult, aged at least 6 years.
Bone condition: poor.
Skeletal elements preserved: possibly associated cranial vault fragments.
Dentition: 2 M_1 antimeres.
Aging criteria: dental attrition.
Pathology and morphological variation: f. c. molare.

TOMB LXX (GRAVE 17), TRENCHES 3 AND 7, UNITS 126 AND 197, N = 2

Individual 126. Subadult female, aged 17–20 years.
Bone condition: fair.
Skeletal elements preserved: skull, right clavicle, scapulae, sternum, left humerus and radius, right

humerus and ulna, hand elements, ribs, vertebrae, pelvis, femora, tibia, fibula, foot elements.
Dentition: set of 28 teeth, missing the R M^3, R I_1, L I_1, I_2.
Aging criteria: skeletal maturation, M3 eruption, dental attrition.
Sexing criteria: cranial morphology, postcranial size, and robusticity.
Pathology and morphological variation: I^2 trait expressed as size reduction (**Fig. 6.43**) (see **Fig. 6.21a** for labial view); crowding and rotation affecting the R I^2 and C^1, and the R C_1; overbite, dental calculus, f. c. molare.

Individual 197. Adult male, >45 years.
Bone condition: fair.
Skeletal elements preserved: cranium, sternum, humeri, radii, left ulna, hand elements, vertebrae, ribs, pelvis, right femur, patella, tibiae, fibulae, foot elements.
Dentition: 21 permanent teeth, including the R I^1–P^3, L I^1, C^1–M^2; R I_1–C_1, M1–M_3; L I_1, C_1, M_1–M_3.
Aging criteria: dental attrition, cranial sutures.
Sexing criteria: skull morphology, postcranial size, and rugosity.
Pathology and morphological variation: linear enamel hypoplasia, overbite; rib injury: two ribs are fused near the tubercle on the left side; osteophytes on the acetabulum.

TOMB LXXI (GRAVE 28), TRENCH 1, UNIT 177, N = 1

Individual 177. Subadult, aged 8 years ± 2 years.
Bone condition: poor.
Skeletal elements preserved: skull with temporals, clavicle, right radius and ulna, femora, tibiae.
Dentition: mixed dentition with 35 teeth, including 13 deciduous (R di^2–dm^2, L di^1, dc^1, dm^2; R and L dc_1–dm_2) and 22 permanent (R I^1–M^2; L I^2–M^2; R I_1, I_2, M_1, M_2; L I_1–C_1, M_1, M_2).
Aging criteria: permanent crown development and eruption (**Fig. 6.44**).
Pathology and morphological variation: slight incisor shoveling, Carabelli's cusps on the M^1s, linear enamel hypoplasia.

TOMB LXXII (GRAVE 24), TRENCH 1, UNIT 162 AND 162a, N = 2

Grave 24 was found to contain the burial of an adult that was mixed with the dentition of a subadult.

Individual 162. Adult, aged 25–35 years.
Bone condition: poor.
Skeletal elements preserved: femora, tibiae, tarsals; fragmentary portion of the pelvis, arms, and ribs.
Dentition: permanent dentition with 6 teeth (L I^1, C^1, P^4–M^3); miscellaneous fragments of mandibular molars.
Aging criteria: dental attrition.
Sexing criteria: none.
Pathology and morphological variation: none.

Individual 162a. Subadult, aged 10 years ± 2.5 years.
Bone condition: poor.
Skeletal elements preserved: non-diagnostic fragments of postcrania.
Dentition: mixed dentition of 36 teeth, including 9 deciduous (R and L di^2, dm^1, dm^2; R dm_1, dm_2; L dm_2) and 27 permanent (R I^1, I^2, P^3–M^3; L I^1, C^1–M^2; R and L I_1–M_2).
Aging criteria: dental development and wear.
Pathology and morphological variation: Carabelli's cusp on M^1s, Carabelli's groove on dm^2s; I^2 trait; linear enamel hypoplasia on L P^4.

TOMB LXXIII (GRAVE 26), TRENCH 1, UNITS 169 AND 169a, N = 2

This grave contained the partial lower skeletons of two adults. The remains are mixed, but it is most parsimonious to attribute the femur and tibiae to the "major" individual, Unit 169, with the bigger feet.

Individual 169. Adult, probable male; no age estimation.
Bone condition: poor.
Skeletal elements preserved: pelvis, left femur, tibiae, fibulae, foot elements.
Dentition: none.
Aging criteria: adult bone morphology.
Sexing criteria: postcranial size and robusticity.
Pathology and morphological variation: none.

Individual 169a. Adult, no specific age or sex estimation possible.
Bone condition: poor.
Skeletal elements preserved: foot elements, possible fibular fragments.
Dentition: none.
Aging criteria: adult bone morphology.

Sexing criteria: none.
Pathology and morphological variation: none.

TOMB LXXIV (GRAVE 29), TRENCH 4, UNITS 210, 230, AND 231, N = 3

This is a grave with three adults in the extended position with intertwined arms. Individual 210 lacks the right leg except for the proximal femur. The arms of the three are intertwined and the elements are overlapping; the left hand of Individual 231 appears to have been below the hip or behind the body.

Individual 210. Adult male, aged 25–35 years.
Bone condition: fair.
Skeletal elements preserved: skull, clavicles, scapulae, humeri, radii, ulnae, hand elements, vertebrae, ribs, pelvis, femora, left patella, left tibia, fibula, and tarsals.
Dentition: complete permanent dentition.
Aging criteria: dental attrition.
Sexing criteria: skull morphology, pelvic morphology, postcranial dimensions.
Pathology and morphological variation: caries, n = 6; I^2 trait; linear enamel hypoplasia; dental calculus; dental crowding and rotation of L C_1 and P_3; Carabelli's cusp; f. c. molare.

Individual 230. Adult male, aged 35–45 years.
Bone condition: fair.
Skeletal elements preserved: skull, clavicles, scapulae, humeri, radii, ulnae, hand elements, vertebrae, ribs, pelvis, femora, tibiae, fibulae, foot elements.
Dentition: 26 permanent teeth, including R I^1–P^3, M^2; L I^1–P^4, M^2; R I_1–M_3, L I_1–P_4, M_2–M_3.
Aging criteria: dental attrition, cranial sutures.
Sexing criteria: cranial morphology, postcranial size.
Pathology and morphological variation: caries, n = 23, antemortem losses, n = 6 possible; abscesses, n = 3 possible; linear enamel hypoplasia, I^2 trait, dental calculus, crowding of the mandibular incisors, rotation of the P^3s, hyper-eruption, f. c. molare, possible osteophytes on vertebrae and lipping on the glenoid fossa.

Individual 231. Adult male, aged 20–25 years.
Bone condition: fair.
Skeletal elements preserved: skull, clavicles, scapulae, humeri, radii, ulnae, hand elements, vertebrae, ribs, pelvis, femora, patellae, tibiae, fibulae.
Dentition: complete permanent dentition.
Aging criteria: dental attrition, bone development.
Sexing criteria: cranial morphology, postcranial robusticity and measurements.
Pathology and morphological variation: linear enamel hypoplasia, incisor shoveling, I^2 trait, Carabelli's cusps, f. c. molare, dental crowding and rotation of mandibular anterior dentition and the R P_3, overbite; Poirier's facet on the femoral neck, subtrochanteric flattening.

Tomb LXXIV (Grave 29) holds three adult males, with the older individual in the middle. The two flanking males appear very similar, so some comparison is informative. The two dentitions do have some important differences. Individual 231 has a much larger, robust dentition with very prominent canines, P_4 molarization, and an M_3 that is larger than the M_2. Individual 210 has molar reduction and similar-sized molars; 231 has I^2 reduction with folded edges and Carabelli's cusps on the M^1s. Individual 210 also has the I^2 trait, but it is not as extreme and does not involve infolding of the borders; the Carabelli's cusp is much smaller and the M^3 reduction is much more asymmetrical; 210 has dental decay and linear enamel hypoplasia, along with slightly more advanced attrition. It is not possible to evaluate 231 for linear enamel hypoplasia.

TOMB LXXV (GRAVE 33), TRENCH 1, UNIT 223, N = 1

Individual 223. Adult male, aged 30–40 years.
Bone condition: poor.
Skeletal elements preserved: skull, clavicles, scapulae, humeri, radii, ulnae, hand elements, vertebrae, ribs, pelvis, femora, patellae, tibiae, fibulae, foot elements.
Dentition: 19 permanent teeth, including the R I^1–I^2, M^2; L I^1, P^3–P^4; R I_1–M_3, L I_1–I_2, M_1–M_3.
Aging criteria: dental attrition.
Sexing criteria: skull and pelvic morphology, postcranial dimensions.
Pathology and morphological variation: caries, n = 5; abscesses, n = 2; linear enamel hypoplasia, malocclusion and crowding of L M_2 and M_3, I^2 trait, dental calculus; right talus has extension of the posterior articular surface for the tibia.

TOMB LXXVI (GRAVE 16), TRENCH 4, UNIT 125, N = 1

Individual 125. Adult, no sex estimation.
Bone condition: poor.
Skeletal elements preserved: cranium with occipital, possible ulna.

Dentition: one tooth, probably L M^3.
Aging criteria: bone morphology.
Sexing criteria: none.
Pathology and morphological variation: none.

Tomb LXXVII (Grave 18), Trench 4, Unit 139, N = 1

Individual 139. Adult male, mid-age.
Bone condition: fair.
Skeletal elements preserved: skull, left scapula, humeri, radii, ulnae, hand elements, vertebrae, pelvis, femora, and tibiae.
Dentition: 4 permanent teeth, including the L P^3–P^4, L C_1–P_3.
Aging criteria: dental attrition.
Sexing criteria: cranial morphology.
Pathology and morphological variation: linear enamel hypoplasia, dental crowding and rotation of the L C_1; antemortem losses, $n = 1$; abscesses, $n = 1$; injury to the frontal; enthesiopathy on the left femur lesser trochanter.

Individual 139 has a depressed circular area, 15.2 × 15.2 mm with the greatest depth of 4.5 mm, on the frontal, clearly a healing injury, over the right eye to the midline (see **Fig. 6.14**). The bone is deflected inward with possible drainage tracks running roughly parallel to the frontal crest. The frontal crest area is thickened and lumpy. There appears to be much less remodeling of the endocranial surface. Still, healing took place and the individual survived the initial trauma.

Tomb LXXVIII (Grave 5), Trenches 2, 3, and 7, Unit 51, N = 1

Individual 51. Adult female, aged 20–25 years.
Bone condition: very poor.
Skeletal elements preserved: skull, clavicles, humeri, radii, ulnae, hand elements, vertebrae, pelvis, femora, tibiae, fibulae.
Dentition: 26 permanent teeth, lacking the R and L I^2; R I_2, P_3; L I_1, M_1.
Aging criteria: dental attrition.
Sexing criteria: postcranial size and rugosity.
Pathology and morphological variation: linear enamel hypoplasia.

Tomb LXXIX (Grave 6), Trench 1, Unit 54, N = 1

Individual 54. Adult, no sex estimation, aged 20–25 years.
Bone condition: poor.
Skeletal elements preserved: femora, tibiae, pelvis.
Dentition: 5 isolated teeth, including the R M^2 or M^3, L C_1, M_1–M_3.
Aging criteria: dental attrition.
Sexing criteria: none.
Pathology and morphological variation: none.

Tomb LXXX (Grave 4), Trenches 4 and 5, Unit 32, N = 1

Individual 32. Adult male, aged 20–30 years.
Bone condition: poor.
Skeletal elements preserved: skull, clavicles, scapulae, humeri, radii, ulnae, hand elements, vertebrae, ribs, pelvis, femora, tibiae, fibulae, foot elements.
Dentition: 30 permanent teeth, lacking the R and L I^2.
Aging criteria: dental attrition.
Sexing criteria: skull morphology, postcranial measurements, and robusticity.
Pathology and morphological variation: remodeled periostitic reactive region on left femur (**Fig. 6.45**).

The medial surface of the left femur, approximately 80 mm below the lesser trochanter, has an irregular bump (approximately 13 mm wide × 29.3 mm long) with a smooth, completely remodeled surface.

Tomb LXXXI (Grave 1), Trench 3, Unit 14, N = 1

Individual 14. Adult male, aged 18–22 years.
Bone condition: poor.
Skeletal elements preserved: skull, fragmentary arm long bones, femora, tibiae, fibula, foot elements.
Dentition: 22 permanent teeth, including the R P^3–M^2, L I^2–M^3; R P_3–M_3, L C_1–M_3.
Aging criteria: dental attrition and root completion.
Sexing criteria: cranial morphology, postcranial size, and robusticity.
Pathology and morphological variation: linear enamel hypoplasia, I^2 trait, numerous wormian bones.

Tomb LXXXII (Grave 9), Trench 4, Units 85, 105, and 106, N = 3

Individual 85. Adult, possible male, aged 26–32 years.
Bone condition: poor.
Skeletal elements preserved: femora.
Dentition: 20 permanent teeth, including R C^1, M^2–M^3; L P^4–M^3; L I_1–M_3, R C_1, P_4–M_3.

Aging criteria: dental attrition.
Sexing criteria: large, rugose femora; large dentition.
Pathology and morphological variation: none.

Individual 105. Young adult, no sex estimate.
Bone condition: poor.
Skeletal elements preserved: occiput, left temporal, and parietals.
Dentition: none.
Aging criteria: none.
Pathology and morphological variation: none.

Individual 106. Possible male, aged 26–32 years.
Bone condition: poor.
Skeletal elements preserved: cranial vault fragments, mandible, femoral diaphyses.
Dentition: 19 teeth, including the R M^1–M^3, L P^3, P^4, M^2, M^3; R C_1–M_3, L P_3–M_3.
Aging criteria: dental attrition.
Sexing criteria: moderately large and rugose femora.
Pathology and morphological variation: linear enamel hypoplasia seen on R C_1, not observable on other teeth due to enamel condition; rotated R P_3.

Tomb LXXXIII (Grave 7), Trench 4, Unit 60, 60a, 60b, and 60c, N = 4

Grave 7 is a complicated grave with evidence for the dentition of at least four individuals.

Individual 60. Subadult, aged 15 years ± 3 years.
Bone condition: very poor.
Skeletal elements preserved: skull fragments, possible scapula, humeri, right radius and ulna, nonidentifiable leg fragments, hand phalanx.
Dentition: 14 teeth, including R C^1, P^3, M^1, M^2; L P^3, M^1, M^2; R P_4–M_2; L P_3–M_2.
Aging criteria: dental attrition.
Sexing criteria: none.
Pathology and morphological variation: metopic suture (**Fig. 6.46**).

The cranial vault fragments include one small portion preserving the metopic suture/coronal suture junction at bregma. The metopic portion is fused endocranially but patent exocranially. The coronal suture is patent throughout. The active metopic portion is obviously not indicative of a very young age for Individual 60, but represents a morphological variant.

Trench 4, Unit 60, Isolated Teeth Designated 60a, b, and c

Near Grave 7, but not initially viewed as part of it, were scattered 29 teeth. These were grouped according to morphology and wear into three individual dentitions and 15 other isolated teeth. Some may belong with the dental sets, but the matches are too tentative to include here. All together, these teeth consisted of:

15 isolated teeth—not associated with any individual
Individual 60a n = 7 teeth
Individual 60b n = 4 teeth
Individual 60c n = 3 teeth

Individual 60a. Young adult; no further age or sex determination possible.
Bone condition: very poor.
Skeletal elements preserved: none.
Dentition: 7 isolated teeth, including R M^1, L M^1, M^2; R M_1, M_2, L P_3, P_4.
Aging criteria: dental attrition.
Sexing criteria: none.
Pathology and morphological variation: L M^2 enamel pearl, f. c. molare.

Individual 60b. Adult, mid-older age; no further age or sex determination possible.
Bone condition: very poor.
Skeletal elements preserved: none.
Dentition: 4 isolated teeth, including R P^4, L M; L P_3, P_4.
Aging criteria: dental attrition.
Sexing criteria: none.
Pathology and morphological variation: none.

Individual 60c. Adult, over 45; no sex estimation possible.
Bone condition: very poor.
Skeletal elements preserved: none.
Dentition: Right M^1, maxillary P, mandibular M.
Aging criteria: dental attrition.
Sexing criteria: none.
Pathology and morphological variation: none.

This dentition, with its extreme sloping and cupped wear, represents one of the oldest individuals from a prehistoric burial at Lofkënd.

Tomb LXXXIV (Grave 2), Trench 1, Units 18, 38, and 38a, N = 3

Individual 18. Adult male, aged 23–26 years.
Bone condition: fair.

Skeletal elements preserved: skull, clavicles, scapulae, humeri, radii, ulnae, hand elements, vertebrae, ribs, pelvis, femora, tibiae, fibulae, foot elements.
Dentition: 15 permanent teeth, including the L P^3, M^2–M^3; R I_1–C_1, M_1–M_3; L I_1, C_1–M_2.
Aging criteria: pubic symphysis, dental attrition.
Sexing criteria: pelvic morphology, postcranial size, and rugosity.
Pathology and morphological variation: linear enamel hypoplasia, enamel extensions on 5 molars.

The dentition, which is large with bulging crowns, has several notable features: 5 of the molars have enamel extensions; the L M^3 has 4 roots and a deep, broad furrow extending down the buccal crown surface; the L P_3 is double-rooted with developed lingual cusplets. With the exception of strong linear enamel hypoplasia on the canines, the dentition is healthy.

Individual 38. Adult male, aged 35–45 years.
Bone condition: poor.
Skeletal elements preserved: skull, left clavicle, scapulae, humeri, radii, ulnae, hand elements, vertebrae, pelvis, femora, tibiae, fibulae, foot elements.
Dentition: 13 permanent teeth, including the R C^1; R I_1–P_4, M_2; L I_1–P_3, M_1, M_3.
Aging criteria: dental attrition, cranial sutures, osteoarthritis.
Sexing criteria: skull morphology, postcranial measurements, and rugosity.
Pathology and morphological variation: caries, n = 8; antemortem tooth losses, n = 3 possible (**Fig. 6.47**); linear enamel hypoplasia; R P_4 and L P_3 rotated 90°; osteoarthritis of lumbar vertebrae, sacrum, acetabulum, and elbow joint surfaces; right elbow dislocation. (See **Fig. 6.13** for views of the elbow dislocation.)

The right distal humerus of Individual 38 is characterized by damage to the olecranon fossa and formation of a new joint surface inferior to it, on the posterior surface (see **Fig. 6.13a–c**). This secondary fossa is not as deep as the normal fossa, but a clear pit for the olecranon process was formed. The trochlea and capitulum dorsal margins have lipping and irregular bone growth. Anteriorly, the original surface along the lateral border of the capitulum has eburnation, indicating loss of articular cartilage. Superior to the capitulum and trochlear surfaces, the alteration of the joint is apparent from the roughened, lipped surfaces for the flexed lower arm. Corresponding changes are seen in the proximal radius and ulna. This suggests a long-term adaptation to the condition.

Individual 38a. Subadult, aged 2–5 years.
Bone condition: very poor.
Skeletal elements preserved: femur, tibia, innominate (**Fig. 6.48**).
Dentition: none.
Aging criteria: tibia size.
Pathology and morphological variation: none.

TOMB LXXXV (GRAVE 10), UNIT 90, UNIT 145 AND 145a, AN INFANT IDENTIFIED IN THE LAB, N = 3

Individual 90. Adult female, mid-age.
Bone condition: fair.
Skeletal elements preserved: skull.
Dentition: none.
Aging criteria: edentulous, but unfused cranial sutures.
Sexing criteria: skull morphology, postcranial size, and rugosity.
Pathology and morphological variation: edentulous; expansion of diploe on frontal, wormian bones.

The skull of Individual 90 has great variance in the vault thickness, with thin parietal and occipital portions but a significantly thicker frontal. The latter is so porous that large pock-like holes dot the surface. While there is taphonomic damage to the frontal surface, the bone is still thicker, and the diploe appears to be expanded.

Individual 145. Adult, possible male; no specific age estimate.
Bone condition: poor.
Skeletal elements preserved: cranium; left humerus, radius, and ulna; right ulna, hand elements, vertebrae, pelvis, femora, tibiae, fibulae, foot elements.
Dentition: 4 permanent teeth, the R I_1, M_2; L I_2, M_2.
Aging criteria: none.
Sexing criteria: cranial morphology, postcranial morphology.
Pathology and morphological variation: none.

Individual 145a. Subadult, <1 year.
Bone condition: good.
Skeletal elements preserved: zygomatic bone and a vertebral centrum.
Dentition: none.

Aging criteria: bone morphology and development.
Pathology and morphological variation: none.

Modern Graves

TOMB LXXXVI (GRAVE 22), TRENCH 1, UNIT 156, N = 1

Individual 156. Adult male, aged >45 years.
Bone condition: excellent.
Skeletal elements preserved: nearly complete skeleton, including ossified throat cartilages; lacking only some delicate cranial elements and hand phalanges.
Dentition: 23 permanent teeth, including the R I^1–M^3, L C^1, M^1, M^2; R I_1–C_1, M_2; L I_1–M_3; (L M_1–M_2 roots only).
Aging criteria: pubic symphysis, auricular surface, dental attrition, cranial sutures.
Sexing criteria: pelvic and skull morphology.
Pathology and morphological variation: caries, n = 5; antemortem losses, n = 9–10; linear enamel hypoplasia; extreme accumulation of dental calculus; rotation of L P_4; cranial asymmetry where the digastric groove is deeper on the right, and the left occipital condyle has a double facet; flattening above lambda possibly associated with early sagittal suture fusion; multiple infraorbital foramina; sternal manubrium is asymmetrical and unfused to the blade; osteophytes on vertebrae (especially C6, T11, T12, lumbars), distal right radius, distal left femur, distal right fibula; Schmorl's nodules on 11 vertebrae, the largest being on the thoracics.

TOMB LXXXVII (GRAVE 8) ANIMAL (SEE CHAPTER 16.1)

TOMB LXXXVIII (GRAVE 3), TRENCH 1, UNIT 21, N = 1

Individual 21. Subadult, neonate, birth ± 2 months.
Bone condition: good.
Skeletal elements preserved: complete skeleton.
Dentition: 18 deciduous tooth germs, lacking the L dc^1, dm^2.
Aging criteria: deciduous crown formation.
Pathology and morphological variation: none.

TOMB LXXXIX (GRAVE 11), TRENCH 1, UNIT 96, N = 1

Individual 96. Subadult, neonate, aged birth ± 2 months.
Bone condition: good.
Skeletal elements preserved: skull with all elements present, complete skeleton with measurable long bones (**Fig. 6.49**).
Dentition: deciduous dentition, 8 germs (R di^2, dm^1; L di^1, dc^1, dm^1; R di_2, dm_1; L di_1).
Aging criteria: crown formation.
Pathology and morphological variation: ridge on distal humerus.

The left humerus has a rugose ridge along the medial supracondylar ridge on the anterior surface (**Fig. 6.50**). It has a regular, rather than a reactive, appearance.

TOMB XC (GRAVE 14), TRENCH 1, UNITS 109 AND 110: PERINATES

Grave 14 contained the remains of two small infants interred together. It is very likely that they were twins. (See **Figs. 3.300, 3.301a–b, 6.1**.)

Individual 109. Subadult, probable perinate.
Bone condition: fair.
Skeletal elements preserved: skull fragments with little of the vault but the right mandible; clavicle, scapulae, right arm (humerus, radius, ulna, and hand), left arm (humerus and radius), ribs, vertebrae, pelvis, femora, tibiae.
Dentition: none.
Aging criteria: size and development of bones.
Pathology and morphological variation: none.

Individual 110. Subadult, perinate.
Bone condition: good.
Skeletal elements preserved: skull with most portions preserved, including the mandible, clavicle, left scapula, right humerus, radius and ulna, hand elements, vertebrae, ribs, pelvis, left femur, tibia, and fibula.
Dentition: 7 tooth buds.
Aging criteria: tooth formation.
Pathology and morphological variation: none.

TOMB XCI (GRAVE 15), TRENCH 1, UNITS 122 AND 128: PERINATES

Grave 15 contained the burials of two young infants, Individuals 122 and 128. The 122 skeleton is more complete, and the elements are substantially larger than those of Individual 128.

Individual 122. Subadult, aged birth ± 2 months.
Bone condition: fair.

Skeletal elements preserved: skull with major elements preserved, including the auditory ossicles and tympanic ring; left clavicle, scapulae, humeri, radii, ulnae, and hands; ribs, vertebrae, pelvis, femora, tibiae, fibulae.
Dentition: tooth germs, including maxillary R di^1, di^2, and a molar fragment; R and L di_1 and di_2.
Aging criteria: dental development.
Pathology and morphological variation: none.

Individual 128. Subadult, preterm?
Bone condition: poor.
Skeletal elements preserved: skull with some diagnostic elements, including the mandible, scapula, arm and hand elements, vertebrae, ribs, tibiae, fibulae.
Dentition: none.
Aging criteria: bone size and development.
Pathology and morphological variation: none.

Tomb XCII (Grave 23), Trench 1, Units 159 and 159a, N = 2

Individual 159. Adult male, aged >50 years.
Bone condition: excellent.
Skeletal elements preserved: complete skeleton, including ossified throat cartilages.
Dentition: 13 permanent teeth, including the R I^1, C^1–P^3, M^1–M^2; L C^1–P^3; R C_1–P_3, L I_2–P_4.
Aging criteria: pubic symphysis, auricular surface, dental attrition, cranial sutures.
Sexing criteria: pelvic and skull morphology (**Fig. 6.10**).
Pathology and morphological variation: caries, n = 11; antemortem losses, n = 11; linear enamel hypoplasia; tooth rotation due to antemortem losses; cranial asymmetry with the mandibular fossa and mastoid process on the right slightly dorsal/medial and fusion of the occipito-temporal suture on the right side; nasal injury; osteophyte development on cervical, thoracic, and lumbar arches; Schmorl's nodules on L1 and L2; bifurcate spines on T10–T12; osteophytes on right metacarpal 1; on the hallux, the proximal articular surface of the left proximal phalanx has a pit measuring 2.5 × 3.5 mm.

The cranium has asymmetry, most notably of the occipital, basicranium, and the temporal bones. The relative positioning of the mandibular fossa reflects this; the right side is more distal and medially positioned. This is distinct from any postmortem damage. There is also asymmetry in the nasal region (**Fig. 6.51**). The left nasal bone is thickened, there is an extra nasal foramen, and the internasal suture is deflected to the left. The perpendicular plate of the ethmoid deviates to the right. These changes reflect trauma to the region and subsequent healing.

Individual 159a. Subadult, perinate, aged birth ± 3 months.
Bone condition: poor.
Skeletal elements preserved: skull (temporal and mandible fragments).
Dentition: none.
Aging criteria: bone development.
Pathology and morphological variation: none.

Tomb XCIII (Grave 19) Animal
(see Chapter 16.1)

Tomb XCIV (Grave 25), Trench 1, Unit 165, N = 1

Individual 165. Subadult, aged 6 months ± 3 months.
Bone condition: fair.
Skeletal elements preserved: skull with most elements represented in good condition—left clavicle, right scapula, humeri, ulnae, hand, vertebrae, ribs, pelvis, femora, tibiae, fibulae.
Dentition: mixed dentition of 22 teeth, with a full set of 20 deciduous teeth and 2 permanent (germs of R and L M_1) teeth.
Aging criteria: development of the dm2s and M_1s.
Pathology and morphological variation: possible small reactive lesion on the right frontal (**Fig. 6.52**).

Tomb XCV (Grave 21), Trench 1, Unit 153, N = 1

Individual 153. Subadult, perinate, aged birth ± 2 months.
Bone condition: poor.
Skeletal elements preserved: skull with major elements preserved; right clavicle, scapula, humerus, radius, ulna, and hand; left arm fragments; ribs, vertebrae, pelvis, tibiae.
Dentition: deciduous dentition with 15 teeth, including R di^1, di^2, dm^1; L di^1, dm^1; hemi mandibles with di_1–dm_2.
Aging criteria: dental development.
Pathology and morphological variation: none.

Tomb XCVI (Grave 20), Trench 1, Unit 148

Individual 148. Subadult, perinate, aged birth ± 2 months.
Bone condition: poor.

Skeletal elements preserved: skull fragments, including petrous portions and auditory ossicles, left clavicle and scapula, arm fragments, vertebrae, ribs, pelvis, left femur, and tibia.
Dentition: fragment of molar crown.
Aging criteria: bone size and comparison with other specimens.
Pathology and morphological variation: none.

TOMB XCVII (GRAVE 39), TRENCH 1, UNITS 250 AND 250a, N = 2

Individual 250. Adult male, aged >50 years.
Bone condition: good.
Skeletal elements preserved: almost complete skeleton.
Dentition: edentulous.
Aging criteria: pubic symphysis, auricular surface, cranial sutures.
Sexing criteria: pelvic and skull morphology.
Pathology and morphological variation: antemortem loss of all teeth; healed cribra orbitalia; Inca bone and wormian bones; asymmetrical clavicles, with the shorter right displaying pitting, lipping, and ridging on the sternal end; two healed right rib fractures (**Fig. 6.53**); osteophytes, some extensive, throughout the vertebral column; osteophytes on patellae; eburnation on the distal radii and carpals; osteophytes on metacarpal 4; osteophytes and eburnation on the phalanges, especially the distal phalanges; osteophytes on the calcanei; slight lipping on the glenoid fossae.

Individual 250a. Adult, aged mid-age–older.
Bone condition: very poor.
Skeletal elements preserved: cranial vault fragments, hand phalanges.
Dentition: 7 permanent teeth, including the R C^1, M^2; L I^2–P^3; R I_1, L I_2.
Aging criteria: dental attrition.
Sexing criteria: none.
Pathology and morphological variation: linear enamel hypoplasia.

TOMB XCVIII (GRAVE 36), TRENCH 1, UNIT 238, N = 1

Individual 238. Subadult, aged 9 months ± 3 months.
Bone condition: very good.
Skeletal elements preserved: skull with delicate elements preserved, clavicles, sternum, scapulae, humeri, radii, ulnae, hands, vertebrae, ribs, pelvis, femora, tibiae, fibulae (**Fig. 6.54**).
Dentition: mixed dentition of 23 teeth, including a full deciduous set except for the R di^1 and the 4 M1 crowns; most teeth still in bone.
Aging criteria: dental eruption and crown formation.
Pathology and morphological variation: none.

TOMB XCIX (GRAVE 45), TRENCH 1, UNIT 277, N = 1

Individual 277. Adult female, aged 25–35 years.
Bone condition: very good.
Skeletal elements preserved: complete skeleton with ossified throat cartilages.
Dentition: complete permanent dentition.
Aging criteria: pubic symphysis, auricular surface, dental attrition.
Sexing criteria: pelvic and skull morphology.
Pathology and morphological variation: caries: n = 10; one small abscess; dental calculus; periodontal disease; linear enamel hypoplasia; wormian bones at pterion, Inca bone, lambdoidal wormian bones; infraorbital suture; extra suture from asterion across the occipital, bilateral; T1 and T2 have asymmetrical superior/inferior articular facets, with cupped articulation on the right; small Schmorl's nodes on the lower thoracics and lumbars; right distal phalanx of the thumb is crushed—the joint surface of the proximal phalanx looks normal.

TOMB C (GRAVE 48), TRENCH 1, UNITS 302 AND 302a, N = 2

Individual 302. Adult female, aged 45–60 years.
Bone condition: very good.
Skeletal elements preserved: complete skeleton with preserved cartilages.
Dentition: 14 permanent teeth, some only roots: R I^2, P^3–P^4, M^2; L I^1–I^2, P^4; R I_1–P_3, L I_2–P_3 .
Aging criteria: pubic symphysis, auricular surface, dental attrition, cranial sutures.
Sexing criteria: pelvic and skull morphology.
Pathology and morphological variation: caries: n = 8, antemortem losses: n = 16, secondary malocclusion due to tooth losses; possible healed cribra orbitalia, infraorbital suture (**Fig. 6.24**), bilateral parietal depressions but no surface porosity (**Fig. 6.55**),

wormian bones, temporomandibular joint deterioration with porosity on the articular eminences and the mandibular condyles; most joint surfaces have degenerative changes, including lipping and eburnation on the cervical vertebra, increased porosity of the thoracic vertebra, lipping on the glenoid fossae, lipping on the elbow joint surfaces, spicules on the patellae, slight lipping on hand and foot elements; several sites of muscle insertion and origin on the humerus, radius, and clavicles are irregular with a bubbly appearance.

Individual 302a. Subadult, aged 14–18 years.
Bone condition: fair.
Skeletal elements preserved: left distal fibula and right navicular.
Dentition: none.
Aging criteria: epiphyseal fusion, postcranial size.
Sexing criteria: none.
Pathology and morphological variation: none.

From Tumulus Fill

The following individuals were recovered from the tumulus fill and not assigned specific grave numbers or skeletal unit numbers. They are identified by the trench, sedimentary unit, and date on which they were excavated.

Trench 1, Unit 7, 6/7/04

Subadult, infant.
The remains of this infant consist of cranial vault, represented by two fragments, and a right femoral shaft.

Trench 1, Unit 9, 25/6/04

Subadult, infant.
A portion of the greater wing of an infant sphenoid was identified from the fill of this unit.

Trench 1, Unit 39, 10/7/04

Subadult, fetus.
Bone condition: good.
Aging criteria: bone morphology and size.
Skeletal elements preserved: cranium.
Dentition: none.
Remains identified included fetal cranium with vault portions, the petrous of the temporal, a sphenoid greater wing, and the sphenoid body.

Trench 1, Unit 39, 8/7/04

Fetal humerus, possibly the same individual as other Unit 39 material above.

Trench 2, Unit 593, 1/7/04 and 8/7/04 (previous Unit 40)

Subadult, aged 6 years ± 2 years.
Bone condition: poor.
Skeletal elements preserved: subadult limb fragments, possible radius and fibula.
Dentition: 3 teeth from a mixed dentition, including R dm^1, L M^1, L M_2.
Aging criteria: dentition, bone size.
Pathology and morphological variation: none.

Trench 2, Unit 594, 06/28/05 (previous Unit 202)

Young adult; no sex determination possible.
Bone condition: poor.
Skeletal elements preserved: cranial vault fragments, right mandibular corpus, and long bone fragments.
Dentition: 6 teeth, including a L C_1 and mandible fragment with R I_2–M_1.
Aging criteria: dental attrition.
Sexing criteria: none.
Pathology and morphological variation: f. c. molare, linear enamel hypoplasia, dental calculus.

Trench 2, Unit 595, 6/27/05

Subadult, aged 6 months ± 3 months.
Bone condition: good bone quality.
Skeletal elements preserved: cranial vault, clavicles, maxilla, mandible.
Dentition: deciduous mandibular molars: R dm_1, R and L dm_2.
Aging criteria: dental development.

This burial, from a non-grave context, is recognized as a separate skeletal unit of infant remains found with a bronze object comprised of a disk and rings. There is metallic staining on the mandible inferior and interior aspects, the dm_2 interior, the clavicle, and the vertebra. The dm_1s look possibly burned, as does one small bone fragment that is gray-black, although this discoloration may just be a result of the staining. There is no clear contextual association between the infant remains and the bronze object such that it can be argued that the infant was buried with the object. Instead, it is likely that the infant burial was disturbed and the bronze object was redeposited near it. The porosity of the infant bone would enhance the potential for metallic staining.

TRENCH 2, UNIT 235, 19/07/05 (BELOW GRAVE 42)

Adult, possible male.
Bone condition: good.
Skeletal elements preserved: right humerus, ulnae, hand elements, vertebra, ribs, pelvis, left first metatarsal.
Dentition: R C_1, L P_4.
Aging criteria: dental attrition, bone morphology.
Sexing criteria: bone size and robusticity.
Pathology and morphological variation: strong linear enamel hypoplasia on the canine, multiple bands.

APPENDIX 1
SUMMARY OF RESULTS OF DNA ANALYSIS OF ANCIENT HUMAN BONE FROM THE LOFKËND TUMULUS

Laura Menez

With contributions by John K. Papadopoulos

INTRODUCTION

John K. Papadopoulos

One of the aims from the outset of the project in 2004 was to test, using DNA analysis, for any consanguinity or relationship between individuals buried in the tumulus. To this end, a pilot project was initiated with Alan Cooper of the Henry Wellcome Ancient Biomolecules Centre at the University of Oxford. Laura Menez, at the time one of the doctoral students in the center, joined the Lofkënd project for two weeks in July 2004, and oversaw the sampling of the excavated skeletons and conducted the DNA analysis in Oxford. A summary of the results of this analysis is presented below. Despite a stringent set of protocols and various levels of sterility under which the samples were taken, the ancient bone was contaminated with modern DNA. For a variety of reasons, the results were confusing and not accurate enough to point to any satisfactory pattern or to be reproducible. The decision to sample the majority of graves from the beginning of the project for DNA analysis, usually with multiple samples from each burial, was not fruitful. Indeed, the discussion of shared dental characteristics (Chapter 6) provided much stronger evidence of likely kin relationships of the individuals buried in the Lofkënd tumulus.

In hindsight, a better approach would have been to target, with very specific questions in mind, individual graves after their excavation, focusing on bone and teeth that were likely to produce usable DNA. For example, the three fragmentary individuals in Tomb I (Grave 64) of Phase I, as well as the three individuals interred together in Tomb LXXIV (Grave 29, Phase Va), or the various individuals found in Tomb XXI (Grave 55, Phase II), Tomb XLVIII (Grave 52, Phase III), and Tomb LXX (Grave 17, Phase Va), could yield potentially interesting results. The fact that these tombs cover various phases of the prehistoric period of use of the tumulus allows testing for population continuity diachronically. Such results have been successfully obtained in the case of the DNA analysis of the Kamenicë tumulus by Todd Fenton and David Foran of Michigan State University (Bejko, Amore, and Aliu 2006:319–322). Thus far, it has been possible only to isolate and analyze mitochondrial DNA, but this is present in very low quantities and usually quite degraded (Bejko, Amore, and Aliu 2006:321). Nevertheless, and in spite of the limitations, robust results have been obtained, particularly of the relatedness of individuals from several contexts in the Kamenicë tumulus, including those within the so-called great circle (Murray 2006), those within the monumental grave structure (Rennick 2005), and from eight sets of double burials covering the entire period of use of the tumulus at Kamenicë (Clemmer 2005). Although the majority of individuals in the great circle do not appear to have been maternally related, those from the other contexts were related (Bejko, Amore, and Aliu 2006:322).

Similar results are still possible for the individuals buried at Lofkënd, and such analyses, together with future work on stable isotopes from both human and animal bone samples from the region of Lofkënd and Apollonia (Chapter 7), can potentially yield important conclusions. As is so often the case, this report is only preliminary.

AIMS, METHODS, AND RESULTS OF THE DNA ANALYSIS

Laura Menez

The aims and objectives of the DNA analysis as originally conceived were twofold: first, to establish relationships between buried individuals using mitochondrial and, if possible, nuclear DNA; and second, to develop a method for sampling human remains from an archaeological context that was both effective and practical for fieldwork. Samples of saliva were voluntarily provided from virtually all those involved in the excavations and this DNA would serve as an exclusion control for DNA extracted from excavated bone. That is, if a DNA sequence that looked like that of a fieldworker was extracted from ancient bone, it could be excluded as a modern contaminant rather than be erroneously included as DNA endogenous to the bone.

There were four levels of sterility under which the samples were collected. Level 1 was a completely non-sterile sampling, and Level 4 represented the maximum sterility possible. Level 2 involved the sampler wearing gloves and a mask but not using instruments that had been cleaned with ethanol (**Fig. A1.1**). Level 3 involved the wearing of gloves and the use of cleaned tools but not the wearing of a facemask. Level 4 entailed the wearing of gloves and a facemask, as well as the use of tools that had been cleaned with ethanol. The idea behind the varying levels of sterility was to determine and monitor the minimum amount of protective clothing to be worn to maintain sterility of the samples and, at the same time, minimize the necessary equipment and protocols. Multiple samples were taken from individual skeletons, with varying levels of sterility. Typically, a combination of samples from the long bones and teeth of an individual in a tomb was taken.

Following preparation in the laboratory, the samples were powdered using a freezer mill, microdismembrator, or similar homogenization equipment. The samples were then decalcified for up to 72 hours in ethylenediaminetetraacetic acid (EDTA) and were extracted twice, either with equal volumes of phenol or once with phenol and once with a mixture of phenol and isoamyl alcohol. Once extracted from the sample, DNA was amplified using polymerase chain reaction (PCR), which was carried out on heating blocks at an annealing temperature of 62° C for 45 cycles. Each PRC that produced a band when viewed on an agarose gel was cloned; eight colonies were selected for each and sequenced. The sequences thus obtained were analyzed using Seqman and Se-AL. Each variation contributed to categorizing that individual as a certain haplotype.

TABLE A1.1 Summary of results of DNA analysis of ancient human bone from the Lofkënd tumulus

Sample name	Tomb & unit no.	Extract	PCR band?	Mutations	Haplotype
ALB BS 1	LXXXI (1), Individual 14	SG 106	No		
ALB BS 2	LXXXIV (2), Individual 18	SG 107	No		
ALB BS 3	LXXX (4), Individual 32	SG 108	Yes	16192, 16256, 16270, 16304	U5a1
ALB BS 4	LXXXIV (2), Individual 38	SG 109	Yes	16192, 16256, 16270, 16304	U5a1
ALB BS 5	LXXIX (6), Individual 54	SG 110	Yes	16260	U?
ALB BS 6	LXXVIII (5), Individual 51	SG 112	No		
ALB BS 8	LXXXII (9), Individual ?	SG 113	No		
ALB BS 9	LXXXII (9), Individual 83	SG 114	No		
ALB BS 10	LXXXII (9), Individual 83	SG 115	No		
ALB BS 11	LXXXII (9), Individual 83	SG 116	Yes	16224, 16311	K
ALB BS 12	LXXXIX (11), Individual 96	SG 118	Yes	16126, 16163, 16186, 16189, 16294	T1
ALB BS 13	LXXXIX (11), Individual 96	SG 119	No		
ALB BS 14	LXXXV (10), Individual 90	SG 120	No		
ALB BS 15	LXXXV (10), Individual 145	SG 121	Yes	16192, 16256, 16270, 16304	U5a1
ALB BS 16	XC (14), Individual 107	SG 122	Yes	16192, 16256, 16270, 16304	U5a1
ALB BS 17	XC (14), Individual 109	SG 124	Yes	Clones did not grow	

Continued on the next page

TABLE A1.1 (*continued*). Summary of results of DNA analysis of ancient human bone from the Lofkënd tumulus

Sample name	Tomb & unit no.	Extract	PCR band?	Mutations	Haplotype
ALB BS 18	XC (14), Individual 110	SG 125	No		
ALB BS 19	XC (14), Individual 110	SG 126	Yes	16362	?
ALB BS 20	LXX (17), Individual 126	SG 127	No		
ALB BS 21	LXVIII (13), Individual 136	Tooth			
ALB BS 22	XCIII (19), Sheep	SG 129	Yes	?	
ALB BS 23	XCIII (19), Sheep	SG 130	Yes	CRS	
ALB BS 24	LXXVII (18), Individual 139	SG 131	Yes	16192, 16256, 16304	H?
ALB BS 25	LXVII (12), Individual 171	SG 132	No		
ALB BS 26	LXXVII (18), Individual 139	SG 133	No		
ALB BS 27	LXXXV (10), Individual 90	SG 134	Yes	CRS	
ALB BS 28	LXVII (12), Individual 100	SG 136	Yes	16192, 16256, 16304, 16343	U3
ALB BS 29	LXVII (12), Individual 171	SG 137	No		
ALB BS 30	LXXXI (1), Individual 14	Tooth			
ALB BS 31	LXXXIV (2), Individual 18 or 38	Tooth			
ALB BS 32	LXXXIV (2), Individual 38	SG 138	No		
ALB BS 33	LXXXIV (2), Individual 18	SG 139	Yes	CRS	
ALB BS 34	LXXXVIII (3), Individual 21	SG 140	Yes	????	
ALB BS 35	LXXVIII (5), Individual 51	Tooth			
ALB BS 36	LXXVIII (5), Individual 51	SG 142	Yes	16126, 16294	
ALB BS 37	LXXIX (6), Individual 54	SG 143	Yes	16192, 16256, 16304	H
ALB BS 38	Tumulus fill, tooth	Tooth			
ALB BS 39	Tumulus fill, tooth	Tooth			
ALB BS 40	Tumulus fill	SG 144	Yes	16192, 16256, 16304	H
ALB BS 41	Tumulus fill	SG 145	No		
ALB BS 42	Tumulus fill	SG 146	No		
ALB BS 43	Tumulus fill	SG 147	No		

As already noted, the ancient bone was quite contaminated with modern DNA for a variety of reasons, such as exposure of the site and practical limitations in controlling sterility. The results were less than satisfactory and were not reproducible enough to be accurate (see summary in **Table A1.1**).

CHAPTER 7

RESULTS OF STABLE-ISOTOPE ANALYSES OF HUMAN BONE SAMPLES FROM LOFKËND

Brian N. Damiata and John Southon

INTRODUCTION

Following the decision to initiate a radiocarbon dating project that would be useful for determining the chronology of the Lofkënd tumulus (Damiata and Southon, Chapter 4; Damiata et al. 2007–2008), the analyses of stable isotopes was included as part of the project. The purpose of the analyses was to obtain information on the paleodiet of this area of Albania. This was envisaged as a piece of a larger, ongoing study that would compare the diet of the inhabitants of the Greek colony of Apollonia (see Chapter 6) with that of an established population in the hinterland of the colony (not only Lofkënd, but other Illyrian populations), and potentially the diet of the metropolis or mother-city of the colony—in this case, ancient Corinth. This chapter summarizes the results of carbon (C) and nitrogen (N) stable-isotope analyses obtained from a total of 25 measurements. The analyses were performed together with the accelerator mass spectrometry (AMS) dating in two batches in January and June 2008.

The methodology for the selection and treatment of the samples was the same as that described for the human bone that underwent AMS dating. Aliquots of the dried collagen were analyzed for C and N stable isotopes using a Fisons NA-1500NC elemental analyzer equipped with a Delta-Plus CFIRMS stable-isotope mass spectrometer.

RESULTS AND DISCUSSION OF THE STABLE-ISOTOPE ANALYSES

The results of the stable-isotope analyses are summarized in **Table 7.1**. The data are reported in the conventional δ-notations that are referenced to the PDB standard for C (Craig 1957) and the AIR standard for N (Mariotti 1983). The reported $\delta^{13}C$ and $\delta^{15}N$ values were measured to a precision of <0.1‰. Also included in the table are data concerning the sex and age of the skeletal remains as determined through morphological examination by Lynne Schepartz (see Chapter 6).

The collagen in bone continually undergoes remodeling and replenishment (Boskey 1999). In particular, its isotopic composition represents the last few years of a person's life (Bell, Cox, and Sealy 2001), often assumed to be the last 10 years. The purity of extracted collagen for a given sample is typically evaluated using three criteria: the C:N ratio, the collagen yield, and the weight percent (wt %) concentrations of C and N (Ambrose 1990; Ambrose and Norr 1992).

The mostly widely used criterion for identifying contamination and/or diagenetic alteration is the C:N ratio. Modern collagen has an atomic ratio of 3.21. Values within an empirically derived range of 2.9–3.6 are commonly accepted for archaeological studies (DeNiro 1985), although Hedges (2000) restricts the range to 2.8–3.3. Values above 3.4 may indicate contamination with carbon-rich substances such as humic acid (Kennedy 1988). Lastly, the Oxford radiocarbon laboratory uses a range of 3.1–3.5 for accepting samples (van Klinken 1999). The measured values reported here are 3.17–3.42, upon conversion from wt % ratio to the more commonly reported atomic ratio (see footnote in **Table 7.1**).

The second criterion for evaluation of the purity of collagen is yield. Modern defatted bone can

TABLE 7.1 Stable-isotope results for collagen extracted from human bone samples

Sample ID	Tomb	Element[a]	Sex	Age	Collagen yield (%)	$\delta^{13}C$ (‰)	$\delta^{15}N$ (‰)	N (wt %)	C (wt %)	C:N[b] (wt %)
LB1E	LXXXI	Lt. tibia	♂	18–22 yr	2.9	–18.1	7.5	14.6	41.7	2.85
LB2B	LXXXIV	Lt. humerus	♂	35–45 yr	4.6	–18.5	6.7	15.3	42.2	2.75
LB4C	LXXX	Rt. humerus/ radius	♂	20–30 yr	4.3	–18.3	7.1	14.9	42.3	2.83
LB5A	LXXVIII	Rt. femur	♀	20–25 yr	0.5	–18.8	6.9	14.6	40.9	2.80
LB9B	LXXXII	Lt. femur	♂?	26–32 yr	0.8	–18.5	7.2	13.9	38.8	2.80
LB10A	LXXXV	Rt. femur	♀	Mid-aged	1.8	–18.3	7.2	15.0	41.6	2.77
LB13A	LXVIII	Long bone			8.4	–18.0	7.1	15.2	42.6	2.79
LB17A	LXX	Rt. radius			0.7	–18.7	7.3	13.0	37.8	2.89
LB22A	LXXXVI	Rt. hand	♂	>45 yr	10.0	–12.4	8.9	15.9	43.3	2.72
LB26A	LXXIII	Rt. tibia	♂?		7.9	–18.3	7.1	15.4	45.0	2.93
LB27A	LXIX	Lt. femur			2.2	–18.2	6.9	15.0	41.6	2.78
LB29A	LXXIV	Rt. humerus			1.1	–18.1	7.0	15.4	43.6	2.83
LB29B	LXXIV	Lt. tibia			2.7	–18.3	6.7	15.3	42.8	2.80
LB29C	LXXIV	Lt. tibia			2.9	–18.6	8.2	14.3	41.6	2.91
LB29F	LXXIV	Rt. humerus			5.1	–18.3	7.0	16.1	44.6	2.76
LB30A	LXV	Lt. tibia/fibula	♀	18–23 yr	0.8	–18.4	6.7	15.0	41.8	2.78
LB31A	LXVI	Humerus		15 ± 2 yr	2.5	–17.8	6.6	14.8	41.9	2.83
LB33B	LXXV	Lt. tibia	♂	30–40 yr	1.7	–18.4	7.0	14.3	40.4	2.83
LB38A	LXXV	Rt. humerus	♂	35–45 yr	0.5	–17.9	7.7	14.5	41.3	2.85
LB39A	XCVII	Mixed phalanges	♂	>50	6.4	–12.5	8.7	15.6	42.5	2.72
LB45A	XCIX	Rt. hand	♀	25–35 yr	18.0	–11.8	8.2	16.9	46.8	2.78
LB48A	C	Rt. hand	♀	45–60 yr	17.0	–12.9	8.3	15.9	43.4	2.73
LB49A	XIII	Rt. hand	♀	25–35 yr	2.8	–18.5	6.5	15.4	43.3	2.81
LB60E	XLV	Lt. tibia			6.0	–17.7	8.5	15.8	43.7	2.77
LB91B	II	Rt. tibia	♂	>45 yr	2.1	–19.1	7.4	14.9	42.2	2.83

[a] Lt., left; Rt., right.

[b] Atomic C:N ratios are higher than those calculated from weights of C and N. The former are calculated directly from the number of atoms of each sample. To convert from wt % C:N ratios to atomic C:N ratios, multiply by 14/12 (i.e., atomic masses of nitrogen and carbon), or 1.16667.

yield as much as 25 wt %, but well-preserved archaeological samples are typically 0.8–3.5 wt %. The measured yields reported here are 0.5–18.0 wt %.

The third criterion is the wt % concentrations of C and N. Modern collagen is over 43 wt % C and 16 wt % N. The Oxford radiocarbon laboratory sets the range for intact collagen at 26–43 wt % and 11–16 wt % for C and N, respectively, for Western European bone samples (van Klinken 1999). The measured values reported here are 37.8–46.8 wt % and 13.0–16.9 wt %, respectively. Thus, by all of the established criteria, the extracted collagen for these samples was of acceptable quality, with no appreciable degradation or contamination.

Stable-isotope analyses can be used to infer paleodiet (see Katzenberg 2000; and Ambrose 1993 for overviews). It is generally assumed that the isotopic composition of collagen mainly reflects the intake of protein when dealing with high-protein diets. In particular, carbon isotopic values are used to differentiate the intake of C_3 and C_4 plants as well as to identify terrestrial and marine components when C_4 plants are absent. Examples of C_3 plants include wheat, barley, rice, root crops, legumes, vegetables, nuts, honey, and most fruits, whereas C_4 plants include millets, maize, sorghum, and sugar cane. Nitrogen isotopic values are used to distinguish the intake of plant and animal proteins by assessing the trophic level of an organism (i.e., its position within the food chain).

Given the paucity of data and particularly the lack of any significant faunal data to describe the regional ecosystem's food web (see Chapter 16.1), interpretations of the isotopic results are tentative and based on comparison with similar studies by Papathanasiou (2003) and Richards and Hedges (2008). The former dealt with six Neolithic Greek sites that included three coastal sites. The latter was concerned with Neolithic, Minoan, and Mycenaean sites. The upper diagram in **Figure 7.1** depicts the distribution of stable isotopes in a generalized food web containing terrestrial and marine components, while the lower one gives the results of the present study with the incorporation of data from the previous studies.

The isotopic values for the Bronze Age and Iron Age samples are similar and indicate terrestrial dietary protein derived mainly from C_3 sources. Intake may include some dairy products and meat but little or no marine sources, as interpreted from the values for $\delta^{13}C$ in general and the relatively depleted values for $\delta^{15}N$ as compared to the diet with an interpreted marine component (Box E). Compared to the Neolithic (Box A) and Minoan (Box F) diets, the Lofkëndian diet shows less protein from dairy and meat (i.e., relatively depleted in $\delta^{15}N$) or possibly more consumption of legumes or legume-consuming carnivores. The diet was probably heavy on the consumption of legumes, including lentils and vetch. Such diets are typical for Late Bronze Age and Early Iron Age sites in Greece (e.g., Kroll 1983; Flint-Hamilton 1999; Petroutsa and Manolis 2010). The modern-age samples indicate consumption of C_4 plants, most likely maize.

Acknowledgment

We thank Xiaomei Xu for performing the stable-isotope measurements and Will Beaumont and Kiyo Beverly for their technical assistance.

Part III
Analyses of Materials from the Tumulus

CHAPTER 8
The Prehistoric Burial Customs

Lyssa C. Stapleton

Introduction

The excavation of the Late Bronze and Early Iron Age burial mound at Lofkënd in Albania was undertaken between 2004 and 2008. During that time, the entire tumulus was excavated, and 100 tombs containing over 150 individuals were uncovered. Eighty-five of these tombs were prehistoric (Chapter 3.1), and 13 dated to a period of reuse of the tumulus in the eighteenth to nineteenth century (Chapter 3.2). The remaining two tombs contained animal, rather than human, burials (Chapters 3.2, 16.1). The prehistoric tombs contained a total of 129 individuals, most of which were inhumations, interred in both single and multiple graves.

In this chapter, I explore aspects of the prehistoric tombs from the Lofkënd tumulus, including their construction and the orientation and disposition of both the tombs and the human remains. Normative burial treatments are defined and conclusions drawn about the magnitude of variability among tombs generally and between normative and wealthy tombs in particular. I also consider a broad range of possible explanations for the unusual demography of the tumulus population and discuss several possible interpretations for patterns in the deposition of non-normative grave goods.

One of the most intriguing aspects of this cemetery, a disparity between grave offerings in the tombs of children and adolescent women and those of the rest of the burial population, developed as the excavation was carried out. This group of comparatively wealthy prehistoric graves is treated as a focus group in this chapter.

Six tombs stood out from the rest by virtue of the type, location, and placement of grave goods deposited in them, as well as the age and sex of the occupant(s) (Tombs XVII [72], XVIII [73], XXI [55], XXVIII [77], XLVIII [52], and LXX [17]). In five of these tombs, several smaller items accompanied a principal object and, like it, were found on or around the cranium of the deceased. These five tombs contained an adolescent female, a child, or an infant adorned with a bronze headband (conventionally referred to as diadem) or other head ornament (see also Papadopoulos 2010a:35, 50–51). The sixth (Tomb XLVIII [52]) had fewer grave goods, but the types of objects and their positions on the skeleton allowed this tomb to be included in the group of wealthiest burials (for related burials in other parts of the Mediterranean, including the Aegean, Italy, and Cyprus, see Stampolides 2012).

Lofkënd mortuary practices, particularly those applied to individuals in the focus group, reflect some social role played by the deceased in life, one that was significant enough to be portrayed in death. While this chapter comprises what I hope is a complete analysis of the mortuary practices at Lofkënd, the six atypical tombs necessitated a focus on questions of gender, social roles, and age. I ask three questions designed to elicit a variety of interpretations for the complex data produced by the Lofkënd tumulus: What are normative burial practices at the Lofkënd tumulus, and how great is the variation among them? How do the six conspicuously wealthy tombs containing subadults differ in ways other than quantity and quality of grave goods from the other tombs in the tumulus? What social behaviors, structures, or roles are expressed in the mortuary practices of these six tombs, and what, if any, relationship can be established between the mortuary evidence and existing theories of rites of passage and social status?

Mortuary studies attempt to gain an understanding of social structure through the interpretation of funerary ritual. Many such studies have focused on the reconstruction of hierarchical social structures, based primarily on tomb characteristics and wealth distribution (Binford 1971; Goldstein 1981; Jørgensen 1987; Saxe 1970; Tainter 1975, 1977). Burials, however, are more than a projection of hierarchical social constructs; they express "a wide range of social messages" that can include "rank, status, age, gender, occupation or social role, physical anomalies or infirmities, cause of death . . . relationship to the principal deceased person (in the case of multiple burials), and behavior prior to death" (Arnold 2001:213). Initial observations showed that the wealthiest tombs at Lofkënd belonged to young women and children. This convergence of wealth, sex, and age appears to represent a socially significant role expressed in mortuary ritual.

Sex and age are integral characteristics that shape most rites of passage—critical life events—such as the onset of puberty, marriage, pregnancy, childbirth, and death (beginning with the seminal work of van Gennep 1908 [1960]; see further Garland 1985:38–47; Huntington and Metcalf 1979; Liston and Papadopoulos 2004:25; Metcalf and Huntington 1991; Morris 1987:29–36). Death forces a transformation of the social role of the deceased within the community, thereby making the funeral a pivotal ritual, one that serves as both a rite of passage and an expression of the status of the deceased, their hierarchical rank, and social persona as perceived by their surviving family and community (Arnold 2001:211, 213–212; Barrett 1990:182; Morris 1987:110; O'Shea 1996:185–186; Pader 1982:42–43; Saxe 1970:4–7, 12; see also Hertz 1907 [1960]:86; Parker Pearson 1982; Ucko 1969). The funeral, however, is a ceremony created by the survivors, and the choice of where, how, and with what the deceased is to be buried falls to them: "Mortuary symbolism is thus employed by mourners concerned not simply with the proper treatment of the dead, but also with the reallocation of rights and duties amongst themselves" (Barrett 1990:182).

For the qualitative portion of this study, I use standard methods of mortuary analysis, including an examination of the physical structures of the tombs and the individuals interred in them, as well as a comparative analysis of the grave goods. The use of componential analysis, and the relationship among different characteristics of the graves, determine which mortuary practices are normative and which, if any, are non-normative. The size of the tomb, investment of labor in its construction, the elaborateness of the grave marker, preparation and position of the body, and the number and type of grave goods are all quantifiable attributes (Binford 1971: 6–29; Goldstein 1981:54–59; Morris 1987:110–119; Saxe 1970: 38–63; Tainter 1977: 110–114). The physical characteristics of a tomb can be examined as components that "partition evidence into types" (Morris 1987: 111); therefore, the mortuary assemblage from Lofkënd contains patterned data, which can reveal some of the underlying social structures guiding mortuary practices. In order to discuss the range of social structure, including rank, status, social personae, and social roles that can be represented through mortuary variability (Arnold 2001: 213; Morris 1987:110–111; Saxe 1970:30–40), I have considered ethnographic as well as cross-cultural parallels and presented several possible explanations for the greater wealth in the tombs of the focus group and the unusual demographic pattern produced by the entire cemetery population at Lofkënd. Even factoring in the comparatively small number of tombs at the site and the difficulties inherent in ethnographic analogy and cross-cultural comparison (Peregrine 2001; Wylie 1985) or in reconstructing social organization from mortuary evidence (Pader 1982:62–64; O'Shea 1984), two explanations suggest themselves immediately—human sacrifice and social rank based on matrilineal or uxorilocal residence patterns—and these will be discussed and assessed.

The Tumulus, Tombs, and Treatment of the Body

The shape of the tumulus prior to excavation was an irregular ovoid earth outcrop, with the bedrock platform on which it lay ultimately 25 m north-south and just under 20 m east-west (see **Figs. 1.3–1.4**, and

2.8, with full details in Chapter 2). The southern end of the tumulus fell away sharply due to the natural escarpment there; this was the most heavily eroded area, and tombs located in this area were among the most damaged (**Fig. 1.8**). It is in the more gradually sloped northern end of the tumulus that the 13 graves that were radiocarbon dated to the early nineteenth century were located, all found on the eastern side of the tumulus, head to the west, facing east (Tombs LXXXVI [22], LXXXVIII [3], LXXXIX [11], XC [14], XCI [15], XCII [23], XCIV [25], XCV [21], XCVI [20], XCVII [39], XCVIII [36], XCIX [45], and C [48]). **Figure 8.1** is a plan of the tumulus showing all of the tombs in the tumulus and highlighting the modern tombs as well as the focus group.

The modern tombs were clearly distinguishable from prehistoric burials because of the extraordinarily good condition of the skeletal material, the fully supine position of the body, the location of the tombs at the northeastern end of the tumulus, their east–west orientation, and, in many cases, the stone lining of the tomb and placement of stones above it. These tombs are not included in this analysis (they are fully discussed in Chapter 3.2), but it is important to note that the creation of these tombs sometimes disturbed earlier burials. In some cases, the modern graves were dug to a greater depth within the tumulus than nearby prehistoric burials. In fact, this method of grave digging characterizes the stratigraphy of the entire tumulus; burials at lower levels sometimes post-date those at higher levels (see Chapters 2 and 4).

Radiocarbon dates place the earliest phase of use in the fourteenth century BC, which corresponds to the palatial phases of the Aegean Late Bronze Age (see Chapter 4). The tumulus continued to be used through the Early Iron Age (for the Late Bronze and Early Iron Age in Albania, see, among others, Davis et al. 2003–2004:6; Galaty 2002:113; Harding 1992; Korkuti and Petruso 1993:706). The prehistoric use of the tumulus may have continued until the late ninth century or just into the early eighth century BC (Chapter 4).

Tomb type

All tombs at Lofkënd were pit burials. Larger tombs were generally rectangular in shape, while smaller tombs tended to be ovoid. In many cases, disturbance by later grave digging, the reopening of tombs for additional interments, and the natural erosion of tombs near the surface or slope of the tumulus drastically affected the preservation of the original grave cut. Of the 85 prehistoric tombs at Lofkënd, 36 retained at least part of their original cut. Preservation of the cut also appeared to relate to whether the tomb contained multiple individuals; 22 of the 36 tombs with preserved cuts contained only one interment. The presence of 11 tombs with multiple interments and only partially intact cuts may indicate that the location of a tomb was marked or otherwise recognizable from the surface of the tumulus during the period of use and that this facilitated the reopening of the tomb.

Multiple tombs and order of interment

Twenty-six of the 85 tombs at Lofkënd contained multiple individuals; 58 were single interments. In many cases, these individuals were not interred simultaneously.

Frequently, the order in which individuals were interred in a tomb was not clear. In some tombs, the occupants were interred simultaneously; Tomb LXXIV (29) consists of an unusual, simultaneous, multiple interment (**Figs. 3.257–3.258**). More often, however, the earlier occupant of the tomb was moved aside to make room for a later interment, making it possible in many instances to determine the order in which individuals were buried. This is well illustrated in the case of Tomb LXX (17), one of the wealthy tombs containing an adolescent female (**Figs. 3.241–3.242**).

All but one of the six tombs in the focus group contained multiple inhumations. The three containing adolescent females (XXI [55], XLVIII [52], and LXX [17]) also held adults but no children (**Figs. 3.63–3.64, 3.154–3.155, 3.241–3.242**). In Tomb LXX (17), the previously inhumed remains of an older adult male were clearly pushed aside for the interment of the adolescent female. Tomb XLVIII (52) contained three individuals; the first to be interred (an adult of indeterminate sex) was completely disturbed when the tomb was reopened for the two later burials. These two individuals—the adolescent female and a young adult male—were probably interred simultaneously. According to the excavator, the male may have been slightly disturbed by the burial of the adolescent female, who was clearly interred last. In Tomb XXI (55), which contained an adolescent female and a young adult male, the position of the male could not be determined, due to disturbance and the poor condition of the skeleton. Only cranial fragments remained from the adolescent female, making the order of interment impossible to determine. The remaining three tombs in this group contained the remains of children or infants. Tomb XVII (72) held two children (**Figs. 3.50–3.51**), two children and an infant occupied Tomb XVIII (73) (**Figs. 3.55–3.56**), and Tomb XXVIII (77) contained a single infant (**Figs. 3.88, 3.89c**). The overall condition of these subadult graves precludes determination of any exact order of interment due to the poor preservation of the bones. The preservation of subadults is generally poorer than that of adults, due to a lesser degree of calcification and articulation of the bone. The condition of the three wealthy infant and child tombs was relatively typical of infant and child tombs throughout the tumulus.

Very few biological relationships among individuals in multiple tombs and throughout the tumulus could be determined, but some dental and other traits are shared (see Chapter 6 and especially **Table 6.14**). Twenty-nine of the skeletons in the tumulus were in too fragmentary a state to be sexed (see the discussion on demographics below; although DNA samples were taken in the 2004 season, no definitive results were obtained; see Papadopoulos, Bejko, and Morris 2007:113; and Appendix 1). There may have been familial relationships between the occupants of multiple tombs, in particular between those reburied in Tomb I, and among the the three males in Tomb LXXIV (29). Any evidence for familial relationships is based on biological traits (see Chapter 6), and there appears to be no convention for burying children with parents or siblings (i.e., children were as likely to be interred with adult males as females, not all graves containing women contained men, and not all child graves contained adults).

Tomb dimensions

There is very little consistency in tomb dimensions (**Table 8.1**). Erosion, the reopening of tombs, and the partial destruction of existing tombs by the construction of new ones naturally affected the preservation of the grave cuts, making original dimensions difficult to determine.

During excavation, determination of the length, width, and depth of a tomb was dependent on the existence of a clear tomb cut. Length and width could not be accurately determined for seriously disturbed tombs. Depth was based on the excavator's determination that the floor of the tomb was reached and is recorded as height above sea level.

Four of the six wealthy tombs retained evidence of their original cut and displayed a range of dimensions (Tombs XVII [72], XVIII [73], XXVIII [77], and XLVIII [52]). There is no apparent similarity in the size of these tombs. Tomb XLVIII (52), which contained two adults, was the largest of the four, and the smallest was Tomb XXVIII (77), which held a single child. None of these six tombs was particularly deep. Tomb XVII (72), containing a child and an infant, was 0.078 m deep, and the others range from 0.35 to 0.10 m in depth, again displaying no consistency in dimensions.

The other tombs in the tumulus show the same range of variability. The mean length of the 35 tombs with some intact cut (including the four from the focus group) is 1.52 m, but the standard deviation is close to 0.50 m. The widths were equally inconsistent; their mean is 0.82 m with a standard deviation of 0.27 m. Multiple graves were not consistently larger than single graves; rather, it is likely that many graves that were eventually reopened for additional interments had originally been created for single individuals. Tomb XII (88) (**Fig. 3.37**) had its north end enlarged either at the time of its original excavation or when it was reopened for a second interment. If this enlargement took place during a reopening of the tomb, it would be a good example of how the original cut of the tomb could be modified or completely destroyed by later use.

Children more frequently occupied tombs with dimensions that were less than 1.50 m in length (10 of the 17 tombs of this size belonged to children), but nearly half of these smaller tombs also contained adults. There is no correlation between tomb size and the sex of the interred.

The mean depth of the tombs at Lofkënd is 0.22 m with a standard deviation of 0.16 m. The actual floor of a tomb frequently could not be identified due to the absence of a distinguishable difference between grave fill and tumulus fill. As a result, depth measurements may be somewhat arbitrary in certain cases. Tombs that preserved poor evidence of the original cut retained little information in terms of measurable dimensions (apart from the skeleton itself).

The largest tomb by far was Tomb I (64). Located near the center of the tumulus, this tomb measured 2.20 m by 1.20 m, and was 1.02 m deep. The only other graves that approached this depth (Tomb

TABLE 8.1 Analytic table of all tombs and their primary characteristics

Tomb	Skeleton	Sex	Age	Tomb length	Skeleon length	Tomb width	Tomb depth	Low point of grave (elevation)	Compass orientation	Degree	Body position (Fig. 8.3)	Multiple/single	Order of interment	Associated stones
(1) LXXXI	—	—	—		—		0.085	109.076			—	S	—	N
	14	M	18-22	—	0.70 m	—	—	—				—		N
(2) LXXXIV	—	—	—	1.92	—	0.70	0.150	108.830	SE-NW	110 (SE)	—	M	—	N
	38a	C	2-5 yr	—		—	—	—				—		N
	38	M	35-45	—	0.65 m	—	—	—				—	1	N
	18	M	23-26	—	1.5 m	—	—	—	E	95	a	—	2	N
(4) LXXX	—	—	—		—		0.173	109.014			—	S	—	N
	32	M	20-30	—	0.75 m	—	—	—	E	89	f	—		N
(5) LXXVIII	—	—	—	2.20	—	1.53	0.082	108.775	SE-NW	100 (SE)	—	S	—	N
	51	F	20-25	—	1.3 m	—	—	—	SE	105	a	—		N
(6) LXXIX	—	—	—		—		0.043	108.872			—	S	—	N
	54	I	20-25	—		—	—	—	SE	130		—		N
(7) LXXXIII	—	—	—		—		0.180	108.830			—	S	—	N
	60	I	15 ± 3 yr	—	0.80 m	—	—	p	E	80		—		N
	60a	I	Young adult	—		—	—	—				—		N
	60b	I	Mid-old adult	—		—	—	—				—	—	N
	60c	I	>45	—		—	—	—				—		N
(9) LXXXII	—	—	—	1.10	—	0.55	0.240	108.559			—	M	—	N
	105	I	Young adult	—		—	—	—				—		N
	106	M?	26-32	—		—	—	—				—		N
	85	M?	26-32	—	1.22?	—	—	—	E	90		—		N
(10) LXXXV	—	—	—		—		0.340	108.571			—	M	—	N
	145a	C	<1 yr	—		—	—	—				—		N
	90	F	Mid-adult	—	1.01 m	—	—	—	NW	270	a	—	2	N
	145	M?	Adult	—	0.75 m	—	—	—	E	90		—	1	N
(12) LXVII	—	—	—		—		0.366	108.133			—	M	—	N
	171	F	23-27	—	0.96 m	—	—	—				—		N
	100	M	>35	—	0.65 m	—	—	—	NE	75		—		N
(13) LXVIII	—	—	—	1.40	—	0.67	0.335	108.304	NW-SE	320-340 (NW)	—	M	—	N
	103	F	Mid-age adult	1.50	0.34 m	0.81	—					—	3/4	N
	103a	I	20-30	—		—	—					—	3/4	N
	149	I	20-25	—		—	—					—	1	N
	136	I	20-25	—		—	—		NW	305	a	—	2	N
(16) LXXVI	—	—	—		—		0.190	108.439			—	S	—	N
	125	I	Adult									—		N
(17) LXX		—	—	—		—	0.351	108.189			—	M	—	N
	126	F	17-20	—	1.02 m	—	—		SW	250	a	—	2	N
	197	M	>45	—		—	—					—	1	N
(18) LXXVII	—	—	—		—		0.149	108.469			—	S	—	N
	139	M	Mid-age adult	—	1.22 m (?)	—	—		NE	80	f	—		N
(24) LXXII	—	—	—		—		0.112	108.378			—	M	—	N
	162a	C	10 ± 2.5 yr	—		—	—					—		N
	162	I	25-35	—	1.56 m	—	—		SE	120	a	—		N
(26) LXXIII	—	—	—	0.90	—	0.85	0.120	108.089			—	M	—	N
	169a	I	Adult	—		—	—					—		N
	169	M?	Adult	—	0.75 m	—	—		NW	120?		—		N
(27) LXIX	—	—	—		—			108.512			—	M	—	N
	551	C	6 ± 2 yr	—	0.20	—	—					—	1	N
	551a	C	6 yr	—		—	—					—		N
	174	I	20-25	—	0.58	—	0.100		E	90	f	—	2	N

Continued on the next page

TABLE 8.1 (*continued*). Analytic table of all tombs and primary characteristics

Tomb	Skeleton	Sex	Age	Tomb length	Skeleon length	Tomb width	Tomb depth	Low point of grave (elevation)	Compass orientation	Degree	Body position (Fig. 8.3)	Multiple/single	Order of interment	Associated stones
	230	M	35-45	—	1.83 m	—	—	—	NE	40-50	i	—	1	Y
	231	M	25-35	—	1.66 m	—	—	—	NE	40-50	i	—	1	Y
(30) LXV		—	—	—	—	—	0.060	108.251			—	S	—	N
	213	F	18-23	—	1.02 m	—	—	—	SE	130	f	—		N
(31) LXVI		—	—	—	—	—	0.150	108.139				S	—	N
	216	F?	15 ± 2 yr	—	1.48 m	—	—	—	NE	70	c	—		N
(32) LXII		—	—	—	—	—	0.059	108.111			—	S	—	N
	220	I	Adult	—		—	—	—				—		N
(33) LXXV	—	—	—		—		0.112	108.153			—	S	—	N
	223	M	30-40	—	1.37 m	—	—	—	E	90	a	—		Y
(34) LXI	—	—	—		—		0.080	108.006			—	S	—	N
	225	I	Adult	—		—	—	—				—		N
(35) LXIII	—	—	—		—		0.083	108.076			—	S	—	N
	234	F?	>45	—	1.10 m	—	—	—	NE	70	a	—		N
(37) LVIII	—	—	—		—		0.094	107.919			—	S	—	N
	243	I	15 ± 3 yr	—		—	—	—	NE	20		—		N
(38) LIX	—	—	—		—		0.182	107.849			—	S	—	N
	247	M	35-45	—		—	—	—	NE	50		—		N
(40) LIV	—	—	—		—		0.107	107.811			—	S	—	N
	252	I	Mid-age adult	—	1.10 m	—	—	—	SW	240-245	h	—		N
(41) XLVII	—	—	—		—		0.218	107.777			—	S	—	N
	256	F	18-21	—	1.25 m	—	—	—	E	100	h	—		Y
(42) XLVI	—	—	—		—		0.138	107.739			—	S	—	N
	263	F	20-25	—	1.35 m	—	—	—	SE	140	h	—		Y
(43) LVI	—	—	—		—		0.203	107.806			—	M	—	Y
	269	F	20-25	—	1.35 m	—	—	—	E	80	f	—	2	N
	296	M	20-25	—	1.25 m	—	—	—	E	80		—	1	N
(44) LX	—	—	—		—		0.120	107.700			—	S	—	N
	273	I	Young adult	—		—	—	—				—		N
(46) L	—	—	—		—		0.123	107.825			—	S	—	N
	284	C	4±1 yr	—	1.00	—	—	—	E	80		—		N
(47) XXII	—	—	—		—		0.140	107.594			—	S	—	N
	294	M?	Mid-age adult	—	0.45	—	—	—	NE	80	h?	—		N
(49) XIII	—	—	—		—		0.253	107.520			—	S	—	N
	305	F	25-35	—	1.20 m	—	—	—	NE	60	f	—		N
(50) XX	—	—	—	2.00	—	1.00	0.171	107.598	E-W		—	S	—	Y
	308	F	20-25	—	1.61 m	—	—	—	E	100	a	—		Y
(51) XLIX	—	—	—		—		0.139	107.661			—	S	—	N
	315	C	3 ± 1 yr	—		—	—	—	NE	75		—		N
(52) XLVIII	—	—	—	1.63	—	0.63	0.309	107.424	SE-NW		—	M	—	N
	318	F	15 ± 3 yr	—	1.41 m	—	—	—	SE	110	a	—	3	N
	372	I	Mid-age adult	—		—	—	—				—	1	N
	366	M	20-25	—	1.41 m	—	—	—	SE	110?	f	—	2	N
(53) LV	—	—	—		—		0.125	107.634			—	S	—	N
	321	C	7 ± 2 yr	—	0.75 m	—	—	—	SE	120-130	f?	—		N
(54) XIX	—	—	—	2.20	—	1.08	0.454	107.340	SW-NE		—	S	—	Y
	323	F	Mid-age adult	—	1.34 m	—	—	—	SW	245	e	—		Y
(55) XXI	—	—	—		—		0.114	107.488			—	M	—	N
	341	F?	15 ± 3 yr	—		—	—	—	NE			—	2	N
	326	M	20-25	—	1.50	—	—	—	NE	50	f	—	1	N
(56) XXIII	—	—	—		—		0.142	107.632			—	S	—	Y
	329	C	8 ± 2 yr	—	1.0 m	—	—	—	NE	80	f	—		Y
(57) XLI	—	—	—		—		0.128	107.556			—	S	—	N
	331	I	20-30	—		—	—	—	NW	335		—		N

Table 8.1 (*continued*). Analytic table of all tombs and primary characteristics

Tomb	Skeleton	Sex	Age	Tomb length	Skeleon length	Tomb width	Tomb depth	Low point of grave (elevation)	Compass orientation	Degree	Body position (Fig. 8.3)	Multiple/single	Order of interment	Associated stones
(58) LVII	—	—	—		—		0.172	107.451			—	S	—	N
	336	I	Mid-age adult	—		—	—	—	W			—		N
(59) XLII	—	—	—	2.20	—	1.15	0.527	107.449	NE-SW		—	M	—	Y
	339	C	4 ±1 yr	—	0.81 m	—	—	—	SW	245	h	—	2/3	Y
	340	C	2 yr ± 8 mo	—	0.66 m	—	—	—	SW	245	h	—	2/3	Y
	338	F	20-25	—	1.22 m	—	—	—	SW	240?		—	1	Y
(60) XLV	—	—	—	1.50	—	0.70	0.383	107.509	SE-NW		—	M	—	Y
	352	F	30-40	—	0.55 m (?)	—	—	—				—	1	N
	345	M	>55	—	1.40 m	—	—	—	SE	105	a	—	3	N
	351	M	18-25	—	1.40 m	—	—	—	SE	105	a	—	2	N
(61) LXIV	—	—	—	1.60	—	0.95	0.569	107.460	SE-NW		—	S	—	Y
	347	C	4 ± 1 yr	—		—	—	—	SE-NW	290?		—		Y
(62) XLIII	—	—	—		—		0.078	107.556			—	M	—	Y
	349	C	18 mo ± 6 mo	—	0.50 m	—	—	—	SE?	125-130		—		N
	356	C	1 yr ± 4 mo	—	—	—	—	—				—		N
	371	C	2 yr ± 8 mo	—	—	—	—	—				—		N
(63) LIII	—	—	—		—		0.140	107.446			—	M	—	N
	359a	C	5 ± 1.5 yr	—		—	—	—				—		N
	359	I	25-30	—	0.71 m	—	—	—	SE	145	h	—		N
(64) I	—	—	—	2.20	—	1.30	1.020	106.729	NW-SE		—	M	—	N
	362	M	25-30	—		—	—	—				—	1	N
	362a	M	18-23	—		—	—	—				—	1	N
	362b	M	>45	—		—	—	—				—	1	N
(65) XLIV	—	—	—		—		0.272	107.364			—	M	—	N
	394	F	18-25	—		—	—	—				—	1	N
	365	M	35-45	—	1.35 m	—	—	—	E	75	b	—	2	N
(66) XXXIX	—	—	—	1.75	—	0.75	0.116	107.303	NW-SE		—	S	—	N
	369	F	25-35	—	1.23	—	—	—	SE	140	d	—		N
(67) XL	—	—	—		—		0.231	107.349			—	S	—	N
	383	M?	>30	—	0.45	—	—	—	W	285		—		N
(68) XVI	—	—	—	1.43	—	0.77	0.280	107.300	SE-NW		—	S	—	Y
	386	C	6 ± 2 yr	—	0.78 m	—	—	—	SE	105	h	—	—	Y
(69) LII	—	—	—	1.40	—	0.58	0.135	107.389	SE-NW		—	S	—	N
	390	M	15 ± 3 yr	—	1.35 m	—	—	—	SE	125	d	—		N
(70) XXX		—		—		—	0.159	107.019				S		N
	398	C	9 mo ± 3 mo	—	.45 m	—	—	—				—		N
(71) XIV	—	—	—	1.26	—	0.59	0.225	107.005	N-S		—	S	—	N
	405	C	6 mo ± 3 mo	—	0.40 m (?)	—	—	—	S	180		—		N
(72) XVII	—	—	—	1.20	—	0.78	0.078	107.324	SE-NW		—	M	—	N
	408a	C	4 ± 1 yr	—		—	—	—				—		N
	408	C	7 ± 2 yr	—	0.90 m	—	—		E	93	f	—		N
(73) XVIII	—	—	—	1.24	—	0.73	0.175	107.174	E-W		—	M	—	N
	412	C	4 ± 1 yr	—		—	—	—	E	90		—		N
	412a	C	3 ± 1 yr	—		—	—	—				—		N
	421	C	5 ± 1.5 yr	—		—	—	—				—		N
(74) XXVI	—	—	—	1.60	—	0.84	0.107	107.237	SE-NW		—	S	—	N
	416	F	>45	—	1.36 m	—	—	—	SE	125	f	—		Y
(75) XXXVI	—	—	—		—		0.125	107.139			—	S	—	N
	420	I	Adult	—	0.45 m	—	—	—				—		N
(76) XXXVII	—	—	—		—		0.556	107.163			—	S	—	N
	424	I	Adult	—	0.40 m	—	—	—				—		N
(77) XXVIII	—	—	—	0.90	—	0.27	0.104	107.009	NE-SW	—	—	S	—	N

Continued on the next page

Table 8.1 (*continued*). Analytic table of all tombs and primary characteristics

Tomb	Skeleton	Sex	Age	Tomb length	Skeleon length	Tomb width	Tomb depth	Low point of grave (elevation)	Compass orientation	Degree	Body position (Fig. 8.3)	Multiple/single	Order of interment	Associated stones
	429	C	3 ± 1 yr	—		—	—	—	NE?	70	—	—	—	N
(78) LI	—	—	—	0.40	—	0.56	0.128	106.937			—	S	—	N
	434	M?	Young adult	—		—	—	—	E?	100?		—		N
(79) XXXVIII	—	—	—		—		0.160	107.229	NW-SE		—	M	—	N
	447	F?	Adult	—		—	—	—	—			—	1	N
	439	M	30-40	—	1.32 m	—	—	—	NW	290	c	—	1	N
(80) XV	—	—	—	1.02	—	0.60	0.301	107.164	SW-NE		—	S	—	N
	444	C	6 mo ± 3 mo	—	0.40 m	—	—	—	SW	200		—		N
(81) III	—	—	—	1.90	—	1.15	0.190	107.080	S-N		—	S	—	Y
	454	M	>35	—	1.17	—	—	—	SE	150	h	—		N
(82) XXVII	—	—	—	1.85	—	1.16	0.410	106.399	NE-SW		—	M	—	N
	461	F?	>35	—		—	—	—	NE	37?		—	1	N
	457a	I	18-25	—		—	—	—				—		N
	457	I	>35	—		—	—	—	NE	37?	a	—	2	N
(83) XXIX	—	—	—	1.05	—	0.55	0.164	107.135	NW-SE	—	—	S	—	N
	469	C	6 mo ± 3 mo	—	0.71 m	—	—	—	SE	140	i	—		N
(84) XXXV	—	—	—	1.20	—	0.64	0.405	106.890	SE-NW	—	—	S	—	N
	472	M	23-27	—	1.40 m	—	—	—	SE	141	a	—		N
(85) XXIV	—	—	—	1.40	—	1.00	0.149	106.660	N-S		—	S	—	N
	478	C	3 ± 1 yr	—		—	—	—				—		N
(86) XXXI	—	—	—	2.10	—	0.75	0.257	106.553	SE-NW		—	S	—	Y
	489	I	>40	—	1.24 m	—	—	—	S	180		—		N
(87) XXXIV	—	—	—	1.20	—	0.63	0.160	106.734	SE-NW		—	M	—	N
	504	M	Adult	—		—	—	—				—		N
	498	M	45-55	—	1.25 m	—	—	—	SE	132	a	—		N
(88) XII	—	—	—	1.40	—	0.60	0.291	106.619	N-S		—	M	—	N
	501	C	1 yr ± 18 mo	—	0.14 m	—	—	—	SE	—		—	1	N
	495	C	9 ± 2 yr	—	1.03 m	—	—	—	NW	336	g	—	2	N
	499	F	20-25	—	1.35 m	—	—	—	SE	150	d	—	1	N
(89) XXXII	—	—	—	2.00	—	0.50	0.330	106.510	N-S		—	M	—	Y
	524	M	18-25	—		—	—	—				—	1	Y
	508	M	20-30	—	1.47 m	—	—	—	SE?	150	a	—	2	Y
(90) XXV	—	—	—	1.10	—	1.10	0.109	106.650			—	S	—	N
	512	C	6-10 yr	—		—	—	—				—		N
(91) II	—	—	—		—		0.200	106.450			—	S	—	Y
	516	M	>45	—	1.46 m	—	—	—	NW	340	g	—		Y
(92) XXXIII	—	—	—		—		0.240	106.310			—	S	—	N
	523	F	16 ± 2 yr	—	1.18	—	—	—	SE	150	a	—		N
(93) XI	—	—	—		—		0.170	106.510			—	S	—	N
	540	C	9 mo ± 3 mo	—	0.20	—	—	—				—		N
(94) IX	—	—	—		—		0.110	105.930			—	S	—	N
	547	F?	Adult	—	0.75	—	—	—	S	170-175		—		N
(95) X	—	—	—		—		0.170	106.060			—	S	—	N
	554	F	30-40	—		—	—	—	NW	340		—		N
(96) V	—	—	—	1.54	—	0.55	0.200	105.240	SE-NW		—	S	—	Y
	569	F	20-25	—	1.25	—	—	—	S	185	j	—		N
(97) VI	—	—	—	1.37	—	0.80	0.107	106.323	SE-NW		—	S	—	N
	573	C	3 ± 1 yr	—	0.15	—	—	—				—		N
(98) IV	—	—	—		—		0.322	104.128			—	S	—	Y
	581	M	20-25	—		—	—	—	S	170	a	—		N
(99) VII	—	—	—	1.68	—	1.33	0.720	105.410	NW-SE		—	S	—	N
	583	C	7 ± 2 yr	—	0.68	—	—	—	SE	160		—		N
(100) VIII	—	—	—	1.33	—	0.88	0.539	105.251	NE		—	S	—	Y
	591	C	3 ± 1 yr	—		—	—	—	NE	40	h	—		N

VII [99], IV [98], VIII [100], XXVII [82], XXXVIII [79]) were, for the most part, near the steep scarp on the southeast side of the tumulus (see **Fig. 8.1**). Their locations on the southeast slope required that the excavators dig a deep pit, since the effects of erosion often resulted in portions of the skeleton moving to lower elevations and some remaining at higher elevations. The recorded depth of these tombs, therefore, does not necessarily reflect their original depth.

ORIENTATION AND POSITION OF THE BODY

Compass orientations are listed only when skeletons appeared to be in their original burial position within the tomb **(Fig. 8.2)**. Notes on orientation record the relationship of the vertebrae to the cranium, and compass readings were taken from the line of the vertebrae when those bones were intact. The position of the skeleton was recorded only when the skeleton was sufficiently intact to show its original posture.

Four of the six individuals in the focus group (skeletons 318 from Tomb XLVIII [52], 408 from Tomb XVII [72], 412 from Tomb XVIII [73], and 429 from Tomb XXVIII [77]) were oriented to the northeast or southeast (**Fig. 8.2**). The adolescent female (skeleton 126) from Tomb LXX (17) was oriented at 250° southwest. The condition of the bones of the adolescent female (skeleton 318) from Tomb XXI (55) was too poor to determine the exact orientation, but she was generally aligned with the male in the same grave, with her cranium to the northeast. Five of the six individuals follow the same general pattern displayed by the rest of the cemetery population, predominantly oriented to the northeast or southeast.

Of the 129 skeletons in the tumulus (including those in the focus group), 82 were intact enough that their approximate orientations could be determined. Sixty-four of the 82 had their crania between 0° and 180° (northeast or southeast); the remaining 18 were oriented between 200° and 359° (southwest or northwest). Of the 30 male skeletons for which orientations could be determined, 3 were not oriented between northeast and southeast. One (skeleton 439) was interred in Tomb XXXVIII (79), a multiple tomb containing one of only two cremations in the tumulus. The identification of sex for the second male skeleton (skeleton 383 from Tomb XL [67]) with this orientation was tentative, due to its extremely fragmentary condition. The third (skeleton 516 from Tomb II [91]) was confidently identified as a male. Skeleton 516 varied only slightly from others in the tumulus by virtue of having been interred lying on his side rather than on his back. The remaining individuals not oriented to the northeast or southeast were females, children, or infants.

Determining the orientation of the tombs was based on the presence of complete or partially intact cuts. For tombs that lacked an intact cut, orientation was based on the skeleton it contained. Fifty-three tombs were oriented either north to south or between 0° and 180°, 18 were oriented between 200° and 359°. The remaining 12 tombs had no discernable orientation. The four wealthy tombs that retained cuts (Tombs XVII [72], XVIII [73], XXVIII [77], and XLVIII [52]) shared the same orientation as the skeletons they contained.

Body position

Among the Lofkënd tombs as preserved, the preferred body position for burial was one in which the torso lay supine with the arms bent at the elbow and folded over the chest or abdomen, often with the fingers intermingled and the legs flexed to some degree (**Table 8.1; Fig. 8.3a–j**). The female skeletons in two of the wealthy tombs (XLVIII [52] and LXX [17]) exemplify this position (see **Fig. 8.3a–b)**. The body position could not be determined for the individuals in the other four wealthy tombs, due to the poor condition of the skeletons.

Of the 57 intact skeletons in their original positions, 39 (approximately 2/3) were buried in some version of the preferred position described above. The variations on this burial position are shown in **Figure 8.3a–f**. Variation occurred in the placement of the arms; some were folded at the elbow, with one hand near a shoulder (**Fig. 8.3d**). Occasionally, skeletons had one arm bent across the torso and the other straight down along one side of the torso (**Fig. 8.3e**). Often one or both of the distal arm bones were missing, making complete identification of the position difficult (**Fig. 8.3f**). Degree of flexion in the legs also varied, as in **Figure 8.3b** and **8.3c**. The extremely flexed position shown in **Figure 8.3c** is discussed further below. In addition, 12 individuals were buried in a flexed position on their sides: seven on the right side (**Fig. 8.3g**) and four on the left (**Fig. 8.3h**). In five cases, the deceased was interred in a fully supine (torso and legs supine) position (**Fig. 8.3i**):

the three males in Tomb LXXIV (29) (see **Figs. 3.257–3.258**) and two infants, one from Tomb XV (80) and one from Tomb XXIX (83). One individual (Tomb V [96]) was placed with her torso prone, her head facing west and her legs flexed (**Fig. 8.3j**).

In at least two cases (Tombs XXXVIII [79] and LXVI [31]), the legs were tightly flexed, so much so that the body must have been placed in the grave with the feet below the hips, forcing the legs into a very tightly bent position (Papadopoulos, Bejko, and Morris 2007:124). **Figures 3.122–3.123** and **3.124a** show Tomb XXXVIII [79], which is the best example of this tightly flexed position and may have been facilitated by securing the body by tying or bundling prior to burial when there was still some degree of movement in the limbs (Papadopoulos, Bejko, and Morris 2007:124; see also Lewartowski 2000:22). The generally small dimensions of tombs at Lofkënd may be the result of a preference for a flexed burial position, although the converse may also be true: the excavation of smaller graves was desirable, requiring the corpse to be interred in a flexed position.

Associated stones

Twenty-one tombs either contained stones or were in some way demarcated by the placement of stones (see **Tables 8.1** and **8.2**). None of the six tombs in the focus group had associated stones.

Stones were placed in a variety of positions, some inside the cut of the grave and apparently directly on top of the skeleton and others outside the grave cut, but, in some instances, bordering or otherwise relating to the tomb. Frequently, the stones were placed on top of the skeleton, sometimes on the cranium, but just as often over other areas of the body. In Tomb LXXIV (29) (**Figs. 3.257–3.258**), stones were placed on top of the cranium of each of the three individuals interred. In most cases, the placement and association with the tomb or skeleton appears intentional. Certain post-depositional processes such as erosion, settling, and the breakage of bone due to the weight of overlying soil (and the stones themselves) makes it difficult to determine how much fill would originally have separated these stones from the deceased. **Table 8.2** shows which tombs had stones and, when applicable, which skeletons within these tombs were found with stones placed near or on top of them. When stones were associated with a tomb but not inside it, they are noted as being outside the cut.

NORMATIVE BURIAL PRACTICES: DO THEY EXIST AT LOFKËND?

Of the various physical characteristics of the 85 prehistoric burials, only the orientation of the skeleton appears to correlate with other mortuary variables. In all but three cases, male skeletons were oriented with their crania to the east (**Fig. 8.2**). One of the three males whose cranium was not in an eastern orientation was skeleton 439 from an unusual tomb, also containing the cremation of an adult female (Tomb XXXVIII [79]). The second of the three, skeleton 383, found in Tomb XXXIX (66), could only be tentatively identified as male. In the third case, the deceased, skeleton 516 from Tomb II (91), an individual with considerable pathologies, was placed in an unusual position within the grave, lying on his right side, his hands folded beneath the cranium. This is the only skeleton in the tumulus with the hands placed in such a position.

An overall preference for a northeast or southeast orientation is evident at Lofkënd, but this was more strictly enforced for males than for females, infants, and children. The correlation, however, is not a strong one, due to the wide range of compass orientation between northeast and southeast (0–180°). No other variable in the tumulus shows any clear association with either sex.

While there is considerable variability among the prehistoric tombs, the key diagram (**Fig. 8.4**) shows that it contained two unusual ("non-normative") graves. Both are cremations. In Tomb XXX (70), the cremated remains of an infant were sparse and scattered, the result of disturbance and erosion. In contrast, the cremated remains of the female in Tomb XXXVIII (79), for whom the age could not be determined, were interred with an inhumed adult male who, as noted above, was placed in the grave in a northwest orientation with his legs folded tightly under him (**Figs. 3.122–3.123, 3.124a**). The two were interred simultaneously; the inhumation showed no signs of having been disrupted by the reopening of the tomb, and the cremated remains were confined to a discrete area south of the pelvis and right femur of the inhumation. The boundaries of the cremated bone were distinguished and enclosed by a line of darker soil. This boundary may indicate that they were originally held in some type of organic container, most likely a cloth or leather sack, which had completely decomposed. This conclusion

TABLE 8.2. Stones associated with prehistoric graves

Tomb	Skeleton	Age	Sex	Location of stone	Elevation
(29) LXXIV	210	29-35	M	Atop cranium	108.353
	210			Atop right arm	108.314
	210/230	35–45		Atop left arm/torso	108.309
	210/230/231	25–35		East of crania	108.399
	230	35–45	M	Atop pelvis	108.328
	230			Atop knees	108.277
	230/231	35–45/20–30	M	Atop crania	108.308
(33) LXXV	223	30–40	M	Atop/among lower torso	—
	223			Below right arm	—
	223			Southeast of cranium	—
(41) XLVII	256	18–21	F	Atop feet	108.044
(42) XLVI	263	20–25	F	Atop pelvis	—
	263			Atop lower legs	107.930
(43) LVI	269/296	20–25	F/M	North of flexed legs	107.895
(50) XX	308	20–25	F	East of feet	107.975
	308			East of feet	107.803
	308			West of knees	107.731
	308			West of pelvis	107.936
	308			Atop lower torso	107.927
	308			West of left upper arm	107.916
	308			East of right upper arm	107.943
	308			Atop and southeast of right shoulder	107.763
(54) XIX	323	Mid-age adult	F	Southeast end of tomb (southeast of feet)	107.978
	323			Southeast end of tomb (southeast of feet)	107.758
	323			South end of tomb (southwest corner)	108.023
	323			Southeast end of tomb (southeast of feet)	108.020
	323			Atop (flexed) lower legs	107.696
	323			Just west of grave cut (northwest corner)	107.692
(56) XXIII	329	8 yr ± 2 yr	C	Northeast "corner" of grave	107.791
	329			Northwest of torso/right arm	107.745
	329			Southwest "corner" of grave	107.86
	329			Southeast "corner" of grave	107.69
(59) XLII	338	20–25	F	Southwest of cranium	107.547
	339, 340	4 ± 1 yr/2 yr ± 8 mo	C/I	Southwest of cranium	107.635
	339			Atop pelvis	107.642
				Southeast "corner" of cut	107.828
(60) XLV	345/351/352	>55/18–25/30–40	M/M/F	Southeast of crania, outside grave cut	107.892
(61) LXIV	347	4 yr ± 1 yr	C	17 stones within the grave cut, above the skeleton	
				Stone at highest elevation	107.727
				Stone at lowest elevation	107.572
(62) XLIII	349	18 mo ± 6 mo	I	Stone southeast of cranium?	107.698
(68) XVI	386	6 ± 2	C	Northeast of cranium	—
(74) XXVI	416	>45	F	Above cranium	107.56
(81) III	454	>35	M	Northeast corner of grave outside cut	107.332
	454			Northeast corner of grave outside cut	107.459
	454			Northern end of grave outside cut	107.488
	454			Northern end of grave outside cut	107.389
	454			Northwest corner of grave outside cut	107.443
	454			Northwest corner of grave outside cut	107.445

Continued on next page

Table 8.2 *(continued)*. Stones associated with prehistoric graves

Tomb	Skeleton	Age	Sex	Location of stone	Elevation
(85) XXIV	478	3 ± 1 yr	C	South of cranium (position of body unknown)	106.992
	478			Southeast of cranium (position of body unknown)	106.866
(86) XXXI	489	>40	?	Below and slightly west of femurs	106.74
	489			East of lower torso	106.775
	489			East of lower torso	106.749
	489			Southeast of lower torso	106.867
(89) XXXII	524	18–25	M	Atop knee of diarticulated skeleton	106.854
	508	20–30	M	East of knees	106.603
	508/524	18–25		Northwest "corner" of grave outside cut	106.909
(91) II	516	>45	M	Atop knees/lower legs	107.035
	516			Atop pelvis/lower torso	106.76
	516			Stone to the west	107.244
(96) V	569	20–25	F	Stone east of lower legs	105.452
	569			Stone northeast of feet	105.425
	569			Stone northeast of feet	105.448
	569			Northwest corner of grave outside cut	106.186
(98) IV	581	20–25	M	Entire grave covered with 25+ stones	
				Stone at highest elevation	104.613
				Stone at lowest elevation	104.945

is supported by the presence of a band of similar gray/black organic soil resembling a cut that defined the entire edge of the grave.

Cremation is often a more elaborate method than inhumation for disposing of a corpse. More time and labor is usually dedicated to cremation. The process includes the gathering of wood, the laying out of the body, the tending of the pyre, and the collection of the remains (Kurtz and Boardman 1971: 73–74; Morris 1987:153–154; Papadopoulos 2005: 246–248, 382–385; Parker Pearson 1999:6–7). Added to this expenditure of labor is the transport of the cremated remains from the site of the pyre to the site of interment. Both cremations at Lofkënd were secondary, and the funeral pyres were not constructed at the site of the grave. Perhaps something distinguished the individuals in Tombs XXX (70) and XXXVIII (79) from the other members of the population buried at Lofkënd (though in neighboring Greece in the Early Iron Age, the distinction between cremation and inhumation was not as important as scholars once thought: see Papadopoulos 2005:382–385, 393–395), but whatever the reasons for cremation, these two individuals were nevertheless interred in the tumulus.

The key diagram **(Fig. 8.4)** illustrates how different Tombs XXX (70) and XXXVIII (79) are from the remainder of the prehistoric tombs; it also shows the frequency of several characteristics of tomb construction and treatment of the skeleton, and establishes which combination is most common at Lofkënd. There is a considerable amount of diversity in tomb construction and treatment of the body in this cemetery. Small clusters of tombs with similar characteristics, however, can be seen at *a1-b1-c1-d1-e5/e9* on **Figure 8.4**, indicating that the most common burial treatment (represented by 10 tombs) is inhumation in a single tomb, with no associated stones and no grave goods. All of the tombs in this group, however, lacked discernable orientation. Since the largest group shown in the key diagram comprises only 10 tombs, a normative burial treatment at Lofkënd cannot be defined. The most that can be said is that each tomb in the tumulus is unique, but there was a consistent practice of burial in pit tombs: general preferences in the orientation of the skeleton, and the majority of bodies placed with the torso supine, legs flexed, and arms folded over the torso.

The key diagram also shows that five of the six wealthy tombs cluster together at *a1, b2, c1, and d4* but are then separated by their orientations. Tomb XXVIII (77) is separated from the other five by virtue of being the only single inhumation in the group. This grouping in the key diagram also shows that, while the type and location of ornaments are the primary link among these tombs, they are all multiple inhumations without associated stones.

On the basis of the evidence presented above, what, if anything, can be determined about the structure of the society of the people who buried their dead at Lofkënd? Some diversity is to be expected in a cemetery that was in use for hundreds of years, and while these mortuary practices appear to lack homogeneity when compared to other Bronze and Iron Age mortuary treatments in the Aegean, they exhibit considerable consistency.

MORTUARY VARIABILITY IN THE BALKANS AND THE AEGEAN

Burial tumuli are known in Serbia, the Former Yugoslav Republic of Macedonia, Bosnia, Croatia, and Montenegro, regions which overlap territories of the ancient Illyrians, and many have been excavated in Albania since the 1950s (see Chapter 17; see also Bejko 2002a; Hammond 1967a:201–426; Hammond 1982:625; Harding 1992:17–18; Wilkes 1992:8, 41, 47–48). In the Aegean, they are common in both the Bronze and Early Iron Ages (see, among others, Pelon 1976:73–152). Inhumation in tumuli is almost ubiquitous as a mortuary treatment in Bronze and Iron Age Albania. The practice apparently appears by the third millennium BC (Harding 1992:17–18; and, most recently, Amore 2010) and continues through the middle of the fourth century BC (Hammond 1967a:405; Wilkes 1992:45; Chapter 17).

In Albania, as at Lofkënd, the grave type within the tumulus can vary among pit graves—cist graves, and at one site (the Pazhok tumulus), wooden coffins (Hammond 1967a:351; Prendi 1982:235; see also Bejko 2002a:174). While the practice of inhumation is clearly the preferred method of disposal, Lofkënd is not the only tumulus that contained cremations (cf. Hammond 1967a:229, 310; Hammond 1982:630; Prendi 1982:235; see also Wilkes 1992:45). Most tumuli in the region were constructed over a central burial, as at Lofkënd, but unlike at Lofkënd, the perimeter of central graves was frequently outlined with a ring of stones, and stones were often piled in cairns or covered the tumulus itself (**Figs. 17.2, 17.3, 17.6**; Bejko 2002a:158; Hammond 1967a: 387, 405, 1982:625–629; Wilkes 1992:41; see also Bejko, Fenton, and Foran 2006). While the methods of tumulus construction and grave type vary across Albania and the Balkans, there is long-term continuity to be found in the consistent use of the tumulus as a monument to the dead. Moreover, when compared to Greek, and especially Aegean, mortuary practice—during the same time that the Lofkënd tumulus was in use—Greek practices show greater variability.

Burial practices on the Greek mainland between ca. 1550 BC, the advent of the Mycenaean period, and ca. 700 BC, which represents the beginning of the Archaic period, were in constant flux (Morris 1987, 1992:103–203; Snodgrass 1971:141–197; Taylour 1983:65; Whitley 1991:23–34). During the Mycenaean period, the form and structure of graves and tombs included single cist graves, multiple shaft graves covered by mounds, chamber tombs, and tholos tombs. (It is important to note, however, that there was considerable overlap in use, and not all types were used in all regions: see Hood [1960], Pelon [1976], Taylour [1983:65–84], Cavanagh and Mee [1998:135–136], and Voutsaki [1995:58–63].) Inhumation was more common, but cremations were not unknown (Kurtz and Boardman 1971:25). The apparent collapse of the Mycenaean civilization and the return to the use of pit and cist graves, the reuse of existing burial monuments, and the increased popularity of cremation have all been attributed to both foreign invaders and internal factors following political collapse. These arguments have been discussed by many scholars (e.g., Cavanagh and Mee 1984:61, 1998:135; Kurtz and Boardman 1971:24–25; Morris 1992:69; Whitley 1991:6–8). Mortuary practices continued to shift along with changes in vase painting style in the Geometric period, increases in the size of cemetery populations, and variations in the elaborateness of tombs apparently linked to class status (Kurtz and Boardman 1971:49–67; Morris 1989:301–302, 1992; Snodgrass 1980:18–25; Whitley 1991); and the Archaic period experienced another shift in popularity of cremation (Kurtz and Boardman 1971:71–76).

The reasons for the long-term consistency in some Albanian mortuary practices—especially when compared to Greece—will not be addressed here, but the contrast between the two regions highlights the uniformity of tumuli use in Albania. Greek mortuary analyses have frequently been used to support theories about the structure of society, such as the reasons for the Mycenaean collapse and the rise of the *polis* in the Archaic period. At Lofkënd, very few patterns are apparent in the tomb structure and body placement, but within Albania the practice of inhuming the dead within a tumulus and around a central burial is fairly consistent.

TUMULI AND CENTRAL BURIALS: IS THERE EVIDENCE OF ORGANIZATION BY LINEAL DESCENT AT LOFKËND?

While nothing is known directly about the social organization of the people who interred their dead in the Lofkënd tumulus, both Hammond (2000:345) and Galaty (2002) have proposed that the prehistoric Illyrians of Albania were a clan-based, tribal system (see further Douzougli and Papadopoulos 2010:9–14, 66–67). In his discussion of the Illyrian tribal system, Galaty (2002:116) presents the possibility that Illyrian tribes marked the boundaries of their territories with tumuli. The visible location of the Lofkënd tumulus (and indeed the visibility of tumuli in general) may have lent itself to that purpose (see also Papadopoulos 2006:75, 78; Chapter 20). The three adult males interred in Tomb I (64) appear to have been moved from another location and reburied on this natural ridge into what is, in all probability, the first burial of this tumulus (this is fully discussed in Chapter 21). The reasons for exhumation and reburial can be many and varied. The relocation of skeletal material is often related to territoriality and the desire to be associated by kinship and descent to the remains. In such cases, proximity to the bones reinforces the lineal descent rights of other members of the society. The establishment and maintenance of both a physical and spiritual connection between a familial or social group and the remains of an important ancestor can support claims to territory as well (Earle 1990:74; Parker Pearson 1999:114–115, 130, 138); secondary inhumations can also be interred in existing burial sites or into those that have fallen into disuse. The intention, however, is the same: to establish a connection to the previous occupants of the tomb or cemetery for the purpose of reinforcing some claim to status or territory (Barrett 1990:183–185, 1994:127).

The circumstances of Tomb I (64) provide support for the argument that the tumulus belonged to a single community or subgroup, probably organized by lineal descent (Goldstein 1981:54). This conclusion is reinforced by the secondary nature of the burial, the central location of the tomb (see **Figs. 2.8, 4.1–4.2, 8.1**), the low elevation of the tomb within the tumulus, and radiocarbon dates that place it among the earliest burials in the tumulus (see Chapter 4). The first characteristic indicates that the skeletal remains were significant enough to be transported and reburied, implying a claim by the persons constructing the tumulus of a relationship to the deceased and perhaps to the surrounding land. The depth of the grave itself and its central location in what was to become the final form of the tumulus also suggest that it was the first grave established at the site (see Chapter 21).

Nothing in the physical construction of the tombs or the treatment of the bodies distinguishes the six wealthy tombs from other prehistoric tombs in the tumulus, and very little can be concluded regarding "normative" practices based on the variables above. The nature of the tumulus itself and its similarity to other tumuli in the region, however, may provide some evidence of social organization based on lineal descent in a tribal or clan structure.

DEMOGRAPHICS

Apart from Tomb I and two cremations, the most conspicuously anomalous tombs at Lofkënd were the six wealthy tombs containing adolescent women and children. The evidence presented above shows the diversity of the mortuary practices at Lofkënd—no physical characteristic of tomb construction or of skeletal position provided additional evidence to distinguish the six wealthy tombs from others in the tumulus.

The tumulus was completely excavated, yielding 129 prehistoric skeletons (i.e., 127 inhumations and two cremations), which represent the complete prehistoric population of the tumulus. Some losses due to the destruction of some graves and skeletal remains through both environmental and human post-depositional processes, however, cannot be disregarded. Physical anthropological analysis can identify the sex of only adolescent and adult skeletons; children and infants cannot be sexed. The demographic tables in this chapter, therefore, include an "indeterminate" ("I") category that is used for adult or adolescent skeletons for which sex could not be determined. Skeletons for which the identification of sex was probable were included in either the male or female group (eight skeletons in the tumulus were identified as most likely male and six as most likely female) (**Fig. 8.5** and **Table 8.3**).

The population of the tumulus was not only small (only 129 interments over a period of about 600 years) but also incomplete. The occupants of the tombs in the focus group belong to three of four demographic groups that were partially excluded from burial at Lofkënd; these include adolescent females, adolescent males, children (aged >3 years), and infants (**Fig. 8.6** and **Table 8.3**). In other words, it appears that only adults were systematically interred in the tumulus and that most members of the community who died prior to reaching adulthood were buried elsewhere or disposed of in some other way.

For the purposes of this study, the 129 individuals will be discussed using two age groupings:

1. The biological age grouping is based on the biological age of the skeleton. The intervals imposed to create these groups were established by Lynne Schepartz (Chapter 6) and are as follows:
   Infant = 0–2 years
   Child = 3–12 years
   Adolescent = 13–18 years
   Young adult = 19–34 years
   Mid-adult = 35–45 years
   Older adult = >45+ years
   (Adult = 18–45+ years)

The distribution of the mortuary population by biological age group is shown in **Figure 8.6** and **Table 8.3**. Biological age groupings are used to establish total numbers by age and sex for demographic calculations.

2. The social age grouping was designed to correspond with patterns in mortuary treatment discerned through an analysis of the population by biological age. It is intended to emphasize the significance of adolescence in this population. The social age groups are as follows and are shown in **Figure 8.7** and **Table 8.4**:
   Infant/child = 0–14 years
   Adolescent/young adult = 15–25 years
   Adult/older adult = 26+ years

During adolescence, an individual is balanced between the dependence of childhood and the responsibilities of adulthood. In certain societies, childhood often ends when the ability to reproduce begins; and in the social groups formed for this study, the "adolescent/young adult" category is based on the age range during which females would be most likely to experience marriage and childbirth. The "adult/older adult" category encompasses ages during which women would still be fertile but after the time when important rites of passage relating to marriage and reproduction would have taken place.

Adolescence is a critical age in pre-modern societies that lose a high percentage of women of reproductive age to the risks of childbirth. Studies of fertility, demographics, and reproductive age in preindustrialized societies often estimate that the female reproductive age begins around age 15 and ends near age 45 (Acsádi and Nemeskéri 1970:175, 177; Hewlett 1991:3). In her study of representations of children in fifth-century Athens, Lesley Beaumont (2000:40–41) identifies three stages of life preceding adulthood: infancy (approximately 0–3 years), prepubertal childhood (approximately 3–13 years), and pubertal youth (13–adulthood). Nancy Demand (1994:10) states that in Classical Greece, menarche was estimated to occur at age 14 and that

TABLE 8.3 Distribution of skeletons by biological age and sex

Age category	No. of burials	% of burials	F	M	I
Infant (0–2)	9	7			
Child (3–12)	26	20			
Adolescent (13–18)	8	6	4	1	3
Adult (19–34)	43	34	12	18	12
Mid-adult (35–45)	22	17	6	11	5
Old adult (>45)	9	7	2	5	2
Adult	12	9	3	3	7
Total	129	100			

F, female; I, indeterminate; M, male.

TABLE 8.4 Distribution of skeletons by social age and sex

Social ages	#
Female adult/older adult (26+ years)	10
Male adult/older adult (26+ years)	22
Indeterminate adult/older adult (26+ years)	8
Female adolescent/young adult (15–25 years)	15
Male adolescent/young adult (15–25 years)	13
Indeterminate adolescent/young adult (15–25 years)	14
Infant/child (0–14 years)	35

this was the time that marked the separation between childhood and readiness for marriage. Renee Pennington (2001:183) reports that among the !Kung, hunter-gatherers of the Kalahari, reproductive age begins as late as 19 years of age. However, Barry Hewlett's (1991:8–9) demographic analysis of "active" hunter-gatherers, sedentary hunter-gatherers, horticulturalists, and pastoralists/agro-pastoralists found a remarkable similarity among the mean fertility, birth, and infant mortality rates of hunter-gatherers and horticulturists/pastoralists—the mean age for when reproduction capacity begins was around 15 years. Since physical anthropological, cultural anthropological, and historical sources appear to agree that 15 is approximately the age most commonly associated with initial reproductive capacity, I have used it in the social age group as an estimate for the early age of reproduction.

Pre-industrial agricultural societies have been found to have high death rates for infants and children—ranging from 30% to 50% (Morris 1987: 57–58; Ubelaker 1989; White and Folkens 2000; see also Paine and Boldsen 2002). Morris's (1987: 57–58) estimates are based on available Iron Age evidence from multiple Greek sites from which he calculates that roughly 45.2–51.8% of children died before 9 years of age (see also Angel 1945; McDonald and Hope Simpson 1969; Young 1939, 1942, 1949, 1951). Hewlett (1991:8–9) determined a somewhat lower number (38%) for horticulturalist/pastoralist societies. However, this number was for children under 12 years and would make the number even higher if calculated for children aged less than 9 years, as in Morris's study.

At Lofkënd, 27% of the mortuary population died before age 12 (see **Fig. 8.5** and **Table 8.3**). A small number of additional infant and child burials can be accounted for by factoring in scattered remains found in the tumulus fill, but these would comprise only 5–10 individuals (Chapter 6). In a normal demographic profile, at least 60% of the deceased would be infants or children (Chapter 6). This indicates that the remains of some infants and children were disposed of in some way other than interment in this tumulus. It also raises an interesting question about why certain infants and children were buried here and others were not, a topic that will be discussed further below.

The dearth of adolescents in the tumulus is even more pronounced than that of infants and children. Only 6% of the mortuary population falls between the ages of 13 and 18 (see **Fig. 8.6** and **Table 8.3**). The number of adolescent females should be much higher; death in childbirth was far more common in pre-modern societies than it is today (Acsádi and Nemeskéri 1970:186). In a prehistoric population with a normal distribution, many more females from this age group should be present in the mortuary population.

In all, only eight adolescents were found at Lofkënd: the three females in the wealthiest tombs, an individual tentatively identified as an adolescent female (Tomb LXVI [31]), one tentatively identified adolescent male (Tomb LXXVI [16]), and three whose sex could not be determined (Tombs XXXIII [92], LVIII [37], and LXXXIII [7]). (The identification of sex in adolescent skeletons can be difficult due to incomplete development in diagnostic bones.) There are even fewer adolescent males than females (see **Table 8.3**); however, the presence of many young adult males in the tumulus demonstrates that the absence of adolescent males may be solely dependent on the general exclusion of that age group.

The size of the living population that buried their dead in the Lofkend tumulus must be considered when gauging the size of the burying group at any given time. Calculations of population size generally depend on information for fertility, subsistence modes, and death rates, as well as evidence from households or settlements. Although nothing is known about the subsistence or settlement patterns of the group that interred its dead at Lofkënd, it has been argued that rates of fertility, population growth, life expectancy, and mortality in a community closed to migration are reasonably constant (Acsádi and Nemeskéri 1970; Wood et al. 1992). The limited data about the subsistence mode of this population, the small skeletal sample from Lofkënd, and the necessity of a wide margin of error within our age groupings, however, make the construction of complete mortality or life tables impossible (Hoppa 2002; Konigsberg and Buikstra 1995; Ubelaker 1989; Wood et al. 2002). The average age at death for the Lofkënd population can be roughly determined using individuals that survived childhood (12+ years) and leaving out skeletons that can be classified only as "adult" (Morris 1987:58); this calculation results in an average age at death of around 25 years. The structure of our biological age groups caused some distortion in this calculation, since the number of years spanned by the lowest and highest age within each group varies. This increases the margin of error, because an average age for each group must be calculated to produce a total average.

Without robust information on life expectancy, estimates of population size are necessarily crude. One formula to calculate population size, proposed and utilized by Morris (1987:74), is appropriate for burying groups in which subadults are not fully represented. This method multiplies the total number of skeletons by the average adult age at death (in this case 25). This total is then divided by the length of time the cemetery was in use. Morris (1987:74) then doubles the total to arrive at a rough figure representing the adult living population. Using the Lofkënd data, the total population is 5.5 persons for a usage span of 600 years.

If accurate, this calculation tells us that the tumulus was by no means a communal cemetery used by a large settlement. Rather, the implication is that the population of the tumulus belonged to some smaller group, such as a family, membership in which afforded them the right to be interred in this prominent location. It may also be that a nomadic or transhumant pastoralist population used the tumulus when they passed through the area, though for a variety of reasons this is highly unlikely (see Epilogue).

The prehistoric population of Albania has frequently been characterized as semi-nomadic and transhumant (Hammond 2000:346–47; but see also Halstead 1987a:80). Galaty (2002:111) also considers the Illyrians to have been semi-nomadic pastoralists with summer and winter grazing territories. Unfortunately, very little archaeological evidence has been compiled to support these models of prehistoric subsistence (Wilkes 1992:126–27; Hammond 2000:346; Galaty 2002:109; Papadopoulos, Bejko, and Morris 2007:110; see also the Mallakastra Regional Archaeology Project [MRAP] Interim Report 1998 [*http://river.blg.uc.edu/mrap/MRAP_en.html]*). Recent reanalysis of the issue reveals that groups in this region as well as farther south into Epirus exploited a range of environments and established both temporary and permanent use sites (Douzougli and Papadopoulos 2010:10). Subsistence based on sedentary mixed farming with some herding is considered to be a more likely lifestyle in the prehistoric Balkans (see Chapter 1 and Epilogue; see also Cherry 1988: 21–22; Douzougli and Papadopoulos 2010:9–14; Halstead 1987a:79, 1990:64). This system would require a portion of the population to follow herds for part of the year, or more locally on a daily basis, a practice that could impact the structure of the burial population as well as the social organization of the living and would also explain the physical robusticity of the population. However, stable isotope analysis of the human skeletons (Chapter 7) does not indicate significant levels of animal protein (whether derived from meat or milk) consumed by the population buried at Lofkënd. A mixed subsistence base including legumes and other plant protein seems more likely; thus, evidence for diet does not support a pastoral lifestyle.

Discussion: The Case for Exclusion— Social Deviancy and Early Rites of Passage

Were the occupants of the tumulus at Lofkënd part of an elite social group or excluded from a more desirable burial place elsewhere based on a perception of otherness? There are too few subadults of both sexes in the demographic profile of the Lofkënd tumulus. This suggests that at Lofkënd, a portion of the subadult population was consistently excluded from burial for at least 600 years. The adult profile, however, is relatively complete, provided that the living population was small.

A perception of physical or social otherness within a society can inspire special treatment of the dead (Hubert 2000:2–3; see also Bradley 1995:vii), and mortuary evidence sometimes suggests exclusion based on physical appearance or disability. The perception that the differential treatment of individuals or groups as deviant can be deciphered from the archaeological record is predicated on the idea that social personae are portrayed in mortuary ritual (Shay 1985:221). Differential treatment as a result of deviancy is difficult to distinguish in the archaeological record but, building on the sociological literature, Talia Shay devised a set of hypotheses to define and identify deviancy archaeologically. These hypotheses were then tested with anthropological and ethnographic data (Shay 1985:222). Many sociologists define deviancy as "any extreme conduct that elicits explicit sanctions from the people of a group, who consider it to threaten them or to produce ambiguity regarding the limits of conduct" (Shay 1985: 222; cf. Erikson 1966:6, 1975:13; Lofland 1969:17). Shay's first hypothesis proposed "that among the social dimensions reflected by burial customs, we can expect to find a variety of behaviors in life and different circumstances surrounding deaths that were regarded as being deviant by different societies" (Shay 1985:223). The second hypothesis states that because deviant behavior would have "breached the normal relationships within their community," the social personae reflected in the mortuary evidence of deviant graves would not reflect their "normal social identities" (Shay 1985: 226). Persons whose deviant behavior was "badly evaluated" by their community would have a "shallow social persona" (Shay 1985:226) represented in their funeral. This would include poor or sparse funerary components (Shay 1985:226). Deviancy that was positively received by the community would result in a more complex funerary treatment representing "greater influence, authority, or power" (Shay 1985:226). The third hypothesis is borrowed from Saxe (1970:118–119) and claims that simple societies will react to "volitional and nonvolitional" (Shay 1985:228) deviancy in the same way. This is not the case, however, in a more complex society where "law and medicine are already differentiated, crime and illness are controlled separately" (Shay 1985:228).

When tested, the first hypothesis confirmed that "mortuary practices reflect, among other things, nonhomogeneously defined deviant actions and circumstances among different societies" (Shay 1985: 223). This finding is further refined by the results of testing Shay's second hypothesis. This showed that "the social personae of deceased deviants can be shallow or complex" (Shay 1985:226). Negatively perceived deviant behavior included murder, sorcery, execution for a crime, disease, mental illness, and death in childbirth. Positively evaluated deviancy could include many of the same behaviors, such as death in childbirth, but also included being struck by lightning, being a holy man or woman or other religious person, and dying in battle (Shay 1985:227). The third hypothesis could not be tested but is confirmed by the data obtained for the previous two hypotheses—smaller groups "do not distinguish between volitional and nonvolitional forms of deviancy" (Shay 1985:228).

Building on the work of Shay, other scholars have observed instances of what may be described as "social deviancy" in the archaeological record. For example, mortuary evidence sometimes suggests exclusion based on physical appearance or disability. An excellent example is the Early Iron Age interment of an adult male found in an abandoned and partially filled well in the area that was to become the Classical Athenian Agora (Little and Papadopoulos 1998). This grave was unusual not only for its location but also for the contracted position of the skeleton, a posture not common in Athenian burial practice during the period (Little and Papadopoulos 1998:379). The remains showed signs of injury: cranial fractures and a broken backbone. While the man had recovered from both of these injuries, he remained crippled and possibly mentally impaired as a result (Little and Papadopoulos 1998:385–392). Little and Papadopoulos concluded that there was a clear parallel between the method of burial (inhumation in a well) and the social standing of the deceased: his injuries and impairments may have destined him for social marginality and excluded him from normal burial rites (Little and Papadopoulos 1998:398).

At Lofkënd, the most distinctive mortuary treatments were limited to a segment of the population that was partially excluded from the cemetery. The unusual demographic pattern at the tumulus, the small total number of graves, and the wealth of the six tombs in the focus group raise several interesting questions about inclusion and exclusion.

Both the smaller size of infant graves and the differential preservation due to lesser degrees of calcification in less developed skeletons (which increases the risk of loss through post-depositional processes) could have contributed to their low number at Lofkënd. It is unlikely, however, that post-depositional processes alone are responsible; if this were the case, the number of child and infant graves would have increased proportionally at the lower depths of the tumulus, where graves were less disturbed. Another explanation is that children and infants had not yet achieved a status that entitled them to the same mortuary treatments as older members of the community. Differential mortuary treatment for infants and children is well summarized by Eleanor Scott (1999) (see also Baxter 2004:97–98; Binford 1971:21–22; Garland 1985:82–86; Houby-Nielsen 2000:152; Liston and Papadopoulos 2004:25; Morris 1987:18–19, 105; Oakley 2003:177; Papadopoulos 2000:111). These sources indicate that the age of 3 years was often the critical stage at which a child could receive the same burial rites as older individuals. Christopher Carr (1995:156) found that the age distinction between adult and child is one of the two most significant factors determining burial practices (the other is hierarchical rank) (also see Crawford 2000:173). In fact, children are often buried outside of public areas rather than in cemeteries. Lewis Binford (1971:21–22) argued that burial under house floors suggested infants and young children had not yet joined the public sphere of society. Frano Prendi (1982:235) notes that during the Neolithic and Early Bronze Age at the site of Maliq in southern Albania, infants were buried under house floors.

In ancient Greece, the first burial distinctions between infants and older individuals appeared in Athens during the Submycenaean and Protogeometric periods (Houby-Nielsen 2000:152; Morris 1987: 18–19). In the Submycenaean period, infants were buried in pits built of small stones, whereas adults were buried in cist tombs. In the Protogeometric, they were inhumed, while the majority of adults were cremated (Houby-Nielsen 2000:152–153; Morris 1987: 18–19; Oakley 2003:177). Little and Papadopoulos (1998) examined skeletal remains deposited in wells in the Aegean and found the remains of one child and approximately 450 fetuses, neonates, or infants within a partially filled well of the Hellenistic period in the Athenian Agora (Papadopoulos 2000: 110; Rotroff, Little, and Snyder 1999). Papadopoulos (2000) observed that the human remains in this Hellenistic well correspond to the overall pattern of Greek burial custom by emphasizing the importance of a child's third birthday. It was at this age that boys would participate in the *choes* ceremony during the Athenian Anthesteria festival (Hamilton 1992). This ceremony was named for a pottery vessel called a *chous* (plural *choes*) that is shaped like a miniature wine jug. These vessels are decorated with conventionalized images of children (predominately boys) wearing amulets and frequently wreaths and are associated with ritual/festival foods (Demand 1994:188; Ham 1999:205; Hamilton 1992:83–114). The ritual took on a particular importance during the last quarter of the fifth century and the first quarter of the fourth century BC when Athens suffered large losses in the Peloponnesian War and from a plague; it was during this period that there was the greatest production of *choes* vessels (Ham 1999:201). *Choes* was the ceremony that marked the important transition point when young boys would be introduced into the family clan (Demand 1994:7, 10; Papadopoulos 2000:111) and, as such, marked an observable, public transition from one social role to another. It was a rite that allowed the community to enact ritually—and thereby ensure—the perpetuation of their society through their offspring.

Neither exclusive use by a small community nor preferential use by a larger community can completely explain the demography at the Lofkënd tumulus. Any interpretation must consider social behaviors and structures that could lead to the exclusion of both male and female adolescents as well as infants and children. Without evidence for alternate disposal practices used by the same community, it is impossible to know why a portion of the population was excluded from burial at Lofkënd. Deviancy does not always elicit a negative response, and exclusion is not always manifested in the unfavorable treatment of the deviant, but it is unlikely that such a high proportion of the subadult population would be considered deviant. Rather, the population of the tumulus may be a case for inclusion rather than exclusion. The inclusion of just a few adolescents, infants, and children at the Lofkënd tumulus may have been due to some special status or role that they played or were destined to play in the community and/or the special circumstances of their birth and death.

DISCUSSION: THE CASE FOR INCLUSION—HUMAN SACRIFICE, ASCRIBED, ACHIEVED, AND UNACHIEVED STATUS, AND MATRILINY

The inclusion of some infants, children, and adolescents in the tumulus at Lofkënd may be just as significant as the exclusion of others. They may have been included due to the status of their family or male relative, or their presence may indicate a status that they were expected to achieve themselves. Three of the tombs in the focus group contained adolescent females and adult males. Whatever the relationships between young women and the adult males—father, uncle, brother, husband, husband-to-be—it is likely that there is a link between these relationships, the age of the female, and the wealth of the tomb.

The contemporaneous burial of husband and wife or concubine is often taken as evidence of human sacrifice or suttee (voluntary suicide of a widow or concubine). Indeed, wealth accompanying a female is usually attributed to the status of a male in such cases. The role of women as grave offerings has been documented in a number of different cultures including India (Davies 1981:100, 103), Bronze Age

China (Davies 1981:104–105; Jiang 2004:124; Jiao 2001:57), the Scythians (Herodotos 4.71; Hughes 1991:44; see also Jones-Bley 2000:129–130; Lincoln 1991:13, 202), and the Inca and pre-Inca cultures of South America (Davies 1981:261; Donnan 1995:150–151; Sutter and Cortez 2005:526; Verano 2008:1048, 1051).

Dennis Hughes (1991:45) provides a set of criteria for determining the likelihood that an individual died as a result of suttee:

> 1) The sex (and age) of the two persons should be determined, not merely from the grave goods but from physical anthropological analysis; 2) The simultaneity of the burials should be established or at least shown to be probable; 3) There should be physical evidence of the violent death of the woman; 4) There should be some indication of the special nature—or what has been called the "ritual character"—of the burial, such as the subordination of the woman's burial to the man's by its position, or a distinction between the two persons regarding the amount of grave gifts or manner of burial; 5) If "suttee" is to be shown to be a custom, then we should expect a certain frequency of man–woman burials in a given cemetery or in tombs of the same type in the same region and period; we might also expect some uniformity in the manner of burial; 6) If it was only the boon of kings, nobles, or warriors to be accompanied in the afterlife by their wives or concubines, then their burials should be clearly distinguishable from those of ordinary citizens.

The extremely small number of graves at Lofkënd containing adolescent females and adult males makes it difficult to argue either for or against a case of human sacrifice; in any event, the existing evidence does not satisfy Hughes's parameters. One tomb contained the simultaneous burial of a male and female (Tomb XLVIII [52]: see **Figs. 3.154–3.155**), but in Tomb LXX (17) (**Figs. 3.241–3.242**), where the skeleton of the male was pushed aside completely, the interments were clearly not simultaneous. No evidence of violent death could be detected in any of the graves or skeletons of this group. If any subordination existed in a multiple burial, it may well have been the subordination of the male whose remains were, in several cases, moved aside for or located below the skeleton of the female, though it is equally possible that this was not subordination but simply a product of who died first. Finally, the most "clearly distinguishable" tombs at Lofkënd are those of adolescent females, children, and infants; male tombs are not significantly different or clearly distinguished from those of the rest of the cemetery population.

The three wealthy child and infant graves (Tombs XVII [72], XVIII [73], and XXVIII [77]) also complicate the argument for suttee. The main difference between these and the three tombs containing adolescent girls is the absence of adult males. Age seems to be the significant factor here; in the focus group, only adolescent females with head ornaments were buried with adult males. This may indicate that regardless of the similarity in ornaments worn by the occupants of the tombs in the focus group, the reasons for their presence are different. Infanticide and the sacrifice of children are common in the mythologies of many cultures, including ancient Greece and India, as well as in Judaism and Christianity (Davies 1981; Hughes 1991; Papadopoulos 2000:97). Where clear archaeological evidence does exist for the sacrifice of children, however, the act of sacrifice and the location of the remains imply a ritual behavior dedicated to a god or gods and aimed at eliciting a specific result. These sacrifices are most often found in ritual locations rather than in cemeteries (Covey 2008: 827; Tierney 1989:30–31; Wilson et al. 2007:16456). If all six of the individuals wearing head ornaments died as a result of suttee, each should be buried with a husband, father, or some type of high-status male. If infants and children were sacrificed for some other reason, it seems more likely that they would be buried in a location that was linked with the purpose or deity to whom they were sacrificed, rather than in a cemetery of a particular social group or community.

Human sacrifice is not a convincing explanation for the relationship between the adult males and adolescent females in Tombs XXI (55), XLVIII (52), and LXX (17). It is significant that more objects appear in the tombs of women and children at Lofkënd than in those of men, but this evidence must be viewed in combination with both the demographic evidence and the fact that the wealth of the tomb is associated with the female (or child) rather than a male.

The Early Iron Age burial of the "Rich Athenian Lady" in the Athenian Agora was originally excavated in 1967 but was reexamined by Papadopoulos and Liston in 2004. In her osteological analysis of the cremated remains, Liston identified fetal bones, an indicator that the woman was pregnant when she died or had given birth soon before her death (Liston and Papadopoulos 2004:19). The contents of the tomb made it perhaps the wealthiest Early Iron Age grave ever found in Athens (Liston and Papadopoulos 2004:12). Based on this evidence, the authors concluded that it may have been her state of pregnancy or recent maternity that entitled her to this elaborate mortuary treatment.

While the tomb of the Rich Athenian Lady is named after the adult occupant, it is also the tomb of an infant, albeit a fetus or neonate. The social status of the child could be viewed as ascribed and based on that of the mother (cf. Langdon 2008:64). This would indicate that the mother's death before, during, or just after pregnancy afforded her a special status that was extended to the child. The infants and children who were given burial rites at the Lofkënd tumulus may also have been the offspring of women who held a special status, played a significant social role, or who died in some transitional state such as childbirth.

At Lofkënd, there are seven graves containing at least one adult and one child. Two (Tomb XII [88] and Tomb XLII [59]) held the remains of both an adult female and two children. The other five tombs contained children with an adult male (Tomb LXXXIV [02]), adults of both sexes (Tomb LXXXV [10]), or adults whose sex could not be determined (Tomb LXIX [27], Tomb LXXXIII [07], and Tomb LIII [63]). No discernible biological traits are shared between the adults and children in these tombs, but traits are shared between children (Chapter 6).

The analysis of skeletal traits, mtDNA analysis, and strontium isotope analysis are some of the methods used to reconstruct group genetic variance as well as post-marital residency patterns (Konigsberg and Buikstra 1995; Mooder et al. 2005, 2006; see also Appendix 1 [following Chapter 6] for a report on the unsuccessful extraction of DNA at Lofkënd). These types of analyses can help to determine the genetic relationships among members of a single community or multiple communities and, therefore, establish patterns of social organization such as matrilocal or patrilocal residency. An association among wealth, tomb type, status, and matrilineal affinity was shown through mtDNA at the Neolithic cemetery at Lake Baikal, Siberia (Mooder et al. 2005:631). Mooder et al. determined that the occupants of certain graves in one discrete cluster (Cluster 2) within the cemetery were related through the maternal line. Furthermore, these graves were the wealthiest in the cemetery; they were all single interments (of both males and females) and were located in an elevated position above the other graves, which tended to be poorer and contained multiple individuals (Mooder et al. 2005:623). The study also identified more subadult burials within this group than in other areas of the cemetery (Mooder et al. 2005: 623). While the wealthiest graves at Lake Baikal did not belong exclusively to women, mtDNA showed that the subgroup that evidently represents the elite of this community was clearly related by matrilineal descent (Mooder et al. 2005:631). The higher number of subadults in this cluster may also indicate that the usual taboos applied to their burial did not apply to higher-ranking family groups.

The pattern seen at Lake Baikal is not unique. Isotopes of strontium, carbon, and oxygen from tooth enamel recovered at Khok Phanom Di in Thailand support conclusions of a matrilocal residence pattern during certain periods (Bentley et al. 2007; Hage and Marck 2003; for stable isotope analysis of human bone samples from the Lofkënd, see Chapter 7). The matrilineal society represented in the cemetery at Khok Phanom Di shares a number of characteristics with Lofkënd, such as the wealth of women's graves and their association with the graves of infants (Bentley et al. 2007; Parker Pearson 1999: 119). As in the cemeteries at Lake Baikal and Lofkënd, poorer graves were also present, indicating that not all members of the lineage possessed the same status or role (Mooder et al. 2005:623; Parker Pearson 1999:119). Unlike the children and infants at Lofkënd, however, those at Khok Phanom Di were buried in the same grave with the richly appointed females and were probably their offspring. In contrast, neither the adolescent nor adult females at Lofkënd were typically interred with infants or children (except for Tombs XII [88] and XLII [59]), and the tombs in the focus group belonged to both adolescents and children but not adults. The high-status burials of the Khok Phanom Di children occurred by virtue of the adult females' rank within the matriliny. At Lofkënd, the only apparent relationship between the three rich tombs of infants and children and any other burials in the tumulus was represented in the three tombs of adolescent females also in the focus group—and that relationship is based solely on the similarity of the grave goods rather than physical proximity or consanguinity. The Khok Phanom Di example provides a strong argument for attributing the high-status burial of infants to their mother's rank within the matriliny. This interpretation, however, should be tempered by the possibility that the manner of burial and associated wealth may also be indicative of the future role of the child and not to the status of the mother (Baxter 2004:96; Oakley 2003:177; Pader 1982:61–62; see also Langdon 2008:64; Liston and Papadopoulos 2004: 29–31) and would, therefore, represent both an ascribed status as well as one that is unachieved. This discussion can also be applied to the evidence from Lofkënd where, in contrast, the three wealthy child tombs (XVII [72], XVIII [73], and XXVIII [77]) are not associated with adults. The social function that their wealth played in the mortuary ritual cannot be directly attributed to an adult family member. The possibility that it is instead related to an unachieved status or role of their own must be explored.

The poetry of Homer may provide a closer (both geographically and chronologically) example of matriliny or, more accurately, uxorilocality. Margalit Finkelberg, as well as Kenneth Atchity and Elizabeth Wayland Barber, have argued that the women of Homer's Greece were the key to both descent and property inheritance (Atchity and Barber 1987; Finkelberg 1991; see also Pomeroy 1975:19, 23). Two often-cited examples are Helen of Sparta and Penelope, the wife of Odysseus. Helen's husband Menelaos succeeded her father as king of Sparta by virtue of his marriage to her. Neither of her two brothers made a claim to the throne. In fact, they helped to choose her husband—and thus their father's successor—and then left home to find wives (Finkelberg 1991:13). Perhaps even more telling is Menelaos's behavior after carrying his wife back from Troy. He did nothing to punish her for her adultery; rather, they appear to have ruled in harmony after their return to Sparta (Atchity and Barber 1987:20). It seems that Menelaos had no recourse: his kingdom and the kingly descent of Sparta came through Helen (Atchity and Barber 1987:20).

Penelope's story as related in the *Odyssey* is similar: she fended off suitors until Odysseus returned to his throne in Ithake, but she had both a son and a father-in-law in residence during her husband's absence. Why didn't they take the throne in Odysseus's stead? Her son, Telemachos, appears to have had no expectation of sitting on his father's throne, but the suitors had every expectation of gaining it through a union with Penelope (Finkelberg 1991: 306–307).

In a matrilineal society, property and descent pass through the female line, and males must marry outside the community or move back to their mothers' residences. If a matrilineal social organization existed in Late Bronze Age Albania, it could, at least in part, explain the shortage of adolescent males in the burial population at Lofkënd. The Navajo people of North America provide a useful ethnographic analogy. When a Navajo girl reaches the age at which she is eligible for marriage, she participates in the Kinaaldá ceremony, during which her responsibilities as an adult woman are explained to her (Stone 2000: 133). This ceremony is a celebration of her transition into adulthood. The Navajo practice a matrilocal form of matriliny in which the groom moves to the residence of the bride's family, bringing a payment of bride wealth in the form of livestock (Aberle 1961:124; Stone 2000:131). The bride and groom then usually reside in a camp limited to members of the clan and consisting of only a few residential structures (hogans) (Aberle 1961:109–110; Stone 2000:128–129). This modern example illustrates several points that are relevant to the evidence from Lofkënd: (1) the transition from adolescent to adult is an important rite of passage which impacts the entire community; (2) while bride wealth is paid, it is in the form of livestock, wealth that may be undetectable in the grave; and (3) Navajo matrilineal groups live in small communities that could produce a sparse burial population.

While descent, property, and perhaps title are passed through the female line in a matrilineal society, matriliny is not necessarily matriarchy, and as the Homeric examples illustrate, kingship and the right to rule may be restricted to males. In fact, the presence of three (related) males in the central tomb at Lofkënd may further support a matrilineal/matrilocal burial pattern by emphasizing the importance of the male role in that community.

Only four adolescent women were interred at Lofkënd, and three of the four were buried with wealth. In order to reconcile this demographic pattern with a matrilocal or matrilineal model, a substantial theoretical leap must be made: the six individuals in wealthy tombs were females in the direct line of descent, and their death represented a distinct loss to the community. This loss and the social significance of their age at death were symbolically represented by the wealth of their graves. Other adolescent females were not interred in the tumulus because they were not in the direct line of descent. Adolescent males married outside their mother's family and were therefore buried in the cemetery of their wives' family. The adult males in the Lofkënd tumulus were either from the lineage group or had married into it. This explanation, unfortunately, cannot explain the presence of a fourth adolescent female buried without wealth. The demography of the Lofkënd tombs could conceivably be attributed to a matrilineal, matrilocal society in which male exogamy is practiced and lineal descent is through the female line. This is, however, an imperfect and, at this point, highly conjectural explanation.

GRAVE GOODS

Each tomb at Lofkënd appeared to be unique—no two had identical dimensions, nor was there ever an exact repetition of other variables such as body position or grave goods. The most distinctive group of tombs comprise the focus group, and they were characterized by the age and sex of their occupants, quantity and type of grave goods they contained, and placement of ornaments on or around the cranium of the deceased.

Less than half of the 85 prehistoric tombs contained any grave goods, but of the 27 females buried in the tumulus, 17 were associated with grave goods, whereas only 13 of the 38 males in the tumulus were buried with material wealth. The mean number of goods per tomb, as well as per person, was one. By contrast, the wealthier tombs contained up to 11 objects, consisting of pottery, bronze, and iron ornaments as well as beads of various materials.

The six wealthy tombs are illustrated in **Figures 3.50–3.51**, **3.55–3.56**, **3.63–3.64**, **3.88–3.91**, **3.154–3.156**, and **3.241–3.242**, and the grave goods found in them are illustrated in **Figures 3.52–3.54**, **3.57–3.58**, **3.65–3.73**, **3.92–3.96**, **3.157–3.159**, and **3.243–3.250**. While some of these tombs contained grave goods that could not be definitively associated with the individual wearing the head ornaments (Tombs XVII [72], XVIII [73], XXI [55], XLVIII [52]), the bulk of the grave goods were firmly associated with a female, child, or infant, and were found on or around the crania of the deceased. Two of these tombs (XXI [55] and LXX [17]) contained adolescent women wearing bronze headbands. Two additional headbands were found in Tombs XVII (72) and XVIII (73) adorning the heads of a 7-year-old and a 3-year-old, respectively. Tomb XXVIII (77) contained a wheel-shaped ornament of bronze, placed on the cranium of a 6-year-old child. The last tomb in this group (Tomb XLVIII [52]) contained the only electrum objects found in the Lofkënd tumulus. These ornaments, a pair of electrum foil disks incised or embossed with concentric circles, were located against either temple of an adolescent female and may have been worn as earrings or, more likely, were attached to a headband or other ornament of organic material otherwise lost. While this young woman was not accompanied by the profusion of bronze ornaments that characterized the other wealthy tombs, the rarity of gold or electrum, the age of the deceased, and the location of the disks on the cranium warrants the inclusion of this tomb in the group of wealthy tombs.

The existence of a correlation between adolescent women and children and head ornaments is tangentially supported by similar relationships that appear between age and sex and other object and material types.

While the group of wealthier tombs was distinctive within the tumulus, **Figure 8.8** graphically illustrates that, in general, women and children received more grave goods than men (this chart excludes skeletons of indeterminate sex). **Table 8.5** shows all skeletons from the tumulus, their age at death, and the grave goods associated with them. Particularly noticeable are the low percentage of adolescent females (age 11–18) in the population and the greater wealth of their tombs. Females between the ages of 25 and 45 have a relatively even distribution of grave goods per individual, as do females in the 18–25 year category. By contrast, males in every age category tend to have fewer goods per person. Infants and children, like adolescent females, have a very high number of grave goods, but these are distributed more evenly throughout the population. This distribution is also represented in **Figure 8.9**, which sorts the same data also using social age categories but includes skeletons that could not be sexed.

The following discussion refers to **Figures 8.10** through **8.15**, which treat objects as types sorted by material of composition, type of object, or, in the case of pottery and bone pins, both. The four most common materials used for objects at Lofkënd are iron, bronze, bone, and ceramic. The only patterning of object location within the tumulus was that of

TABLE 8.5 Analytic table of skeletons and grave goods

Tomb	Skeleton	Age	Sex	Biological Age Group	Pottery	Fibulae				Pins				Bronze, Varia						Beads								Ornaments		Miscellaneous				
						Iiron	Bronze Spectacle	Bow Bronze	Bimetallic	Iron	Bone	Bronze	Bimetallic	Diadem	Earring	Ring	Disk	Disk iron	Spectacle Pendant bronze	Iron	Carnelian	Stone	Amber	Glass	Pottery	Bronze	Spindle whorl bead/button	Bronze	Gold	Knife/ Dagger iron	Implement / weapon iron	Other iron	Other bronze	Total grave goods
(1) LXXXI	14	18–22	M	Young adult																														0
(2) LXXXIV	38a	2–5	C	Child																														0
	38	35–45	M	Mid-adult																										1				0
	18	23–26	M	Young adult																														1
(4) LXXX	32	20–30	M	Young adult						1										1												1		3
(5) LXXVIII	51	20–25	F	Young adult						1																								1
(6) LXXIX	54	20–25	I	Young adult																														0
(7) LXXXIII	60	15 ± 3 yrs	I	Adolescent	1																													1
	60b	35–45	I	Mid-adult																														0
	60c	>45	I	Older adult																														0
	60a	19–34	I	Young adult																														0
(9) LXXXII	105	19–34	I	Young adult																														0
	85	26–32	M?	Young adult																														0
	106	26–32	M?	Young adult																														0
(10) LXXXV	145	Adult	M	Adult																														0
	145a	<1 yr	C	Infant																														0
	90	35–45	F	Mid-adult																														0
(12) LXVII	100	>35	M	Mid-adult																														1
	171	23–27	F	Young adult																							1							0
(13) LXVIII	103	35–45	F	Mid-adult	1													1																2
	136	20–25	I	Young adult			1			1																								0
	149	20–25	I	Young adult																														2
	103a	20–30	I	Young adult																														0
(16) LXXVI	125	adult	I	Adult																													1	1
(17) LXX	126	17–20	F	Adolescent	1	1	1			1			1	1		2								1				2						11
	197	>45	M	Older adult																														0
(18) LXXVII	139	35–45	M	Mid-adult																														0
(24) LXXII	162a	10 ± 2.5 yrs	C	Child																														0
	162	25–35	I	Young adult																														0

Table 8.5 (*continued*). Analytic table of skeletons and grave goods

Tomb	Skeleton	Age	Sex	Biological Age Group	Pottery	Fibulae				Pins				Bronze, Varia						Beads								Ornaments		Miscellaneous				
						Iiron	Bronze Spectacle	Bow Bronze	Bimetallic	Iron	Bone	Bronze	Bimetallic	Diadem	Earring	Ring	Disk	Disk iron	Spectacle Pendant bronze	Iron	Carnelian	Stone	Amber	Glass	Pottery	Bronze	Spindle whorl bead/button	Bronze	Gold	Knife/ Dagger iron	Implement / weapon iron	Other iron	Other bronze	Total grave goods
(26) LXXIII	169	Adult	M?	Adult																														0
	169a	Adult	I	Adult																														0
(27) LXIX	551a	6 yrs	I	Child																														3
	551	6 ± 2 yrs	C	Child			1	1							1																			1
	174	20–25	I	Adult						1																								0
(28) LXXI	177	8 ± 2 yrs	C	Child																														0
(29) LXXIV	230	35–45	M	Mid-adult																														0
	210	20–30	M	Young adult																														0
	231	25–35	M	Young adult																														0
(30) LXV	213	18–23	F	Young adult		1																												1
(31) LXVI	216	15 ± 2 yrs	F?	Adolescent																2														2
(32) LXII	220	Adult	I	Adult																														0
(33) LXXV	223	30–40	M	Mid-adult																														0
(34) LXI	225	Adult	I	Adult																														0
(35) LXIII	234	>45	F?	Older adult	1					2																								3
(37) LVIII	243	15 ± 3 yrs	I	Adolescent			1			2																						1		4
(38) LIX	247	35–45	M	Mid-adult																														0
(40) LIV	252	35–45	I	Mid-adult																														0
(41) XLVII	256	18–21	F	Young adult																														0
(42) XLVI	263	20–25	F	Young adult	1																													1
(43) LVI	269	20–25	F	Young adult	1																													1
	296	20–25	M	Young adult					1																									1
(44) LX	273	19–34	I	Young adult																														0
(46) L	284	4 ± 1 yr	C	Child						1																								1
(47) XXII	294	35–45	M?	Mid-adult																														0
(49) XIII	305	25–35	F	Young adult																														0
(50) XX	308	20–25	F	Young adult																														0
(51) XLIX	315	3 ± 1 yrs	C	Child																														0

TABLE 8.5 (*continued*). Analytic table of skeletons and grave goods

						Fibulae				Pins				Bronze, Varia						Beads								Ornaments		Miscellaneous				
Tomb	Skeleton	Age	Sex	Biological Age Group	Pottery	Iiron	Bronze Spectacle	Bow Bronze	Bimetallic	Iron	Bone	Bronze	Bimetallic	Diadem	Earring	Ring	Disk	Disk iron	Spectacle Pendant bronze	Iron	Carnelian	Stone	Amber	Glass	Pottery	Bronze	Spindle whorl bead/button	Bronze	Gold	Knife/ Dagger iron	Implement / weapon iron	Other iron	Other bronze	Total grave goods
(52) XLVIII	318	15 ± 3 yrs	F	Adolescent	1																								2					3
	372	35–45	I	Mid-adult																														0
	366	20–25	M	Young adult	1															1														0
(53) LV	321	7 ± 2 yrs	C	Child	1	1			1											3		1												7
(54) XIX	323	35–45	F	Mid-adult																														0
(55) XXI	341	15 ± 3 yrs	F?	Adolescent										1			1			3				1								1	1	8
	326	20–25	M	Young adult	1			1																								1	1	4
(56) XXIII	329	8 ± 2 yrs	C	Child						1																								1
(57) XLI	331	20–30	I	Young adult																														0
(58) LVII	336	35–45	I	Mid-adult																														0
(59) XLII	339	4 ± 1 yr	C	Child																														0
	340	2 yrs ± 8 mo	C	Child																														0
	338	20–25	F	Young adult																								2						2
(60) XLV	352	30–40	F	Mid-adult																														0
	345	>55	M	Older adult																											1			1
	351	18–25	M	Young adult																														0
(61) LXIV	347	4 ± 1 yrs	C	Child																														0
(62) XLIII	371	2 yrs ± 8 mo	C	Child																														0
	349	18 mo ± 6 mo	C	Infant																														0
	356	1 yr ± 4 mo	C	Infant																														0
(63) LIII	359a	5 ± 1.5 yrs	C	Child																														0
	359	25–30	I	Young adult	1										2		1		1		1			3										9
(64) I	362b	>45	M	Older adult																														1
	362a	18–23	M	Young adult																														0
	362	25–30	M	Young adult							1																							0
(65) XLIV	365	35–45	M	Mid-adult						1																								1
	394	18–25	F	Young adult						1																								1

TABLE 8.5 (*continued*). Analytic table of skeletons and grave goods

						Fibulae				Pins				Bronze, Varia						Beads								Ornaments		Miscellaneous				
Tomb	Skeleton	Age	Sex	Biological Age Group	Pottery	Iiron	Bronze Spectacle	Bow Bronze	Bimetallic	Iron	Bone	Bronze	Bimetallic	Diadem	Earring	Ring	Disk	Disk iron	Spectacle Pendant bronze	Iron	Carnelian	Stone	Amber	Glass	Pottery	Bronze	Spindle whorl bead/button	Bronze	Gold	Knife/ Dagger iron	Implement / weapon iron	Other iron	Other bronze	Total grave goods
(66) XXXIX	369	25–35	F	Young adult	1		1																											2
(67) XL	383	>30	M?	Mid-adult																														0
(68) XVI	386	6 ± 2 yrs	C	Child																														0
(69) LII	390	15 ± 3 yrs	M?	Adolescent																														0
(70) XXX	398	9 mo ± 3 mo	C	Infant																														0
(71) XIV	405	6 mo ± 3 mo	C	Infant	1																													1
(72) XVII	408a	4 ± 1 yr	C	Child																														7
	408	7 ± 2 yrs	C	Child	1									1	2																			0
(73) XVIII	421	5 ± 1.5 yrs	C	Child																														2
	412a	3 ± 1 yr	C	Child																														0
	412	4 ± 1 yr	C	Child										1			1																	0
(74) XXVI	416	>45	F	Older adult	1																													1
(75) XXXVI	420	Adult	I	Adult																														0
(76) XXXVII	424	Adult	I	Adult																														0
(77) XXVIII	429	3 ± 1 yr	C	Child				1											1			1		6				2						11
(78) LI	434	19–34	M?	Young adult																														0
(79) XXXVIII	447	Adult	F?	Adult																														0
	439	30–40	M	Mid-adult																												1		1
(80) XV	444	6 mo ± 3 mo	C	Infant	1																													1
(81) III	454	>35	M	Mid-adult																														0
(82) XXVII	461	>35	F?	Mid-adult																														0
	457	>35	I	Mid-adult								1																						1
	457a	18–25	I	Young adult																														0
(83) XXIX	469	6 mo ± 3 mo	C	Infant																														0
(84) XXXV	472	23–27	M	Young adult																												1		1
(85) XXIV	478	3 ± 1 yr	C	Child																														0

TABLE 8.5 (*continued*). Analytic table of skeletons and grave goods

						Fibulae				Pins				Bronze, Varia						Beads								Ornaments		Miscellaneous				
Tomb	Skeleton	Age	Sex	Biological Age Group	Pottery	Iiron	Bronze Spectacle	Bow Bronze	Bimetallic	Iron	Bone	Bronze	Bimetallic	Diadem	Earring	Ring	Disk	Disk iron	Spectacle Pendant bronze	Iron	Carnelian	Stone	Amber	Glass	Pottery	Bronze	Spindle whorl bead/button	Bronze	Gold	Knife/ Dagger iron	Implement / weapon iron	Other iron	Other bronze	Total grave goods
(86) XXXI	489	>40	I	Older adult																							2							2
(87) XXXIV	504	Adult	M	Adult																														1
	498	45–55	M	Older adult						1																								0
(88) XII	495	9 ± 2 yrs	C	Child																														0
	501	1 ± 18 mo	C	Infant																														0
	499	20–25	F	Young adult							1																							1
(89) XXXII	524	18–25	M	Young adult																														2
	508	20–30	M	Young adult																										1	1			0
(90) XXV	512	6–10 yr	C	Child																														0
(91) II	516	>45	M	Older adult																														0
(92) XXXIII	523	16 ± 2 yrs	I	Adolescent																														0
(93) XI	540	9 mo ± 3 mo	C	Infant																														0
(94) IX	547	Adult	F	Adult																														0
(95) X	554	30–40	F	Mid-adult																														0
(96) V	569	20–25	F	Young adult							1																							1
(97) VI	573	3 ± 1 yr	C	Child								1																						1
(98) IV	581	20–25	M	Young adult							1																							1
(99) VII	583	7 ± 2 yrs	C	Child								1																						1
(100) VIII	591	3 ± 1 yr	C	Child																						1								1

bone pins, which were all recovered from the southeastern portion of the tumulus (Tombs I [64], IV [98], V [96], and XII [88]) (see **Fig. 8.1**), and among the earliest burials in the tumulus.

Pottery is the most common type of object found at Lofkënd (there were 17 pots found as grave offerings), followed by iron pins (14), glass beads (10), and iron beads (10). No other object type appears more than 10 times. Young adults of both sexes received a relatively high number of grave goods though not as high as adolescent females, infants, and children.

When sorted by age and sex, the pottery displays a pattern that indicates a preference for use in female and child graves (**Figs. 8.10–8.11**; **Table 8.6**). In each of the male graves containing pottery (Tombs XXI [55], XLVIII [52], and LVI [43]), there was a female in the grave as well, making it difficult to determine whether pots were ever intended as grave goods for males. Most tombs contained only one pot, but two (Tombs XVII [72] and XLVIII [52]) contained two.

In most tombs, the pots were located to one side of the cranium. In Tombs XLVIII (52), LXIII (35) (associated with the adolescent female), XXVI (74), LIII (63), and LV (53), the pot was placed to the right of the cranium (although in Tomb LIII [63] it may have been against the forehead). In Tombs XXI (55), XXXIX (66), XLVI (42), and LVI (43), the pot was placed to the left of the cranium. In Tomb XXI (55), the pot has been associated with the adolescent female skeleton but could be associated with the male. It was closer to the female cranium than to the male but was slightly below rather than directly next to it. In Tomb LVI (43), the pot was closer to the female cranium and slightly below the male cranium, which was disturbed, and has therefore been associated with the female. The pots in Tombs XLVIII (52), LXVIII (13) (associated with the young adult male), XIV (17), and XVII (72) were found near cranial fragments, but on which side of the cranium they were placed cannot be determined. In the remaining three tombs containing pottery, the pots were found at some distance from the cranium (Tombs XV [80], the second pot from Tomb XVII [72], and LXX [17]).

While there is a clear preference for placement of pottery near the cranium, there is no correlation with age or sex and no preference for which side of the cranium the pottery was placed. The placement of the pot may, however, be related to the type of pot, though the number of pots is too small to support this observation. In Tomb LXX (17), a stemmed goblet was found placed above the upper femurs/pelvis of the adolescent female (**Figs. 3.241–3.242a**). This pot shape is unique in the tumulus (see Chapter 9). Tomb XV (80) contained an infant and a baby feeder, a ceramic type that is also unique at Lofkënd (**Figs. 3.47, 9.22, 9.56**; no. **9/160**).

Iron pins (**Figs. 10.15–10.18**), the second most common object type in the tumulus, do not appear to have an association with the sex of the deceased but may have some correlation with age. **Figure 8.12** shows iron artifacts sorted by social age but separating infants into a separate group. Iron pins appeared in the tombs of children, adolescent females, young adult males, young adult females, mid-older adult males, and mid-older adult females, but *never* in the tombs of infants. Bronze objects, including a bronze pin, were found in infant tombs, so any restriction must have been related to material rather than function.

Whereas iron is found with individuals in all age groups, with the exception of infants, bronze appears primarily in the tombs of young individuals. Bronze finds tend to decrease as age increases. This distribution is apparent when sorted by either biological or social age. **Figure 8.13** shows bronze finds by social groups. The highest occurrences are in child, infant, and adolescent female tombs, respectively.

The final object type that can be grouped by material but also by function is the bone pin. The pins recovered from the tumulus can be seen in **Figure 10.21**. These were found in only four tombs (Tombs I [64], IV [98], V [96], and XII [88]), which number among the earliest burials in the tumulus. Tomb I (64) was the central (and initial) tomb in the tumulus (see Chapter 21). As already noted, Tomb I (64) was a secondary burial, and the disturbed nature of the skeletal material makes it impossible to associate the pin with any of the three or more individuals buried there. Tomb XII (88) was a multiple burial; the bone pin is clearly associated with the remains of the young adult female, found on her torso (**Fig. 3.37**). Tomb V (96) contained the remains of a young adult female, and in this case the pin was found on the cranium. The final pin comes from Tomb IV (98), also a single inhumation but, notably, a young adult male. The small number of pins makes any discussion of association with adults or young adults statistically irrelevant, but it is interesting that the four pins were all found in tombs excavated at a relatively low elevation in the tumulus. The radiocarbon dates from

TABLE 8.6 Distribution of whole pottery vessels found in burials

Skeleton #	Tomb	Trench	Single/multiple inhumation	Age	Sex	Age group	Object numbers	Pottery
60	LXXXIII (7)	4	Single	15 ± 3 years	I	Adolescent	**10/16** (P047)	1
103/103a	LXVIII (13)	2	Multiple (quadruple)	Mid-age/20–30	F/I	Mid-adult	**10/4** (P077)	1
126	LXX (17)	3	Multiple (double)	17–20	F	Adolescent	**10/15** (P082)	1
234	LXIII (35)	4	Single	>45	F?	Older adult	**10/7** (P166)	1
263	XLVI (42)	2	Single	20–25	F	Young adult	**10/93** (P223)	1
269	LVI (43)	4/1	Multiple (double)	20–25	F	Young adult	**10/95** (P228)	1
318	XLVIII (52)	1	Multiple (triple)	15 ± 3 years	F	Adolescent	**10/8** (P277)	1
366	XLVIII (52)	1	Multiple (triple)	20–25	M	Young adult	**10/96** (P276), **11/94** (SF288)	1
321	LV (53)	1	Single	7 ± 2 years	C	Child	**10/14** (P256)	1
341	XXI (55)	4	Multiple (double)	15 ± 3	F	Adolescent	**10/91** (P253)	1
359	LIII (63)	1	Multiple (double)	25–30	I	Mid-adult	**10/94** (P322)	1
369	XXXIX (66)	1	Single	25–35	F	Mid-adult	**10/3** (P304), **11/16** (SF315)	1
405	XIV (71)	4	Single	6 months ± 3 months	C	Infant	**10/159** (P303)	1
408	XVII (72)	1	Multiple (triple)	7 years ± 2 years	C	Child	**10/1** (P350), **10/90** (P422)	2
416	XXVI (74)	1	Single	>45	F	Older adult	**10/5** (P326)	1
444	XV (80)	2/6/1	Single	6 months ± 3 months	C	Infant	**10/160** P362	1

F, female; I, indeterminate; M, male.

these four tombs show that they are among the earliest in the tumulus, dating from the fourteenth to twelfth centuries BC (Chapter 4).

Fibulae (shown in **Figs. 10.5–10.13**) do not appear with individuals above the age of 25, an age which, interestingly, coincides with the end of young adulthood in the social age groups used in this study. It is also possible that they are associated only with females and children; the two male tombs containing fibulae also contained females (Tombs XXI [55] and LVI [43]). In both of these cases, the fibulae were most likely associated with the male skeleton, but it is still possible that they were interred with the female (**Fig. 8.14**).

Beads are common but appear most often in the wealthiest graves (those of adolescent females and infants) and somewhat less frequently with children (**Fig. 8.15**). Iron beads appear most often with adolescent females, whereas glass beads are more common in child graves. The only beads found with males (Tombs XLVIII [52] and LXXX [4]) were those of iron, the most common bead material.

Male adolescents (of which there were none positively identified, although four adolescent skeletons could not be sexed) appeared to have received no grave goods. The correlation between males and iron implements, particularly weapons, is so straightforward that it does not warrant graphic representation. It cannot be considered a particularly strong correlation, due to the paucity of finds, but all four iron implements from the tumulus were found in association with male skeletons. One iron knife or dagger was found with a male aged over 35 (Tomb LXXXIV [2]), and the tip of another knife was found with a young adult male (Tomb XXXVIII [79]). A spearhead was found with a male above the age of 55 (Tomb XLV [60]), and an iron projectile point was found with a young adult male (Tomb XXXII [89]).

The tip of the iron knife in Tomb XXXVIII (79) and the two arrowheads in Tomb XXXII (89) are unlikely offerings, but rather the remnants of the weapons that killed the deceased (see Chapter 21).

Discussion: The Case for Inclusion

Many more patterns can be detected in the grave goods from Lofkënd than those derived from the physical characteristics of the grave or skeleton. The figures and tables presented above illustrate relationships between age and sex and specific object types and materials. As in other aspects of the unusual demographic profile at Lofkënd, the use of grave goods separates the adult from the subadult population. While several parallels can be drawn between object types and age and/or sex, one of the best-defined correlations is between ornaments placed on the crania and adolescent females, children, and infants.

The three adolescent female tombs in the focus group would certainly prompt some scholars to attribute the wealth of the grave to the status of the associated males—to whom they were likely related by ties of kinship or marriage. Mortuary analyses have frequently focused on reconstructions of hierarchical ranking systems, an approach that tends to privilege male-dominated elite competition while neglecting the multitude of social roles played by all members of a society (Arnold 2002; Arnold and Wicker 2001:vii– ix; Brück 2006:73; Chapman 1977:24; Jørgensen 1987; Randsborg 1980; Rega 1997; Tainter 1978:110; Weglian 2001; Wylie 1992). Mortuary analysis is singularly useful for gender-based analyses of social organization because the juxtaposition of biological sex and mortuary ritual is often clear in the construction of the grave, treatment of the body, or the choice of grave goods (Liston and Papadopoulos 2004; O'Gorman 2001; Stalsberg 2001).

It is often argued that when rank is achieved, the means by which it is obtained is open only to males, but in societies organized by hereditary systems, high status can be conferred on women (Wason 1994:98). Paul Wason notes that a pattern of "high-status" male burials without similar female burials represents a system of inherited prestige and achieved status (Wason 1994:99). In contrast, according to Wason, a cemetery where wealthy burials belong only to females represents a society based on lineal descent, such as a matriliny. The wealth of the six individuals at Lofkënd may represent part of their dowry or bride wealth or an indicator that they held a higher status than others interred in the tumulus. Grave goods, however, should not be taken solely as an indicator of high status. Bride wealth is frequently paid in the form of livestock or other property (Goody 1973:11–12; Russell 1998a:144; Stone 2000:131), which would not be visible in a tomb. The ornaments found in the tombs at Lofkënd, then, may hold symbolic rather than economic value (for further discussion on value, see Papadopoulos and Urton 2012).

Many components of the funerary ritual leave no evidence for the archaeologist. A funerary procession, the number of mourners, feasting, as well as the organic objects that were part of the tomb or were interred with the deceased and completely degraded, are ritual elaborations that are often archaeologically invisible but were indicators of the status of the deceased. The grave goods preserved in the six wealthy tombs at Lofkënd were primarily ornaments rather than utilitarian objects. Their placement on or around the head may have conveyed something to the community about the wearer. A classic case in point is Tomb LXX (17), a young female wearing a bronze headband and various dress fasteners—two fibulae and a pin—on her upper torso (**Fig. 8.16**).

In Martin Wobst's study of the communicative properties of objects, he states: "[I]t is not surprising that human populations should avail themselves of the option to transmit messages in the artifact mode, and that artifact form should be utilized to carry a variety of messages" (Wobst 1977:322). Wobst found that a "headdress, under these circumstances, is singularly appropriate to take on messages of group affiliation, because it is potentially visible to any member of a given social group . . ." (Wobst 1977: 332–333). He argues that objects, particularly those worn on the body, can communicate information ranging from social status, rank, group affiliation, and class affinity to messages of pre- and pro-scription (Wobst 1977:323; see also O'Shea 1996:188). Timothy Earle (1990) has examined grave goods from burials of both the Olmec of Mexico and Hawaiian chiefdoms. Mortuary and other archaeological material allowed him to support his argument that status in complex chiefdoms was conveyed by garments, headgear, and, in the tomb, by grave goods, many of which were items of personal adornment (Earle 1990). He states that less complex, local groups identify with specific "tracts of land" and that often "an origin myth specifies a location from which the group's ancestors sprang . . ." (Earle 1990:74). Local styles and their role in communicating group affiliation develop as a result of a need to prove that members of a community have a claim to and connection with the land and to restrict the access of non-members (Earle 1990:75).

In his work on the Maros cemeteries of Bronze Age Hungary, John O'Shea found that head ornaments were common in all the cemeteries but were almost exclusively worn by adults (O'Shea 1996: 209–211). While both adult males and females wore headdresses, only females wore beaded sashes, and their placement on the body was apparently dependent on the age of the deceased (O'Shea 1996: 211–212). O'Shea concluded, on the basis of the spatial arrangement of males with headdresses, that they were hereditarily ascribed and represented "major social or political positions" (see also Rega 1997). For females, however, he argues that the headdresses represented a different type of social position (O'Shea 1996:264). This was based on the lack of headdresses in the graves of older women (O'Shea 1996:264). The spatial patterning and specific location of sashes within the grave, on the other hand, indicated a distribution along hereditary lines (O'Shea 1996:264). The differing location of the sashes (worn by adult women but placed alongside adolescent women) may have indicated that the adolescent females died prior to achieving an age or rite of passage that would allow them to display the symbol of status that the sash represented (O'Shea 1996: 265; Rega 1997).

While Wobst's examples refer to items worn by the living, Earle's and O'Shea's examples illustrate that grave goods can convey the same types of messages as the ornaments and garments of a living member of a community (cf. **Fig. 8.16**). These items, although selected by the living, are thought to represent the social standing and personae of the deceased (O'Shea 1996:188; Pader 1982:34–38; see also Ucko 1969; Parker Pearson 1982). This concept is aptly described by John Barrett (1990:182): "It is the mourners who are the active participants in the funeral ritual, and the practices are among those which continually bring the social system into being." The difficulty for the archaeologist is interpreting the social meaning of these objects in the context of the culture that created them.

Ancient Greece again supplies an interesting cross-cultural comparison, especially for head ornaments, with an example of headgear and hair treatments communicating social messages. Lesley Beaumont (2000:41) notes that the "pubertal maiden phase" is iconographically identifiable by distinctive clothing and hairstyles and can be distinguished from adult womanhood, which is associated with different hairstyles and sometimes headdresses or veils. Ellen Davis (1986:41) argued that social and religious practices are evident in the artistic conventions of the frescoes in the Bronze Age settlement of Akrotiri on the island of Thera. The hairstyles depicted in these paintings are indicative of age and life stages and thus, to some degree, role and status within a society. Of particular interest are the hairstyles of three young women depicted in the frescoes from Xeste 3: the shaved heads typical of younger children are absent. Instead, these figures have abundant hair but have had their forehead locks cut (Davis 1986:401). In addition to the hairstyle change, the women in this fresco wear a "thin fillet of cloth . . . across the forehead" (Davis 1986:401), which closely resembles a headband or diadem (**Fig. 8.17**). (One additional figure in the Xeste 3 frescoes is younger in age and also wears a fillet [Davis 1986:406].) Davis contends that the shaving and growth of locks of hair on specific areas of the head was linked to rites of passage marking different stages of youth and the transition between adolescence and adulthood (Davis 1986:401). The new haircut, the addition of the cloth fillet, and the artistic representation of these young women as nubile youths imply that their hairstyle and head ornaments relate to marriage or the eligibility for marriage (Davis 1986: 402). Also depicted in the Xeste 3 frescoes is a goddess figure placed in a separate panel above the young women. The relationship between the goddess and the women may be associated with a Classical Greek practice of cutting a bride's hair as a premarital offering to Artemis (Davis 1986:403; Dillon 2002:215; Rehm 1994:29; see also Vérilhac and Vial 1998:287). During the same period, the hair of a Spartan bride was cut off close to the skull (Dillon 2002:215).

Married to Death

In Classical Megara, the cutting of the bride's hair served as an offering to Iphinoë, a virgin maiden (*parthenos*) who died before marriage (Dillon 2002: 215; Dowden 1989:2–3; van Gennep 1908 [1960]:

167). This event was a rite of passage marking the transition from girlhood to womanhood. It differs from the Theran ritual in that the offering was to a human being rather than a goddess, and the ceremony was performed at a tomb rather than a temple (Dowden 1989:3). The offering to Iphinoë conflates death with the loss of maidenhood—a common theme in Greek mythology (Dowden 1989:3).

The cutting of hair is only one symbol of the transition from maiden to woman in Archaic and Classical Greece. John Oakley and Rebecca Sinos (1993:8, 12, 16) refer to the important symbolism of the wreath in ancient Athenian weddings (see also Rehm 1994:14). A conventional way of depicting the bride in Greek vase painting is as a woman holding or wearing a wreath or crown or seated with one above her—the "wedding was the time for the most extravagant adornment of a woman's life" (Oakley and Sinos 1993:16). The wreath was a convention that symbolically communicated both the status of the bride and her participation in an important rite of passage in which she transitioned into womanhood. It signaled her readiness to become a mother, a social role that was important to the entire community.

In many cultures, the emphasis placed on the wedding as a rite of passage reflects the importance of the reproductive capacity of the younger generation for the perpetuation of society. Just as raising children to adulthood is critical for the survival of the community, the death of a young man or woman marks a loss of their reproductive potential and that of the community (Garland 1985:86; Pader 1982:43; Stone 2000:133).

It was the loss of potential that inspired the Archaic Greeks to regard the unmarried dead with special sympathy (Garland 1985:87). In fact, death prior to marriage was considered a disruption of the natural order (Ferrari 2003:36; Pomeroy 1975:62; see also Demand 1994:121). When a young girl died in Archaic Greece, the funeral rite might replace the marriage rite in a ritual not unlike the symbolic loss of maidenhood represented by offerings to Iphinoë: the girl became a "bride of Hades" or "married to death" (Demand 1994:14; Ferrari 2003:27, 35; Oakley and Sinos 1993:6; Rehm 1994:6, 29). The items and rituals intended for her wedding and, therefore, symbolic of her transition from girlhood to womanhood were now symbolic of her unfulfilled life. During her *prothesis* (lying in state), she was bathed and laid out in her wedding clothes (Garland 1985:25). Objects that would have been associated with her wedding celebration, such as the *loutrophoros* or "bath-carrier," a long-necked amphora used to carry water for the bride's bath, accompanied her to the grave (Oakley and Sinos 1993:6; Stears 1998:114, 119). Rush Rehm contends that these vessels were used exclusively as grave goods in the tombs of young people on the cusp of marriage (Rehm 1994: 27; for maiden graves in the Early Iron Age, see Langdon 2008; Papadopoulos 2010a). A maiden's funeral was accompanied by feasting just as her wedding would have been, and, finally, she underwent a change in residence and joined a new community in Hades, just as she would have done when she transitioned from the home of her parents to that of her new husband (Rehm 1994:12; see also Ferrari 2003: 27–28).

The mortuary ritual is a renegotiation of the roles and social personae of the deceased. In death, individuals can no longer fulfill their destined social roles, and in the case of the unmarried dead, this loss is particularly poignant. "Birth, social puberty, marriage, fatherhood (or motherhood), advancement to a higher class, occupational specialization and death" are events that require a readjustment or redefinition of social position (van Gennep 1908 [1960]:2–3). The biological maturation of the body is frequently connected to specific rites of passage, as is the sex of the individual (see Derevenski 2000:9). For a female in ancient Greece, the end of childhood marked a transition to marriageable status during which she was neither child nor adult woman. It was marriage and, subsequently, childbirth that conveyed the status of adulthood (Beaumont 2000:45). The Lofkënd tumulus produced four young women in this transitional age range between childhood and adulthood. Three of them had greater grave wealth than that contained in most tombs in the tumulus, and that wealth was chiefly in the form of ornaments that may have been parts of a headdress constructed of perishable organic material woven of plant fiber or wool or leather. Like lost pieces of a puzzle, we also may never know details about the textiles that made up the complete headdress, though the numerous fibulae and pins associated with Tomb LXX [17] may provide a glimpse into the complex clothing worn to the grave (**Fig. 8.16**; see also Chapter 12 on the textile pseudomorphs). Similar tombs were found belonging to children and infants. The significance of the age of the young women allows the drawing of parallels among these tombs, the importance of objects in conveying social messages, the symbolic significance of head and hair treatments in Archaic and Classical Greece, and the wedding and the funeral in those societies.

Although a marriage in ancient Greece did not take place until a girl had reached sexual maturity (coinciding with the stage of pubertal youth discussed above), betrothal could be arranged when she was very young (Oakley and Sinos 1993:10). Objects that they would have been too young to use often accompanied the children of wealthier Greeks to their graves. It has been noted that when children have object-rich graves or are buried with adult grave goods, the most common interpretation is that the mortuary treatment reflects the status of their parents (Baxter 2004:96; see also Pader 1982:59, 61–62). In the case of the infants and children in wealthy tombs at Lofkënd, another interpretation seems more appropriate—that these were "signs of the adults they would have been rather than . . . objects actually belonging to the deceased" (Oakley 2003:177).

CONCLUSIONS

In the introduction to this chapter, I hypothesized that the mortuary practices at Lofkënd, and specifically those represented in a group of six wealthier tombs, captured a social role played by the deceased in life, one that was significant enough to be portrayed in death. The mortuary analysis of the entire Lofkënd tumulus determined the degree to which these tombs differed from others.

There is no definitively normative tomb type at Lofkënd. The most that can be said about normative mortuary behavior is that there was a general preference for inhumation as well as for the northeast to southeast orientation of skeletons. Burial within a tumulus, often built over a central grave, however, is practiced consistently throughout the region. If the population were sedentary, the location of a tumulus may mark territorial boundaries and perhaps a community organized by lineal descent. This evidence supports existing theories that the Illyrians were organized in clan-based groups (Hammond 2000: 345; Galaty 2002; see also Earle 1990:74 and Douzougli and Papadopoulos 2010).

While there is no normative burial pattern at Lofkënd, the six tombs in the focus group are distinctive (see **Fig. 8.4**). The wealth of these six tombs, however, may be less remarkable than the location of the grave goods around the crania of the deceased and the significant age of the young females. The unusual demographics at Lofkënd are difficult to explain, and the theory of matrilocal marriage practices presented here may expend too much effort attempting to explain both the demography and the wealth of the focus group tombs.

The question that was the most difficult to approach, that of special social statuses, roles, and rites of passage represented in mortuary ritual, proved to be the most productive. While there is no definitive evidence proving that these young women and children held a special status, the examples provided by the Lake Baikal project, Khok Phanom Di, and Early Iron Age to Classical Greece contributed several possible social behaviors and structures that, whether or not they existed at Lofkënd, provided a range of possible interpretations. These include matriliny or matrilocal or uxorilocal marriage practices, rank, the significance of certain treatment of the hair, the communicative properties of headgear, and the special status afforded to females who died prior to marriage and motherhood.

Many interpretations of the social meaning of mortuary ritual, including those of Wason mentioned earlier, conclude that wealthy female burials are a result of ascribed status. I would argue that the evidence from Lofkënd, combined with the cross-cultural information presented here, suggests that the reproductive capacity of the younger generation—a critical factor to the survival of a community—formed the foundation of a highly significant social role. The wealth of the six noteworthy tombs at Lofkënd may represent particular issues raised in the community by the loss of these young women and children. If, as I have proposed, these mortuary practices relate to potential future roles of wife and mother, and perhaps bearer of a lineage, it follows that the latter role within their clan or tribe was ascribed, whereas the former was an achieved status. The ornaments in these tombs could have effectively conveyed the message that the young women who wore them were ready to exchange girlhood for womanhood, but as grave goods they would have symbolized a loss of potential. The presence of such ornaments in the graves of subadults can be interpreted as symbolic of a status they would have achieved much later in life.

CHAPTER 9

THE POTTERY FROM THE TOMBS AND TUMULUS FILL

Seth Pevnick and Esmeralda Agolli

INTRODUCTION

This chapter presents all of the inventoried pottery from the excavation of the Lofkënd tumulus, including the complete and nearly complete vessels found as kterismata in graves, as well as many of the sherds and fragmentary pieces found within topsoil and tumulus and grave fill. Readers familiar with publications of handmade prehistoric and protohistoric pottery from contemporaneous cemeteries and tumuli in surrounding regions of Albania as well as in northern Greece and southern Italy will recognize many important similarities with the Lofkënd pieces, particularly in terms of complete vessels. But because no contemporaneous settlement site has yet been excavated in the region, and because publication of fragmentary pieces remains somewhat limited, scholarly knowledge of a great portion of Late Bronze Age (LBA) and Early Iron Age (EIA) pottery remains incomplete. It is partially in hopes of remedying this deficiency—and of offering a better idea of the full range of fabrics and vessel types in use in LBA/EIA Albania—that it was decided to publish here a much fuller account than usual of fragmentary pieces. Thus, nearly all pieces, both intact and fragmentary, that can be classified in terms of fabric and at least general vessel type or shape are included in the catalogue, which makes up the bulk of this chapter. The catalogue also includes many fragments that can be classified only by fabric, probable vessel size, and preserved portion of vessel (e.g., handle, rim, or base); while some of these pieces may derive from known types, many probably also derive from types not known from any intact pieces, either at Lofkënd or elsewhere. Aside from a short final section devoted to the few non-prehistoric sherds found at Lofkënd, and an appendix (Appendix 2) on the handful of Corinthian sherds found, this chapter focuses on the prehistoric pottery. (Unless otherwise noted, all dates are BC.)

After a section on classification and terminology, a brief typology of the LBA/EIA repertoire at Lofkënd is set forth. Because this typology is based primarily on complete or nearly complete vessels found as grave goods from securely dated tombs at Lofkënd, we have been able to arrange the types chronologically within each fabric group. Fragments from uncertain types within each fabric group are also discussed, but here—because nearly all such fragments were found in topsoil or tumulus fill—dating relies primarily on comparanda from other sites. Following the typology are sections on decorative types and motifs, and, finally, on the intra-tumulus distribution of the Lofkënd pottery.

CLASSIFICATION AND TERMINOLOGY

As with any catalogue and typology of handmade pottery, most of the distinctions drawn here between Lofkënd types are to be classified as etic rather than emic (see Rice 1987:283–285 for a summary of the debate over "real" and "artificial" types). The divisions have been made by means of both objective and subjective criteria, corresponding first to fabric and color, then to vessel size and type, as explained below.

The most basic distinction used here is that of fabric, with fine, semi-coarse, and coarse pieces distinguished from one another primarily on the basis of the size and abundance of inclusions and air pockets, but also of wall thickness and vessel size. For each piece, the catalogue entry includes a description of fabric, with the basic type followed by a listing of the inclusions and air pockets visible to the naked eye with the aid of a handheld magnifying glass. These are further described in terms of frequency, with the terms "abundant," "moderate," or "occasional" adopted here to correspond roughly to frequencies of ca. 20% or more, 10–20%, and 10% or less, respectively, with "very abundant" or "very occasional" used to indicate greater or lesser frequencies (cf. illustrations in Munsell Soil Color Charts [rev. ed. 2000]:9–10). Inclusions are identified where possible (e.g., mica, limestone, quartz, pebble), but are often described simply in terms of basic color—white, black, gray, and so on, with red inclusions that are clearly not pebbles identified as grog. By far the most common inclusion type at Lofkënd is mica, which is generally more silver than gold in color and which actually seems to be found at higher frequencies in fine fabric than in semi-coarse and coarse. The increasing size and frequency of all other inclusions, as well as air pockets and blowouts, can be used to distinguish between the fabric types, with the coarsest fabric type having the greatest size and frequency, the finest fabric the least. Although subjective, these distinctions are internally consistent, made in every case by the two authors (together sometimes with the excavation co-directors); for those that were more difficult to classify, the description reflects this difficulty (e.g., by listing the piece as "fine to semi-coarse" [**9/2** (P212), **9/22** (P074), **9/23** (P278)]). Since coarser fabrics tend generally to be used for larger pieces, vessel size and wall thickness sometimes also aid in fabric classification, particularly for more fragmentary pieces. Similarly, fine fabric tends to be less dense than coarse, and especially dense or heavy fabric is noted where applicable (e.g., in the pithos fragment **9/305** [P011]). But these are certainly not hard and fast rules, as indicated by the presence of several quite large pieces of fine fabric (e.g., **9/41** [P275], **9/88** [P001]) and of several small semi-coarse vessels (e.g., **9/160** [P362], **9/162** [P079]). As with any study of pottery, the identification of fabric at Lofkënd remains subjective.

Fabric color, given as far as possible for both surfaces (exterior and interior) and core in terms of the Munsell Soil Color Charts (rev. ed. 2000), forms the basis for a further division, between two main types of fine fabric—sometimes called simply "light" (for fine light) and "dark" (for fine dark). Color is also listed for semi-coarse and coarse pieces, but in those fabrics, much more color variation occurs, even within individual pieces, and therefore seems to be of less importance. The distinction between fine light and fine dark fabric is most obvious on the complete vessels, nearly all of which can be classified as fine fabric (with the exceptions of two that are semi-coarse: **9/159** [P303] and **9/160** [P362]). Generally speaking, the fine light fabric is yellowish red or reddish yellow (most commonly 5 YR in hue), but it can be as light as yellow (e.g., **9/16** [P047]) and as dark as red (e.g., **9/7** [P166]); the dark fabric is usually very dark gray or black (e.g., **9/95** [P228], **9/96** [P276]), though often mottled with reds and browns (e.g., **9/91** [P253]). Both fabrics are nearly always smoothed, and while the light fabric (and occasionally even semi-coarse) is sometimes burnished, dark fabric is more frequently burnished to a high sheen. Similarly, only the light fabric (and semi-coarse fabric of similar colors) is ever painted or incised, while only the dark fabric is decorated with fine plastic ribbing.[1] Moreover, while there are some general similarities in vessel size and shape between the two fine fabrics, each has its own characteristic forms—globular one-handled vessels with everted rims in light fabric (Types 2a and 2b), sharply delineated biconical forms in dark fabric (Types 4 and 5).

Still, although color, vessel shape, and surface treatment can all be used, at least to some extent, to distinguish between these two fabric groups, two major difficulties remain. First, because of the reddish colors seen in patches on and just beneath the surface of some dark fabric pieces, one wonders whether these apparently different fabrics might derive originally from the same or similar clay, with

[1] It is important to note that this distinction is particular to Lofkënd, and while often true elsewhere, cannot be universally maintained. Thus, at some Korçë basin sites in southeastern Albania, matt-painted decoration occurs on dark fabric (e.g., Drenovë, in Lera 1987b:242–243), while light fabric pieces from other regions sometimes feature ribbing, a type of decoration that we have not seen on light fabric (e.g., Eggebrecht 1988: cat. 50 [Tirana NHM 1130]; the ribbing here is similar to that on **9/90** [P422], rather than finely spaced as on **9/95** [P228], **9/96** [P276], **9/98** [P032], and **9/99** [P050]). In addition, there is a group of highly burnished semi-coarse pieces, often mottled in color, that seem to derive from Bronze Age vessels, as discussed below.

dark colors being produced by a reducing firing atmosphere, lighter ones by an oxidizing one. Although this question cannot be answered without further scientific analysis, it is worth bearing in mind, particularly in those cases in which it is especially difficult to distinguish between the two fabrics. The other difficulty encountered in distinguishing between dark and light fabrics arises when heavy encrustation covers the original surface. This encrustation (which also appears on semi-coarse and coarse fabrics) can range in color from white calcium carbonate to black manganese,[2] with associated soil often adding a brownish or grayish hue. It also sometimes takes on a high sheen (and may even include mica), giving the impression of a burnished surface. Where encountered in significant amounts, encrustation is noted in the catalogue entry.

Within their respective fabric groups (i.e., fine light, fine dark, semi-coarse, and coarse), pieces are further divided, where possible, according to vessel type, as outlined in the brief typology presented below. Because of the relatively small size and fragmentary nature of the ceramic repertoire at Lofkënd, very few complete types can be identified, particularly in semi-coarse and coarse fabric. Likewise, many small fragments cannot be assigned to specific types and are therefore grouped first by fabric and then by the specific portion of the vessel that they preserve—rim, handle, wall, or base, further divided into rim types, handle types, and so on. These pieces are further categorized where possible by original vessel size (as indicated by the extant remains)—small, medium, or large, with a few exceptional pieces classified as very small and very large—according to the rough (and partially overlapping) guidelines set forth in **Table 9.1** (all tables for this chapter placed at the end of the chaper, before Appendix 2)

It must be stressed that these size groupings are particular to the vessels and fragments found at Lofkënd and may not be appropriate for other bodies of material. Moreover, one must not confuse the size of the fragment itself with the size of the original vessel; by looking at the thickness and curvature of even a very small wall fragment, for example, one can usually make a rough approximation of the size of the original vessel, whether very small, very large, or somewhere in between. The resulting groupings, then, link together fragments that seem to derive from similar parts of vessels of similar size. Further links are occasionally made between groups of fragments and particular vessel types, but this is not usually possible. The resulting groupings, by which pieces are arranged in the catalogue, are laid out in tabular form in the respective conspecti for each fabric group (**Tables 9.9–9.13**); these are further explained below in the typology section.

Two other potential means of grouping, by so-called "open" versus "closed" shape, and by decoration type, proved ineffective, though for very different reasons. The "open-closed" distinction, though often used in ceramic studies, turned out to be more confusing than helpful at Lofkënd. While the distinctions between these terms would seem to be fairly obvious—an open shape permitting easy access to its contents, a closed one concealing or protecting them (as suggested by the alternative terms "unrestricted" and "restricted" used by some archaeologists [e.g., Sinopoli 1991:60])—in practice the distinction often proves elusive, particularly when working with small fragments deriving from uncertain or unknown types. Moreover, while most of the complete vessels found at Lofkënd seem to be small or medium-size drinking vessels, suggesting that they should be classified as open, their fairly constricted necks might be considered by some as characteristic of closed vessels. The arbitrary nature of the "open-closed" distinction for these pieces is made clear by the application of two commonly used formulae for distinguishing between open and closed shapes (**Table 9.2**); the first compares the mouth diameter with the maximum diameter, and the second the mouth diameter with the height from base to rim.[3]

---

[2] Black manganese encrustation can also be mistaken for matt-painted decoration as well as bitumen, but we have done our best to avoid such misidentification; see Chapter 16 for more on bitumen at Lofkënd. Manganese encrustation has been identified on selected sherds with X-ray fluorescence (XRF) (by Vanessa Muros), and calcium carbonate encrustation by its reaction to dilute 10% hydrochloric acid (by John Foss).

[3] Rice (1987:212) uses a different distinction for classification of the vessel mouth, if not for the vessel itself: "If [the mouth opening] is equal to or greater than the maximum diameter … it is described as an unrestricted orifice. If it is less than the maximum diameter, it is called a restricted orifice." Rice (1987:217) also makes the following interesting comment in this regard: "The principle of classifying vessel shapes by ratios of height to diameter is clearly a good one."

By the first measure, nearly all of the intact or nearly intact Lofkënd vessels (with the exceptions of the large coarseware amphora **9/259** [P283] and the small biconical kantharoi **9/1** [P350], **9/95** [P228], and **9/96** [P276]) are classified as open shapes, while by the second measure, nearly all (except **9/14** [P256] and the fragmentary profile **9/6** [P073]) are classified as closed. Although the total number of vessels adds up to 19 in both cases, it should be noted that Formula B is inconclusive for **9/94** (P322), while **9/92** (P364), due to its fragmentary state, cannot be assessed by means of Formula A. **Figure 9.1** shows how easily the terms "open" and "closed" can be manipulated.

In other words, for the pottery under consideration here, the open-closed distinction is largely arbitrary; the selection of one formula over the other simply gives the appearance of objectivity to a truly subjective distinction. For this reason, we refrain from using the distinction as an integral part of our classificatory scheme. Still, the terms can be useful when considering the presumed function of the pieces, particularly when shape and surface treatment provide additional clues. Thus, when the terms are used below, they should be understood not as corresponding to any specific numerical formula, but as indications of apparent vessel function, with open vessels being used primarily for drinking and eating, and closed ones for pouring or storing.

Distinction by decoration type can also be misleading, especially when working with many fragmentary pieces, since an undecorated sherd could derive from any portion of a completely undecorated vessel, as well as from an undecorated portion of a decorated vessel. In addition, decoration types span multiple fabric groups, with matt-painted and incised/punched decoration found on both fine light fabric and semi-coarse vessels, and plastic decoration on all four fabrics. Thus, all pieces are classified first by fabric, size, and fragment variety, and only then—where applicable—grouped together by decoration type; where not noted, pieces are undecorated. Abbreviations have been kept to a minimum.

Most of these are self-explanatory, but the differences among PL, PH, and PW must be clarified. Where possible, in every case in which precise orientation is clear, PH has been used. In many other cases, only a general orientation can be determined, allowing for an accurate measurement of width but not height; in these cases, PW and PL are used, with the latter referring to the greatest preserved length along the axis perpendicular to the width.

Finally, in addition to the four fabric groups of LBA/EIA pottery, a very brief fifth section, entitled "Other Fabrics," has been included at the end of the catalogue to describe the 12 pieces of different (probably non-prehistoric) fabric found in the tumulus. Unlike the prehistoric pottery, all of which is handmade, some of these "other" pieces are wheelmade (mostly Corinthian: discussed in Appendix 2). These were found primarily in topsoil and very late fill units, and seem to postdate the active ancient use of the tumulus, as discussed briefly below. Certain non-diagnostic pieces of handmade LBA/EIA pottery that were originally catalogued due to their bitumen coating are listed in the section on bitumen in Chapter 15; where found on pieces that could be classified, the presence of bitumen is noted in the catalogue entry.

Typology

The nature of handmade pottery argues against the high levels of precision and standardization found in wheelmade and moldmade pottery, both within types and on individual pieces. Thus, even on a single very fine piece, for example, one can find significant formal variation, from the thickness of the walls to the roundness of the rim or the profile of the lip (e.g., **9/95** [P228]; overhead photo 2487). Likewise, even two pieces made on the same day by the same potter may not be entirely identical. Although this often makes it difficult or impossible to assign small fragments to particular types, it is nonetheless worthwhile to create a typology that accounts for all of the complete and nearly complete vessels found at Lofkënd, and that also indicates some of the possible types from which the more fragmentary pieces are derived. Indeed, without any previously published typology of prehistoric Albanian pottery on which to build, the Lofkënd corpus, with its relatively limited size and range, actually serves as an ideal starting point, providing an explicitly provisional typology on which future researchers can (and must) build. Following these caveats, this section provides brief explanations and discussions of the types identified at Lofkënd.

These types are discussed here by fabric group, in roughly the same order in which they are listed in the conspecti and catalogue below. Within each fabric group, the types have been arranged in roughly chronological order, based on the overall chronological phases of the tumulus, so that Type 1 appears earlier in the typology and earlier in the tumulus chronology than Type 2, and so on. The tumulus chronology, based on the vertical and horizontal stratification of tombs, together with secure AMS ^{14}C radiocarbon dates from human bone and charcoal, is presented in Chapter 4. Although no ceramic types can be assigned to the earliest phase of Lofkënd burials (Phase I), multiple types belong to each of the following phases (II–Va–b).

The few fragments that can be safely assigned to a given type are similarly numbered. Because the majority of the fragments cannot be confidently attributed to a particular phase, however, the order of numbering has been done largely for convenience rather than chronology, proceeding within each fabric group through all of the handles, and then rims and lips, bodies and necks, and bases, respectively, from the smallest original vessel size to the largest. The conspecti should help to clarify this arrangement, while the accompanying commentary discusses chronology where relevant. Finally, although nearly all of the Lofkënd types sometimes bear matt-painted and/or plastic decoration, such decoration is discussed in terms of the typology only where it seems truly characteristic of a given type; more explicit discussion of decorative schemes is reserved for the following section of this chapter.

Fine light fabric

All of the complete and nearly complete light fabric vessels can be classified as small or medium in size, and most seem to be drinking vessels (and therefore can be considered open in shape [see above for discussion, and below for specific exceptions]). Types 1, 2, and 5 are fairly common, finding numerous parallels at contemporaneous sites, while Types 3 and 4 are quite unusual. None of the types has an accepted standard name within the English-language literature, and although descriptors are attached here to each type, these are either too lengthy or not sufficiently specific to stand alone as names. Following the descriptions of these five types and their specific representatives at Lofkënd, some of the more fragmentary pieces are discussed in similar terms, occasionally as possible additional fragments of the known types, but more usually as representatives of less certain types not fully preserved here. See the individual catalogue entries for additional details.

Type 1—Biconical kantharos (Lofkënd Phase II)

Although seen in only a single vessel at Lofkënd (**9/1** [P350]), the biconical kantharos Type 1 is strikingly similar in shape—with flat base, sharply delineated biconical profile, and conical neck—to Lofkënd fine-dark fabric Type 5 (cf. especially **9/95** [P228]). Outside of Lofkënd, the shape has parallels to the north—in the Mati-Glasinac culture (Bodinaku 1982:72)—as well as to the south—at Kamenicë (Bejko forthcoming: vessel 91/Q489), Prodan (Aliu 1984: pl. VII, 61; pl. III, 28), and Barç (Andrea 1985: pl. XV, Tomb 162, 1). The unpainted body fragment **9/2** (P212), although worn and not well fired, is almost certainly from a similar vessel (note that the Prodan and Kamenicë examples are also unpainted).

Type 2—One-handled drinking vessel (Lofkënd Phases II–Va)

The Lofkënd type most commonly seen at contemporaneous sites in the surrounding region, particularly to the south (e.g., Kamenicë, Korçë, Luaras), the one-handled drinking vessel labeled "Type 2," is characterized by a flat base, rounded lower body, single high-swung handle, and everted rim. At Lofkënd, every vessel of this type also features matt-painted decoration, and many have some form of plastic decoration as well, although similar pieces known from other sites are undecorated. Type 2a (**9/3** [P304], **9/4** [P077], **9/5** [P326]) can be distinguished from Type 2b (**9/7** [P166], **9/8** [P277]) on the basis of the contour of the lower body, with the latter more hemispherical, the former biconical (though still rounded). Within these subdivisions, further differences are apparent, particularly with regard to neck and handle shape. Thus, **9/7** (P166) can be distinguished from **9/8** (P277), for example, not only by the plastic projections and smaller overall size of the latter piece, but also by its proportionally shorter neck. Likewise, the flaring cylindrical neck of **9/3** (P304) differs from the more conical necks of **9/4** (P077) and **9/5** (P326). The strutted (*biforata*) handle of **9/5** (P326) also sets this vessel apart, not only at Lofkënd, where no other piece features such a handle, but also in the larger region, since nearly all other vessels with strutted handles are unpainted. The four projections spread across the front of this piece are also unusual, since such projections normally occur in groups of three (Aliu, personal communication). These differences in shape do not appear to have any real chronological significance, as Type 2a pieces were deposited in tombs of Phases II (**9/5** [P326]), III (**9/3** [P304]), and V (**9/4** [P077], dated by AMS ^{14}C to 876 ± 46 BC), while Type 2b pieces were found in tombs of Phases III (**9/8** [P277]) and IV (**9/7** [P166]). Perhaps more important, then, is the clear evidence that this common ceramic type remained popular for several centuries, throughout much of the active use life of the tumulus. Moreover, with absolute dating for **9/4** [P077] and stratigraphically secure relative dating of the other pieces, the evidence from Lofkënd seems to confirm (and perhaps extend slightly earlier) the eleventh–ninth century dates often assigned to similar types at other sites (e.g., Barç 1 [Andrea 1985], Luaras [Aliu 2004], Kamenicë [Bejko forthcoming]).

The body, neck, and rim of the nearly complete profile **9/6** (P073) all suggest that it derives from a Type 2a vessel, but without any preserved handle this remains uncertain. Many of the even more fragmentary pieces may also belong to this type, and are discussed below.

Type 3—One-handled hemispherical cup or dipper (Lofkënd Phase IV)

This very unusual Lofkënd type, the one-handled cup or dipper, is represented by only a single piece, **9/14** (P256) and finds no precise parallels outside of Lofkënd. Although of similar fabric and decoration to the other complete vessels at Lofkënd, **9/14** (P256) is atypical in shape, its hemispherical body rising from a flattened base and terminating abruptly at the point of maximum diameter with a simple, flat-lipped rim. The extremely open body and unusually raised horizontal handle could function equally well as either drinking vessel or dipper, while the relatively broad lip might be better suited for the latter purpose. A similar piece from Patos was found outside of any grave context (Korkuti 1981: pl. IX,1), making dating difficult. N.G.L. Hammond (1971: 233–234, with additional references) refers to a similar shape found in tombs of both Albania and Greece in the Middle Bronze Age as a dipper, but both the rim and handle differ greatly from ours, and the date is certainly earlier. Likewise, a great number of impasto ware cups of similar (but not identical) shape have been found in the southeastern necropolis of Sala Consilina in the Campanian Salerno province (cf. especially Schale H1f and H4a [Kilian 1970:98, 107–108, Beilage 9.1, 10.1]), dating mostly from the ninth through late eighth centuries (i.e., Sala Consilina IA, B, C, and IIA, B, C [Beilage 23]). Fragments with a similar handle have also been found at Luaras (Aliu 2004: pl. XXV, 279).

Type 4—Stemmed goblet (Lofkënd Phase Va)

The so-called "stemmed goblet" **9/15** (P082) also seems to be unique, both at Lofkënd and in the greater region (Papadopoulos, Bejko, and Morris 2007:118). The most similar known piece was found at Luaras (Aliu 2004: pl. IV, 64), but that vessel has a high-swung handle more similar to those of Lofkënd Type 2. Indeed, the handle on **9/15** (P082) is perhaps its most unusual feature, for although technically a vertical strap handle, it is very tightly curved, with no vertical rise, much like a loop handle. Unlike a true loop handle, however, here the loop is incomplete, its curve completed by the vessel exterior. This distinction is made apparent by contrast with the Lofkënd body fragment **9/82** (P305), which preserves the concave portion of a loop handle attached directly to the vessel exterior (probably from a large amphora with loop handles). One other Lofkënd handle fragment worth mentioning in this regard is **9/40** (P279), which preserves a tightly curved portion of a vertical strap handle, either from a handle similar to that of **9/15** (P082) or from the apex of a high-swung vertical strap handle (as on many Type 1 vessels). The other unusual feature of **9/15** (P082) is its combination of stem and raised, hollowed foot, for which we have found no precise parallels; the flaring foot of the vessel from Luaras mentioned above is actually more similar to the fragmentary foot **9/86** (P108 + P388) found at Lofkënd, for neither of these rises nearly so high as that of **9/15** (P082).

Type 5—Globular kantharos (Lofkënd Phase Vb)

Found in fragmentary form in two graves of later date, as well as in tumulus fill, the globular kantharos Type 5 cannot be described in quite so much detail as the other types. Still, Type 5 is clearly distinguished from the chronologically earlier Type 1 kantharos by its globular body and short cylindrical neck. The most nearly complete specimen, **9/16**

(P047), bears unusual high-swung handles that switch from ring at lower body to vertical strap at rim; also unusual in comparison to other complete vessels at Lofkënd is the complete absence of painted or plastic decoration, and the slightly lighter, yellowish color of the fabric. Given these irregularities, it may be that the more fragmentary pieces **9/17** (P078) and **9/18** (P198), both of which exhibit shorter vertical strap handles and a redder hue to their fabric, are in fact more representative of a globular kantharos type in use at the time. Because of their more fragmentary state, however (neither one preserving either base or second handle), one cannot be certain about this type; it seems safer to label **9/16** (P047) as Type 5a, **9/17** (P078) and **9/18** (P198) Type 5b. Secure dating of similar pieces at Kamenicë seems to confirm the late date (Bejko forthcoming: Q104 [eighth–seventh century]). Comparable pieces have also been found at Luaras (Aliu 2004: pl. XXII, 250–257), Rehovë (Aliu 2012: pl. XXXV, 401), and Vitsa Zagoriou, where it is especially common (Vokotopoulou 1986:10).

Perhaps also connected with this type—though even more fragmentary and with more of an S-profile to its body—are the fragments **9/19** (P386) and **9/20** (P290 + P221), which are probably all from a single very small kantharos with globular lower body, vertical strap handles, and everted rim. Given its poor preservation, this piece can be only provisionally labeled Type 5c, in the hopes that a more complete specimen will be unearthed elsewhere (cf. an equally fragmentary piece from Kamenicë, also found recently in topsoil [Bejko forthcoming: 1/199]).

Uncertain fragments and additional types

Many of the remaining fine light fabric fragments from Lofkënd may well derive from the five types described above, but specific attributions are difficult to make. In addition to the few fragments mentioned above in connection with specific types, the most distinctive pieces are the fragments of everted rims with flat or rounded lip (**9/9** [P020], **9/10** [P213], **9/11** [P004], **9/12** [P131], **9/13** [P202], and several similar but uncatalogued pieces), many of which seem likely to derive from small to medium vessels of Type 2. For instance, **9/43** (P220), **9/44** (P013), and **9/45** (P173), which are all of similar size and shape, might also derive from Type 2 vessels but are too fragmentary to justify such an attribution. Moreover, since similar rims appear on very different types of vessels elsewhere (e.g., amphorae at Vitsa; Vokotopoulou 1986:359–360), circumspection is warranted. Since four of the five types have vertical strap handles, usually rectangular in section, it is not possible to assign any of the handle fragments of these categories to specific vessel types. Still, it is likely that at least some of the handles from small or medium-size vessels derive from one or another of the identified types. Likewise, some of the matt-painted body fragments from small and medium-size vessels may be from known types, but none is sufficiently well preserved to allow a confident attribution.

Perhaps more interesting, many fragments cannot possibly have come from any of the known Lofkënd types. Certainly some, such as the incised body fragments **9/54** (P133) and **9/55** (P093 + P096 + P247 + P370), must derive from small vessels at least similar to the types known from grave offerings; these may even have been grave offerings themselves before being disturbed by later burials and/or erosion of the tumulus. But many other pieces clearly derive from very different types, including both smaller pieces probably predating the main phases of use of the tumulus, and larger ones likely contemporaneous but not deemed appropriate for use as grave offerings. Thus, the three spur handles **9/29** (P006), **9/30** (P064), and **9/33** (P080) are characteristic of LBA types (cf. Korkuti 1981: pl. XI, 11; Aliu 2004: pl. XXVI, 282). Although the precise vessel shape to which such handles were once attached remains uncertain, spur handles diminish in popularity in the EIA (Aliu, personal communication). Similarly, the unusual matt-painted vertical handle **9/26** (P107) seems to derive from an LBA shape (cf. Prendi 1977–1978: pl. IX, 54). Perhaps most common on unpainted kantharoi, the handle also resembles those of the matt-painted vessel from the Barç tumulus now in the Tirana National Historical Museum (TNHM 1020 [Eggebrecht 1988:192, no. 34]; cf. also Aliu 2012: pl. XL, 427, 429 for a similar LBA piece from Rehovë). Not all unusual handles derive from early periods, however. The horizontal handle **9/35** (P287), for example, finds a parallel in the eighth or seventh century from Kamenicë (Bejko forthcoming: Q843).

As for potentially larger types, the small rim/spout fragment **9/51** (P163) seems to derive from a handmade jug with cutaway neck, an EIA shape familiar from sites in northern Greece and Macedonia (cf. Papadopoulos 2005:469–471). No other specific types can be connected with sherds of fine light fabric, but several large matt-painted lower-body sherds (**9/74** [P252], **9/75** [P132 + P324]), together with the large base fragments **9/87** (P097 + P237 + P285), **9/88** (P001), and **9/89** (P232), attest to the existence of large closed shapes, perhaps amphorae (cf. pieces from Kamenicë [Bejko forthcoming: Q407, Q740] and tumulus 1 of Barç [Andrea 1985: pl. II, Tomb 10, pl. VI, Tomb 45]). Additional sherds preserve portions of the upper bodies of such vessels, with both vertical strap handles (**9/83** [P255 + P112], **9/84** [P030]) and loop handles (**9/82** [P305]), while the fragment of a very large vertical strap handle (**9/41** [P275]) is a further indication of large types. Comparable fragments found at Tren have been dated to the first phase of the EIA (Korkuti 1971: pl. IX). A triangular-shaped projection or ledge lug (**9/71** [P127]) may derive from a similar vessel, but probably one of smaller dimensions. All of these pieces are of similar fabric to the complete vessels described above, and should be of EIA date. Fragments of additional large matt-painted vessels were also found at Lofkënd in semi-coarse fabric (see below).

Fine dark fabric

As with fine light fabric, all of the complete vessels of fine dark fabric come from tombs, are small to medium in size, and seem to be drinking vessels, open in shape. Although fewer in number than those in light fabric, the dark fabric vessels still display a range of forms, with only one type represented by more than a single example (i.e., Type 5: **9/95** [P228] and **9/96** [P276]). Types are again numbered in a roughly chronological order, although with so few complete vessels, the precise chronological range of each type must remain uncertain. Two-handled vessels occur more frequently in dark fabric than in light fabric, but in several instances the one- and two-handled dark fabric shapes are quite similar to one another.

Type 1—One-handled drinking vessel (Lofkënd Phase II)

The one-handled drinking vessel Type 1 (**9/90** [P422]) is characterized by a slightly hollowed, raised base, nearly hemispherical lower body, and flaring conical neck. Much more rounded and less squat than the other dark fabric, one-handled drinking vessel (Type 4 below), Type 1 is a very familiar EIA/LBA shape. The high-swung vertical strap handle, roughly triangular in section, and subtle *kanellure* (ribbed) decoration on the lower body are similar to those of two-handled vessels from Pazhok (Eggebrecht 1988:50 [TNHM 1130]) and Patos (Korkuti 1981: pl. IX, 2). The overall shape is similar to that of a much cruder vessel of semi-coarse fabric also found at Lofkënd in a Phase II tomb (SC Type 2—**9/159** [P303]). Further support for such an early date comes in the form of vessels with similar decoration from Gajtan, Kukës, Mat, and Maliq (Prendi 1977–1978:14–15).

Type 2—Globular kantharos with flat base and strap handles (Lofkënd Phase II)

Decidedly less well preserved (and perhaps also less finely made, with neither painted nor plastic decoration) than the other dark fabric types, the globular kantharoi Type 2 are distinguished from their biconical brethren primarily by their rounded lower bodies, but also by their flaring cylindrical necks and wide, flat bases. Type 2a (**9/91** [P253]) is more squat, with a rounded biconical lower body, than Type 2b (**9/92** [P364]), which features only a slight curve in its S-profile, but the similarities between these two pieces far outweigh the differences. No fragments from Lofkënd can be definitively assigned to these types, but the shapes are known from contemporaneous sites of the tenth and ninth centuries (e.g., for **9/91** [P253], cf. vessels from Luaras [Aliu 2004: pl. XXII, 253] and Rehovë [Aliu 2012: pl. XXXV, 401]; for **9/92** [P364], cf. a vessel from the Macedonian settlement of Vardarski Rid [Papazovska 2005: pl. I, 7]).

Type 3—Globular kantharos with rounded bottom and spur handles (Lofkënd Phase III)

As its name indicates, this type, known from the single example **9/93** (P223), is distinguished from the previous one by its rounded base and spur handles as well as its very short neck. The squared spur handles, which are often cited elsewhere as Late Bronze Age characteristics (e.g., Prendi 1977–1978: pl. VIII), find no parallels at Lofkënd within this fabric group and are only roughly similar to the light fabric spur handles mentioned above (**9/29** [P006], **9/30** [P064], **9/33** [P080]); much more similar are the semi-coarse squared spur handles **9/168** (P116), **9/169**

(P119), **9/189** (P088) and perhaps the more fragmentary piece **9/188** (P076). Vessels of similar shape (including the handles) have been found at Luaras as well as Rehovë, where Aliu dates them to the fourteenth–twelfth centuries (Aliu 2004: pl. II, Tomb 12, 26; pl. XXVI, 283; Aliu 2012: pl. XXVI, 344 [Tomb 265]). Andrea (1985: pl. VII, Tomb 50, 2) assigns a later date to a similar piece from Barç (eleventh–eighth centuries).

Type 4—Squat one-handled biconical drinking vessel (Lofkënd Phase III)
Wide and squat in shape, this biconical drinking vessel, **9/94** (P322), shares some features with both light fabric (LF) and dark fabric (DF) pieces. The high-swung, strutted handle, for example, is much more common among light fabric pieces, while the *kanellure* (ribbed) decoration is characteristic of dark fabric. The biconical shape, meanwhile, is seen in both fabrics at Lofkënd (cf. **9/1** [P350] [LF Type 1], **9/95** [P228], **9/96** [P276] [DF Type 5]), although neither of those types is so squat as this, nor with such a sharply everted rim. The combination of all these elements is unique, not only at Lofkënd but elsewhere as well.

Type 5—Biconical kantharos (Lofkënd Phases III–IV)
The only dark fabric type with more than a single representative (**9/95** [P228], **9/96** [P276]), the biconical kantharos Type 5 is similar to both its light fabric counterpart (**9/1** [P350]) and the dark fabric Type 4 discussed above, although neither of those types features such sharp carination in its lower body. Despite their nearly identical forms, the two vessels found at Lofkënd actually differ slightly in size as well as in handle shape and in direction or pattern of ribbing; **9/95** (P228) features ring handles roughly rhomboidal in section (cf. handle fragment **9/97** [P123]) and diagonal ribbing, while **9/96** (P276) has more ovoid ring or strap handles and straight ribbing (cf. ribbed body fragment **9/98** [P032]). The similarly ribbed body fragment **9/99** (P050) is less sharply carinated but is still probably from this vessel type. Since tumulus stratigraphy suggests that these vessels come from different chronological phases (**9/95** [P228] in Phase IV, **9/96** [P276] in Phase III), one wonders whether the slight differences between them may have chronological significance. Outside of Lofkënd, vases of similar form tend to be of light fabric, and usually matt-painted rather than ribbed (e.g., from Rehovë [Aliu 2012: pl. XVIII, 226, Tomb 207] and Korçë [Andrea 1985: pl. XXXV, 3, Tomb 126]). Perhaps the closest parallels for the Lofkënd pieces come from Barç 1 (Andrea 1985: pl. XXX, Tomb 58) and Prodan (Aliu 1984: pl. VII, Tomb 61), though both have been dated later than at Lofkënd (eighth–sixth and eighth–seventh centuries, respectively).

Uncertain fragments and additional types
As in fine light fabric, so also in fine dark fabric might some fragmentary pieces derive from known types. But here confident attributions are even more scarce, and only those mentioned above in connection with Type 5 seem at all secure. Among the other types, only the handles of Types 3 and 4 are especially distinctive; comparanda for the former have been described above, while the latter finds no dark fabric parallels. The body fragments **9/100** (P049 + P063) and **9/101** (P137)—together with the smaller non-joining piece **9/102** (P051) and a number of small, uncatalogued sherds (P162, P415B, P410)—may derive from a type related to Type 1, for they are of a similar fabric to **9/90** (P422) and also show a similar globular lower body with slightly offset flaring conical neck. Even with no preserved base, however, two significant differences are clear: the fragments show none of the *kanellure* decoration seen on **9/90** (P422), and the fragmentary handle does not rise as sharply as that on the complete vessel. Since the fragments do not join, one cannot be certain whether this type had one handle or two; the former seems more likely.

Most of the other noteworthy fragments seem to come from types not known from complete vessels at Lofkënd. The fragments **9/103** (P134) and **9/104** (P233), for example—which are of similar shape and fabric and possibly even non-joining pieces of the same vessel—preserve a type of globular kantharos with vertical strap handles barely rising above the rim. Although the base is not preserved, the unusually low handles distinguish this type from Lofkënd Type 2, and may even suggest a significantly earlier date, with parallels from Maliq dating as early as the Middle Bronze Age (i.e., Maliq Phase IIIc [Prendi 1977–1978: pl. IV, 49]). Also noteworthy and possibly of early date are the fragmentary handles **9/114** (P027), **9/121** (P307), **9/123** (P339), and **9/122** (P361), all of which derive from wide strap or ribbon handles of highly burnished fabric. This fabric is similar to (but finer than) that of several semi-coarse handle fragments thought to be from Bronze Age vessels (e.g., **9/176** [P289], **9/183** [P062], **9/190** [P098]), and may also be from vessels of earlier date; compare again parallels from Maliq (kantharoi with high-swung handles dated to Maliq Phase III, d2–d3 [Prendi 1977–1978: pl. IX, 54]). Several of the vertical strap handles (e.g., **9/116** [P357A], **9/117** [P055], **9/118** [P056], **9/119** [P229]) preserve a tighter bend than that seen on any of the Lofkënd types and may derive from vessels more similar to the light fabric or semi-coarse types, in which such handles (often on larger vessels) are more common; thus, similar types from Pazhok have been compared with the so-called Devollian pottery of EIA date (Bodinaku 1982: pl. II, 2, 3; cf. also Andrea 1996: pl. IV, 11 from Zagorë). The fragment **9/107** (P142), which may belong in this group of sharply bent vertical strap handles, might also be part of a lug, another feature not seen on any of the preserved Lofkënd types. The ridge along the vertical strap handle **9/110** (P103) is unusual for a small vessel of this fabric (cf. the larger coarseware handles **9/274** [P135] and **9/275** [P238] for similar forms at Lofkënd; cf. Hoti 1982a: pl. VI, Tomb 10, and Hoti 1986: pl. IX, 6 for parallels at Bardhoc and Kënetë, respectively). Also unusual is **9/120** (P021), which appears to preserve part of a horizontal handle curving down and away from the rim.

Fewer fragments from large vessels are preserved in dark fabric than in any other fabric group; most notable here are the everted rim fragments **9/153** (P151) and **9/154** (P169), both from large open vessels. Although the profiles of these pieces roughly resemble that of the complete vessel **9/93** (P223), both are from much larger vessels and may actually be from a slightly earlier period (cf. Andrea 1985: pl. IX, varri 66—twelfth–eighth centuries BC [dated LBA by Bejko (personal communication)]; Korkuti 1971: pl. VIII.3 [Tren, EIA, but in LBA shape]; Andrea 1985: pl. IX, Tomb 70 [eleventh– eighth centuries]; Aliu 2012: pls. XL, 429; Prendi, Lera, and Touchais 1996). One final noteworthy piece is **9/126** (P191), which is not only of an unusual fabric somewhere between fine dark and fine light, but also preserves a shallow curve that might be attributed to either a shallow bowl or a spout—in either case, a form not seen in any of the complete types described above (cf. also **9/127** [P089], though this is less unusual). Shallow bowls of EIA date have been found elsewhere (e.g., at Kamenicë [Bejko forthcoming: 294/2084]), but most small spouted vessels are of somewhat later date (e.g., Durić, Glišić, and Todorović 1975 (Romaja): pls. 54, 58, 64).

Semi-coarse

Fragments of semi-coarse fabric account for the largest percentage of inventoried pottery at Lofkënd, despite the fact that only two complete vessels (**9/159** [P303], **9/160** [P362]) and a very fragmentary third (**9/161** [P201]) survive. Unlike the two fineware fabric groups, which have been subdivided by the color of the clay matrix (itself perhaps a result of firing technique, as mentioned above), the semi-coarse fabric group includes pieces in a relatively wide range of colors. In fact, some semi-coarse pieces resemble those known from finer fabric, both in shape, color, and surface treatment—with some of the lighter pieces, for example, bearing matt-painted decoration, and several darker ones highly burnished. In addition to their fabric, however, these semi-coarse pieces also differ from fine-light and fine-dark fabric pieces by their larger size, which itself may have necessitated the use of coarser fabric; in other words, semi-coarse fabric may in many cases derive from the same clay as finer fabric, whether with additional temper added or fewer impurities removed from the raw material.

Type 1—One-handled drinking vessel (Lofkënd Phase II)

The semi-coarse one-handled drinking vessel Type 1 is represented by only a single example at Lofkënd (**9/159** [P303]), although the general shape is not unfamiliar either at Lofkënd or at contemporaneous sites. Although certainly most similar at Lofkënd to dark fabric Type 1 (**9/90** [P422]), it can be easily distinguished from its finer cousin by the rounded, irregular base, the absence of plastic decoration, and the generally cruder construction. Likewise, the coarser fabric and crude construction, as well as the thick handle and flat rim, distinguish it from the finer light fabric Types 2a and 2b (with thinner handle and everted rim). Because of the irregularities and asymmetries in form, precise comparanda are difficult to find, but the piece can be profitably compared with similar EIA vessels from Kamenicë (Bejko forthcoming: 74/1971, 34/2118) and Barç (Andrea 1985: pl. I, Tomb 4).

Type 2—Baby feeder (Lofkënd Phase II)

The semi-coarse baby feeder Type 2 is a very unusual type represented by just a single example at Lofkënd (**9/160** [P362]). Both its shape and its findspot (just above the infant burial Tomb XV [Grave 80]) suggest that "baby feeder" is indeed an apt name, but one cannot be certain about its function. Like Type 1, Type 2 is asymmetrical and crudely made, with little apparent attention given to surface treatment. A roughly comparable piece at Luaras has been assigned an LBA date (Aliu 2004: pl. XVII, 216, Tomb 182), but that from Barç 1 comes closer to the Lofkënd example in terms of both shape and date (Andrea 1985: pl. X, Tomb 89), and Bronze Age examples from the Aegean have been collected by Pomadere (2007). Similar vessels from later periods have been found at other sites as well (e.g., Aliu 1984: pl. IX 113, p. 63; Jubani 1983: pl. VII, Tomb 1, 68; pl. IX, Tomb 1, 112).

Uncertain fragments and additional types

As mentioned above, the vast majority of semi-coarse material comes in the form of fragments. Aside from **9/166** (P120) and **9/182** (P035), two fragments of vertical strap handles similar in form to Type 1 (though perhaps slightly larger in size), no fragments can be confidently attributed to either of the types above. The many joining fragments of **9/161** (P201) provide an important indication of another semi-coarse vessel type, probably medium in size and open in shape (most likely another drinking vessel), with flat, very slightly raised base, globular body, and vertical strap handle (or handles, though only one is preserved), ovoid in section. Unfortunately, these fragments do not preserve enough of the original vessel to allow a more specific reconstruction of the type, or listing of reliable parallels from other sites. Most of the other surviving fragments of semi-coarse fabric derive from vessels at least as large as **9/161** (P201), and one must look beyond the surviving types to understand the semi-coarse pottery from Lofkënd.

A number of the semi-coarse fragments, especially highly burnished spur handles (**9/173** [P314], **9/174** [P205], **9/176** [P289], **9/190** [P098]) and horizontal handles (**9/184** [P099], **9/185** [P245], **9/186** [P344], **9/187** [P345]), seem to be of LBA date, perhaps even predating the active life of the tumulus. Less finely finished handles of similar shape (e.g., **9/170** [P140], **9/172** [P401], **9/189** [P088], **9/181** [P111], **9/168** [P116], **9/188** [P076], **9/171** [P203], **9/175** [P293]) might also be of Bronze Age date, but could just as easily be later continuations of earlier forms (as seen, for example, on the nearly complete fine dark fabric vessel **9/93** [P223]; cf. also the fragmentary handle of EIA date from the settlement of Symizë [Lera 1992: pl. II, 4]). Likewise, widely flaring rims with flat lips (e.g., **9/222** [P017], **9/224** [P040], **9/225** [P053]) seem to derive from large vessels that might date as early as EBA (cf. Jubani 1995: pl. XII, 10), while slightly flaring rims from large open forms (**9/232** [P377], **9/233** [P380]) find LBA parallels (Koçi 1991: pl. II, 30–31). With so little preserved, however, one cannot be at all certain about either the original vessel forms or the date; EIA parallels can be found for other pieces from this same group (cf. **9/227** [P366] with Andrea 1995: pl. I, 10; cf. **9/232** [P377], **9/233** [P380] with Prendi, Lera, and Touchais 1996: fig. 10, 14, 15). Similarly, the large body fragment **9/244** (P288) finds parallels from the Early Bronze Age (e.g., Tumuli 2 and 3 at Shtoj [Jubani 1992: pl. V, 10]), but does not preserve enough of the original vessel to allow confident dating. Semi-coarse rim fragments also find parallels across time periods, from at least LBA (cf. **9/214** [P069] with Andrea 1985: pl. XVI, Tomb 172, 2) to EIA (cf. **9/219** [P083] with Andrea 1985: pl. XIV, Tomb 150, 2; cf. **9/210** [P208] with Andrea 1995: pl. II, 23; cf. **9/211** [P310] with Ylli 1988: pl. I, 2).

The most noteworthy remaining fragments are matt-painted pieces such as **9/220** (P168), which not only preserves familiar EIA painted motifs, but also has a form roughly similar to those of the fine light fabric Type 2, albeit with a small irregularity in its everted rim—either the rise toward a handle in a larger, coarser version of LF Type 2, or else a rudimentary spout for a pouring vessel; for a similar shape from Kamenicë painted with different motifs, see Bejko (forthcoming: Q143). Other semi-coarse rims, including one with matt-painting (**9/221** [P057]), are widely flaring with flat lip, evidence of vessels much larger than those fully preserved. Also noteworthy are the body and neck fragments of large matt-painted pieces, especially the unusual flaring neck **9/247** (P196), which finds an EIA parallel at Tren (Korkuti 1971: pl. IX). Finally, the matt-painted base fragments **9/251** (P105) and **9/254** (P335), similar in form to the fine light fabric bases **9/88** (P001) and **9/87** (P097 + P237 + P285), respectively, again underline the use of semi-coarse fabric to create larger, coarser versions of fine light fabric types, particularly in later periods (cf. Bejko forthcoming: Q1154, Q44/59, dated eighth–seventh centuries). In addition to these matt-painted pieces, a few projections or lugs (**9/255** [P302], **9/256** [P117], **9/257** [P214]), unfortunately preserved without any trace of the vessel walls to which they were once attached, suggest that plastic decoration similar to that seen on several of the intact fine light fabric pieces was also used on semi-coarse vessels. The one incised semi-coarse piece (**9/237** [P209]) is anomalous, as discussed in the following section on incised and punched decoration.

Coarseware

Like semi-coarse fabric, coarseware includes a wide range of different colored clay and tends to be used for larger vessels not deposited in tombs. Unlike semi-coarse fragments, however, few coarseware pieces even resemble smaller types seen in finer fabrics. No coarseware pieces are matt-painted, for example, and only a few preserve plastic decoration; of these, perhaps three (**9/305** [P011], **9/311** [P360], **9/312** [P087]) can be plausibly compared with the plastic decorations seen in fineware, while the remainder must be from large coarse vessels of types discussed in this section. Although only one piece (**9/259** [P283]) is preserved to the extent necessary to classify it as a known type, several fragments provide important information about otherwise lost types.

Type 1—Biconical amphora with vertical neck

The nearly complete vessel **9/259** (P283) is discussed elsewhere (Chapter 15, **15/12**) with regard to its unusual find context and the large amount of bitumen found within and around it. But the piece is equally important for the information it provides about an otherwise unknown ceramic type—the biconical amphora with vertical neck, or coarseware Type 1. Although about one quarter of the vessel is still missing (or preserved only in non-joining fragments), only the precise shape and number of handles and form of the toe or foot are lacking; the biconical body, rounded shoulder, and tall vertical neck with everted rim and rounded lip are quite apparent. The toe or foot, which has a rough and slightly rounded underside, is smaller in size than similar amphorae from other sites and finds no precise parallels at other sites.

Uncertain fragments and additional types

Among the coarseware fragments, **9/290** (P224 + P282 + P411) is most notable, preserving much of the upper profile of a large or very large open vessel with S-curve beneath flat rim with external projection. Another rim fragment (**9/291** [P244]) is likely from the same vessel but offers no further information on its form (and see also similar rim fragments from Prodan [Aliu 1984: pl. X, 136] and Pazhok [Bodinaku 1982: pl. II, 14]). Other rims of similar form from very large vessels (**9/299** [P145], **9/300** [P065]) also find parallels outside of Lofkënd, with dates ranging from MBA (at Cetush [Korkuti and Bunguri 1996: pl. III, 2]) to the Archaic period (at Kënetë; see Hoti 1986: pl. V, Tomb 10). Two different base fragments of similar fabric (**9/315** [P281], **9/316** [P325]) provide possible indications of the form of the base for this vessel. Likewise, the unusual spur handle **9/265** (P136) is of similar fabric and could also be associated with this type, although a slight projection from **9/290** (P224 + P282 + P411) might represent part of a mostly lost attachment for a very different type of handle, no longer preserved.

Other coarseware fragments include a variety of vertical strap handles, often quite large but otherwise similar to those preserved in other fabrics. Two such handles with pronounced central ridges (**9/274** [P135], **9/275** [P238]) find EIA parallels at the settlement of Zagorë in northeastern Albania (Andrea 1996: pl. XIII, 15). Coarse rim fragments, on the other hand, tend to differ from those seen in finer fabrics, with many more flat and flaring rims more appropriate for large, thick-walled vessels. Although too little is preserved of these vessels to assign them to a particular type, they are generally compatible with an EIA date (cf. **9/285** [P154] with material from Bujan [Andrea 1995: pl. II, 23], **9/302** [P313] with pieces from Barç [Andrea 1985: pl. XXI, Tomb 19] and Prodan [Aliu 1984: pl. X, 135]). A single stratigraphic unit at the settlement of Belsh also provides comparative material suggesting a late date (perhaps too late) for many Lofkënd coarseware fragments (cf. Ceka 1977–1978: pls. V. 1–5, VI.1– 4 [sixth century]).

More interesting are a group of unusual rim fragments with sloping walls and partially preserved circular mouths (**9/286** [P230], **9/287** [P231], **9/288** [P176], **9/289** [P161]), which seem to preserve portions of hole-mouthed jars or similar pieces. General comparanda for such vessels can be found in Mycenaean pottery (Mountjoy 1993: figs. 19, 21, 22, 77), and more specific pieces in the Middle Helladic layers at Kastanas in Macedonia (e.g., Hochstetter 1982: pl. 2.4). Since the mouth is the most distinctive portion of these vessels, it is not surprising that other parts of such pots have not been identified. Also distinctive are some very large wall fragments with relief decoration (e.g., **9/310** [P261], **9/313** [P333], **9/314** [P387 + P408]), which almost certainly derive from pithoi or other very large, coarse vessels, and may date as early as the Early Bronze Age, based on parallels from Macedonia (Hammond 1972: fig. 8a–f; Hochstetter 1984) and Maliq (Prendi 1977–1978: pl. III.1, 3 [Maliq Phase IIIa–b = EBA]); for possible later dates, see EIA finds from Tren (Korkuti 1971: pl. 10) and Rehovë (Aliu 2012: pl. I, 28, Tomb 9). Several additional large wall fragments (**9/305** [P011], **9/306** [P141], and two smaller, uncatalogued fragments [P390, P396]) probably also derive from pithoi but are less distinctive and therefore more difficult to date. Among base fragments, **9/318** (P354) may also derive from a pithos (perhaps EIA—see Andrea 1985: pl. XIII, Tomb 122), while the bases **9/317** (P100) and **9/319** (P309) are less massive and may derive from open vessels, perhaps quite late in date (cf. Bejko forthcoming: 1/167, 34/385 [eighth–sixth centuries]).

Finally, there are several unusual coarseware fragments worthy of mention. The fragment **9/320** (P090) preserves two parallel surfaces and a rounded edge, either from a vertical strap handle or an uneven rim, while **9/321** (P180) preserves three finished edges, the outermost bearing one and a half finger impressions, either from a lug handle (on a pithos) or the rim of a large open basin. Six additional pieces (**9/323** [P146], **9/324** [P147], **9/325** [P226], **9/326** [P383], **9/327** [P153], **9/328** [P327]) form a loose group, characterized by a single, straight, finished edge and either unusual curvature or no curvature at all. Although they share a roughly similar red, coarse fabric, they differ in size, shape, and state of preservation; one is from grave fill, and the others from tumulus fill, but all must be ancient. Many preserve large sections with little to no curvature, making it difficult to envision them as pieces of vessels. Although this might suggest to some that they could be tiles, without any known LBA–EIA tile comparanda, this cannot be proven. More likely possibilities include tripod legs (cf. Papadopoulos 2005:479–81); very large vessels in which small sections would not preserve much curvature; or cooking or kiln furniture with straight or flat components (e.g., *pyraunos* [cf. Hochstetter 1984:155–164, pl. 45.9, 10, pl. 59.6, pl. 65.3, pl. 102.5; Morris 2009–2010:52, fig. 41; also found in Albania at Maliq: Andrea 2005–2006: pl. III.7]).

Other fabrics

In addition to the prehistoric pieces discussed above, 12 sherds and groups of joining sherds of non-prehistoric fabrics were also found at Lofkënd, all from topsoil or neighboring units of tumulus fill. All are quite small, and little more can be said about them than what is already listed in the catalogue entries. The misfired and reduced sherds (**9/337** [P340], **9/338** [P301], **9/339** [P338], **9/340** [P139]) might be prehistoric but are noticeably different from the vast majority of pieces found at Lofkënd, including the many pieces of fine dark fabric that are also reduced. Because they cannot be identified with much certainty, these four pieces do not add much to our understanding of the tumulus or the people who created it. Of more interest are the single early modern sherd and the various wheelmade Greek imported sherds. The former (**9/341** [P273]) is apparently a bit of so-called "Painted Ware from Grottaglie and/or Corfu," very late in date (probably nineteenth or twentieth century AD; see catalogue entry for references).[4] This late date and the topsoil findspot of the sherd could be compatible with the modern burials of the tumulus, but the piece is probably unrelated to the active use of the burial mound. The wheelmade Greek sherds, on the other hand, seem to date from the late Archaic or early Classical periods, shortly after the tumulus itself went out of use. Most noteworthy are the numerous badly worn sherds seemingly derived from a single Corinthian kotyle or skyphos, perhaps of the sixth century (**9/332** [P175], **9/333** [P257], **9/334** [P328], **9/335** [P372], **9/336** [P384]; cf. Blegen, Palmer, and Young 1964:106–108). Since this is about the time that Greek colonization begins in this region, such finds are not completely out of place, but one wonders how exactly they ended

[4] This pottery takes its name from Grottaglie, the main production center in Apulia, and from Corfu, where Italian potters made imitations. We must thank Joanita Vroom for taking the time to examine a drawing and photo of this piece and provide us with her expert opinion.

up in the tumulus soil, even if only around the outer edges—did people continue to visit the burial mound even after it went out of active use? (For further discussion, see Appendix 2, Epilogue.)

DECORATIVE TYPES AND MOTIFS

Decoration of pottery at Lofkënd occurs on complete vessels deposited as kterismata as well as on fragments of many different types found in grave fill, tumulus fill, and topsoil. The most recognizable decorative techniques include matt-painting and mastoi (small projections) on fine light and semi-coarse wares, and *kanellure* (fine parallel ribbing) on fine dark fabric. Incised and punched decoration also occurs, but only on a few fragments, almost entirely of fine light fabric (it is also rare at Sovjan in southeast Albania; see Lera, Oberweiler, and Touchais 2011: 46). This section provides both quantitative and qualitative analyses of all means of ceramic decoration found at Lofkënd.

Quantitative analyses

Given the relatively small size of the Lofkënd corpus, quantitative analysis must be pursued with caution, particularly for the complete vessels. Nearly 90% of these whole vessels, for example, bear some type of decoration, but the type and frequency of decoration correlates more with the type of fabric than with any other factor. Thus, while the two complete semi-coarse vessels (**9/159** [P303] and **9/160** [P362]) are undecorated, all of the complete fine light kterismata bear matt-painted decoration, many with mastoi as well. Similarly, four of six fine dark fabric vessels (67%) are decorated with *kanellure*, and none is painted.

A greater variety of decoration is found on sherds from the tumulus fill, although decorated sherds remain far less common than undecorated ones in every fabric group. Fragments of fine fabric tend to follow the decorative types seen on the complete vessels (**Fig. 9.2**), with matt-painted decoration found on light fabric but not dark, and *kanellure* decoration found on dark but not light. In addition, a number of fragments of fine light fabric bear incised decoration. Decoration is also found in the semi-coarse category, primarily in the form of matt-painting (**Fig. 9.3**). Plastic decoration appears less frequently, as illustrated below. Still less decoration is encountered in the coarseware, with only a few sherds from large vessels bearing plastic decoration (**Fig. 9.4**).

Qualitative analyses

The matt-painted decoration

Matt-painted decoration occurs at Lofkënd on both fine light and semi-coarse fabric, but never on fine dark fabric or coarse pottery. Motifs consist entirely of rectilinear (rather than curvilinear) elements, and there is a distinct preference—seemingly regardless of vessel size—for horizontal bands around the neck and handles, and hatched or solid pendent triangles on the shoulders and lower body. In terms of chronology, matt-painted decoration can be definitively associated with Phases II–Va on the basis of its appearance on kterismata from well-stratified tombs at Lofkënd. More precise dates can be assigned to each of the intact vessels of fine-light fabric, all of which bear matt-painted decoration. Unfortunately, the same cannot be said for fragments, which constitute the largest body of matt-painted material from Lofkënd, including many motifs not seen on complete vessels. Thus, where motifs occur on both complete vessels and fragments, the latter can be given a rough date by analogy to the former. But for matt-painted motifs found only on fragments from topsoil and tumulus fill, it is not possible to give a date more precise than the eleventh through ninth or perhaps eighth centuries (i.e., Lofkënd phases III–Va). Although imprecise, this wide range is generally compatible with dates given from both complete vessels at Lofkënd and comparable pieces from contemporaneous sites. The remainder of this section reviews the matt-painted pottery from Lofkënd, in chronological order where possible.

The earliest matt-painted patterns are those found on **9/1** (P350), dated to Phase II (twelfth–eleventh centuries). The hatched and unhatched zigzags, the lattice bands, and the pseudo-Maltese cross are all unparalleled at Lofkënd, on either sherds or complete vessels.

With Phase III (eleventh–tenth centuries), much more common patterns appear, with firmly dated examples found on complete vessels, and similar motifs on fragments from the fill. Particularly noteworthy are the pairings of horizontal bands at shoulder or neck above pendent triangles—usually hatched—stretching toward the base. Horizontal bands, and occasionally wavy lines or zigzags, also occur on neck and lip. Groups of bands, sometimes cross-hatched, are common on handles. All of these patterns continue on complete vessels into Phases IV and Va (late tenth–earlier eighth centuries), with similar motifs found on fragments from the tumulus fill that cannot be so precisely dated; such fragments likely derive from vessels produced in Phases III–Va but could have originated in earlier or later periods as well. As **Tables 9.3** and **9.4** indicate, various combinations of vertical and horizontal bands and zigzags, as well as hatched pendent triangles, occur on complete vessels from tombs as well as on sherds from the tumulus fill, while a number of other motifs are restricted to one context or the other.

In addition to general vagaries of survival, these differences in the findspots of the various motifs are perhaps most attributable to vessel size, with large vessels—whether fine or semi-coarse in fabric—able to accommodate much larger pendent triangles (and perhaps rectangles in some cases [e.g., **9/247** (P196)]) with a variety of types of filling ornament—from the hatching seen commonly on smaller vessels to cross-hatching (e.g., **9/84** [P030], **9/247** [P196]) or lozenges (e.g., **9/248** [P234]). Such designs are not very common elsewhere, but where known, they are found placed vertically on the necks of medium- to large-sized vessels of semi-coarse fabric (cf. fragments from Barç [Andrea 1985: pl. IV, 4, 265] and Tren [Korkuti 1971: pl. IX]). Similarly, the upright triangles seen on fragments of larger vessels of fine light fabric from Lofkënd (**9/80** [P284], **9/84** [P030], **9/72** [P058 + P391], **9/83** [P255 + P112]) find parallels on large vessels at other sites. The receding hatched triangles seen at Lofkënd (**9/245** [P423], **9/246** [P249]) find no precise parallels. Cross-hatched pendent triangles are present on only two sherds, both belonging to medium- to large-sized vessels of fine light (**9/84** [P030]) and semi-coarse fabric (**9/245** [P423]). As with the hatched pendent triangles found on several complete vessels of fine light fabric, these are located on the lower body, although the vessel shape is notably different. This type of design is common on the so-called "Devollian ware," which appears at Maliq in the Late Bronze Age (Phase III, levels D1–D3 [Prendi 1985a: 186, fig. 2]).

Regional considerations

The matt-painted decoration from the Lofkënd tumulus is unique from a regional perspective, comprising the only such material yet known from this region of central Albania. Strikingly similar to the above-mentioned "Devollian Ware" of southeastern Albania (near the eponymous Devoll River and its surrounding valley, as well as the numerous tumuli of the Korçë basin and Kolonjë plateau, and the sites of Maliq, Tren, Barç 1, Kamenicë, Luaras, Rehovë, etc.), the matt-painted motifs found at Lofkënd also find parallels in northwestern Greece (e.g., at Vitsa [Vokotopoulou 1986] and Liatovouni [Douzougli and Papadopoulos 2010]) and southern Italy (e.g., at Salento). Probably the best comparanda for the Lofkënd matt-painted pottery, however—in terms of both matt-painted motifs and vessel form—derive from sites within the Korçë basin of southeastern Albania. It should be noted that such close parallels are partly the result of a good record of excavation and publication from this region; future excavations elsewhere may well provide additional comparanda. Especially important is the material from the first phase of the Barç 1 tumulus, dated to the eleventh–eighth centuries. Some 12 vessels from Barç display the combination noted above as a Lofkënd preference, with horizontal bands on the neck and pendent triangles on the lower body (Andrea 1985: pls. I–XI, 262–272). Also as at Lofkënd, here the preference is for rectilinear rather than curvilinear ornament. With a larger repertoire at Barç, however, several additional vessel shapes and painted motifs are present. Neither large amphorae nor double and triple matt-painted vessels, for example, survive complete at Lofkënd (but see above for fragments that may derive from large amphorae, including, e.g., **9/74** [P252], **9/75** [P132 + P324], **9/82** [P305], **9/83** [P255 + P112], **9/84** [P030], **9/87** [P097 + P237 + P285], **9/88** [P001], and **9/89** [P232]).

Other than the handle fragments **9/36** (P066) and **9/107** (P142), for which no reconstruction is certain, there are no Lofkënd fragments that can be plausibly connected with a double vessel. Similarities with material from the tumulus of Kamenicë are also present, though to a lesser degree. Approximately 10 vessels dating from the tenth–ninth centuries, for example, show the distinctive combination of horizontal bands and pendent triangles (e.g., Bejko forthcoming: Q001, Q476, Q1517; Agolli 2009, 2014). In the accompanying figure, a few representative examples from these sites are placed alongside those from Lofkënd in order to underline the similarities (**Fig. 9.5**).

At sites located farther from the Korçë basin, similarities are also present, though to a lesser degree.

Finds from several late prehistoric cemeteries in northwestern Greece, such as Vitsa, Liatovouni, or Boubousti, for example, exhibit parallels with those from Lofkënd. These relate primarily to the technique and style of decoration rather than to vessel form. Thus, amphorae and jugs with cutaway necks found at the Early Iron Age cemetery of Vitsa (eleventh–ninth centuries), for instance, are painted with horizontal bands and pendent triangles similar to those seen at Lofkënd. Likewise, certain vessels from the Liatovouni cemetery display painted motifs like those discussed above. A few selected pieces are illustrated here to show these similarities (**Fig. 9.6**). In both of these cases, however, one finds similarities with Lofkënd on only a very small proportion of the excavated material; most of the vessels from these sites are quite different in form and decoration from those excavated at Lofkënd (Vokotopoulou 1986; Douzougli and Papadopoulos 2010; Heurtley 1926–1927).

Close comparanda from southern Italy are still less common, consisting of only a few specific designs, such as the pseudo-Maltese cross, solid pendent triangles, or horizontal bands found at Salento in the late ninth century. Not only are these later in date than most of the painted material from Lofkënd, but they may in fact be imports rather than local products (Yntema 1990:56). Because local production of matt-painted pottery in southern Italy seems to be generally later in date than the earliest such pottery from southeastern Albania, it would seem that the direction of influence was from east to west, with matt-painted pottery developing first in Albania and then spreading, whether through the movement of pots, potters, or a combination of the two (Kilian 1974:260; Yntema 1990:57).

If scholars generally agree that matt-painted decoration developed on the eastern side of the Adriatic Sea, however, there is no consensus on its precise place of origin. For Prendi (1985), Korkuti (1969), and Andrea (1985, 2005–2006), the Devoll Valley is the site of origin (hence the name "Devollian"), while other scholars see the core of this tradition in western Macedonia (e.g., at Boubousti, eponymous site of Heurtley's [1926–27] "Boubousti ware"). Psaraki and Andreou (2010), for example, claim that during the late phase of the Middle Bronze Age and the Late Bronze Age, no matt-painted pottery style was imported from outside Macedonia. For Vokotopoulou (1986:364–366), "the Northwestern matt-painted ware" is a continuation of the Middle Helladic matt-painted tradition of central and southern Greece, appearing first in the region of the Axios River valley (known as the Vardar farther north) and in the area of Naousa and western Chalkidike. To fill in the missing links across both time and space, however, Vokotopoulou must rely on the migrations of the so-called "Dorians" mentioned by Herodotos (and maintained by early scholars such as Heurtley).

Such a "Dorian" link is specifically refuted by Bouzek (1985), who points out the minority of painted wares as compared to their unpainted counterparts in the entire repertoire of LBA/EIA pottery. More recently, other scholars have also deviated from the discussion of the Doric migrations, though still connecting the emergence of the matt-painted ware to the north with the earlier traditions of Mycenenan pottery. The geometric designs, for example, are considered regional imitations derived from contacts with the Mycenaean world (Hänsel 1989; Gimatzidis 2010). Likewise, according to Tartaron (2004:85–86), the similarities between LBA/ EIA Devollian and Boubousti ware may be interpreted by means of regional interactions dating back to the Neolithic period, with east–west river valleys allowing relatively easy accessibility and facilitating cultural connections among southeastern Albania, Epirus, and Macedonia (and perhaps even extending to present-day Thessaloniki). Most recently, Horejs (2007a) has argued persuasively that the matt-painted pottery of the north was neither a purely indigenous Late Bronze Age invention nor a product of people migrating from central or southern Greece (see also Horejs 2007b:218–286 for a useful overview of matt-painted pottery in the north Aegean and in the Balkans). Rather, in her words, following "long, extant and apparently continued contact" with those regions, potters in the north adopted and adapted "[t]echniques, ornaments and compositions . . . to local traditional pottery. And so a specific north Greek category of pottery emerged at the beginning of the Late Bronze Age, whose origin is located in Middle Helladic central and southern Greece" (Horejs 2007a).

But how does the matt-painted evidence from the Lofkënd tumulus fit into this discussion? With AMS ^{14}C dates showing that matt-painted pieces in Lofkënd date as early as those found anywhere else (i.e., twelfth-eleventh centuries), the conventional chronology must be revisited. For the AMS ^{14}C applications could considerably change the existing relative chronologies, perhaps helping to resolve the prolonged scholarly debates over the origin of the matt-painted pottery. At the same time, petrographic analysis might help to determine whether **9/1** (P350) and other early matt-painted pieces from Lofkënd were locally produced or imported from the south (although it is clear in any case that there was interaction at this very early date between Lofkënd and areas much farther south, whether the Korçë basin or the Devoll Valley). A single protagonist in the development of matt-painted decoration seems unlikely, and it is hoped that future research—into regional ceramic variations as well as the exchange of other goods and materials—will help to explain the development of this complex phenomenon.

Plastic Decoration

Plastic decoration at Lofkënd occurs on all four types of fabric: fine light, fine dark, semi-coarse, and coarse. Because the types of decoration differ according to fabric, each is treated separately below (**Tables 9.5, 9.6**).

Plastic decoration in fine light fabric

With a single exception (**9/71** [P127] below), all plastic decoration found on fine light fabric at Lofkënd occurs in combination with matt-painted designs. In each of these cases (**9/3** [P304], **9/5** [P326], **9/6** [P073], **9/8** [P277], and **9/14** [P256]), the plastic decoration consists of several mastoi, or projections, placed near the point of maximum diameter. These are usually ovoid or roughly rectangular in section, and located at the front of the vessel, directly opposite the handle. The mastoi are not always evenly spaced. Although each of these instances is on a complete or nearly complete vessel of Phases III–IV (eleventh–ninth centuries), there are too few examples to draw any meaningful chronological conclusions. Already noted above is the unusual set of four, rather than the usual three, mastoi on **9/5** (P326). Because the mastoi do not seem to serve any identifiable function, they are classified here as decorative (and perhaps even symbolic, if one follows the meaning of their name and associates them with female breasts). Parallels for similar matt-painted vessels with mastoi have been found at the tumuli of Barç and Kamenicë in southeastern Albania (Andrea 1985: tab. V, V 34, tab. X, Tomb 81; Bejko forthcoming: Q2156, Q001, Q171), as well as in the cemeteries of Vitsa (Vokotopoulou 1986:46, 6, 5945/T 180) and Liatovouni (P8092; for the matt-painted pottery from Liatovouni, see Douzougli and Papadopoulos 2010:44–45, figs. 17–18). A single triangular projection is preserved on an unpainted body sherd deriving from a medium-sized vessel (**9/71** [P127]) and may be a mastos, though its shape is more like that of a functional ledge lug seen on fragments from larger vessels of coarser fabric (e.g., **9/312** [P087], **9/311** [P360], **9/304** [P189], **9/257** [P214]).

Plastic decoration in fine dark fabric

Plastic decoration in fine dark fabric is restricted almost entirely to so-called *kanellure* decoration, or fine parallel ribbing, which occurs on four complete vessels from tombs (DF Type 1 [**9/90** (P422)], DF Type 4 [**9/94** (P322)], and DF Type 5 [**9/95** (P228) and **9/96** (P276)]), together with two sets of body sherds (also DF Type 5 [**9/98** (P032), **9/99** (P050)]) (**Figs. 9.16–9.18**). The earliest of these, **9/90** (P422), dates to Lofkënd Phase II (twelfth–eleventh centuries) and has more widely spaced grooves or ribbing, unlike those found in tombs of later date. Comparanda have been found at several Late Bronze Age sites to the south or southeast of the Lofkënd tumulus (e.g., Pazhok [Bodinaku 1982: 49– 101], Kastanas [Hänsel 1985:238]), but this type of decoration may derive originally from the Late Bronze Age Urnfield culture in central Europe.

Kanellure decoration continues into the Early Iron Age with Lofkënd DF Types 4 and 5 (in Phases III–IV [eleventh–ninth centuries]). The parallel ridges remain below the necks of the vessels but become narrower, deeper, and more vertical, never sloping as much as on Type 1. Although no precise parallels for these vessels have been found elsewhere, vessels with a similar type of ribbed decoration have been found in the tumuli of Bujan (Andrea 1995: pl. 1, Tuma 1), Myç-Has (Bela 1990:pl. III, Tomb 6, 36, pl. XVI, 269, Tuma 8, 296), Bardhoc (Hoti 1982: pl. X, 7), and Shtoj (Jubani 1992: pl. 2, Tomb 3, pl. IV, Tomb 6]; in these regions of northern Albania, this decorative feature is considered an indispensable element of the pottery (Bodinaku 1982: 72). In a recent study of the pottery from Maliq, Andrea (2005–2006:5–57) divides *kanellure* decoration into two versions: vertical and horizontal. In her view, the vertical version—which is also encountered at Lofkënd—is the more popular. Andrea does not offer any theory regarding the origin of this style, nor cite any regional preference within Albania. Despite the fragmentary evidence, vertical *kanellure* decoration seems to become popular primarily during the Early Iron Age, both in southern Albania, at the sites of Maliq and Tren, and in northern Albania, at the settlement of Zagorë (pottery stratum II) (Andrea 2005–2006:17–18, pls. 13, 14; Prendi 1966:267, pl. XVII, 18; Korkuti 1971:40, pl. 10). Quite contrary to the matt-painted decoration discussed above, the *kanellure* decoration found at Lofkënd indicates a connection with regions to the north, perhaps as far as present-day Kosovo.

A second type of plastic decoration found on fine dark fabric occurs on two separate wall fragments from mid-sized vessels of uncertain shape, each with a single, small, circular projection. The first can be described as a nipple (**9/155** [P421]), the second as a flat pellet lug or perhaps a misproduction (**9/156** [P316]). To our knowledge, neither finds a precise parallel outside of Lofkënd.

Plastic decoration on semi-coarse fabric

Unlike plastic decoration on fineware, which occurs on complete vessels and can therefore be easily connected with specific types, semi-coarse plastic decoration is confined to just three fragmentary projections. All are from tumulus fill, and none preserves any vessel wall. Two (**9/255** [P302] and **9/256** [P117]) are round in shape, similar in form to light fabric mastoi, while a third (**9/257** [P214]) is more triangular. Although perhaps functional lugs, these find comparanda on numerous EIA amphorae (e.g., Aliu 2004: fig. 51, amphorae a, 95, fig. 42, p. 85; pl. XX, 238; Korkuti 1984: pl. 4, 41, p. 58), suggesting that they should be dated to Phases III–IV.

Plastic decoration on coarseware

As with semi-coarse fabric, plastic decoration on coarseware occurs only on fragments, all from topsoil or tumulus fill. These can be divided roughly into two types: projections and relief decoration. In addition to one mastos-like projection (**9/304** [P189]), there is another round projection of different form (**9/310** [P261]), as well as two triangular projections (**9/311** [P360] and **9/312** [P087]) similar in form to the semi-coarse fragment mentioned above (**9/257** [P214]). The two fragments with relief decoration (**9/313** [P333] and **9/314** [P387 + P408]) are more distinctive, each bearing a horizontal band of applied clay with finger or tool impressions—perhaps once rope patterns. As noted above, similar pieces have been dated to periods anywhere from EBA to EIA, and far too little is preserved here to allow a more specific date. According to Tartaron, the so-called finger-impressed decoration (K II type) is typical of the late prehistoric pottery in southern Europe (Tartaron 2004:72), but more specifically Albanian comparanda are found at Maliq (Andrea 2005–2006:9, pl. IV, 4), Prodan (Aliu 1984: pl. XI, 137), and Zagorë (Andrea 1996: pl. III, 16).

Incised and punched decoration

Incised and punched decoration occurs at Lofkënd almost entirely on fragments of fine light fabric found in topsoil and tumulus fill (**Table 9.7**). As mentioned above, however, the large number of joins found among such fragments suggests that at least one incised and punched vessel may once have been deposited as a grave good, and then damaged during a later burial. Before looking at the combined use of incised and punched decoration, we consider the independent use of each of these decorative techniques.

Only two fragments from Lofkënd bear incised decoration alone, but both are quite interesting, with Neolithic comparanda. The first of these, **9/85** (P346), is a small wall fragment broken on all sides. It bears five incised concentric arching lines, all but one of which continue beyond the breaks. Since it is the outermost line that has a visible endpoint, the original motif may have been a spiral, though uneven concentric semicircles (or perhaps even incomplete concentric circles) are also possible. The closest parallels for such a motif date to the Neolithic period (at Barç [Lera 1987a: pl. XVI, 13; pl. XVIII, 21] and Tren [Korkuti 1971: pl. IV]), although the context and fabric of the Lofkënd piece do not seem compatible with such an early date (unless its findspot within the fill of Tomb III [Grave 81] is evidence of careful curation of an ancient piece). The other incised fragment, **9/37** (P342), seems to derive from an unusual loop or ring handle, otherwise unattested at Lofkënd. Its incised decoration consists of three roughly parallel zigzag bands on the upper surface. These are aligned with the apparent curvature of the handle, and broken at either end. A similar design from Neolithic Kolsh was published by Korkuti (1983a:69, pl. XX, 13).

Punched decoration is equally rare at Lofkënd, again with only two small fragments preserved, both from small vessels of fine light fabric quite red in color. Each was found in tumulus fill. The first, **9/52** (P246), preserves portions of two hourglasses, broken all around; these could also be triangles or lozenges connected at apices. The second fragment, **9/53** (P414), is even smaller, preserving just five punched dots, arranged in a row horizontally but broken above and at either end so that the original design remains uncertain. Due to the very limited state of preservation, no reliable parallels can be given for this group, although the fabric and probable vessel size and shape are at least generally compatible with an EIA date.

Four fragments (or groups of joining fragments), all from small vessels of fine light fabric, bear a combination of incised and punched decoration. The first two cases (**9/54** [P133] and **9/55** [P093 + P096 + P247 + P370]) comprise a large number of joining fragments found in disparate units of tumulus fill, possibly all from a single vessel. In both sets of fragments, the decoration runs around the shoulder of the vessel: a single horizontal band of punched diagonal teardrops above incised receding pendent triangles. Although LBA parallels exist for the decoration (from the Vardarski valley [Videski 2005: vol. I, T. II. 9–10, p. 106]), the fabric, vessel shape, and similarity to EIA matt-painted motifs all suggest a later date, with comparable pieces known from tumuli at Barç (Andrea 1985: pl. XIII, Tomb 125.1) and Kamenicë (Bejko forthcoming: 234/550), as well as at the settlement of Sovjan (Lera et al. 2011: fig. 10). In contrast to Andrea, who suggests a broad chronological range from the twelfth to the eighth centuries, Bejko places analogous pieces within the first phase of the Early Iron Age, from the tenth to ninth centuries. Two additional joining fragments, **9/56** (P371), preserve a similar punched and incised motif, but with hatched pendent triangles in place of receding ones. The vessel type and placement of the decoration also resemble those of **9/54** (P133) and **9/55** (P093 + P096 + P247 + P370), although here the affinities with EIA matt-painted motifs are even stronger. Finally, **9/57** (P375) shows an incised zigzag, probably vertical, with interstices irregularly dotted. Because it is broken all around, the full extent of this decoration remains uncertain, but as preserved it has no known parallels, either at Lofkënd or other sites.

Only a single fragment with incised decoration was found at Lofkënd outside the fine light fabric group: the semi-coarse wall fragment **9/237** (P209). Found in topsoil, this piece bears incised decoration on its interior surface, roughly in the shape of the ligature Æ. Although the incision clearly continues beyond the break, suggesting that it occurred before the fragment was broken, it also appears to have been scratched through the encrustation, a possible indication that it occurred more recently. The unusual shape and placement of the incision (on the interior surface), as well as its topsoil findspot, also suggest that this may not be ancient decoration.

Distribution

The distribution of all finds within the tumulus is discussed in more detail elsewhere (Chapter 2), but a few comments about the distribution of pottery, in particular, are worth making. First, nearly all of the complete vessels and nearly complete vessels or profiles found at Lofkënd are small to medium in size, of fine fabric, and found in grave contexts, where they were presumably deposited as grave offerings. Those few that were not found in graves (e.g., **9/6** [P073], **9/92** [P364]), including those pieced together from multiple fragments found in different units (e.g., **9/100** [P049 + P063], **9/20** [P290 + P221], **9/55** [P093 + P096 + P247 + P370]), were almost certainly once offerings as well, disturbed from their original contexts by digging for subsequent burials (in much the same way that many loose fragments of human bone were encountered within tumulus fill and topsoil). Of course, the large number of unique fragments and fragments for which no joins can be found clearly indicate that much fragmentary pottery must have been brought to the tumulus, most probably in soil from settlement sites (as also suggested by the presence in tumulus fill of stone tools and bits of impressed fire-hardened clay from wattle-and-daub architecture; cf. Papadopoulos 2006: 81–83, and Chapter 14). Furthermore, as has been noted above, despite the fact that nearly all of the identifiable ceramic types at Lofkënd occur in fine fabric (whether light or dark), the total number of coarse and semi-coarse fragments actually slightly outnumbers the total quantity of fine pieces. With regard to findspots, non-grave contexts (topsoil and tumulus fill) far outnumber grave contexts in the amount of pottery that they contained, while fine light fabric and fine dark fabric were of course overrepresented among grave offerings. More surprisingly, fragments of fine dark fabric were significantly underrepresented within grave fill contexts, where only four such fragments were found. Fine light fabric, on the other hand, appears underrepresented within tumulus fill and overrepresented elsewhere. Finally, and not surprisingly, the few pieces of wheelmade and modern pottery classified above as "other" were found only in topsoil units and the tumulus fill units around them. **Table 9.8** and the graphs in **Figure 9.7** should help to clarify these distributions; the total number of 396 inventoried pieces includes several diagnostic but unexceptional pieces not ultimately included in the catalogue, and counts as single any groups of joined pieces, even those originally inventoried separately—hence the disparities between these figures and totals presented elsewhere for inventoried and catalogued pottery.

CONCLUSIONS

The ceramic material found at Lofkënd is remarkably varied, particularly given the relatively small size of the corpus. It demonstrates many important links with surrounding regions and helps to clarify the chronology of LBA/EIA pottery at contemporaneous sites. Of the four Lofkënd fabric groups, the two different types of fine fabric are most distinctive and suggest that the Lofkënd tumulus was uniquely located at a sort of crossroads of material culture. The fine light fabric, which often features matt-painted decoration and mastoi like those seen on so-called Devollian ware, demonstrates links primarily to the south, from southern Albania to northern Greece and southern Italy. The fine dark fabric, on the other hand, which is highly burnished and sometimes features *kanellure* decoration, finds parallels mostly to the north. Although much of the finest pottery from Lofkënd—and certainly the most precisely datable material—comes in the form of complete vessels deposited in tombs, many more types are represented by fragmentary pieces from the tumulus fill. This evidence, like that found at contemporaneous sites, suggests that many larger types have been nearly lost from the material record. Although the precise details of these types remain uncertain, it is hoped that the illustrations and descriptions published here will allow links to be made with fragmentary material elsewhere, leading ultimately to a fuller understanding of LBA/EIA pottery in this region. Likewise, although the Lofkënd corpus omits many ceramic types seen at contemporaneous tumuli (e.g., double or triple vessels), it is hoped that the typology provided here will serve as a springboard to future publications, giving all archaeologists working in these areas a common frame of reference within which to discuss past and future finds. (See **Tables 9.9–9.13** for conspecti of each fabric group.)

CATALOGUE

Fine Light Fabric

Unless otherwise noted, matt-painted lines measure 0.2–0.3 cm in width.

9/1 (P350) **Figs. 9.8**, **9.36** (Drawing 77.1; Photos 3199–3206)
Kantharos, FL Type 1.
Tomb XVII (72) (SU 1.408).
H (rim): 7.4 cm; H (max): 9.4 cm; D (rim ext): 5.5–7.0 cm; D (mouth int): 5.6 cm; D (max): 10.5 cm; Th (handles): 0.8; Th (walls): ca. 0.3–0.6 cm; W (handles): 1.5 cm; W (handles at rim): 2.4 cm; W (handles at shoulder): 3.4 cm.
Small kantharos, nearly completely preserved; several chips from surface of lower body, excavation damage to top of one handle. Shape basically as **9/95** (P228), with slight differences: handles here nearly rectangular in section, no outward curve in rise from shoulder; conical neck less sharply angled, creating wider rim, here ovoid in plan; lower body–shoulder junction not quite so sharp. Exterior surface smoothed and burnished, resulting in irregular black lines; moderate encrustation in interior.
Matt-painted decoration: On neck, at top and bottom, identical bands of lattice patterns, consisting of three horizontal lines with vertical hatches. At top of shoulder, two horizontal lines run around vessel, stopping at either side of handles; double zigzag, vertically hatched, runs beneath; at base of each handle, modified Maltese cross, with central rhomboid. On upper surface of rim, five sets of 12 to 13 radiating lines, poorly preserved. Just above rim, extending to handle tops, four zigzag lines run across entire width of handle. All lines very fine, measuring less than 0.2 cm in thickness.

Fabric: Hard; abundant grog inclusions (<0.3 cm), moderate black inclusions; occasional mica sparkles and air pockets, most very small (<0.1 cm). Surface 5 YR 6/6 (reddish yellow); isolated 2.5 YR 6/8 (light red) patch on lower body; core 10 YR 5/1 (gray).

Comparanda: Cf. Mati-Glasinac culture for similar forms.

9/2 (P212), **Fig. 9.36** (Photos 2680–2681)

Shoulder/Body Fragment, probably FL Type 1.

Tumulus Fill (SU 2.202).

PH (max): 2.7 cm; PW (max): 2.8 cm; Th: 0.7–1.0 cm; D (max, ext): ca. 15 cm.

Single fragment preserving carinated junction of shoulder and lower body from small biconical vessel. Surfaces smoothed but uneven; heavy encrustation.

Fabric: Fine to semi-coarse (not well-fired), with moderate mica, small white inclusions, and grog. Surfaces 7.5 YR 5/8 (strong brown), core 7.5 YR 4/0 (dark gray).

Comparanda: **9/1** (complete vessel P350).

9/3 (P304), **Figs. 9.8**, **9.36** (Drawing 71.1; Photos 3063–3069)

One-Handled Vessel, FL Type 2A.

Tomb XXXIX (66) (SU 1.0369).

H (rim): 11.5–11.8 cm; H (max): 14.3 cm; H (projections): 2.3–2.5 cm; D (base): 5.2 cm; D (max): ca. 11.5 cm; D (rim int): 8.1 cm; D (rim ext): 10.5–8 cm; Th (rim): 0.5 cm; Th (handle): 1.7–9 cm; W (handle): 2.2 cm; W (handle at rim): 3.4 cm; W (handle at body): 3.7 cm; W (projections at body): ca. 1.0 cm.

Nearly intact complete vessel; damage limited to vessel surface just below handle. Flat base, rising to conical lower body, curving in to tall, flaring cylindrical neck and everted rim with unevenly rounded lip; asymmetry most apparent on rim. Vertical handle curves gently from shoulder, rising above rim; handle roughly rectangular in section, slightly rounded on exterior, unusually thick in section; vessel walls also unusually thick. Surface slipped and smoothed, with encrustation everywhere, especially across one side of vessel.

Plastic decoration: Three vertical projections opposite handle, protruding ca. 0.5 cm from vessel surface, tapering away from body; projections nearly evenly spaced along shoulder at point of maximum diameter, separated by 2.8 and 2.5 cm, respectively; leftmost projection raised 0.4 cm above other two.

Matt-painted decoration: Two horizontal lines just below rim, disappearing on either side of handle; lines not completely parallel, touching at some points, separated by up to 0.3 cm in others. Around neck, three horizontal zigzags, consisting of six or seven strokes each; remnants of three or four similar zigzags on other side of vessel. Just above shoulder, two additional horizontal lines encircle vessel up to either side of handle; lower line supports 13 pendent triangles, each filled with four to six parallel diagonals; on most triangles, two outermost lines come together to form a single line stretching down to base of vessel. On handle exterior, four bands of five horizontal lines. On upper surface of rim, three groups of radiating lines, varying in number from three to five, not well preserved.

Fabric: Sandy, soft; abundant white inclusions (0.3–1.0 cm), moderate quartz and black inclusions, moderate mica, abundant air pockets (<0.5 cm). Surface 7.5 YR 6/6 (reddish yellow); core 10 YR 6/3 (pale brown).

9/4 (P077), **Figs. 9.8**, **9.37** (Drawing 19; Photos 507–510)

One-Handled Vessel, FL Type 2A.

Tomb LXVIII (13) (SU 2.0103).

H (max): 15.5 cm; H (rim): 11.4–12.0 cm; D (base): 5.5 cm; D (max): 12 cm; D (rim int): 7.5–8.0 cm; D (rim ext): 11.3 cm; Th (walls): 0.6–0.9 cm; Th (handle): 1.4 cm; W (handle): 2.4 cm; W (handle at body): 3.3 cm; W (handle at rim): 4.6 cm.

Fully preserved vessel, not completely uniform in shape. Rounded biconical lower body with flat bottom, conical neck, and everted rim. Handle, roughly rectangular in section, joins body just above point of maximum diameter, rises high and curves back down to join rim. Some surface damage, including chips missing from rim, top of handle, and from body.

Matt-painted decoration: very poorly preserved on only one side of vessel; two horizontal lines just beneath swelling of rim. Beneath these lines, two sets of three parallel vertical lines, still less preserved. Very faint traces of another horizontal line (possibly two) just above shoulder, possible traces of vertical decoration (pendent triangles?) beneath. On handle, two faded groups of three horizontal lines on exterior surface above shoulder; unclear vertical decoration above rim.

Fabric: Hard, sandy, with black and white inclusions (including lime) occasional, mica rare; air pockets abundant across exterior and interior surfaces, up to 0.6 cm. Exterior surface slipped and burnished, color 10 YR 7/3 or 8/3 (very pale brown); chipped surface areas and core 7.5 YR 6/6 (reddish yellow). Encrustation minimal on one side, heavy on other, especially concentrated within loop of handle, around shoulder and neck beside handle.

Dated 876 BC ± 46 years with AMS ^{14}C.

9/5 (P326), **Figs. 9.9, 9.37** (Drawing 88; Photos 3446–3453)

One-Handled Vessel, FL Type 2A.

Tomb XXVI (74) (SU 1.0416).

H (rim): 11.9; H (max): 15.8; H (projections): 2.0–2.7 cm; D (base): 5.0–5.2 cm; D (max): ca. 14.0 cm; D (rim): 10.3 cm; D (mouth int): 7.7 cm; D (handle strut): ca. 1.0 cm; Th (handle): 0.7 cm; Th (rim): 0.5 cm; Th (walls): 0.4–0.7 cm; W (handle): 2.7 cm; W (handle at body): 4.4 cm; W (handle at rim): 4.1 cm; W (projections): ca. 1.0 cm.

Nearly intact one-handled vessel; spalling on one side of lower body, slight ancient damage to rim interior and small part of lower exterior; excavation damage to upper, interior portion of handle. Flat, roughly circular base, with rounded biconical lower body curving out and then back in to meet conical neck, which flares out to everted rim. High-swung, slightly asymmetrical vertical strut handle rises from just above point of maximum diameter; diagonally placed strut, nearly circular in section, meets neck exterior; handle roughly rectangular in section. Surface smoothed; white and gray encrustation scattered, especially on interior.

Plastic decoration: Four vertical projections on lower body, opposite handle, extending ca. 0.5 cm from vessel surface. Projections uniform in width, but not height or spacing (ca. 2.0 cm between center two projections, 2.5 cm between outer and inner projections).

Matt-painted decoration: Two horizontal lines just beneath rim and just above shoulder; pendent triangles, not fully preserved, descend from beneath lowermost horizontal line, terminating in sharp points just above base of vessel; pendent triangles slightly narrower between projections. Two sets of horizontal lines (numbering seven and eight, respectively) with diagonal hatching on handle exterior, above and below strut; beneath lower set of lines, a single pendent triangle, with three diagonal lines branching out from either side.

Fabric: Medium; many very small black inclusions, mica, possibly grog; many small air pockets (<0.5 cm) on both interior and exterior. Surface 7.5 YR 6/6 (reddish yellow); damaged area (2.5 YR 5/6—red); core not visible.

9/6 (P073), **Figs. 9.9, 9.38** (Drawing 23; Photos 668–673)

One-Handled Vessel, FL Type 2A.

Ceramic Deposit within Tumulus Fill (SU 1.091 within 1.039).

H (max preserved): 12.5 cm; D (rim, projected): <13 cm; D (base, projected): ca. 6 cm; Th (walls): 0.4–1.2 cm.

Nine joining fragments preserving full profile; one non-joining rim fragment. Body curves outward to point of maximum diameter, with very small ridge at shoulder; above ridge, conical neck flares out to slightly everted rim. Less than half of base preserved. Small section of rim preserved, quite irregular. No trace of handle extant. Damage to exterior surface, both modern and ancient.

Plastic decoration: Vertical ovoid projections, tapering to tip. One projection preserved intact, one broken but reattached, a third no longer extant. Positions and spacing between projections somewhat irregular.

Matt-painted decoration: Very poorly preserved. Three horizontal lines on neck, traces of paint just above shoulder, possibly two horizontal lines. Traces of three pendent triangles, preserved best between two extant projections; traces of vertical lines or triangles below projections.

Fabric: Soft, with moderate number of mica sparkles and black and white inclusions, ranging in size from very small to 0.3 cm in length. Occasional air pockets visible on exterior and interior surfaces, up to 0.7 cm in length. Exterior and interior surfaces 5 YR 6/8 (reddish yellow); core 2.5 Y 7/2 (light gray); encrustation on interior and exterior. Surface smoothed.

9/7 (P166), **Figs. 9.9, 9.38** (Drawing 47.1; Photos 2471–2477)

One-handled Vessel, FL Type 2B.

Tomb LXIII (35) (SU 1.234).

H (rim): 12.3–12.9 cm; H (max) 16.0 cm; D (base): 5.7 cm; D (max): ca. 13 cm; D (rim ext): 10.3 cm;

D (rim int): 7.2 cm; Th (walls): 0.3–0.5 cm; Th (handle): 1.5 cm; W (handle): 2.5 cm; W (handle at rim): 4.0 cm; W (handle at shoulder): 3.7 cm.

Nearly complete vessel reconstructed from 23 joining fragments, with conical neck, rounded lower body, flat base, and everted rim with rounded lip. Clear line of demarcation created by uneven junction between bottom of neck and top of lower body, which projects very slightly (<0.2 cm). Vertical strap handle, rectangular to ovoid in section.

Matt-painted decoration: Over entire exterior surface. Two wavering horizontal lines run around neck, just above point of minimum diameter. Just beneath neck-body junction, single horizontal line, with 14 pendent triangles extending below, nearly to base; triangles filled with five to eight parallel diagonals, none completely preserved. Pendent triangles even in size and spacing, but one much more narrow opposite the handle. On handle exterior, three bands of horizontal lines; fourth band partially preserved just above rim. Single set of radiating diagonals partially preserved on rim beside handle attachment.

Fabric: Soft, sandy, with abundant black and red inclusions, possibly grog (0.3 mm in diameter); moderate white inclusions, occasional mica. Air pockets (0.5 cm in length) common in both exterior and interior. Surface slipped and burnished, 5 YR 5/6 (yellowish red), with isolated red patches (2.5 YR 5/8) on and around handle; core 5 YR 6/1 (gray).

9/8 (P277), **Figs. 9.9**, **9.38** (Drawing 74.1; Photos 2750–2756)

One-Handled Vessel, FL Type 2B.

Tomb XLVIII (52) (SU 1.0318).

H (max): 11.5 cm; H (rim): 9.5 cm; D (rim): 9.5 cm; D (mouth int): 6.6 cm; D (base): 4.8 cm; D (max): ca. 11.0 cm; Th (handle): 0.8 cm; Th (rim): ca. 0.5 cm; Th (walls): ca. 0.5 cm; W (handle at body): 3.1 cm; W (handle at rim): 4.5 cm; W (handle): 2.2 cm.

Fully preserved one-handled vessel. Round, flat base, rounded hemispherical lower body curving up and outward to shoulder, then inward to conical neck, flaring out to everted rim with rounded lip. Single, vertical strap handle rises from body slightly askew before curving back down to join rim; handle roughly rectangular in section. Surface smoothed, not burnished; some spalling and small cracks on exterior surface.

Plastic decoration: Three rounded conical projections on front of vessel, immediately opposite handle at point of maximum diameter; projections evenly spaced, uniform in size (ca. 1.5-cm base diameter, 0.5-cm projection).

Matt-painted decoration: Two poorly preserved horizontal lines run around neck of vessel; four sets of parallel radiating lines on upper rim surface; best preserved in one set of five near handle-rim junction, elsewhere fragmentary. On handle at and above junction with rim, five parallel horizontal lines.

Fabric: Medium; many black inclusions (<0.3 cm) and mica, moderate number of small white inclusions (0.1–2 cm). Small air pockets on interior and exterior (<0.6 cm). Exterior and interior surfaces 7.5 YR 6/6 (reddish yellow); core not visible; many spots of white and gray encrustation, heavier on interior than exterior.

Comparandum: **9/3** (P304; similar shape and decoration).

9/9 (P020), **Figs. 9.11**, **9.39** (Drawing 5.7; Photos 934–937)

Rim and Neck Fragments (Probable FL Type 2).

Topsoil (SU 2.002).

PH (max): 6.5 cm; PW (max): 7.0 cm; D (rim ext): 12–14 cm; D (rim int): 7–8 cm; Th (rim): 0.5 cm; Th (wall): 0.5–0.7 cm.

Six joining fragments of conical neck and everted rim of closed vessel; rim not quite horizontal, lip rounded. Walls thicken slightly at lower end and surely continued before curving into base, but precise shape and size of original vessel uncertain. Fragments worn, with chips, scratches, and other damage to both surfaces; no encrustation.

Fabric: Soft, sandy fineware with abundant grog (<0.2 cm) and mica sparkles; occasional very small black inclusions and air pockets (<0.1 cm) in core and on both surfaces. Exterior and interior surfaces 5 YR 7/6 (reddish yellow); core 5 YR 6/1 (light gray) to 6/2 (pinkish gray).

Comparanda: Cf. rims **9/48** (P048), **9/12** (P131), **9/13** (P202), plus several uncatalogued frr (P177**,** P263, P264, P348, P425); matt-painted rims **9/10** (P213), **9/47** (P258); complete vessels (**9/7** (P166), **9/8** (P277), **9/5** (P326), **9/6** (P073). Everted rims with rounded lips are common outside of Lofkënd;

compare to amphorae from Vitsa (Vokotopoulou 1986:359–360, drs. 32–44; grave context, ninth–eighth century [but origins of form go back to the Bronze Age]) and Kamenicë (Bejko forthcoming: Q735, from Grave 67, eighth–seventh centuries).

9/10 (P213), **Figs. 9.11**, **9.39** (Drawing 84.2; Photos 2399–2401)
Rim Fragment (Probable FL Type 2).
Tumulus Fill (SU 1.279).
PH (max): 3.9 cm; PW (max): 2.0 cm; Th: 0.5 cm.
Small portion of neck and rim from small open vessel with S-shaped profile, broken just above shoulder; nearly vertical neck rises to slightly everted rim and rounded lip, preserving bit of rise to handle. Surfaces smoothed, once burnished, with scattered encrustation; exterior matt-painted.
Matt-painted decoration (exterior): Just below rim, three horizontal bands.
Fabric: Fine, abundant mica, very rare black inclusions. Surface and core 5 YR 6/8 (reddish yellow).
Comparanda: **9/19** (P386; similar size and shape, more fully preserved); **9/6** (P073; only other Lofkënd example with three horizontal bands below rim).

9/11 (P004), **Figs. 9.11**, **9.39** (Drawing 2.4; Photos 563–566)
Rim and Neck Fragment (Probable FL Type 2).
Surface Collection.
PH (max): 1.9 cm; PW (max): 4.5 cm; PL (max): 3.1 cm; Th: 0.5 cm; D (rim ext): ca. 9 cm(?).
Single fragment of everted, squared rim (with flat lip) of small vessel, perhaps one-handled drinking (Type 2). Surfaces smoothed, now very worn; encrustation on core.
Fabric: Fine light, with abundant red and brown inclusions (grog; <0.3 cm), moderate mica. Exterior surface 7.5 YR 6/4 (light brown), interior surface 5 YR 6/6 (reddish yellow).
Comparanda: **9/9** (P020), **9/12** (P131), **9/13** (P202), **9/48** (P048); plus uncatalogued P177, P263, P264, P348, P425.

9/12 (P131), **Figs. 9.11**, **9.39** (Drawing 39.1; Photos 2227–2228)
Rim and Neck Fragments (Probable FL Type 2).
Tumulus Fill (SU 1.070).
PH (max): 4.5 cm; PW (max): 6.0 cm; D (rim ext, approximate): 20–22 cm; D (rim int, approximate): 17–18 cm; Th (rim): 0.5–0.6 cm; Th (wall): 0.6–0.9 cm.
Two joining fragments of neck and everted rim, with rounded lip. Exterior and interior surfaces smoothed, now quite worn; light encrustation over core and nearly all surfaces, heavier near lower edges; perhaps once painted.
Fabric: Soft, sandy, semi-fine, with few inclusions; occasional mica, grog (<0.1 cm), and white inclusions (<0.2 cm). Surfaces 5 YR 6/6 to 6/8 (reddish yellow) where visible, core as surface, with isolated dark patch 5 YR 4/2 (dark reddish gray).
Comparanda: **9/235** (P084; semi-coarse); **9/220** (P168); similar shape, smaller in size—**9/10** (P213), **9/47** (P258); similar shape, smaller size, no matt-paint—**9/11** (P004), **9/9** (P020), **9/48** (P048), **9/13** (P202); plus uncatalogued P177, P263, P264, P348, P425.

9/13 (P202), **Fig. 9.11** (Drawing 46.3; Photos 2422–2424)
Rim Fragment (Probable FL Type 2).
Tumulus Fill (SU 2.217).
PL (max): 4.1 cm; PW (max): 3.3 cm; Th (wall): 0.9 cm; Th (rim): 0.5 cm.
Single fragment of everted, rounded rim, probably from small one-handled drinking vessel (Type 2). Surfaces almost entirely covered with heavy encrustation.
Fabric: Sandy, red fabric, difficult to see beneath encrustation: moderate mica and very small black inclusions. Surface and core 2.5 YR 4/8 (red).
Comparanda: **9/11** (P004), **9/9** (P020), **9/48** (P048), **9/12** (P131), plus uncatalogued P177, P263, P264, P348, P425.

9/14 (P256), **Figs. 9.10**, **9.39** (Drawing 55.1; Photos 2596–2603)
One-Handled Vessel, FL Type 3.
Tomb LV (53) (SU 1.0321).
H (rim): 5.4 cm; H (max): 8.9 cm; D (rim): 9.4 cm [=D (max)]; D (mouth int): 8.0 cm; D (base): 4.0 cm; Th: 0.8 cm.
Nearly complete one-handled vessel, reconstructed from 12 pieces; damage heaviest opposite handle, at rim, projections, and exterior surface. Flat base, open hemispherical body, simple flat-lipped rim,

horizontal loop handle, set at sharp angle, roughly rectangular in section. Interior surface smoothed, exterior smoothed, slipped, burnished, and painted; heavy encrustation everywhere.

Plastic decoration: Opposite handle, three horizontal projections protrude 1.0 cm from vessel surface, tapering from 1.0 × 1.5 cm at junction with body to 0.3 × 0.8 cm at tip; projections begin 1.0 cm below rim, at maximum diameter of vessel; middle projection not fully preserved.

Matt-painted decoration: Not fully preserved, partially obscured by encrustation. Upper rim surface painted with at least seven groups of parallel lines, evenly spaced; two to six lines remain in each group. Six groups of four parallel lines on both exterior and interior of handle, with third and fourth groups meeting as opposed diagonals at apex of exterior. Two horizontal lines run just beneath rim on exterior, best preserved below handle. Beneath lower horizontal line, below handle, three groups of three solid pendent triangles with elongated apices; traces of similar decoration on either side of and beneath frontal projections. On top of one projection, three lines parallel to vessel surface; similar traces on other two.

Fabric: Hard, sandy; moderate mica and grog, white inclusions (<0.3 mm) rare; limited core visibility. Air pockets (up to 0.3 mm) more visible in interior. Surface and core 5 YR 6/6 (reddish yellow).

Comparanda: Handle/shape—Andrea 1985:57–58, pl. LVII.5.

9/15 (P082), **Figs. 9.10**, **9.40** (Drawing 24.1; Photos 503–506)

Stemmed Goblet, FL Type 4.

Tomb LXX (17) (SU 3.126).

H (max): 15.5–15.8 cm; D (rim ext): 13.0 cm; D (rim int): 9.9 cm; D (base): 8.1 cm; D (max): ca. 14–15 cm; Th (rim): 0.6 cm; Th (handle): 1.1 cm; W (handle): 2.1 cm.

Intact vessel, with everted, nearly horizontal rim, cylindrical neck, and rounded lower body atop hourglass stem and concave disk foot (hollow underside). Vertical strap handle makes small loop from shoulder to just below rim. Pronounced ridge at junction between shoulder and neck. Some damage to vessel surface, including chips on rim, foot, and stem, as well as black stains (manganese?) on lower body, and minor encrustation across exterior surface, heavier within vessel.

Matt-painted decoration: Very poorly preserved. Two horizontal lines just below rim; two zigzag bands on neck, at top and bottom, stopping on either side of handle. Just beneath ridge, single horizontal line above at least five pendent triangles, each filled with parallel diagonals. Zigzag line runs around upper rim surface.

Fabric: Soft, sandy, no large temper; moderate black inclusions (<0.2 cm in size), rare white inclusions and mica, possibly fine grog inclusions. Air pockets abundant, mostly quite small (<0.4 cm). Exterior and interior surfaces smoothed, lightly burnished; surface color 5 YR 6/8 (reddish yellow), core not clearly visible.

Comparandum: **9/40** (P279; handle).

9/16 (P047), **Figs. 9.10**, **9.40** (Drawing 11.1; Photos 511–514)

Kantharos, FL Type 5A.

Tomb LXXXIII (7) (SU 4.059).

H (max preserved): 11.5 cm; H (rim): 8.8 cm; D (body, max): ca. 10.5 cm; D (rim): 6.7–7.7 cm; D (base): 3.7–4.2 cm; Th (walls): 0.4–0.6 cm.

Base and parts of body, neck, rim, and handles, reconstructed from some 27 joining fragments; profile not fully preserved. Irregularly shaped flat base. Body rises gently to maximum diameter just below handles, then inward and back out at neck, rising farther (on single preserved side) to meet upper portion of handle. Extant profile suggests high-swung vertical handles, ovoid at body, rectangular in section at rim junction. Rim and mouth not fully preserved, probably ovoid in plan, widest at handle attachments.

Fabric: Medium hardness, fine to semi-coarse, with occasional black inclusions, very sparse mica sparkles. Many air pockets visible, especially on exterior surface, measuring up to 0.7 cm in length. Interior and exterior surfaces 10 YR 7/6 (yellow), largely covered by encrustation 5 YR 8/2 (pale yellow); core not easily visible, appears similar to surface. Surface smoothed.

9/17 (P078), **Figs. 9.10**, **9.40** (Drawing 20.1; Photos 515–518)

Kantharos, FL Type 5B (?).

Tumulus Fill (SU 4.0079).

H (max preserved): 14.3 cm; Th (handle): 1.2 cm; Th (neck): 0.5 cm; Th (bottom): 1.0 cm; Th (body): 0.7 cm; W (handle): 2.8 cm; W (handle at body): 4.9 cm.

One large fragment preserving most of profile, plus one non-joining fragment of similar fabric and thickness. Bottom of large fragment is rounded, but does not necessarily include original base of vessel, making determination of original orientation difficult; flat base most likely. Remains of prefiring concentric stroke marks on lower interior surface, probably finishing tool marks; these suggest original diameter greater than that shown in drawing. One small vertical strap handle fully preserved, raised slightly above everted rim, joined crookedly to body; handle roughly rectangular in section. Chip missing from neck exterior, and two smaller chips from lower interior.

Fabric: Soft, with many very small (0.1–2 cm) black inclusions and mica sparkles, moderate white inclusions. Many air pockets (0.1–0.9 cm) on both interior and exterior. Color varies slightly: 5 YR 5/8 (yellowish red) to 6/8 (reddish yellow) on exterior, lighter on interior (5 YR 7/8—reddish yellow); core 2.5Y 6/2 (light brownish gray). Dark encrustation scattered uniformly across both interior and exterior; exterior surface smoothed.

9/18 (P198), **Figs. 9.10, 9.40** (Drawing 57.1; Photos 2727–2729)

Kantharos Handle and Body Fragments, FL Type 5B(?)

Tumulus Fill (SU 4.205).

PL (max): 8.4 cm; PW (max): 8.3 cm; W (handle): 2.2 cm; Th (wall): 0.6 cm; Th (handle): 1.1 cm.

Two joining fragments of small to medium-sized two-handled vessel, preserving entire strap handle, roughly rectangular in section, with attachments at rim and at point of maximum diameter on globular lower body; horizontal ridge at lower handle attachment, setting off neck from body. Precise size, shape, and original orientation uncertain—could be more or less open than drawing indicates. Surfaces smoothed but covered with heavy encrustation.

Found with non-joining fragment of similar fabric and size.

Fabric: Fine and sandy, red fabric, with abundant white inclusions and air pockets, moderate black inclusions and mica. Surface and core 5 YR 6/8 (reddish yellow).

Comparanda: **9/28** (P005), **9/22** (P074), **9/27** (P240 + P251), **9/32** (P031), **9/40** (P279); uncatalogued P187, P274, P385, P393.

9/19 (P386), **Figs. 9.10, 9.41** (Drawing 96.1; Photos 3464–3465)

Rim Fragment (FL Type 5C?).

Topsoil (SU 6.002).

PH (max): 6.5 cm; PW (max): 7.1 cm; Th (wall): 0.3–0.4 cm; Th (rim): 0.3 cm; D (max, approximate): 8 cm.

Single fragment preserving nearly entire profile of very small vessel, with globular lower body curving into short vertical neck and everted rim; rim preserves partial rise, probably to join handle. Stump of broken handle attachment on shoulder, roughly rectangular in section. Base not preserved. Uncertain whether this was single or multiple vessel (since these are often small in size), one handle or two, spout or no spout.

Matt-painted decoration: Two parallel lines beneath rising rim; two horizontal lines running around upper shoulder, with solid pendent triangles extending beneath, five to one side of handle stump, one to other.

Fabric: Medium, sandy, fine; abundant mica sparkles, occasional black and white inclusions (<0.2 cm) and air pockets (<0.6 cm). Encrustation over all surfaces. Exterior, interior, and core (where visible) 7.5 YR 6/6 (reddish yellow) to 5/6 (strong brown).

Comparanda: Shape **9/16** (P047), **9/17** (P078); Kamenicë (Bejko forthcoming:1/199 [topsoil/looting remains]).

9/20 (P290 + P221), **Figs. 9.11, 9.41** (Drawing 66.3; Photos 3269–3270) (Drawing 49.2; Photos 2574–2575, 2516–2517 2005 [P221]) (New Photos 3996–3998 [P290 + P221]; 4003 [P290 + P221 beside P386])

Wall/Handle Fragment.

Tumulus Fill (Units 4.286, 4.204).

PH (max): 3.0 cm; PW (max): 5.5 cm; PW (handle): 1.5 cm; Th (wall): 0.3 cm; Th (handle): 0.7 cm.

Two joining fragments from very small vessel with S-shaped profile, preserving convex shoulder with stem and lower attachment of vertical strap handle, bit of turn to neck above. Surfaces smoothed, but worn, exterior painted, once burnished; scattered encrustation.

Matt-painted decoration: Two nearly horizontal lines converging from either side toward handle attachment (perhaps once running all the way around vessel below neck); at left, two solid pen-

dent triangles emerge from lower line. Below handle attachment, horizontal line above four nearly vertical lines, perhaps legs of pendent triangles; above handle, stray mark.
Fabric: Fine, with abundant mica and small (>0.1cm) black and red inclusions; one larger white quartz pebble in handle (<0.4 cm). Surfaces 5 YR 4/4 (reddish brown), core 5 YR 4/1 (dark gray).
Comparanda: **9/19** (P386; nearly identical size, shape, painted decoration; very likely from same vessel—two-handled—despite slight difference in color).

9/21 (P113), **Fig. 9.11** (Drawing 38.4; Photos 2150–2151)
Vertical Strap Handle Fragment.
Tomb LXXV (33) Fill (SU 1.222).
PL (max): 2.3 cm; PW (max): 3.5 cm; Th: 1.2 cm.
Single fragment of vertical strap handle, flattened oval in section, preserving no curve or attachments to vessel. Surface smoothed.
Fabric: Fine, red fabric, with moderate mica, no other inclusions. Surface and core 5 YR 5/6 (yellowish red).
Comparanda: **9/18** (P198), **9/22** (P074), **9/27** (P240 + P251), **9/32** (P031), **9/28** (P005), **9/40** (P279); uncatalogued P187, P274, P385, P393.

9/22 (P074), **Fig. 9.11** (Drawing 15.6; Photos 797–799b)
Vertical Strap Handle Fragment.
Tumulus Fill (SU 4.079).
PL (max): 5.1 cm; PW (max): 4.0 cm; W (handle): 2.6 cm; Th: 1.0 cm.
Two joining fragments of vertical strap handle, roughly rectangular in section, broken at lower attachment and at upper curve returning to body; vessel walls not preserved. Surface smoothed, very worn; encrustation especially heavy on handle interior and over breaks.
Fabric: Sandy, fine to semi-coarse, with abundant mica and white inclusions, moderate black inclusions. Surface 5 YR 6/6 (reddish yellow), core 5 YR 6/2 (pinkish gray).
Comparanda: **9/28** (P005), **9/18** (P198), **9/27** (P240 + P251), **9/32** (P031), **9/40** (P279); uncatalogued P187, P274, P385, P393.

9/23 (P278), **Fig. 9.41** (Photos 2714–2716)
Vertical Strap Handle Fragment.
Tomb LXVIII (35) Skeleton (SU 4.234)—not associated with grave goods.
PL (max): 2.5 cm; PW (max): 2.8 cm; Th: 1.0 cm.
Single fragment of vertical strap handle, roughly rectangular in section, preserving no curve or attachments to vessel. Surface smoothed, very worn.
Fabric: Fine to semi-coarse, very worn; abundant mica, moderate black inclusions and grog. Surface and core 5 YR 5/6 (yellowish red).
Comparanda: **9/18** (P198), **9/22** (P074), **9/28** (P005), **9/32** (P031), **9/40** (P279), **9/27** (P240 + P251); uncatalogued P187, P274, P385, P393.

9/24 (P418), **Fig. 9.41** (Photos 3516–3517)
Vertical Strap Handle Fragment.
Tomb V (96) Fill (SU 2.563)—not associated with grave goods.
PL (max): 2.3 cm; PW (max): 3.1 cm; Th: 0.9 cm.
Single fragment of vertical strap handle, roughly rectangular in section (though section not fully preserved), swelling at base for lower vessel attachment; vessel wall not preserved. Surface smoothed, now worn; encrustation on interior.
Fabric: Fine light fabric; moderate mica and black inclusions (grog?). Surface 5 YR 6/6 (reddish yellow), core 5 YR 6/1 (gray).
Comparanda: **9/18** (P198), **9/22** (P074), **9/28** (P005), **9/32** (P031), **9/40** (P279), **9/27** (P240 + P251); uncatalogued P187, P274, P385, P393.

9/25 (P373), **Figs. 9.11, 9.41** (Drawing 100.3; Photos 3351–3352)
Handle Fragments.
Tomb XXXII (89) Fill (SU 2.507).
PL (max): 2.7 cm; PW (max): 3.3; PW (handle): 1.9 cm; Th: 0.4 cm.
Two joining fragments (from modern break) of ribbon handle, thin rectangular in section, broken at upper curve and attachment to rim. Surfaces smoothed and burnished, exterior painted; scattered encrustation.
Matt-painted decoration: Four horizontal bands.
Fabric: Very fine, moderate mica, occasional fine white inclusions and very fine air pockets. Surfaces 5 YR 6/4 (light reddish brown), core 5 YR 5/1 (gray).

9/26 (P107), **Figs. 9.11**, **9.41** (Drawing 49.1; Photos 2546–2255)
Handle/Body Fragment.
Topsoil (SU 1.141).
PH (max): 8.7 cm; PW (max): 6.8 cm; Th: 0.3–0.5 cm.
Six joining fragments of a high-swung vertical handle, probably from small kantharos. Handle ovoid in section at base, switching to pinched, vertical orientation at top; impression on interior surface near lower attachment. Very little of flat lip preserved, rim sloping down from handle. Surfaces smoothed, exterior painted; heavy encrustation. (Two additional non-joining fragments with same inventory number probably not from same vessel.)
Matt-painted decoration: Beneath handle root, three roughly horizontal lines above top of upright hatched triangle with six diagonal bands parallel to right leg; on either side of handle, portion of a poorly preserved pendent triangle, beneath horizontal line at right; just below rim, poorly preserved horizontal line.
Fabric: Fine, sandy fabric, with abundant mica, occasional black inclusions, and very small air pockets. Exterior and interior surfaces 2.5 YR 6/6 (light red), core mostly as surface, but 10 YR 6/1 (gray) in thick sections.
Comparanda: Barç kantharos (Eggebrecht 1988:192, no. 34 [Tirana NHM 1020], thirteenth–twelfth centuries BC); unusual to find this handle type on a painted vessel (S. Aliu, personal communication).

9/27 (P240 + P251), **Figs. 9.11**, **9.41** (Drawing 62.1; Photos 2696–2699)
Vertical Strap Handle and Wall Fragments.
Tumulus Fill (Units 1.279, 1.240).
PL (max): 6.8 cm; PW (max): 3.8 cm; W (handle): 1.7cm; Th (wall): 0.6 cm; Th (handle): 0.8 cm.
Three joining fragments preserving portion of convex shoulder and flaring neck, with lower attachment of vertical strap handle, ovoid in section; handle rises up and curves back toward body, upper attachment not preserved. Exterior surface smoothed, perhaps once burnished, interior surface more rough; scattered encrustation throughout.
Fabric: Fine red fabric, moderate mica, rare white inclusions. Surface and core 5 YR 6/8 (reddish yellow).
Comparanda: **9/28** (P005), **9/18** (P198), **9/22** (P074), **9/32** (P031), **9/40** (P279); uncatalogued P187, P274, P385, P393.

9/28 (P005), **Figs. 9.12**, **9.42** (Drawing 2.5; Photos 569–572)
Rim and Vertical Strap Handle Fragment.
Surface Collection.
PL (max): 5.0 cm; PW (max): 4.2 cm; W (handle): 3.6 cm; Th: 0.5–0.7 cm.
Single fragment preserving attachment of vertical strap handle, roughly rectangular in section, to everted rim, neck, and shoulder of small, thin-walled vessel, probably open in shape. Surfaces smoothed, perhaps burnished on exterior; heavy encrustation.
Fabric: Fine red fabric, with abundant mica and black inclusions (pebbles?), moderate white inclusions. Surface and core 5 YR 5/8 (yellowish red).
Comparanda: **9/18** (P198), **9/22** (P074), **9/27** (P240 + P251), **9/32** (P031), **9/40** (P279); uncatalogued P187, P264, P274, P385, P393.

9/29 (P006), **Figs. 9.12**, **9.42** (Drawing 2.6; Photos 567–568, 636–637)
Flaring Spur Handle fragment.
Topsoil (2003 Surface Collection).
PL (max): 2.9 cm; PW (max): 3.7 cm; W (spur tip): 3.6 cm; Th: 0.6–1.1 cm.
Single fragment preserving complete tip of flaring spur handle, rectangular in section; broken at thickening for attachment of lower handle element. Surfaces smoothed, now worn, with scattered encrustation; small chips and damage to trapezoid corners.
Fabric: Medium, sandy, with abundant mica, many very small (<0.1 cm) black and white inclusions, occasional grog; air pockets (<0.4 cm) visible especially in upper surface. Surfaces and core 5 YR 5/6 (yellowish red) to 2.5 YR 5/6 (red).
Comparanda: **9/30** (P064).

9/30 (P064), **Figs. 9.12**, **9.42** (Drawing 14.2; Photos 983–988)
Horned Spur Handle Fragment.
Tumulus Fill (SU 2.033).
PL (max): 4.4 cm; PW (max): 2.3 cm; W (spur tip): 2.3 cm; Th: 1.0–1.7 cm.
Single fragment preserving complete tip of horned spur handle, together with considerable section of upper handle element, ovoid in section; broken here and just below attachment of lower element, probably also ovoid in section. Surfaces very smooth, but nearly completely covered by encrus-

tation; damage from excavation just above lower break.
Fabric: Difficult to see beneath encrustation, but appears very fine, with mica, few other inclusions, all very small; surfaces and core 7.5 YR 6/6 (reddish yellow), encrustation 2.5Y 6/2 (light brownish gray).
Comparanda: **9/29** (P006).

9/31 (P114), **Figs. 9.12, 9.42** (Drawing 37.5; Photos 2192–2193)
Vertical Strap Handle Fragment.
Tumulus Fill (SU 1.070).
PH (max): 4.8 cm; PW (max): 3.2 cm; Th: 0.7–1.0 cm.
Fragment of vertical strap handle, roughly rectangular in section, preserving part of vertical rise and angular bend, no attachments to vessel. Surfaces smoothed but worn, exterior painted, slightly damaged, perhaps once burnished; scattered encrustation.
Matt-painted decoration: Above lower break, five slightly irregular horizontal lines; at top, two lines converge in V-shape, probably preserving lower part of motif lost to upper break.
Fabric: Fine, sandy fabric, with moderate mica and black inclusions, occasional air pockets. Surfaces 7.5 YR 6/6 (reddish yellow), core 10 YR 6/1 (gray).
Comparanda: Complete vessels **9/5** (P326), **9/7** (P166).

9/32 (P031), **Fig. 9.12** (Drawing 6.4; Photos 993–996)
Vertical Strap Handle Fragment.
Tomb LXXXI (1) Fill (SU 7.029).
PL (max): 4.9 cm; PW (max): 3.1 cm; Th: 1.2 cm.
Single fragment preserving lower portion of vertical strap handle, roughly rectangular in section, broken at lower attachment and curving sharply upwards; upper attachment, vessel wall not preserved. Surface smoothed, now worn.
Fabric: Fine, sandy, red fabric, with abundant mica and white inclusions, moderate black inclusions, few air pockets (<0.6 cm). Surface 2.5 YR 5/8 (red), core 2.5 YR 4/3 (reddish brown).
Comparanda: **9/28** (P005), **9/18** (P198), **9/22** (P074), **9/27** (P240 + P251), **9/32** (P031), **9/40** (P279); uncatalogued P187, P274, P385, P393.

9/33 (P080), **Figs. 9.12, 9.42** (Drawing 18.3; Photos 997–999d)
Pierced Flaring Spur Handle Fragment.
Tumulus Fill (SU 4.104).
PL (max): 5.2 cm; PW (max): 4.0 cm; W (spur tip): 2.7 cm; Th: 0.7–1.0 cm.
Single fragment (bagged with two non-joining, nondiagnostic sherds of same fabric) of pierced flaring spur handle, broken on upper side on either side of piercing, on lower side just beneath attachment to spur; upper side plano-convex in section, with tapered sides, lower side ovoid, spur tip rectangular. Piercing not quite centered, handle slightly asymmetrical in plan. Surfaces smooth, very worn, with scattered encrustation.
Fabric: Soft, semi-fine, sandy, with occasional mica and grog, infrequent small black, red, and white inclusions, all very small (<0.1 cm); very small air pockets (<0.1 cm) on surfaces and in core. Surfaces 5 YR 5/6–5/8 (yellowish red), core 5 YR 4/1 (very dark gray) to 3/1 (black).

9/34 (P125), **Figs. 9.12, 9.42** (Drawing 39.3; Photos 2688–2689)
Pierced Handle Fragment?
Tumulus Fill (SU 1.070).
PL: 1.6 cm; PW (max): 4.5 cm; Th: 0.9–1.4 cm; D of pierce: 1–2 cm.
Small fragment of unusual shape, with sharply curved profile and partially preserved circular piercing; perhaps from pierced handle. Surfaces smoothed, exterior painted; scattered encrustation.
Matt-painted decoration: Beside piercing, two parallel lines, perhaps traces of others.
Fabric: Fine, abundant mica, occasional black inclusions. Surfaces 5 YR 6/8 (reddish yellow), core 7.5 YR 7/1 (light gray).
Comparanda: Aliu 2004:201, pl. XXI, 246 (open vessel with horizontal handle, triple pierced, perhaps similar to original shape of this piece?).

9/35 (P287), **Figs. 9.12, 9.42** (Drawing 66.2; Photos 2978–2979)
Horizontal Handle Fragment.
Tomb LIII (63) Fill (SU 1.358).
PL (max): 4.4 cm; PW (max): 2.7 cm; Th: 1.4 cm.
Single fragment preserving outermost edges of horizontal handle, pierced at center and probably originally three-sided; irregular ovoid in section. Surface once smoothed, now very worn, partially encrusted.
Fabric: Fine to semi-coarse, darker upper surface, redder core and underside; moderate mica and small black inclusions. Exterior 2.5 YR 5/8 (red) to

2.5 YR 4/2 (weak red) where not covered by encrustation; core 2.5 YR 5/8 (red).
Comparanda: Aliu 2004: pl. XXVI, 286.

9/36 (P066), **Fig. 9.42** (Photos 765–768)
Strutted Handle Fragments.
Tumulus Fill (SU 1.070).
PL (max): 5.5 cm; PL (smaller fr): 2.2–2.6 cm; Th: 0.7–1.4 cm; PW (max): 2.2 cm; PW (large handle): 0.7–1.4 cm.
Three fragments, two joining from strutted handle, one from additional handle, plus two non-joining wall fragments; probably all from same vessel, shape unclear. Largest fragment ovoid in section, flattened at edge, with beginning of ovoid strut projecting at slight angle from convex curve of handle. Original shape and orientation uncertain: flattened end either from pointed handle or from juncture with body, either double vessel or single vessel with strutted handle.
Non-joining handle fragment, nearly circular in section, preserves part of a corner, perhaps from a horizontal (horned?) handle. Fabric and surface very similar to larger handle fragment, but not clear what vessel type could include both of these handle types.
All surfaces smoothed, with minimal encrustation on parts of two sides and over section of broken strut. Black coloring preserved in patches on all exterior surfaces, including flattened end; could be clay slip gloss, matt-painting, or organic residue.
Fabric: Soft fineware, with abundant mica sparkles, occasional black inclusions, perhaps grog, and one quartz pebble, <0.1 cm in diameter. Occasional air pockets, mostly very small (<0.3 cm). Surfaces 5 YR 4/6 to 5/6 (yellowish red), core 7.5 YR 6/2 (pinkish gray).
Comparanda: Aliu 2004:206, pl. XXVI, 283, 291 (for pointed handles); double vessel (pl. XXXV, 35); single vessel with strutted handle (pl. XXXIII, 25).

9/37 (P342), **Figs. 9.12, 9.42** (Drawing 70.3; Photos 3078–3079)
Handle Fragment.
Topsoil (SU 2.002).
PL (max): 2.3 cm; PW (max): 2.5 cm; Th (max): 1.5 cm.
Small fragment from loop or ring handle, roughly triangular in section, perhaps set horizontally. Surfaces smoothed, incised.
Incised decoration: On upper surface, three roughly parallel zigzag bands.
Fabric: Fine to semi-coarse, sandy, with abundant mica and black and white inclusions.
Comparanda: Korkuti 1983a:69, pl. XX, 13 (Neolithic settlement of Kolsh).

9/38 (P236), **Fig. 9.42** (Photos 2564–2565)
Ring Handle Fragment.
Tumulus Fill (SU 4.204).
PL (max): 5.9 cm; PW (max): 5.5 cm; Th (wall): 0.5 cm; D (handle): 1.7 cm.
Single fragment preserving convex wall of vessel with attachment and lower portion of ring handle, oval in section, almost certainly vertical in orientation. Original shape uncertain, but ridge just above handle likely from lower neck. Handle exterior surface once burnished, vessel interior smoothed, now covered by heavy encrustation; black patches (manganese? possibly matt-paint?) scattered on vessel exterior, especially just above handle attachment.
Possible matt-painted decoration.
Fabric: Fine, red fabric, with abundant mica, very small black inclusions, few white inclusions. Surface 5 YR 5/6 (yellowish red), core GLEY2 7/5PB (light bluish gray).

9/39 (P172), **Figs. 9.12, 9.43** (Drawing 86.2; Photos 2344–2346)
Ring Handle Fragment.
Tumulus Fill (SU 1.039).
PL (max): 4.5 cm; D (handle): 1.9 cm; Th (handle): 1.4 cm.
Single fragment of ring handle, oval in section, curved in profile, preserving no attachment to vessel. Surfaces smoothed, with some fine cracks.
Fabric: Fine, sandy fabric, with abundant mica, white and black inclusions. Surface 5 YR 5/6 (yellowish red), core 5 YR 3/1 (very dark gray).

9/40 (P279), **Fig. 9.12** (Drawing 95.3, Photos 2825–2827)
Vertical Strap or Loop Handle Fragment.
Tomb XLVIII (52) Fill (SU 1.317).
PL (max): 4.2 cm; PW (max): 2.6 cm; Th: 0.9 cm.
Single fragment of vertical strap handle, rectangular in section, preserving uppermost, semi-circular curve, but no attachments to vessel body; tightness

of curve also suggests loop handle. Surface smoothed, especially on exterior; exterior painted; scattered encrustation.
Matt-painted decoration: at and above break, three fragmentary horizontal bands.
Fabric: As **9/27** (P240 + P251).
Comparanda: **9/18** (P198), **9/22** (P074), **9/27** (P240 + P251), **9/32** (P031), **9/28** (P005); for tightly curved strap handle, cf. **9/15** (P082); for loop handle attachment, cf. **9/82** (P305); uncatalogued P187, P274, P385, P393.

9/41 (P275), **Figs. 9.12**, **9.43** (Drawing 103, Photos 2703–2705)
Vertical Strap Handle Fragment.
Topsoil (SU 1.278).
PH (max): 3.8 cm; PW (max): 4.4 cm; Th: 1.2 cm.
Single fragment of large vertical strap handle, rectangular in section, preserving 90° bend to horizontal, but no attachments. Surfaces smoothed, exterior burnished.
Fabric: Fine light fabric, moderate mica, few very small black pebble inclusions, few air pockets (<0.2 cm). Surface and core 2.5 YR 5/8 (red).
Comparanda: Aliu 2004:184, pl. IV, 60 (fill of Grave 38, ninth–eighth century); **9/18** (P198), **9/22** (P074), **9/27** (P240 + P251), **9/28** (P005), **9/32** (P031), **9/40** (P279); uncatalogued P274, P385, P393; cf. also **9/277** (P394) for similar size and shape in coarseware.

9/42 (P159), **Fig. 9.12** (Drawing 42.2; Photos 2365–2366)
Rim Fragment.
Topsoil (SU 4.201).
PH (max): 1.2 cm; PW (max): 1.5 cm; Th: 0.4 cm.
Small fragment from very thin-walled open vessel, preserving vertical neck tapering to narrow lip; precise shape uncertain. Surfaces smoothed, now worn.
Fabric: Fine, sandy, light fabric, with abundant mica and very fine black and white inclusions, few very small air pockets. Surface and core 7.5 YR 6/6 (reddish yellow).
Comparanda: Uncatalogued P164, P331.

9/43 (P220), **Fig. 9.12** (Drawing 45.8; Photos 2374–2375)
Rim Fragment.
Tumulus Fill (SU 4.204).
PL (max): 2.3 cm; PW (max): 2.3 cm; Th: 0.5–0.6 cm.
Single rim fragment with rounded lip above very shallow groove; precise shape and orientation uncertain. Surfaces smoothed, worn; scattered encrustation.
Fabric: Fine to semi-fine, sandy, with abundant mica and fine to small white (limestone?) inclusions (<0.3 cm). Surfaces and core 5 YR 4/6 (yellowish red).
Comparanda: **9/44** (P013), **9/45** (P173); uncatalogued P330.

9/44 (P013), **Fig. 9.12** (Drawing 3.4; Photos 581–584)
Rim Fragment.
Topsoil (SU 1.001).
PL (max): 2.1 cm; PW (max): 2.6 cm; Th: 0.6 cm.
Single fragment from small vessel, preserving bit of turn from neck to everted rim, with tapered lip. Surfaces smoothed, worn, encrustation over core.
Fabric: Fine, sandy fabric, abundant mica, fine black, white, red (grog?) inclusions. Exterior surface 7.5 YR 4/3 (brown), interior 5 YR 4/4 (reddish brown), core not visible.
Comparanda: **9/43** (P220), **9/45** (P173); uncatalogued P330.

9/45 (P173), **Fig. 9.12** (Drawing 86.1; Photos 2347–2348)
Rim Fragment.
Tumulus Fill (SU 1.039).
PL (max): 2.2 cm; PW (max): 2.3 cm; Th: 0.5 cm.
Single fragment from rim of small thin-walled vessel with vertical wall, slight S-curve, and rounded lip. Surfaces smoothed, now worn.
Fabric: Fine, with abundant mica and small black inclusions (pebbles <0.2 cm), occasional white inclusions. Surface and core 2.5 YR 5/6 (red).
Comparanda: **9/43** (P220), **9/44** (P013); uncatalogued P330.

9/46 (P268), **Fig. 9.43** (Photos 2670–2671)
Rim Fragment.
Tomb LXIV (61) Fill (SU 2.259).
PH (max): 1.5 cm; PW (max): 3.0 cm; Th: 0.6 cm.
Small fragment of short everted rim with tapered lip, probably from small, open vessel. Surfaces smoothed, now worn, with heavy encrustation.
Fabric: Fine, soft, red fabric, with abundant mica, moderate very small black and white inclusions,

few red and brown inclusions (grog, <0.2 cm). Surface and core 2.5 YR 6/8 (light red).
Comparanda: Uncatalogued P280, P329.

9/47 (P258), **Figs. 9.12, 9.43** (Drawing 60.2; Photos 3999–4000)
Rim Fragment.
Tumulus Fill (SU 1.279).
PH (max): 3.8 cm; PW (max): 3.1 cm; Th: 0.8 cm.
Small fragment of mid- to large-sized vessel with nearly vertical neck and slightly flaring rim; very small bit of tapered lip preserved. Surfaces smoothed and burnished, exterior slipped and painted.
Matt-painted decoration: Checkerboard pattern of lozenges, extending beyond breaks at right and bottom, bound at left by four vertical lines, above by three slightly thicker horizontal lines running just below rim.
Fabric: Fine to semi-fine, with abundant black inclusions and moderate mica. Surfaces 7.5 YR 7/6 (reddish yellow), core 7.5 YR 6/1 (gray).
Comparanda: **9/235** (P084; same shape, different motif); Bejko forthcoming: Kamenicë 1/3 (probably same shape, different motif); **9/83** (P112 + P255; perhaps non-joining fragment from same vessel).

9/48 (P048), **Figs. 9.13, 9.43** (Drawing 13.2; Photos 620–623)
Rim, Neck, and Shoulder Fragments.
Topsoil (SU 1.009).
PH (max): 4.1 cm; PW (max): 3.5 cm; PL (max): 6.3 cm; Th: 0.5 cm.
Two joining fragments of thin-walled vessel, preserving diagonal section of sloping neck/shoulder and slightly flaring narrow rim; precise shape and orientation uncertain because so little of actual rim survives (small drinking cup, or something larger?). Surfaces smoothed, encrustation over core.
Fabric: Fine, red fabric, with abundant mica and very small black inclusions; few very small white inclusions and air pockets (<0.2 cm). Exterior surface 5 YR 5/8 (yellowish red), core 7.5 YR 6/2 (pinkish gray).
Comparanda: Bejko forthcoming: Kamenicë Q771, eighth–seventh century (tumulus fill); Q2163, eleventh–ninth century (dedication deposit).

9/49 (P389), **Fig. 9.43** (Photos 3488–3489)
Rim Fragment.
Tumulus Fill (SU 5.550).
PH (max): 1.1 cm; PW (max): 4.3 cm; Th: 0.9 cm; D (ext): ca. 19 cm.
Small fragment preserving outer edge of widely flaring rim with rounded lip; interior surface matt-painted. Surfaces smoothed; scattered encrustation, heavier on edge.
Matt-painted decoration: Base of cross-hatched triangle.
Fabric: Fine to semi-coarse, with abundant mica, white and black inclusions, occasional grog. Surface 2.5 YR 5/8 (red), core 2.5 YR 5/1 (reddish gray).
Comparanda: **9/221** (P057; similar shape, better preserved, semi-coarse).

9/50 (P395 = **15/21**), **Figs. 9.13, 9.43** (Drawing 96.2; Photos 3494–3495)
Rim Fragments.
Topsoil (SU 5.001).
PH (max): 3.1 cm; PW (max): 4.2 cm; Th: 0.6–1.0 cm.
Three joining fragments of medium to large open vessel; neck flares out to wide, flat rim with rounded external projection. Surfaces smoothed, now very worn, with scattered black accretions—probably bitumen (or manganese) rather than matt paint.
Fabric: Fine to semi-fine, red, very worn; abundant mica, few very fine black and white inclusions and air pockets (<0.5 cm). Surface and core 2.5 YR 5/8 (red) to 6/8 (light red).
Comparanda: Similar shape, coarser fabric, larger size: **9/279** (P003), **9/300** (P065), **9/299** (P145), **9/290** (P224 + 282 + P411), **9/291** (P244), **9/292** (P317).

9/51 (P163), **Fig. 9.13** (Drawing 105, Photos 2838–2839)
Rim/Spout Fragment.
Tumulus Fill (SU 4.204).
PH (max): 1.8 cm; PW (max): 2.5 cm; Th: 0.4–0.5 cm.
Single, thin-walled fragment of nearly vertical, cylindrical neck, flaring very slightly to narrow rim; perhaps spout from jug with cutaway neck. Surfaces smoothed, perhaps once burnished, now very worn, with heavy encrustation.
Fabric: Fine light fabric; encrustation obscures much of fabric—some mica sparkles visible. Exterior and core 5 YR 6/8 (reddish yellow).
Comparanda: For possible shape (spout of handmade jug with cutaway neck), cf. Papadopoulos 2005: pls. 345–354.

9/52 (P246), **Figs. 9.13, 9.43** (Drawing 64.2; Photos 2572–2573)
Wall Fragment with Punched Decoration.
Tumulus Fill (SU 1.279).
PL (max): 1.9 cm; PW (max): 4.0 cm; Th: 0.3 cm.
Single fragment of convex wall, probably from small vessel, perhaps closed in shape. Surfaces smoothed, exterior punched, perhaps once burnished; heavy encrustation.
Punched decoration: Fragmentary hourglasses (or triangles or lozenges connected at apices).
Fabric: Very fine, with occasional mica. Surfaces 10 R 5/8 (red), core 10 R 4/1 (dark reddish gray).

9/53 (P414), **Figs. 9.13, 9.43** (Drawing 97.1; Photos 3787–3788)
Wall Fragment with Punched Decoration.
Tumulus Fill (SU 5.536).
PL: 2.1 cm; PW: 1.9 cm; Th: 0.6 cm.
Small wall fragment from small vessel; shape and orientation uncertain. Surfaces smoothed, interior worn.
Punched decoration: Row of five dots, perhaps horizontal.
Fabric: Fine, abundant mica, many small black inclusions. Exterior surface 2.5 YR 4/2 (weak red), interior surface and core 10 R 3/1 (dark reddish gray).

9/54 (P133), **Figs. 9.13, 9.43** (Drawing 64.4/5; Photos 3416–3418, 3425–3426; 3849–3850 [shows latest join])
Shoulder and Neck Fragments.
Tumulus Fill (SU 1.070).
PH (max): 4.0 cm; PW (max): 9.8 cm; Th: 0.4–0.5 cm; D (neck): ca. 10 cm; D (max): ca. 15 cm.
Seven joining fragments from shoulder and neck of small closed vessel with globular body curving in to roughly vertical neck; precise shape and orientation uncertain. Surface smoothed, worn; scattered encrustation.
Incised decoration: Running around shoulder, single horizontal band of punched diagonal teardrops above incised receding pendent triangles.
(Eleven additional non-joining fragments, very worn and without incised decoration, found nearby [Unit 1.279], perhaps from same vessel.)
Fabric: Fine, with moderate mica and fine black inclusions. Surfaces and core 2.5 YR 6/8 (light red).
Comparanda: **9/55** (P093 + P096 + P247 + P370—possibly from same vessel); for similar shape and decoration, Andrea 1985:39, pl. XIII, Vari 125, 1, dated twelfth–eighth (tenth–ninth centuries BC [Bejko, personal communication]); for similar motifs on Late Bronze Age sherds, cf. Papazovska (2005): vol. I, 106, T. II.9–10; for similar shape, although larger, cf. **9/5** (P326); for similar shape in dark fabric, cf. **9/100** (P063 + P049), **9/101** (P137).

9/55 (P093 + P096 + P247 + P370), **Figs. 9.13, 9.44** (Photos 3847–3848)
Body Fragments.
Tumulus Fill (Units 1.070, 2.078, 1.279, 6.181).
PL (max): 4.2 cm; PW (max): 10.1 cm; Th: 0.6 cm.
Four joining body fragments from shoulder of small closed vessel with incised decoration, found in different stratigraphic units. Exterior surface smoothed, P096 very worn; scattered encrustation.
Incised and punched decoration: Running around shoulder, single horizontal band of punched diagonal teardrops above incised receding pendent triangles.
Fabric: Soft, sandy fineware, with moderate, fine black inclusions and occasional mica and tiny air pockets. Surfaces 5 YR 6/6 (reddish yellow); core varies from surface color to 5 YR 6/2 (pinkish gray).
Comparanda: **9/54** (P133: perhaps slightly smaller, but possibly from same vessel; very similar fabric, differently worn).

9/56 (P371), **Figs. 9.13, 9.44** (Drawing 114.1; Photos 3328–3329)
Neck/Shoulder Fragment.
Topsoil (SU 2.002).
PL: 4.6 cm; PW: 6.6 cm; Th: 0.5 cm.
Two joining fragments from small vessel with convex shoulder and vertical or flaring neck, preserving small portion of attachment for handle of uncertain type. Exterior surface smoothed and incised, interior uneven.
Incised and punched decoration: Roughly horizontal band of punched dots/dashes encircling upper shoulder; portions of two incised hatched pendent triangles below punched band.
Fabric: Fine, with abundant mica and occasional black inclusions. Surface and core 2.5 YR 5/8 (red).
Comparanda: **9/54** (P133); **9/55** (P093 + P096 + P247 + P370; similar motif, incised/punched slightly deeper); Bejko forthcoming: Kamenicë 234/550, for similar incised pendent and hatched triangles, but without punched dots; for punched dots on

larger vessel, cf. Bodinaku 1981:257, pl. 1. Cf. also Korkuti 1983a:69, pl. XX, 9 (Kolsh, Middle Neolithic); Korkuti and Andrea 1974:pl. XIII, 4 (Cakran, Middle Neolithic); Durić, Glisić, and Todorović 1975:41, Tomb 15 (dark fabric).

9/57 (P375), **Figs. 9.13, 9.44** (Drawing 114.2; Photos 3394–3395)
Wall Fragment.
Topsoil (SU 8.201).
PL (max): 1.9 cm; PW (max): 4.0 cm; Th: 0.3 cm.
Small fragment of slightly convex wall with sharp curve on interior just before break—perhaps from neck/ shoulder junction of small closed vessel, but shape and orientation uncertain. Surfaces smoothed, heavy encrustation; exterior decorated, interior uneven.
Incised/punched decoration: Incised zigzag, with interstices irregularly dotted.
Fabric: Fine, with very few visible inclusions (much obscured by decoration and encrustation). Surfaces and core 5 YR 5/6 (yellowish red).

9/58 (P351), **Figs. 9.13, 9.44** (Drawing 72.3; Photos 3207–3208)
Neck/Shoulder Fragment.
Tumulus Fill (SU 1.440).
PH (max): 1.8 cm; PW (max): 3.5 cm; Th: 0.4 cm; D (neck, interior): ca. 7 cm.
Small fragment broken from convex shoulder and vertical neck of small, thin-walled vessel, probably open in shape. Exterior surface smoothed and painted, perhaps slipped; interior not preserved.
Matt-painted decoration: Upper shoulder zone diagonally hatched, bound by two horizontal lines below, one above.
Fabric: Very fine, with abundant mica and occasional black inclusions. Surface and core 5 YR 6/6 (reddish yellow).

9/59 (P262), **Figs. 9.13, 9.44** (Drawing 60.3; Photos 3781–3782)
Wall Fragment.
Topsoil (SU 1.278).
PL: 1.8 cm; PW: 3.1 cm; Th: 0.3 cm.
Single fragment from globular lower body of small vessel; shape uncertain. Surfaces smoothed, exterior painted; heavy encrustation.
Matt-painted decoration: Lower portion of hatched pendent triangle.
Fabric: Very fine, with abundant mica. Surface and core 5 YR 6/8 (reddish yellow).
Comparanda: **9/60** (P092), **9/61** (P071).

9/60 (P092), **Figs. 9.13, 9.44** (Drawing 29.1; Photos 769–772)
Wall Fragment.
Tumulus Fill (SU 4.086).
PH (max): 3.3 cm; PW (max): 3.0 cm; Th: 0.5 cm.
Single fragment from convex shoulder of small vessel, broken just below neck. Surfaces smoothed, exterior burnished, interior worn.
Matt-painted decoration: Upper shoulder zone diagonally hatched, bound by two horizontal lines below, one above; farther down, partially preserved upper portions of two pendent triangles, filled with combination of hatched and solid motifs.
Fabric: Same as **9/59** (P262), **9/61** (P071).
Comparanda: **9/59** (P262), **9/61** (P071).

9/61 (P071), **Fig. 9.44** (Photos 628–631)
Body Fragments.
Tumulus Fill (SU 1.039).
PL × PW (max): 4.3 × 1.8 cm; 3.4 × 3.0 cm; 4.4 × 2.1 cm; 1.1 × 1.6 cm; Th: 0.6 cm.
Seven body fragments from small vessel of uncertain shape: three joining sherds probably from lower neck; two joining sherds preserving lower attachment for vertical strap handle, roughly rectangular in section; single sherd from slightly concave shoulder, broken at junction with neck; and single indeterminate sherd. Surfaces smoothed, exterior burnished and painted.
Matt-painted decoration: three joining sherds: three roughly parallel bands; neck/shoulder fragment: two horizontal bands around neck, zigzag on shoulder.
Fabric: Same as **9/59** (P262), **9/60** (P092).
Comparanda: **9/59** (P262), **9/60** (P092).

9/62 (P019), **Figs. 9.13, 9.44** (Drawing 4.5; Photos 525–528)
Shoulder Fragment.
Topsoil (SU 4.004).
PL (max): 2.7 cm; PW (max): 2.3 cm; PTh (max): 0.5 cm.
Single fragment from shoulder of small vessel, broken just below turn to neck. Exterior surface smoothed and painted, interior completely lost.

Matt-painted decoration: Two thick horizontal lines above upper portion of hatched pendent triangle.
Fabric: Fine, with abundant mica and occasional black inclusions. Exterior surface and core 2.5 YR 6/8 (light red).

9/63 (P416), **Fig. 9.44** (Photos 3819–3820)
Lower Body Fragment.
Tumulus Fill (SU 6.559).
PH (max): 4.3 cm; PW (max): 8.6 cm; Th: 0.6–0.7 cm; D (max): ca. 15 cm.
Single fragment from globular lower body of small vessel, perhaps LF one-handled vessel Type 1 or LF kantharos Type 2. Surfaces smoothed but worn, damaged; scattered encrustation.
Fabric: Fine, sandy fabric, with abundant mica, moderate fine black and white inclusions, occasional grog and air pockets (<0.2 cm). Surfaces and core 5 YR 6/6 (reddish yellow).
Comparanda: Fragmentary profiles: **9/17** (P078), **9/6** (P073); intact vessels: **9/5** (P326), **9/7** (P166).

9/64 (P259), **Fig. 9.44** (Photos 2709–2711)
Shoulder Fragment.
Tomb XLII (59) Fill (SU 2.287).
PL (max): 4.3 cm; PW (max): 4.3 cm; Th: 0.5–0.7 cm.
Single fragment from rounded shoulder of small globular vessel (perhaps LF one-handled Type 1), broken just above sharp turn to (conical?) neck. Exterior surface smoothed but worn, interior uneven; heavy encrustation.
Fabric: Fine to semi-fine, with abundant mica and very fine black and white inclusions, occasional small air pockets (<0.2 cm). Surface and core 5 YR 6/6 (reddish yellow).

9/65 (P012), **Figs. 9.13**, **9.44** (Drawing 3.5; Photos 577–580)
Wall Fragment.
Topsoil (SU 1.001).
PH (max): 2.0 cm; PW (max): 2.7 cm; Th: 0.8 cm.
Small fragment from wall of small vessel, perhaps approaching rim. Surfaces smoothed, exterior probably once painted, now very worn; much of interior surface lost; heavy encrustation on breaks.
Matt-painted decoration: Five very faint parallel vertical lines.
Fabric: Fine, with abundant mica, moderate fine white, and red (grog) inclusions, occasional black inclusions and air pockets. Exterior surface 5 YR 5/8 (yellowish red), interior surface and core as exterior where visible.

9/66 (P250), **Fig. 9.45** (Photos 2686–2687)
Wall Fragment.
Tumulus Fill (SU 1.240).
PH (max): 3.3 cm; PW (max): 2.1 cm; Th: 0.5–0.8 cm.
Small wall fragment, probably from small vessel. Surfaces smoothed, exterior painted, now very worn; heavy encrustation.
Matt-painted decoration: Remnants of two perpendicular lines.
Fabric: Fine, with moderate mica and small red (grog) inclusions, occasional fine white inclusions and air pockets. Surfaces and core 5 YR 5/8 (yellowish red).

9/67 (P296), **Fig. 9.45** (Photos 2832–2833, 3261–3262)
Wall Fragment (Shoulder?).
Tumulus Fill (SU 1.377).
PL (max): 2.8 cm; PW (max): 4.1 cm; Th: 0.4–0.5 cm.
Single fragment from wall of very fine, thin-walled vessel, perhaps from shoulder-neck junction; precise shape and orientation uncertain. Both surfaces smoothed and highly burnished, covered with scattered encrustation; modern scratches on interior.
Fabric: Slightly different fabric from usual fine light at Lofkënd: very fine, with moderate, small black and brown inclusions (possibly grog), occasional mica and white inclusions. Surface and core 2.5 YR 6/8 (red).

9/68 (P165), **Fig. 9.13** (Drawing 42.4; Photos 3551–3552)
Wall Fragment.
Tumulus Fill (SU 4.204).
PL (max): 2.6 cm; PW (max): 2.9 cm; Th: 0.8 cm.
Single fragment from wall of small- to mid-sized vessel, shape uncertain. Surfaces smoothed, exterior perhaps once burnished; encrustation scattered, heavy over breaks.
Fabric: Fine light fabric, perhaps burned or incompletely reduced in firing; abundant mica, moderate black and red inclusions. Surfaces 5 YR 4/4 (reddish brown), core not visible.

9/69 (P106), **Figs. 9.13, 9.45** (Drawing 35.3; Photos 2138–2139)
Neck/Shoulder Fragment.
Tumulus Fill (SU 1.039).
PH (max): 2.6 cm; PW (max): 2.3 cm; Th: 0.7–0.9 cm.
Small fragment from turn of shoulder to neck of medium-sized vessel. Surfaces smoothed, worn; exterior burnished and painted.
Matt-painted decoration: Two horizontal bands beneath "V" shape, probably from either zigzag or apex of pendent triangle.
Fabric: Very fine, with abundant mica, occasional black inclusions (<0.2 cm). Surfaces 7.5 YR 6/6 (reddish yellow), core 7.5 YR 6/1 (gray).

9/70 (P272), **Fig. 9.45** (Photos 2576–2577)
Wall Fragment.
Tumulus Fill (SU 1.240).
PL (max): 1.5 cm; PW (max): 1.9 cm; Th: 1.0 cm.
Very small wall fragment. Surfaces smoothed, exterior burnished and painted, interior almost completely lost; heavy encrustation on interior surface and core.
Matt-painted decoration: Three nearly parallel lines.
Fabric: Very fine, abundant mica, occasional red inclusions. Surfaces and core 5 YR 6/8 (reddish yellow).

9/71 (P127), **Figs. 9.14, 9.45** (Drawing 41.1; Photos 2173–2175)
Wall Fragment.
Tumulus Fill (SU 1.039).
PH (max): 4.8 cm; PW (max): 4.2 cm; Th (wall): 0.7 cm; Th (wall + projection): 2.7 cm.
Single wall fragment of medium-sized vessel with triangular-shaped projection or ledge lug projecting from slightly convex shoulder; wall flares in and back out above lug, broken below rim. Surfaces smoothed, now very worn; heavy encrustation.
Fabric: Sandy, fine to semi-fine, with abundant mica, occasional small red (grog) and white inclusions. Surfaces 5 YR 5/8 (yellowish red), core mostly as surface, with small dark area 5 YR 3/1 (very dark gray).
Comparanda: For similar projections, semi-coarse and coarse: **9/257** (P214), **9/304** (P189), **9/311** (P360), **9/312** (P087).

9/72 (P058 + P391), **Figs. 9.14, 9.45** (Photos 4132, 4134)
Wall Fragments (FL Type 1a?).
Tumulus Fill (Units 6.542 + 1.070).
PL (max): 10.1 cm; PW (max): 10.4 cm; Th: 0.4–1.1 cm.
Five joining wall fragments, mostly from globular shoulder and lower body of medium-sized vessel, probably closed in shape, preserving very small portion of conical neck just below break. Surfaces smoothed, exterior painted, perhaps burnished; scattered encrustation, heavier over breaks.
Matt-painted decoration: Thick horizontal line just below neck, supporting vertical lines above, perhaps from hatched upright triangles; below line, portions of three pendent triangles (central nearly complete), hatched with thick lines.
Fabric: Fine, with abundant mica, moderate small red inclusions, occasional white (limestone) and black inclusions, very few air pockets. Surface and core 2.5 YR 6/8 (light red).
Comparanda: **9/6** (P073).

9/73 (P356), **Fig. 9.14** (Drawing 99.3; Photos 3247–3248)
Body Fragment.
Tomb XXXV (84) Fill (SU 1.471).
PL (max): 4.3 cm; PW (max): 3.8 cm; Th: 0.6–0.8 cm.
Fragment from wall of mid-sized vessel; precise shape and orientation uncertain. Surfaces smoothed, exterior perhaps once painted, interior uneven; heavy encrustation.
Matt-painted decoration: Very faint scattered black, perhaps parallel bands, perhaps manganese encrustation.
Fabric: Fine, with abundant mica, moderate fine white inclusions. Exterior surface 5 YR 5/8 (yellowish red), interior 5 YR 6/8 (reddish yellow), core not visible.

9/74 (P252), **Figs. 9.14, 9.45** (Drawing 86.5; Photos 2554–2555)
Wall Fragment.
Tumulus Fill (SU 1.240).
PL (max): 6.7 cm; PW 4.8 cm; Th: 0.3 cm.
Single wall fragment, probably from lower body of mid-sized or large vessel. Exterior surface smoothed, burnished, and painted; interior surface completely lost; heavy encrustation.
Matt-painted decoration: Lower portions of two elongated pendent triangles, probably hatched.
Fabric: Very fine, with abundant mica. Exterior surface and core (so far as visible) 7.5 YR (brown).
Comparanda: For shape and decoration, cf. Andrea 1985: pl. II, V10, 5; pls. VI; Tomb 45, 1; pl. XXIX,

Tomb 35, 2 (Barç); Bejko forthcoming: Kamenicë 280/1343, 1/17; 376/2277; Mikulčič 1966: pl. VII, a (Živojno).

9/75 (P132 + P324), **Figs. 9.14, 9.45** (Drawing 31.2; Photos 3742–3743)
Wall Fragments.
Tumulus Fill (Units 1.070 + 1.368).
PL (max): 6.1 cm; PW (max): 12.0 cm; Th: 0.6 cm; PD (max): ca. 22 cm.
Two joining fragments from wall of large vessel, probably from lower body, perhaps amphora or similar closed shape. Surfaces smoothed, exterior decorated and highly burnished (with visible grooves); scattered encrustation.
Matt-painted decoration: Three groups of three nearly vertical bands, probably apices of elongated inscribed pendent triangles.
Fabric: Fine, with abundant mica and small black inclusions, occasional white pebbles (<0.4 cm). Exterior surface 2.5 YR 6/8 (light red), interior surface 7.5 YR 5/4 (brown), core 10 YR 4/1 (dark gray).
Comparanda: **9/89** (P232).

9/76 (P431), **Figs. 9.14, 9.45** (Drawing 98.3, Photos 3746–3747)
Wall Fragment.
Tumulus Fill (SU 1.070).
PL (max): 4.4 cm; PW (max): 5.2 cm; Th: 0.9 cm.
Single wall fragment from medium-size or large vessel, perhaps closed in shape. Surfaces smoothed, exterior burnished and painted.
Matt-painted decoration: Small portion of pendent triangle, preserved five lines roughly parallel, two of them join left leg of triangle.
Fabric: Same as **9/78** (P104), **9/79** (P193).

9/77 (P195), **Figs. 9.14, 9.46** (Drawing 49.5; Photos 2514–2515)
Wall Fragment.
Tumulus Fill (SU 4.203).
PL (max): 3.5 cm; PW (max): 3.5 cm; Th: 0.7 cm.
Single wall fragment preserving shoulder of medium-sized vessel of uncertain shape. Surfaces smoothed, exterior slipped and painted; heavy encrustation.
Matt-painted decoration: Small portion from middle of pendent triangle, including part of right leg.
Fabric: Same as **9/78** (P104), **9/79** (P193 + P353 + P115 + P167), but exterior 5 YR 5/8 (yellowish red), interior 7.5 YR 5/4 (brown), core 5 YR 4/1 (dark gray).

9/78 (P104), **Figs. 9.14, 9.46** (Drawing 40.3; Photos 2167–2168)
Wall Fragment.
Topsoil/Surface Find (SU 1.001).
PL (max): 4.0 cm; PW (max): 4.8 cm; Th: 0.9 cm.
Single wall fragment, probably from shoulder of medium or large closed vessel. Surfaces smoothed, exterior painted; scattered encrustation.
Matt-painted decoration: Lower portion of hatched pendent triangle.
Fabric: As **9/76** (P431), **9/79** (P193 + P353 + P115 + P167).

9/79 (P193 + P353 + P115 + P167), **Figs. 9.14, 9.46** (Photos 4129, 4131)
Shoulder, Neck, and Rim Fragments.
Topsoil and Tumulus Fill (Units 1.141, 1.070, 1.440).
PH (max): 12.0 cm; PW (max): 6.8 cm; Th: 0.6–1.0 cm.
Five joining fragments (plus surface chip) from upper body of medium or large vessel with S-profile, perhaps closed in shape; convex shoulder curves in to concave neck, set off by slight ridge, rising to narrow, tapered lip, scarcely preserved. Exterior surface smoothed and painted, but now damaged; perhaps once slipped and burnished; interior smoothed, perhaps burnished at top, also damaged, uneven; scattered encrustation, exterior perhaps slipped, air pockets (<0.6 cm).
Matt-painted decoration: On neck, checkerboard pattern of lozenges, bound by three horizontal lines just below rim, three horizontal lines along ridge of neck-shoulder junction; on shoulder, beneath lowest line, extending beyond breaks, two hatched pendent triangles.
Fabric: Fine, with abundant mica and small black inclusions, moderate red (grog) and white inclusions, occasional air pockets, some large (<0.6 cm). Surfaces 5 YR 6/6 (reddish yellow), core 5 YR 4/1 (dark gray).
Comparanda: **9/76** (P431), **9/78** (P104); Korkuti 1971: pl. IX (Tren settlement, Early Iron Age).

9/80 (P284), **Figs. 9.14, 9.46** (Drawing 48.2; Photos 3274–3275)
Neck/Shoulder Fragment.
Tumulus Fill (SU 4.203).
PL (max): 4.8 cm; PW (max): 3.4 cm; Th: 0.9 cm.

Small fragment preserving neck and shoulder of medium- to large-sized vessel with convex lower body and concave neck. Surfaces smoothed, exterior burnished and painted; scattered encrustation.
Matt-painted decoration: Two horizontal lines along ridge dividing upper and lower bodies; above, lower portion of single upright hatched triangle; below, upper portion of much larger, hatched pendent triangle, with bit of second triangle at left edge.
Fabric: Fine to semi-fine, abundant mica, moderate white (limestone, <0.5 cm) inclusions, and occasional black inclusions. Exterior surface 5 YR 5/4 (reddish brown), interior surface and core 2.5 YR 5/8 (red).
Comparanda: Korkuti 1981:41, pl. IV, Tomb 27 (double vessel found in rich Early Iron Age cremation burial at Patos); Videski 2005: pl. VI, 3 (Vardarski Rid, Late Bronze Age); Vokotopoulou 1986: vol. 2, pl. 336:α.

9/81 (P367), **Fig. 9.46** (Photos 3289–3291)
Wall Fragment.
Tomb XXXV (84) Fill (SU 1.471).
PL (max): 2.2 cm; PW (max): 3.4 cm; Th 0.8 cm.
Single wall fragment from vessel of uncertain shape and size. Surfaces smoothed but worn, exterior painted; scattered encrustation.
Matt-painted decoration: Vertical line, poorly preserved.
Fabric: Very fine, abundant mica, occasional black inclusions. Surfaces 2.5 YR 6/8 (light red), core 5 YR 7/1 (light gray).

9/82 (P305), **Figs. 9.15, 9.46** (Drawing 91.1; Photos 3384–3386)
Body Fragment.
Tumulus Fill (SU 1.377).
PL: 10.2 cm; PW (max): 6.5 cm; Th: 1.0–0.6 cm.
Single fragment of large vessel, probably closed in shape, preserving portions of lower neck and upper shoulder, with broken attachment for loop handle, ovoid in section. Exterior surface smoothed and painted, interior uneven; very heavy encrustation.
Matt-painted decoration: On neck, above handle attachment, six horizontal lines, supporting four vertical lines, probably base of upright triangle. On shoulder, upper portions of two pendent triangles, one on either side of handle; beneath handle, three additional horizontal lines above indistinct motif: small solid pendent triangles or top of cross-hatched area.
Fabric: Fine to semi-fine, heavy, with abundant mica and white (limestone) inclusions (<0.3 cm). Surfaces 2.5 YR 5/8 (red), core 2.5 YR 5/1 (reddish gray).
Comparanda (unpainted): **9/40** (P279; similar handle); cf. also similar handle from Barç (Andrea 1985: pl. XLVI, 1, 2, 3 [Barç Tumulus 2, Tombs 9, 10, twelfth–eighth centuries]).

9/83 (P255 + P112), **Figs. 9.15, 9.46** (Drawings 37.1 [P112], 60.1 [P255]; Photo 4135)
Body/Neck/Handle Fragments.
Tumulus Fill (Units 1. 279, 1.070).
PH (max): 11.2 cm; PW (max). 14.2 cm; PL (handle): 4.3 cm; PW (handle): 3.7–5.9 cm; Th (walls): 0.7–1.2 cm; Th (handle): 1.5 cm.
Five joining fragments from conical lower neck, globular upper shoulder, and vertical strap handle, flattened oval in section, of large vessel, perhaps closed in shape; precise orientation uncertain. Surfaces smoothed, exterior painted but damaged (in excavation?), interior slightly uneven; encrustation heavy in places.
Matt-painted decoration: At neck-shoulder junction, three horizontal bands, interrupted by handle attachment; emerging from lowest line, and from attachment, upper portions of four hatched, pendent triangles; from upper line, on either side of handle, groups of roughly vertical lines, probably hatched, upright triangles. On lower handle exterior, four roughly parallel horizontal bands.
Fabric: Fine, with abundant mica and small black and red inclusions, occasional air pockets (<0.4 cm). Surfaces 5 YR 6/6 (reddish yellow), core 5 YR 6/1 (gray).
Comparanda: **9/47** (P258).

9/84 (P030), **Figs. 9.15, 9.46** (Drawing 7.1; Photos 545–558)
Body and Handle Fragments.
Tumulus Fill (SU 4.035).
PL (max): 5.6 cm; PW (max): 14.5 cm; PW (handle): 4.6 cm; D (shoulder): ca. 30 cm; Th: 0.6–1.2 cm; Th (handle): 0.9 cm.
Two joining body fragments (plus small chips) from shoulder of large vessel, probably closed in shape, preserving attachment and lower portion of verti-

cal strap handle, flattened oval in section. Exterior surface smoothed, burnished, and painted, interior uneven; scattered encrustation.

Matt-painted decoration: Five horizontal bands at base of handle, above top of very poorly preserved cross-hatched pendent triangle; to right, on shoulder, middle section of cross-hatched pendent triangle.

Fabric: Fine, with abundant mica, black and white inclusions, very occasional air pockets. Surfaces 2.5 YR 4/6 (red), core 2.5 YR 4/1 (dark reddish gray).

Comparanda: Vokotopoulou 1986:336, α; Korkuti 1971: pl. IX (Tren; same decoration, on rim).

9/85 (P346), **Figs. 9.15, 9.47** (Drawing 76.2; Photos 3294–3295)

Wall Fragment.

Tomb III (81) Fill (SU 2.453).

PL (max): 3.1 cm; PW (max): 3.9 cm; Th: 0.9 cm.

Small wall fragment, probably from large vessel. Surfaces smoothed, exterior incised, perhaps once burnished, interior cracked.

Incised decoration: Five roughly concentric arching lines, from a spiral or receding circles or semicircles.

Fabric: Fine to semi-coarse, abundant mica, many small black inclusions. Surface and core 10 R 5/8 (red).

Comparanda: Lera 1987a/1: pl. XVI, 13; pl. XVIII, 21 (Late Neolithic settlement at Barç; very similar incised semicircles); Korkuti 1971: pl. IV (Neolithic Tren).

9/86 (P108 + P388), **Figs. 9.15, 9.47** (Drawing 114.6; Photos 3789–3790)

Base Fragments.

Tumulus Fill (Units 1.070, 6.550).

PH (max = Th [max]): 2.1 cm; PW (max): 7.5 cm; D (base): ca. 8 cm (not fully preserved).

Two joining fragments from flaring foot with flat underside preserving bit of concave vessel floor. Surfaces smoothed but very worn, covered by heavy encrustation.

Fabric: Fine, with abundant mica, moderate black and red (grog) inclusions, occasional white inclusions and air pockets (<0.4 cm). Surfaces and core 5 YR 5/6 (yellowish red).

9/87 (P097 + P237 + P285), **Figs. 9.15, 9.47** (Drawing 106.3; Photos 3776–3779)

Base Fragments.

Tumulus Fill (Units 4.286, 7.192).

PH (max): 5.5 cm; PW (max): 10.7 cm; Th (walls): 1.0 cm; D: 9.5 cm.

Seven joining fragments of flaring foot with raised underside, carinated lower edge, and narrow resting surface; concave vessel floor rises up and out, joining flaring walls of base to create lower body of uncertain shape, now mostly lost (perhaps amphora?). Surfaces smoothed, now extremely worn, completely lost in places (and still powdery to the touch), with varying degrees of encrustation.

Fabric: Soft, semi-fine fabric, abundant mica, moderate small black and white inclusions (<0.2 cm), few air pockets (<0.2 cm). Surface 5 YR 5/8–6/8 (red-light red), core 5 YR 6/1 (gray-light gray).

Comparanda: Bejko forthcoming: Kamenicë 294/698, 294/710, 344/804.

9/88 (P001), **Figs. 9.15, 9.47** (Drawing 2.1; Photos 551–554)

Base Fragment.

Surface Collection.

PH (max): 2.4 cm; PW (max): 7.0 cm; Th: 0.8–1.6 cm; D (ext): ca. 6–7 cm.

Single fragment from round, hollowed base of large vessel, with flaring lower walls; vessel floor not preserved. Underside smoothed but cracked and uneven, all other surfaces missing; worn.

Fabric: Fine, with abundant mica, small black inclusions, moderate air pockets (<1.0 cm), occasional grog. Surfaces 2.5 YR 5/8 (red), core 2.5 YR 5/1 (reddish gray).

9/89 (P232), **Fig. 9.15, 9.47** (Drawing 53.1; Photos 2489–2492)

Base/Lower Wall Fragments.

Tumulus Fill (SU 1.281).

PH (max): 6.0 cm; PW (max): 13.0 cm; Th: 0.5–0.8 cm; D (base): ca. 6–7 cm.

Five joining fragments (from modern break) of medium to large vessel, perhaps closed in shape, preserving portion of globular lower body and bit of flat, roughly circular base. Surfaces smoothed, exterior burnished and painted, now damaged, with scattered encrustation; heavy encrustation on interior.

Matt-painted decoration: Six sets of converging vertical lines, probably elongated apices of pendent triangles.

Fabric: Fine, with abundant mica, occasional red inclusions. Surfaces 7.5 YR 6/6 (reddish yellow), core 5 YR 5/8 (yellowish red).

Comparanda: **9/75** (P132 + P324; for painted decoration); **9/317** (P100; similar shape, coarse fabric).

Fine Dark Fabric

9/90 (P422), **Figs. 9.16, 9.48** (Drawing 96.3; Photos 3544–3551)

One-Handled Vessel, FD Type 2.

Tomb XVII (72) (SU 5.408 [in baulk]).

H (rim): 8.8 cm; H (max): 11.1 cm; D (rim): 7.0–7.6 cm; D (base): 4.7–5.1 cm; D (max): 10.6 cm; D (min): 6.0 cm; Th (walls): ca. 0.6 cm; Th (handle): 1.1 cm; Th (rim): 0.4 cm; W (handle): 2.3 cm.

Complete one-handled vessel. Rounded hemispherical body rises from slightly concave base, curving in at shoulder to conical neck, then flaring out to flattened rim. Shoulder set off from neck by incised horizontal line running around vessel; line ends on either side of handle, rises to higher level across from it. Handle roughly triangular in section, slightly concave at top; rises up and away from body before curving back down to join rim. Surface smoothed and burnished; encrustation scattered across all surfaces, more abundant within. Slight damage to exterior surfaces of rim, top of handle, and body.

Plastic decoration: Sharply angled diagonal ribbing runs around body just beneath incised horizontal line, stopping ca. 2.0–2/5 cm from base; ribs spaced evenly, ca. every 2.5 cm, in very low relief, perhaps a result of burnishing rather than impression.

Fabric: Hard to medium, with abundant mica, fine black, white, and reddish inclusions (possibly grog); shiny pebble inclusions, black and red, up to 0.4 cm; abundant air pockets, especially on exterior, up to 0.5 cm. Exterior and interior surfaces mostly 5 Y 2.5/1 (black), some (misfired?) areas as light as 10 YR 4/3 (brown); core not visible.

Comparanda: **9/159** (P303; similar shape in cruder fabric).

9/91 (P253), **Figs. 9.16, 9.48** (Drawing 56.1; Photos 2588–2596)

Kantharos, FD Type 4.

Tomb XXI (55) (SU 4.326).

H (rim): 9.9 cm; H (max): 11.9 cm; D (rim): 7.9 cm; D (mouth int): 6.8 cm; D (max): 11.9 cm; D (base): 5.4–6.0 cm; Th (rim): 0.5 cm; Th (handle): 1.1 cm; Th (body): 0.3 cm.

Fragmentary vessel, largely reconstructed, with additional non-joining chips and fragments; one handle, large section of shoulder and neck wall (ca. 7 × 5 cm), many smaller chips missing. Profile very uneven, especially below shoulder. Flat base, squat, convex-concave body narrowing to nearly cylindrical neck and slightly flaring rim. Surviving handle irregular ovoid in section, continuing profile of rim, curving down almost vertically to join body just above shoulder. Both surfaces smoothed, exterior burnished; no surface decoration.

Fabric: Hard, sandy; moderate number of mica sparkles, white (quartz) and black (grog) inclusions; occasional air pockets (up to 0.3 cm). Exterior and interior surfaces vary widely in color, from 2.5 YR 5/8 (red) to 7.5 YR 2.5/1 (black), due to uneven firing; core similar in places to surface colors, also 2.5 YR 4/3 (reddish brown).

9/92 (P364), **Figs 9.16, 9.49** (Drawing 81.1; Photos 3312–3314)

Fragmentary Kantharos, FD Type 5.

Tumulus Fill (SU 2.493).

H (max): 13.2–13.5 cm; H (rim): 8.4–9.2 cm; D (rim ext, estimated): ca. 9 cm; Th (rim): 0.7 cm; Th (wall): 0.4–0.7 cm; Th (handle):1.1–1.9 cm; W (handle at body): ca. 3.5 cm; W (handle at rim): ca. 3.5 cm.

Twelve joining fragments preserving full profile and ca. 1/4 circumference of vessel; three additional non-joining fragments. Very upright profile, slight S-curve: flat base curves out to nearly vertical lower body, then back to gently flaring cylindrical neck. High-swung vertical handle rises from body at point of maximum diameter, stretching far above body and returning to rim; handle varies in section, rectangular at sides, roughly triangular at top. Surface smoothed and burnished, especially away from handle; encrustation on handle exterior and interior, vessel interior; no surface decoration.

Fabric: Hard, sandy; abundant inclusions: black, white, mica, pebble (<0.6 cm); occasional air pockets (<0.5 cm). Exterior and interior surfaces 7.5 YR 4/1 (dark gray), 10 YR 5/2 (grayish brown), interior also 7.5 YR 4/2 (brown); core 5 YR 5/6 (yellowish red) to 3/1 (very dark gray).

9/93 (P223), **Figs. 9.16, 9.49** (Drawing 78.1; Photos 2478–2483)
Kantharos, FD Type 6.
Tomb XLVI (42) (SU 2.263).
H (rim): 8.2 cm; H (max): 9.9 cm; D (rim): 8.4 cm; D (max): 9.5 cm; D (neck int): 7.2 cm; Th (handle at rim): 1.0 cm; Th (handle at body): 1.5 cm; Th (body): 0.3–0.5 cm; W (handle at rim): 2.6 cm; W (handle at tip): 2.1 cm; W (handle at shoulder): 2.4 cm.
Body and one handle reconstructed almost completely from 19 fragments.
Rounded base and globular lower body, tapering inward at neck beneath slightly flaring rim. Surviving vertical strap handle sharply angled, ovoid in section; rises in nearly straight line from rim to pointed tip, then curves outward to join body at shoulder. Small hole (ca. 1.3 × 0.7 cm) in shoulder beside attachment for missing handle; chips missing from rim and much of vessel surface. Surface once smoothed, maybe burnished; now very poorly preserved.
Fabric: Hard, sandy, abundant mica inclusions, moderate number of red (grog?) and white (quartz?) inclusions (up to 0.4 cm); occasional air pockets. Exterior surface 7.5 YR 2.5/1 (black), with one lighter area (encrustation?) 10 YR 6/3 (pale brown); core 2.5 YR 4/6 (light reddish brown).

9/94 (P322), **Figs. 9.17, 9.50** (Drawing 79.1; Photos 3189–3198)
One-Handled Vessel, FD Type 1.
Tomb LIII (63) (SU 1.0359).
PH (max): 7.5 cm; H (rim): 6.8 cm; D (base): 4.2–4.5 cm; D (max): 10.9 cm; D (rim): 9.8–10.0 cm; Th (walls): 0.3–0.6 cm; Th (handle): 0.6 cm; D (handle strut): ca. 1.0 cm; W (handle at body): 3.7 cm; W (handle): 2.4 cm; W (handle fragment): 2.2 cm.
Nearly complete vessel, reconstructed from some 40 fragments; missing top of handle, fragments from rim, neck, and lower body; damage to mended fragments on bottom and one side, with parts of exterior surface not preserved. Flat, ovoid base rising to sharply delineated biconical lower body, nearly vertical conical neck, and widely flaring, everted rim. Vertical handle curves up and out from point of maximum diameter, with nearly horizontal strut stretching toward broken rim; handle not preserved above strut, but must have continued up before curving back toward rim, as indicated by one non-joining fragment; strut nearly circular in section, handle irregular ovoid. Surface smoothed and highly burnished, especially on upper sections of both interior and exterior.
Fabric: Hard, sandy; many very small white inclusions, moderate mica; single visible black inclusion (ca. 0.2 cm); few visible air pockets, very small (<0.2 cm). Exterior and interior surfaces 2.5 Y 2.5/1 (black); spots of 5 Y 7/3 (pale yellow) on handle and shoulder; core 5 YR 5/6 (yellowish red) to 5 Y 5/1–2 (gray-olive gray).
Plastic decoration: Vertical ribbing across entire preserved shoulder zone, ca. 0.3 cm between ridge crests. Five very shallow and roughly circular impressions (ca. 0.2–0.5 cm in diameter) on exterior portion of non-joining handle fragment.
Comparanda: Profile and fabric more similar to dark fabric kantharoi **9/95, 9/96** (P228, P276) than to one-handled vessels, although no other vessel from Lofkënd is so wide and squat; plastic decoration most similar to **9/96** (P276), but more closely spaced.

9/95 (P228), **Figs. 9.17, 9.51** (Drawing 51.1; Photos 2484–2488, 2526)
Kantharos, FD Type 3.
Tomb LVI (43) (SU 4.267).
H (rim): 8.6–8.9 cm; H (max): 10.6–10.8 cm; D (rim ext): 6.5 cm; D (neck int): 4.4 cm; D (max): 10.8 cm; D (base): 3.7 cm; Th (rim): 0.4 cm; Th (handles at body): 1.5 cm; Th (handles, top): 1.1 cm; Th (handles, rim): 1.1 cm; Th (walls): ca. 0.5 cm; W (handles): 2.1 cm wide; W (handles, top): 1.3 cm; W (handles, rim): 2.5 cm.
Nearly complete vessel reconstructed from 18 joining fragments (and additional chips). Flat base rising to sharply delineated biconical lower body, conical neck, and everted, slightly ovoid rim. Asymmetry apparent especially in lower body and shoulder zone. Rounded handles (rhomboid in section) rise vertically, with slight inward curvature, then curve sharply back toward rim. Chips and fragments missing mainly from one side of vessel, especially shoulder and rim. Surface smoothed over entire exterior, burnished on neck and rim.
Plastic decoration: Ribbing on shoulder zone; shallow (0.1–0.2 cm deep) diagonal striations, 2.1–2.7 cm long, spaced 0.3–0.5 cm between ridge crests. Two sets of parallel diagonal ridges, beginning at handles and oriented opposite to one another,

converge to create two concentric, triangular indentations; junction of opposed diagonals slightly off-center on both sides of vessel, but lost due to damage on one side.

Fabric: Sandy, hard; abundant mica sparkles, occasional grog, other inclusions difficult to see due to dark fabric. Moderate number of air pockets (<0.3 cm). Exterior and interior surfaces 5 Y 2.5/1 (black); several small patches of reddish brown (5 YR 5/4) on one side of rim and at junction of shoulder and lower body; core difficult to see, but where visible, very similar to surfaces; light spot at junction of one handle with body due to dirt and encrustation.

Comparanda: **9/96** (P276; similar shape, fabric, decoration); **9/98** (P032), **9/99** (P050).

9/96 (P276), **Fig. 9.17**, **9.51** (Drawing 59.1; Photos 2743–2749)

Kantharos, FD Type 3.

Tomb XLVIII (52) (SU 1.0366).

H (rim): 8.9–9.2 cm; H (max, handles): 11.1, 11.4 cm; D (rim ext): 6.8–7.3 cm; D (mouth int): 5.8 cm; D (max): 12.3 cm; D (base): 4.0 cm; Th (rim): 0.4 cm; Th (handles at body): 1.5 cm; Th (handles, top): 1.0 cm; Th (handles, rim): 1.2 cm; Th (walls): 0.3–0.5 cm; W (handles): 2.1 cm wide; W (handles, top): 1.2 cm; W (handles, rim): 1.9 cm.

Nearly complete vessel, reconstructed from many fragments. Shape as P228, except for handles—here ovoid in section. Chips and missing fragments from all parts of vessel, especially lower body and one handle. Surface smoothed over entire exterior, burnished on neck and rim.

Fabric: Sandy, hard; abundant mica sparkles, occasional grog; other inclusions difficult to see due to dark fabric. Moderate number of air pockets (<0.3 cm). Exterior and interior 10 YR 3/1 (very dark gray); one small black patch (5 Y 2.5/1) at junction of shoulder and lower body, just beside handle; several very small lighter areas (5 YR 6/6) around rim, near tops of handles, and on neck; core 5 YR 6/6 (reddish yellow).

Plastic decoration: Ribbing across entire shoulder zone between handles; very shallow (<0.1 cm deep) radiating ridges, 2.2–2.5 cm long, evenly spaced 0.4 cm between ridge crests. Ridges nearly vertical, following shoulder profile, but curve slightly in some sections, with no discernible pattern between straight and curved lines. Parts of decoration lost to chips in shoulder surface.

Comparanda: **9/95** (P228; similar shape, fabric, decoration); **9/98** (P032), **9/99** (P050).

9/97 (P123), **Figs. 9.18**, **9.51** (Drawing 39.5; Photos 2156–2157)

Handle Fragment.

Topsoil (SU 1.141).

PL (max): 2.5 cm; PW (max): 1.6 cm; Th: 1.2 cm.

Single fragment of vertical loop or strap handle, rhomboid in section; broken at mid-handle on both sides. Exterior surface smoothed, especially on exterior; worn underneath, with encrustation covering breaks.

Fabric: Fine, medium hardness, with abundant mica, few very small white inclusions, single black pebble. Exterior 2.5 YR 4/6 (red), interior 2.5 YR 2.5/1 (reddish black).

Comparanda: **9/95** (P228).

9/98 (P032), **Figs. 9.18**, **9.51** (Drawing 6.4; Photos 749–752)

Body Fragments.

Interface Topsoil/Tumulus Fill (SU 4.011).

PH (max): 2.4 cm; PW (max): 6.0 cm; Th: 0.5–0.9 cm; D (max, ext): 17 cm.

Two joining body fragments from widest point of biconical vessel, preserving carination; lower section smoothed, perhaps once burnished, now worn; interior surface rougher.

Plastic decoration: Parallel ribbing on upper section.

Fabric: As **9/96** (P276).

Comparanda: **9/99** (P050); complete vessels **9/95** (P228), **9/96** (P276).

9/99 (P050), **Figs. 9.18**, **9.51** (Drawing 7.5; Photos 753–757)

Body Fragment.

Tumulus Fill (SU 1.007).

PH (max): 2.2 cm; PW (max): 2.9 cm; Th: 0.5–0.7 cm.

Single fragment from handle zone. Interior surface perhaps smoothed (but very little preserved).

Plastic decoration: Parallel ribbing.

Fabric: As **9/96** (P276), but interior lighter in color: 10 YR 5/1 (gray).

Comparanda: **9/98** (P032); complete vessels **9/95** (P228), **9/96** (P276).

9/100 (P049 + P063), **Figs. 9.17, 9.52** (Drawing 14; Photos 4027–4029 [P049 + P063], 725–728 [P049], 3562–3563 [P063])
One-Handled Vessel Fragments, FD Type 2.
Tumulus Fill (Units 3.068, 1.070, 2.078?).
PH (max): 7.0 cm; D (max): ca. 12 cm; Th (handle): 0.9 cm; Th (wall): 0.6–0.7 cm; PW (max): 11.1 cm; PW (handle): 3.2 cm.
Four joining fragments, found in different stratigraphic units, from small vessel with globular lower body and conical neck, preserving attachment and lower portion of vertical strap handle, roughly rectangular in section, just above neck-shoulder junction. Exterior surface smoothed and burnished, interior smoothed; encrustation scattered across interior and exterior surfaces, heavy in places. (Two additional non-joining fragments possibly from same vessel.)
Fabric: Hard, sandy; abundant mica and small (<0.3 cm) black, white, and red (grog) inclusions; air pockets especially in interior surface (<0.4 cm). Exterior and interior surfaces both very dark, varying from 10 YR 5/2 (grayish brown) to 3/1 (very dark gray); core 10 YR 6/3 (pale brown).
Comparanda: **9/101** (P137; possibly same vessel, non-joining), **9/102** (P051), **9/159** (P303), **9/90** (P422); **9/105** (P181) and **9/109** (P235): handle frr of similar fabric, shape; **9/91** (P253; similar handle on intact vessel). Cf. also uncatalogued P162, P410, P415B.

9/101 (P137), **Figs. 9.17, 9.52** (Drawing 31.1; Photos 3564–3565)
One-Handled Vessel Fragments, FD Type 2.
Tumulus Fill (Units 1.070, 2.078).
PH (max): 7.0 cm; D (max approximate): ca. 12 cm; Th (handle): 0.9 cm; Th (wall): 0.6–0.7 cm; PW (max): 11.1 cm; PW (handle): 3.2 cm.
Two joining wall fragments preserving rounded lower body and junction with conical neck. Exterior surface smoothed and burnished, interior smoothed; fire-clouding on neck exterior, encrustation scattered across interior and exterior surfaces, heavier toward bottom.
Fabric: Hard, sandy, fine fabric; abundant inclusions, including very small (<0.2 cm) black, white, and mica; black and white pebbles (<0.5 cm); possibly grog; few air pockets, more in interior surface (<0.3 cm). Exterior and interior surfaces both very dark, varying from 10 YR 5/2 (grayish brown) to 3/1 (very dark gray); core 10 YR 6/3 (pale brown).
Comparanda: **9/100** (P049 + P063; possibly same vessel, non-joining), **9/90** (P422), **9/102** (P051), **9/159** (P303); uncatalogued P162, P410.

9/102 (P051), **Figs. 9.18, 9.52** (Drawing 10.7; Photos 588–591)
Shoulder/Neck Fragment.
Tumulus Fill (SU 1. 007).
PL (max): 3.9 cm; PW (max): 2.5 cm; Th: 0.5 cm.
Single fragment from convex shoulder of small vessel, preserving bit of turn to neck. Surfaces smoothed, once burnished, now worn; light encrustation.
Fabric: Fine, with abundant mica, occasional fine white inclusions. Exterior 2.5 YR 4/3 (reddish brown), interior and core 2.5 YR 3/1 (dark reddish gray).
Comparanda: **9/100** (P049 + P063), **9/101** (P137); uncatalogued P162, P410, P415B.

9/103 (P134), **Figs. 9.18, 9.52** (Drawing 107; Photos 4016–4019)
Body, Neck, Rim, and Handle Fragments.
Tumulus Fill (SU 2.239).
PH (max): 6.4 cm; PW (max): 7.3 cm; W (handle): 2.2 cm; Th: 0.4–0.6 cm; D (min [neck int]): ca. 9 cm.
Two sets of joining fragments (mostly from modern breaks) from small vessel; three preserve bit of upper shoulder, vertical neck, and rim, with vertical strap handle, rectangular in section; six preserve small section of globular lower body. Handle rises just above slightly everted rim with tapered lip, mostly lost. Surfaces smoothed, exterior and lower interior once burnished, now worn; scattered encrustation. (Five additional fragments, including one from neck, one from rim, and three from body joining one another [from modern break], do not join above groups and may not belong.)
Fabric: Fine dark, with abundant mica, small white and red (grog) inclusions, occasional larger white quartz (<0.4 cm) and red pebble (<0.2 cm) inclusions. Surfaces mostly 10 YR 2/1 (black), with some lighter areas: 5 YR 4/2 (dark reddish gray) and 7.5 YR 5/3 (brown); core 10 YR 2/1 (black), with very thin reddish layer just below surface in places.
Comparanda: **9/104** (P233; possibly same vessel, non-joining).

9/104 (P233), **Figs. 9.18, 9.52** (Drawing 107; Photos 4007–4012)
Shoulder, Neck, Rim, and Handle Fragments.
Tumulus Fill (SU 2.297).
PH (max): 6.6 cm; PW (max): 7.9 cm; Th: 0.4–0.5 cm.
Six joining fragments (from modern break) of small vessel, preserving parts of upper shoulder, with vertical neck, rim, and vertical strap handle, as **9/103** (P134). Surfaces smoothed, once burnished, now worn; light encrustation. (Non-joining fragment, probably from flaring neck, may not belong.)
Fabric: As **9/103** (P134).
Comparanda: **9/103** (P134; possibly same vessel, non-joining).

9/105 (P235), **Fig. 9.18** (Drawing 84.3; Photos 2562–2563)
Vertical Strap Handle Fragments.
Tumulus Fill (SU 4.204).
PL (max): 6.6 cm; PW (max): 5.5 cm; W (handle): 2.4 cm; Th: 0.7 cm.
Seven joining fragments of vertical strap handle, rectangular in section, preserving junction with thin-walled body; one non-joining fragment. Vessel probably closed, small to medium size, but precise shape uncertain. Surface smoothed and burnished.
Fabric: Fine dark fabric, with abundant mica, very small black inclusions, few white inclusions and air pockets (up to 0.5 cm); traces of encrustation. Exterior 10 YR 3/2 (very dark grayish brown); core as exterior, but also lighter in places: 5 YR 5/6. (yellowish red).
Comparanda: **9/100** (P049 + P063), **9/109** (P181) for similar fabric, shape on handle fragment; **9/91** (P253) for similar handle; possibly **9/93** (P223; spur handle on nearly intact vessel).

9/106 (P242), **Fig. 9.52** (Photos 2636–2637)
Vertical Strap Handle Fragment.
Tumulus Fill (SU 2.288).
PL (max): 4.5 cm; PW (max): 4.4 cm; W (handle): 2.4 cm; Th: 1.1 cm.
Single fragment of vertical strap handle, ovoid in section, preserving part of turn to vertical portion and part of junction with body. Surface smoothed, perhaps once burnished, now very worn.
Fabric: Fine dark fabric, with abundant mica, white inclusions (<0.4 cm), very small air pockets, few black and red inclusions (perhaps grog). Surface and core 5 YR 4 + /3 (reddish brown).
Comparanda: **9/108** (P043), **9/100** (P049 + P063), **9/105** (P235); uncatalogued P241.

9/107 (P142), **Fig. 9.18** (Drawing 42.8; Photos 2386–2389)
Looped Handle Fragments.
Tumulus Fill (SU 4.204).
PL (max): 3.0 cm; PW (max): 2.6 cm; Th: 0.8 cm.
Two joining fragments (modern break) of tightly curved strap, rectangular in section; original shape uncertain, possibly small lug—vertical or horizontal—or apex of vertical strap handle, perhaps even from double vessel. Exterior surface smoothed and burnished, with small ridges on one exterior side; no plastic decoration.
Fabric: Fine dark, medium hardness, with few very small black, white, and red (grog?) inclusions. Exterior and core 10 YR 2/1 (black).

9/108 (P043), **Fig. 9.18** (Drawing 10.3; Photos 624–627)
Vertical Strap Handle Fragment.
Tumulus Fill (SU 1.039).
PL (max): 3.9 cm; PW (max): 1.7 cm; Th: 0.8 cm.
Single fragment of small vertical strap handle broken at both ends, preserving sharp curve, but no attachment to vessel rim or body; oval in section, thickening slightly near lower break, perhaps for connection with body. Surfaces smoothed but now very worn; no traces of burnishing, heavy encrustation in places.
Fabric: Fine dark fabric, with abundant mica and very small white inclusions. Exterior 2.5 YR 5/6 (red), core 2.5 YR 2.5/1 (reddish black).
Comparanda: Uncatalogued P148 (similar shape, fabric).

9/109 (P181), **Fig. 9.18** (Drawing 85.1; Photos 2341–2343)
Vertical Strap Handle Fragment.
Tumulus Fill (SU 1.070).
PL (max): 2.8 cm; PW (max): 2.5 cm; Th: 1.1 cm.
Single fragment of vertical strap handle, rectangular in section, not preserving junction to body or rim, but with scar from spur or strut attachment on underside. Exterior handle surface burnished, interior smoothed.

Fabric: Fine dark fabric, with abundant mica, red inclusions (grog?), moderate encrustation. Exterior 10 YR 2/1 (black), core as surface with lighter-colored layer 5 YR 4/6 (yellowish red).

Comparanda: **9/105** (P235) and **9/100** (P049 + P063; similar fabric and shape on handle fragment); **9/91** (P253) for similar handle; possibly **9/93** (P223; spur handle on nearly intact vessel).

9/110 (P103), **Fig. 9.18** (Drawing 34.2; Photos 2146–2147)

Vertical Strap Handle Fragment.

Tumulus Fill (SU 4.129).

PL (max): 3.3 cm; PW (max): 2.0 cm; Th (max): 0.9 cm.

Single fragment of vertical strap handle, roughly triangular in section, with raised ridge running parallel to long edges of exterior, slightly off-center; broken at curve near top. Surface smoothed; scattered encrustation.

Fabric: Fine dark fabric, with abundant mica and black inclusions, moderate white inclusions. Surface and core 5 Y 2.5/1 (black).

Comparanda: **9/116** (P357A), **9/118** (P056), **9/119** (P229).

9/111 (P041), **Figs. 9.18, 9.52** (Drawing 8.6; Photos 3744–3745)

Rim Fragment.

Tumulus Fill (SU 2.041).

PH (max): 2.3 cm; PW (max): 3.0 cm; Th: 0.4–0.5 cm; D (rim ext): 13 cm.

Small fragment of small or mid-sized vessel, probably open in shape, with upper wall flaring slightly to rim with flat lip. Surfaces smoothed, once burnished, now worn (especially exterior); slight encrustation on exterior; small black spots, perhaps bitumen, on interior.

Fabric: Fine dark fabric, abundant mica, moderate fine black and white inclusions, few very small black, red, and white pebble inclusions (0.1 cm). Exterior 2.5 Y 4/2 (dark grayish brown), core 7.5 YR 4/4 (brown to dark brown) to 4/0 (dark gray).

9/112 (P334), **Fig. 9.53** (Photos 3160–3161)

Rim/handle Fragment.

Tumulus Fill (SU 2.395).

PL (max): 2.6 cm; PW (max): 2.5 cm; Th: 0.5 cm.

Two joining fragments (from modern break) of slightly flaring rim with rounded lip (or handle?); vessel shape uncertain, probably small or medium in size. Surfaces smoothed, perhaps once burnished, now worn.

Fabric: Fine; moderate mica, fine black and white inclusions, few small pebbles (<0.3 cm). Surface mottled 10 YR 5/2 (grayish brown), 5/3 (brown), 2/1 (black); core 2.5 Y 5/3 (light olive brown).

Comparanda: Fabric: **9/152** (P045 + P046), **9/125** (P126), **9/128** (P204), **9/137** (P215), **9/142** (P248), **9/130** (P295), **9/112** (P334). In this group, the difficulty of distinguishing between dark and light fabric comes to a head; most of these pieces are burnished and at least partially reduced, giving the appearance of dark fabric, but often some of the dark color is due also to encrustation, and where this is worn off, the reddish yellow color of light-colored fine fabric is visible. On some pieces (e.g., **9/140** [P044], **9/137** [P215], **9/133** [P171]), the color of one side might be classified as fine light fabric, that of the other fine dark. Without elemental analysis, one cannot be certain, but it seems at least possible that the two fine fabric groups—dark and light—are in fact the same clay, simply treated differently in the firing and finishing processes.

9/113 (P211), **Fig. 9.53** (Photos 2442–2443)

Rim Fragment.

Tumulus Fill (SU 2.202).

PL (max): 3.9 cm; PW (max): 3.1 cm; Th: 0.7 cm.

Single fragment from small or mid-sized vessel with tapered lip; precise shape and orientation uncertain. Surfaces smoothed, now worn, especially just below lip on exterior; lip damaged on interior.

Fabric: Abundant mica and white inclusions (<0.5 cm), occasional black and red inclusions (some grog) and air pockets. Surfaces 10 YR 3/3 (dark brown), core 7.5 YR 3/1 (very dark gray).

Comparanda: **9/124** (P299), **9/127** (P089), **9/138** (P199), **9/146** (P067), **9/147** (P149); uncatalogued P152, P341, P374, P419.

9/114 (P027), **Fig. 9.19** (Drawing 5.6; Photos 908–911)

Wide Strap Handle Fragment.

Interface Topsoil/Tumulus Fill (SU 4.011).

PL (max): 3.5 cm; PW (max): 3.1 cm; Th: 0.7 cm.

Single fragment preserving only one edge of wide strap handle, probably at rim junction; original shape and orientation uncertain.

Fabric: As **9/122** (P361).
Comparanda: **9/121** (P307), **9/122** (P361), **9/123** (P339).

9/115 (P398), **Fig. 9.53** (Photos 3478–3479)
Handle Fragment.
Tumulus Fill (SU 5.535).
PH (max): 4.7 cm; PW (max): 4.0 cm; Th: 1.3–1.7 cm.
Single fragment of vertical strap handle, rounded triangular in section; broken at junction with body and mid-handle. Exterior surface smoothed and burnished, interior smoothed; entire exterior now very worn, much burnishing lost. No plastic or painted decoration.
Fabric: As **9/90** (P422), with core visible mostly black but as light as 2.5 YR 4/6 (red), and some areas of convex side not fully black—7.5 YR 5/2 (brown).
Comparanda: **9/167** (P094B; very similar shape and fabric, classified as semi-coarse with rim **9/206** [P094A]); **9/90** (P422; similar handle shape, but smaller, on intact vessel).

9/116 (P357A), **Fig. 9.19** (Drawing 103; Photos 3226–3227)
Vertical Strap Handle Fragment.
Tumulus Fill (SU 1.440).
PL (max): 6.0 cm; PW (max): 4.6 cm; PW (handle): 3.5 cm; Th: 1.3–1.7 cm.
Single fragment of vertical strap handle, triangular to ovoid in section, broken just above attachment to body and at 90° bend near top. Surface highly burnished on handle exterior, smoothed within. (Found together with uncatalogued P357B, non-joining shoulder fragment (7.8 × 9.0 cm) of same fabric, probably same vessel.)
Fabric: Fine to semi-coarse, heavy, dark fabric—highly burnished, finely shaped, but with abundant large red (grog <0.4 cm) and white (lime?) inclusions; fine cracks visible in surface. Surface mottled 5 YR 5/3 (reddish brown) to 2/1 (black), core 10 R 4/2 (weak red) to 10 YR 2/1 (black).
Comparanda: **9/117** (P055), **9/118** (P056), **9/119** (P229); LBA form?

9/117 (P055), **Fig. 9.19** (Drawing 12.3; Photos 924–927)
Vertical Strap Handle Fragment.
Interface Topsoil/Tumulus Fill (SU 4.011).
PL (max): 2.7 cm; PW (max): 2.8 cm; Th: 0.7 cm.
Single fragment of vertical strap handle, rectangular in section, broken at 90° bend near top; no attachments preserved. Smoothed, not burnished.
Fabric: Fine to semi-coarse, now covered by heavy encrustations. Moderate mica and black inclusions, few white inclusions and air pockets (<0.4 cm). Surface 10 YR 4/3 (brown) where visible.
Comparanda: **9/116** (P357A), **9/118** (P056), **9/119** (P229).

9/118 (P056), **Fig. 9.19** (Drawing 12.6, Photos 2012–2017)
Vertical Strap Handle Fragment.
Tumulus Fill (SU 3.045).
PL (max): 3.2 cm; PW (max): 2.9 cm; Th: 0.9 cm.
Single fragment of vertical strap handle, rectangular in section, broken at 90° bend near top; no attachments preserved (as **9/117** [P055], but slightly larger portion preserved). Smoothed, not burnished, scattered encrustation.
Fabric: Fine to semi-coarse, dark fabric, with many white inclusions, moderate black inclusions and mica, few tiny air pockets. Surface and core 10 YR 2/1 (black).
Comparanda: **9/116** (P357A), **9/117** (P055), **9/118** (P056), **9/119** (P229).

9/119 (P229), **Fig. 9.53** (Photos 2624–2626)
Vertical Strap Handle Fragment.
Tumulus Fill (SU 1.279).
PL (max): 3.9 cm; PW (max): 2.8 cm; Th: 1.0 cm.
Single fragment of vertical strap handle, ovoid in section, broken at 90° bend near top, and farther below; no attachments preserved (as P055, P056, but still larger portion preserved). Smoothed, not burnished, heavy encrustation.
Fabric: As **9/118** (P056), but with few larger black inclusions (<0.3 cm).
Comparanda: **9/116** (P357A), **9/117** (P055), **9/118** (P056)

9/120 (P021), **Figs. 9.19, 9.53** (Drawing 4.4; Photos 596–599)
Rim and Horizontal Handle Fragment.
Tumulus Fill (SU 1.007).
PL (max): 5.7 cm; PW (max): 4.2 cm; Th (handle): 1.5 cm; Th (wall): 0.8 cm.

Single fragment preserving rim, upper wall, and part of strap handle, oval in section; handle moves down from rim, but curves around, probably attached horizontally. Original shape uncertain, but definitely open; wall curvature uneven, precluding estimation of original diameter.

Fabric: Dark, semi-coarse, hard, with many black and white inclusions, abundant mica, few air pockets (<0.2 cm); some small cracks visible in surface. Exterior 2.5 YR 4/2 (weak red), core 2.5 YR 4/8 (red) to 3/1 (dark reddish gray).

9/121 (P307), **Fig. 9.19** (Drawing 70.1; Photos 2944–2945)

Wide Strap Handle and Rim Fragment.

Tumulus Fill (SU 4.286).

PL (max): 7.1 cm; PW (max): 8.2 cm; PW (handle): 4.8 cm; Th: 0.5–0.8 cm.

Two joining fragments (from modern break) preserving straight rim and wide strap (or ribbon) handle of thin-walled open vessel, probably two-handled. Precise shape and orientation uncertain, but handle, with small ridge along either side, begins to curve away from wall at break. Both sides smoothed and burnished.

Fabric: As **9/122** (P361).

Comparanda: **9/122** (P361; similar fabric, handle, different shape); **9/190** (P098; similar handle-rim junction on coarser fabric, with shorter, spur handle—though **9/121** [P307] probably not a spur handle); cf. also **9/123** (P339).

9/122 (P361), **Fig. 9.53** (Photos 3245–3246)

Wide Strap Handle and Rim Fragment.

Tumulus Fill (SU 2.474).

PL (max): 7.0 cm; PW (max): 5.1 cm; PW (handle): 3.8 cm; Th: 0.5–0.8 cm.

Fragment preserving portion of rim and handle of thin-walled, open vessel, with wide strap (or ribbon) handle rising from sharply everted, rounded rim; only one edge of handle clearly preserved. Surface roughly smoothed on exterior, highly burnished within; encrustation scattered on all surfaces.

Fabric: Fine light fabric, with abundant mica and very small black inclusions; white inclusions rare. Exterior 5 YR 5/6 (yellowish red), core 7.5 YR 4/1 (dark gray).

Comparanda: **9/121** (P307; similar handle on different shape); **9/123** (P339; similar fabric, shape).

9/123 (P339), **Fig. 9.53** (Photos 3213–3214)

Vertical Strap Handle Fragment.

Tumulus Fill (SU 1.377).

PL (max): 4.4 cm; PW (max): 5.4 cm; W (handle): 3.8 cm; Th: 1.0 cm.

Three joining fragments of wide strap handle, broken at attachment with rim, and also preserving bit of convex curve of lower body; perhaps from large kantharos. Exterior surface very worn, but once burnished both inside and out; encrustation especially over broken edges.

Fabric: Fine dark fabric, with abundant mica, moderate black inclusions, few air pockets (<0.4 cm). Surface 2.5 YR 3/1 (dark reddish gray) to 5/6 (red); core not visible.

Comparanda: **9/122** (P361), **9/121** (P307), **9/114** (P027).

9/124 (P299), **Fig. 9.53** (Photos 2888–2889)

Rim Fragment.

Tomb XLIV (65) Fill (SU 4.364).

PL (max): 1.8 cm; PW (max): 2.2 cm; Th: 0.6 cm; D (rim ext): ca. 10–12 cm

Single fragment from small vessel with narrow, tapered lip, perhaps shallow cup or bowl. Surfaces smoothed, exterior worn, uneven; interior, lower break possibly burned.

Fabric: Fine, with abundant mica and small black inclusions, occasional white inclusions. Exterior 7.5 YR 4/3 (brown), core and interior 7.5 YR 3/1 (very dark gray).

Comparanda: **9/127** (P089), **9/113** (P211), **9/138** (P199), **9/146** (P067), **9/147** (P149); uncatalogued P152, P341, P374, P419.

9/125 (P126), **Fig. 9.19** (Drawing 39.4; Photos 2148–2149)

Rim Fragment.

Tumulus Fill (SU 1.039).

PL (max): 3.4 cm; PW (max): 1.8 cm; Th: 0.6 cm.

Single fragment of flaring rim with tapered lip from small, thin-walled vessel. Surface smoothed, exterior perhaps once burnished.

Fabric: Fine dark fabric, with abundant mica, few very fine white and black inclusions. Surface mostly 2.5 YR 3/1 (dark reddish gray) with some spots as light as 2.5 Y 6/2 (light brownish gray), others as dark as 5 YR 2.5/1 (black); core black.

Comparanda: **9/111** (P041); uncatalogued P412.

9/126 (P191), **Fig. 9.19, 9.53** (Drawing 83.2; Photos 2218–2219)
Rim Fragments.
Tumulus Fill (SU 4.201).
PL (max): 4.3 cm; PW (max): 4.9 cm; Th: 0.6 cm; D (rim ext): 8–10 cm.
Three joining fragments (two from modern break) from small, thin-walled vessel with shallow, uneven rim and rounded lip, perhaps from shallow bowl, or spout from larger vessel. Surface very worn and uneven, with heavy encrustation.
Fabric: Difficult to decide here between fine light and fine dark fabric—reddish color similar to matt-painted fabric, but slightly darker, with encrustation giving even darker appearance. Abundant mica and black inclusions (mostly very fine, few pebbles <0.3 cm), moderate white inclusions. Surfaces 2.5 YR 5/6 (red), core 2.5 YR 3/1 (dark reddish gray).
Comparanda: Similar lip and wall thickness, not so shallow: **9/127** (P089), **9/113** (P211), **9/138** (P199), **9/147** (P149), **9/124** (P299) **9/146** (P067). See bowls from Kamenicë (Bejko forthcoming: 294/2084; 350/817; 404/954), small spouted vessels from Romaja (Durić, Glišić, and Todorović 1975: pls. 54, 58, 64 [sixth century?]). Cf. also uncatalogued P152, P341, P374, P419.

9/127 (P089), **Fig. 9.19, 9.53** (Drawing 25.2; Photos 733–736)
Rim Fragment (or Spout?).
Tumulus Fill (SU 1.067).
PL (max): 3.5 cm; PW (max): 4.2 cm; Th: 0.6 cm; D (rim ext): 10–12 cm.
Single fragment from small, thin-walled vessel with irregularly curved rim and rounded lip, perhaps flaring toward spout. Surfaces smoothed, perhaps once burnished; heavy encrustation.
Fabric: Fine, with abundant mica, moderate black inclusions (including pebbles <0.2 cm), occasional white inclusions and red pebbles. Surfaces 2.5 YR 4/4 (reddish brown), core 2.5 YR 3/1 (dark reddish gray).
Comparanda: **9/113** (P211), **9/124** (P299), **9/126** (P191), **9/138** (P199), **9/146** (P067), **9/147** (P149); uncatalogued P152, P341, P374, P419.

9/128 (P204), **Fig. 9.19** (Drawing 76.4; Photos 2418–2419)
Rim Fragment.
Tumulus Fill (SU 2.202).
PH (max): 2.1 cm; PW (max): 2.8 cm; Th: 0.6 cm.
Two joining fragments (from modern break) of slightly flaring rim with flat lip; vessel shape uncertain, probably small in size. Surfaces smoothed, interior burnished, exterior worn.
Fabric: Fine; moderate mica, fine black, white, and red inclusions, few small black, gray, and red pebbles (and grog?; <0.3 cm). Surface mottled 5 YR 5/4 (reddish brown), 4/2 (dark reddish gray), 2.5/1 (black); core 5 YR 3/1 (very dark gray) to 2.5/1 (black).
Comparanda: **9/129** (P160), **9/130** (P295), **9/131** (P023), **9/137** (P215) for shape; **9/112** (P334), **9/125** (P126), **9/142** (P248), **9/152** (P045 + P046) for fabric.

9/129 (P160), **Fig. 9.19** (Drawing 42.3; Photos 2363–2364)
Rim Fragment.
Topsoil (SU 4.201).
PL (max): 1.2 cm; PW (max): 2.1 cm; Th: 0.4 cm.
Small fragment from small, thin-walled vessel with flaring rim and nearly flat lip. Surface smoothed, worn; encrustation over core.
Fabric: Fine, with abundant mica and white inclusions, occasional black inclusions. Surfaces 2.5 YR 4/3 (reddish brown), core not visible.
Comparanda: **9/128** (P204), **9/130** (P295), **9/131** (P023), **9/137** (P215).

9/130 (P295), **Fig. 9.54** (Photos 2883–2884)
Rim Fragment.
Topsoil (SU 1.375).
PL (max): 1.4 cm; PW (max): 2.5 cm; Th: 0.5 cm.
Very small rim fragment with flat lip; vessel shape uncertain. Surfaces smoothed and burnished.
Fabric: Fine, with abundant mica, few small red and gray inclusions (<0.4 cm). Surface mottled 5 YR 5/3 (reddish brown) to 3/1 (very dark gray); core 5 YR 3/1.
Comparanda: **9/112** (P334), **9/125** (P126), **9/128** (P204), **9/129** (P160), **9/131** (P023), **9/137** (P215), **9/142** (P248), **9/152** (P045 + P046).

9/131 (P023), **Fig. 9.19** (Drawing 5.2; Photos 604–607)
Rim Fragment.
Tumulus Fill (SU 1.007).

PL (max): 1.8 cm; PW (max): 2.1 cm; Th: 0.5 cm.
Single fragment from small, thin-walled open vessel, with slightly flaring rim and flat lip. Surfaces smoothed, exterior slightly uneven.
Fabric: Sandy, with abundant mica, moderate small black inclusions, occasional very small white pebble inclusions. Surface and core 5 YR 2.5/1 (black).
Comparanda: **9/128** (P204), **9/129** (P160), **9/130** (P295), **9/137** (P215).

9/132 (P150), **Fig. 9.19** (Drawing 45.3; Photos 2376–2378)
Rim Fragment.
Topsoil (SU 1.141).
PL (max): 2.5 cm; PW (max): 2.2 cm; Th (neck): 0.6 cm; Th (rim): 0.3 cm.
Single fragment from small, thin-walled vessel, preserving sharply carinated rim and bit of tapered lip. Surfaces smoothed, interior burnished, scattered encrustation, heavy over breaks.
Fabric: Fine, with abundant mica, moderate fine black and white inclusions, few black and gray pebble inclusions (<0.2 cm) and air pockets (<0.2 cm). Exterior 5 YR 4/4 (reddish brown), interior 2.5 Y 5/3 (light olive brown), core 2.5/1 (black).
Comparanda: Unique shape at Lofkënd? For fabric, cf. especially **9/141** (P206), **9/133** (P171).

9/133 (P171), **Fig. 9.19** (Drawing 64.6; Photos 2415, 2417)
Rim Fragments.
Tumulus Fill (SU 1.039).
PH (max): 2.8 cm; PW (max): 4.2 cm; Th: 0.4–0.6 cm; D (rim ext): ca. 10 cm.
Two joining fragments (modern break) of sharply everted rim and rounded lip of small vessel, probably open in shape. Surfaces smoothed, perhaps once burnished.
Fabric: Fine, with moderate mica and fine black, white, and red pebble inclusions. Exterior surface 7.5 YR 5/4 (brown), with patches as dark as 2.5/1 (black); interior surface and core 4/1 (dark gray) to 2.5/1 (black).
Comparanda: **9/152** (P045 + P046); cf. also **9/215** (P337; similar shape, semi-coarse).

9/134 (P068), **Figs. 9.19, 9.54** (Drawing 16.3; Photos 2022–2025)
Rim Fragment.
Tumulus Fill (SU 2.071).
PL (max): 2.1 cm; PW (max): 2.5 cm; Th: 0.4 cm; D (rim ext): ca. 9–11 cm.
Three joining fragments (modern break) of small, thin-walled vessel with neck curving in to narrow mouth with rounded lip. Surface smoothed, worn.
Fabric: Fine, with abundant mica and black inclusions, occasional white inclusions. Surfaces and core 2.5 YR 4/4 (reddish brown).
Comparanda: **9/16** (P047).

9/135 (P406), **Fig. 9.19** (Drawing 94.3; Photos 3502–3503)
Rim Fragment.
Tumulus Fill (SU 2.532).
PH (max): 4.3 cm; PW (max): 2.4 cm; Th: 0.8 cm.
Single fragment from small globular vessel, preserving S-curve from upper body to neck and everted rim with damaged lip (perhaps rounded); orientation uncertain. Surfaces smoothed, encrustation lightly scattered, heavy over breaks.
Fabric: Fine, with abundant mica and white inclusions, occasional grog, black pebble inclusions, and air pockets (<0.4 cm). Exterior surface 7.5 YR 5/4 (brown), interior surface 7.5 YR 2.5/1 (black).
Comparanda: Cf. **9/136** (P124), **9/93** (P223; complete vessel); cf. also fragments from larger vessels with similar rim, fabric: **9/144** (P343), **9/145** (P404), **9/149** (P033), **9/150** (P194), **9/151** (P197), **9/153** (P151), **9/154** (P169).

9/136 (P124), **Fig. 9.20** (Drawing 39.2; Photos 2690–2691)
Rim Fragment.
Tumulus Fill (SU 1.070).
PH (max): 2.8 cm; PW (max): 2.2 cm; Th: 0.8 cm.
Single fragment with slightly everted rim and thin, rounded lip, probably from small globular vessel. Surfaces smoothed, exterior perhaps once burnished, interior damaged, missing in places; scattered encrustation.
Fabric: Fine, with abundant mica, moderate white inclusions, single air pocket on exterior, below lip (0.3 cm). Surface and core 7.5 YR 2.5/1 (black).
Comparanda: **9/135** (P406), **9/93** (P223; complete vessel); cf. also frr from larger vessels with similar rim, fabric: **9/144** (P343), **9/145** (P404), **9/149** (P033), **9/150** (P194), **9/151** (P197), **9/153** (P151), **9/154** (P169).

9/137 (P215), **Fig. 9.20** (Drawing 45.9; Photos 2404–2405)
Rim Fragment.
Tumulus Fill (SU 4.204).
PL (max): 2.5 cm; PW (max): 2.5 cm; Th: 0.4 cm.
Two joining fragments (from modern break) of rim with flat lip; rim orientation and vessel shape uncertain, small to medium in size. Surfaces smoothed, exterior burnished; scattered encrustation.
Fabric: Fine; moderate mica, fine black and white inclusions, few small white inclusions (<0.2 cm). Interior surface mottled 5 YR 4/4 (reddish brown) to 3/2 (dark reddish brown), exterior and core mostly 2.5/1 (black).
Comparanda: **9/128** (P204), **9/129** (P160), **9/130** (P295), **9/131** (P023); cf. also **9/112** (P334), **9/125** (P126), **9/142** (P248), **9/152** (P045 + P046) for fabric.

9/138 (P199), **Fig. 9.20** (Drawing 45.11; Photos 2392–2393)
Rim Fragment.
Tumulus Fill (SU 4.205).
PL (max): 3.5 cm; PW (max): 2.7 cm; Th: 0.6 cm.
Single fragment from small to mid-sized vessel with convex upper wall and rounded lip; precise shape uncertain. Surface smoothed.
Fabric: Fine, with abundant mica and white and black inclusions (some pebbles <0.2 cm), occasional red (grog) inclusions. Exterior surface 5 YR 4/3 (reddish brown), interior and core 5 YR 2.5/1 (black), with lighter interior patch 2.5 Y 6/3 (light yellowish brown).
Comparanda: **9/113** (P211), **9/124** (P299), **9/127** (P089), **9/138** (P199), **9/146** (P067), **9/147** (P149); uncatalogued P152, P341, P374, P419.

9/139 (P007) **Fig. 9.20** (Drawing 3.2; Photos 2030–2033)
Rim Fragment.
Topsoil (SU 2.002).
PL (max): 1.5 cm; PW (max): 1.7 cm; Th: 0.4 cm; D (rim ext): ca. 8–10 cm.
Small fragment of thin-walled vessel with everted rim and rounded lip, probably open in shape. Surfaces smoothed.
Fabric: Fine, with abundant mica, occasional grog. Surfaces 5 YR 4/3 (reddish brown), core 5 YR 3/1 (very dark gray).

9/140 (P044), **Fig. 9.20** (Drawing 10.2; Photos 2018–2020)
Rim Fragment.
Tumulus Fill (SU 2.041).
PL (max): 2.8 cm; PW (max): 3.6 cm; Th: 0.3–0.4 cm.
Single fragment of small, thin-walled vessel preserving short, everted rim with tapered lip; precise shape and orientation uncertain. Surfaces smoothed, exterior perhaps burnished; scattered encrustation.
Fabric: Fine, with moderate mica, fine black and white inclusions, moderate small pebbles (<0.4 cm). Interior surface 7.5 YR 6/4 (light brown), exterior 3/1 (very dark gray), core mostly as exterior.
Comparanda: P190 (for fabric and shape); P206, P248 (for shape, slightly different fabric).

9/141 (P206), **Fig. 9.20** (Drawing 62.6; Photos 2672–2673)
Rim Fragment.
Tumulus Fill (SU 4.203).
PH (max): 2.7 cm; PW (max): 3.6 cm; Th: 0.6 cm.
Single fragment of everted rim with nearly flat lip, flaring out from vertical neck. Surfaces smoothed, now covered with heavy encrustation.
Fabric: Difficult to see beneath encrustation: moderate mica and fine white inclusions; surface perhaps as light as 5 YR 5/6 (yellowish red); core not visible.
Comparanda: **9/142** (P248; similar fabric, shape); **9/140** (P044; similar shape, slightly different fabric); cf. also uncatalogued P190.

9/142 (P248), **Fig. 9.20** (Drawing 64.1; Photos 2568–2569)
Rim Fragment.
Tumulus Fill (SU 1.279).
PH (max): 2.8 cm; PW (max): 3.8 cm; Th: 0.3–0.4 cm.
Single fragment of everted rim and rounded lip from thin-walled open vessel. Surface smoothed, scattered encrustation, heavy over breaks.
Fabric: Fine dark fabric, moderate mica and fine black and white inclusions; few air pockets (<0.4 cm). Surface and core 5 YR 3/1 (very dark gray).
Comparanda: **9/141** (P206; similar fabric, shape); **9/140** (P044; similar shape, slightly different fabric); cf. also uncatalogued P190.

9/143 (P109), **Fig. 9.20** (Drawing 39.6; Photos 2140–2141)
Rim Fragment.
Tumulus Fill (SU 1.070).

PL (max): 2.0 cm; PW (max): 2.0 cm; Th: 0.4 cm.
Single fragment of everted rim and rounded lip from thin-walled open vessel. Surface smoothed, exterior burnished, scattered encrustation.
Fabric: Fine, with abundant mica, occasional very small black inclusions. Surface 2.5 YR 4/4 (reddish brown), core 2.5 YR 3/1 (dark reddish gray).

9/144 (P343), **Fig. 9.20** (Drawing 66.6, Photos 3076–3077)
Rim Fragment.
Tumulus Fill (SU 4.286).
PH (max): 2.7 cm; PW (max): 3.2 cm; Th: 0.7 cm.
Single fragment of everted rim with rounded lip, probably from small- to medium-sized open vessel. Surface smoothed, perhaps once burnished; scattered encrustation, heavier on breaks.
Fabric: Fine, with abundant mica and red (grog?) and white inclusions. Exterior 7.5 YR 4/3 (brown), interior and core 7.5 YR 2.5/1 (black).
Comparanda: **9/135** (P406), **9/136** (P124), **9/145** (P404), **9/149** (P033), **9/150** (P194), **9/151** (P197), **9/153** (P151), **9/154** (P169).

9/145 (P404), **Figs. 9.20, 9.54** (Drawing 95.1; Photos 3510–3511)
Rim Fragment.
Tumulus Fill (SU 5.535).
PH (max): 3.4 cm; PW (max): 2.8 cm; Th: 0.7 cm.
Single fragment of everted rim with rounded lip, probably from small to medium-sized open vessel. Surface smoothed, exterior perhaps once burnished; encrustation on neck exterior and breaks.
Fabric: Fine; moderate mica and red (grog and pebble) and very small white inclusions. Surfaces 2.5 YR 5/8 (red), core 2.5 YR 2.5/1 (reddish black).
Comparanda: **9/135** (P406), **9/136** (P124), **9/144** (P343), **9/149** (P033), **9/150** (P194), **9/151** (P197), **9/153** (P151), **9/154** (P169).

9/146 (P067), **Fig. 9.20** (Drawing 16.1; Photos 642–645)
Rim Fragment.
Tumulus Fill (SU 1.047).
PH (max): 3.3 cm; PW (max): 4.5 cm; Th: 0.8 cm.
Single fragment from mid-sized open vessel with flaring rim and tapered lip; precise shape and orientation uncertain. Surfaces smoothed, perhaps once burnished; exterior very worn, heavy encrustation over interior and breaks.
Fabric: Fine to semi-coarse, with abundant mica and black inclusions (some pebbles <0.2 cm); three air pockets in exterior (<0.6 cm). Exterior surface 7.5 YR 4/2 (brown), interior 5 YR 7/3 (pale yellow); core 2.5 Y 6/1 (gray).
Comparanda: **9/113** (P211), **9/124** (P299), **9/127** (P089), **9/138** (P199), **9/147** (P149); cf. also uncatalogued P152, P341, P374, P419.

9/147 (P149), **Fig. 9.20** (Drawing 40.7; Photos 2198–2199)
Rim Fragment.
Tumulus Fill (SU 2.239).
PL (max): 3.0 cm; PH (max): 2.5 cm; PW (max): 2.9 cm; Th: 0.6 cm.
Single fragment from medium-sized vessel with flaring rim and rounded lip; precise shape and orientation uncertain. Surfaces smoothed, now very worn; scattered encrustation, heavier on exterior.
Fabric: Fine dark, with abundant white and red (grog?) inclusions, moderate mica and black inclusions. Interior surface 7.5 YR 4/3 (brown), core 7.5 YR 4/1 (dark gray).
Comparanda: **9/113** (P211), **9/124** (P299), **9/127** (P089), **9/138** (P199), **9/146** (P067); cf. also uncatalogued P152, P341, P374, P419.

9/148 (P192), **Fig. 9.20** (Drawing 84.4; Photos 2222–2224)
Rim Fragments.
Tumulus Fill (SU 1.070).
PL (max): 4.2 cm; PW (max): 3.2 cm; Th: 0.5 cm.
Two joining fragments (from modern break) of thin-walled, mid-sized open vessel; S-curve of neck leads to everted rim with rounded or tapered lip, now almost entirely lost. Interior surface smoothed, exterior very worn, original surface perhaps lost; lightly scattered encrustation.
Fabric: Fine, with abundant mica and very small white inclusions. Interior surface and core 7.5 YR 2.5/1 (black), small lighter patch on interior 5 YR 4/4 (reddish brown).

9/149 (P033), **Fig. 9.20** (Drawing 6.3; Photos 959–962)
Rim Fragment.
Tumulus Fill (SU 2.010).
PH (max): 4.5 cm; PW (max): 5.3 cm; Th: 0.8 cm; D (rim ext): 15 cm.

Single fragment from mid-sized open vessel, with S-curve from neck to slightly everted rim with rounded lip. Surfaces smoothed and burnished; encrustation covering breaks.
Fabric: Fine dark fabric, with abundant mica, occasional black and white inclusions and very small air pockets. Exterior mottled: 7.5 YR 4/4 (brown) and 2.5 Y 4/3 (olive brown), with large black patch (7.5 YR 2.5/1); core not visible.
Comparanda: **9/135** (P406), **9/136** (P124), **9/144** (P343), **9/145** (P404), **9/150** (P194), **9/151** (P197), **9/153** (P151), **9/154** (P169).

9/150 (P194), **Fig. 9.20** (Drawing 46.1; Photos 2420–2421)
Rim Fragments.
Tumulus Fill (SU 4.203).
PH (max): 2.3 cm; PW (max): 5.2 cm; Th: 0.6 cm; D (rim ext): 14 cm.
Two joining fragments from mid-sized open vessel, with nearly vertical neck flaring to everted rim with rounded lip. Surfaces smoothed, exterior burnished, chipped in places; encrustation over break.
Fabric: Sandy, with abundant mica, small white pebble inclusions (<0.3 cm), and occasional grog. Surface and core 7.5 YR 2.5/1 (black).
Comparanda: **9/135** (P406), **9/136** (P124), **9/144** (P343), **9/145** (P404), **9/149** (P033), **9/151** (P197), **9/153** (P151), **9/154** (P169).

9/151 (P197), **Fig. 9.20** (Drawing 45.5; Photos 2406–2407)
Rim Fragment.
Tumulus Fill (SU 1.067).
PL (max): 3.2 cm; PW (max): 4.9 cm; Th: 0.4–0.5 cm.
Single fragment preserving bit of everted rim with rounded lip; shape and orientation uncertain—perhaps S-profile, open shape, probably mid-sized. Surface smoothed, interior perhaps burnished (shiny encrustation); encrustation over interior and breaks.
Fabric: Abundant mica, white and red inclusions. Exterior 2.5 YR 4/4 (reddish brown), interior 2.5 Y 8/3 (pale yellow), core 2.5 YR 3/1 (very dark gray).
Comparanda: **9/135** (P406), **9/136** (P124), **9/144** (P343), **9/145** (P404), **9/149** (P033), **9/150** (P194), **9/151** (P197), **9/153** (P151), **9/154** (P169).

9/152 (P045 + P046), **Figs. 9.20, 9.54** (Drawing 10.5; Photos 781–784)
Shoulder, Neck, and Rim Fragments.
Tumulus Fill (SU 2.006).
PH (max): 4.1 cm; PW (max): 4.8 cm; Th: 0.3–0.4 cm; D (rim ext): ca. 10–12 cm.
Two joining fragments from shoulder, neck, and rim of mid-sized, thin-walled vessel with sharply everted rim and nearly flat lip with shallow groove beneath. Surfaces smoothed but cracked, exterior burnished; encrustation, especially on interior and core.
Fabric: Fine, with moderate mica and fine white and black inclusions, few small bits of lime (<0.3 cm) and very thin air pockets. Exterior 10 YR 7/2 (light gray) to 2/1 (black).
Comparanda: **9/133** (P171); cf. also **9/215** (P337; similar shape, semi-coarse).

9/153 (P151), **Fig. 9.20** (Drawing 61.3; Photos 2642–2643)
Rim Fragments.
Tumulus Fill (SU 4.204).
PH (max): 5.5 cm; PW (max): 5.3 cm; Th: 0.3–0.8 cm; D (rim ext): ca. 25 cm.
Seven joining fragments from large open vessel, with S-curve from neck to everted rim with rounded lip. Surfaces smoothed, perhaps once burnished; little encrustation over breaks.
Fabric: Fine dark fabric, with abundant mica and small white inclusions, moderate grog, occasional air pockets (<0.3 cm). Exterior surface 5 YR 4/2 (dark reddish gray), interior surface and core 7.5 YR 2.5/1 (black)
Comparanda: **9/135** (P406), **9/136** (P124), **9/144** (P343), **9/145** (P404), **9/149** (P033), **9/150** (P194), **9/151** (P197), **9/154** (P169).

9/154 (P169), **Figs. 9.20, 9.54** (Drawing 42.7; Photo 2546)
Rim and Neck Fragments.
Tomb LIV (40) Fill (SU 2.254).
PH (max): 5.5 cm; PW (max): 6.7 cm; Th: 0.4–0.6 cm; D (rim ext): ca. 20 cm.
Five joining fragments from large open vessel, with S-curve from neck to everted rim with flat/tapered lip (irregularity not visible in drawing). Surfaces smoothed, exterior once burnished, interior uneven; light encrustation.

Fabric: Fine dark fabric, with abundant mica and white and black inclusions, moderate grog. Exterior surface 5 YR 5/4 (reddish brown), interior surface and core 7.5 YR 2.5/1 (black).
Comparanda: **9/135** (P406), **9/136** (P124), **9/144** (P343), **9/145** (P404), **9/149** (P033), **9/150** (P194), **9/151** (P197), **9/153** (P151); cf. also Andrea 1985: pl. IX, Tomb 66 (twelfth–eighth centuries BC [or Late Bronze Age? (Bejko, personal communication)]); Korkuti 1971: pl. VIII, 3 (Tren, EIA, but originally LBA shape). Cf. also Prendi, Lera, and Touchais 1996: fig. 8, 242 (similar shape, matt-painted fabric).

9/155 (P421), **Figs. 9.20, 9.54** (Drawing 114.5; Photos 3817–3818)
Wall Fragment.
Tumulus Fill (SU 6.562).
PL (max): 3.9 cm; PW (max): 3.5 cm; Th (wall): 0.6–0.7 cm; Th (wall and projection): 1.4 cm; D (projection): 0.9–1.7 cm.
Small wall fragment, probably from mid-sized vessel; surfaces smoothed and burnished, encrustation around breaks.
Plastic decoration: Single circular projection (nipple).
Fabric: Fine, with abundant mica, and occasional red inclusions. Exterior 5 YR 4/3 (reddish brown), 10 YR 2/1 (black).
Comparanda: **9/156** (P316).

9/156 (P316), **Figs. 9.20, 9.54** (Drawing 66.5; Photos 2953–2954)
Shoulder Fragment.
Fill of Tomb I (64) (SU 1.361).
PH (max): 5.1 cm; PW (max): 6.5 cm; Th 0.4–0.7 cm; Th (wall and projection): 0.9 cm; D (max): 12–13 cm; D (projection): 1.2 cm.
Single fragment from convex shoulder of medium-size vessel, preserving bit of turn to neck; precise shape uncertain, perhaps open. Surfaces smoothed and burnished, interior uneven.
Plastic decoration (?): Small circular projection, quite worn; perhaps flat pellet lug, perhaps misproduction?
Fabric: Fine, with abundant mica and white inclusions, occasional black (pebble) inclusions (<0.3 cm). Exterior 7.5 YR 4/3 (brown), interior and core GLEY 12.5/N (black).
Comparanda: **9/155** (P421); Bodinaku 1982: pl. III, 4, 5 (LBA).

9/157 (P210), **Fig. 9.20** (Drawing 46.4; Photos 2444–2445)
Shoulder Fragment.
Tumulus Fill (SU 4. 203).
PL (max): 3.1 cm; PW (max): 3.4 cm; Th: 0.6–0.8 cm.
Single fragment from convex shoulder of mid-sized vessel, broken at curve to neck or rim; precise shape and orientation uncertain. Surfaces smoothed; scattered encrustation around breaks.
Fabric: Fine, with abundant mica and white inclusions, occasional red inclusions. Surface and core 2.5 Y 2.5/1 (black).

9/158 (P349), **Fig. 9.54** (Photos 3296–3297)
Shoulder or Base Fragment.
Tomb XXVII (82) Fill (SU 4.456).
PL 3.4 cm; PW 3.8 cm; Th 0.5–0.7 cm.
Single fragment from mid-sized vessel, preserving curve of ca. 45°, either from carinated shoulder or uneven base and lower wall. Surfaces smoothed but worn, exterior perhaps once burnished; interior slightly uneven; heavy encrustation on exterior and breaks.
Fabric: Fine to semi-coarse, with abundant mica, occasional air pockets. Surfaces 5 YR 5/4 (reddish brown), core not visible.

Semi-Coarse Fabric

9/159 (P303), **Figs. 9.21, 9.55** (Drawing 73.1; Photos 3140–3146)
Crude Light Fabric One-Handled Vessel, SC Type 1.
Tomb XIV (71) (SU 4.405).
H (rim): 7.7 cm; H (max):10.7 cm; D (rim): 7.6 cm; D (mouth int): 6.3 cm; D (max): ca. 9.5 cm; Th (handle, max): 1.8 cm; Th (wall): 0.4 cm.
Complete one-handled vessel, crudely made; several cracks (due to uneven drying and excessively large temper) around rim and on handle exterior. Globular body rises from rounded base to conical neck and flattened rim; vertical handle, irregular ovoid in section, rises unevenly from crooked rim and curves back, joining body just above point of maximum diameter. Rounding of rim, body, and handle irregular throughout. Surface smoothed, not burnished; encrustation occasional on exterior,

abundant within. No plastic or painted decoration.
Fabric: Hard, semi-coarse; large grog and white inclusions (<0.5 cm); mica sparkles and air pockets (<0.3 cm) throughout. Exterior surface 2.5 YR 5/4 (reddish brown), 5 YR 4/4 (reddish brown); rim and interior 5 YR 5/1 (gray); core not visible.
Comparanda: Korkuti 1981: pl. XIV, 3.

9/160 (P362), **Figs. 9.21, 9.55** (Drawing 80.2; Photos 3315–3322)
"Baby Feeder," SC Type 2.
Tumulus Fill (SU 6.240) perhaps associated with Tomb XV (80).
H (rim): 5.2–5.7 cm; H (max): 7.2 cm; D (rim, short side): 3.4 cm; D (rim, spout to handle): 4.5 cm; D (max): 6.0 cm; D (spout int): 0.3 cm; Th (walls): 0.6–0.8 cm; Th (rim): 0.3 cm.
Very small baby feeder, nearly complete; much of spout, chip from one side of lower body now missing. Rounded body, tapering slightly up to ovoid rim. Spout-hole placed 0.5 cm below rim, which rises slightly in this area. Crudely made; uneven base, rim, and profile. Exterior body surface smoothed, handle rough; no surface decoration.
Fabric: Hard, sandy, semi-coarse; abundant mica and black inclusions, moderate white inclusions (including quartz, <0.6 cm), occasional grog. Abundant air pockets (<0.7 cm). Exterior and interior surfaces 5 YR 6/6 (reddish yellow); 2.5 Y 2.5/1 (black) fire clouds scattered across exterior; interior covered with encrustation and root marks.

9/161 (P201 = **15/10**), **Figs. 9.21, 9.56** (Drawing 113; Photos 2509–2513, 3851–3853 [showing joined base])
Base, Body, and Handle Fragments.
Tumulus Fill (Units 1.281, 1.291, 2.202, 7.202).
PH (base frr): 2.4 cm; PW (base frr): 11.2 cm; D (base): ca. 7.5 cm; Th (base): 1.2 cm; D (body frr): ca. 15 cm; PW (max, body frr): 8.9 cm; W (handle): 2.9–4.4 cm; PH (max, body frr): 5.6 cm; Th (wall): 0.8 cm; Th (handle): 1.3 cm.
Many fragments of similar fabric, including two joining groups and many more non-joining, found across three stratigraphic units. First group, of 10 joining fragments, preserves much of flat, very slightly raised base, rising to widely flaring vessel walls; body shape uncertain. Second group, of four joining fragments, preserves globular body of mid-sized open vessel, with attachment for vertical strap handle, ovoid in section; small ridge above attachment seems to set off neck from lower body, suggesting this is lower handle attachment. All surfaces very worn, those of second group coated in many places, including over breaks, with bitumen (i.e., not from active use life of vessel).
Fabric: Soft, sandy, semi-coarse, orange fabric, with abundant mica, very fine black and white inclusions, moderate black, brown, and white pebbles (<0.6 cm) and small air pockets (<0.2 cm). Surfaces 10 YR 5/6 (yellowish brown), core 10 YR 4/4 (dark yellowish brown).

9/162 (P079 = **15/4**), **Fig. 9.56** (Photos 3454–3455)
Body and Handle Fragments.
Tumulus Fill (SU 4.104).
PL (max): 6.1 cm; PW (max): 5.0 cm; W (handle): 2.8 cm; Th (wall): 1.1 cm; Th (handle): 1.2 cm.
Seven fragments of small to mid-sized coarseware vessel, probably open in shape. Largest fragment preserves part of inner and outer vessel walls, with attachment and lower portion of vertical strap handle, irregular ovoid in section; five additional fragments, three joining, preserve other parts of handle, including sharp bend for return to rim. Last fragment of same thickness and fabric, but position on vessel uncertain; this fragment also has hole partially bored into interior surface. Surfaces smoothed, now very worn; bitumen lining vessel interior opposite handle attachment.
Fabric: Soft, sandy fabric, with abundant mica, white, brown, and black inclusions (<0.5 cm), few air pockets (<0.2 cm) and cracks. Surfaces 5 YR 3/1 (very dark gray) to 2/1 (black), with thin layers 5 YR 4/4 (reddish brown) around 5 YR 4/2 (dark reddish gray) core.
Comparanda: **9/236** (P059).

9/163 (P008), **Figs. 9.22, 9.56** (Drawing 3.3; Photos 785–788)
Vertical Strap Handle Fragment.
Tumulus Fill (SU 2.006).
PL (max): 2.4 cm; PW (max): 2.3 cm; Th: 1.0 cm.
Single fragment of vertical strap handle, ovoid in section, preserving slight curve, no attachments to body. Surface smoothed, perhaps once burnished, but now very worn, partially lost. Double diagonal line incised on underside.

Fabric: Dark, sandy, semi-coarse fabric, with abundant mica and small red and white moderate black inclusions and small air pockets. Surface and core 2.5 Y 3/1 (very dark gray).

Comparanda: **9/177** (P037), **9/197** (P143), **9/191** (P399), **9/166** (P120), **9/198** (P320); cf. also uncatalogued P308, P403.

9/164 (P029), **Figs. 9.21, 9.56** (Drawing 6.1; Photos 2522–2523)

Body and Handle Fragment.

Tomb LXXX (4) Fill (SU 4.031).

PL (max): 4.5 cm; PW (max): 7.6 cm; PW (handle): 5.1 cm; Th: 0.5–1.1 cm; Th (handle): 1.0 cm.

Two joining fragments of rounded lower vessel body, medium in size, with lower attachment of vertical strap handle, ovoid in section. Exterior surface smoothed and burnished, interior rough, suggesting a closed shape; now very worn, with scattered encrustation, scratches, and gouges.

Fabric: Semi-coarse, with abundant mica and white inclusions, and few red inclusions (probably grog). Surface and core mottled, from 2.5 YR 5/4 (reddish brown) to 7.5 YR 4/1 (dark gray).

Comparanda: **9/165** (P347), **9/178** (P039), **9/180** (P158), **9/182** (P035), **9/192** (P157), **9/193** (P182), **9/194** (P110), **9/195** (P368), **9/196** (P216); cf. also uncatalogued P018, P183, P402.

9/165 (P347 = **15/19**), **Fig. 9.57** (Photos 3307–3309)

Body and Strap Handle Fragment.

Topsoil (SU 4.201).

PH (max): 2.8 cm; PW (max): 5.1 cm; PW (handle): 3.1 cm; Th (wall): 0.6 cm; Th (handle): 1.3 cm.

Single fragment of convex vessel wall with lower attachment of vertical strap handle, ovoid in section. Surfaces smoothed, perhaps once burnished, now worn; interior partially lined with bitumen.

Fabric: Semi-coarse, sandy, with abundant mica, small white and red inclusions; few larger inclusions and air pockets (<0.3 cm). Exterior surface 2.5 YR 3/4 (dark reddish brown) to 2.5/0 (black); interior surface 2.5 YR 2.5/2 (very dusky red); core 2.5 YR 2/0 (black), with thin layer near surface 2.5 YR 4/6 (red).

Comparanda: **9/164** (P029), **9/178** (P039), **9/180** (P158), **9/182** (P035), **9/193** (P182), **9/194** (P110), **9/195** (P368), **9/196** (P216); cf. also uncatalogued P018, P183, P402.

9/166 (P120), **Fig. 9.22** (Drawing 38.1; Photos 2162–2164)

Vertical Strap Handle Fragment.

Tumulus Fill (SU 2.202).

PL (max): 4.2 cm; PW (max): 2.3 cm; Th: 1.4 cm.

Single fragment of vertical strap handle, oval in section, preserving uppermost section of curve but no attachments to vessel. Surface smoothed, now covered with heavy encrustation.

Fabric: Semi-coarse, originally reddish in color, but neither surface nor core visible beneath encrustation; abundant mica and nondescript inclusions. Surface 5 YR 5/8 (yellowish red).

Comparanda: **9/163** (P008), **9/177** (P037), **9/197** (P143), **9/191** (P399), **9/198** (P320); cf. also uncatalogued P308, P403.

9/167 (P094B), **Figs. 9.22, 9.59** (Drawing 106.2; Photos 2004–2007)

Handle Fragment.

Tumulus Fill (SU 4.086).

PH (max): 4.8 cm; PW (max): 3.8 cm; Th: 1.1–1.8 cm.

Single fragment of vertical strap handle, rounded triangular in section; broken at junction with body and mid-handle. Exterior surface smoothed and burnished, interior smoothed; entire exterior now very worn, much burnishing lost. (Found and initially catalogued with non-joining rim fragment **9/206** [P094A].)

Fabric: As **9/206** (P094A), but without visible air pockets.

Comparanda: **9/90** (P422), **9/115** (P398).

9/168 (P116), **Fig. 9.22** (Drawing 38.5; Photos 2154–2155)

Squared Spur Handle Fragment.

Tumulus Fill (SU 2.202).

PL (max): 2.9 cm; PW (max): 2.7 cm; W (spur tip, approximate): 2.2 cm; Th (max): 1.4 cm; Th (spur tip): 0.3 cm.

Single fragment of squared spur handle, preserving only flattened tip of spur and parts of upper and lower surfaces, diverging sharply from tip but broken before separation of upper and lower elements; squared spur slightly askew in plan. Surfaces roughly smoothed, corners worn.

Fabric: As **9/181** (P111).

9/169 (P119), **Fig. 9.22** (Drawing 38.2; Photos 2160, 2160b, 2161)
Squared Spur Handle Fragment.
Tumulus Fill (SU 2.202).
PL (max): 3.9 cm; PW (max): 3.1 cm; W (spur tip, approximate): 2.2 cm; Th (max): 1.6 cm; Th (strap, max): 0.8 cm.
Single fragment of squared spur handle, preserving tapered end of spur and much of upper surface; fragment broken beneath spur, at attachment for lower element, and probably near rim attachment of upper element. Upper strap plano-convex in section, slightly asymmetrical in plan, with one side flaring out toward rim attachment while other remains straight. Surfaces once smoothed, now worn, especially at corners, and covered by heavy encrustation.
Fabric: Difficult to discern beneath encrustation, appears to be sandy, semi-coarse, 7.5 YR 7/6 (yellowish red) where surface is worn and chipped; core not visible.

9/170 (P140), **Figs. 9.22, 9.57** (Drawing 41.3; Photos 2190–2191)
Pierced Squared Spur Handle Fragment.
Tumulus Fill (SU 1.070).
PL (max): 4.1 cm; PW (max): 3.6 cm; W (spur tip): 2.1 cm; Th (max): 1.6 cm; Th (strap, max): 0.8 cm; D (piercing, approximate): 1.5 cm.
Single fragment of pierced squared spur handle, broken above on either side of piercing, below at attachment of lower handle element; spur tapered at tip, diverging to upper and lower elements, both probably rectangular in section. Side walls flare out slightly toward bottom, upper strap flares out toward rim, flattening to ovoid in section, slightly thicker on one side of piercing than other. Surface smoothed on exterior, not on underside; worn at spur tip, corners chipped.
Fabric: Dark, hard, semi-coarse, sandy with many inclusions; abundant mica, very small (<0.1 cm) black, white, and red inclusions. Occasional very small (<0.2) air pockets in core and on surfaces, one large pit (0.4 cm) on surface at upper corner, uncertain whether air pocket, chip, or otherwise.
Publication: Papadopoulos, Bejko, and Morris 2007: fig. 28.

9/171 (P203), **Fig. 9.22** (Drawing 45.2; Photos 2408–2410)
Flaring Spur Handle Fragment.
Tumulus Fill (SU 2.202).
PL (max): 5.0 cm; PW (max): 3.5 cm; W (spur): 2.6 cm; Th (spur tip): 0.5 cm; Th (upper strap, max): 0.6 cm; Th (lower strap, max): 1.1 cm; Th (max): 1.9 cm.
Single fragment of flaring spur handle preserving flattened spur tip, much of upper element, irregular flattened ovoid in section, and attachment for lower element, ovoid in section. Upper section tapers in from spur tip before flaring widely toward rim, not preserved. All surfaces smoothed, upper perhaps once burnished; lower surface scratched and worn, encrustation scattered on all surfaces, heavy on core.
Fabric: Dark, semi-coarse, with many inclusions: abundant mica, occasional red, white, and black inclusions, mostly very small. Large black bit (pebble?) loose within pocket in exterior surface (0.4 cm), white pebble (0.3 cm) in core at lower attachment; additional air pockets abundant in all surfaces and core, mostly very small, some up to 0.4 cm. Surface 5 YR 5/6 (yellowish red), core not visible.
Publication: Papadopoulos, Bejko, and Morris 2007: fig. 28.

9/172 (P401), **Fig. 9.57** (Photos 3498–3499)
Flaring Spur Handle Fragment.
Tumulus Fill (SU 5.536).
PL (max): 4.7 cm; PW (max): 3.4 cm; Th (max): 1.4 cm.
Single fragment of flaring spur handle, most similar in shape to **9/265** (P136), preserving tip of outer, vertical element, broken at one corner of spur, at junction of two handle straps, and at some distance above shoulder attachment; exterior strap roughly rectangular in section, no trace of inner strap. Surfaces smoothed, perhaps once burnished; encrustation scattered across all surfaces, heavy on broken spur.
Fabric: As **9/170** (P140).

9/173 (P314), **Figs. 9.22, 9.57** (Drawing 70.4; Photos 3093–3094)
Rounded Spur Handle Fragment.
Tumulus Fill (SU 4.286).

PL (max): 4.3 cm; PW (max): 3.6 cm; Th (strap, max): 0.8 cm; Th (max): 1.4 cm.

Single fragment of rounded spur handle fragment, broken at widening of upper element, perhaps near rim, and at attachment of lower element; upper strap flat on top, plano-convex in section, lower roughly rectangular. Spur tapered at tip, asymmetrically rounded in plan. Upper and lower surfaces of spur highly burnished, with wear and slight chipping around edges; underside of upper strap smoothed, not burnished, where lower element once covered it.

Fabric: As **9/190** (P098), with large air pocket (1.3 cm) preserving fiber impression (burned out temper) in underside.

9/174 (P205), **Fig. 9.57** (Drawing 45.1; Photos 2397–2398)

Horned Handle Fragment.

Ceramic Deposit (SU 2.227) within Tomb LVIII (37) Fill.

PL (max): 2.9 cm; PW (min): 2.1 cm; PW (max = spur): 3.9 cm; Th (max): 1.3 cm.

Single fragment of horned handle, ovoid in section, broken relatively close to horned end; break shows no sign of separate attachments, and thickness remains constant, so uncertain whether this was part of spur handle or ornament atop another handle type. Tip and upper surface highly burnished, sides and lower surface smoothed; horn tips slightly worn, chip from upper exterior surface, probably sustained during excavation.

Fabric: As **9/190** (P098).

Publication: Papadopoulos, Bejko, and Morris 2007: fig. 28.

9/175 (P293), **Fig. 9.57** (Photos 2892–2894)

Lug or Spur Handle Fragment.

Tumulus Fill (SU 2.380).

PL (max): 3.3 cm; PW (max): 1.6 cm; Th (tip): 0.8 cm; Th (max): 1.9 cm.

Small fragment of lug or spur handle, preserving squared spur with flattened tip, rectangular in section. Fragment flares widely away from spur, and seems to be diverging into two distinct elements of spur handle, both ovoid in section and broken near same position; shape could also derive from lug handle broken very near attachment to body. Surface very worn, with light areas of encrustation; once smoothed, perhaps burnished.

Fabric: As **9/171** (P203).

9/176 (P289), **Fig. 9.22** (Drawing 66.1; Photos 2822–2824)

Pierced Lug or Squared Spur Handle Fragment.

Tumulus Fill (SU 1.377).

PL (max): 5.4 cm; PW (max): 3.5 cm; Th: 0.4–1.1 cm.

Single fragment of pierced lug or spur handle, broken on either side of piercing, along long side, and at short end; only one finished edge clearly preserved. Orientation uncertain: long side may attach to vessel, forming roughly triangular horizontal handle or lug, with piercing at center, or short broken end is from tip of squared spur, with pierced upper strap; lug more likely due to complete absence of attachment scar for lower element of spur handle, but in either case, less than half of original now preserved. Upper surface highly burnished, lower smoothed, even burnished near finished edge, rough elsewhere.

Fabric: As **9/190** (P098).

Comparanda: **9/179** (P038).

9/177 (P037), **Fig. 9.22** (Drawing 7.3; Photos 616–619)

Vertical Strap Handle Fragment.

Tomb LXXXIV (2) Fill (SU 1.017).

PL (max): 5.2 cm; PW (max): 3.0 cm; Th: 1.0–1.1 cm.

Single fragment of vertical strap handle, rectangular in section and very slightly curved, preserving no attachments. Surfaces smoothed, now very worn, heavily encrusted. Large scratches and gouges on exterior surface.

Fabric: Semi-coarse; abundant mica and black inclusions, one very large white pebble inclusion. Color more representative of encrustation than ceramic: 2.5 Y 4/1 (dark gray).

Comparanda: **9/163** (P008), **9/166** (P120), **9/191** (P399), **9/197** (P143), **9/198** (P320); cf. also uncatalogued P308, P403.

9/178 (P039), **Fig. 9.21** (Drawing 8.5; Photos 612–615)

Body and Handle Fragment.

Tumulus Fill (SU 1.007).

PL (max): 4.6 cm; PW (max): 5.3 cm; PW (handle): 3.6 cm; Th: 0.5 cm.

Single fragment preserving biconical body profile of lower vessel wall, with attachment and lower portion of vertical strap handle, nearly rectangular in section. Surface smoothed, not burnished.

Fabric: Semi-coarse fabric, with abundant mica and black inclusions (<0.7 cm), moderate white inclusions and small air pockets. Exterior surface 5 YR 5/6 (yellowish red), core and interior surface 2.5 Y 2.5/1 (black).

Comparanda: **9/164** (P029), **9/180** (P158), **9/182** (P035), **9/192** (P157), **9/193** (P182), **9/194** (P110), **9/195** (P368); cf. also uncatalogued P183, P402.

9/179 (P038), **Fig. 9.22** (Drawing 8.4; Photos 777–780)

Pierced Handle Fragment.

Tomb LXXX (4) Fill (SU 4.031).

PL (max): 4.2 cm; PW (max): 3.1 cm; Th: 1.2 cm.

Single fragment, nearly flat, with one rounded, finished edge, broken on three other sides, one preserving part of pre-firing piercing. Perhaps from lower section of pierced handle (vertical strap?), broken just above lower attachment to body. Surface perhaps once smoothed, now very worn, cracked and flaked off on part of one side.

Fabric: Sandy, semi-coarse, with abundant mica, small black and red (grog) inclusions, few very small air pockets. Surfaces 2.5 YR 5/6 (red), core 5 YR 2.5/1 (black).

Comparanda: Vokotopoulou 1986: Sch. 4.2089/T40; Aliu 1984: pl. II, Tomb 13.24, eighth–seventh centuries BC; Aliu 2004: pl. XXVI, 206, nos. 281, 284.

9/180 (P158), **Fig. 9.22** (Drawing 61.2; Photos 2656–2657)

Body and Strap Handle Fragment.

Tumulus Fill (SU 1.070).

PL (max): 3.6 cm; PW (max): 3.2 cm; PW (handle): 2.4 cm; Th: 0.5 cm.

Two joining fragments (from modern break) of convex vessel wall with lower attachment of vertical strap handle, rounded rectangular in section. Exterior surface smoothed, interior rough.

Fabric: As **9/192** (P157).

Comparanda: **9/164** (P029), **9/178** (P039), **9/182** (P035), **9/192** (P157), **9/193** (P182), **9/194** (P110), **9/195** (P368), **9/196** (P216); cf. also uncatalogued P183, P402.

9/181 (P111), **Fig. 9.22** (Drawing 40.2; Photos 2144–2145)

Fragment of Vertical Strap Handle.

Tumulus Fill (SU 1.070).

PL (max): 3.4 cm; PW (max): 4.1 cm; Th (max): 1.7 cm; Th (handle, max): 1.1 cm.

Single fragment of plano-convex strap handle, broken in three places: at flaring proximal end near attachment to vessel; down center of strap; and farther from vessel junction, where part of attachment for strut is preserved. Attachment scar could also be for lower element of spur handle (so that preserved element joins rim), but appears more likely to be for strut (with preserved element joining body). Surface incompletely smoothed.

Fabric: Dark, semi-coarse, sandy, crumbly, with many inclusions: abundant mica, red, white, black inclusions, pebble, grog, mostly very small, few in core up to 0.5 cm; few air pockets in core and surfaces (<0.2 cm). Surfaces 5 YR 4/2 (dark reddish gray) to 4/6 (yellowish red), core 5 YR 2.5/1 (black).

9/182 (P035), **Fig. 9.23** (Drawing 8.2; Photos 989–992)

Vertical Strap Handle Fragment.

Topsoil (SU 8.022).

PL (max): 4.7 cm; PW (max): 2.8 cm; Th: 1.5 cm.

Single fragment of vertical strap handle, ovoid in section, preserving part of upward curve of handle and lower widening for attachment to body. Surface smoothed, covered by heavy encrustation.

Fabric: Semi-coarse, with abundant mica, moderate red and white inclusions, few very small air pockets. Surface mottled 10 YR 5/1 (gray) to 6/4 (light yellowish brown); core grayish, but hidden by encrustation.

Comparanda: **9/164** (P029), **9/178** (P039), **9/180** (P158), **9/192** (P157), **9/193** (P182), **9/194** (P110), **9/195** (P368), **9/196** (P216); cf. also uncatalogued, P183, P402.

9/183 (P062), **Figs. 9.23, 9.57** (Drawing 14.3; Photos 639–641)

Body and Handle Fragments.

Tumulus Fill (SU 1.047).

PL (max): 7.4 cm; PW (max): 7.2 cm; Th: 0.4 cm.

Eight joining fragments of lower body of thin-walled, open vessel, preserving attachment and lower portion of vertical strap handle, probably

ovoid in section. All surfaces smoothed, once burnished, now worn.
Fabric: Semi-coarse.
Comparanda: **9/164** (P029), **9/178** (P039), **9/180** (P158), **9/182** (P035), **9/192** (P157), **9/193** (P182), **9/194** (P110), **9/195** (P368); cf. also MBA kantharos from Maliq (now Tirana AM 5661 [Eggebrecht 1988:192, no. 33]), plus uncatalogued P183, P402.

9/184 (P099 = **15/5**), **Figs. 9.23, 9.58** (Drawing 34.1, 35.1; Photos 2213–2215)
Wishbone Handle Fragment.
Ceramic Deposit within Tumulus Fill (SU 4.206—includes also **9/274** [P135], **9/265** [P136]).
PH (max): 7.2 cm; PW (max): 8.2 cm; Th: 2.0–3.7 cm.
Single fragment preserving much of large horizontal wishbone handle, pierced at center, perhaps from large open shape. Broken at join to body and at both extremities; original orientation uncertain. Surfaces smoothed, with some traces of bitumen on lower exterior (perhaps where fingers once grabbed?), but generally worn. Scattered encrustation, especially at breaks.
Fabric: Hard, sandy, semi-coarse, with many small inclusions: abundant mica, occasional grog and pebble inclusions (<0.3 cm): occasional very small air pockets. Surface mottled: 5 YR 5/6 (yellowish red), through reddish brown to 5 YR 3/2 (dark reddish brown); core and remaining burnished areas of exterior 2.5 Y 2.5/1 (black).
Comparanda: None found with single attachment spreading to two arms—the opposite more common (e.g., Hammond 1972: fig. 8*i*–*l*, EBA–MBA); see also **9/186** (P344).

9/185 (P245), **Figs. 9.23, 9.58** (Drawing 84.1 [underside]; Photos 2556–2557)
Square Handle Fragment.
Topsoil (SU 2.002).
PL (max): 7.3 cm; PW (max): 2.9 cm; Th: 1.9 cm.
Single fragment preserving squared-off end of horizontal handle, oval in section, broken at one corner and just beyond other; attachments to body, original orientation not preserved. Large chip missing from underside of one corner (see **Fig. 9.23**); elsewhere surface smoothed but heavily covered with encrustation.
Fabric: Semi-coarse, medium hardness, with abundant small, dark inclusions, possibly grog (all <0.2 cm). Exterior and core 2.5 YR 5/8 (red) to 4/4 (reddish brown) where not covered with 10 YR 7/2 (light gray) encrustation. One long, thin grain (0.8 cm) embedded in the upper surface is magnetic—either iron in the clay matrix or an iron pseudomorph deriving from neighboring material (or corrosion of neighboring material) in the soil.

9/186 (P344), **Figs. 9.23, 9.58** (Drawing 72.1; Photos 3082–3083)
Forked Handle Fragment.
Tumulus Fill (SU 4.441).
PL (max): 8.2 cm; PW (max): 4.5 cm at join to body, 9.4 cm at greatest preserved span; Th: 1.3–1.9 cm.
Three joining fragments preserving part of V-shaped forked handle from large vessel, probably open. Broken at junction with body and at both prongs, one perhaps preserving a second join (to additional part of handle?); original shape uncertain, but orientation clear, with prongs spreading up and away from vessel body. Surface very worn, smoothed more on underside (where hands once grabbed) than on that facing vessel body.
Fabric: As **9/184** (P099), but less evidence of burnishing, and core not so dark: here 5 YR 4/3 (reddish brown) to 4/1 (dark gray).
Comparanda: Vokotopoulou 1986: pl. 57 (for other forked handles, but these seem to have the opposite orientation—forked end attached to vessel); cf. also **9/184** (P099).

9/187 (P345), **Figs. 9.22, 9.58** (Drawing 70.2; Photos 3091–3092)
Pierced Handle Fragment.
Tumulus Fill (SU 1.377).
PL (max): 4.8 cm; PW (max): 4.9 cm; Th: 1.6 cm at body junction, tapering to tip.
Single fragment preserving most of horizontal handle fragment, with hole pierced near junction with body (now partially lost) and again at rounded tip; gouge (pick mark?) on underside. Surface smoothed and burnished, especially on convex side.
Fabric: Semi-coarse, with moderate number of small black inclusions and grog, occasional small (<0.2 cm) air pockets. Exterior 5 YR 5/4 (reddish brown); core 5 YR 4/1 (dark gray) to 3/1 (very dark gray).
Comparanda: Aliu 2004: pl. XXVI, 281 (for attachment at belly/shoulder), pl. XXI, 246 (for attachment above rim).

9/188 (P076), **Fig. 9.22** (Drawing 17.3; Photos 975–978)
Spur Handle Fragment.
Tumulus Fill (SU 2.078).
PL (max): 2.7 cm; PW (max): 2.1 cm; Th: 0.6–1.0 cm.
Single fragment of semi-coarse spur handle, probably rounded; precise type uncertain due to breaks at tip and down center of handle. Surfaces smoothed and worn.
Fabric: Hard, with many black, red, and white inclusions (including quartz), mostly very small, one pebble 0.2 cm; occasional mica. Surface 10 YR 6/3 (pale brown) to 7.5 YR 5/4 (brown); core 7.5 YR 2.5/1 (black).

9/189 (P088), **Fig. 9.22** (Drawing 25.6; Photos 789–792)
Squared Spur Handle Fragment.
Tumulus Fill (SU 4.130).
PL (max): 3.6 cm; PW (max): 2.6 cm; W (spur tip): 2.3 cm; Th: 0.5–2.1 cm.
Single fragment of squared spur handle, preserving only flattened, rectangular tip and junction of upper and lower elements, probably ovoid in section. Surfaces smoothed.
Fabric: Medium, sandy, with many inclusions; abundant mica, very small (<0.2 cm) black, red, and white inclusions, air pockets (<0.1 cm) rare on surfaces, occasional in core (<0.2 cm). Surfaces 7.5 YR 5/4 (brown) to 5/6 (strong brown); core 7.5 YR 2.5/1 (black).

9/190 (P098), **Figs. 9.23, 9.58** (Drawing 30.2; Photos 2179–2180)
Flaring Spur Handle Fragment.
Tumulus Fill (SU 1.039).
PL (max): 7.5 cm; PW (max): 5.9 cm; W (spur tip): 4.1 cm; Th (tip): 0.8 cm; Th (handle): 0.7–1.7 cm.
Single fragment of flaring spur handle, preserving entire upper handle element, rectangular in section, and attachment scar of lower element, triangular in section. Spur terminates in slightly upraised rectangular tip with flaring corners, and flares much more widely at proximal end to join rim, probably with rounded lip; fragment extends just beyond rim attachment, suggesting closed upper body with everted rim. Upper surface and outer edges burnished, now worn in places, lower surface smoothed; encrustation heavy on core, scattered elsewhere.
Fabric: Dark, semi-coarse, medium-hard with many inclusions; abundant mica, black, red (grog?), and white (quartz?) inclusions (<0.3 cm, mostly much smaller); air pockets in core and surfaces, mostly long (<1.1 cm) and thin, perhaps from grain used as temper. Surface color varies widely, from 5 YR 5/4 (reddish brown) to 3/1 (very dark gray); core 5 YR 4/6 (yellowish red) to 2.5/1 (black).
Comparanda: Uncatalogued P379.
Publication: Papadopoulos, Bejko, and Morris 2007: fig. 28.

9/191 (P399), **Fig. 9.58** (Photos 3474–3475)
Vertical Strap Handle Fragment.
Tumulus Fill (SU 5.548).
PL (max): 5.5 cm; PW (max): 3.4–4.0 cm; Th: 1.0–1.4 cm.
Single fragment of vertical strap handle, rounded rectangular in section, broken mid-strap and just above lower attachment to wall (preserving thickening of lower handle, but none of vessel wall). Surfaces smoothed, traces of burnishing on exterior, very worn all around, heavy encrustation over both breaks, one intact edge.
Fabric: Semi-coarse, sandy, with abundant mica, very small black and white inclusions (<0.2 cm). Surface 5 YR 5/6 (yellowish red).
Comparanda: **9/163** (P008), **9/166** (P120), **9/177** (P037), **9/197** (P143), **9/198** (P320); cf. also uncatalogued P308, P403.

9/192 (P157), **Fig. 9.23** (Drawing 41.5; Photos 2220–2221)
Body and Strap Handle Fragment.
Topsoil (SU 4.201).
PL (max): 3.3 cm; PW (max): 4.0 cm; PW (handle): 2.9; Th (wall): 0.6 cm; Th (handle): 1.0 cm.
Single fragment of convex vessel wall with lower attachment of vertical strap handle, flattened oval in section. Surfaces smoothed, exterior probably once burnished, now very worn.
Fabric: Semi-coarse, with abundant mica, very few white and red inclusions and air pockets. Surface 7.5 YR 5/4 (brown), core 10 YR 2/1 (black).
Comparanda: **9/164** (P029), **9/178** (P039), **9/180** (P158), **9/182** (P035), **9/193** (P182), **9/194** (P110),

9/195 (P368), **9/196** (P216); cf. also uncatalogued P018, P183, P402.

9/193 (P182), **Fig. 9.58** (Photos 2429–31, 2621–2623)
Body and Strap Handle Fragment.
Tumulus Fill (SU 1.039).
PL (max): 3.2 cm; PW (max): 3.9 cm; PW (handle): 3.4 cm; Th (wall): 0.5 cm.
Two joining fragments of vessel wall with lower attachment and small lower portion of vertical strap handle, flattened oval in section. Exterior surface smoothed, interior rough.
Fabric: Semi-coarse, with abundant mica and white inclusions (<0.5 cm), moderate red inclusions and air pockets (<0.4 cm). Surface and core 7.5 YR 2.5/1 (black).
Comparanda: **9/164** (P029), **9/178** (P039), **9/180** (P158), **9/182** (P035), **9/192** (P157), **9/194** (P110), **9/195** (P368), **9/196** (P216); cf. also uncatalogued P018, P183, P402.

9/194 (P110), **Fig. 9.23** (Drawing 39.8; Photos 2142–2143)
Body and Strap Handle Fragment.
Tumulus Fill (SU 1.070).
PL (max): 1.9 cm; PW (max): 3.9 cm; PW (handle): 3.3 cm; Th: 0.8 cm.
Single small fragment preserving vessel wall with lower attachment of vertical strap handle, flattened oval in section. Exterior handle surface smoothed and burnished, other surfaces smoothed.
Fabric: As **9/193** (P182).
Comparanda: **9/164** (P029), **9/178** (P039), **9/180** (P158), **9/182** (P035), **9/192** (P157), **9/193** (P182), **9/195** (P368), **9/196** (P216); cf. also uncatalogued P183, P402.

9/195 (P368), **Fig. 9.58** (Photos 3302–3303)
Vertical Strap Handle Fragment.
Tumulus Fill (SU 1.440).
PL (max): 3.6 cm; PW (max): 4.1 cm; Th (handle): 1.1 cm.
Single fragment of vertical strap handle, ovoid in section, preserving lower attachment with very small section of inner vessel wall. Exterior surface smoothed and burnished, other surfaces smoothed only.
Fabric: As **9/193** (P182), but lighter in color: 2.5 YR 5/8 (red) to 10 YR 2/1 (black).
Comparanda: **9/164** (P029), **9/178** (P039), **9/182** (P035), **9/192** (P157), **9/180** (P158), **9/193** (P182), **9/194** (P110), **9/196** (P216); cf. also uncatalogued P183, P402.

9/196 (P216), **Fig. 9.58** (Photos 2640–2641)
Vertical Strap Handle Fragment.
Tumulus Fill (SU 1.240).
PL (max): 3.0 cm; PW (max): 3.6 cm; Th: 1.3 cm.
Single fragment preserving lowermost portion of vertical strap handle, flattened ovoid in section, but missing actual attachment to vessel wall. Surface smoothed, not burnished.
Fabric: As **9/194** (P110).
Comparanda: **9/164** (P029), **9/178** (P039), **9/180** (P158), **9/182** (P035), **9/192** (P157), **9/193** (P182), **9/194** (P110), **9/195** (P368); cf. also uncatalogued P183, P402.

9/197 (P143), **Fig. 9.23** (Drawing 83.3; Photos 2566–2567, 2630–2631)
Vertical Strap Handle Fragments.
Tumulus Fill (SU 1.070).
PL (max): 6.0 cm; PW (max): 3.8 cm; Th: 1.0 cm.
Two joining fragments (from modern break) of vertical strap handle, ovoid in section and very slightly curved, preserving no attachments. Surfaces smoothed, exterior once burnished; scattered encrustation, more abundant on interior surface.
Fabric: As **9/177** (P037), with abundant mica and black inclusions, one very large air pocket (0.4 cm); surface color more visible with less encrustation: 7.5 YR 4/4 (brown).
Comparanda: **9/163** (P008), **9/166** (P120), **9/177** (P037), **9/191** (P399), **9/198** (P320); cf. also uncatalogued P308, P403.

9/198 (P320), **Fig. 9.59** (Photos 2898–2899)
Vertical Strap Handle Fragment.
Tumulus Fill (SU 4.286).
PL (max): 4.8 cm; PW (max): 3.7 cm; Th: 1.0 cm.
Two joining fragments of wide strap handle, ovoid in section, broken mid-handle and just above attachment to body. Surfaces smoothed, not burnished.
Fabric: Semi-coarse, abundant mica and black inclusions, moderate white inclusions, and moderate

very small air pockets. Surface 7.5 YR 4/3 (brown), core 10 YR 2/1 (black).
Comparanda: **9/163** (P008), **9/166** (P120), **9/177** (P037), **9/191** (P399), **9/197** (P143); cf. also wide strap handles of finer fabric: **9/114** (P027), **9/121** (P307), **9/122** (P361), **9/123** (P339), plus uncatalogued P308, P403.

9/199 (P091), **Figs. 9.24, 9.59** (Drawing 25.3; Photos 737–740)
Rim Fragment.
Tumulus Fill (SU 1.067).
PL (max): 2.4 cm; PW (max): 2.8 cm; Th: 0.9 cm.
Very small fragment, probably from rim of small vessel with worn, tapered lip; precise shape and orientation uncertain—could even derive from strap handle? Surface smoothed, very worn.
Fabric: Semi-coarse, sandy, with abundant mica, black and white inclusions, moderate red inclusions, few air pockets (<0.2 cm). Exterior 2.5 YR 2.5/1 (reddish black), interior 2.5 YR 5/4 (reddish brown); core similar to exterior where visible.

9/200 (P218), **Fig. 9.24** (Drawing 45.7; Photos 2384–2385)
Rim Fragment.
Topsoil (SU 2.002).
PL (max): 3.1 cm; PW (max): 3.0 cm; Th: 1.1 cm.
Single fragment from mid-sized vessel, probably open in shape, with slightly everted rim, rounded lip. Surfaces smoothed, interior and lip damaged; heavy encrustation on exterior.
Fabric: Semi-coarse, with abundant mica and red inclusions (grog; <0.4 cm), occasional black inclusions. Exterior surface 5 YR 5/3 (reddish brown), interior 2.5 YR 5/6 (red), core 5 YR 2.5/1 (black).
Comparanda: Uncatalogued P420; cf. also **9/208** (P178; thinner wall, more everted).

9/201 (P054), **Figs. 9.24, 9.59** (Drawing 12.4; Photos 920–923)
Rim Fragment.
Interface Topsoil/Tumulus Fill (SU 4.011).
PL (max): 4.4 cm; PW (max): 4.5 cm; Th: 0.9–1.1 cm; D (rim ext): ca. 9 cm.
Single fragment preserving shoulder, neck, and everted rim, with rounded lip, of medium-sized vessel, probably closed in shape; very slight groove seems to set off neck from shoulder. Surfaces smoothed, now very worn, covered with encrustation.
Fabric: Semi-coarse, red fabric, with abundant white and gray pebble inclusions (<0.4 cm), moderate mica, very fine black and white inclusions, and air pockets (<0.2 cm). Surface 2.5 YR 6/6 (light red).

9/202 (P155), **Fig. 9.24** (Drawing 42.6; Photos 2225–2226)
Rim Fragment.
Tomb LXXV (33) Fill (SU 1.222).
PL (max): 3.2 cm; PW (max): 2.5 cm; Th: 0.8 cm.
Fragment of small open shape, preserving bit of shoulder-neck curve and slightly everted rim with rounded lip. Surfaces likely once smoothed, now very worn, completely missing in places.
Fabric: Semi-coarse, red fabric, with abundant mica, very small white, black, and brown inclusions, and air pockets (<0.3 cm). Surface and core 5 YR 5/6 (yellowish red).

9/203 (P086), **Figs. 9.24, 9.59** (Drawing 25.1; Photos 900–903)
Rim Fragment.
Tumulus Fill (SU 4.086).
PH (max): 3.8 cm; PW (max): 3.4 cm; Th: 0.6 cm.
Single fragment from mid-sized vessel with upper wall curving in to everted rim with rounded lip. Surfaces smoothed, exterior slightly uneven, with damage to lip; heavy encrustation on interior.
Fabric: Dark, semi-coarse, with abundant mica and white inclusions (pebbles <0.3 cm), occasional red pebbles (<0.2 cm). Surfaces and core 7.5 YR 2.5/1 (black).
Comparanda: **9/204** (P417), **9/205** (P405), **9/206** (P094A); cf. also uncatalogued P427.

9/204 (P417), **Fig. 9.59** (Photos 4001–4002)
Rim Fragment.
Tumulus Fill (SU 6.559).
PH (max): 4.1 cm; PW (max): 2.6 cm; Th: 0.6 cm.
Single fragment from upper wall of mid-sized vessel preserving irregularly curved everted rim and rounded lip, perhaps rise to handle attachment or spout. Surfaces smoothed but now very worn, interior uneven; scattered encrustation.
Fabric: Semi-coarse, with abundant mica and white inclusions, occasional black inclusions. Exterior surface 5 YR 4/4 (reddish brown), interior surface as exterior, with black spots (7.5 YR 2.5/1), core 5 YR 5/1 (gray).

Comparanda: **9/203** (P086), **9/205** (P405), **9/206** (P094A); cf. also uncatalogued P427.

9/205 (P405), **Fig. 9.24** (Drawing 94.4; Photos 3512–3513)
Rim Fragment.
Tumulus Fill (SU 5.535).
PH (max): 3.4 cm; PW (max): 2.8 cm; Th: 0.3–1.0 cm; D (rim): ca. 15–20 cm.
Single fragment of mid-sized vessel, probably open in shape, with everted rim and rounded, uneven lip. Surface smoothed but very uneven, heavy encrustation over break and on interior surface.
Fabric: Sandy, semi-coarse, with abundant mica and white inclusions, moderate red inclusions (grog?). Surface 5 YR 4/4 (reddish brown), core 7.5 YR 2.5/1 (black).
Comparanda: **9/203** (P086), **9/204** (P417), **9/206** (P094A); cf. also uncatalogued P427.

9/206 (P094A), **Figs. 9.24, 9.59** (Drawing 29.5; Photos 2004–2007)
Rim Fragment.
Tumulus Fill (SU 4.086).
PH (max): 3.5 cm; PW (max): 4.7 cm; Th: 0.7 cm
Single fragment from mid-sized vessel, probably open in shape, with everted rim and rounded lip. Surfaces smoothed, perhaps once burnished, with scratches on neck; now very worn, scattered encrustation, heavy over breaks. (Found with **9/167** [P094B], non-joining vertical strap handle of similar fabric, but burnished, perhaps slightly finer.)
Fabric: Semi-coarse to fine, with abundant small white inclusions, moderate mica, occasional red (grog?) inclusions and air pockets (<0.5 cm). Surfaces 5 YR 4/3 (reddish brown), core 7.5 YR2.5/1 (black).
Comparanda: **9/203** (P086), **9/204** (P417), **9/205** (P405); cf. also uncatalogued P427.

9/207 (P321), **Fig. 9.59** (Photos 2980–2981)
Rim Fragment.
Topsoil (SU 2.002).
PL (max): 2.9 cm; PW (max): 3.7 cm; Th: 0.8 cm.
Two joining fragments (from modern break) of open vessel with slightly everted rim and rounded lip. Surfaces smoothed but slightly uneven (especially exterior), perhaps once burnished, now worn; heavy encrustation.
Fabric: Semi-coarse, with abundant mica and white inclusions, several very large angular pebbles (<0.9 cm), moderate black inclusions. Core very uniform, 5 YR 5/6 (yellowish red), perhaps preserving original surface color (now obscured by darker multicolored encrustation).

9/208 (P178), **Fig. 9.24** (Drawing 45.10; Photos 2370–2371)
Rim Fragment.
Tumulus Fill (SU 1.070).
PL (max): 2.0 cm; PW (max): 2.6 cm; Th: 0.7 cm.
Small fragment from everted rim with uneven rounded lip, probably of mid-sized vessel. Surfaces smoothed, with heavy encrustation on exterior.
Fabric: Semi-coarse, sandy, with abundant mica, white pebbles (<0.3 cm), and black inclusions. Exterior surface 5 YR 4/4 (reddish brown), interior and core 5 YR 2.5/1 (black).
Comparanda: **9/200** (P218; less everted, thicker walls); cf. also uncatalogued P420.

9/209 (P060), **Fig. 9.24** (Drawing 13.3; Photos 967–970)
Rim Fragment.
Tumulus Fill (SU 3.045).
PH (max): 3.7 cm; PW (max): 3.8 cm; Th: 0.6–0.9 cm.
Two joining fragments from upper body of mid-sized open vessel; wall narrows toward neck, curving slightly before flaring out to everted rim with rounded lip and flat upper surface. Surfaces smoothed, outer lip worn.
Fabric: Dark, sandy, semi-coarse, abundant mica and black pebble inclusions (<0.4 cm), occasional white and red (grog) inclusions (<0.4 cm). Surface and core 2.5 YR 2.5/1 (reddish black).
Comparanda: **9/210** (P208).

9/210 (P208), **Fig. 9.24** (Drawing 45.4; Photos 2379–2380)
Rim Fragment.
Topsoil (SU 4.201).
PL (max): 2.1 cm; PW (max): 2.2 cm; Th: 0.8 cm.
Single fragment from vertical wall of mid-sized vessel with flat lip, very small external ridge. Surfaces smoothed, interior slightly uneven.
Fabric: As **9/209** (P060).
Comparanda: **9/209** (P060).

9/211 (P310), **Fig. 9.24** (Drawing 48.4; Photos 2972–2974)
Rim Fragment.

Topsoil (SU 2.003).
PL (max): 2.7 cm; PW (max): 4.0 cm; Th: 0.7–0.8 cm; D (rim ext): ca. 15 cm.
Single fragment from mid-sized vessel, probably open in shape, preserving shoulder with curve to everted rim, nearly flat lip. Surfaces smoothed, but slightly uneven; scattered encrustation, heavier on interior.
Fabric: Semi-coarse, sandy, with abundant red (grog?) and black inclusions, moderate mica. Surfaces 5 YR 4/4 (reddish brown), core 5 YR 2.5/1 (black).

9/212 (P072), **Fig. 9.24** (Drawing 16.5; Photos 947–950)
Rim Fragment.
Tumulus Fill (SU 2.078).
PH (max): 3.1 cm; PW (max): 5.5 cm; Th: 0.6–1.0 cm; D (rim ext): 12 cm
Two joining fragments (from modern break) of mid-sized vessel, probably open in shape, with everted rim and flattened, tapered lip; part of handle attachment and scratches or tool marks preserved on exterior surface. Both surfaces smoothed, slightly worn; scattered encrustation.
Fabric: Dark, semi-coarse, with abundant mica, several white angular inclusions (0.4 cm), occasional red inclusions. Surface 5 YR 4/4 (reddish brown), core 5 YR 2.5/1 (black).

9/213 (P186), **Fig. 9.24** (Drawing 46.2; Photos 2676–2677)
Rim Fragment.
Tumulus Fill (SU 1.070).
PL (max): 2.3 cm; PW (max): 1.6 cm; Th: 0.7–0.9 cm.
Small fragment from mid-sized vessel with everted rim and tapered lip; precise shape and orientation uncertain. Surfaces smoothed, encrustation over one break.
Fabric: Semi-coarse, with abundant mica and small white inclusions, moderate black inclusions (<0.3 cm), occasional grog, three air pockets (<0.4 cm). Surface 5 YR 3/3 (dark reddish brown), core 7.5 YR 2.5/1 (black).

9/214 (P069), **Figs. 9.24, 9.59** (Drawing 16.2; Photos 741–744)
Rim Fragment.
Tumulus Fill (SU 1.070).
PH (max): 3.1 cm; PW (max): 3.3 cm; Th: 0.8 cm; D (rim ext): ca. 15–20 cm.
Two joining fragments (from modern break) of everted rim with flat lip, probably from medium-sized open vessel. Surfaces smoothed, heavily encrusted.
Fabric: Semi-coarse, with moderate mica and black and white inclusions (<0.3 cm). Surface 7.5 YR 5/4 (brown) where visible, core 7.5 YR 3/1 (black).
Comparanda: **9/230** (P413), **9/231** (P323).

9/215 (P337), **Fig. 9.59** (Photos 3157–3159)
Rim, Neck, Shoulder Fragment.
Tumulus Fill (SU 4.286).
PL (max): 4.2 cm; PW (max): 3.3 cm; Th: 0.6 cm.
Single fragment preserving S-curve of convex lower body, narrow neck, and everted rim, broken below lip. Preserved surface smoothed (perhaps slipped?), interior with horizontal lines, perhaps tooling marks, but much of surface now flaked off; encrustation scattered, heaviest where surface is missing.
Fabric: Semi-coarse, with abundant red, black, and brown inclusions, mostly small, one pebble 0.7 cm. Surface 7.5 YR 3/1 (very dark gray), core 7.5 YR 4/3 (brown).
Comparanda: Similar shape, finer fabric: **9/152** (P045 + P046), **9/133** (P171).

9/216 (P121), **Fig. 9.24** (Drawing 40.5; Photos 2169–2170)
Rim Fragment.
Topsoil (SU 1.141).
PL (max): 3.6 cm; PW (max): 5.2 cm; Th: 0.9 cm.
Two joining fragments preserving bit of curve from neck to widely flaring rim, rounded lip. Surfaces once smoothed, now quite worn, with several chips missing; encrustation on exterior.
Fabric: Semi-coarse, red fabric, with abundant mica, very small white and black inclusions, and air pockets (<0.2 cm); few larger white and gray pebble inclusions (<0.3 cm). Surface and core 5 YR 6/6 (reddish yellow).

9/217 (P318), **Fig. 9.60** (Photos 2890–2891)
Rim Fragment.
Tumulus Fill (SU 2.395).
PL (max): 2.1 cm; PW (max): 3.3 cm; Th: 1.0 cm.
Small fragment from mid-sized vessel with flaring rim and flat lip; small ridge just below lip on exterior. Surface smoothed, exterior slightly uneven; encrustation over core.

Fabric: Semi-coarse, with abundant mica and black and red (grog) inclusions, occasional white inclusions. Exterior surface 2.5 YR 5/6 (red), interior 2.5 YR 3/1 (dark reddish gray), core not visible.

9/218 (P156), **Fig. 9.26** (Drawing 42.5; Photos 2357–2358)
Rim Fragment.
Topsoil (SU 4.201).
PL (max): 3.3 cm; PW (max): 1.3 cm; Th (wall): 0.3 cm; Th (rim): 1.0 cm; D (rim ext): ca. 20 cm.
Single fragment, probably from open shape, preserving rim with flat lip, squared in section, above very small portion of thin vertical wall; difficult to imagine such a thin handmade wall supporting such a wide rim. Surfaces smoothed, now worn.
Fabric: Semi-coarse, brick red, with abundant mica, very fine black, white, and brown inclusions, and moderate brown and red inclusions (grog; <0.3 cm) and air pockets (<0.3 cm). Exterior surface and core 2.5 YR 5/8 (red), interior surface 5 YR 6/6 (reddish yellow).

9/219 (P083), **Fig. 9.26** (Drawing 21.2; Photos 646–649)
Rim and Upper Wall Fragments.
Tumulus Fill (SU 1.067).
PL (max): 6.2 cm; PW (max): 3.7 cm; Th: 0.6 cm.
Six joining fragments (plus one non-joining chip) of mid- to large-sized vessel preserving (roughly vertical?) upper wall leading directly to uneven, rounded lip. Surfaces smoothed, now worn and damaged; scattered encrustation.
Fabric: Semi-coarse, with abundant mica and black inclusions, occasional white inclusions. Surfaces 10 YR 4/3 (brown), core 7.5 YR 4/1 (dark gray).
Comparanda: **9/232** (P377), **9/233** (P380).

9/220 (P168), **Figs. 9.25, 9.60** (Drawing 48.1; Photos 2497–2501)
Rim, Neck, and Upper Body Fragments.
Tumulus Fill (SU 4.204).
PH (max): 9.2 cm; PW (max): 10.0 cm; Th: 0.7–1.2 cm; D (rim ext): 11 cm.
Four joining fragments of mid- to large-sized vessel, including portions of upper body, neck, and nearly half of everted rim with rounded lip; small irregularity preserved in rim, perhaps rudimentary pouring spout or beginning of rise for handle. Surfaces smoothed, exterior painted.
Matt-painted decoration: Two nearly horizontal lines below rim, diverging at right, just above continuous zigzag; farther down, bit of another zigzag, poorly preserved.
Fabric: Sandy, semi-coarse, with moderate mica, grog (<0.6 cm) and white inclusions (<0.5 cm), occasional small air pockets. Surfaces 2.5 YR 6/6 (light red) to 5 YR 6/6 (reddish yellow), core 5 YR 5/1 (gray).
Comparanda: **9/7** (P166; for spout/handle?).

9/221 (P057), **Figs. 9.25, 9.60** (Drawing 17.2; Photos 537–540, 550)
Rim and Neck Fragment.
Tumulus Fill (SU 4.048).
PH (max): 4.4 cm; PW (max): 11.0 cm; Th (wall): 0.8–1.2 cm; Th (rim): 0.8 cm; D (ext, approximate): 25 cm; D (int, approximate): 18 cm.
Rim and neck fragment from large vessel, probably open in shape, with nearly vertical neck flaring widely to rim with flat lip. Surface smoothed, now quite worn, completely flaked off in places; interior painted, exterior perhaps once burnished; heavy encrustation, especially on rim interior.
Matt-painted decoration: On interior rim surface, now largely lost and/or obscured; single line just below inner (lower) edge of rim, beneath two sets of evenly spaced parallel lines radiating out to lip; possible second horizontal line above radiating lines, along outer edge of rim.
Fabric: Soft, sandy, semi-coarse with moderate white, black, and mica inclusions, mostly quite small (<0.2 cm); occasional air pockets (including two ca. 0.6 cm long on interior rim surface). Exterior and interior surfaces 2.5 YR 5/6 (red), core (beneath encrustation) 2.5 YR 6/0 (gray).
Comparanda: **9/216** (P121), **9/222** (P017), **9/223** (P095), **9/226** (P075); cf. also **9/49** (P389; FL, matt-painted).

9/222 (P017), **Fig. 9.25** (Drawing 4.3; Photos 963–966)
Rim Fragment.
Topsoil (SU 4.004).
PL (max): 5.3 cm; PW (max): 3.9 cm; Th: 1.0–1.3 cm; D (rim ext): ca. 25 cm.
Small fragment of widely flaring rim with nearly flat lip, from large vessel, perhaps open in shape. Surfaces smoothed, now very worn, completely missing in places; scattered encrustation.

Fabric: Semi-coarse, sandy, with abundant mica, moderate fine black, red, and white inclusions, few air pockets (<0.6 cm) and larger white (limestone) inclusions (<0.2 cm). Surfaces 5 YR 5/8 (yellowish red), core 2.5 Y 6/2 (light brownish gray).
Comparanda: **9/216** (P121), **9/221** (P057), **9/223** (P095), **9/226** (P075); cf. also **9/49** (P389; FL, matt-painted).

9/223 (P095), **Figs. 9.25, 9.60** (Drawing 29.4; Photos 2000–2003)
Rim and Neck Fragment.
Topsoil (SU 7.186).
PH (max): ca. 3 cm (precise orientation uncertain); PW (max): 6.8 cm; Th (wall): 0.9–1.2 cm; Th (rim): 0.5 cm; D (rim ext, approximate): 24–26 cm; D (rim int, approximate): 19–21 cm.
Single fragment from widely flaring rim, with flat lip, of large vessel, broken at turn to neck; precise orientation and vessel shape uncertain. Surface smoothed, now very worn; perhaps once slipped.
Fabric: Sandy, semi-coarse, with abundant mica and small black inclusions (<0.3 cm); occasional small white inclusions (including quartz, <0.2 cm). Exterior surface 2.5 YR 6/6 (light red), interior 5 YR 6/4 (light reddish brown), with some poorly fired areas much darker (smudging? fire-clouding?); core 5 YR 6/1 (gray).
Comparanda: **9/216** (P121), **9/222** (P017); cf. also **9/221, 9/226** (P057, P075; FL, matt-painted), and Koçi 1991: pl. III, 59 (Himarë, EIA).

9/224 (P040), **Fig. 9.25** (Drawing 8.3; Photos 773–776)
Rim Fragment.
Tomb LXXX (4) Fill (SU 4.031).
PH (max): 6.1–6.6 cm; PW (max): 5.6 cm; D (rim, approximate): 30 cm (depends partly on orientation); Th (rim): 0.9 cm; Th (wall): 0.8–0.9 cm.
Single fragment of neck and rim with flat lip; neck probably close to vertical, but precise orientation uncertain. Surfaces smoothed, exterior perhaps once burnished; fine horizontal lines preserved on exterior and interior; encrustation on core.
Fabric: Medium-hard; abundant inclusions, mica, fine black and white (including quartz; <0.3 cm); air pockets visible in all surfaces (<0.3 cm). Exterior 7.5 YR 5/4 (brown) to 5/6 (strong brown); interior and core as exterior, also ranging to 7.5 YR 3/1 (very dark gray).
Comparanda: **9/228** (P267); **9/225** (P053; thinner wall); cf. also coarseware rims **9/282** (P397), **9/283** (P081), **9/284** (P227), and uncatalogued P184.

9/225 (P053), **Fig. 9.25** (Drawing 12.5; Photos 916–919)
Rim Fragment.
Tumulus Fill (SU 1.047).
PL (max): 5.1 cm; PW (max): 2.2 cm; Th: 0.6 cm.
Single fragment from mid-sized vessel with nearly vertical upper wall flaring to flat lip. Surfaces smoothed, exterior perhaps once burnished; interior cracked, heavy encrustation on exterior.
Fabric: Semi-coarse, with abundant mica and black inclusions, occasional white inclusions. Surfaces 7.5 YR 5/4 (brown), core 7.5 YR 2.5/1 (black).
Comparanda: **9/224** (P040), **9/228** (P267); cf. also coarse rim **9/284** (P227) and uncatalogued P358.

9/226 (P075), **Figs. 9.25, 9.60** (Drawing 18.1; Photos 2008–2011)
Rim, Neck, and Wall Fragments.
Tomb LXXXII (9) Fill (SU 4.084).
PH (max): 3.0 cm; PW (max): 4.9 cm; Th (wall): 0.7– 1.0 cm; Th (rim): 0.7 cm; D (rim ext, approximate): 15–18 cm; D (rim int, approximate): 10 cm.
Two pairs of joining fragments from widely flaring rim, with uneven, rounded lip, of large vessel, probably open in shape; neck close to vertical. Exterior and interior surfaces crudely smoothed, now extremely worn; heavy encrustation, especially on neck interior. Shallow, irregular grooves on neck interior of one fragment, perhaps tooling marks, have collected encrustation to give false appearance of matt-painted lines.
Fabric: Medium, sandy; abundant inclusions, black and white (<0.2 cm), mica, grog (<0.75 cm); air pockets over all surfaces (<0.6 cm). Exterior and interior surfaces 7.5 YR 6/6 (reddish yellow); core 7.5 YR 6/6 to 10 YR 6/2 (light brownish gray).
Comparanda: **9/221** (P057), **9/223** (P095).

9/227 (P366), **Fig. 9.26** (Drawing 114.3; Photos 3298–3299)
Rim Fragment.
Tomb I (64) Fill (SU 1.361).
PL (max): 4.2 cm; PW (max): 4.8 cm; Th: 1.2 cm.
Single fragment of large vessel with wide, flaring rim and narrow rounded lip; horizontal ridges on rim

exterior. Surfaces smoothed, lightly scattered encrustation, heavier on breaks.
Fabric: Sandy, semi-coarse, with abundant small black and white inclusions (including pebbles <0.3 cm), and occasional mica. Surface 5 YR 4/4 (reddish brown), core 10 YR 2/1 (black).
Comparanda: Uncatalogued P381, P352.

9/228 (P267), **Fig. 9.60** (Photos 2668–2669)
Rim Fragment.
Tomb LXIV (61) Fill (SU 2.259).
PL (max): 3.5 cm; PW (max): 4.3 cm; Th: 1.0 cm.
Single fragment of large vessel with slightly flaring rim, flat lip with slight external projection. Surfaces smoothed, interior perhaps once burnished; lip damaged, encrustation over breaks.
Fabric: Sandy, semi-coarse, with abundant black and white inclusions (<0.5 cm), moderate mica, two very small air pockets in interior. Surfaces 5 YR 6/6 (reddish yellow); core not visible.
Comparanda: **9/224** (P040), **9/225** (P053), **9/284** (P227); cf. also uncatalogued P358.

9/229 (P122), **Figs. 9.26, 9.60** (Drawing 39.7; Photos 2158–2159)
Rim Fragment.
Topsoil (SU 1.141).
PH (max): 4.3 cm; PW (max): 4.6 cm; Th: 1.2 cm.
Single fragment of large vessel with everted rim and narrow rounded lip. Surfaces smoothed, interior worn and damaged, exterior missing just below lip, uneven toward bottom.
Fabric: Dark, semi-coarse, abundant mica and red (grog) inclusions, moderate small white inclusions. Patch on interior surface 5 YR 3/3 (dark reddish brown), remaining surface and core 7.5 YR 2.5/1 (black).
Comparanda: **9/227** (P366); cf. also uncatalogued P352, P381.

9/230 (P413), **Fig. 9.26** (Drawing 97.2; Photos 3783–3784)
Rim Fragment.
Tumulus Fill (SU 5.548).
PH (max): 2.8 cm; PW (max): 6.1 cm; Th: 0.7 cm; D (rim ext): ca. 20 cm.
Single fragment of large open shape with everted rim and flat lip with rounded external projection. Surfaces smoothed, perhaps once burnished; now worn, finely cracked.
Fabric: Semi-coarse, with abundant mica, moderate black inclusions, few red (grog?) and white inclusions (<0.3 cm). Surface 5 YR 5/6 (yellowish red) to 2.5/1 (black); core black (where visible).
Comparanda: **9/214** (P069), **9/231** (P323).

9/231 (P323), **Fig. 9.60** (Photos 3080–3081)
Rim Fragment.
Topsoil (SU 2.003).
PH (max): 3.8 cm; PW (max): 5.3 cm; Th: 0.8 cm; D (rim ext): ca. 30 cm.
Single fragment of large open shape with everted rim and flat lip. Exterior surface rough, with horizontal finger or tool marks; interior smooth, perhaps once burnished.
Fabric: Semi-coarse, abundant mica, red and white inclusions (<0.3 cm). Surfaces 7.5 YR 4/2 (brown), core 4/1 (dark gray).
Comparanda: **9/214** (P069), **9/230** (P413).

9/232 (P377), **Fig. 9.26** (Drawing 100.2; Photos 3400–3401)
Rim and Upper Wall Fragment.
Tumulus Fill (SU 1.399).
PL (max): 7.4 cm; PW (max): 4.6 cm; D (rim ext, approximate): 26–30 cm; Th (wall): 0.7–0.9 cm; Th (rim): 0.6 cm.
Single fragment preserving rim and upper wall of large open vessel (bowl?); convex upper wall beneath shallow S-curve and slightly everted rim, tapering slightly to rounded lip. Exterior surface smoothed and burnished, traces of tooling marks on upper exterior; interior smoothed around rim, much more worn below; modern chip from rim exterior; encrustation scattered over all surfaces, heavier on interior.
Fabric: Soft, sandy, many inclusions: mica, very small black inclusions, occasional grog, and one very large (0.6 cm) gray pebble in core; abundant very small air pockets (<0.1 cm) on surfaces, slightly larger (<0.2 cm) in core. Surfaces and core 2.5 YR 6/6 (light red) to 5/6 (red).
Comparanda: **9/233** (P380); cf. also **9/219** (P083; similar fabric, thinner, more vertical walls) and **9/281** (P239), **9/282** (P397), **9/301** (P185): similar everted rim, but flat lip.

9/233 (P380), **Figs. 9.26, 9.60** (Drawing 100.1; Photos 3398–3399)
Rim and Upper Wall Fragment.
Tumulus Fill (SU 2.520).
PH (max): 6.8 cm; PW (max): 6.0 cm; Th: 0.7 cm.
Single fragment preserving rim and upper wall from large open vessel; irregular convex upper wall flares very slightly to uneven rounded lip, now partially damaged. Exterior surface smoothed but uneven, missing at lower right exterior; interior rougher.
Fabric: Semi-coarse, with abundant mica and white and black pebble inclusions (<0.2 cm); one air pocket (0.4 cm) on interior. Surfaces 2.5 YR 5/8 (red), with two black patches on exterior; core 7.5 YR 2.5/1 (black).
Comparanda: **9/219** (P083), **9/232** (P377).

9/234 (P016), **Figs. 9.26, 9.61** (Drawing 3.6; Photos 529–532, 549)
Rim Fragment.
Topsoil (SU 1.009).
PH (max): 4.6 cm; PW (max): 4.9 cm; Th: 1.1 cm.
Rim fragment of large vessel, probably open in shape; rim vertical or very slightly everted, with flat lip, but twisted in section, perhaps for handle attachment or spout. Surfaces once smoothed, now very worn and damaged, especially interior; painted on both sides.
Matt-painted decoration: On exterior, six or seven parallel, diagonal bands, bound above by nearly horizontal band just below rim—perhaps upper part of hatched pendent triangle. On interior, roughly parallel with lip, parts of two or three nearly horizontal bands, thicker than usual. (Found together with smaller, non-joining, chip of similar fabric, preserving bits of matt-paint, perhaps three bands.)
Fabric: Sandy, semi-coarse, with abundant mica, moderate air pockets, and occasional small black and white inclusions. Surfaces 2.5 YR 5/8 (red) or 5 YR 5/8 (yellowish red); core 7.5 YR 6/2 (pinkish gray).

9/235 (P084), **Figs. 9.26, 9.61** (Drawing 21.4; Photos 533–536)
Rim and Neck Fragment.
Tumulus Fill (SU 1.070).
PL (max): 7.0 cm; PW (max): 6.1 cm; Th: 0.7–1.2 cm; D (rim ext): ca. 15 cm.
Rim and neck fragment of large open vessel, with concave neck, slightly everted rim, and tapered lip. Surfaces smoothed, perhaps slipped, exterior painted; scattered encrustation, with modern damage on exterior, ancient damage to lip.
Matt-painted decoration: Poorly preserved, beginning just below rim. Two horizontal lines above and below band of small upright solid triangles; farther below, three vertical lines.
Fabric: Sandy, semi-coarse, medium hardness, with moderate mica and black inclusions, air pockets (<0.4 cm). Exterior and interior surfaces 10 YR 6/6 (brownish yellow); core as surface, with central areas 10 YR 6/2 (light brownish gray).
Comparanda: **9/48** (P048), Bejko forthcoming: Kamenicë Q771, eighth–seventh century (tumulus fill); Q2163, eleventh–ninth century (dedication deposit).

9/236 (P059), **Fig. 9.27** (Drawing 13.5; Photos 979–992)
Carinated Shoulder Fragments.
Topsoil and Tumulus Fill (Units 2.002 and 2.066).
PH (max): 2.5 cm; PW (max): 4.1 cm; Th 0.8 cm.
Two joining fragments (found in different stratigraphic units) from carinated shoulder of small or mid-sized vessel, with bit of lower handle attachment preserved. Surfaces smoothed but very worn, damaged in places.
Fabric: Semi-coarse, with abundant mica and small black (0.2 cm) and red (0.6 cm) inclusions, occasional white inclusions. Surfaces and core 10 YR 2/1 (black), small patch exterior 2.5 YR 5/6 (red).
Comparanda: **9/162** (P079).

9/237 (P209), **Fig. 9.61** (Photos 2402–2403)
Wall Fragment.
Topsoil (SU 1.070).
PL (max): 3.6 cm; PW (max): 2.3 cm; Th: 0.8 cm.
Small fragment from wall of small or mid-sized vessel. Surfaces completely covered by heavy encrustation; unusual lines scratched on interior surface (through encrustation), roughly in shape of ligature Æ, continuing beyond break.
Fabric: Semi-coarse, with abundant mica and white and red inclusions. Surfaces 7.5 YR 4/3 (brown), core 10 YR 3/1 (very dark gray).

9/238 (P028), **Fig. 9.27** (Drawing 5.5; Photos 912–915)
Wall Fragment.
Interface Topsoil/Tumulus Fill (SU 4.011).
PH (max): 3.7 cm; PW (max): 3.4 cm; Th: 0.9–1.5 cm.

Single wall fragment from mid-sized vessel, preserving part of shoulder with lower portion of vertical strap handle attachment. Surfaces smoothed, but damaged; heavy encrustation.
Fabric: Semi-coarse, moderate mica, many black inclusions (0.2 cm) and occasional white inclusions. Exterior 2.5 YR 5/8 (red), core and interior 2.5 YR 3/1 (dark reddish gray).

9/239 (P070), **Fig. 9.27** (Drawing 16.4; Photos 793–796)
Shoulder Fragment.
Tumulus Fill (SU 3.064).
PL (max): 3.7 cm; PW (max): 4.6 cm; Th 0.4–0.6 cm.
Small fragment from shoulder of medium-size vessel, broken just above turn to neck; rim type uncertain. Surfaces smoothed, exterior perhaps once burnished; small patch of encrustation on exterior, scattered elsewhere.
Fabric: Semi-coarse, with abundant mica and black and white inclusions.
Exterior surface 7.5 YR 5/4 (brown), interior 5 YR 4/3 (reddish brown), core GLEY 1 2.5/N (black).

9/240 (P207), **Fig. 9.27** (Drawing 46.5; Photos 2446–2447)
Neck and Shoulder Fragment.
Topsoil (SU 2.002).
PL (max): 4.7 cm; PW (max): 4.2 cm; Th 0.6–0.9 cm.
Neck and shoulder fragment from mid-sized vessel with S-profile below convex shoulder and below everted rim. Surfaces smoothed, exterior perhaps once burnished; heavy encrustation and modern damage to both surfaces.
Fabric: Semi-coarse, with abundant mica and black and white inclusions. Surfaces 10 YR 6/3 (pale brown), core GLEY 1 5/1 (greenish gray).

9/241 (P034), **Figs. 9.27, 9.61** (Drawing 7.2; Photos 521–523, 549)
Shoulder Fragments.
Topsoil (SU 8.022).
PL (max): 5.3 cm; PW (max): 4.0 cm; Th: 0.5–0.9 cm.
Two joining fragments from shoulder of mid- to large-sized vessel, probably closed in shape. Surfaces smoothed, now very worn, exterior painted; heavy encrustation on exterior surface and fractures.
Matt-painted decoration: Lower portion of pendent triangle, solid, hatched, or cross-hatched (drawing more clear than sherd).
Fabric: Soft, sandy semi-coarse, with abundant mica sparkles, occasional black inclusions and air pockets. Exterior and interior surfaces 2.5 YR 5/8 (red), core 2.5 YR 6/0 (gray).
Comparanda: Korkuti 1971: pl. IX for variety of M/P triangles—dated eleventh–ninth centuries BC.

9/242 (P222), **Fig. 9.27** (Drawing 83.4; Photos 2427–2428)
Shoulder Fragments.
Tumulus Fill (SU 4.204).
PL: 3.7 cm; PW 4.6 cm; Th 0.4–0.6 cm.
Two joining fragments from convex upper shoulder of medium-size or large vessel, broken just above turn to neck or rim; precise shape and orientation uncertain. Surfaces smoothed and worn, with damage to exterior, perhaps at point of attachment for handle.
Fabric: Sandy, semi-coarse, with abundant mica and white and black inclusions. Surfaces 2.5 YR 5/6 (red), core GLEY 1 3/N (very dark gray).
Comparanda: Uncatalogued P426.

9/243 (P243), **Fig. 9.27** (Drawing 86.4; Photos 2548–2549)
Wall Fragment.
Tomb LVI (43) Fill (SU 4.269), adhered to ribs.
PL (max): 7.5 cm; PW (max): 5.3 cm; Th: 1.0–1.3 cm.
Single wall fragment from large vessel, perhaps from lower body, broken just above swelling for base? Surfaces once smoothed, now uneven, worn, damaged; light encrustation.
Fabric: Heavy, sandy, semi-coarse, with very abundant mica, abundant small black, white, and gray inclusions, moderate black pebbles (<0.3 cm), occasional air pockets. Surfaces and core 5 YR 5/6 (yellowish red).

9/244 (P288), **Figs. 9.27, 9.61** (Drawing 76.1; Photos 2807–2808)
Wall Fragments.
Topsoil (SU 4.201).
PH (max): 12.5 cm; PW (max): 17.0 cm; Th: 0.9–1.3 cm; D (max, ext): ca. 21–22 cm.
Five joining fragments (from both ancient and modern breaks) preserving portion of globular lower body and rounded shoulder of large semi-coarse vessel, broken just above small ridge (or indentation?) at base of neck; original shape uncertain,

probably open. Exterior surface smoothed, perhaps once burnished, but now very worn and damaged; ridges visible on interior, perhaps once slipped; heavy encrustation on exterior and breaks, scattered on interior.
Fabric: Heavy, semi-coarse fabric, with abundant mica, moderate small black and white inclusions and air pockets (<0.8 cm). Surfaces 5 YR 6/8 (reddish yellow), core 5 YR 4/1 (dark gray).

9/245 (P423), **Figs. 9.27, 9.61** (Drawing 98.4; Photos 3772–3773)
Wall Fragments.
Tumulus Fill (SU 4.034).
PL (max): 4.2 cm; PW (max): 9.8 cm; Th: 1.2 cm; D (ext): ca. 25 cm.
Two joining fragments (from modern break) from lower body of large vessel, perhaps closed in shape. Surfaces smoothed, exterior slipped and painted; heavy encrustation on interior.
Matt-painted decoration: Middle portions of two pendent triangles—on left crossed-hatched, on right receding hatched.
Fabric: Semi-coarse; mica, many black pebbles (<0.5 cm), occasional white inclusions. Exterior surface 2.5 YR 5/8 (red), interior 5 YR 6/8 (reddish yellow), core 5 YR 5/1 (gray).

9/246 (P249), **Figs. 9.27, 9.61** (Drawing 52.2; Photos 2584–2585)
Wall Fragment.
Tumulus Fill (SU 1.240).
PH (max): 5.8 cm; PW (max): 10.5 cm; Th: 0.9–1.1 cm.
Single fragment (plus surface chips) from lower body of large vessel of uncertain shape. Surfaces smoothed, exterior painted, perhaps slipped; scattered encrustation.
Matt-painted decoration: At left, three roughly vertical lines, perhaps from hatched, elongated pendent triangle; at right, lower portion of pendent receding, hatched triangle.
Fabric: Semi-coarse, with abundant mica and red (grog?) inclusions (<0.4 cm). Surfaces 10 R 5/8 (red), core 10 R 4/1 (dark reddish gray).
Comparanda: Prendi 1966:pl. XVIII, motif, p. 267).

9/247 (P196), **Figs. 9.27, 9.62** (Drawing 52.1; Photo 2502)
Neck Fragments.
Tumulus Fill and Topsoil (Units 1.279, 4.201).
PL: 11.3; PW (max): 7.1 cm; Th: 1.3 cm.
Two joining fragments from flaring neck of large vessel, broken below rim and just above shoulders, probably closed in shape. Surfaces smoothed and slipped, exterior painted and damaged (in antiquity).
Matt-painted decoration: Nearly rectangular cross-hatched area, bound below by two irregular horizontal bands; upper portion not preserved.
Fabric: Semi-coarse, with abundant mica, black pebbles (<0.2 cm), and white (<0.2 cm) inclusions, few air pockets (<0.4 cm). Surfaces and core 5 YR 7/6 (reddish yellow).
Comparanda: Korkuti 1971: pl. IX; cf. also Andrea 1985: pl. I, Tomb 5, 1, p. 262 (sherd from grave fill, twelfth–eighth centuries).

9/248 (P234), **Figs. 9.28, 9.62** (Drawing 48.2; Photos 2582–2583)
Neck Fragment.
Tumulus Fill (SU 1.240).
PH (max): 6.5 cm; PW (max): 4.6 cm; Th: 1.3 cm.
Fragment from nearly vertical, flaring neck of large vessel. Surfaces smoothed, exterior slipped and painted; heavy encrustation.
Matt-painted decoration: Checkerboard pattern of lozenges covering entire exterior.
Fabric: Heavy, with abundant mica, black and red (grog) inclusions, moderate air pockets. Surfaces 2.5 YR 5/8 (red), core 2.5 YR 4/1 (dark reddish gray).
Comparanda: Korkuti 1971: pl. IX); Hammond 1972: fig. 14e.

9/249 (P170), **Figs. 9.28, 9.62** (Drawing 53.2; Photos 2493–2496)
Wall Fragment.
Tumulus Fill (SU 1.070).
PL (max): 8.3 cm; PW (max): 14.3 cm; Th: 1.1–1.6 cm.
Single wall fragment from lower body of large or very large matt-painted vessel, probably closed in shape. Exterior surface smoothed, slipped, burnished, and painted; interior uneven; scattered encrustation.
Matt-painted decoration: Two groups of converging vertical lines (seven left and six right), probably from hatched pendent triangles.
Fabric: Semi-coarse, abundant mica and black inclusions, white limestone appearing on exterior (<0.6 cm). Exterior surface 7.5 YR 6/8 (reddish yellow),

interior surface 2.5 YR 6/8 (red), core 5 YR 6/1 (gray).

9/250 (P429), **Figs. 9.28, 9.62** (Drawing 98.1; Photos 3994–3995)
Wall Fragment.
Tumulus Fill (SU 4.130).
PH (max): 2.9 cm; PW (max): 2.5 cm; Th: 0.7–0.9 cm.
Small wall fragment from vessel of indeterminate shape and size. Surfaces smoothed, exterior painted, perhaps slipped; scattered encrustation.
Matt-painted decoration: Upper left-hand portion of hatched pendent triangle, with upper horizontal line above two vertical lines and diagonal to left.
Fabric: Semi-coarse, with abundant mica, small white and black (0.2 cm) inclusions. Exterior 7.5 YR 6/4 (light brown), interior 2.5 YR 5/8 (red).

9/251 (P105), **Fig. 9.28** (Drawing 40.6; Photos 2202–2203)
Base and Lower Body Fragments.
Topsoil (SU 1.001).
PH (max): 3.7 cm; PW (max): 6.3 cm; PTh: 0.7–1.2 cm; D (base): ca. 10–12 cm.
Single fragment from lower body and slightly raised and hollowed base of medium-size or large vessel; vessel floor concave, only minimally preserved; walls of lower body flare out slightly from base. Exterior surface smoothed, burnished, and painted, perhaps slipped; interior very worn; both surfaces damaged, partially missing.
Matt-painted decoration: Two pairs of roughly parallel vertical lines, perhaps elongated apices of pendent triangles.
Fabric: Semi-coarse; abundant mica, small black and white inclusions, and air pockets (<0.4 cm, especially in interior). Surfaces 5 YR 5/8 (yellowish red), core 5 YR 6/1 (gray).

9/252 (P174 = **15/9**), **Fig. 9.28** (Drawing 106.4, Photos 2507–2508)
Base Fragment.
Tumulus Fill (SU 1.070).
PH (max): 7.5 cm; PW (max): 11.6 cm; Th (base): 1.9 cm; Th (wall): 0.8–1.3 cm; D (base): ca. 11 cm.
Single fragment from large vessel, preserving small portion of flat, raised base, together with concave inside floor and rise of flaring wall. Surfaces smoothed, covered with heavy encrustation; bits of bitumen over encrustation and breaks (i.e., not indications of use).
Fabric: Semi-coarse, heavy, red fabric, difficult to see under encrustation: moderate mica, fine black and white inclusions, few small air pockets (<0.2 cm). Surfaces 5 YR 6/8 (reddish yellow), core 5 YR 7/2 (pinkish gray).
Comparanda: **9/259** (P283), **9/318** (P354), **9/319** (P309): similar shape, coarse.

9/253 (P335), **Fig. 9.28** (Drawing 90.2; Photos 3215–3216)
Lower Wall Fragments.
Topsoil (SU 2.002).
PL (max): 9.6 cm; PW (max); 5.3 cm; Th: 1.0–2.0 cm.
Large fragment from lower body of large or very large vessel, broken just above foot of raised, hollowed base and concave vessel floor. Surfaces smoothed, exterior slipped and painted; heavy encrustation.
Matt-painted decoration: Very poorly preserved traces of vertical lines, perhaps lower portions of pendent triangles.
Fabric: Semi-coarse, with abundant mica and red inclusions (grog; <0.3 cm), occasional small white inclusions. Surfaces 2.5 YR 5/8 (red), core 5 YR 5/1 (gray).
Comparanda: **9/254** (P430).

9/254 (P430), **Figs. 9.28, 9.62** (Drawing 99.2; Photos 3774–3775)
Base and Lower Body Fragments.
Tumulus Fill (SU 5.558).
PL (max): 12.4 cm; PW (max): 7.2 cm; Th: 1.0–2.2 cm.
Two joining fragments from lower body of very large vessel, probably closed in shape, broken just above foot of raised, hollowed base; part of concave vessel floor and raised underside preserved. Exterior surface smoothed and painted, now very worn and damaged, interior uneven; scattered encrustation.
Matt-painted decoration: Very poorly preserved; on wall, lower portions of two pendent triangles, perhaps cross-hatched; around base, perhaps two horizontal bands.
Fabric: Semi-coarse to coarse, abundant mica, many black inclusions, occasional white pebbles (<0.7 cm). Surfaces 5 YR 6/8 (reddish yellow), core 5 YR 5/1 (gray).
Comparanda: **9/253** (P335).

9/255 (P302), **Fig. 9.62** (Photos 3098–3100)
Projection/Lug.
Tumulus Fill (SU 4.286).
PH (max): 1.4 cm; PW (max): 1.7 cm; Th: 0.9 cm.
Rounded projection, fully preserved, with no trace of originally attached wall or vessel (probably small). Surfaces smoothed, but worn; heavy encrustation.
Fabric: Semi-coarse to fine, abundant mica, occasional black and white inclusions. Surface and core 10 YR 4/1 (dark gray).
Comparanda: **9/256** (P117), **9/257** (P214).

9/256 (P117), **Fig. 9.28** (Drawing 38.3; Photos 2152–2153)
Projection/Lug.
Tumulus Fill (SU 2.202).
PH (max): 2.1 cm; PW (max): 2.8 cm; Th: 2.2 cm.
Rounded projection, fully preserved, with no trace of originally attached wall or vessel (probably medium to large in size). Surfaces mostly smoothed but worn, rough underside; light encrustation. Surfaces smoothed and uneven.
Fabric: Semi-coarse, with abundant mica and black (<0.2 cm) and red inclusions. Surface 5 YR 4/4 (reddish brown), core 2.5 YR 2/1 (black).
Comparanda: **9/255** (P302), **9/257** (P214).

9/257 (P214), **Fig. 9.62** (Photos 2615–2616)
Projection/Lug.
Tumulus Fill (SU 1.279).
PH (max): 2.6 cm; PW (max): 3.4 cm; Th: 1.8 cm.
Triangular projection, fully preserved, with no trace of originally attached wall or vessel. Exterior surface smoothed but worn, with finger impressions around base; interior (underside) rough; heavy encrustation.
Fabric: Semi-coarse, with abundant mica, occasional black and white inclusions. Surface and core 10 YR 7/6 (yellow).
Comparanda: **9/255** (P302), **9/256** (P117).

9/258 (P269), **Fig. 9.62** (Photos 2684–2685)
Handle Fragment?
Tumulus Fill (SU 1.240).
PL (max): 2.6 cm; PW (max): 2.3 cm; Th: 1.2 cm.
Single fragment of unusual shape: flat, with two finished edges, one straight, the other rounded, perhaps from pre-firing piercing; probably a handle. Surfaces smoothed, very worn, scattered encrustation.
Fabric: Red, semi-coarse, sandy, with abundant mica and black inclusions, moderate air pockets (<0.3 cm), occasional white inclusions. Surface and core 5 YR 7/8 (reddish yellow).

Coarse Fabric

9/259 (P283 = **15/12**), **Figs. 9.29, 9.63, 15.4** (Drawing 92.1; Photos 2738–2742 [sherds with bitumen], 2763–2768 [sherds with dirt and bitumen]; 3219– 3225a [full vessel, post-conservation]; 3361–3363 [full vessel with bitumen])
Biconical Amphora with Vertical Neck, Coarseware Type 1.
Ceramic Deposit within Tumulus Fill (SU 4.280), near Tomb LXIII (35).
H: 41 cm; H (neck): 14 cm; Th (wall): 0.9–1.1 cm; Th (base): 1.7 cm; D (max, at handles): 33 cm; D (neck ext): 18 cm; D (rim ext): 20 cm; D (base): ca. 8 cm; D (handles): ca. 3 cm (set 8 cm apart).
Large amphora, nearly completely reconstructed from many fragments (preserving complete profile, complete neck, ca. 75% of body). Biconical body flares from narrow toe-like base with rough and slightly rounded underside to rounded shoulder and tall vertical neck with everted rim and rounded lip. Attachments for horizontal ring handle set at point of maximum diameter, just beyond midpoint of body; attachments for second handle probably appeared on opposite side, now lost.
Surfaces smoothed, exterior highly burnished, now worn. Black smears or streaks, apparently bitumen, applied to exterior on upper shoulder and lower neck. Vessel found smashed, containing soil with bitumen flecks and patches; pool of cooled bitumen conforms to shape of inside floor.
Fabric: Dark coarseware, with abundant mica (gold in color) and very fine white and black inclusions, moderate larger white inclusions (pebbles <0.8 cm). Exterior surface mottled 7.5 YR 4/2–4/4 (brown to dark brown), interior surface 2.5 YR 5/4 (reddish brown), core 7.5 YR 2/0 (black).

9/260 (P024), **Fig. 9.63** (Drawing 5.1; Photos 608–611)
Vertical Strap Handle Fragment.
Tumulus Fill (SU 1.007).
PL (max): 3.1 cm; PW (max): 3.2 cm; Th: 1.0–1.1 cm.
Single fragment of vertical strap handle, roughly rectangular in section, preserving part of curve,

but no attachments; crudely made, with rough surfaces, now very worn, with scattered encrustation.

Fabric: Coarse red fabric, with dark gray core and abundant mica and red, white, and black inclusions, some quite large (1.0 cm). Surface 2.5 YR 4/6 (red) to 4/2 (weak red), core 2.5 YR 2.5/1 (reddish black).

Comparanda: **9/262** (P052), **9/264** (P061), **9/272** (P400); cf. also semi-coarse **9/162** (P079), **9/177** (P037), **9/191** (P399), **9/197** (P143), **9/198** (P320).

9/261 (P036), **Fig. 9.30** (Drawing 8.1; Photos 989–992)

Vertical Strap Handle Fragment.

Tumulus Fill (SU 2.023).

PL (max): 3.5 cm; PW (max): 4.7 cm; W (handle): 2.8 cm; Th: 1.4 cm.

Fragment of vertical strap handle, ovoid in section, broken just above attachment, probably above rim. Surfaces rough, with scattered encrustation.

Fabric: Coarse, red fabric, with abundant inclusions: mica, red, white, and black inclusions, quite large pebbles (1.0 cm). Surface 5 YR 4/3 (reddish brown), core 5 YR 3/1 (very dark gray).

9/262 (P052), **Fig. 9.30** (Drawing 12.2; Photos 592–595)

Wall and Vertical Strap Handle Fragment.

Tumulus Fill (SU 1.007).

PH (max): 5.2 cm; PW (max): 3.8 cm; PW (handle): 2.6 cm; Th (max): 3.0 cm; Th (handle): 1.6 cm; Th (wall): 1.2 cm.

Single fragment preserving wall with attachment and lower portion of vertical strap handle, irregular ovoid in shape; wall convex, but precise orientation uncertain. Surface smoothed, not burnished.

Fabric: Coarse, sandy fabric, abundant mica, black and white inclusions. Exterior 10 YR 5/4 (yellowish brown), core 7.5 YR 3/1 (very dark gray).

Comparanda: **9/260** (P024), **9/264** (P061), **9/272** (P400); cf. also semi-coarse **9/177** (P037), **9/191** (P399), **9/197** (P143), **9/198** (P320).

9/263 (P432) Not Illustrated

Wall/Handle Fragment.

Ceramic Deposit (SU 2.227) within Tomb LVIII (37) Fill.

PL: 5.1 cm; PW: 4.1 cm; Th: 1.6 cm.

Wall fragment preserving body and lower attachment of vertical strap handle, oval in section; shape uncertain, but perhaps closed, with wall roughly vertical. Surfaces smoothed, but worn, cracked.

Found together with 25 non-joining pieces of similar fabric.

Fabric: Coarse, sandy, abundant mica, many angular white (0.4–06 cm), black (0.5 cm), and occasional red (0.4 cm) pebbles. Many small black, white, and red inclusions. Surfaces 10 R 4/6 (red), core 2.5 YR 3/1 (dark reddish gray).

9/264 (P061), **Figs. 9.30, 9.63** (Drawing 15.5, Photos 632–635)

Vertical Strap Handle Fragments.

Tumulus Fill (SU 1.047).

PL (max): 7.3 cm; PW (max): 4.0 cm; PW (handle): 2.4 cm; Th: 1.0 cm.

Seven joining fragments (from modern break) of convex body with attachment and lower portion of vertical strap handle, oval in section, very poorly preserved. Exterior surfaces smoothed, interior quite rough; no evidence of burnishing.

Fabric: Dark, coarse; abundant black inclusions (pebbles up to 0.8 cm), moderate white inclusions. Dark core 7.5 YR 3/1 (very dark gray) sandwiched between 5 YR 5/6 (yellowish red) layer just beneath 5 YR 5/4 (reddish brown) surface.

Comparanda: **9/260** (P024), **9/262** (P052), **9/272** (P400); cf. also semi-coarse **9/177** (P037), **9/191** (P399), **9/197** (P143), **9/198** (P320).

9/265 (P136), **Figs. 9.30, 9.63** (Drawing 41.2; Photos 4013–4015)

Flaring Spur Handle Fragments.

Ceramic Deposit within Tumulus Fill (SU 4.206—includes also **9/184, 9/274** [P099, P135]).

PL (max): 10.9 cm; PW (max): 5.1 cm; W (handle): 3.2–4.3 cm; W (spur tip): 4.1 cm; Th (wall): 0.7–1.2 cm; Th (max): 3.0 cm; H (aperture): 3.5 cm; W (aperture): 2.5 cm; D (at junction of body and handle): ca. 15 cm.

Flaring spur handle, nearly completely reconstructed from some 15 fragments, broken from large coarseware vessel at both shoulder and rim; profile partially preserved, showing closed lower body curving sharply into shoulder, then up and out to slightly flaring rim. Precise orientation uncertain, but handle rises nearly vertically from shoulder, obliquely from rim, joining sharply at flattened rectangular top, with flaring corners and slight surface indentations just beneath top; both upper and lower elements roughly rectangular in section. Surfaces smoothed, with faint parallel lines (tooling or burnishing marks) and small areas of burnishing visible in places; chips missing from many areas, surface especially worn at left exterior edge, light encrustation scattered, especially at handle tip. No plastic or painted decoration.

Fabric: Crumbly coarseware, many inclusions, red, white (including quartz), black, mostly very fine (<0.2 cm). Surfaces and core 5 YR 3/3 (dark reddish brown) to 2.5/1 (black).

Publication: Papadopoulos, Bejko, and Morris 2007: fig. 28.

9/266 (P128), **Fig. 9.30** (Drawing 37.1; Photos 2196–2197)

Circular Handle Fragment.

Tumulus Fill (SU 1.039).

PH (max): 4.5 cm; PW (max): 2.8 cm; Th: 2.1 cm.

Single fragment preserving outer edge and bit of inner edge of circular, horizontal handle, broken at both sides before join with body; generally round in section, flatter underneath. Original shape and orientation uncertain, surface smoothed.

Fabric: Coarse, medium hardness, with very abundant inclusions (<0.5 cm): white, black, and quartz pebbles; grog. Exterior 7.5 YR 5/4 (brown), core 7.5 YR 3/1 (very dark gray).

9/267 (P363), **Fig. 9.63** (Photos 3234–3235)

Handle Fragment.

Topsoil (SU 2.002).

PL (max): 3.5 cm; PW (max): 3.1 cm; Th: 2.3 cm.

Single fragment of coarse ceramic, probably from horizontal semi-circular handle.

Surface smoothed, preserving two roughly parallel surfaces, as well as circular inner piercing (ca. 2.5 cm diameter) and bit of circular outer edge (ca. 9.5-cm diameter); less than one-fourth of circle preserved.

Fabric: Very coarse, dark, sandy fabric, with abundant mica, black and white pebbles (<0.3 cm), moderate red inclusions (<0.4 cm). Surface 7.5 YR 5/4 (brown), core 7.5 YR 3/1 (very dark gray).

Comparanda: For more complete horizontal handles of similar shape, cf. Kamenicë P178, P179 (unpublished, from topsoil); cf. also similar horizontal handle from tripod cauldron (Papadopoulos 2005: pl. 389, fig. 179e, T123–125), and spindle whorls for similar shape, fabric (but smaller size [e.g., **11/10** [SF045]).

9/268 (P219), **Fig. 9.64** (Photos 2634–2635)

Vertical Strap Handle Fragment.

Tumulus Fill (SU 4.207).

PH (max): 2.3 cm; W (handle): 3.3 cm; PW (max): 4.5 cm; Th (handle): 1.3 cm; Th (max): 3.1 cm.

Single fragment preserving attachment and lower portion of vertical strap handle, roughly rectangular in section; interior surface lost. Surfaces smoothed, scattered encrustation.

Fabric: Coarse, red fabric, with dark gray core and abundant mica and red, white, and black pebble inclusions (<0.3 cm); one void (from pebble or air pocket) nearly 1 cm. Surface 5 YR 5/8 (yellowish red), core 5 YR 3/1 (very dark gray).

9/269 (P118), **Fig. 9.30** (Drawing 38.6, Photos 2188–2189)

Vertical Strap Handle Fragment.

Tumulus Fill (SU 2.202).

PL (max): 3.9 cm; PW (max): 5.5 cm; Th: 1.5 cm.

Single fragment of large vertical strap handle, flattened oval in section, preserving no curve or attachments. Surfaces smoothed, cracked.

Fabric: Coarse, with abundant mica and red, white, and black inclusions (including pebbles <0.4 cm). Surface 5 YR 5/8 (yellowish red) to 7.5 YR 5/4 (brown), with spots dark like core, 5 YR 3/1 (very dark gray).

9/270 (P179), **Fig. 9.64** (Photos 2349–2350)

Vertical Strap Handle Fragment.

Topsoil (SU 2.002).

PL (max): 3.0 cm; PW (max): 2.7 cm; Th: 1.0 cm.

Single fragment of vertical strap handle, flattened oval in section, preserving no curve or attachments. Upper surface smoothed, underside rough.

Fabric: Coarse to semi-coarse fabric, with abundant inclusions: red, white (including quartz), and black inclusions (<0.2 cm). Surface 5 YR 4/3 (reddish brown), with black patches like core—5 YR 2/1 (black).

9/271 (P392), **Fig. 9.30** (Drawing 95.4; Photos 3482–3483)

Vertical Strap Handle Fragment.

Tumulus Fill (SU 5.536).
PL (max): 2.5 cm; PW (max): 4.5 cm; Th: 1.2 cm.
Two joining fragments of vertical strap handle, flattened oval in section, preserving no curve or attachments. One surface smoothed, the other rough; both covered with fine cracks.
Fabric: Dark, coarse fabric, with abundant red, gray, and white inclusions, moderate pebbles (<0.5 cm) and mica, few air pockets (<0.3 cm). Surfaces and core 5 YR 3/1 (very dark gray).

9/272 (P400), **Fig. 9.64** (Photos 3725–3726)
Vertical Strap Handle and Body Fragment.
Tumulus Fill (SU 5.536).
PL (max): 4.7 cm; PW (max): 3.7 cm [=W(handle)]; Th: 1.7 cm.
Single fragment of exterior wall with attachment and lower portion of vertical strap handle, ovoid in section and diverging greatly from body (perhaps neck of amphora?); interior wall very worn, not fully preserved. Surfaces smoothed, now worn, scattered encrustation.
Fabric: Coarse, dark, sandy fabric, with moderate mica and very small black and white inclusions. Exterior surface 7.5 YR 6/6 (reddish yellow), interior/core 7.5 YR 2.5/1 (black).
Comparanda: **9/260** (P024), **9/262** (P052), **9/264** (P061); cf. also semi-coarse **9/177** (P037), **9/191** (P399), **9/197** (P143), **9/198** (P320).

9/273 (P332), **Fig. 9.64** (Photos 3153–3154)
Vertical Strap Handle Fragment.
Tumulus Fill (SU 2.431).
PL (max): 3.8 cm; PW (max): 5.0 cm; Th: 1.5 cm.
Single fragment of vertical strap handle, roughly triangular in section (slightly rounded exterior, central ridge on interior), broken just above lower attachment. Exterior surface smooth, perhaps once burnished, interior rough; both surfaces worn, scattered encrustation.
Fabric: Dark, coarse fabric, with abundant mica, fine black, white, and gray inclusions, few pebbles and air pockets (<0.3 cm). Surface and core 10 YR 3/2 (very dark grayish brown).
Comparanda: Typical of northern Mati region (Korkuti, personal communication).

9/274 (P135), **Fig. 9.30, 9.64** (Drawing 39.9; Photos 2181–2183)
Vertical Strap Handle Fragment.
Ceramic Deposit within Tumulus Fill (SU 4.206—includes also **9/184, 9/265** [P099, P136]).
PL (max): 4.7 cm; PW (max): 4.0 cm; Th: 0.8–1.6 cm.
Single fragment of vertical strap handle, flattened oval in section, with external central ridge, broken just above lower (body) attachment and beyond sharp bend to horizontal; no vessel walls preserved. Surfaces smoothed, exterior once burnished; scattered encrustation.
Fabric: Dark, coarse fabric, with abundant mica, black, red, and white inclusions, fine cracks in core and side surface. Exterior mottled, but mostly 10 YR 2/1 (black) where not covered by encrustation; core 2.5 YR 3/2 (dusky red) to 4/3 (reddish brown).
Comparanda: **9/275** (P238; similar fabric and shape); cf. also **9/110** (P103; similar shape, smaller, FD).

9/275 (P238), **Fig. 9.30** (Drawing 61.5; Photos 2627–2629)
Vertical Strap Handle Fragment.
Topsoil (SU 4.201).
PL (max): 6.0 cm; PW (max): 4.2 cm; Th: 1.2–1.8 cm.
Single fragment of vertical strap handle, rectangular in section, with external central ridge, broken mid-handle below, beyond 90° turn to horizontal above; no attachments preserved. Surfaces smoothed, once burnished exterior, now worn.
Fabric: Dark reddish, coarse fabric, with abundant mica, red and white inclusions (<0.5 cm), moderate black inclusions, small air pockets and cracks. Exterior 2.5 YR 5/6 (red), interior 2.5 YR 4/3 (reddish brown); core not visible.
Comparanda: **9/274** (P135; similar fabric and shape); cf. also **9/110** (P103; similar shape, smaller, FD).

9/276 (P365), **Fig. 9.64** (Photos 3310–3311)
Vertical Strap Handle Fragment.
Tumulus Fill (SU 2.380).
PH (max): 7.3 cm; PW (max): 7.6 cm; Th (wall): 1.1–1.4 cm; Th (max): 2.7 cm; Th (handle): 2.1 cm.
Single wall fragment of very large coarseware vessel, with attachment for vertical strap handle (or ledge lug?), flattened oval in section. Surfaces smoothed, exterior badly worn, encrusted.
Fabric: Coarse, crumbly fabric, with abundant small black, white, and red (grog) inclusions (<0.2 cm),

moderate mica. Exterior surface 2.5 YR 5/8 (red), interior 5 YR 5/3 (reddish brown), core 5 YR 3/1 (very dark gray).

9/277 (P394), **Fig. 9.30** (Drawing 94.5; Photos 3471–3472)
Vertical Strap Handle Fragment.
Tumulus Fill (SU 6.541).
PH (max): 3.9 cm; PW (max): 5.6 cm; Th: 1.4 cm.
Single fragment of large vertical strap handle, rectangular in section, preserving 90° bend to horizontal, but no attachments. Surface smoothed.
Fabric: Coarse, red fabric, abundant mica, black pebbles, and white inclusions (<0.5 cm), few small air pockets in exterior. Exterior 2.5 YR 5/6 (red), core 2.5 YR 3/1 (dark reddish gray).
Comparanda: **9/41** (**P275**; same size and shape, burnished FL).

9/278 (P376), **Fig. 9.64** (Photos 3731–3732)
Coarseware Rim Fragment.
Tumulus Fill (SU 5.279).
PH (max): 3.4 cm; PW (max): 4.2 cm; Th: 0.9 cm.
Single fragment preserving neck and rim of open coarseware vessel; neck flares out to flat rim with thick external projection. Exterior surface rough, interior smoothed, core covered by encrustation.
Fabric: Coarse, sandy fabric, abundant mica and black pebble inclusions (<0.6 cm), few air pockets in exterior (one 0.7 cm). Surface 5 YR 5/6 (yellowish red).
Comparanda: **9/279** (P003), **9/280** (P217), **9/290** (P224 + P282 + P411), **9/291** (P244), **9/293** (P022), **9/294** (P026), **9/298** (P291), **9/299** (P145), **9/300** (P065).

9/279 (P003), **Figs. 9.31**, **9.64** (Drawing 2.3; Photos 559–562)
Rim and Neck Fragment.
Surface Collection (2003).
PH (max): 5.1 cm; PW (max): 5.0 cm; Th: 0.9–1.1 cm.
Single fragment preserving straight flaring neck and wide flat rim with rounded external projection; outward angle to upper body just preserved. Surfaces smoothed, now very worn, especially exterior; encrustation on interior surface.
Fabric: Soft, coarse fabric, very porous, with abundant air pockets (mostly very small, but up to 0.5 cm) and very small black, white, and red (grog) inclusions; moderate mica; fine cracks on exterior. Exterior surface 7.5 YR 4/3 (brown), interior surface 10 YR 6/3 (pale brown); core as surface, but darkening at lower break to 10 YR 2/1 (black).
Comparanda: **9/298** (P291), **9/299** (P145), **9/300** (P065); cf. also **9/290** (P224 + P282 + P411), **9/291** (P244): similar shape, slightly darker, less porous fabric.

9/280 (P217), **Fig. 9.64** (Photos 2830–2831)
Rim Fragments.
Tumulus Fill (SU 2.202).
PH (max): 2.9 cm; PW (max): 3.9 cm; Th: 1.5 cm.
Two joining fragments preserving rounded lip, with external projection; wall appears to thicken below lip, but difficult to ascertain original shape—perhaps no flaring neck. Surfaces smoothed, covered with heavy encrustation.
Fabric: Coarse, sandy fabric, with abundant black and white inclusions, moderate mica. Surface 2.5 YR 5/8 (red), core not fully visible, but appears similar to surface.
Comparanda: **9/278** (P376), **9/279** (P003), **9/290** (P224 + P282 + P411), **9/291** (P244), **9/293** (P022), **9/294** (P026), **9/298** (P291), **9/299** (P145), **9/300** (P065).

9/281 (P239), **Fig. 9.31** (Drawing 66.4; Photos 2654–2655)
Rim Fragment.
Topsoil (SU 1.278).
PL (max): 4.6 cm; PW (max): 4.0 cm; Th: 0.9 cm.
Single fragment from large coarseware vessel, with nearly vertical neck, flaring out to rounded lip. Surfaces smoothed, with scattered encrustation.
Fabric: Coarse, sandy, with abundant mica and small black inclusions, occasional larger black inclusions (<0.5 cm) and very small red and white inclusions, few air pockets (<0.4 cm). Surfaces 2.5 YR 5/6 (red), core 2.5 YR 5/1 (reddish gray).
Comparanda: **9/301** (P185), **9/282** (P397); cf. also uncatalogued P184.

9/282 (P397), **Fig. 9.31** (Drawing 114.4, Photos 3785–3786)
Rim and Neck Fragment.
Tumulus Fill (SU 5.536).
PH (max): 5.4 cm; PW (max): 5.2 cm; Th: 1.1–1.3 cm; D (rim ext): ca. 15 cm.

Single fragment of large vessel with nearly vertical neck curving out slightly to rim with flat, wide lip with irregular, rounded edges. Surface smoothed, now worn, with heavy encrustation, especially on interior surface and core.

Fabric: Dark, coarse, sandy, with abundant mica, small black, white, and red (grog?) inclusions; few very small air pockets. Surface mostly 7.5 YR 4/3 (brown), with dark patch 2.5/1 (black); core mostly covered, but similar.

Comparanda: **9/281** (P239), **9/301** (P185).

9/283 (P081), **Fig. 9.31** (Drawing 18.4; Photos 503–506)

Rim Fragment.

Tomb LXVIII (13) Fill (SU 2.102).

PH (max): 2.6 cm; PW (max): 4.4 cm; Th: 1.0–1.5 cm; D (rim ext): ca. 25 cm.

Single fragment from flaring rim, with flat lip, of large vessel, probably open in shape. Surfaces smoothed but very worn; ancient damage to outer lip; scattered encrustation.

Fabric: Coarse, very sandy, with abundant mica and very small black, white, and red (grog?) inclusions, occasional air pockets (<0.3 cm). Rim and exterior 2.5 YR 4/6 (red), interior 2.5 YR 4/2 (weak red) to 3/2 (dusky red), core 2.5 YR 3/1 (dark reddish gray).

Comparanda: **9/282** (P397; no raised lip), **9/284** (P227); cf. also semi-coarse **9/224** (P040), **9/225** (P053), **9/228** (P267).

9/284 (P227), **Fig. 9.31** (Drawing 62.1; Photos 2648–2649)

Rim Fragment.

Tomb LIX (38) Fill (SU 4.246).

PH (max): 3.0 cm; PW (max): 3.0 cm; Th: 0.8–0.9 cm.

Single fragment of wall, probably vertical, flaring out very slightly to flat lip; rounded interior, slightly projecting exterior. Surfaces smoothed, with rough horizontal lines visible on both sides; surfaces now worn, slight chipping from interior and rim surfaces, light encrustation over all surfaces, heavier on interior and core.

Fabric: Coarse, sandy, with abundant very small black inclusions, occasional mica; few very small air pockets (<0.1 cm). Surface color 10 YR 5/4 (yellowish brown) to 4/4 (dark yellowish brown), but largely hidden by encrustation.

Comparanda: **9/285** (P154); cf. also semi-coarse **9/224** (P040), **9/225** (P053), **9/228** (P267), uncatalogued P358.

9/285 (P154), **Fig. 9.31** (Drawing 85.3; Photos 2520–2521)

Rim Fragment.

Tumulus Fill (SU 1.070).

PH (max): 3.3 cm; PW (max): 2.3 cm; Th: 1.3 cm.

Single fragment from vertical upper wall or neck of large coarseware vessel, flaring out slightly to flat lip with very slight rounded exterior projection. Interior surface smoothed, exterior rough, with letter "A" incised by modern tool.

Fabric: Coarse, red fabric, sandy, with abundant mica, black and white pebble inclusions, few air pockets in exterior. Surface and core 2.5 YR 5/8 (red).

Comparanda: **9/284** (P227); cf. also semi-coarse **9/224** (P040), **9/225** (P053), **9/228** (P267), uncatalogued P358.

9/286 (P230), **Fig. 9.31** (Drawing 85.5; Photos 2552–2553)

Fragment.

Tumulus Fill (SU 4.204).

PL (max): 6.4 cm; PW (max): 5.8 cm; Th: 1.4 cm; D (mouth): ca. 5 cm.

Nearly flat fragment preserving portion of gently sloping upper wall (shoulder) and nearly circular mouth of large coarseware vessel, perhaps hole-mouthed jar; upper (exterior) surface and lip smoothed, underside rough, with excess clay beneath lip.

Fabric: Red coarseware, abundant mica and small black inclusions. Surface and core mostly 2.5 YR 5/8 (red).

Comparanda: **9/287** (P231), **9/288** (P176), **9/289** (P161); for hole-mouthed jar, see Mountjoy 1993: figs. 19, 21, 33, 77; Hochstetter 1984: Taf 2.4; cf. also Bejko forthcoming: Kamenicë Unit 292, Q635 (tumulus fill; finer fabric, painted).

9/287 (P231), **Fig. 9.31** (Drawing 86.3; Photos 2432–2433)

Fragment.

Tumulus Fill (SU 4.204).

PL (max): 6.0 cm; PW (max): 3.1 cm; Th (wall): 0.8 cm; Th (lip): 1.3 cm.

Single fragment preserving portion of downward sloping upper wall (shoulder), with bit of circular mouth, perhaps from hole-mouthed jar. Lip and surfaces smoothed, now worn; scattered encrustation.

Fabric: Semi-coarse, abundant mica, few very small white and black inclusions. Exterior 7.5 YR 6/6 (reddish yellow), core 7.5 YR 2.5/1 (black).

Comparanda: **9/286** (P230), **9/288** (P176), **9/289** (P161); for hole-mouthed jar, see **9/286** (P230) above.

9/288 (P176), **Fig. 9.64** (Photos 2658–2659)
Fragment.
Tumulus Fill (SU 1.070).
PL (max): 3.7 cm; PW (max): 3.6 cm; Th: 1.2 cm; D (mouth): ca. 5 cm.
Single fragment preserving portion of downward sloping upper wall (shoulder), with nearly circular mouth of large coarseware vessel, perhaps hole-mouthed jar. Upper (exterior) surface smoothed, lip and underside rough, now all worn; encrustation especially on upperside and lip.
Fabric: Pale coarseware, sandy, with abundant mica and very fine black and white inclusions. Surface and core 5 YR 5/8 (yellowish red).
Comparanda: **9/286** (P230), **9/287** (P231), **9/289** (P161); for hole-mouthed jar, see **9/286** (P230) above.

9/289 (P161), **Fig. 9.65** (Photos 2660–2661)
Fragment.
Tumulus Fill (SU 2.202).
PL (max): 2.5 cm; PW (max): 3.7 cm; Th: 1.2–1.5 cm.
Small, roughly flat fragment preserving a single indentation or finished edge. Surfaces and edge rough, encrustation over breaks. Perhaps hole-mouthed jar, but too little preserved to be sure—maybe hole from strainer or cooking vessel?
Fabric: Coarse, red; abundant mica and very small black and white inclusions. Surfaces 2.5 YR 5/8 (red), core 2.5 YR 3/1 (dark reddish gray).
Comparanda: **9/286** (P230), **9/287** (P231), **9/288** (P176); for hole-mouthed jar, see **9/286** (P230) above.

9/290 (P224 + P282 + P411), **Figs. 9.31, 9.65** (Drawing 105, Photos 2518–2519 [P224], 2694–2695 [P282]; 3554–3555 [P224 + P282 + P411]
Rim, Neck, Body Fragments.
Tumulus Fill (Units 2.202 [P224]; 1.279 [P282]; 6.556 [P411]).
PH (max): 10.6 cm; PW (max): 16.5 cm; Th (lip): 1.3– 1.5 cm; Th (wall): 0.8–1.1 cm; diameter (rim, approximate): 28–38 cm (very irregular curvature).
Five joining fragments preserving rim, neck, and partial body profile of large or very large coarseware vessel. Convex body curves in to straight flaring neck, wide flat rim with rounded exterior projection. Exterior surface smoothed, perhaps once burnished; interior surface roughly smoothed, traces of horizontal tooling marks on neck, blackened below. Ancient chips from rim exterior, excess clay not trimmed from interior.
Fabric: Hard, sandy, coarse, with abundant mica and red (grog?), white, and black inclusions, mostly very small; also pebbles (<0.3 cm); few air pockets in exterior and interior (<0.6 cm). Interior surface 5 YR 3/1 (very dark gray), exterior varying, 5 YR 3/1 to 7.5 YR 5/4 (brown); core 5 YR 3/1 where not covered by encrustation.
Comparanda: Rim **9/291** (P244; no join, likely same vessel); bases **9/315** (P281), **9/316** (P325): no joins, but one or other perhaps from same vessel; cf. also **9/279** (P003), **9/292** (P317), **9/298** (P291), **9/299** (P145), **9/300** (P065), and **9/50** (P395; similar shape, FL).

9/291 (P244), **Fig. 9.32** (Drawing 50.3; Photos 2492–2493)
Rim, Neck, Body Fragments.
Tumulus Fill (SU 2.202).
PL (max): 9.9 cm; PW (max): 5.5 cm; Th (lip): 1.4 cm; Th (wall): 0.8 cm.
Nine joining fragments of rim, neck, and small bit of upper body of large coarseware vessel; shape and surface treatment as **9/290** (P224 + P282 + P411).
Fabric: As **9/290** (P224 + P282 + P411).
Comparanda: **9/290** (P224 + P282 + P411; no join, likely same vessel); bases **9/315, 9/316** (P281, P325; no joins, but one or other perhaps from same vessel); cf. also **9/279** (P003), **9/292** (P317), **9/298** (P291), **9/299** (P145),**9/300** (P065).

9/292 (P317), **Fig. 9.65** (Photos 2900–2903)
Rim and Neck Fragments.
Tumulus Fill (SU 2.403).
PH (max): 4.1 cm; PW (max): 4.5 cm; Th (lip): 1.7 cm; Th (neck): 0.9 cm.
Three non-joining fragments of straight flaring neck and wide flat rim with rounded exterior projection;

curve and tapering to upper body just preserved. Surface treatment as **9/290** (P224 + P282 + P411).

Fabric: As **9/290** (P224 + P282 + P411, but parts of exterior surface and lip considerably lighter in color: 5 YR 5/6 (yellowish red [though darkening toward bottom to 2.5/1—black]).

Comparanda: **9/290, 9/291** (P224 + P282 + P411, P244; no joins, perhaps same vessel); bases **9/315, 9/316** (P281, P325; no joins, but one or other perhaps from same vessel); cf. also **9/279** (P003), **9/298** (P291), **9/299** (P145), **9/300** (P065).

9/293 (P022), **Fig. 9.31** (Drawing 5.3; Photos 600–603)
Rim Fragment.
Tumulus Fill (SU 1.007).
PH (max): 2.0 cm; PW (max): 2.9 cm; Th: 1.2 cm.
Single fragment preserving flat rim with very slight external projection; lip angle suggests flaring neck, but not fully preserved. Surfaces smoothed, scattered encrustation, especially on breaks.
Fabric: Dark, reddish, coarse fabric, abundant gray, black, and white inclusions (some quartz, some very large [1.0 cm]). Exterior surface 2.5 YR 3/2 (dusky red), lip and interior 2.5 YR 4/4 to 4/6 (reddish brown to red).
Comparanda: **9/278** (P376), **9/279** (P003), **9/280** (P217), **9/290** (P224 + P282 + P411), **9/291** (P244), **9/294** (P026), **9/298** (P291), **9/299** (P145), **9/300** (P065).

9/294 (P026), **Fig. 9.31** (Drawing 5.4; Photos 904–907)
Rim Fragment.
Interface Topsoil/Tumulus Fill (SU 4.011).
PH (max): 2.8 cm; PW (max): 3.4 cm; Th: 0.9–1.2 cm.
Single fragment preserving flat rim with very slight external projection; lip angle suggests flaring neck, but not fully preserved. Surfaces smoothed, exterior shiny with encrustation.
Fabric: Dark, coarse fabric, moderate mica and red inclusions (<0.5 cm), few air pockets. Surface and core 7.5 YR 4/1 (dark gray).
Comparanda: **9/278** (P376), **9/279** (P003), **9/280** (P217), **9/290** (P224 + P282 + P411), **9/291** (P244), **9/293** (P022), **9/298** (P291), **9/299** (P145), **9/300** (P065).

9/295 (P085), **Figs. 9.32, 9.65** (Drawing 21.3; Photos 971–974)
Rim Fragment.
Tumulus Fill (SU 4.129).
PH (max): 2.9 cm; PW (max): 5.1 cm; Th (wall): 1.3; Th (lip): 2.0 cm.
Single fragment of flaring wall of large coarseware bowl, with flat rim set off by indented curve on interior, slightly rounded projection on exterior. Lip smoothed, perhaps burnished; interior surface smoothed; exterior rough, with tooling marks, excess clay; encrustation, especially on exterior and breaks.
Fabric: Coarse, heavy fabric, with abundant small black, white, and red inclusions, moderate white pebbles (<0.5 cm). Surfaces 5 YR 4/4–4/6 (reddish brown-yellowish red), core 5 YR 3/2 (dark reddish brown).

9/296 (P014), **Figs. 9.32, 9.65** (Drawing 4.2; Photos 585–587)
Neck/Rim Fragment.
Topsoil (SU 1.001).
PH (max): 4.2 cm; PW (max): 4.2 cm; Th: 1.0–1.5 cm.
Single fragment of large vessel with neck flaring sharply to wide rim; lip not preserved, orientation uncertain. Surfaces smoothed, exterior worn; scattered encrustation.
Fabric: Coarse, heavy fabric, with abundant mica, red (grog), white, and black inclusions, occasional pebbles and air pockets (<0.4 cm). Interior surface 5 YR 4/6 (yellowish red), exterior 5 YR 3/2 (dark reddish brown), core 5 YR 2.5/1 (black).
Comparanda: **9/302** (P313).

9/297 (P428), **Fig. 9.65** (Photos 3752–3753)
Rim Fragment.
Tumulus Fill (SU 5.548).
PH (max): 5.7 cm; PW (max): 4.4 cm; Th (wall): 1.2–1.5 cm; PTh (rim): 2.5 cm.
Single fragment of large coarseware vessel, preserving S-curve of upper wall flaring to everted rim with flat upper surface; outer lip edge not preserved. Exterior wall and upper rim surfaces smoothed, interior surface rougher, with tool marks and excess clay; scattered encrustation, especially on breaks.
Fabric: Coarse, red, with abundant small red and brown inclusions (grog? <0.2), moderate mica, black, red, and brown pebbles (<0.8 cm), and air

pockets (0.3 cm). Exterior surface 5 YR 4/6 (yellowish red), interior surface and core 5 YR 5/6 (yellowish red).

9/298 (P291), **Figs. 9.32, 9.65** (Drawing 62.5; Photos 2815–2817)
Rim Fragment.
Tomb XLIV (65) (SU 4.364).
PH (max): 4.7 cm; PW (max): 4.4 cm; Th (wall): 1.2–1.3 cm; Th (rim): 1.4–1.5 cm; diameter (rim, approximate): 30–40 cm.
Single fragment of flaring neck and wide, flat rim with small indentation beneath external projection, from very large vessel; slight curvature above lower break may begin curve to convex lower body. Ancient chip from rim exterior, heavy encrustation over all surfaces; exterior and interior smoothed.
Fabric: Very hard; few inclusions visible beneath encrustation: mica, black and white (<0.1 cm); air pockets in exterior (<0.1 cm). Surfaces 7.5 YR 6/6 (reddish yellow) where visible, mostly covered by encrustation 7.5 YR 4/1 (dark gray) to 2.5 Y 7/2 (light gray).
Comparanda: **9/279** (P003), **9/290** (P224 + P282 + P411), **9/291** (P244), **9/292** (P317), **9/299** (P145), **9/300** (P065).

9/299 (P145), **Fig. 9.32** (Drawing 50.2; Photos 2530–2531)
Rim Fragment.
Tumulus Fill (SU 2.202).
PH (max): ca. 6.0 cm; PW (max): 5.1 cm; Th (wall): 0.7–1.3 cm; Th (rim): 1.1–1.2 cm; diameter (rim ext, approximate): 34–37 cm.
Single fragment, shape as **9/298** (P291), with more pronounced external projection and neck tapering before curving out to thinner wall. Surface smoothed, more finely, with fine lines (tooling marks?) visible, on exterior; clay not fully smoothed on rim interior, slightly uneven on exterior. Encrustation heavy on all surfaces.
Fabric: Very hard; abundant mica inclusions, fewer small black and white (<0.2 cm); air pockets (<0.3 cm) in all surfaces. Exterior surface 7.5 YR 4/1 (dark gray) to 4/4 (brown), interior surface 2.5 Y 7/3 (pale yellow) to 5/3 (light olive brown), core as surface, also as dark as 7.5 YR 2.5/1 (black).
Comparanda: **9/298** (P291), **9/300** (P065); cf. also similar rims of darker fabric **9/279** (P003), **9/290** (P224 + P282 + P411), **9/291** (P244), **9/292** (P317).

9/300 (P065), **Fig. 9.32** (Drawing 13.4; Photos 729–732)
Rim Fragment.
Tumulus Fill (SU 1.070).
PH (max): 6.0–6.5 cm; PW (max): 4.8 cm; Th (wall): 0.7–0.8 cm; Th (rim): 1.2–1.3 cm; diameter (rim ext, approximate): 30–40 cm.
Single fragment preserving flat rim with rounded external projection, flaring neck curving gently outwards to shoulder. Neck profile slightly uneven, with very shallow ridges, perhaps due to coil construction. Surfaces smoothed, perhaps once burnished, now covered by heavy encrustation; modern damage to rim.
Fabric: Medium; few inclusions visible beneath encrustation: mica, black and white pebbles (<0.3 cm); few air pockets (<0.3 cm). Color difficult to discern beneath heavy encrustation (10 YR 6/4—light yellowish brown); surfaces and core 5 YR 5/6 (yellowish red).
Comparanda: **9/298** (P291), **9/299** (P145); cf. also similar rims of darker fabric, **9/279** (P003), **9/290** (P224 + P282 + P411), **9/291** (P244), **9/292** (P317)

9/301 (P185), **Fig. 9.32** (Drawing 46.7; Photos 2674–2675)
Rim Fragment.
Tumulus Fill (SU 1.070).
PL (max): 4.4 cm; PW (max): 4.2 cm; Th: 1.0–1.5 cm.
Single fragment from very large coarseware vessel, preserving irregular rim with rounded lip, perhaps slightly everted from vertical neck/wall. One surface smoothed, the other rough and irregular, but precise shape and orientation uncertain.
Fabric: Coarse, sandy fabric, with abundant white, black, and gray inclusions (<0.3 cm), moderate mica. Surfaces 2.5 YR 5/6 (red), core 10 YR 6/2 (light brownish gray).
Comparanda: **9/281** (P239), **9/282** (P397).

9/302 (P313), **Fig. 9.32** (Drawing 94.1; Photos 2904–2905)
Rim Fragment.
Tumulus Fill (SU 1.377).
PH (max): 6.3 cm; PW (max): 4.9 cm; Th: 1.0–1.6 cm.
Single fragment from large vessel of uncertain

shape, probably open, with neck flaring sharply to wide rim and narrow, rounded lip; wall thickens considerably from neck to rim. Surfaces smoothed but worn and damaged; much of lip lost to modern break.
Fabric: Dark, coarse, with abundant mica and red (grog and pebbles), brown, and white inclusions (<0.3 cm), occasional air pockets (<0.3 cm). Interior surface 5 YR 4/4 (reddish brown) to 3/1 (very dark gray); interior surface and core 5 YR 2.5/1 (black).
Comparanda: **9/296** (P014).

9/303 (P336), **Fig. 9.32** (Drawing 95.2; Photos 3151–3152)
Rim Fragments.
Tumulus Fill (SU 4.286).
PL (max): 7.0 cm; PW (max): 6.8 cm; Th: 1.1 cm.
Six joining fragments from large coarseware vessel with widely flaring neck, everted rim, and rounded lip; surfaces smoothed but damaged, exterior worn; scattered encrustation.
Fabric: Sandy, semi-coarse, with abundant mica and small black, white, and red inclusions. Surfaces 2.5 YR 5/6 (red), core 2.5 YR 3/1 (dark reddish gray).

9/304 (P189), **Fig. 9.32** (Drawing 76.3; Photos 2353–2354)
Wall Fragment.
Tumulus Fill (SU 1.070).
PH (max): 3.5 cm; H (projection): 1.5; PW (max): 3.0 cm; W (projection): 2.2 cm; Th (wall): 0.9 cm; Th (wall + projection): 1.8 cm.
Small wall fragment of medium or large coarseware vessel with ovoid projection, not fully preserved. Exterior surface smoothed but worn; interior uneven; both surfaces cracked.
Fabric: Coarse, sandy, with abundant mica, white pebbles (<0.3 cm), and small red and black (<0.2 cm) pebble inclusions. Surfaces and core 5 YR 5/6 (yellowish red).
Comparanda: **9/311** (P360), **9/312** (P087); cf. also semi-coarse **9/257** (P214).

9/305 (P011), **Fig. 9.32, 9.65** (Drawing 3.1; Photos 573–576)
Pithos Wall Fragment.
Topsoil (SU 1.001).
PL (max): 8.1 cm; PW (max): 7.3 cm; Th: 1.6–2.5 cm.
Single wall fragment of very large, thick-walled coarseware vessel (pithos), with finger groove in exterior surface perhaps marking division between shoulder and neck. Exterior surface smoothed, scattered encrustation, heavier on exterior and core.
Fabric: Very heavy, coarse red fabric, unique at Lofkënd, with abundant mica, fine black inclusions, white, gray, and red pebbles (or grog[?], <0.6 cm), moderate air pockets (<0.5 cm). Surface and core mottled 2.5 YR 4/6 (red) to 3/4 (dark reddish brown).
Comparanda: **9/306** (P141), uncatalogued P390, P396.

9/306 (P141), **Fig. 9.32** (Drawing 42.1; Photos 2204–2205a/2205b)
Wall Fragment.
Topsoil (SU 2.002).
PL (max): 7.2 cm; PW (max): 4.6 cm; Th: 2.8–3.2 cm.
Single fragment from wall of very large, thick-walled coarseware vessel, possibly pithos. Surfaces smoothed, interior slightly uneven, with light encrustation; encrustation heavier elsewhere.
Fabric: Heavy, coarse to semi-coarse fabric, with moderate inclusions: mica, small black and white grains and pebbles (<0.5 cm), grog (<0.3 cm). Exterior surface 5 YR 5/6 (yellowish red), interior 5 YR 5/4 (reddish brown), core 5 YR 4/1 (dark gray) to 2.5/1 (black).
Comparanda: **9/305** (P011), uncatalogued P390, P396.

9/307 (P306), **Fig. 9.65** (Photos 3096–3097)
Wall Fragments.
Tumulus Fill (SU 1.387).
PH (max): 10.2 cm; PW (max): 20.2 cm; Th: 1.2–1.5 cm; PD (max): >50 cm.
Thirteen joining and eight non-joining fragments from wall of very large coarseware vessel, probably open in shape. Exterior surface rough and very worn, interior smoothed but uneven; scattered encrustation, heavy in places.
Fabric: Coarse, crumbly fabric, with abundant mica and fine black and white inclusions, moderate red (grog?) inclusions. Exterior surface 5 YR 5/6 (yellowish red), interior surface 5 YR 4/2 (dark reddish gray), core 5 YR 2/1 (black).

9/308 (P010), **Fig. 9.32** (Drawing 4.1; Photos 939–942)
Wall Fragment.

Topsoil (SU 3.003).
PL (max): 6.4 cm; PW (max): 5.0 cm; Th: 1.0–1.2 cm.
Single fragment from wall of large coarseware vessel. Surfaces smoothed, worn, scratched; encrustation heavy on exterior surface and core.
Fabric: Coarse, with abundant mica, fine black and white inclusions, moderate air pockets (<0.4 cm) and pebble inclusions, some quite large (<1.2 cm). Exterior surface 5 YR 6/6 (reddish yellow), interior surface 2.5 YR 5/8 (red); core not fully visible.

9/309 (P042), **Fig. 9.33** (Drawing 10.1; Photos 3821–3822)
Wall Fragment.
Tumulus Fill (SU 1.017).
PL (max): 7.2 cm; PW (max): 3.8 cm; Th: 1.5–1.7 cm.
Single fragment from wall of large coarseware vessel. Surfaces smoothed, exterior worn, uneven, pitted (or perhaps punched decoration [roughly triangular]?); heavy encrustation over core.
Fabric: Coarse, with abundant small black inclusions, moderate mica, grog, and dark pebble inclusions. Surfaces and core 5 YR 5/6 (yellowish red).

9/310 (P261), **Figs. 9.33, 9.66** (Drawing 61.1; Photos 2666–2667)
Wall Fragment.
Tumulus Fill (SU 1.279).
PH (max): 5.2 cm; PW (max): 7.5 cm; Th (wall): 0.9 cm; Th (wall + projection): 1.7 cm; D (max): ca. 35 cm; D (projection): 3.3 cm.
Wall fragment of large coarseware vessel with uneven round projection beneath small groove (the latter perhaps unintentional). Surfaces smoothed, interior perhaps once burnished, exterior rough.
Fabric: Hard, sandy; many inclusions, mica, black, white (<0.3 cm); air pocket cracks visible in interior. Surface 2.5 YR 5/6 (red), core 10 YR 2/1 (black).
Comparanda: Andrea 1985: pl. I, Tomb 6, 2.

9/311 (P360), **Fig. 9.66** (Photos 3228–3229)
Wall Fragment.
Tumulus Fill (SU 1.399).
PH (max): 4.0 cm; H (projection): 2.2 cm; PW (max): 4.4 cm; W (proj): 3.7 cm; Th (wall): 1.3 cm; Th (wall + projection): 3.3 cm.
Small wall fragment of large coarseware vessel with roughly triangular projection or lug flaring up and out from wall at about 45° angle. Surfaces worn, uneven.
Fabric: Coarse, sandy; abundant mica, black and white pebble inclusions (<0.4 cm). Surface 2.5 YR 5/6 (red), core 5 YR 2.5/1 (black).
Comparanda: **9/312** (P087); cf. also **9/71** (P127; similar projection in LF) and complete vessels with projections from Luaras (Aliu 2004: fig. 51, amphorae a, p. 95; fig. 42, p. 85; pl. XX, object 238, p. 200) and Patos (Korkuti 1981: pl. 4, 41, p. 58).

9/312 (P087), **Figs. 9.33, 9.66** (Drawing 25.5; Photos 759–762)
Wall Fragment.
Tumulus Fill (SU 1.067).
PH (max): 5.1 cm; PW (max): 5.0 cm; Th (wall): 1.2 cm; Th (wall + projection): 2.6 cm.
Small wall fragment of large coarseware vessel with roughly triangular projection or lug jutting straight out from wall. Surfaces uneven.
Fabric: Coarse, heavy, sandy, with abundant mica, white pebbles (<0.7 cm) and small red pebbles; hole in fracture from very large inclusion (ca. 1.6 cm), now lost. Exterior surface 2.5 YR 4/8 (red), interior 2.5 YR 3/2 (dusky red), core 10 YR 2/1 (black).
Comparanda: **9/304** (P189), **9/311** (P360), **9/257** (P214; semi-coarse).

9/313 (P333), **Fig. 9.33** (Drawing 72.2; Photos 3155–3156)
Wall Fragment.
Tumulus Fill (SU 2.431).
PH (max): 4.4 cm; H (band): 1.5 cm; PW (max): 5.2 cm; Th (wall): 1.2 cm; Th (wall + band): 2.0 cm.
Triangular fragment from wall of very large coarse vessel. Surfaces smoothed, interior perhaps once burnished; scattered encrustation.
Fabric: Coarse, with abundant mica, black pebbles (<0.3 cm), and red inclusions (possibly grog [<0.4 cm]). Exterior surface 2.5 YR 5/6 (red), interior surface and core 7.5 YR 2.5/1 (black).
Plastic decoration: Part of raised horizontal band of applied clay, now damaged—perhaps once rope pattern or finger-impressed band.
Comparanda: **9/314** (P387 + P408).

9/314 (P387 + P408), **Figs. 9.33, 9.66** (Drawing 95.5; Photos 3462–3463)
Wall Fragment.

Topsoil (Units 2.002, 6.002).
PH (max): 4.8 cm; PW (max): 7.7 cm; Th: 1.3–2.2 cm.
Two joining wall fragments from body, probably near midpoint, of very large, coarse vessel. Exterior surface smooth, with plastic decoration; interior rougher, less even; scattered encrustation.
Fabric: Hard, dark red, coarse, with abundant mica and very fine black and white inclusions, moderate larger, red inclusions (grog? <0.5 cm) and white pebbles (<0.3 cm). Exterior surface 5 YR 5/6, interior surface and core 5 YR 4/6 (yellowish red).
Plastic decoration: Horizontal band of applied clay with finger or tool impressions.
Comparanda: **9/313** (P333), **9/321** (P180). Many EBA pithoi with applied bands, but more often with round pock marks rather than finger impressions (e.g., Hammond 1972: fig. 8a–f; Hochstetter 1984; Schicht 9, 13, 8; Prendi 1977–1978: pl. III.1, 3); for later date (EIA), cf. Korkuti 1971: pl. 10 (Tren); Aliu 2012: pl. I, 28, Tomb 9 (eleventh–tenth centuries).

9/315 (P281), **Figs. 9.33, 9.66** (Drawing 61.4; Photos 2809–2812)
Base Fragment.
Tomb XIX (54) Fill (SU 2.333).
PH (max): 2.8 cm; PW (max): 9.0 cm; D (base): ca. 9–10 cm; Th (base, min): 1.7 cm.
Two joining fragments (from modern break) of roughly circular base, with nearly flat underside, slightly splayed foot. Fragment just preserves gently angled rise of vessel wall, both interior and exterior. Surfaces smoothed, exterior more finely; encrustation unevenly scattered.
Fabric: Medium, sandy; many inclusions, mostly very fine, black, white, mica, possibly grog; air pockets (<0.3 cm) visible in core, underside, interior. Exterior surface variable, 5 YR 5/6 (yellowish red) to 3/1 (very dark gray); underside 7.5 YR 5/4 (brown) to 4/1 (dark gray); interior and core 5 YR 4/1 (dark gray).
Comparanda: **9/290** (P224/282/411; no join, perhaps from same vessel), **9/291** (P244), **9/292** (P317); **9/316** (P325; different base form, same fabric).

9/316 (P325), **Figs. 9.33, 9.66** (Drawing 67.3; Photos 2950–2952)
Base Fragment.
Tumulus Fill (SU 1.377).
PH (max): 5.5 cm; PW (max): 9.2 cm; D (base): ca. 7 cm; Th (base): 1.2 cm; Th (wall): 0.9–1.1 cm.
Single fragment of large coarseware vessel preserving part of round base with slightly raised underside, concave floor, and flaring lower wall. Surfaces smoothed, exterior more finely; light encrustation on all surfaces and core.
Fabric: Hard, sandy; many inclusions, mica, black, white, red pebbles (<0.6 cm); air pockets (<0.4 cm) visible in core, interior, much smaller in exterior and underside. Core 2.5 YR 5/8 (red) to 5 YR 2.5/1 (black); exterior and underside 7.5 YR 4/2–5/4 (brown), interior 10 YR 5/3 (brown) to 3/1 (very dark gray).
Comparanda: **9/319** (P309), uncatalogued P433; cf. also **9/290** (P224 + P282 + P411); no join, perhaps from same vessel, **9/291** (P244), **9/292** (P317); **9/315** (P281; different base form, same fabric).

9/317 (P100), **Figs. 9.33, 9.66** (Drawing 34.3; Photos 2206–2208, 4020–4022)
Base Fragment.
Tumulus Fill (SU 2.202).
PH (max): 5.6 cm; PW (max): 10.0 cm; Th: 1.0 cm; D (base): ca. 10 cm (but very irregular).
Two joining fragments (modern break) of flat, roughly circular base of large coarseware vessel, probably open in shape; wall rises at low angle from base. Surfaces uneven, but smoothed, very worn.
Fabric: Coarse, sandy, abundant black inclusions (<0.7 cm), moderate mica, few white inclusions, air pockets, and cracks in surfaces and core. Exterior surface 5 YR 6/6 (reddish yellow), interior surface covered with encrustation, core 7.5 YR 2.5/1 (black).
Comparanda: **9/89** (P232; similar shape, FL).

9/318 (P354), **Fig. 9.66** (Photos 2958–2960)
Base Fragment.
Tomb I (64) Fill (SU 1.361).
PH (max): 4.5 cm; PW (max): 5.2 cm; Th: 2.4–2.8 cm.
Single fragment preserving small portion of flat base, probably round, together with sloping inside floor and rise of flaring wall. Surfaces very rough, heavy encrustation.
Fabric: Very coarse, heavy fabric; abundant black inclusions, fewer white pebbles, perhaps red inclusions difficult to see under encrustation. Exterior

5 YR 6/8 (reddish yellow), core 5 YR 3/1 (very dark gray).
Comparanda: **9/317** (P100; similar shape, thinner); **9/316** (P325), **9/319** (P309): similar fabric, slightly raised.

9/319 (P309), **Fig. 9.33, 9.67** (Drawing 67.1; Photos 2955–2957)
Base Fragment.
Tumulus Fill (SU 1.377).
PH (max): 3.6 cm; PW (max): 4.4 cm; Th (wall): 1.1 cm; Th (base): 1.8 cm; D (base): ca. 9 cm.
Single fragment preserving small portion of flat, raised base with bit of vessel floor from large coarseware vessel; wall rises sharply up from base. Surfaces rough, encrustation over breaks.
Fabric: Coarse, sandy fabric with abundant black and white pebbles (<0.8 cm), rare mica and air pockets. Surface 2.5 YR 4/6 (red), interior 5 YR 5/6 (yellowish red).
Comparanda: Uncatalogued P433; **9/259** (P283; bitumen pot, with smaller base of similar shape and fabric); **9/316** (P325; for similar fabric and form, but with raised underside).

9/320 (P090), **Figs. 9.33, 9.67** (Drawing 25.4; Photos 955–958)
Vertical Strap Handle or Rim Fragment.
Tumulus Fill (SU 1.070).
PL (max): 5.9 cm; PW (max): 3.3 cm; Th: 1.6 cm.
Three joining fragments (from modern break) preserving two parallel surfaces, very slightly curving, and one rounded edge; either from vertical strap handle, ovoid in section, with no preserved attachments, or from uneven rounded rim of very thick-walled vessel with vertical upper walls. Lighter colored side (interior, if rim) smoothed, other side left rough.
Fabric: Coarse, dark fabric, with abundant mica and red, white, and black inclusions, (<0.3 cm), few small air pockets (<0.2 cm). Surfaces 7.5 YR 4/2 (brown/dark brown) to 10 YR 5/4 (yellowish brown), core 7.5 YR 3/0 (very dark gray).

9/321 (P180), **Figs. 9.34, 9.67** (Drawing 85.2; Photos 2338–2340)
Pithos Lug or Basin Rim Fragment.
Topsoil (SU 1.070).
PL (max): 4.5 cm; PW (max): 2.8 cm; Th: 2.1 cm.
Single fragment, nearly square in section, probably from large coarseware vessel, with at least three finished edges, outermost bearing one and a half finger impressions. Orientation must be horizontal, but could derive from lug handle (on pithos) or from rim of large open basin.
Fabric: Coarse, red fabric, with abundant mica, and red, white, and black inclusions (<0.2 cm), moderate white and gray pebbles (<0.5 cm) and air pockets (<0.5 cm). Surface mottled 5 YR 5/6 (yellowish red), 4/4 (reddish brown) to 3/2 (dark reddish brown); core 7.5 YR 3/2 (very dark gray).

9/322 (P260), **Fig. 9.34** (Drawing 63.2, Photos 2692–2693)
Fragment.
Tumulus Fill (SU 1.279).
PL (max): 9.2 cm; PW (max): 6.2 cm; Th: 1.3–2.0 cm.
Irregularly shaped ceramic fragment, preserving a single rounded edge, smoothed concave surface, and very rough, uneven convex surface; unevenly fired. Possibly rim and upper wall from very large basin, bowl, or mortar.
Fabric: Coarse, red fabric, abundant mica and black inclusions, moderate black and white pebble inclusions (<0.5 cm), few air pockets. Exterior 5 YR 4/6 (yellowish red), core 7.5 YR 2.5/1 (black).
Comparanda: **9/326** (P383).

9/323 (P146), **Fig. 9.34** (Drawing 87.1, Photos 2544–2545)
Fragments.
Tumulus Fill (SU 2.228).
PL (max): 7.8 cm; PW (max): 6.5 cm; Th: 1.2 cm.
Two joining fragments of irregularly shaped flat object, preserving a single flat, finished edge, mostly straight but curving out before break. One surface smoothed, the other rough, with indentations, perhaps finger marks, running parallel to finished edge; additional set of grooves running perpendicular lost at break. Encrustation on smooth surface. Finished edge resembles rim, but flat shape does not correspond to any known vessel type—perhaps a tile or tripod leg?
Fabric: Coarse, sandy fabric, abundant mica and tiny black and white inclusions, few air pockets. 5 YR 4/6 (yellowish red).
Comparanda: **9/324** (P147), **9/325** (P226), **9/326** (P383).

9/324 (P147), **Fig. 9.34** (Drawing 62.4; Photos 2632–2633)
Fragment.
Tumulus Fill (SU 4.204).
PL (max): 4.9 cm; PW (max): 4.4 cm; Th: 1.2–1.5 cm.
Single fragment of flat, nearly square-shaped object, broken all around; one side "corrugated" with irregular finger strokes, the other uneven; heavy encrustation.
Fabric: Heavy, coarse, with moderate mica, small black inclusions, possible grog, and air pockets, occasional dark pebbles (<0.5 cm). Surface and core 5 YR 5/8 (yellowish red).
Comparanda: **9/322** (P260), **9/323** (P146), **9/325** (P226), **9/326** (P383).

9/325 (P226), **Fig. 9.34** (Drawing 62.2; Photos 2646–2647)
Fragment.
Tomb LIX (38) Fill (SU 4.246).
PL (max): 3.9 cm; PW (max): 5.0 cm; Th: 1.4 cm.
Single fragment of flat, triangularly shaped object, with one roughly finished, nearly flat, but curving edge. One surface smooth, but with excess clay around edge, other surface rough, now worn and encrusted.
Fabric: Sandy, coarse, with abundant mica and small black, white, and red pebble inclusions (<0.2 cm). Exterior and core 2.5 YR 4/8 (red).
Comparanda: **9/322** (P260), **9/323** (P146), **9/324** (P147), **9/326** (P383).

9/326 (P383), **Fig. 9.67** (Photos 4004–4005)
Fragment.
Tumulus Fill (SU 2.487).
PL (max): 5.3 cm; PW (max): 5.5 cm; Th: 1.5 cm.
Single fragment of flat, triangularly shaped object, with one roughly finished, nearly flat edge; one surface smoothed, the other more rough; scattered encrustation, especially on edge. No visible curvature suitable to any known vessel type—perhaps from a tile or tripod leg (but no tapering as in other tripod legs [cf. Papadopoulos 2005: fig. 67d, 157k; T123-3:179c, pl. 386] and no known comparanda for EIA tiles).
Fabric: Coarse, light fabric, hard texture; abundant mica and black and white inclusions (<0.5 cm), moderate air pockets (<0.4 cm). Surface 2.5 YR 5/6 (red), core 2.5 YR 2.5/1 (reddish black).
Comparanda: **9/322** (P260), **9/323** (P146), **9/324** (P147), **9/325** (P226).

9/327 (P153), **Fig. 9.34** (Drawing 85.4; Photos 2520–2521)
Fragment.
Tumulus Fill (SU 1.070).
PL (max): 6.2 cm; PW (max): 4.9 cm; Th: 1.1 cm.
Single fragment of irregularly shaped object, nearly flat, with one finished edge; one surface smoothed and burnished, now partly covered by encrustation, the other surface smoothed. Both edge and profile present very slight double curve, but not enough to place them on any known vessel type.
Fabric: Coarse, sandy fabric, abundant mica and black and white pebbles (<0.3 cm; more visible on unburnished surface), few air pockets (<1 cm). Surface 5 YR 5/6 (reddish yellow), core 7.5 YR 3/1 (very dark gray).

9/328 (P327), **Fig. 9.67** (Photos 3182–3185)
Fragments (A–D).
Tumulus Fill (SU 4.286).
PL (A, max): 13.3 cm; PW (A, max): 11.6 cm; Th (A): 0.7–1.7 cm; D (B, max): 6.8 × 6.4 cm; Th (B): 1.1–1.4 cm; D (C, max): 5.0 × 5.7 cm; Th (C): 0.9–1.4 cm; D (D, max): 4.5 × 5.7 cm, Th (D): 0.9–1.2 cm.
A—Eight joining fragments of irregular shape, with convex exterior and one nearly straight, finished edge, with uneven, rounded lip, rising slightly at one end. Unusual curvature could be from very large vessel, with finished edge a rim, perhaps rising toward handle; also perhaps from tripod leg or *pyraunos* skirt, with finished edge once vertical. Exterior surface smoothed but uneven, with many fine cracks and large darkened areas, whether from reduced firing atmosphere or from burning during use or after deposition; interior surface also damaged.
B—Two joining fragments of similar fabric to above, nearly flat, with no finished edges, but very slight curve above shallow groove, perhaps from upper shoulder of very large vessel.
C—Single wall fragment, also of similar fabric, but perhaps from smaller vessel, broken all around.
D—Wall fragment of large vessel of different fabric, broken all around; definitely does not belong with above.
Fabric: A, B, C: Coarse, crumbly, with moderate mica, black, white, gray, and red inclusions (<0.6 cm);

surfaces mottled: exterior 7.5 YR 5/4 (brown) to 2.5/1 (black), interior 5 YR 5/6 (yellowish red) to 3/1 (very dark gray), core 3/1. D: Coarse to semi-coarse, hard, with abundant mica, small black and white (quartz and pebble) inclusions; surfaces 2.5 YR 5/6 (red), core 10 YR 6/3 (pale brown).

Comparanda: **9/322** (P260), **9/323** (P146), **9/324** (P147), **9/325** (P226), **9/326** (P383). Cf. also **9/307** (P306) for large wall fragments from very large vessel of coarse crumbly fabric. For *pyraunos*, see Hochstetter 1984: pl. 45.9, 10, pl. 59.6, pl. 65.3, pl. 102.5.

9/329 (P369), **Fig. 9.67** (Photos 3300–3301)
Fragment.
Tumulus Fill (SU 1.440).
PL (max): 5.1 cm; PW (max): 3.4 cm; Th: 1.4 cm.
Single fragment of irregular shape, very nearly flat, perhaps from wall of large coarseware vessel. Exterior surface smoothed, with two post-firing incised furrows (ca. 0.1–0.3 cm deep), not parallel, not necessarily ancient; additional damage around edges; interior surface smoothed.
Fabric: Coarse, with abundant mica and small black, white (limestone), and red (grog) inclusions, moderate pebbles (<0.5 cm). Exterior 5 YR 4/3 (reddish brown), interior 10 YR 3/2 (very dark grayish brown), core 5 YR 5/6 (yellowish red) to 5 YR 2/1 (black).

Other Fabrics—Non-Prehistoric

9/330 (P002), **Figs. 9.35, 9.68** (Drawing 2.2; Photos 555–558)
Base Fragment.
Pre-excavation Surface Collection.
PH (max): 1.3 cm; PW (max): 3.8 cm; Th: 0.2–0.9 cm; D (base): 6–7 cm.
Single fragment from wheelmade base with articulated foot, raised underside, and slightly concave vessel floor; lower walls not preserved. Surfaces smoothed, now very worn. Attic skyphoid cup?
Fabric: Very fine, with occasional very small black inclusions and air pockets.
Exterior surface 10 YR 8/4 (very pale brown), core and interior surface 2.5 YR 5/8 (red).

9/331 (P424), **Fig. 9.35** (Drawing 97.3; Photos 4023–4024)
Rim Fragment.
Topsoil (SU 2.002).
PH (max): 2.8 cm; PW (max): 2.8 cm; Th: 0.3 cm.
Rim fragment of open vessel, with vertical wall rising at steep angle to narrow, tapered lip, now damaged. Surfaces smoothed, interior slightly scratched and pitted; light encrustation.
Fabric: Very fine, with occasional mica. Exterior surface 10 YR 8/4 (very pale brown) to 7.5 YR 7/6 (reddish yellow), interior surface and core 5 YR 6/6 (reddish yellow).

9/332 (P175), **Fig. 9.35** (Drawing 45.6; Photos 2367–2369)
Wall Fragment.
Tumulus Fill (SU 1.070).
PL (max): 2.3 cm; PW (max): 1.7 cm; Th: 0.4–0.6 cm.
Small wall fragment from small, open vessel, perhaps wheelmade Corinthian kotyle. Surfaces smoothed, very worn; fine ridges on interior; scattered brown encrustation.
Fabric: Very fine, with very few visible inclusions tiny air pockets; surfaces and core 2.5 Y 7/2 (light gray).
Comparanda: **9/333** (P257), **9/334** (P328), **9/335** (P372), **9/336** (P384); for Corinthian kotylai, see Blegen, Palmer, and Young 1964:106–108.

9/333 (P257), **Figs. 9.68, A2.1a-b** (Photo 3364)
Corinthian Kotyle Fragments (Rim, Body, and Handle).
Tumulus Fill and Topsoil (Units 1.240, 1.278, 1.279).
Rim fragments: PH (max) × PW (max): 3.1 × 5.2 cm; 2.3 × 4.2 cm; body frr: D (max): 7.1 × 3.0 cm; 3.9 × 2.0 cm; 3.3 × 2.3 cm; 2.9 × 2.5 cm; 3.6 × 1.0 cm; Th: 0.3–0.5 cm; D (rim): ca. 12 cm; D (handle): 0.8–0.9 cm.
Twelve fragments from Archaic Corinthian kotyle (wheelmade), found across three stratigraphic units over 2 days, including two joining pairs from rim, seven from lower body (of which three join), and one handle fragment; no portion of the base remains. Painted decoration is almost completely lost; handle once completely black or brown, scattered brown or black paint (dull) on both surfaces of all other pieces; one lower body fragment preserves parts of six rays or vertical bands, emerging from incised horizontal line below, perhaps just above missing foot. Double incised horizontal band preserved on three joining and one non-joining body fragment. Heavy encrustation across all surfaces.

Fabric: Very fine, with occasional mica, fine black inclusions, and air pockets. Surfaces and core 10 YR 8/3–8/4 (very pale brown).

Comparanda: **9/332** (P175), **9/334** (P328), **9/335** (P372), **9/336** (P384); for Corinthian kotylai, see Blegen, Palmer, and Young 1964:106–108. Note that ray-based Archaic kotylai often have animal or pattern motifs on body, not seen here; also no trace of usual patterns beneath rim.

9/334 (P328), **Fig. 9.35** (Drawing 67.5; Photos 3180–3181)
Wall Fragment.
Tumulus Fill (SU 2.431).
PL (max): 3.0 cm; PW (max): 3.6 cm; Th: 0.5–0.6 cm.
Single wall fragment from small or medium-size vessel, perhaps wheelmade Corinthian kotyle. Dark red, diagonal band painted on exterior, fine ridges on interior; scattered encrustation, heavy on breaks.
Fabric: Fine, with moderate mica, occasional fine brown inclusions. Exterior surface 7.5 YR 8/6 (reddish yellow), interior surface 10 YR 8/6 (yellow), core 10 YR 7/1 (light gray).
Comparanda: **9/332** (P175), **9/333** (P257), **9/335** (P372), **9/336** (P384); for Corinthian kotylai, see Blegen, Palmer, and Young 1964:106–108.

9/335 (P372), **Fig. 9.68** (Photos 3366–3367)
Wall Fragment.
Tumulus Fill (SU 1.509).
D (max): 1.3 × 1.1 cm; Th: 0.4 cm.
Very small wall fragment, perhaps from wheelmade Corinthian kotyle. Surfaces smoothed, exterior damaged; heavy encrustation.
Fabric: Fine, with occasional mica, fine brown inclusions. Surfaces and core 2.5 Y 8/3 (pale yellow).
Comparanda: **9/332** (P175), **9/333** (P257), **9/334** (P328), **9/336** (P384); for Corinthian kotylai, see Blegen, Palmer, and Young 1964:106–108.

9/336 (P384), **Fig. 9.68** (Photos 4025–4026)
Wall Fragment.
Topsoil (SU 1.375).
D (max): 1.7 × 0.9 cm; Th: 0.3 cm.
Very small body fragment from small, thin-walled vessel, probably wheelmade. Surfaces smoothed, with very fine ridges; scattered encrustation on surface, heavy on core.
Fabric: Very fine, few fine black inclusions. Exterior surface 7.5 YR 7/6 (reddish yellow), interior 2.5 YR 6/8 (light red), core not visible.
Comparanda: **9/332** (P175), **9/333** (P257), **9/334** (P328), **9/335** (P372), for other small, wheelmade sherds.

9/337 (P340), **Fig. 9.68** (Photos 3070–3071)
Rim Fragment.
Tumulus Fill (SU 2.403).
PH (max): 2.3 cm; PW (max): 3.3 cm; Th: 0.7 cm.
Single fragment from rim of small vessel, tapering to narrow lip; no curvature preserved, so shape and orientation unclear. One surface smooth, the other rough and uneven, both worn, now heavily encrusted.
Fabric: Surface 2.5 Y 7/3 (pale yellow), core 5/1 (gray) where visible.

9/338 (P301), **Fig. 9.69** (Photos 2836–2837)
Base Fragment?
Topsoil (SU 2.002).
PL (max): 3.2 cm; PW (max): 1.9 cm; Th: 0.8 cm; D (base, ext): 13 cm.
Single fragment, probably from disk foot (but maybe uneven rim or even crooked strap handle), with raised underside, convex upper surface. Surfaces smoothed, now very worn, heavy encrustation.
Fabric: Fine to semi-fine, with abundant mica, few other visible inclusions. Surface GLEY2 4/10B (dark bluish gray).
Comparanda: **9/339** (P338; similar fabric).

9/339 (P338), **Fig. 9.69** (Photos 3177–3179)
Rim Fragment.
Tumulus Fill (SU 4.286).
PL (max): 2.4 cm; PW (max): 1.8 cm; Th: 0.5 cm.
Single fragment from small vessel with convex lower body, narrow lip, and everted rim. Surfaces very worn, especially on exterior; scattered encrustation. Gray color suggests misfiring.
Fabric: Fine, with abundant mica, moderate small black and white inclusions; one pebble (0.3 cm). Surface and core GLEY2 5/5PB (bluish gray).
Comparanda: **9/338** (P301; similar fabric).

9/340 (P139), **Fig. 9.69** (Photos 2165–2166)
Lower Body Fragment.
Tumulus Fill (SU 1.070).
PL (max): 5.5 cm; PW (max): 5.3 cm; Th: 0.8–1.1 cm.

Single fragment probably from lower body of globular vessel, very worn, with many blisters and air pockets on exterior, perhaps from overfiring/vitrification; possibly a production discard or waster. Interior surface smoothed, now very worn, with heavy encrustation; scattered encrustation on exterior.

Fabric: Difficult to discern original fabric; sandy, semi-coarse, with abundant white inclusions, moderate red (grog) inclusions, occasional mica. Densely packed air pockets on exterior probably result of misfiring rather than original fabric; air pockets result in very light weight. Exterior and interior surfaces 5 YR 6/6 (reddish yellow), 10 YR 4/1 (dark gray).

9/341 (P273), **Figs. 9.35, 9.69** (Drawing 60.4; Photos 2712–2713)

Wall Fragment (from Cup?).

Topsoil (SU 1.278).

PL (max): 2.8 cm; PW (max): 3.1 cm; Th: 0.5 cm.

Single fragment of thin-walled, glazed, wheelmade vessel, probably a cup, with scar for handle attachment on exterior; early modern painted ware from Grottaglie and/or Corfu (nineteenth–twentieth centuries AD). White glaze on both sides, partially flaked off; exterior includes bits of blue, purple, yellow, and orange decoration, perhaps florals and tendrils.

Fabric: Fine light fabric, with very few visible inclusions. Surface and core 10 YR 8/3 (very pale brown).

Comparanda: Vroom 1998:138–142, figs. 4–6; Vroom 2003:182–184, figs. 6.15, 6.38 and 6.47; Vroom 2005:184–185.

ACKNOWLEDGMENTS

A full draft of this chapter was completed in summer 2009, with some revision and updating of bibliographic references in late 2011 and in 2013; a preliminary report on the pottery was presented at the AIA Annual Meeting in 2009 (for an abstract, see www.archaeological.org) and posted at the *Aegeo-Balkan Prehistory* website (*http://www.aegeobalkan-prehistory. net/article.php?id_art=18*). Our most sincere thanks are due to Lorenc Bejko, Sarah Morris, and John Papadopoulos for entrusting us with the study and publication of this material, and for their substantial assistance throughout the process. They have saved us from many errors, but all that remain are our own.

TABLE 9.1 Lofkënd ceramic vessel size chart (cm)

Size	Th (wall)	D (max)	H (max)
Very small	ca. 0.3–0.6	<8	<8
Small	ca. 0.4–0.6	ca. 8–12	ca. 8–12
Medium	ca. 0.4–1.0	ca. 12–18	ca. 12–24
Large	ca. 1.0–1.5	ca. 18–36	ca. 24–48
Very large	>1.5	>36	>48

TABLE 9.2 Commonly used formulae for distinguishing between open and closed shapes

	Formula A	Formula B
Open	D (mouth) ≥ 50% D (max)	D (mouth) > Ht(rim)
Closed	D (mouth) < 50% D (max)	D (mouth) < Ht(rim)

Table 9.3 Idealized version of matt-painted motifs found on tomb pottery

Motifs	Rim/neck	Shoulder	Handle	Number of occurrences	Phase/date	Fabric/type
Hatched zigzag				**9/1** (P350)	Phase II LBA /EIA: 12th–11th c. BC	Fine light/ kantharos
Pseudo Maltese cross				**9/1** (P350)	Phase II LBA /EIA: 12th–11th c. BC	Fine light/ kantharos
Lattice band				**9/1**(P350)	Phase II LBA /EIA: 12th–11th c. BC	Fine light/ kantharos
Zigzags				**9/1**(P350)	Phase II LBA /EIA: 12th–11th c. BC	Fine light/ kantharos
Horizontal bands				**9/4** (P077), **9/3** (P304), **9/5** (P326), **9/7** (P166), **9/19** (P386)	Phase III LBA /EIA: 11th–10th c. BC	Fine light/ one-handled vessels
Multiple horizontal bands				**9/3** (P304), **9/5** (P326), **9/7** (P166), **9/8** (P277), **9/14** (P256)	Phase III LBA /EIA: 11th–10th c. BC	Fine light/ one-handled vessels
Hatched pendent triangles		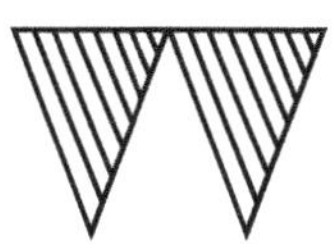		**9/5** (P326), **9/6** (P073), **9/7** (P166), **9/15** (P082)	Phase III LBA /EIA: 11th–10th c. BC	Fine light/ one-handled vessels, stemmed goblet
Hatched pendents & elongated apex				**9/3** (P304)	Phase III LBA /EIA: 11th–10th c. BC	Fine light/ one-handled vessel
Cross-hatched rectangle				**9/5** (P326), **9/19** (P386)	Phase III LBA /EIA: 11th–10th c. BC	Fine light/ one-handled vessels
Solid triangles		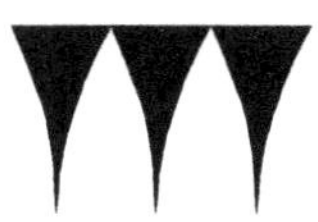		**9/14** (P256)	Phase III LBA /EIA: 10th–9th c. BC	Fine light/ one-handled vessel
Wavy line				**9/3**(P304), **9/15** (P082)	Phase III LBA /EIA: 9th–8th c. BC	Fine light/ one-handled vessel, stemmed goblet

TABLE 9.4 Idealized version of matt-painted motifs found on fragments not from tombs

Motifs	Rim/neck	Shoulder	Handle	Number of occurrences	Phase/date	Fabric/type
Zigzag				**9/220** (P168)	Phase II LBA /EIA: 12th–11th c. BC	Fine light/ semi-coarse
Horizontal bands				**9/220** (P168), **9/235** (P084), **9/19** (P386)	Phase II LBA /EIA: 12th–11th c. BC Phase III LBA /EIA: 11th–10th c. BC	Fine light/ semi-coarse
Multiple horizontal bands				**9/25** (P373), **9/31** (P114), **9/34** (P125), **9/40** (P279), **9/10** (P213), **9/47** (P258), **9/82** (P305), **9/84** (P030)	Phase III LBA /EIA: 11th–10th c. BC	Fine light/ semi-coarse
Vertical bands				**9/235** (P084), **9/220** (P168)	Phase III LBA /EIA: 11th–10th c. BC	Semi-coarse/ uncertain
Upright triangles				**9/80** (P284), **9/82** (P305)	Phase III LBA /EIA: 11th–8th c. BC	Fine light/ uncertain
Solid lozenges				**9/248** (P234)	Phase III LBA /EIA: 11th– 8th c. BC	Semi-coarse/ uncertain
Cross-hatched rectangle				**9/247** (P196)	Phase III LBA /EIA: 11th–10th c. BC	Semi-coarse/ uncertain
Horizontal lines hatched diagonally				**9/58** (P351)	Phase III LBA /EIA: 11th –10th c. BC	Fine light/ uncertain
Hatched pendent triangles		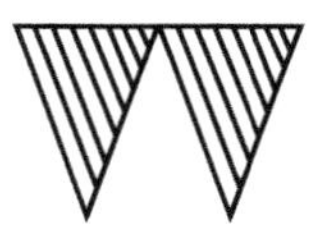		**9/26** (P107), **9/59** (P262), **9/62** (P019), **9/65** (P012), **9/74** (P252), **9/75** (P132+P324) **9/76** (P431), **9/77** (P195), **9/78** (P104), **9/80** (P284), **9/82** (P305), **9/89** (P232), **9/234** (P016), **9/241** (P034), **9/248** (P234), **9/250** (P429), **9/251** (P105), **9/254** (P430)	Phase III LBA /EIA: 11th–10th c. BC	Fine light, semi-coarse/ uncertain
Cross-hatched pendent triangles				**9/84** (P030)	Phase III LBA /EIA: 11th–8th c. BC	Fine light/ uncertain
Receding hatched & hatched pendent triangles				**9/246** (P249)	Phase III LBA /EIA: 11th–8th c. BC	Semi-coarse/ uncertain
Receding hatched & cross-hatched pendent triangles				**9/245** (P423)	Phase III LBA /EIA: 11th–8th c. BC	Semi-coarse/ uncertain

TABLE 9.5 Idealized version of plastic decoration on tomb pottery

Motifs	Shoulder	Handle	Number of occurrences	Phase/date	Fabric/type
Four projections			**9/5** (P326)	Phase II LBA/EIA 12th–11th c. BC	Fine light/ Type 2a
Three projections	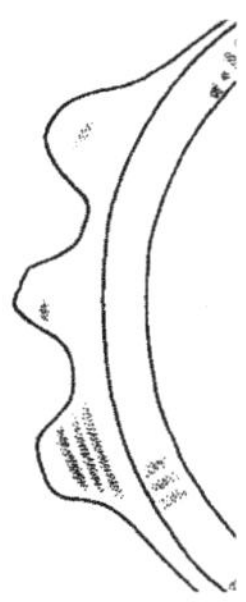		**9/3** (P304), **9/8** (P277), **9/14** (P256), **9/6** (P073)?	Phase III LBA/EIA 11th–8th c. BC	Fine light/ Type 3
Ribbing: diagonal striations			**9/90** (P422)	Phase II LBA/EIA 12th–11th c. BC	Fine dark/ Type 1
Punched dots		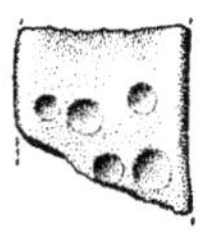	**9/94** (P322)	Phase II LBA/EIA 12th–11th c. BC	Fine dark/ Type 4
Sharply angled diagonal ribbing	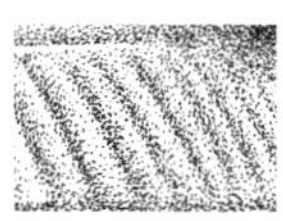		**9/94** (P322), **9/95** (P228), **9/96** (P276)	Phase III+IV LBA/EIA 11th–9th c. BC	Fine dark/ Types 4, 5

TABLE 9.6 Plastic decoration found in the tumulus fill

Motifs	Shoulder	Rim	Number of occurrences	Phase/date	Fabric/type
Vertical ribbing	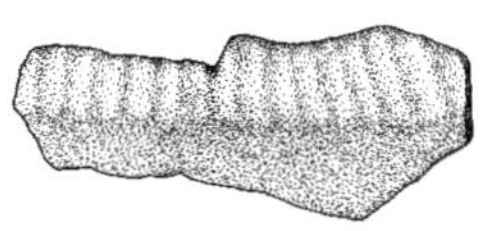		**9/98** (P032), **9/99** (P050)	Phase IV EIA late 10th–9th c. BC	Fine dark/ Types 4, 5?
Finger impressions		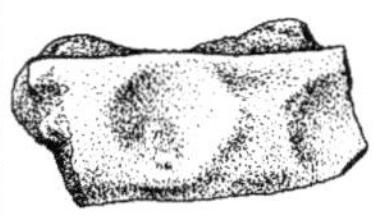	**9/327** (P153), **9/314** (P387+ P408)	Phase V EIA late 8th c. BC	Coarse/ uncertain

TABLE 9.7 Idealized versions of incised and punched decoration found on fragments from tumulus fill

Motifs	Rim/neck	Shoulder	Handle	Number of occurrences	Phase/date	Fabric/type
Zigzag band				**9/37** (P342)	Phase II LBA/EIA 12th–11th c. BC	Fine, semi-coarse/ uncertain
Concentric circle				**9/85** (P346)	Phase II LBA/EIA 12th–11th c. BC	Fine, semi-coarse/ uncertain
Row of dots				**9/53** (P414)	Phase II LBA/EIA 12th–11th c. BC	Fine/uncertain
Punched hourglass				**9/52** (P246)	Phase II LBA/EIA 12th–11th c. BC	Fine/uncertain
Incised pendent triangles				**9/54** (P133)	Phase IV LBA/EIA 10th–9th c. BC	Fine/uncertain
Hatched pendent triangles				**9/56** (P371)	Phase IV LBA/EIA 10th–9th c. BC	Fine/uncertain
Vertical zigzag with dotted interstices				**9/57** (P375)	Phase II LBA/EIA 12th–11th c. BC	Fine/uncertain

TABLE 9.8 Lofkënd pottery by fabric and find context

Fabric Type	Find context				
	Topsoil	*Tumulus fill*	*Grave fill*	*Grave offering*	*Total*
Fine light	25	62	17	9	113
Fine dark	11	57	4	6	78
Semi-coarse	21	87	9	2	119
Coarse	11	54	9	0	74
Other	6	6	0	0	12
Total	74	266	39	17	396

TABLE 9.9 Fine light fabric conspectus

Type	Size	Complete (or nearly)	Handle(s)	Rim and lip	Body and neck	Base
1	Small	**9/1**[b]		Sharp biconical: **9/2**		
2a	Medium	**9/3**[a,b], **9/4**[b], **9/5**[a,b], **9/6**[a,b]		Everted, rounded/flat; cf. **9/9** (including neck), **9/10**[b], **9/11**, **9/12**, **9/13**		
2b	Small/medium	**9/7**[b], **9/8**[a,b]				
3	Small	**9/14**[a,b]				
4	Medium	**9/15**[b]				
5a	Small/medium	**9/16**				
5b		**9/17**, **9/18**				
Uncertain type	Very small	S-profile (Type 5c?): **9/19**[b]		Vertical/tapered: **9/42**	S-profile, with vertical strap-rectangle (Type 5c?): **9/20**[b]	
	Small		Vertical strap/ovoid: **9/21** Vertical strap/rectangle: **9/22**, **9/23**, **9/24** Ribbon: **9/25**[b] Other: **9/26**[b]		**9/52**[c], **9/53**[c], **9/54**[c], **9/55**[c], **9/56**[c], **9/57**[c], **9/58**[b], **9/59**[b], **9/60**[b], **9/61**[b], **9/62**[b], **9/63**, **9/64**, **9/65**[b], **9/66**[b]	
			Vertical strap/ovoid: **9/27** Vertical strap/rectangle: **9/28**			
	Medium		Small/medium: spur: **9/29**, **9/30** Medium: vertical strap/rectangle: **9/31**[b], **9/32** Spur: **9/33** Pierced: **9/34** Horizontal: **9/35** Strutted(?): **9/36** Medium/large: Horizontal ring/loop: **9/37**[c] Ring: **9/38**, **9/39** Vertical strap/rectangle-loop?: **9/40**[b]	Small/medium: everted/flaring rim: **9/43**, **9/44**, **9/45** Medium: vertical/flaring, tapered lip: **9/46**, **9/47**[b] Incurving?: **9/48** Widely flaring: **9/49**[b]	Small/medium: **9/67**, **9/68** Medium: **9/69**[b], **9/70**[b], **9/71**[a], **9/72**[b], **9/73**[b] Medium/large: **9/74**[b], **9/75**[b], **9/76**[b], **9/77**[b], **9/78**[b], **9/79**[b], **9/80**[b], **9/81**[b] Large: **9/82**[b], **9/83**[b], **9/84**[b], **9/85**[c]	Medium/large: **9/86**
	Large		Vertical strap/rectangle: **9/41**	Medium/large: flaring rim, flat lip, external projection: **9/50** Rim/spout, jug with cutaway neck(?): **9/51**		Flaring foot: **9/87** Hollowed: **9/88** Flat: **9/89b**

[a] Plastic decoration.
[b] Matt-painted.
[c] Incised/punched.

TABLE **9.10** Fine dark fabric conspectus

Type	Size	Complete (or nearly)	Handle	Rim and lip	Body and neck	Base
1	Small	**9/90**[a]				
2a	Small	**9/91**				
2b	Small/medium	**9/92**				
3	Small	**9/93**				
4	Small	**9/94**[a]				
5	Small	**9/95**[a], **9/96**[a]	Vertical ring-round/rhomboidal: **9/97**		Biconical: **9/98**[a], **9/99**[a]	
Uncertain	Small	**9/100**, **9/101** **9/103**, **9/104**	*Small/medium* Vertical strap-rectangle: **9/105** Vertical strap-oval: **9/106**	Tapered: **9/124** Flaring, tapered: **9/125** Shallow bowl/spout: **9/126**, **9/127** Flaring, flat: **9/128**, **9/129**, **9/130**, **9/131** Sharply carinated, tapered: **9/132** Sharply everted: **9/133** Narrow mouth, rounded lip: **9/134** Everted, rounded: **9/135**, **9/136** *Small/medium* Rounded/tapered lip: **9/111**, **9/112**, **9/113** Flaring, flat: **9/137** Rounded/tapered lip: **9/138** Everted, rounded/tapered: **9/139**, **9/140**, **9/141**, **9/142**, **9/143**, **9/144**, **9/145**	**9/102**	
	Medium		Looped/lug: **9/107** Vertical strap-oval: **9/108** Vertical strap-rectangle: **9/109** Vertical strap-sharp bend, ridge: **9/110**	Flaring, tapered: **9/146** Flaring, rounded: **9/147** Everted, rounded/tapered: **9/148**, **9/149**, **9/150**, **9/151** Sharply everted: **9/152**	**9/155**[a], **9/156**[a], **9/157**	
	Medium/large		Wide strap/ribbon: **9/114** Vertical strap-triangular: **9/115** Vertical strap-sharp bend: **9/116**, **9/117**, **9/118**, **9/119**	Everted, rounded/tapered: **9/153** Everted, irregular lip: **9/154**	**9/158** (uncertain shape: shoulder or base)	
			Horizontal/ovoid: **9/120** Wide strap/ribbon: **9/121**, **9/122**, **9/123**			

[a] Plastic decoration.

Table 9.11 Semi-coarse fabric conspectus

Type	Size	(Nearly) complete	Handle	Rim and lip	Body and neck	Base	Uncertain/other
1	Small	**9/159**					
2	Very small	**9/160**					
Uncertain	Small			Rounded/tapered: **9/199**			Lug? **9/255**[a]
	Small/medium		Vertical strap-ovoid: **9/162**		Carinated: **9/236** Wall: **9/237**[c]		
	Medium	**9/161**	Vertical strap-ovoid: **9/163, 9/164, 9/165** Vertical strap-oval: **9/166** Vertical strap-triangle: **9/167** Squared spur: **9/168, 9/169** Pierced, squared spur: **9/170** Flaring spur: **9/171, 9/172** Rounded spur: **9/173** Horned: **9/174** Lug/spur: **9/175, 9/176** (pierced)	Slightly everted, rounded: **9/200** Everted, rounded: **9/201, 9/202, 9/203, 9/204, 9/205, 9/206, 9/207, 9/208** Everted, round, flat top: **9/209, 9/210, 9/211** Everted, flat/tapered: **9/212, 9/213, 9/214** Sharply everted: **9/215** Flaring, rounded: **9/216** Flaring, flat: **9/217** Flat, squared lip: **9/218**	Shoulder: **9/238, 9/239** Shoulder/neck: **9/240**	*Medium/large* Slightly raised and hollowed: **9/251**[b] *Large* Flat, raised: **9/252** Raised, hollowed: **9/253**[b] Very large Raised, hollowed **9/254**[b]	
	Medium/large		Vertical strap-rectangle: **9/177, 9/178** Vertical strap-pierced: **9/179** Vertical strap-flat oval: **9/180** **9/181** (with strut?) Vertical strap-ovoid: **9/182, 9/183** Horizontal wishbone: **9/184** Squared horizontal: **9/185** Forked horizontal: **9/186** Pierced horizontal: **9/187** Spur: **9/188** Squared spur: **9/189** Flaring spur: **9/190**	No neck, uneven lip: **9/219** Everted/rounded: **9/220**[b] *Large* Widely flaring, flat: **9/221**[b], **9/222, 9/223, 9/224, 9/225** Widely flaring, founded: **9/226, 9/227** Flat lip, uncertain neck/rim: **9/228** Everted, rounded: **9/229** Everted, flat: **9/230, 9/231** Open shape—slightly flaring, rounded: **9/232, 9/233** Vertical/slightly everted, flat: **9/234**[b] Vertical/slightly everted, tapered: **9/235**[b]	Shoulder: **9/241**[b], **9/242** *Large* Body: **9/243, 9/244, 9/245**[b], **9/246**[b] Flaring neck: **9/247**[b], **9/248**[b] *Very large/large* **9/249**[b] Uncertain size **9/250**[b]		Lug?: **9/256**[a] *Uncertain size* Lug?: **9/257**[a] Pierced? **9/258**
	Large		Vertical strap-rectangle: **9/191** Vertical strap-flat oval: **9/192, 9/193, 9/194, 9/195, 9/196** Vertical strap-ovoid: **9/197, 9/198**				

[a] Plastic decoration.
[b] Matt-painted.
[c] Incised/punched.

TABLE **9.12** Coarseware conspectus

Type	Size	Nearly complete	Handle(s)	Rim and lip	Body and neck	Base	Uncertain
1	Large	**9/259**					
Uncertain type	Medium/ large		Vertical strap-rectangle: **9/260** Vertical strap-ovoid: **9/261, 9/262, 9/263** Vertical strap-oval: **9/264**	Flat rim, external projection: **9/278**	Wall, ovoid projection: **9/304**[a]		
	Large		Spur handle: **9/265** Horizontal handle: **9/266, 9/267** Vertical strap-rectangle: **9/268** Vertical strap-flat oval: **9/269, 9/270, 9/271** Vertical strap-ovoid: **9/272** Vertical strap-triangle: **9/273** Vertical strap-ridge: **9/274, 9/275**	Flat rim, external projection: **9/279** Rounded, external projection: **9/280** Flaring, rounded: **9/281** Flaring, flat: **9/282, 9/283** Vertical, flat: **9/284, 9/285** Hole-mouthed jar?: **9/286, 9/287, 9/288, 9/289** *Large/very large* Flat rim, external projection: **9/290, 9/291, 9/292, 9/293, 9/294** Flat rim, external and internal projection: **9/295** Flaring, no lip: **9/296, 9/2**		Raised, slightly hollowed: **9/315, 9/316** Flat: **9/317, 9/318** Flat, raised: **9/319**	Handle? Rim? **9/320**
	Very large		Vertical strap-rectangle: **9/276** Vertical strap-90° bend: **9/277**	Flat rim, external projection: **9/298, 9/299, 9/300** Flaring, rounded/tapered: **9/301, 9/302, 9/303**	Pithos wall: **9/305** Pithos wall(?): **9/306** Wall: **9/307, 9/308, 9/309** (punched?) Wall, round projection: **9/310**[a] Wall, triangular projection: **9/311**[a], **9/312**[a] Wall, relief band: **9/313**,[a] **9/314**[a]		Pithos lug? Large basin rim?—**9/321**[a] Uncertain size Rim? **9/322** Tile/tripod leg/pyraunos? **9/323, 9/324, 9/325, 9/326, 9/327, 9/328** Wall (incised?) **9/329**

[a] Plastic decoration.

TABLE **9.13** Non-prehistoric fabrics

Fabric type	Catalogue#
Greek imported	**9/330** (Attic?); **9/331** (Greek?),**9/332, 9/333, 9/334, 9/335, 9/336** (Corinthian)
Uncertain, reduced	**9/337, 9/338, 9/339**
Misfired/vitrified	**9/340**
Early modern	**9/341**

Appendix 2
Illyria Capta: Corinthian Wheelmade Pottery from the Lofkënd Tumulus

Sarah P. Morris

As Chapter 9 has demonstrated, the fill of the tumulus, and indeed its topsoil, was dominated by local (Illyrian) handmade pottery of the Late Bronze and Iron Ages, much of it contemporary or older than the tumulus burials, with very few wheelmade sherds or imports from beyond Illyria. Only scarce surface finds (e.g., **9/330** [P002] (Attic?), **Fig. 1.13**) indicate later historical (Greek or Roman) periods, and even medieval and early modern finds were rare (e.g., a single fragment of early modern porcelain, **9/341** [P273]) beyond pieces of modern tile and nail (Chapter 1, **Table 1.1**). In more recent centuries, this prominent hill attracted shepherds, farmers, and soldiers, who littered the tumulus with gun cartridges (28 were recovered from topsoil in all four trenches, along with four bullets (see Appendix 3, following Chapter 10) and other debris such as modern glass, plastic, and the like (Chapter 1, **Table 1.3**, **Fig. 1.16a–b**).

The latest ancient finds, however, were a handful of wheelmade sherds that make a striking impression as imported Greek ceramics, unmistakably of Corinthian clay, and at least two had the form of the kotyle, or straight-sided cup, common in Archaic and Classical Greece (**Table A2.1**). None survives in more than a few fragments, and those that do (**9/333** [P257]) were recovered from various different units of topsoil and disturbed tumulus fill across two trenches, as if broken and scattered long ago (**Figs. A2.1–A2.2**). The sherds were too small and fragile to clean thoroughly, mend, or restore in drawings: traces of an incised rosette and parallel rays on lower body sherds (no foot fragments were found) help date them to the Archaic Corinthian period, somewhere in the sixth century BC. This places them long after the last burial in the tumulus (no later than 800 BC; see Chapters 3 and 4), two or more centuries later, but they offer a tantalizing link to long-term history in the region.

The city of Corinth (in a legend later revised, during the Peloponnesian War, to include Kerkyra as co-founder: Stocker 2009:266–285) established a city named for the god Apollo at the mouth of the Aoös/ Vjösë River and near that of the Seman (some 30 km northwest of the Gjanicë valley), sometime in the late seventh or sixth century BC (Stocker and Davis 2006). At a more proximate distance, the prominent hilltop citadel of Margëlliç, which dominates the Gjanicë valley from the north, is visible from the mound and lies only a few kilometers away (**Figs. 1.5a–c, 20.6**). Inhabited for at least a short time in the Bronze Age, this site produced Corinthian imports since the seventh century in the form of Type A amphorae, and may be the closest locale (along with Mashkjezë, near Cakran) in regular contact with Greek maritime traffic along the Adriatic (Stocker and Davis 2006:87, citing Ceka 1983a, 1985b, and 1986a; Stocker 2009:329–352). Yet surface finds of Corinthian finewares, anywhere outside the necropolis (see below) and cult centers (Davis et al. 2006), are rare in the Apollonia area, including material collected in the MRAP regional survey (none were found west of the Gjanicë River valley). Beyond Apollonia, only Margëlliç (possibly a transshipment center for Corinthian amphorae or their contents: Davis et al. 2007:18–19; Stocker 2009:323–324) produced early Archaic material and Archaic finewares, including Corinthian painted sherds like the ones from the Lofkënd tumulus (Stocker 2009: 835–847), always in association with indigenous material. This recalls the unusual circumstances of Grave 45 in Tumulus 9 at Apollonia, the burial of a young woman with Illyrian bronze jewelry (bracelets, spectacle fibulae) but also a pair of Corinthian kotylai (one Protocorinthian, the other closer to the date of the Lofkënd fragments: Amore 2010:154; Amore and Dimo 2010:363–364, nos. 1/29–30).[1] Such shapes were also common in tumulus fill at Apollonia (Amore and Dimo 2010: 364), but Apollonia Grave 45 offers a striking juxtaposition of Illyrian artifacts (in bronze) and imported Corinthian pottery (Amore 2010: 693–694). At least one family brought together material markers of indigenous and imported cultures to commemorate the death of a young woman (Illyrian bride of a Corinthian colonist? a female with a mixed dowry of local and imported possessions?). To find Corinthian sherds at another Illyrian burial tumulus, however removed from the nearest settlement (Margëlliç?), marks a different encounter between native burial customs and intrusive Greek objects, one that may have led to their fusion at Apollonia.

---

[1] Two examples closer in style and date to the Lofkënd example(s) from a slightly later burial, Grave 60, a primary cremation with many ceramic offerings, are probably Attic (Amore and Dimo 2010:363–365, nos. 1/31–32; the same grave also held Lakonian kraters and black-glazed pottery, for an overall date in the late sixth century). "Grave" 62 (a mud brick structure with burned animal bones) held another pair of Corinthian kotylai, dated to the Middle Corinthian period (Amore and Dimo 2010:363–366, nos. 1/33–34).

TABLE A2.1 Corinthian pottery, Lofkënd tumulus (**Fig. 9.68**: **9/333**, **9/335**, **9/336**; **Fig. 9.69**: **9/339**)

Inventory/Catalog #	Trench	Unit	Date	Description
9/331 (P424)	2	2 (topsoil)	10/7/2007	Rim: kotyle?
9/332 (P175)	1	70 (tumulus fill)	1/7/2005	Wall (kotyle?)
9/333 (P257)	1	240 (tumulus fill)	19/7/2005	1 rim, 3 body
	1	278 (topsoil, below 1 m)	22/7/2005	1 rim, 2 body (including rays)
	1	279 (tumulus fill)	22/7/2005	1 handle?
9/334 (P328)	2	431 (dark soil in NE corner)	5/7/2006	Wall fragment, painted (dark red)
9/335 (P372)	1	509 (dark unit west of topsoil; from baulk)	20/7/2006	Wall (kotyle?)
9/336 (P384)	1	375 (topsoil, start of 2006 excavations)	21/7/2006	Body sherd
Probably not Corinthian				
9/339 (P338)	4	286 (tumulus fill)	5/7/2006	Wall, rim

Most Illyrian tumuli of the Early Iron Age (including those at Apollonia, later reused by Greeks) were abandoned by the middle of the eighth century BC, along with settlements, which either disappeared (as at Kraps; see Stocker 2009:828–830) or contracted severely (as at Margëlliç; see Stocker 2009: 830). This disruption was accompanied by an important shift toward new and more easily defensible sites (or, in the case of Margëlliç, toward more easily defended areas of the same site), many first settled in the later seventh century BC (including perhaps Apollonia)[2], but others as late as the sixth century BC (Ceka 2011). Some speculate that in the intervening century or so, many Illyrian tribes may have moved to Italy, possibly disrupted or displaced by new Balkan arrivals (Stocker 2009:834 [citing Hammond 1992; Lamboley 1987; Lomas 2000; Prendi 1982]). New sites in the area near the Lofkënd tumulus include Klos, Mashkjezë, and Gurëzezë, not previously occupied by newly arrived Greeks, although imported Greek sherds first appear there. Whether these were permanent settlements remains an open question: close investigation of the Grunas fortress in the Shala valley (Galaty, Lafe, and Tafilica 2011; Galaty et al. 2013) indicates an initial phase of massive stone terracing and imported clay for packing, floors, and platforms (ca. 800 BC), followed by a later phase of largely extramural settlement (ca. 800–400 BC), before abandonment with some minor seasonal reuse. Few other protohistoric fortified sites have yielded enough stratified material for secure sequences of occupation (Ceka 2011; Gjipali 2011; Lima 2013).

One can only speculate that, long after the tumulus covered its last native burials, its prominent dome attracted visitors to this site, whether Greek or Illyrian, who left sherds of Greek cups at the site, whether as a casual loss or a deliberate offering to the dead. Unlike at Apollonia, such visitors were not inspired to reuse the same mound for their own burials, as evidently happened in the Kryegjatë cemetery (e.g., Tumulus 10; see Amore 2010). Whoever these visitors to Lofkënd were, as they dropped or deposited Corinthian vessels, they left signs of contact with a new, intrusive culture that was to change the face of Illyrian history and material culture forever. At the same time, they or others carried away, to sites like Apollonia, indigenous practices such as tumulus burial, such that, as in Horace's famous phrase for the Roman capture of Greece, new arrivals were converted to some of the native customs of the culture they had invaded.

Comparanda (*kotyle*/*skyphos*)

Apollonia: Bereti et al. 2007:130, fig. 55.6

Apollonia necropolis: Amore and Dimo 2010:363–372, "Kotyle" (esp. 1/31–1/40, Middle Corinthian)

Apollonia, Bonjaket sanctuary: Davis et al. 2011:211–212 (Trench 16T: two kotylai, 16-3 and 16-11, ca. 550 BC; not illustrated)

Corinth: Weinberg 1943, #337–354 (Middle Corinthian); Palmer in Blegen, Palmer, and Young 1964:104–108; Amyx and Lawrence 1975:73–78, 209, pl. 39 (Late Corinthian), An 31, An 41, pl. 67 (beginning of Late Corinthian), An 58, pl. 66 (Middle Corinthian)

Olympia: Heiden 2000:190, no. 69, pls. 57, 66 (Early Corinthian, from well filled ca. 600 BC), H 7.3 cm, D 9.3 cm; p. 191, no. 77, pls. 59, 66 (Late Corinthian, from well under stadium, North Wall), H 7.9 cm, D 11.7 cm; no. 80, pls. 59, 67 (Late Corinthian), H 8.5 cm, D 12.1 cm

---

2 Ceka (2011:649–650) claims "une donnée stratigraphique sûre" for dating Apollonia's earliest wall to the final quarter of the seventh century, citing Balandier, Koço, and Leinhardt (2007: 167–169, fig. 76), but they make no such claim—in fact, they call for sondages to obtain dates for the walls (pp. 185–186). Rather, Ceka compares the masonry of the acropolis walls at Apollonia to walls at Margëlliç and Mashkjezë, dated by Protocorinthian and Early Corinthian sherds, respectively.

Chapter 10

Objects of Terracotta, Metal (Gold/Electrum, Bronze, Iron, and Bimetallic), Semi-Precious Stone, Faience, Glass, and Worked Bone

John K. Papadopoulos and Rovena Kurti
With contributions by Vanessa Muros and Yannis Mylonas

Introduction

The small finds other than pottery are presented in this chapter. The material includes that deposited in tombs as well as that encountered in the tumulus and in the topsoil. We begin with the objects believed to be ancient based on their context, followed by those considered ancient on the basis of style, fabric, and feel that were encountered in topsoil. The survey of ancient objects is followed by the modern (ancient and modern material picked up on the surface prior to excavations is presented separately in Chapter 1). Individual classes of object are presented in a typology, with full details of their chronology and their cultural affinities provided in the introductory remarks of each category.

The chapter begins with terracotta objects, specifically, with the terracotta spindlewhorls, beads, or buttons, which follow on from the pottery presented in Chapter 9. As the majority of the terracottas are spindlewhorls, beads, or buttons, determining whether individual objects were items of personal ornament or implements/tools was not possible, particularly for those pieces that were not encountered in tombs. In contrast, the remainder of the ancient material is presented under the following two broad categories: jewelry and other personal ornament items, and tools/weapons. Within each of these categories, the material is presented according to the type of object (e.g., fibulae, pins, pendants, disks/bosses, spiral coils, rings, headbands [diadems], beads, spearhead, arrowheads, knives, and so on), the material of which it was made (gold/electrum, bronze, iron, bimetallic, stone, faience, glass, bone), and, wherever possible, according to type. A draft catalogue of selected pieces of jewelry was initiated by Rovena Kurti and completed by John Papadopoulos, together with the comparanda from Albania, the Aegean, other parts of Europe, and, wherever possible, the ancient Near East. The typology of the fibulae and pins was also prepared by Papadopoulos. The sections on the faience and glass were prepared by Papadopoulos and Vanessa Muros, while the overview of the spent gun cartridges and bullets (Appendix 3) was prepared by Papadopoulos and Yannis Mylonas.

At least two tombs (XXXVI and LXIX) yielded small and fragmentary shells that were found to be naturally perforated (see Chapters 3 and 16.2). In both tombs the shells were minuscule (less than 1 cm in length), and there was never more than one in each of the two graves. Both fragments preserve the operculum of land gastropods (see **Fig. 3.119a–b**). Although naturally perforated objects could have served as amulets or ornaments, and are often found with women and children in other parts of Europe in the Early Iron Age (see Pauli 1975), the two Lofkënd examples were almost certainly naturally perforated fragments of shell that were not utilized as beads or other ornaments. The position of the shells in each tomb is noted in Chapter 3, but there is no discussion of them in this chapter, as they are not considered items of personal ornament but rather as naturally occurring in the fill of each tomb.

Conspectus of the Small Finds[1]

I. **Terracotta objects**
1. Terracotta spindlewhorls, beads, or buttons: **10/1–10/8**
2. Terracotta spindlewhorls, beads, or buttons formed from broken fragments of pottery: **10/9**
3. Terracotta, other (possible loomweight?): **10/10**

II. **Jewelry and other personal ornament items**
1. Objects of gold/electrum: ear or head ornaments: **10/11, 10/12**
2. Fibulae
   a. Bronze fibulae
      Spectacle fibulae
         Type I.1.a: **10/13**
         Type I.1.b: **10/14**
         Type I.1.c: **10/15, 10/16, 10/17**
         Type I.1.d: **10/18**
      "Cassibile" fibulae
         Type I.2.a: **10/19, 10/20**
         Type I.2.b: **10/21**
   b. Iron fibulae
      Type II.1: **10/22**
      Type II.2: **10/23, 10/24**
   c. Bimetallic fibulae
      Type III.1: 10/25, 10/26
3. Pins
   a. Bronze pins
      Type I.1: **10/27**
      Type I.2: **10/28**
      Type I.3: **10/29**
      Bronze pin shafts of unidentified type: **10/30, 10/31, 10/32**
   b. Iron pins
      Type II.1: **10/33, 10/34, 10/35, 10/36, 10/37, 10/38, 10/39, 10/40, 10/41**
      Type II.2: **10/42, 10/43**
      Type II.3: **10/44**
      Type II.4: **10/45**
      Fragmentary iron pin shafts: **10/46, 10/47, 10/48, 10/49**
   c. Bimetallic pins
      Type III.1: **10/50**
      Type III.2: **10/51**
   d. Bone pins
      Type IV.1: **10/52**
      Type IV.2: **10/53**
      Type IV.3: **10/54**
      Type IV.4: **10/55**
      Bone pin shaft of unknown type: **10/56**
4. Pendants
   a. Small bronze double-spiral ("spectacle") pendants: **10/57, 10/58**
   b. Wheel pendants: **10/59**
5. Disks/bosses
   a. Bronze: **10/60, 10/61, 10/62, 10/63**
   b. Iron: **10/64**
6. Spiral coils
   a. Small bronze coils (beads): **10/65, 10/66, 10/67**
   b. Other bronze spirals (earrings or hair rings?): **10/68, 10/69, 10/70**
   c. Iron spiral coils: **10/71, 10/72**
7. Bronze rings
   a. Plain rings, including possible finger rings: **10/73, 11.74, 10/75, 10/76, 10/77, 10/78, 10/79**
   b. Earrings: **10/80, 10/81, 10/82, 10/83**
8. Headbands (diadems)
   a. Bronze: **10/84, 10/85, 10/86, 10/87**
9. Length of copper/copper alloy wire: **10/88**
10. Beads
   a. Tubular iron beads
      Type I: **10/89, 10/90, 10/91, 10/92, 10/93, 10/94, 10/95, 10/96, 10/97, 10/98, 10/99**
      Type II: **10/100, 10/101**
   b. Stone beads: **10/102, 10/103, 10/104**
   c. Faience bead: **10/105**
   d. Glass beads
      Beck Type I.C.1.a and related: **10/106, 10/107, 10/108, 10/109**
      Beck Type I.B.1.a and related: **10/110, 10/111, 10/112, 10/113**
      Beck Type II.C.1.a/XXIII.A.2.a: **10/114**
      Beck Type VII.D.1.a/XLVII.A.7.a: **10/115**
      Fragmentary glass beads of uncertain type: **10/116, 10/117**

III. **Tools/weapons**
1. Iron spearhead: **10/118**
2. Iron arrowheads: **10/119, 10/120**
3. Iron knives: **10/121, 10/122, 10/123**
4. Worked bone **10/124**
5. Unidentified objects of copper/copper alloy: **10/125, 10/126, 10/127, 10/128, 10/129, 10/130, 10/131**

[1] Remnants of textiles or textile pseudomorphs noted in the catalogue entries are fully discussed in Chapter 12.

6. Unidentified iron objects: **10/132, 10/133, 10/134**

IV. **Modern objects and fragments**
1. Copper, zinc, tin, and lead alloy: **10/135**
2. Iron: **10/136, 10/137, 10/138, 10/139, 10/140, 10/141, 10/142, 10/143, 10/144**
3. Modern iron and other objects and fragments

TERRACOTTA OBJECTS

Spindlewhorls, Beads, or Buttons

Eight terracottas may be classed as spindlewhorls, beads, or buttons, all of biconical or related form (**Fig. 10.1**). Of these, four were found in tumulus fill (**10/3** [SF 169], **10/4** [SF 168], **10/5** [SF 065], **10/8** [SF 138]), and one in topsoil (**10/7** [SF 293]), none of which can be dated precisely. Only three derive from tombs: **10/6** (TLXVII-1), which dates to the late Phase Va (ninth or early eighth century BC), and two from Tomb XXXI (**10/1, 10/2**), which can be assigned to Phase II (twelfth–eleventh centuries BC). The example from Tomb LXVII was associated with a female burial, the two in Tomb XXXI with an individual of indeterminate sex.

Such small pierced terracottas are usually variously classified as spindlewhorls, beads, or buttons (cf. Barber 1991:39–788; Heurtley 1939:165, fig. 35p; 203, fig. 67l–ee; 213, fig. 83l–o; 231, fig. 104a–l; 240, fig. 112h; McDonald, Coulson, and Rosser 1983:287; Popham, Sackett, and Themelis 1979–1980:83; Wells 1983a:80), designations that cover the range of likely possibilities. The three examples found in tombs share a similar placement in the tomb: that in Tomb LXVII was found west of the westernmost femur of the female inhumation, and the two from Tomb XXXI were found along the northeast face of what appeared to be the left femur of the deceased. Unless displaced, the position of the three terracottas in tombs makes it difficult to see them as beads or buttons actually worn by the deceased at the time of death, though they may originally have served such a function in life. That similar terracottas served as beads or buttons is clear from various contexts in Greece. At least one appears to have been worn as a bead in Tomb 10 in the Early Iron Age cemetery at Torone (Papadopoulos 2005:553–554). In Mycenaean Greece, related terracotta or stone examples are sometimes interpreted as buttons; for example, Mylonas (1948:73, with n. 67 for a list of examples) notes: "That the bodies were laid dressed is indicated by the discovery of a great number of buttons (often called spindlewhorls) . . . attached to the clothes." Alternatively, Iakovidis (1977:113–119) interprets the Mycenaean examples as dress weights (see also Iakovidis 1969–1970, vol. 2:277–281; Furumark 1972b:89, with references; Barber 1991; Rehak 1996).

In Albania, similar terracottas are known from several prehistoric tumuli, both from tombs and tumulus fill, including Burreli (Kurti 1977–1978:173, pl. I, Tomb 2, no. 9; 174, pl. II, Tomb 4, no. 4; 186, pl. XIV, Tomb 76, no. 5), Patos (Korkuti 1981:41, pl. IV, Tomb 27), Shtoj (Jubani 1992:49, pl. II, Tomb 10, no. 5), Shkrel (Jubani 1995:82, pl. VII, Tumulus II, no. 1; 85, pl. X, Tumulus V fill, nos. 8–9; 86, pl. XI, Tumulus VI, no. 5), Kënetë (Jubani 1983:126, pl. VI, no. 62 [tumulus fill]), and Prodan (Aliu 1984:62, pl. VIII, no. 84 [tumulus fill]), although details of their precise position in the grave are often lacking. They are also known from Neolithic and Bronze Age settlement sites (e.g., Andrea 1996:42, pl. I, nos. 4–10; 47, pl. VI, nos. 8–10, 12–20; Gjipali 1995:51, pl. XII, no. 7; Jubani 1994:99, pl. I, no. 1; cf. also Ugolini 1942:183–184, fig. 188, no. 3). This basic type of biconical spindlewhorl, bead, or button is standard in northern Greece during the Bronze and Early Iron Ages (in addition to the examples cited in Heurtley 1939, see, among others: Andronikos 1969:260, pl. 116ζ; Andronikos et al. 1988:169, no. 87; Carington Smith and Vokotopoulou 1988:368, no. 5; Casson 1923–1925:21, fig. 10; Vokotopoulou 1969a: pl. 255γ; Wardle 1980: pl. 22:e) and is common elsewhere in the Bronze Age and much later contexts (for bibliography, see Papadopoulos 2005:567–568, n. 2).

10/1 (SF 389, TXXXI-2), **Figs. 3.105b, 10.1** (Sheet 98.5)
Terracotta Spindlewhorl, Bead, or Button.
T86-2 (SU 2.0489).
H: 0.027; D: 0.023–0.024; D (perforation): 0.005; Wt: 11.3 g.
Intact, except for minor chipping, especially around one aperture. Surface much worn.
Fabric: As **10/2** (SF 388), fired close to light yellowish brown (10 YR 6.4).
Shape and surface smoothing as SF 388, but more regularly biconical, with top and bottom around perforation flat.

Cf. Bartoloni et al. 1980:162, pl. 54, no. 10; Blinkenberg 1931:139, pl. 14, no. 377; Papadopoulos 2005: pl. 450c, i, no. 55.

10/2 (SF 388, TXXXI-1), **Figs. 3.105a, 10.1** (Sheet 98.6)
Terracotta Spindlewhorl, Bead, or Button.
T86-1 (SU 2.0489).
H: 0.025; D: 0.023; D (perforation): 0.004; Wt: 9.1 g.
Complete except for minor chipping at one point on body and at both apertures. Surface much worn.
Fabric: Semi-coarse with moderate small to medium white, light-colored, and some dark inclusions (1.0–2.0 mm), and some fine silvery mica. Surfaces fired close to light yellowish brown and pale brown (10 YR 6/4–6/3).
Form roughly biconical, but a little irregular and slightly more rounded than SF 389. Surfaces smoothed but much worn.
Cf. **10/1** (SF 389); Casson 1923–1925:21, fig. 10 (center); Koukouli-Chrysanthaki 1992:410, fig. 94, ΣII, 1γ; Papadopoulos 2005: pl. 450b, h, T46-3.

10/3 (SF 169), **Fig. 10.1** (Sheet 57.3; Photo 2717)
Terracotta Spindlewhorl, Bead, or Button.
Tumulus Fill (SU 2.0239).
H: 0.030; D: 0.033–0.034; D (perforation): 0.006; Wt: 22.7 g.
Intact; cracked at one point but not broken; surfaces somewhat worn.
Fabric: Fine, with only the occasional very small white and light-colored inclusion erupting onto the surface and some fine silvery mica. Surfaces fired close to pale brown and brown (10 YR 6/3–5/3); slightly blackened at one point.
Form biconical, consisting of two more or less equal parts; top and bottom around perforation flat. Surface polished smooth, but only a faint sheen preserved here and there.
Cf. Aliu 1984:62, pl. VIII, no. 84; Benac and Čović 1956: pl. X, no. 11; Boardman 1967:235, fig. 155, no. 511; Heurtley 1939:231, fig. 104:f (Late Bronze Age); Koukouli-Chrysanthaki 1992:410, fig. 94, ΣII, 1α; Papadopoulos 2005: pl. 450b; Truhelka 1904: pl. XIV, no. 3; Vokotopoulou 1986: fig. 60β, inv. 5400.

10/4 (SF 168), **Fig. 10.1** (Sheet 57.4; Photo 2721)
Terracotta Spindlewhorl, Bead, or Button.
Tumulus Fill (SU 1.0039).
H: 0.022; D: 0.032; D (perforation): 0.005; Wt: 11.1 g.
Reconstructed from three joining fragments preserving almost complete spindlewhorl, bead, or button, except for chips on one side and at one of the apertures; surfaces worn.
Fabric: Fine to semi-fine, with only the occasional small to very small white inclusion and a dusting of fine mica. Core fired black; surfaces fired close to brown and strong brown (7.5 YR 5/4–5/6). Surface blackened in parts.
Sharply biconical, composed of two equally symmetrical parts. Surface where best preserved burnished, in parts with a slight sheen.
Cf., among others, Koukouli-Chrysanthaki 1992:410, fig. 94, ΣII, 1β; Papadopoulos 2005: pl. 450c; Vokotopoulou 1986: fig. 60β, inv. 2449.

10/5 (SF 065), **Fig. 10.1** (Sheet 10.6; Photo 682)
Terracotta Spindlewhorl, Bead, or Button.
Papadopoulos, Bejko, and Morris 2007:132, fig. 26c.
Tumulus Fill (SU 1.0007).
H: 0.023; D: 0.025; D (perforation): 0.006–0.007; Wt: 10.4 g.
Intact, except for minor chipping around perforation; surface rather worn.
Fabric: Fine to semi-fine, with few visible impurities, although with the occasional small white inclusion and a fine dusting of surface mica. Fabric fired close to strong brown (7.5 YR 5/6) and yellowish red (5 YR 5/6).
Form biconical but asymmetrical. Surface smoothed.
Cf. Heurtley 1939:231, fig. 104:c, i (Late Bronze Age); Koka 2012:252, pl. VI, Tumulus 2, Tomb 10, no. 5; Kurti 1977–1978:173, pl. I, Tomb 2, no. 9.

10/6 (SF 130, TLXVII-1), **Figs. 3.228, 10.1** (Sheet 29.6; Photo 686)
Terracotta Spindlewhorl, Bead, or Button.
Papadopoulos, Bejko, and Morris 2007:132, fig. 26c.
T12-1 (SU 3.0100).
H: 0.022; D: 0.027; D (perforation): 0.002 (at center); Wt: 10.4 g.
Single fragment preserving most of spindlewhorl, bead, or button, heavily chipped on one side and at both apertures. Surface rather worn.
Fabric: Semi-coarse, with moderate small white and light-colored inclusions (0.5–1 mm) and a little fine silvery mica. Core fired dark gray to black;

surface fired close to yellowish red and reddish yellow (5 YR 5/6–6/6).

Form biconical, more symmetrical, and a little more pushed down, than **10/5** (SF 065). Surface roughly smoothed. Unusually small perforation at center, suggesting a bead or button rather than spindlewhorl.

Cf. **10/5** (SF 065); cf. also Bunguri 1994:99, pl. I, nos. 1–2; Heurtley 1939:231, fig. 104:e (Late Bronze Age); Jubani 1983:126, pl. VI, no. 62; Kurti 1977–1978:173, pl. I, Tomb 2, no. 9; 186, pl. XIV, Tomb 76, no. 5).

10/7 (SF 293), **Fig. 10.1** (Sheet 98.7; Photo 2906)
Terracotta Spindlewhorl, Bead, or Button.
Topsoil (SU 4.0201).
H: 0.021; D: 0.029; D (perforation): 0.006; Wt: 7.3 g.

Single fragment preserving about one-half of spindlewhorl, bead, or button; surfaces much worn and blackened in parts due to soil action.

Fabric: Semi-fine to semi-coarse with occasional small white and light-colored inclusions (0.5–1 mm) and a little mica. Core and surfaces mainly fired close to pale brown and brown (10 YR 6/3–5/3).

Sharply biconical, consisting of two equal parts; top and bottom around perforation flat. Surface smoothed, but much worn.

Cf. Koukouli-Chrysanthaki 1992:410, fig. 94, ΣΙΙ, 1δ; Von der Osten 1937:199, fig. 198, c 16.

10/8 (SF 138), **Fig. 10.1** (Sheet 29.7; Photo 694)
Large, Fragmentary, Terracotta Spindlewhorl, Bead, or Button.
Papadopoulos, Bejko, and Morris 2007:132, fig. 26c.
Tumulus Fill (SU 1.0067).
H: 0.029; D: 0.051; D (perforation): 0.008–0.009; Wt: 23.6 g.

Single fragment preserving almost one-half of spindlewhorl, bead, or button. Surface worn, especially near center and at apertures.

Fabric: Semi-coarse, with moderate medium to large white and light-colored inclusions (0.5–1.0, 1.0–2.0 mm) and a little fine mica. Surface blackened at one point; elsewhere core and surface fired close to yellowish red (5 YR 5/6).

Form biconical, consisting of two more or less equal parts. Larger, more symmetrical and more pushed down than **10/5** (SF 065) and **10/6** (SF 130). Surface where less worn burnished, with a slight luster where best preserved.

Among others, cf. Heurtley 1939:231, fig. 104:h (Late Bronze Age); Kurti 1977–1978:174, pl. II, Tomb 4, no. 4; Themelis 1969:172, fig. 9, no. 19); Truhelka 1904: pl. XIV, nos. 4–5; cf. also Bartoloni et al. 1980:162, pl. 54, no. 15.

Spindlewhorl, Bead, or Button Formed from Broken Fragments of Pottery

The solitary example of a terracotta spindlewhorl, bead, or button formed from a broken fragment of pottery, **10/9**, although ancient, cannot be dated more precisely. The idea of crudely forming a broken pottery sherd into a disk and drilling a hole in its center for use—for whatever purpose—is a common phenomenon cross-culturally. It occurs frequently in Albania in Neolithic settlement sites (e.g., Gjipali 1995:51, pl. XII, no. 9 [Rashtan]; Gjipali 1999–2000:69, pl. XVI, no. 11 [Rajcës]; Korkuti 1982a:113, fig. 9, no. 1; 145, pl. XVII, no. 20 [Vashtemi]), as it does in Bronze Age and Early Iron Age burial tumuli (e.g., Aliu 1984:62, pl. 8, no. 80 [Prodan]; Bela 1990:122, pl. IV, no. 56; 127, pl. IX, no. 125 [Myç-Has]; Jubani 1995:85, pl. X, Tumulus IV, no. 1 [Shkrel]; Koka 2012:290. pl. XLI, Tumulus 10, no. 70 (Shtoj); Kurti 1977–1978:186, pl. XIV, Tomb 76, no. 4 [Burreli]; cf. Jubani 1992:52, pl. V, Tomb 12, no. 3 [Shtoj]). Similar pieces are often referred to elsewhere as spindlewhorls (e.g., Davidson 1952:175, pl. 177, no. 1215; see further the discussion in Papadopoulos 2005:556; Popham, Sackett, and Themelis 1979:83–84, pl. 65a–h; pl. 145, P Tomb 36, no. 6; pl. 189, T Tomb 36, no. 2), although many in Early Iron Age Greece are also thought to have served as "counters," or "gaming pieces" (*pessoi*) (Wells 1983b:228, nos. 516–518; for further discussion and alternative uses, see D'Onofrio 2007; Papadopoulos 2002). They are very common in prehistoric Macedonia from the Early Neolithic period through the Early Iron Age (Heurtley 1939:139, fig. 7k–l [Early Neolithic]; 203, fig. 67gg–jj [Early Bronze Age]; 213, fig. 83u [Middle Bronze Age]; 231, fig. 104m–p [Late Bronze Age]; 240, fig. 112i–j [Early Iron Age]), as they are elsewhere in Greece (e.g., Rahmstorf 2003:413, figs. 16–17, various examples from Late Bronze Age Tiryns), and in other parts of the Balkans (e.g., Hiller and Nikolov 1997: pl. 70, nos. 1, 5 [Karanovo, Neolithic]; pl. 172, no. 1 [Karanovo, Bronze Age]).

10/9 (SF 413; originally P129 + P130), **Fig. 10.2** (Sheet 37.6; Photo 2194)
Terracotta Spindlewhorl, Bead, or Button Formed from Broken Fragment of Pottery.
Tumulus Fill (SU 1.0039).
H (Th of sherd): 0.010; D: 0.041–0.042; D (perforation): 0.008–0.010; Wt: 12.5 g.
Reconstructed from two joining fragments preserving a little more than one-half of object.
Fabric semi-coarse, with occasional small white and light-colored inclusions (0.5–2.0 mm) and moderate fine mica. Core and interior surface fired close to reddish brown (5 YR 5/3), exterior surface closer to reddish brown (5 YR 5/4), in places blackened.
Discoid spindlewhorl, rather than bead or button, formed from a broken fragment of semi-coarse pottery by crudely chipping around the edges and drilling a central hole from the underside. Surface on exterior roughly smoothed, but more carefully than interior.

Other (Possible Loomweight?)

Since its discovery, **10/10** (**Fig. 10.3**) has been variously referred to as either a "terracotta spindlewhorl, bead, or button," or else a "handle fragment" of a semi-coarse to coarse pot. In drawing it, Ilir Zaloshnja reconstructed it as a possible handle fragment, and the possibility that it was a handle cannot be dismissed. The fragment, however, is not like any other handle from Lofkënd, and it is, at the same time, a most unlikely spindlewhorl, bead, or button. Although fragments are, as Sparkes and Talcott (1970:2) warn, "dangerous allies," we thought it best to present it here as a problem piece, in the hope that a convincing parallel might be found. The possibility that the fragment derives from a loomweight, rather than a spindlewhorl, bead, or button, should not be overlooked, particularly as there are, as far as we know, no clear loomweights from the assemblage of materials from Lofkënd, and the presumed importance of textiles and weaving faces an uncomfortable dearth of appropriate weights for the loom. If a loomweight, then it is different from the large and heavy pyramidal weights known in Early Iron Age Greece (Papadopoulos 2005:555–556, with full discussion and references), Italy (e.g., Peroni and Trucco 1994a:530– 531, fig. 153, no. 93A), and the central Balkans (e.g., Truhelka 1904:35–36, figs. 12–13, 15, pls. X–XII).

10/10 (SF 045), **Fig. 10.3** (Sheet 10.4; Photo 680)
Fragment of Unidentified Terracotta, either a Pot Handle or Fragment of a Possible Loomweight.
Tumulus Fill (SU 3, baulk cleaning).
PL: 0.034; H (same as Th): 0.015; Wt: 8.6 g.
Single fragment preserving unknown portion of handle or terracotta object; surfaces worn.
Semi-coarse to coarse fabric, with several moderate white and light colored inclusions (0.5–2.0 mm) and a little mica. Clay core fired black; surfaces variously fired from very dark grayish brown to grayish brown (10 YR 3/2 to 4/2).
Small perforation, impossible to measure accurately, and different when viewed from either side of the fragment, but it seems too small for a normal handle. If a handle, then the section is elliptical, with the top and bottom flat. Surfaces only very roughly finished on all sides.
For prehistoric loomweights in Albania of a form not unlike **10/10** (SF 045), cf. Korkuti 1982a:114, fig. 10, center and right; Korkuti 1983a:26, fig. 16a, c; see also Bunguri 1993:63, pl. II, no. 8.

JEWELRY AND OTHER ITEMS OF PERSONAL ORNAMENT

Objects of Gold/Electrum (Ear or Head Ornaments)

The two following ornaments, **10/11** and **10/12** (**Fig. 10.4**), are of electrum rather than gold. Both were found in situ, one on either side of the cranium of the female in Tomb XLVIII, at or near the position of the ears. The tomb is assigned to Phase III (eleventh–tenth centuries BC). Their position in the grave indicated that they were earrings, but how they were attached remains an issue. Alternatively, they may have served as ornaments of some other item of personal decoration, conceivably even elements of a headband or diadem, made of some organic material otherwise lost (for leather headbands with attached thin bronze or copper sheeting, see Gimbutas 1965:254, fig. 163, nos. 30–31; and 599, fig. 420, the headband associated with the so-called Abashevo woman at Vilovatovo in central Russia). In Classical Greek, there are various words for such head ornaments, such as *stephane* (στεφάνη) or *tainia* (ταινία) (and cf. κρήδεμνον, Doric κράδεμνον), as opposed to *diadem* (διάδημα), which was normally worn by royalty (see Papadopoulos 2010a, with full discussion and references).

The two are identical, circular, slightly domed disks of gold/electrum foil, with their edges on the underside folded down as if to attach the foil and hold it in place to an organic material, otherwise lost. The repoussé decoration comprises an embossed dot at center, with three concentric rings surrounding it on the inner portion of disk; these rings, which define small ridges on the exterior, are further decorated with short diagonal strokes, resembling a rope pattern. The outer portion of each disk is decorated with fine incised diagonal strokes arranged in five rings, defining two rows of a herringbone pattern, with an additional inner row of diagonal strokes.

The closest Early Iron Age parallels known to us, of more or less contemporary date, include a disk of gold foil from Tomb XI in the Fortetsa cemetery near Knossos and two from a Subprotogeometric II–III grave at Lefkandi; the more recent excavations at the Toumba cemetery at Lefkandi have yielded additional gold disks (at least seven from Tomb 63 [Late Protogeometric]), which, although illustrated, are not yet properly published (see Popham and Lemos 1996: pl. 131). In describing the Fortetsa piece, which has a diameter of 0.037 m, Brock writes: "Sides bent down. Decoration of impressed compass-drawn circles and Vs within a line. Dots around edge" (Brock 1957:22, pl. 13, Tomb XI, no. 193). Although the form and size of the Fortetsa disk is precisely that of the two Lofkënd examples, the decoration is very different, and it was found with the cremated remains believed to be those of a woman (Brock 1957:19). The Lefkandi disks are only fractionally smaller (D: 0.032), and appear to be broken and frayed; the better preserved one is described as a plain "disk of gold foil with central hole for attachment, folded over at the edges—the cover for an object of perishable material. Two fragments, the second probably from another similar example" (Popham, Sackett, and Themelis 1979–1980:191, pl. 189, Toumba Tomb 36, no. 17). Of the disks from these three sites, those of Lofkënd are probably the earliest. Of slightly later date, Late Geometric or Archaic, are several gold disks from the "aire sacrificielle" north of the sanctuary of Apollo Daphnephoros at Eretria (Huber 2003:53–54, pls. 45, 120–121, nos. O 75–79; O 82–89, 92). Some of these are described as gold "disques" (O 75–79), others as "appliqués à trous de fixation" (O 82–85), and as "appliqués en relief" (O 86–89, O 92). All are decorated with some form of impressed motifs; of the disks, the largest has a diameter of 0.032–0.033 (O 75), the smallest 0.015; of the various appliqués, O 86, with its edges folded over in the same manner as **10/11** (SF 290) and **10/12** (SF 291), has a diameter of 0.020 m; whereas O 89 has a preserved length of at least 0.035 m, and O 92 a diameter of 0.034 m. The closest of the Eretria gold foil disks and appliqués in terms of size and decoration to those from Lofkënd is perhaps O 92, which is decorated with four stamped concentric circles.

The Lofkënd, Lefkandi, and Fortetsa disks, together with those from Eretria, bear a general likeness to the well-known and variously decorated gold "runde Goldplättchen" from Grave Circle A at Mycenae (Karo 1930–1933: pls. XXVIII–XXIX; note also similar gold disks from the Kuppelgrab at Volos-Kapakli in Thessaly: Avila 1983b:28, fig. 5, nos. 4–7), which, with diameters of between 0.051 and 0.085 m, are larger than those of the Early Iron Age. Closer in shape and size are the smaller gold "Knöpfe" from Shaft Graves IV and V (Karo 1930–1933: pls. LIX–LX, LXII–LXV), especially several from Grave IV that are decorated with incised concentric circles (Karo 1930–1933:85, pl. LIX). The smallest of these (D: 0.011–0.023 m) are smaller than the Lofkënd, Lefkandi, and Fortetsa disks, but some of the more elaborately decorated are closer in size (D: 0.029–0.037 m). The majority of these were thought to have been sewn onto garments of one form or another (Karo 1930–1933:183; cf. Dakoronia 2003:346, fig. 1). Similar gold disks are well known in the Archaic period. In describing the eight gold disks from Perachora (D: 0.014–0.030 m), Payne noted that they were made of thin, beaten gold, carelessly cut into a roughly circular shape, with one or more holes near the rim, "and were obviously made to be sewn on to dresses —a clear survival, or revival, of Mycenaean tradition" (Payne 1940:74, pl. 18, nos. 1–3, 5–7, 25–26, with reference to Karo 1930–1933). On the basis of their decoration, Payne assigned at least one of the Perachora disks to the end of the Geometric period, citing parallels from Ephesos (cf. Hogarth 1908: pl. 8, nos. 3, 19, 27–29; also pl. 10, nos. 7, 12–13, 23) and Crete, and concluded that it "seems likely that it was in Ionia that this Mycenaean tradition survived" (Payne 1940:74). Together, the Lofkënd, Lefkandi, and Fortetsa disks help bridge the gap between the Bronze Age and the Early Iron Age types. In other parts of Europe, related ornamental gold disks, often referred to as "plates," are known from Transylvania and Hungary (Gimbutas 1965:57, fig. 21, nos. 8–20; 191, fig. 123, the latter after Popescu 1944).

Farther east, two related gold "discoid foils" with a convex surface are known from Ashdod—a little larger (ca. 6 cm in diameter) than the examples from Lofkënd—with comparanda from Megiddo, Tel Dan, Tell el-'Ajjul, and Kāmid el-Lōz, mostly of Late Bronze Age date (Dothan and Ben-Shlomo 2005:126–127, fig. 3.38). The Ashdod examples differ from the Lofkënd ornaments in that at least one has remnants of bronze adhering to the perimeter of the gold foil. In discussing them, the authors suggest that the Ashdod examples may have served as ornaments for ceremonial bronze sword pommels, or else were attached to some sort of garment, perhaps as gold-plated bronze buttons; they also note that this type of object seems to indicate Aegean affinities (Dothan and Ben-Shlomo 2005:127).

The possibility that the Lofkënd electrum disks were attached to some type of head ornament, rather than to earrings, is bolstered, for example, by a fragmentary gold headband or diadem from the Archaic and Classical votive deposit at the temple of Apollo Alaios at Crimisa in southern Italy, which has two attached gold rosettes (Orsi 1933:87–88, fig. 49). In a similar vein, a small gold diadem from the shaft graves of Grave Circle A at Mycenae has three variously shaped gold attachments dangling from the diadem, one in the center and one near each terminal (Karo 1930–1933: pl. XXXIX, nos. 231, 236–237). If worn around the head, the two side attachments would have been more or less at the position of the ears. Although very different from the foil disks of Lofkënd, and considerably later, the attachments on the Crimisa diadem, like those on the much earlier headband/diadem from Mycenae, at least show that gold attachments to diadems were used in antiquity, even though they do not provide precise parallels for the Lofkënd disks. It is noteworthy that the two gold foil disks from Lefkandi already cited were found in the same tomb as a gold diadem, although it is clear from the plan of the grave that the disks were not found in direct association with the diadem (see Popham, Sackett, and Themelis 1979–1980:191, pl. 164).

10/11 (SF 290, TXLVIII-3), **Fig. 10.4** (Sheet 54.2; Photo 2613)
Gold/Electrum Foil Ear or Head Ornament.
Papadopoulos, Bejko, and Morris 2007:124, fig. 16.
T52-3 (SU 1.0318), right side of cranium.
D: 0.038; Wt: 1.1 g.
Complete; torn on one side; condition otherwise good.
Circular, slightly domed, disk of gold/electrum foil, with edges on underside folded down as if to attach the foil and hold it in place, to an organic material, otherwise lost. Repoussé decoration, comprising an embossed dot at center, with three concentric rings surrounding it on inner portion of disk; these rings, which define small ridges on the exterior, are further decorated with short diagonal strokes, resembling a rope pattern. Outer portion of disk decorated with fine incised diagonal strokes arranged in five rings, defining two rows of a herringbone pattern, with an additional inner row of diagonal strokes.
Cf. **10/12** (SF 291). Found in situ over the region of the ear on both sides of the cranium of the female in Tomb XLVIII (52), the ornaments are either gold/electrum foil attached to earrings made of an organic material not preserved, or parts of an organic headdress, resembling a diadem, otherwise not preserved.

10/12 (SF 291, TXLVIII-4), **Fig. 10.4** (Sheet 54.3; Photo 2613)
Gold/Electrum Foil Ear or Head Ornament.
Papadopoulos, Bejko, and Morris 2007:124, fig. 16.
T52-4 (SU 1.0318), left side of cranium.
D: 0.038; Wt: 0.9 g.
Condition as **10/11**, but torn on two sides, as shown.
Shape and incised decoration as **10/11** (SF 290).

Fibulae

The Lofkënd fibulae are presented below in three categories according to the material from which they were made: bronze, iron, and bimetallic (bronze and iron). As various scholars have established, fibulae first appear in Europe in the course of the later second millennium BC, and it is generally accepted that fibulae derived from pins, specifically from the so-called "eyelet" pins that were secured with thongs or threads of organic material (Alexander and Hopkins 1982:401; and, generally, Alexander 1973a, 1973b). Undet was the first to suggest, in 1889, that the substitution in the eyelet pins of bronze for the organic material led to the development of fibulae (Undet 1889; Alexander and Hopkins 1982; for good examples of eyelet pins, see Montelius 1895: pl. 24, no. 5; pl. 38, no. 1; Åberg 1935:34, figs. 60–62; Gimbutas 1965: pl. 56, no. 4; Alexander 1973b:217, fig. 1; 220, fig. 3).

When looking at the Lofkënd fibulae collectively, what is immediately striking is the variety of the types present (in both bronze and iron, as well as the bimetallic), and the fact that there are several idiosyncratic types that are, to date, only known or best represented at Lofkënd, including a distinctive iron fibula that one of the authors (Papadopoulos 2010b) has dubbed the "Lofkënd type." These are discussed in detail below, but it is worth noting here that even in the case of well-established fibula types, there are important differences between the Lofkënd fibulae and those from other nearby sites. For example, the large bronze spectacle fibula (Type I.1a) is unique to Lofkënd, as are the iron fibula of Type II.1 and the bimetallic Type III.1, whereas at the nearby and largely contemporary tumulus at Patos, the most common type of fibula is a plain arched fibula (see Korkuti 1981:55, pl. XVIII, nos. 4–8; 44, pl. VII, Tomb 68), which is not found at Lofkënd.

Bronze fibulae

There are only two broad types of bronze fibulae at Lofkënd, the spectacle fibula (Type I.1) and what is often referred to as the "Cassibile" type after the site in Sicily (Type I.2). Both types are further subdivided into variants.

Spectacle fibulae

Designated Type XIV ("Agrafes en spirales") by Blinkenberg (1926:253–262) and conventionally referred to as "Plattenfibeln" in German (e.g., Sundwall 1943:170–176) and as "spectacle" fibulae in English (Alexander 1965; Benton 1950, 1952), the type is well represented in many parts of Europe, especially in central, eastern, and southeastern Europe (e.g., Bader 1983:41–71, pls. 5–23, nos. 25–127, esp. pls. 11–22; Betzler 1974: esp. pls. 20–71; Gedl 2004: pls. 42–48; Gergova 1987: pls. 18, 19; Glogović 2003: pls. 14–34; Laux 1973; Novotná 2001: pls. 14, 15; Říhovský 1993: esp. pl. 12; Vasić 1999: pls. 8–22).

Four subtypes are here presented, each defined by the manner in which the spiral loops are connected. Type I.1a (**10/13**; **Fig. 10.5**), which is rare, has a figure-eight connecting loop in the center consisting of two triple-coiled spirals; Type I.1b is the same, but with double-coiled connecting spirals; Type I.1c has single-coiled connecting spirals, whereas the small Type I.1d has no connecting spirals.

We know of no parallel for Type I.1a in Albania. Elsewhere in Europe, this type of spectacle fibula, with its triple connecting loop, is exceedingly rare. Alexander (1965:9, ill. 2) lists the type with a double loop (his Type Ig, our Type I.1b), but none with a triple loop. The only parallel we are aware of is a fragmentary fibula from Slovenia (Mason 1996:36, fig. 20, no. 7).

Type I.1b (**10/14**; **Fig. 10.6**) is rather more common. There are a number of published examples from Albania (including Aliu 1984:62, pl. VIII, no. 65 [same as 66, pl. XII, no. 9]; Andrea 1976a:144, pl. II, no. 2; Andrea 1985:262, pl. I, Tomb 1, no. 1; 267, pl. VI, Tomb 46, no. 1; pl. LVI, nos. 2–3; Prendi 1975:131, pl. II, no. 13), and they are especially common at the Luaras tumulus in southeast Albania (Aliu 2004: pl. III, Tomb 17, no. 29; pl. III, Tomb 19, no. 31; pl. X, Tomb 108, no. 130; pl. X, Tomb 109, no. 133; pl. XI, Tomb 128, nos. 156–157), and the type is well known in other parts of Europe, including Austria, Hungary, and the former Yugoslavia (see esp. Alexander 1965:9, 11, 21 ill. 2, Type Ig).

In contrast, Type I.1c (**10/15–10/17**; **Fig. 10.7**) is ubiquitous in many parts of Europe, including Albania (e.g., Aliu 1995:141, pl. II, Tomb 11, no. 20 [same as pl. IX, no. 1]; Aliu 2004: pl. I, Tomb 6, no. 7; pl. II, Tomb 13, no. 27; pl. III, Tomb 21, no. 30; pl. III, Tomb 26, nos. 39–40; pl. VI, Tomb 65, no. 82; pl. VIII, Tomb 89, nos. 114–115; pl. IX, Tomb 104, no. 123; pl. IX, Tomb 105, nos. 125–126; pl. X, Tombs 108, 109, nos. 131, 138; pl. XI, Tomb 122, no. 141; pl. XV, Tomb 171, nos. 197–198; Aliu 2012: pl. XI, Tomb 160, nos. 146–147; cf. pl. XXII, Tomb 222, nos. 266, 270; Andrea 1976a:144, pl. II, no. 1; Andrea 1985:268, pl. VII, Tomb 50, no. 1; pl. LVI, no. 4; and Andrea 1981a:226, pl. II, no. 5; Bodinaku 1982:95, pl. VI, no. 14 [and color pl. B]; 99, pl. X, Tumulus 7, Tomb 6, no. 14 [same as 101, pl. XII, no. 12]; Koka 1985:249, pl. I, no. 11; Koka 2012:252, pl. III, Tumulus 1, no. 38; Korkuti 1981:43, pl. VI, Tomb 43; Kurti 1983:105, pl. III, Tomb 13, no. 3; 107, pl. V, Tomb 17, no. 1), Greece (e.g., Blinkenberg 1926:257, fig. 303, XIV 2e; Brock 1957:54, pl. 37, Tomb X, no. 558; Casson 1923–1925, pl. III, no. 2a–b; Chrysostomou 2008:39, fig. 17 [two examples]; Coldstream and Catling 1996: fig. 171, Tomb 292, f.3; Dawkins et al. 1906–1907:84, fig. 20a; 113, fig. 3b; Dawkins 1929: pl. LXXXI [three examples, upper right]; pl. LXXXII, nos. c–d, h, m; pl. CXXXIII, a; Felsch 2007: pl. 8, nos.

506, 508; pls. 31–32, nos. 503–508; Furtwängler 1890: pl. XXI, no. 359; Heurtley 1939:240, fig. 112:o; Kilian-Dirlmeier 2002: pl. 40, nos. 580–582; Onasoglou 1981:18–19, pl. 17β; Philipp 1981: pl. 21, no. 1070; Radt 1974:124–126, fig. 38, nos. 4–13; Reichel and Wilhelm 1901:53, fig. 83; Rhomiopoulou 1971:39, fig. 2 [three examples]; Rhomiopoulou and Kilian Dirlmeier 1989:94, fig. 6, no. 8; 95, fig. 7, nos. 1–2, 17, 21; 108, fig. 17, nos. 9, 14, 17; 109, fig. 18, nos. 6, 14; 118, fig. 29, nos. 1–2, 5–7, 31; Rhomiopoulou and Touratsoglou 2002:48–49, M 1051–1052 [Tomb 6]; 65, α.α [Tomb 31]; Savvopoulou 2004:315, fig. 12 [top and middle]; Vokotopoulou 1986: fig. 109α–β, ια–ιβ, inv. 2357/ T113; inv. 2261/T103; inv. 2323/T46; pl. 211β; pl. 240γ; pl. 247β–γ; pl. 248β [upper right]; Waldstein 1905:240, pl. LXXXV, esp. no. 818), the Adriatic generally (see Lo Schiavo 1970: pl. XIII:17; pl. XXX:2), the central Balkans, including Serbia, Kosovo, Bosnia Herzegovina, and the Former Yugoslav Republic of Macedonia (FYROM) (e.g., Garašanin 1954: pl. XX, nos. 1–4; pl. XXVI, no. 6; pl. XLVII, no. 2; Vasić 1999: pl. 14, nos. 190–194; pl. 15, nos. 199–200; pls. 16–22 [mainly with wire circular in section, some rhomboidal]; Vukmanović and Radojčić 1995:15–16, no. 3), Slovenia (e.g., Mason 1996:15, fig. 2, no. 8; 15, fig. 3, no. 5; 33, fig. 19, no. 9; 36, fig. 20, no. 8;), the Italian peninsula (e.g., de la Genière 1968: pl. 31, no. 7; von Eles Masi 1986: pl. 42, nos. 628–629), and in other parts of Europe (e.g., Åberg 1935:119, fig. 201 [third row, left]; 150, fig. 245 [bottom]; Kilian 1975b: pl. 36, no. 16; pl. 43, no. 22; pl. 54, no. 10; pl. 65, no. 2).

For Type I.1d (**10/18**; **Fig. 10.8**), we know of only a few parallels in Albania (two examples from the tumulus at Prodan: Aliu 1984:62, pl. VIII, nos. 66–67; 66, pl. XII, no. 8 [both from tumulus fill]; one from the Luaras tumulus: Aliu 2004:132, Type III; pl. III, Tomb 18, no. 33; and another from Dardanisë: Jubani 1985:220, pl. I, no. 5), but they are well represented elsewhere in the Balkans and the Aegean, with the wire both circular and rhomboidal in section (e.g., Åberg 1935:49, figs. 81–83; Benac and Čović 1956: pl. II, no. 13; pl. IV, no. 3; Benac and Čović 1957: pl. II, no. 15; pl. VI, nos. 4, 6, 8; pl. IX, nos. 5–8; pl. XXX, no. 8; pl. XXXIV, no. 3; pl. XXXV, nos. 5–8; pl. XXXVII, no. 15; pl. XXXVIII, no. 4; pl. XL, no. 10; pl. XL, no. 8; Blinkenberg 1926:256, fig. 302, XIV 1a; Casson 1923–1925: pl. III, no. 2c; Coldstream and Catling 1996: fig. 159, Tomb 75, f.35; Furmánek, Veliačik, and Vladár 1999:78, fig. 34, no. 26; pl. 37:b [larger than the example from Lofkënd]; Garašanin 1954: pl. XVIII, nos. 6–7; pl. XXVI, no. 5; pl. XLVII, no. 3; Kilian 1975b: pl. 37, nos. 11–15; pl. 42, no. 10; pl. 43, nos. 14, 16; pl. 52, no. 5; pl. 54, no. 6, pl. 61, no. 8; pl. 67, nos. 11–12; pl. 70, no. 1; pl. 73, no. 7; Mason 1996:16, fig. 4, no. 7; 18, fig. 6, no. 3; Perdrizet 1908:112, fig. 404bis; Peristeri 2004:261, fig. 9 [right]; Popović and Vukmanović 1998:83, fig. 61; 123, pl. 1, no. 17; Vasić 1999: pl. 8, nos. 101–109; pl. 9, esp. nos. 114, 116–122, 130–133, of which nos. 114, 121–122, and 130–133 are closest to the solitary example from Lofkënd; pl. 10, various examples, esp. nos. 134–135; pl. 11, esp. nos. 146, 150, 151, and 157; Savvopoulou 2004:315, fig. 12 [bottom]; Truhelka 1904:143, fig. 82; pl. XL, no. 10; pl. LXXXI, no. 1; cf. also Gimbutas 1965: pl. 23, no. 3b [from the Jenišovice hoard, Bohemia]).

Type I.1.a

10/13 (SF 107, TLXX-3), **Figs. 3.246, 10.5** (Sheet 26.1; Photo 711)

Bronze Spectacle Fibula, Type I.1a.

Papadopoulos, Bejko, and Morris 2007:121, fig. 12b.

T17-3 (SU 3.0126).

L: 0.172; H (same as D spiral loops): 0.069–0.071; H (interconnecting loops): 0.067; Wt: 205.3 g.

Complete; pin very slightly deformed and broken at juncture with spiral loop; good dark green/black patina.

Large spectacle fibula, made of continuous length of bronze wire, circular in section, formed into two spiral loops, with a figure-eight connecting loop in the center, which consists of two triple-coiled spirals. The terminal of one spiral loop (coiled eight times) was bent over and hammered flat to form the catchplate. The terminal of the other loop (coiled seven times) continues into the pin, which is circular in section and tapers slightly toward the point.

Such a triple-coiled figure-eight loop is unique in Albania and rare elsewhere in Europe.

Type I.1.b

10/14 (SF 170, TLVIII-1), **Figs. 3.195, 10.6** (Sheet 87.2; Photo 2440)

Spectacle Fibula, Type I.1b.

T37-1 (SU 2.0243).

L: 0.080; H (same as D spiral loops): 0.032–0.035; H (interconnecting loops): 0.039; Wt: 28.9 g.

Almost complete, except for point of pin, broken; pin and parts of both spiral loops slightly deformed. Green patina.

Small to medium-size spectacle fibula, made of continuous length of bronze wire, circular in section, formed into two spiral loops, with a figure-eight connecting loop in the center, which consists of two double-coiled spirals. The terminal of one spiral loop (coiled four times) was bent over to form the catchplate. The terminal of the other loop (also coiled four times) continues to form the pin, which is circular in section, though misformed.

Possible fabric pseudomorphs.

Cf. especially Alexander 1965:9, pl. 2, no. 3; Aliu 1984: 62, pl. VIII, no. 65 (Prodan, tumulus fill).

Type I.1.c

10/15 (SF 106, TLXVIII-3), **Figs. 3.233, 10.7** (Sheet 27.1; Photo 709)

Bronze Spectacle Fibula, Type I.1c.

Papadopoulos, Bejko, and Morris 2007:119, fig. 10.

T13-3 (SU 2.0102).

L: 0.114; H (same as D spiral loops): 0.048, 0.049; H (interconnecting loops): 0.028; Wt: 72.6 g.

Intact; good dark green patina.

Spectacle fibula, made of continuous length of bronze wire, rhomboidal in section, formed into two spiral loops, with a figure-eight connecting loop in the center, which consists of two single-coiled spirals. The terminals of both spiral loops are bent slightly and taper toward a point; the terminal of one spiral loop is bent to form the catchplate, which is a plain clasp. The pin is made separately of bronze wire, circular in section, tapering toward a point. The pin terminal opposite the point is hammered flat and looped around the terminal of one of the spiral loops in order to hold the pin in place.

10/16 (SF 315, TXXXIX-2**), Figs. 3.130; 10.7** (Sheet 75.8; Photo 3167)

Spectacle Fibula, Type I.1c.

T66-2 (SU 1.0369).

L (not including pin): 0.091; L (with pin): 0.093; H (same as D spiral loops): 0.042, 0.040; H (interconnecting loops): 0.022; Wt: 28.5 g.

Intact; good dark green patina.

Spectacle fibula, made of continuous length of bronze wire, rhomboidal in section, formed into two spiral loops (with six to seven coils), with a figure-eight connecting loop in the center, which consists of two single-coiled spirals. The terminal of one spiral loop tapers slightly toward a blunt point; the terminal of the other spiral loop is bent over to form the catchplate, which is a plain clasp. The pin is made separately of bronze wire, circular in section, tapering toward a point. The pin terminal opposite the point is hammered flat and looped around the terminal of one of the spiral loops in order to hold the pin in place.

10/17 (SF 434, TLXIX-2), **Figs. 3.238, 10.7** (Sheet 102.10; Photo 3618)

Spectacle Fibula, Type I.1c.

T27-2 (SU 5.0551).

L: 0.073; H (same as D spiral loops): 0.032, 0.030; H (interconnecting loops): 0.020; Wt: 17.8 g.

Reconstructed from three joining fragments, one preserving all of fibula, the other two the pin. Well preserved; good green patina.

Spectacle fibula, made of continuous length of bronze wire, rhomboidal in section, formed into two spiral loops (each with six coils), with a figure-eight connecting loop in the center, which consists of two single-coiled spirals. The terminal of one spiral loop is bent over to form the catchplate, which is a plain clasp. The terminal of the other loop continues to form the pin, which is circular in section and tapers to a well-preserved point.

Fiber pseudomorphs visible around start of pin.

Cf. **10/15** (SF 106) and **10/16** (SF 315). See also Lo Schiavo 1970: pl. XIII:17 (Prozor); pl. XXX:2.

Type I.1.d

10/18 (SF 157), **Fig. 10.8** (Sheet 40.1; Photo 2356)

Small Bronze Spectacle Fibula, Type I.1d.

Papadopoulos, Bejko, and Morris 2007:132, fig. 26a.

Topsoil (SU 4.0004).

L: 0.040; H (same as D spiral loops): 0.018; Wt: 4.4 g.

Single fragment preserving all of fibula except for greater part of pin; excellent condition, with good dark green patina.

Small spectacle fibula, made of continuous length of bronze wire, circular in section, formed into two spiral loops. The terminal of one spiral loop (coiled four times) is bent to form the catchplate, which is a plain clasp. The terminal of the other loop (also coiled four times) continues to form the pin, which is not preserved.

Cf. Garašanin 1951:64; Slakovic-Duric 1965:542, pl. 1, no. 7.

"Cassibile" fibulae

There are three examples from Lofkënd of the distinctive type of fibula that we have conventionally referred to as "Cassibile" type (**10/19**, **10/20**, **10/21**; **Figs. 10.9**, **10.10**), named after the site in Sicily excavated by the great Paolo Orsi (see Orsi 1899:117–146, esp. 137–138, pl. XIII, nos. 6–7, and cf. also nos. 1–4). In discussing this characteristic fibula, Orsi (1899:137) writes: "ultima emanzione la serpeggiante ad occhio," and the type is often referred to in the Italian literature as "fibula serpeggiante" (e.g., Lo Schiavo 1983–1984:135, fig. 47, no. 2), and as "Schlangenfibeln" in German (e.g., Philipp 1981:287–289, nos. 1031–1045; Sundwall 1943:136–169), with Klaus Kilian specifically referring to the type as "sizilischen Schlangenfibeln" (Kilian 1970:332, pl. 9, I, no. 3), and Juliette de la Genière as "fibule à arc serpentant de type 'sicilien'" (de la Genière 1968:315, pl. 31, no. 5). Although very common on the Italian peninsula (e.g., Åberg 1930:59, fig. 157 [Cumae]; Bartoloni et al. 1980:173, pl. 65, no. 2 [Cassibile, Tomb 3]; Blinkenberg 1926:197–204, fig. 221, cf. no. 2; Kilian 1970: pl. 32, III, no. 7a; pl. 61, I, no. 1b; pl. 155, II, no. 7a; pl. 202, I, no. 6a; Maxwell-Hyslop 1956:131, fig. 2; 136, fig. 4; Montelius 1895: pl. XVI, nos. 221, 230; pl. XVII, no. 236; Montelius 1904: pl. 177, no. 2; Sundwall 1943:136–156, esp. 143, figs. 208, 210, 211; 150, figs. 225–226; 151, fig. 227; Müller-Karpe 1959: various examples, pls. 1–2, 4, 6, 8, 16, 18, 22, 27, 58, 85; esp. pl. 27, B, no. 1, and C, no. 2 [both from Tarquinia]; pl. 85, D, no. 13 [Bismantova]; see also 198, fig. 32, no. 7; 208; Peroni and Trucco 1994b:684, pl. 130, esp. 8–9, 11–12; also 739, pl. 152, no. 7; 751, pl. 157, no. 6), the type is closely connected with the Balkans (Bietti Sestieri 1973:423–424; cf. Vasić 1999: pl. 25, no. 285). Referring to them specifically as "Cassibile" fibulae, Bietti Sestieri (1973:424) cites examples from Enkomi on Cyprus, and Vrokastro, Kavousi, and Kydonia on Crete (cf. Blinkenberg 1926:55, fig. 27, I, 12a; Boardman 1961:121, pl. XLV, no. 529; Coldstream and Catling 1996: fig. 158, Tomb 45, f.4), as well as Lapithos (Cyprus), Hama, and Megiddo in the eastern Mediterranean.

The three examples from Lofkënd all have incised decoration, although they vary somewhat in shape, to such an extent that we have distinguished two sub-types. The first (Type I.2a), **10/20** (TXXVIII-3) and **10/19** (TLXIX-1), are long and relatively slender in terms of their height (**Fig. 10.9**), whereas the second, **10/21** (TXXI-2), is shorter in length but proportionately taller in height (**Fig. 10.10**). Chronologically, the two examples from Tombs XXI and XXVIII (**10/21** and **10/20**) are both relatively early in the Lofkënd sequence, both belonging to Phase II (twelfth–eleventh centuries BC), while **10/19** (TLXIX-1) is late, the tomb assigned to Phase Va (ninth–earlier eighth century BC, with an AMS ^{14}C calibrated date of 863 ± 44 BC).

Elsewhere in Albania, there is a well-known Cassibile fibula, similar to our **10/21** (SF 251), from the nearby tumulus of Patos (Korkuti 1981:44, pl. VII, Tomb 67 [same as 55, pl. XVIII, no. 9]; with further discussion in Bodinaku 1984:49–50, 57, pl. I, no. 1; Korkuti 1985:96, fig. 3; 102, no. 3), where the type is specifically referred to as Cassibile.

Type I.2.a

10/19 (SF 431a, TLXIX-1), **Figs. 3.237, 10.9** (Sheet 102.4; Photo 3597)

Large Bronze Fibula, Type I.2a, "Cassibile" Type, with Incised Decoration.

T27-1 (SU 5.0551).

L: 0.157; H: 0.021; Wt: 19.5 g.

Complete, pin broken in center; good dark green patina.

Large bow fibula, with a spring (two turns) at the apex of the bow, made of thick bronze wire, circular in section, rising at an angle to form a broad, roughly triangular but asymmetrical shape. Terminal hammered flat and folded over to form a long catchplate. Spring (two turns) continues to form pin, circular in section and tapering to a well-preserved point encased in the catchplate. Bow on either side of spring at apex decorated with fine incision, as shown, consisting of fine diagonal strokes defining groups of herringbone pattern.

Cf. **10/20** (SF 337), **10/21** (SF 251); also Åberg 1930: 59, fig. 157.

10/20 (SF 337, TXXVIII-3), **Figs. 3.94, 10.9** (Sheet 69.4; Photo 3136)

Bronze Fibula, Type I.2a, "Cassibile" Type, with Incised Decoration.

T77-3 (SU 4.0429).

L: 0.094; H: 0.018; Wt: 6.7 g.

Complete, but perhaps repaired; good green/dark green patina.

Bow fibula, with a spring (one turn) at the apex of the bow, made of bronze wire, circular in section, rising at an angle to form a roughly triangular but asymmetrical shape. Terminal hammered flat and folded over to form catchplate. The opposite terminal is slightly bent, with the actual terminal apparently broken, resembling a broken spring. Pin, mostly circular in section, tapering to a well-preserved point encased in the catchplate; the opposite end of the pin is hammered flat, resulting in a more rectangular section, with the terminal wrapped around what may originally have been the broken spring of the fibula. It is likely that this represents a repair to the fibula. Bow on either side of spring at apex decorated with fine incision, as shown, consisting of zigzags and chevrons.

Cf. **10/19** (SF 431a).

Type I.2.b

10/21 (SF 251, TXXI-2), **Figs. 3.66, 10.10** (Sheet 54.1; Photo 3530)

Bronze Fibula, Type I.2b, "Cassibile" Type, with Incised Decoration.

T55-2 (SU 4.0326).

L: 0.089; H: 0.031; Wt: 8.1 g.

Intact, except for minor chipping to edge of catchplate; well preserved, with good green patina.

Bow fibula, with a spring (one turn) at the apex of the bow, made of bronze wire, circular in section, except for the springs, where the section is square, rising at an angle to form a roughly triangular but asymmetrical shape. Terminal hammered flat and folded over to form catchplate. Spring (one turn) continues to form pin, circular in section and tapering to a well-preserved point; pin slightly bent. Bow on either side of spring at apex decorated with fine incision, as shown, consisting of zigzags and chevrons.

Cf. **10/20** (SF 337) and **10/19** (SF 431a); also Åberg 1930:10, fig. 10; cf. 13, fig. 20; 53, fig. 144; Korkuti 1981:44, pl. VII, Tomb 67; 55, pl. XVIII, no. 9.

Iron fibulae

There are only three iron fibulae from the tumulus, belonging to two types, all from tombs that are relatively late in the series (Phases IV and Va). The first, Type II.1, of which we know no good parallel anywhere in Europe, is represented by one example (**10/22**; **Fig. 10.11**); Type II.2, which has been dubbed the Lofkënd type (Papadopoulos 2010b), is represented by two examples, **10/23** (TLV-2) and **10/24** (TLXV-1) (**Fig. 10.12**), both assigned to Phase IV. Although there are a number of related iron and bronze fibulae in the Balkans, there are, to date, only a few published examples that can be assigned to the same type: what should be a fragmentary catchplate from Tumulus I at Kënetë in the northeast Albania along the east bank of the Drini i Zi (Drin) River, excavated by Bep Jubani, and a fragmentary fibula from the Molossian cemetery at Liatovouni, excavated by Angelika Douzougli (for the cemetery at Liatovouni, near Konitsa, see Douzougli 1996; Douzougli and Papadopoulos 2010). The example from Kënetë, although said to recall a fibula, was not recognized as such (Jubani 1983:84, 123, pl. III:29); that from Liatovouni was originally classified as an "unidentified iron attachment" (Papadopoulos 2010b: 242, 245, fig. 8). These examples are now joined by another, almost complete, fibula from Mojsinje in western Serbia dating to the ninth/eighth century BC (see Nikitović, Stojić, and Vasić 2002:105, pl. III:1, XVI:2). The fact that examples have been found as relatively far afield as Lofkënd, in the river valley of the Aoös in northwest Greece, and at Kënëte in northeast Albania and in western Serbia, may suggest that the type is more common than is currently established.

Type II.1

10/22 (SF 110, TLXX-4), **Figs. 3.247, 10.11** (Sheet 30.30; Photo 3766)

Iron Fibula, with Arch Made of Continuous Coil, Type II.1.

Papadopoulos, Bejko, and Morris 2007:120, fig. 12d.

T17-4 (SU 3.0126).

L (bow): 0.098; PL (pin shaft): 0.063; Wt: 31.4 g.

Almost complete; arched bow intact, with most of catchplate and part of spring preserved; most of pin preserved (small missing portion of pin shaft may be the section corroded and now adhering to arch of bow, PL: 0.021). Fibula heavily corroded.

The arched bow is made of one piece of iron wire coiled to form a tubular bow of dense coils. Catchplate heavily corroded, so it is difficult to reconstruct its precise form; roughly rectangular as preserved. Spring, one turn, continues from the coil of the bow. Pin shaft circular in section, tapering toward point, which is not clearly preserved.

Pseudomorphs of textile present.

Type II.2

10/23 (SF 261, TLV-2), **Figs. 3.183, 10.12** (Sheet 82.1; Photo 3147)

Large Arched Iron Fibula with Two Springs and Large Lunate Catchplate, Type II.2.

Papadopoulos 2010b:239, fig. 4.

T53-2 (SU 1.0321).

L (arch, spring to spring): 0.119; L (including catchplate): 0.126; H (arch); 0.047; Wt (all frr): 28.9 g.
Reconstructed, as shown, in two preserved groups of fragments, plus ten small to minuscule non-joining fragments and chips. All fragments corroded. Including the non-joining fragments, the fibula is more or less complete.
Arched bow, circular in section; spring (two turns), developing into pin, which is also circular in section, tapering toward sharp point, well preserved (in catchplate). Bow at opposite end connected to center of catchplate by a second spring (one and one-half turns). Large lunate or crescent-shaped catchplate, hammered flat, with lower edge upturned to form lip in order to accommodate pin.
Textile pseudomorphs present on many fragments.
Cf. **10/24** (SF 162), but with more elongated, and curved, catchplate. The only close parallels are the fragmentary fibulae from Kënëte, Liatovouni, and Mojsinje referred to previously. For smaller iron fibulae of related, but not the same type from the Glasinac region of Bosnia, cf. Benac and Čović 1956: pl. XXXV, 10 (Planje, Tum. I, Gr. 3); Benac and Čović 1957: pl. XI, 27 (Gosinja planina, Tum. I, Gr. 1); pl. XII, 18 (Podlaze, Tum. LXXXVI).

10/24 (SF 162, TLXV-1), **Figs. 3.222, 10.12** (Sheet 44.1; Photo 2736)
Large Arched Iron Fibula with Two Springs and Large Lunate Catchplate, Type II.2.
Papadopoulos 2010b:241, fig. 6.
T30-1 (SU 1.0213).
L (as preserved, including catchplate): 0.150; Wt (all frr): 47.6 g.
Reconstructed from various joining fragments, as shown, plus 28 non-joining fragments and chips; including all fragments, the fibula is more or less complete, but heavily corroded.
Arched bow, circular in section, and slightly thicker at apex of arch; spring (one preserved turn, probably originally two), developing into pin, which is also circular in section, tapering toward point, but less well preserved than SF 261. Bow at opposite end connected to center of catchplate by spring, only partially preserved; there appears to be an additional iron backing at the center of the catchplate to reinforce the juncture (there is no similar backing on **10/23** [SF 261]). Large lunate catchplate, as SF 261, but heavier, and a little more triangular than crescent-shaped, with upturned lip to accommodate pin.
Cf. **10/23** (SF 261) and comparanda from Kënëte, Liatovouni, and Mojsinje cited previously.

Bimetallic fibulae

Two examples only, both of the same type and both from tombs dating to Phase IV (late tenth–ninth century BC), are made of both iron and bronze. The fibulae are essentially figure-eight-shaped, composed of two circular disks of iron sheet; the edges of the smaller **10/26** (TLVI-2) are decorated with repoussé dots, barely visible in the corrosion; the iron disks of the larger **10/25** (TLV-3) are plain (**Fig. 10.13**). The only other difference between the two is that the one-piece supporting band which forms the pin and catchplate of **10/25** is of bronze, attached to the underside of the iron disks by means of bronze rivets, which form the conical bronze bosses on top. In contrast, the one-piece supporting band of **10/26** is of iron rather than bronze, and evidently attached to the iron disks by means of iron rivets, the conical tops of which are covered with bronze. The pins of both fibulae are of iron, circular in section.

Although the type bears a general likeness to the figure-eight fibulae, whether of bronze or iron, with decorated plates of ivory or bone (e.g., Benton 1953: pl. 63, nos. C.48–C.51; Blinkenberg 1926:262–279, 284–285; Dawkins et al. 1906–1907:84, fig. 20c; Dawkins 1929: pls. CXXXII–CXXXIII; Deonna 1938: pl. LXXXVI, various examples; Dunbabin 1962:433–437, pls. 183–185, with full list of comparanda from sites in mainland Greece, the islands, as well as Asia Minor, south Italy, and Cyrene; Hogarth 1908: pl. 32, nos. 1–12; pl. 36, nos. 1–32, 36–37; cf. also Kilian 1975b: pl. 26, no. 1; Kilian-Dirlmeier 2002:94, pl. 93, no. 1457, and the attachments 1458–1459; for the type in Albania, see Aliu 1994:42, 61, pl. II, Tomb 2, nos. 9–10 [two examples of the sixth century BC, Borovës]; Aliu 1985:280, pl. II, no. 25 [seventh–fifth centuries BC, Kolonjë region]), they are very different. We know of no similar fibula in Albania.

A related form of bimetallic fibula is known in Thessaly, which is essentially a normal "spectacle" fibula made of continuous iron wire, but with the addition of bronze bosses on the front; there are four from Marmariane (Heurtley and Skeat 1930–1931:35–37, fig. 37, nos. 13–15; Kilian 1975a:148–149, pl. 58, nos. 1711–1714) and one from Platykampos (Kilian 1975a: no. 1714a; Theochari 1966:45–46, fig. 11:7, no. 3); the bronze bosses of one of the Marmariane examples is plated with gold leaf (Heurtley and Skeat 1930–1931:35–37, no. 15); Heurtley refers to a similar "spectacle" fibula in bronze from central Europe (Åberg 1931:91, fig. 180 [Schrozhofen]), and the type is well represented at Mlad and Gosinja planina in Bosnia Herzegovina (Benac and Čović 1956: pl. XXIX, no. 1; Benac and Čović 1957: pl. VIII, nos. 14–16; pl. IX, nos. 14–15).

Perhaps the nearest parallels to the two Lofkënd fibulae are six related fibulae from Vitsa Zagoriou; the six were found in pairs in three tombs, the earliest assigned to the end of the ninth century BC (four examples), the latest to the period ca. 750–720 BC (Vokotopoulou 1986:309–310, fig. 108α–ε). Unlike the iron fibulae from Marmariane already referred to, and a related iron fibula from Vergina (Andronikos 1969:227; cf. also Kilian 1973a:7, fig. 2, nos. 1–4, esp. no. 4), the Vitsa fibulae were made of sheet iron disks, with a series of bronze bosses (see esp. Vokotopoulou 1969a: pl. 99α, inv. 2381/T 21, fig. 108γ; pl. 223γ–δ, inv. 2438, fig. 108α–β). The primary difference between the Vitsa and Lofkënd fibulae is that whereas the latter have a single bronze boss at the center of each disk, the Vitsa examples have multiple bronze disks, arranged in a line of three at the center of each disk, with a further boss on either side of these, at the edges, and another at the center (cf. also Dawkins et al. 1906–1907:84, fig. 20b, 113, fig. 3g).

Type III.1

10/25 (SF 262, TLV-3), **Figs. 3.184a–b, 10.13** (Sheet 83.1; Photos 3378–3379)
Bimetallic (Iron and Bronze) Figure-Eight Fibula, Type III.1.
T53-3 (SU 1.0321).
L: 0.084; D (individual disks): 0.041–0.042; D (bronze bosses): 0.011–0.012; L (supporting bronze band forming pin and catchplate): 0.069; PL (pin): 0.047; Wt (all frr): 18.0 g.
Reconstructed from various joining fragments preserving greater parts of both iron disks, all supporting bronze band forming the pin and catchplate, and much of the pin (preserved in two or three fragments); 13 additional non-joining fragments clearly represent missing parts of both disks; including these fragments, the fibula is almost complete. Iron elements heavily corroded; bronze elements much better preserved, with a good green patina.
Figure-eight fibula, with two iron disks. One-piece supporting band of bronze attached to the underside of the disks by means of bronze rivets, with bronze conical bosses on top; supporting band is slightly thinner at the center, wider at the point where the rivets are attached. One terminal tapers to a point and is bent back to form the catchplate; at the opposite terminal, a strip of iron overlays the bronze, and this continues, without a looped spring, to form the pin. The pin is circular in section and tapers toward a point, which is preserved.
Cf. **10/26** (SF 231). The two are very similar, except that SF 232 has a bronze rather than an iron supporting band, and lacks the punched decoration of dots around the edges of the disks of SF 231.

10/26 (SF 231, TLVI-2), **Figs. 3.190, 10.13** (Sheet 90.1; Photo 3532)
Bimetallic (Iron and Bronze) Figure-Eight Fibula, Type III.1.
T43-2 (SU 4.0269).
PL: 0.064; L (original, est.): 0.068; D (individual disks): 0.037; D (bronze bosses): 0.010; L (supporting iron band forming pin and catchplate): 0.050; PL (pin): 0.042; Wt (all frr): 14.2 g.
Reconstructed from various joining fragments preserving most of fibula, including one complete disk and one missing fragments/chips, most of supporting band, and greater part of pin; ten additional non-joining fragments and chips. Iron heavily corroded; bronze bosses somewhat better preserved.
Figure-eight fibula, with two iron disks, the edges of which are decorated with a band of repoussé dots, barely visible in the corrosion but evidently punched from the underside. One-piece supporting band of iron attached to underside of disks by means of two iron rivets, with bronze conical bosses on the top; this supporting band is thinner at the center, wider at the point where the rivets are attached. One terminal continues, without a looped spring, to form the pin; the opposite terminal is bent to form the catchplate; both of these terminals are broken. Pin circular in section.
Cf. **10/25** (SF 262), which is slightly larger.

Pins

Pins, whether dress or hair pins, are relatively common at Lofkënd, and they are found in the graves of both men and women, as well as children/infants (see Chapter 8). The contextual evidence for each pin is discussed more fully under the individual types. In comparison to the iron pins, which are relatively numerous, pins made of bronze are not common. There are only three complete examples, all different types. Two of the bronze pins, **10/27** and **10/28** (**Fig. 10.14**), are early, both dating to Phase I (fourteenth or thirteenth century BC), and both contemporary with the Mycenaean Palatial period in the Aegean; the third, **10/29**, is slightly later, but still early, dating as it does to Phase II (transitional Late Bronze to Early Iron Age, twelfth or eleventh century BC). In addition to complete bronze pins, there are fragments from an additional three pins, all encountered in the tumulus fill and not precisely datable (**10/30**, **10/31**, **10/32**). The total number of iron pins that can be assigned typologically is 13 (nine Type II.1, two Type II.2, one Type II.3, and one Type II.4). There are, in addition, four fragmentary iron pin shafts that cannot be assigned typologically (**10/46–10/49**), three of them from tombs. The iron pins, together with the two bimetallic pins and four bone pins, will be discussed in more detail below.

A critical problem in the archaeology of Albania has been the long-standing belief that dress pins essentially appear in the period of transition from the Bronze Age to the Iron Age. A related phenomenon has plagued the appearance of the dress pin in the Aegean. Indeed, a good deal of the literature on the dress pin in the Early Iron Age has been fixated on the origin and use of dress pins, and especially the evident change in fashion, as it is assumed to have accompanied the introduction of the peplos (see Desborough 1972:295; Hood and Coldstream 1968:214). Much of the evidence, however, cited in favor of this view seems rather pressed (e.g., Desborough 1972: 295; cf. Hägg 1967–1968). A number of scholars have argued that pins were introduced either from central Europe or northern Italy, or else from the east (see, among others, Bouzek 1985:165–167; Desborough 1964:53–54, 1972:296–298; Hood and Coldstream 1968:214–218; Hood, Huxley, and Sanders 1958–1959:235–237), while other scholars prefer to see a more local development—and therefore continuity—from the metal and bone/ivory pins of the Late Bronze Age (see esp. Deshayes 1966:204–207; Kilian-Dirlmeier 1984a:80–83). The presence, however, of pins in Mycenaean and earlier Bronze Age contexts in the Aegean began to raise doubts about influences or the introduction of pins from the north or east, and as early as 1971, Snodgrass (1971:227–228) wrote:

> These observations, taken with the fact that Mycenaean tombs had quite commonly contained small pins of bone, ivory and even bronze; with the lack of analogous Central European examples which show any chronological priority; and with the general absence of the earlier types from Greek sites north of Athens, suggest that the origins of the straight pin in Greece need to be reconsidered. In any case, the development of iron versions of these pins in Greece, rather before the middle of the eleventh century, takes the question, from then on, far beyond the reach of any explanation in terms of European influence.

More to the point, both Erwin Bielefeld (1968: C 38–40) and Imma Kilian-Dirlmeier (1984c) have argued authoritatively against the introduction of pins by invaders from the north. The antiquity of the use of dress pins in the Aegean needs to be emphasized, for pins interpreted as dress pins by Christos Tsountas at Chalandriani are Early Cycladic in date (third millennium BC) (Tsountas 1899:101–102); there are bronze pins in Middle Helladic burials at Corinth (Shear 1930:408; cf. Blegen, Palmer, and Young 1964:7–8, Grave 2, nos. 5, 10), and there are similar Middle Helladic and Late Helladic bronze pins from tombs in the West Cemetery at Eleusis (Mylonas 1975: vol. III, pl. 50, χ6, pl. 51, Μπ4–χ23 [Middle Helladic], pl. 51, Λπ16–χ22; Ζπ6–χ12 [Late Helladic]); Minoan examples are listed by Sinclair Hood (Hood and Coldstream 1968:214); and there are the large and elaborate pins from the Shaft Graves at Mycenae thought to be dress pins and evidently associated with the burials of males (Karo 1930–1933:173–174, esp. pl. 18, nos. 245–247). It is worth adding that pins are a prominent feature of the objects of personal use of the Early and Middle Cypriot Bronze Age, continuing into the Late Bronze and Early Iron Age (see Catling 1964:69–74, figs. 5–6 [for Early and Middle Cypriot pins]; 237–239 [for the Late Cypriot pins]). As we shall see, the *Rollenkopfnadel*, or rolled-head pin, enjoys a venerable prehistory going back to the Chalcolithic period at various sites in the Aegean and continuing through the Early, Middle and Late Bronze Age and well into the Early Iron Age. Consequently, Paul Jacobsthal's (1956:1) statement that the "history of the Greek pin does not begin before the later twelfth century" no longer stands.

In Albania the early date of many of the pins of Lofkënd, some as early as the fourteenth century BC, is now supported by the discovery of bone pins at Sovjan, two from stratum 5 c1 and one from stratum 5 c2 (Lera et al. 2011:49, fig. 23; Touchais and Lera 2007:147, pl. 30:g). More importantly, the levels from which the Sovjan pins derive have been dated by AMS ^{14}C to the later part of the fifteenth century BC, contemporary with Late Helladic II in the Aegean and with the closing stages of Maliq III d1 (stratum 5 c2), and to the broad period from ca. 1400–1000 BC (contemporary with Late Helladic IIIA to IIIC Late/Submycenaean in the Aegean, and with Maliq III d2), for the pins for stratum 5 c1 (for the absolute chronology, see Lera et al. 2011:49–51, with figs. 24–25). Two of the Sovjan pins are therefore contemporary with the earliest of those from Lofkënd, and one is even earlier.

Bronze pins

The solitary example of Type I.1 from Lofkënd was found in Tomb VI (**10/27**), an inhumation of an infant that can be dated to the Late Bronze Age (Phase I) (**Fig. 10.14**). This basic type, with a small rounded head and incised decoration on the upper shaft is common in Albania (e.g., Aliu 1984:57, pl. III, Tomb 16, no. 30, p. 62, pl. VIII, nos. 74, 76; Aliu 2007: pl. lxii, Type IA3, cf. Types IIA3 and IIB1; Bodinaku 2001–2002:82, pl. VI, Tomb 13, no. 4; 85, pl. IX, Tomb 15, nos. 4–5; Ceka 1975:158, pl. IV, Tomb 4; Kilian-Dirlmeier 1984b: esp. 100, pl. I, no. 7; 101, pl. II, nos. 17, 23, 24, 26; Prendi 1975:131, pl. II, nos. 2, 4 [cf. also no. 5]; cf. Aliu 2012: pl. XLIV, no. 471), particularly at nearby Patos, where it is the most common type of bronze dress pin (Korkuti 1981:40, pl. III, four examples, decorated and undecorated, from Tombs 16, 17, 20, 22; 41, pl. IV, Tomb 26; 42, pl. V, four examples from Tombs 28, 33, 34, 38; 43, pl. VI, Tomb 58; 44, pl. VII, two examples from Tombs 59 and 68; 46, pl. IX, no. 13 [tumulus fill]; of these, the closest to the Lofkënd pin is 44, pl. VII, Tomb 59); it is worth adding that the upper shafts of several of the pins of this type from Patos are not decorated. Related types, sometimes with a slight swelling on the upper shaft, and occasionally with a square rather than conical head, are fairly well represented in the Luaras tumulus in southeast Albania (Aliu 2004: pl. I, nos. 3, 5; pl. III, no. 34, pl. IV, no. 45; pl. XV, no. 189; cf. also pl. XIII, nos. 166–167; see also Aliu 2007). In discussing the chronology of this type of pin in Albania, Prendi (1975:117) notes that it appears in tombs of the Protogeometric and Geometric periods, but it can now be traced back at least to the Late Bronze Age, and not to the latest phases of the period.

Although common in Albania, this type is also found in central Europe, particularly in Germany (e.g., Müller-Karpe 1959: pl. 170, E, no. 3 [Reipersdorf]; pl. 193, nos. 24–26 [Roseninsel]).

This basic form of bronze pin, usually undecorated, can be traced back to at least the Early Bronze Age in both the Aegean (e.g., Blegen 1928:183–184 [same as Kilian-Dirlmeier 1984a:18, pl. 1, no. 8], which is more biconical; Lamb 1936:166, fig. 48a, various examples; Kilian-Dirlmeier 1984a:41, pl. 2, no. 78 [Shaft Grave period]) and Anatolia (e.g., Von der Osten 1937:195, fig. 195, esp. d 2706, e 645).

Type I.1

10/27 (SF 438, TVI-1), **Figs. 3.24, 10.14** (Sheet 101.1; Photo 3559)
Bronze Dress Pin, Type I.1, with Small Rounded Head and Incised Decoration.
T97-1 (SU 2.0573).
L: 0.127; D (head): 0.005–0.006; Wt: 6.5 g.
Intact; good green/dark green patina.
Shaft circular in section, tapering toward well-preserved point, and slightly thinner in section at juncture with head. Small, rounded, slightly domed, head. Upper part of shaft decorated with fine incised diagonal strokes, as shown, defining zigzags and chevrons.
Cf. Aliu 1987: T164, pl. VII, 142; Andrea 1985: T: 7, pl. II, no. 2, T: 36, pl. V, no. 2, T: 49, pl. VII, no. 1 (dated in the Early Iron Age, eleventh to the first half of the eighth century) and with the decoration in some cases covering more of the shaft than in **10/27**; see Andrea 2009–2010:271, pl. II, Tomb 19, no. 22 (Shuec); Prendi 1957: fig. 2c; for a series of related pins dated to the eleventh century from the tumuli at Apollonia, see Amore 2010:655–656, nos. 13.1, 13.2, 13.5, 13.6.

The one example of Type I.2 from Lofkënd was found in Tomb VII (**10/28**), the inhumation of a child aged 7 (± 2) years at death, evidently worn on the right shoulder. The tomb can be securely dated to the earliest phase (I) of the tumulus, belonging to the fourteenth or thirteenth century BC.

This type of pin with concave disk head is not common in Albania (Aliu knows of one similar pin from Barç found by Andrea [personal communication]), although it is worth noting a related pin, made of iron (see Prendi 1975:118, 131, pl. II, no. 6). Related types are found elsewhere in Europe: in Italy, for example, a number of pins are not unlike **10/28**, although none has a head as proportionately large or as concave as the example from Lofkënd (cf. Müller-Karpe 1959: pl. 104, nos. 57, 74 [both from Peschiera]). A related type is particularly popular in the region of the northwest Alps, as well as in Germany, Austria, and Hungary, but these are invariably decorated on the upper shaft, and only rarely are the heads as concave as the Lofkënd example (among many, see esp. Beck 1980: pl. 4, no. 7; pl. 12, no. 5; pl. 28, nos. 2–3 [no. 2 is particularly close]; pls. 31–32, various examples; pl. 37, no. 5; David 2002: pl. 342, no. 1; Holste 1953:32, fig. 2, esp. nos. 1, 4; 56, fig. 4, nos. 1, 15; 64, fig. 5, no. 1; pl. 5, no. 1; pl. 9, no. 2; Willvonseder 1937: pl. 31, no. 6; pl. 42, no. 4; pl. 43, no. 1 [with hole at upper shaft]; pl. 49, no. 1; pl. 53, no. 1). A close parallel, dating to the eighth or seventh century BC, was found at Assiros Toumba in central Macedonia (Wardle and Wardle 2000:673, fig. 6, SF 1208).

The size of the head and the prominent concavity of **10/28** (TVII-1) is similar to the pin of a Late Bronze Age fibula from Sicily (Blinkenberg 1926:44, fig. 7; Sundwall 1943:158, fig. 243, D IV α a 2), as well as a series of bronze pins with prominent concave disk head and a spur on the upper shaft from the Lusatian cemetery at Kruszyniec in Silesia (e.g., Gimbutas 1965:289, fig. 195, nos. 10–11).

Type I.2

10/28 (SF 440, TVII-1), **Figs. 3.27, 10.14** (Sheet 101.4; Photo 3552)

Bronze Dress Pin, Type I.2, with Concave Disk Head.

T99-1 (SU 2.0583).

L: 0.175; D (head): 0.024; Wt: 26.1 g.

Intact; good dark green/black patina.

Large dress pin, with shaft, circular in section, tapering toward point and flaring out toward head. Large disk head, with concave upper face.

The solitary example of Type I.3 with rolled head—the classic *Rollenkopfnadel* or *Rollennadel*—in bronze from Lofkënd was found in Tomb XXVII and can be dated to Phase II, which is transitional Late Bronze to Early Iron Age (twelfth or eleventh century BC). It has a length of 0.128. The shaft is circular in section, tapering to a point; the upper part of the shaft was hammered flat, resulting in a rectangular section, and rolled to form the distinctive, T-shaped hollow head. Although there is only one example in bronze at Lofkënd, the type in iron is by far the most popular pin in the Lofkënd tombs (see below). Elsewhere in Albania, the type is one of the most common (see, among others, Aliu 2007: pl. lxii, Type V; Andrea 1985:271, pl. X, Tomb 84, nos. 1–2; 288, pl. XXVII, Tomb 27, no. 1; Kilian-Dirlmeier 1984b:102, pl. III, nos. 42–46; as well as 102, pl. III, nos. 39–40 with incised decoration [both from Patos, same as Korkuti 1981:41, pl. IV; pl. IX, no. 7]; Prendi 1975:131, pl. II, no. 1; Andrea 1976a:133, 143, pl. I, no. 3). In the tumulus at Kuç i Zi, the type is found in both bronze (Andrea 1976b:223, pl. VIII, Tomb 52, no. 2) and iron (Andrea 1976b:220, pl. V, Tomb 27, no. 1; 225, pl. X, Tomb 80, no. 1). One of the bronze pins of this type at Patos is especially close to ours from Lofkënd (cf. Korkuti 1981:45, pl. VIII, Tomb 74).

This basic type of pin enjoys a venerable prehistory in Greece. The earliest published example, in almost pure copper, comes from Chalcolithic Sitagroi in northeast Greece; with a preserved length of 0.080 m, the pin was found in Sitagroi Phase III, which is broadly dated to 3800–2700 BC (Elster and Renfrew 2003:305, fig. 8.1f; pl. 8.2a [SF 880]). There is a similarly early example from Lerna, Phase IVB, which is Early Bronze Age, about the same size (L: 8.05 cm) as the pin from Sitagroi (Kilian-Dirlmeier 1984a:25, pl. 1, no. 20); there is another from the prehistoric settlement under the Heraion on Samos (Milojcić 1961:53, no. 13 [same as pl. 50, no. 6]); three examples from Early Bronze Age Thermi on Lesbos (Lamb 1936:167, 178, fig. 48, pl. XXV, nos. 32.15, 32.35, 32.46), and one from Phylakopi on Melos (Cherry and Davis 2007:413–414, fig. 10.4, no. 707). Middle Bronze Age examples are known from Palamari on Skyros (Parlama et al. 2010:289, fig. 13 [middle]). Elsewhere in Greece, the type is attested in the Late Bronze Age (Felsch 2007:269, pl. 22, nos. 203–204 refers to an example from Kalapodi as Mycenaean; Kilian-Dirlmeier 1984a: pl. 5, nos. 146–151 illustrates three Late Helladic examples from Mycenae [see also Tsountas 1888: pl. 9, no. 25], one from Pylos [LH IIIB2], and one from Nichoria, the latter [same as McDonald et al. 1975:118, pl. 27: e [LH II or IIIA] stylistically the closest to the Lofkënd example; Kyparissis 1919:117, fig. 32, no. 1; there is also a Late Bronze Age example from Aiani, Karamitrou-Mentessidi 2008:74, fig. 115, third from right). The type is also attested in the Early Iron Age

(Submycenaean Athens: Kraiker and Kübler 1939: 40, SM Grave 85 [L: 0.096 m]; McDonald 1972:263, pl. 51:g [Protogeometric Nichoria]; Müller-Karpe 1962:87, fig. 5, no. 1; Wide 1910:29–30, fig. 13, two examples from the Submycenaean graves on Salamis [L: 0.165 and 0.125 m]; Andronikos 1969: pl. 95, P III η; Radt 1974:126, pl. 38, no. 26 [Protogeometric Vergina]; see also Desborough 1952:152). An example of a *Rollennadel* was found in Kastanas Level 14b, which is transitional Late Bronze Age/Early Iron Age (Hochstetter 1987:31, pl. 4, no. 26 [same as pl. 27, no. 8]. Another, found in Lefkandi Palia Perivolia Tomb 21, is Subprotogeometric II (Popham, Sackett, and Themelis 1979–1980:245, pl. 136, P Tomb 21, no. 11), and still another was found at Thessalian Philia (Kilian-Dirlmeier 2002: pl. 8, nos. 51–56 [with reference to a Subprotogeometric example from Halos]). The most recent overview of the type in Greece is by Reinholdt (2008:123–124, pls. 9–10, nos. 005–008; 128, pl. 14l 135–137, pls. 21–23).

By the Geometric and Archaic period, the type is found at many sites in Greece in both bronze and iron (Kilian-Dirlmeier 1984a:206–207, pl. 84, nos. 3383–3407, illustrates examples from Olympia [same as Philipp 1981:88–93, pls. 35–36, nos. 268–294], Corinth, Bassai [Kourouniotis 1910:326, fig. 50, top right, no. 4], Mantineia in Arcadia, and cites examples from sites outside the Peloponnese; see also Waldstein 1905:240, pl. LXXXIV, nos. 811–812, which are probably rolled-head pins rather than the broken pins of fibulae). The type is found at Isthmia from the Archaic through Roman periods (Raubitschek 1998:48, pl. 34, nos. 184–188). In dealing with numerous examples of this type from Emporio on Chios, Boardman (1967:223) writes: "it is remarkable that most of the straight pins from the archaic deposits are 'roll pins,' a type not hitherto met in Ionia, and rare enough in the rest of Greece" (Boardman 1967:223–224, fig. 145, nos. 377–382). With the examples already cited, and numerous more from Archaic and Classical sites, including, among others, Tegea (Dugas 1921:376–377, fig. 39, nos. 108–109), Halae (Goldman 1941:418, fig. 61, nos. 1–2), Dodona (Evangelides 1935:242, pl. 22β, nos. 4, 14), Olynthos (Robinson 1941:363–364, pl. CXV, nos. 1755–1762), Karphi (Pendlebury et al. 1937–1938:106, pl. 28:3, no. 381 [M. 10]; pl. 29, nos. 471, 565, 306, 503), Knossos (Coldstream 1973:146–148, figs. 34–35, pl. 90, nos. 127–129), as well as elsewhere on Crete (Benton 1939–1940:58, pl. 32, no. 35; Boardman 1961:32–35, fig. 14D), Aigina (Furtwängler, Fiechter, and Thiersch 1906: pl. 114, nos. 17–18 [plain and with twisted shaft]), and Thera (Dragendorff et al. 1903:302, fig. 490a), the type is not as rare as Boardman supposed.

This type is very common in the central and northern Balkans (see Furmánek, Veliačik, and Vladár 1999:36, fig. 10, no. 23; 87, fig. 39, no. 2; Lo Schiavo 1970:463, pl. XXXVI:17, with listed parallels from Kompolje [Tomb 169], Prozor, Osor, Jablanc, and Jezerine [Tomb 41-323]; Mason 1996:15, fig. 2, no. 1; Truhelka 1904: pl. XXXVIII, no. 24), as it is on the Italian peninsula, especially in northern and central Italy (e.g., Åberg 1932:47, fig. 63 (second from left); 62, fig. 109, no. 10; Montelius 1895: pl. 7, no. 8 [Peschiera]; pl. 91, no. 4 [Villanova]; Montelius 1904: pl. 221, no. 7 [Chiusi]; Müller-Karpe 1959: pl. 56, A, no 10 [Pianello]; pl. 59, K, no. 2 and pl. 71, C, left [Bologna, San Vitale, Graves 461 and 260]; pl. 88, no. 11 [Cremona]), but is also found on Pithekoussai (Macnamara 2006:271, fig. 2, no. 7), and in central Europe north of the Alps (Beck 1980: pl. 23, no. 9; Childe 1929: pl. XI, no. C1; David 2002: pl. 172, no. 6; pl. 353, nos. 2–3; Gimbutas 1965:115, fig. 75, nos. 10, 16; 277, fig. 184, no. 4; 289, fig. 195, no. 5; 295, fig. 201, no. 3; 417, fig. 271, no. 3; Holste 1953:41, fig. 3, no. 2; Müller-Karpe 1959: pl. 116, no. 3 [Haidin]; pl. 118, no. 3 [Marburg]; pl. 164, no. 4 [Pfeffingen]; pl. 166, C, no. 3 [Reismühl]; pl. 193, nos. 28–31 [Roseninsel]; pl. 196, A, no. 8 [Ellmosen]; pl. 202, C, no. 8 [Höfen]; pl. 205, B, no. 12 [Essfeld]; von Sacken 1868: pl. XVI, nos. 2–3 [with both plain and twisted shaft]). The type is also found in Georgia (Gimbutas 1965:514, fig. 350, no. 11).

In northern Italy, particularly in the Bologna region (Savena and San Vitale), there is a related Late Bronze Age pin with a crook, bent back, before the rolled head proper (e.g., Jacobsthal 1956: figs. 352–353; Müller-Karpe 1959: pl. 58, F, no. 1 [Bologna, Savena, Grave 528]; pl. 65, F, no. 4 [Bologna, San Vitale, Grave 652]; pl. 73, E, no. 1 [Bologna, Savena, Grave 146]; pl. 73, Q, no. 1 [Bologna, Savena, Grave 92]; Randall-MacIver 1924: pl. 2, no. 6). Another related type is the rolled-head pin with part of the shaft twisted, found on either side of the Adriatic (e.g., Alexander 1964:175, fig. 9, no. 10; cf. also no. 13 [Yugoslavia]; Kilian 1970: pl. 152, II, no. 2 [Sala Consilina]).

There has been a good deal of discussion about the origin of this type of pin. Jacobsthal, in his seminal study, concluded that these "pins originate in the Early Bronze Age in Europe and the Near East, and survive into the Iron Age" (Jacobsthal 1956:122, with full discussion of the type on 122–123, figs. 350–351, 354–356, 360–361). Other scholars prefer an eastern origin. Hetty Goldman, for example, stated: "The type is very old, going back to the prehistoric Bronze Age. It is sometimes referred to as of Cypriote origin but, as a matter of fact, it has a very wide distribution throughout Anatolia and is also found in Mesopotamia. One was found in Thera in an archaic grave and quite a number in the crematorium of Gorica" (Goldman 1940:421, with references); the type is well known on Cyprus (e.g., Dikaios 1969: pl. 163, nos. 3, 13; Gjerstad et al. 1935: 35, pl. IX, 1, pl. CLII, 6; Catling 1964:238, fig. 22, nos. 23–24), and in Anatolia (e.g., Emre 1978: 119–120, figs. 123–124 [Type d], dated to the second millennium BC, with additional comparanda cited from Boğazköy, Alacahöyük, Beycesultan, Elbistan-Karahöyük, and at the Karum at Kanesh; Goldman 1956:285, fig. 430, nos. 181–190; fig. 431, nos. 191–199 [Tarsus, Early Bronze II–Late Bronze II]; Goldman 1963:375, fig. 175, nos. 58–61 [Iron Age Tarsus]; Lohmann 2007:144, fig. 33, PA 4d5–02 [seventh century BC]; Schalk 2008:191, 225, figs. 12:a–e, 35, nos. 266–303 [Type IX]; Schliemann 1880:564, nos. 1231, 1234, 1236–1237; 586, no. 1351 [Troy]; Schmidt 1902:253, nos. 6370, 6386, 6395; Von der Osten 1937:196, fig. 196, nos. e 907, d 1769, e 699, e 662, d 1563, e 1102 [Alishar Hüyük, of which d 1563 is closest to the Lofkënd pin]). The type is found as far east as northern Baktria (modern Afghanistan: Kaniuth 2006:119, nos. 214–215, Type F-5, with reference to parallels from Central Asia, Iran, and Pakistan). An eastern origin was championed by Catling (1964:238), who concluded that they "are clearly of Near Eastern origin," citing examples from Byblos, Gezer, Megiddo, Atchana, and Tarsus, among other sites. Catling also noted, as did Boardman a few years later, that the form rarely appears in the Aegean (Catling 1964:238). The numerous examples cited above from Greece show that the type is not as rare in the Aegean as it was once assumed. Indeed, the examples in Greece, the Italian peninsula, and central Europe, coupled with the fact that the type is found as early in Greece as it is in the east, if not earlier, does not give priority to an eastern origin, and we would concur with Jacobsthal's judicious statement that the type traces its origins to Europe and the Near East.

Type I.3

10/29 (SF 354, TXXVII-1), **Figs. 3.87a–b, 10.14** (Sheet 80.1; Photo 3324)
Bronze Dress Pin, Type I.3, with Rolled Head.
T82-1 (SU 4.0457).
L: 0.128; W (head): 0.012; Wt: 5.9 g.
Intact; good green patina.
Shaft mostly circular in section, tapering to a well-preserved point: upper part of shaft below head hammered, resulting in a rectangular section; the uppermost part hammered flat and rolled to form the distinctive T-shaped, hollow, head.
Cf. similar iron pins: **10/33** (SF 171), **10/35** (SF 390), **10/37** (SF 046).

Bronze pin shafts of unidentified type

10/30 (SF 105), Not Illustrated
Bronze Fragment, Unidentified Object, Conceivably Terminal of Rolled-Head Pin or Casting Waste.
Tumulus Fill (SU 2.0117).
PL (presumed head): 0.017; PH: 0.011; Wt: 1.1 g.
Single fragment, heavily corroded.
T-shaped, with horizontal element resembling rolled head of pin; vertical element conceivably pin shaft, circular in section.

10/31 (SF 227), Not Illustrated
Fragment of Bronze Pin Shaft.
Tumulus Fill (SU 4.0204).
PL: 0.044; Wt: 2.2 g.
Single fragment, broken on both sides, preserving portion of pin shaft; good dark green patina.
Shaft circular in section, tapering toward one end. Small notch on shaft near one end.

10/32 (SF 249), Not Illustrated
Fragment of Bronze Pin Shaft and Point.
Tumulus Fill (SU 1.0279); found in Fill immediately south of Tomb C (48).
PL: 0.024; Wt: 1.3 g.
Single fragment preserving small portion of pin shaft and point; good dark green patina.
Shaft circular in section, tapering toward point, which is hammered slightly flat. Perhaps from fibula rather than dress pin?

Iron pins

In comparison to bronze, iron pins have been much less studied, both in Albania and elsewhere, and even establishing a typology can be difficult owing to the state of preservation of the iron and the normally heavy corrosion products. The iron pins of Lofkënd can be divided into four types, the most common of which is Type II.1 with rolled head (nine examples), followed by Type II.2 with a plain disk head (two examples), and one example of each of Types II.3 and II.4.

Type II.1

The most common of the prehistoric iron pins from Lofkënd is that with rolled head (Type II.1) (**Fig. 10.15**). There are as many as seven examples of the type from tombs at Lofkënd (**10/33** [TLVIII-2], **10/34** [TLXIII-2], **10/35** [TXXXIV-1], **10/36** [TLXXX-1], **10/37** [TLXXVIII-1], **10/38** [TLXVIII-4], and perhaps also **10/41** [TLXX-7]). The earliest is **10/35** (Phase III), the latest **10/37** and **10/36** (Phase Vb); two belong to Phase IV (**10/33** [TLVIII-2] and **10/34** [TLXIII-2]), and two to Phase Va (**10/38** [TLXVIII-4] and **10/41** [TLXX-7]). Two more examples—one from tumulus fill **10/39** (SF 66), another from topsoil **10/40** (SF 18)—cannot be precisely dated.

We have already discussed the type in bronze. Iron examples of the type are fairly well represented in Albania (e.g., Andrea 1976b:220, pl. V, Tomb 27, no. 1; 225, pl. X, Tomb 80, no. 1; Kurti 1977–1978:160, 178, pl. VI, Tomb 26, no. 3; 163, 182, pl. X, Tomb 50, no. 4; 163, 183, pl. XI, Tomb 53, no. 3 [Burreli]; perhaps two more pins, Hoti 1982a:43, pl. VIII, Tomb 19, and one of the iron pins from Patos, Korkuti 1981:40, pl. III, Tomb 17, may be of this type), as they are in Greece (e.g., Felsch 2007:282, pl. 27, nos. 395, 403–404 [Geometric Kalapodi]; Heurtley and Skeat 1930–1931:36–37, fig. 15, nos. 17–18 [Protogeometric Marmariane]; Rhomiopoulou and Touratsoglou 2002:89, no. M 1074 [fifth century BC Mieza]). Chronologically, the closest iron parallels to the example of this type from Tomb XXXIV (**10/35**, Phase III: eleventh–tenth centuries BC) are the two Protogeometric examples from Marmariane. In describing the two Thessalian examples, Heurtley and Skeat (1930–1931:36–37) clearly state that they are iron pins "with pierced cylindrical head" (the type with T-shaped or mallet- or hammer-shaped head is very different: e.g., Waldstein 1905:215–216, pl. LXXX, nos. 352–383, Type e; Dugas 1921:378, fig. 40, no. 132). Iron rolled-head pins are common at Vitsa Zagoriou in Epirus; the site has yielded seven examples, found in six tombs, all of females, and all dating to the first half of the eighth century BC (Vokotopoulou 1986:307–308, fig. 114ι–ιε, fig. 115δ–ε).

10/33 (SF 171, TLVIII-2), **Figs. 3.196, 10.15** (Sheet 90.3; Photo 2216)
Fragmentary Iron Dress Pin, Type II.1, with Rolled Head.
T37-2 (SU 2.0243).
PL: 0.101; L (head): 0.011; Wt: 8.3 g.
Reconstructed from three joining fragments preserving greater part of pin, except for missing lower shaft and point; seven additional non-joining fragments; heavily corroded.
Shaft mainly circular in section; upper part of shaft below head hammered, resulting in more rectangular section; uppermost shaft hammered flat and rolled to form the distinctive T-shaped head, although head is not as wide as other examples, such as **10/37** (SF 046).
Possible textile pseudomorphs.
Cf. **10/35** (SF 390), **10/37** (SF 046).

10/34 (SF 254, TLXIII-2), **Figs. 3.216, 10.15** (Sheet 89.2; Photo 2536)
Iron Dress Pin, Type II.1, with Rolled Head.
T35-2 (SU 4.0234).
PL: 0.134; L (head): 0.012; Wt: 7.8 g.
Seven joining fragments preserving almost complete pin except for very tip of point; heavily corroded.
Shaft mainly circular in section, somewhat bent, tapering toward point, which is not preserved; upper part of shaft below head hammered, resulting in more rectangular section; uppermost shaft hammered flat and rolled to form the distinctive T-shaped head.
Textile pseudomorphs, herringbone.

10/35 (SF 390, TXXXIV-1), **Figs. 3.113, 10.15** (Sheet 89.3; Photo 3383)
Fragmentary Iron Dress Pin, Type II.1, with Rolled Rim.
T87-1 (SU 1.0498).
PL: 0.122; L (head): 0.013; Wt (all frr): 5.9 g.
Reconstructed from seven fragments, joining to form three groups of fragments that do not clearly join, plus two additional non-joining fragments; all fragments heavily corroded.
Shaft mainly circular in section, slightly curved, tapering to well-preserved point. Upper part of shaft below head hammered, resulting in more

rectangular section; uppermost shaft hammered flat and rolled to form the distinctive T-shaped head, although head is not as wide as other examples such as SF 046.
Textile pseudomorphs.
Cf. especially **10/33** (SF 171).

10/36 (SF 423, TLXXX-1), **Figs. 3.272, 10.15** (Sheet 101.3; Photo 3456)
Fragmentary Iron Dress Pin, Type II.1, with Rolled Head.
T4-1 (SU 5:0032).
L: 0.129; W (head): 0.010; Wt: 6.5 g.
Completely reconstructed from six joining fragments; heavily corroded; shaft bent/misformed at one point.
Pin shaft mostly circular in section, tapering toward a fairly well-preserved point; opposite end hammered flat and rolled to form the distinctive T-shaped head.
Some fiber/textile pseudomorphs.

10/37 (SF 046, TLXXVIII-1), **Figs. 3.267, 10.15** (Sheet 110.1)
Iron Dress Pin, Type II.1, with Rolled Rim.
T5-1 (SU 2.0051).
L: 0.140; L (head): 0.018; Wt: 8.4.
Completely reconstructed from four joining fragments; heavily corroded.
Shaft mainly circular in section, tapering to well-preserved point; upper part of shaft below head hammered, resulting in more rectangular section; uppermost shaft hammered flat and rolled to form the distinctive T-shaped head.
Significant remains of textile pseudomorphs.

10/38 (SF 116, TLXVIII-4), **Figs. 3.234, 10.15** (Sheet 28.1; Photo 402)
Iron Dress Pin, Type II.1, with Rolled Head.
T13-4 (SU 2.0136).
PL: 0.121; L (head): 0.010; Wt: 6.6 g.
Reconstructed from six joining fragments, heavily corroded, almost complete, except for very tip of point; plus one small non-joining fragment.
Shaft mainly circular in section, slightly bent; upper part of shaft below head hammered, resulting in more rectangular section; uppermost shaft hammered flat and rolled to form the distinctive T-shaped head, although head is not as wide as other examples such as SF 046.
Textile pseudomorphs in two corrosion bits, Z-twist.
Cf., among others, SF 046, SF 390.

10/39 (SF 066), **Fig. 10.15** (Sheet 102.5)
Fragmentary Iron Dress Pin, Type II.1, with Rolled Head.
Tumulus Fill (SU 2.0040).
PL: 0.107; Wt: 9.4 g.
Three main fragments, probably originally joining, preserving all of head and significant portions of shaft of pin; three additional non-joining fragments, little more than corrosion products. All fragments heavily corroded.
Very small portion of uppermost shaft, immediately below head, hammered and rolled to form head; remainder of shaft circular in section, tapering significantly toward point, which is not preserved.

10/40 (SF 018), **Fig. 10.15** (Sheet 89.1)
Fragment of Iron Dress Pin, Type II.1, with Rolled Head.
Topsoil (SU 2.0002).
PL: 0.027; Wt: 1.2 g.
Single fragment, heavily corroded, preserving all of head and small portion of upper shaft of pin.
Upper shaft hammered and more or less rectangular in section, rolled to form head. Lower preserved shaft circular in section.

10/41 (SF 414, TLXX-7), **Figs. 3.250, 10.15** (Sheet 110x.2)
Fragmentary Iron Dress Pin, Perhaps Type II.1.
T17-7 (SU 3.0126).
PL (both non-joining frr): 0.067 (0.035+ 0.032); Wt (both frr): 2.8 g.
Two non-joining fragments, one preserving point of pin, the other what appears to be either the head or a substantial area of corrosion. Both fragments heavily corroded.
Shaft circular in section, tapering to well-preserved point. One fragment preserves either the head, conceivably but not certainly of a Type II.1 pin, with rolled head, or simply corrosion.
Textile pseudomorphs.

Type II.2
In comparison to Type II.1, the two examples of this type from Lofkënd are both relatively early: TXXIII-1 belongs to Phase II, with an AMS ^{14}C date of 1070 ± 59 BC, and TXLIV-2 to Phase III (**Fig. 10.16**). Type II.2 has an essentially plain head that is only slightly, if at all, articulated from the shaft. Two related iron pins from Tumulus 6 at Shtoj in the Shkodër

region of northern Albania are probably of this type (Koka 1990:34, 64, pl. II, Tomb 3, no. 22; 35, 66, pl. IV, Tomb 5, no. 52; cf. also 65, pl. III, Tomb 4, no. 50), and it is worth noting that in the case of some of the pins from the nearby tumulus at Patos, the form of the head was unclear due to corrosion and their general state of preservation, and it is not impossible that one or two of these are of this type (see Korkuti 1981:41, pl. IV, Tomb 25; 45, pl. VIII, Tomb 70). A pin from Kastanas in central Macedonia assigned to Level 4 (seventh or sixth century BC) is probably of this type (Hochstetter 1987:31, pl. 4, no. 22).

10/42 (SF 260, TXXIII-1), **Figs. 3.78, 10.16** (Sheet 69.2; Photo 2814)
Iron Dress Pin, Type II.2, with Disk Finial.
T56-1 (SU 4.0329).
L: 0.142; D (head): 0.012; Wt: 18.8 g.
Reconstructed into two groups of six fragments, which now do not clearly join among themselves, but which did originally, plus four non-joining fragments/chips. All fragments heavily corroded.
Shaft slightly bent, circular in section, and comparatively thick at juncture with head, tapering to sharp point, which is well preserved. Pin head articulated from shaft by groove, partly obliterated by corrosion, but clear enough. Plain disk head, only slightly greater in diameter than upper part of shaft. Additional iron on top of disk resembling a dome almost certainly corrosion and not part of original pin.
Textile pseudomorphs on lower part.
Cf. **10/43** (SF 305); cf. also Andrea 2009–2010:276, pl. VII, Tomb 61, no. 82.

10/43 (SF 305, TXLIV-2), **Figs. 3.145, 10.16** (Sheet 89.4; Photo 3285)
Iron Dress Pin, Type II.2, with Disk Finial.
T65-2 (SU 4.0394).
PL: 0.143; D (max, head): 0.010; Wt (both frr): 16.6 g.
Two fragments, very heavily corroded, probably originally joining, preserving all or most of pin, except the point.
Shaft circular in section, and comparatively thick at juncture with head, tapering toward opposite end. Although heavily corroded, the pin head appears to be articulated, at one point, from shaft by groove, much obliterated by corrosion. Plain disk head, only slightly greater in diameter than upper part of shaft.
Retains excellent textile pseudomorphs. Z-twist cordage visible in two areas on the longer fragment (b). Single piece of cord visible approximately 5.5 cm from break and 3 cm from tip. At tip (non-break end) knotted and possibly braided strands of Z-twist cord are visible.
Type as **10/42** (SF 260).

Type II.3

Little can be said conclusively about **10/44** (TLXIII-3), as the pin is heavily corroded and the head is largely obscured by corrosion (**Fig. 10.17**). Nevertheless, the form of the head as preserved, the clear articulation between the head and the shaft, and the fact that the upper shaft immediately below the head is square in section, whereas the remainder of the shaft is circular in section, all seem to suggest that the head, or at least part of it, was intentional and not just the result of corrosion. Although it may seem premature to assign this one problematic pin its own type, what is fairly clear is that the form of the head is such that it cannot belong to one of the other Lofkënd iron pin types, and as such distinguishing it typologically makes sense.

The pin derives from a tomb assigned to Phase IV (late tenth–ninth century BC), and it was one of two iron pins in the grave, one each found on the left and right shoulders, respectively. The precise form of the head remains difficult to determine on account of the corrosion products. As preserved, the head almost resembles an animal head, but this is highly unlikely. Although animal-head pins are known from antiquity (e.g., Jacobsthal 1956: figs. 261–263) they are not only very different from **10/44** (TLXIII-3), but considerably later in date. Rather than read too much into the form of the head as it survives, it might be better to consider the head as a sphere or globe that was clearly articulated from the shaft. The tumulus burials at Kënetë have yielded a number of iron pins with bulbous, rounded heads that are not unlike **10/44** (e.g., Jubani 1983:121, pl. I, Tumulus I, Tomb 3, no. 9; 124, pl. IV, Tumulus II, Tomb 2, no. 34; 128, pl. VIII, no. 84 [tumulus fill]; cf. Hoti 1981:217, pl. III, nos. 1–2), and there are related iron pins elsewhere (e.g., Felsch 2007:280, pl. 27, no. 377). It is even possible that these iron pins ultimately derive from a bronze prototype with bulbous head known from Kënetë and Kuç i Zi (see Kilian-Dirlmeier 1984b:103, pl. IV, nos. 55–56), where the head is rather differently articulated from the shaft than the Lofkënd example, with a concave upper element.

10/44 (SF 253, TLXIII-3), **Figs. 3.217, 10.17** (Sheet 69.1; Photo 2586)
Iron Dress Pin, Type II.3, with Spherical Head (Resembling an Animal Head).
T35-3 (SU 4.0234).
PL: 0.140; PL (head): 0.016; Wt: 7.3 g.
Reconstructed from seven joining fragments, almost complete, except for point; heavily corroded.
Pin shaft mainly circular in section, tapering toward point, which is not preserved; upper shaft, below head, square in section. The precise form of the head remains difficult to determine on account of the heavy corrosion. As preserved, it resembles an animal-head finial, but this is far from clear.
Textile pseudomorph.
The form of the head is such that it cannot belong to one of the other Lofkënd iron pin types.

Type II.4
The solitary example of this type of iron pin comes from Tomb LVIII, which is assigned to Phase IV (late tenth–ninth century BC) (**Fig. 10.18**). On account of the poorly preserved state of the cranium of the deceased, the decision was taken in the field to block-lift the skull, and **10/45** was found in the lab during the process of cleaning the cranium. The pin could, therefore, have been either a dress pin, worn on the upper torso very near the head (cf. Kilian-Dirlmeier 1984b:106, pl. VII:1–3), or else a hair pin (cf. Kilian-Dirlmeier 1984b:106, pl. VII:4).

The terminal of **10/45** is bent back onto itself to form the head of the pin; this is clearly intentional and does not appear to be accidental. The general appearance recalls a modern crochet hook—or crochet needle—which is a type of needle with a hook at one end used to draw thread through knotted loops. The context of the piece near the cranium of the deceased would suggest that it was, rather, a dress or hair pin, although **10/45** may have been used for other functions. In any case, we know of no close parallel for **10/45** from Albania, Greece, or the Balkans more generally.

10/45 (SF 226, TLVIII-3), **Figs. 3.197, 10.18** (Sheet 101.5; Photo 2535)
Iron Dress Pin, Type II.4, with Bent-Back Head.
T37-3 (SU 2.0243).
PL: 0.147; Wt: 9.8 g.
Four joining fragments preserving almost complete pin except for tip of point; heavily corroded.
Pin shaft circular in section, tapering toward point, which is not preserved. Pin originally tapering also toward point at the opposite end, which is bent back onto itself to form the head.
Textile pseudomorphs visible. The twist (Z) and the weave (plain) are preserved.
For a bronze pin of related form from a late Classical/Hellenistic farmhouse near Edessa, see Chrysostomou 2008b:93, fig. 3.

Fragmentary iron pin shafts

10/46 (SF 304a–c, TXLIV-1), **Fig. 3.144** (Photo 2986)
Fragmentary Shaft of Iron Dress Pin.
T65-1 (SU 4.0365).
PL (frr a+b): 0.073; PL (fr c): 0.036; Wt (all frr): 7.9 g.
Two joining and one non-joining fragments, all heavily corroded, preserving greater part of shaft of pin, but nothing clearly of the head or point. It is possible, if not likely, that all three fragments originally joined, but that the junctures are now obscured by the corrosion.
Shaft circular in section.
Possible textile pseudomorphs on the pin, but further examination is required.

10/47 (SF 155, TLXIX-4), **Fig. 3.240** (Photo 3592)
Fragmentary Shaft of Iron Dress Pin.
T27-4 (SU 1.0174).
PL: 0.067; Wt: 4.1 g.
Reconstructed from three joining fragments preserving lower portion of shaft of pin, including the poorly preserved point; heavily corroded.
Pin shaft circular in section, tapering toward point.
Presence of pseudomorphs of textile braided pattern.

10/48 (SF 225, TL-1), **Fig. 3.164** (Photo 2531)
Fragmentary Shaft of Iron Dress Pin.
T46-1 (SU 4.0283).
PL (joining frr): 0.076; Wt (all frr): 7.2 g.
Main piece reconstructed from four joining fragments, plus numerous additional non-joining fragments and chips, preserving portion of shaft of pin; all fragments heavily corroded.
Shaft evidently circular in section, tapering toward point at one end; opposite end thicker. Conceivably, but not certainly of Type II.2, with disk finial, but the corrosion is too great to tell.

Textile pseudomorphs visible on several fragments.

10/49 (SF 085), Not Illustrated
Iron Pin Fragments, Conceivably Modern.
Tumulus Fill (SU 2.0066).
PL (joining frr): 0.071; Wt (all frr): 6.1 g.
Two joining fragments and one non-joining one, preserving portion of shaft of possible pin; much corroded.
Main shaft circular in section, tapering slightly toward one preserved end; opposite end articulated, but broken. Original form unclear.

Bimetallic pins

Type III.1

The large bimetallic pin **10/50** is one of the finest metal objects encountered in the tumulus (**Fig. 10.19**). The pin was found in situ above the upper right humerus of the female buried in LXX, and it was thus worn on the upper right torso, a worthy counterpart to the large spectacle fibula worn by the deceased on the other side (**10/13** and see **Fig. 8.16**). Although it is possible that **10/50** was not a clothing fastener but rather a scepter/rod—similar objects are known from other parts of Europe where they have been interpreted as marking social status on special occasions by being carried or held (see, e.g., some of the Dürrnberg graves, Penninger 1972– 1978)—this appears less likely given the position of the piece on the deceased. The fact that the pin was completely engulfed by textile pseudomorphs (see Chapter 12) further suggests that this was a pin worn by the deceased, not a scepter/rod placed on the body of the young woman. The tomb dates to Phase Va and is therefore ninth to earlier eighth century BC. The fragmentary **10/51** should be the bronze head from a dress pin similar to **10/50**. The overall form of **10/50** is essentially an elaborated version of the earlier long bronze pins with knobbed head and incised decoration (such as Andronikos 1969:234, fig. 74, esp. KVIIα; Kilian-Dirlmeier 1984b:101–102, pls. II–III, nos. 15–16, 18, 21–22, 27–31), but with the shaft made separately of iron and subsequently attached.

As a category, bimetallic pins have been little studied. Although there are, for example, bimetallic pins from Protogeometric Theotokou in Thessaly that are roughly contemporary to, or a little earlier than, **10/50** (Wace and Thompson 1912:213), with shafts of iron and heads of bronze, there are none as elaborately decorated as **10/50** (or, for that matter, **10/51**). In its overall appearance, **10/50** is, especially in the manner of its decoration, reminiscent of large bronze spindles that are a feature of the indigenous cultures of southern Italy. This type of spindle is known in the indigenous cemetery at Sala Consilina (Kilian 1970: pl. 158, I, mo. 1n; Beil. 16, Type U4a), a similar spindle was dedicated at the early Greek sanctuary on the Timpone Motta at Francavilla Marittima in northern Calabria (Papadopoulos 2003:125, fig. 155a–b, no. 441), and a spindle with incised decorated shaft (and disks) comes from Pantano di Cleto (Luppino 1982:76, pl. 9, no. 8; for a related spindle, but without the incised decoration, see Orsi 1926: 63–64, fig. 7 [left]; note also another spindle from Osteria dell'Osa: Bietti Sestieri 1992:197, fig. 8.15 [center]). But these are spindles made entirely of bronze, not bimetallic dress pins. Without close parallels, **10/50** is a welcome addition to the corpus of Early Iron Age pins from southeastern Europe.

10/50 (SF 111, TLXX-5), **Figs. 3.248a–b, 10.19** (Sheet 28.2; Photos 3802–3803)
Bimetallic (Bronze and Iron) Dress Pin, Type III.1.
Papadopoulos, Bejko, and Morris 2007:120, fig. 12c.
T17-5 (SU 3.0126).
PL (entire pin): 0.278; L (bronze head): 0.065; D (conical head): 0.012; L × W (projecting prongs): 0.023 × 0.023; D (lower molding): 0.013; D (iron shaft): 0.007–0.012; Wt: 46.0 g.
Almost complete, except for tip of point. Iron shaft heavily corroded, reconstructed from five fragments. Bronze head patinated, preserved complete, but with tips of two prongs slightly damaged.
Bronze head consists of shaft, circular in section, surmounted by conical finial. About 0.015 m below the head is an articulated projection, with four prongs defining a cross. Lower portion of bronze shaft, which tapers slightly toward the top, is defined by two projecting fillets, the lower of the two at juncture with iron pin shaft; the shaft between the two fillets is decorated with fine incision, as shown, consisting of vertical chevrons; the upper part of the shaft below the head was probably similarly decorated, though this is now barely visible. Iron shaft circular in section, tapering toward the point, which is not preserved.
Minerally preserved remains of textile on iron pin—largest section near where iron pin attaches to copper alloy handle—fibers seem to be in a Z-twist

and in a plain weave. Fibers and weave visible in other areas along the pin.

Type III.2

10/51 (SF 067), classified as Type III.2 and found in tumulus fill (**Fig. 10.20**), is listed here for comparison to **10/50**. Although it resembles some of the "geschlossene Bommeln" published by Kilian-Dirlmeier (1979: esp. 60, pl. 22, no. 375, said to be from Chalkidike, now in the Benaki Museum, Athens; cf. also Vokotopoulou 1986: pl. 12β, fig. 115β, inv. 4949), at one end of which the bronze splays to form a loop for suspension, it is almost certainly a bimetallic pin, with a bronze head and iron shaft, since the lower end of the bronze shaft preserves the connection with the iron pin shaft, and there are significant traces of iron corrosion on the bronze head.

A related bronze element, very similar in form, and referred to as one of two "Rippenfibeln," was found in Arareva gromila Grave 1 (Benac and Čović 1957:79, pl. XXXXI, no. 6; Alexander 1964:173, fig. 8, no. 2), and another, very similar, described as a bronze "Schmucknadelkopf" has long been known from Donja Dolina (Truhelka 1904:72, fig. 49, pl. LXXXI, no. 34). These two parallels, the closest to our pin head, both come from Bosnia Herzegovina.

10/51 (SF 067), **Fig. 10.20** (Sheet 15.2; Photo 386)
Fragmentary Bronze Pin Head of Bimetallic (Bronze and Iron) Dress Pin, Type III.2.
Tumulus Fill (SU 2.0040).
PL: 0.047; D (bead with projections): 0.014; Wt: 10.8 g.
Single fragment preserving portion of bronze head of bimetallic pin; upper shaft broken, finial not preserved; original lower shaft preserves connection with iron pin shaft. Significant traces of iron corrosion on bronze head.
Main lower portion of bronze pin head shaft circular in section and ridged (15 ridges); this main portion of the shaft is surmounted by a large bead with four small conical projections; the bead is almost square when seen from above, due to the projections. The bead is surmounted by a small section of ridged shaft (three ridges), which tapers toward a point, the uppermost part of which is broken. The minuscule portion of the iron pin shaft preserved at the lower end of the bronze pin head is circular in section.
Cf. Andrea 1985: pl. XI, Tomb 97, no. 2.

Bone pins

There are only four complete or near-complete bone pins from tombs in the tumulus, together with a fragmentary example, **10/56** (SF 412), from Tomb I (**Fig. 10.21**). All five pins are early, deriving from tombs that can be assigned to Phase I of the tumulus (fourteenth–thirteenth century BC). Bone pins contemporary to, and even earlier than, those of Lofkënd are known from stratified levels at Sovjan (Touchais and Lera 2007:148, pl. 30:g; Lera et al. 2011:49, fig. 23). In terms of function, the context of the pin in Tomb V (**10/53**) suggests that it served as a hair pin rather than a dress pin; the pin in Tomb IV (**10/54**) was found on the torso of the deceased, immediately below the cranium, between the ribs and right scapula, and was probably worn as a dress pin; in the case of Tomb XII (**10/55**), the pin appeared to belong with the female, rather than the infant, and although this was not absolutely clear, the pin could hardly have served as a hair pin, and must have been used to fasten clothing. The disarticulated bone in Tomb I was such that context provided no information as to the function of the two pins in the grave (**10/52, 10/56**). Interestingly, the pins of Tombs I and IV were associated with males, those in Tombs V and XII with females. Bone pins were, therefore, not gender-specific at Lofkënd. In addition to serving as hair or dress pins, bone pins in some cultures may have been used as "ear ornaments" worn through a piercing in the ear (see Yphantidis 2006: pl. 35, no. 6).

Although bone pins are not uncommon in Albania (e.g., Aliu 2004: pl. XI, Tomb 131, nos. 148–151; pl. XVI, Tomb 174, no. 204; Lera et al. 2011:49, fig. 23), they have been little studied as a group. The seminal study of bone pins in southern Greece, specifically the Peloponnese, is Imma Kilian-Dirlmeier's 1984 *Prähistorische Bronzefunde* volume (Kilian-Dirlmeier 1984a). Kilian-Dirlmeier traces the prehistory of the bone pin from the Early Bronze Age into the Mycenaean period and beyond as part of her study of bronze pins (Kilian-Dirlmeier 1984a:28–36, 51–65). Since her study, additional bone pins have been noted from Asine in the Argolid, some of them Middle Bronze Age in date (Nordquist 1987:159, fig. 19; Krzyszkowska 1996:87, 90, fig. 2, no. 13), Tiryns (e.g., Kilian 1981:179, fig. 33b, esp. top row, second from left; Kilian 1982:416, fig. 33), and at least one from the island of Aigina (Walter and Weisshaar 1993:294–295, fig. 5); they have also been found in Mycenaean tombs in the area of the later Athenian Agora (Immerwahr 1971:217, pls. 50, 77, Tomb XXI-15

[LH III]; 190, pl. 40, Tomb VII-32 [LH II–III]) and in the Late Bronze Age levels at Assiros Toumba in central Macedonia (Wardle 1980:253, pl. 22:d); additional Middle Bronze Age examples, at least five, are now known from Megali Magoula, Galatas in Troizinia (see Konsolaki-Yiannopoulou 2010:76, fig. 4). By the seventh century BC, there is an astonishing array of bone pins of various types from the Archaic Artemision at Ephesos (Hogarth 1908:187–189, pls. 33–34).

In terms of type, all four of the more complete bone pins from Lofkënd are unique. Type IV.1 from Tomb I, which terminates in a plain head that was simply rounded off, finds close parallels in the Bronze Age, if not earlier (e.g., Lamb 1936:202, pl. XXVII, nos. 22–23 [Early Bronze Age Thermi]; Goldman 1956:308, fig. 438, nos. 40, 62; Nordquist 1987:114, 159, fig. 19, no. 66 [Middle Helladic Asine]; Taylour and Janko 2008:417–418, fig. 9.1, no. 6003 [Middle Helladic Agios Stephanos]; Heurtley 1939:230, fig. 230r [Late Bronze Age Macedonia]; Evely 2006:295, fig. 5.14, no. 2 [Late Helladic IIIC Lefkandi]; Protonotariou-Deïlaki 2009:522, pl. Γ34, no. 6, bottom [Late Helladic Argos]), into the Early Iron Age (e.g., Hochstetter 1987: pl. 14, nos. 1, 3–7 [Kastanas, Levels 16, 12–9, which range in date from the Late Bronze Age through the early stages of the Early Iron Age]; Batziou-Eustathiou 1999:122, fig. 15, nos. BE 5999, BE 5997, BE 5998), as well as in the historic era (e.g., Davidson 1952:287, pl. 120, no. 2385 [first or second century AD]; Deonna 1938: pl. LXXXIV, 717, nos. 1–13, esp. second from left; sixth and seventh from right; Dusenbery 1998:1013, 1015, nos. S158–S159, S238–S241 [both Roman]). Similar plain-headed pins in bronze are known from Cyprus (Early Cypriot II–Middle Cypriot III), where they are said to be the most numerous class in the Cypriot Bronze Age (Catling 1964:70, fig. 5, no. 12); there is one from Troy (Schmidt 1902:288–289, esp. no. 7867), and there is a related bronze type from Bosnia Herzegovina (Benac and Čović 1956: pl. XX:16 [Bandino Brdo, Tum. IV, "*älteres Stratum*"]). The almost complete example from Tomb I (**10/52**) can be dated to the fourteenth century, if not slightly earlier, on the basis of ^{14}C AMS dating, and it is worth adding that the bone pin fragment from the same tomb (**10/56**) is probably of the same type. Close parallels to **10/52** (TI-1) were found in the tumulus burials at Pogoni in Epirus, near the Albanian frontier (Andreou 1982:59, fig. 14; Andreou and Andreou 1999:85, fig. 36, and fig. 33).

Type IV.2 (**10/53**) terminates in a small disk finial and has elaborate incised decoration on the upper part of the shaft. There is a similar bone pin from Amantia in Albania (Anamali 1972:163, pl. VI, no. 5 [far left]), which is later than the example from Lofkënd, and it is impossible to determine from the published photograph whether the pin is decorated. A similar bone or ivory pin, decorated with two incised and framed latticed bands on the upper shaft, was found in Lapithos on Cyprus, Tomb 602 (Gjerstad et al. 1934:40, pl. LIX, no. 40), and a related pin, undecorated, is known from the sanctuary of Demeter at Knossos (Coldstream 1973:168–170, fig. 43, no. 315).

The fragmentary Type IV.3 (**10/54**) is the smallest and most finely carved of the bone pins from Lofkënd. We have referred to the type as having a beaded head, and elsewhere in the literature, pins of this type, whether of bone or bronze, are referred to as "vase-headed" (e.g., Catling 1964:239; Alexander 1964:167–169, fig. 6, nos. 3–4 [Type II]) or with a "pomegranate" head (Jacobsthal 1956:185–200, fig. 48; Goldman 1963:375, fig. 175, no. 63); related examples are occasionally referred to as "poppy-headed" (e.g., Duru 2008:180, fig. 362). A particularly close parallel was found in Tomb 131 in the Luaras tumulus (Aliu 2004:56–57, pl. XI, no. 149; cf. no. 148), and there is a related bronze example from the Korçë basin (Andrea 1976a:143, pl. I, no. 5). Elsewhere, related examples are known from Middle Helladic Eutresis and Agios Stephanos in Greece (Goldman 1931:213– 214, fig. 284, nos. 5, 7; Taylour and Janko 2008:418, fig. 9.1, no. 6008), Early Iron Age Yugoslavia (Alexander 1964: 167–169, fig. 6, esp. nos. 3–4) and Cyprus (Gjerstad et al. 1934:187, pl. XLII, 3, Lapithos Tomb 403, no. 5; 263, pl. LVII, 4, Lapithos 429, no. 25; 331, pl. LXIII, 1, Agios Jakovos Tomb 8m no. 13; 483, pl. LXXVIII, 1, Enkomi Tomb 3, nos. 240–241), and in a Hellenistic or earlier context at Corinth (Davidson 1952:282, pl. 118, no. 2291 [labeled as ivory]).

The final pin type, the small Type IV.4, with square head, finds no exact parallel. There is a related Early Bronze Age example from Lerna that is proportionately longer and with a less distinctly square head (Kilian-Dirlmeier 1984a:28, fig. 2, B2), another from Thermi with a splaying head (Lamb 1936: pl. XXVII, no. 29), and a related type with a pierced or drilled hole (Kilian-Dirlmeier 1984a:28, fig. 2, B3–B4), but these are far removed from the Lofkënd example both in form and date. A Late Bronze Age pin from Tarsus is a little closer in shape, but has incised decoration in the form of a Z on the square head (Goldman 1956:308, fig. 438, no. 65; cf. also no. 64). At the other end of the time scale, **10/55** bears a general likeness to a series of bone styli. As Davidson (1952:185) states, the two essential requirements of the stylus—a point and a flat, blunt end for erasure—are "fulfilled by many instruments." Davidson further notes that styli are generally made of bronze or bone (Davidson 1952:185). Among historical examples, perhaps the closest to ours are some of the bone "styli" of the late sixth and early fifth centuries BC found both below and above the temple area at Halae (Goldman 1940:425–426, fig. 78, nos. 1–12, esp. nos. 9–12), although context alone does not provide clear evidence that these were styli as opposed to pins, whether hair or dress pins. Related examples, most often referred to as styli, are common in the Hellenistic and especially the Roman period (e.g., Davidson 1952:185–187, pls. 83–84, esp. nos. 1348, 1365; also 285, pl. 119, no. 2336; cf. Blinkenberg 1931:150, pl. 16, nos. 422–423; Corbett 1949:340, pl. 101, nos. 131–132; Dawkins 1929:242, pl. CLXXII, no. 5; Deonna 1938: pl. LXXX, various examples; Dunbabin 1962:445–447, pl. 189, nos. A357–A373, esp. A366; Perdrizet 1908:162–163, fig. 679; see also Waldstein 1905:353, pl. CXL, no. 85). In his discussion of the so-called bone styli from Perachora, Dunbabin (1962:445–446) writes: "One cannot say for certain that they were not used for toilet purposes such as the spreading of ointment or even as pins. . . . They probably had several uses: Galen [XII.865] says that teeth may be extracted with a stylus." Elsewhere in the Balkans, bone pins with flat, rectangular heads are found occasionally, such as the example from Lepenski Vir (Srejović and Babović 1981: 95, no. 296).

Type IV.1

10/52 (SF 411, TI-1), **Figs. 3.5, 10.21** (Sheet 102.9; Photo 3387)

Bone Pin, Type IV.1, with Plain Round Head.

T64-1 (SU 1.0361).

PL: 0.155; D (max): 0.005; Wt: 3.8 g.

Reconstructed from six joining fragments, almost complete, except for very tip of point and minor chipping of shaft. Surface pitted at points, but with good polished surface where better preserved.

Comparatively long pin, with shaft circular in section, tapering toward point, which is not preserved. Pin terminates in a plain, rounded head.

Cf. **10/56** (SF 412). Among others, cf. Heurtley 1939:230, fig. 230r (Late Bronze Age); Kilian 1981: 179, fig. 33b (top row, second from left) (Late Bronze Age Tiryns); Nordquist 1987:114, 159, fig. 19, no. 66 (Middle Helladic Asine).

Type IV.2

10/53 (SF 437, TV-1), **Figs. 3.21, 10.21** (Sheet 101.2; Photo 3556)

Bone Pin, Type IV.2, with Small Disk Finial and Incised Decoration.

T96-1 (SU 2.0569).

PL: 0.157; D (max, upper shaft): 0.008; Wt: 4.5 g.

Reconstructed from six joining fragments, almost complete, except for tip of point; small portion of the head missing. Surface rather worn.

Comparatively long pin with shaft round in section, tapering toward point, which is not preserved. Uppermost part of the shaft widest, surmounted by short neck tapering toward top, with small, flat-topped disk finial. Uppermost 0.055 m of the shaft bears incised decoration in the form of four registers, each with three parallel rows of incised horizontal zigzags, the registers separated from each other by three incised lines, with three incised lines at top and bottom of decorated zone.

Cf. Gjerstad et al. 1934:40, pl. LIX, Lapithos Tomb 602, no. 40.

Type IV.3

10/54 (SF 443, TIV-1), **Figs. 3.18, 10.21** (Sheet 101.6; Photo 3625)

Small Bone Pin, Type IV.3, with Beaded Head.

T98-1 (SU 2.0581).

PL (all frr): 0.068; D (head, max): 0.006; Wt (all frr): 0.8 g.

Three joining fragments preserving portion of shaft, plus two additional joining fragments, probably but not clearly joining with those of the shaft, preserving the entire head and upper portion of shaft; surface pitted and worn.

Comparatively thin shaft, circular in section; it appears that the shaft is thinnest immediately below the head, becoming thicker toward the midsection before tapering toward the point of the pin, which is not preserved. The head is offset from the shaft by four small projections (cf. the more prominent projections on the bimetallic pin, **10/50**), surmounted by a bead of piriform shape, itself surmounted by a small cylindrical neck and disk finial, not unlike the finial on **10/53** (SF 437). Together, the bead and disk finial resemble a small vase, even a small pomegranate or poppy.

Cf. Aliu 2004: T:131, pl. XI, no. 149; T:141, pl. XVI, no. 204 (probably with slightly longer neck); cf. also the bead resembling a small vase framed by a disk instead of little spikes T:131, pl. XI, no. 148.

Type IV.4

10/55 (SF 404, TXII-1), **Figs. 3.39, 10.21** (Sheet 80.4; Photo 3355)
Small Bone Pin, Type IV.4, with Square Head.
T88-1 (SU 2.0499).
L: 0.084; L × W (head): 0.012 × 0.013; Wt: 1.8 g.
Reconstructed from two joining fragments, complete; surface pitted and a little worn, but with polish still preserved on the shaft and one side of the head.
Shaft round in section, tapering to well-preserved point. Head carved from the same piece of bone as the shaft, almost square in shape, roughly ovoid in section, with one side slightly flatter than the other. Grooves along shaft and head, initially thought to be possible decoration, probably the result of post-depositional damage.

Bone pin shaft of unknown type

10/56 (SF 412, TI-2), **Fig. 3.6** (Photo 3389)
Fragmentary Bone Pin Shaft, Undetermined Type.
T64-2 (SU 1.0361).
PL: 0.059; D (max): 0.005–0.006; Wt: 1.3 g.
Reconstructed from four joining fragments preserving small portion of shaft of pin; surface pitted and rather worn.
Shaft: circular in section, tapering slightly, as preserved, toward one end.
Cf. **10/52** (SF 411) and perhaps from the same type; cf. also Kilian-Dirlmeier 2002: pl. 71, no. 1130; note also the fragment from the Mycenaean tholos at Nichoria, McDonald et al. 1975: pl. 27:e.

Pendants

Small bronze double-spiral ("spectacle") pendants

There are only two examples of the small double-spiral pendant ornament so common in Late Bronze and Early Iron Age Europe, one from Tomb LIII (**10/57**), the other from Tomb XXVIII (**10/58**) (**Fig. 10.22**). The latter comes from a tomb that can be assigned to Phase II (twelfth–eleventh centuries BC), the former to Phase III (eleventh–tenth centuries). Made of continuous bronze wire, the spirals frame a drawn-up loop designed for suspension, and such ornaments are often found in situ suspended as parts of earrings (such as the gold pendants of this form from Subprotogeometric Lefkandi: Popham, Sackett, and Themelis 1979–1980: pl. 173, Toumba Tomb 13, nos. 17–17; pls. 231b, 221f), necklaces (e.g., Gimbutas 1965:109, fig. 71, nos. 1, 3 [latter part of a necklace with blue glass beads from Middle Bronze Age Russia]), fibulae (e.g., Bartoloni et al. 1980:175, pl. 67, no. 6; von Eles Masi 1986: pls. 24–25, nos. 397B and 401; pls. 26–27, nos. 405–406; pl. 63, no. 861; pl. 73, no. 966; pl. 84, no. 1043;), headbands/diadems (e.g., Gimbutas 1965:598–599, fig. 420), or attached to a textile. There is also a double-spiral ornament of precisely this form from Nagybátony in northern Hungary that served to attach the two ends of a bronze belt (Gimbutas 1965:295, fig. 201, no. 19).

The two pendants are similar to one another, and both are made of a continuous length of bronze wire, mostly circular in section, tapering slightly toward the center of each spiral. Pendant **10/57** was formed into two spiral coils/loops, with four or three to four coils, respectively, while **10/58** was formed into two spirals, each with six coils. The wire between the two spirals on both pendants was drawn up to form a loop. Pendant **10/58** differs slightly from **10/57** in that the loop appears to have been reinforced, or conceivably repaired, with a thin strip of bronze hammered flat and wrapped around the loop.

In Albania, similar ornaments are known from the tumuli at Patos (Korkuti 1981:41, pl. IV, Tomb 27 [two examples]; same as 55, pl. XVIII, nos. 1, 3) and Prodan (Aliu 1984:62, pl. VIII, no. 68). Comparanda from other parts of the Aegean and Europe are given in the catalogue entries below.

10/57 (SF 300, TLIII-4), **Figs. 3.174, 10.22** (Sheet 68.6; Photo 3053)
Bronze Spectacle Ornament, Pendant.
T63-4 (SU 1.0359).
L: 0.028; H: 0.020; D (individual spirals): 0.014, 0.012; Wt: 1.7 g.
Two joining fragments preserving complete ornament; good green/dark green patina.
A continuous length of bronze wire, mostly circular in section, tapering slightly toward the center of each spiral, formed into two spiral coils/loops (four and four–three coils, respectively), the wire between the two spirals drawn up to form a loop.
Cf. **10/58** (SF 335). For Greece, see, among others, Andronikos 1969:256, fig. 91; Radt 1974:134, pl. 40, nos.

17–18 (Vergina); Kilian 1975a: pl. 78, nos. 66– 67 (Pherai, Enodia Sanctuary); Petropoulos 1987–1988: pl. IA, fig. 13 (Ano Mazaraki); Rhomiopoulou 1971:40, fig. 3 (three examples, from Spelaion near Grevena in western Macedonia); note also the gold pendants, clearly part of earrings, from Lefkandi: Popham, Sackett, and Themelis 1979–1980: pl. 173, Toumba Tomb 13, nos. 17–18; pls. 231b, 221f (Subprotogeometric II). For the central Balkans, see Alexander 1972:89, fig. 51, IIIc; Benac and Čović 1956: pl. XLVI, no. 3; pl. XLVII, nos. 1–2; Benac and Čović 1957: pl. I, nos. 1, 5–9 (the Glasinac examples are larger than that from Lofkënd); Garašanin 1954: pl. XLVII, no. 1 (six examples); Lo Schiavo 1970: pl. II: 4 (two examples from Kompolje, Tomb 222), pl. III:5 (Kompolje, Tomb 263), pl. XI:8 and XXXVII:16 (Prozor). For Italy, see Åberg 1932:47, fig. 63 (top right); also 15, fig. 8 (right); Bartoloni et al. 1980:189, pl. 81, no. 8 (Golasecca); 189, pl. 81, no. 2 (Allumiere, Poggia La Pozza); Kilian 1970: pl. 269, no. 4; Montelius 1904: pl. 147, no. 18 (Novilara [Pesaro]); Papadopoulos 2003:70–71, fig. 90c–d, no. 179; Stoop 1987:22–24, figs. 3–5; Zancani Montuoro 1980–1982: pl. XXXII, no. 2; pl. XXXV, no. 7; cf. also Warneke 1999:167, fig. 81; Beil. 1, no. 29. For Slovakia, see Furmánek 1980:7–11, pls. 2–3, nos. 5–61 ("Brillenanhänger"); Furmánek, Veliačik, and Vladár 1999:36, fig. 10, no. 16; 85, fig. 37, nos. 32, 34; 87, fig. 39, no. 16. For Hungary, see Kovács 1977:34, no. 7; 39, no. 7. For Germany and Austria, see Gimbutas 1965:254, fig. 163, nos. 3–7 (Straubing); 289, fig. 195, nos. 12–13; and 462, fig. 305, no. 4 from northern Carparthia; Holste 1953: pl. 12, no. 16 (large); pl. 25, no. 2 (numerous examples in situ); Müller-Karpe 1959: pl. 180, C, nos. 1–2 (Riegsee). For Russia, see Gimbutas 1965:111, fig. 73, no. 3.

10/58 (SF 335, TXXVIII-1), **Figs. 3.92, 10.22** (Sheet 75.5; Photo 3249)
Bronze Spectacle Ornament, Pendant.
T77-1 (SU 4.0429).
L: 0.035; H: 0.028; D (individual spirals): 0.017–0.018; Wt: 3.4 g.
Four joining fragments, much corroded, preserving complete ornament/pendant.
A continuous length of bronze wire, mostly circular in section, tapering slightly toward the center of each spiral, formed into two spiral multiple coils/loops (six coils), the wire between the two spirals drawn up to form a loop. This loop appears to have been reinforced, conceivably repaired, with a thin strip of bronze hammered flat and wrapped around the loop.
Cf. **10/57** (SF 300) and parallels cited there. Particularly close are the following: Åberg 1935:74, fig. 140 (top right); 76, fig. 143 (right); Holste 1953:92, fig. 11, no. 8; Müller-Karpe 1959: pl. 180, B, no. 5 (Riegsee).

Wheel pendants

Although such small wheels can derive from small votive chariots or wagons (such as Heilmeyer 1994: pl. 72:7, no. 72), the form and context of **10/59** (**Fig. 10.23**) suggests that it served as an item of jewelry, such as the many later examples from the Enodia sanctuary at Pherai (Kilian 1975b: pl. 79, esp. nos. 23–31, which more clearly served as pendants). The fullest discussion of these is by Imma Kilian-Dirlmeier, who traces the various types of *Radanhänger* from Late Helladic IIIC through the later stages of the Geometric period (Kilian-Dirlmeier 1979:16–29, pls. 4–10). Pendant **10/59** belongs to the type with two or more wheels that are connected by four spokes. The type is well-known in Greece (e.g., Olympia, Delphi, Pherai), but also in FYROM, particularly at the site of Vranje (Kilian-Dirlmeier 1979:29, pl. 10, nos. 149–153). Other types of wheel pendants are common in Greece (e.g., Agios Panteleimon [Patelli], Argive Heraion, Chalkidike, Dedeli, Delphi, Ithake, Lousoi, Olympia, Pherai, Philia, Sparta, Tegea, Thebes, Thermon, Vergina), as well as at Suva Reka in Kosovo, Brazda near Skopje and Radanje (FYROM) (see Kilian-Dirlmeier 1979: pls. 4–9). Various types of wheels or wheel pendants are also common throughout Europe in the "Urnenfelderzeit" (see, among others, Müller-Karpe 1959: pl. 21:a, no. 6 [Cumae]; pl. 118, no. 22 [Marburg]; pl. 183, various examples, from Grünwald, Grave 1), and during the Archaic period in southern Italy (Papadopoulos 2003:126–127, fig. 156, no. 443, with full discussion of South Italian types).

In Albania, there are various examples of multiple wheel pendants connected by spokes, including the examples from Katundas near Berat (Braka 1987:47, pl. III:8–9, esp. no. 9), from Hamallaj near Dürres (Hoti 1993:126, 131, pl. III:1), and Klaus Kilian also lists a wheel pendant from Kuç i Zi (Kilian 1975b: 133). An overview of wheel pendants, together with a discussion of connections on both sides of the

Adriatic, is provided by Korkuti (1985:96–97, figs. 4–5; 102, pl. I, nos. 6–10).

10/59 (SF 336 + SF 343, TXXVIII-2), **Figs. 3.93, 10.23** (Sheet 75.6; Photo 3169)
Wheel Pendant.
T77-2 (SU 4.0429).
D (outer circle): 0.052–0.055; H (including loop): 0.063; Wt (all frr): 6.0 g.
Wheel reconstructed from eight joining frr; pendant loop reconstructed from four joining fragments, plus two non-joining fragments of wheel. Fragments preserve all of loop and most of wheel, except for small portion of outer and less of inner circles; two of the spokes broken, with small parts missing. All fragments corroded, but condition quite good, considering the thinness of sheet bronze. The possibility that the piece had a backing of organic material was noted in the field but could not be confirmed.
Wheel pendant made of thin sheet bronze, cut to form inner and outer circles, connected by four spokes, with strip of the same sheet bronze, though fractionally thicker, rolled over to form loop for suspension. The outer circle is less wide than the inner circle, and the spokes are slightly wider than both circles.
Clearly a wheel pendant, this piece is related to, but is not precisely the same as, the numerous bronze wheel pendants found in Greece and other parts of the Balkans, for which, see especially, Kilian-Dirlmeier 1979: pls. 4–9 (various examples); of these, pl. 10, nos. 151–153 (two from Vranje and one from Pherai) are the closest; for Albanian comparanda, see especially Braka 1987:47, pl. III:9; Hoti 1993:126, 131, pl. III:1.

Disks/Bosses

Bronze

There are four bronze disks or bosses from the tumulus (**Fig. 10.24**). Two of these, **10/60** (TXVIII-2) and **10/63** (TXXI-5), were almost certainly associated with the bronze headbands found in the tombs and worn by the deceased. A third, **10/62** (TLIII-2), was found on the upper torso of the deceased, immediately to the northwest of the cranium, and was therefore probably worn as pendant or bead. The fourth, **10/61**, was found in topsoil, so little can be said about its function or date. Three of the disks have perforations in the center: that on **10/60** is large, whereas those on **10/61** and **10/62** are small. In the case of **10/63**, there is a small embossed dot at the center rather than a perforation. Apart from **10/61**, which cannot be dated more precisely than Late Bronze or Early Iron Age, two of the disks/bosses, **10/60** and **10/63**, are assigned to Phase II (twelfth–eleventh centuries BC), while **10/62** belongs to Phase III (eleventh–tenth centuries BC). As **10/60** and **10/63** were associated with bronze headbands, they are best discussed as part of the larger object of which they were decorative elements. Item **10/61** stylistically belongs with **10/60**, as both are decorated with repoussé decoration. On **10/60**, the decoration consists of a ring of dots more or less halfway between the edge and the perforation, with the area thus enclosed by the ring decorated with four arches of dots defining a cross. The edge of **10/61** was decorated with three parallel rings of small repoussé dots, punched from the underside.

In many respects, **10/62** differs from the other bronzes of this category in several respects. Unlike **10/60** and **10/61**, it is undecorated, but the bronze seems different in its appearance and feel to the other bronze disks/bosses, being rather more substantial and clearly cast rather than cut into shape. Its exterior surface was finished smooth, whereas the surface of the underside was only roughly finished, with tooling marks visible all over. Such a distinctive finish is a characteristic feature of many of the indigenous bronzes of Early Iron Age and Archaic South Italy (cf., among others, Papadopoulos 2003:112–117, figs. 139–145, nos. 413–415 [the so-called "dischi composite"]).

10/60 (SF 318, TXVIII-2), **Figs. 3.58, 10.24** (Sheet 75.3; Photo 3089)
Perforated Bronze Disk, with Repoussé Decoration.
Papadopoulos 2010a:40, fig. 8.
T73-2 (SU 4.0410).
D: 0.047–0.050; Wt: 10.7 g.
Complete, but slightly split at one point, with additional cracks around perforation; slightly warped. Both surfaces corroded.
As preserved, the disk is flat but was probably originally slightly domed; edge a little irregular, as if cut rather than cast. Comparatively large perforation at center, also rather irregular. Repoussé decoration consists of a single ring of dots approximately halfway between edge and perforation; the area

enclosed by this ring is further decorated with four arches of dots, roughly equidistant from one another, defining a cross of sorts.

10/61 (SF 163a–d), **Fig. 10.24** (Sheet 54.2; Photo 3707)
Small Perforated Bronze Boss, with Repoussé Decoration.
Tumulus Fill (SU 2.0202).
D: 0.033; H: 0.007; Wt (all frr): 2.5 g.
Reconstructed from various joining fragments, almost complete, except for small portion of body; with three additional non-joining fragments. Thin sheet bronze, extremely fragile; green/dark green patina.
Thin sheet bronze disk formed into a small domed boss, with small central perforation. Edge of boss decorated with three parallel rings of small repoussé dots, punched from the underside.

10/62 (SF 297, TLIII-2), **Figs. 3.172, 10.24** (Sheet 68.4; Photo 2990)
Bronze Disk, Perforated in the Center (Button or Small Boss).
T63-2 (SU 1.0359).
D: 0.033; H: 0.004–0.005; Wt: 6.1 g.
Intact; green patina, different from other bronzes from Lofkënd.
Sheet bronze more substantial than **10/61** (SF 163a–d), formed into domed boss, with small central perforation. Exterior surface finished smooth; in comparison, the surface of the underside is roughly finished, with tooling marks visible all over. This distinctive finish is a characteristic feature of many of the indigenous bronzes of South Italy; cf., among others, Papadopoulos 2003:112–117, figs. 139–145, nos. 413–415.

10/63 (SF 259, TXXI-5), **Figs. 3.69, 10.24** (Sheet 102.12)
Fragmentary Small Bronze Boss.
Papadopoulos 2010a:42, fig. 11.
T55-5 (SU 4.0326).
PL: 0.031; D (est.): 0.032–0.035; Wt (all frr): 2.1 g.
Main fragment reconstructed from several joining frr preserving about one-third of boss, plus numerous non-joining fragments and chips (20+); bronze heavily corroded.
Small circular boss, slightly domed, made of thin sheet bronze. Small embossed dot at center. Perhaps associated with the bronze headband, **10/86** (TXXI-4).

Iron

The solitary example of an iron disk/boss, **10/64** (**Fig. 10.25**), was found in Tomb LXVIII, which is assigned to Phase Va (ninth–earlier eighth century BC). It was found in situ on the north side of the cranium of one of the deceased interred in this grave. More conical than domed, the form of **10/64** is not unlike that of **10/63** in one respect, namely, that rather than being perforated, there was a small central depression at the apex. A small fragment of iron was also encountered in the tomb and thought to be a possible finial associated with **10/64**, though it does not clearly join and its association with the disk/boss remains uncertain.

10/64 (SF 092, TLXVIII-2), **Figs. 3.232, 10.25** (Sheet 18.2; Photo 3791)
Iron Boss.
T13-2 (SU 2.0103).
H (boss): 0.021; D (boss): 0.047–0.049; PH (finial): 0.021; Wt (all pieces): 19.2 g.
Two joining fragments preserving complete boss, plus one non-joining fragment preserving (small?) portion of possible finial. All pieces corroded.
Small conical boss, with small central depression at apex. Small non-joining segment of possible finial evidently circular in section. Since the presumed finial does not clearly join, it is unclear whether it belongs with the underside or exterior of the boss, but it was found in Tomb LXVIII on the underside of the boss.
Textile pseudomorphs.

Spiral Coils

Small bronze spiral coils (beads)

There are three examples of this small type of bronze spiral coil (**Fig. 10.26**), all of which are diminutive: one from Tomb VIII (**10/65**), belonging to Phase I (fourteenth–thirteenth centuries BC) and one of the earliest tombs in the tumulus, perhaps originally connected with the left hand of the infant, which was largely not preserved; and two examples from Tomb XVII (**10/66** and **10/67**), which is also relatively early (Phase II: twelfth–eleventh centuries BC). Given their context in Tomb XVII, these two examples are conceivably beads or else head ornaments associated either with the headband **10/84**

(SF 349) or the earrings **10/80**, **10/81** (SF 313a, c). It is even possible that **10/66** and **10/67** may have been hair ornaments, used to fasten the ends of braids (Bettina Arnold, personal communication).

This type of coil, in various sizes, is ubiquitous throughout Europe. Similar bronzes in Albania have been published from various sites (e.g., Aliu 1995:141, pl. II, Tomb 11, no. 18 [Psar]; Aliu 2004: pl. II, nos. 13–15; pl. VIII, nos. 99–102; pl. XXVIII, no. 327 [Luaras]; Aliu 2012: pl. I, Tomb 10, nos. 17–18; pl. VII, Tomb 136, no. 114; pl. XI, Tomb 160, nos. 157–158; pl. XIII, Tomb 170, no. 181; pl. XLVI, nos. 503–504; Andrea 1976b:221, pl. VI, Tomb 34, no. 4; 225, pl. X, Tomb 68, no. 1; 227, pl. XII, Tomb 116, nos. 1–2 [Kuç i Zi]; Andrea 1985:272, pl. XI, Tomb 97, no. 4; Andrea 1995:117, pl. II, Tumulus 6, Tomb 1, no. 16 [Bujan]; Andrea 2009–2010:272, pl. III, Tomb 31, nos. 44–45; 279, pl. X, no. 23; Bodinaku 1981:259, pl. III, no. 4 [Permet]; Bodinaku 2001–2002:83, pl. VII, Tomb 13, no. 8; 84, pl. VIII, Tomb 23, no. 4 [Dukat]; Kurti 1976:247, pl. IV, no. 6 [Mat]; Kurti 1977–1978:179, pl. VII, Tomb 32, no 8; 184, pl. XII, Tomb 58, no. 2; Kurti 1983:103, pl. I, Tomb 4, no. 3; 106, pl. IV, Tomb 22, no. 13 [Burreli]).

The type is very common in different parts of Greece, particularly in the north (e.g., Andronikos 1969:225–227; Radt 1974:124, pl. 38, nos. 1–3; Rhomiopoulou and Kilian-Dirlmeier 1989:95, fig. 7, nos. 11–14; 26–28; 108, fig. 17, nos. 4, 11–12; 109, fig. 18, nos. 7, 22–23, 28–39; 118, fig. 29, nos. 12–18; 119, fig. 30, nos. 1–9 [Vergina]; Kilian 1975a: pl. 70, nos. 43–50 [Thessalian Pherai]; Kilian-Dirlmeier 2002: pl. 175, nos. 2969–2975 [Philia in Thessaly]; cf. Kilian 1975b: pl. 55, nos. 4–5; pl. 67, no. 2; Hochstetter 1987: pl. 4, no. 13 [Kastanas]; Rhomiopoulou 1971:38, fig. 1 [four examples from Spelaion near Grevena]; Vokotopoulou 1986: fig. 111α–δ [Vitsa, all 850- 800 BC]; Koukouli-Chrysanthaki 1992:412, fig. 99, no. 1 [Thasos]; Felsch 2007:335. pl. 45, no. 1506 [Kalapodi]), but is also known farther south (e.g., at Late Helladic Mycenae: Xenaki-Sakellariou 1985: pl. 3, no. 2418).

In other parts of the Balkans, they were recorded early on at Donja Dolina by Truhelka (1904:143, fig. 81; pl. L, no. 39; pl. LIII, no. 24; pl. LVIII, nos. 5, 7, 14–15; pl. LX, nos. 33–34; pl. LXXVI, no. 10; pl. LXXVII, nos. 5, 18). Elsewhere in the Balkans, they are common in Serbia (e.g., Garašanin 1954: pl. XVIII, no. 8; pl. XXV, no. 4; Popović and Vukmanović 1998:123, pl. 1, no. 8; 126, pl. 4, no. 7; 128, pl. 6, nos. 8–10; 129, pl. 7, nos. 12–13; 131, pl. 9, no. 9; 132, pl. 10, no. 7; cf. 140, pl. 18, nos. 4–5), in Bosnia Herzegovina (e.g., Benac and Čović 1956: pl. IV, no. 2; pl. V, nos. 13–14; pl. VI, nos. 6–9; pl. IX, no. 12; pl. XIV, no. 8; pl. XV, nos. 7–8; pl. XVI, no. 14; pl. XXI, no. 13; pl. XXII, nos. 14–15; pl. XXX, no. 6; pl. XXXV, nos. 6–8; pl. XL, no. 3; pl. XLI, nos. 3–6; pl. XLII, nos. 4–5, 9, 12; Benac and Čović 1957: pl. II, nos. 11–13; pl. XI, nos. 14–16; pl. XXI, no. 9; pl. XXV, nos. 21–26; pl. XXXIII, no. 22; pl. XXIV, no. 51; pl. XXXIX, nos. 2, 24), Slovakia (e.g., Furmánek, Veliačik, and Vladár 1999:35, fig. 9, nos. 5–6; 36, fig. 10, no. 14; 75, fig. 31, nos. 23, 26; 85, fig. 37, nos. 24, 26; 93, fig. 42, nos. 6–7), and in Slovenia (e.g., Teržan 1995: pl. 107, nos. 122–157, pl. 165, no. 6).

They are plentiful in Italy, especially in the Emilia, around Bologna, at San Vitale and Savena (e.g., Müller-Karpe 1959: pl. 62, A, no. 6; B, no. 5; pl. 64, B, no. 3; C, no. 4; pl. 66, C, no. 8; D, nos. 4–5; F, no. 6; pl. 68, K, no. 5; pl. 69, A, no. 6; pl. 70, A; pl. 77, K, no. 4; pl. 78, N, no. 4; J, no. 2; L, no. 2; T, no. 1; pl. 79, F, no. 7; pl. 80, E, no. 3; cf. Warneke 1999:40, fig. 11, no. 139), although the type is found elsewhere in Italy as well (e.g., Kilian 1970: numerous examples on pls. 34, 43, 55, 68, 71, 87, 97, 99, 101, 113, 128, 136, 139, 147–148, 157, 178, 185, 196, 203, 214, 237, 242, 251, and 254 [Sala Consilina]; Montelius 1904: pl. 144, no. 13 [Novilara, Tomb 5]; Papadopoulos 2003:71–72, fig. 92e–h, nos. 185–186 [Francavilla Marittima], with reference to further Italian comparanda; Peroni and Trucco 1994a:572, pl. 120, no. 10; Peroni and Trucco 1994b:686, pl. 131, no. 15; 740, pl. 153, nos. 5–7 [Sibaritide]), Austria and the eastern Alps (e.g., Müller-Karpe 1959: pl. 111, C, nos. 4, 7–8 [Maria Rast, Grave 89]; Von Sacken 1868: pl. XVII, no. 18; Willvonseder 1937: pl. 38, no. 5; pl. 48, no. 4), Germany (e.g., Holste 1953: pl. 12, nos. 14–15; Müller-Karpe 1959: pl. 118, no. 18 [Marburg]; pl. 124, D, no. 7 [Mixnitz]; pl. 180, B, no. 12 [Riegsee]; pl. 186, C, no. 8 [Unterhaching]), Hungary (Childe 1929:277, fig. 149, nos. 10–11; David 2002: pl. 140, no. 17; pl. 156, no. 7; pl. 181, nos. 9–16; pl. 185, nos. 5–12; pl. 187, no. 11; pl. 231, nos. 15a–c; pl. 257, nos. 7a–g; pl. 264, no. 6; pl. 265, nos. 10a–d; pl. 271, nos. 3–4; pl. 275, no. 10; pl. 321, no. 8; Kovács 1977:34, no. 11; 39, nos. 14–15), and elsewhere in Europe (e.g., Gimbutas 1965:41, fig. 11, no. 9; 252, fig. 162a, no. 1; 254, fig. 163, nos. 26–29; 257, fig. 166, no. 2; 269, fig. 176, no. 10). Similar spiral coils are also common in gold (e.g., Åberg 1932:76, fig. 154, bottom left; fig. 155, bottom center).

In discussing the common bronze coils from Vergina, which are in many cases considerably longer than the examples from Lofkënd, Andronikos (1969: 225–227, fig. 66) referred to them as "σύριγγες," and went on to discuss the famous passage in *Iliad* 17.52 ("πλοχμοὶ θ' οἳ χρυσῷ τε καὶ ἀργύρῳ ἐσφήκωντο"), together with the reference in the scholiast on *Iliad* 18.402, and Eustathios on the same passage ("οἱ δὲ χρυσᾶς εἶναι σύριγγας, ὡς οἷον σωληνίσκους, αἷς πλόκαμοι περιέχονται"), with such small bronze spiral coils in mind. Perhaps the earliest attempt to link the references in Homer to specific Early Iron Age spiral ornaments—and certainly the *locus classicus* on the Homeric references to such objects—goes back to Helbig, who had in mind larger spiral rings, bracelets, and armbands (Helbig 1887:242–245), some of them closer to the somewhat wider spiral coils, such **10/68** from Lofkënd.

10/65 (SF 444, TVIII-1), **Figs. 3.30, 10.26** (Sheet 102. 11; Photo 3607)
Minuscule Bronze Spiral Coil (Bead).
T100-1 (SU 2.0591).
L: 0.005; D: 0.003; Wt: >0.1 g.
Intact; good green patina.
Minuscule spiral coil bead formed of thin bronze wire, plano-convex in section.
Cf. **10/66** and **10/67**.

10/66 (SF 313b = 348, TXVII-5), **Fig. 10.26** (Sheet 68.8b)
Small Bronze Spiral Coil (Bead or Head Ornament).
Papadopoulos 2010a:39, fig. 5b.
T72-5 (SU 1.0408).
L: 0.019; D: 0.004–0.005; Wt: 0.7 g.
Two joining fragments preserving complete spiral coil; good green/dark green patina.
Thin bronze wire, circular in section, coiled to form spiral.
Cf. **10/67** (SF 313d).

10/67 (SF 313d, TXVII-6), **Figs. 3.54, 10.26** (Sheet 68.8d; Photo 3055)
Small Bronze Spiral Coil (Bead or Head Ornament).
Papadopoulos 2010a:39, fig. 5d.
T72-6 (SU 1.0408).
L: 0.018–0.019; D 0.004; Wt: 0.9 g.
Intact; good dark green patina.
Thin bronze wire, circular in section, coiled to form spiral.
Cf. **10/66** (SF 313b).

Other bronze spirals (earrings or hair rings?)

The three following bronzes are related to the small bronze spiral coil beads discussed above, but slightly larger, and may have served as earrings (**Fig. 10.27**). One of them, **10/70** (TLIII-5), was found in the lab during the cleaning of the cranium, which had been block-lifted, and must have served either as an earring, a hair ring, or fastener (for which, see Blegen 1937:41, fig. 65, no. 13 [Middle Helladic Grave XVII]; Papadopoulos 2005:559–560, fig. 67f, pl. 461a–b), or, conceivably and less likely, a bead suspended around the neck. The two examples from Tomb XLII (**10/68** and **10/69**) are contextually interesting: they were the only grave goods in the tomb, found immediately to the west and northwest of the southern skeleton in the tomb. Although neither **10/68** or **10/69** could be directly associated with any one of the three individuals in the tomb, it is highly possible that both were in their original positions, that is, that the ornaments were worn by the adult, assuming that she was originally in the central portion of the grave, and remained where they were once the cranium and the remainder of her skeleton was pushed to the north in order to accommodate the two infants. The possibility that they were associated with one or both of the infants cannot be ruled out categorically, but it seems less likely. Consequently, both objects are best associated with the adult female and, if so, must have been worn as earrings. All three spirals are from tombs assigned to Phase III (eleventh–tenth centuries BC).

The two spirals from Tomb XLII (**10/68, 10/69**) are almost certainly the same, although one is more fragmentary than the other. Elsewhere in Albania, similar spiral ornaments have been published from Myç-Has (Bela 1990:130, pl. XII, Tomb 6, no. 175; cf. 130, pl. XII, Tomb 12, no. 202), Psar (Aliu 1995:144, pl. V, no. 50), Luaras (Aliu 2004: pl. II, Tomb 14, no. 25; pl. VI, Tomb 61, no. 18; pl. XV, Tomb 167, nos. 193–194), and Permet (Bodinaku 1981:258, pl. II, no. 13); from tombs in the Korçë basin (Andrea 1985:265, pl. IV, Tomb 22, no. 1; 267, pl. VI, no. 5; 289, pl. XXVIII, Tomb 34, no. 4); and there are several more from the Rehovë tumulus (Aliu 2012: pl. I, Tomb 10, nos. 19–20; pl. II, Tomb 56, nos. 34–36; pl. XI, Tomb 160, nos. 156, 161–162; pl. XXI, Tomb 221, no. 260; pl. XXII, Tomb 222, nos. 284, 286). Similar spiral ornaments are common in Late Bronze and Early Iron

Age Greece (e.g., Dawkins et al. 1906–1907:116, fig. 6c; Dawkins 1929: pl. LXXXVa [Sparta]; Pendlebury et al. 1937–1938: pl. 28:1, nos. 162–163, M. 8, [Karphi]), Italy (e.g., Peroni and Trucco 1994b:740, pl. 153, no. 4), as well as elsewhere in Europe (e.g., Åberg 1932:59, figs. 99, 101; 65, fig. 113, no. 8; Gimbutas 1965:271, fig. 178, no. 9).

The primary difference between the two spirals from Tomb XLII and the one from Tomb LIII is that in the former the wire is tightly coiled, whereas in the latter it is more open, with far fewer turns. As such, **10/70** more closely resembles a series of spirals that are likely hair rings or fasteners (see esp. Papadopoulos 2005:559-60, nos. T7-6, and cf. T10-8, T69-4, and T102-6). This type of personal ornament, as Desborough (1972:304–305; cf. Higgins in Popham, Sackett, and Themelis 1979–1980:220) notes, is mostly found in central mainland Greece with evident northern connections, and examples are fairly common both in bronze and gold (e.g., Higgins 1980:89–91 [examples from Athens, Argos, Homolion, Agrinion]; cf. Batziou-Eustathiou 1984:79, fig. 5 [gold]; Doumas and Marangou 1978:213, no. 74 [Skyros]; Kilian 1975a: pl. 70, nos. 33–41; Koukouli-Chrysanthaki 1992:409, 411, fig. 97, pl. 355, esp. nos. 8–12 [Thasos]; Kraiker and Kübler 1939:85, fig. 4, left; Kübler 1943: pl. 39; Philipp 1981: pl. 42, nos. 547–556; Popham, Sackett, and Themelis 1979–1980:220 [gold]; Ruppenstein 2007:24, 229, Beil. 11, no. 18 [gold]). The type is quite common in Macedonia in both metals, especially at Vergina (Andronikos 1969:240, fig. 80; 241, fig. 82 [bronze]; 259, fig. 99 [gold]; see also Casson 1923–1925:25 [referred to as "finger rings" of wire of several convolutions]). In Albania, similar spirals are found at Luaras (Aliu 2004: pl. XV, Tomb 168, no. 192; cf. the larger examples, such as pl. XXVIII, nos. 322, 324). TLIII-5 may also be related to a series of spirals found in southern Italy (e.g., Papadopoulos 2003:75, fig. 97, nos. 201–202 [Francavilla Marittima]).

10/68 (SF 267, TXLII-1), **Figs. 3.138, 10.27** (Sheet 111; Photo 3768)
Fragmentary Bronze Spiral Ornament, Perhaps Earring.
T59-1 (SU 2.0338).
PH (main fr): 0.011; D: 0.010; Wt (all frr): 0.08 g.
Single fragment preserving portion of spiral ornament, plus 16 minuscule non-joining fragments; heavily corroded.
Continuous thin bronze wire, circular in section, coiled to form a spiral ornament. The position of this piece, together with **10/69** (SF 265), indicated that both were conceivably earrings worn by the adult female; alternatively, they could have served as beads or some other item of personal ornament.
Cf. **10/69** (SF 265).

10/69 (SF 265, TXLII-2), **Figs. 3.139, 10.27** (Sheet 117.5; Photo 3711)
Fragments of Bronze Spiral Ornament (or Five Separate Small Rings).
T59-2 (SU 2.0338).
PL (largest fr): 0.007; D (est.): 0.010; Wt (all frr): 0.2 g.
Some 26 fragments, all much corroded, some probably but not clearly joining, preserving portion of spiral ornament, or else five smaller rings, as drawn.
Spiral ornaments as **10/68** (SF 267).

10/70 (SF 307, TLIII-5 **Figs. 3.175, 10.27** (Sheet 68.7; Pl. 3059)
Small Bronze Spiral Coil.
T63-5 (SU 1.0359).
D (max): 0.010; Wt: 0.2 g.
Reconstructed from two joining fragments, preserving portion of spiral coil; corroded.
Thin bronze wire, circular in section, formed into a spiral coil, as shown; it is unclear how much of the original object is not preserved. Conceivably an earring or part of a head ornament.
Cf. especially Papadopoulos 2003:75, fig. 97, nos. 201–202; Truhelka 1904: pl. L, no. 31; pl. LI, no. 12.

Iron spiral coils

The two iron coils presented here (**Fig. 10.28**) are best seen in the context of the bronze spirals discussed above, but made of iron (for related spiral coils in bronze, "spiraline," see Papadopoulos 2003: 71, fig. 92a–b). The small **10/72** dates to Phase II (twelfth– eleventh centuries BC) and is among the earliest of the iron objects from the tumulus, the larger **10/71** is assigned to Phase IV (late tenth–ninth centuries). Of the two, **10/71**, although resembling the iron fibula of Type II.1 (**10/22**), cannot be a fibula on account of a clasp-like fragment that is probably part of it. Resembling a linked chain, the object is most likely part of necklace that was worn around the neck of the deceased (the fragments of **10/71** were found associated with the block-lifted cranium). What is perhaps most interesting is that the fragmentary **10/72** may very well belong to the same necklace as the other tubular beads from Tomb XXI (see **10/91** and **10/92**), and this bolsters the interpretation of both iron coils as parts of necklaces worn by the deceased.

10/71 (SF 172, TLVIII-4), **Figs. 3.198, 10.28** (Sheet 103.2; Photo 2434)
Fragments of Iron Coil Resembling Linked Chain.
T37-4 (SU 2.0243).
PL (as drawn): 0.085; Wt: 7.7 g.
Five joining and two non-joining fragments, all heavily corroded, preserving portion of iron coil.
Continuous length of iron wire forming at least seven coils. Small non-joining fragment may derive from a clasp; another preserves a very small portion of thin iron wire.
A slight curvature of the fragments as drawn suggested the possibility of a fibula (cf. the iron fibula, **10/22** [SF 110] with arch made of continuous coil, Type II.1), but this is unlikely, given the clasp-like non-joining fragment. Associated with the cranium of the deceased in Tomb LVIII, the object is more likely part of a necklace worn around the neck. If the coil does derive from a fibula, then it belongs to a type with looped arch first noted in bronze at Hallstatt; see Von Sacken 1868: pl. XIII, no. 12.

10/72 (SF 264, TXXI-9), **Figs. 3.72, 10.28** (Sheet 103.3; Photo 3687)
Fragmentary Iron Coil.
T55-9 (SU 4.0341).
PL: 0.024; Wt: 1.3 g.
Single fragment, much corroded, preserving portion of iron coil.
Continuous length of iron wire forming six preserved coils. Perhaps part of the same iron coil attached to the iron tubular bead **10/91** (TXXI-7, SF 257b).
Cf. **10/91** (TXXI-7, SF 257b).

Bronze Rings

Plain bronze rings, including possible finger rings

Not one of the bronze rings presented here is demonstrably a finger ring (**Fig. 10.29**). Two were found associated with the bronze headband of the adolescent female in Tomb LXX (**10/73, 10/74**). Another (**10/79**) was not found in situ but rather during the process of cleaning the skeletal remains in the lab. Although it was reasonably clear that **10/79** was associated with the male burial in the tomb (SU 326), the preserved state of the human remains, coupled with the fragmentary nature of the ring, was such that it could not be determined whether the ring was worn. Four additional bronze rings were encountered in topsoil or in tumulus fill, so they provide no contextual evidence as to their function (**10/75–10/78**). Of the rings found in tombs, the earliest is **10/79**, which is assigned to Phase II (twelfth–eleventh centuries BC), whereas the two from Tomb LXX (**10/73** and **10/74**) date to Phase Va (ninth–earlier eighth centuries BC).

Among the rings presented here, the form and size of **10/79** is such that it may have been a finger ring, although this cannot be established with certainty. Made of thin sheet bronze hammered flat, the ring, if complete, would have had overlapping terminals, allowing it to be adjusted to the size of the finger on which it was worn. Rings of this type are fairly standard in Early Iron Age tombs in the Aegean. In Athens, for example, they are well represented (see Kraiker and Kübler 1939:33–34, 85, fig. 3 [right)]; Müller-Karpe 1962:86, fig. 4, no. 12; cf. p. 84, fig. 2, no. 13 [both Submycenaean]; Parlama and Stampolidis 2000:45, nos. 13–14; Ruppenstein 2007:24, Beil. 11, Gr. 136, no. 15; 28, Beil. 13, Gr. 143, no. 8). There are nine rings of this type from tombs at Lefkandi, six of which are Submycenaean, two Early Protogeometric, and one, much smaller than the rest, dating to Subprotogeometric I (Popham, Sackett, and Themelis 1979–1980:248). Elsewhere in the Aegean, rings of this type are common in, among other places, Aigina (Furtwängler, Fiechter, and Thiersch 1906: pl. 116, esp. no. 38), the Argolid (Daux 1957: 663–664, fig. 55; Kokkou-Vyridi 1977:177, fig. 4, no. E 1969; Verdelis 1963:7, fig. 3, various examples), Elis (Eder 2001:92, pl. 14a, nos. e–i), as well as elsewhere in the Peloponnese (e.g., McDonald, Coulson, and Rosser 1983: esp. 300, fig. 5.10 [Nichoria]; Philipp 1981:141–142, pl. 42, nos. 515–517 [Olympia]; cf. also Taylour and Janko 2008:446, pl. 52, nos. 7011–7012 [Agios Stephanos]), central Greece (Felsch 2007:300–303, pl. 37, nos. 621–622, 630, 661 [Kalapodi]; Vlachogianni 2000: 397, fig. 28 [Ellopia]), Thessaly (Arachoviti 1994:132, 134, fig. 11, no. BE 8655; Batziou-Efstathiou 1984:76, fig. 1; Kilian-Dirlmeier 2002:12, pl. 8, no. 59, cf. no.

64), Macedonia (Andronikos 1969:238–241, fig. 80 [Vergina]; Hochstetter 1987:35, pl. 5, nos. 1–4 [Kastanas]; Papadopoulos 2005:559, fig. 70, T11-1 [Torone]), Epirus (e.g., Vokotopoulou 1986:312–314, fig. 31, Type γ), Crete (Brock 1957:71, 199, no. 795; Coldstream and Catling 1996:557; Rethemiotakis and Englezou 2010: pl. 55, ill. 141), and the Dodekanese (e.g., Morricone 1978:84–85, fig. 78; pp. 166–167, fig. 297); similar rings are also known on Cyprus (e.g., Karageorghis 1983:138, pl. 78, no. 33 [Tomb 62]; 219, pl. 143, nos. 60, 71 [Tomb 76]).

Although **10/74** resembles the common closed ring with a section that is plano-convex which is common in the Aegean and over much of Europe, this was not a finger ring but an ornament associated with the bronze headband from the same tomb. Among the rings encountered in topsoil or tumulus fill, a few may conceivably be finger rings (e.g., **10/75**, **10/77**), on account of their shape and size, but several, like **10/73** and **10/76**, are clearly not finger rings. In the Aegean, such rings are referred to as *krikoi*—κρίκος in the singular (sometimes κίρκος) in Classical Greek, as opposed to δακτύλιος (used specifically for a finger ring or signet)—and are found in various shapes and sizes. They are especially common dedications at sanctuaries throughout mainland and insular Greece (see Papadopoulos 2003:156, n. 380 for full references). In an overview of early rings in the Aegean, Dakoronia (1989) has argued that such rings were dedicated for the value of the bronze, and she raised the possibility that such rings may have represented recognized items of value or weights in a pre-monetary system. Moreover, Greek literature from as early as Homer is replete with references to *krikoi*, and their various functions: bronze rings could have served as fasteners for a horse's breast-band or to a carriage pole, as eyelet-holes in sails, as curtain rings, nose-rings, as links in a chain, and so on (see *LSJ* sv. κρίκος). Identical rings are found in great numbers in southern Italy, Sicily, and beyond, both in colonial Greek sanctuaries as well as indigenous cemeteries (see Papadopoulos 2003:156, n. 382, for a full listing).

10/73 (SF 091d, TLXX-2a), **Figs. 3.245a, 10.29** (Sheet 102.1; Photo 3615)
Bronze Ring.
Papadopoulos 2010a:46, fig. 14.
T17-2a, associated with Bronze Headband, **10/87** (TLXX-2).
D: 0.021; Wt: 1.4 g.
Complete ring reconstructed from two joining fragments; corroded.
Plain ring; circular to ovoid in section.

10/74 (SF 091e, TLXX-2b), **Figs. 3.245b, 10.29** (Sheet 102.2; Photo 3613)
Bronze Ring, Resembling Finger Ring.
Papadopoulos 2010a:46, fig. 14.
T17-2b, associated with Bronze Diadem, TLXX-2.
D: 0.024; Wt: 3.1 g.
Intact; good dark green patina.
Plain ring, plano-convex in section.
Cf. Wace and Thompson 1912:212, fig. 147n.

10/75 (SF 163e), **Fig. 10.29** (Sheet 54.2; Photo 3600)
Bronze Ring.
Tumulus Fill (SU 2.0202).
D: 0.024; H: 0.002; Wt: 1.0 g.
Reconstructed from two joining fragments, complete. Good dark green patina.
Plain ring, thin band, plano-convex in section.

10/76 (SF 163f), **Fig. 10.29** (Sheet 54.2; Photo 3597)
Bronze Ring.
Tumulus Fill (SU 2.0202).
D: 0.025; H: 0.002; Wt: 2.4 g.
Intact; good dark green patina.
Plain band, section in places elliptical, elsewhere faceted or lozenge-shaped, indicating that the ring was cast in two-part mold, clearly preserving the line around the inner circumference where there was a misalignment of the mold.
Cf. Papadopoulos 2003:107–108, fig. 134o–p, no. 382 (Francavilla Marittima).

10/77 (SF 163g), **Fig. 10.29** (Sheet 54.2; Photo 3599)
Bronze Ring.
Tumulus Fill (SU 2.0202).
D: 0.022; H: 0.002; Wt (all frr): 0.7 g.
Reconstructed from five joining fragments, almost complete, much corroded.
Plain, thin band, circular in section, thicker on one side. Probably originally a continuous band.
Cf. **10/78** (SF 163h).

10/78 (SF 163h), **Fig. 10.29** (Sheet 54.2; Photo 3598)
Bronze Ring.
Tumulus Fill (SU 2.0202).
D: 0.022; H: 0.001; Wt (all frr): 0.2 g.

Now two joining fragments, extremely thin and fragile, preserving greater part of ring.
Plain, very thin band, circular in section.
Cf. **10/77** (SF 163g).

10/79 (SF 292a, TXXI-10), **Figs. 3.73, 10.29** (Photo 3710)
Fragmentary Ring of Thin Sheet Bronze.
T55-10 (SU 4.0326).
PL: 0.020; D (est.): 0.024; H: 0.007; Wt (all frr): 0.5 g.
Main piece reconstructed from three joining fragments preserving about one-third of ring, plus five non-joining fragments; all fragments heavily corroded.
Ring made of thin sheet bronze; conceivably finger ring.

Earrings

The four rings presented here (**Fig. 10.30**) share a similar form, made of a thickish bronze wire, more rhomboidal than circular in section, tapering toward pointed overlapping terminals. In addition, all four rings were found in contexts that would confirm, or at least suggest, that they were worn to the grave as earrings. The two bronze earrings in Tomb XVII (**10/80** and **10/81**) were found overlying the bronze headband in the tomb and were originally thought to be part of it, but the nature of both the rings and headband was such that this was unlikely, and **10/80** and **10/81** were probably worn by the deceased separate from the headband. In the case of Tomb LIII, the bronze ring **10/82** was found under the upper torso, in the area of the sternum of the deceased, and the context would indicate that it was either worn as an earring or as a hair-tie. The earring in Tomb LXIX (**10/83**) was found by the right side of the cranium of the child (SU 551) at the southwest corner of the tomb, and it can only have been worn as an earring.

Chronologically, the four earrings, despite their small number, cover most of the period of the use of the tumulus. The earrings in Tomb XVII date to Phase II (twelfth–eleventh centuries BC), that in Tomb LIII to Phase III (eleventh–tenth centuries BC), while the latest, **10/83**, is assigned to Phase Va (ninth–earlier eighth centuries BC). This basic type of earring is particularly common in central mainland Greece and Macedonia, where it is found in the Early Iron Age in both bronze and gold (see Desborough 1972:304–305; Higgins 1980:89–91; Kraiker and Kübler 1939:85, fig. 4 [left]; Kübler 1943: pl. 39; Popham, Sacket, and Themelis 1979–1980:220; see also Kilian 1975a: pl. 70; Philipp 1981: pl. 42, nos. 549–556), and related earrings are also common in South Italy during the Early Iron Age and Archaic periods (Guzzo and Carrara 1981:447, fig. 4, Tomb 1, no. 4 [left]; 450, fig. 6, Tomb 2, nos. 3–4; Pasqui 1888: pl.XV, fig. 12 [Torre Mordillo]; Ruffo 1994–1995:99, fig. 64, no. 441; Zancani-Montuoro 1974–1976: pl. XXI, no. 14; Zancani-Montuoro 1977–1979: pl. Xb, no. 6). They are also common in many others parts of the Balkans, including Albania, and Europe (see comparanda under **10/80**).

10/80 (SF 313a, TXVII-3), **Fig. 10.30** (Sheet 68.8a)
Bronze Earring.
Papadopoulos 2010a:39, fig. 5a.
T72-3 (SU 1.0408).
H: 0.029; D (max): 0.029; Wt: 5.6 g.
Intact; good dark green patina.
Earring formed of bronze wire, more rhomboidal than circular in section, thicker on one side, and tapering toward overlapping terminals.
Cf. **10/81** (SF 313c) and **10/82** (SF 299). Cf., among others, Andrea 2009–2010:272, pl. III, Tomb 28, no. 36; 276, pl. VII, nos. 85–88; 279, pl. X, no. 25 (Shuec); Benac and Čović 1956: pl. XVII, no. 9 (referred to as "Bronzering [Finger ring? Ohrring?])"; Furmánek, Veliačik, and Vladár 1999: 75, fig. 31, no. 22 (Slovakia); Müller-Karpe 1959:52, pl. 207, A5 (Fuchsstadt); Tsountas 1908: pl. 4, no. 6, from a Mycenaean tomb at Sesklo).

10/81 (SF 313c, TXVII-4), **Figs. 3.54, 10.30** (Sheet 68.8c; Photo 3055)
Bronze Earring.
Papadopoulos 2010a:39, fig. 5a, no. c.
T72-4 (SU 1.0408).
H: 0.025; D (max): 0.025; Wt: 3.7 g.
Intact; good dark green patina.
Earring formed of bronze wire, circular to rhomboidal in section, thicker toward bottom, and tapering toward points at both terminals, which are overlapping.
Cf. **10/80** (SF 313a).

10/82 (SF 299, TLIII-3), **Figs. 3.173, 10.30** (Sheet 68.9; Photo 3987)
Bronze Earring.

T63-3 (SU 1.0359)
H: 0.034; D: 0.031; Wt: 4.3 g.
Intact; good green/dark green patina.
Earring formed of bronze wire, rhomboidal in section, thicker toward bottom, and tapering toward points at both terminals, which are overlapping.

10/83 (SF 431b, TLXIX-3), **Figs. 3.239, 10.30** (Sheet 102.3; Photo 3601)
Bronze Earring.
T27-3 (SU 5.0551).
L (max same as D): 0.024; Wt: 4.2 g.
Intact; slightly corroded; green patina.
Thick bronze wire, elliptical to rhomboidal in section, tapering toward both terminals, which overlap considerably.

Headbands (Diadems)

The bronze headbands (often referred to as diaems) of Lofkënd are among the most characteristic items of personal ornament in the richest graves of the tumulus, all of them of young females (**Figs. 10.31–10.34**). Their significance, in terms of funerary custom and the notion of "marriage to death" are considered more fully in Chapter 8. There are four bronze headbands from burials, the three from Tombs XVII (**10/84**), XVIII (**10/85**), and XXI (**10/86**) the earliest, all three more or less contemporary and assigned to Phase II (twelfth–eleventh centuries BC). The contextual information regarding each of the headbands is presented in detail in Chapter 3, but it is important to note here the other objects found in association with the headbands that have already been discussed. In Tomb XVII, what were identified in the field as two bronze "earrings" (**10/80** and **10/81**), together with two small spiral coils (**10/66** and **10/67**), were found overlying the headband and were originally thought to be part of it, although as noted above, these may well have been worn by the deceased separate from the headband. The circumstances of Tomb XVIII were such that it was difficult to determine which of the individuals interred in the grave was associated with the bronze disk, **10/60**, although it seems reasonable to conclude that it was part of bronze headband **10/85**. Another bronze boss, **10/63**, this one more fragmentary, also appears to have associated with the headband in Tomb XXI, **10/86**.

The term "diadem" should be used conventionally, for although we may never know by what word the ancient Illyrians of Lofkënd referred to their bronze headbands, Greek literature provides several terms by which these distinctive items of personal ornament were known (see Papadopoulos 2010a:50–51). The most common term is στεφάνη, which generally means anything that surrounds or encircles the head for defense or ornament (*LSJ* sv. στεφάνη); the word is attested in both Homer and Hesiod. There is also the word ταινία, which most often refers to a band, fillet, and especially a headband, worn as a sign of victory (whether athletic, military, or even a beauty contest); it can also be a breastband worn by young girls, or else an abdominal band or even a bandage (*LSJ* sv. ταινία). Another term, κρήδεμνον (Doric κράδεμνον), refers to a woman's headdress or veil, and this term, too, is attested in Homer (*LSJ* sv. κρήδεμνον; Papadopoulos 2010a:50). It is ironic, therefore, that the term by which most bronze headbands are identified in the modern literature, διάδημα—diadem—is the least appropriate or least likely to have been used in antiquity, especially in early Greece. In several ancient sources, a διάδημα (as in διαδέω, to bind on either side) is a band or fillet round the *tiara* (τιάρα) worn by the Persian king (see esp. Xenophon, *Cyropedia* 8.3.13; Plutarch 2.488d). A διάδημα was also worn by Alexander the Great and his successors, and by kings generally. Indeed, the word is not common—if at all attested—before the Persian Wars (for the diadems worn by Alexander, see *LSJ* sv. διάδημα, and esp. Arrian, *Anabais* 7.22.2; Herodian Grammaticus, *Herodiani technici reliquiae* 1.3.3; and see further Papadopoulos 2010a:50–51).

In Albania, bronze headbands are known in burials from a number of tumuli, all of them decorated with various motifs, whether incised or repoussé, including Patos (Korkuti 1981:44, pl. VII, Tomb 68; also the fragmentary examples: 44, pl. VII, Tomb 63), Burreli (Kurti 1983:91, 96, 103, pl. I, Tomb 4, no. 19 [deriving from one of the richest tombs in the tumulus]), a number from Barç Tumulus I (Andrea 1985: 23, pl. IV, Tomb 19, no. 2; 25–26, pl. V, Tomb 34, no. 2; 32–33, pl. IX, Tomb 78, no. 1; cf. the thin sheet bronze 34, pl. X, Tomb 98, no. 1; and the "bracelet" 52, pl. XXII, no. 2), several from the tumuli at Shtoj (Jubani 1992:27, 49, pl. II, Tomb 8, no. 3; Koka 1990:64, pl. II, Tomb 3, no. 8; 35, 66, pl. IV, Tomb 5, no. 53 [the latter with incised motifs including some found on matt-painted pottery]; Koka 2012:274, pl. XXV, Tumulus 6, Tomb 3, no. 8; pl. XXVII, Tumulus 6, Tomb 5, no. 53), and Shtikë (Aliu 1996:63, 76, pl. IV, no. 1); elements of a bronze headband virtually identical to those on **10/87** have been recently published from Tumulus I at Shuec (Andrea 2009–2010:277, pl. VIII, Tomb 64, no. 95). Furthermore, bronze headbands continue to be used in later burials in Albania, as the magnificently decorated bronze example from Tomb 20 in Tumulus I in the necropolis of Apollonia testifies (Mano 1971:142, pl. XLIII, no. 1).

In the Balkans north of Albania, bronze headbands/diadems are well represented in Bosnia Herzegovina, at Donja Dolina (Truhelka 1904: pl. XL, no. 1), and in the various graves of the Glasinac region, including ones that can be assigned to the later stages of the Bronze Age and to the Early Iron Age (Benac and Čović 1956:55, pl. XXIV, no. 2 [Podlaze, Tumulus LXXXX, Grave 1]; note also the incised decorated band from Mlad, Tumulus X, Grave 9:57, pl. XXIX, no. 4; Benac and Čović 1957:66, pl. VI, no. 12 [Ilijak, Tumulus IV, Grave 1]; 67, pl. X, no. 11 [Rusanovići, Tumulus XXV, Grave 2]; 69, pl. XIII, no. 7 [Ilijak, Tumulus III, Grave 2]; 69, pl. XV, no. 14 [Ilijak, Tumulus III, Grave 9]). Farther north, bronze headbands, variously referred to as a "Stirnband" or more simply as a "Bronzeband," are well known in a number of funerary contexts (e.g., Åberg 1932:49, fig. 67 [top]; 81, fig. 169 [left], fully discussed on p. 78). A copper headband from Vyčapy-Opatpvce in Slovakia, with pointillé decoration, is assigned to the Early Bronze Age (Gimbutas 1965:40, fig. 10, no. 26), though the absolute date remains something of a problem. Copper and bronze headbands with a backing of leather are known from Late Bronze Age Straubing and elsewhere (see Gimbutas 1965:254, fig. 163, nos. 30–31; cf. 599, fig. 420).

In neighboring Epirus and Greek Macedonia, bronze headbands very similar to those of Lofkënd are known at Vitsa Zagoriou (Vokotopoulou 1986:152–153, fig. 3; fig. 108ζ, pl. 245α, inv. 2350, from the richly furnished female inhumation, Tomb 113), in the tumulus burials at Pogoni, which should date to the period of transition between the Late Bronze and the Early Iron Age (Andreou and Andreou 1999a:81, figs. 12, 14–15), Dodone (Evangelides 1935:235–236, fig. 9, nos. 53–55; pl. 23α, no. 1; pl. 24α, no. 23), and Vergina (Andronikos 1969:251–254, fig. 88; Radt 1974:132–133, pl. 40, no. 1), the Vitsa and Vergina examples dating to the ninth century BC, as well as one from an Early Iron Age burial at Spelaion near Grevena (Rhomiopoulou 1971:38, fig. 1); elsewhere in western Greece, there is a repoussé-decorated piece of sheet bronze from Ithake that is perhaps from a belt rather than a headband, with very similar decoration to the headband from Vitsa (Benton 1953:352, pl. 69, nos. E.243, E.243a; same as Kilian 1975b: pl. 16, no. 12). Bronze headbands, whether decorated or plain, are known from other sites in Greece (e.g., Dakoronia 2006:502, fig. 26.12, from east Lokris, with attached double ax).

It is worth adding that headbands in the Aegean found in situ in tombs go back to the Early Bronze Age, such as the celebrated silver headband from Grave 14 at Dokathismata on the island of Amorgos (Tsountas 1898:154, pl. 8, no. 1; also illustrated in Åberg 1933:72, fig. 130). In describing the tomb, Tsountas notes that among the human remains, only the cranium with the headband was preserved, and it was behind the cranium that the silver pin, surmounted by an animal, was found (Tsountas 1898:154; for the pin, see pl. 8, no. 66; Papadopoulos 2010a:47–49, fig. 15), perhaps a hairpin; it is even possible that the small pin may have been used to fix the headband in place. Be that as it may, the fact that only the cranium survived recalls many of the Late Bronze and Early Iron Age tombs at Lofkënd, and it may well be that Tomb 14 at Dokathismata was that of a female child. Headbands continue into later stages of the Bronze Age, as the many gold examples from the shaft graves at Mycenae (Karo 1930–1933: pls. XXXVI–XXXVII, nos. 232–235; pl. XXXVIII, nos. 219, 286–287; pl. XXXIX, nos. 231, 236–239; 70, pl. LXXII, no. 219; pl. XLV, nos. 292–293, 311; cf. also some of the "Goldbänder" and "Armbänder," 124, fig. 45, nos. 638, 654; pl. XLIII, nos. 255, 257) and the Aigina Treasure (Evans 1892–1893:210–211, nos. L–M) testify. There is also a gold headband/diadem in the Middle Bronze Age shaft grave at Kolonna on Aigina, published by Imma Kilian-Dirlmeier, who also provides a distribution map of gold headbands in Early, Middle, and Late Bronze Age contexts in the Aegean and Anatolia (Kilian-Dirlmeier 1997:19, fig. 6, no. 9; 54–57, fig. 26; pls. 1, 3).

For the Classical Greek world, one of the earliest studies of bronze headbands/diadems, all of them with incised decoration, was Adolf Furtwängler's seminal 1890 publication of the bronzes and other small finds from Olympia (Furtwängler 1890:46–48, pls. XVIII–XIX, nos. 297–316). In his brief analysis, Furtwängler noted a number of early terracotta figurines, of Geometric or Early Orientalizing date, of human figures wearing headbands/diadems, many of them with incised decoration (including Furtwängler 1890: p. XV, nos. 264, 265; pl. XVII, no. 290). Two of these are clearly female (Furtwängler 1890: no. 265, 290), but in the case of the third, which has both a headband decorated with incised zigzag and a belt with similar zigzag (Furtwängler 1890: no. 264), gender was not clear; there are no breasts and no clear genitalia and, as such, a young girl is probably what is indicated. In his discussion of the headbands, Furtwängler (1890:46) wrote:

> Wie die primitiven Bronze- und Terrakottastatuetten lehren, war ein breites Diadem mit linearer Verzierung (vergl. 264, 265, 290) ein charakteristischer Teil der Frauentracht, den ja auch die ältere Poesie mit den beliebten Beiworten καλλιστέφανος εὐστέφανος χρυσάμπυξ und dergleichen an den Frauen hervorhebt. Zahlreiche kleinere und grössere Diademe aus dünnen Bronzeblech mit geometrischen Ornamenten haben sich in den tiefsten Schichten um die Altäre in Olympia gefunden, die offenbar selbständige Weihegaben waren. Sie fanden sich mit den bereits besprochenen primitiven Tier- und Menschenfiguren zusammen und gehören ohne Zweifel derselben ältesten Epoche an wie diese. Besonders häufig waren sie in der Gegend des Pelopions.

The Olympia headbands/diadems were clearly dedications at the sanctuary at Olympia, but their context provides little evidence for their use, though one might speculate that given Furtwängler's connection of these objects with women, they were probably dedications to Hera rather than to Zeus. Be that as it may, many of the Olympia headbands are decorated with incised zigzags, not the same as, but not unlike, those on **10/87** (Furtwängler 1890: pl. XVIII, nos. 297–302), whereas others are decorated with repoussé dots (Furtwängler 1890: pl. XVIII, nos. 303–309; pl. XIX, no. 310–312, 316), sometimes defining circles, semicircles, or running spirals. A number of related sheets of decorated hammered bronzes are classified as "Armringe" (Furtwängler 1890: pl. XXII, nos. 381–382; pl. XXIII, no. 380), and one or two of them may conceivably be headbands. One of these is decorated with opposed diagonals (no. 381), another with meander (no. 382), and a third with zigzag (no. 380). In a similar vein, one or two of the so-called belts from Hallstatt, particularly some of the smaller and more fragmentary examples, might be from headbands (see esp. Kilian-Dirlmeier 1972: pl. 49, nos. 660, 679; pls. 52–53, nos. 549, 553; some of the pieces on pl. 48), so too some of the fragmentary pieces from Emporio on Chios classified as belts (Boardman 1967:220–221, fig. 143, no. 323; cf. also some of the "bracelets," esp. 212–214, pl. 87, nos. 261–265).

In addition to the pieces already cited, bronze headbands, including miniatures, are well represented among the votives in a number of Greek sanctuary sites (e.g., Bather 1892–1893:251, fig. 20 [Athens, Acropolis]; Bouzek 1982: fig. 7, no. 1 [Troizen]; Kilian-Dirlmeier 2002:68–69, pl. 66, nos. 1024–1033 [Athena-Itonia sanctuary near Philia in Thessaly, though not all are certainly headbands]; Kourouniotis 1910:324–25, fig. 47 [Bassai]; Orsi 1933:97, fig. 58 [silver diadem, from the sanctuary of Apollo at Crimisa]; Payne 1940:181, pl. 81, nos. 2–5 [Perachora]; cf. also the bronze mask with slits for eyes, 181, pl. 81, no. 1; Reichel and Wilhelm 1901:56, figs. 102–104 [Lousoi]; Waldstein 1905:266–267, pl. XCIX, nos. 1590–1599 [Argive Heraion]).

Related gold headbands/diadems (for which, see generally, Greifenhagen 1970: pl. 3, nos. 1–5; pl. 4, nos. 1–3; pl. 6, nos. 4, 6; pl. 11, nos. 5–7; Greifenhagen 1975: pl. 2, nos. 6–8; pl. 3, nos. 1–6; pl. 74, no. 7; Higgins 1980:96–97; Kourouniotis 1913: esp. 290–296, various examples, pls. XIV–XVI; Ohly 1953; Reichel 1942; Szilgágyi 1957) are well known from both Greek sanctuary sites (Payne 1940:185 [Perachora]; Orsi 1933: 87–88, fig. 149 [Crimisa]; cf. the gold/electrum "strips" in Hogarth 1908:109, pl. IX, nos. 29, 52 [Ephesos]) and from tombs (esp. in Athens: Alexandri 1968:23, fig. 3 [Athens, Kriezi Street, Middle Geometric]; Brückner and Pernice 1893:109–111, fig. 7; 125–127, fig. 24 [Athens, Kerameikos]; Droop 1905–1906:91–92, fig. 12 right [Athens, Kynosarges, Geometric]; Kübler 1954:190, 238, pl. 158, M 111, Grave 43 [Athens, Kerameikos]; Schlörb-Vierneisel 1966:7–8, Beil. 13, no. 6 [Athens, Kerameikos]; Stavropoullos 1965:78, pl. 44α, β [two headbands, Athens, Kavalotti Street, Geometric]; but also at Eretria: Bérard 1970:36, pls. B1, B2, no. 14.4, fully discussed on pp. 36–45; and at Lefkandi on the island of Euboia: Popham, Sackett, and Themelis 1979– 1980:188, pl. 187, Toumba Tomb 33, nos. 6–7; 190, pls. 189, 232a, Toumba Tomb 36, no. 2 [Subprotogeometric II–III]. Headbands are also common in tombs on Cyprus: Gjerstad et al. 1934:250, pl. LV, 2 [Lapithos, Cyprus, Tomb 425, nos. 7–8, Cypro-Geometric II]; also the diadems and gold bands: pl. LXXVIII, 3, Enkomi Tomb 3, nos. 94–95, 111, 150–151, 156–158, 160, 198, 235–236 [Late Cypriot II]; pl.

LXXX, 2, Enkomi Tomb 8; pl. LXXXI, 2, Enkomi Tomb 10, nos. 57–58; pl. LXXXVII, 1, Enkomi Tomb 17; pl. CXLVI, CXLVII, no. 1]); cf. the gold foil headband and diadem-like object from the Danish oak-coffin graves: Randsborg and Christensen 2006:18, fig. 8). There is also, at Eretria, a related object in a non-funerary context, found in the goldsmith's workshop hoard (Themelis 1983:163, fig. 13, no. 14).

Bronze

10/84 (SF 349, TXVII-2), **Figs. 3.53, 10.31** (Sheet 93; 11.31a–b; Photo 3331)
Bronze Headband.
Papadopoulos 2010a:38, fig. 4a–b.
T72-2 (SU 1.0408).
L (all frr lined up together): 0.495; L (main fr at front same as D of headband): 0.210; W: 0.015; Wt: 23.0 g.
Reconstructed from seven joining fragments, complete. Surface corroded at points, but generally well preserved, with a good dark green patina.
Headband made of thin strip of hammered sheet bronze. The upper and lower edges of the strip are decorated by an incised horizontal line that runs around the diadem. For the greater part of the length of the diadem, there are two additional incised lines between the two near the edges, resulting in four parallel incised lines. At one point near the center of the front of the diadem, there are two vertical incised lines running between the two horizontal lines near the edges, defining a small area or metope of incised herringbone decoration, with a portion of additional hatching, almost certainly part of a herringbone pattern, to the right. There appear to be additional incised vertical and diagonal lines, some of which are perhaps scratches, on the exterior face near one of the terminals.
The system of tying the two terminals of the headband is interesting. As preserved, the terminals are tied together by remnants of textile preserved in the corrosion. The textile itself, which is mineralized, is a rectangular strap or cord but with no clearly visible weave or thread (depleted?). What is particularly interesting is that the terminal on the side opposite the mineralized textile does not join with either of the ends on the other side (both of which appear to be smooth), and it is possible that the end was simply inserted into the two strips of tied bronze. Whether this was the original binding system, or an expedient because the headband had broken, is uncertain, but it is one way of securing a larger headband/diadem on the head of a child.

10/85 (SF 317, TXVIII-1), **Figs. 3.57, 10.32** (Photo 4128)
Fragmentary Bronze Headband.
Papadopoulos 2010a:40, fig. 7.
T73-1 (SU 4.0412).
L (all frr lined up together): 0.444; L (longest preserved group of joining frr close to D of headband): 0.140; W: 0.018; D (rivet heads): 0.006; Wt (all frr, including conservation backing): 10.9 g.
Reconstructed from numerous joining and non-joining fragments, preserving all or greater part of headband. Some fragments better preserved than others, but mostly extremely corroded and fragile.
Headband made of thin strip of hammered sheet bronze. No visible incised decoration. The two terminals were connected by two bronze rivets, well preserved.
For the basic form, cf. **10/84** (SF 349).

10/86 (SF 255, TXXI-4), **Figs. 3.68, 10.33** (Sheet 115; Photo 4127)
Fragmentary Bronze Headband.
Papadopoulos 2010a:42, fig. 10.
T55-4 (SU 4.0341).
L (all frr lined up together): 0.495; L (longest preserved group of joining frr): 0.146; W: 0.026; Wt (all frr, including conservation backing): 23.4 g.
Reconstructed from numerous joining and non-joining fragments, preserving all or greater part of headband. A few fragments slightly better preserved than others, but most pieces corroded and very fragile.
Headband made of thin strip of hammered sheet bronze. As with other headbands/diadems, the width remains more or less constant, but at one point the joining fragments appear to define a slightly greater width, almost as if the strip of bronze develops into a disk. This point does not appear to be associated with the fastening of the terminals; rather, it may well represent the center of the headband, worn in the middle of the head. It is precisely at this point where highly fugitive traces of possible, but very poorly preserved and fine, incised decoration were noted.
Cf. **10/85** (SF 317) and, for the basic form, **10/84** (SF 349).

10/87 (SF 091 + SF 089, TLXX-2), **Figs. 3.244, 10.34** (Sheet 104; Photo 4124)
Fragmentary Bronze Headband.
Papadopoulos 2010a:44, fig. 13a–b.
T17-2 (SU 3.0126).
PL (all main joining or nearly joining frr): 0.586; W (diadem band): 0.025; H (hooks, max): 0.045; W (hooks, at lower terminals): 0.058; W (hook, at upper connection, max): 0.038; D (rivet head): 0.005; Wt (all frr, including chips): 41.4 g; Wt (small piece of iron): 0.4 g.
Reconstructed as shown, from numerous joining and non-joining fragments, preserving greater part of headband; extremely fragile and corroded, with incised decoration only visible under raking light. Fragment "a" refers to all the fragments of the diadem, except for the two pieces ("b" and "c") preserving the hooks. The various components of the headband on the drawing are labeled "A" through "F" for the purposes of describing the incised decoration.
Headband made of hammered strip of sheet bronze; the terminals connected on fragment "c" ("A" on the drawing) by means of two rivets. The neatly articulated terminals of the headband are framed on either side by a hook, each hook connected by three rivets. The bronze rings **10/73** (SF 091d) and **10/74** (SF 091e) may have been originally associated with the hooks. The entire headband is decorated with fine incision, as shown. For the main part of the headband, the decoration consists of three registers of finely incised multiple horizontal zigzags, each register separated, as well as framed above and below, by a row of fine repoussé dots. For a portion of the band of the headband (labeled "B" on the drawing), the upper and lower registers are decorated with hatched dogtooth instead of the multiple zigzags. The three registers of multiple zigzags on the fragment labeled "C" on the drawing are framed, probably on either side, by hatched dogtooth set vertically and framed by repoussé dots. Similar registers of horizontal multiple and vertical zigzags and hatched dogtooth appear to frame a series of circles on fragments "A," "C," "E" and "F" on the drawing, each formed of fine repoussé dots, except, perhaps, for the circle on fragment "A," which is very poorly preserved and barely visible (it, too, may have been formed by repoussé dots, but too little survives to be certain). All the circles, except perhaps for that on fragment "F," are further decorated with hatching set in groups, as shown, but are too poorly preserved to determine the precise design.
Although headbands with incised and repoussé decoration are known from other contemporary or near contemporary sites in Albania, the only close parallel is the fragmentary headband from Tomb 64 in Tumulus I at Shuec in the Korçë basin (Andrea 2009–2010:277, pl. VIII, Tomb 64, no. 95).

Length of Copper or Copper Alloy Wire

The following piece was found in topsoil (**Fig. 10.35**), so context provides no clue to its date or even whether it is ancient or modern. Although the piece may well be modern, the fact remains that very similar lengths of wire are known from a number of prehistoric sites in Europe (e.g., Müller-Karpe 1959: pl. 183, no. 24 [Grünwald, Grave 1]; cf. David 2002: pl. 233, no. 20:a–j), and there are related lengths of thin strip bronze wire from the Archaic sanctuary at Francavilla Marittima, the extramural sanctuary of Sybaris, in northern Calabria (see Papadopoulos 2003:135, fig. 170a–f).

10/88 (SF 316), **Fig. 10.35** (Photo 3118)
Length of Copper/Copper Alloy Wire.
Topsoil (SU 2.0003).
PL (max, as twisted): 0.038; Th: 0.001–0.002; Wt: 3.0 g.
Thin copper or copper alloy wire twisted, as shown; both terminals appear to be cut. Condition quite good.
Thin wire, circular in section.
Context does not determine whether this wire is ancient or modern.

Beads

Several objects already discussed may well have served as beads—such as the small bronze spiral coils, which were conceivably beads (**10/65–10/67**), and, of course, the terracotta spindlewhorls, beads, or buttons (**10/1–10/9**)—although this could not be established with certainty.

Tubular iron beads

There are a total of 13 tubular iron beads from the tumulus (**Figs. 10.36–10.37**), 11 deriving from five tombs (**10/89, 10/101** [TLXVI-1, TLXVI-2]; **10/90, 10/100** [TLXXX-2, TLXX-3]; **10/91, 10/92** [TXXI-6,

TXXI-7]; **10/94** [TXLVIII-5]; and **10/95**, **10/96**, **10/97**, **10/98** [TLV-4–TLV-7]), one from topsoil (**10/99** [SF 124]), and another from tumulus fill (**10/93** [SF 126]). The two examples from Tomb XXI are the earliest, dating to Phase II (twelfth–eleventh centuries BC); the solitary example from Tomb XLVIII belongs to Phase III (eleventh–tenth centuries BC); the beads from Tombs LXVI and LV are Phase IV (late tenth–ninth centuries BC); the two examples from Tomb LXXX belong to the latest of the prehistoric burials in the tumulus (Phase Vb). The two beads from topsoil and tumulus fill cannot be precisely dated. The chronological distribution of the iron beads extends, therefore, through most of the period of the use of the tumulus except for the very earliest phase.

The tubular iron beads from Lofkënd belong to one of two basic types: Type I (**10/89–10/99**) are both the most numerous and include the earliest of the iron tubular beads, though the type is long-lived, extending from Phase II into the latest Phase Vb. Although similar to Type II, they differ from them by having a prominent central rib; although poorly preserved, three examples (**10/97** [SF 266B], **10/98** [SF 266A], **10/99** [SF 124]) are probably of Type I. The prominence of the central rib can vary (**10/90** has the most prominent rib), but in all cases it is reasonably clear. There are only two clear examples of Type II (**10/100** [SF 13] and **10/101** [SF 164a]); both are plain tubes, **10/100** has more or less straight sides, but spreading in diameter toward one side; **10/101** has concave sides; both lack the prominent central rib of Type I. In terms of size, the smallest beads have a length of 0.017 m, the largest 0.025 m, and in terms of diameter they range between 0.012 and 0.022 m.

Elsewhere in Albania, tubular iron beads, particularly of Type I, are well represented at a number of Early Iron Age sites, including Katundas (Braka 1987:43, 47, pl. III, nos. 1–3) and Shtoj (Koka 1990:34, 64, pl. II, Tomb 3, no. 9; Koka 2012:274, pl. XXV, Tumulus 6, Tomb 3, 13–14 examples of both types, clearly part of a necklace). In dealing with the examples from Katundas, Braka (1987:43) writes: "Rruazat e hekurit (varri 11) jane gjithsej 32 copë, 16 copë bikonike me diameter 1,5 cm, 15 copë kanë gjatësinë 2,2 cm, ndërsa njëra 2,4 cm. Kjo e fundit është e vendosur në qendër të gjerdamit dhe ndan dy rruazat e gurit të shndrithëm (Tab. III, 1,2). Inventari i këtij varri na ndihmen për përcaktimin e drejtimit veri-jug të varreve dhe për datimin e tij." At Shtoj, there were some 14 examples, with a length of 0.022–0.034 m and a diameter of 0.015–0.019 m, all evidently part of a necklace from Tomb 3 (Koka 1990:34, 64, pl. II, no. 9; 73, pl. XI, no. 3); two smaller tubular iron beads from Tomb 4 have a length of 0.015 and a diameter of 0.010 m. In describing these distinctive beads, Koka (1990:64) compares them to human vertebrae. There is also a closely related bronze cylindrical bead, similar to our Type I, among the finds from the Luaras tumulus (Aliu 2004: pl. XV, Tomb 171, no. 199).

In Early Iron Age Greece, this type of tubular iron bead is rare, but a related type in bronze, which usually served as a finger ring and is often found in situ worn by the deceased, is well represented in the Argolid, especially at the Argive Heraion (Waldstein 1905: pl. XCI, nos. 1487–1488, 1496–1498), in the Late Bronze and Early Iron Age burials at Mycenae (e.g., Danielidou 2008: pl. 62:β; Desborough 1955:242, no. 2, pl. 49f, inv. 54-162: three in all, two with a raised ridge round the middle), Argos (Kritsas 1972: pl. 134:α), and in one of the Geometric tombs at Tiryns (Verdelis 1963:44, fig. 14, Beil. 19, Gr. XXV:3; L: 0.037 m; D: 0.025 m). Another example from an Early Iron Age tomb at Mycenae was made of iron sheet coated in bronze (Desborough 1954:263, pl. 45, inv. 53-621: L: 0.020 m; D: 0.021 m). A similar finger ring, in bronze, was discovered in situ worn on one of the fingers of the hand of the deceased in a Geometric tomb at Mycenae excavated in 2008 by Christofilis Maggidis (personal communication). A related type of bronze ring is also known among the Geometric and Archaic votive finds from Sparta (e.g., Buschor and von Massow 1927:35, fig. 17, no. 6; Raftopoulou 1998:134, fig. 12.15 [center right, various examples]); there is also an example from a Geometric tomb at Traganas in east Lokris (Onasoglou 1981:20, pl. 18β [middle]).

Type I

10/89 (SF 164b, TLXVI-2), **Figs. 3.225b, 10.36** (Photo 3697)
Tubular Iron Bead, Type I.
T31-2 (SU 4.0214).
L: 0.020; D: 0.012; Wt: 2.0 g.
Intact; heavily corroded.
Cylindrical bead, thinner than SF 164a, flaring toward both edges and with prominent central rib.

Very clear traces of cord preserved on interior.
Cf. **10/90**, **10/91**, **10/92**, **10/93**, **10/94**, **10/95**, **10/97**.

10/90 (SF 424, TLXXX-3), **Figs. 3.274, 10.36** (Sheet 102.6; Photo 3983)
Fragmentary Tubular Iron Bead, Type I.
T4-3 (SU 5 0032).
PL: 0.018; D (max): 0.016; Wt: 3.9 g.
Single fragment, heavily corroded, preserving over one-half of bead.
Cylindrical bead, flaring slightly toward preserved end, with very prominent central rib.
Fiber pesudomorph from inside.
Cf. **10/89** (SF 164b).

10/91 (SF 257b, TXXI-7), **Figs. 3.70b, 10.36** (Sheet 46.8; Photo 3705)
Tubular Iron Bead, Type I, with Attached Iron Coil.
T55-7 (SU 0341).
L (bead without iron coil): 0.018; PL (with iron coil): 0.024; D: 0.015; Wt: (both frr): 2.8 g.
Bead complete; iron coil fragmentary, one portion still attached to bead, plus one other non-joining fragment; all pieces heavily corroded.
Cylindrical bead flaring slightly toward both edges and with prominent central rib. Iron coil made of thin wire, attached at one end, as shown.

10/92 (SF 257a, TXXI-6), **Figs. 3.70a, 10.36** (Sheet 46.8; Photo 3703)
Tubular Iron Bead, Type I.
T55-6 (SU 4.0341).
L: 0.017; D: 0.013; Wt: 2.3 g.
Complete, but very heavily corroded; broken through at one point.
Cylindrical bead flaring slightly toward both edges and with prominent central rib.

10/93 (SF 126), **Fig. 10.36** (Photo 3915)
Fragment of Tubular Iron Bead, Type I.
Tumulus Fill (SU 1.0067).
L: 0.017–0.018; PW 0.012; D (est.): 0.015; Wt: 1.4 g.
Single fragment, corroded, preserving almost one-half of bead.
Fragment from tubular bead; it is unclear from the corrosion whether the bead is a plain cylinder or has a central swelling on exterior.

10/94 (SF 288, TXLVIII-5), **Figs. 3.159, 10.36** (Photo 3681)
Tubular Iron Bead, Type I.
T52-5 (SU 1.0366).
L: 0.025; D: 0.016; Wt: 4.2 g.
Almost complete, except for chipping at one edge; much corroded.
The largest of the Lofkënd tubular iron beads; cylindrical, flaring toward both edges and with prominent central rib.

10/95 (SF 287a, TLV-6), **Figs. 3.185c, 10.36** (Sheet 110.4; Photo 3699)
Tubular Iron Bead, Type I.
T53-6 (SU 1.0321).
L: 0.019–0.020; D: 0.013; Wt: 1.4 g.
Intact; corroded.
Cylindrical bead flaring slightly toward both edges, with central rib not as prominent as on other beads.

10/96 (SF 287b, TLV-7), **Figs. 3.185d, 10.36** (Sheet 110.5; Photo 3701)
Tubular Iron Bead, Type I.
T53-7 (SU 1.0321).
L: 0.018; D: 0.014; Wt: 2.9 g.
Intact; very heavily corroded.
Cylindrical bead flaring slightly toward both edges and with prominent central rib. Seam clear on one side, establishing that it was rolled to form the tubular bead. Although other similar beads preserve possible traces of a seam (cf. **10/101** [SF 164a]), this is the clearest example.

10/97 (SF 266b, TLV-5), **Figs. 3.185b, 10.36** (Sheet 110.6; Photo 3944)
Tubular Iron Bead, Type I.
T55-5 (SU 1.0321).
L: 0.019; D: 0.013–0.015; Wt: 3.3 g.
Complete: but heavily corroded.
Cylindrical bead flaring slightly toward both edges and probably with central rib, though this is unclear due to corrosion.

10/98 (SF 266a, TLV-4), **Figs. 3.184a–b, 10.36** (Sheet 110.7; Photo 3955)
Fragmentary Tubular Iron Bead, Type I.
T55-4 (SU 1.0321).
PL: 0.016; D: 0.015; Wt: 4.0 g.
Reconstructed from two joining fragments, preserving most of bead, except for much of one edge; heavily corroded.

Cylindrical bead flaring slightly toward both edges and with prominent central rib.

10/99 (SF 124), **Fig. 10.36** (Sheet 110x.3; Photo 3754)
Fragmentary Tubular Iron Bead, Type I.
Topsoil (SU 4.0004).
PL: 0.024; D: 0.015; Wt: 2.3 g.
Eight joining fragments preserving greater part of tubular iron bead; heavily corroded and somewhat misformed.
Cylindrical bead, probably a plain cylinder.
Textile pseudomorphs.

Type II
10/100 (SF 013, TLXXX-2), **Figs. 3.273, 10.37** (Sheet 18.5; Photo 3609)
Tubular Iron Bead, Type II.
T4-2 (SU 4.0031).
L: 0.025; D 0.018–0.022; Wt: 11.3 g.
Completely reconstructed from 18 joining fragments; corroded.
Plain tubular bead.
Cf. **10/101** (SF 164a).

10/101 (SF 164a, TLXVI-1), **Figs. 3.225a, 10.37** (Photo 3695)
Tubular Iron Bead, Type II.
T31-1 (SU 4.0214).
L: 0.018; D: 0.016; Wt: 4.1 g.
Intact; split but not broken; heavily corroded.
Plain cylindrical bead, flaring ever so slightly toward both edges. What is considered a split is a possible, but unclear, seam (cf. **10/96**).
Traces of cord preserved on interior.
Cf. **10/100** (SF 013).

Stone beads

There are only three beads of semi-precious stone from the tombs of the Lofkënd tumulus, all made from a similar crypto-crystalline quartz, either carnelian or sardonyx (**Fig. 10.38**). All three are confirmed as beads on account of their position in the tomb. The stone bead in Tomb XXVIII (**10/102**) was found together with other beads of glass and various bronzes in close association with the cranium of the deceased. The stone bead in Tomb LIII (**10/103**) was one of three roughly spherical beads—the others made of glass or glass paste—that were found in line in situ on the lower torso of the deceased, immediately above the pelvis. The location of the beads in the grave would suggest that rather than a necklace worn around the neck, this was perhaps a string of beads worn around the hips, or else placed at the hips. The stone bead in Tomb LXX (**10/104**) was found immediately adjacent to the front of the skull, and would have been worn by the deceased around the neck. Despite their small quantity, the three stone beads represent a broad time span, as they date to Phases II (**10/102**), III (**10/103**), and Va (**10/104**). In terms of their shape, the small spherical bead from Tomb LXX (**10/104**) accords with Beck's (1928: pls. II–III [between 4 and 5]) Type I.C.1.a; **10/102** and **10/103** are generally similar to Beck's Type I.B.1.a, though **10/102** is somewhere between Beck's Types I.B.1.a and I.B.1.b.

As for the stone, the material of all three is probably carnelian or else sard or sardonyx. As Higgins (1980:36) explains: "When the same stone is cut horizontally, as in the cutting of cameos, it is known as *onyx* or (if one layer is sard) *sardonyx*." The latter is normally an orange-red variety of chalcedony, similar to, but darker than, carnelian (which is also known as cornelian), and classed by some scholars as a variety of carnelian. In his definition of "cornelian (sard)," Higgins (1980:37) writes:

> The ancient *sardius*. A reddish or brownish chalcedony. Most periods. Some authorities use the term *cornelian* for the red varieties, *sard* for the brown, but the distinction is frequently hard to draw, and here the term *cornelian* is used for both varieties.

Among the 375 beads from the Late Bronze Age tombs at Perati in east Attica, there are 75 made of sard/carnelian and, like Higgins, Iakovidis was loath to distinguish between the two (Iakovidis 1969–1970:vol. 2, 302, prefers to classify the beads as "σάρδιος/κορβαλίνης," i.e., sard/carnelian). Carnelian or sard beads of globular or roughly spherical shape are ubiquitous throughout the Aegean in the Late Bronze Age, and are, in fact, found in most periods (in addition to the examples from Perati already noted [nine of the 75 sardonyx/carnelian beads listed by Iakovidis are spherical and a further 18 are flattened spherical], see esp. Higgins 1980:64, 75–76, with references on 205–210; see further, among others, Kavvadias 1912:263, fig. 42; 265, fig. 46 [left and center]; also 263, figs. 40:α-β; Marinatos 1932:42, pls. 14–15 [various examples]; Souyoudzoglou-Haywood 1999:42). Similar beads are found in the Early

Iron Age Aegean, though not in the same quantities as in the Late Bronze Age. There are, for example, a number of spherical and globular beads from Early Iron Age tombs at Knossos that are published as carnelian (e.g., five spherical carnelian beads [D 0.008] from the Fortetsa tombs, Brock 1957:22, no. 194:b [Tomb XI]; for more carnelian beads from Fortetsa, see Brock 1957:100, nos. 1153–1165 [Tomb I]; 97, no. 1116 [Tomb II]; see also Coldstream and Catling 1996:622 [Tomb J]); there is also the necklace of beads from Vergina thought at first to be of amber, but later diagnosed as sard (Andronikos 1969:254, fig. 89; Coldstream 1977:45; Radt 1974: pl. 40, no. 40).

The bead typology given in the catalogue entries below follows that established by Beck (1928: pls. II–III), as do the perforation types (see Beck 1928:51–52, pl. IV).

10/102 (SF 340, TXXVIII-6), **Figs. 3.95d, 10.38** (Sheet 75.1; Photo 3057)
Rounded Bead of Crypto-Crystalline Quartz (Carnelian or Sardonyx).
T77-6 (SU 4.0429).
H: 11–12 mm; D: 17 mm; D (perforation): 1.98–3.62 mm; Wt: 5.1 g.
Complete except for minor chipping; surface pitted and a little worn, but condition generally good.
Reddish brown colored stone.
Bead rounded, more pushed down than spherical in shape, approaching biconical; flatter version of **10/104**; perforation slightly wider at one end (perforation Type III: 1.98–3.62 mm).
Cf. **10/103** SF 296; for the type, see Beck 1928: Type I.B.1.a, resembling I.B.1.b.

10/103 (SF 296, TLIII-7), **Figs. 3.176b, 10.38** (Sheet 68.3; Photo 2842)
Rounded Bead of Crypto-Crystalline Quartz (Carnelian or Sardonyx).
T63-7 (SU 1.0359).
H: 10 mm; D: 16 mm; D (perforation): 1.92– 2.86 mm; Wt: 3.4 g.
Complete, except for very minor chipping around perforation; surface only very slightly pitted.
Reddish brown colored stone, with darker bands.
Bead rounded, more pushed down than spherical in shape, approaching biconical; perforation Type III (1.92 mm–2.86 mm).
For the type, cf. **10/102** (SF 340), and especially Beck 1928: Type I.B.1a.

10/104 (SF 113, TLXX-6), **Figs. 3.249, 10.38** (Sheet 10.9; Photo 3854)
Spherical Bead of Crypto-Crystalline Quartz (Carnelian or Sardonyx).
Papadopoulos, Bejko, and Morris 2007:120, fig. 12e.
T17-6 (SU 3.0126).
H: 8 mm; D: 8 mm; D (perforation): ca. 2 mm; Wt: 0.7 g.
Complete except for very minor chipping; surface only very slightly pitted.
Stone reddish brown.
Small spherical bead; perforation Type III (1.58–2.16 mm).
For the type, see Beck 1928: Type I.C.1.a.

BEADS OF FAIENCE AND GLASS

John K. Papadopoulos and Vanessa Muros

Full details of the faience, glass, and stone beads have been tabulated by Vanessa Muros and presented in **Table 10.1**.

Faience bead

There is only one faience bead from the tumulus (**Fig. 10.39**), **10/105**, from Tomb LIII, assigned to Phase III (eleventh–tenth centuries BC). Made of a whitish-yellow matrix, with a light blue-green glaze, which is much worn, the bead is roughly cylindrical and thus accords with Beck's (1928: pls. II–III) Type I.D.1.b. The surface, however, is decorated with a distinctive corrugation, consisting of a series of impressed horizontal and vertical lines forming a grid. The type accords with Beck's (1928:27, fig. 23) Group XXV, nos. A.4 and A.5 (the former is a cut faience or glass granulated bead from Egypt, Nineteenth Dynasty; the latter a faience granulated bead from the Twenty-third Dynasty). Faience beads of this type are occasionally found in Albania, with examples having been recorded from the Barç and Pazhok tumuli (see catalogue entry for details).

Although in Egypt faience beads are known from as early as the predynastic period (4th millennium BC), and in Mesopotamia even earlier, in the 5th millennium BC, it is likely that the faience in Albania, together with the glass beads, derives from Syria or Cyprus (for the antiquity of faience beads, see Panagiotaki 1997:303 [with references]). Among the earliest examples of faience beads on the Greek mainland are those recently published from the

Early Bronze Age graves at Agios Mamas in Chalkidike (Pappa 2010:386–388 ([three faience beads from the pithos inhumation, Tomb 6]; also pl. 7, nos. 1–2; pl. 8, no. 3; see also 401–403 for a discussion of the history of faience, including the earlier finds of Walter Heurtley in Macedonia), which now join the Early Minoan examples long known (for Early Minoan faience, see Foster 1979:34, 56–60; Panagiotaki 1997; see also Panagiotaki 2000:154– 157; Panagiotaki 2001).

10/105 (SF 298, TLIII-9), **Figs. 3.177, 10.39** (Sheet 68.1; Photo 2850)
Cylindrical Faience Bead, corrugated.
T63-9 (SU 1.0359).
H: 17.26 mm; D: 8.92 mm; D (perforation): 1.86–1.78 mm; Wt: 1.2 g.
Complete except for minor chipping at one end; surfaces much worn and slightly pitted.
Whitish yellow faience interior, with light blue-green glaze, much worn. Roughly cylindrical, almost melon-shaped bead, with distinctive corrugated decoration consisting of a series of impressed horizontal and vertical lines forming a grid. Perforation Type III.
For the type, see Beck 1928: Type I.D.1.b/ XXV.A.4–5. Cf. Andrea 1985:34, 271, pl. X, V83:7 (Barç); Bodinaku 1982:T 18, pls. VI:13, X, VI:16 (Pazhok).

Glass beads

There are 12 glass beads from the tumulus (**Figs. 10.40–10.43**). The earliest derive from tombs that are assigned to Phase II (twelfth–eleventh centuries BC) (TXXI-8 [**10/114**], and the six beads from Tomb XXVIII [**10/106, 10/108, 10/112, 10/115–10/117**]). There are two beads, TLIII-8 (**10/107**) and TLIII-6 (**10/109**), that can be assigned to Phase III (eleventh-tenth centuries BC), and one, TLV-8 (**10/113**) that is Phase IV (tenth–ninth centuries BC). Two other beads, **10/110** and **10/111**, were found in topsoil and cannot be dated. In **Figures 10.40–10.43**, we have grouped the glass beads by type, following, once more, the typology established by Beck. **Figure 10.40** assembles the beads that are largely spherical (Beck Type I.C.1.a). Of these, **10/106** and **10/107** are true spheres; **10/108** is also spherical, but a little flatter than **10/106**; the opaque orange-brown glass of **10/108** is decorated with an opaque white glass horizontal band at its midpoint, which Beck (1928: 46–48) classifies as Group XLVII (zone, striped, wave, and chevron beads). In the case of **10/109**, the opaque reddish brown glass is decorated with dots of white glass, which corresponds to Beck's (1928:41–46) Group XLVI (spot beads and eye beads and pendants with circular eyes).

The beads assembled in **Figure 10.41** are rounded but somewhat flatter than those that are spherical, and they accord with Beck's Type I.B.1.a. In the case of **10/112**, there are possible but unclear traces of fluting, while the rounded form of **10/113** is almost biconical. In the Aegean, the cemetery at Perati has produced numerous glass beads and provides many parallels for the beads from Lofkënd, some of which are roughly contemporary or later. Among the 103 glass or glass-paste (υαλόμαζα) beads from Perati assembled by Iakovidis, no fewer than 55 are spherical, and virtually identical to those of Lofkënd (Iakovidis 1969–1970: vol. 2, 302–305, fig. 128, no. 1, and table, p. 302; cf. Xenaki-Sakellariou 1985:292, Types 1 [σφαιρικός] and 2 [στρογγυλός]). The slightly more flattened, but still rounded form of **10/110–10/113** is also paralleled by a number of glass beads from Perati, especially Iakovidis Shape 2 (αρτόσχημοι), and also perhaps his Shape 5 (αμφικωνικαί) (Iakovidis 1969–1979: vol. 2, 305, fig. 128, nos. 2 and 5; for earlier Mycenaean spherical and flattened spherical glass beads, see, among many others, Kyparissis 1919:117, fig. 31; Wace 1932: pl. XX, no. 13 [Tomb 523]; pl. XXXIV, nos. 37:a–d [Tomb 517]). There is also a spherical bead from Submycenaean Grave 143 in the Athenian Kerameikos (Ruppenstein 2007:232, pl. 33). Spanning as they do the transition from the Late Bronze Age to the Early Iron Age, the examples from Perati and Athens are as early as the earliest of Lofkënd beads, and herein lies the importance of the glass beads from our tumulus. For although glass beads are commonly found in Illyrian graves of the Early Iron Age and the Archaic period, few are as precisely dated as the examples from Lofkënd.

As for the material, although there is some evidence for the production of European glass, especially in the form of beads, as early as the eleventh–ninth centuries BC, particularly at the site of Frattesina in northern Italy (see esp. the evidence presented in Henderson 1988; for further discussion on glass beads in European contexts, see Haevernick

1981, 1983, 1987), it seems more reasonable to conclude that the majority of the glass beads from Lofkënd, like those of the Aegean, were imported from Phoenicia or North Syria or Cyprus (cf. Kilian-Dirlmeier 2000:161, who considers the faience and glass beads from Early Iron Age Athens to be imported) or else Egypt (for the similarities between Egyptian and Mycenaean glass, see esp. Nikita and Henderson 2006; Walton et al. 2009). For the Late Bronze Age, the evidence from the fourteenth century BC shipwreck at Uluburun is unequivocal, with Canaanite jars filled with glass and faience beads, which highlight the quantities of such items shipped from their likely place of manufacture in the Levant to Mycenaean Greece (e.g., Pulak 2001:41–42; Ingram [2005:121–122] estimates that about 9,500 glass and 75,000 faience beads had been on board the vessel) in addition to the glass ingots. Within the Aegean, there is now good evidence to suggest the manufacture of glass beads at Methone in Pieria in the Early Iron Age (Manthos Bessios, Director of the KZ′ Ephorate of Prehistoric and Classical Antiquities of Pieria, personal communication).

Beck Type I.C.1.a and related

10/106 (SF 342, TXXVIII-8), **Figs. 3.96b, 10.40** (Sheet 110.10; Photo 3127)
Spherical Glass Bead.
T77-8 (SU 4.0429).
H: 14.1 mm; D: 13.12 mm; D (perforation): 4.46 mm; Wt: 1.5 g.
Reconstructed from three joining fragments preserving about two-thirds of bead; very poorly preserved, extremely fragile, with surfaces much pitted.
Opaque yellow-brown glass. Bead of spherical form; probably perforation Type IV(?).
For the type, see Beck 1928:Type I.C.1.a.

10/107 (SF 295, TLIII-8), **Figs. 3.176c, 10.40** (Photo 2858)
Fragmentary Spherical Glass Bead.
T63-8 (SU 1.0359).
PL (largest fr): 0.011; Wt (all frr): 0.5 g.
Shattered into numerous fragments (18+; illustrated are the five best-preserved fragments, including two pieces each reconstructed from two joining fragments), preserving all or most of bead.
Opaque golden yellow glass. Bead clearly spherical but too fragmentary to determine precise type, original dimensions, perforation, or any decoration.
For the type, see Beck 1928: Type I.C.1.a.

10/108 (SF 341, TXXVIII-7), **Figs. 3.96a, 10.40** (Sheet 10.9; Photo 3391)
Spherical Glass Bead.
T77-7 (SU 4.0429).
H: 10.2 mm; D: 10.51 mm; D (perforation): 3.54–3.90 mm; Wt: 1.5 g.
Reconstructed from three joining fragments, complete, but poorly preserved, with surface much pitted.
Opaque orange-brown glass, decorated with opaque white glass horizontal band at center of bead. More or less spherical bead, but a little flatter than **10/106** (SF 342) on top and bottom. Perforation Type III.
For the type, see Beck 1928: Type I.C.1.a/ XLVII. A.1.a.

10/109 (SF 294, TLIII-6), **Figs. 3.176a, 10.40** (Sheet 68.2; Photo 2847)
Spherical Glass Bead.
T63-6 (SU 1.0359).
H: 12.76 mm; D: 14.33 mm; D (perforation): 3.48–3.24 mm; Wt: 1.6 g.
Intact; but poorly preserved, with surface much pitted.
Opaque reddish brown glass, decoration with dots of white opaque glass on main body of bead and white glass band on one end near perforation. Bead spherical. Perforation Type III.
For the type, see Beck 1928: Type II.C.1.a/ XLVI.A.2 (cf. XLVI.A.2.b.3 and XLVI.A.2.d).

Beck Type I.B.1.a and related

10/110 (SF 125), **Fig. 10.41** (Sheet 10.8)
Rounded Glass Bead.
Topsoil (SU 4.0004).
H: 06.36–8.9 mm; D: 10.48 mm; D (perforation): 4.42–4.69 mm; Wt: 0.9 g.
Intact, but split at one point; surface iridescent, somewhat worn and fragile, but generally quite good.
Translucent yellow-green glass bead of rounded form, more pushed down than spherical, rather irregular, being higher on one side than the other. Perforation Type III/VI.a.
For the type, see Beck 1928: Type I.B.1.a. Cf. Liatovouni T20-11–T20-16.

10/111 (SF 450), **Fig. 10.41** (Photo 3890)
Fragmentary Rounded Glass Bead.
Topsoil (SU 4.0201).

Table 10.1 Description of the beads

No.	Material	Color	Decoration	Bead type[a]	Perforation type[a]	Bead dimensions (max)[b]	Perforation diameter	Weight
10/102	Quartz (carnelian/ sardonyx?)	Reddish brown		I.B.1.a	III	H = 11.01 mm; D = 17.3 mm	1.98–3.62 mm	5.1 g
10/103	Quartz (carnelian/ sardonyx?)	Reddish brown with darker bands		I.B.1.a	III	H = 9.74 mm; D = 16.05 mm	1.92–2.86 mm	3.4 g
10/104	Quartz (carnelian/ sardonyx?)	Reddish brown		I.C.1.a	III	H = 7.62 mm; D = 8.40 mm	1.58–2.16 mm	0.7 g
10/105	Faience	Light blue-green glaze; whitish yellow interior	Series of impressed horizontal and vertical lines forming a grid	I.D.1.b/XXV.A.5	III	H = 17.26 mm; D = 8.92 mm;	1.86–1.78 mm	1.2 g
10/106	Glass	Opaque yellow-brown		I.C.1.a	IV?[c]	H = 14.10 mm; D = 13.12 mm	4.46 mm	1.5 g
10/107	Glass	Opaque golden yellow		Too fragmentary	Too fragmentary	Too fragmentary	Too fragmentary	0.5 g
10/108	Glass	Opaque orange-brown	Opaque white horizontal band	I.C.1.a/XLVII.A.1.a	III	H = 10.20 mm; D = 10.51 mm	3.54–3.90 mm	1.5 g
10/109	Glass	Opaque reddish brown	Impressed white opaque spots and line	II.C.1.a/XLVI.A.2	III	H = 12.76 mm; D = 14.33 mm	3.24–3.48 mm	1.6 g
10/110	Glass	Transparent yellow-green		I.B.1.a	III/VI.a	H = 6.36–8.90 mm[d]; D = 10.48 mm	4.42–4.69 mm	0.9 g
10/111	Glass	Transparent blue-green		I.B.1.a	III/ VI.b	H = 7.9 mm: D = 9.72 mm	6.52–6.84 mm	0.4 g
10/112	Glass	Translucent blue-green	3 or 4 ribs?	II.B.1.a/XXIII.A	III/VI.b	H = 5.41 mm; D = 7.16 mm	4.12 mm	0.2 g
10/113	Glass	Translucent dark green		III.B.1.a	III/VI.a	H = 3.32 mm: D = 5.5 mm	1.82–1.88 mm	>0.1 g[e]
10/114	Glass	Opaque white	5 vertical ribs	II.C.1.a/XXIII.A.2.a	III?[f]	H = 11.22 m; D = 11.56 mm	3.24–3.32 mm	0.6 g
10/115	Glass	Dark green glass	Opaque white waves or chevrons	VII.D.I.a/XLVII.A.7.a	III	L = 16.52 mm; D = 14.56 mm; H = 10.40 mm	4.18–4.22 mm	3.4 g
10/116	Glass	Opaque golden yellow	Opaque white spots	XLVI.A.2[g]	Too fragmentary	Too fragmentary	Too fragmentary	0.6 g
10/117	Glass	White opaque glass with translucent light blue-green glass		Too fragmentary	Too fragmentary	Too fragmentary	Too fragmentary	0.4 g

[a] Beck 1928. For the bead shape, two types are given if the bead is decorated. For the perforation, two types are given if the diameter of the perforation is particularly large.

[b] Maximum dimensions are given: length (L) = distance between the two ends; diameter (D) = maximum width of the bead in cross-section (transverse); and height (H) = distance between convex side and flat side. In the catalogue, H refers to the dimension of the bead along the axis of the perforation.

[c] A part of the perforation of the bead is damaged and therefore the classification is only an approximation based on what is preserved.

[d] For this bead, two lengths are given because of the pronounced height difference across one end of the bead.

[e] The bead is very light and the scale used measures only to a tenth of a gram. Therefore, the weight of this bead falls between 0.0 and 0.1 g.

[f] The perforation is damaged toward one end, and therefore the shape is an approximation based on the preserved section of the perforation. The dimensions given for the perforation diameter are also based on the preserved length of the perforation.

[g] Although the bead is too fragmentary to provide a type based on shape, it could be classified according to the decoration preserved on the fragments.

H: 7.9 mm; D: 9.72 mm; D (perforation): 6.52–6.84 mm; Wt: 0.4 g.
Single fragment preserving about one-half of bead. Surface iridescent, worn, and somewhat pitted.
Translucent yellow-green glass bead of rounded form, more pushed down than spherical. Perforation Type III/VI.b.
For the type, see Beck 1928: Type I.B.1.a.

10/112 (SF 338, TXXVIII-4), **Figs. 3.95c, 10.41** (Sheet 75.4)
Small Glass Bead.
T77-4 (SU 4.0429).
H: 5.41 mm; D: 7.16 mm; D (perforation): 4.12 mm; Wt: 0.2 g.
Complete but split at one point; surface iridescent, poorly preserved, fragile, and somewhat pitted.
Translucent blue-green glass bead, rounded in shape; possible, but unclear traces of fluting. Perforation Type III/VI.b.
For the type, see Beck 1928: Type II.B.1.a/XXIII.A.2.

10/113 (SF 283, TLV-8), **Figs. 3.186, 10.41** (Sheet 110.8; Photo 3763)
Small Glass Bead.
T53-8 (SU 1.0321).
H: 3.32 mm; D: 5.5 mm; D (perforation): 1.82–1.88 mm; Wt: >0.1 g.
Intact; condition good.
Translucent dark green glass. Small rounded bead, approaching biconical. Perforation Type III/VI.a.
For the type, see Beck 1928: Type III.B.1.a.

Beck Type I.C.1.a/XXIII.A.2.a.

10/114 (SF 258, TXXI-8), **Figs. 3.71, 10.42** (Sheet 110.11; Photo 3864)
Spherical Fluted Glass Bead.
T55-8 (SU 4.0341).
H: 11.22 mm; D: 11.56 mm; D (perforation): 3.24–3.32 mm; Wt: 0.6 g.
Main piece reconstructed from three joining fragments, plus many (7+) non-joining fragments and chips, preserving about two-thirds of bead. Poorly preserved, fragile, with surfaces somewhat pitted.
Opaque white glass. Bead spherical and fluted, with five vertical ribs. Perforation type unclear due to damage.
For the type, see Beck 1928: Type II.C.1.a/XXIII.A.2.a.

Beck Type VII.D.1.a/XLVII.A.7.a

10/115 (SF 339, TXXVIII-5), **Figs. 3.95a–b, 10.42** (Sheet 75.2; Photos 3119, 3124)
Elliptical-Shaped, Plano-Convex Glass Bead with Marvered Trails.
T77-5 (SU 4.0429).
H: 16.52 mm; D (max): 14.56 mm; W: 10.4 mm; D (perforation): 4.18–4.22 mm; Wt: 3.4 g.
Intact, comparatively well preserved, with surface only partly worn.
Rod-formed, dark green glass with marvered trails of opaque white glass resembling waves and chevrons. Bead essentially scaraboid in form: elliptical in shape and plano-convex in section. Perforation Type III.
For the type, see Beck 1928: Type VII.D.I.a/XLVII.A.7.a.

Fragmentary glass beads of uncertain type

10/116 (SF 344, TXXVIII-9), **Figs. 3.96c, 10.43** (Photo 3128)
Fragmentary Glass Bead.
T77-9 (SU 4.0429).
PL (largest fr): 0.008; Wt (all frr): 0.6 g.
Shattered into 13 fragments, preserving most, if not all, of bead; surfaces much worn.
Opaque golden yellow glass decorated with opaque white spots. Too fragmentary to determine original dimensions or perforation type.
Cf. **10/109** (SF 294). Although the bead is too fragmentary to determine type based on shape, it can be classified according to the preserved decoration as in Beck 1928: Type XLVI.A.2 (cf. XLVI.A.2.b and XLVI.A.2.d).

10/117 (SF 351, TXXVIII-10), **Figs. 3.96d, 10.43** (Photo 3129)
Fragmentary Glass Bead.
T77-10 (SU 4.0429).
PL (largest fr): 0.011; Wt (all frr): 0.4 g.
Shattered into six fragments, preserving portion of bead.
White opaque glass with translucent light blue-green glass. Too fragmentary to determine bead type, original dimensions, perforation, or decoration.

TOOLS/WEAPONS

Iron Spearhead

The solitary spearhead was found with the older adult male in Tomb XLV (**Fig. 10.44**). It conforms to Snodgrass's (1964:125, 127–128, fig. 8b) Type M, which is generally a plain, smallish type, with a flat blade and with the socket hammered into a tube. It is only known in iron and may well be a development of Snodgrass's bronze spearhead of Type K in iron (see Snodgrass 1964:125; for the type in Italy, see Peroni and Trucco 1994b:751, pl. 157, no. 4 [Torre Mordillo]; for Bosnia Herzegovina, see Truhelka 1904: pl. LXIII, no. 3; pl. LXIV, no. 4; pl. LXXIII, nos. 5, 7–8 [Donja Dolina]). The earliest example listed by Snodgrass (M1) comes from an Early Protogeometric tomb at Knossos (Boardman 1960:133, pl. 39, Tomb IV, no. 8 [L: 0.205]), a date that accords nicely with the Lofkënd example. Indeed, an AMS ^{14}C date from human bone from the individual with which our spearhead was found indicates a date in the tenth century BC, which would make the Lofkënd spearhead among the earliest examples of this type. Snodgrass (1964:128) notes that this basic form of spearhead continued to be in wide use in later periods, with Archaic examples known from Isthmia (Broneer 1958:35, pl. 17c) and Mykale in Asia Minor (Lohmann 2007:153, fig. 36, PA 6b3-01), and even later examples from Trebeniste (Vulić 1933: 475–478, fig. 17a–b) and the Polis Cave on Ithake (Benton 1934–1935:69–70, fig. 18i, no. 3 [L: 0.142 m]), the latter in an apparently Hellenistic level. Since Snodgrass's (1964) seminal study, several important typologies of iron spearheads in Greece have appeared, especially those from Vitsa Zagoriou in Epirus (Vokotopoulou 1986: figs. 100–102, various examples, esp. pl. 92β, inv. 4960, fig. 91κε; pl. 117α, inv. 5960, fig. 91ιστ [all Type M]; see also Vokotopoulou 1985) and Kalapodi in ancient Phokis (Schmitt 2007: pls. 67–90), as well as the numerous examples from Liatovouni near Konitsa in Epirus (Douzougli and Papadopoulos 2010:41, fig. 15).

Similar iron spearheads are known at a number of sites in Albania, including, among others, Kuç i Zi (Andrea 1977–1978:150, pl. III, Tomb 9, nos. 3–4), Kruma (Jubani 1982:156, 188, pl. VI, no. 36 [three examples, esp. center and right]), Kënet (Hoti 1986: 54, no. 8; 64, pl. V, Tomb 17 [left]), Plasa (Aliu 1988: 208, pl. I, no. 5), Shtoj (Koka 1990:63, pl. I, Tomb 1, no. 6; Tomb 2, no. 7; 64, pl. II, Tomb 3, no. 24; 66, pl. IV, Tombs 6, 7, 9, nos. 54, 57, 60; 67, pl. V, Tomb 11, no. 67; 68, pl. VI, Tomb 13, no. 76; Koka 2012: numerous examples, pls. V–IX, XIII–XIV, XVI, XIX– XX, XXII, XXIV–XXV, XXVII–XXIX, XXXVI, XLII–XLIV; also Koka 1985:250, pl. II, no. 4), Myç-Has (Bela 1990:119–136, pls. I–III, VII–XI, XIII– XVIII, various examples), Psar (Aliu 1995:142, pl. III, Tomb 7, no. 25 [found together with an iron sword]), and Rehovë (Aliu 2012: pl. III, Tomb 72, no. 48; pl. IX, Tomb 156, nos. 127, 128; pl. X, Tombs 158–159, nos. 133, 141; pl. XIII, Tomb 163, no. 176; pl. XLIII, nos. 449–453). Iron spearheads of other types are known from other sites in Albania (e.g., Jubani 1992:48–49, pls. I–II, Tomb 1, nos. 5–7; Tomb 2, nos. 6–10; 50–51, pls. III–IV, Tomb 5, nos. 5–7; Tomb 6, nos. 10–12), and related spearheads continue into later phases of the Early Iron Age (e.g., Aliu 1994: pls. I, III, VII–XIV, XVI–XVIII, XXIV, including some Type M; Bunguri 1989) into the Classical and Hellenistic periods (e.g., Kurti 1987:108, pl. V, Tomb 1, no. 9; 113, pl. X, Tomb 17, nos. 1–2). Tumulus I at Shuec in the Korçë basin has yielded a variety of iron spearheads of Early Iron Age date (Andrea 2009–2010:270, pl. I, Tomb 6, no. 6; 271, pl. II, Tomb 11, nos. 16–17; 272, pl. III, Tomb 22, nos. 27–28, Tomb 29, no. 42; 274, pl. V, Tombs 42 and 43, nos. 60–61; 276, pl. VII, Tomb 60, no. 79, Tomb 68, no. 89; as well as several found in the tumulus fill, outside of tombs, such as 279, pl. X, nos. 28–31). A useful overview of bronze weapons, including spearheads, in the Late Bronze Age in Albania and Epirus is presented by Soueref (1989).

10/118 (SF 263, TXLV-1), **Figs. 3.148a–b, 10.44** (Sheet 103.1; Photos 3722, 3723)
Iron Spearhead.
T60-1 (SU 1.0345).
PL: 0.182; L (shaft): 0.078; D (shaft at terminal): 0.024; PW (blade, max): 0.039; Th (blade): 0.005; Wt (all frr): 132.1 g.
Reconstructed from fragments preserving most of spearhead, except for point of blade; there are an additional 31+ fragments, plus over 100 minuscule chips, all non-joining, or not clearly joining. Spearhead very heavily corroded, but clearly deposited into tomb intact.
Medium-sized spearhead, with circular shaft, the lower end of which was rolled to form a tube, with the seam visible near the terminal; traces of wood preserved in the corrosion on shaft interior (see

Chapter 12). No preserved fastening hole(s) or reinforcing rings. Comparatively slender blade, the precise form of which is unclear due to corrosion, but probably leaf-shaped; no central rib, resulting in a flat, thin blade, tapering toward point, which is not clearly preserved.

Cf. Type M1 of the iron spearheads from Vitsa (Vokotopoulou 1986: figs. 100–102; pl. 92β, inv. 4960, fig. 91κε; pl. 117α, inv. 5960, fig. 91ιστ; Liatovouni, T31-3 [inv. 7934] [unpublished]; cf. also Popović and Vukmanović 1998:137, pl. 15, no. 7; Schmitt 2007: esp. pls. 78–79, nos. 143, 152–153).

Iron Arrowheads

There are only two arrowheads from the Lofkënd tumulus, both iron, and both from Tomb XXXII (**Fig. 10.45**), where they were the only material associated with the deceased (**10/119**, **10/120**). The circumstances of the arrowheads within the tomb, and the possibility that they were used to kill the occupant of the grave, are discussed elsewhere (Chapters 3, 21). Here we are concerned with the typology of the arrowheads and their cultural affinities.

The tomb in which the Lofkënd arrowheads were found is assigned to the later stages of Phase II (twelfth–eleventh centuries BC) and must be roughly contemporary with the period of transition from the Bronze Age to the Iron Age in Greece (Submycenaean—Early Protogeometric). Arrowheads, whether of bronze or iron, or even stone, are exceedingly rare in the Late Bronze and Early Iron Age of Albania. The only published arrowheads in Albania prior to the historic period that we are aware of are two examples from the Drino valley in southern Albania (Budina 1975:385, pl. III, nos. 1–2) and an iron arrowhead from Pazhok Tumulus 7, Tomb 22, which was the only find in the tomb (Bodinaku 1982:68, 94, pl. V, no. 1); more recently, a solitary possible iron arrowhead has been published from Tumulus I at Shuec (Andrea 2009–2010:273, pl. IV, Tomb 37, no. 51). The two arrowheads from Lofkënd Tomb XXXII (89) are, therefore, a welcome addition to the list of known examples from Albania.

Both of the Lofkënd arrowheads belong to Snodgrass's (1964:145, fig. 9) Type 2, which is described as a "barbed and tanged arrowhead, without boss" (Snodgrass 1964:148). In his overview of the type, Snodgrass (1964:148) notes that "its importance lies in the fact that it spans the almost empty period between Mycenaean and seventh-century Greece," and he goes on to note the solitary Protogeometric arrowhead from the Athenian Kerameikos (Avila 1983a:146, pl. 54, no. 1074; Kübler 1943:27, 35, pl. 38, Grave 28, inv. M34; Müller-Karpe 1962:92, fig. 10, no. 7) and the four arrowheads found at Vergina that were known to him at the time (see now Andronikos 1969:272–273, and see further below); the same basic type with a long tang is known at Karphi (Pendlebury et al. 1937–1938: pl. 29:1, nos. 672, 552, 454), Fortetsa (Brock 1957:54, 202, pl. 171, no. 574 [Tomb X]), and Arkades (Levi 1931: pl. XI, various examples) on Crete, dating between the eleventh and seventh centuries (Snodgrass 1964:148). The majority of the examples of the type listed by Snodgrass are of iron, though later examples in bronze are noted from Chios and Delos (Snodgrass 1964:148). This basic form of arrowhead derives from a Bronze Age prototype in bronze classified as Type VIb and dated from Late Helladic II to IIIC and into later periods by Buchholz (1962:11, fig. 7, Type VIb; 19, fig. 12, nos. n–z, α–β) and as Class 2a by Avila (1983a:105, pl. 27, nos. 724A–Q), especially the examples from Langada on Kos (for which see further Morricone 1965–1966:269–271, figs. 303–306; cf. also Blegen 1937:116, fig. 263, no. 4; fig. 336, no. 3; fig. 461, no. 4).

As has been often noted in the modern literature, the abundance of arrowheads and the evidence they provide for the practice of archery in the Late Bronze Age (for which see Avila 1983a; Buchholz 1962) is matched by their rarity in the Early Iron Age in Greece (see Popham, Sackett, and Themelis 1979–1980:256–257; Snodgrass 1964:144, 148; 1971:274–275). The discovery, however, of 15 Early Iron Age iron arrowheads from Vergina by Andronikos (1969: 272–273), of further examples by Petsas (1961–1962: 227, fig. 10, Grave LXV ΒΔ; cf. Radt 1974:140), and an additional six arrowheads found together in a tomb in the Malamas field (Tumulus Γ, Grave 1) at Vergina (Rhomiopoulou and Kilian-Dirlmeier 1989:114, fig. 25, nos. 2–7), coupled with the publication or mention of previously overlooked examples (Avila 1983a: 146–147, appendix 5; see further Alexandri 1968:22; Béquignon 1937:52 [Tombs 85–86]; Desborough 1952:133; Papadopoulos 2005:130, 562, 819, fig. 112c, T56-3; Vokotopoulou 1986: fig. 107β, inv. 5286; Weinberg 1948: 206, pl. 72, no. B 10; and two iron objects previously classified as a javelin or small chisel, and a pin or nail in Blegen 1952:289 that have been reinterpreted as arrowheads: see Papadopoulos and

Smithson forthcoming), has added significantly to the evidence for the practice of archery in the Early Iron Age. The later, Archaic and Classical period, evidence for archery in the Greek world, including representations of archers and archery in Greek art, has been collected in a number of specialized studies (Bulanda 1913, which also gathers comparative evidence from Egypt, Arabia, Ethiopia, Babylonia, as well as the Hittite and Skythian material; Schaumberg 1910; Snodgrass 1964:141–156; Wrede 1916: 362–367).

In Greece, the site that has arguably provided the most significant evidence for archery in the Early Iron Age is Lefkandi. This includes what was interpreted as a "quiverful" of iron arrows from the Late Protogeometric Tomb T26 (nos. 19a–j), the schematic drawing of archers on the Middle Protogeometric hydria from Tomb S51 (no. 2), and the possible remains of a composite bow made of horn or antler from T Pyre 1 (not precisely dated), although the latter could represent the remnants of a lyre instead (Popham, Sackett, and Themelis 1979–1980:256). In describing the Lefkandi arrowheads, Catling (in Popham, Sackett, and Themelis 1979–1980:256–257) writes that the type consists of "flat arrow plates ... evidently slotted together into a split shaft tip and probably bound in place by thread or gut." This is exactly how the Lofkënd arrowheads would have been attached to the shaft of the arrow.

10/119 (SF 410, TXXXII-1), **Figs. 3.108a, 10.45** (Sheet 82.2; Photo 3396)
Iron Arrowhead.
T89-1 (SU 2.0508).
PL: 0.042; PW 0.017; Wt (all frr): 1.6 g.
Reconstructed from three main fragments, almost complete; the small missing portions of one of the barbs and the smaller portion on one side of the point are probably represented in several fragments (originally three but subsequently fragmented into smaller pieces) that are difficult to join. All fragments heavily corroded.
Arrowhead formed of relatively thin piece of iron, with point, tang, and barbs flat. It is unclear whether the tang, as preserved, is intact or broken; barbs on both sides of tang, as shown.
This was one of two only arrowheads at Lofkënd; the other, **10/120** (SF 409), is from the same tomb.

10/120 (SF 409, TXXXII-2), **Figs. 3.108b, 10.45** (Sheet 82.3; Photo 3382)
Fragmentary Iron Arrowhead.
T89-2 (SU 2.0508).
PL: 0.031; PW 0.018; Wt: 1.7 g.
Single fragment, heavily corroded, preserving greater part of point, but nothing of the tang or barbs.
Arrowhead as **10/119** (SF 410), with point relatively flat in section, but a little wider. Form of tang and barbs unknown.

Iron Knives

There are only three knives from the tumulus, two of which are certainly ancient, all three fragmentary (**Fig. 10.46**). The best-preserved and almost complete knife comes from Tomb LXXXIV-1 (**10/121**), one of the very latest of the prehistoric burials in the tumulus, with a calibrated AMS ^{14}C date of 821 ± 24 BC. The fragmentary tip of the iron knife from Tomb XXXVI-II (**10/122**) was found associated with the male inhumation in the tomb, and its fragmentary state, coupled with its location in situ, suggests that it was the tip of a knife that was used to kill the individual (this is more fully discussed in Chapters 3, 8, and 21); the tomb is assigned to Phase III (eleventh–tenth centuries BC). A third fragmentary iron blade, **10/123** (SF 230), was found in topsoil near Tomb LX, but not clearly associated with the burial. Judging by its form and heaviness, this knife is almost certainly modern, but it is listed here for comparison. Little can be said, therefore, about TXXXVIII-1 (**10/122**) or SF 230 (**10/123**). Although often associated with male burials, iron knives are equally common in the graves of women, not least at Vitsa and Liatovouni in Epirus (see Vokotopoulou 1986:297–299; Douzougli and Papadopoulos 2010:40, 43–45, fig. 16), as well as in Athens (e.g., Young 1949: 297, pl. 72, no. 31; von Freytag gen. Löringhoff 1974). This has a lot to do with the fact that they are multipurpose tools; as Catling (in Popham, Sackett, and Themelis 1979–1980:257) so nicely put it, such knives are "ideal general purpose implements that would be equally useful for cutting food, cleaning game, whittling, pruning and—conceivably—shaving."

The long and slender form of TLXXXIV-1 (**10/121**), with the cutting edge on the concave side, accords with Type 1 in the typology of iron knives from Vitsa Zagoriou, with slightly curved blade and triangular haft (Vokotopoulou 1986:297–299, fig. 27, Type 1). As Vokotopoulou has shown, this is the earliest of the Vitsa types, beginning ca. 850 BC and continuing through the early fifth century BC (Vokotopoulou 1986:298, fig. 27). In the typology of iron knives from the North Cemetery at Knossos, **10/121** (TLXXXIV-1) may be compared to Type D knives, which have a "marked curvature in both the cutting edge and back, so that the former is concave" (Coldstream and Catling 1996:587, fig. 178, Tomb 296, f.66); these range in date from Submycenaean through Protogeometric and Geometric, into the Orientalizing period. Although slightly curved, this type is very different from the lunate, sickle-shaped blades so common in Greek contexts (see esp. Kron 1998; cf. Rhomaios 1929:213–216, figs. 20–23).

The type of iron blade, very slightly curved, with the cutting edge on the inner (concave) side, finds numerous parallels throughout the Aegean, especially in the north (Macedonia, Epirus, and Thessaly, for which see, among others, Andronikos 1969:266–269, esp. 268, fig. 104, nos. AA1, K1β; 269, fig. 105; Casson 1923–1925: pl. II, no. 1c; Kilian 1983:138, fig. 7, no. 5; Koukouli-Chrysanthaki 1992:404–408, fig. 90 [bronze]; fig. 91 [iron]; Papadopoulos 2005:817, fig. 198d; Savvopoulou 1988:229, fig. 13 [top row]; Wace and Droop 1906–1907:323, 326, fig. 12i; Wace and Thompson 1911–1912:26–27, fig. 15, no. 4), but is also very common in Athens (e.g., Blegen 1952: 281, fig. 3, pl. 75c, nos. 4–5; Kübler 1954: pl. 166, inv. M 54, M 80, M 96; cf. also M 120; Müller-Karpe 1962:109, fig. 27, nos. 5, 7; cf. 111, fig. 29, no. 4; Thompson 1947:196, fig. 1, pl. 41, no. 2 [bottom]; von Freytag gen. Löringhoff 1974: pl. 5, 14, nos. 47–48; Young 1939:104, fig. 73, no. XI 6; Young 1949:297, pl. 72, no. 31), and elsewhere (e.g., Hood, Huxley, and Sandars 1958–1959:255, fig. 32, Tomb VII, no. 12 [with bronze rivets]; Popham, Sackett, and Themelis 1979–1980: pl. 6.7, P31.7, pl. 246h).

This basic type in bronze with the cutting edge on the concave side can be traced back to the Bronze Age (see esp. Andronikos 1969:268, n. 6; Sandars 1955: 178, fig. 2, nos. 3–4; Vokotopoulou 1969b). Among the objects from the tumulus fill was a fragment of thin sheet bronze, **10/126**, which is classified as unidentified, but may conceivably have been the tang of a bronze blade. This is presented here only for comparison.

Iron knives of similar form are common in Albania, with published examples from Patos (Korkuti 1981:45, pl. VIII, Tomb 76), Kuç i Zi (Andrea 1976b: 220, pl. V, Tomb 28, no. 2; 223, pl. VIII, Tomb 52, no. 1, and Tomb 57, no. 1; 226, pl. XI, Tomb 97, no. 4; 229, pl. XIV, no. 14), Kënetë (Hoti 1981:212, 217, pl. III, nos. 3–4; Hoti 1986:63, pl. IV, Tomb 12, left; cf. 68, pl. IX, no. 4; Jubani 1983:81, 121, pl. I, Tumulus I, Tomb 3, no. 10), Pazhok (Bodinaku 1982:67, 99, Tumulus 7, Tomb 15, V/10), Burreli (Kurti 1983:107, pl. V, Tomb 19, no. 5), Shtoj (Koka 1990:64, pl. II, Tomb 3, no. 25), Myç-Has (Bela 1990: esp. 129, pl. XI, Tomb 8, no. 151), Borovës (Aliu 1994:33, Type I), and Shuec (Andrea 2009–2010:275, pl. VI, Tomb 57, no. 75; 276, pl. VII, Tomb 68, upper right, Tomb 61, no. 83; 279, pl. X, nos. 26–27); related knives of bronze are also well represented (see esp. Aliu 2004:190, pl. X, Tomb 111, no. 137; 193, pl. XIII, no. 169). Iron knives similar to **10/121** are found elsewhere in Europe (e.g., Foltiny 1961:289–290; Randall-MacIver 1927: pl. 6, no. 5 [Este]; Truhelka 1904: esp. pl. XXXVII, no. 13; pl. XLVI, no. 1; pl. LIV, no. 23 [Donja Dolina]; with further discussion of the chronology in Snodgrass 1962).

10/121 (SF 019, TLXXXIV-1), **Figs. 3.284, 10.46** (Sheet 21.1; Photo 3439)
Iron Knife.
T2-1 (SU 1.018).
PL: 0.177; W (max): 0.016; Wt: 25.1 g.
Almost complete; main portion of blade reconstructed from six joining fragments; tip of point and tip of tang preserved, but not joining; four additional non-joining fragments. All fragments heavily corroded.
Blade, only slightly lunate-shaped, tapering to point, with cutting edge on the concave side; convex edge thicker and duller. Tang also tapers to sharp point, preserving small portion of wood pseudomorph, indicating wooden haft (see Chapter 12).
Cf. especially the iron knives from Vitsa: Vokotopoulou 1986:297–299, fig. 27, Type 1; and from the North Cemetery at Knossos on Crete: Coldstream and Catling 1996:587, fig. 178, Tomb 296, f.66.

10/122 (SF 370, TXXXVIII-1), **Figs. 3.126a–b, 10.46** (Sheet 75.5; Photos 3149, 3150)
Tip of Iron Knife Blade.
T79-1 (SU 1.0439).
PL: 0.039; Wt: 1.2 g.
Single fragment, much corroded, preserving complete point and portion of blade of iron knife.
Fragment, as preserved, triangular in shape, with blade tapering to a sharp point. Too little survives to determine the cutting side of the blade. The fragment may conceivably be from the haft of a knife rather than the point, though this is less likely.
Cf. **10/121**.

10/123 (SF 230), **Fig. 10.46** (Photo 3952)
Fragmentary Iron Blade, Probably Modern.
Found in topsoil: (SU 2.0274) near Tomb LX.
PL: 0.071; PW (max): 0.023; Wt (all frr): 15.3 g.
Two joining fragments, corroded, preserving portion of blade and smaller portion of haft. Three additional non-joining fragments, much corroded.
Thin, but heavy blade, with central supporting ridge and cutting edges on both sides, tapering to a point; haft, which appears to have been hammered flat, articulated from blade by a T-shaped element; most of haft not preserved.
The blade does not look ancient.

Worked Bone

Very little can be said with certainty about the solitary fragment of worked bone from Lofkënd, **10/124**, but given the dearth of worked bone from the tumulus—which is not surprising, as ours is not a settlement context—we thought it best to include this one object (**Fig. 10.47**). Its context within the tumulus fill is such that it should be ancient. Its uncanny resemblance to an arrowhead may be purely fortuitous, but that it tapers to a point is clear enough, and the fact that its surface is highly polished is telling. If a tool, as seems likely, it most closely resembles an awl.

10/124 (SF 347), **Fig. 10.47** (Sheet 68.5; Photo 3062)
Worked Bone Point Fragment.
Tumulus Fill (SU 4.0286).
PL: 0.024; PW 0.009–0.010; Wt: 0.8 g.
Single fragment, broken laterally and at one end; condition good.
Fragment clearly worked, arrow-shaped as preserved, but broken, tapering toward point at one end. Original surface highly polished either from use or by intent. The fragment as preserved is too broad to be a pin; if a tool, it most resembles an awl.

Unidentified Objects of Copper/Copper Alloy

The objects presented here (**Fig. 10.48**) are mainly of bronze and all are considered to be ancient, either on account of their context, found in undisturbed tumulus fill, or else by their form. Most derive from tumulus fill, but a few were encountered in topsoil. None can be dated with any precision. Among the fragments, **10/125** preserves a portion of a shaft of a bronze pin, probably from a fibula rather than a dress pin; alternatively, it may be a needle, an item not uncommon in the Balkans, both in the Bronze and Early Iron Ages (see, among others, Benac and Čović 1956: pl. IX:13 [Sjevrska, Tum. VI, Gr. 3]; pl. XVI:4 [Ilijak, Tum. XII, Gr. 1; pl. XXXIX:17 [Taline, Tum. XIX, Gr. 8]). The roughly triangular form of **10/126**, as preserved, is conceivably the tang of a bronze blade, but this cannot be established. The object **10/127** resembles a sprue—also referred to as "dross"—being the waste from casting in the lost-wax technique. The sprue can be either the hole through which the metal is poured into the gate and from which it goes into a mold, or it can refer to the waste piece cast in this hole. Such sprues are occasionally even found among votive offerings in early Greek sanctuaries, including Olympia (Furtwängler 1890:153, no. 976 [not identified as a sprue but as a form of "andere Tiere") and the extramural sanctuary of Sybaris at Francavilla Marittima (Papadopoulos 2003:139– 140, esp. no. 529; Stoop 1980:178, 189, fig. 41). Although encountered in topsoil, the small but relatively thick sheet of bronze, **10/128**, which preserves an edge and is also folded over, certainly looks ancient rather than modern. The minuscule and heavily corroded pieces of copper/copper alloy, **10/129–10/131**, are typical of other small pieces of bronze or copper encountered in the tumulus fill (see inventoried pieces listed below).

10/125 (SF 044), **Fig. 10.48** (Sheet 15.1)
Bronze Pin Fragment, Perhaps from Fibula; Alternatively, a Needle.
Tumulus Fill (SU 1.0047).
PL: 0.046; Wt: 0.6 g.
Single fragment, broken at both ends, preserving small portion of pin shaft; good dark green/black patina; corrosion products at both broken ends.
Thin pin shaft, circular in section. Shaft seems too thin for a normal dress pin; perhaps originally from a fibula, alternatively a needle.

10/126 (SF 212), **Fig. 10.48** (Photo 2449)
Fragment of Thin Sheet Bronze, Unidentified Object.
Tumulus Fill (SU 1.0070).
PL: 0.023; Wt: 0.5 g.
Single fragment, broken at one end; good dark green patina.

Small fragment of sheet bronze, roughly triangular, tapering toward point. Conceivably the tang of a bronze blade.

10/127 (SF 408), **Fig. 10.48** (Photo 3990)
Bronze Fragment, Unidentified Object, Resembling Sprue or Dross.
Tumulus Fill (SU 4.0204).
PL (max): 0.014; Wt: 2.4 g.
Single fragment, corroded.
In form, this piece resembles a sprue, sometimes referred to as "dross," representing waste from casting in the lost-wax technique.
Cf. discussion in Papadopoulos 2003:139.

10/128 (SF 393), **Fig. 10.48** (Photo 3349)
Fragment Sheet Bronze.
Topsoil (SU 2.0003).
L × W (max): 0.028 × 0.023; Wt: 11.5 g.
Single fragment, broken on three sides but preserving a clear edge on the fourth. Good dark green patina.
Thick sheet of bronze, with clear edge on one side; opposite side folded over. Object unidentified, but the bronze looks ancient.

10/129 (SF 104), **Fig. 10.48** (Photo 398)
Bronze Fragments, Unidentified Object.
Tumulus Fill (SU 4.0035).
PL: 0.017; Wt: 1.2 g.
Two minuscule fragments, heavily corroded.

10/130 (SF 378), **Fig. 10.48** (Photo 3284)
Minuscule Fragment of Bronze.
Tumulus Fill (SU 2.0474).
D (max): 0.004; Wt: >0.1 g.
Much corroded.
Minuscule fragment; conceivably, but not certainly, part of a small spiral coil.

10/131 (SF 134), **Fig. 10.48** (Photo 2132)
Bronze Fragment, Unidentified Object.
Tumulus Fill (SU 2.0118).
PL: 0.009; Wt: 0.2 g.
Miniscule fragment, heavily corroded.

Inventoried pieces of copper/copper alloy
SF 415, Tumulus fill (SU 2.0287); PL: 0.012; Wt: >0.1g.
SF 419, Tumulus fill (SU 2.0078); PL: 0.010; Wt: 0.5g.
SF 420, Tumulus fill (SU 2.0078); PL: 0.003; Wt: >0.1g.
SF 421, Tumulus fill (SU 2.0078); PL: 0.009; Wt: 0.4g.
SF 426, Tumulus fill (SU 5.0536); PL: 0.012; Wt: 0.7g.

Unidentified Iron Objects

The three following objects (**Fig. 10.49**) are considered to be ancient on account of their context, but their fragmentary and heavily corroded state impedes more detailed comment. Two of the pieces are conceivably fragments from blades (**10/132, 10/133**), and one is plausibly the fragmentary shaft of a pin (**10/134**).

10/132 (SF 256, TXXI-3), **Figs. 3.67, 10.49** (Sheet 117; Photo 3680)
Fragments of Sheet Iron.
T55-3 (SU 4.0326).
PL × PW (largest fr): 0.017 × 0.018; Wt (both frr): 1.0 g.
Three non-joining fragments; heavily corroded.
Thin pieces of sheet iron, hammered flat; unidentified object, conceivably, but not clearly, a blade?

10/133 (SF 384, TXXXV-1), **Figs. 3.116, 10.49** (Photo 3282)
Iron Fragments, Unidentified Object.
T84-1 (SU 1.0470).
PL (largest fr: 0.018; Wt (all frr): 0.6 g.
Three non-joining fragments, heavily corroded.
Fragments conceivably, but not clearly, from a blade.

10/134 (SF 217), **Fig. 10.49** (Photo 2526)
Fragment of Iron Pin.
Tumulus Fill (SU 4.0204).
PL: 0.018; Wt: 0.4 g.
Single fragment, broken at both ends, preserving small portion of possible iron pin; much corroded.
Small section only of shaft, circular in section.

Modern Objects and Fragments

Copper, Zinc, Tin, and Lead Alloy

Although encountered in the tumulus fill, the following object was found at the edge of the tumulus and is either medieval or modern on account of its metallography.

10/135 (SF 333 **Fig. 10.50** (Sheet 80.3; Pl. 3239)
Unidentified Ornament, Made of Alloy of Copper, Zinc, Tin, and Lead; Medieval/Modern.
Tumulus Fill: (1.0377).
PL: 0.047; PW 0.014; Wt: 4.0 g.
Single fragment, preserving perhaps one-half of original ornament; preservation otherwise good.
As preserved, roughly crescent-shaped, but originally probably eye-shaped, with two small holes, only partially preserved, for suspension/attachment. Embossed floral decoration, as shown, on exterior; underside finished smooth.

Inventoried piece of copper alloy
SF 245 (Tomb C, SU 1.0302) PL: 0.010; Wt: 0.2 g. Found in a modern grave that also yielded a fragmentary copper alloy coin.

Iron

10/136 (SF 098), **Fig. 10.51** (Sheet 28.3; Photo 306)
Iron Nail Head, Conceivably Modern.
Tumulus Fill (SU 1.0112).
PL: 0.028; W (head): 0.024; Wt: 5.6 g.
Single fragment preserving all of head and portion of shaft; corroded.
Shaft: hammered flat, rectangular in section, tapering toward lower end, significantly wider at juncture with head. Head composed of sheet iron, hammered flat, folded around upper shaft.
Similar: SF 432 (SU 6:0002): L: 0.034; W (head): 0.022; Wt: 3.8 g.

10/137 (SF 010), **Fig. 10.51** (Photo 289)
Fragmentary Small Iron Stud.
Topsoil (SU 2.0002).
PL: 0.010; D (head): 0.009; Wt: 0.9 g.
Single fragment, corroded, preserving all of head and portion of shaft of iron stud.
Small, hemispherical or dome-shaped head; thin central shaft, evidently circular in section.
The piece is probably modern, but it is not unlike the numerous small studs made of bronze and common in the Late Bronze and Early Iron Ages.
Cf. Liatovouni T59-15 (inv. 8546), Type α (unpublished).

10/138 (SF 289), **Fig. 10.51** (Photo 3951)
Small Iron Rivet, Probably Modern.
Topsoil (SU 1.0278).
H: 0.011; D (head): 0.010–0.018; L (shaft): 0.009; Wt: 2.0 g.
Intact; much corroded.
Short shaft, rectangular in section at juncture with head, tapering sharply to point; elliptical head.

10/139 (SF 103), **Fig. 10.51** (Photo 308)
Iron Nail, Modern.
Fill of Tomb LXVIII (intrusive) (SU 2.0102).
L: 0.038; Wt: 1.9 g.
Twisted out of shape; corroded.
Modern nail.

10/140 (SF 086), **Fig. 10.51** (Photo 2200)
Fragment of Iron Spike, Probably Modern.
Topsoil (SU 2.0002).
PL: 0.066; Wt: 16.9 g.
Single fragment, broken at both ends, preserving portion of shaft of spike; corroded.
Shaft square in section, tapering slightly toward one end.
Cf. **10/136** (SF 98).

10/141 (SF 306), **Fig. 10.51** (Photo 2993)
Fragments Iron Alloy: Modern Bottle Cap.
Topsoil (SU 2.0002).
PL (largest fr): 0.020; Wt (all frr): 2.0 g.
Nine fragments, all probably joining, but much corroded, preserving greater portion of bottle cap.
Cap flat on top, with scalloped edges.

10/142 (SF 114), **Fig. 10.51** (Photo 2128)
Fragments of Sheet Iron.
Topsoil (SU 2.0002).
PL × PW (largest fr): 0.021 × 0.016; Wt (all frr): 1.2 g.
Four non-joining fragments; corroded.
Thin pieces of sheet iron, hammered flat; unidentified object, perhaps modern.

10/143 (SF 228), **Fig. 10.51** (Photo 2525)
Fragments of Iron Sheet.
Topsoil (SU 2.0003).
PL (largest fr): 0.023; Wt (all frr): 0.8 g.
Three non-joining fragments, corroded, preserving small portion of iron sheet.
Thin iron sheet, hammered flat; the largest fragment preserves an edge, which is evidently curved.

10/144 (SF 301), **Fig. 10.51** (Photo 3281)
Fragments of Iron Sheet (Modern).

Topsoil (SU 2.0002).
PL (largest fr): 0.021; Wt (all frr): 1.4 g.
Six fragments, not clearly joining, preserving small portion of iron sheet; heavily corroded.
Iron sheet hammered flat; one fragment has edge folded over.

Inventoried pieces of iron sheet or strips

SF 060, Topsoil (SU3.0064); L: 0.015; Wt: 0.8 g.
SF 090, Topsoil (SU 4.0004); PL: 0.020–0.023; Wt: 8.1 g.
SF 109, Tumulus fill (SU 1.0039); PL: 0.020; Wt: 0.6 g.
SF 165, Topsoil (SU 1.0141); L (as bent): 0.075; W (max): 0.012; Wt: 14.9 g.
SF 392, Topsoil (SU 2.0003); PL × PW 0.020 × 0.013; Wt: 1.4 g.
SF 416, Topsoil (SU 2.0002); PL × PW (largest fr): 0.023 × 0.020; Wt: 1.3 g.
SF 418, Tumulus fill (SU 2.0040); L × W (max): 0.012 × 0.009; Wt 0.3 g.

Modern Iron and Other Objects and Fragments

Unless otherwise specified, fragments of glass refer to modern glass.

Inventoried

SF 001 SU 1.0001 Fragments (8+) of unidentified iron object (7.5 g).
SF 002 SU 1.0001 Single fragment. Point of iron spike or nail (2.6 g).
SF 004 SU 2.0002 Fragments (4) iron sheet (0.7 g).
SF 374 SU 2.0002 Fragments (3) iron sheet (0.5 g).
SF 401 SU 1.0375 Fragment bottle glass (14.1 g).
SF 435 SU 6.0002 Fragments (2) iron sheet (2.7 g).

Not inventoried

TR 1 SU 1.0001: Three fragments glass (3.4 g).
TR 1 SU 1.0009: One small iron stud (0.7 g); two fragments sheet iron (1.1 g); nine fragments of glass (9.9 g).
TR2 SU 2.0002: Two fragments iron sheet (1.6 g); fragment of iron sheet (0.2 g); fragment of iron button (0.7 g); two fragments glass (2.3 g); fragment glass (3.0 g); minuscule fragment glass (>0.1 g); thin hollow white plastic tube (PL: 0.047; D: 0.004; Wt: 0.3 g); minuscule fragment blue glass (0.010 × 0.009; >0.1 g).
TR2 SU 2.0003: Small piece of plastic (>0.1 g); two fragments glass (1.5 g); two fragments glass (including one jar rim) (18.5 g); 14 fragments glass (32.0 g); two fragments glass (11.0 g); fragment iron sheet (1.4 g); four fragments glass (4.1 g); fragment glass (1.0 g); minuscule fragment blue plastic (PL: 0.007; Wt: >0.1 g).
TR2 SU 2.0006: Two fragments of iron sheet (0.4 g); two fragments of iron sheet (0.5 g); gun cartridge, spent shell (L: 0.059; D: 0.012; Wt: 13.1 g); gun cartridge, spent shell (L: 0.054; D: 0.011; Wt: 10.0 g).
TR2 SU 2.0010: Gun cartridge, spent shell (L: 0.052; D: 0.011; Wt: 9.6 g).
TR3 SU 3.0003: Five fragments of iron, unidentified and heavily corroded (1.7 g).
TR3 SU 3.0008: Two fragments of glass (2.3 g).
TR3 SU 3.0064: Two fragments of glass (2.6 g).
TR4 SU 4.0004: Fragment sheet iron (0.5 g); cigarette filter (used); two fragments glass (9.2 g).
TR4 SU 4.0201: Two fragments of glass (4.3 g).
TR8 SU 8.0201 Fragment of glass (1.5 g).

Experimental Archaeology: A Note on the Making of Fibulae

John K. Papadopoulos

In a small exercise of experimental archaeology, Lorenc Bejko, Surya Lela, and other members of the Lofkënd team attempted to make fibulae following some of the standard prehistoric types found at the site (**Fig. 10.52**). Unlike their ancient counterparts, the modern fibula makers did not have to start from scratch. Instead, they purchased copper wire, ready made. Their challenge was thus to attempt to replicate, as accurately as possible, the form of the prehistoric and protohistoric fibulae. They did not attempt to decorate any of the modern fibulae with incised decoration, following the decoration seen on some of the ancient fibulae. In **Figure 10.52**, the fibula on the top right replicates the well-known Lofkënd Type I.2 bronze fibula, often referred to as "Cassibile" type (**10/19–10/21**), especially **10/21**; that on the bottom right replicates in copper the idiosyncratic Lofkënd Type II.1 iron fibula, **10/22**, of which there was only a single example from the site. The two fibulae on the left are variants of **10/22**. Other fibula types that were made include the ubiquitous "spectacle" fibulae. Even with the use of modern wire, the making of a spectacle fibula can be quite challenging, and extra care was needed in the initial stages of the making process to ensure that both of the spirals and their connecting loops were symmetrically laid out. Although with practice anyone could become a competent fibula maker, the challenge was really in the entire *chaîne opératoire*, and by purchasing their copper wire, the experimental archaeologists at Lofkënd did not have to prospect for the metal ore, smelt it, and prepare it in a form that could be used. Moreover, by using purchased copper, they did not have to experiment with different alloys of bronze, nor did they attempt to work with iron, especially for the making of the larger Type II.2 iron fibulae. Nevertheless, the experiment was worth the effort, and replicas of various different types of prehistoric and protohistoric jewelry made by the members of the Kamenicë team can be found in the visitors' building next to the conserved Kamenicë tumulus in southeast Albania.

APPENDIX 3
MODERN GUN SHELLS AND BULLETS ON THE SURFACE OF THE LOFKËND TUMULUS

John K. Papadopoulos and Yannis Mylonas

INTRODUCTION

The surface and topsoil levels of the tumulus contained a good many spent shells of gun cartridges, together with the occasional bullet. These were carefully collected, their position on or in the tumulus recorded, cleaned, conserved, photographed, and catalogued. A draft catalogue was prepared by Papadopoulos, who recorded the dimensions of the shells and bullets, and who also arranged the material into a preliminary typology. In the case of the cartridge shells, the typology was based on the inscriptions, or head stamps, found at the bottom end of the cartridge shells. Photographs of the material were subsequently shown to Yannis Mylonas, a military uniforms and weapons expert at the War Museum in Athens, who was able to identify most of the shells and also some of the bullets. With the exception of a solitary shell that dates to 1913, all of the material presented in this appendix belongs to the years of World War II, and represents the material record on the ground of one of the many skirmishes of the Greco-Italian war fought in Albania and Epirus in 1940 (**Fig. A3.1**). Located as it was close to the main road linking northwest Greece with Albania, the Lofkënd tumulus was not the site of a major battle, but the shells and bullets found provide testimony of the type of ammunition that was standard issue for both the Greek and the Italian armies, with several other types of ammunition that may have been used by Albanian partisans. All of the shells and bullets found at the site are catalogued in this appendix.

A total of 28 cartridge shells were recorded from the Lofkënd tumulus. In contrast, there were only four bullets. Cartridge shells do not fall far from the place where the gun was fired, so all of the shells were fired from guns held by soldiers or other people stationed at the tumulus. In contrast, the number of bullets fired at the tumulus and which made their way into the topsoil or fill of the mound was small indeed. Given the relatively broad variety of cartridge shells and bullets, which must represent at least five different types of gun, it is highly unlikely that the weapons from which the cartridges were fired—all manufactured before World War II or in the early years of the war—were relics kept by local villages who used them for hunting from the tumulus in the post-war period. What is perhaps most interesting is that both Greek and Italian rounds of ammunition were fired from and/or toward the tumulus. In terms of the distribution of the material on the surface and upper levels of the tumulus, there was no clear pattern, as the material was found in all four sectors of the mound.

SPENT SHELLS OF GUN CARTRIDGES

The diagnostic shells belongs to three main series, one being the standard ammunition of the Greek army, the second and third of the Italian army. There is a solitary shell that is Yugoslavian, which is thought to have been used by Albanian partisans. These various series are distinguished by their head stamps. A few of the shells are so corroded or poorly preserved that their head stamps are illegible; these are presented separately. Also catalogued separately are the bullets.

The Greek Series

The ten examples of Greek cartridge shells are identified by their head stamps with the letters Ε Σ on the left and right, which stand for Ελληνικός Στρατός (Greek Army), with additional letters at the top of the head stamp, often as a monogram: ΕΠΚ. This was an abbreviation for ΕΛΛΗΝΙΚΟ ΠΥΡΙΤΙΔΟΠΟΙΕΙΟ ΚΑΛΥΚΟΠΟΙΕΙΟ (often abbreviated as ΠΥΡ-ΚΑΛ), which is usually translated as the Greek Powder and Cartridge Company in English (or Poudrerie et Cartoucherie Hellénique in French). At the bottom end of the head stamp appears the date of manufacture. Up to 1938, the date was written in full, but beginning in

1939, the date was abbreviated to the last two digits of the year, or at least this was the case for the ΠΥΡ-ΚΑΛ shells found at Lofkënd.

Founded in 1874, the company was the main producer of ammunition and explosives in Greece. These shells were from cartridges of the Mannlicher-Schönauer M1903, 6.5-mm caliber rifle, which was the standard issue of the Greek Army from 1907 to 1941 (**Fig. A3.2**). The Mannlicher-Schönauer M1903 was fabricated in Austria by Steyr only for the Greek army, so this type of rifle is normally referred to as Greek. As the shells of this series are all dated, they are presented in their chronological order (**Figs. A3.3–A3.4**).

A3/1 (SF 405), **Figs. A3.3–A3.4** (Photos 3368–3369)
Gun Cartridge (Spent Shell).
Topsoil (SU 2.0002).
L: 0.054; D (base and body): 0.011; D (top): 0.008; Wt: 9.8 g.
Head stamp (top): Monogram ΕΠ (rather than ΕΓ); (left): E; (right): vacant; (bottom): 1913.

A3/2 (SF 072), **Figs. A3.3–A3.4** (Photos 2119–2120)
Gun Cartridge (Spent Shell).
Topsoil (SU 4.0004).
L: 0.054; D (base and body): 0.011; D (top): 0.008; Wt: 9.8 g.
Head stamp (top): Monogram ΕΠΚ; (left): E; (right): Σ; (bottom): 1936.

A3/3 (SF 014), **Fig. A3.4** (Photos 2123–2124)
Gun Cartridge (Spent Shell).
Topsoil (SU 1.0009).
L: 0.054; D (base and body): 0.011; D (top): 0.008; Wt: 9.4 g.
Head stamp (left): E(?); (bottom): 1936.

A3/4 (SF 042), **Fig. A3.4** (Photos 2125–2126)
Gun Cartridge (Spent Shell).
Topsoil (SU 1.0009).
L: 0.052; D (base and body): 0.011; D (top): 0.008; Wt: 9.5 g.
Head stamp (top): *Vacant* (or illegible?); (left): E; (right): *Vacant* (or illegible); (bottom): 1937.

A3/5 (SF 057), **Figs. A3.3–A3.4** (Photos 2099–2100)
Gun Cartridge (Spent Shell).
Topsoil (SU 1.0002).
L: 0.054; D (base and body): 0.011; D (top): 0.008; Wt: 9.7 g.
Head stamp (top): Monogram ΕΠΚ (though resembling ΕΞΚ); (left): E; (right): Σ; (bottom): 1938.

A3/6 (SF 311), **Fig. A3.4** (Photos 2913–2914)
Gun Cartridge (Spent Shell).
Topsoil (SU 4.0286).
L: 0.054; D (base and body): 0.011; D (top): 0.008; Wt: 9.6 g.
Head stamp (top): Monogram ΕΠΚ?; (left): E; (right): Σ; (bottom): 39.

A3/7 (SF 380), **Fig. A3.4** (Photos 3241–3242)
Gun Cartridge (Spent Shell).
Topsoil (SU 2.0002).
L: 0.054; D (base and body): 0.011; D (top): 0.008; Wt: 10.3 g.
Head stamp (top): Monogram ΕΠΚ; (left): E; (right): Σ(?); (bottom): 40.

A3/8 (SF 381), **Figs. A3.3–A3.4** (Photos 3236–3237)
Gun Cartridge (Spent Shell).
Topsoil (SU 1.0399).
L: 0.054; D (base and body): 0.011; D (top): 0.008; Wt: 10.1 g.
Head stamp (top): Monogram ΕΠΚ (though resembling ΕΞΚ); (left): E; (right): Σ(?); (bottom): 40.

A3/9 (SF 394), **Fig. A3.4** (Photos 3433–3434)
Gun Cartridge (Spent Shell).
Topsoil (SU 1.0375).
L: 0.057; D (base and body): 0.012; D (top): 0.009; Wt: 10.1 g.
Head stamp (top): Monogram ΕΠΚ(?); (left): E; (right): Σ; Bottom: 40.

A3/10 (SF 430), Not Illustrated
Gun Cartridge (Spent Shell).
SU 6.0002.
L: 0.054; D (base and body): 0.011; D (top): 0.008; Wt: 10.0 g.
Head stamp (top): Monogram + K; (left): E; (right): Σ; (bottom): 40.

The Italian Series

There are two series presented here. The first, with the characteristic SMI head stamp, was identified by Mylonas; the second was first identified by Papadopoulos after further research and subsequently and independently confirmed by Mylonas. There are eight Italian cartridge shells that are identified by their distinctive head stamp: at the top are the letters SMI, which stand for Società Metallurgica Italiana. At the bottom of the head stamp are three numbers, the lowest 924, the highest 937; these are abbreviations of the year of manufacture, 924 = 1924, 937 = 1937, and so on. The earliest of the series, **A3/11** has a five-pointed star to the right and left. By 1927, if not earlier, these stars disappeared, at least from the head stamps on the cartridge shells from Lofkënd. Cartridges of this type were used for the Mannlicher-Carcano M1891 rifle, which was produced from 1892 to 1945 in both rifle and carbine versions; it was used by most Italian troops in World War I and by Italian and some German forces in War World II (**Fig. A3.5**). The company, founded in 1886 and based in Florence, continues today, having changed its name in 2006 to KME Group Spa. During World War II, the company was a government contractor.

SMI head stamps

A3/11 (SF 070), **Figs. A3.6–A3.7** (Photos 2117- 2118)
Gun Cartridge (Spent Shell).
Topsoil (SU 4.0004).
L: 0.052; D (base and body): 0.011; D (top): 0.008; Wt: 9.6 g.
Head stamp (top): SMI; (right and left): *; (bottom): 924.

A3/12 (SF 003), **Figs. A3.6–A3.7** (Photos 2103- 2104)
Gun Cartridge (Spent Shell).
Topsoil (SU 2.0002).
L: 0.052; D (base and body): 0.011; D (top): 0.008; Wt: 12.2 g.
Head stamp (top): SMI; (bottom): 927.

A3/13 (SF 402), **Figs. A3.6–A3.7** (Photos 3353–3354)
Gun Cartridge (Spent Shell).
Topsoil (SU 1.0375)
L: 0.052; D (base and body): 0.011; D (top): 0.008; Wt: 9.7 g.
Head stamp (top): SMI; (bottom): 927.

A3/14 (SF 284), **Fig. A3.7** (Photos 2856–2857)
Gun Cartridge (Spent Shell).
Topsoil (SU 2.0002).
L: 0.052; D (base and body): 0.011; D (top): 0.008; Wt: 9.9 g.
Head stamp (top): SMI; (bottom): 928.

A3/15 (SF 303), **Figs. A3.6–A3.7** (Photos 2917–2918)
Gun Cartridge (Spent Shell).
Topsoil (SU 2.0002).
L: 0.052; D (base and body): 0.011; D (top): 0.008; Wt: 9.6 g.
Head stamp (top): SMI; (bottom): 931.

A3/16 (SF 302), **Fig. A3.7** (Photos 2915–2916)
Gun Cartridge (Spent Shell).
Topsoil (SU 2.0002).
L: 0.052; D (base and body): 0.011; D (top): 0.008; Wt: 9.6 g.
Head stamp (top): SMT or SMI; (bottom): 936.

A3/17 (SF 073), **Fig. A3.7** (Photos 2113–2114)
Gun Cartridge (Spent Shell).
Topsoil (SU 4.0011).
L: 0.052; D (base and body): 0.011; D (top): 0.008; Wt: 9.6 g.
Head stamp (top): SMI; (bottom): 937.

A3/18 (SF 372), **Fig. A3.7** (Photos 3175–3176)
Gun Cartridge (Spent Shell).
Topsoil (SU 1.0375).
L: 0.052; D (base and body): 0.011; D (top): 0.008; Wt: 9.8 g.
Head stamp (top): SMI; (bottom): 937.

A.A head stamps

This series is marked by the distinctive A.A head stamp, the letters A.A (the two separated by a full stop) appearing at the bottom of the head stamp; at the top is a code, either C-37 or C-40 (**Figs. A3.8–A3.9**). The C stands for Capua and is an abbreviation of Pirotechnia di Capua, which during the Second World War was part of the Italian government arsenal. The number should be the year of manufacture, with 37 as an abbreviation for 1937, 40 for 1940, dates that would fit perfectly. If this identification is correct, then we have the initials of two known inspectors who worked for the Pirotechnia di Capua—Aldo and Adamo—who always stamped their initials together to produce A.A.

There are six cartridge shells with the A.A head stamp, three dating to 1937 (two of which are certain, one probably 1937), and three dating to 1940, though the stamp on one of these is also not very clear. The only difference between the C-37 and C-40 shells is that in the former the C-37 is at the top and the A.A at the bottom, whereas in the latter the C-40 is at the bottom, the A.A at the top.

A3/19 (SF 386), **Figs. A3.8–A.3.9** (Photos 3370– 3371)
Gun Cartridge (Spent Shell).
Topsoil (SU 1:.0375).
L: 0.059; D (base and body): 0.012; D (top): 0.009; Wt: 13.4 g.
Head stamp (top): C-37; (bottom): A.A.

A3/20 (SF 015), **Figs. A3.8–A3.9** (Photos 2101–2102)
Gun Cartridge (Spent Shell).
Topsoil (SU 1.0002).
L: 0.059; D (base and body): 0.012; D (top): 0.009–0.010; Wt: 16.0 g.
Head stamp (top): C-37; (bottom): A.A.

A3/21 (SF 160), **Fig. A3.9** (Photos 2115–2116)
Gun Cartridge (Spent Shell).
Topsoil (SU 4.0201).
L: 0.059; D (base and body): 0.012; D (top): 0.009; Wt: 13.7 g.
Head stamp (top): C-37?; (bottom): A.A.

A3/22 (SF 250), **Fig. A3.9** (Photos 2109–2110)
Gun Cartridge (Spent Shell).
Topsoil (SU 2.0002).
L: 0.052; D (base and body): 0.011; D (top): 0.008; Wt: 10.0 g.
Head stamp (top): A.A; (bottom): C-40.

A3/23 (SF 308), **Figs. A3.8–A3.9** (Photos 3372–3373)
Gun Cartridge (Spent Shell).
Topsoil (SU 1.0375).
L: 0.052; D (base and body): 0.011; D (top): 0.008; Wt: 10.2 g.
Head stamp (top): A.A; (bottom): C-40.

A3/24 (SF 391), **Fig. A3.9** (Photos 3374–3375)
Gun Cartridge (Spent Shell).
Topsoil (SU 2.0003).
L: 0.052; D (base and body): 0.011; D (top): 0.008; Wt: 10.0 g.
Head stamp (top): A.A; (bottom): C-40?

Shell from Yugoslavian Cartridge

The head stamp of this cartridge shell was, for a while, something of an enigma (**Fig. A3.10**). In recording the stamp, Papadopoulos read, beginning at 9 o'clock, T B 3, with the number 32 at the bottom. In searching for this type of head stamp, Mylonas quickly determined that the number 3 at 3 o'clock was not a number, but the Cyrillic letter "Z": З. The three Cyrillic letters would be equivalent to VYZ. The cartridge shell was thus identified as Yugoslavian, but from an unidentified factory. This sort of cartridge shell was often used by partisans, so it is possible that this was from the gun of an Albanian fighter, though Albanian partisans usually used Italian Carcano and German Mauser rifles.

A3/25 (SF 108), **Fig. A3.10** (Photos 2121–2122)
Gun Cartridge (Spent Shell).
Topsoil (SU 1.0002).
L: 0.057; D (base and body): 0.012; D (top): 0.009; Wt: 1.9 g.
Head stamp (top): T; (left): B (on its side); (right): 3 (on its side); (bottom): 32.

Illegible/Uncertain

Little can be said with certainty about the following cartridge shells (**Fig. A3.11**). The dimensions of **A3/26** accord best with the Greek series, manufactured by PYR-KAL, so the shell is probably Greek. The dimensions of **A3/28** suggest that it is an Italian cartridge shell, with similar dimensions recorded both from the SMI and A.A C-40 series. The length of **A3/27** is fractionally smaller than all of the other shells, but it is closer to the Italian shells as opposed to the Greek or Yugoslavian shells.

A3/26 (SF 422), **Fig. A3.11** (Photos 3538–3539)
Gun Cartridge (Spent Shell).
SU 6.0141.
L: 0.054; D (base and body): 0.011; D (top): 0.008; Wt: 10.2 g.
Head stamp: Illegible.

A3/27 (SF 229), **Fig. A3.11** (Photo 2532)
Gun Cartridge (Spent Shell).
Topsoil (SU 2.0003).
L: 0.048; D (base and body): 0.011; D (top): 0.009; Wt: 7.9 g.
Head stamp (top and bottom): Illegible.

A3/28 (SF 439), Not Illustrated
Gun Cartridge (Spent Shell).
SU 1.0375.
L: 0.052; D (base and body): 0.011; D (top): 0.008; Wt: 9.6 g.
Head stamp: Illegible.

BULLETS

In comparison to the cartridge shells, there were very few bullets (**Fig. A3.12**). Of the three certain bullets, **A3/29** is of a type that was fired from a French-made Lebel rifle that was commonly used by the Greek army in World War II. The overall length and diameter of **A3/29** is precisely that of bullets fired from such Lebel rifles. The dimensions of **A3/30** accord well with bullets fired from an Italian Breda machine gun. The dimensions of **A3/31** were not diagnostic enough for identification. Item **A3/32** is clearly a Greek bullet.

A3/29 (SF 211), **Fig. A3.12** (Photo 2361)
Iron Bullet.
Topsoil (SU 2.0002).
L: 0.028; D (max): 0.008; Wt: 6.5 g.

A3/30 (SF 350), **Fig. A3.12** (Photo 3238)
Iron Bullet.
Topsoil (SU 2.0002).
L: 0.035; D (max): 0.008; Wt: 12.0 g.

A3/31 (SF 428), Not Illustrated
Iron Bullet.
SU 6.0002.
L: 0.033; D (max): 0.008; Wt: 15.3 g.

A3/32 (SF 442), Not Illustrated
Copper Alloy Bullet.
SU 2.0002.
L: 0.033; D (max): 0.008; Wt: 13.3 g.
Head stamp (top): Monogram + K; (left): E; (right): Σ; (bottom): 40.

Chapter 11

Analytical Studies of the Metal Objects from Lofkënd

Vanessa Muros and David A. Scott

Introduction

The study of the metallic finds excavated from the tumulus at Lofkënd provides a unique opportunity to examine a number of objects from the same context and time period, adding to the body of data published for metal artifacts from Albania as well as southeastern Europe. This study, which utilizes optical microscopy, portable X-ray fluorescence spectroscopy (pXRF), X-ray diffraction (XRD) analysis, and metallographic examination, provides only preliminary results from the work undertaken thus far on the Lofkënd metals. Additional studies are ongoing, but these initial results help to provide a better understanding of the metallurgical techniques used in this area during the late Bronze Age to Early Iron Age.

Methodology

A total of 162 metal artifacts were excavated at Lofkënd. These range from copper alloys, silver-copper alloys, gold, and iron to composite objects made from a copper alloy and iron. This particular study focused on analysis of the copper alloy objects (including the composite objects of bronze and iron), the silver-copper alloys, and the gold/electrum objects.

Optical microscopy

Prior to undertaking any sampling or analysis, all metallic artifacts in this study were examined using a binocular microscope (7–40x magnification). This helped to identify areas for analysis, as well as inform on condition, technology and manufacture, and potential sample sites.

Portable X-ray fluorescence spectroscopy (pXRF)

Portable X-ray fluorescence spectroscopy (pXRF) was used to provide a general understanding of the alloy composition of 70 copper alloy and gold artifacts excavated at Lofkënd and to guide sampling and further analysis. Results of the analysis were used to provide qualitative data to allow for comparison of the objects with each other and to look for any overall differences in composition that could be detected by pXRF. Because the analysis was not done on a polished metal surface but rather on a corroded surface, quantitative analysis could not be undertaken. Instead, when looking at the data acquired for each object and comparing them across types, relative peak height was used to provide a method to determine possible differences in composition. Because the condition of each object varied according to the amount of corrosion present and degree of mineralization, which in turn would affect the elements present in those surface layers and in what quantity, the relative peak height was used to provide very cursory information on possible differences that could guide future research and sampling, as well as analysis using other instrumentation.

Objects were analyzed using a Bruker AXS TRACeR III-V portable X-ray fluorescence analyzer (rhodium X-ray tube and Si-PIN diode detector). Readings were taken from the surfaces of the artifacts, on an area about 4–5 mm in diameter, which had been cleaned of soil and thicker corrosion layers but still retained a patina and some corrosion. The coating that had been applied during conservation treatment to each object, consisting of a corrosion inhibitor (benzotriazole or BTA) and an acrylic resin (Paraloid B-48N), was not removed prior to analysis (see Chapter 5). Multiple areas were analyzed on each object, especially if the object was made of different components (e.g., fibula with a separate pin) or a visual difference was observed in the thickness or shape/manufacture of different sections of the object.

The pXRF data were acquired using three acquisition parameters. A setting with no filter run at 40 kV/1.6 μA under vacuum was used to include a broad range of elements. A 1-mm copper (Cu) filter, with the spectra acquired at 15 kV/15 μA under vacuum, was used to focus on lighter elements. A filter composed of 1-mm titanium (Ti) and 12-mm aluminum (Al), with acquisition parameters set at 40 kV and 1.6 μA and no vacuum, was used to focus on heavier elements. Spectra were acquired for 5 minutes and interpreted using Bruker's S1pXRF software (version 3.8.3).

Metallographic examination

Twenty-three copper alloy objects of the 87 excavated were sampled for metallographic examination. Determining which objects would be sampled was based on several criteria: results of the pXRF analysis; condition of the object (whether there were existing damaged areas or breaks that could be sampled rather than sampling an intact object) and the object type. All samples were mounted and prepared using previously published methods (Scott 1991). Samples were taken with a jeweler's saw and then embedded using Buehler's Epoxicure resin. They were then prepared by grinding using various grits of sandpaper with a final polishing stage using 6-micron and 1-micron Buehler Metadi Diamond Suspension on Buehler Mastertex polishing cloths. The samples were examined using a Nikon Epiphot metallographic microscope under both reflected and cross-polarized light, first in their unetched condition and then following etching using alcoholic ferric chloride (Scott 1991:72). Samples were photographed using either a Nikon D70 or Nikon D300 digital SLR camera. The sizes of these polished samples were too small for analysis using pXRF in order to obtain any quantitative compositional information. However, future analysis will be conducted using other analytical techniques to obtain these data.

X-ray diffraction (XRD)

X-ray diffraction (XRD) analysis was used to identify corrosion products and accretions on the surfaces of 24 objects in order to provide information on the condition of the artifacts and the burial environment. Samples were taken from the surface of the object using a scalpel. The samples were mounted on a glass spindle and analyzed using a Rigaku R-Axis Spider X-ray diffractometer. XRD spectra were recorded at 50 kV/40 mA using a Cu-Kά target for 900 seconds. XRD data were processed and matched against reference spectra from the International Center for Diffraction Data (ICDD) files using the JADE, v8.2 software from Materials Data Inc.

RESULT OF pXRF ANALYSIS

Bronzes

Based on the results of the pXRF analysis of the copper alloy objects at Lofkënd, the majority of these objects are tin bronzes, most of which are leaded (**Table 11.1**). These alloys are typical of the late Bronze Age and Early Iron Age in Europe and the Aegean (Craddock 1976, 1977; Scott et al. 2003:171; Tylecote 1976:29). For several of the pieces, it was difficult to determine whether lead was present in the alloy without the use of another analytical technique, and therefore it is possible that these objects are simply tin bronzes. In a few examples, tin was difficult to detect, and it may be that in those cases, the objects may be primarily copper with an extremely small amount of tin present.

Elements such as arsenic (As), silver (Ag), antimony (Sb), and nickel (Ni) were detected and attributed to trace elements in the ores used (Craddock 1988:323; Patterson 1971; Scott et al. 2003:184). The appearance of these elements in the alloy suggests the possible use of copper sulfide or Falherz ores which were commonly used during the Bronze Age (Craddock 1988:324; Patterson 1971:11). Phosphorus was found on several of the objects analyzed and thought to be due to the burial environment (see discussion of XRD results below). Other elements found, such as silica (Si), calcium (Ca), aluminum (Al), iron (Fe), titanium (Ti), sulfur (S), and chlorine (Cl), were attributed to the burial environment

TABLE 11.1 pXRF results showing elements attributed to alloy composition (major and trace elements)

No.	Description	Elements attributed to alloy composition (major and trace elements)
10/11 (SF 290)	Ear or head ornaments	Au, Ag, Cu
10/12 (SF 291)	Ear or head ornaments	Au, Ag, Cu
10/13 (SF 107)	Spectacle fibula—pin	Cu, Sn, Pb, As
10/13 (SF 107)	Spectacle fibula—spiral	Cu, Sn, Pb, As
10/14 (SF 170)	Spectacle fibula	Cu, Sn, Pb? 1
10/15 (SF 106)	Spectacle fibula—area of spiral	Cu, Sn, Pb, As
10/15 (SF 106)	Spectacle fibula—center of pin	Cu, Sn, Pb, As
10/16 (SF 315)	Spectacle fibula—coiled section	Cu, Sn, Pb, As?
10/16 (SF 315)	Spectacle fibula—pin section	Cu, Sn, Pb,As?, Ag?, Ni?
10/17 (SF 434)	Spectacle fibula—pin section	Cu, Sn, Pb, Ni?
10/17 (SF 434)	Spectacle fibula—coils	Cu, Sn, Pb, Ni?
10/18 (SF 157)	Spectacle fibula	Cu, Sn, Pb
10/19 (SF 431a)	"Cassibile" fibula—pin section	Cu, Sn, Pb, As?, Ag?, Ni?
10/19 (SF 431a)	"Cassibile" fibula—bow/spring section	Cu, Sn, Pb, As?, Ag?, Ni?
10/20 (SF 337)	"Cassibile" fibula with incised decoration—area of decoration	Cu, Sn, Pb, As?, Ag?
10/20 (SF 337)	"Cassibile" fibula with incised decoration—pin	Cu, Sn, Pb, As?
10/20 (SF 337)	"Cassibile" fibula with incised decoration—clasp	Cu, Sn, Pb, As?, Ag?
10/21 (SF 251)	"Cassibile" fibula with herringbone design—center of pin	Cu, Sn, Pb, Ni?
10/21 (SF 251)	"Cassibile" fibula with herringbone design—clasp	Cu, Sn, Pb, As?
10/21 (SF 251)	"Cassibile" fibula with herringbone design—coil on bow	Cu, Sn, Pb
10/25 (SF 262)	Bimetallic fibula, figure-8 shape—rivet	Cu, Sn, Pb?
10/25 (SF 262)	Bimetallic fibula, figure-8 shape—bronze strip to which iron disks are attached	Cu, Sn, Pb?, Ni?
10/26 (SF 231)	Bimetallic fibula figure-8 shape—rivet	Cu, Sn, Pb, As
10/27 (SF 438)	Pin with herringbone decoration	Cu, Sn, Pb, Ni?, Ag?
10/28 (SF 440)	Pin with concave head—head	Cu, Sn, Pb
10/28 (SF 440)	Pin shaft	Cu, Sn, Pb
10/29 (SF 354)	Pin with rolled head—section of head	Cu, Sn, Pb, As?
10/29 (SF 354)	Pin with rolled head—section below head	Cu, Sn, Pb, As?
10/29 (SF 354)	Pin with rolled head—center of pin shaft	Cu, Sn, Pb, As?
10/30 (SF 105)	Unidentified fragment—metal waste?	Cu, Sn, Pb, As
10/31 (SF 227)	Pin fragment pointed with a notch (end of pin)	Cu, Sn, Pb, As
10/32 (SF 249)	Pin fragment with rounded edge	Cu, Sn, Pb, As, Ni?
10/50 (SF 111)	Bimetallic pin—copper alloy head	Cu, Sn, Pb, As
10/51 (SF 067)	Bimetallic pin head—protrusions on shaft head	Cu, Sn, Pb, As?
10/51 (SF 067)	Bimetallic pin head—threaded portion (with some iron from pin)	Cu, Sn, Pb, As?
10/57 (SF 300)	Spectacle pendant	Cu, Sn, Pb, Ag, As, Ni
10/58 (SF 335)	Spectacle pendant	Cu, Sn, Pb
10/59 (SF 336)	Wheel-shaped pendant	Cu, Sn, Pb?, As?
10/59 (SF 343)	Curved fragment	Cu, Sn, Pb?, As?

Continued on next page

Table 11.1 (*continued*). pXRF results showing elements attributed to alloy composition

No.	Description	Elements attributed to alloy composition (major and trace elements)
10/60 (SF 318)	Perforated disk with punched circular decoration	Cu, Sn, Pb
10/61 (SF 163a–d)	Perforated disk with punched circular decoration	Cu, Sn, Pb
10/62 (SF 297)	Perforated disk	Cu, Sn, Pb, Ag, As, Ni
10/63 (SF 259)	Perforated disk with punched circular decoration	Cu, Sn, Pb?, As?
10/65 (SF 444)	Spiral ornament or bead	Cu, Sn, Pb, Ni?, Ag?
10/66 (SF 313b)	Spiral ornament or bead	Cu, Sn, Pb
10/67 (SF 313d)	Spiral ornament or bead	Cu, Sn, Pb, Ni?
10/68 (SF 267)	Spiral ornament or bead	Cu, Sn, Pb?
10/69 (SF 265)	Spiral ornament or bead	Cu, Sn, Pb
10/70 (SF 307)	Coiled wire or ring	Cu, Sn, Pb, Ag, Sb, Ni
10/73 (SF 091d)	Ring (smaller, reconstructed)	Cu, Sn, Pb, As?, Ag?
10/74 (SF 091e)	Ring (larger, intact)	Cu, Sn, Pb, As?, Ag?
10/75 (SF 163e)	Ring (thin, reconstructed)	Cu, Sn, Pb, As, Ni?
10/76 (SF 163f)	Ring (flat, intact)	Cu, Sn, Pb, As, Ag?, Ni
10/77 (SF 163g)	Ring (thick, fragmentary)	Cu, Sn, Pb, As, Ag?
10/78 (SF 163h)	Ring (thin, fragmentary)	Cu, Sn, Pb, Ag, Ni, Sb
10/79 (SF 292a)	Ring	Cu, Sn, Pb?, As?
10/80 (SF 313a)	Earring	Cu, Sn, Pb, As?, Ni?
10/81 (SF 313c)	Earring	Cu, Sn, Pb, Ni?
10/82 (SF 299)	Earring	Cu, Sn, Pb, Ag, As, Ni
10/83 (SF 431b)	Coiled wire or ring	Cu, Sn, Pb, As?
10/84 (SF 349)	Headband	Cu, Sn, Pb?
10/85 (SF 317)	Headband	Cu, Sn, Pb, As?
10/86 (SF 255)	Headband	Cu, Sn, Pb
10/87 (SF 089)	Headband	Cu, Sn, Pb
10/87 (SF 089)	Headband—rivet	Cu, Sn, Pb
10/87 (SF 091a)	Headband—S-shaped decorative element	Cu, Sn, Pb, As?, Ag?
10/87 (SF 091a)	Headband—S-shaped decorative element (rivet)	Cu, Sn, Pb, As?, Ag?
10/87 (SF 091b)	Headband—S-shaped decorative element	Cu, Sn, Pb, As?, Ag?
10/87 (SF 091b)	Headband—S-shaped decorative element (rivet)	Cu, Sn, Pb, As?, Ag?
10/125 (SF 044)	Pin fragment	Cu, Sn, Pb, As
10/126 (SF 212)	Unidentified fragment with finished edge (blade?)	Cu, Sn, Pb?, As?
10/127 (SF 408)	Unidentified fragment	Cu, Sn, Pb, As?
10/129 (SF 104)	Unidentified fragment	Cu, Sn, Pb, As?
10/130 (SF 378)	Unidentified fragment	Cu, Sn, Pb?, As?
10/131 (SF 134)	Unidentified fragment	Cu, Sn, As?
10/135 (SF 333)	Modern ornament with raised vegetal design	Cu, Zn, Pb, Sn, Ag?, Ni?
TXCII-1 (SF 153)	Coin	Cu, Ag, Pb, As?
TXCIV-1 (SF 133)	Coin	Cu, Ag, Pb?
TXCVIII-1 (SF 167)	Modern clasp	Cu, Zn, Pb
TC-1 (SF 246)	Coin (fragmentary)	Cu, Sn, Zn, Pb?, Ni
Tomb C (SF 245)	Unidentified fragment	Cu, Sn, Pb, As
SF 415	Unidentified fragment	Cu, Sn, Pb?

TABLE 11.1 (*continued*). pXRF results showing elements attributed to alloy composition

No.	Description	Elements attributed to alloy composition (major and trace elements)
SF 417	Unidentified fragment	Cu, Sn, Pb, As?
SF 419	Unidentified curved fragment	Cu, Sn, Pb?
SF 420	Unidentified circular fragment (rivet?)	Cu, Sn, Pb, As?, Ni?
SF 421	Unidentified fragment with slight curve	Cu, Sn, Pb?
SF 426	Unidentified fragment	Cu, Sn, Pb, As?

Note: For some of the elements detected, the peak height was small and only slightly taller than the background. Because of the limitations of pXRF, it was difficult to determine whether the element was present. These elements have been marked with a question mark (?). Other analytical techniques should be used to determine whether the element is present.

(**Table 11.2**). Although iron can be an indicator of the type of ore used and the smelting process (Craddock and Meeks 1987), in the case of the Lofkënd metals its presence was thought to be due to the incorporation of soil into the corrosion products. Further studies of the alloy and ores used would be required to differentiate whether iron was present as a trace element in the alloy. A similar issue arose with the detection of sulfur and its attribution as an element in the alloy or from the surface as part of the corrosion. Because the presence of sulfur-containing corrosion products was detected using XRD, the detection of sulfur for now is thought to be due to the burial environment, until further analysis can be conducted to determine its origin more precisely.

Five of the objects analyzed had very different compositions in comparison to the majority of copper alloy objects analyzed, and clearly belong to recent centuries. Three of the objects (**TXCVIII-1** [SF 167: modern clasp], **Tomb C-1** [SF 246: Ottoman coin], **10/135** [SF 333: modern ornament]) contained zinc, and all are modern (see Chapters 1, 3, and 10). Two coins (**TXCIV-1** [SF 133]; **TXCII-1** [SF 153]) analyzed were silver-copper alloys. These objects come from burials dated to 1801 ± 153 AD based on recent radiocarbon dating of organic materials (Damiata et al. 2007–2008; Chapter 4) and thus date to the modern use of the tumulus. The majority of the graves excavated at Lofkënd date from the fourteenth century BC to the ninth or early eighth century BC, placing the main use of the tumulus starting in the Late Bronze Age, or Mycenaean era in Greece, through the Early Iron Age (Papadopoulos 2010b:236).

The pXRF data not only allowed for a comparison of the general composition of artifacts across object types (see below), but it also allowed for general patterns in alloy types used at Lofkënd to be observed (**Fig. 11.1**). Based on the initial pXRF results of the analysis of the copper alloy, silver-copper alloy, and gold artifacts, it appears that there were several alloys from different ore types being used at the site. The majority of the objects appear to be made from a copper-tin-lead alloy with traces of arsenic (total of 23 artifacts), followed by copper-tin-lead alloys (19 objects) with no additional minor or trace elements detected attributed to the alloy or ore (**Fig. 11.2**). From the data collected, it appears that only one of the copper alloy objects did not contain lead (**10/131** [SF 134]) and only one contained antimony (**10/78** [SF 163h]). The three zinc artifacts analyzed were each made using a slightly different alloy. Again, because the pXRF analyzed only the surface, which was affected by corrosion, it is possible that certain elements may not have been detected in all the samples or areas enriched in certain elements. Further analysis would help in better understanding the types of alloys and ores used at Lofkënd and will be included in future work on this material.

DISCUSSION OF pXRF RESULTS BY OBJECT CATEGORY

Gold

Ear or head ornaments (**10/11** [SF 290], **10/12** [SF 291]: **Fig. 10.4**)

Two gold objects, ear or head ornaments made of gold foil (**10/11** [SF 290] and **10/12** [SF 291]), were also analyzed using pXRF. Because standards and calibration files for various gold alloys had been

TABLE 11.2 pXRF results showing elements attributed to corrosion or burial deposits

No.	Description	Elements attributed to corrosion or burial environment
10/11 (SF 290)	Ear or head ornaments	Si, Ca, Fe
10/12 (SF 291)	Ear or head ornaments	Si, Ca, Fe
10/13 (SF 107)	Spectacle fibula—pin	Al, P, S, Cl, Fe
10/13 (SF 107)	Spectacle fibula—spiral	Al, P, S, Fe
10/14 (SF 170)	Spectacle fibula	Al, P, Cl, Ca, Ti, Fe
10/15 (SF 106)	Spectacle fibula—area of spiral	Al, S, Ca, Fe
10/15 (SF 106)	Spectacle fibula—center of pin	Al, Si, S, Ca, Fe
10/16 (SF 315)	Spectacle fibula—coiled section	Al, Si, Fe
10/16 (SF 315)	Spectacle fibula—pin section	Al, Si, Fe
10/17 (SF 434)	Spectacle fibula—pin section	Al, Si, Ca, Fe
10/17 (SF 434)	Spectacle fibula—coils	Al, Ca, Ti, Fe
10/18 (SF 157)	Spectacle fibula	Al, Si, S, Cl, Fe
10/19 (SF 431a)	"Cassibile" fibula—pin section	Al, Cl, Fe
10/19 (SF 431a)	Spectacle fibula—pin	Al, Si, P, Fe
10/20 (SF 337)	"Cassibile" fibula with incised decoration—area of decoration	Al, Si, Cl, Fe
10/20 (SF 337)	"Cassibile" fibula with incised decoration—pin	Al, Si, Fe
10/20 (SF 337)	"Cassibile" fibula with incised decoration—clasp	Al, Si, Fe
10/21 (SF 251)	"Cassibile" fibula with herringbone design—center of pin	Al, Si, Cl, Fe
10/21 (SF 251)	"Cassibile" fibula with herringbone design—clasp	Al, Si, P, Fe
10/21 (SF 251)	"Cassibile" fibula with herringbone design—coil on bow	Al, Si, Cl, Fe
10/25 (SF 262)	Bimetallic fibula, figure-8 shape—rivet	Si, K, Ca, Fe
10/25 (SF 262)	Bimetallic fibula, figure-8 shape—bronze strip to which iron disks are attached	Al, Si, K, Ca, Fe
10/26 (SF 231)	Bimetallic fibula figure-8 shape—rivet	Si, Ca, Fe
10/27 (SF 438)	Pin with herringbone decoration	Al, Si, Ca, Fe
10/28 (SF 440)	Pin with concave head—head	Al, Si, Ca, Ti, Fe
10/28 (SF 440)	Pin shaft	Al, Si, Ca, Fe
10/29 (SF 354)	Pin with rolled head—section of head	Al, Ca, Ti, Fe
10/29 (SF 354)	Pin with rolled head—section below head	Al, Si, P, Cl, Ca, Fe
10/29 (SF 354)	Pin with rolled head—center of pin shaft	Al, Si, Ca, Fe
10/30 (SF 105)	Unidentified fragment—metal waste?	Al, Si, Ca, Fe
10/31 (SF 227)	Pin fragment pointed with a notch (end of pin)	Al, P, Fe
10/32 (SF 249)	Pin fragment with rounded edge	Al, Si, Fe
10/50 (SF 111)	Bimetallic pin—copper alloy head	Al, Si, Cl, Ca, Fe
10/51 (SF 067)	Bimetallic pin head—Protrusions on shaft head	Al, Si, Ca, Fe
10/51 (SF 067)	Bimetallic pin head—threaded portion (with some iron from pin)	Al, Si, Ca, Fe
10/57 (SF 300)	Spectacle pendant	Al, Si, P, Fe
10/58 (SF 335)	Spectacle pendant	Al, Si, K, Ca, Ti, Fe
10/59 (SF 336)	Wheel-shaped pendant	Al, Si, S, Ca, Fe
10/59 (SF 343)	Curved fragment	Al, Si, P, S, Ca, Fe

TABLE 11.2 (*continued*). pXRF results showing elements attributed to corrosion or burial deposits

No.	Description	Elements attributed to corrosion or burial environment
10/60 (SF 318)	Perforated disk with punched circular decoration	Al, Si, Ca, Fe
10/61 (SF 163a–d)	Perforated disk with punched circular decoration	Al, Cl, Fe
10/62 (SF 297)	Perforated disk	Al, Si, Fe
10/63 (SF 259)	Perforated disk with punched circular decoration	Al, P, Ca, Ti, Fe
10/65 (SF 444)	Spiral ornament or bead	Al, Si, Mn, Fe
10/66 (SF 313b)	Spiral ornament or bead	Al, S, Ca, Fe, Ti
10/67 (SF 313d)	Spiral ornament or bead	Al, Cl, Ca, Fe
10/68 (SF 267)	Spiral ornament or bead	Al, Si, Fe
10/69 (SF 265)	Spiral ornament or bead	Al, Si, Cl, Fe
10/70 (SF 307)	Coiled wire or ring	Al, Si, Fe
10/73 (SF 091d)	Ring (smaller, reconstructed)	Al, Cl, Ca, Ti, Fe
10/74 (SF 091e)	Ring (larger, intact)	Si, Ca, Ti, Fe
10/75 (SF 163e)	Ring (thin, reconstructed)	Al, Si, Ca, Fe
10/76 (SF 163f)	Ring (flat, intact)	Al, Si, P, Fe
10/77 (SF 163g)	Ring (thick, fragmentary)	Al, Fe
10/78 (SF 163h)	Ring (thin, fragmentary)	Si, Cl, Fe
10/79 (SF 292a)	Ring	Al, Si, S, Fe
10/80 (SF 313a)	Earring	Al, Cl, Fe
10/81 (SF 313c)	Earring	Al, Si, Fe, Ti
10/82 (SF 299)	Earring	Al, Si, Cl, Ti, Fe, Mn
10/83 (SF 431b)	Coiled wire or ring	Al, P, Cl, Fe, Ca
10/84 (SF 349)	Headband	Al, Ca, Fe
10/85 (SF 317)	Headband	Al, Si, Cl, Fe
10/86 (SF 255)	Headband	Al, Si, Fe
10/87 (SF 089)	Headband	Al, P, Cl, Ca, Fe
10/87 (SF 089)	Headband-rivet	Al, Si, P, Cl, Ca, Ti, Fe
10/87 (SF 091a)	Headband—S-shaped decorative element	Al, Si, Ca, Ti Fe
10/87 (SF 091a)	Headband—S-shaped decorative element (rivet)	Al, Si, Ca, Fe
10/87 (SF 091b)	Headband—S-shaped decorative element	Al, Si, Cl, Ca, Ti, Fe
10/87 (SF 091b)	Headband—S-shaped decorative element (rivet)	Al, Si, S, Ca, Ti, Fe
10/125 (SF 044)	Pin fragment	Al, Si, S, Fe
10/126 (SF 212)	Unidentified fragment with finished edge (blade?)	Al, Si, Fe
10/127 (SF 408)	Unidentified fragment	Al, Si, Ti, Fe
10/129 (SF 104)	Unidentified fragment	Al, Si, Ci, Ca, Fe
10/130 (SF 378)	Unidentified fragment	Al, Si, Ca, Ti, Fe
10/131 (SF 134)	Unidentified fragment	Al, Si, Fe
10/135 (SF 333)	Modern ornament with raised vegetal design	Al, Si, Fe
TXCII-1 (SF 153)	Coin	Al, Si, Ca, Fe
TXCIV-1 (SF 133)	Coin	Al, Si, S, Ca, Fe

Continued on next page

TABLE 11.2 (*continued*). pXRF results showing elements attributed to corrosion or burial deposits

No.	Description	Elements attributed to corrosion or burial environment
TXCVIII-1 (SF 167)	Modern clasp	Al, Ca, Fe
TC-1 (SF 246)	Coin (fragmentary)	Al, Si, Ca, Ti, Fe
Tomb C (SF 245)	Unidentified fragment	Al, Si, S, Fe
SF 415	Unidentified fragment	Si, Fe
SF 417	Unidentified fragment	Al, Ca, Fe
SF 419	Unidentified curved fragment	Al, Si, Ca, Ti, Fe
SF 420	Unidentified circular fragment (rivet?)	Al, Si, Ca, Ti, Fe
SF 421	Unidentified fragment with slight curve	Al, Si, Cl, Ti, Mn, Fe
SF 426	Unidentified fragment	Al, Si, Ti, Mn, Fe

established with the particular instrument used, these objects were analyzed quantitatively to try to approximate the alloy composition. Based on the results (approx. 85% Au, 14% Ag, 0.6% Cu), the composition falls within the range of objects previously identified from the Mediterranean as being made from electrum (Karydas 2007:421–425) or within the limits of a slightly silver-rich native gold deposit. However, further analysis should be conducted using other techniques to obtain additional compositional data to confirm these initial findings.

Bronze, including bimetallic jewelry (bronze and iron)

Spectacle fibulae (**10/13** [SF 107], **Fig. 10.5; 10/14** [SF 170], **Fig. 10.6; 10/15** [SF 106], **Fig. 10.7; 10/16** [SF 315], **Fig. 10.7; 10/17** [SF 434], **Fig. 10.7; 10/18** [SF 157], **Fig. 10.8**)

Six spectacle fibulae were found at Lofkënd. The fibulae were similar in appearance but differed in terms of their method of manufacture. The wire used to make the fibula was either circular (**10/13** [SF 107], **10/14** [SF 170], **10/17** [SF 434], and **10/18** [SF 157]) in section, or square (**10/15** [SF 106] and **10/16** [SF 315]). The fibulae with the square wire had a separate pin attached. The pins with the circular wire appear to be made from one piece of wire, with one end fashioned into a clasp and the other shaped to be the pin.

Whether the fibula was made from one piece of wire or a separate segment was used for the pin, the compositions were similar. All were leaded bronzes, but with relatively small peaks for both tin and lead. The addition of only a small amount of tin and lead may have allowed the metal to remain soft and malleable enough to allow for easily working the wire into the elaborate coil shape of the fibula. It is interesting that lead is often found in these spectacle ornaments, since the general view of bronze-working is that lead should be absent or very low in hammered sheetwork because the presence of lead can potentially result in lamination or delamination of layers of hammered bronze. No differences were noted in elemental composition on **10/15** (SF 106) and **10/16** (SF 315) between the coiled area of the fibula and the separate pin fragment. As with many of the other copper alloy objects analyzed from Lofkënd, the possible presence of arsenic was detected in all the fibulae, except **10/14** (SF 170) and **10/18** (SF 157). A small peak for nickel was detected in fibulae **10/16** (SF 315) and **10/17** (SF 434). Silver may be present in **10/16** (SF 315) as well.

Bow fibulae, "Cassibile" Type, I.2a and I.2b (**10/19** [SF 431a], **Fig. 10.9; 10/20** [SF 337], **Fig. 10.9; 10/21** [SF 251], **Fig. 10.10**)

Three bow fibulae, each decorated with a herringbone design on the area of the bow, were analyzed. Each fibula was slightly different in appearance and size. Fibulae **10/19** (SF 431a) and **10/21** (SF 251) were each made from one piece of metal. Fibula **10/20** (SF 337) had a separate pin that was attached to the end of the bow. Despite these differences in appearance, all three seemed to be leaded bronzes of very similar composition. Each fibula was analyzed in three areas (the pin, clasp, and bow), and there were no observable differences in the major elements detected or their relative peak heights across the objects. Some small differences did appear in the minor or trace elements detected. Arsenic was found in the three artifacts, but silver appeared to be present in **10/19** (SF 431a) and **10/20** (SF 337). Nickel was found in **10/21** (SF 251).

Bimetallic fibulae, Type III.1 (**10/25** [SF 262] and **10/26** [SF 231], **Fig. 10.13**)

Two bimetallic fibulae in the shape of a figure-eight, were analyzed. The decorative parts of the objects were made of two disks of iron fixed to a copper alloy strip by two copper alloy rivets. The pins were also iron and were made as separate pieces. The copper alloy was leaded bronze. The lead peak on **10/25** (SF 262) was quite small, and therefore further analysis is necessary to determine whether that element was present. Arsenic was found in the rivets of **10/26** (SF 231) and nickel in **10/25** (SF 262) as trace elements.

Pins, and fragment of pin shaft (**10/27** [SF 438], **10/28** [SF 440], **10/29** [SF 354]: **Fig. 10.14, 10/125** [SF 044], **Fig. 10.48**)

Three of the pins discussed here were found within graves of the tumulus (**10/27–10/29**). The fragmentary pin shaft **10/125** (SF 044), from tumulus fill, is a thin, cylindrical bronze object that appears to be a pin, but it is not clear if it is the pin of a fibula or part of a dress or hair pin because it is damaged at either end. It has been included in this discussion of the other pins found.

All the pins appear to be leaded tin bronzes. In comparing the different spectra acquired from each object, there seem to be differences in relative peak height and the minor or trace elements detected. Pin **10/125** (SF 044) seems to have larger peaks for tin than the other pins. The pin with the rolled head, **10/29** (SF 354), had a larger lead peak than the others, whereas **10/28** (SF 440) had very small peaks for both tin and lead in comparison to the other pins. The pin with the herringbone decoration, **10/27** (SF 438), contained traces of nickel and silver. Pin **10/29** (SF 354) could possibly contain silver. The pin shaft and head of each pin (except for **10/125** [SF 044]) were analyzed with the pXRF, and no differences in general alloy composition were noted across the object.

Pin fragments (**10/31** [SF 227] and **10/32** [SF 249])

Two fragments were found from two different copper alloy objects that resemble the shaft of pins. The objects were made of a round wire that tapers to a point. Both fragments are made from leaded tin bronze, but **10/32** (SF 249) seems to contain more tin and lead based on peak height; **10/32** also seems to contain some traces of arsenic. Fragment **10/31** (SF 227) may contain some arsenic, but the peak appears to be too small to make this exact determination; another technique should be used to verify the presence of this element.

Bimetallic pins, Types III.1 and III.2 (**10/50** [SF 111]; **10/51** [SF 067])

Two bimetallic pins made of a leaded bronze head/handle with an iron pin were analyzed. Only the pin head/handle of **10/51** (SF 067) was preserved. Arsenic was present in **10/50** (SF 111) and possibly in **10/51**.

Small double spiral "spectacle" pendants (**10/57** [SF 300], **Fig. 10.22; 10/58** [SF 335], **Fig. 10.22**)

Two double spiral "spectacle" pendants were found in tombs. Pendants **10/57** (SF 300) and **10/58** (SF 335) are very similar in appearance, consisting of a thin wire twisted to make two coils connected by a loop from which it could be hung or attached to another ornament. The loop portion of **10/58** (SF 335) appears to have a raised decoration.

Both ornaments are made of a leaded tin bronze alloy, but differ slightly in the relative peak height of the elements. Pendant **10/58** (SF 335) appears to have very small peaks for tin and lead, suggesting a small amount of these elements in the alloy. Pendant **10/57** (SF 300) has a much larger lead peak, as well as some arsenic, silver, and nickel. The presence of these elements perhaps suggests the ore used came from a copper, arsenic, silver and nickel mineralization belt (**Fig. 11.3**). This is likely true in the case of other finds from Lofkënd that contain these same elements. Pendant **10/57** (SF 300) was found with an ovoid ring (**10/82** [SF 299]), both of which have similar elemental compositions, suggesting that they could have been made from the same batch of metal alloy.

Wheel-shaped pendant (**10/59** [SF 336 + SF 343], **Fig. 10.23**)

A wheel-shaped pendant, **10/59** (SF 336), was found in one of the graves at Lofkënd, located on the skull of an individual in association with several other copper alloy artifacts and beads. Next to the pendant was a flat, curved strip of metal (SF 343) thought possibly to attach to the end of the pendant to allow it to be hung. Both metallic objects were analyzed and were very similar in composition. The pieces are both tin bronze. It is possible that there is some lead present in both artifacts, but the peaks for this element were extremely small, so its presence should be confirmed with another technique.

Disks/bosses with central perforation (**10/60** [SF 318]; **10/61** [SF 163a–d]; **10/62** [SF 297]; **10/63** [SF 259]: **Fig. 10.24**)

Four circular disks, each with a central perforation, were excavated. Each disk, however, differed slightly in appearance from the others. Somewhat conical in shape, **10/61** (SF 163a–d) was decorated along the edge by a row of punched or incised circles. Only half of **10/63** (SF 259) is preserved, but it is curved in appearance and has punched or incised decoration as well. Disk **10/62** (SF 297) is slightly curved in section, its concave surface containing several ridges, running in different directions, thought possibly to be made from irregularities or markings from the mold in which the object was cast (see results of metallographic examination below). Disk **10/60** (SF 318) was flatter and larger than the other two disks. It had an incised or punched circular decoration along the edge.

Despite the differences in shape and size, the general compositions of the decorated disks were similar. All were tin bronzes with some lead present. The lead peak for **10/63** (SF 259) and **10/60** (SF 318) was very small, making it difficult to determine whether lead was actually present. By comparison, **10/62** (SF 297) had a larger lead peak and contained silver, arsenic, and possibly nickel.

Spiral ornaments or beads (**10/65** [SF 444], **10/66** [SF 313b], **10/67** [SF 313d]: **Fig. 10.26**; **10/68** [SF 267], **10/69** [SF 265]: **Fig. 10.27**)

Five coiled wire beads or ornaments were analyzed and found to be made of leaded tin bronze. Two of them, **10/69** (SF 265) and **10/68** (SF 267), are similar in appearance and were found together. Another two, **10/66** (SF 313b) and **10/67** (SF 313d), are similar in appearance and each was found in association with an ovoid ring (**10/80** [SF 313a] and **10/81** [SF 313c], respectively). The fifth, **10/65** (SF 444), is also made from a coiled wire, but the wire is wider and flatter in section than that used for the other two ornaments/beads.

Despite differences in appearance, all the ornaments are similar in composition, with small peaks of tin and lead. The small addition of these elements likely imparted strength to the alloy while keeping the metal malleable enough for it to be easily shaped and coiled. Ornaments **10/68** [SF 267] and **10/69** [SF 265] had the smallest peaks for lead of the five analyzed. Ornaments **10/66** (SF 313b) and **10/67** (SF 313d) are very similar in composition, as are the rings they were found with. Ornament **10/65** (SF 444) could possibly contain traces of nickel and silver in the alloy, but further analysis would be required to be certain.

Coiled wire or ring (**10/70** [SF 307], **Fig. 10.27**; **10/83** [SF 431b], **Fig. 10.30**)

These two objects consisted of a piece of leaded bronze wire that was twisted into a coil. **10/70** (SF 307) contains very small peaks for what looks like silver, nickel, and arsenic, and possibly antimony, all trace elements found in copper ores.

Rings (**10/73–10/74** [SF 091d–e], **10/75–10/78** [163e–h]: **Fig. 10.29**; **10/79** [SF 292a], **Fig. 10.29**)

Seven circular rings were found in the tumulus, each located on areas near the upper body of the buried individuals; **10/79** (SF 292a) was found in the lab during the process of cleaning block-lifted skeletal remains. All the rings, except for **10/79**, appear to be made from a ternary alloy of copper-tin-lead. The relative peak heights of the rings differed from object to object, even of those within the same burial context. Rings **10/73** (SF 091d) and **10/77** (SF 163g) had the smallest peaks for lead of the group; **10/78** (SF 163h), on the other hand, had the most intense peaks for tin and lead, followed by **10/75** (SF 163e). The peaks for lead in **10/79** are very small, and it is not clear if the element is actually present.

Minor amounts of arsenic were found in all rings, except for **10/78** (SF 163h). For **10/79**, the peaks were extremely small, and it was not clear if arsenic is present. Silver and nickel were also likely present in several of them. Ring **10/78** (SF 163h) may contain antimony and is the only copper alloy object with such a prominent peak for that element.

Earrings (**10/80–10/81** [SF 313a and 313c]; **10/82** [SF 299], **Fig. 10.30**)

Three ovoid earrings were analyzed, which appear to be made from leaded tin bronze. The pXRF results of **10/80** and **10/81** (SF 313a and SF 313c) were very similar compared to that of **10/82** (SF 299). Earrings **10/80** and **10/81** both had small peaks for lead and tin. They also possibly contained arsenic and nickel as trace elements. Similar in size and appearance, both were found associated with small spiraled wire ornaments (**10/66** and **10/67** [SF 313b and d]). Earring **10/82** (SF 299), which was similar in appearance to the other two and only slightly larger, had a much higher peak for lead and tin. It appears that a slightly different alloy composition was used in the manufacture of this particular earring.

Headbands (**10/84** [SF 349], **Fig. 10.31; 10/85** [SF 317], **Fig. 10.32; 10/86** [SF 255], **Fig. 10.33; 10/87** [SF 089 + SF 091], **Fig. 10.34**)

Four headbands, all of relatively similar composition, were found. They all seem to be tin bronzes, possibly leaded. However, the peaks for lead are extremely small in two of the headbands (**10/87** [SF 089] and **10/84** [SF 349]). The tin peaks appear relatively small as well in all these artifacts, which may suggest a composition primarily of copper.

Headband **10/87** (SF 089) had some attached decorative elements (SF 091) consisting of a series of flat, curved pieces of metal held in place with round rivets. pXRF analysis of the decorative elements, including the rivets, showed a composition similar to that of the headband. Traces of arsenic and possibly silver were detected in these added elements, which were not seen in the areas analyzed on the main section of the headband. At this time, further conclusions cannot be drawn regarding these trace elements without additional analysis of all sections of this headband.

Unidentified copper alloy fragments (**10/30** [SF 105]; **10/126** [SF 212], **10/127** [SF 408], **10/129** [SF 104], **10/130** [SF 378], **10/131** [SF 134]: **Fig. 10.48**; SF 415, 419, 420, 421, and 426: all from tumulus fill; SF 245 (found in modern Tomb C)

Several copper alloy fragments were found at the tumulus which could be either scraps of metal waste or fragments from an artifact that could not be identified. All the pieces appear to be made from tin bronze, and several contain small amounts of lead (**10/30** [SF 105], SF 245, **10/127** [SF 408], SF 420, SF 426). The other fragments analyzed may have lead, but the peak was too small to make that determination. The majority of the pieces also appeared to contain possible arsenic as a trace element.

Modern coins (**TXCII-1** [SF 153], **Fig. 3.308a–b; TXCIV-1** [SF 133], **Fig. 3.315a–b; TC-1** [SF 246], **Figs. 3.329a, 3.330**)

Two of the coins found are silver-copper alloys (**TXCII-1** [SF 153] and **TXCIV-1** [SF 133]), which may contain very small traces of lead. **TC-1** (SF 246) is of a very different composition. This coin was found to contain copper, tin, and zinc, along with possible traces of lead and nickel.

Modern clasp (**TXCVIII-1** [SF 167]; **Fig. 3.325a–b**)

A brass clasp, containing lead, was excavated from a modern infant burial.

Modern ornament with raised vegetal design (**10/135** [SF 333], **Fig. 10.50**)

The object appeared to be a disk-shaped fragment with a raised design on one surface, and the design appeared organic and vegetal. The disk was comprised of copper, tin, zinc, and lead, similar to coin **TC-1** (SF 246) but with a larger lead peak. Traces of silver and nickel were detected.

METALLOGRAPHIC EXAMINATION

Twenty-three copper alloy objects were sampled for metallographic examination in order to provide general information on the techniques employed in the manufacture of metal artifacts at Lofkënd (**Table 11.3**). An attempt to take representative samples was made, but because samples could only be taken from broken areas or the ends or edges of objects, the microstructure observed could be specific only to the area sampled. However, based on overall observations of the microstructure in relation to its shape and form, it was possible to obtain a general idea of working techniques that could be extrapolated across the entire object and the collection of copper alloy artifacts at this site.

TABLE 11.3 Summary of metallographic examination

Object no.	Description	Metallographic structure	Other observations
10/13 (SF 107)	Spectacle fibula	Heavily worked and annealed; small grain size	No cuprite corrosion visible in sample
10/14 (SF 170)	Spectacle fibula	Cold worked and annealed; small grains	Some cuprite but not a continuous layer over metallic surface
10/17 (SF 434)	Spectacle fibula	Cold worked and annealed; heavily worked; small grains	Some cuprite but not a continuous layer over metallic surface
10/18 (SF 157)	Spectacle fibula	Cold worked and annealed	No cuprite corrosion visible in sample
10/19 (SF 431a)	"Cassibile" fibula	Cold worked; flattened and aligned inclusions due to working	Some cuprite but not a continuous layer over metallic surface
10/28 SF 440)	Pin with concave head	Cold worked and annealed	Some cuprite but not a continuous layer over metallic surface
10/30 (SF 105)	Unidentified fragment (metal waste?)	Cast	
10/31 (SF 227)	Pin fragment pointed with a notch (end of pin)	Cold worked and annealed; small grains	Some cuprite but not a continuous layer over metallic surface
10/51 (SF 067)	Bimetallic pin	Head cast around pin with some cold working; both dendrites and grains present; coring present	
10/57 (SF 300)	Spectacle pendant	Cold worked and annealed; small grains	Corrosion shows Liesegang banding
10/59 (SF 336)	Wheel-shaped pendant	Cold worked; flattened and aligned inclusions due to working	Corrosion shows Liesegang banding
10/61 (SF 163a–d)	Perforated disk with punched circular decoration	Cold worked; flattened and aligned inclusions due to working	Some cuprite but not a continuous layer over metallic surface
10/62 (SF 297)	Perforated disk	Cold worked and annealed; flattened and aligned inclusions due to working	
10/66 (SF 313b)	Spiral ornament or bead	Cold worked and annealed	
10/70 (SF 307)	Coiled wire or ring	Cold worked and annealed	
10/73 (SF 091d)	Ring (smaller, reconstructed)	Cast with some localized cold working	Corrosion shows Liesegang banding
10/75 (SF 163e)	Ring (thin, reconstructed)	Heavily cold worked and annealed; more heavily worked toward edges; grain size varies	No cuprite corrosion visible in sample
10/77 (SF 163g)	Ring (thick, fragmentary)	Cold worked; made by folding a strip and then hammering to round shape; more heavily worked on side opposite fold	Some cuprite but not a continuous layer over metallic surface
10/78 (SF 163h)	Ring (thin, fragmentary)	Cold worked and annealed; small grains	
10/84 (SF 349)	Headband	Cold worked and annealed; heavily worked	Corrosion shows Liesegang banding
10/86 (SF 255)	Headband	Cold worked; flattened and aligned inclusions due to working	
10/87 (SF 089)	Headband	Cold worked and possibly annealed; heavily worked	
10/135 (SF 333)	Modern ornament with raised vegetal design	Cast with localized working and annealing at edges	

The metalworking techniques employed are fairly similar among the artifacts sampled. The majority of objects seem to have been cold worked and annealed after casting, with annealing used as the final stage of production. The objects appear to be heavily worked due to their small grain size and evidence of strain lines. The grain size tends to be uniform across many of the objects sampled. No sample retained a purely cast structure, except one thought to be casting waste (**10/30** [SF 105]). Those samples that still retained a dendritic structure also showed evidence of cold working, especially around the edges.

Rings were the only object type that showed notable variation in metalworking techniques. Four distinct methods were used to manufacture each of the rings (**10/73** [SF 091d], **10/75** [SF 163e], **10/77** [SF 163g], **10/78** [SF 163h]):

1. Cast with some localized working
2. Cold working and annealing, with heavier working at the edges creating smaller grains toward the edges
3. Folding a strip of metal, then cold working and annealing to make a rounded shape
4. Heavy cold working and annealing producing a very small and uniform grain size

It is not clear why these particular objects show differences in methods of manufacture. Unfortunately, little could be said about the date of the rings, as three of the four were encountered either in topsoil or in tumulus fill (**10/75**, **10/77**, **10/78**). Further study of object composition and a comparison of the manufacture of these types of artifacts with those in the region may shed some light on why such variation is observed.

Many of the samples examined seem to contain blue-gray inclusions, thought to be indicative of copper sulfide present due to the particular ore used. When viewed under polarized light, these small inclusions appear dark rather than red, which is typical for copper sulfide inclusions. Although lead was detected via pXRF analysis on some of the samples, lead was observable in the structure of only a few of the samples (**10/30** [SF 105], **10/51** [SF 067], **10/135** [SF 333]). Tin was found in every sample examined using pXRF, but no $\alpha+\delta$ eutectoid phase was visible in the microstructure. This could be due to the fact that only a small amount of tin was used for the alloy and therefore no eutectoid was formed. Further analysis of the samples would be helpful in better understanding the metal composition in order to correlate that data with what is observed in the metallographic examination.

Metallographic examination also allowed observation of corrosion layers and the extent of that corrosion into the metallic structure. Evidence of intergranular corrosion was seen. On many of the samples, however, the corrosion was found on the surface and only slightly extended into the metallic core. An interesting observation was the lack of cuprite, or copper(I) oxide (Cu_2O), on three of the samples (**10/13** [SF 107], **10/18** [SF 157], **10/75** [SF 163e]), or only the formation of a few discrete areas of cuprite across the metallic surface. This is unusual since the formation of cuprite as one of the first corrosion layers would be expected on copper alloy objects from a terrestrial site. Green corrosion products, generally dark green or greenish brown in color, were found to sit directly on the metal surface and even extend into the metallic core. Findings from the analysis of selected corrosion products and discussion of the corrosion structure are found later in this chapter (Corrosion Studies section).

Description of metallographic examination

Spectacle fibula (**10/13** [SF 107], **Fig. 10.5**)

A sample was taken from the break edge where the pin attaches to one of the coiled sections of the fibula. The wire used for the fibula is circular in section. The structure appears to be composed of very small grains with strain lines along the edges from heavy cold working. Many of the grains exhibit twinning, and therefore annealing may have been the final stage of production. Blue-gray copper sulfide inclusions are visible throughout the sample from the copper ore used.

Examination of the visible corrosion layers shows no formation of the red cuprite layer that usually overlies the metal core. In this sample, there is only a dark green corrosion layer directly overlying the uncorroded metal. In areas where there is intergranular corrosion along the edges, cuprite is visible, and the green corrosion layer extends into the grain structure of the sample.

Spectacle fibula (**10/14** [SF 170], **Fig. 10.6**)

The sample was taken from the end of the clasp. The area was shaped by cold working and annealing.

Small grains are visible and many of these exhibit twinning. Strain lines are visible within some of the grains, suggesting heavy cold working took place. Small blue-gray inclusions are visible within the sample, and these are thought to be copper sulfide from the copper ore used.

In examining the corrosion, the cuprite layer does not appear to be continuous over the entire surface, and in some sections there is no cuprite between the metal and the green corrosion layer on the surface. There is intergranular corrosion occurring within the grain boundaries, and in some areas, only this green corrosion extends into the sample.

Spectacle fibula (**10/17** [SF 434], **Fig. 10.7**)

A sample was taken from a break edge of the wire near a coiled area. The sample shows that the pin has a structure of very small grains with straight twin lines, indicating cold working with annealing used as the final step. Strain lines are visible at the edges of the sample, indicating heavy cold working. There is corrosion along the strain lines. Corrosion is also visible along the grains at the edge of the sample. There are blue-gray inclusions in the alloy thought to be copper sulfide. The metal is porous, as is evident from the small black holes visible in the section (**Fig. 11.4**).

The fibula is in good condition, and from the sample taken, the corrosion on the object is not extensive. No cuprite layer is visible in section when examined using polarized-light microscopy. Instead, a dark green or green-brown corrosion has formed on the metallic surface.

Small spectacle fibula (**10/18** [SF 157], **Fig. 10.8**)

The sample was taken from a broken area of the wire. The wire is round in section. The structure is composed of grains, slightly larger in size than seen on the other objects examined. Twinning is visible on some of the grains, indicating that the metal was cold worked and annealed. The wire was likely cast and then hammered to create the spiral design. Small blue-gray inclusions of copper sulfide are visible throughout the sample.

The sample does not appear to be heavily corroded but does show some intergranular corrosion toward the edges of the wire. This sample does not show signs of the formation of cuprite but instead has a dark green-colored corrosion product that has formed directly over the metal surface.

"Cassibile" fibula (**10/19** [SF 431a], **Fig. 10.9**)

A sample was taken from the center of the pin near a break. The structure is comprised of twinned grains indicating the use of cold working and annealing. Strain lines are evident throughout the entire sample, showing a heavy amount of cold working. Small blue-gray inclusions are visible and thought to be copper sulfide present due to the type of copper ore used. Some of these inclusions appear elongated due to hammering.

Examination of the corrosion layers showed some unusual green corrosion forming on the surface. Cuprite has formed over areas of the metal surface, but it is often not present as a continuous layer. On some sections of the surface, there is a dark green corrosion layer, some of which is located directly over the metal. This corrosion sometimes extends into the metal along areas of intergranular corrosion between the grain boundaries. The green corrosion appears much darker toward the metal core and then gets lighter as it moves toward the exterior of the sample.

Pin with concave head (**10/28** [SF 440], **Fig. 10.14**)

The sample examined was taken from a break edge at the tip of the pin. The structure is composed of grains with twin lines. This shows that the pin was made by cold working and then annealing. Strain lines and the grain size suggest that the pin was heavily worked. Intergranular corrosion is visible throughout the sample. Intragranular corrosion is also evident. Cuprite corrosion is visible along some areas of the surface, but the layer is not continuous. In a few spots, green corrosion sits directly on the metal surface.

Unidentified fragment, possibly metal waste (**10/30** [SF 105])

The sample was taken from a copper alloy object that could not be identified based on its shape. It was small in size, broken, and appeared to have a section that was rolled or bent. Initially it was thought that the fragment had come from an object such as a pin or disk. Examination of its structure revealed no clear dendrites or grains (**Fig. 11.5**). Some long, thin metallic inclusions are visible that appear to be copper in a corroded matrix. There are also some large gray-silver inclusions that could be lead. The sample does not appear to come from a manufactured object but is thought possibly to be metal waste material.

Pin fragment (**10/31** [SF 227])

The sample was taken from the break edge of the pin fragment, and not the preserved tip. The structure of the pin is composed of small grains, many of which are twinned. This indicates that the pin was made through cold working and annealing. There are some strain lines visible, showing that it was heavily worked. Blue-gray copper sulfide inclusions are present. The inclusions are round but appear elongated in areas due to the amount of hammering required to shape the pin.

It appears as if very little cuprite has formed on the metal surface. The cuprite layer is also not consistent over the entire surface of the object. In sections, green corrosion sits directly over the metal core, and where there is intergranular corrosion, the green corrosion seems to extend into the center of the pin.

Bimetallic pin (**10/51** [SF 067], **Fig. 10.20**)

Only the head of the pin is preserved from this object, but in cross-section a piece of an iron circular pin is clearly seen inserted into the handle. A sample was taken from the lower end of the head, which also retained part of the central iron pin.

The copper pin head has a dendritic structure from casting, with some coring present. It appears to have been cast directly over the iron pin (**Fig. 11.6a**). In some areas, the dendrites appear slightly deformed and resemble grains. There are also many strain lines (**Fig. 11.6b**). These features suggest heavy cold working of the piece after casting. This may mean that the head, which may have been cast onto the iron pin, did not fit well and had to be hammered in order to fit on end of the pin. Small blue-gray inclusions are visible and thought to be copper sulfide inclusions from the copper ore used. Small, gray speckled lead inclusions are visible. Corrosion of the surface has extended into the head of the pin, evident by the intergranular corrosion seen between the arms of the dendrites and the corrosion along the strain lines.

Double spiral "spectacle" pendant (**10/57** [SF 300], **Fig. 10.22**)

A sample was taken from one of the ends of the wire used to make the coiled design. The wire is rectangular in section with rounded, outer edges. It is slightly flatter and longer on one side when viewed in section. The structure is made of small grains, with strain lines showing extensive cold working. Some of the grains are twinned, indicating that the piece was also annealed. Small blue-gray copper sulfide inclusions are visible.

There seems to be intergranular corrosion of the structure along the edges of the sample. Not much cuprite has formed, but it is visible in some areas. Some of the corrosion appears layered, suggesting formation due to the Liesegang phenomenon (cf. **Fig. 11.9**). This banded corrosion is formed due to the diffusion of components from the metal and the burial environment which react and then periodically precipitate out of solution as insoluble corrosion products (Scott 2005, 1985).

The thickness of the corrosion layer is not even across the surface, and there is more surface corrosion on one side of the sample than the other. This may be because the object has a spiral design and the less corroded surface corresponds to the edge of the wire that is closest to an adjacent wire, which is more protected and not as exposed to the elements as the other faces of the wire.

Wheel-shaped pendant (**10/59** [SF 336 + SF 343], **Fig. 10.23**)

A sample was taken from a broken edge of one of the spokes of the wheel **10/59** (SF 336). The sample is corroded, but some of the grain structure has been preserved within the corrosion. There are also a few small metallic grains still present, showing that the pendant was made through cold working. Within the center of the sample, elongated inclusions are visible which have been flattened due to hammering that took place to make the pendant **(Fig. 11.7)**.

The wheel pendant is very corroded, with little or no metal preserved. The corrosion on the surface appears to be layered in a way suggesting formation due to the Liesegang phenomenon.

Disks/bosses with central perforation (**10/61** [SF 163a–d], **Fig. 10.24**)

A sample was taken from a fragment that could not be joined to the disk. The disk was very corroded and no metal remained. However, the microstructure, consisting of small metallic grains, was preserved in the corrosion layers. Inclusions in the alloy appeared flattened and were arranged in a linear pattern due to the hammering that was used to form the thin disk.

Disks/bosses with central perforation (**10/62** [SF 297], **Fig. 10.24**)

A sample was cut from the edge of the disk. Examination showed a structure made of twinned grains, indicating that the disk was made through alternating stages of annealing and cold working. The surface of the disk appears smooth on the exterior and has ridges that appear almost as tool marks on the interior surface. However, no differences in the structure across the sample were noted. One possibility is that the ridges seen on the interior are due to the mold that was used initially to cast the piece.

Blue-gray inclusions are visible, some of which have become elongated due to heavy hammering of the disk. These inclusions are thought to be copper sulfide, which is present because of the particular ore used. In section, the sample is more heavily corroded toward the outer edge of the disk, with some intergranular corrosion present toward the surface.

Spiral ornament or bead (**10/66** [SF 313b], **Fig. 10.26**)

The sample was taken from the end of the wire used to make the coil or spiral design. The wire was made through cold working, evident from the small grain size. Twinned grains are present, and it seems that the ornament was shaped through a combination of cold working and annealing. Small, round, blue-gray copper sulfide inclusions are present.

Examination of the corrosion layers on the sample shows very little cuprite formed directly on the metallic surface or within the corrosion layers. In certain areas, a blue-green corrosion layer has formed over the metallic surface and extends into areas of intergranular corrosion.

Small spiral (earring or hair ring) (**10/70** [SF 307], **Fig. 10.27**)

The sample was taken from one end of the wire. The wire is circular in cross-section. The sample is very corroded, but small grains are visible in the center, indicating that the ornament was made through cold working. Some of the grains show twinning due to annealing. Strain lines are visible.

Ring (**10/77** [SF 163g], **Fig. 10.29**)

A sample was taken from the break edge of one of the fragments of the ring. Examination of the ring using the binocular microscopic (5–45x) showed a ridge along the circumference with an opening along it, suggesting that the ring was made through folding a strip or thin sheet of metal. This fold is visible in the metallographic section (**Fig. 11.8**). The ring shows a structure made of small grains, indicating that it was cold worked, with some of the grains exhibiting twinning due to annealing. The grain size varies across the sample, with larger grains visible near the fold where the ring was not as heavily worked. Strain lines are visible along some of the grains. Small, round, blue-gray inclusions visible throughout the sample are thought to be copper sulfide, which derives from the copper ore used.

In some areas, this sample exhibits the typical corrosion layers found on many archaeological objects where cuprite is formed first and then overlaid with other corrosion products, including what are thought to be copper carbonates. However, the cuprite layer is not continuous over the entire surface, and in some sections there is no cuprite between the metal and the green corrosion layer. In addition, some of the green corrosion does not exhibit the typical color of the commonly found copper carbonate malachite, but is slightly brown or greenish brown in areas.

Ring (**10/73** [SF 091d], **Fig. 10.29**)

The sample was taken from a break edge. The ring appears to have been cast, because a dendritic structure is present (**Fig. 11.9**). In the center of the sample near one edge, there seems to be an area where the dendrites show deformation and strain from cold working. Next to these dendrites, grains are visible, and it seems that the ring was locally worked in this area. There are many strain lines in this area as well.

The piece of the ring sampled seems to be corroded through much of the section. There seems to be preferential α-phase corrosion. There is also intergranular corrosion in the locally worked area. The corrosion on the surface of the ring shows the formation of cuprite with an overlying green corrosion product that is not the typical color of the expected copper carbonate malachite but is browner in tone. The corrosion shows the formation of the Liesegang phenomenon observable as banded or layered corrosion.

Ring (**10/75** [SF 163e], **Fig. 10.29**)

A sample was taken from one of the broken edges of the ring. The sample shows a microstructure made up entirely of grains showing that it was cold worked after casting. Twinning is present and so the piece was also annealed during the working process. The grains vary in size, with smaller grains found toward the edges, indicating that it is more heavily worked in those areas **(Fig. 11.10)**. Strain lines are also present. The grains on the edges of the sample are not well defined and show signs of stress cracking. This is due to the extensive cold working that took place in those areas. Corrosion is found within these cracks as well as across the strain lines and grain boundaries. Blue-gray inclusions, thought to be copper sulfide, are visible. Those inclusions located toward the edges have been slightly elongated due to the heavy cold working. The ring that was sampled appears slightly ovoid and flat in section. It is possible that the cast wire or strip was rectangular in section or only slightly rounded, and heavy working of the edges was required to make the ringer rounder.

This sample does not have any cuprite on the surface but instead has a green-colored corrosion product that formed directly over the metallic surface.

Ring (**10/78** [SF 163h], **Fig. 10.29**)

A sample of the ring was taken from a break edge. The ring is circular in section and has a structure composed of small twinned grains, indicating the ring was cold worked and annealed after casting. Blue-gray inclusions, thought to be copper sulfide, appear elongated due to the shaping of the ring through hammering.

Headband (**10/84** [SF 349], **Fig. 10.31**)

A sample was taken from a fragment of the headband that could not be reattached. The sample was very corroded, but a few metallic grains were visible. Some of the grain structure was also visible in the corrosion. Strain lines were observed within some of the grains, indicating that the metal was heavily worked to make the thin, flat headband. Some twinning was preserved in the structure, suggesting the use of annealing during manufacture. Corrosion on the headband surface has a slightly layered appearance formed by the Liesegang phenomenon **(Fig. 11.11)**.

Headband (**10/86** [SF 255], **Fig. 10.33**)

This headband was completely mineralized and barely any metal remains in the core. A few small metal grains, some with strain lines, are visible, indicating the headband was made through heavy cold working, likely from hammered bronze sheet. The grain structure is also preserved in the corrosion. Some inclusions in the sample appear elongated due to the hammering required to make the thin strip of metal used for the headband.

Headband (**10/87** [SF 089 + SF 091], **Fig. 10.34**)

This headband was completely mineralized and no metal remains in the core (**Fig. 11.12**). A small fragment from the main section of the headband (SF 089) was taken for metallographic examination since it is sometimes possible for the grain structure to be preserved in the corrosion layers. Examination of the cross-section shows that the headband is very corroded, with a few metallic grains visible. It appears as if some of the grains, which are small in size, exhibit strain lines or twinning from extensive cold working and annealing. Etching the sample did not provide any further information on the structure.

Examination of the sample under cross-polarized light showed both cuprite corrosion and what is thought to be green copper carbonate corrosion (malachite). Areas of a white corrosion product are visible and thought possibly to be tin oxide, but further analysis is required.

Modern ornament with raised vegetal design (**10/135** [SF 333], **Fig. 10.50**)

The sample examined was taken from a broken edge of the ornament. The object has primarily a dendritic structure, indicating that it was not heavily worked after casting. Grains are visible at the edges, some with strain lines, and this may be due to working done to finalize the shape or better define the decoration on the concave surface. Some of the grains toward the edges are twinned, indicating a mixture of annealing and cold working as the final stages of production. Large gray speckled inclusions of lead are visible. Within these areas of lead there are white inclusions where some other material has recrystallized within it. One possibility is that the white inclusions are tin oxide, but analysis would be required in order to clearly identify this material **(Fig. 11.13)**.

Corrosion Studies: Results of XRD Analysis and Metallographic Examination

XRD analysis was carried out on 24 samples of corrosion products or other surface deposits/accretions to learn more about the condition of the objects and the burial environment (**Table 11.4**). Prior to sampling, the objects were examined using a stereomicroscope (7–40x magnification). The majority of the samples analyzed were corrosion products from copper alloy objects. Corrosion products ranged in color from light green to blue-green to dark green. Many of the objects exhibited surface corrosion after cleaning that was dark green/green-black or brown-green in color. The dark green/green-black corrosion was a relatively smooth corrosion layer. The iron objects were covered with corrosion products in colors ranging from orange to brown. Deposits of a soft white material, or a hard white somewhat translucent accretion, were found on many of the objects, especially the iron artifacts.

The results of the XRD analysis of the copper alloy artifacts showed the presence of typical corrosion products expected for archaeological objects from terrestrial sites: cuprite (Cu_2O), malachite ($CuCO_3{\cdot}Cu[OH]_2$), and in one case azurite (2 $CuCO_3{\cdot}Cu[OH]_2$) (Scott 2002). Cassiterite (SnO_2) was found on six of the samples analyzed (**Fig. 11.14**). This was likely formed due to high levels of carbon dioxide in the sandy soil of the tumulus in the presence of moisture, which contributed to the preferential corrosion of copper and the formation of tin oxide (Gettens 1970; Scott 1994:16–21). Quartz was also found in some of the samples, but this was to be expected, since some of the corrosion incorporated material from the burial environment.

The samples analyzed from the two iron objects, which consisted of soft, powdery, white accretions and harder, white, semi-translucent accretions, were composed of calcite ($CaCO_3$). These deposits on the surface of the iron objects are the result of the alkaline, calcareous-rich soils the tumulus was made from (Papadopoulos et al. 2007:140–144).

In addition to these expected corrosion products, several other corrosion products, some of which are not commonly found on archaeological bronzes or are perhaps under-reported, were discovered. A discussion of those results follows.

Chlorides

Atacamite, a copper trihydroxychloride ($Cu_2[OH]_3Cl$), was found on seven of the objects sampled. Although often associated with arid or highly saline environments, atacamite has been found to form where high chloride and copper concentrations occur in pits on the surface of an object (Scott 2002:131). The presence of bronze disease was visually identified by the conservators on the project, and the presence of chlorides in the corrosion detected in the field using a silver nitrate microchemical test (Odegaard, Carroll, and Zimmt 2005:108–109).

Another copper chloride corrosion identified was connellite ($Cu_{19}[SO_4]Cl_{14}[OH]_{32}{\cdot}3H_2O$), a copper chloride sulfate hydroxide hydrate, found on a bronze coiled wire or ring (earring) (**10/83** [SF 431b]) (**Fig. 11.15**). Connellite has been reported on a few archaeological examples (Muros, Parry, and Whyte 2002:3–5; Scott 2000:48–49) and is generally found as bright blue, needle-like crystals growing on the surface of the object. No needle-like corrosion was found on the ring, but the corrosion sampled was blue-green in color. Based on the XRD analysis, the sample was made up of a mixture of corrosion products that also included malachite, brochantite ($Cu_4SO_4[OH]_6$), atacamite, and cuprite, all of which are corrosion products previously found in association with connellite (Scott 2002:139–141).

Although not reported as a common corrosion product on archaeological artifacts, connellite could occur more frequently than expected. Studies have shown that small amounts of sulfate ions in solution with chloride ions can allow for the formation of connellite (Pollard, Thomas, and Williams 1990). This more common occurrence is further supported by the stability diagram for connellite showing its formation at chloride ion levels that can be present in most natural environments at a pH of 5 to 8 (Scott 2000:49). Studies of the Lofkënd tumulus soil have found that the pH falls within this range (Foss, personal communication; Muros, Lewis, and Bardho 2006; O'Grady and Bardho 2005; see further Chapter 16.4), thus providing conditions suitable for the formation of this particular corrosion product.

Phosphates

The presence of sampleite, $NaCaCu_5(PO_4)4Cl{\cdot}5H_2O$, and libethenite, $Cu_2(PO_4)(OH)$, was found on two of

Table 11.4 Summary of XRD results

Sample no.	Object	Sample description	XRD results
10/14 (SF 170)	Spectacle fibula	Dark green corrosion	Cuprite Malachite
10/17 (SF 434)	Spectacle fibula	Globular, compact green corrosion	Brochantite Calcite Malachite
10/18 (SF 157)	Spectacle fibula	Green corrosion	Malachite
10/19 (SF 431a)	"Cassibile" fibula	Dark green corrosion	Cuprite Malachite
10/22 (SF 110)	Iron fibula	White powdery material	Calcite
10/25 (SF 262)	Bimetallic fibula, figure-8 shape	Green-blue corrosion	Azurite Malachite
10/28 (SF 440)	Pin with concave head	Green corrosion	Calcite Cuprite Malachite
		Lighter green corrosion	Brochantite Cuprite Malachite
10/30 (SF 105)	Unidentified fragment—metal waste?	Green corrosion	Calcite Malachite Quartz
10/31 (SF 227)	Pin fragment, pointed with a notch (end of pin)	Green corrosion	Cuprite
10/32 (SF 249)	Pin fragment with rounded edge	Green corrosion	Cassiterite Cuprite Malachite Quartz
10/57 (SF 300)	Spectacle pendant	Dark green corrosion	Atacamite Calcite Cuprite
10/59 (SF 336)	Wheel-shaped pendant	Green corrosion	Atacamite Brochantite Sampleite
10/61 (SF 163a–d)	Perforated disk with punched circular decoration	Green corrosion	Hydroxyapatite
10/62 (SF 297)	Perforated disk	Green corrosion	Calcite Cassiterite Quartz
10/63 (SF 259)	Perforated disk with punched circular decoration	Compact corrosion within punched decoration	Atacamite Brochantite Cuprite Malachite
		Green corrosion on surface	Brochantite Cuprite Malachite

Continued on next page

TABLE **11.4** (*continued*). Summary of XRD results

Sample no.	Object	Sample description	XRD results
10/66 (SF 313b)	Spiral ornament or bead	Green corrosion	Cassiterite Cuprite Malachite
10/73 (SF 091d)	Ring (smaller, reconstructed)	Dark green corrosion	Atacamite Cassiterite Cuprite
10/75 (SF 163e)	Ring (thin, reconstructed)	Dark green corrosion	Cassiterite
10/77 (SF 163g)	Ring (thick, fragmentary)	Green corrosion	Cuprite Malachite
10/84 (SF 349)	Headband	Green corrosion	Atacamite Cuprite Malachite
10/83 (SF 431b)	Coiled wire or ring	Blue-green corrosion	Atacamite Brochantite Calcite Connellite Cuprite Malachite
10/86 (SF 255)	Headband	Green corrosion	Libethenite Sampleite
10/87 (SF 089)	Headband	Dark green corrosion	Atacamite Cuprite Malachite
10/118 (SF 263)	Iron spear	White powdery corrosion	Calcite Quartz
		Yellow-brown compact corrosion	Calcite Goethite
10/135 (SF 333)	Modern ornament with raised vegetal decoration	Dark green-black corrosion	Cassiterite Cuprite Zinc oxide

the bronze objects. A wheel-shaped pendant (**10/59** [SF 336 and SF 343]) contained sampleite, along with atacamite and brochantite. A bronze headband (**10/86** [SF 255]) contained both sampleite and libethenite **(Fig. 11.16).** The occurrence of sampleite has been reported in objects from Egypt, Turkey, Ecuador, and Peru (Fabrizi et al. 1989; Scott 2002:240–246). The presence of sampleite together with libethenite on a bronze and iron bed from Western Turkey has recently been published (Scott and Maish 2010:12–15).

The stability diagrams for both sampleite and libethenite show that high levels of phosphate are required, and the presence of sodium, calcium, and chloride, in the case of sampleite, for these corrosion products to form (Scott and Maish 2010:16–17). Chloride ions in the soil have already been identified due to the formation of atacamite, connellite, and likely paratacamite on the copper alloy objects. As for the presence of phosphorus and calcium, the likeliest source is the burial context of the objects, that is, graves. The two objects on which sampleite and libethenite were found were grave goods placed on the skulls of individuals. Dissolution of the bone during burial could have contributed to the high amounts of phosphorus and calcium needed for the formation of this corrosion. The presence of hydroxyapatite ($Ca10(PO4)_6(OH)_2$) within the corrosion layers on a bronze perforated disk (**10/61** [SF 163a–d]) supports the suggestion of dissolved components of bone being incorporated into the corrosion layers of the artifacts. pXRF showed a small peak of phosphorus for many of the objects analyzed, which could also suggest the presence of this element within the corrosion on other objects due to its proximity to bone. Sampleite could be found more frequently in future studies, if XRD analysis is conducted on corrosion taken from bronze artifacts found within grave contexts.

Sulfates

Another corrosion product found on five of the objects analyzed, which is not commonly reported on recently excavated archaeological objects, is brochantite ($Cu_4SO_4(OH)_6$) (**Fig. 11.17**). Brochantite is more commonly found on outdoor sculpture in polluted environments where there are high levels of sulfur dioxide in the air (Graedel et al. 1987:641). However, brochantite has been reported on bronzes from archaeological sites (Balasubramaniam et al. 2002:5; Ingo et al. 2006), especially from areas where there is a high presence of sulfur due to pollution within the burial environment (Nord, Mattsson, and Tronner 2005).

When this corrosion product was identified on the Lofkënd pieces, it was initially thought that the sulfur source was from the soil. Soil studies undertaken at the tumulus and surrounding areas, however, did not find high amounts of sulfur or deposits of sulfur-containing minerals or materials (J.E. Foss, personal communication; see also Chapter 16.4). It is proposed that the most likely source for the sulfur ions needed to form brochantite at Lofkënd would be the decomposition of organic material in the burial environment, such as wool textiles, pseudomorphs of which have been found on many of the metallic finds (see Muros, Chapter 12), and the bodies within the graves. Microorganisms in the soil break down the textiles and bodies within the graves after burial, releasing sulfur contained in the proteins of these organic materials (Waksman and Starkey 1931:168). Most of the sulfur liberated would have been in the form of sulfides and formed this type of corrosion initially on the objects, such as seen with bronzes corroding in anaerobic conditions underwater and in waterlogged soils (Duncan and Ganiaris 1987:109; Schweizer 2005). In the case of the aerobic environment at Lofkënd, a secondary process involving oxidation could have taken place to convert the sulfides to sulfates (Schweizer 2005:46; Waksman and Starkey 1931:170–172), which may have allowed for the formation of brochantite. There are also microorganisms that produce sulfates directly as a by-product of metabolic activity (Costello 1969; Waksman and Starkey 1931:168), which also could have contributed to the formation of brochantite.

Lack of cuprite formation

The final unusual corrosion phenomenon found on three (**10/13** [SF 107], **10/18** [SF 157], **10/75** [SF 163e]) of the metallographic samples examined was the lack of any cuprite formation (**Fig. 11.18**). On about eight other samples examined, the cuprite layer was discontinuous and only visible in small discrete areas. Since the bronzes were excavated from a terrestrial environment where moisture and oxygen were present, cuprite would be expected as one of the first corrosion layers formed. In the case of three of the objects, no cuprite was observed in the samples examined, and only a dark green or green corrosion layer was visible sitting directly on the metallic surface or extending into cracks or between grains within the center of the sample.

The reason for the lack of cuprite could not be definitively determined, but some possibilities are suggested. The Lofkënd tumulus is composed of calcareous-rich sandy soils that are well draining (Foss, personal communication; Chapter 16.4; Papadopoulos, Bejko, and Morris 2007:140–144), allowing for the movement of oxygen and water. Therefore, an anaerobic environment is not the cause for the lack of cuprite formation. The creation of a microenvironment within each of the graves, in discrete areas of each grave, or on individual artifacts, however, due to the decomposition of organic material in the burial environment, could be a likely reason. In examinations conducted on patina formation on outdoor sculpture in a sulfur-rich atmosphere, in particular sulfur dioxide, the conversion of cuprite to brochantite has been observed (Graedel 1987:760–761; Livingston 1991:1404; Scott 2002:149). Due to the high amounts of sulfur gases that could be present in the graves due to the decomposition of organic material, a similar reaction could have taken place.

The rate at which this reaction occurs is influenced by pH, where acidic conditions can cause dissolution of the cuprite to produce more copper ions for the reaction (Graedel 1987:761) or influence the growth of cuprite (Scott 2002:149). Although the pH of the Lofkënd soils is alkaline, areas with low pH could be created due to the activity of microorganisms breaking down the human remains and other organic materials in the grave. It is possible that acids released during the decomposition of these materials or through bacterial-metabolic reactions not only contribute to the production of certain corrosion products (Caneva et al. 2008:154–156; Costello, Nugari, and Salvadori 1969), but could cause acidic conditions along certain areas of the object's surface that could affect the formation of cuprite or promote its conversion to brochantite.

There are other possible diagenetic processes that could result in the absence or patchy presence of cuprite observed at this site, such as an alteration of the burial environment due to changes that took place subsequent to the original burial. Since we expect cuprite to form adjacent to the metallic substrate, it may have originally formed in the burial but then was possibly subject to depositional alteration as a result of changing burial conditions in the tumulus. The presence of brochantite and copper phosphate salts argues for a variety of burial environmental parameters that may be implicated in the general absence of cuprite on these artifacts.

FUTURE WORK

Future examination and analysis of the Lofkënd metals is planned to provide more information on ancient metallurgy in this region during the Late Bronze and Early Iron Ages, as well as help to explain some of the condition issues observed on the sampled artifacts. Further studies of the corrosion layers, including the use of scanning electron microscopy with energy dispersive spectroscopy (SEM-EDX), will be performed to obtain compositional data on the 22 sampled metals. This analysis may also help to understand better the corrosion processes that these artifacts underwent, especially for those that did not show cuprite formation. Samples were taken of 35 metals for lead isotope analysis to help determine the source of metal ores used in this area. This analysis is currently ongoing, with the results to be published once completed.

ACKNOWLEDGMENTS

The authors would like to thank Alma Bardho, a conservator on the Lofkënd Archaeological Project, for her help with the pXRF analysis of the metallic finds and the sampling of the copper alloy artifacts for metallographic analysis. They would also like to thank those colleagues who assisted with the interpretation of some of the analytical data and read over early drafts of this chapter.

Chapter 12

Textiles and Other Mineralized Organic Remains at Lofkënd

Vanessa Muros

Introduction

Artifacts made from organic materials generally do not survive in the archaeological record because they require specific burial conditions for their preservation. Often information about these objects is obtained not from the objects themselves but from impressions or *pseudomorphs* (the replacement of organic materials with minerals or metallic corrosion products) found on other artifacts. At Lofkënd, no organic materials such as cloth, wood, or cordage have been preserved in the burials, but evidence of these materials has been found as pseudomorphs on several of the metallic finds, providing technological information about the objects and people who used this tumulus. This report documents the textile and wood pseudomorphs found on the Lofkënd objects and attempts to provide a preliminary interpretation of the pseudomorphs in order to identify the organic materials and any technological information that can be obtained from the scant remains (for textiles and pseudomorphs from the Aegean and Cyprus, see Barber 1991:174, n. 12; for the Ukraine and the former U.S.S.R., see Barber 1991:144, n. 7; for central Europe, see, especially, the work of Hans-Jürgen Hundt 1959, 1960, 1961, 1962, 1967, 1969, 1974).

Examination and Documentation Methods

As part of the routine examination of artifacts conducted by the project conservators, all iron and copper alloy finds brought to the lab were examined for the presence of organic pseudomorphs using a stereomicroscope (7–45x magnification). Objects that would have been attached to cloth or textile, such as pins and fibulae, beads that would have been strung together, and blades or other tools/weapons that would have had handles were found in many cases to contain some trace of a previously existing organic material. When pseudomorphs were identified, they were documented in written form, using a description and sketches, as well as through photography, using a digital SLR camera for both macro- and microphotography. Among the 160 iron and copper alloy objects treated in the conservation laboratory, pseudomorphs were found on 35.

The pseudomorphs found were often in a poor state of preservation. Since the organic component of the material has been replaced with iron or copper corrosion products, the pseudomorphs were often powdery, brittle, and friable. On many of the objects, most of the organic material was gone and only very small areas contained any traces of cloth, cordage, or wood. When pseudomorphs were present, they were often obscured by soil and corrosion that could not be removed due to the fragility of the pseudomorphs. Areas of cloth or textile pseudomorphs were often abraded or damaged.

All the textile or fiber pseudomorphs found are recorded and described in **Table 12.1**, listing when possible the direction of spin, the angle of spin, the diameter of the yarns (in both directions), estimated thread counts (in both directions), and the weave structure. For pseudomorphs in better states of

Table 12.1 Description of textile and fiber pseudomorphs found on Lofkënd metallic objects

No.	Object	Weave pattern	Yarn spin/ thread twist/ ply	Spin angle[a]	Yarn/ thread width	Thread count/cm	Angle of crossing elements	Other observations
10/14 (SF 170)	Cu alloy spectacle fibula	Too poorly preserved to determine						Globular and linear corrosion pattern in areas possibly suggesting a textile
10/17 (SF 434)	Cu alloy spectacle fibula	Too poorly preserved to determine						Fibers found wrapping around pin at point of attachment to fibula coil
10/22 (SF 110)	Iron fibula with coiled arch	Too poorly preserved to determine	Z-spun; single	15–20° (loose)	0.3–0.5 mm		90°	
10/23 (SF 261)	Iron arched fibula	Plain(?) weave	Z-spun; single ply	20° (loose)	0.4–0.5 mm (in both directions)	26 (direction 1) by 19 (direction 2)	85–90°	Only a few yarns preserved, making accurate measurements of diameter and thread count difficult
		V-shaped elements, chevrons?	Spin not preserved; single ply		0.3–04 mm		85°	
10/24 (SF 162)	Iron arched fibula	Plain weave?			0.5–0.6 mm (direction 1); 0.6–0.7 mm (direction 2)	24 (direction 1) by 22 (direction 2)		Globular and linear corrosion pattern in certain areas may suggest a textile
10/25 (SF 262)	Bimetallic fibula, figure-8 shape	Too poorly preserved to determine	Z-spun; single ply		0.3–0.4 mm			
10/26 (SF 231)	Bimetallic fibula, figure-8 shape	Twill?	No spin preserved; single		0.5–0.8 mm			
		Cord? Thread?	No spin preserved; single ply		0.7–0.9 mm			Two yarns preserved on pin seem to wrap around shaft and cross over each other
10/33 (SF 171)	Iron dress pin with rolled head	Plain weave?	Z-spun, single	15–20° (loose)	0.3–0.4 mm (direction 1); 0.3–0.5 mm (direction 2)	24 (direction 1) by 17 (direction 2)	80–90°	
		Plaited(?) element (figure-8 shape)	No spin preserved; single		0.6–0.8 each yarn			
10/34 (SF 254)	Iron dress pin	Plain weave or oblique interlacing	Z?-spun; single ply	20–25° (medium)	0.3–0.4 mm (direction 1); 0.4–0.5 mm (direction 2)	43 (direction 1) by 28 (direction 2)	75–80°	
		Twill (herringbone?)	No spin preserved				30–40°	Width of individual yarns cannot be measured due to poor preservation
10/35 (SF 390)	Iron dress pin	Balanced plain weave or oblique interlacing?	Z-spun; single ply	30–40° (medium)	0.4–0.5 mm	13 (direction 1) by 13 (direction 2)	85–90°	

[a] Spin angle, 0–20° loose; 25–40° medium; 45° tight.

Table 12.1 (*continued*). Description of textile and fiber pseudomorphs found on Lofkënd metallic objects

No.	Object	Weave pattern	Yarn spin/ Thread twist/ ply	Spin angle[a]	Yarn/ thread width	Thread count/cm	Angle of crossing elements	Other observations
10/36 (SF 423)	Iron dress pin	Too poorly preserved to identify			0.3–0.4 mm			
10/37 (SF 046)	Iron dress pin with rolled head	Faced plain weave (weft faced?)	Z-spun; single	25–30° (medium)	0.3–0.5 mm (faced yarns); 0.7–1 mm (direction 2)	25 (faced yarns) by 10 (direction 2)	90°	
		Thread or cord	Z-spun; Z-twist; 2- ply		Thread: 0.7–0.8 mm; yarns: 0.3–0.4 mm			
		2/1 twill(?) (V-shaped elements)	Spin not preserved; single		0.5 mm		85°	
10/38 (SF 116)	Iron dress pin	Too poorly preserved to determine	Z-spun; single	25–40° (medium)	0.4–0.5 mm			Series of yarns found that seem to wrap around pin near center
10/40 (SF 018)	Iron dress pin with rolled head	Too poorly preserved to determine			0.5–0.6 mm			Series of yarns that seem to wrap around shaft of pin
10/41 (SF 414)	Iron dress pin	Too poorly preserved to determine	No spin preserved		0.9–1 mm			Fibers and yarns wrapped around a section of the pin
10/42 (SF 260)	Iron dress pin	Weft-faced plain weave	Z-spun; single ply	25–35° (medium to tight)	0.6–0.7 mm (wefts); 0.7–0.9 mm (warps)	22 (wefts) by 18 (warps)	45°	Selvedge preserved
10/43 (SF 305)	Iron dress pin with disk finial	2/2 twill?	Z-spun; single ply		0.5–0.6 mm (both directions)		75–80°?	2 weft elements seem to cross over 2 warps, but preservation very poor
10/44 (SF 253)	Iron pin with spherical head	Plain weave or oblique interlacing	Z-spun; single ply	30–40° (medium)	0.3–0.4 mm (both directions)	30 (direction 1) by 23 (direction 2)	85–90°	
10/45 (SF 226)	Iron dress pin with bent-back head	Faced plain weave (weft faced?)	Z-spun; single	30° (medium)	0.3–0.4 mm (faced yarns); 0.5–0.8 mm (direction 2)	35 (faced yarns) by 15 (direction 2)	90°	
		Ovoid element near tip	Z-spun single	35–45° (medium to tight)	0.6–0.7 mm		45°	
10/46 (SF 304a-c)	Iron dress pin	Plain weave	Z?-spun; single ply	10–15° (loose)	0.5–0.7 mm (wefts and warps)	21 (wefts) by 18 (warps)	90°	Selvedge preserved

[a] Spin angle, 0–20° loose; 25–40° medium; 45° tight.

Continued on next page

TABLE 12.1 (*continued*). Description of textile and fiber pseudomorphs found on Lofkënd metallic objects

No.	Object	Weave pattern	Yarn spin/ thread twist/ply	Spin angle[a]	Yarn/thread width	Thread count/cm	Angle of crossing elements	Other observations
10/47 (SF 155)	Iron dress pin	2/2 twill?	Z-spun; single	10–20° (loose)	0.45–0.7 mm		85°	
		Plaited cord? (V-shaped elements)	No spin preserved		V-shaped element 0.7–0.8 mm; yarns-0.4–0.5 mm		10–25°	
10/48 (SF 225)	Iron dress pin	2/1 or 2/2 twill?	Z-spun; single	30° (medium)	0.5–0.7 mm		85–90°	
10/50 (SF 111)	Bimetallic pin (Cu alloy and Fe)	Balanced plain weave	Z-spun; single	20–30° (medium)	0.6–0.7 mm (direction 1); 0.6–0.8 mm (direction 2)	20 (direction 1) by 13 (direction 2)	90°	
10/63 (SF 259)	Cu alloy disk or boss	Yarn	Z-spun; single ply	35–45°	1.5–1.8 mm?			Single yarn found among fragments of bronze disk
10/71 (SF 172)	Iron coil	Too poorly preserved to determine	Z-spun; single	30–40° (medium)	0.6–0.7 mm		80–85°?	Too little of textile is preserved with only one or two crossing elements
10/83 (SF 431b)	Cu alloy earring	Too poorly preserved to determine						Mineralized fibers found wrapping around lower portion
10/84 (SF 349)	Cu alloy headband	Strap or cord			L: 78.7 mm; W: 3.6 mm; T: 1.7 mm			
10/89 (SF 164b)	Iron tubular bead	Cord	Z-twist, 2- or 3-ply	25–40° (medium)	Thread: 2.1–2.3 mm; yarn: 1.4–1.6 mm			
10/90 (SF 424)	Iron tubular bead	Cord?			1.5 mm			Fiber pseudomorphs found on interior but not clear if cord or cloth
10/91 (SF 257b)	Iron tubular bead with attached iron coil	Too poorly preserved to determine			0.7 mm?			Mineralized fibers toward one end extending into the interior; not clear if cord or cloth
10/92 (SF 257a)	Iron tubular bead	Too poorly preserved to determine			0.5 mm?			
10/99 (SF 124)	Iron tubular bead	Oblique weave pattern	S?-spun; single	15–20° (loose)	0.2–0.4 mm (wefts); 0.4–0.6 mm (warps)	28 (wefts) by 17 (warps)	80°	
		Balanced plain weave?	Z-spun, single	35–40° (medium)	0.3-0.5 mm	20 (direction 1) by 19 (direction 2)	90°	
		Cord	Z-spun, Z-twist, 2- ply	25–30° (medium)	Thread: 0.5–0.7 mm; yarn: 0.2–0.4 mm			

[a] Spin angle, 0–20° loose; 25–40° medium; 45° tight.

preservation, a more detailed description is found below. When weave structures are given, these are offered as tentative identifications due to the condition of the pseudomorphs and how little of the material has been preserved. Factors such as distortion of the textile during burial and poor or selective preservation of individual elements or certain areas of the cloth made a more definitive identification difficult. These factors also made understanding the cloth or textile in its entirety impossible (i.e., What was the cloth used for or made into? Were several weave types used to make one cloth? Was it dyed or decorated?). Poor preservation also affected the identification of weft and warp elements, as well as the determination of thread count. When the warp and weft yarns could not be identified, the diameter is given as "direction 1" and "direction 2." The few thread counts that could be provided are extrapolated from the small section of the remaining woven yarn pseudomorphs and again are reported as directions 1 and 2 (Julie Unruh, personal communication).

Weaving Terms and Techniques

The term *fiber(s)* is used in this report to describe the individual elements or components of fibrous materials from a plant or animal (Emery 1994 [1966]:4). They are also referred to sometimes as *ultimates*. The earliest fibers used for weaving were from plants, with linen, or flax, being one of the more common fibers used (Barber 1991:10). Animal hair was introduced as a textile fiber after the domestication of animals and was most commonly found as wool from sheep. Although linen and wool are the more predominant fibers in the archaeological record, evidence shows that other plant materials, such as nettle (in northern Europe) and hemp, and animal hairs other than wool were also used to make cloth (Barber 1991:30, 197). Archaeological remains of textiles found at sites in central and southern Europe dating from the Bronze and Iron Ages indicate that linen and sheep wool were predominantly used (Barber 1991:9–34). The use of these materials changes over time, however, with flax (linen) being the fiber primarily used in the Bronze Age, and wool and flax in the Iron Age (Jørgensen 2003:61–70). Based on this evidence, the likely fibers used for the textiles at Lofkënd were flax and wool.

Fibers are often spun on a spindle to create a long continuous strand to be used for weaving, referred to as a *yarn*. The fibers can be spun either in an S-direction (following the slope of the letter "S") or a Z-direction (following the slope of the letter "Z") (Emery 1994 [1966]:11). Archaeological evidence has shown that certain regions and cultures preferred certain spin directions for their yarns. For example, in Egypt, where linen was extensively used, the plant fiber was generally spun in the S-direction, following the natural twist of the linen fiber (Barber 1991:66). In Europe, Z-spin was preferred for both plant and animal fibers (Jørgensen 2003:70).

Several yarns can be twisted together, again either in an S- or Z-direction, to create a *plied yarn* or *thread*. The number of plies corresponds to the number of yarns that have been twisted together. If it is a single yarn, it is referred to as a "single(s)" (Barber, personal communication). When two yarns are twisted together, it is two-ply and so on. In the description of the Lofkënd material, the *spin* direction refers to the direction the bundle of fibers has been turned to create a yarn, and *twist* refers to the direction several spun yarns have been turned to create a thread.

Woven fabrics consist of a *warp*, which is the element put under tension during weaving so that it is stable, and a *weft*, the active element that is passed back and forth through the warp to create a woven textile (Emery 1994 [1966]:74). When cloth is made on a loom, a *selvedge* (closed edge of the textile) is created when the weft is one long continuous yarn and is passed back around the warps at the end of a row to be woven into the next one (Barber 1991:9). Examination of the edges and other features of a textile, such as additions to or changes in the weave pattern at the edges of the textile, can provide insight into weaving techniques, and sometimes indicate the types of looms used (Barber 1991:126–144; Hoffman 1964:152–183). For example, in warp-weighted looms, the top and sides of the textile are selvedges (with the top often containing a heading band or starting edge). When a horizontal ground loom is used, all four sides of the cloth have closed edges. In the examination of the Lofkënd textiles, two objects were found (**10/42** [SF 260] and **10/46** [SF 304a–c], both iron dress pins) that contained textile pseudomorphs of one side that was clearly a selvedge.

Archaeological evidence indicates that warp-weighted looms were used in Europe to make textiles as early as the sixth millennium BC (Barber 1991:93). In its simplest form, the loom was set upright, usually tilted back at an angle, with the warps hanging from a top bar with weights at the bottom of each warp to pull them tight (Wild and Rogers 2003:13–14). Although it cannot be known for certain, the common use of the warp-weighted loom in areas such as Greece and central Europe during the Bronze and Iron Ages make it a likely candidate for the type of loom used to weave the Lofkënd textiles.

One of the earliest weaves found is *plain* or *tabby* weave, where the weft is woven over one yarn and under one in both directions. When the same numbers of threads are woven per centimeter in each direction, the weave is known as a *balanced plain weave*. A weave can still be considered balanced even if the diameter of the yarns in each direction varies, as long as the thread count is the same in each direction (Barber, personal communication). If the threads are densely packed with more yarns or threads per cm in one direction, it is known as a *faced plain weave*. The weave can be warp- or weft-faced. With many of the Lofkënd pseudomorphs, when a plain weave is preserved, it is difficult to determine which element is the warp and which is the weft without the presence of a selvedge. In these cases, the weave is described simply as "faced." Pseudomorphs of this weave pattern are found on many of the Lofkënd iron artifacts.

Another common weave found in Europe, as early as the second millennium BC, is *twill* (Barber 1990:186). In creating the twill pattern, the weft goes over and under the warp in succession. In the next row, the weaver offsets this pattern by one warp thread, which forms diagonals on the cloth. Twill weaves are recorded with a number designating how many wefts are passing over the number of warps (e.g., 2/2 twill is where the weft passes over two and under two warps).

Evidence from central Europe indicates twill weave was found as early as the Neolithic and had spread across Europe, including into the northern Balkans, during the Iron Age (Jørgensen 2003:67). Very little textile evidence has survived from the Iron Age in southern Europe (Barber 1991:197). With the discovery of textiles from the Hallstatt culture in central Europe, a culture that commonly used twill weave in their textiles, and evidence of the influence of this weaving technology as far west as Spain (Jørgensen 2003:70), it is possible that twill patterns were commonly woven into southern Albania textiles, as they were over much of Europe during this time. Due to the poor preservation of the pseudomorphs on many of the Lofkënd objects, it is not clear whether twill weave was common. The pseudomorphs on only one object (**10/47** [SF 155]) could be definitively identified as a 2/2 twill weave. Several other objects with textile pseudomorphs have crossed elements that seem to suggest a twill weave or a herringbone pattern, but the exact determination of the weave cannot be made. For these examples, tentative twill identifications are given.

In addition to the techniques described above, textiles and other items such as cordage can be made through *interlacing*, *plaiting*, and *twining*, which do not necessarily require a loom (Emery 1994 [1966]: 60–64). In interlacing, all the elements are active, being woven under and over each other. In this technique, elements cross at oblique angles, and the edges can, as in oblique interlacing, show elements turning at different angles than those seen on a plain weave. With plaiting, all the elements are active and each element interlinks with the others (Barber 1991:5; Emery 1994 [1966]:60–61). On some of the textile pseudomorphs found at Lofkënd, it is possible that techniques such as oblique interlacing were used.

FIBER IDENTIFICATION: METHODOLOGY AND RESULTS

In the field, small samples were taken of the pseudomorphs from 18 objects and sent to the conservation laboratories of the UCLA/Getty Conservation Program for examination, in the hope of identifying the organic material found on the Lofkënd metals. Samples taken were first examined under transmitted and polarized-light microscopy (5–50x magnification) for identification of the fibers. Due to the degree of mineralization of the pseudomorphs, individual fibers and their diagnostic characteristics were not preserved, and therefore the fibers could not be identified. Since microscopy could not be used, alternative methods for fiber identification were investigated that could possibly determine whether the fibers were plant (cellulosic) or animal (proteinaceous) in origin, even if a more specific identification could not be determined.

Fourier transform infrared spectroscopy (FTIR) has been utilized for the identification of fibers with some success (Garside and Wyeth 2003; Stuart 2007), and this technique was first used to identify the general type of fiber preserved on the Lofkënd metal artifacts. A portion of the samples taken from each object was analyzed using a Perkin-Elmer Spectrum One instrument equipped with a solid-state attenuated total reflectance (ATR) sample stage. Spectra were recorded in the 4000–550 cm^{-1} region, with a resolution of 4.0 cm^{-1}, and matched against the UCLA/Getty Conservation Program's IR database, as well as the spectral database of the Infrared and Raman Users Group (IRUG). No results were obtained from the FTIR analysis of the pseudomorphs, suggesting that either the sample size was too small or not enough organic material remained to be detected by the instrument.

Microchemical tests previously used for the identification of pseudomorphs to determine whether the fibers were cellulose or protein based were performed on the remaining samples (Anheuser and Roumeliotou 2003). Because of the small sample size, only the Herzberg stain for cellulose and the Naphthol blue-black stain for proteins could be carried out. Each test was performed twice to verify results, especially if a positive result was obtained. If sample size did not permit both tests to be repeated using new samples, the protein microchemical test was repeated on the sample tested with the Herzberg stain after rinsing with deionized water.

Of the 19 samples tested, only 6 gave positive results (**Table 12.2**). The majority of the samples tested negative for both cellulose and protein, suggesting

TABLE **12.2** Results of microchemical testing on textile pseudomorphs

No.	Object	Cellulose	Protein	Results
10/22 (SF 110)	Iron fibula with coiled arch	—	—	
10/23 (SF 261)	Iron fibula	—	+?	
10/33 (SF 171)	Iron dress pin with rolled head	+?	+?	Samples taken from object seem to test positive for both cellulose and protein—results difficult to interpret and contamination possible
10/33 (SF 171)	Iron dress pin with rolled head	—	+?	
10/34 (SF 254)	Iron dress pin	—	+?	Protein test may be positive—difficult to interpret
10/35 (SF 390)	Iron dress pin	—	—	
10/37 (SF 046)	Iron dress pin with rolled head	—	—	
10/42 (SF 260)	Iron dress pin	—	—	
10/45 (SF 226)	Iron dress pin with bent back head	—	+	
10/46 (SF 304c)	Iron dress pin	—	—	
10/47 (SF 155)	Iron dress pin	—	+?	Protein test possibly positive—difficult to interpret
10/48 (SF 225)	Iron dress pin	—	+?	
10/50 (SF 111)	Bimetallic pin (Cu alloy and iron)	—	—	
10/63 (SF 259)	Thread found with Cu alloy disk/boss	—	+	
10/84 (SF 349)	Copper alloy headband	—	+	
10/89 (SF 164b)	Iron tubular bead	—	+	
10/99 (SF 124)	Iron tubular bead	—	+	Textile found on exterior
10/99 (SF 124)	Cord inside iron tubular bead	+	—	Sample looked woody—not clear if it is a contaminant or actually part of the cord
10/101 (SF 164a)	Cord inside iron tubular bead	—	+?	
10/118 (SF 263)	Iron spearhead	—		
10/118 (SF 263)	Organic material from spear socket	—	—	

that the pseudomorphs were too mineralized and not enough organic material remained in the sample (Anheuser, personal communication). This meant that a negative result did not necessarily exclude that fiber type as being present in the pseudomorph. Interpretation of tests performed using Naphthol blue-black was often difficult because several of the pseudomorphs did not turn blue as expected. They became slightly green-colored or only faintly blue in areas. Formation of blue crystals underneath or on top of samples, possibly due to drying of the reagent, also made interpretation of the results confusing. For these samples where the result was unclear, only a tentative identification is given.

Of the six positive results obtained, five were for protein, indicating the possible use of animal fiber in the textile. One sample of a pseudomorph taken from what was thought to be a cord from the inside of an iron tubular bead (**10/99** [SF 124]) tested positive for cellulose. Under transmitted light microscopy (50–200x magnification), the sample looked woody in appearance. Another sample from **10/99** (SF 124) was taken to verify the initial positive test result for cellulose. However, the second sample tested negative and did not resemble the previous sample under the microscope. It is not clear if the woody-looking material was actually used for the cord or if it is another organic material, possibly from the burial environment, that became incorporated into the corrosion on the interior of the tube.

SUMMARY AND CONCLUSIONS

The descriptions and images of the pseudomorphs presented provide a preliminary interpretation of the textile remains at Lofkënd and some insight into the technology of weaving taking place during the Late Bronze and Early Iron Ages in southern Albania. Although the textile remains found are extremely small and in poor condition, some technological information concerning common weaving techniques can be obtained.

The majority of the spun fibers are spun in the Z-direction, and the yarns used for weaving tend to be singles. Plain weave seems to be the most common weave found on all the metal objects, and the weave tends to be faced (**Table 12.1**). On two of the pseudomorphs examined (on **10/42** [SF 260] and **10/46** [SF 304a–c], both iron dress pins), a selvedge is preserved for the plain woven textiles, allowing the weave to be identified as weft-faced plain weave. There is also evidence of twill weaving on some of the pieces (**Table 12.1**). Several of the pseudomorphs preserved show differences in the diameters of the yarns in different directions. Clearly a loom was used at Lofkënd for the manufacture of most of the textiles, if not all.

Thread count was extrapolated for several of the pseudomorphs showing a plain weave or oblique interlacing pattern. The average thread count was 25 threads/cm by 18 threads/cm. Four of the thread counts (on **10/24** [SF 162], **10/35** [SF 390], **10/46** [SF 304a–c], and **10/99** [SF 124]) suggest a balanced plain weave. The yarns were densely packed on three of the artifacts (**11/34** [SF 254], **11/37** [SF 046], and **11/45** [SF 226]). The preserved weave structure on two of these artifacts (**10/37** [SF 046] and **10/45** [SF 226]) was identified as possibly weft-faced plain weave, with all three showing many more threads in one direction (the weft) than the other (warp).

Finally, two weave patterns distinct from that found on the other objects examined also were found on several of the objects. Two iron pins (**10/35** [SF 390] and **10/44** [SF 253]) had the remains of a textile made with an open weave pattern where the elements crossed over and under each other as in plain weave. An iron tubular bead (**10/99** [SF 124]) had pseudomorphs that seem to have the remains of two textiles lying on top of each other, one resembling the open weave seen on the pins; the other seems to be a more complex weave that cannot, at the moment, be clearly interpreted.

Of the fibers subjected to microchemical testing, most of the fibers tested that yielded a positive result were protein-based, which could suggest a predominance of animal fibers used for the textiles. However, more testing needs to be done to identify the fibers found on the textiles in order to draw more reliable conclusions about the types of fibers used for textiles/clothing at Lofkënd, especially due to the limitations of microchemical testing and the possibility for contamination of the pseudomorphs during burial.

It is the hope that through the documentation and publication of these objects, further information can be elucidated from the pseudomorphs that have been preserved to provide a more concrete interpretation of the remains as well as provide more information for experts and scholars on Albanian textiles of this time period, for which little information currently exists.

DESCRIPTION OF SELECTED TEXTILE REMAINS

Iron Objects

Fibulae

10/24 (SF 162) Arched fibula

The remains of the pseudomorphs of some yarns were found at the end of the pin, but no twist was visible. Some of these seem to wrap around the pin, but no weave or other structure is identifiable. Similar untwisted fiber ultimates and yarns are found on the spring section of the fibula.

In examining a fragment containing a section of the spring and bow, an unusual corrosion pattern was found. Visible on the surface was iron corrosion that had formed as small raised circles with a globular appearance (**Fig. 12.1**). Some of this corrosion formed linear patterns. This corrosion has a structure similar to a textile and is thought to perhaps follow the weave pattern or individual yarns. Similar types of corrosion were found on a few of the copper alloy objects.

Using the pseudomorphs in this area, estimates of yarn diameters were determined. In direction 1, the "yarns" or woven elements measure approximately 0.5–0.6 mm in diameter and 0.6–0.7 mm in the other direction. The thread count was extrapolated from what was preserved and estimated to be 24 threads/cm (direction 1) by 22 threads/cm (direction 2). It is possible that this was a balanced plain weave textile.

Pins

10/33 (SF 171) Dress pin with rolled head

Fragments of preserved textile are found on several areas on the lower section of the pin near the tip and toward the center of the pin. The textile is poorly preserved in this area, and for most of the yarns, only the lower half is preserved. This makes interpretation of the weave structure difficult.

In one area, there appears to be a series of "bumps" preserved in the corrosion that in raking light resembles a plain weave (**Fig. 12.2**). The angle at which the yarns cross ranges between 80 and 90°. Approximate measurements were taken of yarn diameters in this section of preserved textile, and it appears that there are slightly different yarn diameters used for the elements in each direction. In direction 1, the yarns measure 0.3–0.4 mm, and the yarns in direction 2 average 0.3–0.5 mm. The thread count appears to be 24 threads/cm (direction 1) by 17 threads/cm (direction 2). The difference in thread count in each direction suggests this weave is a faced plain weave. On top of this section of plain weave, there is a yarn that appears to be lying over two or three yarns and positioned in a straight line (**Fig. 12.2**). It is difficult to determine whether this one yarn actually passes below any of the other yarns. It is not clear how the single yarn fits into the plain weave pattern and whether it could be a string/cord attached or related to some other part of the textile, such as a sewn stitch (Barber, personal communication).

Most of the yarns in this area do not seem to have preserved their spin, and in some cases there appears to be Z-spun yarns with a few that are S-spun. It is difficult to say for certain whether yarns spun in two different directions were used for this textile, as has been found with certain decorative weaving techniques or preparation of the warp and weft of wool cloths (Barber 2008; Broholm and Hald 1940:126). Another possibility for the different spin observed is that what remains of the yarns are negative casts of spun yarns, which would appear to have a reverse spin (Julie Unruh, personal communication). The fibers that are spun seem to be spun loosely (15–20°) and are singles.

On another section of the pin closer to the head, several crossed yarns are visible that make a figure-eight pattern (**Fig. 12.3**). There seem to be two threads used to make this element, but it is possible there is a third yarn within the figure-eight that is obscured by corrosion. What is interesting about this thread is that the yarn diameter is 0.6–0.8 mm, larger than the yarns used for the plain weave textile found on the same object. This may support the idea that this is a solitary element, like a cord.

10/35 (SF 390) Dress pin

A weave pattern has been preserved in the textile pseudomorphs located near the tip of this pin. The pattern seems to have a single yarn that passes over and under another yarn as in a plain weave (**Fig. 12.4**). The yarns cross at angles slightly less than 90°, creating diamond- or ovoid-shaped voids, as well as an overall angular direction to the weave. It is similar in some respects to the weave pattern seen on **10/44** (SF 253), but the crossing of the yarns and their spacing are more regular on this pin. The estimated thread count is 13 threads/cm in both directions. Because no selvedge has been preserved, it is difficult to determine whether the cloth was woven as a balanced plain weave or was made through interlacing, creating, for example, a plain, oblique interlaced pattern. All yarns are Z-spun, at a medium to tight angle (30–40°) and are singles.

10/37 (SF 046) Dress pin with rolled head

Textile pseudomorphs were found over several areas of the pin. The yarns appear to be singles and Z-spun. There are many yarns that no longer retain their spin, but those that do seem to be spun at 25–30° (medium range). The yarns used for the cloth are about 0.7–1 mm thick.

Near the central section of the pin, there is a small section of well-preserved textile pseudomorphs showing the yarns, which cross at 90° and were woven using plain weave (**Fig. 12.5**). There is a difference in diameter between the yarns running in each direction. The horizontal yarns visible in **Figure 12.5** range from 0.3 to 0.5 mm (direction 1). The vertical yarns (direction 2) that are not visible in the image are about 0.7 to 1 mm in diameter. The estimated thread count for the textile is 25 threads/cm (direction 1) by 10 threads/cm (direction 2). This makes the weave faced. Since there is no selvedge, the direction to which it is faced cannot be determined, but based on the appearance and the straight direction of the vertical yarns, a suggestion of "weft-faced" has been made.

In some areas, the edge of this textile fragment resembles a twined element rather than the selvedge seen on the sides of a plain woven fabric (**Fig. 12.6**). It is not clear if this appearance is due to distortion of or damage to the edge. Other possibilities for the appearance of this edge could include a decorative edge or twined cord added to the textile, or a starting edge (or heading band) of the cloth. Examples of heading bands with twined or plaited cords passed through the warps of a plain weave textile have been found on archaeological textiles (Broholm and Hald 1940; Burnham 1965:172). However, due to the amount of soil and corrosion in this area, further interpretation could not be done, and it is still unclear as to what is occurring at the edges of the textile.

Below this area where the textile edge is preserved, there is a single two-ply, Z-twisted thread that does not seem to be connected to any of the woven sections (**Fig. 12.7**). The spin on the individual yarns used to make this thread are not so well preserved, but on one yarn it looks as if it may have been spun in the Z-direction. The yarns used for this thread are also much finer than those used for the plain weave. The approximate thickness of the yarns is 0.3–0.4 mm, and they make a thread about 0.7–0.8 mm. It is unclear why this fiber is located in this area or what its function may have been.

Near the preserved section of a faced plain weave is another area with textile pseudomorphs that are not as well preserved. The yarns have been damaged and only the lower half of them remains. What remains appears to be a V-shaped or chevron pattern where the elements cross at about 85° (**Fig. 12.8**). Due to the deteriorated nature of the pseudomorph, it is difficult to see clearly how the warp and weft cross each other. In one area, it appears as if there are two elements crossing over a single one, as seen in herringbone or "zigzag" twill weave (Emery 1994 [1966]:94–105). Other areas, however, appear to have only one element crossing over and under the other, as in plain weave. To complicate the interpretation further, two of the yarns seem to be S-spun (yarn diameter is about 0.5 mm), although most do not have a spin preserved in this area. The weave pattern is tentatively identified as 2/1 twill since it cannot be clearly identified. Because this is a different weave pattern from the pseudomorphs found on other areas of the pin, this could suggest that two different textiles were present or that it was a textile made using different weave patterns.

On the same pin fragment as the V-shaped elements, near the break edge toward the center of the pin, there is a series of fibers or yarns that appear to wrap around the circumference of the pin (**Fig. 12.9**). No spin or weave pattern is visible. Interpretation of why the pseudomorphs look this way has not been possible. It is not clear if this is a thread or strand that was wrapped around the pin. Another explanation could be that this is the result of distorted or poorly preserved textile pseudomorphs. It could also be possible that these pseudomorphs represent the appearance of the cloth after the pin pierced through it. Whatever the case, these similar "wrapped" yarns or fibers are found on other pins from the tumulus, such as **10/34** (SF 254), **10/38** (SF 116), **10/40** (SF 018), **10/41** (SF 414), and the pin of fibula **10/24** (SF 162).

10/42 (SF 260) Dress pin

Near the tip of the pin, there is a fragment of a well-preserved textile pseudomorph (**Fig. 12.10**). The yarns are spun in the Z-direction, at 25–35° (medium to tight range). The weave was identified as a plain weave and also has a selvedge preserved (identified by the white arrows). The diameter of the yarns differs slightly in each direction. The weft threads are 0.6–0.7 mm in diameter and the warps are 0.7–0.9 mm. The estimated thread count is 22 wefts/cm and 18 warps/cm, making this a weft-faced plain weave.

10/44 (SF 253) Dress pin with spherical head (resembling an animal head)

On one area near the center of the pin, there are the remains of a textile with yarns that are singles and spun in the Z-direction at roughly 30–40° (medium to tight spin). Some of the yarns cross at an angle slightly less than 90°, and the weave pattern creates a triangular-shaped void between some of the warps and wefts (**Fig. 12.11**). Crossed yarns with triangular voids are also found nearby in one other area. Examination of the individual yarns did indeed show that in areas it is possible to see one yarn crossing over and under another, though the appearance of the weave made the determination of the technique difficult.

Near the woven yarns described above, originating from one end of the pseudomorphs with the triangular voids, was a feature that appeared almost like a cord or strand. Upon closer examination, it resembles the broken or cut edges of the plain weave textiles found on **10/37** (SF 046) and **10/45** (SF 226), suggesting a plain weave pattern for this textile (**Fig. 12.12**). However, it cannot be known for certain whether the entire textile was plain weave or if different techniques or the remains from different textiles are found on this pin. For example, oblique interlacing (Emery 1994 [1966]:62–63) produces a weave where yarns travel over one and under the other as in plain weave, and the yarns cross at angles less than 90°. The estimated thread count is 30 threads/cm by 23 threads/cm. Without preservation of more of the textile, including a selvedge, identification of all the techniques used for producing this cloth cannot be known for certain.

An unusual knob or circular protrusion is visible on one side of the pin, directly below the head of the pin (**Fig. 12.13**). This area initially appeared to be just an area of voluminous iron corrosion, but after closer examination, the protrusion appears to have striations around it. These have created a highly linear pattern of corrosion that seems to wrap around the raised area. Towards the top of this knob there are impressions of yarns that appear to be spun in the Z-direction. Although the nature of these striations and this protrusion cannot be determined, it is possible that its appearance is due to the mineralization of a textile fragment in this area.

10/45 (SF 226) Dress pin with bent-back head

Visible on the central section of the pin and toward the tip are two areas with textile pseudomorphs preserved. The cloth preserved on the central section is in much better condition and appears to be a faced plain weave (**Fig. 12.14**). The yarns running in the horizontal direction in the figure (direction 1) measure 0.3–0.4 mm. The yarns running perpendicular to these (direction 2) are 0.5–0.8 mm in diameter. The thread counts were estimated as 35 threads/cm (direction 1) by 15 threads/cm (direction 2), supporting the designation of the weave as faced. The yarns used were singles and Z-spun, with the spin at about 30° (medium range).

A damaged edge of the textile shows the "over and under" crossing of the weft over the straighter warps that appear ovoid in section. Similar to the pseudomorph found on **10/37** (SF 046), this textile could have possibly been weft-faced.

At the very end of the pin, at the tip, there are some yarns that run partially around the pin and then form a pattern where the warp and weft yarns appear to cross at about 45° angles, creating a triangular or ovoid space (**Fig. 12.15**). This appears to be different from the other weave patterns seen on the pin where faced plain weave was found. It is also different in diameter (0.6–0.8 mm). It is possible that these yarns are similar to the "figure-eight" element seen on pin **10/33** (SF 171) (**Fig. 12.3**).

10/46 (SF 304a–c) Dress pin

This object consists of three fragments, two of which have been joined together (A and B). Fragments A and C contain pseudomorphs of a textile.

On fragment A, near the join edge with fragment B, there is a small area of textile preserved which has a plain weave (**Fig. 12.16**). A portion of the textile has been pushed up due to the formation of corrosion underneath, but there appears to be a selvedge at the end of this raised section.

Measurements of the warp and weft elements show that they are similar in diameter, measuring about 0.5–0.7 mm. The estimated thread count is 21 wefts/cm by 18 warps/cm and is a plain weave. The majority of the yarns do not retain their spin, except on fragment C, where there are several yarns preserved that show a Z-spin. The yarns on this fragment appear loosely spun, at about 10–15°.

10/47 (SF 155) Dress pin

Preserved on this pin are the pseudomorphs of a textile pattern on the central section of the pin and another on the flat break end, where different weaving techniques may have been employed. Many of the yarns are poorly preserved and only the lower halves remain. Most of these yarns do not retain their spin, and some appear to turn slightly in an S-direction, although it is not clear whether this represents the true spin or just some type of distortion that occurred during burial or deterioration. On a few areas of the pin, where whole yarns are preserved, they are clearly Z-spun with a spin only about 10–20°. All the yarns found are singles.

On the central section of the pin, the weave pattern is preserved as a series of interlocking V-shapes, with the warp and weft yarns crossing at about 85° (**Fig. 12.17**). A very small section of this weave remains, but there are two yarns, probably the wefts, crossing over two yarns that seem to be warps. Due to the number of and shape made by the yarns, it is possible that this textile is a 2/2 twill. However, with so little preserved, it is very difficult to say for certain.

Near the break edge, the yarns preserved also make a series of V-shapes but cross at a more acute angle (ranging from 10 to 25°) (**Fig. 12.18**). Only one small area of this pattern is preserved, and therefore it is difficult to determine whether this weave structure extends across a larger piece of cloth or whether this is a decorative section or some kind of cordage. Based on what remains, the pseudomorph could represent some type of plaited strip or cord.

10/50 (SF 111) Bimetallic (bronze and iron) pin

The pseudomorphs found in two areas on this pin represent the best preserved examples found on any of the metallic artifacts from the Lofkënd tumulus (**Fig. 12.19**). The yarns used are Z-spun singles. The degree of spin is approximately 20–30° (medium range), though there are yarns where the spin is much looser or not preserved. The warp and weft yarns seem to cross at 90°. The yarn diameter ranges from 0.6 to 0.7 mm in direction 1 and 0.6 to 0.8 mm in direction 2. The estimated thread count is 20 threads/cm for direction 1 and 13 threads/cm in direction 2, making this a faced, plain weave textile.

Beads

10/99 (SF 124) Tubular bead

On a tubular-shaped object thought to be a bead, textile pseudomorphs were preserved on most of the exterior surface. Yarns were found spun in two directions, and two overlying sections of the pseudomorphs each seem to have different weave structures, possibly suggesting two different textiles were preserved on this object.

On two areas of the tube, pseudomorphs with an unusual pattern were observed (**Fig. 12.20**). This weave structure has several yarns running at about an 80° angle over thicker yarns that are straight and run parallel to each other. The yarns running at an angle are in poor condition, and in many cases the whole yarn is not preserved. The preservation of the straighter yarns is better, and it appears that the fibers were spun in an S-direction. The straighter yarns, which may possibly be the warp, measure approximately 0.4–0.6 mm in diameter as opposed to the angled yarns, possibly the weft, that are 0.2–0.4 mm in diameter. Thread counts were estimated at 28 wefts/cm by 18 warps/cm.

In between two sections of the pattern described above, below the unusual weave, there is an area where a different weave structure seems to be preserved (**Fig. 12.21**). The yarns used are singles, Z-spun, at an angle of about 20° (medium range). The pattern resembles the weave structure seen on **10/35** (SF 390) (**Fig. 12.18**) where the crossing yarns create an open weave of ovoid- or diamond-shaped elements. A few yarns do appear as if they cross over and under each other as in a plain weave pattern, with yarns crossing at around 90°. The estimated thread count is 20 threads/cm by 19 threads/ cm, making it a balanced plain weave.

The relationship between the two different textile patterns observed on this object is unclear. The current thought is that the two different patterns could be due to the preservation of two different pieces of woven cloth.

Bronze Objects

No pseudomorphs of woven cloth were found on any of the bronzes. However, a few objects do have some evidence of fibers or yarns preserved on the objects, such as the pin of spectacle fibula **10/17** (SF 434), which has some unspun yarns wrapped around it (**Fig. 12.22**).

As stated previously, some of the bronze objects had unusual corrosion patterns that were somewhat linear and globular in appearance, which mimics a textile (**Fig. 12.23**). As with the iron fibula **10/24** (SF 162), this corrosion pattern could not be attributed to any particular weave pattern.

10/63 (SF 259) Perforated disk/boss

Although no textile pseudomorphs were found on the artifact, a single Z-spun yarn was found among the fragments of the disk when excavated (**Fig. 12.24**). The yarn, which is colored green from copper corrosion products, is very well preserved and was spun tightly (35–45°). The diameter of this yarn is about 1.8 mm, much thicker than any of the yarns found on any of the textile pseudomorphs. The diameter of the yarn is similar to the cords found on the iron beads (**10/89** [SF 164b] and **10/101** [SF 164a]) discussed in the section below. It is not clear what the relation of this yarn is to the disk or boss, but perhaps due to its diameter it is a fragment of cordage.

CORDS AND STRAPS

Several objects excavated from the tumulus were found with pseudomorphs from what has been identified as cordage or straps based on their appearance and the type of object they were found on.

Iron Objects

Beads

10/99 SF 124 Tubular bead

Examination of the interior of the tube revealed what looked like a small cord running halfway down the interior surface (**Fig. 12.25**). The cord appears to be made of at least two yarns. The fibers seem to have been Z-spun at about 25–30° (medium range), possibly to make a two-ply, Z-twisted thread. The diameters of the individual yarns are about 0.2–0.4 mm and are twisted to make a 0.5–0.7-mm thread.

10/89 (SF 164b) and **10/101** (SF 164a) Tubular beads

These two beads, found in close proximity to each other, both contain the remnants of a cord preserved on their interior. The cord runs the entire length of each bead and extends slightly out of one end of **10/89** (SF 164b). The cord on both objects is mineralized and covered with thick corrosion in areas, obscuring some features. Based on the similarities of the cordage and the proximity of the two beads to each other, these beads could have been strung together prior to burial.

The cord inside **10/101** (SF 164a) is not as well preserved as **10/89** (SF 164b) (**Fig. 12.26**). The cord is covered with a smooth layer of iron corrosion. The individual fibers are not visible and the spin cannot be determined. It appears that more than one yarn was used to make a Z-twisted cord, but it is not clear if it is two or three plied. The individual yarns in the cord measure about 1.2–1.3 mm and have been twisted to make a 2–2.3-mm thread.

The cord inside **10/89** (SF 164b) is better preserved, but corrosion again obscures the direction of spin and the number of yarns used to make the cord (**Fig. 12.27**). Like the cord in **10/101** (SF 164a), it appears to be made from two- or three-ply yarns that were Z-twisted. The cord preserved in this bead is similar in diameter to that in **10/101** (SF 164A), where the threads are about 1.4–1.6 mm in diameter twisted to make a 2.1–2.3-mm thread.

Bronze Objects

Headbands

10/84 (SF 349) Headband

This headband is made from a strip of bronze sheet that overlaps at the ends to form a closed shape. At the point where the two metal pieces overlap, there is a series of pseudomorphs in the shape of a flat strap or cord that wraps around the headband at angles (**Fig. 12.28**). On the front of the headband, a section of this strap seems to have been preserved intact, providing its thickness (1.7 mm) and width (3.6 mm). The straps seem to have extended across the entire section of overlapping metal, an area about 78.7 mm long.

On one edge of the headband, a rounded mass of fibers is visible (**Fig. 12.29**). Here it looks as though there may be some non-mineralized fibers still preserved. The fibers appear white and relatively flat, some with rounded, slightly pointed ends. Under examination with a binocular microscope (45x magnification), no features are visible on the surface of the fibers. It is not clear what this bundle of fibers represents, but it is possible this could be where the strap was tied off to make a knot.

Wood Pseudomorphs

Pseudomorphs of wood were found only on three iron objects. On two of the objects, the wood corresponds to the area where the object is thought to have had a handle or would be the point of attachment to something wooden. In the third instance, the function of the object is unknown and therefore the purpose of the wood is unclear. For this latter object, it is possible that the wood pseudomorphs are the remains of some woody plant in the burial environment, but it is still included in this description since the material is found on a large portion of the object. In all cases, the wood was too mineralized to allow for identification of the wood species or of any other diagnostic features.

10/121 (SF 019) Iron knife

A small fragment of minerally preserved wood was found on the area that corresponds to the tang and was the material used for the handle (**Fig. 12.30**). The grain of the wood is visible.

10/118 (SF 263) Iron spearhead

Wood pseudomorphs were found on the interior of the socket of the spear (**Fig. 12.31**). The spear was set onto a wooden shaft or handle so that it could be used.

10/133 (SF 384) Unidentified fragments

Although the object that these iron fragments came from could not be identified, they contain pseudomorphs of what is thought to be wood. The preservation of the pseudomorphs is not very good, but the linear, striated appearance of the material resembles the grain of wood (**Fig. 12.32**).

A section of the two largest fragments of **10/133** (SF 384) also contained preserved pseudomorphs of a material with what looked like a cellular structure (**Fig. 12.32**). It resembles either the cell structure of wood or cancellous bone. Conservators consulted with physical anthropologists and a paleobotanist on the project to help make the determination. Although initially thought be cancellous bone, the linear, fibrous-looking areas were thought to more closely resemble wood. What the function of the wood was in relation to the object is unknown.

Acknowledgments

The author would like to the thank the conservators of the Lofkënd Archaeological Project, Sarah Cleary, Alma Bardho, Caitlin O'Grady, and Allison Lewis, for their initial identification of pseudomorphs on many of the objects described in this chapter. The author is very grateful to those colleagues and experts, especially Elizabeth Barber and Julie Unruh, who provided much needed guidance, suggestions, and edits.

CHAPTER 13

THE LITHIC ARTIFACTS

Jamie D. Aprile

INTRODUCTION

Chipped stone artifacts were recovered from both graves and fill contexts in the Lofkënd tumulus, as well as in topsoil, and they date to a wide range of periods including the Middle Paleolithic, Upper Paleolithic, Mesolithic, Neolithic, and Bronze Age. Finding this type of material was not completely unexpected, since lithic artifacts have been reported from many tumuli in Albania, including those at Vajzë (Prendi 1957), Pazhok (Bodinaku 1982), Shkrel (Jubani 1995), and Apollonia (Amore 2010:621–626; Bodinaku 2001–2002), although in the earlier studies, only a very small sample (usually between 2 and 10 examples) of the lithic artifacts were described in the text and illustrated. Prendi (1957:92), and more recently Jubani (1995: 72) asserted that the chipped stone represented intrusions in the tumulus fill because the objects typologically matched earlier periods than the burials. In the case of the Lofkënd tumulus, this appears to be the case as well; however, multiple depositional processes likely account for their appearance.

Ninety-four tools (including cores) were identified and catalogued, while an additional 495 pieces of debitage were collected, for a total of 589 chipped stone objects. Forty-one tools and 158 debitage fragments, approximately 34% of the total collection, were recovered from surface and topsoil contexts. Forty-four tools and 244 debitage fragments, approximately 49% of the assemblage, were found in tumulus and pit fills. An additional nine pieces of debitage were recovered from excavation of the soft, sandy bedrock below the tumulus, hinting at the source for some of these objects in the local soils. Eight graves in the tumulus yielded a total of nine tools, while 84 fragments of debitage were recovered from 42 graves; thus, only approximately 16% of the total lithic artifacts in the collection were recovered from tomb contexts. Based on the following analysis of these tools and the soils in which they were recovered, these artifacts most likely arrived in the graves through the soils used to construct the mound, the action of building it, or repeated interment activities.

Despite the large number of chipped stone objects, their presence does not provide definitive chronological information regarding the construction of the tumulus. Over half (57%) of the tools catalogued could not be assigned a reliable date, and many were diagnostic to a range of periods sometimes extending over several thousand years. Because of this, these objects do not represent pieces that can confidently be used to date the tumulus or any individual burial within it. They more likely represent a range of dates encompassing both the original deposition of the soils later used to construct the mound and the span of time during which the tumulus was used for burials. Objects that date to the Neolithic and/or Bronze Age, such as the blades and denticulated sickle elements, are of a standard, highly functional shape that could have been manufactured and used either before or after the introduction of metal tools. These objects could be contemporaneous with the construction of the tumulus in the Late Bronze Age. Given the appearance of daub (see Papadopoulos, Chapter 14) and the relative absence of faunal bones and carbonized floral remains (see Marston, Chapter 16.1) in the tumulus fill, it is difficult to determine whether the soils used to construct the tumulus were intentionally brought from settlement contexts. The character of sites found in the intensive survey of the vicinity and the distribution of lithic artifacts in the tumulus suggest that tools of the Neolithic/Bronze Age periods arrived in the mound through interment activities or settlement soil packages, while those from the Paleolithic and Mesolithic were collected with local clay-rich soils used to stabilize the tumulus.

Paleolithic and Mesolithic Tools and Flakes

The 18 Paleolithic and Mesolithic objects recovered from the tumulus most likely arrived there in local soils used to construct the tumulus. Ten of the artifacts (approximately 56%) were found in surface and topsoil units, and eight (approximately 44%) were from tumulus fill contexts. Because the soil volume of the tumulus fill is much greater than that of the topsoil, this indicates that there were relatively more tools of the Paleolithic and Mesolithic in the topsoil than in the rest of the tumulus. The archaeological survey conducted around the tumulus (see Aprile, Chapter 18) provided ample evidence that the area was occupied during the Paleolithic period, although the assemblage has not yet been fully studied and catalogued. Lithic artifacts were recovered in survey tracts within 200 m of the mound, and many others were found diffusely scattered over the landscape as well as in several dense clusters labeled as sites. These surface finds almost certainly resulted from disturbance and mixing in surface sediments caused by erosion, agricultural use, and terracing activities, but the appearance of Site S004 eroding from the cut-slope along a terrace clearly indicates the existence of buried deposits from the Paleolithic period in the vicinity.

There was some correspondence between the survey finds, the tumulus lithic artifacts, soils in the vicinity, and tumulus construction. The lithic finds from the archaeological survey generally came from soil unit 1 zones (see Foss, Chapter 16.4), shale-derived soils with a high clay content. Foss reports that this parent material was collected from nearby the tumulus and placed over the mound and in bands within the fill to help stabilize the eroded Miocene sandstone component of the tumulus fill, which was also available in the vicinity, including on the promontory where the tumulus was built. Eight of the 13 identifiable Middle and Upper Paleolithic/ Mesolithic tools and flakes were recovered from the surface (**13/82**, **13/89**), topsoil units (**13/83**, **13/23**, **13/90**, **13/93**), and the fills of Tombs LVI (43) (**13/22**) and LXXVIII (13) (**13/88**), which were both partially overlain by topsoil. Two other Mesolithic objects, a perçoir/scraper and a core rejuvenation flake (**13/34** and **13/78**, respectively), were also recovered from the topsoil of Trench 2. These objects most likely arrived in these contexts with the soil brought to create the cap of clay-rich soil used to consolidate the mound and prevent erosion.

The tumulus fill units where Paleolithic and Mesolithic objects were recovered, however, had some variation in their soil contents. Several identified stratigraphic units of more clay-rich soils appear to be associated with bands of Foss's soil unit 1 laid in the tumulus fill to stabilize the sandy soil used to construct much of the mound. SU 1.0039 contained two objects (**13/19**, **13/36**) and displayed a similar color and proportion of clay to sand compared to the topsoil. During excavation, this unit appeared to be an old tumulus surface that created a narrow band between the modern topsoil and the sandy, lighter-colored tumulus fill of SU 1.0007 and 1.0070 in Trench 1. This surface could have been buried during reuse of the tumulus in that sector in historical times, since the site has not been stable for long enough to allow natural horizonation to occur (Foss, Chapter 16.4). Thus, the artifacts found in this context are also likely a result of ancient tumulus construction activities that brought locally available clay-rich soil to the mound. SU 2.0118 also consisted of more clay-rich soil in the midst of the tumulus fill, so the Paleolithic artifact (**13/92**) from this unit most likely arrived in the mound from nearby soil unit 1 deposits. A Mesolithic perçoir (**13/31**) was also recovered from SU 1.0067, another soil unit 1 clay-rich stabilizing deposit in the tumulus fill; however, this unit also yielded a Neolithic/Early Bronze Age blade. A Mesolithic bifacial arrowhead with a tranchet edge (**13/2**) was recovered from another clay-rich deposit in the fill, SU 2.0228. This is a rather unusual piece, similar to ones found in Upper Mesolithic and Neolithic levels at Franchthi Cave in Greece (cf. Perlès 1990: fig. 16, nos. 12–13), although those may have different manufacturing techniques (C. Runnels, personal communication).

The remaining Paleolithic artifacts from the tumulus fill appeared in three soil deposits from stratigraphically lower parts of the tumulus. The lighter-colored, sandier fill appears to have been derived from soil unit 2, generally a yellowish brown sandy loam that eroded from local soft sandstone outcrops, including the one on which the tumulus rests. Stratigraphically, the Paleolithic artifacts from three sandy fill units may thus be associated with the earliest construction activities for the mound, when early objects that were already present in the landscape could have been more easily mixed into the tumulus fill. SU 8/2.0204 (**13/91**) was a characteristic example of this type of fill and was deposited directly over the bedrock at the base of the mound. Two other stratigraphic units, SU 2.0532 (**13/94**) and 2.0474 (**13/81**), were deposited over graves and pits that had been dug down into the bedrock. No Mesolithic objects occurred in these lower deposits.

The diagnostic Paleolithic and Mesolithic artifacts catalogued from the tumulus tended to be found stratigraphically closer to the bedrock, in the topsoil, and in zones of clayey soil in the tumulus fill. This pattern suggests that the older tools in the mound arrived there in local soils used to construct the mound, and not through purposeful inclusion. It also suggests that more Mesolithic material may be identified in the survey collection during future analyses. This does not preclude the possibility of mixing between tumulus deposits originating in soil units 1 and 2. Because of the relatively limited distribution of very early artifacts in the mound and the identification of those objects on the surface in the vicinity, it appears to be the case that Paleolithic and Mesolithic objects arrived in the tumulus accidentally rather than as purposeful inclusions with symbolic significance. The same cannot be said for Neolithic and Bronze Age lithic artifacts, although the relative density of material between the topsoil and tumulus fill follows a similar pattern.

NEOLITHIC AND BRONZE AGE TOOLS

Many tools diagnostic to the Neolithic and Bronze Age were found in the tumulus, but the origin of these objects remains unclear. Artifacts from these periods have not yet been identified in the preliminary analysis of the Lofkënd survey assemblage, so it appears that those found in the tumulus did not arrive there through gathering soil from the immediate surroundings. This does not completely rule out the possibility of a short- or long-term prehistoric settlement in the area. As indicated by the soil survey, erosion has been a significant factor in shaping the modern landscape and surface deposits in the vicinity. The appearance of Paleolithic objects on the surface may indicate that cultural deposition from later periods simply has not survived in situ. Furthermore, it is possible that the nearby modern communities of Gjinoqara and Ngrançija were built atop small, local prehistoric settlements, and relatively recent construction activities could have destroyed or obscured their presence. Non-agricultural land was not surveyed within the villages, so the survey data cannot provide certainty regarding the presence or absence of prehistoric settlements. The total lack of lithic blades recovered from the survey strongly suggests, however, that the community that buried people in the tumulus did not live and work in the immediate vicinity.

The distribution of diagnostic Neolithic and Bronze Age tools was noted across different types of soils in the mound. A total of 22 objects dating to the Neolithic and/or Bronze Age were identified in the assemblage. Of those, 2 (9%) were found in graves, 9 (41%) were recovered from topsoil, and 11 (50%) were found in the tumulus fill units. Considering that the greater volume of soil in the tumulus lies in the fill, this indicates that tools of the Neolithic and Bronze Age were concentrated in a greater density on the surface and in topsoil. Tools of these periods are not more closely associated with soil unit 1 or 2 deposits in the tumulus as was the case with the older Paleolithic and Mesolithic tools. They were frequently recovered from mixed or sandy stratigraphic units, suggesting that they were deposited in the tumulus during the course of interment activities.

The Neolithic objects identified in the tumulus came from a variety of contexts, but relatively few were found outside of the topsoil and tumulus fill. Only two pieces were recovered from a grave and a pit, both near the bottom of the mound. A truncation with lateral retouch (**13/14**) was the only piece found in a tomb in the fill of Tomb XXXI (86). A Neolithic bladelet with use-wear (**13/15**) was also found deep in the tumulus inside a pit dug into the soft bedrock in SU 2.0474, along with a Paleolithic bec (**13/81**, mentioned above) and a non-diagnostic core (**13/65**). Since these two Neolithic objects were found near the bottom of the tumulus, they could be associated with relatively early construction and interment activities. Their stratigraphic position deep in the tumulus indicates that the inclusion of artifacts not present in the landscape in the immediate vicinity began early in the history of the mound.

In the clayey topsoil, Neolithic and Bronze Age objects were found in many of the same units as Paleolithic and Mesolithic artifacts. On the surface, a retouched Neolithic blade segment with polish (**13/3**) was found. In the topsoil, both complete and broken blades (**13/9**, **13/12**, **13/13**) dating to the Neolithic were recovered. Three additional objects found in the topsoil could not be assigned a date more specific than Neolithic/Early Bronze Age. Blades in that category were recovered from Trenches 3 (**13/4**) and 4 (**13/5**). A Neolithic/Early Bronze Age denticulate sickle element (**13/32**) was also found in the topsoil of Trench 1, while Neolithic and/or Bronze Age core-shaping elements (**13/75**, **13/76**) were recovered from the topsoil in Trenches 4 and 6 (baulk).

Ten artifacts dating to the Neolithic and Bronze Age were also recovered from tumulus fill contexts of both clay-rich and sandy soils. One tool was recovered from the more clay-rich deposit SU 1.0067, including a Neolithic/Early Bronze Age broken blade (**13/6**), along with the Mesolithic perçoir (**13/31**) discussed above. Several tools were recovered from contexts in Trench 2 that were highly variable in their color and texture. A Neolithic bladelet with a scraper and use-wear (**13/16**) as well as a Neolithic/Bronze Age core-shaping element (**13/77**) were recovered from the fill unit SU 2.0202, which was variable yet generally more clayey than sandy. A large Late Neolithic/Bronze Age blade (**13/10**) was found in SU 2.0288, another variable deposit consisting of clay-rich and pure sand deposits. A Neolithic retouched blade with polish (**13/11**) was recovered from SU 2.0297, a variable deposit similar to 2.0288, with zones of clay and pure sand. In the case of these variable Trench 2 stratigraphic units, it is impossible to tell whether the diagnostic lithic artifacts were recovered from a more clayey or sandy zone within the defined context, but the degree of variation might suggest that these stratigraphic units were disturbed repeatedly over the course of the use-life of the mound for burials.

Two stratigraphic units from Trench 4 near the top of the tumulus also produced Neolithic/Bronze Age artifacts. A Neolithic/Early Bronze Age denticulate sickle element (**13/33**) from SU 4.0035 and a Neolithic/Bronze Age core-shaping element with use-wear (**13/74**) from SU 4.0079 were both found in deposits with a relatively even mix of clay and sand, suggesting they may have been mixed deposits of soil units 1 and 2—not unexpected, given that the two soil units appeared stratigraphically higher in the tumulus, where more frequent disturbance would occur during interments.

In sandy contexts, a Neolithic/Early Bronze Age broken blade (**13/7**) appears in sandy tumulus fill in SU 1.0070, while a blade segment with use-wear (**13/8**) appeared in the similar deposit SU 4.0204. Finally, a broken blade with a utilized notch (**13/17**) was found in the sandy yet dark-colored SU 4.0286 near a portion of the single-course stone "wall" that stretched between Trenches 1 and 4 (see Chapter 2).

Neolithic and Bronze Age diagnostic artifacts most likely arrived in the tumulus deposits through the activities of the people building the mound and interring members of their community. They were not found on the surface in the vicinity except on the mound itself. The lack of blade cores indicates that the tools arrived fully formed and were abandoned, although the presence of core-shaping elements indicates that a limited number of blades might have been produced at the mound from prepared cores during temporary use of the locale for burials. The very limited number found in graves suggests that they were not intended as grave goods but rather reflect a by-product of construction or interment activities. Blades and denticulates, especially those with polish, could have been used to cut grasses and other plants that had grown over the top of the tumulus between interments (cf. the tumulus surface between seasons of excavation and after its mud brick reconstruction when thick grasses and weeds grew over the mound; see Chapter 22). Because the community that buried people at Lofkënd did not live in the immediate vicinity, they may have had to travel some distance and spend time at the location during funeral activities, producing some limited and ad hoc tools and other fugitive habitation remains. The detritus remaining from those camps may then have been incorporated into rebuilding the tumulus either intentionally or inadvertently after excavating a grave and performing funeral rituals. This is further supported by the presence of a greater density of tools found in the topsoil units.

NON-DIAGNOSTIC TOOLS

A total of 54 tools catalogued from the assemblage were not diagnostic enough to be assigned to a specific chronological period. Of these, 23 (approximately 43%) were found in surface and topsoil contexts. The remaining 31 tools (approximately 57%) were recovered from tumulus and grave fills; removing the 6 tools (approximately 11% of the total of non-diagnostic tools) found in five graves leaves 25 tools in tumulus fill contexts (approximately 46% of the total non-diagnostic tools). This follows the general pattern of greater artifact density in the topsoil compared to tumulus fills observed for datable lithic objects.

Many of these tools could be morphologically described and functionally identified in the catalogue, but a number of factors prevented them from being chronologically recognizable in mixed deposits. Those factors included breakage of diagnostic elements and the generally non-diagnostic quality of some examples of functional categories of tools used for millennia, such as perçoirs and scrapers. All of the cores found in the tumulus belong to this category of non-diagnostic tools. They were amorphous and multidirectional with apparently little care taken to predetermine the shape of flakes produced, suggesting that they were used ad hoc or as part of a tradition very difficult to detect in a mixed deposit.

Surface and topsoil contexts held a variety of non-diagnostic tools. The surface collection at the tumulus yielded a core fragment (**13/59**) and flake with a notch and distal retouch (**13/37**). Topsoil units in Trench 1 produced eight catalogued tools, including three cores or core fragments (**13/61**, **13/69**, **13/71**), two worked flakes (**13/44**, **13/49**), one perçoir (**13/30**), one scraper (**13/27**), and one broken tool (**13/38**). In Trench 2, non-diagnostic tools included four cores (**13/57**, **13/53**, **13/72**, **13/64**), two perçoirs (**13/28**, **13/29**), and three scrapers (**13/18**, **13/24**, **13/25**). Only three tools were recovered from the topsoil of Trench 4, including one core (**13/62**), one scraper (**13/21**), and one notch tool (**13/86**). Baulk Trench 7 also produced a core (**13/60**).

Six non-diagnostic tools were found in tomb fills in the tumulus. In Tomb LXXIV (2), one core (**13/56**) was found in a clayey fill. In Tomb XLII (59), two retouched flakes (**13/47**, **13/39**) were recovered from another clayey fill that also contained calcareous material most likely derived from local bedrock deposits. Tomb I (64) produced a scraper with a notch formed on an exhausted core (**13/79**) from its fill. A chert fragment with utilized edges (**13/85**) was recovered from Tomb III (81) in an equally clayey and sandy fill located near the edge of the tumulus. Tomb XXXII (89) produced another combination tool with a scraper and notch (**13/35**) shaped on a flake in a slightly sandier fill unit also close to the edge of the tumulus. Tombs XLII (59) and LXXXIV (2) were located near topsoil units, close to the surface in the case of Tomb LXXXIV (2), and near the west edge of the tumulus in the case of Tomb XLII (59). Tombs containing tools were generally associated with the base and edges of the tumulus, where clayey soil from the topsoil and soil unit 1 stabilizing layers would have been more likely to introduce lithic material during interment activities. This could have been the result of tools arriving in these deposits both from the landscape and from the agency of the community using the tumulus.

The 25 non-diagnostic tools from tumulus fill contexts also displayed variety. About half of these were recovered from variable soil contexts in Trench 2 containing patches of clay and sand in dark brown to yellowish and even red-brown colors. Tools found in these stratigraphic units included four cores (**13/54**, **13/58**, **13/67**, **13/73**), five worked flakes (**13/41**, **13/42**, **13/43**, **13/45**, **13/46**), one scraper (**13/80**), and a fragment with a notch and denticulate (**13/84**). In a pit dug into bedrock, a core (**13/65**) was found. Clayey or variable stratigraphic units also produced tools in Trench 4, where a retouched flake (**13/50**) was recovered, and in baulk Trench 8, where a core (**13/55**) was found. In more sandy, yellowish to olive brown tumulus fill contexts, 11 non-diagnostic tools were recovered. Trench 1 produced several, including a core (**13/68**), a utilized flake (**13/51**), a utilized chunk (**13/87**), and a scraper (**13/20**). In a pit dug into bedrock in Trench 1, one utilized flake (**13/40**) was found. In Trench 4, non-diagnostic tools included cores (**13/63**, **13/70**), worked flakes (**13/48**, **13/50**), and a scraper (**13/26**). Baulk Trench 6 yielded another core (**13/66**). The prevalence of tools found in Trench 2 is not necessarily an important pattern. Early in the excavation process, Trenches 2 and 3 were combined into a single area. This suggests that the non-diagnostic tools were relatively evenly distributed between the north and south portions of the tumulus fills.

Debitage

The debitage in the tumulus was classified into four basic morphological categories: cortical flakes, non-cortical flakes, chunks, and shatter. Objects assigned small find (SF) numbers in the field that were found to be debitage were tallied with the counts and weights of the SU bulk finds of these types of objects. A grand total of 495 pieces of debitage totaling 2.023 kg were recovered from the tumulus, including 104 cortical flakes (0.696 kg), 95 non-cortical flakes (0.247 kg), 242 chunks (1.058 kg), and 54 pieces of shatter (0.022 kg). Forty-one graves contained debitage, including 12 cortical flakes, 20 non-cortical flakes, 31 chunks, and 21 pieces of shatter, for a total of 84 pieces. Three non-cortical flakes and six chunks were recovered from excavations of the soft bedrock below the tumulus. Pit fills (SU 2.0474, 2.0543) contained one cortical flake, one non-cortical flake, and two chunks. Topsoil contexts in the tumulus contained 37 cortical flakes, 30 non-cortical flakes, 75 chunks, and 16 pieces of shatter for a total of 158 pieces of debitage. Tumulus fill contexts contained 53 cortical flakes, 40 non-cortical flakes, 124 chunks, and only 17 pieces of shatter, for a total of 234 pieces of debitage.

Cortical flakes and chunks generally denote earlier stages of the manufacturing process, but because the number of cortical and non-cortical flakes is relatively similar, the larger number of chunks does not necessarily indicate that cores were being directly utilized at the mound for the purpose of producing tools. This pattern could have easily resulted from random fractures that occurred when stone or metal tools were used to collect soil to construct the mound or to excavate for interments. Only approximately 11% of the total debitage was comprised of pieces of shatter. This could indicate that those smaller pieces were very frequently missed during modern excavation, or that more refined flint knapping was not being conducted at the site and the majority of lithic fragments that appear in the mound were the result of natural deposition from soils in the surrounding vicinity, which was a focus of Paleoithic and possibly Mesolithic occupation. Considering the fact that lithic collection was generally good and yielded a large assemblage, it is reasonable to suggest the latter is true. Patterning in the debitage distribution when contexts are sorted by clayey and sandy texture mirrors the patterning found in the Paleolithic/Mesolithic tools, supporting this interpretation. Debitage occurring in the sandier fill units could thus have arrived in the tumulus both through mixing between soil units 1 and 2 and through interment activities at the mound in ancient times. The larger number of chunks suggests natural deposition of these pieces is more likely, as shatter would be more likely if tools were being produced on site or fractured during excavation.

When the tumulus contexts are sorted between clayey and sandy textures, the frequency patterns observed among the Paleolithic and Mesolithic tools are supported. In the clayey contexts, including one bedrock context, all of the topsoil contexts, 24 graves, one pit fill, and 25 tumulus fill contexts, a total of 345 pieces of debitage (totaling 1.478 kg) were collected. Seventy-four cortical flakes, 75 non-cortical flakes, 160 chunks, and 36 pieces of shatter were found in these contexts. In the sandy contexts, including two bedrock contexts, the fill of 15 graves, one pit fill, and 24 tumulus fill contexts, a total of 125 pieces of debitage (totaling 0.433 kg) were collected. Several contexts were not able to be assigned to clayey or sandy soil textures, but they represent a very small number of the total pieces of debitage. Approximately 70% of the pieces of debitage were recovered from the tumulus topsoil and fills with a more clayey texture, while about 26% were found in contexts with a sandy texture. This suggests that the debitage more likely arrived in the tumulus from the surrounding soils of the area that were used to build the mound. Those found in the sandier-textured contexts may have been deposited with the later tools through the process of interment or through slight mixing with the clayey soils that would result from repeated excavations and redepositions.

Summary

The lithic tool assemblage from the tumulus at Lofkënd most likely arrived within the mound through two depositional processes. There is a higher density in both diagnostic and non-diagnostic stone tools in the tumulus topsoil than in the tumulus fills. Patterning between finds of Paleolithic and Mesolithic stone tools in the texture and color of the deposits, in addition to the fact that Paleolithic stone tools are known to occur nearby in the landscape, suggests that these objects arrived in the tumulus in the soils from the vicinity used to build the mound. The lack of Neolithic and Bronze Age tools and blade cores in the vicinity and their patterning in the topsoil and tumulus fills indicate that those tools may have been brought to the site as functional objects for the people constructing the mound and conducting interment activities.

CATALOGUE

The purpose of this catalogue is to provide a detailed record of the stone tools, cores, and debitage recovered from the Lofkënd tumulus. Because it encompasses a collection of lithics essentially out of context from a variety of time periods and from tool technologies mixed together in the Late Bronze Age to Early Iron Age tumulus fill, the identifications and descriptions of the formal qualities of the finds are based on basic known lithic types and comparisons with other known and dated collections whenever possible. Because of the nature of the assemblage, no statistics were run to attempt to identify modes in the measured attributes of the collection, as would usually be conducted in a lithic analysis aimed at creating a site-based formal artifact typology (cf. the Bordian method, Debénath and Dibble 1994:174).

Methods

A number of analytical tools were used in cataloging the assemblage. All measurements of length, width, and thickness were taken using a 6-inch dial caliper (Spi). Weights were measured using a table-top digital scale (Ohaus HS-120) unless the weights were greater than 120 g, in which case a 500-g suspended scale (Homs) was used. Low-power magnification was achieved through the use of a 20 x 18-mm jeweler's loupe. Colors were judged and recorded using a Munsell Soil Color Chart (1994 revised edition) under indirect natural light.

Polish and patina are indicated in the catalogue when they were observable by the naked eye or under 20x magnification. When polish is mentioned, it refers to the chemical process of binding silica from grasses and wood onto the surface of the lithic artifact during use (Andrefsky 2005:196). It is characterized by a glossy surface, ripples and flow zones, pits wiped to a comet shape, a lack of overall abrasion, scratches polished away, and a smoothing of the peaks of flake scars (Witthoft 1967). Patina is a chemical weathering process that alters the surface of the chert depending on the specific environmental conditions to which the piece was subjected. It can result in a darker surface, a white or bluish white surface, or a glossy surface (Luedtke 1992:108–109). Patina can be differentiated from polish by the lack of smoothing along the peaks of flake scars.

The pieces are discussed in the catalogue using a consistent orientation. When a drawing is published for an object, the description follows the left-hand image unless noted otherwise. When there is no drawing, one can assume that the worked end of the piece is pointing up and the dorsal side is facing the viewer unless noted otherwise. In the case of flake tools with multiple worked margins, the distal end is up with the dorsal side facing the viewer.

Classification Scheme

The tools in this collection were categorized using a macroscopic morphological typology based on one devised by Andrefsky (2005: fig. 4.7, see also fig. 1). All pieces believed to be intentionally modified or used are classified as tools and catalogued in detail. The non-modified flakes and other fragments are separated as debitage. These items are presented in tabular format (**Table 13.1**) with counts and weights in four categories: cortical flakes (displaying cortex on the dorsal surface), non-cortical flakes (displaying no cortex), chunks, and shatter. Flakes had to display a bulb of percussion and ventral surface with no flake scars other than eraillures. The angular shatter was divided by size into pieces larger than 1 cm (chunks) and smaller than 1 cm (shatter). These debitage categories are generally used as a rough measure of the proportion of early (cortical flakes and chunks) to later stage (non-cortical flakes and shatter) chipping in the assemblage.

The tools in the assemblage were further classified according to known pseudo-functional[1] types that occur in the Balkans (**Fig. 13.1**). Because only two bifaces were identified, those pieces were simply presented under the biface category (A). The flake

[1] I use the phrase "pseudo-functional" because the labels applied to these tools sometimes, but not always, refer to suites of morphological characteristics that have an assumed functional quality, such as scraping hides. Use of these types follows more closely the Bordian method of lithic classification and was used to connect these tools to traditional classification schemes commonly used in the region.

TABLE 13.1 Lithics summary chart

Trench and unit	Context type	Number of cortical " flakes	Cortical flake weight (g)	Number of non-cortical flakes	Non-cortical flake weight (g)	Number of chunks	Chunk weight (g)	Number of shatter fragments	Shatter fragment weight (g)	Inventoried pieces
Surface collection										SF 8/23*, SF 9*, SF 5*, SF 6*, SF 7/22*
1.0001	Topsoil	4	15.2	5	19.5	6	25.2	2	1.2	SF 29*, SF 31, SF 33, SF 190*, SF 206*, SF 207, SF 32*
1.0007	Tumulus fill	2	28.1			3	11	2	0.9	SF 25, SF 56, SF 94, SF 148
1.0009	Topsoil	3	14.3	4	10.9	10	22.5	1	0.3	SF 50*, SF 26*, SF 52*, SF 88, SF 96, SF 97
1.0017	Fill of Grave 2			2	5.9	1	1.9			SF 49
1.0039	Tumulus fill	3	24.4	4	24.9	2	40.8			SF 119*, SF 34, SF 41*, SF 68, SF 102, SF 143
1.0047	Tumulus fill			1	8.8	1	0.8			SF 47
1.0053	Fill of Grave 6			2	1.6			1	0.2	SF 43
1.0067	Tumulus fill	4	9.2	1	1.6	7	46.1			SF 129*, SF 62, SF 63, SF 144*, SF 198
1.0070	Tumulus fill	3	15.8	1	1.7	7	34.9	1	0.7	SF 123*, SF 199*, SF 139*, SF 59*, SF 75, SF 82, SF 120, SF 200
1.0095	Fill of Grave 11					1	3.1			
1.0108	Fill of Grave 14							2	0.2	
1.0120	Fill of Grave 15					1	1.1			
1.0127	Tumulus fill							1	0.7	
1.0141	Topsoil	1	11.7			3	51.2			SF 150*
1.0158	Fill of Grave 23							1	0.5	
1.0213	Fill of Grave 22	1	0.8							
1.0222	Fill of Grave 33					1	0.8			
1.0240	Tumulus fill			1	2.4	2	5.7			SF 279
1.0278	Topsoil	2	11.9							SF 236*, SF 242*, SF 233, SF 239

* Asterisk indicates pieces chosen for inclusion in final catalogue. Pieces without asterisks were included with debitage in the analysis, yet field designation was maintained in records and collection storage

Table 13.1 (*continued*). Lithics summary chart

Trench and unit	Context type	Number of cortical flakes	Cortical flake weight (g)	Number of non-cortical flakes	Non-cortical flake weight (g)	Number of chunks	Chunk weight (g)	Number of shatter fragments	Shatter fragment weight (g)	Inventoried pieces
1.0358	Fill of Grave 63					2	3.4			
1.0399	Tumulus fill	1	14.7			2	4.7			SF 398, SF 399
1.0440	Tumulus fill	1	12.7			1	1.7			
1.0497	Fill of Grave 87	1	5.3							
2.0000	Out of context	1	8.5	1	0.5	4	48.5			SF 327, SF 329
2/7.0002	Topsoil									SF 400
2.0002	Topsoil	11	48.1	12	18.8	20	52.4	5	2.8	SF 36*, SF 69*, SF 320*, SF 322*, SF 323*, SF 361*, SF 192*, SF 214*, SF 215*, SF 235, SF 220*, SF 35, SF 183, SF 184, SF 186, SF 187, SF 328, SF 368, SF 375, SF 379, SF 397, SF 441, SF 448*, SF 449*, SF 453*
2.0003	Topsoil	4	7.4	2	5.8	13	26.1			SF 324*, SF 278*, SF 237, SF 369
2.0006	Topsoil					2	3.7			
2.0017	Fill of Grave 2									SF 132*
2.0024	Tumulus fill					2	1.9			
2.0037	Tumulus fill	2	32.5							SF 58
2.0040	Tumulus fill	1	1.7							
2.0066	Fill of Grave 13 (upper)			1	1.4					SF 99
2.0071	Tumulus fill					1	2.2			
2.0078	Tumulus fill			3	7.4	1	4.9			SF 140, SF 147*
2.0097	Fill of Grave 13	1	8.9							SF 131
2.0102	Fill of Grave 13			1	0.9					SF 121*, SF 141
2.0117	Tumulus fill					1	4			SF 118*
2.0202	Tumulus fill	7	61.8	6	11.6	18	102.1	1	0.2	SF 195*, SF 196, SF 174, SF 175, SF 176, SF 177, SF 191, SF 197*, SF 201, SF 203, SF 204, SF 205, SF 216, SF 222, SF 223, SF 224, SF 232*, SF 241*

* Asterisk indicates pieces chosen for inclusion in final catalogue. Pieces without asterisks were included with debitage in the analysis, yet field designation was maintained in records and collection storage

Continued on next page

TABLE 13.1 (*continued*). Lithics summary chart

Trench and unit	Context type	Number of cortical flakes	Cortical flake weight (g)	Number of non-cortical flakes	Non-cortical flake weight (g)	Number of chunks	Chunk weight (g)	Number of shatter fragments	Shatter fragment weight (g)	Inventoried pieces
2.0226	Tumulus fill					1	4.5			SF 180 with SF 181* (joining) from SU 2.0229
2.0228	Tumulus fill									SF 182*
2.0235	Tumulus fill			1	0.5	2	12.5			SF 271
2.0239	Tumulus fill									SF 194*
2.0242	Fill of Grave 37	1	11.9							
2.0253	Tumulus fill					1	0.9			SF 179
2.0254	Fill of Grave 40	1	1.1			1	2.5			
2.0257	Fill of Grave 41					1	0.9			
2.0259	Fill of Grave 61			1	2.2					SF 234
2.0272	Fill of Grave 44					1	5.9			
2.0287	Fill of Grave 59					1	3.5			SF 240*, SF 280*
2.0292	Fill of Grave 54					2	2.1			SF 238, SF 272
2.0295	Fill of Grave 47							1	0.8	
2.0297	Tumulus fill			1	6.1	4	32			SF 247*, SF 248, SF 270, SF 277
2.0370	Fill of Grave 68			1	2.7			1	0.2	SF 325
2.0379	Bedrock			1	0.5	3	10.1			
2.0380	Tumulus fill	3	4.8	2	1.4	2	4			SF 326*, SF 371*
2.0382	Fill of Grave 67					1	0.9			
2.0395	Fill of Grave 81			1	0.4	1	59.3			SF 345, SF 362*, SF 363
2.0397	Fill of Grave 70					1	1.3			
2.0403	Bedrock			1	0.5	1	2.6			
2.0423	Fill of Grave 76			1	0.7	2	5.4	3	0.5	
2.0431	Tumulus fill	1	21.2	3	10.5	12	49.7	1	0.3	SF 359* SF 364*, SF 365, SF 366*, SF 367
2.0432	Fill of Grave 78			1	0.9					
2.0453	Fill of Grave 81					1	1.7	3	0.4	2.6

* Asterisk indicates pieces chosen for inclusion in final catalogue. Pieces without asterisks were included with debitage in the analysis, yet field designation was maintained in records and collection storage

Table 13.1 (*continued*). Lithics summary chart

Trench and unit	Context type	Number of cortical flakes	Cortical flake weight (g)	Number of non-cortical flakes	Non-cortical flake weight (g)	Number of chunks	Chunk weight (g)	Number of shatter fragments	Shatter fragment weight (g)	Inventoried pieces
2.0474	Pit fill			1	0.2	2	2.7			SF 385*, SF 373*, SF 376, SF 377*
2.0475	Fill of Grave 75 (upper)	1	0.7	1	0.6	3	5.5			
2.0481	Fill of Grave 86					1	0.6			SF 396*
2.0487	Tumulus fill	1	2.1	2	6.1	1	0.5	1	0.4	SF 403
2.0491	Fill of Grave 88					1	1.3			
2.0507	Fill of Grave 89			1	1.6					SF 451*
2.0510	Fill of Grave 90			1	1.3					
2.0522	Tumulus fill	1	4.5							
2.0525	Bedrock			1	0.2	2	2.9			
2.0532	Tumulus fill	1	5.1			6	29.7	1	0.5	SF 447*
2.0543	Pit fill	1	11.9							SF 427
2.0568	Fill of Grave 86			1	0.1					
2.0569	Fill of Grave 96							1	0	
2.0577	Tumulus fill			1	0.1					
2.0578	Fill of Grave 99	1	5.8					1	0.3	
2.0586	Fill of Grave 100	1	0.4	2	3.8					
3.0003	Topsoil					1	1			
3.0008	Topsoil					1	3.4			SF 20*, SF 27
3.0045	Tumulus fill	2	14.1			1	0.9	2	1.6	SF 53
3.0064	Topsoil					1	0.7			
3.0068	Tumulus fill	4	36.1							SF 95
3.0069	Tumulus fill					2	5.3			SF 61
3.0092	Tumulus fill	1	0.9			1	1.4			
4.0004	Topsoil	3	21.7	1	0.9	1	8.8			SF 152*, SF 87*

* Asterisk indicates pieces chosen for inclusion in final catalogue. Pieces without asterisks were included with debitage in the analysis, yet field designation was maintained in records and collection storage

Continued on next page

TABLE 13.1 (*continued*). Lithics summary chart

Trench and unit	Context type	Number of cortical flakes	Cortical flake weight (g)	Number of non-cortical flakes	Non-cortical flake weight (g)	Number of chunks	Chunk weight (g)	Number of shatter fragments	Shatter fragment weight (g)	Inventoried pieces
4.0011	Topsoil			3	29.9	4	28	3	1.4	SF 51*, SF 21, SF 28*, SF 30, SF 55, SF 71, SF 74
4.0035	Tumulus fill			2	7.3	4	7			SF 64*, SF 100
4.0048	Tumulus fill	1	2.2							SF 54*, SF 101
4.0057	Tumulus fill	1	3.5				2.2			SF 76
4.0059	Fill of Grave 7			1	1.7	2	18.4	2	0.5	SF 81
4.0079	Tumulus fill									SF 83*
4.0086	Tumulus fill			3	4.3	9	33.3			SF 122, SF 142, SF 151*
4.0087	Fill of Grave 9							1	0.2	
4.0104	Tumulus fill									SF 145*
4.0115	Tumulus fill			1	1.4	1	2			
4.0129	Tumulus fill			3	20.5	2	13.1			SF 112, SF 115, SF 146
4.0130	Tumulus fill	1	5.3			4	7.3			
4.0131	Tumulus fill					1	1.8			
4.0138	Fill of Grave 18			1	1					SF 117
4.0201	Topsoil	3	22.4	1	0.9	7	51.8	2	1.1	SF 209*, SF 185, SF 269*, SF 178, SF 193*, SF 275, SF 276
4.0204	Tumulus fill	3	15.8			7	35.4			SF 274*, SF 210*, SF 188, SF 189, SF 221, SF 281, SF 292b*
4.0207	Tumulus fill			1	1.6			1	0.1	SF 208
4.0209	Fill of Grave 29							1	0	
4.0268	Fill of Grave 43									SF 213*
4.0286	Tumulus fill	2	10.2	1	6.2	4	5.9	2	0.6	SF 353*, SF 243, SF 273, SF 334
4.0400	Fill of Grave 71	1	2.3	1	1.6	2	5			SF 312, SF 314
5.0004	Topsoil							1	0.3	
5.0025	Topsoil			1	2.1					
5.0026	Tumulus fill	1	15.9							SF 93

* Asterisk indicates pieces chosen for inclusion in final catalogue. Pieces without asterisks were included with debitage in the analysis, yet field designation was maintained in records and collection storage

Table 13.1 (*continued*). Lithics summary chart

Trench and unit	Context type	Number of cortical flakes	Cortical flake weight (g)	Number of non-cortical flakes	Non-cortical flake weight (g)	Number of chunks	Chunk weight (g)	Number of shatter fragments	Shatter fragment weight (g)	Inventoried pieces
5.0140	Fill of Grave 18	2	7.3			1	1.7			
5.0179	Tumulus fill			1	2.6					
5.0268	Fill of Grave 43					1	1			
5.0278	Topsoil					1	3.4			SF 454*
5.0375	Topsoil					1	4.7			
5.0535	Tumulus fill							3	2.5	
6.0002	Topsoil							1	0.3	SF 436*
6.0541	Tumulus fill							1	0.4	
6.0542	Tumulus fill	1	41.2							SF 425*, SF 429
7.0002	Topsoil	1	4.6							
7.0027	Topsoil									SF 40*
7.0126	Fill of Grave 17	1	0.3							
7.0186	Topsoil	1	3.5							
7.0529	Tumulus fill							1	0.5	
7.0530	Tumulus fill	1	9.7							
8.0129	Tumulus fill					1	6.7			SF 282*
8.0204	Tumulus fill	1	9.1			2	20.2	1	0.6	SF 452*
8.0207	Tumulus fill	1	22.2							SF 406
	Grand total:	104	695.7	95	247.3	242	1058	54	21.6	

* Asterisk indicates pieces chosen for inclusion in final catalogue. Pieces without asterisks were included with debitage in the analysis, yet field designation was maintained in records and collection storage

and core tools were subdivided into several different categories to provide a more clear organization for the catalogue. Pieces that were clearly formed on a flake blank were placed under the flake tool category. They were subdivided into blades (B1), scrapers (B2), perçoirs (B3), denticulates (B4), multitools (B5), microliths (B6), and other flake tools (B7). Category B7 included retouched and utilized flakes that did not fit into the other types. Pieces worked on amorphous or bifacially worked blanks, which were not bifacially formed tools, were classified as core tools. These were subdivided into cores (C1), core-shaping elements (C1a), scrapers (C2), becs (C3), multitools (C4), and other core tools (C5). Similar to B7, the pieces in C5 included retouched and utilized objects that did not fit into other types. A separate section of the catalogue (D) is reserved for a small collection of Middle Paleolithic flakes and tools recovered from the tumulus.

Raw Material

The raw material used for creating chipped stone tools is known as chert, a rock composed mostly of microcrystalline quartz (a silicate mineral) and a wide variety of impurities. The common usage of the term "flint" in Europe to refer to microcrystalline silicates seems to arise from usage in Great Britain, where flint is defined as the high-quality, dark-colored rock that erodes from chalk deposits, while chert is a lighter-colored, lower-quality material (Luedtke 1992:5–6). Rather than referring to all good-quality, workable silicates by the name given to a single type, this work will use the term "chert" to refer to microcrystalline silicate rocks used in the manufacture of stone tools. Geologists, petrologists, mineralogists, and archaeologists use various terms (chert, jasper, chalcedony, agate, carnelian, etc.) for silicates that vary visibly, chemically, and structurally, but they tend to use those terms in different ways that often depend on their level of analysis (Luedtke 1992:6). Since this study uses only macroscopic investigation techniques and thus cannot detect details of crystalline structure or chemical composition, "chert" will be used as a catch-all term for silicate rocks that can be chipped to form tools.

The creation of a chert typology for this assemblage posed problems similar to those facing the author when attempting to categorize the tools within the assemblage as a whole. Lithic analysis is usually based on considering the structure of an entire group of tools and flakes from a single site dating to a relatively limited time period. The Lofkënd tumulus lithic assemblage is composed of tools from a very wide variety of technological facies and time periods that defies traditional analytical techniques. As such, this typology is purely descriptive and does not purport to identify any chert types in the geological sense (Luedtke 1992:6). Chert can display a wide range of variability in its visible properties that do not necessarily reflect the chemical variability more characteristic of its origin (Luedtke 1992:76); thus, chert collected a few meters apart from the same source *could* look rather different in color, texture, and inclusions. Lacking a geological survey of the region for outcrops and an analysis of the range of variability in local chert sources, this typology provides only a comprehensive list of the observed sets of macroscopic characteristics found in this collection.

The chert typology presented here to categorize the raw materials used to make tools and cores found in the Lofkënd tumulus is organized around three major attributes of the stone: texture, inclusions, and color (**Table 13.2**). Each type thus has three components in the name referring to where it falls in the range of variation observed in those categories. Judgments regarding each piece were made subjectively by the author. Individual Munsell readings were not considered in the process; the pieces were holistically compared and grouped. Munsell

TABLE 13.2 Schematic of the chert typology

Texture	Grainy (1)		Smooth (2)		
Inclusions	Mostly small visible fossil inclusions (a)	Mostly large visible fossil inclusions (b)	Mixed large and small fossil inclusions (c)	Isolated large fossil inclusions (d)	No visible fossil inclusions (e)
Color	Numbered consecutively with lowercase Roman numerals in the order identified within each texture/inclusion group. Color numbers refer to actual colors. See **Table 13.3** for Munsell readings for each type				

TABLE 13.3 Munsell readings for each chert type

Chert type	Munsell color
1ai	10 YR 6/1; 2.5 Y 6/2; 5 Y 6/1
1aii	2.5 Y 7/1
1aiii	10 YR 6/2; 10 YR 7/2
1aiv	10 YR 6/3; 2.5 Y 6/3
1av	10 YR 4/2; 10 YR 5/1
1avi	5 YR 4/3; 5 YR 4/4; 7.5 YR 4/3
1bi	10 YR 8/1
1bii	10 YR 7/2; 2.5 Y 7/3
1biii	10 R 6/4, 10 YR 6/4; 10 YR 5/3; 10 YR 6/3; 10 YR 6/4; 10 YR 6/6; 10 YR 7/3; 2.5 Y 6/4
1biv	10 YR 7/2
1bv	10 R 5/2; 10 R 6/3
1ci	10 YR 6/2
1cii	10 YR 6/4
1di	10 YR 6/4; 2.5 Y 6/2; 2.5 Y 6/3; 2.5 Y 7/3 to 2.5 Y 7/6
1dii	10 YR 5/1, 10 YR 5/3; 10 YR 5/4; 10 YR 6/2; 2.5 Y 5/1
1diii	10 R 4/2; 10 R 5/3; 10 YR 7/2 to 2.5 YR 5/2; 2.5 YR 7/2; 5 YR 7/3
1ei	10 YR 5/3
2ai	10 YR 5/2
2aii	2.5 Y 6/2; 2.5 Y 6/4
2di	10 YR 6/2 to 10 YR 8/1
2ei	10 YR 6/4; 10 YR 7/4; 2.5 Y 6/3
2eii	10 YR 7/4
2eiii	10 YR 4/4; 10 YR 5/4; 10 YR 5/6; 10 YR 6/6
2eiv	10 YR 4/4; 10 YR 5/3; 10 YR 5/4
2ev	10 YR 5/3; 2.5 Y 4/2
2evi	10 YR 4/4
2evii	10 YR 4/4; 10 YR 6/4; 5 YR 5/2; 7.5 YR 5/2; 7.5 YR 5/4
2eviii	2.5 YR 3/4; 5 YR 3/4
2eix	Gley chart 2, 7/1
2ex	10 YR 2/1; 10 YR 3/2; 10 YR 6/4 to 10 YR 5/1; 2.5 Y 2.5/1; 2.5 Y 4/3
2exi	2.5 YR 6/4
2exii	10 R 4/4
2exiii	5 YR 6/4

ranges for each type were assembled following division (**Table 13.3**). The number of types proposed in this study is relatively large, but, as stated above, these types are not to be read as representative of individual chert sources or geological chert types. In the interests of clearly recording the variation in the collection, the author chose to separate types as often as necessary to reflect recurring sets of observed characteristics.

A separate section of the raw material typology was set aside to describe pink, red, and sooty black colored cherts, which are possibly burned or heat-treated to improve the mechanical properties of the rock for knapping. Heat treatment has been shown experimentally to alter color, often but not always to a pink or reddish hue, at temperatures of about 200° C to 500° C (Luedtke 1992:100–101). Because of the lack of known local variation mentioned above and the paucity of stone tool assemblages excavated and published for this area, it is nearly impossible to tell which of the reddened cherts might have been heat-treated intentionally, and similar results can occur inadvertently when chert is accidentally thrown into a normal fire. Heat treatment is very rare in Old World lithic assemblages as an intentional alteration (C. Runnels, personal communication), which suggests that the heat-related color changes are more likely accidental than intentional. Also, the collection recovered from the tumulus fill may represent a wide variety of primary contexts; therefore, no discussion will be undertaken of patterning in the heat treatment of raw materials in this assemblage. Several pieces bear signs of damaging burning as well, but this effect as well as the heat treatment could also have been caused quite recently in surface and topsoil finds by the agricultural practice of burning stubble in fields. Several of the reddish-colored pieces, especially the dark red microblades (**13/15, 13/16, 13/17**) and single very large blade (**13/10**), were assumed to be made of a red variety of unaltered chert and thus were categorized with the regular smooth cherts.

The best-known chert from the Gjanicë River valley is a honey-brown variety identified at a Paleolithic hunting station at Kraps west of Lofkënd. That material has not been published, however, other than as a brief summary with no Munsell readings or color illustrations. Based on firsthand identification of honey-brown chert collected from the Lofkënd Survey Project by Muzafer Korkuti, Type 2eiii is most likely the same chert type as the Kraps honey-brown, which was also found in the form of finished tools at Kryegjatë B (Runnels et al. 2004:13). Types 2eiv, 2ev, and 2evi may be variations on that type, with color banding and a slightly more brown to grayish hue. As at Kryegjatë, most of the artifacts from the Lofkënd tumulus made of this very translucent yellowish brown chert were tools, although several were not of a well-defined type.

Raw material types

The catalogued tools assigned to each chert type are listed at the end of the description for reference. That information is repeated in the catalogue entry.

1ai Grainy, very small, fine fossiliferous inclusions with a few large fossil inclusions; generally, pale gray to light brownish gray. Translucent. **13/31, 13/8, 13/40.**

1aii Grainy, very small, fine fossiliferous inclusions with a few large fossil inclusions; pale gray to white, banded, with the darker bands matching the color of type 1ai. Translucent. **13/72.**

1aiii Grainy, very small, fine fossiliferous inclusions with very few large fossil inclusions; light yellowish brown to gray, banded. Yellowish white pitted cortex, cortical iron oxide staining. Translucent. **13/65, 13/93.**

1aiv Grainy, with fewer very small, fine fossiliferous inclusions than types 1ai–1aiii; grayish brown and white immediately beneath white pitted cortex. Some light white mottling. Translucent. **13/74, 13/7.**

1av Grainy, very small, fine fossiliferous inclusions; darker grayish brown with lighter and darker bands. Translucent. **13/57, 13/6.**

1bi Grainy, densely packed, large fossiliferous inclusions with a dull, pocked surface; white. Opaque. **13/13.**

1bii Grainy, densely packed large fossiliferous inclusions with a dull, pocked surface; very light, yellowish brown with light yellow to white pitted cortex and some iron oxide staining below the surface. Opaque. **13/78** (with voids), **13/42, 13/20, 13/24.**

1biii Grainy, densely packed, large fossiliferous inclusions with some smoother, less grainy areas and some voids; yellowish brown to light yellowish brown with yellowish brown pitted cortex. Opaque. **13/54, 13/22, 13/23, 13/94, 13/68, 13/56, 13/61, 13/80, 13/77.**

1biv Grainy yet almost smooth, with medium-large fossiliferous inclusions and some voids; very pale brown with yellowish brown staining and deep, large pits on the cortex. Opaque. **13/73.**

1ci Grainy, mixed large and small fossiliferous inclusions; grayish brown with light white mottling. Very similar in color to type 1aiv. Translucent. **13/11, 13/12.**

1cii Grainy, mixed large and small fossiliferous inclusions with some smoother, lighter-colored areas; yellowish brown. Opaque. **13/71.**

1di Grainy, isolated, sometimes clustered large fossiliferous inclusions; light gray to yellowish brown. White pitted cortex, yellowish brown iron oxide staining below cortex. Slightly translucent. **13/91,**

13/79, 13/55, 13/90, 13/88, 13/21, 13/37, 13/53, 13/92, 13/87.

1dii Grainy, isolated, sometimes clustered large fossiliferous inclusions; darker grayish brown with some light white mottling and small fossils sometimes visible. Uneven distribution of inclusions in the pieces. Slightly translucent. **13/43, 13/63, 13/58, 13/70, 13/39.**

1ei Slightly grainy, poor quality with many fractures; yellowish brown. Opaque. **13/60.**

2ai Smooth, very tiny fossiliferous inclusions with very few large fossils; pale yellow. Translucent. **13/47.**

2aii Smooth, very tiny fossiliferous inclusions with very few large fossils; yellowish brown. Slightly translucent. **13/82, 13/86.**

2di Smooth, very slightly grainy in limited areas with a few large fossil inclusions and scattered dark specks; white to very light gray. Slightly translucent. **13/1.**

2ei Smooth, almost creamy in texture with dark, thin lines on one piece. Some slight lighter and darker diffuse mottling; pale brown. Translucent only at the thinnest margins. **13/89, 13/36, 13/84, 13/29.**

2eii Smooth with dark wavy bands and lighter and darker diffuse mottling; light yellowish brown. Slightly translucent. **13/48.**

2eiii Smooth, no inclusions; yellowish brown. Translucent. **13/2, 13/9, 13/34, 13/83, 13/3.**

2eiv Smooth, no inclusions; grayish brown, lighter yellowish brown, and yellowish brown with mottling and white, chalky cortex. Translucent. **13/46, 13/62, 13/64, 13/59.**

2ev Smooth, no inclusions; grayish brown with a band of lighter brown below iron oxide staining below the white, chalky cortex. Translucent. **13/52, 13/27.**

2evi Smooth, no inclusions; dark brown. Translucent. **13/45, 13/33.**

2evii Smooth, no inclusions; light yellowish brown with moderate to heavy white mottling. Translucent. **13/19, 13/14, 13/41, 13/4, 13/18.**

2eviii Smooth, no inclusions; dark red. Opaque. **13/10, 13/16, 13/15, 13/17.**

Burned or potentially heat-treated cherts

1avi Slightly grainy; pink to pinkish tan. Translucent. **13/50, 13/49, 13/81.**

1bv Grainy with densely packed large fossiliferous inclusions; pink to grayish pink. Opaque. **13/69, 13/66.**

1diii Grainy, isolated large fossiliferous inclusions; medium pink to yellow. Opaque. **13/26, 13/28, 13/44, 13/38, 13/35.**

2eix Smooth, no inclusions; bluish gray with white mottling. Many spalls and fractures. Opaque. **13/85.**

2ex Smooth, no inclusions, waxy; yellowish brown where not altered to a smoky gray or black. Whitish, chalky, pitted cortex. Translucent only at the thinnest margins. **13/51, 13/76, 13/32, 13/75, 13/5.**

2exi Smooth, no inclusions, waxy; pink. Translucent. **13/30.**

2exii Smooth, waxy, heavily fractured with many spall scars; dark red. Translucent. **13/67.**

2exiii Smooth, waxy, heavily fractured; orange-red to tan with dark lines around the fractures. Translucent. **13/25.**

Catalogue of Lithic Artifacts

A. Bifaces

13/1 (SF 152), **Figs. 13.4, 13.8**
Biface Preform
Topsoil (SU 4.0004).
L (max): 32.0 mm; W (max): 20.0 mm; Th (max): 8.9 mm; L, left abraded edge: 21 mm; L, right abraded edge: 17 mm; Wt: 5.8 g.
Raw material: Mostly smooth chert with few large fossiliferous inclusions.
Color: 10 YR 6/2, light brownish gray, on the used half of the biface; 10 YR 8/1, white, on the unused half.
Condition: Broken on one end. It is unclear whether this is the proximal or distal end of the tool preform, but it is most likely the proximal end. For the sake of clarity, the shaped end of the tool will be referred to as "distal" (pointing up in the drawing), while the broken end will be referred to as "proximal."
Description: A bifacially worked preform with heavy abrasion along half of the remaining right and left margins. Cross-section is roughly lenticular but is angular and uneven. For clarity of discussion, this tool can be divided into two zones. The first zone is the distal half of the tool, which exhibits light retouch and use-wear abrasion. The second zone is the proximal half of the tool, which exhibits no use-wear and no retouch. The two zones will be discussed separately below.

In the first zone, moderate retouch with feathered terminations on the dorsal surface on the right margin show that this portion of the piece was shaped for use after the second portion was no longer shaped or used. The right margin curves on to the distal end. Light retouch on the ventral surface on the straight left margin with feathered terminations indicates the preparation of that edge as well. Both of those shaped edges were very heavily used and show significant abrading and multiple, overlapping step fractures. The very tip of the piece shows evidence of light use. The edge angles in the first zone of approximately 30° for the left margin and 20° for the right indicate that this side of the piece was worked further into the production process (Andrefsky 2005: table 7.7) than was the second half.

In the second zone, the tool preform was left largely unmodified. The edge angles are approximately 50° for the left margin and 30° for the right margin. On the ventral side of the piece (in **Fig. 13.4 [13/1]**, the drawing on the right), a single flake scar extending across the full width of the piece with a deep step in the center may have been the flawed flake that ended the bifacial production process.

The width/thickness ratio (Andrefsky 2005: table 7.7) of 2.6 for the first zone and 2.2 for the second zone are both indicative of a Stage 2 Edged Biface. According to Johnson's (1989:124) classification system using a different approach, because the lateral margins have been worked and all visible cortex has been removed from the tool, the piece would be considered a Preform I.

Date: Neolithic?

13/2 (SF 182), **Fig. 13.3**

Bifacial Arrowhead with Tranchet Edge

Tumulus fill (SU 2.0228).

L (max): 27.1 mm; W (max): 12.0 mm; Th (max): 3.5 mm; L, tool in proposed orientation as barb: 24.0 mm; Wt: 1.2 g.

Raw material: Smooth, fine chert; translucent.

Color: 10 YR 4/4, dark yellowish brown. The translucency makes the color appear lighter in certain lighting conditions.

Condition: Complete tool.

Description: Bifacial arrowhead broken to form a tranchet edge. The blank was likely a flake, but it could have been a blade with a trapezoidal cross-section. The tranchet edge is unworked and mostly undamaged, while the other margins are all worked. On the dorsal side, shallow, short retouch scars with feathered terminations extend continuously around the curved margin. On the ventral side, minute flake scars with stepped terminations predominate, with a few larger, very shallow flake scars especially near to the shaped end. Similar examples from northern contexts in Epiros date to the Mesolithic (e.g., Preveza and Thesprotia), while examples from southern Greece occur in Upper Mesolithic and Neolithic contexts (e.g., Franchthi).

Date: Mesolithic or Neolithic.

Comparanda: Perlès 1990: fig. 16 nos. 12–13, fig. 27.

B. Flake tools

B1. Blades

13/3 (SF 5), **Figs. 1.15, 13.4, 13.8**

Retouched Blade Segment with Polish

Surface find (2003 season).

L (max): 25.7 mm; W (max): 14.5 mm; Th (max): 12.6 mm; L, left margin: 15.9 mm; W (max) polish, left margin, dorsal side: 2.1 mm; W (max) polish, left margin, ventral side: 1.2 mm; W (max) polish, right margin, dorsal side: 2.8 mm; W (max) polish, right margin, ventral side: 2.2 mm; Wt: 1.5 g.

Raw material: Smooth, fine chert with a flaw at the distal end; translucent.

Color: 10 YR 5/4, yellowish brown; reddened at the distal end near flaw.

Condition: Broken perpendicular to the longitudinal axis on both ends; this was most likely a feature of the manufacture. Distal end broken again at approximately 50° angle from a blow to the center on the dorsal side.

Description: Most likely a sickle blade. Trapezoidal cross-section. Two worked edges. Both of the straight, parallel blade edges show unimarginal retouch on the dorsal side. Polish on the dorsal and ventral sides of both edges, but the polish and retouch both are greater on the right margin. Continuous retouch flake scars have been smoothed by the polish on the right margin. Clustered retouch flake scars on the left margin have feathered terminations.

Date: Neolithic?

Comparanda: Korkuti 1983a: pl. 2.16 (Kolsh).

13/4 (SF 20), **Fig. 13.4**
Broken Retouched Blade
Topsoil (SU 3.0008).
L (max): 16.3 mm; W (max): 13.7 mm; Th (max): 3.7 mm; L, worked edge: 38 mm; Wt: 1.0 g.
Raw material: Fine, smooth chert.
Color: 10 YR 4/4, dark yellowish brown, with white inclusions at proximal end.
Condition: Broken perpendicular to the longitudinal axis at the distal end.
Description: Broken blade with the striking platform and left margin retouched into a single, continuous convex working edge and used. Trapezoidal cross-section. Unimarginal dorsal retouch on proximal end and left margin, although abrasion and small random flake scars are evident from use on the right margin, proximal end, and the ventral side along the entire working edge. Continuous retouch on the proximal end shows hinged flake terminations, while clustered retouch on the right margin shows feathered flake terminations. The proximal retouch was conducted first in forming the tool, followed by the retouch on the right margin. The uppermost feathered termination right margin retouch scar, which aided in creating the rounded tip on this blade tool, overlies one of the hinged, proximal end retouch scars.
Date: Neolithic or Early Bronze Age?

13/5 (SF 28), **Fig. 13.4**
Bladelet Segment with Wear
Topsoil (SU 4.0011).
L (max): 8.8 mm; W (max): 7.9 mm; Th (max): 1.9 mm; L, worn edge: 7.1 mm; Wt: 0.1 g.
Raw material: Fine chert; slightly translucent at blade edge.
Color: 2.5 Y 2.5/1, black.
Condition: Broken perpendicular to the longitudinal axis on one end, and broken at a 10° angle on the other end. The patina on the tool appears worn away in except in a few small places, leaving a dull surface.
Description: Tiny blade segment with no discernible force ripples. Triangular cross-section. With the angled break up, the wavy, uneven right margin shows signs of wear in the form of abrasion on the dorsal side that may have been partly polished away, but the surface wear removed any signs of luster or reflectivity. The straight left margin appears very sharp and unused.
Date: Neolithic or Early Bronze Age?

13/6 (SF 129), **Fig. 13.8**
Broken Blade
Tumulus fill (SU 1.0067).
L (max): 25.2 mm; W (max): 12.6 mm; Th (max): 3.3 mm; Wt: 1.1 g.
Raw material: Fine, waxy chert with few tiny white specks; completely translucent.
Color: 10 YR 4/2, dark grayish brown.
Condition: Broken on a 10° angle to the longitudinal axis at the distal end.
Description: Broken blade. Well-preserved bulb of percussion on the ventral proximal end. The left and right margins are very slightly convex. There is some irregularity in the cross-section, which, considering also the lack of significant use-wear, suggests this could be a core-reshaping flake. The cross-section is trapezoidal; however, the central flake scar on the dorsal surface is to the right of center and irregular in width. Hinged flake scars are on the dorsal proximal end. Very slight use-wear is evident in small chips along the right and left margins, although the edges are still sharp. No retouch.
Date: Neolithic or Early Bronze Age?

13/7 (SF 139), Not illustrated
Broken Blade
Tumulus fill (SU 1.0070).
L (max): 18.3 mm; W (max): 15.6 mm; Th (max): 6.6 mm; Wt: 1.9 g.
Raw material: Fine chert with tiny white specks and very little white mottling; translucent.
Color: 10 YR 6/3, pale brown.
Condition: Broken at the distal end perpendicular to the longitudinal axis. Also broken on the right-hand side. Right-side break shows a single, clear flake scar, which removed approximately one-quarter of the remaining blade. It is unclear whether the right-side break occurred before or after the distal break.
Description: Proximal end of a broken blade. The bulb of percussion is missing due to an eraillure flake scar. Notching in the platform at the point of percussion and beside it suggests preparation prior to flake removal with a punch, necessitated by the 50° angle of the striking platform from the ventral surface. The blade was most likely originally triangular in cross-section. There is evidence

of another facet just below the striking platform on the right dorsal side, which suggests that the piece could have been trapezoidal prior to the right-side break, although the center flake scar would have been skewed far to the right. This may have been a core-reshaping flake used to remove a spur near the striking platform left by previous blade removals. The straight left margin shows minor chipping from possible use, but no retouch.

Date: Neolithic or Early Bronze Age?

13/8 (SF 210), **Fig. 13.4**

Blade Segment with Lateral Use-Wear

Tumulus fill (SU 4.0204).

L (max): 15.3 mm; W (max): 15 mm; Th (max): 3.8 mm; Wt: 1.0 g.

Raw material: Grainy chert; translucent.

Color: 2.5 Y 6/2, light brownish gray.

Condition: Broken at both the proximal and distal ends. The graininess of the chert caused some hinging along the margin of the left flake scar on the dorsal side, as shown in **Figure 13.4 (13/8)**.

Description: Blade segment with use-wear on the right margin. The left margin is straight with only slight chipping. The right margin is very uneven and chipped. There is no evidence of retouch. The blade segment is trapezoidal in cross-section.

Date: Neolithic or Early Bronze Age?

13/9 (SF 220), **Fig. 13.4**

Broken Blade

Topsoil (SU 2.0002).

L (max): 20.1 mm; W (max): 11.0 mm; Th (max): 3.0 mm; Wt: 0.6 g.

Raw material: Fine, smooth chert; completely translucent.

Color: 10 YR 5/6, yellowish brown. The translucency affects the color reading to some extent.

Condition: Broken at the distal end perpendicular to the longitudinal axis.

Description: Broken blade with use-wear along both the right and left margins. The left margin is very slightly convex with evidence of use in the form of tiny, irregular chips that dulled the edge. The right margin is uneven but more or less straight with a small notch from a single flake scar. The blade is trapezoidal in cross-section, although the flake scar in the center on the dorsal side leading up from the striking platform is more pronounced than the underlying flake scar, which makes the remaining distal end also trapezoidal.

Date: Neolithic?

13/10 (SF 244), **Fig. 13.4**

Large Broken Blade

Tumulus fill (SU 2.0288).

L (max): 52.5 mm; W (max): 26.2 mm; Th (max): 7.1 mm; L, left margin: 56 mm; L, right margin: 42 mm; Wt: 12.0 g.

Raw material: Smooth chert with tiny white inclusions that cause very shallow fracturing as force travels through the piece during knapping; completely opaque.

Color: 5 YR 3/4, dark reddish brown.

Condition: Broken at the distal end perpendicular to the slightly curving longitudinal axis.

Description: Heavy, thick blade with the longitudinal axis curving about 15° to the right 20 mm from the break. Triangular cross-section. Very pronounced, well-preserved bulb of percussion on the ventral face. No evidence of use on the broken edge. The left margin is straight until the axial curve forms a convex edge near the distal end, and it shows evidence of either very minor retouch or use-wear in the form of a few scattered, small flake scars on both the dorsal and ventral sides. The left margin is very smooth, suggesting the piece has been ground smooth on that edge, except for the last 5 mm before the distal break, which shows sharp use-wear. The right margin is slightly concave, with very heavy use-wear in the form of multiple, overlapping stepped and feathered fractures primarily on the dorsal side. The striking platform appears to have been prepared with multiple, overlapping, small flake scars. Patina on all surfaces of the tool.

Date: Late Neolithic or Bronze Age?

Comparanda: Sordinas 1970: fig. 7 (Sidari, Bronze Age, Level A); Lera 1988: pl. 1.12 (Dërsnik, Late Neolithic).

13/11 (SF 247), **Fig. 13.4**

Retouched Blade Segment with Polish

Tumulus fill (SU 2.0297).

L (max): 40.7 mm; W (max): 13.6 mm; Th (max): 5.8 mm; L, worked edge: 40.7 mm; W (max), polish from left margin, dorsal side: 4.5 mm; W (max), polish from left margin, ventral side: 6.3 mm; Wt: 3.1 g.

Raw material: Grainy, slightly fossiliferous chert; somewhat translucent.
Color: 10 YR 6/2, light brownish gray, with many small white fossil inclusions.
Condition: Broken at both the proximal and distal ends.
Description: Blade segment retouched only on the straight left margin. Trapezoidal in cross-section on the proximal end; triangular in cross-section on the distal end. Continuous bimarginal retouch along the left margin; unimarginal retouch continues onto the proximal end (pointing up in **Fig. 13.4 [13/11]**) on the dorsal surface only. Both feathered and smoothed retouch flake scars. Retouch appears to have occurred at different times during the use-life of the piece. Flake scars with differing degrees of polish occur discontinuously along the worked edge. In its final form, the tool appears somewhat denticulated, but it is unclear whether that was its original form or a result of the extensive retouching and use over the course of time. The straight right margin is sharp and unmodified, with only tiny chips from the thinnest part of the blade edge.
Date: Neolithic.
Comparanda: SF 453; Korkuti and Andrea 1974: pl. 1 (Cakran).

13/12 (SF 453), **Fig. 13.4**
Blade
Topsoil (SU 2.0002).
L (max): 30.4 mm; W (max): 13.8 mm; Th (max): 5.2 mm; Wt: 1.9 g.
Raw material: Grainy, slightly fossiliferous chert. Somewhat translucent.
Color: 10 YR 6/2, light brownish gray, with small fossil inclusions and a few large fossil inclusions.
Condition: Broken unevenly at the distal end.
Description: Blade with slight use-wear on the straight left margin. No retouch. Flake scars on the right dorsal surface are uneven and form a small peak in the trapezoidal cross-section of the proximal end of the flake. Triangular cross-section at the distal end.
Date: Neolithic.
Comparanda: SF 247; Korkuti and Andrea 1974: pl. 1 (Cakran); see also Korkuti 1974.

13/13 (SF 269)**, Figs. 13.4, 13.8**
Broken Blade with Use-Wear
Topsoil (SU 4.0201).
L (max): 24.5 mm; W (max): 22.8 mm; Th (max): 5.5 mm; Wt: 4.4 g.
Raw material: Dense, fossiliferous chert; completely opaque.
Color: 10 YR 8/1, white, with brown specks.
Condition: Broken at both ends. Hinging on the proximal break.
Description: Thick, broken blade. The straight left margin shows signs of use in the form of abrasion and small, discontinuous scars on the dorsal and ventral surfaces. The uneven right margin shows heavier signs of use. Continuous step fractures extend 15.3 mm along the center of the blade edge on the dorsal side, with abrasion on either end. The ventral surface has been heavily abraded and shows only faint remnants of flake scars.
Date: Neolithic.
Comparanda: Prendi and Aliu 1971: pls. 1-1, 5 (Kamnik).

13/14 (SF 396), **Figs. 13.4, 13.8**
Truncation with Lateral Retouch
Tomb XXXI (86) fill (SU 2.0481).
L (max): 34.1 mm; W (max): 16.1 mm; Th (max): 4.7 mm; L, worked scraper edge: 22 mm; L, retouched lateral margin: 33 mm; Wt: 3 g.
Raw material: Smooth, fine chert; translucent.
Color: 7.5 YR 5/4, brown, with diffuse white inclusions.
Condition: Complete, with shiny, lustrous patina.
Description: Multiuse tool based on a blade blank. Four flake scars on the dorsal surface; irregular in cross-section. Two small flake scars at the proximal end mark the removal of the striking platform, possibly intentionally. The slightly convex left lateral margin below the truncation shows evidence of use in the form of abrasion and tiny flake scars on both the dorsal and ventral sides. Part of the distal truncation is shaped by lamellar retouch into a narrow point. A tiny burin-like spur was also formed by retouch on the distal end of the blade. The straight right margin was retouched on the ventral surface with pressure flaking.
Date: Neolithic.
Comparanda: Korkuti and Andrea 1974: pls. 1.7, 1.8 (Cakran); Korkuti 1974: pl. 1 (Cakran).

13/15 (SF 373), Not illustrated
Bladelet with Use-Wear
Pit fill (SU 2.0474).
L (max): 13.4 mm; W (max): 6.6 mm; Th (max): 1.5 mm; Wt: 0.2 g.

Raw material: Smooth, fine chert; translucent partially because of thinness.
Color: 5 YR 3/4, dark reddish brown.
Condition: Broken at the distal end.
Description: Very small bladelet fragment. Triangular cross-section. Abrasion on the left and right margins.
Date: Neolithic.
Comparanda: **13/16** (SF 241), **13/17** (SF 353).

13/16 (SF 241), Not illustrated
Bladelet with Use-Wear and End Scraper
Tumulus fill (SU 2.0202).
L (max): 24.7 mm; W (max): 10.0 mm; Th (max): 2.5 mm; Wt: 0.8 g.
Raw material: Smooth, fine chert; slightly translucent.
Color: 2.5 YR 3/4, dusky red, with lighter pink cortex at the distal end.
Condition: Complete.
Description: Bladelet with plunging flake termination. Abrasion on both the right and left margins. Marginal use-wear is greater on the right side than on the left. Step fractures on the lighter pink cortex at the distal end forms a scraper. Triangular cross-section at the distal end and trapezoidal in the rest of the piece.
Date: Neolithic.
Comparanda: **13/15** (SF 373), **13/17** (SF 353).

13/17 (SF 353), Not illustrated
Blade with Utilized Notch
Tumulus fill (SU 4.0286).
L (max): 22.6 mm; W (max): 8.8 mm; Th (max): 3.1; L of notch along longitudinal axis: 3.6; depth of notch from right margin: 2 mm; Wt: 0.7 g.
Raw material: Smooth, fine chert; cortex on dorsal surface. Translucent at the margins.
Color: 5 YR 3/4, dark reddish brown, with 5 YR 8/3, pink, cortex.
Condition: Broken at the distal end.
Description: Small blade with a notch on the right margin. Bulb of percussion with eraillure flake scar. Trapezoidal in cross-section. The left margin is very slightly convex with two separate tiny flake scars on the dorsal surface and some abrasion along the edge. The right margin is straight with very slight wear along the thinnest edge of the blade. The notch is on the right margin 5.1 mm from the striking platform and shows the multiple, random stepped flake scars evident of use.
Date: Neolithic or Early Bronze Age.
Comparanda: **13/15** (SF 373), **13/16** (SF 241).

B2. Scrapers

13/18 (SF 36), Not illustrated
Fragment of a Scraper
Topsoil (SU 2.0002).
L (max): 11.4 mm; W (max): 20.3 mm; Th (max): 9.9 mm; L, preserved worked scraper edge: 14.9 mm; Wt: 3.2 g.
Raw material: Smooth chert with numerous tiny white specks. Major fracture plane running through the remaining piece perpendicular to the worked edge. The fracture plane in this chunk of chert affected the breakage pattern described under "Condition" (below). Translucent only at the edges.
Color: 5 YR 5/2, reddish gray.
Condition: Broken on clean planes on five of the six sides of the chunk. The angle of the very flat lateral breaks parallel the fracture plane visible in the piece. The other breaks on the dorsal, ventral, and proximal ends are also very flat. Patina on the dorsal and ventral surfaces as well as on the worked distal edge indicate that these surfaces are old, while the breaks on the left and right margins and proximal end are younger. The piece was thus originally larger.
Description: Fragment of a scraper on a piece of chert with strong fracture planes. Unimarginal steep retouch on the dorsal side at an approximately 70° angle at the scraper edge, which curves back gently onto the flat dorsal surface. The continuous retouch flake scars, which shaped the working edge, have feathered terminations. The slightly convex distal margin shows abrasion from use.
Date: Not diagnostic.

13/19 (SF 119), **Fig. 13.3**
Broken Multifaceted Scraper
Tumulus fill (SU 1.0039).
L (max): 22.7 mm; W (max): 12.6 mm; Th (max): 6.3 mm; L, left scraper edge: 14.9 mm; L, right scraper edge: 15.6 mm; L, distal scraper edge: 19 mm; Wt: 2.2 g.

Raw material: Smooth chert; slightly translucent.
Color: 7.5 YR 5/2, brown, with heavy white mottling.
Condition: Broken at the proximal end. Lustrous patina on all surfaces except the break.
Description: Scraper with three discernible working edges. The cross-section of the flake blank is triangular. The straight left scraper edge is shaped by short, continuous flake scars with feathered terminations on the dorsal surface. The edge angle is 80°. Use-wear along the edge. This edge is interrupted by a small peak formed by a bump in the flake blank, which was modified by a single flake scar on the ventral surface. The use-wear from the left margin is continuous up to this peak. The distal end of the blank is shaped into a pointed scraper with lamellar retouch on the dorsal surface. The edge angle on the distal scraper is 70°, and it also shows abrasion from use-wear. The right margin scraper is distinguished from the pointed distal scraper by the end of the lamellar retouch. Shorter, more uneven retouch flakes with feathered and stepped terminations created a jagged edge. The edge angle is uneven and changes from approximately 80° to 50° along the right margin. This object could possibly be an unfinished Neolithic unifacial point, but the quality of the lamellar retouch near the pointed end suggests that the steep edge angle was intentional and the object is a scraper.
Date: Upper Paleolithic? Possibly Neolithic.

13/20 (SF 123), Not illustrated
Side Scraper with Light Distal Use-Wear
Tumulus fill (SU 1.0070).
L (max): 19.3 mm; W (max): 26.4 mm; Th (max): 7.0 mm; L, worked scraper edge: 13.5 mm; L, used distal edge: 27.0 mm; Wt: 4.2 g.
Raw material: Grainy, fossiliferous chert; translucent only at the thinnest edge.
Color: 2.5 Y 7/3, pale yellow, with a small area of red at the striking platform consistent with iron oxide staining from cortical weathering on similar pieces.
Condition: Mostly complete flake blank. Bulb of percussion missing from an eraillure flake scar. May be broken on the unworked right margin.
Description: Side scraper on a flake with light use-wear on the thin distal margin. The straight but slightly jagged scraper edge is shaped by short, shallow flake scars with feathered terminations on the dorsal surface. The edge angle is approximately 85°. The distal end of the flake displays a smooth, sharp, even convex angle from a good feather termination. That sharp edge was used and consequently slightly chipped and abraded, but not retouched.
Date: Not diagnostic.

13/21 (SF 209), **Fig. 13.6**
Inverse Side Scraper
Topsoil (SU 4.0201).
L (max): 31.8 mm; W (max): 33.3 mm; Th (max): 7.9 mm; L, worked scraper edge: 23.5 mm; Wt: 5.6 g.
Raw material: Grainy, fossiliferous chert; completely opaque.
Color: 2.5 Y 7/3, pale yellow, to 2.5 Y 7/6, yellow. The color grows more yellow as it gets closer to the cortical surface at the striking platform.
Condition: Complete flake tool.
Description: Side scraper with all retouch on the ventral face (use the right-hand drawing of the ventral surface in **Fig. 13.6** [**13/21**] to orient left and right). The main scraper edge is on the left margin of the flake. Continuous feathered retouch scars extend the length of the slightly concave edge. A flake scar on the dorsal surface creates an edge angle of approximately 65°. The distal edge of the flake was broken off and retouched slightly in a small cluster of flake scars also on the ventral surface to form an additional slightly concave scraping edge with an approximate edge angle of 90°. The retouch scars on this facet terminate at a 90° angle to the ventral surface.
Date: Not diagnostic.

13/22 (SF 213), **Figs. 13.2, 13.8**
Unguiform End Scraper
Tomb LVI (43) fill (SU 4.0268).
L (max): 33 mm; W (max): 32 mm; Th (max): 15.5 mm; L, scraper edge: 44 mm; Wt: 12.5 g.
Raw material: Grainy, fossiliferous chert; completely opaque.
Color: 10 R 6/4, pale red, where heated on ventral side; 10 YR 6/4, light yellowish brown where not heated on dorsal side.
Condition: Complete artifact, although burning detached three large spalls from the ventral surface and altered the color heavily. Some black staining or sintering along the scraper edge and on the distal

end of the ventral surface, but the black staining was removed by the spalls in that area. The damage was likely caused by burning after the tool had been made and used.

Description: Unguiform end scraper with thick, steep scraping edge. The retouch scars are primarily on the dorsal side, but at least two flakes were removed from the ventral side as well. The edge angle is approximately 70° along the worked scraper edge. The worked edge is continuously retouched, with large flake scars with feathered terminations where preserved. Many of the shaping flake scars terminate in a peak in the distal center of the piece, which would have provided a platform for gripping the scraper with the thumb and forefinger.

Date: Upper Paleolithic or Mesolithic.

13/23 (SF 320), **Fig. 13.2**
Unguiform Scraper
Topsoil (SU 2.0320).

L (max): 26.9 mm; W (max): 19.6 mm; Th (max): 6.8; L, worked edge/perimeter of piece: 72.0 mm; Wt: 4.2 g.

Raw material: Grainy chert; completely opaque.

Color: 10 YR 6/3, pale brown, with darker yellowish brown staining near a patch of cortex remaining on the distal end.

Condition: Complete tool.

Description: Thick flake with small area of cortex remaining on distal end, with all margins worked at a steep angle for scraping (in **Fig. 13.2 [13/23]**, this piece shows the distal end down and the proximal end up). Teardrop-shaped. Striking platform formed into a point by lateral retouch was heavily utilized but not retouched and could form a tang. The continuous convex working edge begins at the striking platform and extends around the edge of the piece to return to the striking platform. At the distal end, the retouch scars only extend halfway across the thickness of the piece; they exhibit stepped terminations, which indicate they were truncated by the remaining cortex. The lateral retouch scars all end on an angle with the dorsal surface. Three small, isolated shallow use-wear scars are on the ventral distal end; the entire margin of the piece is abraded. The edge angle along the left lateral margin is approximately 50°; along the distal margin 90°; along the right margin 80°.

Date: Upper Paleolithic or Mesolithic.

13/24 (SF 324), **Fig. 13.5**
Lateral Scraper on a Cortical Flake
Topsoil (SU 2.0003).

L (max): 41.0 mm; W (max): 39.0 mm; Th (max): 11.1 mm; L, worked scraper edge: 23.8 mm; Wt: 14.9 g.

Raw material: Grainy, dense chert. Flat, pitted cortical surface 1.5 mm thick; completely opaque.

Color: 2.5 Y 7/3, pale yellow, with darker yellowish brown cortical surface.

Condition: Complete flake tool.

Description: Cortical flake with a retouched and used lateral scraper on the left margin. Continuous retouch on the dorsal side with stepped terminations just below the cortical surface. Very slight use-wear and chipping on the right lateral margin, dorsal side. Possible broken-off perforator or burin element just above the scraper marked by a pointed extension of the cortical surface. The possible broken pointed element was shaped by flaking, which formed the scraper and two flake scars on the distal left lateral margin. Edge angle is approximately 58°.

Date: Not diagnostic.

13/25 (SF 361), **Figs. 13.6, 13.9**
End Scraper on a Very Thick Flake with Cortex
Topsoil (SU 2.0002).

L (max): 35.5 mm; W (max): 55.4 mm; Th (max): 16.1 mm; L, worked and used scraper edge: 38.8 mm; Wt: 44.3 g.

Raw material: Smooth chert with cracking and crazing from heat; completely opaque. Water weathered white pitted cortex 5.8 mm thick.

Color: 5 YR 6/4, light reddish brown, with darker linear inclusions.

Condition: Heavily affected by heating. Cortex extends along the entire left margin. Extreme spalling and fracturing along the thick striking platform on the proximal end and right lateral margin. Two isolated spalls on the dorsal surface. More heavily heat treated on the left side of the piece, dark red in color gradually lightening to the Munsell color identified above.

Description: End scraper on a very thick flake. The flake is wider than it is long, providing a long edge for the scraper, which was worked dorsally and worn into a slightly convex curve. Dorsal retouch scars have feathered terminations. A single ventral

retouch scar, which causes an S-curve in the edge, also has a feathered termination. Edge angle is 70°. The scraper edge shows significant abrasion from heavy use. The heating, which affected the proximal end and right margin so heavily, does not appear to have affected the scraper edge, suggesting the heat treatment happened prior to the creation of the tool and may have been an intentional part of the production process.
Date: Not diagnostic.

13/26 (SF 151), Not illustrated
Scraper on the Proximal End of a Flake
Tumulus fill (SU 4.0086).
L (max): 22.7 mm; W (max): 26.8 mm; Th (max): 5.9 mm; L, scraper edge: 8.5 mm; Th (max) scraper edge: 3.0 mm; Wt: 3.3 g.
Raw material: Smooth, dense chert. Dark speckled inclusions; completely opaque.
Color: 10 YR 7/2, light gray, with 2.5 YR 5/2, weak red, heat-treated portion on right side.
Condition: Complete flake.
Description: Scraper formed on the right proximal margin of a wide, short flake. Ventral retouch with a 90° edge angle. Abrasion along the dorsal proximal edge.
Date: Not diagnostic.

13/27 (SF 32), Not illustrated
Primitive Lateral Scraper on a Cortical Flake
Topsoil (SU 1.0001).
L (max): 40.4 mm; W (max): 25.9 mm; Th (max): 10.1 mm; L, scraper worked edge: 22.5 mm; Wt: 12.4 g.
Raw material: Smooth chert with few fossiliferous inclusions. Rough, pitted cortex. Opaque.
Color: 10 YR 5/3, brown, with light yellowish brown bands just below the cortex.
Condition: Complete cortical flake.
Description: Heavy flake with cortex on the entire dorsal surface. Scraper is shaped by a few continuous dorsal flake scars on the right proximal margin, taking advantage on the right distal margin of the angle of the natural stone at that spot. The shaped part of the scraper in effect continues the natural edge. Edge angle is 90° along both the shaped and natural portions of the scraper edge. Abrasion along the scraper edge. The other flake scar margins around the striking platform are pristine, which aided in identification of this piece as a primitive tool rather than a rolled cortical flake.
Date: Not diagnostic.

B3. Perçoirs (Perforators)

13/28 (SF 69), **Fig. 13.5**
Small Perçoir on a Broken Cortical Flake
Topsoil (SU 20002)
L (max): 20.6 mm; W (max): 18.0 mm; Th (max): 3.4 mm; Wt: 1.2 g.
Raw material: Grainy chert with red, pitted cortex; slightly translucent at the thinnest margins.
Color: 5 YR 7/3, pink.
Condition: Thin flake broken at the proximal end as indicated by waves of force on the ventral surface.
Description: Small perçoir (perforator) on the distal end of a broken cortical flake. Two small chips were removed from the feathered termination of the flake to form two concave notches that created the spur. No retouch. Use-wear on the spur and the thin concave distal margins on either side of it. The left margin is completely unused. The right margin was shaped similar to the distal margin by snapping off bits of the thin edge to form a smaller spur. The right concave margins and spur are also abraded.
Date: Not diagnostic.

13/29 (SF 322), **Figs. 13.5, 13.9**
Perçoir with Lateral Use-Wear
Topsoil (SU 2.0002).
L (max): 22.5 mm; W (max): 14.7 mm; Th (max): 8.1 mm; Wt: 2.2 g.
Raw material: Smooth, creamy chert; completely opaque.
Color: 10 YR 7/4, very pale brown, with a very slightly reddened tip.
Condition: Appears to be complete, but a break on the distal end (opposite the perforator tip) with very light wear suggests the original tool was longer.
Description: Long, thin perçoir tool with light retouch and use-wear along the left margin and heavy use-wear along the right margin. The left margin retouch consists of continuous dorsal flaking, which created two shallow concave notches, each with use-wear in the center. The point is roughly triangular formed by the ventral surface, a single flake scar on the right side, and the retouch on the

left side. The right margin is straight, with an approximately 30° angle at the middle. The dorsal right margin is unretouched. Very heavy abrasion and multiple overlapping step fractures run the length of the right margin, and a cluster of very shallow flake scars with feathered and stepped terminations on the ventral side may have helped alter the angle of the edge from being more convex to straight with an angled turn.
Date: Not diagnostic.

13/30 (SF 29), **Figs. 13.5, 13.9**
Perçoir with Use-Wear
Topsoil (SU 1.0001).
L (max): 20.5 mm; W (max): 16.4 mm; Th (max): 5.1 mm; Wt: 1.9 g.
Raw material: Smooth chert with smooth, dark red cortex.
Color: 2.5 YR 6/4, weak red.
Condition: Flake broken at the proximal end (bottom in **Fig. 13.5 [13/20]**).
Description: Small, roughly triangular broken flake with two small tips shaped on the right lateral margin. The thick left margin is completely covered in cortex. The distal end is shaped on the dorsal surface into a used narrow scraper with a 75° edge angle. Use-wear abrasion from the scraper extends onto the distal edge of the right margin. The right margin begins with a small spur shaped by tiny retouch flakes on the dorsal surface and a single flake scar on the ventral surface, all with feathered terminations. The right margin below the spur is concave and shaped bimarginally by a few flake scars on both the dorsal and ventral surfaces. In the center of the right margin is another spur that could act as a perforator or graver. It was formed primarily by tiny dorsal retouch flakes, which have been almost entirely obliterated by use-wear, although the flakes that shaped the two concave edges on either side of it contributed to the shape. The concave edge below the perforator/graver was bimarginally shaped by a hinged flake on the dorsal side and a feathered flake on the ventral surface. The dorsal edge and the right half of the ventral edge of the proximal break were not retouched, but both were abraded from use.
Date: Not diagnostic.

13/31 (SF 144), **Figs. 13.3, 13.8**
Perçoir
Tumulus fill (SU 1.0067).
L (max): 25.6 mm; W (max): 32.6 mm; Th (max): 10.1 mm; L, notch edge: 14.0 mm; L, perforator edge: 20.0 mm; Wt: 5.9 g.
Raw material: Slightly grainy, fossiliferous chert; translucent.
Color: 5 Y 6/1, gray, with white fossil inclusions occurring frequently throughout.
Condition: Complete flake tool.
Description: Flake shaped into a perçoir on the distal end. In **Figure 13.3 (13/31)**, the normal flake orientation is shifted about 45° counterclockwise to emphasize the perforator point. That triangular point is still relatively sharp but has light use-wear abrasion on the tip and two small retouch scars on the right facet. The notch to the left was formed by a single flake and shows some use-wear in the center. On the left side, a secondary small round point was formed by notches from retouch scars on either side with stepped terminations and an edge angle of approximately 90°. Both the two lateral notches and the rounded spur of the perforator were abraded.
Date: Mesolithic?

B4. Denticulates

13/32 (SF 150), **Figs. 13.4, 13.8**
Sickle Element
Topsoil (SU 1.0141).
L (max): 21.6 mm; W (max): 13.4 mm; Th (max): 3.0 mm; PL, worked denticulate edge: 18.2 mm; Wt: 1.0 g.
Raw material: Smooth, fine chert with cortex on the left beveled margin extending 2 to 2.5 mm into the tool (indicated on **Fig. 13.4 [13/32]** by a thin line of stippling through the left flake scar in the center of the piece on the dorsal side; also shown with a thin stippled line on the ventral side).
Color: 10 YR 2/1, black. Color was altered by heating. Traces of the previous color indicate it was brown or tan, but insufficient area remaining to take a Munsell reading.
Condition: Flake broken on the proximal end perpendicular to the longitudinal axis with spall scars from heating. Dorsal (left margin) and ventral (right margin) surfaces on the distal end damaged by spalling as well. Areas of spalling are indicated in **Figure 13.4** by random stippling. The original length of the piece is obscured by the heat-related breakages.
Description: Cortical flake with polish on the right

margin retouched into a denticulate. The unshaped left margin of the flake with remaining cortex would have provided an excellent place to grip the small flake. The right margin is straight, with a relatively steep angle, approximately 50° from the ventral surface. The dorsal facet of that flake scar was heavily polished. The single flake scars that formed each of the four denticulate notches along the right margin were struck from the ventral surface; the notches were subsequently retouched with very minute pressure flakes from the dorsal surface. The retouch flake scars that created the denticulates lack the same degree of polish evident around them. While the edge of the right margin where it has been removed shows evidence of polish, the retouch flake scars on both the dorsal and ventral surfaces have clearly defined feathered terminations and show no evidence of polish, although the piece overall has a lustrous patina.
Date: Neolithic or Early Bronze Age.

13/33 (SF 64), Not illustrated
Sickle Element
Tumulus fill (SU 4.0035).
L (max): 17.6 mm; W (max): 11.4 mm; Th (max): 3.2 mm; L, left margin: 14.4 mm; L, right margin: 17.6 mm; Wt: 0.8 g.
Raw material: Smooth, fine chert; translucent.
Color: 10 YR 4/4, dark yellowish brown.
Condition: Flake broken at the distal end.
Description: Roughly blade-shaped flake with a utilized edge on the left margin and denticulate edge on the right margin. The left margin is straight and ground down by tiny overlapping flakes on the ventral side. The right margin is straight and abraded from use. Single flakes forming denticulate teeth along the edge are worked on both the dorsal and ventral sides. Two notches are worked on the dorsal side only, two on the ventral side only, and one is bimarginally retouched.
Date: Neolithic or Early Bronze Age.

B5. Multitools

13/34 (SF 323), **Figs. 13.3, 13.8**
Perçoir on an End Scraper
Topsoil (SU 2.0002).
L (max): 15.8 mm; W (max): 20.8 mm; Th (max): 8.9 mm; L, left scraper edge: 14.0 mm; L, right scraper edge: 13.0 mm; Wt: 2.1 g.
Raw material: Smooth chert with darkened staining on the left dorsal surface and left proximal side; completely opaque except at the very thinnest parts of the perforator spurs.
Color: 10 YR 6/6, brownish yellow.
Condition: Complete tool.
Description: Triangular multitool that appears to be a perçoir on an end scraper shaped on a flake. The left distal edge is concave with an edge angle of 70°. Retouch scars appear to be mostly stepped, but very heavy abrasion from use-wear has nearly obliterated them. The right distal margin is retouched by a single large flake into a concave edge like a broad U-shape. Very heavy abrasion and step fracturing along the edge indicate heavy use. The right distal edge angle is 65°. The proximal margin on the dorsal side of the striking platform was also abraded on the right and left sides, with a 5-mm gap between the used areas. Both of those edges have an approximately 90° edge angle. No ventral retouch on the piece. The spurs appear at each point of the triangular shape and show evidence of slight chipping and abrasion from use but no direct retouch to form the points; however, retouch scars that originally formed the perforators may have been obscured by the scraper use-wear.
Date: Mesolithic?

13/35 (SF 451), **Fig. 13.6**
Combination Tool with Scraper and Notch
Tomb XXXII (89) fill (SU 2.0507).
L (max): 27.7 mm; W (max): 17.4 mm; Th (max): 5.5 mm; L, scraper edge: 31.0 mm; L, perforator edge: 23.0 mm; Wt: 2.8 g.
Raw material: Grainy, slightly fossiliferous chert; opaque.
Color: 2.5 YR 7/2, pale red.
Condition: Proximal end of the flake is fractured.
Description: Flake worked on the lateral margins into a scraper multitool with a notch. The left lateral margin is worked on the dorsal surface into a scraper with a sinuous, S-shaped curve. Edge angle is approximately 75°. The right lateral margin is also steeply retouched on the dorsal side only, with an edge angle of approximately 60° at the proximal end. In the center, the edge is worked into a point formed by a notch. The proximal and distal ends are unworked. Abrasion on the left and right margins from use.

Date: Not diagnostic, but possibly Neolithic.

Comparanda: Korktui 1982a: pl. 2.18 (Vashtëmi, Early Neolithic), although it is unclear in the image if there are scrapers on the lateral margins.

B6. Microliths

13/36 (SF 41), **Figs. 1.15, 13.3**
Geometric Microlith
Tumulus fill (SU 1.0039).
L (max): 16.1 mm; W (max): 18.3 mm; Th (max): 5.1 mm; L, scraper edge: 15.2 mm; Wt: 1.1 g.
Raw material: Smooth, fine chert with a large fossil inclusion; translucent only at the thinnest margins.
Color: 10 YR 6/4, light yellowish brown, with lighter yellowish brown staining from cortex on the dorsal surface.
Condition: Complete flake.
Description: Irregular trapeze microlith. Wedge-shaped flake with continuous retouch scars along the straight worked edge. Edge angle is 80°. All retouch worked from the dorsal surface.
Date: Mesolithic.

B7. Other flake tools

13/37 (SF 6), **Figs. 1.15, 13.5**
Flake with Notch and Distal Retouch
Surface find (2003 season).
L (max): 30.4 mm; W (max): 29.5 mm; Th (max): 10.9 mm; L, worked notch: 24.0 mm; L, lateral retouch: 13.6 mm; Wt: 7.4 g.
Raw material: Grainy, slightly fossiliferous chert. Small area of dark, reddish cortex to the right of the striking platform.
Color: 2.5 Y 6/3, light yellowish brown.
Condition: Appears to be complete. Possibly the tip of the distal end of the flake is broken off.
Description: Roughly triangular, thick flake with a notch in the left margin, cortex on the right margin, and distal retouch. The cross-section is also roughly triangular. The left margin is convex before the notch, then concave. The notch appears to have been shaped by a single blow to the ventral surface of the flake, which was interrupted by a flaw in the chert, which resulted in a step fracture inside the notch. Similar steps occur in the dorsal surface of the piece. The right margin is straight and entirely composed of cortical surface. The distal end is convex but was partially straightened by a segment of continuous ventral retouch. Retouch flake scars exhibit both feathered and stepped terminations as a result of the fossiliferous inclusions in the raw material.
Date: Not diagnostic.

13/38 (SF 26), **Fig. 13.5**
Broken Tool with Use-Wear and Light Retouch
Topsoil (1.0009).
L (max): 17.3 mm; W (max): 19.4 mm; Th (max): 5.0 mm; L, worked edge: 34.0 mm; Wt: 1.7 g.
Raw material: Grainy, slightly fossiliferous chert; completely opaque. Shiny, lustrous patina except in the center; convex flake scar of the dorsal surface.
Color: 10 R 5/3, weak red.
Condition: Fragmentary, but unable to estimate how much of the tool is missing. Patina on all breaks indicate the tool was broken in antiquity. The break occurs at a 20° angle to the longitudinal axis as best as can be reconstructed.
Description: Fragment of a tool with use-wear and light retouch along the shaped edge. Could have been a scraper or the distal end of a point formed from a flake. The cross-section is lenticular, or elliptical, a more common cross-section for bifaces than for flake tools, which are usually plano-convex or unifacial. Relatively thick flake. The bimarginally worked edge is convex and extends along the right margin and distal end. On the dorsal side, the continuous retouch is shallow and composed of flake scars with feathered terminations. Shallow, clustered retouch scars with both feathered and stepped terminations extend along the worked edge on the ventral side of the piece. The worked edge shows evidence of use-wear in the form of abrasion. The broken edge was also worked slightly to reshape the left distal corner of the tool that remains. The initial break was a hinged fracture. A single flake was next removed from the proximal end by striking the ventral surface near the center of the diagonal break to prepare a platform to remove an additional flake from the dorsal surface whose scar is obvious on the left margin **Figure 13.5** (**13/38**).
Date: Not diagnostic.

13/39 (SF 280), Not illustrated
Retouched Flake
Tomb XLII (59) fill (SU 2.0287).

L (max): 28.4 mm; W (max): 16.7 mm; Th (max): 4.9 mm; L, retouched edge: 31.0 mm; Wt: 2.1 g.
Raw material: Smooth chert. Some fossiliferous inclusions. Slightly translucent.
Color: 10 YR 5/3, brown, to 10 YR 5/4, yellowish brown.
Condition: Complete flake.
Description: Light retouch on the right margin of a flake forming a convex curved edge. The chipping along the margin creates a slightly serrated edge. All retouch is on the dorsal surface, light abrasion along the edge. The shape of this flake suggests it could be related to **13/2** (SF 182), possibly representing a blank for an unfinished tool of that type.
Date: Not diagnostic.

13/40 (SF 321), **Fig. 13.7**
Utilized Flake
Pit fill (SU 1.0378).
L (max): 56.0 mm; W (max): 34.8 mm; Th (max): 12.1 mm; L, utilized edge: 21.0 mm; Wt: 25.2 g.
Raw material: Grainy, banded chert; relatively translucent.
Color: 10 YR 6/1, gray.
Condition: Complete flake.
Description: Large flake with a hinged termination utilized on the distal half of the right margin. No retouch. Abrasion from use forms a slightly concave edge. The striking platform may have been utilized for scraping as well, but the abrasion on that surface could have been from platform preparation rather than use.
Date: Not diagnostic.

13/41 (SF 147), Not illustrated
Flake with Notch and Retouched Edge on Proximal End
Tumulus fill (SU 2.0078).
L (max): 39.1 mm; W (max): 30.9 mm; Th (max): 12.9 mm; L, proximal worked and used edge: 25.0 mm; L, distal worked edge: 13.0 mm; Wt: 10.0 g.
Raw material: Smooth, fine chert; slightly translucent at the thinnest margins.
Color: 10 YR 6/4, light yellowish brown with white mottling.
Condition: Flake may be broken on the right margin.
Description: Triangular flake with a thick striking platform. On the right dorsal edge of the platform, a concave notch was formed by a step-fractured flake scar and retouched with a small, feathered flake scar. The notch shows use-wear abrasion. In the center of the dorsal edge of the platform, small, continuous feathered termination retouch scars on the platform surface form an S-shaped edge. This edge shows less use-wear than the notched portion of the piece. Very slight retouch on the distal end of the flake gave the right distal margin a convex curve with fine denticulation. This edge also shows slight use-wear.
Date: Not diagnostic.

13/42 (SF 180 + SF 181), Not illustrated
Retouched Flake
Tumulus fill (SU 2.0226 and 2.0229).
L (max): 37.8 mm; W (max): 31.3 mm; Th (max): 5.8 mm; L, retouched edge: 42.0 mm; Wt: 5.6 g.
Raw material: Grainy, fossiliferous chert; somewhat translucent.
Color: 10 YR 7/2, light gray, with white and dark brown inclusions.
Condition: Broken into two pieces, probably during excavation. Very slight patina on the dorsal and ventral surfaces, but no patina on the break.
Description: Flake worked with wide, shallow notches along the left margin curving onto the distal end. The left margin is shaped by six wide chips. The two chips closest to the proximal end form a larger spur between them than the other retouch scars. The distal margin is worn slightly on the ventral surface by scars that span the thickness of the piece. The retouch scars display little to no use-wear. The right margin shows signs of very slight use and chipping along the straight center part of the margin. The proximal end of the right margin cuts back toward the striking platform below the slightly used area and displays no clear evidence of retouch or use. The notches along the margins could be a result of trampling since there is little to no evidence of use.
Date: Not diagnostic.

13/43 (SF 194), Not illustrated
Flake with Two Small Notches Forming a Spur
Tumulus fill (SU 2.0239).
L (max): 34.3 mm; W (max): 40.0 mm; Th (max): 13.8 mm; L, retouched edge: 14.2 mm; Wt: 16.4 g.
Raw material: Slightly grainy, fossiliferous chert; slightly translucent at the thinnest margins.
Color: 10 YR 5/1, gray, with white mottling (possibly cortical weathering).

Condition: Complete flake tool.
Description: Thick flake with some possible cortical weathering on the dorsal right side. The left lateral margin is retouched to form two shallow notches near the distal end that create a small spur between them. The continuous retouch scars span the thickness of the margin, which is the thinnest part of the flake. The rest of the flake is unused and unmodified.
Date: Not diagnostic.

13/44 (SF 190), Not illustrated
Retouched Flake
Topsoil (SU 1.0001).
L (max): 41.4 mm; W (max): 26.2 mm; Th (max): 13.4 mm; L, utilized edge: 20.6 mm; Wt: 11.4 g.
Raw material: Grainy, fossiliferous chert. Heat treated. Only translucent at the thinnest margins.
Color: 10 R 4/2, weak red, with lighter inclusions.
Condition: Complete flake.
Description: Flake with the left lateral distal margin utilized. Very small, shallow retouch flake scars are evident on the ventral surface, but their terminations are indistinct because of the texture of the chert.
Date: Not diagnostic.

13/45 (SF 366), Not illustrated
Utilized Flake with a Possible Perforator
Tumulus fill (SU 2.0431).
L (max): 26.2 mm; W (max): 24.0 mm; Th (max): 7.5 mm; L, utilized edge: 12.7 mm; L, perforator point: 1.1 mm; Wt: 5.3 g.
Raw material: Smooth, fine chert; somewhat translucent.
Color: 10 YR 4/4
Condition: Complete flake.
Description: Used edge on a flake terminating in a possible perforator spur. Left margin forms a convex curve with abrasion along the edge. Very light clustered retouch on the dorsal and ventral surfaces. The spur is shaped by several flake scars on the dorsal surface perpendicular to the spur.
Date: Not diagnostic.

13/46 (SF 232), Not illustrated
Retouched Flake
Tumulus full (SU 2.0202).
L (max): 30.0 mm; W (max): 41.2 mm; Th (max): 11.8 mm; L, right retouched edge: 23.0 mm; L, left retouched edge: 22.0 mm; Wt: 13.4 g.
Raw material: Smooth, fine chert. Some fracturing. Translucent only at the thinnest margins.
Color: 10 YR 4/4, dark yellowish brown, with lighter brown and white mottling.
Condition: Complete flake.
Description: Thick flake with the distal right and left corners retouched and used. The left distal corner was bimarginally retouched into a convex, almost pointed curve. Dorsal retouch was continuous with small feathered retouch scars. Ventral retouch was clustered with a few small, feathered retouch scars. Abrasion along the retouched edge. The right distal corner was also bimarginally retouched into a wider convex curve. Dorsal retouch was scattered, with both feathered and stepped terminations. Ventral retouch was clustered on the right margin with feathered terminations. Abrasion along that retouched edge as well.
Date: Not diagnostic.

13/47 (SF 240), **Fig. 13.6**
Double Notched Retouched Flake with Use-Wear
Tomb XLII (59) fill (SU 2.0287).
L (max): 23.9 mm; W (max): 23.9 mm; Th (max): 4.4 mm; L, proximal retouched edge: 14.0 mm; L, left retouched edge: 16.0 mm; L, right retouched edge: 23.0 mm; Wt: 2.3 g.
Raw material: Smooth chert, slightly waxy; translucent.
Color: 10 YR 5/2, dark grayish brown.
Condition: Complete flake.
Description: Flake with retouch and use on the proximal end and both margins. The proximal end was chipped twice to form a spur. The left shallow notch and the spur are abraded, but the right shallow notch shows only two tiny flake scars as evidence of use. The left margin was continuously retouched along the proximal end, with tiny feathered termination flake scars along the proximal end on the dorsal side in a ragged but straight edge. A single ventral flake scar interrupted that edge, forming a notch in the center of the left margin. Very light abrasion along the left margin. The right margin was continuously retouched along the dorsal side, with tiny flake scars that span the thickness of the margin. On the ventral side, two small flake scars form a tiny spur in the center of the edge. Abrasion along the right margin.
Date: Not diagnostic.

13/48 (SF 292b), **Fig. 13.7**
Unifacially Worked Flake Tool
Tumulus fill (SU 4.0204).
L (max): 34.0 mm; W (max): 14.8 mm; Th (max): 4.5 mm; Wt: 2.8 g.
Raw material: Smooth, fine chert with thin, darker bands; slightly translucent.
Color: 10 YR 7/4, very pale brown.
Condition: Broken twice at the distal end.
Description: Unifacially worked flake. The distal end of the flake is covered with high-quality, shallow pressure-flaking scars creating a smooth, convex surface. On the proximal end of the flake, two large flake scars dominate the surface. All retouch and shaping on the piece occur on the dorsal surface. Light abrasion from use on both the left and right margins where the pressure flaking is preserved, and only very minor chipping along the margins on the proximal end where the large flake scars reach the margins. The quality of the pressure flaking shaping the dorsal surface of this piece is the best of the assemblage.
Date: Not diagnostic.

13/49 (SF 52), **Fig. 13.5**
Flake with Lateral Retouch
Topsoil (SU 1.0009).
L (max): 26.3 mm; W (max): 17.5 mm; Th (max): 5.5 mm; L, retouched edge: 20.4 mm; Wt: 2.4 g.
Raw material: Smooth, waxy, heat-treated chert with isolated inclusions. Whitish smooth cortex at the proximal end of the flake. Translucent.
Color: 5 YR 4/3, reddish brown.
Condition: Distal left corner of the flake may be broken off. Does not affect the retouched edge.
Description: Flake with bimarginal lateral retouch on the right edge. Continuous retouch on the dorsal side with both feathered and stepped terminations. Isolated, scattered retouch scars on the ventral side. Abrasion along the worked margin. Other edges of the piece show very light abrasion and no retouch.
Date: Not diagnostic.

13/50 (SF 54), Not illustrated
Retouched Cortical Flake
Tumulus fill (SU 4.0048).
L (max): 26.1 mm; W (max): 33.7 mm; Th (max): 11.3 mm; L, distal retouched edge: 27.0 mm; L, right retouched edge: 22.0 mm; Wt: 7.2 g.
Raw material: Smooth, slightly grainy chert. Heat treated. Whitish cortex on one facet on the dorsal surface.
Color: 5 YR 4/4, reddish brown.
Condition: Complete flake.
Description: Cortical flake with retouched edges on the distal and right margins. The distal retouch consists of two shallow notches on the right side with a small spur in between them. Abrasion from use in the left notch. To the left of the notches, a very straight edge was retouched by tiny flakes that span the thickness of the margin. On the right margin, a large notch was created by a single large flake and retouched with a single small flake within the notch. Clustered retouch flake scars continue along the right dorsal margin. Light use-wear in the notch.
Date: Not diagnostic.

13/51 (SF 59), **Fig. 13.9**
Utilized Flake with Light Retouch
Tumulus fill (SU 1.0070).
L (max): 33.7 mm; W (max): 20.8 mm; Th (max): 6.8 mm; L, right utilized edge: 24.0 mm; L, proximal utilized edge: 8.8 mm; Wt: 5.8 g.
Raw material: Smooth, fine, waxy chert. Possibly heat treated or burned. Slightly translucent at the thinnest margins.
Color: 10 YR 3/2, very dark grayish brown.
Condition: Flake broken at the distal end.
Description: Flake with two utilized edges with light retouch. The right margin distal end of the flake was lightly retouched with continuous feathered termination flake scars on the dorsal side. The edge terminates in a single flake scar, which created a shallow notch in the edge. Heavy abrasion on the un-notched portion of this worked edge. On the left proximal end, the striking platform was altered into a straight left lateral edge by two shallow retouch flake scars with feathered terminations on the ventral side. Some abrasion along this edge as well, but not as heavy as on the right distal utilized edge. Heavy retouch with stepped fractures on the dorsal surface behind the point of percussion, but this could represent platform preparation rather than a utilized edge.
Date: Not diagnostic.

C. Cores and other non-flake tools

C1. Cores

13/52 (SF 360), **Fig. 13.9**
Multidirectional Core with Cortex
Tumulus fill (SU 1.0377).
L (max): 34.8 mm; W (max): 30.5 mm; Th (max): 22.0 mm; core maximal linear dimension: 4.03 cm; Wt: 27.1 g; Andrefsky core size value: 109.
Raw material: Smooth chert with whitish pitted cortex; opaque.
Color: 2.5 Y 4/2, dark grayish brown, with lighter 2.5 Y 7/4, pale yellow, below the iron oxide stained cortex.
Condition: Complete core.
Description: Multidirectional core worked on three faces, with cortex covering the rest. Edges used as striking platforms. Approximately 11 flake scars.
Date: Not diagnostic.

13/53 (SF 214), Not illustrated
Multidirectional Core with Cortex
Topsoil (SU 2.0002).
L (max): 33.0 mm; W (max): 23.2 mm; Th (max): 16.6 mm; core maximal linear dimension: 4.05 cm; Wt: 15.3 g; core size value: 62.
Raw material: Slightly grainy chert with fossil inclusions. Whitish pitted cortex. Opaque.
Color: 2.5 Y 6/2, light brownish gray, with darker yellowish brown iron oxide staining below the cortex.
Condition: Complete core.
Description: Multidirectional core worked on three faces. Cortex covers half of the rest of the piece, while an area with heavier patination displaying older flake scars covers the rest. Six visible flake scars. Edges used as striking platforms.
Date: Not diagnostic.

13/54 (SF 371), Not illustrated
Multidirectional Core with Cortex
Tumulus fill (SU 2.0380).
L (max): 44.8 mm; W (max): 39.1 mm; Th (max): 28.6 mm; core linear dimension (max): 5.25 cm; Wt: 60.6 g; core size value: 318.
Raw material: Grainy chert with darker yellowish brown pitted cortex; opaque.
Color: 10 YR 6/6, brownish yellow.
Condition: Complete core. Cortex beginning to form over flake scars on one face, but can still be easily chipped off.
Description: Multidirectional core worked on four faces with cortex covering the rest. Approximately six visible flake scars. Edges used as striking platforms.
Date: Not diagnostic.

13/55 (SF 282), Not illustrated
Multidirectional Core with Cortex
Tumulus fill (SU 8.0129).
L (max): 33.9 mm; W (max): 40.7 mm; Th (max): 18.2 mm; core maximal linear dimension (max) 4.63 cm; Wt: 20.1 g; core size value: 93.
Raw material: Slightly grainy chert with whitish pitted cortex. Darker and lighter bands under the cortex.
Color: 2.5 Y 6/3, light yellowish brown.
Condition: Likely broken. One face is convex as a flake would be, rather than concave as would be expected for a core.
Description: Multidirectional core worked on three faces, with the rest covered by cortex. Four visible flake scars, five if one counts the convex surface on one of the worked faces. Edges used as striking platforms.
Date: Not diagnostic.

13/56 (SF 132), Not illustrated
Unifacial Core
Tumulus fill (SU 2.0117).
L (max): 48.1 mm; W (max): 45.8 mm; Th (max): 15.8 mm; core linear dimension (max): 5.34 cm; Wt: 34.3 g; core size value: 183.
Raw material: Very rough, grainy chert. Darker pitted cortex.
Color: 10 YR 6/4, light yellowish brown.
Condition: Possibly broken. One face of the core is flat, with no flake scars.
Description: Unidirectional unifacially worked core. Three flake scars visible on one face, all struck from the same edge of the piece. The face on the opposite side of the core is flat.
Date: Not diagnostic.

13/57 (SF 192), Not illustrated
Multidirectional Core
Topsoil (SU 2.0002).
L (max): 46.6 mm; W (max): 34.4 mm; Th (max):

19.1 mm; core linear dimension (max): 5.19 cm; Wt: 30.4 g; core size value: 158.
Raw material: Slightly grainy chert. Whitish, slightly pitted cortex. Translucent at the margins.
Color: 10 YR 5/1, gray, with lighter and darker banding.
Condition: Complete core.
Description: Multidirectional core worked on three faces. Six visible flake scars struck from at least three edges of the piece.
Date: Not diagnostic.

13/58 (SF 195), Not illustrated
Exhausted Multidirectional Core
Tumulus fill (SU 2.0202).
L (max): 52.1 mm; W (max): 31.4 mm; Th (max): 15.2 mm; core linear dimension (max): 5.21 cm; Wt: 23.8 g; core size value: 124.
Raw material: Grainy, fossiliferous chert. Pitted yellowish white cortex. Opaque.
Color: 2.5 Y 5/1, gray.
Condition: Core is possibly broken on one face, but the edge could be an old flake scar on this exhausted core.
Description: Exhausted multidirectional core worked on two faces with a narrow third worked face remaining. Cortex covers the rest of the piece. Six visible flake scars. One large flake scar on each of the two main faces from the same side of the piece and three small ones from the opposite side on one face. Edges used as platforms.
Date: Not diagnostic.

13/59 (SF 9), **Figs. 1.15, 13.5**
Core Fragment
Surface find (2003 season).
L (max): 28.1 mm; W (max): 25.8 mm; Th (max): 14.1 mm; core linear dimension (max): 3.33 cm; Wt: 12.9 g; core size value: 43.
Raw material: Smooth chert with a few fossil inclusions. Whitish pitted cortex.
Color: 10 YR 5/3, brown.
Condition: Broken. This piece was knocked off a core as a flake, as shown by one convex face with a bulb of percussion.
Description: Core fragment removed from the original objective piece as a flake. Multidirectional. The reshaping was probably necessitated by a very uneven scar on the dorsal side of this core flake, which has a step fracture in the middle and a hinged termination. Total of six visible flake scars. Edges used as striking platforms. The exterior of the core is still partially covered with cortex; the rest is smooth but more heavily patinated than the interior where flakes were detached.
Date: Not diagnostic.

13/60 (SF 40), Not illustrated
Multidirectional Core
Topsoil (SU 70027)
L (max): 32.6 mm; W (max): 40.0 mm; Th (max): 31.5 mm; core linear dimension (max): 4.22 cm; Wt: 46.6 g; core size value: 197.
Raw material: Smooth chert. Whitish pitted cortex. Some of the whitish cortex is worn or flaked away to reveal iron-oxide-stained cortical surface with more patina than the flaked surfaces. Opaque.
Color: 10 YR 5/3, brown.
Condition: Complete core.
Description: Multidirectional core worked on three faces. Edges used as striking platforms. Major step fracturing evident in at least two of the visible flake scars. These failed flakes likely caused the core to be abandoned while it was still relatively large. Approximately nine flake scars total.
Date: Not diagnostic.

13/61 (SF 50), Not illustrated
Core Fragment
Topsoil (SU 1.0009).
L (max): 55.5 mm; W (max): 43.5 mm; Th (max): 23.5 mm; core linear dimension (max): 5.62 cm; Wt: 53.2 g; core size value: 299.
Raw material: Dense, fossiliferous chert with holes. Poor quality for flint knapping. Pale brown pitted cortex.
Color: 10 YR 7/3, very pale brown.
Condition: Broken from the original objective piece as a flake. One face is convex with a bulb of percussion.
Description: On the dorsal side of core fragment, two clear flake scars from the same edge. One possible additional flake scar on that face, but the heavy fracturing obscures its origin. The third worked face is also very heavily fractured. The extremely poor quality of the chert caused it to be abandoned while it was still relatively large.
Date: Not diagnostic.

13/62 (SF 51), Not illustrated
Exhausted Multidirectional Core
Topsoil (SU 4.0011).
L (max): 24.5 mm; W (max): 18.5 mm; Th (max): 15.0 mm; core linear dimension (max): 2.89 cm; Wt: 8.6 g; core size value: 25.
Raw material: Smooth chert, patinated cortex.
Color: 10 YR 5/4, yellowish brown.
Condition: possibly broken on fractured face.
Description: Exhausted core worked on five faces. One of the faces shows greater patination than the rest. Multidirectional. Approximately nine visible flake scars; two of these are more heavily patinated, and one is heavily fractured. Edges used as striking platforms.
Date: Not diagnostic.

13/63 (SF 145), **Figs. 13.7, 13.10**
Exhausted Multidirectional Core
Tumulus fill (SU 4.0104).
L (max): 50.8 mm; W (max): 44.5 mm; Th (max): 15.3 mm; core linear dimension (max): 5.08 cm; Wt: 30.9 g; core size value: 157.
Raw material: Slightly grainy fossiliferous chert with white mottling. Whitish pitted cortex. Translucent at the thinnest margins.
Color: 10 YR 6/2, light brownish gray.
Condition: Complete core.
Description: Exhausted multidirectional core. Edges used as striking platforms. Worked on two faces, the rest covered by cortex. Eleven visible flake scars.
Date: Not diagnostic.

13/64 (SF 278), Not illustrated
Exhausted Multidirectional Core
Topsoil (SU 2.0003).
L (max): 22.0 mm; W (max): 24.7 mm; Th (max): 13.8 mm; core linear dimension (max): 2.99 cm; Wt: 8.2 g; core size value: 25.
Raw material: Smooth chert. Small depression with whitish pitted cortex. Translucent at the thinnest margins.
Color: 10 YR 5/4, yellowish brown.
Condition: Possibly broken on one face.
Description: Exhausted multidirectional core worked on four faces. Approximately eight flake scars, but heavy fracturing on one face obscures flake scars. One face displays a very flat surface, which may be indicative of a fracture along natural cleavage planes. Edges used as striking platforms.
Date: Not diagnostic.

13/65 (SF 385), Not illustrated
Multidirectional Core
Pit fill (SU 2.0474).
L (max): 46.4 mm; W (max): 44.5 mm; Th (max): 22.3 mm; core linear dimension (max): 4.76 cm; Wt: 35.5 g; core size value: 169.
Raw material: Slightly grainy chert with yellowish brown pitted cortex. Opaque.
Color: 10 YR 7/2, light gray, with lighter gray band.
Condition: Complete core on a thick cortical flake.
Description: Multidirectional core worked on four faces, with cortex on half of one of the faces. Edges used as striking platforms. Nine flake scars. Worked on a thick, angular flake with the bulb of percussion preserved on one face.
Date: Not diagnostic.

13/66 (SF 425), Not illustrated
Burned Multidirectional Core
Tumulus fill (SU 6.0542).
L (max): 60.1 mm; W (max): 59.3 mm; Th (max): 46.1 mm; core linear dimension (max): 6.22 cm; Wt: 200.0 g; core size value: 1244.
Raw material: Grainy, fossiliferous chert with pitted whitish and yellowish cortex. Opaque.
Color: 10 R 5/2, weak red.
Condition: Core mostly covered with cortex. Heavily weathered from rolling. Flake scar margins are evenly abraded around the piece.
Description: Burned chert cobble with approximately 11 flake scars visible on the surface. Some of the smaller scars could be from natural flaking during the weathering process. Cortex beginning to form over several of the flake scars. This piece could have been used as a hammerstone, but the overall pitting and wear from weathering prevents distinguishing use-wear of that type. Edges of the natural cobble and previous flake scars were used as striking platforms.
Date: Not diagnostic.

13/67 (SF 359), Not illustrated
Burned Multidirectional Core
Tumulus fill (SU 2.0431).
L (max): 38.3 mm; W (max): 41.5 mm; Th (max):

27.2 mm; core linear dimension (max): 4.52 cm; Wt: 51.4 g; core size value: 232.
Raw material: Smooth chert with curved fracture planes. Whitish pitted cortex. Opaque.
Color: 10 R 4/4, weak red.
Condition: Complete core. Cortex remaining in three small isolated patches.
Description: Burned multidirectional core worked on five faces. Edges on the natural cobble and previous flake scars used as platforms. Approximately 12 visible flake scars, but the fracturing evident along several margins may have obscured others.
Date: Not diagnostic.

13/68 (SF 199), Not illustrated
Multidirectional Core
Tumulus fill (SU 1.0070).
L (max): 57.8 mm; W (max): 51.8 mm; Th (max): 33.3 mm; core linear dimension (max): 5.78 cm; Wt: 89.8 g; core size value: 519.
Raw material: Grainy, dense chert. Yellowish brown pitted cortex. Opaque.
Color: 10 YR 6/4, light yellowish brown.
Condition: Complete core.
Description: Multidirectional core worked on three faces on a cobble. Seven flake scars. Edges of the natural cobble and previous flake scars used as striking platforms. More than half of the cobble is covered in cortex.
Date: Not diagnostic.

13/69 (SF 236), Not illustrated
Burned Multidirectional Core
Topsoil (SU 1.0278).
L (max): 39.6 mm; W (max): 42.0 mm; Th (max): 24.0 mm; core linear dimension (max): 4.97 cm; Wt: 38.3 g; core size value: 190.
Raw material: Grainy, fossiliferous chert. One flake scar displays a fossil imprint of a 8 × 5 mm shell fragment. Whitish and dark brown pitted cortex. Opaque.
Color: 10 R 6/3, pale red.
Condition: Complete core.
Description: Burned multidirectional core worked on three faces. Eleven visible flake scars. Edges used as striking platforms.
Date: Not diagnostic.

13/70 (SF 274), Not illustrated
Unidirectional Core
Tumulus fill (SU 4.0204).
L (max): 28.3 mm; W (max): 53.7 mm; Th (max): 29.2 mm; core linear dimension (max): 5.37 cm; Wt: 57.1 g; core size value: 307.
Raw material: Grainy chert. Yellowish brown pitted cortex. Opaque.
Color: 10 YR 5/3, brown.
Condition: Complete core.
Description: Unidirectional core on a small chert cobble. Four flake scars struck from one cortical edge of the piece. Worked on one long face.
Date: Not diagnostic.

13/71 (SF 242), Not illustrated
Exhausted Multidirectional Core
Topsoil (SU 1.0278).
L (max): 46.7 mm; W (max): 31.4 mm; Th (max): 27.5 mm; core linear dimension (max): 4.67 cm; Wt: 27.2 g; core size value: 127.
Raw material: Fossiliferous chert with lighter grainy mottling; opaque.
Color: 10 YR 6/4, light yellowish brown.
Condition: Complete core.
Description: Exhausted multidirectional core worked on four faces. Roughly pyramid shaped. No cortex remaining. Edges of the piece used as striking platforms. Ten to 12 flake scars visible, but fracturing on one face obscures them.
Date: Not diagnostic.

13/72 (SF 215), Not illustrated
Exhausted Multidirectional Core
Topsoil (SU 2.0002).
L (max): 30.7 mm; W (max): 31.4 mm; Th (max): 24.2 mm; core linear dimension (max): 3.66 cm; Wt: 23.4 g; core size value: 86.
Raw material: Slightly grainy banded chert.
Color: 2.5 Y 7/1, light gray, with slightly darker gray bands.
Condition: Complete core.
Description: Exhausted multidirectional core worked on four faces. Roughly pyramid shaped. Edges used as striking platforms. Approximately 10 flake scars.
Date: Not diagnostic

13/73 (SF 326), **Figs. 13.7, 13.9**
Multidirectional Core
Tumulus fill (SU 2.0380).
L (max): 17.9 mm; W (max): 38.3 mm; Th (max):

29.8 mm; core linear dimension (max): 4.10 cm; Wt: 21.1 g; core size value: 87.
Raw material: Slightly grainy fossiliferous chert. Yellowish pitted and pock marked cortex. Opaque.
Color: 10 YR 7/2, light gray, with yellowish brown iron oxide staining under remnants of cortex.
Condition: Complete core.
Description: Multidirectional core worked on five faces, with a small patch of cortex preserved. One face is unidirectionally worked from the edge of one of the other faces. Edges are used as striking platforms.
Date: Not diagnostic.

C1a. Core-shaping elements

13/74 (SF 83), **Fig. 13.9**
Core-Shaping Element with Use-Wear
Tumulus fill (SU 4.0079).
L (max): 59.5 mm; W (max): 35.3 mm; Th (max): 9.3 mm; L, utilized edge: 40.5 mm; Wt: 17.2 g.
Raw material: Smooth, slightly grainy chert with very thick, pitted cortex.
Color: 2.5 Y 6/3, light yellowish brown, with white as the chert approaches the cortex. Munsell reading taken from the core color of the chert, not the cortex.
Condition: Flake with the bulb of percussion fractured away.
Description: Large cortical flake with the left margin chipped slightly and used. The dorsal retouch flakes with stepped terminations are wide and shallow and form a serrated edge.
Date: Neolithic or Bronze Age?

13/75 (SF 193), Not illustrated
Core-Shaping Element with Use-Wear
Topsoil (SU 4.0201).
L (max): 32.1 mm; W (max): 20.9 mm; Th (max): 6.2 mm; L, utilized edge: 33 mm; Wt: 4.0 g.
Raw material: Smooth, fine chert. White pitted cortical surface with iron oxide staining. Slightly translucent.
Color: 10 YR 6/4, light yellowish brown, with dark gray mottling and a band of 10 YR 5/1, gray, running through the flake parallel to the cortical surface.
Condition: Broken at the distal end.
Description: Utilized cortical flake with a small notch. The flake is wedge shaped, with the narrow left margin used. The right dorsal side of the piece is covered in cortex, which provided backing for the utilized edge. The proximal half of the left margin is convex and shows light abrasion from use. The distal half of the preserved edge displays an uneven concave notch from ventral retouch. The very distal end of the edge displays two small chips that may or may not have been intentional.
Date: Neolithic or Bronze Age.

13/76 (SF 436), Not illustrated
Core-Shaping Element with Use-Wear
Topsoil (SU 6.0002).
L (max): 48.2 mm; W (max): 16.1 mm; Th (max): 5.2 mm; L, utilized edge: 22.9 mm; Wt: 4.8 g.
Raw material: Smooth chert with white pitted cortex. Opaque. Probably burned.
Color: 2.5 Y 4/3, olive brown. Appears to be blackened yellowish brown.
Condition: Flake broken at the proximal end. Striking platform and bulb of percussion missing, but the rest of the flake morphology indicates that it is a flake.
Description: Utilized cutting edge on the right edge of a flake with cortex on 75% of the dorsal surface. Triangular in cross-section. The cortex provides a backing for the utilized edge. Shallow, clustered retouch along the straight right margin with stepped terminations. Light abrasion from use.
Date: Neolithic or Bronze Age.

13/77 (SF 197), Not illustrated
Core-Shaping Element with Use-Wear
Tumulus fill (SU 2.0202).
L (max): 38.3 mm; W (max): 19.4 mm; Th (max): 8.0 mm; L, utilized edge: 26.1 mm; Wt: 6.6 g.
Raw material: Grainy, fossiliferous chert. Lighter, pitted cortical surface with darker brown staining directly beneath; completely opaque.
Color: 10 YR 5/3, brown.
Condition: Complete cortical flake.
Description: Utilized cortical flake. The flake is wedge shaped, with the longitudinal axis of the flake running parallel to the utilized edge. The thick right dorsal side with cortical surface preserved at the distal end provides backing for the edge. The thick right dorsal side also displays two broad flake scars on the proximal half, which also provide backing for the blade. The thin left margin shows abrasion from use along the distal two-thirds of the edge.
Date: Neolithic or Bronze Age.

13/78 (SF 448), **Fig. 13.3**
Core Rejuvenation Flake
Topsoil (SU 2.0002).
L (max): 38.5 mm; W (max): 43.2 mm; Th (max): 16.5 mm; Wt: 20.5 g.
Raw material: Grainy chert, with small voids. Yellowish pitted cortex. Opaque.
Color: 2.5 Y 7/3, pale yellow.
Condition: Complete flake.
Description: Flake with cortex on the dorsal distal portion. Three flake scars on the rest of the dorsal surface. Two of the scars form a wing-like shape on the dorsal surface, with a more heavily patinated scar in the center.
Date: Mesolithic or later.

C2. Scrapers

13/79 (SF 395), **Figs. 13.6, 13.10**
End Scraper with a Notch on an Exhausted Core
Tomb I (64) fill (SU 1.0361).
L (max): 46.3 mm; W (max): 45.5 mm; Th (max): 11.6 mm; L, scraper edge: 11.8 mm; L, shaped point: 7.9 mm; core linear dimension (max) 5.02 cm; Wt: 26.9 g; core size value: 135.
Raw material: Slightly grainy chert with fossil inclusions. Whitish pitted cortex. Opaque.
Color: 2.5 Y 6/2, light brownish gray, and darker yellowish brown staining below cortex.
Condition: Complete tool with use-wear.
Description: Thin exhausted core with numerous flake scars on both surfaces, with one edge reshaped into a scraper with a notch. The notch forms a 95° angle in the margin of the piece. The scraper edge angle is 75°. Retouch scars with feathered terminations along the scraper edge. The retouch scars on the shaped point have stepped terminations. The point formed by the notch is roughly triangular in shape and shows light usewear at the tip, suggesting it may have been used as a perçoir. Light abrasion along the scraper edge as well. One of the cortical edges of the piece may also have been utilized, but the heavy fracturing may have been a feature of the raw material and its use as a core.
Date: Not diagnostic.

13/80 (SF 118), Not illustrated
Broken Scraper on a Chunk
Tumulus fill (SU 2.0117).
L (max): 32.3 mm; W (max): 22.7 mm; Th (max): 7.6 mm; L, preserved scraper: 12.6; Wt: 6.4 g.
Raw material: Grainy, fossiliferous chert. Opaque.
Color: 10 YR 6/4, light yellowish brown.
Condition: Darkened cortical surface on the dorsal side of the piece. Part of the scraper edge is broken off. The piece may be formed on a flake, but the bulb of percussion would be broken off with the upper part of the scraper if that is the case. The discussion will proceed as if the piece is a flake for the sake of clarity of description.
Description: A scraper on an amorphous, roughly flake-shaped piece of chert. With the surmised proximal end of the piece down, the straight scraper edge is on the left ventral distal end of the blank. The edge angle is approximately 85°. The proximal end of the scraper is broken off. A small spur was formed by the end of the scraper retouch at the left distal corner of the piece. Light retouch on the distal end beside the spur. Abrasion along the worked edges.
Date: Not diagnostic.

C3. Becs

13/81 (SF 377), Not illustrated
Bec on a Cortical Chunk
Pit fill (SU 2.0474).
L (max): 28.8 mm; W (max): 20.3 mm; Th (max): 9.2 mm; L, notched utilized edge: 15.0 mm; L, perforator spur: 3.0 mm; Wt: 3.9 g.
Raw material: Slightly grainy, fossiliferous chert; somewhat translucent.
Color: 7.5 YR 4/3
Condition: Unclear whether the chunk was broken off from a flake.
Description: Small notch formed by dorsal retouch next to the pointed end on a roughly triangular chunk creating a bec. Retouch scars are very shallow, with stepped terminations. Abrasion along the S-shaped notched margin.
Date: Paleolithic?

C4. Multitools

13/82 (SF 008/023), **Figs. 1.15, 13.2**
Combination Tool on a Bifacially Worked Piece with Use-Wear Surface
L (max): 22.1 mm; W (max): 20.5 mm; Th (max): 7.9 mm; total L, edge: 72.0 mm; Wt: 3.6 g.
Raw material: Smooth chert with a few scattered,

white fossil inclusions; translucent at the thin edges.

Color: 2.5 Y 6/2, light brownish gray.

Condition: Complete tool. May have been formed on a broken biface or small exhausted core.

Description: Roughly triangular piece with every edge retouched and/or used. Lenticular in cross-section. Flake scars on the both the dorsal and ventral sides of the piece, which appear to be unrelated to the final shape, indicate that this piece may be formed from a broken biface or an exhausted core. Beginning at the top **Figure 13.2** (**13/82**) and proceeding clockwise around the piece (see the left hand drawing for orientation), the pointed tip at the top of the piece was formed by one flake scar from the left margin and one from the right on the dorsal side. Very shallow retouch on the ventral side with feathered terminations reshaped the right margin of the pointed end very slightly. This type of shaping on a point is consistent with Upper Paleolithic borer tools. Just below the pointed tip, the edge continues as a straight scraper edge with use-wear. Dorsal retouch is evident, with hinged and stepped flake scars. The edge angle is 70°. The rest of the right lateral margin continues straight below a 50° angle turn in the margin. This edge is classified as a cutting edge by the shallow angle. Two retouch scars are visible on the dorsal surface with feathered terminations. On the ventral surface, two small retouch scars are clustered together with feathered terminations. Abrasion and tiny overlapping step fractures from use-wear are seen along the edge. The bottom edge of the piece has not been retouched until it approaches the left margin. The ventral edge along the bottom shows evidence of use-wear, but the dorsal edge does not. This end of the tool is the area most likely to reflect that this tool was formed on a broken biface. At the left margin, retouch on the dorsal side forms a sinuous scraper edge, which first curves out to form a convex margin and then curves back in to form a concave curve just below the pointed tip. The edge angle is 80°.

Date: Upper Paleolithic?

13/83 (SF 206), **Fig. 13.2**
Combination Tool on Bifacially Worked Piece
Topsoil (SU 1.0001).

L (max): 25.5 mm; W (max): 18.2 mm; Th (max): 9.8 mm; L, retouched notched edge: 24.0 mm; Wt: 4.1 g.

Raw material: Smooth, fine chert; slightly translucent.

Color: 10 YR 5/4, yellowish brown.

Condition: May be complete. The blank shape is indeterminate.

Description: Retouched notched tool on a roughly triangular piece of chert. Bimarginal retouch along the notched edge (top in drawing of **Fig. 13.2** [**13/83**]). The bimarginally flaked edge is concave with continuous feathered and stepped termination retouch scars along the front side and a cluster of stepped retouch scars on the back. The edge is heavily abraded from use. The rest of the piece shows no signs of use.

Date: Paleolithic?

13/84 (SF 364), **Figs. 13.7, 13.10**
Retouched Fragment with Notch and Denticulate
Tumulus fill (SU 2.0431).

L (max): 21.3 mm; W (max): 44.1 mm; Th (max): 7.1 mm; W, notch: 10.5 mm; L, retouched edge: 17.7 mm; Wt: 4.7 g.

Raw material: Smooth, fine chert; opaque.

Color: 10 YR 7/4, very pale brown, with darker brown stripes.

Condition: Fragment, possibly of a flake. Thick break along one side, as would be expected from a broken flake. No bulb of percussion preserved.

Description: Roughly triangular fragment with retouch along the left side and a notch on the right side. As shown in **Figure 3.7** (**13/84**), with the longest side of the piece and the flat, possible ventral flake surface pointing down, the dorsal retouch along the left margin is irregular yet continuous along the worked part of the edge. The margin is roughly straight, but appears to be a small dentinculate based on the ragged spurs left between the retouch flake scars, which have feathered terminations. The notch on the right margin was created by a single dorsal retouch flake. The interior of the notch shows abrasion. To the right of the notch, light chipping forms another ragged, straight denticulate edge similar to the left margin, with spurs left between the flake scars.

Date: Not diagnostic.

C5. Other Core Tools

13/85 (SF 362), Not illustrated
Fragment with Utilized Edges
Tomb XXXVII (76) fill (SU 2.0423).

L (max): 36.1 mm; W (max): 21.8 mm; Th (max): 6.0 mm; L, straight utilized edge: 18.9 mm; L, concave utilized edge: 19.0 mm; Wt: 5.1 g.
Raw material: Smooth chert, burned, with white mottling. Smooth, dark, pitted cortex on the left dorsal surface.
Color: Gley Chart 2, 7/1; light bluish gray due to burning.
Condition: Burned fragment of chert. Possibly a flake whose bulb of percussion spalled off on the flat ventral side of this piece; because of this possibility, the end with the ventral spall will be referred to as the proximal end of the piece. The distal end is broken into a pointed shape as well, but there is no sign of intentionality or use in the breaks. The piece is fractured on the proximal end into a point as well, but these two breaks also show no evidence of use.
Description: Utilized lateral margins on a burned fragment of chert. The left lateral margin is shaped into a rough denticulate with some retouch in the notches. Continuous retouch flake scars on the dorsal surface terminate in cortex. Scattered retouch or use flake scars visible on the ventral surface as well. The right lateral margin is roughly S-shaped with no retouch, but it has abrasion along the edge and spalls on the ventral side.
Date: Not diagnostic.

13/86 (SF 87), **Fig. 13.6**
Notch Tool
Topsoil (SU 4.0004).
L (max): 31.4 mm; W (max): 19.0 mm; Th (max): 11.4 mm; L, heavily used notched edge: 35.0 mm; Wt: 7.3 g.
Raw material: Smooth chert, few inclusions; opaque.
Color: 2.5 Y 6/4, light yellowish brown.
Condition: Appears to be a complete tool, but the tip of the pointed end may be broken off if it were once sharper.
Description: Wedge-shaped chunk of chert with a wide notch on the right (narrow) margin. The notch is very heavily abraded from use, as is the straight edge just below it. The wedge shape created a very efficient cutting tool because the wide side of the piece provided good leverage for the tool user, although the piece was not retouched to provide that backing for the blade. The left margins also show some abrasion, but not as significant as the notched margin. The pointed tip shows some use-wear around the lateral margins, but may have been snapped off, since the very tip displays a fairly clear, unabraded scar.
Date: Not diagnostic.

13/87 (Bulk collection from SU 1.0377), Not illustrated
Double Notch Tool on a Cortical Chunk
Tumulus fill (SU 1.0377).
L (max): 21.7 mm; W (max): 4.9 mm; Th (max): 5.6 mm; L, notched edge: 12.3 mm; Wt: 1.6 g.
Raw material: Smooth chert with fossiliferous inclusions. Yellowish brown pitted cortex. Translucent only at the thinnest margin.
Color: 2.5 Y 6/2, light brownish gray.
Condition: Blank is possibly a broken flake.
Description: Chunk of chert with two small notches on one margin, which form a small spur between them. Viewing the piece as oriented with the flake scars for the notches visible and pointing up, the two small notches were each formed by a single retouch flake scar with a feathered termination; abrasion in the left notch extending up on the spur between them. An additional concave flake scar to the left suggests that this piece could be a broken denticulate.
Date: Not diagnostic.

D. Middle Paleolithic flakes and tools

13/88 (SF 121), **Fig. 13.2**
Broken Levallois Flake
Tomb LXVIII (13) fill (SU 2.0102).
L (max): 24.4 mm; W (max): 39.7 mm; Th (max): 4.1 mm; Wt: 4.4 g.
Raw material: Slightly grainy fossiliferous chert; partially translucent.
Color: 2.5 Y 6/3, light yellowish brown.
Condition: Flake diagonally broken at the distal end.
Description: Broken Levallois flake. Likely a core-reshaping flake. On the dorsal surface, one relatively shallow central flake scar is flanked by flake scars with parallel margins following the longitudinal axis of the flake.
Date: Middle Paleolithic.

13/89 (SF 007/022), Fig. 13.2
Broken Levallois Flake
Surface find (Mapping Station 4).
L (max): 30.7 mm; W (max): 48.8 mm; Th (max): 6.4 mm; Wt: 7.9 g.

Raw material: Smooth chert; slightly translucent at the thinnest margins.
Color: 2.5 Y 6/3, light yellowish brown.
Condition: Broken diagonally at the distal end.
Description: Broken Levallois flake. One flake scar on the dorsal surface runs parallel along the right proximal margin of the flake, but this scar could be shatter from the removal of this flake, since the bulb of percussion negative from the scar on the dorsal surface is missing. Small area of dark brown, heavily patinated iron oxide staining remains on the right lateral margin.
Date: Middle Paleolithic.

13/90 (SF 449), **Fig. 13.2**
Levallois Flake
Topsoil (SU 2.0002).
L (max): 39.0 mm; W (max): 41.0 mm; Th (max): 7.5 mm; Wt: 8.0 g.
Raw material: Slightly grainy, fossiliferous chert; slightly translucent at the thinnest margins.
Color: 2.5 Y 6/3, light yellowish brown.
Condition: Complete flake.
Description: Flat Levallois flake shaped like a hexagon. Four flake scars on the surface, which terminate near the center of the dorsal side of the flake, as would be expected from a lineal or possibly recurrent Levallois core of roughly round shape. One additional small flake was detached from the dorsal surface within one of the four flake scars, which displays a stepped termination at the thickest center part of the flake where the other flake scars terminate. This small flake scar was likely either from a failed first attempt to remove this flake or shatter from the removal.
Date: Middle Paleolithic.

13/91 (SF 452), Not illustrated
Middle Paleolithic Flake
Tumulus fill (SU 8.0204).
L (max): 30.1 mm; W (max): 31.4 mm; Th (max): 10.4 mm; Wt: 10.9 g.
Raw material: Grainy, slightly fossiliferous chert. Yellowish brown patina on the dorsal surface. Opaque.
Color: 2.5 Y 6/3, light yellowish brown.
Condition: Complete flake.
Description: Roughly square-shaped flake. Very heavy patination on the dorsal surface over three clear flake scars, which was partially removed by a small flake on the dorsal surface behind the striking platform. Joint in the ventral surface from the right distal corner to the center of the left margin changes the angle of the ventral surface by about 30°.
Date: Middle Paleolithic.

13/92 (SF 149), **Fig. 13.8**
Primitive Perçoir
Tumulus fill (SU 2.0118).
L (max): 61.5 mm; W (max): 43 mm; Th (max): 24 mm; total L, worked edges: 59 mm; Wt: 51.4 g.
Raw material: Thin cobble of grainy chert with yellowish white, pitted cortex; completely opaque.
Color: 10 YR 6/4
Condition: Tool appears complete.
Description: Roughly formed perçoir (perforator) on a cobble fragment with cortex completely covering three faces and partially covering a fourth broken face. The point is triangular in shape, formed by three flake scars that are retouched along two edges. Referring to the side of the point with two flake scars as the dorsal surface, the margin to the left of the point was retouched into a deep notch by a cluster of stepped flake scars. The margin to the right of the point was retouched on the ventral side by scattered isolated short flake scars with feathered terminations.
Date: Middle Paleolithic?
Comparanda: Runnels 2004: fig. 7, SF 0028. This depicts a denticulate, but the flaking method producing a large, blocky point on a cobble is similar. Fistani 1989: pl. 2, top, Mousterian (Gajtan).

13/93 (SF 454), Not illustrated
Middle Paleolithic Cortical Flake?
Topsoil (SU 5.0278).
L (max): 50.7 mm; W (max): 34.8 mm; Th (max): 7.1 mm; Wt: 11.3 g.
Raw material: Slightly grainy chert. Brown pitted cortex. Slightly translucent.
Color: 10 YR 6/2, light brownish gray.
Condition: Complete flake. Possibly broken along the right margin.
Description: Possible Levallois flake. Dorsal surface is dominated by a single flake scar whose longitudinal axis runs perpendicular to the axis of this flake. The left and right margins also display nar-

row flake scars, but the right one may be broken off. Cortex on the proximal margin. The flake morphology is not particularly diagnostic; however, the raw material is very similar to that of the other Middle Paleolithic flakes in this collection, and the nature of the flake scars on the dorsal surface suggests that it dates to this period.
Date: Middle Paleolithic?

13/94 (SF 447), **Fig. 13.3**
Middle Paleolithic(?), Cortical Flake
Tumulus fill (SU 2.0532).
L (max): 46.4 mm; W (max): 46.1 mm; Th (max): 11.6 mm; Wt: 21.1 g.
Raw material: Grainy, fossiliferous chert with tiny voids and inclusions. Yellowish, pitted cortex. Opaque.
Color: 2.5 Y 6/4, light yellowish brown.
Condition: Complete flake.
Description: Possible core-shaping flake mostly covered by cortex on the dorsal side. Three other flake scars with darker yellowish brown staining or patination. Broad, flat distal end flares out from a thick striking platform. Although this piece is not particularly diagnostic, it shares the raw material and general size and character of the other Middle Paleolithic flakes and tools in the assemblage.
Date: Middle Paleolithic?

Chapter 14
Daub

John K. Papadopoulos

Introduction

One of the most prominent finds in the tumulus fill, and certainly among the most unexpected, was daub. The numerous pieces of this material were originally entered into the Lofkënd database as "fired clay, not pottery," and were counted and weighed along with pottery, flint, and other finds; a few were selected for inventory, either for their context or on account of their surviving impressions. Many of these pieces and lumps of clay are amorphous, but many preserve reed, rod, or stake impressions, indicating that the clay had been used as an integral component of wattle-and-daub architecture. Such pieces of daub are common in prehistoric settlement sites, but the incidence of so much daub (see **Table 14.1**) in a burial tumulus came as a surprise, requiring explanation.

Elsewhere in Albania, daub has been reported from a variety of settlement contexts, especially for the Neolithic period (see Korkuti 1982a:94, fig. 4, nos. 1–2; Korkuti 1995a:41–43, pl. II, no. 3 [Vashtëmi]; Korkuti and Andrea 1974:53–60 [Cakran]; Lera 1987:28, fig. 3 [Barç]; and perhaps also Kolsh: see Korkuti 1977–1978), but also from the Bronze Age and Early Iron Age levels at Sovjan in the Korçë basin (Lera, Prendi, and Touchais 1996:1008, fig. 7; Prendi, Lera, and Touchais 1996:237, fig. 5, 262; Prendi and Touchais 1999:23, fig. 10). It is worth noting that daub is common as remains of building material elsewhere in the prehistoric Balkans (e.g., Hiller and Nikolov 1997:94–95, pl. XXIII [Karanovo]; Grbić 1960: pl. 5, nos. 1–2 [Porodin]).

Daub was not just an archaeological artifact in Albania and elsewhere in the Balkans, but has continued through various times in the historic era, surviving into the present. In 1904, Carl Patsch recorded a modern wattle-and-daub house in Muzakia (Patsch 1904:134, fig. 105), as did Luigi Ugolini a few decades later (Ugolini 1927:59–60, pl. XCVI, fig. 114), and similar houses still exist in Albania and specifically in the region of Lofkënd (**Figs. 14.1–14.3**), though they are becoming rare. The same type of daub is used in Albania for other purposes, including as a waterproof covering for haystacks (**Fig. 14.4a–b**).

As for the chronology of the Lofkënd daub, there is no way to gauge its date, except that the daub had to be earlier than the contexts in which it was encountered at Lofkënd, including as occasional grave fill, but primarily as tumulus fill. How much earlier is impossible to estimate, since wattle-and-daub architecture was in use from the Neolithic period through the Early Iron Age and beyond.

In the context of the stratigraphy and processes of tumulus formation, the pieces of daub from Lofkënd should be taken together with the other category of prominent material encountered both on the surface and throughout the fill of the Lofkënd tumulus: chipped stone tools (see Aprile, Chapter 13). The combination of chipped stone tools and remnants of wattle-and-daub architecture either suggests that such material was plentiful enough in the nearby landscape to be used as tumulus and grave fill, or it raises the intriguing possibility that those burying the dead intentionally brought material

TABLE 14.1 Uninventoried pieces of daub from the various stratified units at Lofkënd by number and weight

Trench and unit	Context type	No. of frr/pieces	Wt	Inventoried pieces
Surface collection		5	26 g	
SU 1.0007	Tumulus fill	175	232 g	**14/1** (SF 77), **14/2** (SF 37)
SU 1.0009	Topsoil	99	235 g	
SU 1.0020	Grave 3 fill	1	14 g	
SU 1.0037	Tumulus fill	27	123 g	
SU 1.0039	Tumulus fill	395	148 g	**14/6** (SF 128), **14/7** (SF 80)
SU 1.0047	Tumulus fill	65	197 g	**14/4** (SF 78)
SU 1.0053	Tumulus fill	29	70 g	
SU 1.0067	Tumulus fill	351	1012 g	**14/13** (SF 127), SF 166
SU 1.0070	Tumulus fill	786	2591 g	**14/10** (SF 135)
SU 1.0089, 1.0145	Grave 10 fill	9	15 g	
SU 1.0095	Grave 11 fill	2	6 g	
SU 1.0127	Tumulus fill	12	54 g	
SU 1.0129	Tumulus fill	8	?	
SU 1.0141	Topsoil	104	269 g	**14/17** (SF 173)
SU 1.0143	Grave 19 fill	2	1 g	
SU 1.0147	Grave 20 fill	3	10 g	
SU 1.0155	Grave 22 fill	15	47 g	
SU 1.0161	Grave 24 fill	2	6 g	
SU 1.0162	Grave 24 skeleton	4	5 g	
SU 1.0176	Grave 28 fill	2	5 g	
SU 1.0212	Grave 30 fill	7	25 g	
SU 1.0222	Grave 33 fill	11	40 g	
SU 1.0237	Grave 36 fill	2	8 g	
SU 1.0240	Tumulus fill	78	350 g	
SU 1.0278	Topsoil	139	442 g	
SU 1.0279	Tumulus fill	190	789 g	
SU 1.0281	Tumulus fill	40	180 g	
SU 1.0301	Grave 48 fill	15	20 g	
SU 1.0317	Grave 52 fill	2	6 g	
SU 1.0335	Grave 58 fill	1	0.5 g	
SU 1.0344	Grave 60 fill	3	2 g	
SU 1.0358	Grave 63 fill	7	16 g	
SU 1.0361	Grave 64 fill	312	863 g	**14/14** (SF 387)
SU 1.0368	Grave 66 fill	7	14 g	
SU 1.0375	Topsoil 283		465 g	
SU 1.0377	Tumulus fill	313	925 g	SF 355
SU 1.0378	Tumulus fill	19	30 g	
SU 1.0380	Tumulus fill	63	76 g	
SU 1.0387	Ceramic unit in tumulus fill	1	1 g	
SU 1.0389	Grave 69 fill	1	5 g	
SU 1.0399	Tumulus fill, near topsoil	175	481 g	
SU 1.0407	Grave 72 fill	12	60 g	SF 357
SU 1.0415	Grave 74 skeleton	18	40 g	
SU 1.0438	Grave 79 fill	9	18 g	
SU 1.0440	Tumulus fill	375	969 g	SF 358
SU 1.0460	Tumulus fill, near topsoil	73	165 g	
SU 1.0466	Grave 83 fill	3	2 g	
SU 1.0471	Grave 84 fill	50	115 g	
SU 1.0482	Tumulus fill	22	40 g	
SU 1.0497	Grave 87 fill	45	51 g	
SU 1.0500	Tumulus fill	141	283 g	
SU 1.0503	Tumulus fill	103	127 g	
SU 1.0509	Tumulus fill	23	47 g	
SU 1.0518	Tumulus fill	80	171 g	
SU 2	Surface sweeping and scraping 2006	54	78 g	
SU 2.0002	Topsoil	770	1565 g	
SU 2.0003	Topsoil	218	383 g	
SU 2.0005	Tumulus fill	6	21 g	
SU 2.0006	Tumulus fill	33	156 g	
SU 2.0010	Tumulus fill	8	63 g	
SU 2.0023	Tumulus fill	67	245 g	**14/11** (SF 039)
SU 2.0024	Tumulus fill	5	30 g	
SU 2.0033	Tumulus fill	68	245 g	
SU 2.0036	Tumulus fill	39	155 g	
SU 2.0037	Tumulus fill	67	174 g	
SU 2.0040	Tumulus fill	32	124 g	
SU 2.0041	Tumulus fill	8	8 g	
SU 2.0042	Tumulus fill	39	138 g	
SU 2.0050	Tumulus fill	18	27 g	
SU 2.0066	Tumulus fill	31	39 g	

Continued on next page

TABLE 14.1 (*continued*). Uninventoried pieces of daub from the various stratified units at Lofkënd

Trench and unit	Context type	No. of frr/ pieces	Wt	Inventoried pieces
SU 2.0071	Tumulus fill	43	80 g	**14/5** (SF 079)
SU 2.0072	Tumulus fill	139	200 g	
SU 2.0073	Tumulus fill	12	8 g	
SU 2.0075	Tumulus fill	5	5 g	
SU 2.0078	Tumulus fill	318	654 g	
SU 2.0097	Tumulus fill	7	35 g	
SU 2.0102	Grave 13 fill	69	120 g	
SU 2.0117	Tumulus fill	30	48 g	
SU 2.0118	Tumulus fill	21	25 g	
SU 2.0150	Grave 13 east fill	5	8 g	
SU 2.0170	Tumulus fill	16	61 g	
SU 2.0198	Tumulus fill (23)	10	?	
SU 2.0199	Tumulus fill	6		
SU 2.0200	Tumulus fill	40		
SU 2.0202	Tumulus fill	512	1772 g	**14/12** (SF 218), SF 219
SU 2.0217	Tumulus fill	45	77 g	
SU 2.0228	Tumulus fill	29	60 g	
SU 2.0229	Tumulus fill	1	1 g	
SU 2.0235	Tumulus fill	10	62 g	
SU 2.0239	Tumulus fill	81	268 g	
SU 2.0243	Grave 37 skeleton	4	23 g	
SU 2.0244	Grave 37 fill	4	7 g	
SU 2.0252	Grave 40 skeleton	1	35 g	
SU 2.0253	Tumulus fill	63	116 g	
SU 2.0254	Grave 40 fill	10	20 g	
SU 2.0259	Tumulus fill	30	62 g	
SU 2.0266	Grave 42 fill	5	5 g	
SU 2.0273	Grave 44 skeleton	8	8.5 g	
SU 2.0274	Grave 44 fill	4	3 g	
SU 2.0287	Tumulus fill	51	170 g	
SU 2.0288	Tumulus fill	17	135 g	
SU 2.0292	Tumulus fill	21	38 g	
SU 2.0297	Tumulus fill	58	227 g	
SU 2.0330	Grave 57 cut	3	8 g	
SU 2.0370	Grave 68 fill	56	148 g	
SU 2.0379	Tumulus fill	53	89 g	
SU 2.0380	Tumulus fill	111	368 g	SF 309
SU 2.0382	Grave 67 fill	1	1 g	
SU 2.0395	Tumulus fill	105	120	
SU 2.0397	Grave 70 fill	25	25 g	
SU 2.0398	Grave 70 skeleton	11	11.5 g	
SU 2.0403	Tumulus fill	38	58 g	
SU 2.0419	Grave 75 fill	1	10 g	
SU 2.0423	Grave 76 fill	9	25 g	
SU 2.0431	Tumulus fill	425	748 g	
SU 2.0442	Tumulus fill	15	28 g	
SU 2.0446	Grave 80 fill	2	2 g	
SU 2.0453	Grave 81 fill	85	114 g	
SU 2.0458	Tumulus fill/ near bedrock	1	1 g	
SU 2.0474	Tumulus fill	171	316 g	
SU 2.0475	Grave 85 fill	100	175 g	
SU 2.0481	Grave 86 fill	20	45 g	
SU 2.0487	Near topsoil	69	106 g	
SU 2.0491	Grave 88 fill	31	56 g	
SU 2.0505	Grave 88 fill	5	2 g	
SU 2.0507	Grave 89 fill	44	70 g	
SU 2.0510	Grave 90 fill	8	14 g	
SU 2.0520	Tumulus fill	1	2 g	
SU 2.0527	Under Grave 88	1	3 g	
SU 2.0532	Tumulus fill (380)	136	309 g	
SU 2.0533	Tumulus fill (487)	5	10 g	
SU 2.0538	Grave 93 fill	6	10 g	
SU 2.0543	Tumulus fill/pit	108	254 g	
SU 2.0546	Grave 94 fill	5	7 g	
SU 2.0565	Tumulus fill/ near bedrock	10	16 g	
SU 2.0566	Tumulus fill/ near bedrock	12	21 g	
SU 2.0568	Grave 96 fill	2	7 g	
SU 2.0570	Near topsoil	12	28 g	**14/8** (SF 446)
SU 2.0571	Grave 97 fill	4	7 g	
SU 2.0577	Tumulus fill	17	79 g	
SU 2.0578	Grave 99 fill	26	88 g	
SU 2.0580	Grave 98 fill	9	10 g	

TABLE 14.1 (*continued*). Uninventoried pieces of daub from the various stratified units at Lofkënd

Trench and unit	Context type	No. of frr/ pieces	Wt	Inventoried pieces
SU 2.0586	Grave 100 fill	5	25 g	
SU 2.0592	Under Grave 98	12	12 g	
SU 3.0003	Topsoil	5	25 g	
SU 3.0008	Tumulus fill	13	21 g	
SU 3.0045	Under Grave 1	95	270 g	
SU 3.0046	Tumulus fill	14	37 g	
SU 3.0055	Tumulus fill (46)	1	2 g	
SU 3.0061	Tumulus fill	58	72 g	
SU 3.0064	Topsoil	25	46 g	
SU 3.0069	Tumulus fill	13	27 g	
SU 3.0077	Tumulus fill	22	23 g	
SU 3.0092	Tumulus fill	77	115 g	
SU 3.0099	Grave 12 fill	39	61 g	
SU 3.0134	Grave 17 fill	9	25 g	
SU 3.0184	Under Grave 12	9	15 g	
SU 4	Back fill	3	2 g	
SU 4.0004	Topsoil	58 + 10	197 g	
SU 4.0011	Topsoil	194	753 g	**14/3** (SF 38)
SU 4.0015	Tumulus fill	3	11 g	
SU 4.0031	Grave 4 fill	47	341 g	
SU 4.0034	Ceramic unit, tumulus fill	10 (large)	175 g	
SU 4.0035	Tumulus fill	100	278 g	
SU 4.0043	Grave 4 fill	8	9 g	
SU 4.0048	Ceramic unit in tumulus fill	148	498 g	
SU 4.0057	Tumulus fill (35)	45	97 g	
SU 4.0059	Grave 7 fill	15	50 g	
SU 4.0079	Tumulus fill	27	48 g	
SU 4.0084	Grave 9 fill	22	46 g	
SU 4.0085	Grave 9 skeleton	1	4 g	
SU 4.0086	Tumulus fill	232	947 g	
SU 4.0093	Topsoil above Grave 9	3	14 g	
SU 4.0104	Tumulus fill	111	311 g	
SU 4.0115	Tumulus fill	60	150 g	
SU 4.0129	Tumulus fill	72	154 g	
SU 4.0130	Tumulus fill	72	236 g	
SU 4.0131	Tumulus fill	26	35 g	
SU 4.0132	Tumulus fill	9	15 g	
SU 4.0138	Grave 18 fill	32	67 g	
SU 4.0201	Topsoil	323	623 g	
SU 4.0203	Tumulus fill	211	435 g	
SU 4.0204	Tumulus fill	955	2532 g	SF 286
SU 4.0205	Tumulus fill	49	262 g	
SU 4.0207	Tumulus fill	71	69 g	
SU 4.0209	Grave 29 fill	39	58 g	
SU 4.0215	Grave 31 fill	13	43 g	
SU 4.0233	Grave 35 fill	71	69 g	
SU 4.0246	Grave 38 fill	21	49 g	
SU 4.0268	Grave 43 fill	38	26 g	
SU 4.0280	Ceramic unit in tumulus fill	11	44 g	
SU 4.0283	Grave 46 fill	4	28 g	
SU 4.0286	Tumulus fill	736	2,350 g	**14/15** (SF 310), SF 330+331
SU 4.0311	Grave 51 upper fill	12	16 g	
SU 4.0314	Grave 51 lower fill	3	0.5 g	
SU 4.0325	Grave 55 fill	10	5.5 g	
SU 4.0328	Grave 56 fill	18	15.5 g	
SU 4.0364	Grave 65 fill	80	167 g	
SU 4.0365	Grave 65 skeleton	1	1 g	
SU 4.0369	Grave 66 skeleton	6	4 g	
SU 4.0400	Grave 71 fill	77	159 g	
SU 4.0410	Grave 73 fill	12	20 g	
SU 4.0413	Tumulus fill	4	11 g	
SU 4.0421	Grave 73 skeleton	1	11 g	
SU 4.0424	Grave 76 skeleton	4 scraps	<0.1 g	
SU 4.0425	Grave 77 fill	9	10 g	
SU 4.0433	Grave 78 fill	1	1 g	
SU 4.0441	Ceramic unit in tumulus fill	14	20 g	**14/16** (SF 356)
SU 4.0449	Tumulus fill	4	2 g	
SU 4.0456	Grave 82 fill	29	34 g	
SU 4	Baulk cleaning	29	23 g	

Continued on next page

TABLE 14.1 (*continued*). Uninventoried pieces of daub from the various stratified units at Lofkënd

Trench and unit	Context type	No. of frr/ pieces	Wt	Inventoried pieces
SU 5.	Baulk cleaning	9	34 g	
SU 5.0001	Topsoil	66	165 g	
SU 5.0017	Grave 2	1	11 g	
SU 5.0025	Topsoil	2	11 g	
SU 5.0026	Tumulus fill	16	58 g	
SU 5.0086	Tumulus fill (11)	60	155 g	
SU 5.0140	Topsoil	15	20 g	
SU 5.0178	Tumulus fill	10	22 g	
SU 5.0179	Tumulus fill	3	6 g	
SU 5.0203	Tumulus fill	15	119 g	
SU 5.0204	Tumulus fill (86)	14	31 g	
SU 5.0268	Grave 43 fill	7	20 g	
SU 5.0278	Topsoil	56	157 g	
SU 5.0279	Tumulus (70)	55	185 g	
SU 5.0407	Grave 72 fill	8	22 g	
SU 5.0534	Tumulus fill	112	205 g	
SU 5.0535	Tumulus fill	149	252 g	
SU 5.0536	Tumulus fill	289	757 g	
SU 5.0537	Grave 4 fill	10	10 g	
SU 5.0548	Tumulus fill (535)	149	509 g	**14/9** (SF 445)
SU 5.0549	Grave 27 fill	6	24 g	
SU 5.0561	Tumulus fill	28	75 g	
SU 5.0546	Grave 94 fill	5	7 g	
SU 5.0570	Tumulus fill, near topsoil			
SU 6	Baulk cleaning	7	2 g	
SU 6.0002	Topsoil	13	35 g	
SU 6.0070	Tumulus fill	22	62 g	
SU 6.0141	Topsoil	29	75 g	
SU 6.0181	Tumulus fill (70)	5	5 g	
SU 6.0240	Tumulus fill	7	5 g	
SU 6.0279	Tumulus fill (70)	10	13 g	
SU 6.0375	Topsoil	1	2 g	
SU 6.0541	Tumulus, below topsoil	18	85 g	
SU 6.0542	Tumulus fill (541)	83	175 g	
SU 6.0555	Tumulus fill (550)	2	20 g	
SU 6.0556	Tumulus fill (555)	23	43 g	
SU 6.0557	Tumulus fill (556)	1	5 g	
SU 6.0559	Tumulus fill	44	105 g	
SU 6.0560	Tumulus fill	18	36 g	
SU 6.0562	Tumulus fill	15	23 g	
SU 7.0002	Topsoil	8	8 g	
SU 7.0027	Topsoil	52	150 g	
SU 7.0029	Grave 1 fill	2	?	Cf. **14/14** (SF 387)
SU 7.0187	Tumulus fill	2	12 g	
SU 7.0188	Tumulus fill	10	11 g	
SU 7.0189	Tumulus fill	1	1 g	
SU 7.0191	Tumulus fill	8	3 g	
SU 7.0192	Tumulus fill	7	6 g	
SU 7.0193	Tumulus fill	6	35 g	
SU 7.0194	Tumulus fill	20	25 g	
SU 7.0195	Tumulus fill	10	12 g	
SU 7.0196	Tumulus fill	18	22 g	
SU 7.0197	Grave 17 fill	3	18 g	
SU 7.0202	Topsoil	9	14 g	
SU 7.0217	Tumulus fill	8	3 g	
SU 7.0228	Tumulus fill	5	5 g	
SU 7.0529	Tumulus fill	30	35 g	
SU 7.0530	Baulk cleaning	6	25 g	
SU 7.0531	Baulk cleaning	9	27 g	
SU 8.0022	Baulk cleaning	25	43 g	
SU 8.0028	Grave 1 fill	7	30 g	
SU 8.0114	Baulk cleaning	15	83 g	
SU 8.0129	Tumulus fill	34	47 g	
SU 8.0201	Topsoil	69	215 g	
SU 8.0204	Tumulus fill	96	210 g	
SU 8.0207	Tumulus fill (129)	63	110 g	
Totals		15,875	40 kg	

from sites that were not in the immediate vicinity of the tumulus, as well as from sites that were already ancient. The ramifications of these two categories of material on and in the tumulus are discussed more fully in Chapters 2 and 20.

In this chapter, I give an overview of the prehistory of archaeological daub, a detailed contextual quantification of the daub from the Lofkënd tumulus, including a count and weight of all the daub encountered in the mound (**Table 14.1**), as well as a catalogue of some of the more diagnostic and representative pieces of the material.

The Prehistory of Archaeological Daub

In his discussion of the buildings of the earliest phase of the acropolis of Neolithic Sesklo, Christos Tsountas (1908:79–84) described what he referred to as clay-built huts (πηλόκτιστοι καλύβαι) constructed of wattle-and-daub that was used for both the walls and the roofs of the buildings. Tsountas went on to illustrate a representative sample of pieces of daub from Neolithic Sesklo, including pieces with the impressions of fine reeds (Tsountas 1908:79–84, fig. 13), pieces with impressions of larger timbers (Tsountas 1908: fig. 14), as well as pieces of clay with impressions of stakes that also preserved what Tsountas (1908:79–84, figs. 15–16) referred to as the upper edges of a wall; he also illustrated a reconstruction of a Neolithic hut, showing what he considered the position of the various pieces of daub (Tsountas 1908: fig. 17). Similar pieces of daub were also encountered at Neolithic Dimini, and several of these are now in the National Archaeological Museum of Athens (**Fig. 14.5**). It was only two years later that Oscar Montelius illustrated and described similar construction in the Neolithic of Italy in the following terms: "Les parois de ces habitations primitives étaient revêtues d'argile; on en a trouvé des fragments (fig. f), durcis par le feu qui detruisait la cabane" (Montelius 1910:551–552, fig. f).

Some two decades later, Carl Blegen illustrated a number of pieces of daub and labeled them "fragments of clay packing from roof" (Blegen 1928:13, fig. 12). In his discussion of the Early Helladic House of the Pithoi at Zygouries in the Peloponnese, Blegen (1928:13) wrote:

> The roof itself was undoubtedly flat. Some fragments of clay packing and surfacing give an idea of the manner of construction. Resting on the walls... were heavy wooden beams. These were apparently not squared, but small tree trunks or logs left in the round and placed close together. Over these was spread a layer of clay, filling up chinks and leveling the platform. Upon this was laid in turn a row of reeds, running not parallel to the heavy beams but diagonally across them. Several fragments of clay packing were found, bearing on one side the impress of large logs, on the other that of reeds (Fig. 12). Above the reeds, finally, was laid a thick surfacing of clay. Fragments of this also were found preserving on their lower side the impression of the reeds, and smooth on their upper surface. Several other hardened bits bore the impression of reeds both on their top and bottom; where these were employed is not clear.

Soon after, Hetty Goldman, in her discussion of Middle Helladic architecture at Eutresis in Boeotia, noted numerous fragments of wattle-and-daub, which she interpreted as deriving from a roofed shed built to protect storage pithoi, and she envisaged a Middle Helladic roofing system similar to the Early Helladic roof previously noted by Blegen (Goldman 1931:62–63, fig. 71). She illustrated, in splendid drawings by Piet de Jong, two pieces of daub and a reconstruction of the roofing system (**Fig. 14.6**). Goldman also noted that this manner of roofing could be used for a slanting as well as a flat roof and argued that this was the prevalent form of roofing throughout the Bronze Age of the Aegean.

In his discussion of the architecture of Early Neolithic Servia in Macedonia, Walter Heurtley noted that although poorly preserved, enough was found to identify rectangular houses built of wattle-and-daub (Heurtley 1939:52). Rather than the roofing system (as it was outlined by both Blegen [1928] and Goldman [1931]), it was the actual walls of the buildings themselves that were constructed in this manner. He writes: "Most of the fused rubbish was in shapeless lumps, which made it clear that the walls themselves were not made of bricks. But a number of them bore clear impressions of reeds or had been carefully moulded to fit beams" (Heurtley 1939:53), and he went on to describe and illustrate four different examples of such impressions, many of them similar to those found among the pieces of daub at Lofkënd. Indeed, successive levels of Neolithic settlement at Servia consisted of houses built of wattle-and-daub, and Heurtley noted that it was probably easier to demolish and rebuild such a house than to repair it (Heurtley 1939:52–53). A fuller account of the prehistoric building techniques of Neolithic Servia was published by Cressida Ridley and her collaborators, who noted that:

> Although little of the buildings themselves survives above floor level, the destruction fires of Middle Neolithic Phase Four and Late Neolithic Phase Six caused large parts of the clay packing around the timbers of walls and roofs to be burnt hard and thus preserve some traces of the upper structure as impressions in the burnt lumps, though it is not always clear which belong to the walls and which to the roof. (Ridley, Wardle, and Mould 2000:79)

On the basis of surviving pieces of daub, Ridley, Wardle, and Mould (2000:80–90) were able to reconstruct accurately various structures, including one with closely set uprights of larger timbers, some squared (Ridley, Wardle, and Mould 2000:80, fig. 3.2); another with closely packed reeds, thought to be for partition walling (Ridley, Wardle, and Mould 2000: 81, fig. 3.3); and yet others, some with massive timbers, together with split segments of poles (Ridley Wardle, and Mould 2000:82–85, figs. 3.4–3.7). More than this, Ridley, Wardle, and Mould (2000:87–90, figs. 3.8–3.11) were able to differentiate, in certain cases, daub with impressions deriving from roofs as opposed to walls, as well as daub thought to be from the construction of a beamed floor. A similarly detailed study from Early Bronze Age Elmalı-Karataş in Anatolia, indeed one of the fullest accounts of daub walls and roofs published, is presented by Jayne Warner (1994:143–167, pls. 156–161; also pls. 198–201 for modern wattle-and-daub architecture).

It was precisely this type of house built of wattle-and-daub that Gordon Childe (1929:99, with figs. 58–59) described for Neolithic central Europe (see also Montelius 1910:551–552, fig. f, and 552–553, n. 1, with references to earlier literature), a type of construction that was to last in Hungary throughout the Bronze Age (Kovács 1977:21, figs. 1, 3). In Anatolia, the same type of wattle-and-daub architecture enjoyed a venerable prehistory, from the Neolithic period surviving into the modern era (e.g., Duru 2008: 161, figs. 325–326; Eslick 1992:15–16, pl. 98a–f; Goldman 1956:54, fig. 194a; Warner 1994:143–176, pls. 156–161; 198–201), as it also did in Egypt, especially for roofing (e.g., Kemp 2000:93–96, fig. 3b).

In other parts of Greece, daub has been reported from a variety of prehistoric sites (e.g., Bakalakis and Sakellariou 1981: pl. 6b, no. 1 [Paradimi]; Gimatzidis 2010: Beil. 19a [Early Iron Age Sindos/Anchialos]; Hänsel 1989:95, fig. 28; 141, fig. 50; 194, fig. 78, no. 1 [left]; 215, fig. 90 [Kastanas]; Hänsel and Aslanis 2010:66, fig. 31; 109, fig. 54; 137, fig. 72; 153, fig. 82; 169, fig. 89 [Agios Mamas]; Hellström 1987:137–138, fig. 52 [Paradeisos]; Lamb 1936:40 [Thermi]; Pappa 1998:27, fig. 18 [Kastri, Nikiti, Chalkidike]; Stephani 2010:300–302 [various examples from Late Bronze Age Angelochori]; Taylour and Janko 2008: pl. 17 [Ayios Stephanos, Lakonia]; Wardle 1996:42–44 [Nea Nikomedeia]). Few projects, however, have systematically quantified their findings of daub. Pontus Hellström and collaborators noted approximately 30 kg of burned clay, both with and without reed impressions, from the late Neolithic settlement at Paradeisos in Thrace that they interpreted as deriving from clay floors and wattle-and-daub constructions of house walls (Hellström 1987:138). Despite the fact that the area excavated at Paradeisos was small, it is noteworthy that an even smaller burial mound in Albania contained more daub—approximately 40 kg in all—than the portion of a prehistoric settlement site in northern Greece that was excavated. This establishes the fact that daub was not just an incidental component of the tumulus fill at Lofkënd.

In the context of burial tumuli, Lofkënd is not the only burial mound to have yielded daub. At least one fragment, simply noted as a "keramisches Fragment," but clearly daub, is illustrated from Planje, Tumulus I, Grave 3, in Bosnia Herzegovina (Benac and Čović 1956:59, pl. XXXV, no. 11). This, together with the sheer quantity of daub from the fill of the Lofkënd tumulus, makes it likely that daub may have been a more ubiquitous but overlooked component of other Illyrian tumuli. The ramifications of the Lofkënd daub, including the possibility that this material, once sun-dried, may have been deliberately fired for reuse as rubble fill, are considered more fully by Martin-McAuliffe (Chapter 20).

CATALOGUE

The dimensions of the fragments of daub refer to the PL × PW × PTh (in meters). All of the following pieces are single fragments unless otherwise noted.

14/1 (SF 077), **Figs. 14.7, 14.8** (Sheet 32.1; Photos 2038b-c)
Fragment of Daub
Papadopoulos, Bejko, and Morris 2007:135, fig. 30 (top left).
Tumulus fill (SU 1.0007).
0.069 × 0.051 × 0.034; Wt: 79.6 g.
Semi-coarse to coarse clay, with small white and dark-colored inclusions, occasional larger inclusions, and quite a bit of mica; little visible organic material. Fabric on one side close to reddish yellow (5 YR 6/6), reddish gray (5 YR 5/2) on the other.
Flat and even surface on one side; at least one rod impression on side adjacent to flat surface.

14/2 (SF 037), **Figs. 14.7, 14.8** (Sheet 32.2; Photos 2045a, c)
Fragment of Daub
Papadopoulos, Bejko, and Morris 2007:135, fig. 30 (top right).
Tumulus fill (SU 1.0007).
0.054 × 0.047 × 0.035; Wt: 48.3 g.
Fabric and color as **4/1** (SF 077).
Small portion of flat surface preserved on one side; one preserved rod impression on side opposite flat surface.

14/3 (SF 038), **Figs. 14.9, 14.10** (Sheet 33.1; Photos 2040a–d)
Fragment of Daub
Papadopoulos, Bejko, and Morris 2007:135, fig. 30 (bottom right).
Topsoil (SU 4.0011).
0.066 × 0.057 × 0.037; 74.7 g.
Fabric as **4/1** (SF 077); color in the range of reddish yellow (5 YR 6/6) and light red (2.5 YR 6/6) on one side, reddish gray and dark reddish gray (5 YR 5/2–4/2) on the other.
Portion of flat surface preserved on one side, which bears traces of three linear impressions, perhaps of organic material or else tooling/finishing marks. At least two, and perhaps three or more, preserved rod impressions (the two clear rod impressions are on the side opposite the flat surface, one of the other possible impressions is on the side adjacent to it).

14/4 (SF 078), **Figs. 14.9, 14.10** (sheet 36.1; photos 2044a, c)
Fragment of Daub
Papadopoulos, Bejko, and Morris 2007:135, fig. 30 (bottom left).
Tumulus fill (SU 1.0047).
0.053 × 0.049 × 0.030; Wt: 52.8 g.
Fabric and color as **4/1** (SF 077).
Small portion of flat surface preserved on one side; at least one impression of a substantial rod or stake preserved on the side adjacent to the flat surface.

14/5 (SF 079), **Fig. 14.11** (Photos 2046a, b)
Fragment of Daub
Tumulus fill (SU 2.0071).
0.061 × 0.059 × 0.036; Wt: 55.9 g.
Coarse clay with many small to large inclusions, primarily white (some 8.0 mm), but less mica; impressions of organic material and calcareous incrustation all over. Fabric on one side close to reddish yellow (5 YR 6/6), elsewhere closer to pinkish gray and reddish gray (5 YR 6/2–5/2).
Small portion of flat surface preserved on one side; at least two and probably three rod/stake impressions on side opposite flat surface, one of which is from a substantial rod.

14/6 (SF 128), **Fig. 14.11** (Photos 2042a, b)
Fragment of Daub
Tumulus fill (SU 1.0039).
0.065 × 0.053 × 0.036; Wt: 53.4 g.
Coarse clay with numerous small to medium and some large inclusions, primarily white and light-colored, and a dusting of fine silvery mica; impressions of organic material all over. Fabric mostly reddish gray (5 YR 5/2), at one point closer to yellowish red and reddish brown (5 YR 5/6– 5/4).
One very small portion may preserve an original flat surface; at least one rod impression on opposite side, and one or more additional rod impressions here and there.

14/7 (SF 080), **Fig. 14.11** (Photos 2049a, b)
Large Fragment of Daub
Tumulus fill (SU 1.0039).
0.102 × 0.080 × 0.023; Wt: 147.7 g.
Semi-coarse clay with occasional small white and light-colored inclusions and a light dusting of fine silvery mica. Fabric mostly yellowish red (5 YR 5/6), elsewhere blackened.
Substantial original surface on one side, characterized by a curvature, not unlike pottery but different, as the surface undulates following the contours

of the rods/stakes onto which the clay was attached. The fragment preserves two convex areas with a concave area in between. At least one rod/stake impression on opposite side, corresponding with concave area on the surface; possible segments of two additional rod impressions on either side of the central impression. There is, in addition, a deep impression resembling a tear on the side opposite the surface.

14/8 (SF 446), **Fig. 14.12** (Photos 3536–3537)
Fragment of Daub
Tumulus fill, near topsoil (SU 2.0570).
0.051 × 0.034 × 0.029; Wt: 29.7 g.
Coarse clay with many small to large white inclusions, but no mica to speak of. Fabric evenly fired close to reddish yellow (5 YR 6/6).
Although small, the fragment is interesting in that one side preserves a good flat surface, with two very slight undulations/concavities. At least two rod impressions on side opposite surface, and there appears to be an impression on each of the two sides adjacent to the surface.

14/9 (SF 445), **Fig. 14.12** (Photos 3534–3535)
Fragment of Daub
Tumulus fill (SU 5.0548).
0.050 × 0.043 × 0.024; Wt: 33.1 g.
Semi-coarse clay with a few small white and light-colored inclusions, but little or no mica. Fabric evenly fired close to yellowish red (5 YR 5/6), except for minor spots of blackening.
Two substantially preserved rod/stake impressions on one side; no clearly preserved surface, although the side opposite the impressions may well have been the surface.

14/10 (SF 135), **Fig. 14.12** (Photos 2041a, b)
Fragment of Daub
Tumulus fill (SU 1.0070).
0.080 × 0.060 × 0.035; Wt: 90.7 g.
Fabric as **14/9** (SF 445); fabric mostly reddish yellow (5 YR 6/6) except for slight blackening on one side.
Two roughly parallel rod impressions on one side; the opposite side is probably the original, but uneven or poorly preserved, surface.

14/11 (SF 039), **Fig. 14.12** (Photos 2039a, b)
Fragment of Daub
Tumulus fill (SU 2.0023).
0.067 × 0.057 × 0.038; Wt: 69.0 g.
Semi-coarse to coarse fabric with many small to medium and some large inclusions, primarily white and light-colored, and a little fine silvery mica. Fabric close to reddish yellow (5 YR 6/6) on one side, closer to light reddish brown and pinkish gray (5 YR 6/3–6/2) elsewhere, and in parts blackened.
Two roughly parallel rod impressions on one side; the opposite side is probably the original surface.

14/12 (SF 218), **Fig. 14.13** (Photos 2381–2382)
Small Fragment of Daub
Tumulus fill (SU 2.0202).
0.039 × 0.029 × 0.014; Wt: 8.5 g.
Semi-coarse clay, with small to medium white and light-colored inclusions, but only a slight dusting of mica. Fabric, where not blackened, close to light reddish brown and reddish brown (5 YR 6/3–5/3).
One side clearly an original surface, nicely finished flat; opposite side, although small, has at least six small reed/rod impressions, some parallel, others converging.

14/13 (SF 127), **Fig. 14.13** (Photos 2047b, c)
Fragment of Daub
Tumulus fill (SU 1.0067).
0.051 × 0.051 × 0.032; Wt: 29.8 g.
Coarse clay, with small to large white and light-colored inclusions, but very little mica. Fabric where not blackened close to light reddish brown (5 YR 6/4).
One side, which appears to be the surface, preserves at least one, and probably two, rod impressions running more or less parallel but slightly curved. The two adjacent sides bear impressions of more substantial rods running the opposite direction to those on the surface.

14/14 (SF 387), **Fig. 14.13** (Photos 3230, 3233)
Fragment of Daub
Tomb I (Grave 64) fill (SU 1.0361).
0.055 × 0.051 × 0.029; Wt: 46.9 g.
Semi-coarse clay with small white, light-colored, and dark inclusions, and a dusting of fine silvery mica. Fabric mostly reddish yellow (5 YR 6/6), elsewhere closer to light reddish brown and pinkish gray (5 YR 6/3–6/2); one small area blackened.
Three roughly parallel rod/stake impressions on one side, the two at the sides from more substantial

rods/stakes; central portion of opposite side preserves portion of a flat surface.

14/15 (SF 310), **Fig. 14.13** (Photos 2921, 2926)
Fragment of Daub
Tumulus fill (SU 4.0286).
0.063 × 0.059 × 0.047; Wt: 71.2 g.
Coarse clay with many small to large inclusions of various colors, including white, light, and dark; a little fine silvery mica. Clay mostly reddish yellow (5 YR 6/6), elsewhere closer to light reddish brown and pinkish gray (5 YR 6/3–6/2). Several tears here and there.
Impressions from two substantial rods/stakes on one side, roughly parallel, with two smaller impressions at the edge of the same side, and another possible impression near the other edge on the same side. Opposite side may preserve a small portion of the original flat surface.

14/16 (SF 356), **Fig. 14.14** (Photos 3116–3117)
Fragment of Daub
Ceramic unit (SU 4.0441) in tumulus fill.
0.064 × 0.058 × 0.030; Wt: 47.2 g.
Semi-coarse clay with small to medium primarily white and light-colored inclusions and a very fine dusting of silvery mica. Clay mostly light brown (7.5 YR 6/4).
Three roughly parallel rod impressions clearly preserved on one side, with traces of one or two additional impressions preserved at the edges. Opposite side probably the original surface, though more undulating than flat.

14/17 (SF 173), **Fig. 14.14** (Photos 2394–2395)
Fragment of Daub
Topsoil (SU 1.0141).
0.033 × 0.030 × 0.025; Wt: 13.1 g.
Coarse clay with many small to large white and light-colored inclusions, but only a very fine dusting of silvery mica. Clay mostly reddish yellow and light reddish brown (5 YR 6/6–6/4), elsewhere much blackened.
One side preserves portion of original flat surface; impression of a substantial rod/stake on opposite side, with smaller portion of additional impression at the edge.

Other inventoried fragments of daub

SF 011 SU 3.0003 (Topsoil) 0.039 × 0.030 × 0.026; Wt: 20.6 g
SF 012 SU 1.0001 (Topsoil) 0.038 × 0.026 × 0.022; Wt: 16.5 g
SF 017 SU 1.0007 (Tumulus fill) 0.053 × 0.042 × 0.028; Wt: 47.3 g
SF 166 SU 1.0067 (Tumulus fill) 0.078 × 0.054 × 0.046; Wt: 98.5 g
SF 219 SU 2.0202 (Tumulus fill) 0.073 × 0.062 × 0.040; Wt: 96.0 g
SF 286 SU 4.0204 (Tumulus fill) 0.070 × 0.042 × 0.033; Wt: 54.4 g
SF 309 SU 2.0380 (Tumulus fill) 0.051 × 0.045 × 0.040; Wt: 55.2 g
SF 330+331 SU 4.0286 (Tumulus fill) 0.055 × 0.049 × 0.041; Wt: 75.9 g
SF 355 SU 1.0377 (Tumulus fill) 0.031 × 0.030 × 0.026; Wt: 17.5 g
SF 357 SU 1.0407 (Grave 72 fill) 0.025 × 0.025 × 0.016; Wt: 8.5 g
SF 358 SU 1.0440 (Tumulus fill) 0.071 × 0.053 × 0.025; Wt: 65.2 g

Chapter 15

Bitumen at Lofkënd: Deposits, Sherds, and Containers

Sarah P. Morris

The landscape surrounding the Lofkënd tumulus, and the Gjanicë valley in particular, as it joins the Seman [Apsos] River en route to the Adriatic, is rich in fossil hydrocarbon deposits, with petroleum compounds ranging from heavy fraction (asphalt) to liquid (petroleum) and light (methane gases, naphtha, etc.). The nearby town of Ballsh, first settled in the late antique period as a successor to Byllis and site of a large Early Christian basilica, is the center of modern Albania's petroleum industry, and gives its name to native diesel ("naftë e Ballshit"). Southwest of the Gjanicë valley, across the Mallakastër ridge that separates it from the lower Vjosë and below, as well as across from ancient Byllis, the Shushicë River joins the Vjosë near modern Selenicë. This locale makes Albania famous as home to a native form of asphalt (a natural alloy of 80–85% carbon and 9–10% hydrogen, plus traces of sulfur, nitrogen, and oxygen, which melts at 90–120° C) which is especially pure in valuable subsurface deposits of 86% natural bitumen (asphaltite). Its properties and processing were witnessed and described by early modern travelers (Holland 1815:519–524; Patsch 1904:98), many of whom knew its ancient reputation. In post-antique times, this important mineral was extracted for industrial, cosmetic, and decorative uses abroad (by Venetian, Ottoman, British, French, and Italian interests; Anhegger 1943:174–176), coming into greater demand after the introduction of tar-bound macadam (**Fig. 15.1**) and then waning in favor of petroleum products from the same region in the twentieth century. In recent post-communist times, operations at the rich underground mines have been revived in a French enterprise named for the source (*www.selenicebitumi.com*), which has trademarked its chief product as Selenizza SLN120®. For all of these reasons, a major question for the Lofkënd team was the potential role of this form of hydrocarbon in the ancient economy.

In antiquity, this area was renowned for its bitumen mines, especially among the Romans, who exploited the local combination of timber and mineral resources for building and repairing ships (Cabanes 2002; Morris 2006). Roman coins showed the personification of the ancient Aoös as a river god supervising the transport of a giant log, so closely was the area identified with its forests (Gjongecaj and Picard 2007:105). Modern speculation makes the co-presence of natural asphalt, near rivers that led to a well-timbered hinterland, one initial attraction for Corinthians to the Illyrian coast, as it may have supported their early shipbuilding industry, long before the Romans made it famous (Cabanes 2002; Morris 2006).

The high quality of *Illyrica pix* was praised in Roman times (Ovid, *Ars amatoria* 2.658, *Epistulae ex Ponto* 4.14.45), with both Apollonia (which may mean the Selenicë source) and Dyrrhachium (modern Durrës) mentioned for their springs that produce this material (Vitruvius, *De architectura* 8.3.8). This substance and its surface manifestations (in lumps, pools, springs, or deposits both liquid and flaming) attracted the special attention of ancient scientists and compilers of scientific wonders (*thaumata*) (Aelian, *Varia historia* 13.16; Ampelius 8.1; Cabanes 2007; Dioscorides, *Materia medica* 1.83–84; Pliny, *Naturalis his-*

toria 2.237, citing Theopompus; cf. Aristotle [auctor ignotus], *De mirabilius auscultationibus* 833d, 842b 19l; Stocker 2009:56–60). Ancient sources also report one particular aspect that made it a natural wonder, a famous oracle linked to a fiery source of gases at Nymphaeum, probably emissions of methane co-present with tar and oil (Caesar, *Bellum Civile*; Dio Cassius, 41.45; Plutarch, *Sulla* 27; Strabo, 7.5.8, from Poseidonios; Pliny, *Naturalis historia*, 2.106.228, 3.23.145). By the third century, this site was celebrated on the reverse of local coins, showing three eponymous Nymphs as dancing female figures near a flaming cauldron (**Fig. 15.2**) (Gjongecaj and Picard 2007:85, fig. 20, nos. 8–9; Head 1911:314).

The precise location of this oracle remains uncertain: ancient remains were noted 200 years ago on the banks of the Vjosë River near the Selenicë mines (Holland 1815:523), but have disappeared today, and no longer indicate either the site or its ancient toponym (Amore 2010:18–19; Cabanes 2007). Meanwhile, geological studies of the Greek oracle at Delphi claim to confirm ancient reports (in Strabo, Plutarch, Pausanias) of natural gases at the site as the source of inspiration for its priestesses, with ethylene a likely suspect (de Boer et al. 2001; Hale et al. 2003; Lehoux [2007] doubts this theory). This has renewed attention in the oracle reported at Nymphaeum as another possible candidate for such psychotropic intoxication. But without a secure location or comparable procedures reported (according to Dio Cassius, 41.45, incense was thrown on the flames at Nymphaeum to obtain divine replies to a question), conclusive results are not possible.

As discussed in an earlier study (Morris 2006), bitumen was widely popular in Old World prehistory as an adhesive (mortar) between courses of mud bricks, a binding agent added to straw and clay in manufacturing bricks, or in selective applications to render built surfaces waterproof (Forbes 1936; Moorey 1994:332–335; Stol 2012). It also served as a binding and sealing agent for ceramics or for hafting lithics since the Old Stone Age, primarily in the Near East (Boeda et al. 1998). Mixed and heated with other materials such as clay, minerals, crushed matter, and organic ingredients, it also formed a pliable matrix or mastic for manufacturing artifacts such as beads, plaques, figurines, and vessels (Connan and Dechesne 1996; Serpico 2000:456). Closer to Albania, it was used to decorate pottery, as a layer for inlaid straw or incised motifs, in the Carpathian basin (Hungary) in the Tisza culture, or Early Copper Age (Raczky 1987:68, fig. 8), but also for repairing Classical vessels in South Italy (Reyer-Völlenklee 2005). But in the pre- or proto-history of Albania, its sole reported use so far has been to repair Neolithic sherds from Cakran, not far from Lofkënd (Korkuti 1995a:145; Morris 2006:99–100, fig. 3), and no local lithics show signs of hafting to wood with bitumen.

In Classical times, fossil bitumen was frequently used to line local Greek and Roman transport and storage vessels. A number of the pithoi reused for burials in the necropolis of Apollonia, identified as local and Corinthian types of the sixth and fifth centuries BC, were coated with bitumen, in some cases on the exterior as well as inside the cavity of the vessel (Amore 2010:582–588);[1] recent research by an Albanian-German team in the theater of Apollonia turned up ca. 1,000 Hellenistic and Roman sherds with traces of bitumen, also found on those from the "amphora wall" (Fiedler and Döhner 2013; Lahi and Fiedler 2010). Roman amphora fragments collected over a wider area in the MRAP survey were also lined with bitumen (Stocker 2009:526, 527). Presumably this substance improved their watertight properties for storing liquids such as wine, rather than marking them as primary containers for tar in liquid form (reused in walls and burials). Pine resins were added to wine, probably as preservatives and later as flavorants, since the earliest specimens collected and analyzed by archaeologists (from Haji Firuz Tepe, in sixth-millennium Iran: McGovern 2003:64–73; Morris 2008:117), creating a mixture still in use as modern Greek *retsina*, but fossil resins were never involved. A comprehensive study of such vessels in Albania, and of possible local industries in the extraction and distribution of asphalt, remains to be performed.

---

[1] Samples of bitumen from about 15 vessels were collected in 2006 for possible future analysis; meanwhile, residue analysis performed at Millsaps College on samples of bitumen-lined pottery from Apollonia was unsuccessful in identifying oil, red wine, or white wine (or indeed any fatty acids), perhaps because hydrocarbons repel residues, or due to earlier excessive washing of sherds (Michael Galaty, personal communication). Fiedler and Döhner (2013:133, n. 6) mention pithoi from the necropolis from Apollonia but do not cite Amore 2010.

Given all these expectations surrounding bitumen in the Lofkënd environment, it was not surprising to find asphalt in evidence on material from the tumulus, and on the survey.

Twenty ceramic examples listed below represent sherds found in tumulus excavations (plus one from final baulk removal), all from handmade pots with traces of bitumen (asphalt), applied presumably as lining or waterproofing to interior surfaces (**Figs. 15.3, 15.5**). A few also show shiny patches on exterior surfaces (**15/5** [P099], **15/7** [P138], **15/10** [P201], **15/11** [P254], **15/16** [P297], **15/20** [P355, **15/21** [P395]), which may indicate a form of decoration and/or waterproofing; some of these may date to the Bronze Age (**15/5** [P099]), and possibly were coated in a slip mixed with bitumen to add luster. The fragments are too small or worn to determine original size, shape, and appearance of the vessels, but most indicate shapes and fabrics older than the burials. Others were found with patches or splashes of bitumen that even cover breaks (**15/9** [P174]), as if in contact with a liquid form of bitumen after breakage (rather than deliberately applied for repairs). Most came from fill layers of the tumulus or topsoil, while a few (noted in bulk finds, not inventoried below) made their way into the fill of Early Iron Age graves (e.g., in Tombs LXVII, LXXXI; see Chapter 3), but none are from the deeper levels and earliest graves of the Late Bronze Age, reached in the final season (Phases I–II). The fact that bitumen was commonly found on sherds from the systematic survey in the area (Chapter 18), and in informal surface collections, suggests that the natural environment around the tumulus may have also brought many artifacts into contact with this mineral, without human agency. However, as reported above, none of the stone tools from the tumulus (Chapter 13) or surface survey (Chapter 18) showed traces of bitumen as adhesives for fastening to wooden shafts, nor did any of the fired fragments of building material (Chapter 14) offer evidence for the use of bitumen or its compounds as a mortar (not surprising, as no true masonry mud bricks were found).

In addition to casual finds of redeposited sherds with tarred surfaces, the excavation of the Lofkënd burials also produced a highly unusual discovery. A large vessel (**9/259** [P283]) was found smashed in situ in the vicinity of Tomb LXIII, but not close enough to be clearly associated or called a grave good: it lay in a deposit (SU 280) 20 cm below and north of the burial, identified as part of tumulus fill (**Fig. 15.4**; see also **Figs. 3.214**). Lying on its side, it rested against a large stone at its east end, which may have shattered it or broken its fall: the stone was found over the mouth of the vessel but seems too large and heavy to have served as its lid. The condition of the vessel on discovery suggests that it was dropped or abandoned, then crushed after deposition by the accumulation of earth in the tumulus, with its uppermost exposed surface, the side of the vessel, largely removed by subsequent burial activity (e.g., SU 207).

First in its importance, this vessel preserves the entire profile of a large, coarseware container, unique in the repertoire of Illyrian ceramics (**Fig. 9.63: 9/259**). Given that most surviving material evidence for protohistoric Illyria derives from cemeteries, knowledge of indigenous ceramics is limited to fine wares such as the matt-painted and dark burnished fabrics found as grave goods in the tumulus (see Chapter 9), making this vessel an anomaly and a challenge. In terms of fabric, the bitumen container shows a dark, coarse matrix and smoothed brown surface, perhaps once burnished, common in coarseware sherds from tumulus fill (**Fig. 9.29**). The stub of a single, horizontal belly handle suggests a second mirrored it on the opposite side (**Fig. 9.63**), to give it a shape best described as a belly-handled amphora. While found with bitumen inside the lower cavity, clearly once in liquid form that had cooled and hardened in the shape of the vessel, its original purpose could have been for any contents, whether liquids, foodstuffs such as dry grains or pulses, or other household commodities. Whether it was designed for industrial substances such as bitumen, which have household uses in ancient and contemporary societies (as adhesives, salves, and so on; see Ochsenschlager 1992, 2004; Wendt 2009), is unknown. However, it was deliberately marked around the exterior neck and shoulder join with streaks of bitumen (**Fig. 9.63**), perhaps before firing, which may have been intended to signal its contents.

What makes this shape more intriguing, given its final use as a container for bitumen, is its relevance to a passage in Herodotos (4.195) describing the collection of bitumen on Zakynthos, an Ionian island rich in tar deposits. At this source, tar wells up naturally in lakes, as it does anywhere when gases co-present in hydrocarbon deposits push materials upward.[2] The historian reported the collection of bitumen from pools of water, its transferral to rock-cut basins, and eventually to containers that he called *amphorai*, presumably for shipment to markets and consumers.[3] This makes our discovery at Lofkënd a possible remote ancestor or relative of such containers, with a wide mouth (for filling, and cooling, molten tar?) and two side handles for lifting and carrying, and if not, at least an important addition to the local ceramic repertoire in coarseware vessels.

What purpose it served in proximity to protohistoric burials is more difficult to ascertain and may be partially answered by comparison to the other significant bitumen discovery from the Lofkënd tumulus. At least one burial in the south-central area (Sector/Trench 2: **Figs. 3.61–3.62**), Tomb XX (50) from Phase II, dated to the Late Bronze–Early Iron Ages (twelfth–eleventh centuries BC), contained a skeleton of an adult female (20–25 years of age) with a deposit of bitumen that had dried across the lower left ribs, left arm folded over the pelvic area and edge of the left pelvis (**Fig. 15.6**). The tomb belongs to the second phase of the tumulus, perhaps a century (or more) earlier than Tomb LXIII (35) of Phase IV, or late tenth to early ninth century, and its relationship to the amphora containing bitumen is even more remote. One can only imagine that this irregular patch represents an application of the substance as adhesive or decoration to skin or clothing on the body deposited for burial, perhaps to seal a shroud or bundle (no textile impressions were visible on the fragments retrieved, even under magnification). The location of this irregularly shaped patch means that it was deposited after the body was arranged for burial in the grave, with arms folded across the lower torso. In a burial without grave goods, this curious patch of bitumen calls for special attention.

Bitumen is common in mortuary contexts in ancient Egypt, which imported it from natural asphalts in the Dead Sea (Jordan valley) since 900 BC or perhaps earlier (Peters, Walters, and Moldowan 2005:23–328; Serpico 2000:455, 464–468). Even earlier, bitumen was applied to human crania in postmortem processes, largely in the Pre-Pottery Neolithic period in the Near East (Arensburg and Hershkowitz 1989; Bonogofsky 2011). Ancient authors claimed that bitumen (rather than the pine-tree resin better known in the process) was used in mummification (embalming) in Egypt (Diodorus Siculus, 2.48.6–8, after Hieronymus of Cardia; Lucas and Harris 1962 [1934]:303–308; Serpico 2000:464–468). Procedures involved applying various resins in perfume or ointment form to human skin and its coverings, or resin-soaked linen to cover wound openings for embalming. More recently, patches of bitumen, often on linen, have been identified, primarily in post-Pharaonic mummies (Rullkötter and Nissenbaum 1988). However, nothing about the burial in Tomb XX allows us to confirm the traces of bitumen either as part of a deliberate mortuary practice or postmortem treatment of the skeleton, or, alternatively, to explain its origin in an accident (e.g., from spilled pitch from torches in nocturnal ceremonies or burial rites, or from a jar such as **9/259**). If bitumen were widely used in mortuary preparations or graveside rites in Late Bronze and Early Iron Age Albania, we would expect to see evidence for it associated with a number of other burials. Curiously, a recent discovery at Apollonia of a threshold deposit in a Classical house includes the debris of a transport amphora full of liquid bitumen that may have been poured out as part of a foundation sacrifice (Fiedler and Döhner 2013:131 n. 2), suggesting long-term local use of bitumen for ritual purposes.[4] Meanwhile, the two unusual finds at Lofkënd, both a vessel that held liquid bitumen found in association with protohistoric burials, and the deposit of dried bitumen found in connection with a buried individual, must remain, for the moment, unexplained anomalies in the local mortuary record.

---

2 This corrects earlier misunderstandings of Vitruvius 8.3.8 (Forbes 1936:29, followed by Morris 2006:96) that made Zakynthos a region of Illyria. Our thanks to John Hale for pointing this out to us, as well as to Shari Stocker for putting it in print (2009:34, n. 82).

3 Compare the storage of liquid bitumen in clay containers, in ancient Mesopotamian texts (Stol 2012:50, n. 23), and the presence of bitumen inside large Classical and Hellenistic vessels from Apollonia (Fiedler and Döhner 2013), including amphoras (see below, n. 6).

4 The deposit is described as a "Bauopfer" and closely resembles the "saucer pyres" in Classical Athens and elsewhere (Rotroff 2013). I am grateful to Manuel Fielder for advance discussion of this.

Catalogue of Sherds with Bitumen from Lofkënd (Tumulus Fill, 2004–2006)

15/1 (P 015), **Fig. 15.3a** (Photo 2462)
SU 1.001 (23/06/2004).
Flat body sherd with bitumen on inside surface.
PL (max): 15 mm; PW (max): 11–15 mm; Th: 5–8 mm.
Fabric fine, red; surface smoothed, worn outside, coated with bitumen inside.
Interior 7.5 YR 2.5/1, exterior 5 YR 4/4.

15/2 (P 018), **Fig. 15.3b** (Photo 2458)
SU 2.006 (2004).
Body sherd of open shape preserving lower attachment of vertical strap handle. Traces of bitumen on interior, below handle.
PL (max): 40 mm; PW (max): 27 mm; Th: 8–19 mm; W (handle): 2 cm; Th: 3 mm.
Fabric semi-coarse, dark gray with small red and white inclusions, air pockets (blow-outs). Core dark gray to black (5 YR 2.5/1, 5 Y 3/1).

15/3 (P 025a, b), **Fig. 15.3c** (Photo 2454)
SU 4.011 (28/6/2004).
Two non-joining wall fragments of fine ware shape (cup?); tiny traces of bitumen on inside surfaces. Maximum preserved dimensions: 20 × 22 mm; Th: 6 mm.
Fabric fine, dark-gray core (2.5 YR 3/1) with tiny red and white inclusions, reddish exterior surface (5 YR 6/6).

15/4 (P 079 = **9/162**), **Fig. 9.56**
SU 4.104 (14/7/2004).
Seven fragments, some joining, of coarseware open shape with strap handle angled at top; lined with bitumen.
PW (max): 5.0 cm; PH (max): 2.8 cm; W (handle): 2.8 cm; Th: 1.2 cm; Th (wall): 1.1 cm.
Fabric coarse, gray core inside red-brown surface layers, black surface. Surface smoothed, interior of one body sherd lined with bitumen. Dark gray core (5 YR 4/2) between red-brown layers (5 YR 4/4), black surface (2/1).

15/5 (P 099 = **9/184**), **Figs. 9.23, 9.58**
SU 1.167 (2005).
Handle, wishbone shape, of large open cup or kantharos with convex wall: triangular perforated spur curving up from single lower handle attachment, separating into two spurs attached to (missing) upper wall. PH: 7.2 cm; PW (upper break): 8 cm; W (lower handle attachment): 5.6 cm; Th: 2 cm; D (hole): 1.5 cm (max depth 2 cm).
Heavy semi-coarse fabric, black (2.5 YR 5/1); surface mottled brown, black (5 YR 3/2), very worn. Patches of shiny black preserved on exterior handle below perforation, trace inside handle: heavy burnish, or slip with bitumen? Bronze Age.

15/6 (P 101 + P 102), **Fig. 15.3d** (Photo 2171)
SU 1.141 (22/6/2005 and 21/6/2005).
Two joining sherds of large open shape, interior coated with bitumen; raised relief band (partially broken away) on exterior.
P 101, max dimensions: 29 × 25 mm; Th: 12 mm; P 102, max dimensions: 47 × 32 mm, Th: 13 mm.
Fabric semi-coarse, sandy texture, dark core with small inclusions, fine mica; exterior surface red, smoothed and burnished (worn on P 101); interior coated with bitumen (visible bubbles and blisters).

15/7 (P 138), **Fig. 15.3e** (Photo 2465)
Morris 2006:101, fig. 4.
SU 1.070 (28/6/2005).
Three sherds (two joining), from convex wall of open vessel lined with bitumen; small trace of bitumen on outside of smaller joining sherd.
Max dimensions: 63 × 29 mm (two joining sherds), 37 × 29 mm (single sherd); Th: 5–8 mm.
Layer of bitumen about 1 mm thick, surface blistered and bubbled. Fabric semi-coarse, dark red.

15/8 (P 144), **Fig. 15.3f** (Photo 2468)
SU 2.202 (2/7/2005).
Two joining sherds (recent break) from wall of thick vessel, coated with bitumen inside. Max dimensions: 40 × 60 mm; Th: 11–14 mm.
Fabric semi-coarse, small chip inclusions, fine white grit; inner core brown (7.5 YR 3/2), outer layers red (5 YR 5/6).

15/9 (P 174 = **9/252**), **Fig. 9.28**
SU 1.070 (1/7/2005).
Base (flat underside) and lower wall of large bowl with patches of bitumen adhering to inside floor

and along upper edge of right-hand break. PH: 70 mm; Th (wall): 75 mm; Th (floor): 195 mm; D (base): 110 mm.
Fabric semi-coarse, orange-red clay throughout (5 YR 6/8), fine mica.

15/10 (P 201 = **9/161**), **Figs. 9.21, 9.56, 15.3g**
SU 7.202 (baulk) + 1.291.
Flat base, convex wall and vertical strap handle (ovoid in section) of small closed vessel (squat jug or amphora), very worn, with traces of bitumen inside and out, including on handle break, restored for complete body profile (missing neck and rim, upper join of handle). Joining fragments of handle and wall from SU 7.202 (7/7/05) and SU 1.291 (west deposit of ceramic fragments (12/7/05); joining base fragments from SU 1.291 (east deposit of ceramic fragments (12/7/05); non-joining wall fragment from SU 1.281 (12/7/05).
PH (base to neck, restored): 15 cm; D (estimated max): 19 cm; D (base): 7.5 cm; Th: 1.2 cm; W (max, handle): 4 cm.
Fabric semi-coarse, yellow-brown core (10 YR 4/4) with fine black particles, yellow chips, streaks; surface (under bitumen) light red (5 YR 5/8).

15/11 (P 254), **Fig. 15.3h** (Photos 3276–3277)
SU 1.291 (12/7/05).
Body sherd from large coarse vessel, with bitumen coating on both sides.
Max dimensions: 37 × 48 mm; Th: 12.8 mm.
Fabric coarse; brick-red, fine white grit.

15/12 (P 283 = **9/259**), **Figs. 9.29, 9.63, 15.4**
SU 4.280 (21/7/05) found on side, smashed in situ just north of Tomb LXIII, which runs under baulk between Trenches 1, 2, and 4; filled up to one-third of height with bitumen (cooled against wall in shape of vessel) and loose flecks of bitumen in soil inside.
Large vessel found in fragments, preserving complete profile of coarseware amphora (missing half of body/profile) with vertical neck rising to everted rim, rounded lip; body biconical, narrow base (toe) with rounded underside. Roots of single horizontal ring handle set at midpoint (max D); second handle presumed on opposite side.
H: 41 cm; D (max): 33 cm; D (rim): 20 cm; D (base): 8 cm.
Coarse fabric, brown to dark brown (7.5 YR 2/0), with large white pebbles and inclusions (up to 7 mm), gold mica, fine white and black grit. Surface mottled brown to dark brown (7.5 YR 4/2–4/4); interior face (2.5 YR 5/4, reddish brown), smoothed and burnished. Black streaks and smears of bitumen applied to exterior on lower neck, upper shoulder, and join: to decorate vessel, or mark contents? (cf. Fiedler and Döhner 2013). Contents saved (soil sample 84).

15/13 (P 286), **Fig. 15.5a** (Photo 2854)
SU 4.280 (22/6/06).
Sherd broken from floor of large (open?) vessel with patch of bitumen on interior floor. Max dimensions: 4.7 × 3.7 cm.
Fabric semi-coarse, surface smoothed.

15/14 (P 292), **Fig. 15.5b** (Photo 2834)
SU 4.286 (24/6/06).
Rim fragment from medium-sized open vessel, with flaring wall and flat rim; lined with bitumen.
PH: 3.05 cm; PW: 2.3 cm; Th: 0.7–0.8 cm.
Fine red clay, fine white particles; surface smoothed, interior coated with bitumen. Munsell 2.5 YR 5/8.

15/15 (P 294), **Fig. 15.5c** (Photo 2885)
SU 1.375 (26/6/06).
Wall fragment of large open vessel with flakes of bitumen adhering to interior.
Max dimensions: 1.8 and 2.0 cm; Th: 0.9 cm.

15/16 (P 297), **Fig. 15.5d** (Photo 2895)
SU 2.380 (26/6/06).
Wall fragment coated with black on exterior surface: bitumen or manganese deposit?
Max dimensions: 3.2 × 2.7 cm.
Fabric semi-fine, light red. (2.5 YR 5/8).

15/17 (P 311), **Fig. 15.5e-f** (Photos 2984–2985)
SU 4.286 (27/6/06).
Wall fragment of medium-sized open vessel, lined with bitumen.
Max dimensions: 2.8 × 2 cm; Th: 0.45 cm.
Fabric semi-coarse, sandy; dark core, surface smoothed, coated with bitumen.

15/18 (P 319), **Fig. 15.5g** (Photo 3110)
SU 4.286 (27/6/06).

Small fragment of open vessel lined with bitumen.
Max dimensions: 2.4 × 1.3 cm; Th: 0.95 cm.
Fine reddish clay (2.5 YR 5/8), tiny air holes.

15/19 (P 347 [**9/165**]), **Fig. 9.57**
SU 4.201 (11/7/06).
Wall and lower handle attachment of strap handle (flattened oval in section) set at angle to open vessel wall; interior shows traces of bitumen.
PW: 5 cm; PH: 2.95 cm; W (handle): 3.1 cm; Th: 1.3 cm.
Fabric semi-coarse, sandy, dark core with mica sparkles, inside red surface layer; exterior rough, very worn; now dark brown. Surface smoothed, bitumen patches inside.

15/20 (P 355), **Fig. 15.5h-i** (Photos 3209–3210)
SU 1.440 (11/7/06).
Wall fragment of small to medium vessel with patches of bitumen adhering to exterior.
Max dimensions: PH: 2.6 cm; PW: 3.1 cm; Th: 0.75 cm.
Fabric semi-coarse, dark between red layers; surface smooth, dark: bitumen applied.

15/21 (P 395 [**9/50**]), **Figs. 9.13, 9.43**
SU 5.001 (baulk removal, topsoil level, 29/6/2007).
Three joining rim fragments of open shape: flaring wall, flat rim thickened on exterior with projecting lip. Patches of shiny black layer on exterior wall, top of rim and possibly inside face very worn.
PH: 3.1 cm; PW: 4.2 cm; Th: 0.4 cm.
Fabric semi-fine, reddish core and surface (2.5 YR 6/8).

Special finds: bitumen deposit, Tomb XX (50)
Special Find 252, **Fig. 15.6a-b**
Morris 2006:102, fig. 6 (Photo 1151); Papadopoulos, Bejko, and Morris 2007:129, fig. 23 (Max Farrar).
Tomb XX (Grave 50) 2.308 (skeleton, Individual 308), 19/7/2005.
Inhumation in simple pit (non-rectangular), partially lined with stones. Skeleton in extended supine position, head tilted, left arm crossed across lower torso, legs slightly flexed. Layer of bitumen below and around left pelvis and over left arm, lower left ribs of skeleton (adult female, ca. 20–25 years), broken into many small flakes and chips (2 mm thick), as if dried from a liquid layer. W: 180 g (with soil); after cleaning, 23 g without soil: associated soil sampled (145 g). No grave goods (two scraps of ceramic in fill).

ACKNOWLEDGMENTS

We are grateful to John Hale for his visit to Apollonia and Selenicë in 2008, his interest in this oracle, and enlightenment on Delphi and other oracles as possible sites of chemically induced trances and utterances (for a popular account, see Broad 2006).

CHAPTER 16
ENVIRONMENTAL ARCHAEOLOGY AT LOFKËND

John M. Marston, Evi G. Vardala-Theodorou, and J.E. Foss
With contributions by George Theodorou and Evangelos Veltzelos

CHAPTER 16.1
ENVIRONMENTAL ARCHAEOLOGY AT THE LOFKËND TUMULUS: RESULTS OF ZOOARCHAEOLOGICAL, FLOTATION, AND WOOD CHARCOAL ANALYSES

John M. Marston

INTRODUCTION

The Lofkënd tumulus was not a good candidate for comprehensive environmental archaeological investigation: composed entirely of secondary deposits of sediment from other areas of the neighboring cultural landscape, there was little reason to believe that animal and plant remains pulled from the tumulus fill matrix would be indicative of human–environment interaction contemporaneous with the construction of the tumulus. Despite this assumption, however, a comprehensive collection plan for animal bones, flotation samples, and wood charcoal was included in the excavation plan for three reasons: to teach American and Albanian students methods and strategies of environmental archaeological sampling, to ensure that potentially important classes of information were not missed during excavation, and to expand the meager record of sampling and analysis of archaeological plant and animal remains from excavations in Albania.

Three types of environmental archaeology sampling were conducted at the tumulus: zooarchaeological analysis of animal bone, anatomical analysis of wood charcoal fragments, and sampling of small soil samples for flotation to recover carbonized plant macroremains. Animal bones were the most common of these remains, and the analysis of the zooarchaeological remains was the most informative of these analyses, especially relating to the two animal burials discovered. The wood charcoal analysis was limited in scope but suggests similarities between forest structure today and in prehistory. The flotation analysis, in contrast, yielded few, if any, ancient plant remains. The absence of meaningful quantities of burned seed and charcoal remains from the tumulus is instructive in and of itself, however, indicating that on-site burned offerings or funeral pyres were not a component of burial ritual at Lofkënd.

This chapter presents the results of these three analyses in turn, including full lists of all taxa encountered as tables within this chapter. The full inventory of animal bones identified is listed separately at the end of this section (**Table 16.5**).

ANIMAL REMAINS AND ANIMAL BURIALS

The Lofkënd tumulus contained several hundred fragments of animal bone in addition to human remains from the 100 graves discovered during the excavations. Most of this animal bone consists of small fragments, especially teeth, that were mixed in with the tumulus fill and redeposited from their original context. In addition to these scattered remains were three articulated sheep skeletons that were placed directly over two early modern graves (**Fig. 16.1**). Two are neonatal infant lambs, less than 2 months of age, and all may have been sacrificed as part of the burial ritual (Papadopoulos, Bejko, and Morris 2007:117).

Archaeological animal remains can be used to investigate questions of hunting and herding strategies (Bird et al. 2002; West 2009; Zeder 1991), diet and cuisine (Lightfoot et al. 1998; Stiner et al. 2000), economy (Arbuckle 2012; Choyke and Bartosiewicz 2001; Miller et al. 2009; Mulville and Outram 2005), human impacts on the environment (Grayson 2001; Steadman and Martin 2003), and ritual use of animals (Hamilakis and Konsolaki 2004; Ryan and Crabtree 1995). The assemblage from the Lofkënd tumulus fill, however, is not associated with its original deposition context. Based on the limited data available from the burial tumulus, it is only possible to make general conclusions about human diet and subsistence practices during the period in which the tumulus was constructed; I then turn to focus more specifically on the animal burials.

Comparative zooarchaeological studies from Albania are practically non-existent. Excavations at Konispol Cave in southern Albania have yielded wild and domestic animal remains from the Paleolithic and Neolithic periods (Schuldenrein 1998). The site of Sovjan in eastern Albania has yielded rich zooarchaeological remains from the Late Bronze Age, contemporary with the earliest remains from Lofkënd (Allen 2005:336–337; Gardeisen et al. 2002). This leaves the data from Lofkënd as the only record of animal use during the Early Iron Age in Albania, at least for the time being. The conclusions of this report regarding animal use are tentative and subject to further revision as more zooarchaeological research is undertaken in Albania, but important in the current absence of contemporary studies.

Methods

Stray pieces of bone were collected by hand during excavation and placed in paper bags labeled with the trench and unit numbers. All excavated soils were also screened with 1- or 0.5-cm mesh, and any bone recovered in the screen was bagged together with the hand-collected bone. Physical anthropologists examined all bone in the laboratory and removed the majority of human bone collected from tumulus fill contexts. The small quantity of bone that could not be positively identified as human or animal was so poorly preserved as to be of no use. All burials, human and animal, were recorded separately on grave context sheets and were photographed and drawn. Animal bone mistakenly collected with human burials was bagged by the physical anthropologists and returned to the faunal specialist. The animal burials were wrapped in paper towels and aluminum foil and boxed separately. All faunal specimens were shipped to UCLA for analysis.

All animal remains were analyzed at the UCLA Zooarchaeology Laboratory by the author and Thomas Wake. Justin Lev-Tov (Statistical Research, Inc.) also looked at the remains of the three sheep from burial contexts. Each context bag was emptied on a tray and the bone fragments were sorted by taxon and element. The taxon, element, part, side, and fragment count were recorded in an Excel spreadsheet for each bag. When possible to determine, age class was recorded, as well as cut marks and evidence of working or burning. The animal burials were unwrapped separately and studied in additional detail, although the same information was recorded.

For all animal bone fragments from tumulus fill contexts (not animal burials), the number of identified specimens (NISP) was calculated by counting all separate skeletal element fragments that did not refit. If pieces refit, they were counted as a single fragment. Mandibles/maxilla with teeth were counted as a single element for calculation of NISP. The minimum number of individuals (MNI) was also calculated for each identified taxon following the definition of White (1952, 1953) as modified by Reitz and Wing (1999:196). The MNI is based on repeated instances of the same skeletal element from animals of the same age class. For example, although not more than one instance of each skeletal element of cattle was identified, teeth from juvenile, adult, and old adult individuals were recorded, resulting in an MNI of 3 (**Table 16.1**).

Results

At least 11 vertebrate taxa were identified from the tumulus fill. Land snail shells also were systematically collected separately and are published by Evi Vardala-Theodorou (Chapter 16.2). Fragmentary human remains were also present in the animal bone samples; these fragments were returned to the physical anthropologists. All animal taxa identified are summarized in **Table 16.1**, excluding the sheep burials, which are treated separately later in this

TABLE 16.1 Taxa identified from tumulus fill

Taxon	Common Name	NISP	MNI
Testudo sp.	Tortoise	1	1
Aves	Indeterminate bird	1	
Galliformes	Fowl	1	1
Mammalia	Indeterminate mammal	192	
Canis domesticus	Dog	7	1
Equus sp.	Horse or donkey	1	1
Rattus rattus	Black rat	1	1
Mus cf. *musculus*	Mouse	1	1
Artiodactyla	Indeterminate artiodactyl	3	
Cervidae	Cervid	1	1
Sus scrofa	Pig	26	3
Bovidae	Indeterminate bovid	8	
Bos taurus	Cattle	11	3
Caprinae	Sheep or goat	95	5
Capra hircus	Goat	4	1
Ovis aries	Sheep	1	1

chapter. Bone counts from individual excavation units are presented in **Table 16.5**.

Many taxa are represented by single skeletal elements. This includes a single humerus of a galliform bird, possibly a domestic chicken, and a long bone shaft fragment of a large bird, which might belong to the same taxon. One reptile specimen, a tortoise pygal bone, was identified. Among the mammals, there were only single specimens of rat (incisor), mouse (incisor), horse or donkey (premolar), and a single antler base fragment of a cervid, either European roe deer (*Capreolus capreolus*) or European red deer (*Cervus elaphus*). The only taxa with multiple identified elements were dog, pig, cattle, goat, and sheep. All measurable elements appear to be domestic in size, although the small sample size and fragmentary nature of most specimens prevent conclusive statements about the absence of wild species of these taxa.

Cattle, pig, sheep, and goat made up the majority of identified vertebrate remains: 45% of total NISP, and 92% of specimens identified to family or better. Within each of these taxa, teeth make up two-thirds or more of identified specimens, which is more than twice the rate that teeth appear in the mammalian assemblage as a whole (**Table 16.2**). This is due to two factors. The first is the identification bias inherent in mammal bone identification. Certain elements, such as teeth, are highly identifiable to genus, while others, such as long bone shaft fragments and rib fragments, are rarely identifiable beyond class. The second is that teeth are small and more durable than bone due to their enamel coating, so are often preserved to a greater degree than long, thin bones. The apparent prevalence of teeth in the Lofkënd tumulus fill is the result of this difference in preservation, while the preponderance of teeth among identified domestic animal remains is due to identification biases.

It is not surprising that the majority of faunal remains from Lofkënd are dog, sheep, goat, cattle, and pig, as these species form the core of the Neolithic domesticated animal complex that moved into Europe around 6000 BC (Price 2000). Most of these animals were used for food and for secondary products (wool, hair, and milk; Sherratt 1981). Dog remains form a minor component of the faunal assemblage, as at many contemporary sites (Snyder and Klippel 2003), and do not appear to have been eaten at Lofkënd: there are no cut marks on the bones, which are mostly limb bone fragments, and no evidence of burning on the bones. Since animal remains are randomly mixed into the fill of the tumulus, except for the two early modern burials, their original use or deposition contexts are not clear. Studies of seasonality of culling (Davis 2005), herd management strategies (Zeder et al. 2006), and breed characterization (Ioannidou 2003) require much greater sample sizes and chronological control of stratigraphic deposition than is available with the Lofkënd remains. As such, the following analysis only focuses on general diet and herding strategies evident from these remains in

TABLE 16.2 Percentage of teeth among domestic food animal specimens compared to percentage of teeth among all mammal bone fragments

Taxon	Number of tooth fragments	NISP	% of NISP
Sheep and goat	80	95	84.2
Pig	21	26	80.8
Cattle	9	11	81.8
All mammals	121	351	34.5

Note: Mandible and maxilla fragments, even with teeth, were counted as non-teeth.

the context of comparative studies from Albania and Late Bronze Age and Early Iron Age sites in Greece.

Diet and herding practice in Albanian prehistory

Direct evidence for the butchering and consumption of animals by the prehistoric inhabitants of the Lofkënd area does not exist. The tumulus fill was redeposited from somewhere in the surrounding landscape, and both wood charcoal and pottery (Pevnick and Agolli, Chapter 9) from the fill are mostly contemporary with the burials and thus with the construction of the tumulus, although some of the pottery is earlier. Other material, however, is considerably earlier. The lithics include examples as early as the Paleolithic, whereas the remnants of daub from wattle-and-daub architecture, while not precisely datable, find their closest parallels in Neolithic settlements in Albania, though they are known in Early Iron Age Sovjan (Papadopoulos, Chapter 14). Although it is likely that much of the animal bone is contemporary with the burials, this cannot be established with certainty, and only direct radiocarbon dating of the animal bone will resolve the issue. This said, it is possible, and perhaps likely, that the majority of the animal bones are contemporary with the construction of the tumulus and thus represent animal use in the Late Bronze and Early Iron Ages.

Twelve fragments of animal bone (3% of total vertebrate specimens) show evidence of burning, and none bear any evident cut marks. It is likely that these bones come from undifferentiated domestic trash deposits, as suggested by the presence of charcoal, pottery, and pieces of daub (construction debris) among the fill contexts. The small sample size and tertiary deposition of the animal remains prevents quantitative analysis, except in general terms. Based on the NISP and MNI statistics from the tumulus fill (**Table 16.1**), sheep and goat appear to have been the most common domestic animals in the area during late prehistory. Cattle and pig likely played smaller roles in the herding economy.

Unfortunately, no contemporary zooarchaeological remains have been published in full from Albania. As already noted, animal remains from two sites have been analyzed: Upper Paleolithic-Neolithic Konispol Cave (Schuldenrein 1998) and Neolithic–Bronze Age Sovjan (Allen 2005; Gardeisen et al. 2002). The Konispol Cave remains have not been published but include wild sheep, goat, and pig, as well as domestic sheep and goat from Neolithic levels (Russell 1998b; Schuldenrein 1998:506–507). Domestic sheep, goat, and pig bones were also recovered from the Final Neolithic open-air site at Doliana, just across the border in northwestern Greece (Halstead 2005:47).

Sovjan is the only site in Albania from which animal remains partly contemporary to those from Lofkënd have been analyzed. They are published in preliminary form (Gardeisen et al. 2002) and summarized by Allen (2005:336–337) in her analysis of diet in the Bronze Age levels of the site:

> Zooarchaeological data from Sovjan indicate the use of more than 50 species of fish, as well as eels, wild birds, wildcats, boar, roe deer, and otter that were hunted. In addition, domesticated cattle, sheep, goats and pigs were integrated into the agricultural economy (Gardeisen et al. 2002), but their relative contribution to the economy has not yet been reported. There is preliminary evidence to suggest that domestic animals, especially cattle (adapted to marsh grazing), were increasingly important in the LBA (Late Bronze Age).

While no numerical data on the frequency of these remains is available, it is not surprising that Late Bronze Age Sovjan shares the same domestic species with Lofkënd, since those species are common across Europe and the Near East after the earliest Neolithic. The range of wild animals at Sovjan is much greater than that observed at Lofkënd. Sovjan sits along Lake Maliq and has significant habitat variability among lake, wetland, plain, and mountain, allowing for a more diverse natural animal population. In addition, the contexts excavated at Sovjan were primarily domestic, so more likely to contain food waste, and were waterlogged when excavated, which can improve preservation of organic remains. All deposits at Sojvan were water-screened, in contrast with selected flotation samples from Lofkënd, which resulted in greater recovery of smaller remains during excavation. No fish bones were found at Lofkënd from either hand-collected or water-screened deposits, which is not surprising considering its upland location. Fish clearly played a much greater role in the economy of Sovjan, due to its location on Lake Maliq, as reflected in the diversity of fish species identified at that site.

In neighboring Greece, a much greater amount of zooarchaeological research has been done at Late Bronze and Early Iron Age sites. The data from northwestern Greece, the southern portion of Epirus, are still somewhat limited. Tartaron (2004: 127) notes finds of animal bone at two Bronze Age sites in the region, and mentions that "the recovery of animal bones from excavated sites in Epirus is far more common than has been reported." Animal remains have been published from several sites in mainland Greece with Late Bronze or Early Iron Age levels, including Kastanas (Becker 1986; Reichstein 1979) and Dimitra (Yannouli 1997) in northern Greece, as well as Eleusis (Cosmopoulos et al. 2003), Kalapodi (Stanzel 1991), Tiryns (von den Driesch and Boessneck 1990), Pylos (Nobis 1993), and Nichoria (Sloan and Duncan 1978) in central and southern Greece; there is also some Early Iron Age faunal material in a cemetery, as opposed to domestic, context at Torone (Bökönyi 2005; Ruscillo 2005) in the north Aegean.

Comparison of animal remains from these sites with those at Lofkënd reveals little additional information about contemporary herding practices and diet in the area of the site. Every contemporary site with a reasonable sample size includes remains of domestic goat, sheep, pig, and cattle. Sheep and goat are usually the most common remains in archaeological contexts (e.g., Forstenpointner 2003); this pattern extends to mammal remains from burial contexts at Early Iron Age Torone (Bökönyi 2005: 318–319). These animals appear to have formed the bulk of animal protein in the diet. Examples of extensive utilization of wild animals, such as badgers (Snyder and Klippel 1996), fish (Mylona 2003), and aquatic mammals (Gardeisen et al. 2002), are related to local environmental conditions and are uncommon, although biases in collection techniques may lead to the underrepresentation of fish in many zooarchaeological analyses. Overall, we can conclude only that animal husbandry at Lofkënd was similar to that of other inland areas across the Balkans in the Late Bronze and Early Iron Ages.

Animal burials at Lofkënd

Two incomplete neonatal sheep skeletons and one adult sheep head and forelimbs were found directly overlying Tombs LXXXVI and XCIII (Graves 22 and 23) (**Fig. 16.1**), the two well-built, stone-lined cist graves from the early modern period (one skeleton directly dated to AD 1810 ± 145, Damiata and Southon, Chapter 4.2). The neonatal sheep above Tomb LXXXVI (Grave 22) was numbered Tomb LXXXVII (Grave 8); this skeleton is mostly complete and articulated, suggesting that it was deposited whole, but the skeleton is missing several limb bones and appears to have been disturbed prior to excavation (**Fig. 16.2**). A lower third milk molar shows very slight wear, placing the age at death for this individual at 6–8 weeks. The sheep skeleton above Tomb XCII (Grave 23), however, is very fragmentary: only one metacarpus, one metatarsus, part of the right mandible and maxillary teeth, a few skull fragments, and a single first phalanx survive. This was not identified as an animal burial during excavation but was numbered as Skeleton 112. The remains were concentrated in a small area adjacent to scattered human remains (Skeleton 111), and were clearly disturbed between burial and excavation, as was the human Skeleton 111 (**Fig. 16.3**). Only portions of the cranium of Skeleton 112 were articulated when found. This lamb was also 6 to 8 months old when killed. There are no butchering (or any other) marks on the bones of either lamb, so the cause or method of death is not possible to determine.

The adult sheep burial of Tomb XCIII (Grave 19) also overlies Tomb XCII (Grave 23). Its stratigraphic relationship with Skeleton 112 was unclear, but, due to the difference in preservation, it does not appear to have been deposited simultaneously; these should probably be interpreted as two separate animal burials. Grave 19 contains the entire preserved cranium of a sheep approximately 2 years of age, with its front forelimbs and the left rear forelimb placed atop the cranium (**Fig. 16.4**). These are complete from the metapodial down, although one toe is missing from the hind leg. The cranium was also photographed in situ (**Fig. 16.5**). There are no cut marks visible on these bones, and the degree to which the forelimbs were articulated when deposited is unclear; some of the toes are articulated, others are not.

The position of these animal burials directly over the center of the two cist graves, and the presence of precisely one neonatal sheep directly atop each of the two cist graves, suggests a connection between the human and animal burials. The adult sheep head faces to the northeast, the same direction as the individual interred in Grave 23. Based on these positional similarities, the excavators suggested that the animals were sacrificed as part of the burial ritual (Papadopoulos, Bejko, and Morris 2007:117). Further interpretation of these burials requires comparison with other data on animal sacrifice and burial across the Balkans.

Animal sacrifice and burial in the Balkans

The best known type of sacrifice in southeastern Europe during antiquity is the Classical Greek tradition of so-called "Olympian" sacrifice, attested in a variety of historical sources and iconographically (van Straten 1995). In this ritual an animal was sacrificed and its flesh roasted and consumed by the participants. The fat and bones were burned for the gods, a clever trick devised by Prometheus so that the Olympian gods received only the scent of flesh (Hesiod, *Theogony* 540–557; van Straten 1988). This practice has been identified archaeologically through faunal analysis of burned animal bones from several cult sites in Greece (Forstenpointner 2003). Sheep, goat, pig, and cattle bones, especially those of the haunch, are present in these contexts (Chenal-Velarde and Studer 2003; Forstenpointner 2003). This ritual, however, has little to do with the ritual care of the dead, since the bones are deposited disarticulated in trash heaps.

Related types of Greek sacrifice in which fleshed portions of the animal were cremated include the holocaust, in which the (dead) animal was burned whole, and the *enagizein* sacrifice, an apparently uncommon ritual in which portions of the animal were butchered for consumption and other portions were burned while fleshed (Ekroth 2002; Hughes 2005; Powell 2010). The *enagizein* sacrifice is notable because it was associated with funerary rites and hero cults, and an apparent example has been documented from an Archaic Greek burial at Tumulus 9 of Apollonia, Albania (Amore 2010:173; Powell 2010: 55). This burial, Grave 62, contained a sacrifice of at least two adult sheep, the remains of which were placed in a grave pit (albeit one that never received a human body) along with burned charcoal and fragments of pottery vessels. The bones were heavily burned and were either fleshed or green when burned (Powell 2010:853–855). Two bones show butchery marks, including a scapula and radius from a left forelimb, suggesting that this joint was removed and defleshed prior to burning and deposition (Powell 2010:855). The combination of apparently fleshed and defleshed bone being burned led Powell to interpret this deposit as an *enagizein* sacrifice, one of few ever identified archaeologically (Powell 2010:855). Other burials at Apollonia may have included the same type of animal sacrifice, including Grave 74 in Tumulus 10, which contained fragments of a possible ovicaprine skeleton unearthed in a baulk (Amore 2010:254). These types of deposits might have been included among the later graves at Lofkënd, but no evidence for such sacrifices was recovered during excavations; the animal burials described here are much later in date and show no evidence of butchery or burning.

A broader survey of both ethnographically documented cultures around the world (see review in Wilson 1999) and archaeological remains from prehistoric Europe indicates that animal remains were often physically associated with human burials. The presence of animal parts, including crania, antlers, tusks, mandibles, and worked bone objects, is common in Neolithic burials across the Balkans (Bailey 2000). Burial ritual involving animal remains is also common among the Iron Age cultures of northern Europe (e.g., Grant 1984; Wilson 1999) and has been used to identify Celtic burials in Hellenistic Anatolia (Dandoy et al. 2002). Dog burials are particularly common among animal burials in Late Bronze and Early Iron Age Greece (Day 1984; Snyder and Klippel 1997). These may be intended to accompany the soul of the deceased to the underworld (Day 1984), or they may simply be the hunting dogs of the interred individual (Hamilakis 2003). In none of these cases, however, are the heads or entire bodies of animals placed directly over a human grave.

More relevant to the Lofkënd burials are historical accounts that link the practice of animal sacrifice with religious rites related to the dead throughout southeastern Europe. The Albanian custom of *fërlik* includes the killing and roasting of a pig or lamb on Saint Nicholas Day, which is observed by both Christians and Muslims, when the spirits of the dead are invited into the home for an evening of feasting and merriment (Elsie 2001:89). Balkan peasants performed sacrifices of domestic animals to initiate the construction of a house and as part of funerary ritual for the dead as recently as the last century (Stahl 1976). In this ritual, men slaughter only black animals by cutting the throat, which "takes place at the moment of the lowering of the corpse into the Grave and is related to the burial; it thus consecrates the Grave" (Stahl 1976:447). Chickens, sheep, and cattle are considered "pure" animals appropriate for sacrifices to God, while cats, dogs, goats, and horses are "impure" and tainted by the Devil; impure animals are only sacrificed as a cure for certain diseases such as epilepsy (Stahl 1976: 448–449).

Although the information above suggests that these three animals were sacrificed and deposited whole as ritual offerings associated with the opening of the grave, it is also possible that the sacrifice was instead intended to appease the spirits of the dead whose remains were disturbed when the grave was dug. Tombs LXXXVI and XCII (Graves 22 and 23) disturbed the prehistoric Tombs LXV and LXXXV (Graves 30 and 10), respectively, and the animal sacrifice may have been made in response to that inauspicious accident (Papadopoulos, Bejko, and Morris 2007:117). It is worth adding that although some of the other modern burials were found overlying prehistoric graves, only Graves 22 and 23 disturbed earlier remains so clearly. It also remains unclear why the adult sheep head and forelimbs of Tomb XCIII (Grave 19) were buried above Tomb XCII (Grave 23), possibly disturbing the earlier neonatal burial of Skeleton 112.

The motivation for the probable sacrifice and burial of these animals is not possible to determine, but the physical relationship of sacrificed sheep with human graves is in keeping with historically documented burial rites practiced in Albania. The neonatal lambs were disturbed after burial, that in Tomb LXXXVII (Grave 8) to a small degree and Skeleton 112 to a large degree. These burials led to the interment of the only articulated animal skeletons in the tumulus, and likely comprise the only animal remains deliberately deposited at Lofkënd. The variation in burial practice between the ancient and modern human graves is reflected in the animal remains from the tumulus, giving additional insight into the mortuary customs of Albania's past.

Archaeological Wood Charcoal: Implications for Forest Structure

The Lofkënd tumulus was in use for some 700 years in antiquity, as well as during the early modern period, and during the course of its construction multiple fragments of charred wood became interred with its inhabitants. These pieces of charcoal, scattered haphazardly through the redeposited soils that built the tumulus, were not deliberate grave offerings, nor are they an unbiased representation of the surrounding woodland environment. They are the product of human interaction with, and modification of, a natural landscape, much like the tumulus itself.

Using simple light microscopy, it is possible to identify carbonized wood, usually to the genus level (Schweingruber 1990). Wood identification and consideration of the archaeological and temporal context of its collection and deposition allows us to understand past landscapes and the nature of human interaction with the biotic environment (Asouti and Austin 2005; Dufraisse 2006; Marston 2009; Thiébault 2002). This section summarizes results of wood charcoal analysis at Lofkënd and suggests possible behavioral and ecological interpretations of these remains.

Methods

Charcoal was hand collected in the field by each sector/trench supervisor. Soil samples were collected systematically for flotation to recover macrobotanical remains, including charcoal, but early trials showed a paucity of finds in these samples (see below). No wood charcoal was recovered from the flotation samples. Hand-collected samples were wrapped in aluminum foil, catalogued on site, and shipped to UCLA for analysis.

Most charcoal samples included dirt from the field and very small pieces of charcoal (or even charcoal dust) resulting from post-excavation breakage of the charcoal pieces. To limit the sample to pieces of charcoal likely to be identifiable, all samples were screened through a 2-mm geological sieve, and the resulting fraction larger than 2 mm was weighed and retained for analysis. The transverse section of each piece of charcoal was examined with a 7.5–75x zoom stereomicroscope, and was broken to reveal a fresh section if necessary. Pieces requiring the examination of diagnostic characters in the radial and tangential sections were broken along those planes and examined at 100–400x with an incident-light compound microscope. Each piece of charcoal was recorded and placed with like pieces from the same sample. If multiple types were observed in a sample, all fragments were identified and weighed in bulk by type. The weight of each taxon present in the sample was recorded, as was the number of pieces of charcoal in the sample.

Results

Twenty-four charcoal samples, excavated between 2004 and 2007, were examined in this study, although eight of these contained only soil and no identifiable charcoal fragments. The remaining 16 samples are presented in **Table 16.3**.

Two types of wood charcoal were identified among these remains: oak (*Quercus* sp.) and maple (*Acer* sp.). Several wood fragments were unidentifiable; these may be fragments of bark or knotted wood with irregular cell structures; in one case, the wood was impregnated with clay and could not be broken for detailed observation. No more than one wood taxon was identified in each sample (excepting sample 22, where the unidentifiable wood is likely to be oak bark), which is a product of the sampling methodology. Most often, a single piece of wood charcoal is found in excavation, so the multiple fragments found in each sample are a product of fragmentation during excavation and later handling.

Identification of oak and maple wood was accomplished by comparing microscopic wood anatomy details with comparative specimens and published wood identification manuals (Schoch et al. 2004; Schweingruber 1990). The details observed on the archaeological wood specimens are listed below:

Acer sp. (maple)
X-section: diffuse porous, growth ring indistinct, pores in small groups to short oblique bands (less often solitary), paratracheal parenchyma, rays multiseriate
R-section: spiral thickenings, rays homogeneous, long procumbent ray cells
T-section: rays 2- to 5-seriate, 7–15 cells high

Quercus sp. (oak)
X-section: ring porous (when growth ring boundary visible), dendritic pore bands, uni- and multi-seriate rays, frequent radial splits
R-section: not observed
T-section: not observed

TABLE **16.3** Identified hand-collected charcoal from Lofkënd tumulus

				Oak		Maple		Unidentifiable	
Sample	Trench	Unit	Grave	Count	Weight	Count	Weight	Count	Weight
12	2	202		2	0.89				
13	2	244	37			6	0.39		
14	4	289/290		6	0.93				
15	4	290		3	1.32				
17	4	286				1	2.72		
18	4	286				1	0.72		
20	2	395				1	0.07		
22	2	453	81	4	0.04			5	0.12
23	2	475						1	0.11
27	1	361	64	1	0.74				
28	5	279						3	a
29	5	279		1	2.95				
30	7	531/228				8	16.96		
34	4	329	56	6	0.07				
"c1"	2	2				1	0.59		
"c2"	2	489	86			1	0.76		

[a] This fragment was impregnated with clay, so the charcoal weight could not be recorded.

Oak is easily identified from the transverse section, so it is not necessary to examine the other planes of the wood. It is the only European wood type with thin and thick rays and dendritic pores. Maple can be confused with other diffuse-porous woods, including linden (*Tilia* sp.) and domestic fruit trees (*Prunus* sp.), which also have pore groups, multiseriate rays, and spiral thickenings. Neither *Tilia* nor *Prunus* has paratracheal parenchyma, however, which is unusual in maple wood but present in the samples here. The pattern of pores on these fragments is also closer to that of comparative maple specimens than the other genera.

It is not possible to attempt a spatial analysis of the wood charcoal distribution within the tumulus due to the small number of fragments discovered within the mound. Many of the charcoal fragments were submitted for radiocarbon analysis to date the construction of the tumulus and specific graves (see Damiata and Southon, Chapter 4.2, and Damiata et al. 2007–2008), so it is possible to discuss the chronological distribution of the samples. The four samples that date to the modern period (cal. AD 1800 ± 150) are all maple. Of the two samples dating to the Archaic period (ca. 800–500 BC), both of which come from the same context (Trench 5, SU 279), one is oak and the other indeterminate. The remaining 10 samples date to the Late Bronze and Early Iron Ages, of which three consist of maple, six oak, and one is indeterminate. This may indicate that different wood resources have been available in the area over the use period of the tumulus, but the small sample size belies any definitive conclusions on this matter.

Ecological and behavioral interpretations

Few publications of wood charcoal analysis from archaeological contexts in Albania are available, especially in comparison with its neighbors Greece and Italy. Julie Hansen published the charcoal remains from Upper Paleolithic-Neolithic Konispol Cave in southern Albania (Hansen 1999, 2001), and Susan E. Allen completed a dissertation on the carbonized and waterlogged plant remains from Neolithic–Early Iron Age Sovjan in southeastern Albania, including an analysis of wood charcoal from the site (Allen 2005). Only one other paleoethnobotanical study has been done of Albanian material, published in a report on carbonized seeds from the Neolithic site of Maliq (Xhuveli and Schultze-Motel 1995). Other local paleoenvironmental data are extremely limited as well, with a single set of lake and peat cores published from Lake Maliq, which lies beside the sites of Sovjan and Maliq in southeastern Albania (Denèfle et al. 2000; Fouache et al. 2001).

Given this limited comparative material, other authors have made comparisons with a broad range of published paleoethnobotanical and paleoenvironmental data. Fouache and colleagues compare the Lake Maliq pollen cores with cores from northern Greece (Fouache et al. 2001), Hansen compares Konispol Cave to Franchthi Cave in Greece (Hansen 2001), and Allen references Bronze and Iron Age paleoethnobotanical studies from Greece, Macedonia, and Bulgaria (Allen 2005:12). Since the charcoal samples from Lofkënd are limited in number and variety, they do not warrant such extensive treatment.

In comparison with Konispol Cave and Sovjan, the species diversity at Lofkënd is much lower, most likely due to the much smaller number of samples at Lofkënd. Oak charcoal was recovered from Late Bronze and Early Iron Age habitation levels at Sovjan (Allen 2005) and Paleolithic-Neolithic levels at Konispol Cave (Hansen 2001:422–424). Probable maple charcoal comes from all habitation periods at Konispol Cave (Hansen 2001:422–424). Maple charcoal has not been identified at Sovjan or in the Maliq cores, but maple pollen is an infrequent member of the Maliq pollen profile (Denèfle et al. 2000:428). Maple is a minor component of the contemporary deciduous oak forest at both sites (Allen 2005:127–128; Schuldenrein 1998:507).

The Flora Europaea describes eight species of maple and nine species of oak as native to Albania: *Acer platanoides*, *A. campestre*, *A. tataricum*, *A. pseudoplatanus*, *A. heldreichii*, *A. obtusatum*, *A. hyrcanum*, *A. monspessulanum* (Tutin 1968–1980:238–239); *Quercus coccifera*, *Q. ilex*, *Q. trojana*, *Q. macrolepis*, *Q. cerris*, *Q. petraea*, *Q. robur*, *Q. frainetto*, *Q. pubescens* (Tutin et al. 1993:72–76). Unfortunately, it is not possible to distinguish among individual species of maple or oak from charcoal anatomy (Schweingruber 1990); the small number of samples from Lofkënd does not allow for differentiation among different sub-genera of oak. Such a distinction would be useful, as different species of oak, in particular, occupy different ecological niches. Oak forests, dominated by *Q. cerris* and *Q. pubescens*, extend from the coastal plains of Albania up to the base of beech forests, which begin between 900 and 1,300 m in elevation (Nuttonson 1947:12). A maquis scrub, often dominated by oak, is characteristic of coastal Mediterranean vegetation and present in large areas of western Albania (Markgraf 1932:21–25). *Q. ilex* is the dominant species in this phytogeographic association (Markgraf 1932:25).

Phytogeographic surveys of Albania suggest that Lofkënd lies in the upper elevations of the coastal plain, with a maquis to oak forest vegetation type (Markgraf 1932: map). Maple, which is not a component of the oak maquis, is found as a minor component of oak forests throughout the Balkans (Dida 2003:1; Markgraf 1932:31; Turrill 1929:134). The presence of maple at Lofkënd in both the ancient and modern charcoal samples suggests that the area has been forested, likely with an oak-dominated plant community, throughout the Late Holocene. This conclusion is reinforced by the presence of such forests in the area today and by paleoenvironmental data showing little climate change in Albania over the past 4,000 years (Denèfle et al. 2000; Fouache et al. 2001). The absence of oak among the early modern samples from the tumulus is unlikely to be significant, as oak does grow in the immediate vicinity of the site today (personal observation) and is more likely a sample size effect.

Given the variety of secondary hardwood tree types within the deciduous oak forests of Albania, why is maple the only other component of the tumulus fill? The small sample size precludes any behavioral conclusions regarding this specificity. One possibility is that maple was locally prevalent during the period of time in which soil was deposited in the tumulus. Local dominance of secondary tree taxa has been noted across the Balkans (Turrill 1929:134).

The construction of the tumulus at Lofkënd was the result of deliberate human action resulting from social norms regarding disposal of the dead. The presence of wood charcoal within the fill is entirely unrelated to this ritualized activity and is instead an unintentional marker of one way in which people interacted with the living landscape of Lofkënd. Charcoal fragments present in the soil may have been the result of cooking fires or of forest fires. The scarcity of charcoal within the tumulus precludes further discussion of how it was generated, but it is clear from radiocarbon dating that the charcoal is contemporary with the burial of adjacent human remains (Damiata and Southon, Chapter 4.2; Damiata et al. 2007–2008). While the charcoal tells us little about human actions, it does offer a rare glimpse into the forest ecology of Late Bronze and Early Iron Age Albania. Future excavation and recovery of botanical remains in the Lofkënd region is needed to understand the ecological history of central Albania in more detail.

FLOTATION AT THE LOFKËND TUMULUS

Flotation is a mechanism for separating plant remains from archaeological soil samples using the principle of relative density in a water bath: light plant remains float, while heavier soil, metal, ceramic, and most bone fragments sink. Multiple methods exist, ranging from skimming plant remains from a bucket of water by hand to using an automated pump that forces water through a large barrel or tank (Marston et al. 2014; Pearsall 2000). Flotation samples were systematically collected and floated, both by hand and with a flotation tank, from all contexts at the Lofkënd tumulus, but very few plant remains were discovered. This brief contribution outlines the methods used and scant findings from flotation at Lofkënd. It goes without saying that plant remains would be expected to be considerably higher in settlement contexts.

Methods

Soil samples for flotation analysis were taken from a variety of contexts in the tumulus. The priority was to sample the fill of excavated graves, but samples were also taken from fill deposits to allow for comparison between potential grave offerings and random fill inclusions. Over the four seasons of excavation, 151 soil samples were taken in volumes of approximately 1 to 10 liters, of which approximately 140 samples were floated.

During the first season of excavation (2004), the author floated 12 soil samples, ranging from 1 to 3 liters in size, by hand to establish the viability of flotation at the site. No flotation apparatus was available, so samples were floated in a large bucket holding approximately 20 liters of water. Cheesecloth, with an inconsistent mesh size, was used to capture the light fraction using a tea strainer. The heavy fraction was captured in window-screen mesh with a mesh size of approximately 2 to 3 mm. Results of this flotation were discouraging, as no charcoal or plant remains were discovered, and flotation was discontinued as inefficient until a suitable flotation apparatus for high-volume processing could be procured.

During the final season of analysis in 2008, a flotation tank (converted oil barrel) was available for use, courtesy of the Franco-Albanian expedition at Byllis. The remaining 130 or so samples were floated with this apparatus by undergraduate field-school students under the supervision of Sarah Morris. The apparatus was filled with water from the top with a hose attached to a water pump, and fixed with a spout so that the light fraction could overflow into a screen. The heavy fraction was captured in 0.5-mm plastic mesh, and the light fraction was captured with a 0.5-mm geological sieve or cheesecloth spread over a larger-mesh-size geological sieve. The volume of soil floated was not recorded for any of these samples; however, most were approximately 3 to 5 liters, with some smaller samples from grave or vessel fill contexts.

Heavy fractions from Lofkënd from the hand-floated samples were sorted in aggregate by the author at Apollonia in 2004, as were the light fractions, for which a low-power stereomicroscope was used. The small size of these fractions, especially the light fractions, made size fractionation unnecessary. The barrel-floated samples had larger heavy fractions that were size-fractioned when wet through nested geological sieves and sorted by field-school students at Apollonia. The light fractions were sorted by the author at the UCLA Paleoethnobotany Laboratory; again, no size-fractionation was necessary due to the small size of each sample.

Results

No botanical remains, either charcoal or seed, were recovered from the 12 hand-floated samples sorted in 2004. The remaining samples were examined at UCLA in 2008 and yielded few seed remains. The seeds recovered are summarized in **Table 16.4**, which includes the seeds from heavy fractions sorted out in the field. Charcoal was present in a small number of samples, but no fragments were of sufficient size to be identified. The weight of charcoal from light fractions was generally insignificant (<0.1 g) and not recorded systematically. For analysis of larger wood charcoal fragments recovered from the tumulus, see the previous section.

No identifiable carbonized seeds were recovered from these samples. One carbonized seed fragment that may be a partial pulse seed (pea, vetch, or lentil) was found in one sample; two possible large seed fragments (one carbonized, one uncarbonized) of indeterminate nature were identified during excavation and recorded separately. Eleven uncarbonized seeds found in these samples are likely modern, and include common herbaceous plants native to the area: *Chenopodium album* (goosefoot, Chenopodiaceae/goosefoot family), *Teucrium* sp. (germander, Lamiaceae/mint family), possible *Veronica* sp. (speedwell, Plantaginaceae/plantain family), and unidentified members of Cyperaceae (sedge family) and Fabaceae (pea family). Also identified were four uncarbonized seed pods of the genus *Medicago* (medick, Fabaceae). All of these are common herbaceous plants in the Balkans (Tutin 1968–1980).

Conclusions

Few conclusions can be drawn from these limited results. The seeds identified are common herbs and field weeds across Europe and do not indicate anything particular about the landscape. Additionally, as they are all likely modern, they tell us nothing about the ancient landscape. The carbonized possible pulse fragment and large seed fragment might be ancient, but neither is identifiable. The only conclusion to draw from this analysis is that there are virtually no seed remains preserved within the tumulus, which is not surprising considering the limited quantities of charcoal and animal bone discovered in the fill. The burial ritual used at Lofkënd does not appear to have incorporated the burning of plant remains, and what few seeds and charcoal do appear in the tumulus are likely random components of fill deposits or, in the case of the uncarbonized seeds, modern intrusions.

ACKNOWLEDGMENTS

My thanks go to the directors of the Lofkënd excavations (John Papadopoulos, Sarah Morris, and Lorenc Bejko) for the opportunity to direct a trench at the site and to direct the recovery and analysis of the remains as published here. Thanks also to Naomi F. Miller and Virginia Popper for assistance in the identification of the plant remains and to Thomas Wake and Justin Lev-Tov for assistance in faunal analysis.

TABLE 16.4 Seeds and plant parts recovered from Lofkënd tumulus flotation samples

Soil sample	Trench	Unit	Grave	Fraction	Mesh size	Taxon
86	1	344	60	Heavy	2 mm	
61	1	212		Light		Cf. pulse fragment, carbonized
75	4	311		Light		Chenopodium album, uncarbonized
136	6	141		?	0.25 mm	Unknown seed, uncarbonized
139	6	541		Light		Chenopodium album, uncarbonized
150	2	580	98	Light		4 Medicago pods, uncarbonized
115	2	453	81	Heavy	5 mm	Fabaceae, uncarbonized
106	2	423	76	Light		Cyperaceae, uncarbonized
129	1	509		Heavy	2 mm	Fabaceae, uncarbonized
40	4	124	16	Light		Teucrium, 2, cf. Veronica, uncarbonized
34	1	91		Light		Unknown seed, uncarbonized
30	1	53	6	Light		Chenopodium album, uncarbonized
	3	126	17			Possible seed fragment, carbonized
	4	93	9			Possible seed fragment, uncarbonized

TABLE 16.5 Full inventory of animal bones identified from Lofkënd tumulus

Season	Trench	Unit	Grave	Taxon	Element	Part	Side	Count	No. burned	Age
	1	1		*Ovis/Capra*	Mandible	With p1 and p2	Right	1		Young
2004	1	1		*Testudo*	Carapace	Pygal		1		
2004	1	1		Mammal	Unidentifiable bone	Fr		1		
2004	1	7		Cervidae	Antler base	Fr		1	1	
2004	1	7		*Sus*	Molar			1		
2004	1	9		Mammal	Unidentifiable bone	Fr		2		
2004	1	9		Aves	Long bone	Shaft fr		1		
2004	1	18	2	*Bos*	Premolar/molar	Fr		3		
2004	1	27		*Ovis/Capra*	Molar			1		
2004	1	39		*Ovis/Capra*	Premolar	Fr		1		
2004	1	39		*Ovis/Capra*	Petrosal	Fr		1		
2004	1	39		Mammal	Unidentifiable bone	Fr		3		
2004	1	39		*Ovis/Capra*	Premolar/molar	Fr		3		
2004	1	39		*Ovis/Capra*	Premolar/molar	Fr		2		
2004	1	39		Med//lg mammal	Long bone	Shaft fr		1		
2004	1	47		*Bos*	Upper premolar 3		Left	1		Very young adult
2004	1	47		*Ovis/Capra*	Upper molar	Fr		1		
2004	1	47		*Ovis/Capra*	Upper molar 1		Right	1		Young adult
2004	1	67		*Sus*	Lower incisor 3		Right	1		
2004	1	67		*Ovis/Capra*	Lower molar	Fr		1		
2004	1	67		*Ovis/Capra*	Premolar/molar	Fr		2		
2004	1	67		*Canis*	Humerus	Distal fr	Left	1		

Table 16.5 (*continued*). Full inventory of animal bones identified from Lofkënd tumulus

Season	Trench	Unit	Grave	Taxon	Element	Part	Side	Count	No. burned	Age
2004	1	67		*Canis*	Metatarsus 2	Proximal fr	Right	1		
2004	1	67		Medium mammal	Long bone	Shaft fr		9		
2004	1	67		*Ovis/Capra*	Lower incisor 1		Left	1		
2004	1	67		*Ovis/Capra*	Molar			1		Juvenile
2005	1	67		*Ovis/Capra*	Calcaneus	Fr		1		
2005	1	67		Bovidae	Premolar/molar	Fr		3		
2004	1	70		*Capra*	Trapezoid		Right	1		
2004	1	70		Large mammal	Unidentifiable bone	Fr		1		
2004	1	70		Medium mammal	Long bone	Shaft fr		1		
2004	1	70		Large mammal	Long bone	Shaft fr		2		
2004	1	70		*Ovis/Capra*	Premolar			1		
2005	1	70		Large mammal	Long bone	Shaft fr		8		
2004	1	110	14	*Mus*	Lower incisor		Left	1		
2004	1	111		Artiodactyla	Cranium	Fr		2		
2004	1	126	17	Medium mammal	Rib	Fr		1		
2004	1	126	17	Mammal	Unidentifiable bone	Fr		3	3	
2004	1	127		*Ovis/Capra*	Premolar/molar	Fr		1		
2005	1	278		*Ovis/Capra*	1st phalanx		Right	1		
2005	1	278		*Bos*	Upper molar 1		Right	1		Young adult
2005	1	278		*Ovis/Capra*	Radius	Shaft fr	Right	1		
2005	1	278		*Bos*	Calcaneus	Fr	Left	1		Juvenile
2005	1	278		Large mammal	Long bone	Shaft fr		7		
2005	1	278		*Ovis/Capra*	Mandible	Fr with p3, m1, m2	Left	1		Adult
2005	1	278		Large mammal	Unidentifiable bone	Fr		1		
2005	1	278		Bovidae	Premolar/molar	Fr		2		
2006	1	362	64	*Sus*	Molar	Fr		3		
2006	1	362	64	*Sus*	Podial	Fr		1		
2006	1	362	64	Medium mammal	Long bone	Shaft fr		1		
2006	1	362	64	*Ovis/Capra*	Radius	Shaft fr		1		
2006	1	362	64	*Ovis/Capra*	Rib	Fr		1		
2006	1	362	64	*Ovis/Capra*	Long bone	Shaft fr		2		
2006	1	362	64	*Sus*	Scapula blade	Fr		1		
2005	1	366	52	*Ovis/Capra*	Metapodial	Fr		1		
2005	1	369	66	Canis	Tibia		Left	1		
2006	1	377		*Ovis/Capra*	Metapodial			1		Juvenile
2006	1	377		*Ovis/Capra*	Premolar/molar	Fr		2		
2006	1	378		Mammal	Unidentifiable bone	Fr		4		
2006	1	390	69	Canis	Molar	Fr		2		Juvenile
2006	1	399		*Bos*	Lower premolar 2		Right	1		Adult

Continued on next page

TABLE 16.5 (*continued*). Full inventory of animal bones identified from Lofkënd tumulus

Season	Trench	Unit	Grave	Taxon	Element	Part	Side	Count	No. burned	Age
2006	1	440		*Bos*	Vertebra	Fr		1		
2006	1	440		*Ovis/Capra*	Premolar/molar	Fr		2		
2005	1	318/366	52	*Ovis/Capra*	Molar	Fr		3		
2004	2	2		*Ovis/Capra*	Upper molar 3		Right	1		Old adult
2004	2	2		*Ovis/Capra*	Molar			1		
2004	2	2		Medium/large mammal	Long bone	Shaft fr		10		
2006	2	2		*Ovis/Capra*	Molar	Fr		2		
2006	2	3		*Sus*	Molar	Fr		1		
2004	2	6		*Bos*	Lower molar 2		Right	1		
2004	2	23		*Ovis/Capra*	Upper molar 1		Left	1		
2004	2	40		*Ovis/Capra*	Premolar/molar	Fr		1		
2004	2	42		*Ovis/Capra*	Premolar/molar	Fr		1		
2004	2	71		Medium/large mammal	Long bone	Shaft fr		7		
2004	2	72		Bovidae	Premolar/molar	Fr		1		
2004	2	78		*Ovis/Capra*	Molar	Fr		1		
2004	2	118		*Ovis/Capra*	Premolar/molar	Fr		1		
2004	2	136		Small mammal	Long bone	Shaft fr		1		
2005	2	200		*Ovis/Capra*	Upper molar 1		Left	1		Old adult
2005	2	202		Medium/large mammal	Long bone	Shaft fr		3		
2005	2	202		*Ovis/Capra*	Molar	Fr		1		
2005	2	202		*Ovis/Capra*	Premolar/molar	Fr		4		
2005	2	202		Large mammal	Long bone	Shaft fr		8		
2005	2	202		Large mammal	Rib	Shaft fr		1		
2005	2	202		*Sus*	Upper incisor		Left	1		
2005	2	202		*Sus*	Premolar/molar			3		
2005	2	202		Med/lg mammal	Long bone	Shaft fr		4		
2005	2	202		*Ovis/Capra*	Premolar/molar	Fr		1		
2005	2	202		*Ovis/Capra*	Metapodial	Shaft fr		1		
2005	2	202		Artiodactyla	Rib	Shaft fr		1		
2006	2	231		Mammal	Long bone	Shaft fr		1	1	
2005	2	252	40	Sus	Incisor			1		
2006	2	370		*Ovis/Capra*	Molar	Fr		3		
2006	2	380		*Ovis/Capra*	Molar	Fr		1		
2006	2	420	75	*Ovis/Capra*	Premolar/molar	Fr		2		
2006	2	431		*Ovis/Capra*	Radius	Distal fr		1		
2006	2	431		*Ovis/Capra*	Cranium	Fr		1		
2006	2	431		Medium/large mammal	Unidentifiable bone	Fr		28		
2006	2	431		*Capra*	Premolar/molar			1		
2006	2	474		*Canis*	Premolar			1		

TABLE 16.5 (*continued*). Full inventory of animal bones identified from Lofkënd tumulus

Season	Trench	Unit	Grave	Taxon	Element	Part	Side	Count	No. burned	Age
2006	2	478	85	*Ovis/Capra*	Premolar/molar	Fr		1		
2004	3	8		*Ovis/Capra*	Premolar/molar	Fr		1		
2004	3	45		*Capra*	Lower molar 2		Left	1		Adult
2004	3	64		*Ovis/Capra*	Premolar/molar	Fr		1		
2004	3	68		Large mammal	Long bone	Shaft fr		12		
2004	3	68		Large mammal	Long bone	Shaft fr		10		
2004	3	69		*Sus*	Premolar			1		
2004	3	92		*Ovis/Capra*	Premolar/molar	Fr		1		
2004	3	171		Medium mammal	Vertebra	Fr		1		
2004	4	11		Galliformes	Humerus		Right	1		
2004	4	35		*Bos*	Molar			1		
2004	4	48		*Sus*	Incisor	Fr		1		
2004	4	48		Large mammal	Long bone	Shaft fr		1		
2004	4	59		*Ovis/Capra*	Tibia	Distal fr	Right	1		
2004	4	59		Large mammal	Long bone	Shaft fr		1		
2004	4	59		*Ovis/Capra*	Humerus	Distal fr	Left	1	1	
2004	4	79		*Ovis/Capra*	Molar			1		
2004	4	79		Medium mammal	Long bone	Shaft fr		1		
2004	4	86		Mammal	Unidentifiable bone	Fr		1		
2004	4	86		*Ovis/Capra*	Lower molar	Fr		1		
2005	4	86		*Sus*	Molar			1		Adult
2005	4	86		Medium/large mammal	Humerus	Distal shaft fr		1		
2004	4	115		Mammal	Unidentifiable bone	Fr		1		
2004	4	130		*Ovis/Capra*	Lower molar 3	Fr	Left	1		
2005	4	131		Medium/large mammal	Long bone	Shaft fr		1	1	
2004	4	138		*Capra*	Upper molar 1		Right	1		Adult
2004	4	138		Mammal	Unidentifiable bone	Fr		1		
2005	4	201		Bovidae	Premolar/molar	Fr		1		
2005	4	201		*Ovis/Capra*	Upper molar 1		Left	1		Adult
2005	4	201		*Ovis/Capra*	Upper molar 3		Right	1		Old adult
2005	4	201		*Ovis/Capra*	Premolar	Fr		1		
2005	4	201		Medium/large mammal	Long bone	Shaft fr		1		
2006	4	201		*Sus*	Lower p1			1		
2006	4	201		*Sus*	Maxilla	With p3	Right	1		
2006	4	201		*Bos*	Upper premolar 1		Right	1		Old adult
2005	4	203		*Ovis/Capra*	Premolar/molar	Fr		1		
2005	4	204		Medium/large mammal	Long bone	Shaft fr		4		
2005	4	204		Large mammal	Long bone	Shaft fr		5	1	

Continued on next page

TABLE 16.5 (*continued*). Full inventory of animal bones identified from Lofkënd tumulus

Season	Trench	Unit	Grave	Taxon	Element	Part	Side	Count	No. Burned	Age
2005	4	204		*Ovis/Capra*	Lower molar 1		Left	1		
2005	4	204		Medium/large mammal	Unidentifiable bone	Fr		1	1	
2005	4	204		*Ovis/Capra*	Tibia	Proximal shaft fr	Right	1		
2005	4	204		*Ovis/Capra*	Lower molar 3		Left	1		Adult
2005	4	204		*Sus*	Molar			1		
2005	4	204		Medium/large mammal	Long bone	Shaft fr		5	1	
2005	4	204		Large mammal	Long bone	Shaft fr		1		
2006	4	204		*Ovis/Capra*	Humerus	Distal fr	Left	1		
2006	4	204		Medium/large mammal	Long bone	Shaft fr		1		
2005	4	205		Large mammal	Unidentifiable bone	Fr		1		
2005	4	205		Bovidae	Premolar/molar	Fr		1		
2005	4	205		*Ovis/Capra*	Scapula blade	Fr		2		
2005	4	209		Medium/large mammal	Long bone	Shaft fr		2	1	
2005	4	215	31	*Ovis*	Upper molar 1		Left	1		Adult
2005	4	215	31	*Ovis/Capra*	Premolar/molar	Fr		3		
2005	4	286		*Ovis/Capra*	Upper molar 3		Left	1		Adult
2005	4	286		*Sus*	Premolar/molar	Fr		2		
2006	4	286		Mammal	Long bone	Shaft fr		2		
2006	4	286		*Ovis/Capra*	Premolar/molar	Fr		1		
2006	4	286		*Sus*	Maxilla	With p2, p3, m1	Right	1		Young adult
2006	4	286		Med/lg mammal	Long bone	Shaft fr		2		
2006	4	286		Med/lg mammal	Unident. bone	Fr		10	1	
2006	4	286		*Sus*	Mandible	p3, molar 1	Left	1		Older juv
2006	4	286		Mammal	Long bone	Shaft fr		1		
2006	4	286		*Ovis/Capra*	Premolar/molar	Fr		2		
2006	4	286		*Ovis/Capra*	Lower molar 3		Left	1		
2006	4	286		*Ovis/Capra*	Lower molar 2		Left	2		Young adult
2005	4	329		Rattus	Lower incisor		Left	1		
2006	4	394		*Ovis/Capra*	Tarsal			1		
2006	4	412		*Sus*	Premolar/molar	Fr		1		
2006	4	448		*Sus*	Mandible	Anterior fr with p1, p2, p3, m1	Right	1		Juvenile
2006	4	450		*Sus*	Maxilla		Right	1		Juvenile
2004	4	b.c.		*Ovis/Capra*	Premolar/molar	Fr		1		
2007	5	535		Medium mammal	Long bone	Shaft fr		1		

TABLE 16.5 (*continued*). Full inventory of animal bones identified from Lofkënd tumulus

Season	Trench	Unit	Grave	Taxon	Element	Part	Side	Count	No. Burned	Age
2007	5	536		*Ovis/Capra*	Premolar/molar	Fr		1		
2007	5	536		*Ovis/Capra*	Premolar/molar	Fr		1		
2007	5	536		Large mammal	Long bone	Shaft fr		1		
2007	5	536		*Ovis/Capra*	Lower incisor	Fr	Left	1		
2007	5	548		*Canis*	Tibia	Shaft fr		1		
2007	5	555		*Ovis/Capra*	Humerus	Distal fr	Right	1		
2006	7	2		*Equus*	Lower premolar			1		
2006	8	201		*Ovis/Capra*	Lower molar 3		Left	1		Juvenile
2006	8	204		Medium mammal	Unidentifiable bone	Fr		1		
2006	8	204		*Ovis/Capra*	Lower molar	Fr		1		

CHAPTER 16.2
THE MOLLUSCA REMAINS FROM LOFKËND

Evi G. Vardala-Theodorou

INTRODUCTION

The mollusc material from the archaeological excavations at Lofkënd is presented in this chapter. The excavations conducted between 2004 and 2007 brought to light 100 tombs, many of which yielded various quantities of molluscs; additional shells were recovered in the tumulus fill. The molluscs were carefully collected stratigraphically, and they represent, almost exclusively, land shells belonging to the following species (the name and year of the person identifying the species is given in parentheses for each species): *Pomatias elegans* (Müller, 1774), *Helicella itala* (Linnaeus, 1758), *Xeromunda vulgarissima* (Mousson, 1859), *Lindholmiola girva corcyrensis* (Rossmässler, 1838), *Cernuella virgata* (Da Costa, 1778), *Monacha cartusiana* (Müller, 1774), *Trochoidea pyramidata* (Draparnaud, 1805), *Siciliaria stigmatica* (Rossmässler, 1836), *Strigilodelima conspersa* (Pfeiffer, 1848), *Mastus pupa* (Linnaeus, 1758), *Poiretia delesserti* (Bourguignat, 1852), *Aegopis* cf. *verticillus* (Ferussac, 1822), and *Cerastoderma edule* (Linnaeus, 1758). With the exception of a single brackish-water mollusc, their occurrence is attributed to natural processes that do not involve human agency. As far as we know, this is the first time that such material has been collected in its entirety and recorded and studied from a burial tumulus in Albania.

In addition to providing data about the paleoenvironment (Claassen 1998), molluscs with their shells served humans in a variety of ways in antiquity. Quite apart from providing information on dietary habits, molluscs could be used by peoples in different parts of the world as an item or store of value (molluscs used as coins, as in the cowrie shells of Ghana; see Williams 1997:211, figs. 304–305), and as an item of trade; they could also be used as tools or as jewelry or items of personal ornament. Furthermore, molluscs provide information on coastal conditions, such as types of coastlines, and sea-level fluctuations during the period of deposition of the archaeological material, thus allowing for the reconstruction of paleo- or archaeo-environments (Goodwin et al. 2003). Shells can also provide information on absolute chronology by different methods, such as specimens with thick shells that can by dated by thermoluminescence (TL) (Michael et al. 1998; Zacharias et al. 1999, 2002). In contrast to all of these varied uses of molluscs, the shells recovered from the Lofkënd tumulus only yield information about background environment at the site.

Usually the aim of studying malacological material is to elucidate aspects concerning ancient lifeways and the adaptation of humans to the paleoenvironment. Equally important is the interpretation of the occurrence of mollusca that were accumulated after the deposition of the archaeological material. These remains must be treated separately from those that are part of the archaeological collection and are directly correlated with human activities, as are marine molluscs. Sometimes the occurrence of land species in graves depends on climatic conditions (temperature, humidity), soil chemistry, and so on, and in such cases they are not correlated with human activities. It is also possible to find fossil marine, brackish, or freshwater specimens in burial contexts. These can sometimes be part of the sediment in which a grave was opened.

MATERIAL AND METHODS

The study of the Lofkënd malacological collection was carried out at the Apollonia Archaeological Park. The material was mostly collected either by hand in the field or through dry sieving; smaller quantities were also recovered by water sieving. Molluscs were also collected in their natural habitat during a visit to the excavation site, for comparative studies. Sorting took place and the material was grouped by genus or species. Specimens corresponding to trenches and graves were weighed. Due to the extreme fragility of the small land shells, the collection was not washed. This resulted in soil residues being left inside or on the shells. The percentage of the fragmented material is very small. The systematic study of the material based on notes and photographs taken in the field took place in Athens, Greece, where sampled species were compared directly with specimens at the Goulandris Natural History Museum collection.

A stereoscope Leica MZ8 was used for the exact determination of the species in the lab. Taxonomy was based on that currently valid for European species (Anderson 2005; Bank et al. 2001; Cossignani and Cossignani 1995; Falkner et al. 2001; Štamol 2004; Welter-Schultes 2005; Zenetos et al. 2005; see also Fauna Europaea web site at *www.faunaeur.org*). The material is stored in paper bags in the storage facilities of the excavation at the Apollonia Archaeological Park.

RESULTS

Thirteen mollusca species have been identified. Most specimens are complete. In total, the mollusca from the tumulus weighed 7.835 kg, of which 2.308 kg were collected in topsoil, surface layers or levels near topsoil, and in backfill, while the 3.552 kg correspond to tumulus fill proper, including various deposits, ceramic units, baulk, clay fill, and so on. The rest was collected from 70 graves (not all tombs yielded shells).

The *Helicella itala* is by far the most common land snail, accounting for 5.195 kg (66.3% of the total by weight). It was found in abundance in Tombs XXXI (Grave 86), XXXIX (Grave 66), and XCVII (Grave 39), which represent both prehistoric and modern burials (**Fig. 16.6**). It is also abundant in the area today. During weighing, it was possible to document the presence of some specimens of *Xeromunda vulgarissima* that could not be separated without washing the extremely fragile material.

The next most common species by weight is *Pomatias elegans*. Its total weight is 1.806 kg (23% of the total by weight). *Pomatias elegans* is abundant in Tombs I (Graves 64) and XXXI (Grave 86) (**Fig. 16.7**). The next most abundant species by weight (0.4 kg, about 5% of the total by weight) is *Lindholmiola girva corcyrensis*, followed by *Trochoidea pyramidata* (0.37 kg or 4.7% of the total). The rest of the mollusca are present in very small quantities. Weight distribution per grave is given in **Table 16.6**.

None of the above-mentioned species could have been used for food. The extremely scarce brackish-water type *Cerastoderma edule*, and the freshwater *Unio* sp. (Trench 2, Unit 532, SF 433, mother-of- pearl) fragments could have been transported easily by birds, although one cannot exclude the possibility that they were transported by humans. Fragments of *Helix* have been collected in the intensive surface survey of the region at Track A 0116 (8/7/2008), Team 1, 7, 8, 10, 13 (Chapter 18) and in the tumulus from Trench 1, Unit 9. Although *Helix* is edible, it does not occur in high enough quantities in the samples to be considered as food remains. Some indeterminable fragments of aquatic shell have been collected in Trench 2, Unit 297.

In some graves, the following species were collected in significant amounts (see **Table 16.6** for quantification):

- *Pomatias elegans* in Tombs I (64) and XXXI (86)
- *Helicella itala* and *Xeromunda vulgarissima* in Tombs I (64), V (96), XXXI (86), XXXIX (66), and XCVII (39)
- *Trocoidea pyramidata* in Tomb XXXI (86)

There is no correlation with the orientation of the grave, although some of these burials are on the eastern side of the tumulus. In addition, it was possible to identify the following:

- Freshwater shell fragment (Trench 2, Unit 297; Trench 1, Unit 1)
- Fragment of land shell, indeterminate (Trench 1, Unit 17, Tomb LXXXIV [2])
- Fragment of shell *Helix* sp. (Trench 1, Unit 9), and fragment of *Helix* sp. in (Track A 0 116)
- Operculum of the gastropod *Pomatias elegans* (Trench 5, Tomb LXIX [27] SF 551; Trench 2, Tomb XXXVI [75] fill [from sieve] SF 346)
- Fragment (mother of pearl) of freshwater bivalve (Trench 2, Unit 532, SF 433)

TABLE 16.6 Weight distribution of shells per grave plus tumulus fill in contrast to topsoil and associated deposits

Grave/context	Context	a	b	c	d	e	f	g	h	Total
Topsoil, surface, fill near topsoil, backfill	106 units	582.7	1692.7	43.1	5.4	3.1	1.5		0.2	2328.7
Tumulus fill, including ceramic units, baulk cleaning, and other deposits	112 units	894.0	2378.1	318.5	12.6	23.3	0.5	0.3	2.7	3630
Tomb I	Grave 64	77.3	62.7	6.8	0.4	1.9			1.1	150.2
Tomb III fill	Grave 81	15.6	26.1	5.5	1.1	1.2	0.1		0.1	49.7
Tomb IV	Grave 98	1.9	18.1	0.3						20.3
Tomb V fill	Grave 96	10.3	44.9	0.6	0.8		0.1			56.7
Tomb VI patch of dark soil in fill of grave	Grave 97	3.2	17.4	0.8	0.3					21.7
Tomb VIII	Grave 100	0.8	11.3	0.4	0.5					13
Tomb XII fill	Grave 88	5.1	20	0.9	0.1	0.5			0.1	26.7
Tomb XIII	Grave 49		0.2							0.2
Tomb XIV	Grave 71	3.61	19.3	0.8						23.71
Tomb XV fill	Grave 80		0.1							0.1
Tomb XVI fill	Grave 68	3.6	21.1	0.8	0.7	0.2				25.7
Tomb XVII fill	Grave 72		3.4	0.2						3.6
Tomb XX	Grave 50	1.2	1	0.1						2.2
Tomb XXI fill	Grave 55	0.5	16.4							16.9
Tomb XXII	Grave 47	0.7	4.9	0.1	0.1		0.1			5.9
Tomb XXIII fill	Grave 56	0.4	1.9							2.3
Tomb XXVI	Grave 74		11.2	0.1						11.3
Tomb XXVII	Grave 82	9.8	13,8		0,1					23.7
Tomb XXVIII	Grave 77	1.6	9.2	0.2						11
Tomb XXIX	Grave 83		0.1							0.1
Tomb XXX	Grave 70	11.7	20.2	1.9	0.5					34.3
Tomb XXXI	Grave 86	105.1	239.5	2.7	1.2	0.1	348.6			697.2
Tomb XXXII fill	Grave 89	6.6	11	1.1	0.3		19			38
Tomb XXXIII fill	Grave 92	0.3								0.3
Tomb XXXIV fill	Grave 87		0.3							0.3
Tomb XXXV fill	Grave 84	2.9	9	0.2						12,1
Tomb XXXVIII fill	Grave 79	3.1	16.8	2	0.4					22.3
Tomb XXXIX	Grave 66	1.6	155.5	0.1	0.4	0.1				157.7
Tomb XLIII	Grave 62		0.1							0.1
Tomb XLIV	Grave 65	8.5	17.7		0.1					26.3

Continued on next page

Table 16.6 (*continued*). Weight distribution of shells per grave plus tumulus fill in contrast to topsoil and associated deposits

Grave/context	Context	a	b	c	d	e	f	g	h	Total
Tomb XLVI	Grave 42		0.6		0.2					0.8
Tomb XLVII	Grave 41		1.2							1.2
Tomb XLVIII fill	Grave 52		4.2							4.2
Tomb XLIX upper fill	Grave 51	1.1	4.8							5.9
Tomb L fill	Grave 46		4.6							4.6
Tomb LII fill	Grave 69	0.7	10	0.1						10.8
Tomb LIII fill	Grave 63	0.7	2.8	0.2						3.7
Tomb LIV	Grave 40		0.1							0.1
Tomb LVI fill	Grave 43	1.5	5.9		1.5					3
Tomb LVIII	Grave 37		1.4							1.4
Tomb LIX fill	Grave 38	0.8	13.8	0.1						14.7
Tomb LX	Grave 44	2.59	8.89	0.39						11.87
Tomb LXII	Grave 32	0.5	0.9							1.4
Tomb LXIII fill	Grave 35	0.5	3	0.2						3.7
Tomb LXV fill	Grave 30		0.9							0.9
Tomb LXVI fill	Grave 31		10.6							10.6
Tomb LXVII fill beneath the grave	Grave 12	1.8	2.9	0.6	0.2					5.5
Tomb LXVIII	Grave 13	1.6	12.9	0.4	0.2					15.1
Tomb LXIX fill	Grave 27		1.7							1.7
Tomb LXX fill	Grave 17	2	2.2	0.2						4.4
Tomb LXXI	Grave 28		2							2
Tomb LXXII fill	Grave 24		2.4							2.4
Tomb LXXIV fill	Grave 29	1.7	21.1	1.8	0.2	0.9				25.7
Tomb LXXV fill	Grave 33	0.4	2.4	0.1						2.9
Tomb LXXVI cut	Grave 16		1.6							1.6
Tomb LXXVIII fill	Grave 5		1.7							1.7
Tomb LXXX	Grave 4	0.7	3.6	0.2						4.5
Tomb LXXXI fill	Grave 1	1.4	2.7							4.1
Tomb LXXXV fill	Grave 10		0.6							0.6
Tomb LXXXVI	Grave 22		0.5							0.5
Tomb LXXXVII fill	Grave 8	6.4	13.3	0.1	0.1					19.9
Tomb LXXXIX fill	Grave 11	11.1	3.9	6.9					0.1	22
Tomb XCII fill	Grave 23	4.7	10.1	0.6	1.2					16.6
Tomb XCIII	Grave 19	0.3								0.3
Tomb XCVI	Grave 20		0.7							0.7
Tomb XCVII fill	Grave 39	0.7	14	0.1						14.8
Tomb XCVII cut	Grave 39	12.8	171.1	0.9		0.5				185.3
Tomb XCVIII fill	Grave 36		10.1							10.1
Tomb XCIX fill	Grave 45	0.2	1.2	0.3						1.7
Tomb C fill	Grave 48		0.4	0.1						0.5
Total (grams)		1,806.3	5,194.9	400.4	27.9	31.8	369.9	0.3	4.3	7,835.8

Note: Total weight for each available species: (a) *Pomatias elegans*; (b) *Helicella itala* and *Xeromunda vulgarissima*; (c) *Lindholmiola girva corcyrensis*; (d) *Mastus pupa*; (e) *Monacha cartusiana*; (f) *Trocoidea pyramidata*; (g) *Cernuella virgata*; (h) *Siciliaria stigmatica* and *Strigilodelima conspersa*.

Description of Material Identified at Species Level

1. (A) *Pomatias elegans* (**Fig. 16.8**)
Prosobranch molluscs, dextral, conical, thick and solid. Spire 4.5 to 5 whorls. Last whorl about two-thirds of the shell height. Suture deep, umbilicus open, small; aperture nearly circular. Opercul paucispiral calcareous nucleus subcentral. Outer surface reticulated with spiral threads and radial ribs, gray-violet in color.
Habitat: Widely distributed in shady biotopes, damp sites, and warm climates, usually on calcareous matrix.

2. (B) *Helicella itala* (**Fig. 16.9**)
Disk-rounded and compressed shell. Conical spire with 5.5 to 6.5 flat whorls. Last whorl shortly and sharply descending. Outer surface with fine irregular growth lines. It is characterized by the oblique aperture on the parietal wall and by the edges that come very close together. Umbilicus wide, open, one-third of the shell diameter.
Habitat: Grassy hillsides and fields.

3. (B) *Xeromunda vulgarissima*
Flat shell, 4.5 to 5 whorls rounded at the periphery, suture not very deep, umbilicus wide, more than one-fourth of shell diameter and excentrical, aperture margin sharp, with white internal lip. Umbilicus open, color white, with two brown bands on upper side and three to four narrower bands in the base.
Habitat: Dry, in dunes, chalky; it is a very expansive species and forms large populations; also open habitat (Hausdorf 1988).

4. (C) *Lindholmiola girva corcyrensis*
Shell low discoid, with 6 to 7 flat whorls and deep suture. Apex smooth. In the upper part nearly flat; periphery slightly keeled in its upper part, rounded at the base. Aperture narrow, crescent-shaped, truncated, with white margin, covering a small part of the umbilicus. H: 4–6 mm; W: 9 × 12 mm.
Habitat: Limestones; also fallen leaves in lowland and mountain forests.

5. (G) *Cernuella virgata*
Spherical shell with a little conical spire and 5 to 7 whorls, finely and regularly grooved. Suture shallow. Aperture rounded, with sharp margin, and an internal lip. Umbilicus narrow, partly covered by the reflection of the lip. The shell is usually white with brown spiral bands. H: 6–19 mm; W: 8–25 mm.
Habitat: Dry and open sites, dunes, on calcareous ground.

6. (E) *Monacha cartusiana* (**Fig. 16.10**)
Spherical, compressed shell, thin, translucent, polished at the base. Has 5 to 6.5 convex whorls, with large last whorl; aperture elliptical with prominent white internal lip. Umbilicus narrow, almost covered.
Habitat: Open sunny habitats, near running water, shrubs, and ditches.

7. (F) *Trochoidea pyramidata* (**Fig. 16.11**)
Shell conical, opaque, and finely grooved. Spire with 5 to 6 whorls, slowly growing with deep suture. Last whorl large, rounded, not keeled. Aperture slightly oblique, wider than its height, with sharp margin and white internal lip. Umbilicus narrow. Unicolor white, or with bands or spots. W: 6–15 mm; H: 5–11 mm.
Habitat: These snails form large populations, on plants in warm and dry places, usually in coastal vicinity.

8. (Ha) *Siciliaria stigmatica* (**Fig. 16.12**)
Shell sinistral, medium, conical fusiform, with 10 to 11 whorls, regularly growing and fine striated outer surface. Sutures with papillae. Cervix with weak basal keel and shallow basal furrow; aperture with very weak lip, margin not (or only weakly) connected at parietal side; parietalis reaching margin, columellaris runs inside almost without curve; lower palatalis straight below the center of the columellaris, connected to lunula; lunula dorsal, in its upper section curved, subcolumellaris visible in an oblique view (Welter-Schultes 2005). H: 13–20 mm; W: 3–4 mm.

9. (Hb) *Strigilodelima conspersa*
Shell brown, light brown or with violet hue, densely striated or ribbed, nine to eleven whorls, cervix finely ribbed with curved ribs, aperture rounded with strong white lip, margin connected at parietal side by a sometimes very thin white layer, usually not detached, broad, parietalis not reaching margin, columellaris prominent, almost horizontal and far distant from parietalis, lunula dorsolateral,

strongly curved, subcolumellaris visible in the aperture and usually reaches apertural base. H: 15–20 mm; W: 3.3–4.5 mm.

10. (D) *Mastus pupa* (**Fig. 16.13**)

Shell ovoidal-fusiform, spire with 5 to 6 regularly growing whorls, with shallow sutures. Last whorl one-half of shell height. Outer surface finely striated. Umbilicus open, narrow, slit-like. Aperture semi-ovate, with white lip, angular tooth weak or regular, and white parietal lip, oblique, one-third of the shell height. Dirty yellowish, not very glossy. H: 12–15 mm; W: 4–6 mm.

Habitat: Open, dry, stony, biotopes.

11. (L) *Poiretia delesserti* (**Fig. 16.14**)

Shell with 6 rapidly expanding whorls, separated by fathom-like suture and a large aperture, regularly wide. The solid, not slender shell, is very finely ribbed. Columella strongly curved at the base. H: 34–38 mm; W: 13–15 mm. *Poiretia* is a Mediterranean genus whose species are predacious and live mainly on other gastropods.

Habitat: Same as *Pomatias* (Subai 1980).

12. (K) *Aegopis cf. verticillus* fragment

Shell flattened and round, with a low conical spire and slightly translucent whorls. Outer surface in the upper part striated, with a fine spiral sculpture producing a granular pattern; in the base, smooth and very glossy. Spire with whorls keeled, last whorl rounded. Yellowish brown on upper side, greenish yellow around the umbilicus, which is very deep, wide and open. Six to eight distinct radial color bands. H: 16–17 mm; W: 26–30 mm.

Habitat: Under leaves and between stones on humid soils, in mountain forests of lower altitudes.

13. (M) *Cerastoderma edule* (**Fig. 16.15**)

Shell solid, closed, globular in shape, more rounded than the *C. glaucum*, equivalve, inequilateral. Umbones with beaks in front of the midline. Outer surface with 20 to 25 regular ribs. Hinge with two small cardinal teeth, two anterior lateral and two posterior. Color brown, greenish toward the posterior region, dull. The ribs are not visible from the interior side. L: 20–35 mm (Vardala-Theodorou 1998).

Habitat: Found in brackish waters, mainly estuarine conditions, in mud at the infralittoral zone.

The molluscs presented above belong to the following orders.

Gastropoda

Order: Neotaenioglossa (identified Haller, 1892)

Family Pomatiidae
  Pomatias elegans

Order: Pulmonata (identified Cuvier in Blainville, 1814)

Family Hygromiidae
  Helicella itala and *Xeromunda vulgarissima*
  Cernuella virgata
  Monacha cartusiana
  Lindholmiola girva corcyrensis
  Trocoidea pyramidata

Family Clausiliidae
  Siciliaria stigmatica
  Strigilodelima conspersa

Family Enidae (Buliminidae)
  Mastus pupa

Family Oleacinidae
  Poiretia delesserti

Family Zonitidae
  Aegopis cf. *verticillus*, fragment

Bivalvia

Family Cardiidae
  Cerastoderma edule

Discussion and Conclusions

The analysis of the molluscan material and the composition of samples, the relative frequency of the different species, and the comparison with those still living in the area of the Lofkënd archaeological site, provided us the necessary answers about the presence of the molluscs in the excavation area. Most of the species analyzed belong to land gastropods. Bivalves are extremely rare and found only in fragments. A comparison of our list with the existing living Albanian species (Dhora and Welter-Schultes 1996; Reischütz and Sattmann 1990; Reineck and Singh 1975; Sattmann and Reischütz 1994) shows similar environmental conditions with those prevailing around Lofkënd today. *Helicella itala* occurring in abundance prefers grassy hillside fields, and *Pomatias elegans* prefers shady biotopes in damp sites. The encountered species belong to land species living today in Albania (Guisti et al. 1995; Welter-Schultes 2005). Their occurrence in the graves does not follow any pattern. There is only one brackish specimen (*Cerastoderma edule*), indicating nearby areas of low salinity. *Cerastoderma* could be transported by humans or birds (Reineck and Singh 1975), since it was not found during the soil study of the area. The same species appears in all trenches and in the graves. They cannot be attributed to food remnants. Their ecological environment is almost the same as the present.

ACKNOWLEDGMENTS

The author would like to thank John Papadopoulos and Sarah Morris for giving her the opportunity to visit Lofkënd and work with the mollusca collection. Special thanks are owed to K. Vardinoyanni and F. Welter-Schultes. The fieldwork was financed by the Steinmetz Chair in Classical Archaeology and Material Culture at UCLA. Thanks are also due to Lorenc Bejko, then of the International Center for Albanian Archaeology and the Institute of Archaeology, Tirana, now at the University of Tirana.

CHAPTER 16.3
A NOTE ON THE OCCURRENCE OF FOSSIL WOOD AT LOFKËND

George Theodorou and Evangelos Velitzelos

During archaeological field survey, a few specimens of fossil wood fragments had been collected at the area of the Lofkënd archaeological site (**Fig. 16.16a–b**). Large specimens of fossil wood, if available in the area, could have been easily overlooked as pieces of stone, especially during the process of collection of specimens that could possibly be correlated with human actions. The collected material also included a few Miocene fossil marine specimens, belonging to the Neogene marine fauna of the area (*Pleurotoma* sp. and *Ostrea gryphoides*) (Koskeridou et al. 2005) (**Fig. 16.17**).

The occurrence of small fragments of fossil wood is clearly more interesting than the common marine fossils. This material was analyzed in thin sections (**Fig. 16.16a–b**) prepared at the Department of Historical Geology and Paleontology of the University of Athens. Due to the small dimensions of the samples, only sections covering circumscribed areas could be made. This did not allow determination of species with any certainty. The available fragments could belong to the same geological unit of fossil trees of the western part of north Greece. A geological survey of the area must be carried out in the future in order to locate the place of their origin. This could allow the discovery of a fossil petrified forest and would give the opportunity for a complete study of adequately large sections of tree trunks.

CHAPTER 16.4
SOILS INVESTIGATIONS OF THE LOFKËND ARCHAEOLOGICAL SITE AND SURROUNDING LANDSCAPES

J.E. Foss

INTRODUCTION

The investigation of the soils and landscapes at the Lofkënd archaeological site was initiated in July 2005 and continued in the 2006 and 2008 field seasons. The study consisted of two phases: (1) characterization of the soils at the Lofkënd site and (2) evaluation of the regional distribution of landscapes and soils in the study area. This investigation included identifying the geologic sediments for the parent material of the soils and the morphological characteristics of the soils at the excavation. The major soils occurring at the site were described according to standard methods and nomenclature. The description of the soils included the color, texture, structure, consistency, boundary of horizons, and relative content of carbonates. Based on these characteristics, the various layers were assigned horizon designations. These horizon designations indicate the development and history of the soils.

The regional distribution of soils was determined by traversing the area and observing landscapes and soils in road cuts, through various excavations, and by using an auger to examine soils to depths of 1.5 m where practical. This is a general soil survey; a more detailed study and analysis are needed for specific site characterization.

Soil samples were taken to represent the typical soils at the Lofkënd excavation. Three profiles were selected for laboratory analysis; these analyses included chemical properties (elemental composition) and physical characteristics (texture). Several bulk samples were taken for micromorphology to examine some unknown fibrous materials.

Results and Discussion

Regional soils

According to the geologic map, the site is located in Miocene sediments that are characterized as clays, aleurolites (silt stones), and sandstones. The site itself occurred in sandstone that was loosely cemented by carbonates. The bedrock was easily dug in most cases, although some strongly cemented beds occurred at the site and in the surrounding region. The lithic beds were generally less than 0.5 m in thickness. The bedrock had many thick (>1 cm) bands of calcium carbonate; this provided a method of determining when the excavation encountered the undisturbed bedrock.

Figure 16.18 is a regional soils map of the study area surrounding the Lofkënd archaeological site. Five major soil units were identified, based primarily on the geologic sediments or the parent material of the soils. The variability of landscapes and soil morphology was based strongly on the geology of the area. Under these five major soil units, 12 mapping units were delineated, based primarily on slope and the attendant soil morphology. The major units are described below.

Unit 1. Shale-derived soils

The shale-derived soils are formed in Miocene sediments. The landscapes are mainly gently sloping (3–8%) and strongly sloping (8–16%). **Figure 16.19** shows a typical landscape of the 1B soil mapping unit just north of the Lofkënd site. These soils are the most productive agricultural soils of the upland area.

The shale-derived soils have weathered for long periods and show characteristics such as strong structural development, some leaching of carbonates, and organic matter accumulation. **Figure 16.20** is a photographic profile of the typical clayey soil in Unit 1. The high clay content and the nature of mineralogy result in deep cracking during the dry season (**Fig. 16.21**) and development of slickensides. Slickensides are wedge-shaped structural units common in the Vertisol soil order.

The erosion potential of this unit would be moderate as a result of the clay texture and high organic matter in the surface horizon. Runoff, however, would be moderately rapid because of the low infiltration capacity once the surface soil has become saturated.

The fertility status would be quite high, especially on more moderate slopes, because of the nutrient composition and high water-holding capacity. This soil would be difficult to till at certain moisture levels because of the high clay content.

Unit 2. Sandstone-derived soils

Sandstone-derived soils occur throughout the study area mainly on steeply sloping scarp areas, ridge lines, and sandy knobs (e.g., Lofkënd site). The parent material for these soils is loosely cemented sandstone; the sandstone generally has carbonates as the cementing agent, dominantly fine sands, and contain some lithified rock layers.

The lithified layers are usually less than 0.5 m in thickness. Gravelly sediments were also noted in some areas of this sandstone deposit. Landscapes in Unit 2 range from very steep (30–45%+) to strongly sloping (8–16%). **Figure 16.22** shows the very steeply sloping area on the east side of the study area, and **Figure 16.23** is a landscape with steeply sloping soils (16–30%).

The soils located on the sandstone are generally very weakly developed and contain some organic accumulation in surface horizons where stability of landscapes exists. **Figure 16.24** is a typical example of the eroded slopes found in these mapping units. A Bw horizon (subsoil) is only found in sandstone soils on the more moderate slopes or ridgetop positions.

As evident in the photographs, these soils are highly erodible as a result of the steep slopes and the domination of fine sands that are very susceptible to erosion. The lack of adequate vegetative cover also contributes to eroded landscapes. The agricultural potential of these soils is minimal because of the low water-holding capacity, low organic matter, and low nutrient supply.

Unit 3. Reddish soils developed on sandy-clayey deposits

These soils noticeably contrast with surrounding areas because of their reddish hue. Soils on the more stable slopes have reddish colors (5 YR 4/6), well-developed argillic horizons (clayey), and deeply weathered profiles. **Figure 16.25** shows an area of reddish soils on top of the hill but with eroding side slopes. With the strongly developed and deep profile, these soils have weathered for a considerable amount of time; it is estimated that these landscapes would be at least 30,000 to 40,000 years in age. The reasons for the contrasting characteristics of these soils are likely the result of parent material differences (as contrasted to shale and sandstone) and the stability of landscapes over time. **Figure 16.26** shows an isolated hill of the reddish soils; note the amount of erosion on the side slopes at this location.

The reddish soils are leached of carbonates, and on landscapes with minimal erosion, these soils would be marginally productive. The nutrient status is probably not as high as those soils developed on shale. As noted on the photographs, erosion is a major hazard on these landscapes. The soils would have moderate water-holding capacity.

Unit 4. Soils developed on valley alluvium

Soils developed on local alluvium in small, narrow valleys are mainly Holocene in age. Erosion from upland areas has resulted in accumulation of sediment during cycles of erosion and sedimentation. The landscape of one area is shown in **Figure 16.27**. A profile described in an adjacent valley showed numerous discontinuities indicating a number of sedimentation episodes. The texture was mainly loam and clay loam for the surface horizons but very sandy at depths of 110 to 130 cm.

These soils would be very productive as a result of favorable texture, suitable slopes, and moderate nutrient availability. Flooding would be a hazard in some low-lying areas.

Unit 5. Soils developed on alluvium along the Gjanicë River

The Gjanicë River marks the southern border of the study area associated with the Lofkënd archaeological site. The wider section of the river valley has a floodplain with large accumulations of gravel and sand; thus this system shows some high-energy periods of flooding and deposition. Higher terrace formations were noted in many areas along the river, especially where river migration has left higher-elevation sediments as contrasted to the lower floodplain. **Figure 16.28** shows the productive terrace along the Gjanicë River. A soil profile described on this terrace showed major stratification of sediments and textures ranging from very fine sandy loam to clay loam in the upper profile. At 100 to 150 cm, the texture was either loamy sand or fine sandy loam. The soils were found to contain free carbonates throughout the profile.

The soils on the high terrace (mapping unit 5B) would be very productive, with adequate water-holding capacity and nutrient availability. Some degree of flooding would be the main hazard. Soils on the floodplain (mapping unit 5A), however, would have minimal value for agricultural production as a result of frequent flooding and coarse-textured soils.

LOFKËND SITE

The soils at the Lofkënd archaeological site were developed primarily from calcareous sandstone, but additional clayey sediments had been added to the surface of the tumulus and in various layers throughout the profiles. Because of the erosive nature of the fine sands at the site, this clayey material would be useful in stabilizing the structure and in minimizing susceptibility to erosion. **Figures 16.29** and **16.30** illustrate the soils occurring at the site. The dark-colored surface in the profiles is either a silt loam or silty clay loam with moderate blocky structure. This clayey material is believed to have been placed on top of the tumulus to protect it from erosion. Also, similar material is found in thin bands throughout the burial zones to control erosion temporarily.

In looking for the source area of the fine-textured sediment covering the tumulus and interbedded with the sandstone, soils with clayey surfaces and subsoils were located just north of the site. A traverse was conducted using an auger to examine soils to depths of 1 to 2 m. Approximately 30 m north of the site, shale became the parent material for soils, and this formation continued for at least 120 m north. Thus, there were ample source areas available to obtain the clayey sediments near the tumulus.

In the surrounding area, the more moderate slopes seem to be dominated by shale bedrock. In observations west of the site, shale-derived soils were noted in the cornfield located in a saddle position and up the hill toward the site to perhaps 70 m from the top of the hill. The soils derived from shale are more productive than those weathered from the sandstone; this was evident in observations of the crop production in the region. The clayey soils weathered from shale typically have higher plant-available water content and more plant nutrients than soils weathered from sandstone.

Additional observations in the area west of the site yielded evidence of slope failure occurring at the contact between the sandstone and shale layers. In one area, a shale soil was observed occurring above the sandstone (**Fig. 16.31**), but the profile had been displaced from the slope above. The soil derived from shale had not been disturbed or mixed to any great degree suggesting that it had moved en masse down the slope. This type of landscape process is common in the region.

Tables 16.7 to **16.9** provide the detailed descriptions of soils examined at the site and in the general vicinity of Lofkënd. (See Addendum A for abbreviation definitions, and Addendum B for discussion of

TABLE **16.7** Profile descriptions of soils examined in excavations at Lofkënd (colors are for moist conditions)

Horizon	Depth (cm)	Color	Texture	Structure	Consistence	Boundary	Carbonates
Profile S05AL1[a]							
A1	0–16	2.5 Y 4/3, 4/4	sicl	2fabk	vh dry	gs	vst
A2	16–30	2.5 Y 5/4, 5/3	sicl	2msbk	vh dry	gs	vst
A3	30–43	2.5 Y 5/3, 5/4	sicl	2msbk	vh dry	gs	vst
AB	43–56	2.5 Y 5/4	sicl	2mabk	h dry	gs	vst
C1	56–68	2.5 Y 5/4	sil	1mabk	vh dry	cw	vst
2A	68–83	2.5 Y 4/3, 5/3 (20%)	sicl	1mabk	h dry	cw	vst
2C2	83–96	2.5 Y 5/3, 5/4	l	1cpl, 1fabk	fr moist	cw	vst
3A	96–104	2.5 Y 4/3	sil	1vfabk	h dry	cw	vst
3C3	104–112	2.5 Y 5/4	vfsl	1mpl	h dry	cw	st
4A	112–130	2.5 Y 5/4, 6/4	fsl	2mabk	h dry		st
4C4	130–160	2.5 Y 5/3	vfsl				
4C5	160–180	2.5 Y 6/3	vfsl				
4C6	180–195	2.5 Y 6/3, 6/2	vfsl				
Profile S05AL2[b]							
A1	0–22	2.5 Y 4/3	sil	2fabk	vh dry	gs	st
A2	22–33	2.5 Y 4/3	sicl	2mabk	vh dry	gs	st
A3	33–46	2.5 Y 4/3	sicl	2msbk	vh dry	cs	vst
AB	46–60	2.5 Y 4/3–4/4	sicl	2mabk	vh dry	cs	vst
2BC	60–72	2.5 Y 5/3–5/4	fsl	1msbk-0m	vh dry	cs	vst
2C	72–92	7.5 YR 6/3	vfsl	1mabk	vfr moist	-	st
Profile S05AL3[c]							
Ap	0–27	10 YR 4/3	sicl	1fpr–2msbk	vh dry	cs	st
A1	27–44	10 YR 4/3,-4/2	sicl	2msbk	vh dry	gs	vst
A3	44–60	2.5 Y 4/3–3/3	sicl	2mabk	vh dry	gs	vst
A4k	60–78	2.5 Y 4/3	sicl	3msbk	vh dry	as	vst
2C1	78–93	10 YR 4/4	fsl	1cpl	vh dry	as	vst
3Ab	93–118	2.5 Y 5/3–5/4	fsl	1cpl	h dry	as	st
3C2	118–130	2.5 Y 5/3–5/4 (5%)	fsl	1cpr	fr moist	aw	st
4Ak	130–147	5 Y 4/3	sicl	2fsbk	h dry	as	vst
5C3	147–165	10 YR 5/3–5/4	fsl	1cpl–1mabk	fr moist	gs	st
5C4	165–188	10 YR 5/3–5/4	fsl	1cpl	fr moist		st

[a] Sampled and described July 20, 2005; auger used for description below 130 cm; burial was present at base of profile at 130 cm; three block samples taken of dark colored (stratified) region for micromorphology; shells were common in A1 to A3 horizons.

[b] Described and sampled on July 20, 2005; few to many, fine (<1 mm thickness) calcium carbonate filaments in upper 72 cm; many filament bands (0.5–1.0 cm thickness) in C horizon; shells were common in A2 and A3 horizons.

[c] Described and sampled on July 22, 2005; filaments: A1 2%; A3 5–10%; A4k 30% plus masses of carbonates; 2C1 3%; 3Ab and 3C2 3%; 4Ak 50% filaments and some masses plus thin carbonate coatings on ped faces; 5C3 and 5C4 5% filaments.

Table 16.8 Description of soil profile below Tomb XIX (54) at Lofkënd archaeological site

Horizon	Depth (cm)	Color	Texture	Additional Notes
Profile S05AL5				
1	0–180	Section was removed during excavation		
C1	180–200	2.5 Y 6/3	fsl	Some coarse carbonates
C2	200–215	2.5 Y 6/4, 6/3	fsl	
C3	215–230	2.5 Y 6/3	fsl	Mottles c1f 2.5 Y 5/6
C4	230–250	2.5 Y 6/3, 6/4	fsl	Mottles f1f 2.5 Y 5/6
C5k	250–270	2.5 Y 6/3, 7/3	fsl	Some cementation, carbonate concretions
C6	270–290	2.5 Y 6/3	fsl	Mottles m1d 2.5 Y 5/6
C7	290–310	2.5 Y 5/3, 6/3	sl	Mottles m1d 2.5 Y 5/6 > med sand
C8	310–320	2.5 Y 6/3	sl	Mottles m1d 2.5 Y 5/6 > med sand
C9	320–330	2.5 Y 6/2	sl	Mottles f1d 2.5 Y 5/6

Note: Auger used to sample below Tomb XIX (54), near the south end of site (2 m from edge of escarpment); an estimate was made of the amount of sediment removed during excavation; carbonate concretions present are similar to the "loess puppies" found in the U.S. Midwest, concretions were maximum 1–2 cm in length; soil described and sampled July 22, 2005.

photomicrographs of two samples from the bottom of Tomb XLIX [51].) The profiles consist of either an A horizon (surficial accumulation of organic substances and inorganic sediments) or C horizons (relatively unweathered sediments). The area has not been stable for sufficient time to develop any further horizonation. Also, the calcareous nature of the sandstone tends to inhibit the weathering processes of clay and iron translocation in the soils.

The parent material of calcareous sandstone is characterized as (1) very fine sandy loam grading to sandy loam, (2) color of 2.5 Y 6/2, 6/3 (light brownish gray to light yellowish brown), (3) calcareous with bands (0.5–1.5 cm thickness) of calcium carbonate, and (4) some layers of strongly cemented sandstone occurring at depths below 2 m. Occasional coarse fragments of gravel size were also noted in the profiles; these originated from the sandstone bedrock.

LABORATORY DATA

The laboratory data on the particle size analysis and chemical composition of two profiles sampled at Lofkënd are given in Addendum C, **Tables 16.10** and **16.11**. The particle size analysis shows that the total sand content ranged from 36 to 69% in profiles 1 and 3. Sand content in the parent sandstone ranged from 64 to 88%. The particle size of profiles 1 and 3 were strongly influenced by the addition of clay and silt from the shale-derived soil material that was added to the tumulus. The fine and very fine sand fractions dominated the sand component in profiles 1 and 3. However, discontinuities in profile 3 from 60 to 130 cm had higher coarse and medium sand contents than adjacent horizons. These differences are related to the amount of mixing between the sandstone and more clayey shale-derived soils.

The chemical analysis of profiles 1 and 3 demonstrate the dominance of calcium throughout the profiles. The A3k and 4Ak horizons in profile 3 had higher values of calcium than adjacent horizons. The k subscript indicates the field morphology noted the more obvious accumulation of calcium carbonate such as stronger reaction to acid and more evidence of carbonate stringers and coatings. The k horizons are regions of accumulation of carbonates formed by the translocation process of pedogenesis.

The extractable phosphorus (P) in profiles 1 and 3 averaged 273 mg/kg while the C horizons of profile 5 had an average extractable P value of 119 mg/kg. Thus, the human activity surrounding the tumulus has resulted in additional P being added to the upper 1–2 m of soils in the tumulus. Some accumulation of barium (Ba), cobalt (Co), potassium (K), nickel (Ni), and lead (Pb) were noted in the surface horizons of tumulus soils.

TABLE 16.9 Soil profile descriptions taken along traverses north and west of excavations at Lofkënd archaeological site

Horizon	Depth (cm)	Color	Texture	Additional Notes
Northern traverse				
Auger 1[a]				
Ap	0–15	2.5 Y 5/3	sil	
Bw	15–30	2.5 Y 5/3, 5/4 (10%)	sil-sicl	
C1k	30–50	2.5 Y 5/3	vfsl-si	
C2	50–80	2.5 Y 5/3	vfsl	
C3	80–100	2.5 Y 5/3, 5/4	vfsl	
Auger 2				
Ap	0–25	2.5 Y 5/3, 5/4 (10%)	sicl	Mottles f1d 10 YR 5/8
Bwk1	25–50	2.5 Y 5/3	sicl	Mottles m2d 10 YR 5/8
Bwk2	50–65	2.5 Y 5/3	sicl	Mottles m2d 10 YR 5/8
BCk	65–90	5 Y 5/3	sicl	Mottles m2d 10 YR 5/8; pieces of shale present
C	90–100	5 Y 5/3	sicl	
Auger 3[b]				
Ap	0–25	2.5 Y 5/3	sicl-sic	
Bw	25–50	2.5 Y 5/3	sicl	
Bwk	50–70	2.5 Y 5/3; 5 Y 5/3	sicl	
BCk	70–80	5 Y 5/3	sicl	Mottles c1d 10 YR 5/8
Auger 4[c]				
Ap	0–25	2.5 Y 5/3	sicl	
C1	25–40	2.5 Y 5/3	sil	Mottles f1d 2.5 Y 6/8
C2	40–80	2.5 Y 5/3	sil	Mottles f1d 2.5 Y 5/8
2Bwb	80–90	5 Y 5/3	sicl	Mottles f1d 5 Y 5/8
West profile (300 m west of site)[d]				
Ap	0–20	2.5 Y 3/3, 5/3, 5/4	sicl	
Bw	20–50	2.5 Y 5/3	sicl, fsl	Mixed materials, some rounded gravels
Ck1	50–70	2.5 Y 5/3	sicl, fsl	Mottles f1d 10 YR 5/8; mixed materials
Ck2	70–100	2.5 Y 5/3	sicl, fsl	Mottles m1d 10 YR 5/8
2C3	100–120	2.5 Y 5/3, 6/3	vfsl	Sandy bedrock evident

[a] Greater than 15% calcium carbonate in C1k; platy bedrock encountered in C3; reaction to dilute acid was strong throughout profile.
[b] Pieces of bedrock, silt stone, present in auger.
[c] Site was located in saddle position so deposition of sediment was evident in upper 80 cm.
[d] Profile was located on a hill about 300 m directly west of the archaeological site; tree was located (4 m east) near the auger test; 10% slope, northern aspect. Mottles result from impeded drainage.

Summary

The basic parent material for soils occurring at the Lofkënd archaeological site is weakly cemented Miocene sandstone. However, fine-textured sediment derived from soils weathered from shale was added to the top of the tumulus. This clayey sediment also occurs in thin bands within the tumulus. The addition of the fine-textured sediment from off site is believed to have been added periodically to help prevent erosion of the tumulus during various stages of building. The fine and very fine sands weathered from the calcareous sandstone are very erosive (both by wind and water) and thus would be a problem in stabilizing the landscape during the individual burials as well as for the completed tumulus. The sources of the fine-textured sediment appear to be the clayey soils derived from shale that occur about 30 m north of the site.

Soil morphology at the tumulus generally consists of dark-colored (olive brown to light olive brown) silt loam or silty clay loam A horizons that exhibit blocky structure underlain by fine or very fine sandy loam C horizons. The C horizon color ranges from light olive brown to light brownish gray. The entire soil is calcareous and thus indicates minimal amount of leaching. As a result of the calcareous nature of the parent material and the minimal age and conditions for weathering to take place, the soils at the site show little horizonation except for the depositional episodes associated with the burials.

Agriculture is the main land use in the region, with soils and landscapes having a wide range of capabilities for crop production. There are numerous areas north and west of the site that have very low potential for crop growth as a result of the dominantly sandy textures. The most productive areas are dominated by shale-derived soils; this was obvious in observing plant growth in the immediate area around the tumulus. The terrace soils along the Gjanicë River also provide productive agricultural soils.

Acknowledgments

I would like to thank Michael Timpson for his help in reviewing this manuscript and introducing me to the landscapes of the study area. Thanks also to Sarah C. Sherwood for examining and preparing the photomicrographs (Addendum B) of the opaque material located in one of the profiles.

Addendum A: Abbreviations

Texture (USDA textural classes)	
vfsl	very fine sandy loam
fsl	fine sandy loam
sl	sandy loam
sil	silt loam
sicl	silty clay loam
sic	silty clay
Structure	
Grade:	0 structureless; 1 weak; 2 moderate; 3 strong
Size:	f fine; m medium
Type:	abk angular blocky; sbk subangular blocky; pl platy; m massive
Mottles	
f1d	few, fine, distinct
m2d	many, medium, distinct
c1d	common, fine, distinct
m1d	many, fine, distinct
Consistence	
vh	very hard
h	hard
fr	friable
Boundary	
g	gradual
c	clear
a	abrupt
s	smooth
w	wavy
Carbonates: reaction to HCL acid	
vst	very strong
st	strong

Addendum B: Photomicrographs

Sarah C. Sherwood

Samples 511-1 and 511-2 were sent to Spectrum Petrographics in Vancouver, WA where they were partially embedded in epoxy, mounted on 2 × 3 inch (5 × 7.5 cm) glass slides and ground to 30 microns. The samples were collected from the base of Tomb XLIX (51).

The samples primarily consist of opaque organic material overlying a poorly sorted calcareous fine sand/silt. The photomicrographs focus on this organic material in plane polarized light. The material could be carbonized; however, there is no evidence for burned sediment or ash. The thin sections reveal a clearly expressed structure that should allow specialists in archaeobotany to identify the type of plant material. **Figures 16.32** to **16.35** emphasize the material structure and arrangement with different views and magnifications. A section of the organic fiber (?) appears twisted as one might expect in woven fabric. The fruit in **Figs. 16.36** and **16.37** may also be identifiable and provide insight into the type of organics present in Tomb XLIX (51).

ADDENDUM C: LABORATORY DATA

Appendix C comprises two tables, **6.10** and **6.11**.

TABLE 16.10 Particle size analysis of three profiles at Lofkënd archaeological site (percentages)

Horizon	Depth cm	vc sand	c sand	med sand	fine sand	vf sand	total sand	silt+ clay
SO5AL1-1								
A1	0-16	1.97	4.46	4.13	10.97	18.76	40.29	59.71
A2	16-30	1.87	3.61	4.28	15.89	20.84	46.49	53.51
A3	30-43	2.16	4.94	4.68	13.61	16.39	41.78	58.22
AB	43-56	4.05	7.15	2.72	10.78	16.91	41.61	58.39
C1	56-68	2.23	0.89	3.15	9.96	20.48	36.71	63.29
2Ab	68-83	0.89	0.99	1.27	14.49	25.34	47.98	52.02
2C2	83-96	1.4	1.06	1.2	13.72	26.45	43.83	56.17
3Ab	96-104	1.67	1.71	2.24	10.77	21.53	37.92	62.08
3C3	104-112	0.27	0.49	0.24	13.38	31.39	45.77	54.23
4Ab	112-130	2.38	1.07	1.69	13.91	21.78	40.83	59.17
SO5AL13								
Ap	0-27	3.71	6.49	5.78	11.56	13.24	40.78	59.22
A1	27-44	3.68	3.03	4.67	15.09	14.87	41.34	58.66
A2	44-60	3.03	7.69	8.25	17.17	16.37	52.51	47.49
A3k	60-78	2.01	12.08	4.99	16.48	14.7	50.26	49.74
2C1	78-93	2.24	20.69	9.72	20.2	14.11	66.96	33.04
3Ab	93-118	2.77	9.19	8.49	16.44	18.13	55.02	44.98
3C2	118-130	1.49	13.7	7.52	24.47	21.84	69.02	30.98
4Akb	130-147	5.45	5.27	4.28	2.61	6.85	24.46	75.54
5C3	147-165	3.7	1.94	3.15	27.68	23.14	59.61	40.39
5C4	155-188	1.61	2.55	5.91	31.27	28.2	69.54	30.46
SO5AL5								
C1	180-200	9.54	2.42	5.85	35.8	10.56	64.19	35.81
C2	200-215	3.64	2.86	8.99	39.59	14.45	69.53	30.47
C3	215-230	3.41	2.83	13.82	40.86	12.31	75.23	26.77
C4	230-250	5.63	2.5	3.72	53.91	11.51	77.27	20.73
C5k	250-270	7.37	3.88	3.9	42.11	11.1	68.56	31.64
C6	270-290	2.33	34.25	22.75	20.64	6.03	86	14
C7	290-310	1.37	31.55	13.25	29.52	7.43	83.12	16.86
C8	310-320	0.9	20.49	19.92	34.49	8.3	84.1	15.9
C9	320-330	4.12	35.77	19.92	17.93	5.81	83.55	16.45

Table 16.11 Chemical characteristics of soils at Lofkënd archaeological site

Horizon	Depth (cm)	As	Ba	Ca	Cd	Co	Cr	Cu	Fe	K	Mg	Mn	Na	Ni	P	Pb
SO5AL1																
A1	0–16	4.58	38.83	55288	0.25	2.13	1.56	20.03	412	182	621	228	32	8.9	271	3.59
A2	16–30	4.8	67.98	60303	0.21	1.88	1.85	22.64	407	160	693	271	28	7.51	245	3.05
A3	30–43	5.56	74.32	72814	0.23	1.99	2.46	30.04	402	149	778	315	30	7.62	237	3.03
AB	43–56	5.7	40.46	61594	0.26	1.69	2.19	24.83	493	149	691	254	35	7.03	247	3.34
C1	56–68	5.95	42.52	85659	0.23	2.11	2.81	35.17	358	142	932	351	37	7.21	246	2.55
2Ab	68–83	5.69	28.36	78767	0.22	1.75	2.4	32.05	359	112	944	284	37	6.7	250	2.54
2C2	83–96	6	27.18	85992	0.21	1.57	2.6	34.55	321	88	986	309	39	6.23	230	2.19
3Ab	96–104	5.62	37.78	75258	0.23	1.77	2.78	32.26	438	118	947	308	41	8.61	280	2.85
3C3	104–112	5.96	27.91	88433	0.2	1.17	2.14	35.91	243	59	1065	237	37	4.92	239	1.81
4Ab	112–130	5.08	38.27	72498	0.21	1.1	1.6	29.53	407		926	198	42	7.15	349	2.58
SO5AL3																
Ap	0–27	4.98	133.56	52652	0.28	3.35	2.29	20.82	484	222	634	224	39	11.43	255	4.67
A1	27–44	5.11	89.63	59945	0.26	2.27	2.44	31.41	568	205	722	233	41	10.17	418	3.71
A2	44–60	5.31	46.57	68628	0.25	1.99	2.57	32.29	462	174	740	242	40	8.89	423	2.99
A3k	60–78	5.62	42.38	72566	0.27	2.26	2.63	33.32	485	205	856	223	40	9.43	390	3.51
2C1	78–93	2.77	114.36	31694	0.15	1.74	1.65	5.91	337	117	427	207	31	8.13	160	2.56
3Ab	93–118	5.06	61.6	70572	0.21	2.12	2.82	28.96	351	131	821	239	38	8.17	299	2.4
3C2	118–130	4.96	81.7	73762	0.19	2.16	2.62	27.97	303	86	1014	259	34	7.23	250	1.89
4Ak	130–147	6.76	29.9	108484	0.24	1.13	1.48	43.42	150	160	902	227	45	6.03	22	0.18
5C3	147–165	4.65	42.66	63625	0.19	1.48	2.21	23.81	337	93	929	237	38	7.1	328	1.97
5C4	165–188	4.98	29.42	70279	0.19	1.3	2.19	27.28	268	75	871	225	40	5.97	341	1.68

Part IV

The Tumulus in Its Context

Chapter 17

Research on Tumuli in Albanian Archaeology

Lorenc Bejko

The History of Tumuli Exploration in Albanian Archaeology

More than 65 years ago, Shtjefën Konstantin Gjeçovi (see **Fig. 21.2**), a well-educated Roman Catholic priest, was the first Albanian to excavate an Iron Age cemetery for academic purposes (Gjeçovi 1920). He also provided for the first time in the history of Albanian archaeology a coherent procedure for a systematic excavation. Gjeçovi's excavations of four tumuli near Kuvendi i Toroshanit were carried out very carefully, accompanied by a detailed description of the inventories, including dimensions and position of grave goods, position of the skeletons, and grave types (Gjeçovi 1920:110–111). He certainly knew about the tumuli excavated at Glasinac in the central Balkans at the beginning of the century, and used that information to make a comparative study of the burial customs of Illyrians in different parts of the peninsula. Gjeçovi's classification and dating of the archaeological material, which were based on typological similarities with finds from Glasinac, along with the rest of his activity (excavations, surveys, collections, and so on), make him the first Albanian archaeologist to have a great impact on the future of tumuli studies in the country.

For more than 30 years after Gjeçovi's work, not much was added to the field. During the 1950s, however, large-scale systematic excavations began in a number of tumuli around the country. Among the most important were the excavations of various tumuli in the Mat region (Rrethe Bajze, Kokërdhok, Bushkash, Bruç, Urakë, Shtogj, Perlat, Klos) during the 1952–1960 period (Islami et al. 1955), the tumuli at Vajzë in 1953 and 1954 (Prendi 1957), the tumuli of the Dropulli i Sipërm region (Prendi 1956, 1959) in 1955 and 1956, and tumuli at Mjedë (Shkodër) in 1958 (Islami and Ceka 1965:451–454). During the 1950s, a total of 44 tumuli were excavated, of which about 35 were from the Mati region alone (Prendi 1988:10). These excavations provided a great deal of information about tumulus construction, burial customs and rituals, grave types, funerary artifacts, and the chronology of most of their development (for the location of tumuli, see **Fig. 17.1**).

During the 1960s and early 1970s, the geographic span of excavated cemeteries was extended, with new projects undertaken at Pazhok (Islami and Ceka 1965:445–449), Barç (**Fig. 17.2**) and Kuçi i Zi from 1969 to 1971 (Andrea 1985), Çepunë in 1969 (Budina 1969), Bajkaj in 1970 (Budina 1971), and at Krumë and Çinamak (**Fig. 17.3**) in the Kukës region between 1969 and 1971 (Jubani 1971, 1982, 1983). With these new excavations, much new data was added to the field of tumuli studies in the country. These data were the basis for more generalized studies on a regional scale reflected in the work of Prendi (1974, 1977–1978). It is also important to mention that by this time, the work of Albanian archaeologists and their publications received the attention of foreign scholars (e.g., Hammond 1967a, 1967b, 1974; Wardle 1972). This was of special interest for the future development of tumuli research in the country, because for the first time the results of many years of exploration were put in a wider regional context. Questions such as the origin of the Albanian tumulus burials and their connection with Mycenae and the Ionian islands were also addressed.

During the 1970s and 1980s, other cemeteries were excavated not only in regions already known as rich in tumuli burials, but also from new areas. More tumuli were excavated at Pazhok in 1973 (Bodinaku 1982), Mat (Kurti 1977–1978, 1983), and in the Kukës (Hoti 1981, 1982a) regions, as a continuation of previous excavations. The most important evidence during this period, however, came from cemeteries excavated in previously unknown areas such as Kolonjë, with its tumuli at Prodan (Aliu 1984), Rehovë (Aliu 1987), Psar (Aliu 1995), Shtikë (Aliu 1996), Luaras (Aliu 2004), and the flat cemetery at Borovë (Aliu 1985), and in the Vjosë River valley with the tumuli at Piskovë, Rapckë, and Grabovë (Bodinaku 1977- 1978, 1981, 1988), and the Shkodër area with the tumuli at Shtoj and Shkrel (Koka 1985). Other isolated cemeteries were discovered and excavated during the late 1970s and 1980s at Patos (Korkuti 1981) (**Figs. 17.4–17.5**), Gërmenj (Andrea 1981a), Katundas (Braka 1987), Cerrujë (Andrea 1981b:270), Hamallaj (Hoti 1982b:255), Bujan (Andrea 1984:261, 1986:254), and Mujaj (Bela 1987).

A total of 156 tumuli were excavated all over the country from 1952 to 1987. Their location and number are given in **Table 17.1**. This is certainly a large number of excavated sites, which underlines the substantial investment that was dedicated by Albanian archaeology to the exploration of tumuli. During the 1990s, however, almost nothing was added to the list of excavated sites except for a single tumulus excavated at the necropolis of Apollonia in 1996 by a joint Albanian-French team (Dimo and Fenet 1996).

In the first decade of the twenty-first century, a relative revival of the tumuli exploration was observed, with three important and complex projects organized at Kamenicë (Bejko 2002b; Bejko, Fenton, and Foran 2006) (**Fig. 17.6**), Apollonia (Amore 2010) (**Figs. 17.7–17.8**), and at Lofkënd (this volume).

These new projects were clearly based on novel research questions and sound methodology. With the depth of information that has resulted from a multidisciplinary approach, these recent projects represent an important development in the tradition of tumuli exploration in the country.

The relatively long tradition of the study of tumuli in Albania has been generally characterized by a progressive development of excavation techniques, as well as of the structure of the observed data from the individual burials and overall cemeteries. The major focus of research, however, has been on material culture, relative chronology, and burial customs as important means of expressions of ethnic identity for the prehistoric communities. Discussions of social organization as seen from cemetery data have

TABLE 17.1 Location and number of tumuli excavated in Albania during the 1952–1987 period (after Prendi 1988)

Location	Number of tumuli
Mat	53
Krumë (Kukës)	8
Kënetë (Kukës)	6
Bardhoc (Kukës)	2
Çinamak (Kukës)	28
Mujaj (Kukës)	3
Bujan (Tropojë)	9
Mjedë (Shkodër)	1
Shtoj (Shkodër)	9
Dedaj (Shkodër)	3
Hamallaj (Durrës)	1
Pazhok (Elbasan)	8
Cerujë (Gramsh)	1
Patos (Fier)	1
Piskovë (Përmet)	1
Rapckë (Përmet)	1
Grabovë (Përmet)	1
Vajzë (Vlorë)	4
Dukat (Vlorë)	3
Çepunë (Gjirokastër)	1
Vodhinë (Gjirokastër)	1
Bodrishtë (Gjirokastër)	2
Kakavijë (Gjirokastër)	1
Bajkaj (Sarandë)	2
Barç (Korçë)	2
Kuç i Zi (Korçë)	2
Prodan (Kolonjë)	1
Rehovë (Kolonjë)	1
Total	**156**

been not only limited in number but also grounded on weak theoretical bases. The relationships between social characteristics and forms of burial, so widely debated in the 1960s and 1970s elsewhere, were not really touched on or questioned in the Albanian context. Only occasionally do generalizations on the social status of buried individuals and social organization of various communities appear, based on assumed, but not tested and well-supported principles. Mortuary variability, however, in various regions of the country has only been partially studied. It has been concentrated mostly on the differences of tumuli and grave construction between regions, ignoring the variability of mortuary programs and patterns of social structure of the cemeteries.

Another important shortcoming of the archaeological work of the last century has been the lack of systematic recovery and study of skeletal remains from cemeteries. Only in the 1990s did a more conspicuous interest emerge aimed at more elaborate and systematic demographic studies of skeletal materials (Dhima 1992).

Tumuli Exploration and the Quest for Ethnogenesis of the Illyrians

Tumuli burials have been long identified with one of the most important cultural expressions of the Illyrians in the region (Garašanin 1973; Islami et al. 1955; Jubani 1969; Korkuti 1969).

There are two main issues related to the use of tumuli as an ethnic marker of the Illyrians: (1) the chronology of the early appearance of tumuli in Albania, and (2) the chronology of the development of Illyrian ethnic identity (of which the use of tumuli burials was assumed to be an essential part). As for the first appearance of tumuli, extensive evidence has indicated that they were being used since the Early Bronze Age (Andrea [1985] for Barç; Bodinaku [1977–1978, 1981] for Piskovë; Jubani [1969, 1971, 1974] for Shkrel and Shtoj; Koka [1985] for Shtoj Tumulus 6), but data for relating this with the formation of Illyrian ethnic identity are not always direct and easily identifiable. Several arguments were formulated by Islami and Ceka, Korkuti, and Prendi that see, respectively, the Early, Middle, and Late Bronze Age as the crucial moment when the process of Illyrian ethnogenesis begins. It is not our goal here to explore the theoretical basis behind such highly complicated processes and the validity of the archaeological arguments for them; however, specific categories of material culture and the custom of using tumuli burial always appear as crucial data in such arguments. Tumulus burial did eventually become the dominant burial ritual among the Illyrians throughout the Late Bronze and Early Iron Ages, so an understanding of whether they appeared in the Balkans with the "new tumuli-using population" in the Early Bronze Age, or whether they were adopted by the local population during the Middle or Late Bronze Age, had important implications for the autochthonous (or non-autochthonous) nature of the ethnogenetic process.

The exploration of a large number of tumuli in almost all regions of the country made it possible to compare data from their architecture, internal organization, grave types, and related rituals. In an early generalized study that Jubani presented in 1969, several of the regional trends are noted (Jubani 1969). Similarities and differences observed were used to argue for the overall cultural unity of the Illyrians, underlining at the same time regional characteristics that were also visible in the patterns of pottery distributions and metal artifacts. The basic shared ritual of burying the dead under tumuli was considered not only as characteristic of the Illyrians, but also as a uniting feature of all regional Illyrian tribes. The major features of tumulus construction were also seen as shared elements of all regions, together with the positioning of the burial mounds mostly in and along river terraces. Among the regional differences, Jubani mentions the way of creating the inner tumulus (in the south, by means of a heap of stones, vs. in the north, where only soil was used), the preference for different grave types (in the north, stone-lined; in the south, cist graves), the ritual of broken objects (in the north, objects were broken after the creation of the inner tumulus; in the south, this occurred within the fill of the inner tumulus or immediately above it), and the ritual of pottery breaking that has been observed only in the tumuli of Kukës (Jubani 1969:97). These differences, however, were seen by Jubani and his contemporary colleagues as indicators of different regional "archaeological cultures," but not necessarily as arguments for the lack of homogeneity of the Illyrian population in the territory of Albania.

This topic was later taken up by Prendi in "Unity and diversity in the Illyrian culture of the Iron Age"

(Prendi 1985b), where northern and southern Illyrian cultural groups were described and characterized. Thus, tumulus burials have provided most of the arguments for the chronology and character of the process of ethnogenesis of the Illyrians, but also for the unique and regional differences of their culture.

TUMULI, BURIAL CUSTOMS, AND THE STUDY OF PREHISTORIC SOCIETY

The exploration of tumulus burials has been very important also for the information yielded regarding the social organization of Illyrian prehistoric groups. Based on the structure and stylistic characteristics of grave goods, many contextual assumptions have been made for individual differentiation within Illyrian society. Particular attention has been devoted to rich burials, which have been easily identified with emerging elites within a given social group. Attempts also have been made to identify these social groups represented in tumuli. Many consider them cemeteries of extended families, even if these assumptions are not based on any coherent analysis of the demographic structure of individual tumuli.

In the studies by Jubani, for instance, some important gender differences are underlined between members of the prehistoric communities as represented by characteristics of grave goods (Jubani 1974: 192). Yet Jubani has also attributed differences in grave goods between tumuli at Kukës to the emergence of inequalities within a specific cultural group (Jubani 1974:192). It is generally assumed in Albanian studies of tumuli that a direct relationship existed between the social position of the individual and his or her treatment at death. For Andrea (1985), graves with no or few grave goods reflect indirectly the low social status of the buried individuals, while the rich burials, particularly those with imported pottery and gold and amber ornaments, certainly belonged to the heads of rich families who were of higher social status. This is further seen as an indication of the "existence of a rich tribal aristocracy, not only interested, but also in possession of necessary means of acquisition of these expensive items" (Andrea 1985:210–211). Prendi also observed that age distinctions were important, such as the exclusion of children from formal interment in a community's burial mound during the Late Bronze Age, following instead the Neolithic tradition of interment under house floors (Prendi 1977–1978:23).

The concept of "tribal aristocracy" is present in almost all assessments of social organization. Prendi believed that "there is indirect evidence that since the Middle Bronze Age, within the egalitarian communities had emerged a new social group, the tribal aristocracy that differed from the rest of the population in its better life conditions" (1977–1978:23). Consequently, he attributed to this emerging social group the graves of Vajzë, Vodhinë, Midhe, and Pazhok which were richly equipped with grave goods partially acquired through exchange networks with the Aegean world. Prendi saw an important social change toward the end of the Iron Age when he commented on the appearance of the flat cemetery at Borovë in the Kolonjë region (Aliu 1985, 1994) and the scepter-like objects buried with a few males (Prendi 1988:23). Prendi argued that these particular objects symbolized the authority of a "military aristocracy" and that the Illyrians were moving toward a class-structured society. This new society did not consider tumuli as the most suitable burial custom, or at least tumuli were not capable of reflecting the new social changes. This becomes one reason for the almost total abandonment of tumuli by the end of the sixth century BC (Prendi 1988:23).

There is at least one known site, however, where tumuli continued to be the dominant way of burying the dead even after the sixth century BC, and this is the Greek colony of Apollonia (Amore 2010; Mano 1971). The dominant explanation for this continuity has been linked to the importance of native Illyrians in the social make-up of the new city and the adaptation of native burial customs (Mano 1976).

Even if burial data were used almost exclusively to make inferences on social structures, the relationship between burial and society was never discussed, but only assumed. A rigid Marxist social theory provided the framework for social evolution, and the interpretation of tumuli became part of this picture.

MATERIAL CULTURE AND CULTURAL TRADITIONS OF THE ILLYRIANS AS REVEALED FROM THE STUDY OF TUMULI

One of the main concerns of tumulus studies has traditionally been the material culture recovered from them. The characteristic of cemeteries for providing a series of closed contexts not only offered the opportunity for exploring the material culture, but also created the basis for the construction of relative chronologies. These analyses were of great value in the early phases of the development of archaeology in Albania, particularly when the settlement data were not only limited but could not be based on secure stratigraphies.

In 1974, the newly founded journal *Iliria* published the first general overview by Frano Prendi on the characteristics of the Iron Age in Albania (Prendi 1974:103–130). It is evident from this important article that almost 90% of the data and sites involved were tumulus burials. Fundamental chronological issues of the Iron Age, such as the first appearance of iron objects (and consequently the beginning of the Iron Age) and its phasing, were also argued on the basis of data collected from the excavations of various tumuli. Another similarly important generalization was published by Prendi a few years later on the cultural characteristics and sequence of the Bronze Age (Prendi 1977–1978:5–58). It is evident again that, excepting a few important settlements, such as Maliq, Tren, and Gajtan, the construction of cultural histories of the Bronze and Iron Ages in the country were based on the consolidated tradition of exploration of tumulus burials. From them were derived important chronological markers, a wide range of artifact categories, relative chronologies, and the bases for regional comparisons as well as country-wide generalizations.

Prendi's general works were followed by systematic analyses of material culture (mainly typological) of individual sites by Andrea (1985: Barç and Kuç i Zi), Aliu (1984: Prodan; 1987: Rehovë; 1995: Psar; 1996: Shtikë; 2004: Luaras), Korkuti (1981: Patos), Bodinaku (1982: Pazhok), Jubani (1971: Çinamak; 1982: Krumë; 1983: Kënetë), and others. An accumulation of large amounts of data made possible, and to some degree necessary, the in-depth characterization of separate categories of finds, which was observed from the second half of the 1980s. A series of articles on matt-painted pottery by Andrea (1985) and Bodinaku (1989, 1990b) extended and refined observations advanced previously by Prendi (1974, 1977–1978) and Korkuti (1970) on the origins and historical development of this group of pottery. The earliest appearance of matt-painted decoration on local ware is argued to be established on the basis of a closed context provided by grave 162 of tumulus 1 at Barç (Bodinaku 1989:53). Its changing features were also seen chronologically, based on evidence from closed burial contexts that offered associations with other chronologically sensitive artifacts. Detailed discussions by Bodinaku (1989, 1990b) followed the arguments that Kilian (1972:115–122) and Hochstetter (1982) had advanced on the origins and chronology of the matt-painted pottery in Albania.

Prendi, Korkuti, Andrea, Aliu, Bejko, and particularly Bodinaku classified and provided typological analyses of swords and daggers from tumuli and burial contexts, focusing on their relationships with traditions in the Aegean, the Italian peninsula, and central Europe. Many of the typological specificities of these objects were considered indicators of local adaptations to outside influences, including the local development of bronze metallurgy in Albanian prehistory.

Jewelry was one of the most frequently studied groups of artifacts. Pins, fibulae, bracelets, pendants, diadems, beads, and the like were the focus of specific studies (Kilian-Dirlmeier 1984a, 1984b, 1984c on pins; Aliu and Bejko on fibulae; Papadopoulos 2010a on headbands/diadems), or considered together as components of assemblages (Prendi, Andrea, Aliu, Korkuti, and Bodinaku, among others; see further Chapter 10). This group of objects comprised important evidence for chronological ordering of graves and other chronologically less sensitive artifacts, as well as differentiation of group identities and the definition of regional cultural traditions. Numerous assessments of cultural relationships and influences between local groups and more distant cultures of the region are derived from analyses of jewelry.

In the early 1980s, concerns were raised that generalizations about characteristics of material culture based only on tumuli/cemetery evidence were somewhat distorted. Thus, the dichotomy of cemetery versus settlement material became one of the main constraints for cultural history generalizations, as well as one of the main issues that would redirect future priorities/explorations in the country. Even if never explicitly stated, the concern with this dichotomy indirectly underlined the symbolic use of material culture and the primary importance of context in archaeological interpretation. Both remain central issues for the development of archaeological thought in the country.

ARCHAEOLOGY OF RITUALS

The few studies in traditional Albanian archaeology that focused on prehistoric rituals are dominated by death rituals and specifically tumulus burials. The use of tumuli themselves is frequently considered to be the most obvious ritual that appeared with the beginning of the Bronze Age and dominated the Illyrians' social practices throughout prehistory. In 1969, Jubani discussed a number of observations that indicated material evidence for death and burial rituals (Jubani 1969:91–101), including the way of arranging the central grave, formulation of the inner tumulus, the large circular feature around the central grave, and raising the tumulus through numerous interments. Another group of observations was related to the ways of treating the dead (inhumation versus cremation), types of grave structures used, and orientation of the individuals in burial. Differences and similarities in burial rites were then identified across geographical regions of the country and cross-referenced with other material culture evidence.

Jubani, however, noted some interesting observations of rituals carried out after covering the central grave or at completion of the inner tumulus. This ritual consisted of intentional breaking of complete pots, found in large numbers in fragmentary condition over the central grave or over the pile of stones of the inner tumulus. Jubani relates this ritual of complete pot breaking to the importance of the individual (or his or her social role) in the central grave, or to the beginning a new phase of raising the tumulus. Unfortunately, this evidence has not been confirmed in other excavations and has remained unexplored in subsequent publications.

Another observation appears to recur in most of the excavated tumuli—archaeological finds in the tumulus fill that predate (sometimes by millennia) the process of raising the tumulus itself. Until recently (Papadopoulos 2006; Papadopoulos, Bejko, and Morris 2007), this observation had not been thoroughly described and interpreted, even as it was considered evidence for intentional transportation of soil from a community to the tumulus being built (see discussions at the annual meeting of Albanian Institute of Archaeology [Tiranë 1986]). This recurrent connection of settlement and cemetery through the use of settlement deposits, or even earlier deposits from the surrounding landscape, as a symbolic component of the tumulus fill becomes a ritual that provides a spirit of place, and thus constructs social/historical memory (see Chapters 13, 14, 18, 20).

Finally, an important contribution to the exploration of the ritual dimensions in the context of tumulus burial was given by Nicholas Hammond, who suggested seeking answers to questions of rituals related to tumulus burial in the historical records of Homer in the *Iliad* (Hammond 1976:132). Hammond's argument is that even though tumulus burial ceased in Mycenaean Greece before the end of Late Helladic IIIB, Achilles honors Patroklos by using a burial rite practiced in Albania and farther north:

> They made a pyre a hundred feet in length and breadth, and with sorrowful hearts laid the corpse on top. At the foot of the pyre they flayed and prepared many well-fed sheep and shambling cattle with crooked horns. The great-hearted Achilles, taking fat from all of them, covered the corpse with it from head to foot, and then piled the flayed carcasses round Patroklos. To these he added some two-handled jars of honey and oil, leaning them against the bier; and in his zeal he cast on the pyre four high-necked horses, groaning aloud as he did so. The dead lord had kept nine dogs as pets. Achilles slit the throats of two of them and threw them on the pyre. Then he went on to do an evil thing—he put a dozen brave men, the sons of noble Trojans, to the sword, and set the pyre alight so that the pitiless flames might feed on them (Homer, *Iliad* 23.164–177, after Hammond).

In *Iliad* 23.249–261, the funeral of Patroklos continues:

> They went about the business as the swift son of Peleus had directed. First they put out with sparkling wine all parts of the funeral pyre in which the flames had done their work and the ash had fallen deep. Then, with tears on their cheeks, they collected the white bones of their gentle comrade in a golden vase [*phiale* in Greek], closed it with a double seal of fat, laid it in his hut and covered it with a soft linen shroud. Next they designed his barrow [tumulus; *sema* in Greek] by laying down a ring of stone revetments round the pyre. Then they fetched earth and piled it up inside. [Note also the funeral of Hektor in *Iliad* 24.]

This call to look for "historical" records (or anthropological observations) in order to understand some of the material evidence from tumuli, whether to detect "heroic" practices of burial or simply dimensions of burial ritual, has provided a valuable research avenue.

ENGENDERED TUMULI AND THE ARCHAEOLOGY OF THE HUMAN BODY IN ALBANIA

By focusing on individuals of the remote past, excavations of tumuli have offered a unique opportunity to explore issues of gender, as well as demography, pathology, diet, and many other factors related to the human body. Little of the potential of these explorations to provide data on social conditions and actions of past communities was realized in Albanian archaeology in the second half of the twentieth century. I believe this occurred for two main reasons: (1) research priorities were primarily focused on constructing cultural sequences and ethnic characteristics (craniometric analyses were developed in conformity with these priorities), and (2) lack of expertise and human resources to cope with the scale of prehistoric cemetery explorations.

Albanian archaeology treated social theories of gender only peripherally. Cemeteries provided the opportunity to observe male and female individuals beyond material remains, based on their associations with grave goods "typically characteristic" for males or females. Weapons and jewelry were the most commonly used artifact types for distinguishing gender. The discussion, however, generally ended there. Cemetery populations were composed of males and females who used different material culture to define themselves. The only detailed information on the role and place of women in ancient society was provided by a study of Pierre Cabanes published in *Iliria* in 1983 (Cabanes 1983). It has to do with the somewhat more active role that women played in ancient Epirote society, compared to women in Greek city-states, as far as property rights, decision-making in family matters, and rights to free slaves were concerned. These were considered privileged information that Classical archaeology could retrieve from historical sources and inscriptions, and its influence on the study of prehistoric tumuli was very limited. Similarly, the seriation analysis that Imma Kilian-Dirlmeier (1984b) and Biba Terzan (1984, 1985) conducted with data from Albanian tumuli had little influence. They could be considered the first attempts to define gender systematically through the symbolic use of material culture in burials, and this cast light on those individuals located in the gray area that does not permit a direct affiliation with either of the gender groups. Only in the early third millennium AD was the engendering of tumuli considered an important part of the research agenda as well as an essential dimension of prehistoric social constructions (Bejko 1999–2000; Kurti 2006).

Most of the work on skeletal remains in Albania during the second half of the twentieth century was linked to the work of Alaksandër Dhima (1980, 1983, 1984, 1985a, 1985b, 1992). He studied remains from a large number of cemeteries (Koman, Piskovë, Rrapckë, Barç, Kuç i Zi, Rehovë, Luaras, Pazhok, Liqedh, Dyrrachium, Ballsh, Byllis, Prosek, Brrar, Apollonia, Berat), from nearly all regions of the country, and over exceptionally extended time periods. His early studies focused on physical anthropology in describing the ethnogenesis of Albanians. Studying a representative sample of the modern Albanian population, Dhima identified "main characteristics of the physical typology of modern Albanians" (Dhima 1985a:43–150) and then did the same with the early medieval Albanians, prehistoric Illyrians, and those of the Classical period. These later samples present necessarily a number of problems related to their size, geographical distribution, and missing data. Dhima, however, was able to conduct typological and comparative analyses, which stand at the core of his arguments for the "autochthonous process of formation of the Albanians from the bio-anthropological perspective." The main conclusions of his work include (1) the local anthropological element of both Illyrians and Albanians is comprised of Adriatic, Mediterranean, and, to a lesser degree, Alpine types; (2) the identification of the same anthropological type over a very long time span indicates a common ethnic base; and (3) the continuous increase in the Adriatic type, over other typological elements, makes it the principal converging factor in the formation of Albanians (Dhima 1985a:231–235; 1985b:293–300). Dhima's later studies focused mainly on paleo-demographic profiles of the social groups represented in cemeteries (Dhima 1992). Beyond the identification of age and sex of individuals, he systematically considered epigenetic variations, paleo-pathologies, paleo-demographic indicators, hypothetic assessment of population (settlement) size, and fertility indicators, among other issues (Dhima 1992:263–286). In addition to patterns of death and dying, Dhima's work provided insight into the lives of individuals and communities in the prehistoric past.

LANDSCAPES OF TUMULI AND OTHER RECENT DEVELOPMENTS IN THE STUDY OF PREHISTORIC BURIALS

One of the early discussions of tumuli concerned their place in the environment along with their geographical distribution. Both observations have proved to be useful concepts for understanding the spread of the tumulus phenomenon and its intensity, as well as the relationship that tumuli have had with each other and with other features of the ancient landscape. This dimension of information contained in the location of tumuli has not been fully explored in traditional Albanian archaeology but has appeared as a new direction in recent years. Excavation of the Lofkënd tumulus has certainly become an opportunity to discuss this issue fully (Chapter 20). Aspects such as the visibility of the tumulus from many points of the landscape, and its physical dominance, inter-visibility between tumuli, and visual communication between tumuli and settlements are some of those observations that help reconstruct the active social landscape of past societies. On its own terms, this social landscape is a product of social action, while it plays an active role in shaping the society responsible for it. This dual relationship contains many other social aspects such as ritual, intra- and inter-group relationships, territorial marking, accessibility to natural resources, and so on. Recent discussions along these lines (Papadopoulos 2006; Papadopoulos, Bejko, and Morris 2007) have also provided the theoretical base for decisions to conserve, protect, or reestablish the physical integrity of tumuli in their characteristic landscape (Papadopoulos, Bejko, and Morris 2008, 2009; Chapter 22).

Among other recent trends in the study of tumuli in Albania could be listed the full-scale and systematic recovery and analysis of skeletal material from burials *and* tumulus fill. This practice, adopted since 2000, at Kamenicë, Apollonia Tumuli 9, 10, 11, and Lofkënd, has opened a broad avenue of research into population studies. Beyond paleo-demographic studies and paleo-pathologies, the appearance of DNA and biodistance analysis have already contributed much to the understanding of social dynamics in the past (Bejko, Fenton, and Foran 2006).

When the Lofkënd project began in 2004, it was located in an academic context with a long tradition of tumulus exploration. This volume shows that contemporary trends (cf. Amore 2010) are being consolidated and enriched with new data and methodological and theoretical innovations, all of which have brought new insights to the past.

Chapter 18

An Intensive, Systematic Archaeological Survey of the Landscape around the Lofkënd Tumulus

Jamie D. Aprile

Introduction

The Lofkënd tumulus excavation brought to the fore a number of important and interesting questions about the relationship between the cemetery and the surrounding landscape that could only be addressed by collecting detailed archaeological data from an intensive systematic survey. A number of Early Iron Age tumuli have been excavated in Albania such as those at Pazhok, Barç, and Kamenicë (Chapter 17), but no settlements have been found in association with those sites using extensive survey techniques, which usually identify fortifications, cities, and cemeteries. The Lofkënd Project provided an excellent opportunity to join excavation with survey to understand the immediate surroundings of the tumulus and to identify, as a particular goal, the nature of any Late Bronze or Early Iron Age habitation nearby. Furthermore, aside from learning more about the tumulus specifically, conducting a survey in the Gjanicë River valley has provided important data on the nature of settlement across time within an east-west inland valley far from major sites. This report is a summary of the results of the survey largely conducted in 2008; a fuller report, with the possibility of additional fieldwork, is intended for the future.

Intensive systematic survey in Albania has been conducted in large, historically important regions to gain an understanding of the nature of settlement near heavily studied sites. In contrast, the Lofkënd Survey sampled an area that has received little attention. Three important intensive surveys in Albania conducted in the last ten years have primarily been focused on understanding the hinterlands of the Greek and Roman cities of Butrint, Apollonia, and Dyrrachium (modern Durrës)—all three along the coast—and how they interacted with local populations (Davis et al. 2003; Davis et al. 2003–2004; Lafe 2006; Pluciennik 2004). A major survey in the Shala Valley region of Albania (Galaty et al. 2005, 2011; Galaty et al. 2013) has contributed significantly to our understanding of the later periods of Albanian history as well as the nature of settlement in the Alpine north. Another survey conducted in 2005–2007 in Albania's major southeastern agricultural valley, the Korçë basin (*http://www.icaa.org.al/Anglisht/kobas.html*), has revealed very important data concerning the nature of settlement in this region that serves as a rich food-producing area as well as an important inland connection between Albania and its Balkan neighbors to the south and east. The Gjanicë River valley provides an important route of communication between the coastal plains and the inland valleys and mountain passes. No intensive archaeological surveys have been conducted in such a liminal area, so the data provided from this project will help to fill in initial models of the long-term history of settlement in Albania.

As seen in Chapter 20, the tumulus has played an important role as a landmark in the long-term history of the local landscape. Data collected from the survey is an integral part of that analysis, as it provides a window into both the built environment and the memories of the locality that work in concert in the modern inhabitants to structure their perceptions and use of the landscape today.

Methodology

The methodology used by the Lofkënd Survey was modeled from the techniques developed in Greece by Cherry, Davis, and Mantzourani (1991) for the Northern Keos Survey and improved for modern computerized recording methods by Davis et al. (1997) for the Pylos Regional Archaeological Project (for a timely overview of the past, present, and future of regional survey in the Mediterranean focusing on the Aegean, see Cherry 2003). These methods were also used for the recent surveys in Albania mentioned above near Apollonia, Durrës, and Butrint as well as in the Shala Valley. Methodological similarity among these surveys creates a number of comparable data sets from a variety of projects that will enable future macro-scale regional analysis of archaeological site patterning in the south and southwestern Balkan Peninsula.

Field survey was conducted in a systematic, intensive manner. Using existing field boundaries, the area was divided into tracts that measured, for the most part, approximately 0.5 ha and no more than 1 ha, if possible (**Fig. 18.1**). Field surveyors walked across each tract along transects 15 m apart collecting 100% of the archaeological materials encountered within 1 m to each side of the transect line. In cases where the density of artifacts was too great to collect everything, representative diagnostic pieces were collected and the rest were counted. A global positioning system (GPS) device (Trimble Pathfinder ProXRS) with real-time satellite differential correction provided by Omnistar was used to trace tract and approximate site boundaries as well as any important features.

The area surveyed was approximately 1.7 km wide at the widest point, and 3 km long, and was defined arbitrarily by geographical features that surrounded the tumulus excavation. The boundaries consisted on the west and east of spring-fed streams that drain into the Gjanicë River, the southern boundary. The northern boundary was a high ridge line that divides the Gjanicë River valley from the Seman River drainage to the north. The survey area was subdivided into three areas: Area A, the central area immediately surrounding the tumulus; Area B, the southern area incorporating the river bottom and the hills nearest the river; and Area C, the northern area incorporating the deeply divided uplands leading up to the highest ridge line to the north. The project did not have time to achieve 100% coverage; thus the area was strategically sampled to provide continuous coverage from the river to the upland areas, resulting in a roughly southwest-northeast corridor.

In addition to the intensive survey, this project conducted an extensive survey to visit known sites in the area. These visits consisted of grab samples of diagnostic artifacts and single GPS points to aid in locating those sites in the vicinity of the survey area more accurately. These data will not be considered in future statistical analyses conducted for the intensive systematic survey; they will be used, however, to contextualize and extend the discussion of the final results, and they will be included here to provide a more complete picture of the landscape within which to analyze the tumulus as a landmark.

Results

Following this methodology, eight sites were identified or visited in the survey area. A site, for the purposes of this report, consists of an area containing a high density of artifacts in a generally definable area. Because many of the intensive surveys in Albania have been in the vicinity of cities densely populated in antiquity, they have needed more stringent definitions to separate more dense clusters of artifacts from the pervasive background scatter, but in the area surveyed for this project, the overall density was very thin and clusters were distinct. Wherever higher densities of artifacts occurred compared to the surrounding areas, a site was defined. These definitions are, of course, to a certain extent arbitrary and subject to dispute; however, they do reflect the locations in the landscape where artifact concentrations existed. These boundaries are not meant to be indicative of the size or extent of ancient habitation but rather to represent modern observation of the archaeological record on the surface.

An informal survey of local knowledge regarding archaeological sites in the area was also conducted during the course of field walking. When local villagers and farmers were encountered, we inquired as to what they could tell us about the history of the area and any ancient remains they might know about. In some instances, this information was extremely important in understanding both the formation and current state of a site we identified during the survey. Whenever possible, this information has been included in the description of the site.

Site S001: Mezhdat e Kuqe

This location is a Hellenistic-Roman period burial site consisting of an unknown number of graves on top of a bedrock outcrop. The hilltop was identified as Mezhdat e Kuqe by residents of the nearby village of Gjinoqara and translates roughly to "Red Ridge." The site was heavily disturbed by the construction of terraces for olive trees and unauthorized excavation, some very recent, as evidenced by fresh pick marks in the earth and scattered human bone on the surface. Conversations with local farmers revealed that the site had been known for many years. One man even recounted how he had played with human skulls on that hill as a child. The site boundary was defined by the farthest pockets of tile identified as having washed down-slope from the central, highest point of the site where the remains of graves are located.

The visible archaeological remains on the hilltop included a large, dense scatter of tile, a few remnants of built walls and possible grave cists, and a scatter of human bone clustered near the highest point of the site. Material was collected from a regular pass across Tract C0046, three extensive grab samples (X0001, X0007, X0008), and 10 "dog-leash" collection units with a 1-m radius. The tile mostly consisted of non-diagnostic pieces and Corinthian-style roof tiles. The amount of tile (411 fragments) recovered from the dog-leash collection units scattered around the site suggests that tile graves ("*alla cappuccino*"), which are similar to those found at Apollonia in Tumulus 10 (Amore 2010:101, fig. 5.24), existed in this cemetery area prior to erosion and modern disturbance. Abundant evidence for erosion on the west, east, and south sides of the hill and an immediate decrease in tile density to the north, where there is no erosion, indicate that most likely all of the material originated at the top of the hill and traveled down-slope (**Fig. 18.2**). A heavily disturbed grave cist was identified during survey (**Fig. 18.3**). It consists of two rough stone slabs set upright approximately 40 cm apart; this cist had been truncated by illicit excavation or erosion at some time in the past and appeared to be empty. The other remaining construction on the hilltop is made up of a 30-cm length of a wall consisting of three courses of rough field stones, horizontal tile or brick, and more rough field stones. The original function of this small wall segment could not be identified. Very little pottery was collected during intensive field walking and site collection.

This site was informally collected in 2005 by a visitor to the tumulus excavation and yielded more chronologically diagnostic pieces than were recovered during the 2008 survey. Pottery finds include a West Slope Ware cup rim, two large pithos fragments with possible bitumen residue, and two fragments of African Red Slip Ware. These finds are the best evidence to date the site to the Late Roman period, although the Corinthian tiles found during the intensive collection point to a Hellenistic date as well. Further study is necessary to understand the chronology of this site more clearly.

Site S002: Mashkullorë tumulus

The Mashkullorë tumulus is located approximately 2.3 km across a valley to the northwest of the Lofkënd tumulus, and was already known to Albanian archaeologists prior to this survey. It measures roughly 31 m north to south and 22 m west to east. The longest axis of the tumulus lies at approximately 45° off of magnetic north. The tumulus could be as much as 6 m high, although there is no way to determine the thickness of the archaeological deposits without excavation, especially considering that the excavated tumuli in this region (such as Patos and Lofkënd) were built upon slightly elevated bedrock outcroppings. The surface is covered with river-smoothed pebbles. Two joining fragments of modern glazed pottery were collected as a non-systematic collection in addition to five small chunks of chert as a geological sample. Approximately 20 fragments of modern tile were not collected. No prehistoric archaeological materials were identified on the surface.

Today the tumulus is located in the center of the village of Mashkullorë. An Albanian military surveying pin is set in concrete on the top of the mound, similar to the pin set into the Lofkënd tumulus prior to excavation. A man from the village informed us that in the recent past, a military tower was constructed on the north side of the mound. Another local man remembered the construction of the tower; he informed us that the builders had found graves on the north side of the mound but had avoided the modern cemetery still visible on the south side (**Fig. 18.4**). The foundations for that construction and several others on the north and east sides of the tumulus are visible on the surface. The early modern cemetery—with both marked and unmarked standing grave stones dating as late as the 1960s—covers the south side of the mound and extends to the modern road through the village. When digging drainage ditches in the vicinity of the tumulus, the villagers report finding pottery and skeletons.

Site S003: Belishovë A

This site consists of a scatter of prehistoric stone tools and debitage clustered primarily near a modern, functioning oil derrick at the edge of the floodplain of the Gjanicë River. Artifacts from this site were recovered from Tracts B0013, B0014, B0015, B0015a, and B0005; no dog-leash units were collected for this site. The site boundaries as shown in **Figure 18.1** illustrate the area where the lithic scatter was most dense. The finds from B0005 were located approximately 200 m to the east, with little material in between. Because of the proximity of this site to the oil derrick and the modern highway that was built at some point between the early 1970s and mid-1980s, there is a very real possibility that these artifacts were displaced down-slope during construction. Tracts B0166–B0178 continuing up the hill north of the road in the vicinity of the site did not yield any finds whatsoever, suggesting that the site did not wash down from a higher elevation. Tract B0023 north of the road near B0005, however, did produce a single piece of chert that was not chosen for inventory.

The site can best be characterized as an open-air, short-term settlement site of the Middle Paleolithic. Five cores, two perforators, and one scraper were recovered from the regular intensive collection in a variety of cherts, including the honey-brown type known from the nearby area of Kraps and the Kryegjatë B excavation. The site consists of a diffuse scatter of cores, tools, and debitage (as shown in **Fig. 18.1**), but it may extend farther to the east along the lowest margins of the hills, where several pieces of worked chert were collected, just north of the 5B floodplain soil boundary (as illustrated by John Foss in **Fig. 16.18**).

Site S004: Visokë A

This site consists of a very dense scatter of chert eroding out of a field terrace scarp from a clearly defined bed of cobbles and gravel distinct from the surrounding soil matrix. Tracts B0074 and B0075 lie to the north and south of the eroding edge, respectively, although the scatter is so close to the eroding terrace wall (**Fig. 18.5**) that no material from the site was collected during regular intensive survey. The field walker noticed the material off transect and alerted the author to its presence, at which time it was deemed to be a unique geological and archaeological feature. Soil cores sunk along the terrace scarp as well as 2 m to the south were composed of shale soil indicating a well-developed, old soil profile. The bed appears in the scarp for approximately 7 m. The lithic scatter was very limited in extent, so a single dog-leash unit was collected in the center of the area.

The majority of the stone artifacts recovered from this area were cores (19 pieces) or debitage (24 pieces), none of which were selected for inventory during the initial bulk finds analysis. Several pieces consist of the honey-brown chert known from Kraps and noted also in the Kryegjatë B excavation. A total of 61 pieces of chert were collected from the dog-leash unit, 4 of which were very large (10–15 cm in diameter) tested cores. The bed of cobbles could be an outcrop of favorable chert used in prehistory as a source for collecting raw materials for making tools found at the other prehistoric sites located by this survey and in nearby areas. More detailed analysis of this collection is necessary to clarify its history and function.

Site S005: Gjinoqara A

This site was identified both through archaeological survey and local oral history as an important location in the vicinity of the village of Gjinoqara (**Fig. 18.1**). One man interviewed by Surja Lela identified this spot as the location of a church that had been torn down around 1924 in association with the arrival of Austro-Hungarians, and another mentioned that the oral history of the village recorded a church being located in that area. The site lies on the west edge of an erosional ditch that deeply incises the hillside from north to south, and several pieces were collected from an area where soil had recently slumped down into the ravine. This suggests the possibility of buried archaeological features, although there is no direct evidence.

The site was initially identified by a grab sample collection (X0005) gathered while moving around the edge of the field where the site lies to avoid very dense wheat plants that reduced visibility to zero. That collection included more diagnostic pieces than were recovered from later collection in the site. Two rims of Red Slip Ware, as well as a painted or glazed handle along with a variety of wheelmade coarse wares, kitchen wares, wheelmade semi-coarse wares, and tile were found. Five dog-leash units were collected around the site in addition to a regular tract (C0114) after the wheat had been harvested. More Medieval/Byzantine glazed and incised pottery was recovered from those collections, but more careful study is needed to characterize the history of this site.

Site S006: Ngrançija A

Site S006 is a hilltop grape vineyard and the adjoining field south of the abandoned village school in Ngrançija (**Fig. 18.1**). There are two major components to the site, a Paleolithic stone scatter and a tile and pottery scatter most likely dating to the Late Roman period. Artifacts were recovered from Tracts C0145 and C0146. The lithic scatter observed was very dense and focused in Tract C0145, but many of the pieces of chert were broken cobbles shattered by recent plowing in the vineyard and thus were not collected. The historical component of the site is scattered and diffuse and centered on the north edge of Tract C0145 and southern half of C0146.

The lithic artifacts recovered from Tract C0145 include a retouched and used flake and an end scraper from 162 total lithic pieces collected. Only three good-quality pieces of chert were collected from C0146, suggesting that the recent plowing had a major impact on the visibility of the artifacts on the surface at this location. From the historical component, a variety of wheelmade coarse, semi-coarse, and fineware sherds were recovered from both tracts, in addition to a few fragments of possibly prehistoric handmade coarseware and pithos fragments. Twenty-six pieces of tile and a fragment of daub with reed impressions were also recovered from this site. The most datable piece is the toe of a Roman transport amphora. The abandoned school building contains some reused architectural members that are interesting, considering the nearby presence of these remains (**Fig. 18.6a–b**). The remains collected from this site require further study to identify the items more specifically and to refine the chronology.

Site S007: Belishovë B (upland)

This project visited the upland village of Belishovë to the south of the survey area in search of a reported "proto-urban" period site (Ceka 1983a, 1983c, 1985a, 1990a, 1990b; Lafe 2003). Ols Lafe reported the site to be 3 km west of the city of Ballsh at 508 m ASL. The author visited a site in the middle of the village using directions provided by a local archaeologist; however, the site described here as S007 is almost certainly not the proto-urban site of Belishovë. The site we visited is approximately 4.5 km west-northwest of Ballsh at an elevation of 451 m ASL. We were not able to locate the reported fortification walls; thus we have no information to provide about their location and preservation. We did locate a number of wall foundations in the area near the village school, but they were no more than 30 cm wide, consisting of field stones no larger than 30 cm in diameter, not the large, "cyclopean" or ashlar blocks expected of proto-urban architecture reported by Ceka and Lafe. A villager informed us that more ancient walls had been standing above the surface as recently as the summer of 2007, but they had all been deconstructed to clear the area and build field walls. The modern villagers' field walls are long and over 2 m high in many places, suggesting the extensiveness of the earlier walls that existed in the area.

The other archaeological finds in the area consisted of modern tile and pottery and an extensive early modern cemetery in the vicinity of the village school. Several different styles of grave stones were identified in the underbrush in an area of approximately 1 ha to the north and northwest of the school building. A few pieces of modern pottery and tile were collected as a non-systematic sample. The wall foundations and extensive historic-period cemetery in the area suggest that excavation would likely yield interesting results for understanding the growth and development of the village of Belishovë, including its religious history.

Site S008: Bektashi Teqe and Tyrbe

This site reflects recent historical changes in the region. The location was initially identified by an extremely dense tile scatter (over 300 fragments counted along the four transects through Tract A0086) and concrete foundations from a razed building. Two farmers working in a field nearby informed us that the location had been a Bektashi *teqe* (tekke), a place of worship, which was either destroyed or converted by the Communist regime in 1967, and a stable was built on the spot (labeled *stalla*, stable, on the Albanian Army map in **Figure 18.1**, which was made in the 1980s). Following the collapse of that government in the 1990s, the villagers tore down the stable. Just north of the ruins of the building, a *tyrbe*, or shrine for a Bektashi imam's tomb, about 5 m in diameter, was being rebuilt in Tract A0079. A small early modern to modern cemetery located in Tracts A0083 and A0084 among a well-established olive grove was identified by the same informants as the burial area for the family of the head Bektashi imam.

Most of the finds from these tracts include modern tile and window glass from the destruction of the stable and modern pottery, including two joining fragments of transfer-printed porcelain imported from China. Small quantities of chipped stone were also recovered from these tracts. The concrete foundations of the stable were relatively high compared to the rest of the surface in the area (**Fig. 18.7**), so it is possible that foundations from the earlier structure could be preserved underneath, although there is no direct evidence of this.

OTHER ARCHAEOLOGICAL FINDS AND FEATURES

There were several other archaeological surface features of note in the area that were not considered dense enough to warrant being named sites. To the west of Site S008, a diffuse scatter of Hellenistic tile was recovered from Tracts A0242 and A0243 associated with non-diagnostic sherds of wheelmade fine and coarse wares (**Fig. 18.1**). Early modern and modern cemeteries and graves were also located in the village of Ngrançija, in addition to what may be an isolated grave south of S004 (see labeled areas and the point feature south of S004 on **Fig. 18.1**). An inscribed pithos rim was recovered from the roadside west of S001 and northwest of S005 (see point feature on **Fig. 18.1**). A small, round-shaped stone, possibly a millstone, was also located but not collected from the edge of Tract B0024 (see point feature just north of the major east-west road in **Fig. 18.1**).

Scatters of prehistoric lithic materials were recovered as well. On the gravel field road west of S004 (see the point feature west of S004 on **Fig. 18.1**), a small scatter of chipped chert was noted but was not collected or considered a site because the material could have been transported to the spot with the gravel brought in for the roadbed; alternatively, it could have eroded from the adjacent road cut or the slope above. In Tracts B0139 and B0140, on a hilltop halfway between S004 and S006 (**Fig. 18.1**), a very dense scatter of irregular fragments of chert covered a recently plowed field in a zone of darker grayish brown soil compared to the surrounding yellowish brown soil. A Middle Paleolithic flake (F046) was recovered from Tract B0140, but the rest of the lithic material collected was not diagnostic and had been heavily damaged by plowing.

General patterns in the surface archaeological record across the survey area are shown in **Figures 18.8** through **18.10**. In **Figure 18.8**, the distribution of all the pottery recovered is shown. Pottery is more concentrated in the northern and central portions of the area where more historic-period material was recovered and sites S001, S005, and S006 were identified. The concentrations in the northernmost tracts are likely a result of the proximity of those tracts to several houses of Gjinoqara. Tile (**Fig. 18.9**) was more widely distributed around the survey area, but most was not chronologically diagnostic. The very heavy concentration in a single tract in the north is associated with S001. The very heavy tile scatter in the south-central part of the area stemmed from the destruction of a Communist-period stable shown on maps as late as the 1980s. There is no structure in that area at present; however, several large piles of stone cobbles and the tile scatter among freshly planted olive saplings represent the remains of this building.

Lithic scatters were recovered from many areas of the survey corresponding to geographical features. One scatter lies in close proximity to the floodplain of the Gjanicë River on the lowest slopes above the river. The largest scatter occurs over the relatively gentle hills in the middle of the valley in the vicinity of S004. The abundance of unworked chert in this area, much of it not collected, suggests that this area was a natural source of chert that was utilized as a raw material source during the Paleolithic. No Neolithic or Bronze Age blades were recovered from the survey despite the appearance of those types in the Lofkënd tumulus (Chapter 13), so it is impossible to say if this source was used in later periods. Future study of these finds will consider the patterning of cores and tools in relation to subsistence and raw material procurement strategies.

Conclusion

Although small in scale, this intensive survey provides important additional data about a little-studied area in the archaeological landscape: the Gjanicë River valley connecting the Adriatic coastal plains to the inland mountain valleys and basins of eastern Albania and northern Greece. Compared to other intensive surveys in Albania conducted in the vicinity of Greek colonial cities, the overall artifact scatter observed in this study was very thin. Six new sites were identified, and two other known sites outside of the survey area were visited. Open-air Paleolithic finds recovered in several locations will provide very important new data about the little-known nature of Paleolithic occupation in central Albania and may contribute to knowledge of raw material procurement strategies. A more detailed study of the individual artifacts and patterning across the landscape is forthcoming.

CHAPTER 19

THE THREE-DIMENSIONAL MODEL: DIGGING INTO INFORMATION DESIGN

Christopher Johanson

INTRODUCTION

Multidimensional, digital imaging of archaeological sites remains a practice in its infancy. Robust early adoption and now virtual ubiquity on site and in the lab have led to an extraordinary blossoming of archaeological visualizations that range in scale and scope from simple plans and elevations to interactive, web-based, geographic information system (GIS) applications, and complex three-dimensional, real-time models. Yet rhetorical modes and design priorities that are so obvious as to be taken for granted by scholars who make text-based arguments are largely ignored in the creation of these images. Purpose, intent, and design, thesis, structure, and argument are too often subordinated to the immediate availability of technological skill sets at hand and preconfigured, predetermined outputs. This is the functional equivalent to the PowerPoint effect so eloquently articulated by Tufte (2006), in which the template of the software produces imagery that fails to communicate effectively. Default rendering modes are the norm, and this newly established scholarly tradition hinders productive innovation when the models available to emulate are poorly executed. Either the images have been generated with little critical reflection on design, or they have been produced by default settings of an unthinking computer only to yield a preponderance of noise and "chart junk" (Tufte 1983) —for instance, extraordinarily dense and illegible topographic lines, useless and flawed textures and shadows, or even simple visual artifacts, such as the jagged aliased lines generated by default in most GIS software packages. These visual representations proliferate, built without care and sometimes without purpose, constructed only due to a fleeting connection between archaeological project and skilled technical staff.

The problem is exacerbated by the current lack of theory rigorously applied to the archaeological site report. Burrell (2011:796) underscores the problem in a critical review of the state of the modern site report. After noting the great weight placed on a team to produce a final, permanent record, the variegated forms such reports take, and the impossibility of the entire endeavor when authors are confronted with budget and time restrictions, she calls for focused syntheses that would be "shorter, less expensive, and even—is it possible?—interesting." Yet this lean report, when coupled with "attendant archives" of data, "should still sum up the excavation, 're-create' the strata destroyed, explicate the site's development across its phases, and recount the finds" (Burrell 2011: 796). Even for a critic of the meandering, gargantuan report, the implicit goal of the enterprise is to offer a means to "re-create" what was. This goal will always be unattainable, but technological advances offer the illusion of attainability by seemingly providing a means to record and replay *everything.*

A three-dimensional model of an archaeological site lies at the intersection between a deluge of inputs and the need for well-crafted outputs. It is only as strong as the data that has been input but is virtually illegible if it attempts to represent the totality of all ingested material. It can serve as the source for a simple axonometric, digitally rendered drawing, a front-end graphical user interface to a complex database of stratigraphy and small finds, or even a hyper-realistic rendering of landscape and inhabitants. When the goal of preserving and disseminating as much data as possible confronts this limitless capability of the digital, bounds must be set.

The burial tumulus at Lofkënd presents a particularly difficult subject for digital reconstruction and visualization. It is almost entirely organic, contains complex and overlapping stratigraphy, and houses one of the most difficult finds to model in three dimensions, skeletal remains. The most common representational technique to reveal interior burials of tumuli uses a hand-drawn, three-dimensional sketch or hand-built, three-dimensional physical model with cross-section or quarter-sectional cut. When cross-sections of tumuli, grave mounds, and barrows exist, however, they inevitably reveal a clearly articulated architectural form within. As an example that proves the rule, the more famous cross-sectional cuts of the Great Tumulus at Vergina, the so-called tomb of Philip II, are centered on a buried architectural form, the royal tomb, constructed from ashlar masonry (Andronikos 1984:98–99; Dimacopoulos 1997:17). These representations avoid the difficulties of depicting overlapping burial pits such as those present at Lofkënd. Such reconstructions are more often found at the juncture between academia and popular culture, where perhaps the most successful attempt to visualize complex, overlapping burials is the series of images produced by Ned Seidler for *National Geographic* to illustrate the three-dimensional layering of graves in the tomb of the Lord of Sipan on the north coast of Peru (Alva 1988). In this case as well, the skeletal remains were wrapped in textiles and placed within the tomb in a deliberate order.

The tumulus at Lofkënd comprises one hundred graves, six burial phases, and no monumental form of architecture, even though it is, in and of itself, a monument. One cannot simply build a three-dimensional representation of such a site without a clear articulation of the outputs required. The tendency to preserve and represent everything makes this enterprise even more difficult. In comparison, digital reconstructions of archaeological sites that contain monumental architecture offer various levels of immediate utility, even if the initial conceptualization of the project was somewhat vague. Creating hypothetical walls and roofing enables the analysis of sight lines, acoustics, and vague approximations of urban experiences. At the very least, such models, even when built on no clear theoretical foundation, give scholars a clearer means of understanding spatial relationships on site. To represent a tumulus, the task is not to reconstruct, but to visualize information.

INPUTS

There are possible data-gathering techniques that could offer a degree of flexibility in the production of three-dimensional digital artifacts and the rich data objects that might accompany a future archaeological report centered on the excavation of a tumulus. Each is constrained by classic restrictions for any project, time and money, along with the limitations of current technology.

For a tumulus, rapidly advancing technological affordances offer surprisingly simple and cost-effective methods of documentation. To record the day-by-day progress of excavation, one might perform aerial surveys with geographical positioning system (GPS)-guided, gyro-stabilized, radio-controlled helicopters (Team Tanah Datar 2012) in order to build a daily, comprehensive photographic record. The same technology, programmed to take advantage of the three-dimensional motion of these nimble survey vehicles, also enables the creation of photogrammetrically generated, three-dimensional digital state models of the site (Olson et al. 2013). Rather than take a series of aerial photographs whose location is determined by a two-dimensional grid, one can establish a three-dimensional, GPS-guided path to allow for a daily, replicable survey of the entire, relatively small area. The three-dimensional path would enable closer inspection from multiple angles of baulk walls, trenches, and preserved stratigraphy of the entire tumulus. Coupled with this work, at the end of each day, or after the cleaning of each excavated grave, one might perform a short, calibrated, photographic survey of the grave itself, with an aim to gain complete coverage of the grave and its contents. In each case, the resulting data would again be used to create photogrammetrically generated, three-dimensional meshes of the site. The layers of each photogrammetric survey could then serve as the core of a volumetric visualization tool that would let one slice into the tumulus in much the same way that the early visible human project (Kerr, Ratiu, and Sellberg 1996; Tam 2010; Temkin et al. 2006) enabled the three-dimensional exploration of a similar data set of images.

In a slightly more time-consuming approach, complementary to the photogrammetric study, one could employ targeted laser scans of burial pits and the finds within (Payne 2011). Daily scans of each grave and its contents from multiple angles would eliminate the majority of sight-line occlusions and might offer more precise measurements, but would be slower to implement than a photogrammetric approach. Scans of small finds, however, would offer a relatively rapid method for building a three-dimensional database of point clouds that might later be processed and compared algorithmically. For grave pits and skeletal remains, off-site work flows would need to be developed to scan quickly individual bones and fragments from the skeletal remains. These digital artifacts might subsequently be pieced together following the outline of a larger, grave-scale scan.

If there were money and time, such approaches would enable a digital re-excavation in years to come, but we are still far removed from the day when adequate databases, robust storage and retrieval mechanisms, and effective interfaces will allow for cogent and efficient use of this information. The modern archaeological project already faces a data deluge that can demand significant time and effort to manage and navigate the complex and inefficient data-management solutions available to the project. The site report of the far future may very well present pools of machine-actionable data, but one suspects that the fundamental problem affecting visualization will remain: complexity must be abstracted so that the data are legible and useful.

This project lacked the resources and the time to engage in a sophisticated data collection enterprise. Instead it sponsored an effort in repurposing extant data and adapting to ongoing processes on site. The Experiential Technologies Center (ETC) was brought into the project through a fortuitous opportunity provided by generous funding from the ETC Steinmetz Outreach Program. The program sent graduate students to the field to perform ad hoc digital asset surveys while working briefly but intensely with the data collectors on site to formulate simple, working three-dimensional models.[1] The process encouraged creative thinking and the development of rudimentary design solutions. The results were both the designed maps of the site (**Figs. 2.2** and **2.3**) and a series of study models (**Fig. 19.1a–c**). The first phase of the construction process proceeded as described in the following.

Building the fully textured, real-time model was largely an assembly process. Key to this process was the methodology employed by the archaeological project. Many excavations record survey data in two-dimensional, vectorized form. The X and Y coordinates are embedded in the survey point recorded in a master computer-aided design (CAD) file, but the elevation data are recorded only as an annotation. In anticipation of later three-dimensional representations, Max Farrar, the excavation surveyor, attached three dimensions (X, Y, and Z) to all data collected over the first two field seasons. Therefore, survey points, contour lines, grave-bounding boxes, and drawing-frame data were all defined by points and polylines. The ETC team used a variety of programs and methods to build upon these data to create the three-dimensional, real-time model of the tumulus. Rhino 3D was used to derive a three-dimensional surface mesh from the contour lines in the original CAD file. The textured, three-dimensional graves were built in MultiGen Creator. The surveyed point data from the drawing frames were used to create the base polygon for each grave. The drawings were applied as textures to the polygon by mapping the surveyed points marked on the drawings to the polygon in the model. The polygon was then trimmed to fit the correct orientation of the grave as represented by the drawing. A photograph of each grave was aligned to the drawing and applied to an overlay polygon. The extruded height was derived from the measured distance between the highest and lowest recorded point on each skeleton. The combined model (topographic mesh and individual graves) was transferred into SketchUp to create the print images (Papadopoulos, Bejko, and Morris 2007:137).

The original model was an experiment. Begun first as a data-gathering exercise, it was transformed into an interactive model and envisioned to serve as a graphical front-end interface to the database maintained by the project. The digital model was then geolocated, its origin (the 0,0,0 location in a Cartesian system) given an approximate latitude and longitude. The team experimented with various geographically based visualization and navigation tools

[1] The Steinmetz Outreach Program has sponsored similar projects at the archaeological sites of Nysa and Magnesia in western Turkey. The latter has resulted in a radical rethinking of the archaeological site report entitled Immersive Coordinates: Digital Anatolia (*https://securegrants.neh.gov/publicquery/main.aspx?f=1&gn=HD-51465-11*).

such as Google Earth. Geographical points were taken on site through a crude GPS system and matched to concrete monuments previously installed on site to support the local survey (Papadopoulos, Bejko, and Morris 2007:137). Each of these experiments led to a dead end—the time and effort to build the outputs revealed the limits of what was possible. If more sophisticated data-gathering techniques such as those outlined above had been used, more could have been done. Such techniques were not available to the project due to funding, time, and technology constraints. Even if they had been, however, the fundamental design problem would remain.

INFORMATION DESIGN

Three-dimensional, digital representations have reached a level of sophistication where certain strengths are undeniable. First among them: process is where most of the benefit of a three-dimensional modeling endeavor resides (Favro 2006). Much like the act of drawing once functioned for the analog architect, photographing sites and constructing digital models serve as heuristic tools for interacting with data. Such processes ought to be integrated directly into project work flows rather than relegated to a digital ghetto. As a heuristic device, one can build and sketch in order to test ideas. The strength in the process is the modeling and exploring of ideas (Johanson 2009).

Second, once the process is complete, three-dimensional digital models can communicate specific visual and spatial information in ways superior to text alone. Not all arguments are multidimensional, but some necessitate a deep understanding of spatial relationships (Favro and Johanson 2010:15–16). Good three-dimensional representations, like good maps with carefully designed visual elements and symbology, cut through the complexity of a multiplicity of overlapping data by means of simple abstraction. Although good design should be applied whenever possible, a poorly designed three-dimensional model, when processed with rudimentary rendering technology, can still provide amazingly compelling images. In the multivocal world of archaeology, even these can be useful.

Therefore, visualizing the tumulus at Lofkënd requires a careful consideration of an array of possibilities, all ultimately intended to make the underlying data legible in ways that text and two-dimensional imagery cannot. The tumulus is not a feature ever intended to be experienced through a first-person walk-through as one might experience architectural forms. Certainly, the tumulus affected the experience of surrounding landscape (see Chapter 20), but its internal structure was perceived, from a phenomenological perspective, at most, one open grave at a time. Rather than aim to reconstruct digitally, as one might according to the paradigms of architectural reconstruction, the project team defined four design priorities to guide the final outputs:

1. To provide an overview of the shape of the tumulus as defined by the location of the burial pits
2. To see, at a glance, the significant phases of the tumulus and the reconstructed chronology of the burials
3. To enable rudimentary spatial analysis of the location of the burials
4. To juxtapose the landscape of the fourteenth century BC, prior to the first burial, with that of the fully formed tumulus of the nineteenth century AD.

MAPPING SPACE AND TIME

To navigate a three-dimensional tumulus, one needs a three-dimensional map, not a two-dimensional archaeological plan of overlapping information, but an abstract representation designed to reveal and narrate rather than aggregate. The ETC team coaxed the data into a legible representation of the totality of the burials in context in order to indicate the horizontal and vertical spatial transformation of the tumulus over time, to convey, at a glance, the six distinct phases reconstructed by the archaeological team, and to represent the reconstructed chronology of the burials through visual representation (**Fig. 19.2**). The design is grounded in what Tufte (1983:92) calls "the fundamental principle of good statistical graphics: above all else show the data." **Figure 19.2** aims to present the most amount of information with the least amount of (digital) ink. The design privileges hierarchical sequence as originally theorized by Bertin (2011 [1967]) and modified by Mijksenaar (1997:38–39), wherein sequential position and direction indicate spatial transformation on the horizontal axis and chronology on the vertical. Rows contain each of the six primary phases of the tumulus. The columns contain representations of the graves present during the corresponding phase. The designed graphic conveys spatial relationships between the individual graves along the horizontal axis (column 1), the vertical axes (column 2), before combining the two to expose the three-dimensional interrelationship of each grave (column 3). The current phase is then inserted into an accretive representation that demonstrates the three-dimensional growth of the tumulus over time (column 4). Column 5, to the right of the four main columns, presents the supporting information of distinguishing variables in the form of micro-graphics of each grave. Tomb numbers are embedded in the micro-graphic to indicate the reconstructed chronology of the graves.

The figure serves as a high-resolution, small-scale graph that combines the conclusions of the archaeological project as they relate to chronology, the documentary strategy used by the dig in the form of aggregated grave drawings, and a temporal graph of the shape of the tumulus over time. At a glance, column 5 reveals the relatively consistent activity on site over the centuries, with what might be a slight lull during the Early Iron Age.

Spatial Analysis

The three-dimensional model, when analyzed at a larger scale, also offers simple mechanisms to discover spatial clusters (**Fig. 19.3**). Within the fully manipulable real-time model of the tumulus, animated rotation of the tumulus highlights two distinct spatial footprints. During the first two phases of the tumulus, dating from the fourteenth to the eleventh century, two separate areas for burial were established. One, centered on Tomb I, establishes the center of the mound that would later rise above the surrounding landscape. The other follows the relatively steep, southern slope of the original hill. This second set of burial interventions into the south slope did not contribute to the building of the vertical mound. This spatial disjunction, already discovered without the use of the digital model (see Chapters 2 and 4), is clearly delineated in the interactive model, and here highlighted in large-scale plan and section coupled with oblique view of the tumulus from the south.

Tumulus in Context

As seen in **Figures 19.2** and **19.3**, the recorded data from the individual graves are sufficient to enable the reader to estimate the height and form of the surface of the tumulus during each documented phase. The same data can also support a hypothetical reconstruction of the pre-burial site of the tumulus (**Fig. 19.4**). The maximum heights of the burials from the first phase were used to generate contour lines that approximate the original topography of the fourteenth-century BC landscape. The methodology is crude and hardly representative of scientific reconstruction of landscape. Therefore, rather than operate within the parameters of rigorously defined information visualization techniques, the final, experimental design blends art with science to offer two hypothetical, stylized views of the tumulus, one dating to the fourteenth century BC, the other to the nineteenth century AD. Based on these machine-generated views, the early site was a prominent outcropping in the surrounding landscape but did not establish a prominent visual presence until after human intervention.

Conclusion

The three-dimensional model of the tumulus began as a simple visualization exercise, but transformed over time into a critical engagement with theories of information design, data representation, and statistical graphics. During the process of digital creation, the complex nature of the spatial interrelationships of the grave pits and skeletal remains forced the ETC team to develop creative solutions to visualize the data effectively. Instead of attempting to construct a highly detailed, hyper-realistic, three-dimensional mesh—the traditional path to three-dimensional model building on archaeological sites—the team instead created an explicit set of data abstractions. Theories of information design were marshaled to produce data-rich charts to convey chronology and spatial relationships, to narrate and to argue. Stylized renderings were then developed to represent the purely hypothetical. Although the priorities of archaeological practice might compel one to record and reproduce *all* data from an archaeological project, and computational technology might offer illusory hope that a complete visual record might be obtained, this experiment underscores the need for and utility of a focused, design-forward approach to archaeological visualization where complexity is reduced rather than magnified through the use of deliberate information design.

Chapter 20

Lofkënd as Cultivated Place

Samantha L. Martin-McAuliffe

Context and Outline of Present Study

The first part of this chapter considers in detail the situation of the tumulus: its physiognomy, immediate setting, and relationship with the surrounding topography. A close reading of the burial mound within its terrain is crucial for a number of reasons. On one level, it clarifies the present conditions of the tumulus, thereby creating a well-defined cognitive map, not just of the burial mound itself but also its place. At the same time, a careful analysis illuminates new avenues of discussion and inquiry by enabling us to perceive subtle physical and geographical relationships. This line of investigation will hopefully develop questions about the role of the tumulus beyond its immediate function as a burial site. Although the tumulus is arguably integral to its topography, it is also emphatically the product of human agency. For this reason the present study will examine how this burial mound and its vicinity can be defined as a landscape. The final part of this section proposes that Lofkënd embodies a pre-understanding of place, and furthermore, that place itself is a provision of the construction of the tumulus.

Following the situational study of the tumulus, this chapter will examine the cultivation of place. The second section begins by revisiting the etymology of the word "cultivate," paying particular attention to how it can embody both literal and metaphorical connotations. The discussion proposes that the relationship between the Lofkënd tumulus and its topography is stratified. In this particular place, we see very pragmatic and fundamental expressions of cultivation, such as sustenance, and yet there are also conditions that give rise to more formal associations that have to do with settlement. Beyond this, the burial mound has the capacity to harbor much more figurative meanings that encompass notions of memory and honor. Ultimately, this middle section proposes that the burial mound was cultivated with intention, and that people—inhabitants, users, passers-by—catalyzed a number of conditions that shaped its locale, physical construction, and wider understanding.

The third and final section of this chapter explores how we can understand and describe a burial tumulus as a repository. Most burial places can be referred to as containers for the dead, yet the situation at Lofkënd calls for special consideration. Unlike, for example, a cut-stone mausoleum, the form of the Lofkënd tumulus is directly dependent on, and shaped by, its contents. Put categorically, this burial mound could not be emptied of its contents without also losing its physical integrity. This situation raises many questions, both practical and philosophical, about the nature of boundaries in general. Starting from the tumulus itself, this section will progressively move outward to examine the extent of the mound's influence in its landscape, including its relationship to other tumuli and natural landmarks. The final portion of this chapter expands the discussion of boundaries even further. While the term "boundary" often evokes material qualities in the first instance, it also can embody abstract and metaphorical conditions. The burial tumulus at Lofkënd explicitly relates to the afterlife, and therefore it is imperative that we also study how it attends to the limits and thresholds of time.

Situation

Setting

The prehistoric burial tumulus of Lofkënd sits at the southern tip of a foothill that is ancillary to a ridge of the Mallakastër Hills. This primary ridge runs approximately northwest-southeast, and as it extends past the tumulus, it bends and continues farther south (**Fig. 20.1**). Although much higher in elevation than the hill on which the tumulus is located, this ridge is eclipsed in size by another, more rugged range that rises beyond it in the north and a short distance in the east (**Fig. 20.2**). Extending in a southwesterly direction, the hill of the burial tumulus looks toward the valley of the Gjanicë River, which runs farther to the southwest (**Fig. 20.3**). Immediately beyond the river rises yet another imposing ridge stretching northwest-southeast. This line of mountains effectively divides the Gjanicë from the larger and wider Vjosë River valley.

When studied from the point of view of a 1:100-m contour map, the hill of the burial tumulus takes on the shape of a long, narrow spur (**Fig. 2.3**). The slopes on its south and east sides are a great deal steeper than the terrain to its immediate west and north. This means that both the view from the tumulus and the view toward it can change significantly, depending on the place of observation. At times, the tumulus appears to jut forth impressively from an uneven bluff, such as when it is viewed from the modern village of Ngrançija to the immediate southeast (**Fig. 20.4**). In contrast, when seen from the site of another burial tumulus, that of Mashkullorë in the modern village of Gjinoqara Mashkullorë to the northwest, the Lofkënd tumulus takes the appearance of a protuberance growing atop a moderate incline (**Fig. 20.5**). The greatest distances from which Lofkënd is discernible lie in the west, and the mound at Mashkullorë is only one of a number of known sites in this direction. Much farther west-northwest rises Margëlliç, one of the most visually prominent and historically important sites in the region (**Fig. 20.6**). And, albeit not directly in the line of sight of Lofkënd, the burial tumulus of Patos (**Fig. 22.14**) is a moderate distance (14 km) to the west, and also in the Gjanicë River valley. The varying perspectives onto the tumulus are important to consider, not only because they ask us to question how the mound was seen in antiquity, but also because they remind us that the situation of this burial site is never part of a single, framed landscape; it can be understood from many perspectives. As we shall see, these perspectives are both literal and figurative.

When we situate the Lofkënd tumulus in a broader topographical context, we begin to discern a greater complexity to its position on the foothill. In a very basic way, the tumulus occupies a middle ground, and this can be understood both in terms of vertical elevation and physical distance. While the mound stands at about 357 m above sea level, the ridge to the north reaches well over 400 m at several points. In the opposite direction, approximately 1.5 km south of the tumulus, runs the Gjanicë River, which is less than 150 m above sea level. As such, the burial mound effectively stands as an intermediary between the river at the south and the mountains in the north. Furthermore, we can extend this observation by factoring in the presence of the southern mountain ridge. Considered in tandem, the ridges seem to shape a generous riverine valley; like brackets, they form both the southern and northern visual boundaries for the tumulus.

It also is important to consider the topographical physiognomy of the terrain that lies between the tumulus and the river. Beyond being situated at the edge of a foothill, the tumulus stands on the northeastern edge of an irregular horseshoe, or crescent-shaped basin, that slopes directly into the course of the river (**Fig. 20.1**). In other words, not only is the prehistoric site of Lofkënd located in a valley, but it also stands on the perimeter of a secondary valley—a valley contained within a valley. The western side of this crescent is shaped by a series of plateaus and gentle ridges, which today mainly serve as terraced farm tracts. At its southern tip, a knobbly hill protrudes, and where this hill meets the course of the river, it creates, with the adjacent slope of the southern ridge, one of the narrowest passes in the Gjanicë River valley as a whole (**Fig. 20.7a–b**). From the visual perspective of the tumulus, the passage functions as a kind of natural gateway between this particular section of the valley and the regions farther west. The eastern side of the crescent appears as a relatively tall yet asymmetrical ridge (**Fig. 20.8**). Like the hill of the burial tumulus, it is a great deal steeper at the east than in the west. However, in contrast to the form of the ridge in the west, it widens and tapers outward as it descends toward the Gjanicë. At the point where it meets the river, the basin as a whole grows broader, especially on its southern side, where the corresponding mountain slope is relatively moderate.

Approach

While it is possible to access the Lofkënd tumulus from the north, it is more likely that a visitor will first view and approach the burial mound from the south. Today, one of the principal highways in modern Albania (SH4) runs through the Gjanicë River valley, providing a north-south artery between Durrës on the west coast and the Greek border in the south. However, given its close proximity to the Gjanicë River, this thoroughfare likely marks the natural, and therefore ancient, route-way for the valley as a whole. This means that while the river itself was probably never sizable enough for watercraft, it did serve as a literal course or directional path in a manner similar to the present day. Now, as in antiquity, this route linked the immediate valley with other regions both to the northwest and the southeast.

When traveling along the highway from the northwest, the tumulus first comes into view shortly after navigating the narrow pass—the gateway—in the valley (**Fig. 20.9**). At approximately 1 km farther east, a subsidiary road climbs up the valley and eventually leads to the village of Ngrançija. Just before this smaller road, however, lies a large tract of level farmland between the present course of the Gjanicë and the highway. The visibility of the tumulus from this tract is especially clear: It appears directly to the northeast and seems to cap a low hill, which is effectively the very edge of the crescent-shaped basin (**Fig. 20.10**). Notably, the burial mound is situated in such a way that it appears almost isolated on the skyline. It is not merely on the horizon, but clearly *part of* the horizon. The mountain ridge immediately behind the tumulus in the north is perceptible, yet it does not dominate the mound. Farther east, however, the larger ridge is clearly distinguished and, in fact, looks more imposing from the valley floor than from the tumulus itself. Overall, the view of the tumulus from the river, and thus the principal travel route, is distinct and unobstructed. This perspective should not be underestimated, because it will play a vital role in our discussion of the wider symbolism of the burial mound. Yet it is important to underscore that this is not the only perspective. For example, the views toward the Lofkënd tumulus from the west have already been noted. Given that this is a prehistoric archaeological site, many questions remain about how the burial tumulus at Lofkënd was experienced and viewed by its earliest users. Consequently, it is a challenge to describe and define the tumulus as a landscape in a modern sense. This difficulty must be addressed in full.

The relevance of the term "landscape"

> Landscape is a term which both invites and defies definition.
>
> *Gosden and Head (1994:113)*

Landscape is not a single discrete entity, but a term characterized by exceptionally varied and at times contradictory conditions (Jackson 1984). While the discussion of landscape within the realm of archaeology is now commonplace, it is in no way straightforward or unproblematic (Johnston 1998). This chapter does not seek to advance or dispute any one particular methodology of landscape studies but instead draws upon a range of disciplines that consider landscape in order to address the situation at Lofkënd in a comprehensive manner. First and foremost, it is worthwhile revisiting the concept of landscape with a particular view toward its relevance to ancient sites.

The modern English term "landscape" derives from the Dutch *landschap*, a concept first used by painters in the seventeenth century to refer to their subject matter, specifically a region or territory of land (Haber 1995:38; Hirsch and O'Hanlon 1995:2). Although "landscape" long ago expanded beyond a purely technical usage and now carries a very broad set of associations, it should not be overlooked that its modern definitions maintain strong connections with painting. For example, in the second edition of the *Oxford English Dictionary* (*OED*), landscape is defined as a "picture representing natural inland scenery" and "the background of scenery in a portrait." It is important to underscore how both of these definitions assume that the landscape is framed. The subject matter, whether the topography itself or a man-made object in it, is confined to boundaries that are established first by the author-viewer, and ultimately a formal, physical limit, such as the edge of a photograph or a picture frame. As such, the landscape-as-picture is abstracted, unchanging, and static, both spatially and temporally. Beyond this, we can also observe that it is passive. One or more individuals have chosen—interpreted—a place *as a landscape*. In other words, a landscape is not a predetermined and complete "ready-made" object to which humans become cognizant (Knapp 1999:230) but rather a place marked with human intent. It is a way of seeing the given world that is construed by us (Cosgrove 1985:46; see also Berger 1972; Fitter 1995:8–9). We will return to this specific discussion of landscape in the final section of our study where we consider the nature of boundaries at Lofkënd.

A third definition of landscape in the *OED* provides further insight into how we interpret the natural world in the present day: "view or prospect of natural inland scenery, such as can be taken at a glance from one point of view." Significant to this definition is the idea that a landscape is not only contained and static, but that our own place, our role as viewers, is equally limited or even wholly detached. While it may be possible to provide multiple views onto a landscape, the term "prospect" alludes to something quite pointed and specific. It is no wonder that the pioneers of the English garden movement in the eighteenth century emphasized the potential prospect of a client's garden (Cosgrove 1985; Fitter 1995:8–14; Hunt 1987). They were designing and fabricating a landscape at the behest of a particular physical and social point of view: that of the wealthy, often aristocratic proprietor. An interesting visual parallel is Humphry Repton's trade card, which depicts the "landscape-gardener" standing with surveying tools at an elevated point overlooking a carefully orchestrated lake and woods (**Fig. 20.11**).

Revisiting the conventional definitions of landscape calls attention to how we may unconsciously carry with us preconceived inferences about what a landscape is or can be. Moreover, such an exercise asks that we become more selective about how we use particular terms. One result of this deliberation is the clarification of the relationship between landscape and people. Humans are not the recipients of landscape but instead are pivotal to the very conception of it (Jackson 1984). Ultimately, we can argue that a landscape is something determined and generated by us (Cosgrove 1984), and as such, it can be interpreted from many perspectives, both spatial and metaphorical. We transform land into landscape by literally marking the earth or, even more simply, by "projecting our ideas" and beliefs onto it (Ashmore and Knapp 1999:8). This means that certain conditions must surface before we can have a landscape. Namely, the understanding of place precedes the establishment of landscape (Casey 1993:177). For this reason, it is more straightforward to consider Lofkënd first as a place and then, only later, a landscape. This asks us to take heed of the conditions and possibilities that were in position and subsequently allowed for the construction of the tumulus. In doing so, we start to understand how the formation of Lofkënd as a landscape—if it can be called one—is not in the first instance the building of the tumulus but instead the interpretation of this valley by humans as the potential place for a burial mound (Casey 1993:175–177). The cultivation of place starts earlier than any tangible construction, and moreover, this situation develops because of human commitment.

Tumulus as place

> We do not dwell because we have built, but we build and have built because we dwell, that is, because we are dwellers. *Heidegger (1971:148)*

The human-made tumulus at Lofkënd does not alter the shape of the general topography in a significant way, but it is possible to contend that it situates the terrain by engaging the natural features of the surrounding land (Barrett 1999:255–256). This interpretation of place derives from Martin Heidegger's (1971:152) description of a bridge in his essay, "Building Dwelling Thinking." It is worthwhile quoting the passage at length:

> The bridge swings over the stream. . . . It does not just connect banks that are already there. The banks emerge as banks only as the bridge crosses the stream. The bridge designedly causes them to lie across from each other. One side is set off against the other by the bridge. Nor do the banks stretch along the stream as indifferent border strips of the dry land. With the banks, the bridge brings to the stream the one and the other expanse of the landscape lying behind them. It brings stream and bank and land into each other's neighborhood.

Heidegger contends that the interaction between the man-made bridge and the natural stream, banks, and land is a considered relationship. These elements are not separate, isolated units that are attributed to a preordained landscape, but rather it is our recognition of their relationship and mutuality that establishes the landscape in the first place (Tilley 1994:14–17, 1996:161). Therefore, human embeddedness—what Heidegger called "being-in-the-world"—is intrinsic to how we at first perceive and then define our surroundings as landscapes. This idea can be productively extended. Arguably, Heidegger's bridge does not participate in the definition of landscape only after it has been built, but instead there is a fore-understanding of place that must manifest before any move toward construction. Being attuned to the situatedness of the land is what allows and informs building. Crucially, we begin to formulate our relationships with the land as soon as we begin to inhabit it. Habitation, in this respect, is not confined to the idea of a formal or fixed settlement but instead is generated through dwelling. To Heidegger, this mode of being is existential and presupposes all acts of building: "To build is in itself already to dwell" (Heidegger 1971:146).

Dwelling suggests an attuned or sustained relationship with the land, but it is not limited to an exact period of time or precise terms of participation. It does not, in fact, require permanent residency or any residency at all; it can be merely wandering or passing through a region (Casey 1993:116). The significant point is that place is occupied, even informally, by someone or some group. This perspective has many contexts. Notably, Henry David Thoreau, himself a land surveyor, intrinsically understood how the idea of dwelling was bound by human intention: "Wherever I sat, there I might live, and the landscape radiated from me accordingly. What is a house but a *sedes*, a seat?" (Thoreau 1960:59). Because the act of dwelling is not restricted to a specific period of time, it applies to the present day as much as it does to the past. This is especially significant considering that the tumulus at Lofkënd still remains as a monument of sorts. From this perspective, we can begin to interpret its construction not as an embryonic act of occupation but rather as one facet of a long-standing tradition of dwelling in the Gjanicë River valley. The physical marking of the land and the building of the tumulus effectively confirmed through tangible means a pre-existing contract of engagement with the valley. This suggests, importantly, that neither the tumulus nor its site is neutral—*groundless*. There was a particular value attached to this location and it was chosen with intention.

Although it may be impossible to pinpoint conclusively why a collection of individuals decided to construct the tumulus at Lofkënd, it is tempting to surmise that there was a sort of modification in the nature of dwelling within the Gjanicë valley. This area was surely known (and plausibly used) for generations, but at some point the dwellers altered their relationship with the land. A loose or unbound topological mode of dwelling developed into one of greater permanence and specification that claimed a kind of identity with the terrain. From this point onward, it could be argued that the valley was a more concrete kind of dwelling place that not only maintained a more formalized reciprocity with its occupants but that also started to affect the identity of a group of people and the shape of the region (Ashmore and Knapp 1999:6; Basso 1996:55). The construction of the tumulus, although not a dwelling in a conventional sense, can be seen as a willful determination to cultivate identity and give it permanence, what Casey describes as an "architectonic and cultural concrescence" (Casey 1993:177). Lofkënd thus is understood as a landscape because it has been cultivated by human inhabitation. This does not signify that the meaning of it as a landscape has remained the same. As we shall see, it has probably expressed different things to a broad series of valley inhabitants. However, in each successive phase, its situation as a landscape is a result of the intentions of those who made use of the land (see Crumley 1994:6–8).

Heidegger's discussion of dwelling as a mode of being strives to define how we affirm place. By looking beyond the basic definition of the German verb *bauen* (to build, to construct), we can uncover some alternative meanings that suggest not only the physical act of construction but also dwelling itself (Heidegger 1971:146–147). Heidegger contends that these two activities, building and dwelling, are indivisible, but he also extends their interpretation by drawing a link between their abstract and material characteristics: "[T]his word *bauen* however *also* [emphasis in original] means at the same time to cherish and protect, to preserve and care for, specifically to till the soil, to cultivate the vine" (Heidegger 1971:147). For Heidegger, place exists namely because we dwell in it through cultivation. This situation is precisely what needs to be examined closely and clarified in terms of the tumulus at Lofkënd. We wish to understand the conditions by which the tumulus and its broader topography were cultivated as a place.

Cultivators of Place

A stratified understanding of cultivation

The synergy among place, dwelling, and cultivating is embodied by the Latin verb *colere* (till, tend). In "Building Dwelling Thinking," *colere* is introduced specifically to remind us how the act of building can be interpreted metaphorically in the sense of a preservation of place (Heidegger 1971:147). It must be underscored, however, that *colere* itself is not limited to figurative definitions but also encompasses literal meanings. Because this study is concerned with how people dwell in order to cultivate place, it is essential to understand cultivation as a stratified enterprise, extending from the pragmatic and mundane to the metaphorical and symbolic. For example, the most basic uses of *colere* are primarily agricultural, such as tilling soil and tending land. As Heidegger observed, the term also carries social meanings relating to human habitation, in particular the idea of settling, dwelling, and taking care of something. Beyond this, *colere* can express much wider, symbolic connotations—namely, to honor, revere, or even worship. Cicero notably fused the agrarian and figurative inflections of the term, proposing that the human mind must be cultivated in order "to fruit" (Cicero, *Tusculanae disputationes* 2.5).

Human beings are both propagators and beneficiaries of cultivation. Their physical toil provides nourishment, and this in turn anchors their relationship with the land. Ultimately, this reciprocity with the environment affords both community and reflection. It is possible to see all these meanings embodied within and around the burial mound at Lofkënd. Here the cultivators can be understood as both the living and the dead. Generally, within archaeology, the dead command most of our attention, but it is arguably the living who matter most in this portion of our analysis of Lofkënd.

Lofkënd: sustenance, settlement, and the consecration of place

Evidence of pragmatic cultivation in the vicinity of the Lofkënd tumulus is omnipresent and unmistakable. Today the burial mound is located among the crops of a farmer, and moreover, the valley as a whole provides the resources for basic nourishment: fertile soil, adequate terrain for grazing, and water. Although it is difficult to determine exactly how much the topography of this area has changed since antiquity, there are clues that this region was used as source of sustenance in antiquity (see Chapters 16.1, 16.4, 18). In other words, the living not only cultivated the valley at an early point in time, but they also continue to cultivate the land around Lofkënd in a very practical way by tilling the soil, planting crops, and maintaining pasturelands. Although these uses of the land are typical in everyday terms, they are essential to the establishment of place. Only with a foundational stratum of mundane and ordinary conditions can remarkable and paradigmatic meanings transpire.

Particular topographical characteristics of this portion of the Gjanicë River valley make it ideal for prolonged inhabitation. While the distance between the two main mountain ridges is several kilometers wide, the immediate region of the tumulus is more modest in scale. As described earlier, the burial mound is situated at the edge of a smaller, secondary valley that extends into the river and contains both slopes and floodplains. Furthermore, the Gjanicë originally served as a backbone for the entire region, simultaneously providing nourishment and physical orientation. This contained landscape closely corresponds to the kind of setting that geographers suggest is optimal for communities. As a "diversified ecological niche," the valley could have originally provided the essential amenities for subsistence while still maintaining relationships and connections with more distant groups of people (Tuan 1974:117).

It should not be overlooked that pragmatic cultivation always has the capacity to nurture more symbolic connotations. Farming, even in terms of basic subsistence, implies a commitment to the land as well as attention to a particular locale over a long period of time. Planting crops and keeping livestock literally ensures the provision of nourishment, but it also metaphorically puts down roots for a community (Casey 1993:173–175). Through their continual occupancy and marking of the land, humans also cultivate their inhabitation. In some regard, this is extremely fundamental. Individuals must take care of their land if they want to continue using and dwelling in it. It also means, however, that there is a kind of common, focused interest that builds a shared identity over time. This is community in a very basic sense. In other words, by cultivating the soil, dwellers are also cultivating their awareness of themselves and of their boundaries. They are acting as part of a collective, and hence they develop a common orientation (Tuan 1974). In this respect, size also matters. The burial tumulus may have belonged to a larger pattern of settlement that extended throughout the Mallakastër Hills, but its immediate setting is a well-defined "physiographic unit" that affords all the basic conditions for communal existence. The importance of this situation is twofold: because the valley belonging to the tumulus is modest in size, it is more likely to foster close ties to the land. In turn, these relationships can endure, thereby cultivating longstanding, personal commitments (Tuan 1974:101).

At some point, the meaning of the landscape at Lofkënd surpasses the pragmatism of subsistence and becomes something idealized. We can surmise that this region was developed as a dwelling place prior to the construction of the tumulus, and thus it is likely that early occupants grew accustomed to and identified with a broad and loose combination of features in the valley: its contours and scale, the location and shape of particular fields and slopes, passages, points of access, and views or perspectives. Some of these features are palpable, and others are intangible or changing, but they are all unequivocally genuine attributes. It is not necessarily a challenge to recognize these individual characteristics. Rather, the real difficulty lies in how to encapsulate and perceive them as a whole. In terms of spatial ordering, topographical place is inherently ambiguous. It is nearly always problematic to describe a landscape definitively and to determine the point at which a loose set of characteristics becomes understood as an ordered, delineated place. The answer perhaps can be found not so much in the decipherment of formal boundaries but instead in an awareness of how place can be given orientation. More will be said about the nature of boundaries in relation to Lofkënd in the final section of this chapter. At this point, we wish to reflect on how the mound is essentially an intervention, something that is fabricated by the human hand and imposed onto the landscape. Its placement establishes orientation because it is perceived as a fixed marker; not something that necessarily delimits or circumscribes topographical boundaries, but instead an object that radiates influence. The tumulus does not physically change the ground on which it stands. Instead, the mound changes our interpretation and perception of the surrounding land (Papadopoulos 2006:83; see also Barrett 1999:256; Thomas 1991:30). This situation is crucial and can be understood more fully by looking beyond Lofkënd. Scholars working outside of archaeology have also commented on how inanimate objects may claim dominion over a locale and establish place (Tuan 1977:162; Harrison 2003:19). More than once, Wallace Stevens's poem *Anecdote of the Jar* has been used to illustrate this phenomenon (Stevens 2006:66–67). It is worthwhile quoting a stanza from the poem in order to clarify how it relates to the present discussion:

> I placed a jar in Tennessee,
> And round it was, upon a hill.
> It made the slovenly wilderness
> Surround that hill.

Tuan notes that the jar, like a sculpture, can in certain circumstances transcend its inanimate condition and "incarnate personhood" (Tuan 1977:162). The jar is lifeless and yet it is precisely this quality that transforms it into a "metaphor for human striving" (Tuan 1971:184). It becomes a representation, a symbol for the determination of an individual or a community to create place and establish authority over a particular locale. We can understand the jar in Stevens's poem as analogous to the burial tumulus at Lofkënd in two different yet connected ways. The mound has the capacity to lend tangibility to ephemeral things, such as human intentionality and identity. By literally standing for inhabitation, it can be seen as a form of architecture that "localizes caring" for the land and represents those who dwelled in the valley (Casey 1993:175). Beyond this, the tumulus is everlasting, or it at least transcends the finitude of those responsible for its construction. This suggests that it consecrates not only place but also time. Ultimately, however, we must consider how the situation at Lofkënd is a great deal more complex than Stevens's jar and Tennesseean hill. Far from being an empty vessel, or even an earthen sculpture, the burial tumulus is a receptacle for the dead. The relationship between its contents and physical location deserves to be considered in full.

Tumulus as repository

The Lofkënd tumulus is first and foremost a monument to the deceased. While it is imperative to evaluate its broader context and conditions, our study must eventually return to examining its intrinsic value as a container for the dead. To this end, there are a number of factors associated with burials in general that warrant reappraisal. Tombs, graves, and memorials to the dead are cultural universals. Because they are profound and yet a basic aspect of every community, they are also understood to be common. The tradition implied by graves means that they are typical, and perhaps this is why their deep-seated messages and contributions are frequently overlooked. In a very literal way, a marked burial serves the dead precisely because it is a repository for their remains. That these containers promised accommodation—a home—for the dead is attested by the earliest cultures (Colvin 1992). Apart from being a dwelling, a grave is commonly seen as a mark of respect, for the dead deserve an honorable burial. These conditions both focus on how graves serve the dead—what they offer those who have passed. They are undeniably important, but the gravity of their reverence can divert our attention away from exploring how graves can serve us, the living.

We have no evidence to suggest that people died at the Lofkënd burial tumulus, and thus it is unlikely that this place marks profound—or even particular—moments in people's lives. A more plausible hypothesis is that individuals were interred at Lofkënd as a kind of memorial, a remembrance of the past that specifically benefited the living. Burial tumuli, like monumental tombs, are purposefully visible for our sake, not for the sake of the dead. It is worthwhile mentioning that one Greek word for grave, *sema*, is also translated as "sign" (Harrison 2003:20; Thomas 2007:166). A grave must be visible so that the living may recognize it. As David Lowenthal has pointed out, although memorials refer to those who lay beneath them, "the marking function is no longer consequential once bodies have moldered into dust or have been removed to make way for others. . . . Cemeteries matter less as repositories for the dead than as fields of remembrance for the living; the unmarked grave goes unseen" (Lowenthal 1979:123). However much the tumulus builders may have perceived their act as a form of respect for their ancestors, the mound ultimately provided solace and reassurance for those who survived.

Because it is deliberately visible and durable, the burial mound embodies a twofold promise: it ensures that the living would have a place to recall their ancestors, a kind of *aide-mémoire* that could trigger recollection. Its physical presence guarantees that the living could connect with their forebears in a tangible way. Furthermore, the monumentality of the burial imparts permanency, suggesting that the memories of the dead could be retrieved not merely in the immediate aftermath of interment but also in perpetuity (Bradley 1985:9; Harrison 2003:39). As a veritable "mound of memory," the tumulus contained the literal and metaphorical roots of the community, or communities, who used it (Papadopoulos 2006). It also is important to take into account how the mound was not built at once but rather grew through accretion over time and, crucially, was stratified. This means that the earliest graves may have been understood as a kind of origin or lineal foundation (cf. Barrett 1999:261–262; see also Chapter 21). The characteristics of one particular grave may attest to this interpretation. Tomb I (Grave 64) is one of the most complex burials in the mound, containing both human and animal remains (see Chapters 3.1, 6, 16.1, 21). Moreover, it appears to be both centrally located and perhaps the earliest of all the graves in the tumulus. While the specific identity of the persons interred here is clearly unobtainable, the grave itself nevertheless raises an important issue concerning symbolic function. Whether this grave was in some way primogenitary, thereby embodying a kind of paradigmatic ancestry, will be given further consideration in the final section of this chapter.

We can be more specific about the role of memory at Lofkënd by looking at it from the point of view of the mound's topographical context. The likelihood that the immediate valley surrounding the tumulus generated strong ties of identity and even ownership was discussed earlier in this chapter. It was suggested that the tumulus provided orientation for those who were habituated to this particular area of the Gjanicë River valley. In view of this, the role of the burial can take on a different dimension. It was not only a repository of memories of the dead, a place to commemorate and recall ancestors in general. It was also potentially a container for memories that were directed toward an inheritance of this particular place: the hills, terraces, pasturelands, and river. If this valley was a locus of identification for dwellers—the living cultivators of the land—then we may be able to hypothesize that the dead, as tumulus dwellers, also assumed some of this meaning.

In a way, the dead, although silent and inanimate, are a kind of community. Unlike the living, however, they have a remarkable staying power. Notably, the tumulus does not preserve the past itself but what the past stood for—a particular social collective (Lowenthal 1979:121). In other words, the dead provide reassurance. By dwelling perpetually in a singular place, they assert continuity (Harries 1998:310). If this tumulus does indeed mark a territory that was once controlled by a specific group, then the presence of ancestors attests to that claim. Most importantly, the dead transcend the restrictions that inevitably challenge the living. Because their presence is not finite, they effectively promise to inhabit and oversee the land forevermore. This symbiotic relationship among the tumulus, its contents, and setting may further illuminate the choice of location for the burial. Once the Gjanicë River valley was a locus of sustained cultivation, the tumulus may have acted as a landmark for the immediate region. Moreover, if those who inhabited the valley were also responsible for the construction of the tumulus, then the burial would have been more than a topographical feature that helped establish location. It also would have staked a territorial claim. Because it juts forth from the hill on which it is positioned, the mound can be perceived from considerable distances. This is further emphasized by the fact that it appears more conspicuous and dominating when viewed from the river, which would have been the principal access route into the valley in antiquity. Such a situation deserves close study, for it implies that the tumulus served as a kind of signpost for anyone entering or passing through the region, sending a message that they were within the realm of a claimed territory (cf. Bradley 1998:51; Harries 1998:293; cf. more generally, Tilley 1994; Van Dyke 2003a, 2003b, 2008). The dead, as Robert Harrison has argued, can assert extraordinary authority over the land: "The surest way to take possession of a place and secure it as one's own is to bury one's dead in it" (Harrison 2003:24). It is also worthwhile pointing out that both ancient and modern cultures have used the dead to claim political influence over territories. Harrison describes how Virgil's Aeneas buries the dead at calculated points in his journey in order to ensure that such places could be claimed by Rome in the future (Harrison 2001:401, 2003:27). In a wholly different chronological context, Ralph Waldo Emerson argued in his lecture "The Conservative" that property rights are influenced by reverence for traditional use and ancestral names (Ferguson 1971: 184–200).

From the point of view of the dead, the tumulus can be understood as an expressly intentional repository. The deceased were interred deliberately and for specific aims; and yet they were not, of course, the only objects contained within the mound. To some extent, it is reasonable to argue that the entire mound itself is purposeful because it was wholly fabricated as a monument through human endeavor. However, two particular categories of tumulus fill at Lofkënd deserve further investigation. The excavations brought to light large amounts of chipped stone tools (lithics) and a rough, fired clay material. The latter has been described as daub because much of it retains impressions of plant forms (reeds, twigs, straw, rods, etc.), thus suggesting it was used as a plaster in wattle-and-daub (see Chapter 14). The presence of both the lithics and the daub is perplexing, not entirely due to their appearance but rather to their abundance in such a context. Because these materials were found in unusually large quantities, it is worth speculating on what they may reveal about the occupants and builders of the tumulus.

The situation posed by the lithics has been discussed elsewhere (Chapter 13; see also Papadopoulos 2006:81–83). Most of these finds appear to pre-date the tumulus, and more than a few come from the Middle Paleolithic period. An archaeological survey of the area around Lofkënd has identified concentrations of Middle Paleolithic tools and debitage within the nearby vicinity of the tumulus (Chapter 18). At this time it is impossible to determine exactly why the lithics were deposited in the burial mound, although some may have reached the mound along with specific soils selected for their special properties (clay, compaction, color, density). Their presence, however, has led to further speculation about the relationship of the tumulus builders with their more ancient predecessors. If the stone tools were deliberately included in the burial, then it may be possible to ask whether they were understood to refer to a primordial and venerated past. This recalls another matter previously discussed, that of dominion over place. The physical presence of the dead may have legitimized control over a particular locale, but whether this interpretation can be broadened to include ancestral tools and materials remains a compelling yet open question. This is especially relevant for the case of the daub.

The presence of the daub, like that of the lithics, is enigmatic because of its sheer volume (Chapter 14). Both materials clearly had former uses that were separate from the burial. While the lithics are inherently mobile tools, the physical characteristics of daub imply that it once had a stationary and yet ultimately more ephemeral function. Although we can only speculate on its exact original use, what can be maintained with confidence is that it was expressly a building material. This gives rise to the interesting possibility that these fragments could be the traces of architecture, specifically houses or shelters. Wattle-and-daub is one of the most simple and straightforward building techniques, and it can be used to construct both walls and roofs. It is common to many cultures and can still be seen in use today in Albania (**Figs. 14.1–14.4**). If the daub included within the tumulus fill at Lofkënd had architectural origins, then it would be reasonable to ask whether its presence was also curated and possibly even held symbolic connotations. Although any hypothesis must be understood as conjecture, it is tempting to infer that this material could have been symbolic of a lost, impermanent form of dwelling that was preserved in an eternal dwelling, a house of the dead. This proposition can be extended by questioning whether the daub referenced specific communities rather than simply dwellings in general.

One way of developing our hypothesis would be to question whether the building material had originally composed part of the earthly homes of those interred in the tumulus. In this situation, portions of the dwellings belonging to the deceased would have been collected to accompany them in the afterlife (cf. Gosden and Head 1994). This would mean, therefore, that the architectural traces contained by the tumulus were transported not only from a separate physical location but also from a different period of time. Although their integration into the mound was perhaps intended to benefit the dead, it more plausibly served the living. The physical remains of the dead themselves can shed influence over a particular territory, but the deposition of their homes—a fundamental demonstration of proprietorship—can suggest even greater agency. By including these fragments in the tumulus fill, the ancient inhabitants of the Gjanicë River valley may have been able to confer the authority of their ancestors across both topographical distances and over long spans of time.

A curious attribute of the daub found in the tumulus is its physical condition: it has been hardened through a firing process, and although it cannot be defined as pottery, it appears that it has been preserved in a manner similar to pots. One possible explanation for this may help bolster the hypothesis that the daub once belonged to dwellings, or at least some form of shelter. An analysis of sintered daub excavated from a Neolithic settlement in Calabria, Italy, has shown that the hardened clay material was likely fired purposely by its creators (Shaffer 1993). The study assessed a number of experiments that tested the conditions and thresholds by which daub can sinter. Most of these tests involved igniting sample wattle-and-daub buildings to gauge how long a structure would need to burn in order to sinter large quantities of daub. Notably, such fires failed to produce substantial amounts of fired daub, even when they were allowed to smolder over long periods of time (Shaffer 1993:60–62). It therefore appears more probable that the clay material was deliberately fired in controlled environments at high temperatures. What is especially relevant for this present study is the hypothesis that sintered daub was originally preserved for pragmatic reasons: "[T]he Neolithic residents might have intentionally burned dilapidated wattle and daub structures so as to harden the daub and preserve it for incorporation in the walls of new buildings" (Shaffer 1993:62). Recycled, fired daub would not only lend increased durability to new structures, but it would also reduce the amount of new, fresh daub that was required for building. What this means in terms of the tumulus at Lofkënd specifically is still a matter of debate, but it does provide an attractive interpretation for why the daub had been fired. One way of building on this analysis may be to look at how the daub changed meaning from the point of view of preservation. For example, it is conceivable that this material initially was sintered for reasons of efficiency, perhaps for the construction of shelters, and possibly even for dwellings that were nearby to and contemporary with the tumulus. Later, the daub may have been recycled yet again and permanently preserved as tumulus fill. From this perspective, we can posit that the daub fragments are not simply associated with the deceased themselves. Instead, they have a broader frame of reference. If the daub is interpreted as the remains of architecture, then it may be understood by extension to be the last trace of an all but vanished settlement. Even though such a community was ultimately ephemeral, its permanence was materialized by the construction of the tumulus.

As a repository for the deceased, the Lofkënd tumulus by extension also contained the living community and their land. Through its physical position it gathered and bounded place, validating access to the valley as well as its resources (cf. Buikstra and Charles 1999:207–208). This means, though, that in drawing the landscape into its presence, the burial mound also excluded other people and places. Any container creates a relationship between what is bounded and what is outside—excluded. When examining the burial mound in detail, it can be easy to lose track of the fact that it plays a role within a broader topography. Most importantly, its location and physical appearance allows it to be seen from numerous directions as well as from other ancient sites, in particular the tumulus at Mashkullorë in the immediate northwest. These characteristics raise two important topics that have not been addressed in their entirety. The first entails clarifying the conditions that alert us to the tumulus's role as a kind of cynosure. Beyond this, we must also consider how this burial mound stood in reciprocity with other places.

BOUNDARIES AND BOUNDEDNESS

Between a container and its contents

We may refer to the tumulus at Lofkënd as a gravesite, but it is not simply an area of land, a plot set aside for the deceased. The fundamental and lasting value of this place is wholly contingent upon its exterior appearance. In contrast to the hill on which it stands, the mound swells and bulges, projecting upward and outward from a slope (**Fig. 20.12**). The profile of the mound immediately suggests its role as a kind of vessel, and although it is essentially composed of natural material, its shape discloses that it is human-made. It is exactly this juxtaposition of contours that draws attention to the tumulus and signals its presence, thereby communicating to us that it is architecture. Adolph Loos observed this elemental condition of building and making when he remarked that "[i]f we find a mound in the forest, six feet long and three feet wide, heaped up with a spade in the shape of a pyramid, then we become solemn and something tells us: somebody lies buried here—this is architecture!" (quoted in Harries 1998:292; see also Safran and Wang 1985:104–108). Loos's comment underscores the importance of deliberate construction in our interpretation of architecture. The tumulus at Lofkënd may outwardly look very different from a mausoleum composed of cut stone, yet both are intentional repositories for the dead; neither expresses this essential quality more than the other.

By referring to the Lofkënd tumulus as a repository, we unconsciously acknowledge a binary condition: an interior and exterior. Although we first become acquainted with the burial mound by virtue of its external appearance—its profile and volume—the greater part of our attention concentrates on its contents. The mound was intended to hold the dead in one place, presumably for time eternal, and the excavation subsequently recorded in detail not only the remains of deceased but also their belongings. Within such a focused study, however, it can be easy to overlook how the contents themselves—for example, the bones and associated finds—are not the physical repository, even though they directly inform its shape and size. This situation is more complicated than it initially appears and therefore deserves further consideration.

In the first instance, it seems logical and straightforward to describe the burial mound as a container because it clearly surrounds and retains a number of objects. However, defining the precise nature of this place as a receptacle becomes much more difficult once we try to be specific about the difference between the container itself and what is contained (Casey 1997:50–63). Unlike other repositories of the dead, such as mausolea or catacombs, the tumulus was never hollow, and it was incapable of being emptied without losing its physical integrity. It is also important to remind ourselves that the dead and their associated belongings were not the sole contents of the burial mound. Apart from these finds, the tumulus contained a substantial amount of other material, such as stones, soil, and daub (for the latter, see the above section). In archaeology, the word used to refer to this material is "fill," a term that clearly reinforces the idea that the tumulus has both an inside and an outside. It is relatively unproblematic to describe the fill that is located deep within the tumulus as part of the interior, but as we move closer toward the surface, it becomes increasingly awkward to discriminate between the fill and the outermost layers of the burial. These elements appear to be coextensive. In other words, at the extremity of the tumulus, the surrounded material paradoxically seems to be integral—even identical—to the very thing that does the surrounding (Casey 1997:54–55; see also Aristotle, *Physics* 211b10–14). It is difficult to reach a solution to this philosophical problem, but a compromise might be attained by perceiving the bulk of the fill as separate from the outermost layer of the burial mound. More specifically, the extremity of the tumulus could be interpreted as a discrete composition of soil, stones, grasses, and their roots, which together provide an outer barrier—a kind of crust or membrane—for the repository as a whole. However, it should also be mentioned at this point that the location of some burials further complicates efforts to distinguish between the outside and inside of the tumulus. A number of graves are located not deep within the center of the mound but rather to the southeast, beyond the main area of the bulge. This condition raises important questions about how we interpret and describe the nature of boundaries at Lofkënd in general.

The Lofkënd tumulus has no formal, explicit boundary, but its surface, in marking the transition from the inside to the outside, does form a kind of enclosure. By "holding in" the deceased, it expresses a sense of "boundedness" (Casey 1997:54–55). The shape of the tumulus is a contour—that is, an outline. A contour always serves in the first instance the feature it modifies: a hill, lake, or valley. It is intrinsically something that bounds a place, but by virtue of its inclusiveness, it also serves to keep other things at bay. At Lofkënd, the outer surface of the mound preserves the conditions as well as the contents of the burial. Yet this limit is not one-sided, for as much as it retains the tumulus's contents, it is also the place from which the burial engages with the world. The double-sided nature of boundaries also interested Heidegger, and in "Building Dwelling Thinking," he argues that "[a] boundary is not that at which something stops, but as the Greeks recognized, the boundary is that from which something *begins its presencing*" [italics in original] (Heidegger 1971:154; see also Casey 1997:63). Here boundary is understood as a form of enclosure, a containment that articulates a place or a thing. Within this containment, however, there exists a constant and mutual relationship with the edge. It contributes to the ordering of a place and yet also seems to represent it. Crucially, the boundary is not a third, independent region between two realms but instead a place qualified by a threshold (Eliade 1957:181). The two Greek terms central to Heidegger's argument are *peras* (end, limit, boundary) and *horismos* (boundary, limit). While similar, each word is characterized by a particular set of conditions. *Peras* suggests a close and yet deferential relationship to a thing or place that is contained. In other words, it is primarily focused upon and even included within whatever is limited. Casey uses Proclus to clarify this situation: "[T]he limits surrender themselves to the things they limit; they establish themselves in them, becoming, as it were, parts of them and being filled with their inferior characters" (quoted in Casey 1997: 63). *Horismos*, in comparison, is attuned not only to what it marks out, but also what lies beyond, in the distance, outside of what is contained. It associates, or even negotiates, with the limited place, rather than wholly belonging to it. This can be clarified by considering the word "horizon," which embodies characteristics of a boundary: It is a *skyline* but not a place of ultimate finality. Hans-Georg Gadamer maintained that the horizon, in being constantly in motion, conveys a sense of possibility and imminence (Gadamer 1989:301–307).

Beyond the tumulus

Throughout *Anecdote of the Jar*, Wallace Stevens remains oblique about the possible contents stored within the vessel. It is eventually revealed in the third and final stanza that the jar is "gray and bare," thus suggesting that it is empty or unfurnished of goods. By this point, however, the jar's interior contents are not a significant concern for the reader. In fact, its predicted emptiness is a potent foil to the uncompromising authority that the jar holds over its surroundings (Carroll 1987:36–37). Starting early in the first stanza, Stevens constantly emphasizes the container's commanding external influence through repetition and alliteration of the word "round"

(Leggett 1992:199–209). As a result, the whole poem embodies a metaphorical sense of circumambulation, as if all the lines of the verse wrapped around the hill where the jar was placed (Carroll 1987:36). This further reinforces the locative presence of the vessel and especially its ability to impose order and control over its environs. While it may contain only air, it manages to possess everything beyond it. Finally, we are told by Stevens that it simply "took dominion everywhere." Harrison relates the jar to architecture, not merely since both are man-made, but rather because each "domesticate" place (Harrison 2003:19). The Tennesseean countryside is at first a receptacle for the jar, but the presence of this vessel eventually creates, and then orientates, the landscape.

Because the poem is figurative, it can suggest that the jar holds unbridled authority, even omnipotence, both in time and space. A burial tumulus, however, accords with parameters. An earlier discussion in this chapter surveyed how the Lofkënd tumulus could have played a significant role in the ways a community forged an identity with and claimed possession of the land. A close reading of the burial in its setting suggested that it maintains a strong visual reciprocity with a certain section of the Gjanicë River valley. In overlooking a modest topographical niche, and by virtue of being a burial, the mound figuratively and literally bounded its immediate environment. But the question that has not been addressed in its entirety is how this tumulus could also have kept other things—places and people—out of bounds.

By visually calling attention to itself the Lofkënd tumulus serves as a directional landmark at the same time as it symbolizes a common interest. In other words, it is a focalization of place and memory for certain people. Although this study already considered how the burial may have distinguished the territory of a particular community, it is also crucial to assess, by extension, how the mound also could have differentiated this specific place from other domains, perhaps even the property of other groups. Barbara Bender detects that landscape in general can be dualistic or even polysemous in character: "A place inflected with memory serves to draw people towards it or to keep them away, permits the assertation or denial of knowledge claims, becomes a nexus of contested meaning" (Bender 2002:104). Up to this point, our investigation has been limited to the immediate context of the tumulus, both in terms of its physical location and the individuals who were closely associated with its use. However, there is also potentially a great deal to learn about this place by looking at it from a broader topographical perspective. More specifically, the tumulus has never been nowhere, wholly detached from and irrelevant to the rest of the world. In asking who and what was excluded from its domain, we instinctively begin to widen the context of our investigation, and part of this discussion requires a shift in approach. Although we have considered how the tumulus is a place that *overlooks* a certain region, we must also investigate how it is a site that could have served as a *lookout* for a particular territory.

It is difficult and probably injudicious to delineate the extent of the tumulus's dominion with geometric precision, but the absence of formal boundaries does not necessarily mean that Lofkënd lacked territorial limits. Some topographic features that were described early in this chapter can also be interpreted as boundaries. A relatively straightforward example would be the Gjanicë River, which may well have served as a natural borderline in the south. Closely related to this is the narrow pass or gateway between the river and a moderate peak southwest of the tumulus. The physically restrictive character of this place makes it an ideal boundary. Beyond these two examples, it is difficult to detect the boundaries of the burial mound with any degree of certainty. It is possible that the line of the smaller crescent-shaped basin between the tumulus and river served as a natural perimeter. Beyond this, it would be reasonable to assume that some boundaries, such as field paths or stones, have not survived. Although these observations must be understood as conjecture, they become more substantial when considered in relation to a different kind of evidence that was mentioned at the beginning of this study: other ancient sites, especially nearby tumuli.

There are at least two other known tumuli in the general vicinity of Lofkënd, those of Mashkullorë and Patos. The example of Mashkullorë (**Fig. 18.4**) is especially significant for this present investigation because it has a clear, unimpeded view of the tumulus of Lofkënd. Even though it remains unexcavated, the presence of Mashkullorë triggers important questions about the positioning of burial mounds in the valley. Like Lofkënd, it is situated in the valley between the river and a ridge of the Mallakastër

Hills. The view from the tumulus is also expansive, especially toward the river basin in the southeast. Another similarity between the two that becomes apparent when studying a contour map is the fact that each is situated at the edge of a moderate ridge (**Fig. 20.1**). It is tempting to deduce that these parallels were intentional, as if the tumulus builders had purposely followed similar approaches or requirements for siting their burial mounds. This situation underscores how Lofkënd itself was bounded and qualified by some kind of perimeter. At this point, it is impossible to reach any definite conclusion for these observations, but the similarities between the two tumuli do permit us to think beyond the perception that burials were the domains of isolated communities. Ultimately, the presence of Mashkullorë discloses that Lofkënd was neither a place of unqualified solitude nor a landmark that possessed limitless territorial authority.

While it is plausible to describe Lofkënd and Mashkullorë as neighbors, in reality they are separated by several kilometers of hilly terrain. They maintain an association largely because of their shared sight-line, not because they are immediately accessible or especially close. In fact, their placement in the river valley appears to underscore their value as landmarks. Because they are simultaneously separate and mutually visible, each is able to distinguish and affirm a particular terrain. This relationship seems to suggest a kind of posturing, as if the burial mounds were initially constructed to assert territoriality as well as to prevent encroachment between two similar yet distinct communities. Perhaps this hypothesis could serve as a future point of departure for new studies into the prehistory of the Mallakastër region. It would be especially interesting to consider whether ancient communities used monumental tombs not only to ensure their long-standing dominion over particular areas of land but also to delimit, survey, and defend their territory from that of their neighbors or rivals. This would imply that boundaries between the tumuli had the potential to embody negotiations as well as conflicts between groups of people.

This proposition is noteworthy in light of the fact that regions of Albania call upon the dead to establish boundaries even in the modern period. A conversation between travelers in Ismail Kadare's novel *Broken April* elucidates this tradition: "Those are the *muranës* that the innkeeper mentioned. . . . They served as boundary marks between fields or property lines. . . . That's what the *Kanun* says. 'When a death occurs during a boundary dispute, the grave itself serves as a boundary mark'" (Kadare 2003:98; see also Chapter 21). Even though Kadare's narrative is fictional, it reveals insight into the culture of mountain communities in northern Albania. The people of this region live in strict accordance to an ancient law code, the *Kanun*, which stipulates that the graves of those who perish through family rivalries—blood feuds—take up secondary meanings as boundary markers (Hasluck 1954:95–102; Chapter 21). These graves, called *muranës,* embody both the person and their family. This suggests that there is a long tradition of correlating death, burials, and boundaries in Albania. The grave boundary speaks to an immediate present, of the need to preserve customs and uphold order between clans. But another more metaphorical meaning is simultaneously present in the *muranës*. These markers also allude to boundaries on a vast and occasionally unfathomable scale: the limits of time.

Dialectics of boundaries and time

Earlier in this chapter, the tumulus at Lofkënd was described as an intervention. In formal architectural terms, the mound is added to the land and hence contributes to a landscape. In terms of time, though, the tumulus signifies a different kind of intervention. It changes how we may measure time, and in some cases it can also be interpreted as something that represents a rupture of time. Irrespective of human presence, nature exists in self-renewing cycles. The seasons are the best example of cyclical time, and with their passing they bring life, death, and regeneration. This process continues with the presence of humans, and it can even be argued that it endures despite our attempts to stymie or curb it. Beyond this, we are also responsible for introducing other kinds of time alongside the cycles of nature. Two forms of temporality are particularly relevant to this study: the finite and the eternal.

Whereas the tumulus is a physical intervention, our own presence interrupts the cyclicality of nature and brings to the fore an awareness of our mortality (Harrison 2003:19). For us, time is limited and has absolute boundaries. Because we are constantly confronted with and reminded of the restrictions of time imposed upon us, we tend to see the world linearly. The past is historical and can be measured, divided, and classified, and although the future can in some respects be planned or mapped, it is ultimately non-negotiable. One way of reconciling our inevitable finitude is to mark out instances or intervals of time and attempt to give them permanence. By concretizing specific points or periods, we commit them to memory; we memorialize them. Graves, of course, are probably the most common forms of memorials. While they may stand as solemn, even meditative, testaments to the lifetimes of those who have passed away, they can also embody conflict and struggle. Notably, the *muranës* described in Kadare's novel straddle different modes of time. They mark a rupture in time, the sudden and usually violent death of a kinsman, yet they also testify to a seemingly never-ending struggle between families as much as they memorialize an individual. These grave markers attempt to turn raw, fleeting moments into definable and lasting acts.

Like the *muranës*, ancient burial tumuli are in the first instance practical and conspicuous markers. They indicate through their shape and size the presence of the dead. Beyond this, they could also been seen as efforts to formalize a fragile or uncertain condition. If the Lofkënd and Mashkullorë tumuli represented the territories of rival groups, then their presence would have signified the endeavor to safeguard areas of land. It is important to recall that tumuli can be multivalent. Their construction is an act of reverence for the deceased, and they can help recall or communicate with the dead. Moreover, the dead themselves can be used to legitimize a claim to the land at the same time as they promise to hold dominion over it well beyond the temporal limitations of the tumuli builders. The tumuli exist because of death, and yet they are death-defying.

Harrison describes burials as nexus points, places where cyclical time, human finitude, and eternity coexist: "As the primordial sign of human mortality, the grave domesticates the inhuman transcendence of space and marks human time off from the timelessness of the gods and the eternal returns of nature" (Harrison 2003:23). However much the burial mound refers to human ambitions and finitude, it ultimately is also unmoored from time. It points toward an unfathomably vast temporal context—eternity. We know very little about the religious practices of the tumulus builders, but we can make some observations about the possible sacred connotations of the mound by reconsidering its physical context. One important thing to underscore is that the tumulus was not built at the highest point in the valley. The same is true for the other known tumuli in the vicinity, Mashkullorë and Patos. Instead, they all were situated between the tall ridge of the Mallakastër Hills in the north and the lowest point of the valley, the river. Although it is impossible to reach any conclusion about this with certainty, it is tempting to surmise that the mountaintops were reserved for the divine (Tuan 1974:118). This would mean that the tumulus stood both physically and metaphorically between the domain of the living and the heavens. It therefore can be described as an *axis mundi,* a symbolic center and meeting point of the profane and paradigmatic (Eliade 1954, 1957).

The shape of the burial mound is also significant and necessitates further comment. Sacred architecture often mimics mountains, most certainly in an effort to communicate with the heavens but also to provide a visible focus for a community. One merely has to think of the pyramids rising to the west of the Nile, or the skyline of the medieval town of Chartres, to understand the ability of vertical architecture to "conquer the scattering power of open space" (Harries 1998:184). Despite its relatively modest scale, the Lofkënd tumulus nonetheless embodies centuries of human striving through accretion and reuse.

Another way that the tumulus might have expressed an attempt to harness paradigmatic time was through the worship of a principal grave. Tomb I (Grave 64), mentioned earlier in this chapter, is distinguished both by its central position in the tumulus and its early date (fully discussed in Chapter 21). Although many questions remain about the role of this tomb, its distinctive properties may point to an archetypal meaning (further explored in Chapter 21). If the grave was originally intended to be foundational, then the deceased buried within it could have been valued as primogenitors, the earliest ancestors of the community. Tomb I would have symbolized the very beginnings of a lineage, thereby strengthening later claims to the land itself. Although the identities of those interred in this tomb are lost to us, the special attention they received in death suggests that they may have possessed uncommon status during their lifetimes. Perhaps these individuals were exemplary figures who, over time, became mythicized and metamorphosized into archetypes for the community (Eliade 1954:42). The value of this kind of transformation should not be underemphasized, as it would have played a crucial role in the construction of identity for the early inhabitants of Lofkënd. The mythicization of an important ancestor, or group of forebears, would eventually enable a community to trace their territorial inheritance not just to the distant past but also to the very beginning of time.

We cannot say when the Lofkënd tumulus ceased to represent the specific ambitions of its builders. Even though the burial mound has proven to be remarkably durable in physical terms, the particularities of its original significance have not survived, and it is unlikely that they will ever be recovered. Put categorically, the tumulus has outlived the patronage of its particular culture (Tuan 1977:162). This does not mean, however, that the wider message embodied by the tumulus was terminated with the passage of time. On the contrary, some characteristics of this feature in the Gjanicë River valley suggest that it has assumed new roles and meanings, even in the present day.

Conclusion: Afterlife and the Persistence of Value

Given its rich ancient past, it can be easy to overlook how the tumulus at Lofkënd still plays an active role in the landscape of the Gjanicë River valley. At a very basic level, we can contend that the mound maintained a "presence in human consciousness" even through periods when no new burials were interred: physically, it was difficult to destroy and thus it endured (Bradley 1998:71). It is also worthwhile mentioning that the tumulus was never robbed. The exact reasons for this are open to conjecture, but this condition is highly significant, because it allows us to question whether the burial mound was valued long after the memories of its builders had faded. It is unlikely that we will ever know for certain whether all communities associated with the valley knew that the mound at Lofkënd was a burial. However, we have one intriguing clue that suggests this site maintained value as a sacred place throughout many periods. Below the tumulus, on the very slope of the bluff on which it was built, stands a modern Muslim cemetery (**Fig. 20.13**). When the tumulus is viewed from the village of Ngrancija in the southeast, the scattering of white tombstones that make up this modern cemetery are distinctly visible (see above, **Fig. 20.4**). They appear just below and to the east of the prehistoric burial mound yet clearly above the modern village. The two burial places are physically separate from one another, but their positioning implies that the older, more ancient gravesite atop the bluff corroborates an aura of sacredness and respect. This situation is especially interesting given that it does not seem to be a coincidence: just below the neighboring tumulus of Mashkullorë stands another modern cemetery (**Fig. 20.14**).

An earlier study of Lofkënd addressed the present-day conception of the tumulus by its surrounding community (Papadopoulos 2006). We know for certain that the residents of the villages nearby Lofkënd understood the mound as a burial site prior to the archaeological excavation. However, what they were not cognizant of was its antiquity. Instead, they conceived of it as the final resting place for foreign soldiers who fought in wars that were no longer part of popular memory: either the First World War or the Balkan Wars (Papadopoulos 2006:78). This kind of distance—both in cultural identity and time—is especially significant, because it is precisely the unknowingness of a place that makes it open to interpretation. Further to this, it is crucial to distinguish between the basic function of the tumulus and its specificity. In other words, the monument's meaning as a gravesite persisted over time while the particular importance of its contents waned (see Tuan 1977: 164). Certainly, in the modern period at least, the identity of those interred was irrelevant; what mattered most was that people were buried there in the first place. As such, the collective presence of the dead made the tumulus a kind of relic that "stood simply as a mute testimony to an unchangeable order" (Barrett 1999:263).

Although it is a monument, the Lofkënd tumulus is not monolithic in its meaning. Over the course of its long life span, its significance has changed and also widened. The tumulus builders may well have regarded their effort as an attempt to commemorate their own dead. If this were the case, then the mound at its early stages would have evoked memories of the deceased. Memory, however, is a great deal more difficult to preserve than a tangible monument, and therefore it is unlikely that the recollection of specific individuals could have lasted beyond a few generations (Eliade 1954:42–43; Thomas 2007: 166). But in some situations, when the particular details of individuals grow faint in popular memory, a new form of recollection can take their place. A historical figure may transform into hero or legend, what Mircea Eliade refers to as a "mythical model" (Eliade 1954:43). In this situation, the memory of the past is still very real, yet it is embellished or reworked, often to help strengthen a community in the present as well as for the future (Alcock 2002).

Through archaeology, we can determine with relative precision the periods when Lofkënd was an active burial site. Nonetheless, it is still impossible to estimate how long the mound carried any symbolic links—however tenuous—with its point of origin. What can be said, however, is that at some stage this burial place transformed in meaning. It went from being a site that referred to memories that were rooted in particular people or events, to being a place that evoked the veneration of the deceased more generally. For now, it is difficult to establish how Lofkënd was perceived symbolically in the early modern period. The details of those who reopened the tumulus as a burial site during that time remain enigmatic. But their choice of this place as a burial is compelling. It implies that the tumulus not only signified death but also conveyed a sense of deep-seated reverence. Ultimately, all who used the tumulus were connected through their actions, even though they held little or no knowledge of each other.

Once the archaeological excavations revealed the age of the Lofkënd tumulus, the site gained new significance within the nearby community. In learning of the mound's ancient past, local residents developed not only a wider perspective of this place but also a sense of pride (Papadopoulos, Bejko, and Morris 2008). Although the monument had long been a familiar aspect of their landscape, it had also been interpreted as something foreign prior to excavation (Papadopoulos 2006:78). Now it is seen instead as a tangible reference to the distant antiquity of their own country. And, once again, it is a focal point of identity in the community. This has been significantly enhanced through the reconstruction of the tumulus (Chapter 22). Following the completion of the excavation, the entire mound was rebuilt in its original place and to full scale, largely using mud brick to rebuild the excavation baulks (Papadopoulos, Bejko, and Morris 2008; Chapter 22). It now stands as a monument to the past as well as being an important contribution to the development of archaeological interpretation and heritage in modern Albania.

The construction of the burial tumulus at Lofkënd in antiquity formalized, in physical terms, a well-established human commitment to the Gjanicë River valley. Although many questions remain about the tumulus builders, we can perceive them as valley dwellers. They surely identified with this area and its wider topography regardless of whether they formally resided in the immediate vicinity of the burial mound. These individuals were cultivators of place, and their choice of location for the tumulus reflected their connection with the valley. This was not a casually selected site but rather a place chosen with intention. The living could reconcile their own finitude by creating a place where their ancestors would continue to dwell in the valley in perpetuity. For centuries, the dead at Lofkënd claimed dominion over the landscape. Now, millennia later, the dead are no longer physically present, but they continue to exert influence. Through excavation, details of their deep-rooted and complex history have come to light. Ultimately, it is envisioned that the knowledge of this past will help enrich Albanian cultural identity well into the future.

Chapter 21

The Beginning and the End of the Lofkënd Tumulus and the Prehistory of the *Kanun*

John K. Papadopoulos

Death by Violence at Lofkënd

When looking at all of the material from the Lofkënd tumulus buried with the dead, one feature that stands out is the dearth of weapons (see Chapter 10). The quantity of weapons is paltry: a solitary iron spearhead from Tomb XLV (**10/118**); two iron arrowheads from Tomb XXXII (**10/119**, **10/120**); and a few iron knives (one comparatively well preserved from Tomb LXXXIV [**10/121**] and the tip of a knife from Tomb XXXVIII [**10/122**]; a third iron knife encountered in topsoil is probably modern [**10/123**]).

The solitary spearhead in Tomb XLV was found associated with the older male (one of three individuals buried in this tomb), and the position of the deceased in relation to the spearhead was such that it suggests the deceased may have been holding the missing wooden shaft of the spear in his right hand. Although rare at Lofkënd, this was a pattern familiar in many other contemporary burial grounds, not least Vitsa Zagoriou (Vokotopoulou 1986) and Liatovouni (Douzougli and Papadopoulos 2010), where virtually every male was interred with two or more spears. Several of these graves are of warriors, but the fact that all of the males at Liatovouni were accompanied by two, sometimes more, spears, may reflect hunting practices, as well as defense from wild animals or from groups of bandits (see discussion in Douzougli and Papadopoulos 2010:40–42). But whether for protection from a wild animal or human foe, the spear is a consummate weapon, though it need not necessarily reflect a warrior ethos.

In a similar vein, the almost complete iron knife in Tomb LXXXIV repeats a pattern seen elsewhere in the Balkans; but in nearby Epirus, iron knives—unlike spears, which are invariably associated with adult males—are common in tombs of both men and women (Vokotopoulou 1986:297; Douzougli and Papadopoulos 2010:43). As is noted in Chapter 10, the all-purpose nature of such implements made them indispensable for both sexes (see the comments of Catling in Popham, Sackett, and Themelis 1979–1980:257; Papadopoulos 2005:561–562). So there is no need to see the knife in Tomb LXXXIV as a weapon necessarily. The same, however, is not true for the tip of an iron knife found in Tomb XXXVIII (**10/122**) or the arrowheads in Tomb XXXII (**10/119**, **10/120**).

Both tombs are fully described in detail in Chapter 3. In the case of Tomb XXXII, the two arrowheads were found associated with one of two adult males in the grave, a fully articulated skeleton identified as an adult male aged 20–30 years. The two arrowheads were found on his torso, one to the east, the other to the west; the latter in the area of the abdomen, the former on the mid/right torso. These were the only two arrowheads in the entire tumulus, and their position on the torso of the deceased was such that they could either be interpreted as two arrows placed on—or even held by—the deceased, or else as projectiles that led to the demise of the individual through interpersonal violence. The bone was too fragmentary to preserve any clear evidence for death by arrowheads. Of these two interpretations, the placement of arrowheads in a tomb as tomb "offerings" seems strange indeed, since in most Late Bronze and Early Iron Age burials where arrowheads are found, they are usually found in association with other weapons, especially swords, and often in a quiver containing multiple arrows. The possibility that this male was killed by arrows cannot be easily dismissed.

In the case of the iron knife tip found in Tomb XXXVIII, the circumstances were most unusual. The grave was a double burial of an inhumed adult male, aged 30 to 40 years, and a cremated individual, possibly a female. The male inhumation was well preserved and is described as large and robust (see Chapter 6). Most unusual was the fact that the legs were very tightly flexed, doubled back on themselves. The degree of flexing was not only intentional but suggests that the body was tightly bundled in order to carry the corpse to the tumulus. This interpretation was corroborated by the evidence of black lines noted in situ that were almost certainly the remnants of a textile or skin, perhaps a shroud, that seemed to encase the inhumation, and perhaps also the cremation. As is detailed in Chapter 3, the cremation was secondary; the cremated remains were collected and placed immediately to the southeast of the inhumation and may well have been wrapped separately in textile or leather. This was one of only two cremations in the entire tumulus. But the grave became even more interesting during the process of lifting the skeletal remains: a fragment of an iron knife blade, indeed, the actual tip of the knife, was found immediately below the thoracic vertebrae of the inhumed male. The fragmentary state of the blade, coupled with its position in situ, suggested that the adult male may well have died by being stabbed in the abdomen, side, or back. This combination of features—a tightly flexed and seemingly bundled inhumation within what appeared to be a substantial textile or skin, a secondary cremation interred together with the inhumation, and the possibility of a violent death—suggests that both individuals may have died at some distance from the tumulus and were bundled and transported to their final resting place.

The two males in Tombs XXXII and XXXVIII represent the only examples for which we have enough evidence by way of weaponry to suggest, at least, the possibility of death by violence. In both cases, a violent death could not be corroborated by the study of their skeletal remains, because unless the bone had been hit by the knife or arrows found in each burial, there would be no clear osteological evidence. This said, in both cases the scenario of a violent death remains plausible, if not likely, in terms of all the evidence at hand. But these two tombs are not the only unusual, non-normative burials in the Lofkënd tumulus. Two others, Tombs I (Grave 64) and LXXIV (Grave 29), are unique not only in the context of Lofkënd but also when compared to what is known of both Illyrian and Balkan burial customs more generally. Moreover, one of these tombs is probably the earliest burial at Lofkënd, the other among the latest, although clearly not *the* latest.

THE BEGINNING AND THE END OF THE LOFKËND TUMULUS: (RE)CONSTRUCTING A NARRATIVE ON THE BASIS OF HUMAN REMAINS

The beginning

Every cemetery, every burial ground, has a formal beginning—a time when the decision was made to locate a cemetery and, in this case, an earthen mound, in a particular place, for mortuary "monuments are both the context for funerary rites and their consequence" (Barrett 1990:182). Normal cemeteries are often not very conspicuous at ground level, unless individual grave markers are monumental; in contrast, a tumulus "is deliberately prominent in the landscape" (Boyd 1993:563). Although we may never know why this particular spot was chosen for the site of a tumulus, we at least know approximately when the decision was taken. Among the graves of the earliest Phase I, one is so extraordinary, that it is the only candidate for a foundational event, and its establishment within the landscape fundamentally changed that landscape forever.

Tomb I, which is described in detail in Chapter 3, stands apart from all other graves in the Lofkënd tumulus. It is considerably larger and deeper than any other tomb, and it is located in what was to become—by the end of the period of the prehistoric use of the tumulus—the central part of the mound. Dated by AMS from charcoal encountered in the tomb to 1373 ± 57 cal BC, the tomb is one of the earliest features established at the site, and as the ^{14}C date derives from charcoal, as opposed to human bone collagen, the date could be even earlier, or later. Either way, it is one of only a handful of tombs at the site that can be reasonably dated to the fourteenth century, and perhaps to the late fifteenth century BC. Unusually, the tomb comprised nine arbitrary layers of bone that were recorded and lifted. All the bone was disarticulated, representing only a small portion of the various individuals interred. This was the only burial at Lofkënd that was clearly a secondary inhumation. The human remains include portions of at least three individuals, a count based on the number of mandibles. All three individuals are male, aged 25 to 30 years (Individual 362), 18 to 23 years (Individual 362a), and more than 45 years (Individual 362b); the first two of these males appear to be related (see Chapter 6). The total weight of all the human bone amounted to a mere 2.55 kg, and the preservation of the bone ranged from poor to very good (see Chapter 6). Apart from the mandibles establishing the minimum number of individuals, together with a fragmentary maxilla and an eroded right zygomatic, there were no other cranial elements preserved. Furthermore, the thoracic and lumbar vertebrae, along with the very large and durable femora and tibiae, are not present. Such an element representation is quite unusual, where skulls and the larger long bones are usually the focus of collection for secondary interment (cf. Boyd 1993). In her discussion of the mortuary practices at the third-millennium BC settlement complex at Tell Banat in Syria, Anne Porter (2002) discusses multistage burials in which acts such as the disarticulation of the skeletal remains destroyed the individual identities of the deceased but rendered them cultural artifacts to be used in forming collective identities.

Tomb I was also unusual in that it yielded, commingled among the human remains, fragments of several animals, including a pig or pigs (molar, podial, and scapula blade fragments), sheep/goat(s) (rib, radius, and long bone shaft fragments), and a long bone shaft fragment of a medium-size mammal (see Chapter 16.1). Quantitatively, this was the greatest number of animal bone remains in any one tomb. Although bone fragments were encountered in the fill of several graves, only three burials—Tombs XLVIII (Grave 52), LXVI (Grave 31), and LXX (Grave 17)—had more than one animal bone, and in all three burials, there were no more than two animal bone fragments. Moreover, in all of these graves, the animal bone fragments encountered were incidental to the fill of the tomb; they were not a primary component of funerary ritual. In contrast, the animal bone remains in Tomb I were found commingled with the human remains and thus a full-fledged part of the deposit, whatever their function or meaning.

As with many of the Lofkënd tombs, there were no grave goods, at least none that were noted in situ, and only a small quantity of fragmentary residual pottery in the fill; without AMS dating, the tomb could not be dated. The preliminary sorting of the bone deposited in the tomb by Lynne Schepartz (Chapter 6), however, brought to light not only the animal remains but two bone pins, one almost complete, the other fragmentary. Although all of the bone pins at Lofkënd are associated with tombs of Phase I, those in Tombs I and IV were found in male burials, whereas those in Tomb V and XII were found with females. The fact that a bone pin was found with the male in Tomb IV warns against assuming gender specificity for such items of personal ornament.

The fact that the tomb is now dated among the earliest burials at Lofkënd, and perhaps the earliest, is significant when coupled with all of its distinctive features. Located as it was at the highest point of the bedrock outcrop that was to serve as the base of the tumulus, Tomb I must have been a foundational event. Although one or two of the tombs of Phase I to the southeast may be as early, or even earlier, than Tomb I, the stark contrast between these individual inhumations in contracted postures in simple pits, on the one hand, and the circumstances of Tomb I, on the other, is so marked that it only bolsters the fact that this was the solitary Phase I tomb that was in any significant way different and, I would argue, special. The act of digging the pit for Tomb I in the place where it was located, and depositing the partial and secondary remnants of at least three males, together with the animal bones, is, as a ritual event, not seen in any other burial in Albania of which I am aware.

To what extent the remains of the three individuals deposited in Tomb I represent "ancestors" (cf. Barrett 1988), we may never know, though related circumstances have been recorded in northwest Greece. The burial in Tumulus Γ at Ephyra contained, in addition to the primary inhumation fully extended and articulated, 13 skulls, together with some disarticulated bone (Papadopoulos 1984; Boyd 1993:564). Similarly, one of the cist tombs among the mounds excavated at Merope Pogoni, in northwest Greece and closer to the modern Albanian frontier than Ephyra, had what appears to be a deliberate arrangement of disarticulated human bones, including five skulls (Boyd 1993:565; for the tumulus burials of Pogoni, see Andreou 1981, 1982a, 1982b, 1983, 1985, 1997; Andreou and Andreou 1987).

In the case of Lofkënd Tomb I, the process of intentionally bringing *these* human remains to *this* particular place is charged. As an act, it lays claim to the ground as the burial place of a particular group, probably a single extended family rather than a larger clan (see Epilogue). Furthermore, the choice of such a visible spot in the landscape, on a natural bedrock outcrop on a prominent ridge, seen from kilometers away, laid claim not only to a place of burial but to the broader surrounding territory. This one event, and the construction of the subsequent mound that was heaped around Tomb I and the other early burials, was what transformed this particular space into the place that now belonged to a particular group (cf. Tuan 1977).

Landscapes constitute cultural images that can tell us a lot about the ways social groups locate themselves in their environment, and constructed landscapes doubly so (Daniels and Cosgrove 1988: 1). Indeed, the commemorative framework of landscape, and of the monuments within that landscape, forms the very matrix through which memory works: people derive identity from shared remembrance or social memory (Alcock 2002:1, 183). The establishment of the tumulus at the place we now call Lofkënd focused not only the memory, but the identity of the group that was using it, to one particular spot in the landscape. This place was now their territory. Every member of the community knew that their ancestors were buried in this mound and that they themselves, and their offspring, would be buried in the same place.

The end of the Lofkënd tumulus

From its inception with the burial of Tomb I, there was never really an end to the Lofkënd tumulus. The permanence and durability of the mound, located as it is in the middle of Albania's oil fields at Ballsh, an area vulnerable to serious environmental degradation, survived into the twenty-first century, until a small group of archaeologists tore it apart. Our rebuilding of the tumulus (see Chapter 22), however, has ensured the continuity of the monument, at least for a time. That there was an end of the prehistoric group that used the tumulus is beyond doubt. Sometime around 800 BC (see Chapter 4) the prehistoric burials stop, and apart from items like the fragmentary Corinthian pottery found on or near its surface (Appendix 2), the tumulus was not used as a place of burial until the modern era, sometime around AD 1800 (give or take a century; see Chapter 4). So the question that remains is, what brought about the demise of the group that was burying their dead here for some 600 years during the closing stages of the Late Bronze Age and the Early Iron Age? Demise may not be the right word in this context—the group using the tumulus for so long may have simply moved away on account of any number of economic, political, or natural reasons—but I think that the demise of the group is exactly what happened.

As is further outlined in Chapter 4 and in Appendix 2 (following Chapter 9), the last burial at Lofkënd is separated by some two centuries from the time of the establishment of the Greek colony at Apollonia. The coming of Greeks, or at least the Corinthians of Apollonia, had little if anything to do with the end of the tumulus as a burial ground. As for what precisely happened, we have no direct evidence, but there are hints in the human remains themselves that can help (re)construct a narrative of what may have happened. As was the case with Tomb I, it is the exception, not the rule, that provides the starting point for the end of the Lofkënd tumulus.

Of the late prehistoric burials, one stands out as being so idiosyncratic that it demands some attempt at explanation. Tomb LXXIV (Grave 29) is not the final prehistoric burial in the tumulus; that dubious honor probably belongs to Tomb LXXXV (Grave 10). Indeed, the circumstances of Tomb LXXIV were so extraordinary that, when first encountered, we had no clue as to whether the tomb was ancient or modern. Until AMS dating clarified the matter once and for all, the going theory was that this one grave belonged neither with the prehistoric burials nor with the modern tombs in the northeast sector of the tumulus. Rather, it may have been the grave of soldiers killed in fighting in one or other of the battles of the twentieth century—a scenario inspired by Ismail Kadare's *The General of the Dead Army* (see further Appendix 3)—or perhaps three recently murdered men from a local village that happened to be buried here under mysterious circumstances.

The primary features of Tomb LXXIV are outlined in Chapter 3, and details of the human anthropology of the three individuals are provided in Chapter 6. The three males were interred in fully extended supine positions, with their arms intertwined and other body elements overlapping, such as the left hand of Individual 231 located below his hip or behind his body. The oldest of the three, Individual 230, aged 35 to 45 at death, was in the middle, flanked by younger males, the one to his right (Individual 210) aged 25 to 35 years, the other to his left (Individual 231) aged 20 to 25. I would dearly like to know the relatedness of these three individuals in terms of their DNA (cf. Appendix 1). It is possible, for example, if the age of the central individual is closer to 45 and that of the two younger males in their 20s, that we have the tomb of a father and two sons, but this is only speculation. That the three were buried together is beyond doubt, so all three died at the same time or within days of one another.

Tomb LXXIV was different from all the modern burials in three important respects. First of all, it had a totally different orientation, being northeast-southwest, with heads to the northeast, unlike the standard east-west orientation of the modern burials, whose heads were to the west, facing toward the east. Second, Tomb LXXIV contained three individuals side by side, unlike the modern burials, which were individual interments, apart from the two infants buried together in Tomb XC. Finally, their placement in the tumulus, to the northwest of the mound, was unlike the modern tombs, all of which were located in the northeast sector of the tumulus.

Similarly, the differences between Tomb LXXIV and the other prehistoric burials were marked, the primary difference being that all three individuals were in fully extended supine positions. All the other prehistoric burials at Lofkënd were in a contracted position, whether slightly flexed, like the individual in Tomb XX (Grave 50), or extremely contracted, such as the inhumation in Tomb XXXVIII (Grave 79)—who was probably murdered—or in the fetal positions of the individuals in Tombs XLVI (Grave 42) and XLVII (Grave 41) (see Chapter 8, and particularly **Fig. 8.3**). Moreover, in the majority of prehistoric burials that contained more than one individual, the standard practice was to reopen the grave and inter the later burial, a process that led to the remains of the earlier individual being disturbed, whereas those of the latest individual were usually well articulated. Whatever the reason(s) for the idiosyncratic deposition of the three men in Tomb LXXIV, they found no parallel at Lofkënd or, as far as I know, in any other tomb in Albania (for the typical depositions of prehistoric individuals in Albanian tumuli, see Andrea 1985:252–254, figs. 14–15, 17–19).

I would contend that the establishment of Tomb I marked the formal beginning of the burial ground for the group and that the burial of the three males in Tomb LXXIV represented the beginning of the end of the tumulus. After the interment of Tomb LXXIV, there were fewer than a dozen tombs in the tumulus (i.e., the 10 tombs of Phase Vb, and the last burial of Phase Va, Tomb LXXV). As for the age and sex of the individuals interred in the graves post-dating Tomb LXXIV, nine were adult males and four are of indeterminate sex, but there were only two females, one buried alone in Tomb LXXVIII, the other, in Tomb LXXXV—the very last burial in the tumulus—interred with an adult male.

The three males in Tomb LXXIV could have died in any number of ways, such as accident or pestilence, but the simultaneous interment of two younger adult men and a third, older male cannot be easily explained through mishap or disease alone. A much more compelling interpretation would be that all three died violent deaths; and here it is worth bearing in mind the individuals that may have been murdered in Tombs XXXII and XXXVIII discussed at the beginning of this chapter. I would like to suggest that the pattern of formalized *vendetta* documented in modern northern Albania had its origins in a much older time, and that it is a pattern with a prehistoric past. Feuds are a characteristic feature of many tribal societies, and multiple burials such as Tomb LXXIV are found in many other Iron Age contexts in prehistoric Europe (see especially Oeftiger 1984; for feuds more generally in Europe in later times, see, among others, Netterstrøm and Poulsen 2007; Tuten and Billado 2010; cf. Byock 1982). While it is possible that some form of internecine violence is represented in this multiple burial, suggesting that it might be directly linked to the historical Albanian *vendetta,* this cannot be substantiated by the evidence presented here, though the possibility is worth considering.

In modern times, the story of formalized feuds between families and clans is best told in Ismail Kadare's haunting *Broken April.* The setting is the high plateau in northern Albania—the *rrafsh*—in mid-March. Young Gjorg Berisha lies in wait for the man who killed his brother in their family's blood feud with the Kryeqyqes. The feud itself began 70 years earlier, when an unexpected, unknown guest was killed just on the boundary of the village. Although the murder was not committed by Gjorg's family, they inherited the debt under the law of the *Kanun* (for which see Hasluck 1954; Valentini 1945, 1956). Forty-four murders have already been committed in this feud. In other feuds, hundreds have been killed, and whole villages can fall under the blood law. Houses are pulled down, fields left fallow, graves stand as monuments and boundary markers at the scenes of death. The victim's blood-stained shirt is hung out; when its stains become yellow, the family must avenge the death, or face dishonor.

Kadare's story is a fictional account based on stories of actual events and inspired by the *Kanun*, the text of which tradition ascribes to Lekë Dukagjini (1410–1481) (**Fig. 21.1**). One of the Albanian national heroes of the fifteenth century, Lekë Dukagjini was a contemporary and comrade-in-arms of the greatest of the national heroes of Albania, Gjergj Kastrioti, better known as Skanderbeg (Fox 1989: xvi). As a number of scholars have observed, the *Kanun* derives from the Greek *kanon*—κανών—which in Classical antiquity usually referred to a straight rod or bar, oftentimes a rule, even a measure (Camaj 1989:xiii), and by Byzantine times acquired a legal significance (Fox 1989:xvi). The Albanian text of the *Kanun* was collected and arranged by Shtjefën Konstantin Gjeçovi (1874–1929), a Franciscan priest (**Fig. 21.2**) and a pioneer in the exploration of Illyrian tumuli in Albania (Chapter 17). As Martin Camaj (1989:xiii) explains, the old Albanian term is *doke*, derived from *dukem*, as in "appear," "behave," the plural *doket* referring to a collection of laws that determine how one behaves with acquaintances and strangers. The *Kanun* of Lekë Dukagjini was not the only code of customary law in Albania; in Krujë, Dibër, and Mat, the ancestral stronghold of the Kastrioti family, the *Kanun* of Skanderbeg was in force, while other areas had their own respective legal codes (Fox 1989:xvi). The *Kanun* of Lekë Dukagjini was the one most widely followed, however, especially in the mountains of Lezhë, in Dukagjin, in Shkodër, in Gjakovë, in Kosovo, and elsewhere (Fox 1989:xvii).

The important point is that made by Leonard Fox (1989:xvi), who notes that Albanian customary law evolved over a period of many centuries, well before and after the life of Lekë Dukagjini, and that some of the customs described in the *Kanun* may date back to remote antiquity, perhaps even influenced by ancient Illyrian tribal or clan-based law.

Particularly informative are the sections of the *Kanun* of Lekë Dukagjini that deal with the murderer. In the text compiled by Shtjefën Gjeçovi and translated by Leonard Fox (*Kanuni i Lekë Dukagjini: the code of Lekë Dukagjini*), we read (Chapter CXIX):

> § 844. As soon as a murderer has killed someone, he must inform the family of the victim, in order that there should be no confusion regarding the identity.
>
> § 845. The murderer must send someone to request a truce.
>
> § 846. The murderer, if he is able to do so himself, turns the victim over on his back. If he can, well and good; if not, he must tell the first person he meets to turn the victim over on his back and place his weapon near his head.
>
> § 847. The murderer may not dare to take the victim's weapon. If he commits such a dishonorable act, he incurs two blood feuds.
>
> § 849. The murderer may move around at night, but at the first light of day he must conceal himself.
>
> In Chapter CXXII, which governs the truce:
>
> § 855. The law requires a negotiator for the truce; to agree to a truce is the obligation of an honorable man.
>
> § 856. If the family of the victim has agreed to a truce with the murderer, the latter, even though he is the one responsible for the death, must go to the funeral and accompany the body to the cemetery, and attend the wake.
>
> § 868. It is not only a law, but an obligation for honorable men not to be contemptuous of anyone in the family of the victim, even if he is poor and weak.
>
> As for the cemetery, Chapter II:
>
> §5.1) In the cemetery belonging to a brotherhood or a clan, the dead or murdered of another brotherhood or clan cannot be buried. If

someone does this without the permission of the brotherhood or clan to which the cemetery belongs, the *Kanun* requires that the foreign corpse should be disinterred from the cemetery.

The *Kanun* is by no means a relic of the past: the permanence of the *Kanun* may be seen in its influence on Albanian folk literature, and especially in terms of ethical and aesthetic expression (Camaj 1989:xiv). In Albanian literature, the *Kanun* is a code of honor, best encapsulated in terms such as *besa*—a sacred promise, but also a word of honor, faith, trust, protection, truth—and hospitality, "which involves uncompromising protection of a guest, even one with whom the host is in a state of blood-feud" (Camaj 1989:xiv). The *Kanun* survived the era of communism (Fox 1989:xix), and as recently as the summer of 2011, a story in *The Guardian Weekly* was entitled: "Where revenge is in the blood: a brutal code of retribution is still tearing families apart in Albania," a violent reality that forced many young men to seek sanctuary abroad (Smolar 2011:28). Indeed, on June 15, 2003, the year of my first visit to Lofkënd, a 23-year-old man was murdered at the village of Lofkënd in the Mallakastër Hills. The circumstances of the murder were unclear, but in Skrapar in southeastern Albania, that very same day, two murders shocked the city. On Saturday night, Naim Xhemili (80) murdered Vladamir Guri only a few meters away from the police station. The brother of the victim then murdered Xhemili. The police report laconically stated that the murders had revenge as motives (for the Lofkënd and Skrapar murders, see United Nations Field Security Office, Tirana, Albania, weekly security report, June 15, 2003 [Week 24]).

CODA

We may never know how the three men in Tomb LXXIV met their end, and, in the absence of DNA analysis, we may never know whether they were related. All we know is their gender, their approximate ages at death, and the fact that they died at one time, or within days of one another, and were buried together in the one tomb, which was part of the same ritual event. After them, the community that buried their dead at this tumulus in the Mallakastër Hills passed away in what can only be described as a fairly rapid succession:

- an adult male aged 30–40 in Tomb LXXV (33)
- a mid-age adult male in Tomb LXXVII (18), who sustained an injury to his forehead, above the right eye, though he survived the initial trauma
- adults of indeterminate sex in Tombs LXXVI (16) and LXXIX (6)
- an adult female aged 20–25 years in Tomb LXXVIII (5)
- a male aged 20–30 years in Tomb LXXX (4)
- an adult male aged 18–22 years in Tomb LXXXI (1)
- two likely males aged 26–32 years in Tomb LXXXII (9), together with a third adult of indeterminate sex
- the complex circumstances of Tomb LXXXIII (7), which yielded evidence for the dentition of at least four individuals (full details in Chapter 6)
- Two adults males in Tomb LXXXIV (2), one aged 23–26, another 35–45 years, together with a child/infant aged 2–5 years
- And finally, the two interments in Tomb LXXXV (10), a mid-aged adult female, together with another adult of unclear age, perhaps a male

In earlier periods, in contrast, there was a greater proportion of females, infants, and children. To be sure, there is nothing conclusive in this evidence, but the overall pattern is intriguing, as it brings to the fore the possibility of interpersonal violence in several of the tombs of Phases Va and Vb, and more solid evidence for violent deaths in the case of Tombs XXXII and XXXVIII in Phases II and III of the use of the tumulus.

In terms of a tribal, clan-based society, such violence occurred under prescribed rules or laws which, like Albanian folksongs, were not written—until, of course, a much later time—but sung, passing from mouth to mouth and from generation to generation (Camaj 1989:xv). As Camaj (1989:xiv) notes, Albanian customary law left its mark on the character of the people, especially in their "sense of honor, vengefulness, courage and decisiveness in critical situations, and a feeling of closeness within the family, the brotherhood, and the clan."

Chapter 22

Heritage Management and the Future of the Tumulus

John K. Papadopoulos, Lorenc Bejko, and Sarah P. Morris

Introduction and Soil Morphology[1]

The visibility of the Lofkënd tumulus has been an issue we have continually stressed (Chapters 1, 20, Epilogue), and one that made it such an obvious target of investigation. It was also the visibility and materiality of the tumulus that prompted the decision, from the very inception of the project, to rebuild the mound. We could not eradicate this monument from the landscape. Accordingly, funding was allocated from the beginning of the project for the ultimate rebuilding or reconstruction of the tumulus as close to its original appearance as possible.

Given its dominating position in the landscape, on a ridge on such a naturally imposing spot, it is noteworthy that the mound was essentially intact for well over two and a half thousand years after its period of primary use. Although some erosion of burials was noted prior to excavation around the edges of the tumulus (see **Fig. 1.8**), especially along the steep south side, the tumulus survived largely unscathed. Because the tumulus did not collapse or erode more than it had, or wash away in the periodic storms that can be ferocious in this part of Albania, we suspected early on in our excavations that something in its composition assisted in holding it together. Perhaps this was nothing more than the existence of an underlying structure, such as a ring of stones or other means of soil retention seen in other tumuli (e.g., Andrea 1985:242–243, 245–246, 249–250; Aliu 2004:22–31, 2006:52, fig. 5:4; Bejko, Fenton, and Foran 2006:312–315, figs. 3–5; Chapter 17). Indeed, the excavation of individual tombs in the Lofkënd tumulus was exacerbated by the dense, hard, cement-like quality of the soil. Although a partly preserved curved line of stone was encountered at depth in the northern sector of the tumulus (designated "Wall" 1 in Chapter 2; see also Papadopoulos, Bejko, and Morris 2007:131, fig. 25), this by itself did not hold the tumulus together, and so it was to the composition of the fill of the tumulus that we turned.

In order to explore the composition of the soil more systematically, John Foss and Mike Timpson were invited to collaborate on the project. Their preliminary findings confirmed our suspicions (see Foss and Timpson 2007; Foss, Chapter 16.4): they suggested that a fine-textured sediment obtained off-site was added periodically to help prevent erosion of the tumulus. As determined by coring, this clayey fine-textured sediment derived from soils weathered from shale, located about 30 m north of the site. The basic parent material for soils at Lofkënd is weakly cemented Pliocene sandstone, which weathers into very fine, erosive sands that pose difficulties for stabilizing the earth matrix during individual burials, as well as for the completed tumulus. It was to this parent material that soil was brought in from elsewhere in the landscape, along with clayey sediment deriving from shale

[1] The present chapter is a somewhat expanded version of a paper first published in *Antiquity* (Papadopoulos, Bejko, and Morris 2008), but with more on the making of the mud bricks; a related version, in Albanian, was also published in Papadopoulos, Bejko, and Morris (2009).

added throughout the tumulus to control erosion (for detailed descriptions of the soils at the site and near the tumulus, see Foss and Timpson 2007:140–144, tables 1–3; Foss, Chapter 16.4). That some earth was brought to the site from the greater region of Lofkënd, perhaps even in substantial volumes, is indicated by the presence of Late Bronze and Early Iron Age pottery, by Paleolithic, Neolithic and Bronze Age lithic artifacts, as well as other finds, such as numerous pieces of fire-hardened clay daub with reed, rod, or stake impressions used in wattle-and-daub architecture contemporary with, or predating, the burials (fully quantified in Chapters 2, 9, 13, and 14; see also Papadopoulos 2006:81–83, figs. 6–7; Papadopoulos, Bejko, and Morris 2007:134–135, figs. 29–30).

Stratigraphically, the tumulus was formed over a period of over half a millennium (ca. 1400–800 BC) by a relatively limited number of soil units, largely cultural, which tended to be consistent over large areas of the mound. A more detailed account of the stratigraphy of the tumulus is presented in Chapter 2, but two aspects regarding the fill of the mound are worth noting here. First of all, the process of mounding began fairly early in the creation of the tumulus, at least by the end of the Phase II burials. Second, distinct localized dumps of earth in certain deposits were clarified by the process of backfilling the tumulus at the end of each season and reopening it the following season (see Chapter 2). In our own backfill, clear variations in the soil were the result of separate loads of earth and virtually identical to the ancient stratigraphy of the tumulus as excavated. Thus, what seemed to be confusing stratigraphy was clarified by our backfilling, which revealed different loads of dumped earth as distinct stratigraphic units within a single event, demonstrating how a similar process, repeated over time with different types of soil, gave the tumulus its distinctive stratigraphic profile.

Reconstructing the Tumulus

Since the tumulus was divided into four sectors, each separated by baulks of 0.50 m, the process of backfilling at the end of each excavation season (2004, 2005, and 2006) was straightforward (**Fig. 22.1**). Plastic sheeting was laid out on the unexcavated deposits and along the sides of the baulks, and the soil from the excavation dump was reintroduced to each of the sectors. The basic structure of the tumulus was indicated by the unexcavated baulks, against which the earth was piled; in essence, the baulks served as a framework for the backfill. In the process of backfilling we periodically attempted to compact the earth as much as possible. This method of backfilling proved both inexpensive and effective. On returning to the tumulus the following season, we noted slight subsidence at the interface of the baulks and the backfill, but this was never more than a few centimeters. Moreover, vegetation had quickly taken root in the comparatively looser earth of the backfill, and within a year, a dense array of shrubs, grasses, and flowers effectively stabilized the tumulus and held it together, preventing further erosion (**Fig. 22.2**). Some erosion was noted, especially along the more exposed south side of the tumulus, but this was at the steepest point of the mound and where its height was greatest from the surrounding surface. Elsewhere, particularly to the north, the mound sloped more gently toward the surface of the ridge on which the site was located, and here erosion was either minor or non-existent.

So long as the baulks were in place, the need to conceive of an alternate plan of reconstruction was unnecessary. On account of a variety of factors—including the location of numerous graves, not least of which was the central grave, and to clarify a number of stratigraphic features—it was necessary to remove the baulks (**Figs. 22.3–22.4**). We had originally decided to leave intact a 0.50-m square pillar of unexcavated earth at the highest point of the tumulus and near its center, as a remnant of one section of the stratigraphic sequence, but the location of the central grave (**Fig. 22.3b**) and fear of collapse of the resulting pillar made us abandon this idea. Without the baulks, however, the whole issue of reconstruction was more complex, despite the fact that a three-dimensional digital model of the tumulus had been constructed by Chris Johanson and Itay Zaharovitz (Chapter 19; Papadopoulos, Bejko, and Morris 2007: 135–138, fig. 33).

It was clear that to be successful, any solution had to focus on materials readily available near the site, and we deliberately avoided possible "high-tech" solutions. Given our knowledge of the soils of the Lofkënd region and given our experience in backfilling the tumulus over a period of three successive years, we settled on the low-tech and inexpensive expedient of rebuilding the baulks and using them as a framework for retaining the excavation soil. We explored a number of options as to the material to be used in the rebuilding of the baulks. But in the end, our method was to construct mud bricks made out of the soil of the tumulus and to use these to (re)construct the baulks (although mud bricks were never used in the original monument). The basic idea of our reconstruction project was that over time the mud brick used to rebuild the tumulus would disintegrate and effectively become part of the tumulus, a kind of decayed skeleton holding together the structure of the mound.

Our soil scientist, John Foss, stressed the need for using the fine, clayey sediment that derived from weathered shale, but the local excavation workmen, experienced in the making of mud brick, were sensitive to the different types of soil and had independently rejected the sandy parent soil of the area that had weathered from sandstone. They did, however, prefer a mixture of the shaley and sandy earth, and in the end they essentially used the earth of the tumulus—minus the human-made material—to make their mud bricks.

Four workmen from the nearby villages of Ngrançija and Gjinoqara made approximately 1,200 mud bricks, each measuring 42 × 18 × 17 cm and weighing about 20 kg, over a period of some 10 days in the summer of 2006, using one double mold. The four men worked in teams of two, having first dug two rectangular pits for the mixing of soil, water, and straw. The first team, consisting of the two older and more experienced men, mixed the earth, straw, and water by treading with their feet; the correct consistency was gauged by feel, with the various components added as needed in the process of treading. Once the mixture was ready, they would pile it on a sheet of plastic ready for the other team to take over. The second team of the two younger but also experienced men would throw the mixture into the mold, making sure that it was as compact as possible, smooth it off with wet hands, and then together they would lift the mold, carry it to the drying area, up-end it, and let the bricks dry (the process of lifting the mold and exposing the wet bricks was compared by the workmen to the production of squares of *loukoum* or Turkish delight). Every day or so, the bricks were turned to promote better drying. Under the normal summer conditions, the drying of the bricks was complete in about four days. As there is no water supply at hand on the site, all the water was brought up in large plastic containers by donkey, and the straw used for binding the mud was prepared each day from the wheat and barley chaff from the recently harvested local fields.

Although well-versed in the making of mud bricks, the workmen who made the Lofkënd bricks may well represent the last generation of craftsmen in this part of Albania capable of making mud bricks, since commercially manufactured bricks and concrete are now both readily available and affordable. It is worth noting that among the almost 20 workmen who took part in the excavations of the tumulus over four seasons, these four were the most experienced in the making of mud bricks. In the context of mud brick production in the Near East, Delougaz (1933:6–7; cf. Moorey 1999:305) writes: "brickmaking does not require any special technical knowledge, so that practically every villager does it occasionally. Of course there are some men in every village specially skilled in the making and handling of mudbricks." Delougaz goes on to mention a brick maker in the Diyala region in the early 1930s who could make almost 3,000 mud bricks a day (cited in Moorey 1999:305). When told about such production feats, some of the Lofkënd workmen expressed awe, others skepticism.

Our decision to rebuild the excavation baulks using mud bricks was not initially informed by any broader ethnoarchaeological and experimental archaeology studies, nor was it originally intended to. The decision was a logical expedient, but had further consequences, since little if anything is known on the prehistoric and historical use of mud bricks in Albania before the modern era. The process of recording the making of the mud bricks—both in photographs and film—documents a dying craft in the area.

The entire process of making the mud bricks was captured in a series of photographs taken by Richard MacDonald (**Figs. 22.5–22.8**), who also took video footage over several days of the whole process. An overview of the mud brick works immediately adjacent to the tumulus is shown in **Figure 22.5**. The various phases in the preparation of the mud brick mixture are seen in **Figure 22.6**, including the two rectangular mixing pits, together with the water carried to the site by donkeys (**Fig. 22.6a**); the actual preparation by mixing the soil, water, and straw (**Fig. 22.6b**); the treading of the mixture (**Fig. 22.6c**), and finally the prepared mixture heaped into a pile, smoothed by hand, ready for the mold (**Fig. 22.6d**). The carefully smoothed pile of clay seen in **Figure**

22.6d is a feature of pottery and roof-tile manufacture in other parts of the Mediterranean, depicted in the iconography of Athenian black-figure pottery of the sixth century BC (on the celebrated drinking cup known as Robinson's "kiln" skyphos, see Eisman and Turnbull 1978:395, ill. 1; 396, figs. 1–4; Scheibler 1995: 110–111, fig. 102), and recorded on celluloid in a film dating to 1947 entitled *Triumph Over Time* made by the American School of Classical Studies at Athens depicting tile makers at Corinth making roof tiles (Vogeikoff-Brogan 2007).

The molding of the bricks is seen in **Figure 22.7**: the foreman, Baki Ymeri, with the double mold that made all the mud bricks used in the Lofkënd tumulus (**Fig. 22.7a**); putting the clay into the mold (**Fig. 22.7b**); and lifting the mold with the bricks to the drying area (**Fig. 22.7c**). As the clay dried, the workmen would turn the bricks to expose them to the sun (**Fig. 22.7d**). Once dried, the mud bricks were piled up in as compact a form as possible (**Fig. 22.8**); this was necessary in order to cover them with plastic sheeting in the event of rain, and as it turned out, this precaution was necessary, since several days of heavy rainfall that ensued would have damaged or destroyed the exposed bricks.

Many aspects in the process of making the Lofkënd mud bricks find parallels in the ancient and modern Near East, where the making of mud brick has an older and more venerable antiquity than temperate Europe (for an excellent summary, see Moorey 1999:302–309). First of all, the preferred time for manufacturing mud brick in the Near East is May–June, "immediately after the spring rains, when water would be plentiful and the whole summer lay ahead, if necessary, for drying," whereas July–August was the period for building (Moorey 1999:304). In Albania, May and early June may be wet and not suitable for the making of mud brick, so July represents the best time and accords nicely with May–June in the hotter and drier east. In the Near East, mud bricks were usually produced in rectangular wooden molds, open at the top and bottom, most often singly, but occasionally in twos or threes (Moorey 1999:305). The Lofkënd workmen used one double mold, which differed from the Near Eastern version only in that it was not open at the bottom; the workmen preferred this, as the mixed earth, straw, and water could be better compacted within the mold. In the Near East, chopped straw or dung was commonly used as temper, and in one calculation, some 100 bricks required about 60 kg of straw, or about one-eighth of a hectare of barley (Oates 1990:390; Moorey 1999:305). In the Near East, the bricks are dried with regular turning for a period of time appropriate to the season; for example, in the modern village of Khorsabad, as Loud and Altman (1938:13) recorded, "bricks are left to dry in the summer sun for but a day or two" (Moorey 1999: 305), although in cooler Albania, four days were required. As for size, the Lofkënd mud bricks, as noted, measured 42 × 18 × 17 cm. In the ancient Near East, sizes varied from site to site, and in modern times from brick maker to brick maker. Moorey (1999: 306–307) records some standard ancient sizes, as follows:

59 × 24–29 × 8–9 cm (Baghouz)
81 × 45 × 10 cm (Baghouz)
50–70 × 21–30 × 6 cm (Tell es-Sawwam)
45–42 × 22–24 × 7–8 cm (Uruk, Ubaid period)
20 × 8.5 × 8 cm (Jamdat Nasr, Uruk III)
23 × 9 × 6.5 cm (Jamdat Nasr, Uruk III)
33 × 24 × 6 cm (Telloh)

Considerably larger mud bricks, each weighing approximately 34 kg, were made for the reconstruction of a small stretch of the ancient fortification wall of the Hittite capital at Hattuša (**Fig. 22.9**). As Jürgen Seeher notes, the reconstructed wall itself was 7–8 m tall, with the towers 12–13 m in height, and only some 65 m of the length of the wall were reconstructed, representing about 0.7% of the total length of the wall, which is calculated at about 9 km (Seeher 2008:62, with fig. 13; see also Yildirim and Gates 2007:295–296, 298, fig. 9). The formula used for the Hattuša bricks was 2,700 kg of earth, 100 kg of straw, and about 1,500 liters of water (see Seeher 2008:60, fig. 9, for the forming and drying of the bricks; 60–61, figs. 10–11, for the construction of the wall; and 61–62, figs. 12–13, for the final result; for related experimental studies in Hittite drilling and sawing of stone, see Seeher 2005, 2007).

In nearby Greece, the government policy of not supporting building loans for mud brick houses meant that they are no longer being built, and most of the existing mud brick buildings were constructed in the nineteenth or earlier twentieth century (Walsh 1980:155). One of the most important studies of ancient and modern mud brick construction in Greece remains the unpublished doctoral dissertation of Vicky Ann Walsh, who also embarked on a project of experimental archaeology by building a mud brick house in the village of Episkopi on Cyprus (Walsh 1980:162–167). The dissertation, which also includes a useful discussion of New World and African practices in construction with adobe, contains an important section on mud brick strength tests, which focuses on the sand/silt/clay ratio of the bricks and shows that mud bricks could be used safely in houses up to three stories in height, with heavy beamed roofs.

As noted, some 1,200 mud bricks were made at Lofkënd in the summer of 2006. We had anticipated that the excavations would be finished by the end of the season, with enough time remaining in the final weeks to complete the reconstruction of the tumulus. More experienced archaeologists, however, would have anticipated that the season's end would throw up all sorts of surprises, and the Lofkënd tumulus was no exception. The discovery of numerous tombs in the final weeks of the 2006 season, not least of which was a complex central burial with several disarticulated human and animal remains, coupled with the difficult schedule of Alket Islami, who was slated to take final low-level aerial photography of the site from a paramotor or paraglider (for which see Papadopoulos, Bejko, and Morris 2007:126, 130–131, figs. 24–25; Chapter 2), but did not manage to do so, meant that we effectively ran out of time to complete the reconstruction of the tumulus. Our new quandary was what to do with the bricks. Once again, the baulks came to the rescue. Since we only managed to clear parts of the baulks, enough of their structure remained to be of use, and we decided to pile up the mud bricks against the remaining baulks, cover them with plastic sheeting, backfill the tumulus, and hope for the best.

It was, therefore, with some trepidation that we approached the removal of the backfill at the beginning of the 2007 season. To our delight, the mud bricks had survived the winter and a particularly rainy spring intact with minimal damage (**Fig. 22.10**), and we were able to remove the 2006 mud bricks, complete the excavation, and use them, as originally planned, in the reconstruction of the tumulus. In lifting the 2006 mud bricks, however, we noted that a good proportion, almost one-half, tended to crack or to break in two; this was exacerbated by the built-up moisture from under the plastic sheeting used to distinguish the original deposits of the baulks (and the mud bricks) from the backfill. In comparison to the standard sizes of sun-dried mud bricks in the Near East listed above, the closest in general proportions are those from Jamdat Nasr, although ours were considerably larger and heavier and about twice the size. In hindsight, a slightly shorter brick, measuring 20–30 cm in length as opposed to 42 cm, and 14–18 cm square, would not have cracked or broken so readily.

By the end of the 2007 campaign, the steepness of the exposed bedrock ridge on which the tumulus was built, particularly to the south but also to the east and west, necessitated some form of retention, at least until such time as the undergrowth could take hold and help stabilize the soil. Consequently, in the summer of 2007, we made an additional 878 mud bricks for the rebuilding of the baulks and for use around the edges of the tumulus for retention. The process was precisely that described for the bricks made in 2006, except that we used two slightly smaller molds, measuring 40 × 18 × 14 cm. We preferred not to make significantly smaller bricks but to keep to a more or less standard size that would be better for rebuilding the walls of the baulks. **Figure 22.11** shows the mud bricks built into walls replicating the baulks of the tumulus, with appropriate openings to allow backfilling. It was against these mud brick walls that we piled up the excavation backfill one last time.

As Foss and Timpson noted (in Papadopoulos, Bejko, and Morris 2007:140), and as Foss reinforced in Chapter 16.4, the site is located on sandstone that was loosely cemented by carbonates, and bedrock was easily dug in most cases, though some strongly cemented beds occurred at points. In this final reconstruction of the tumulus, we decided to replace the backfill directly onto the bedrock of the site, without any intervening geo-textiles or plastic sheeting. Enough documentation was implanted into various stages of the backfill to confirm that this was, indeed, a reconstructed tumulus devoid of any ancient material or human remains. This was not to dissuade any would-be tomb robbers. We were confident that the reconstructed tumulus would not be defiled, at least not for some time, as the process of reconstructing the tumulus was not only carried out by the men and women of Ngrançija and Gjinoqara, but witnessed by many more. The excavation and reconstruction of the mound was a part of the continuing story of this place. By reconstructing our tumulus, we found that we could maintain its monumentality and give it added value from the new knowledge of its relevance to the local people.

A parting view of the rebuilt tumulus at the end of the 2007 season is presented in **Figure 22.12**. In 2008, when we returned for a final study season, we saw the tumulus covered with undergrowth, as it had been every year during our excavation (**Figs. 20.12, 22.13**). Instead of the customary wildflowers (see **Fig. 22.2**), however, the wheat and barley chaff used as temper for the mud bricks took root, and the tumulus was covered by a protective growth of mostly barley.

Continuing the Monument

Our decision to rebuild the tumulus was in many ways determined in response to the current practice of the complete removal of tumuli from their landscape (cf. Chapter 17). Hundreds of burial tumuli have been excavated in Albania, among them Barç, Mati (where numerous tumuli have been located and explored), Pazhok, Dukat, Kukës, Kuçi i Zi, Piskovë, Vajzë, Vodhine, Dropull, and Bajkaj, many of which have appeared in readily available overviews of Albanian prehistory (Hammond 1982:624–636; Prendi 1982:189, 211, 214, 216–218, 222, 235; see also Papadopoulos, Bejko, and Morris 2007; Bejko, Chapter 17). The policy of total excavation of these tumuli, however, has led to their eradication from the landscape. Indeed, the nearest contemporary tumulus to Lofkënd, at Patos (Korkuti 1981), was removed entirely from its landscape, and the spot where the mound stood formed the foundations of a modern house (**Fig. 22.14**). Consequently, a landscape once dominated by tumuli no longer exists.

Similar losses have been felt in many other parts of Europe and in Central Asia, thanks to the practice of the total excavation of burial mounds. In previous centuries, by contrast, prominent mounds were often tunneled and trenched and, although much damaged, they remain an integral part of the landscape. A classic case in point is provided by the royal mounds at Gamla Uppsala in Sweden, dating to the fifth and sixth centuries AD, particularly the Eastern—or Aun's—Mound (Anund et al. 1998; Gräslund 2000; Klingmark 2003). In Gamla Uppsala, however, many less prominent mounds have disappeared due to farming practices and quarrying (it was gravel quarrying, and subsequent looting, that threatened some of the tumuli of Apollonia and led to the rescue excavations by the Albanian Rescue Archaeology Unit [ARAU] of the complex of Tumuli 9, 10, and 11 at Apollonia; see Amore 2010:1). Of an estimated 2,000–3,000 mounds at Gamla Uppsala originally, only some 250 barrows remain (**Fig. 22.15**). A related loss of tumuli in the Shkodër region of Albania, due primarily to farming, has been recently recorded by Bejko and Galaty.

Elsewhere, burial mounds have been reconstructed and their management undertaken. The mound at Ladby in Denmark, for example, complete with a ship, was totally excavated and subsequently rebuilt—together with the ship—by the Carlsberg Foundation (Sørensen 2001). Many reconstructions of archaeological sites are fraught with controversy (see various papers in de la Torre 1997), and in the realm of mounded tombs, one of the most controversial is the reconstruction of Newgrange in Ireland. There, a reinforced concrete retaining wall was built by the Office of Public Works in order to keep the restored material in place, and to which quartz and granite were affixed with mortar and metal pins, intended to present what the excavator thought would have been the original drum-shaped mound (Cooney 2006:697; O'Kelly 1979). Later scholars, however, offered different interpretations, both of the stratigraphic sequence of the mound and its shape (dome-shaped, as opposed to drum-shaped), as well as of the role of the quartz and granite layer (for discussion and bibliography, see Cooney 2006; Eriksen 2006). Such an intervention—difficult, if not impossible, to reverse—imposes onto the landscape an interpreted and often controversial vision of a monument.

Our reconstruction of the Lofkënd tumulus was not aimed at an interpreted vision of what the mound may have looked like at any point in its history, but rather was intended to restore the mound as close to its original appearance prior to excavation in 2004 as possible. In this respect, our aims were closer to the management project at Sutton Hoo in Suffolk, southeast England, initiated in 1992, where the various mounds, with the exception of Mound 2, were restored to their 1983 height (Carver 2005:57). In the case of Mound 2, a more ambitious project was launched in 1997, where the mound was reconstructed on the basis of its excavation and of a mathematical equation to determine its original, seventh-century AD height (Carver 1998; 2005:46, plate 15; 57). In addition to consolidation and management of the site, the reconstruction of Mound 2 at Sutton Hoo allowed the observation of its rate of erosion (Carver 2005:57). Much closer to home, a number of mound reconstructions have been documented in various countries of west-central and eastern Europe. The important point here is the ubiquity of such practices in many other parts of Europe.

OTHER TUMULUS RECONSTRUCTIONS IN ALBANIA AND GREECE

The reconstruction of the Lofkënd tumulus represents one of the few monuments of this type to have been rebuilt and restored to its landscape. Elsewhere in Albania, the only other burial tumulus to have received any attention of this sort is the tumulus of Kamenicë in the Korçë basin in southeast Albania. Unlike Lofkënd, the Kamenicë tumulus had a substantial structure of stone rings for the earlier burials of the Early Iron Age, and an even more considerable stone-constructed extension for the later, post–seventh-century BC tombs (Bejko, Fenton, and Foran 2006:312–315, figs. 3–5; all of the earlier burials were excavated, but only a few of the later burials were cleared [it is estimated that the Kamenicë tumulus originally held about 400 graves]). Given the size and nature of the funerary architecture, the decision was taken to prepare the Kamenicë site as an open-air museum accessible to the public (**Fig. 22.16**), and, under the direction of Lorenc Bejko, the site was fenced, the stone-constructed elements of the tumulus were stabilized and conserved, a water channel was dug to divert rainwater away from the archaeological remains, and a visitor center was built, housing an overview of the excavation and a history of tumulus excavations, as well as replicas of the finds from the tumulus (the site was inaugurated and the museum opened to the public in June 2007). Thus prepared, Kamenicë has been a mainstay for visitors in this part of Albania—not least for school groups—and the reconstruction has fared well in the years since its opening to the public.

Perhaps the largest tumulus to have been "reconstructed," and certainly the most elaborate and expensive undertaking of its type, was the construction of the massive crypt and museum, together with the reconstruction of the tumulus above it, at the site of the royal tombs of the late fourth century BC at Vergina in northern Greece, including the tomb considered by some scholars to be that of Philip II, the father of Alexander the Great. This massive undertaking, initiated in 1983 and finally completed in July 1993 when the site and museum were opened to the public, with a total project cost of US $7.02 million, involved the construction of a crypt to cover the substantial stone-built monuments, including the "royal tomb," the "tomb of the prince," the "tomb of the free-standing columns," and the "tomb of Persephone," as well as the Heroön (Dimacopoulos 1997). The reconstructed tumulus-cum-museum at Vergina not only protects the tombs and other ancient constructed monuments, it also houses the extraordinary finds from these rich burials and provides access to thousands of people who visit the site annually.

In contrast to both Vergina and Kamenicë, the dearth of architectural remains at Lofkënd meant that once our tumulus was excavated, the only thing visible was the bedrock ridge on which the mound was built. There was none of the substantial stone-constructed funerary architecture of Kamenicë, nor anything like the spectacular late Classical and early Hellenistic monuments of Vergina. Our challenge was, therefore, a much simpler one and more easily met: to restore a once conspicuous mound to its landscape. Consequently, anyone traveling today on the main road between Fier and Tepelenë—as it is still the main road between Albania and Greece (though perhaps not for long)—will continue to notice the small but prominent mound that has taken the name of Lofkënd, located in the Mallakastër Hills on the north side of the road between Patos and Ballsh.

CONCLUSIONS

In this chapter, an attempt was made to situate the aims of the Lofkënd reconstruction project against the backdrop of several mound reconstruction projects in other parts of Europe, particularly in Scandinavia, the British Isles, and the Balkans. These projects, like ours, are not only part of a wider debate on the preservation, restoration, and reconstruction of archaeological monuments (see, especially, Schmidt 1997; Sullivan 1997), but testimony to the fundamental sea change in archaeology that has come about through the deepening respect for the physical and social consequences of field excavation (Lyons 2003:299). The past is an endangered and contested commodity, and one of the aims of archaeology is to reconcile three goals—scientific research, public access, and long-term preservation—often at odds with one another (Lyons 2003). Cultural resource management demands participation from local community interests at all levels, and the process of engaging in dialogue with stakeholders acknowledges that the concept of archaeological patrimony means different things to different people (Lyons 2003:303). The eventual conservation and display of the site is increasingly seen as an integral part of the design phase. As Nicholas Stanley Price (2003:269) notes:

> In the long history of archaeological excavation in the Mediterranean region, issues such as site preservation, presentation to the public and management have tended to be addressed after the fieldwork has finished. By contrast, contemporary thinking would stress that they need to be considered in advance of any fieldwork, and especially in advance of the use of any destructive technique such as excavation.

In the context of Lofkënd, it is interesting to note what the local inhabitants of the area thought of the mound prior to excavation. When asked what the tumulus represents, most if not all of the local villagers of nearby Ngrançija and Gjinoqara considered the mound as a burial place not for the long dead but for those just beyond the grasp of memory. The tumulus was most often conceived of as a collective grave of foreign soldiers, casualties of the First World War or one or other of the campaigns of the Balkan Wars. Ironically, that which was there the longest was considered by many as something of an import, a monument of a recent foreign culture. The realization that the tumulus contained prehistoric burials did a lot more than awaken a sense of local pride. It brought the abstractions of time and memory down to earth and focused them—once more—on one particular place: a tumulus dominating the local landscape. In antiquity, the mound was a focus of memory and identity. It provided the prehistoric inhabitants of this part of the Gjanicë River valley not only an image of, but an anchor to, their past, as well as offering them a conceptual point from which to approach their future (Papadopoulos 2006:83–84). In a similar vein, the reconstruction of the tumulus restores the commemorative framework of landscape and of this particular monument within its landscape. The local and social benefits of reconstructing the tumulus lie precisely in the fact that the reconstruction satisfies the needs of the present without rendering those of the future impossible.

Epilogue

Sarah P. Morris and John K. Papadopoulos

Epilogue 1
The Lifeways of the Tumulus Builders

John K. Papadopoulos

In Chapter 1, we proposed that to understand the landscape of Lofkënd, one had to understand the nature of the tumulus, since it represented the only clear evidence of the use of the landscape in the Late Bronze and Early Iron Ages in this particular place. We also contextualized the issue against the backdrop of both the establishment of the Greek colonies, especially Apollonia sometime around 600 BC, and the rise of large urban agglomerations, which were well established in the Illyrian landscape by the fourth century BC, at nearby sites such as Klos-Nikaia and Byllis, as well as a little farther to the south and west at the site of Amantia. Lofkënd was also ringed by a series of "proto-urban" sites, like Mashkjezë, Gurëzezë, and Dimal—the chronology and character of which are still poorly understood—as well as the other hilltop site at Margëlliç, which at least boasted some Mycenaean material. We also noted the information gleaned from settlements known in our Classical sources as *komai* (i.e., unwalled villages) farther south, in Greek Epirus. Indeed, the interplay among unwalled villages, colonial metropoleis, and proto-urban centers created a dynamic situation, one that allowed for all sorts of political and social experimentation. Consequently, we speculated that the potential implication for the nature of settlement in this region may be more far-reaching than we originally anticipated. The tumulus, however, ceased to be used as a prehistoric burial place sometime around 800 BC (for the end of the tumulus, see Chapters 4 and 21), some two centuries before the establishment of the Greek colony at Apollonia. Although the intensive survey of the area around the tumulus has filled in many gaps, the nature of settlement in the Late Bronze and Early Iron Ages continues to be elusive (see Chapter 18). What remains to be determined is the mode of life and subsistence that the Late Bronze and Early Iron Age inhabitants of the area around the tumulus adopted, one that permitted them to inhabit this landscape for some 600 years.

As Samantha Martin-McAuliffe stated in her conclusion to Chapter 20, the construction of the burial tumulus at Lofkënd in antiquity physically formalized a well-established human commitment to the Gjanicë River valley. She also noted that despite the fact that many questions remain about the tumulus builders, we can perceive them as valley dwellers and, as such, they identified with this area and its wider topography. More than this, these "individuals were cultivators of place, and their choice of location for the tumulus reflected their connection with the valley" (Martin-McAuliffe, Chapter 20).

There is also the information to glean from the physical study of the human population at Lofkënd (Chapter 6), together with the information for a paleodiet suggested by the stable isotopes (Chapter 7). For example, the biophysical affinities of the Lofkëndis and their relationship to the prehistoric population from Apollonia indicate that although both the Lofkëndis and Apollonians were routinely subject to low levels of stress during childhood, as suggested by their dentition, the Lofkënd population may have been exposed to fewer stresses and stresses of lower intensity (they were also less affected by malaria than the population of Apollonia) (Chapter 6). Furthermore, the differences between the prehistoric and modern populations at both Lofkënd and Apollonia were marked; among other things, changes in the diet related to greater reliance on grains may have resulted in the rise of caries rates for the modern populations. Isotopic values (Chapter 7) for the prehistoric population of Lofkënd indicate that terrestrial dietary protein derived mainly from C_3 sources (e.g., wheat, barley, root crops, legumes, vegetables, nuts, honey, and fruits). Intake may have included some dairy products and meat but little or no marine sources. Compared to the Aegean Neolithic and Minoan diets, the Lofkëndian diet shows less protein from dairy and meat, or perhaps more consumption of legumes or legume-consuming carnivores. In contrast, the modern population indicates a consumption of primarily C_4 plants, probably maize.

Given the small size of the sample, little more can be said of the general health and diet intake of the prehistoric Lofkëndis, but more can be said about the general lifeways of the inhabitants of Lofkënd who buried their dead in the tumulus, and to contextualize these lifeways with what is known from farther south. Indeed, study of the Epirote *komai*, introduced in Chapter 1, particularly those at higher elevations, has spawned the view that the inhabitants of much of northwest Greece and southern Illyria were nomadic stock-breeders (see especially Vokotopoulou 1986:3, 340). Many scholars had assumed a seasonally defined exploitation of the physical environment, one that was imposed by the geomorphology of the region, which favored the development and growth of an economy largely based on animal husbandry. In such a perspective, the agricultural production of cereals, pulses, and legumes must have always been limited in scope and of secondary importance to stock breeding. This point of view was championed by Hammond (1967a, 1982) and Dakaris (1976), and it was also assumed by Vokotopoulou (1986:340; see also Vokotopoulou 1973; Bintliff 1977:12–17; Kilian 1972, 1973b; for further discussion, see Douzougli 1996; Zachos 1997), despite the fact that a number of influential scholars already in the late nineteenth century had comprehensively argued for the shortcomings of such a view (e.g., Hahn 1891; Weber 1894). Stable isotope analysis presented in Chapter 7 also undermines the view of the importance of nomadism.

In a number of important studies, Paul Halstead and John Cherry, working independently, challenged the prevailing view that the mountainous environment and pan-Balkan affinities of the material culture of the Pindos indicated transhumant or nomadic pastoralism, and the same holds true for southern Illyria. Halstead effectively showed that two of the three diverse models of subsistence today—namely, nomadic specialized pastoralism, as practiced by the Sarakatsani, and transhumant diversified pastoralism, following the Vlach model—simply did not exist in later prehistory (Halstead 1987a:79–81; 1990:62–64). Indeed, the appearance of the Vlachs in northern Greece and neighboring lands sometime around the end of the first millennium AD may reflect the colonization of a new economic niche (Halstead 1987a:81; for the Vlachs, see further Avramea 1974). Rather, as Halstead argues, the archaeological evidence for expansion of human settlement in the Pindos from the late second millennium BC is interpreted in terms of sedentary mixed farming, replicating the subsistence strategy that dominated the lowlands of Greece throughout the Neolithic and Bronze Age. Halstead was careful to note that a mixed farming economy does not preclude either seasonal use of distant pastures or differences within and between communities in the relative importance of arable and pastoral farming (Halstead 1990:72; for Neolithic and Bronze Age lowland subsistence strategy throughout Greece, see Halstead 1987b; see further Halstead 1981, 1988, 1989, 2000). Halstead goes on to note that firm palynological evidence for the impact of early farming even in lowland vegetation is notoriously difficult to find and that there is simply no palynological evidence from the top of the Pindos range. Pollen cores, however, from middle-altitude sites suggest that deforestation on the scale of today did not take place until the last few centuries (Halstead 1987a:81, 1990:66; echoed by Cherry 1988:15; for pollen cores in Greece, see Bottema 1974, 1979, 1982; Athanasiadis 1975; Greig and Turner 1974; Turner 1978; Turner and Greig 1975; Willis 1989, 1992).

In a similar vein, Cherry has argued that "specialized pastoralism is an unlikely mode of subsistence in the context of the types of environmental settings and farming systems in early prehistoric Greece and, indeed, is an adaptation that has emerged historically only in specific socio-economic and political circumstances" (Cherry 1988:9). In so doing, Cherry deconstructed two assumptions that have plagued the field—namely, that pastoralism in Greece is largely environmentally defined and, second, that the environment has remained essentially unchanged since antiquity (Cherry 1988: 14; cf. Fotiadis 1995; see further Lewthwaite 1981). In an interesting sideline, Cherry (1988), with reference to the earlier work of Eric Higgs and his colleagues (see Higgs et al. 1967), drew attention to the fact that even students of the Paleolithic in Epirus have come under the sway of ethnography to such an extent that modern pastoralists in northwest Greece have been used as a direct model for seasonal transhumant patterns among Upper Paleolithic hunters at sites such as Asprochaliko and Kastritsa. Similarly, Claudia Chang expressed serious reservations as to the validity of projecting the Sarakatsani model directly onto the Bronze or Early Iron Age (Chang 1994:354–355). In attempting to come to grips with why pastoralism has been so widely assumed as the dominant economic model in the study especially of Early Iron Age Greece, Cherry followed the work of Brent Shaw, by concluding:

> From Homer to Ammianus Marcellinus, the pastoralist is defined simply via logical opposition to the essential criteria of civilization: mobile and without established homes, non-urban, *polis*-less, without properly constituted rules or law-codes, lazy and parasitic (because he does not *work* the land and *harvest* crops, like the farmer), an eater of meat (often *raw human* flesh) rather than grain, and a drinker of milk, not wine. By a confused social syllogism, nomadic pastoralism comes to bear the full stigma of uncivilized barbarity. (Cherry 1988:29; Shaw 1982–1983: especially 12–13)

In his cross-cultural overview of pastoralism, C.R. Whittaker makes two important points: the first is that pastoralism is an ideal type that has never really existed in pure form, not even among the Masai or Berbers; and, second, that pastoralism must always start from agriculture (Whittaker 1988:1, 3). For the latter point, Whittaker follows Fernand Braudel's argument that it was the farmers' determination to plow that opened "up the way for shepherds" (Braudel 1995:179; Whittaker 1988:3). Moreover, as Peregrine Horden and Nicholas Purcell (2000:80–87, 197–200) stress, it is mistaken to regard pastoralism as the characteristic activity of the inhospitable mountains. The advantages for sedentary farmers to keep some animals are many. First of all, animals can be "stored food" against the shortages of vegetable products and, second, in the particular setting of the Mediterranean, animals do not need to compete with humans for land (Horden and Purcell 2000:198; for "social storage," see O'Shea 1981). As Horden and Purcell go on to state: "[M]anaged animals are an incomparably flexible resource. Apart from their flesh (and other post-mortem benefits such as leather, bone, and horn), species domesticated in the Mediterranean . . . have offered milk, clothing, manure, and energy for traction, transportation and water-lifting" (Horden and Purcell 2000:199). Cherry (1988:21) echoes this point: "[M]odest flocks of ovicaprids clean and fertilize the cultivated fields near the settlement and—in the process—convert unusable plant-materials into meat and milk serving as dietary supplements of fall-backs in time of crop failure."

In the specific case of Epirus (a region, unlike the Mallakastër Hills, which is both environmentally diverse and vertically differentiated), Konstantinos Zachos has suggested that, in the case of the plain or basin of Ioannina—the ancient Hellopia (for which see Plutarch, *Pyrrhus* 5, 2)—there may have been no need for a winter-in-the-plains, summer-in-the-mountains schedule, since the ecosystem provided all the resources necessary for year-round habitation (Zachos 1997). A similar conclusion was also argued by Thomas Tartaron for the lower Acheron valley in southern Epirus (Tartaron 2004), and as Angelika Douzougli and Zachos (Douzougli and Zachos 2002) have argued, in many parts of Epirus the combination of lowland plains and upland areas provides an abundance and diversity of resources permitting a mixed economy of farming and localized herding.

The archaeological patterns for the Late Bronze and Early Iron Ages that have recently emerged from Epirote sites such as Vitsa Zagoriou, Pogoni, and Liatovouni, when taken together with the material coming to light in Albania, seem unequivocal. They suggest that from the Neolithic period through the years of the fifth and earlier fourth centuries BC, an economy based on sedentary mixed farming and associated localized herding remained the primary subsistence strategy.

The valley dwellers, as Samantha Martin-McAuliffe casts them, of the Gjanicë River at Lofkënd were no exception. The demography of the tumulus, with just over 150 individuals scattered across some six centuries, suggests a very small population, at best an extended family. Their settlement must have been similarly small, reflecting the small unwalled villages like those of the *komai* of Vitsa and Liatovouni in northwest Greece or the site at Tekke Melan in southern Albania (introduced in Chapter 1). We can well imagine a network of such settlements, all part of a clan or tribe, collectively constructing the walled refuges and meeting places that have come to be known as "proto-urban" centers, despite their lack of any "urban" accoutrements. Indeed, the first fundamental reorganization of this economic, social, and political way of life occurs in the fourth century BC, at which time previously scattered groups are centralized within and around new fortified cities like Klos-Nikaia and Byllis, which represent the earliest evidence for full-fledged urbanism in the region.

If the population that buried their dead at the Lofkënd tumulus were valley dwellers, as seems most likely, then proximity to the Gjanicë River would have been crucial in order to secure year-round water supply for humans, animals, and agriculture. The dynamic nature of this riverine landscape could well have wiped out all vestiges of any small settlement or hamlet located on or near the banks of the Gjanicë. The same riverine landscape in the form of two nearby and roughly parallel rivers, the Vjöse/Aoös and Seman, have left the ancient harbor city of Apollonia today over 26 km from the sea. This was a landscape dominated by rivers, as was Epirus to the south, and humans inhabiting this landscape exploited these rivers to their advantage, from the Paleolithic period well into modern times.

EPILOGUE 2
FROM THE STONE AGE TO THE RECENT PAST: THE CULTURAL BIOGRAPHY OF A LANDSCAPE AND OF AN ILLYRIAN TUMULUS

Sarah Morris and John K. Papadopoulos

The long-term history of a modest mound in south central Albania is not limited to the time span of the burials it covers (Chapter 4), from the Late Bronze Age until the eighth century BC. Rather, its form and contents embrace many experiences, from the palimpsest of a Paleolithic landscape, to a modern perch for soldiers, which sustained and extended the pre- and post-mortem life of its occupants. Just as we have sought to understand the formation and afterlife of the mound as a deliberately cultivated place that creates order in the surrounding landscape (Chapter 20), in this section we aim at a narrative of the excavated mound over time, to situate in the wider history of Illyria.

THE STONE AND BRONZE AGES

Nature played the earliest formative role in this narrow valley, whose river cut steep slopes on both flanks and deposited broad areas of rich loam on either side, as it wound its way to join the Vjosë/Aoös River and reach the sea. During the waning of the last Ice Age, a cooler, wetter climate covered these slopes with dense forests of oak and maple (Chapter 16.1; Fouache 2007:9). During the Middle Paleolithic (ca. 200,000–50,000 BP), at least one small band of hunters frequented the river and its springs, where animals came to feed and drink, and left signs of a small, short-term open-air site on or above the east bank of the Gjanicë River, below the modern village of Belishovë (Chapter 18, Site 003). The kind of stone tools they made and used, and the debris they left from manufacturing and reusing them, turned up in scatters across diverse areas on the eastern slopes of the river, including in the immediate vicinity of the tumulus, and on its surface (Chapter 1). More significantly, 589 chipped stone objects from the Paleolithic through the Bronze Ages made their way into the fill of the tumulus (Chapter 13). This may seem a paltry number, but the quantity of lithic artifacts for a burial tumulus of this size is significant. The chipped stone tools were perhaps collected deliberately or, more likely, having traveled with certain types of soils (especially clays), selected to give a growing mound of burials a stable matrix (Chapters 13, 16.4, and 20). Thus, it may have been the associated materials (compacted and denser soils) that decided the afterlife of such tools and drew them into the formation of a Late Bronze and Early Iron Age mound. Henceforth, the site first chosen for burials around 1400 BC (Chapter 4) carried with it a pedigree of older human habitation and activity in the surrounding environment, spanning back as early as at least 50,000 years, and perhaps considerably earlier, before a family group brought their dead to rebury them on a prominent ridge.

It is difficult to say more about sites and life during the Stone Age and earlier Bronze Ages in this region based on Lofkënd results. The Neolithic period is hard to trace in this area, perhaps because new settlements drew farmers and herders away toward permanent settlement in inland valleys at sites such as Cakran (Korkuti and Andrea 1974). By the Middle Bronze Age, the hilltop site of Kraps (its modern name) on a ridge at the northern end of the Gjanicë valley shows occupation and was evidently an "Illyrian hamlet" during the Late Bronze through Iron Ages (Stocker 2009:584–589, 816–817, Site 038, fig. 7.1). Much larger and longer-lived was the site that dominated the area by the Late Bronze Age, visible from both Kraps and the tumulus site at Lofkënd: Margëlliç (Stocker 2009:595–600, Site 041). Indeed, as settlement and survey at this site have shown, it was typical of other sites in the area in showing occupation from the Late Bronze through the Early Iron Age, albeit with its own necropolis. This leaves the Gjanicë valley still poor in settlements that could have buried their dead at Lofkënd.

THE PERIOD WHEN THE TUMULUS WAS IN USE (1400–800 BC)

At some point in the Late Bronze Age (during the century around and after 1400 BC), a small family group or extended clan claimed the ridge above the river and brought to it a collection of the remains of their own dead. In a fundamental rite establishing a new home for their dead, past and future, they dug a roughly rectangular pit down into bedrock on the very peak, and buried in it the partial remains of at least three men (one mature, and two adults who were closely related; see **Table 6.14**), along with the bones of pigs, sheep, and goats (**Fig. 3.8**, Chapter 16.1). In doing so, they mirrored the same deliberate gestures of other groups of this period, who initiated a new burial ground with the bones of their "ancestors," in at least two sites in southern Epirus (northern Greece, Ephyra and Pogoni: Boyd 1993). Few finds in what we have called "Tomb" I (Grave 64) allow us to pinpoint this moment in time: two bone pins (**10/52, 10/56**) may or may not belong to those interred, and a single scrap of charcoal (dated to 1373 ± 57 BC) may have been an old—or recent—piece of wood when it was swept in with human and animal remains. What remains crucial is how fundamental this activity was in making a space of nature into a place of culture, one still active today.

For all intents and purposes, those who reburied their dead here created a new and permanent locus for family ties across the past and into the future. For shortly thereafter, a series of individuals—men, women, and children (sometimes mother and children together, such as in Tomb XII) were interred on the steep slope to the south and east of this central pit atop the bedrock peak (**Table 4.1**). Tombs II and III date within the same generation (Chapter 3.1, Chapter 4, **Table 4.1**), the first pair in a total of 12 burials over the course of the fourteenth century BC, or perhaps somewhat later. According to the finds selected for their final resting place, adults and children were wrapped in garments fastened with bone (**10/53–10/55**) and bronze (**10/27–10/28**) pins, although some of these pins served as hairpins. The textiles are now lost (pseudomorphs are rare outside iron finds, albeit some were preserved with bronze objects: Chapter 12) and the pins may have been the limit of their funerary finery. One infant was buried perhaps holding a tiny bronze bead (**10/65**, Tomb VIII), but no pots were deposited with the dead, and no other jewelry or ornaments were collected to adorn the dead until the next period of tumulus use (**Fig. 4.1**). Either ceramics were not used in Phase I of the tumulus, they or their contents not considered important, or conversely, they were too valuable to give to the dead, and most of the jewelry in use was functional. The impression given by this first phase of burials is of a modest community, closely related. Subadults and adolescents are conspicuously absent in this early skeletal population, which consists primarily of adults and infants/children; the lives they led were hard and left their mark on their bones and teeth (Chapter 6). Burials and contents form a striking contrast to more spectacular tombs of Late Bronze Age "warriors" buried with sword, armor, and vessels (as at nearby Patos, or at Molossian Liatovouni; see Korkuti 1981; Douzougli and Papadopoulos 2010).

During its second phase of use (Tombs XIV–XXXIII), the mound grew, with burials spreading out farther from the original deposit, especially to the broader, flatter plateau north and west of Tomb I (**Fig. 4.1** and Chapter 19). A higher number of children, young adults, and females were buried in this period, the onset of a pattern that lasts through several phases of burials and is highly unusual by demographic standards for prehistoric populations, where one would expect greater numbers of infants and mature adults (Chapter 6, **Table 6.3**). Whether this reflects higher juvenile death rates coupled with low population replacement due to unusual losses of prime age adults (Chapter 6), or deliberate selection of children and young adults for burial (Chapter 8), is difficult to determine. More multiple inhumations were performed in Phase II (e.g., Tombs XXI, XXVII, XXXII), and the first of two secondary cremations was interred in the tumulus (Tomb XXX, an infant). For the first time, several (young) individuals were provided with ceramic vessels (**9/59–9/60**, perhaps full of food or drink), and the earliest handsome array of gifts accompanied two children (7 and 4 years of age) who were buried together in Tomb XVII. Two vessels and several items of bronze jewelry, including a headband, earrings, and spiral coils, mark the first in a pattern of lavish gifts for the very young (**Figs. 3.50–3.54**). Tomb XVIII similarly paired bronze jewelry with young infants, and when a young female (15 years) joined an older male (parent, brother, spouse?) in Tomb XXI, she or they received lavish bronze and, for the first time, iron jewelry. As discussed in Chapter 8, this phase marks the first period when adults and males received fewer grave gifts (often none) than the youthful dead (e.g., Tomb XXVIII, a child of 3 years crowned with glass and bronze beads) and young women, although not every child was thus endowed (e.g., Tombs XXIII–XXV). One of two adult males in Tomb XXXII was buried with a pair of iron arrowheads. The deposition of the projectiles was such that they could either be interpreted as two arrows placed on, or even held by the deceased, or else as the projectiles that killed the individual (the bone was too fragmentary to preserve any clear evidence for death by arrowheads). These were among the very few weapons that mark this period of transition from the Bronze Age to the Iron Age, dated by a single charcoal sample from Tomb XXIII to ca. 1070 BC. Indeed, weapons were rare in any phase of the tumulus, and comprise a solitary iron spearhead found with the older male in Tomb XLV (Phase III); the few iron knives encountered in tombs are best seen as all-purpose tools rather than full-fledged weapons.

Most of these patterns continue in Phase III (Tombs XXXIV–LIV, Early Iron Age, again dated by a single charcoal sample in Tomb XLV to 953 ± 53 BC), without overt signs of value attached to war and weapons, tools, or traded implements: indeed, spindlewhorls, beads, or buttons were worthy of burial, elsewhere a knife (Tomb XXXV), and in general domestic offerings shape an impression of a community that lavished metal ornaments largely on those who died before achieving full life status (Chapter 8) or a critical *rite de passage*. A second example of cremation at Lofkënd, in Tomb XXXVIII, marks an unusual addition of a cremated individual (probable female) to the burial of an adult male who may have died from a stab wound and was "bundled" (carried in a tightly flexed position), perhaps both transported from a locale of sudden or violent death, and after cremation on a funeral pyre elsewhere, to the peaceful and remote (family or clan) cemetery at Lofkënd. Like the triple burial (Tomb XLV, source of the human collagen sample that yielded an absolute date of 953 ± 53 BC) that contains the only spearhead (and one of the very few true weapons) in the Lofkënd tumulus, unusual circumstances may have brought three family members together in death in both of these graves. Despite the higher numbers of adult males (vs. female) in Phases III, Va, and Vb (**Table 6.4**), almost none were buried with signs of military prowess or other achievements in life: this does not necessarily mean the Lofkënd clan enjoyed peaceful lives, as such insignia may simply not have mattered to their relatives. In death, at least, there is little if any attempt to mark the status of adult males: "big men" are not part of the funerary rituals at Lofkënd. These patterns conform to customs for modern "Illyrians" (Rascians of the Moesia area), as observed by a curious traveler of the early nineteenth century (Shoberl 1827:162 [with comments inserted by Sarah Morris]):

> When a man dies, they dress him up in his clothes, put on his boots, and place near him his tobacco pouch, pipe, knife, fork and other implements which he used when alive. [Then his widow asks him if he wants anything else!]

What remains unknown is where the living population associated with the Lofkënd tumulus resided. The remains of building materials from grave and tumulus fill (fire-affected daub: Chapter 14) are not linked to any particular period of the past; indeed, like the stone tools, they could greatly predate or postdate the times of burial (see Chapter 14).

In terms of physical stature, the prehistoric individuals from the tumulus were taller than their modern counterparts, both those estimated from current populations and those buried in Tombs LXXXVI–C (Chapter 6). Collectively, they show signs of conventional sexual dimorphism (**Table 6.5**), with a greater range of dimorphism among males, as if different tasks and nutrient levels shaped the lives and physiques of individual men. In terms of diet and health, the Lofkënd population suffered from high levels of linear enamel hypoplasia among its adults, both male and female (**Table 6.6**), perhaps from shared exposure to stress (disease and/or periods of poor nutrition) during childhood. Few individuals (one child and five adults: Chapter 6) showed signs of possible cranial porosity in the cribra orbitalia area, but whether those who suffered from anemia, malnutrition, infections, or parasites did not survive for burial at this place, or a majority of the community had escaped these challenges, cannot be confirmed. While some individuals suffered from osteoarthritis and osteophysis of the vertebrae, this largely affected the aged (e.g., Individual 495 in Tomb XLV, Lofkënd's oldest male). Dental caries are relatively high compared to levels in other Iron Age populations, approaching those common for the Neolithic, with higher incidence of caries for men than for women (probably due to a greater number of older males in this population: **Table 6.7**), and there is frequent dental crowding of the kind attributed to a softer diet. Non-metric dental traits strongly suggest close kin relations between members of this community, whether measured in the first grave or in one of the last (**Table 6.14**). In terms of life histories, beyond various injured legs and arms, one individual (139, in Tomb LXXVII) sustained and survived a serious head wound above his right eye (**Fig. 6.14**), and at least one man may have died from arrow wounds (Individual 508, in Tomb XXXII, a relative of the other male in the same burial: **Table 6.14**), but few skeletons show signs of constant battle or multiple injuries.

How does this human community compare to others in the area and to those buried later? What is striking about the Lofkënd population is the contrast in their physical health with the early modern individuals interred in the same tumulus (Chapters 3 and 6–7). All five modern adults (buried ca. 1700–1900? AD) had degenerative joint disease of their thoracic vertebrae, and most of their teeth were carious or lost, possibly because of their age (**Tables 6.7–6.9**). Among those buried at Apollonia in the prehistoric period (60 individuals, Early Bronze through Late Iron Ages, in Tumuli 9, 10, and 11: Schepartz in Amore 2010), older males were even more dominant in numbers, with a higher proportion of young adults than standard for prehistoric populations, as if the same burial selectivity was practiced there as at pre- and proto-historic Lofkënd. In other convergences, the Classical and Hellenistic individuals at Apollonia share traits (such as elongated crania) and other biodistance factors with both prehistoric populations, whereas the modern populations of both Lofkënd and Apollonia tumuli differ from them and resemble more closely modern Balkan individuals. At the same time, individuals buried at Apollonia, both in prehistoric and colonial (Classical and Hellenistic) times, show significantly higher rates of stress, disease, or malnutrition (in bones and teeth): possibly the delta and wetlands at the mouth of the Vjosë/Aoös and Seman rivers exposed local populations to malaria and other maladies. These signs of stress or disease are even higher among the Classical and Hellenistic individuals at Apollonia, whether through changes in lifestyle under Greek colonization, higher density of occupation, or a combination of environmental and social factors.

Overall, those living near Lofkënd were marginally healthier than their Apollonia counterparts when they died, but they were also younger at the time of death. More striking is the decline in health in modern populations at both Lofkënd and Apollonia, reinforced by stable isotope analysis of samples of human bone collagen (Chapter 7) that points to a higher reliance on plants high in C_4 but low in protein and iron (probably maize) in modern times, while prehistoric Lofkëndis probably consumed more legumes (C_3 plants) than dairy products and meat. This undermines the notion of an ancient pastoralist economy based on meat and milk and suggests instead more reliance on cultivated plants as a primary food source in the Late Bronze and Early Iron Ages.

Material culture interred with the dead, essentially metal jewelry and clay vessels, links this community closely with those using the same objects farther north in the Balkans, to the south in Epirus, Macedonia, and Thessaly, and across the sea in Italy, without any precision as to their intra-Illyrian identity. In terms of social patterns, signs of ranking and hierarchy are difficult to detect in the burials, whether in artifacts or in factors of health, and were probably absent in daily life: one senses a community not much larger than an extended family, on the margins of more prosperous and ostentatious societies, but eventually, perhaps, a victim of distant or growing conflict (Chapter 21).

It is during the final phase of its Iron Age use (900–800 BC) that the tumulus seems to accumulate a higher number of burials as well as a rise in double and multiple inhumations: a growth in population, rise in death rates, or increase in group death by violence? One reason could be increased stress or conflict, of the kind that may well have ended the lives of two young men buried together with an older male, in a striking, anomalous way, stretched out in extended positions, parallel and close to each other, without offerings (Tomb LXXIV: Chapter 21). And it is hard not to weigh these statistics against the abrupt end of burial at this site, around or slightly after 800 BC, just about the time when elevated sites are newly fortified in the area (Gurëzezë, Mashkjezë) and in the distant north (Grunas, Shala: for which, see Galaty et al. 2013). Possibly displaced by newly intrusive tribes, to the point of forced migration across the Adriatic to southern Italy (Stocker 2009: summary), the community that buried its dead at Lofkënd no longer did so in the eighth century BC. Whether absorbed into larger (safer?) fortified sites such as Margëlliç, or relocated across the sea bridge to Italy—for which there is, to date, no conclusive evidence—its fate belongs to an internal history that is still poorly understood yet surely indigenous in the origin of its developments and independent of any Greek ambitions in this area.

ILLYRIA CAPTA: THE ADVENT OF CORINTHIANS

By the time an unknown visitor left a few Corinthian kotyle sherds on the surface of the tumulus (Appendix 2, and see **9/332–9/336**, and **Table A2.1**), no one had been buried here for over 200 years, and by now the mound had assumed its role as a local monument. In that capacity, it may well have attracted such a visitor through its distinctive shape, and inspired one of them, or others who viewed these mounds in the Illyrian landscape, to bury their dead in a similar formation, as they did at Apollonia (Amore 2010). In dealing with the numerous tumulus burials in the Kryegjatë valley at Apollonia, Sharon Stocker and Jack Davis (2006:91) ask:

> Did the earliest Greek settlers at Apollonia make a conscious decision to imitate a form of burial that they found in the landscape—one that was not then customary in the areas of southern Greece from which they emigrated? By practicing burial in tumuli, did they intentionally distinguish their mortuary customs from those of other Greek settlements in Albania, in particular Epidamnos-Dyrrachium? Did they view such a form of burial as a heroic reflection of practices familiar to them from Homeric epic?

In so doing, Stocker and Davis go on to cite Ian Morris (1997:558), who emphasized that "the epics and the material record were both produced by real people, acting in pursuit of their own goals in a period of dramatic change.... [I]n the rituals that produced the archaeological record, Iron Age Greeks manipulated material culture as a non-verbal language through which they debated who they were and where they stood relative to each other, the larger east Mediterranean world, and the lost heroic past" (Stocker and Davis 2006:91).

The Corinthians who settled in Apollonia ushered in a new era and level of human occupation in this part of coastal Albania, one that irrevocably changed the surrounding landscape forever. Although some of the tumuli of the Kryegjatë valley at Apollonia clearly date as early as the Early Bronze Age, the majority of burial mounds date from the Archaic through Hellenistic and Roman periods. Whichever way one counts the Apollonian tumuli—Vangjel Dimo (2004:12) counted some 450 tumuli in 1995–1996, whereas Stocker and Davis's (2006:89) fieldwork yielded a more conservative minimum estimate of about 100 tumuli—their number in the Kryegjatë valley alone is staggering and, in terms of what existed before, represents an exponential number of people inhabiting the region. Although a handful of Corinthian sherds at Lofkënd may seem a pale reflection of what was happening closer to the delta of the Vjosë and Seman rivers, it is nevertheless symptomatic of the era of Greek colonization.

While we have proposed several scenarios for the end of our Illyrian tumulus and the rise of novel urban forms, first as Illyrian proto-urban fortresses (Ceka 2011; Lafe 2003) and then as Hellenized cities, the exploration of the Lofkënd tumulus does not make a straightforward and lasting contribution to understanding this evolution. Instead, it offers a tantalizing link between the last of the tumuli and the beginning of life in Greek-style cities, in the form of a handful of sherds. While it is remotely possible that later (Archaic) burials were inserted into the mound and disturbed between the sixth century BC and the modern era of burials, we would expect more substantial remains of disturbed tombs and their artifacts, from the intervening years, if this were true.

THE LOFKËND AREA IN THE GREEK AND ROMAN PERIODS

Little is known about this area during the Classical through Late Roman periods, when our knowledge is limited to surface finds, but two sites lie nearby and document occupation permanent enough for burials. Due north of the tumulus and slightly to the east, and only a few hundred meters away, a bedrock outcrop was chosen for rock-cut cavities to hold inhumations, some covered with tiles, during the Hellenistic and Roman periods (Chapter 18, Site 001, Mezhdat e Kuqe: **Figs. 18.1–18.3**). At nearby Gjinoqara, at least one site yielded Roman through Medieval ceramics, and later became the location of a church torn down in 1924. But for most of the Classical and Hellenistic periods, it appears that Margëlliç remained the largest settlement of the area, and by the Roman period, not only Byllis but a new late-antique Christian community at Ballsh dominated this region.

As for the tumulus itself, material clearly postdating the Corinthian and predating the modern was limited indeed (see **Table 9.13**). Among the inventoried pottery presented in Chapter 9, there is a base fragment that is plausibly Athenian, perhaps from a skyphoid cup (**9/330**), picked up on the surface of the tumulus; another fragment from topsoil may be Greek (**9/331**). The remainder is essentially non-diagnostic, consisting of three fragments of a gray, reduced fabric (**9/337–9/339**), and a fire-damaged sherd (**9/340**), about which very little can be said. Of these, one (**9/338**) derives from topsoil, and the remainder are from tumulus fill and may well be Late Bronze or Early Iron Age. Among the non-inventoried wheelmade undecorated pottery summarized in Chapter 2 (see especially **Table 2.3** and associated discussion), one fragment was noted as plausibly being Hellenistic, but the sherd was too fragmentary and poorly preserved to assign a date with any confidence.

FROM LOFKËND TO LIKOFONE: CHRISTIANS AND MUSLIMS IN THE GJANICË VALLEY

For centuries, the tumulus stood undisturbed by serious interventions, especially after the decline of nearby Byllis and Ballsh (the seat of successive Christian bishoprics through the Middle Ages) and the conversion of most of the Orthodox population to Islam (Chapter 3.2). But a few Christians remained in the area or migrated to it, like the solitary Vlach registered in the Austrian census of the early twentieth century as Eastern Orthodox (Chapter 3.2). A century earlier, local inhabitants, probably Christians who did not share the Muslim cemeteries of their neighbors (at Ngrancija, southeast of the tumulus, or at Mashkullorë: **Fig. 18.4**), chose the north end of the tumulus to inter a few of their dead. Whether they were attracted by the visible mound or its distance from Muslim practices, or were remotely aware of how it harbored the dead of a distant past, must remain a question. Those whom they buried here fit a more predictable demographic profile: infants or neonates, and older adults, few buried with grave goods beyond the brass clasp that fastened an infant garment (**TXCVIII-1**: **Fig. 3.325a–b** [SF 167]) or a coin to pay the ferryman of the dead in three tombs, two adults and one infant (**TXCII-1, Figs. 3.306c–d, 3.308a–b; TXCIV-1, Figs. 3.314c, 3.315a–b; TC-1, Figs. 3.329a, 3.330**). By their position (facing east to Jerusalem) they identify themselves as Christians, but no testimonia survive to give them a more proximate identity or explain the circumstances under which they buried those most vulnerable to mortality (elderly adults and newborn infants) in this place and only over a period of a few years. What helps make these late graves at Lofkënd less anomalous are the many cases of medieval and early modern interventions noted at other prehistoric mounds in southeastern Europe and in Anatolia (see discussion in Chapter 3.2).

Of the material from the surface or topsoil of the tumulus, one fragment (**9/341**) is clearly modern and derives from a vessel of an early modern painted Grottaglie ware from the nearby Greek island of Kerkyra (Corfu). Among the few non-inventoried wheelmade undecorated sherds from the surface or topsoil units of the tumulus summarized in Chapter

2, virtually all are modern but could not be dated more precisely.

DARK CONTINENT: ALBANIA AND EUROPE IN A CENTURY OF VIOLENCE

In its final sustained period of use, the tumulus at Lofkënd became a vantage point for warfare during the twentieth century. The spent cartridges released from at least 30 shots were left on the mound, many finding their way into the surface levels of the prehistoric tumulus (Appendix 3, following Chapter 10), and one or two even farther down. As fate would have it, a Greek born in Albania, who moved across the border to Greece as a youth in the wake of these events, became our consultant on these bullets, thanks to his position at the War Museum in Athens and his generosity with his time and expertise.

The twentieth century marked the liberation of nations like Albania from the Ottoman Empire, its birth as a state following the first world conflict, its experiences under monarchy, communism, and democracy punctuated by the outbreak of the Second World War, which brought Albania into the path of an expansionist neighbor, Italy under the Fascists. The archaeological component of Mussolini's vision sent Luigi Ugolini and his fellows to excavate Butrint and explore Rome's "Trojan" and Augustan predecessor (Gilkes 2003; Ugolini 1927; see also Francis 2005 for the prehistorian Luigi Cardini), but it also sent more explicitly strategic expeditions to recolonize what the Romans occupied as Illyria, in both world wars. The first one haunted our tumulus with local legends of fallen soldiers and the immortal fiction of Kadare (*General of the Dead Army*: Papadopoulos 2006), but it also brought Italian soldiers, or their ordnance, to the tumulus. The Second World War brought Italy to the borders of Greece, where some of the hardest campaigns were fought in deep snow and in the highest mountains (such as Mount Grammos), but it also drew Greece deep into Albanian territory, as far north as Vlorë, and evidently up the Gjanicë valley, as the many spent cartridge shells of standard Greek rifles testify.

POST-COMMUNIST ALBANIA: THE TEKKE AND THE PEAR TREE

In a visit to the site prior to excavations in 2003, Bejko and Papadopoulos chanced upon the local owner of the land on which the tumulus stood, a certain Rrapi Malaj, whose house was located down the slope, immediately to the east of the tumulus. Following discussions with him, we were heartened by his response to and general approval of the project as we had outlined it. Having the approval of the landowner is essential to the success of any archaeological project, and in this case, we assumed, would be assured through an appropriate "rental" fee (better known by the Persian term *baksheesh*), as indeed it was. When we returned with a full team of excavators and specialists the following year, on the very first day of excavations, we encountered not Rrapi Malaj, but Baki Ymeri, who was working the land together with his family. When we approached him to establish his relationship to the land, his response took us by surprise, for he also claimed to be the landowner. Upon further discussion with him and with Rrapi Malaj, it was clear that both gentlemen had genuine claims of ownership. Although we were resigned to a double dose of "rental" fee, we wished to get to the bottom of the situation and understand the nature of both claims, and it was here where the diplomacy and tact of the then-director of the Institute of Archaeology, Academy of Sciences of Albania, Muzafer Korkuti, proved pivotal, as he spanned the Communist and post-Communist eras of his homeland. As it turned out, one owner represented the local religious organization, the *tekke* (see Chapter 18, site S008, **Fig. 18.1**) and had been caretaker of land that belonged to it, despite the fact that the *tekke* was no longer in use, while a farmer claimed ownership due to his cultivation of the land. Their dual claims might eventually be resolved by court decisions in Albania necessary in a post-Communist era during the process of establishing a national land registry. In the meantime, our project honored their dual claims to ownership and therefore compensation, in the spirit of both local cooperation and intellectual commitment to the notion that landscapes, ancient and modern, generate multiple human stakeholders.

A final casualty of our project was the pear tree, grafted from both wild and cultivated species, that clung precariously to the summit of the tumulus before we began excavation (**Fig. 2.5a**: view of tumulus in 2004). It had to be removed during our second season, for which compensation was paid, but this only provoked a new campaign of tree planting by the landowner (Rrapi Malaj) that disturbed at least one prehistoric grave. At that point, owners and archaeologists reached an agreement to suspend further planting, and uprooting, of fruit trees, for the sake of the dead.

After Archaeology: Reconstruction and the Return of a Monument to Its Landscape

Our initial commitment and final obligation involved the restoration of a natural and human-made phenomenon to its critical place in the landscape. As detailed in Chapter 22 and in earlier publications, the same team that deconstructed the tumulus turned to its reconstruction, in the final season, in a process that recycled the debris of a human-built mound, minus its human occupants and their final possessions, into a reconstituted monument (**Fig. 22.12–22.13**). While we cannot return to their final resting place the ancient population that chose to be buried here, a memorial to their lives and afterlives will rise forever in the local landscape.

What we (re)constructed in the landscape was not a burial mound but the visibility of a tumulus. In the *longue durée* of its existence, the mound served as a place of burial for only a relatively short period of time, ca. 1400–800 BC, and again for an even shorter period of time around AD 1800. Within this heap of earth were lodged Paleolithic artifacts as well as bullets and spent gun cartridges from the Second World War, and lots more in between. On its surface, we encountered even more modern garbage (presented in Chapter 1.2). Yet at various times in its existence, this heap of earth served as a focus of shared remembrance, social memory, and even identity. The Lofkënd tumulus provided the Late Bronze and Early Iron Age inhabitants of the Gjanicë valley not only an image of, but an anchor to, their past, as well as offering them a conceptual point from which to approach their future. For the inhabitants surrounding the mound today, the excavated and reconstructed heap of earth represents a very different focus of identity, but it has became part of their—and our—cultural heritage.

Final Thoughts: Reflexive Archaeology

Modern archaeology turned the occupants of a single, extended-family cemetery of the prehistoric era in a remote valley of Albania into citizens of a global age. In this disruptive process, they were exposed by a team in part from a continent unknown in Europe in their lifetimes, analyzed with instruments and methods of the nuclear age, introduced to their Stone Age ancestors whose tools were folded into their final resting places, only to be reburied in museum boxes near their relatives at Apollonia. The last of them was interred before Greeks adopted the alphabet and brought it to Albania (seen early at Apollonia, not least in a Corinthian or Corinthian-style sarcophagus), and all of them are now visible in digital form, as long as the Internet lasts.

For the modern team who explored the tumulus, numerous life passages marked the five years of excavation for its members: six young archaeologists from the United States and Albania became parents of offspring destined for healthier lives than children born in Bronze and Iron Age Illyria; several students earned graduate degrees and many began careers; one landowner celebrated the marriage of his daughter, herself a member of the field team at Lofkënd. Thus, the rites of birth and marriage carried human biographies forward beyond our multiple encounters with the end of life. The excavation directors brought to the tumulus their expectations from related sites (a central warrior grave, double or triple vases, and bronze swords were on order), and we anticipated a single season, two at the most. Five hard-working seasons later, we are grateful for our immersion in the early history of Illyria.

Summary

Tuma Prehistorike e Lofkëndit në Shqipëri

Përmbledhje

KAPITULLI 1.
Hyrje

John K. Papadopoulos, Sarah P. Morris, Lorenc Bejko, dhe Lynne A. Schepartz, me kontribut nga Seth Pevnick dhe Esmeralda Agolli

Rajoni i Mallakastrës dhe Shqipëria jugperëndimore, mbartin shumë varreza tumulare të epokave të Bronzit dhe Hekurit (Kapitujt 1, 17) (**Figs. 1.1, 1.2, 17.1**). Disa janë dëmtuar shumë kohë më parë, të tjera janë ripërdorur si vendvarrime moderne, dhe vetëm pak syresh janë trajtuar në kushte shkencore, si ajo e Patosit rreth 14 km larg Tumës së Lofkëndit (**Fig. 1.10**). Kuptimi ynë sot për pre- dhe proto-historinë e Ilirisë është formuar në një pjesë të mirë nga përmbajtja e këtyre varrezave tumulare (Kapitulli 17), të tilla sa e shkuara e Ilirisë ende shfaqet nga pamjet tërheqëse të këtyre "kësulave të kujtesës" (Kapitulli 20). Përkundër këstaj historie, kodra që mban të njëjtin emër me atë të fshatit më të afërt, Lofkënd, shënon një profil të veçantë në hapësirën përreth (**Figs. 1.3, 1.4**) si dhe një rast të patrazuar për arkeologjinë. Tuma gjithashtu është vendosur pranë një sërë vendbanimesh të epokave të ndryshme: ai neolitik në Cakran, i Bronzit në Margëlliç, të periudhës "protourbane" në Gurëzezë dhe Mashkjezë, të periudhës Klasike/Helenistike në Apolloni, Bylis, Klos-Nikaia, dhe ato Romake e Bizantine në Bylis e Ballsh (**Fig. 1.2**). Në konsideratë të kësaj panorame, vendvarrimi na ftoi të ushtronim një gërmim tërësor, me metoda moderne shkencore rreth një hapësire qartësisht të përdorur nga Epokat e Bronzit në atë të Hekurit (**Seksioni 1.2**), dhe gjithashtu të rishikonim domethënien e kësaj tume dhe simotrave të saj lidhur me lindjen e urbanizimit dhe kompleksitetit shoqëror në Ilirinë antike.

Vendvarrimi dhe kuadri i tij mjedisor dhe historik

Rajoni bregdetar perëndimor i Shqipërisë konfigurohet nga rrjedha dhe depozitimet e lumenjve të mëdhenj të ushqyer nga përrenj që penetrojnë përmes përbërjeve ranore, gëlqerore dhe argjilore (**Fig. 1.1**; Kapitulli 16.4). Lumi i Gjanicës rrjedh në drejtimin juglindje-veriperëndim dhe bashkohet me Lumin Seman duke iu drejtuar detit. Lugina e tij i përket prapatokës së bregdetit që tërhoqi që në krye të herës kolonitë greke (**Fig 1.2**). Në synimin e hulumtimeve në vendbanimet antike të këtij rajoni janë vendosur qytetet Greke (Apollonia) dhe gjithashtu vendbanime afatgjata si ai i Margëlliçit (**Figs. 1.5a–b, 20.6**). Një prej projekteve të para sistematike sipërfaqësore, emërtuar sipas rajonit të Mallakastrës, ku shtrihet edhe tuma shqyrtoi rrethinat e Apollonisë duke zgjeruar njohuritë mbi historinë lokale nga Paleoliti në kohët moderne (*http://river.blg.uc.edu/mrap/MRAP_en.html*).

Zona në drejtimin veri-jug të rrugës automobilistike nga Fieri në Ballsh shfaq fusha të kultivuara përgjatë lumit, shpate të lartë dhe tarraca me të mbjella ose kullota, fshatra të shpërndara dhe aty këtu ndonjë pompë nafte (**Fig. 1.6**; Kapitulli 15). Në kohët prehistorike, kjo zonë duhet të ketë qënë e mbuluar nga pyje të dëndur e të përshtatshëm për rritjen e kafshëve të egra të mëdha, kjo duke iu referuar sasisë së bollshme të veglave të gurit në depozitat antike që gjenden tashmë në sipërfaqe (Kapitujt 13, 18). Nga Epoka e Bronzit deri në periudhën Bizantine dhe përtej, zona ka mbartur vendbanime, varreza dhe faltore, disa nga të cilat të eksploruara në kohët moderne. Në përgjithësi, kjo zonë përmendet nga autorët antikë si territor i një ose më shumë fiseve me përkatësi nga Iliria jugore dhe Epiri.

Hekateu i Miletit (fundi i shek. gjashtë dhe fillimi i pestë p.e.s.) përmend disa fise në Epir dhe Ilirinë jugore (**Fig. 1.7**). Lofkëndi shtrihet në veri të fiseve epirote në Ilirinë jugore; më qartë, në veri dhe lindje të Kaonisë antike (midis lumit Kalamas ose Thyamis në Epir dhe Gadishullit të Akroreauneve, Karaburunit në perëndim të Vlorës, Auloni antik) dhe në veri dhe perëndim të Atitanisë (**Fig. 1.7**). Straboni (7.7.8) quan fise ilire në këtë rajon, Bylinët, Taulantët, Parthinët dhe Brygët; Plini Plak (3.23.145) e vendos Nymfeun antik (Nymphaeum, i njohur për bitumin (Kapitulli 15) në zonën e banuar nga Amantët (Amantes) dhe Bylinët (Buliones). Kjo e vendos Lofkëndin (afër me të dy, Bylisin dhe Nymphaion) në territorin e Bylinëve.

Kontaktet e qarta me Egjeun lidhen me Epokën e Bronzit të vonë në Margëlliç, me qeramikën Mikene, kohë përpara themelimint të kolonisë greko-korintase të Apollonisë rreth 600 p.e.s. (Straboni 8.316, Pseudo-Scymnus 439, Tuqididi 1.26.2,: Pausania 5.22.4). Sidoqoftë periudha kohore mes epokave të Mikenës dhe kolonizimit grek ende mbart një sërë pikëpyetjesh për prapatokën.

Tuma e Lofkëndit shtrihet 25 km në lindje dhe juglindje të Apollonisë rreth 357 m mbi nivelin e detit, në një kreshtë ranore me drejtim nga verilindja në jugperëndim (**Figs. 1.1–1.3, 1.7, 2.1**). Përdorimi prej gjashtë shekujsh me varrimet e hershme (rreth 1400–800 p.e.s.), bashkë me 15 varret moderne në skajin verilindor e shdërroi këtë kreshtë në një kodër të spikatur dhe qartësisht të dukshme prej së largu (**Fig. 1.3**).

Erozioni i vazhdueshëm në anën jugore kishte nxjerrë në pah kocka njerëzore (**Fig. 1.8**). Fragmente të hershme qeramike dhe vegla guri ndesheshin të shpërndara në sipërfaqe pa ndonjë shenjë dëmtimi (**Figs. 1.14, 1.15, Tabelat 1.1, 1.2**). Tuma ka marrë emrin e njërit prej fshatrave modern që shtrihet 3 km në lindje, në një ngrehinë të dukshme (shih. Mashkullorë **Figs. 18.4, 20.13**, dhe **20.14**) dhe Patos, (**Figs. 1.10, 1.11**).

Rajoni i Lofkëndit përfaqëson njërën prej zonave më të pasura arkeologjike të Shqipërisë, rrethuar nga qendra domethënëse protourbane dhe Klasike/Helenistike si Margëlliçi (**Fig. 1.2**) në perëndim dhe veriperëndim, Gurëzeza, dhe Mashkjeza në jug dhe perëndim, Bylisi dhe Klos-Nikaia në jug, dhe Dimali (Kalaja e Krotinës) në verilindje. Vendndodhje të rëndësishme në rajon përfshijnë tumën e Epokës së Bronzit dhe Hekurit në Patos (14 km në perëndim-veriperëndim), gjetjet e Epokës së Bronzit në Drenovë, dhe vendbanimin e Neolitit të mesëm në Cakran. Pra, ngritja modeste në horizontin e luginës së Gjanicës shënonte në fakt një vend të pasur me histori të gjatë përreth tij.

Objektivat e Projektit

Hulumtimet më të fundit në Ballkanin jugor e kanë zhvendosur vëmendjen e kërkimeve arkeologjike të pre-dhe proto-historisë përtej etnogjenezës së ilirëve dhe shqiptarëve (Kapitulli 17), ose origjinës së qyteteve dhe kolonive, për të konsideruar zhvillimin e mënyrave të hershme të jetesës përkundër përmbajtjes së historisë klasike. Vendndodhja e Lofkëndit ndërlidhte kërkimet më të fundit në Iliri dhe Epir (si ato të Kamenicës në Shqipërinë juglindore ose Liatovounit dhe Vitsa Zagoriou në Greqinë veriperëndimore: **Fig. 1.7**), me arteriet antike kulturore dhe historike nga malet e Pindit e deri në Detin Jon. Çështje kryesore i lidhin këto zona me anë të kulturës materiale si qeramika e pikturuar e punuar me dorë, objektet prej metali dhe konteksti i një grupi të ri varresh premtojnë përgjigje të freskëta lidhur me problematikën e kontakteve dhe shkëmbimeve kulturore, mënyrën e jetesës, dimensionet e marrëdhënieve fisnore dhe ndërveprimet kulturore mes tyre. Lidhur me këtë proces, gjetjet sipërfaqësore në tumën e Lofkëndit mbulonin hapësirë të rëndësishme kohore të shekujve të prehistorisë (të pakta nga vëzhgimi sipërfaqsor i MRAP) pas rënies së vendbanimeve të fortifikuara të Epokës së Bronzit të tipit të Margëlliçit (**Fig. 1.5a–b**) deri pak përpara lindjes së qendrave protourbane (Gajtani afër Shkodrës, ose Gurëzeza afër Lofkëndit: **Figs. 1.1, 1.2**). Ne synuam të mbushnim këtë boshllëk me materiale të detajuara, të datuara në mënyrë shkencore që siguronin informacion për shoqërinë njerëzore, mjedisin, praktikat e varrimit, teknologjinë e materialeve dhe modelet e zhvillimit shoqëror në Epokat e Bronzit dhe të Hekurit. Për të gjetur vendbanimin shoqërues bashkëkohor, nisëm një vëzhgim sipërfaqsor përreth zonës, dhe përftuam rezultate krejt të papritura për periudhën prehistorike, kryesisht të epokës Paleolitike (Kapitulli 18). Prioritet kryesor mbeti studimi i lindjes së vendbanimeve më të qëndrueshme që tejkalonin për nga përmasat dhe kompleksiteti fshatrat e përbërë nga rrethi i një familje të zgjeruar dhe sqarimi nëse këto u shfaqën si zhvillim i brendshëm apo si imitime të qyteteve Greke.

Gërmimet në Lofkënd na lejuan të shqyrtonim një sërë fenomenesh të ndërlidhura: lidhja e vendvarrimit nga një popullsi e veçantë; modeli i vendbanimeve në Shqipërinë jug-qëndrore bazuar në blegtori dhe bujqësi (identifikimi i sedentarizmit të kombinuar me bujqësi dhe/ose kullota si mjet kryesor e strategjik jetese); eksplorimi i demografisë dhe marrëdhëniet e përkohshme mes popullsisë së Lofkëndit (fis i bazuar në lidhje familjare) dhe kolonistëve të Apollonisë, një *polis* Greek i krijuar rreth 200 vjet më vonë (Kapitulli 4; Shtojca 2, Kapitulli 9); shqyrtimi i origjinës, datimi dhe natyra e qendrave të reja "protourbane" në prapatokën ilire gjatë Hekurit të zhvilluar (Margëlliç, Gurëzezë, Klos-Nikaia, dhe Mashkjezë): si qyteza të vërteta, strehime të fortifikuara, vendbanime sezonale, apo pika rajonale të përdorura për tregti ose takime? Vrojtimet dhe gërmimet e fundit në Butrint, Kalivo, dhe Kepin e Stillos në Kaoni kanë dhënë vetëm depozita të përziera me qeramikë prehistorike pa datim absolut, por tashmë një fortifikim prehistorik (në luginën e Shalës në Shqipërinë veriore) është e datuar saktësisht në Hekurin e hershëm (datimi me radiokarbon AMS, rreth 800 p.e.s. i kalibruar).

Si edhe në Itali, poliset arkaike greke përgjatë bregut të Ilirisë e Jugut (Apolloni dhe Epidamnos) ndryshojnë nga zhvillimet e vendasve në brendësi të prapatokës, ku varreza në zona të spikatura si Lofkëndi duhet patjetër të jenë mbështetur nga vendbanime bashkëkohore. Sidoqoftë i vetmi instalim me datim në Bronzin e vonë dhe Hekurin e hershëm (Krapsi afër Patosit) ka material të Bronzit të vonë dhe vetëm pak të periudhës së hershme të Hekurit. Gjithashtu vëzhgimi sipërfaqësor i MRAP (përfshi edhe vëzhgimin tonë përreth tumës: Kapitulli 18) ka prodhuar të dhëna të paqarta të vendbanimeve të epokave të Bronzit dhe Hekurit. Tuma 10 e Apollonisë së fundmi dha varre prehistorike me qeramikë të Bronzit të vonë, gjilpëra prej kocke dhe bronzi, një shpatë bronzi dhe thikë dhe një figurinë terrakote në formë violine të periudhës së Bronzit të hershëm, të gjitha pranë varreve Klasike. Përmbajtja e saj shënon një kapitull të ri në prehistorinë e kolonisë Greke, por jo ndonjë të dhënë për vendbanimin. Ne kërkuam të kuptonim këtë tranzicion nga fshatrat e pafortifikuar–komai greke (Tuqididi 1.5, 1.10, 3.94; Herodoti 5.98; Aristoteli, *Poetika* 1448a.36; *Politika* 1261a.28) karakteristik të një *etnosi,* shtet fisnor bazuar në lidhjet familjare–në qytetet e "fortifikuara" dhe raportet e tyre me modelet Greke (si në fushëgropën e Janinës). Në Epir, dy varreza Molose me vendbanimet përkatëse (Vitsa Zagoriou dhe Liatovouni) vazhduan të jenë aktive përgjatë periudhës Klasike ndërsa rajoni fqinj i Korçë-Kolonjës në Shqipërinë juglindore njohu vendbanime të reja në maja kodrash në fund të Epokës së Bronzit dhe Hekurit të hershëm. Ne synuam të kuptonim mënyrësn se si këto tuma dhe qyteza u zhvilluan përpara dhe jashtë qyteteve Klasike, në zonat në veri të Greqisë.

Në mbështetje të këtyre qëllimeve studimore u krijua një ndërmarrje e përbashkët me burime njerëzore, aftësi, përgjegjësi dhe partnerë të rinj nga Shtetet e Bashkuara dhe Shqipëria për përgatitjen e profesionistëve të së ardhmes në shumë disiplina. Studentë dhe specialistë nga një numër institucionesh në të dyja shtetet u bashkuan për punën në terren dhe laborator, ndarjen e aftësive, pajisjeve, dhe eksperiencave. Ne gjithashtu synuam të mbështesnim praktika dhe trajnime në konservim dhe bashkuam konservues të trajnuar nga Shtetet e Bashkuara dhe Shqipëria. Gjatë sezonit të fundit, rishtas u organizua një shkollë terreni me një grup studentësh Amerikanë dhe Shqiptarë me qëllim përvetsimin e teknikave të vëzhgimit sipërfaqësor, identifikimin dhe katalogimin e gjetjeve, analizat e guaskave dhe kockave, fluotacionin dhe klasifikimin e mbetjeve bimore nga kampionet e dheut. Për botimin përfundimtar, materiale kryesore ose çështje të ndryshme iu caktuan studentëve nga Shqipëria dhe Shtetet e Bashkuara, në disa raste duke punuar si bashkë-autorë (Kapitujt 8, 9, 10, 13, 16.1, 18), kjo për të përmbushur synimet tona në trajnimin e gjeneratës së re të arkeologëve në të dyja shtetet.

Metodologjitë

Projekti i Lofkëndit u zotua të mbledhjë dhe analizojë në tërësi materialin e tumës: i gjithë materiali u trajtua në sitën e thatë, nga secili kontekst varri u ruajtën për fluotacion kampione dheu dhe përmbajtjet arkeologjike u numëruan, peshuan dhe dokumentuan çdo ditë në arkivin dixhital të të dhënave GSHASH (Grupi Shqiptar i Arkeologjisë së Shpëtimit, Kapitulli 2, Seksioni 1). Një studiues i dherave hulumtoi jo vetëm matriksin e tumës, me përbërjen dhe formacionin por gjithashtu edhe dherat dhe gjeomorfologjinë e mjedisit përreth (Kapitulli 16.4). Për më tepër, rëndësi të veçantë i kushtuam edhe objekteve jo-artefakte në përmbajtjen e mbushjes së tumës, veçanërisht fragmenteve të shumta të baltës së pjekur nga strukturat e plitharit/qerpçit të padatuara saktë, (Kapitulli 14), e vetmja e dhënë e jona për arkitekturën prehistorike lokale. Dokumentimi vendosi standarte të larta: arkitekti dhe vëzhguesi i terrenit Max Farrar dokumentoi dhe shkarkoi çdo ditë të dhëna nga stacioni total dhe skicoi kontekstin e çdo varri *in situ* për formatin përfundimtar dixhital (Kapitulli 3). Gjatë katër sezoneve, dy artistë të talentuar, dokumentuan në fotografi me ngjyra tumën dhe objektet (2004–2007; Kapitulli 2, Seksioni 6; *http://www.sscnet.ucla.edu/ioa/staff/papadopoulos/ lofkend/photos.html*). Një dokumentim shumë i rëndësishëm i është bërë tumës edhe me fotografi ajrore nga lartësi të ulëta prej Alket Islamit me mjetin e tij fluturues (Kapitulli 2 Seksioni 5).

Bioarkeologjia mbeti një nga komponentët kryesorë të këtij projekti studimor. Lynne Schepartz bashkëpunoi në përcaktimin e një protokolli të detajuar dhe kompleks për zhvendosjen, pastrimin e plotë, klasifikimin dhe analizimin e çdo skeleti. Përpara këtij projekti, analiza sistematike ishin realizuar në varrezat tumulare të Kamenicës dhe Apollonisë, sidoqoftë në Lofkënd patëm rastin të mblidhnim dhe analizonim të plotë popullsinë e tumës (Kapitulli 6). Laboratori Wellcome në Universitetin e Oksfordit kreu analizat e kockave njerëzore në 2004 (Shtojca 1, Kapitulli 6) dhe Brian Damiata mori kampione nga kockat njerëzore, dhëmbët dhe mbetjet prej druri të karbonizuar për analizat AMS në Laboratorin Keck të Universitetit të Kalifornisë në Irvine bashkë me John Southon (Kapitulli 4). Kjo nismë prodhoi datimet e para absolute për varret e Lofkëndit, disave nga Apollonia dhe disa nga të dhënat më të hershme të izotopeve strontium nga lashtësia shqiptare (Kapitulli 7). Mbetje organike u shënuan vetëm në dy varre me djegie me përmbajtje farash, bimësh ose druri. Mbi 250 kg dhé (rreth 1/5 e volumit të mbushjes së tumës), u mblodh dhe iu nënshtrua fluotacionit duke dhënë fillimisht material bimor të periudhës moderne (Kapitulli 16.1). Mbetjet e drurit të karbonizuar ndonëse të rralla dhe shumë herë të shpërbëra, dhanë dy specie kryesore, lis dhe panjë (Kapitulli 16.1) dhe 18 datime absolute radiokarboni (Kapitulli 4.2). Kockat e kafshëve ishin të rralla, me përjashtim të tre vendndodhjeve me mbetje moderne të dhenëve (Kapitulli 16.1). Mbetjet e guaskave të mbledhura sistematikisht ishin gjithnjë gastropodë toke të periudhës moderne (Kapitulli 16.2). Ndonëse këto strategji dhanë dëshmi minimale për mjedisin e lashtë dhe historinë njerëzore, në Shqipëri ato janë praktika relativisht të reja dhe ne mbetëm të kënaqur që vendosëm të tilla metodologji në protokollin e gërmimit.

Një seri analizash të materialeve nga profesorë dhe studentë të Institutit Cotsen të Arkeologjisë në UCLA u përqëndruan në rruazat prej qelqi dhe gjetjet metalike. Një instrument i lëvizshëm me rreze-X na lejoi të identifikonim disa objekte metalesh (vëthët prej "ari" **10/11, 10/12**, doli të ishin elektrum kjo pas aplikimit të një metode që shkatërronte minimalisht kampionin e metalit për analizat e izotopeve të plumbit me IPDMS (induktim i plazmës dyshe mas spektometër). Në total, 162 objekte prej metali u analizuan nga mikroskopi optik pXRF dhe analizat e zbërthimit me rrezet-X, 23 objekte prej aliazhi bakri u ekzaminuan për metalografinë dhe afërsisht 35 tekstile pseudomorfe ndeshur në produktet e korroduara të stolive të hekurit u fotografuan dhe analizuan (Kapitulli 12).

Për të kuptuar historinë afatgjatë të zonës përreth tumës, ne e zgjeruam projektin tonë me një vëzhgim sistematik sipërfaqësor në disa trakte në shpate dhe në luginë (Kapitulli 18), kjo e bazuar në analizat e mëparshme të stralleve të ndeshura në tumë (Kapitulli 13).

Rindërtimi dhe Jeta e Përtejme:
Kodra e kujt? Kujtesa e kujt?

Profili imponues i tumës kërkonte respektin tonë dhe rikthim në formën origjinale. Ne jo vetëm rikrijuam kodrën pas çdo sezoni duke mbushur dhe e kthyer në formën e origjinale (Kapitulli 2, Seksioni

1), por ndërrmorëm edhe rindërtimin përfundimtar pa materialet e tumës. Të njëjtët punëtorë që gradualisht hoqën shtresat e tumës prodhuan rreth 2.000 tulla qerpiçi të përbëra nga dheu i hequr nga tuma, ujë dhe kashtë (Kapitulli 22). Kjo lejoi rindërtimin e plotë të tumës me materialin e saj brenda mjedisit përkatës duke i kthyer komunitetit të sotëm dhe të ardhëm një simbol madhor të së shkuarës së tyre në rrethinat e tyre. Statusi i saj si monument kërkonte më shumë, dhe Samantha Martin-McAuliffe (arkeologe terreni dhe skicuese gjatë dy sezoneve të fundit të punës) me syrin e një arkitekteje historiane krijoi për tumën një pamje elegante dhe imagjinare (Kapitulli 20).

Një tjetër tip rindërtimi ofrohet në Kapitullin 21. Dy varre të dalluar unik si në Lofkënd dhe Shqipëri, u gjetën disi të veçuar, më i hershmi (Varri 1) i tumës dhe një prej më të vonëve (Varri LXXIV). Të dy sugjeronin, ngjarjen e themelimit të tumës, nga komuniteti që prentendonte identitetin e tij në mjedisi e vet, dhe ngjarjen më të vonë (me shumë mundësi e dhunshme, me lënien dhe varrimin e pazakontë të tre meshkujve) që tregon braktisjen përfundimtare.

Përfundimisht teknikat moderne për rikrijimin e varrezës antike sollën pamjet tre-dimensionale në një model 3-D të krijuar nga Qendra për Tekonologjinë Eksperimentale në UCLA (*http://www.sscnet.ucla.edu/ioa/staff/papadopoulos/lofkend/mapping.html*; Kapitulli 19).

Çdo projekt arkeologjik i shekullit të njëzet e një përtej synimeve shkencore, vendos prioritete postmoderne, për trashëgimninë kulturore (*http://www.kamenicatumulus.org/*). Në Lofkënd ne integruam historinë e tumës me biografinë kulturore rajonale (Epilogu), duke sinkronizuar aktivitetin në të, me kolonizimin Korintas (Shtojca 2, Kapitulli 9), varret e hershme moderne (Kapitulli 3, Seksioni 2) dhe përdorimi i vonë nga ushtarë të shekullit të njëzetë (Shtojca 3, Kapitulli 10). Në përshtypjet e vendasve, tuma mbartte mbetje përtej kujtesës së tyre, të të rënëve në një luftë në të shkuarën e largët. Por arkeologët në harmonizim me historinë kulturore dhe etnografinë, teksa spekulonin për para-historinë e tumës së Lofkëndit disa varre kërkuan t'i krahasonin me *Kanunin* Shqiptar ose me kodin feudal të nderit dhe gjakmarrjes (Kapitulli 21). Të varrosurit në tumë, vizitorët e varreve të saj, ripërdoruesit e vendit për të vdekurit e tyre dhe përfundimisht ata që sollën varret në dritën e modernitetit kontribuan në vazhdimësinë e historisë së varrezës (shih Epilogun).

Mbledhja e Materialeve Sipërfaqësore në Tumën e Lofkëndit

Përpara gërmimeve, u ndërrmorën dy raunde të mbledhjes sipërfaqësore mbi dhe përqark tumës, një në 2003 (kufizuar në materialet antike) dhe një mbledhje e plotë në 2004 e të gjithë materialeve, antike dhe moderne. Ky rast klasik i mbledhjes intensive të materialit sipërfaqësor përpara fillimit të gërmimit sistematik është prezantuar në tabela të ndara për qeramikën antike, qerpiç dhe tjegulla moderne çatie, stralle dhe materiale moderne (Kapitulli 1.2). Një pjesë e madhe e materialit modern sipërfaqësor me shumë mundësi ishin lënë nga familja që kishte kultivuar tokën rreth tumës në dekadat e fundit ose ndoshta nga pronarët vendas përgjegjës për Teqenë (Vendndodhja V008), në jugperëndim të tumës (Kapitulli 18, **Fig. 18.1**).

Pjesa më e madhe e materialit qeramik ishte antike–kryesisht e Epokave të Bronzit të vonë dhe Hekurit të hershëm–me gjashtë fragmente të inventarizuara (Kapitulli 9); fragmente të tjera prehistorike ngjajnë me katër grupet e brumit të vendosura për qeramikën e tumës, më tepër me fragmentet e shumta të ndeshura në dheun e mbushjes (**Fig. 1.14a–b**). Materiale të tjera të paraqitura në formë tabele (**Tabela 1.1**), në sasi dhe peshë tregojnë sasinë e materialit prehistorik të dukshëm në sipërfaqen e një tume të pagërmuar.

Strallet e mbledhura në 2003 (pesë fragmente të inventarizuara: **Tabela 1.2, Fig. 1.15**) përfshinin materiale të Epokave të Paleolitit, Mezolitit, Neolitit dhe Bronzit (Kapitulli 14); sidoqoftë stralle kaq të hershme ishin të papritur për një varrezë tumulare Bronzit të vonë dhe Hekurit të hershëm, ato mund të krahasohen me lehtësisht me gjetjet e Paleolitit të hershëm, mesëm dhe të lartë dhe Mezolitit nga Apollonia, në Fier dhe gjetiu në zonën e vëzhgimit të MRAP. Vëzhgimi i mëpasshëm i rajonit nga Aprile (Kapitulli 18) nxorri në dritë një stacion të rëndësishëm gjahtarësh në hapësirë afër Ngraçijes, me nyje të izoluara aktiviteti afër tumës dhe bregut verior të Lumit të Gjanicës.

Të gjitha materialet moderne u mblodhën gjatë dy ditëve të para punës së terrenit në sezonin e vitit 2004 dhe përfshijnë një grup materialesh hekuri, qelqi, plastike, cohe dhe të tjerë (**Tabela 1.3**, **Fig. 1.16a–b**).

Vetëm gëzhojat boshe dhe fragmente këllëfësh hekuri që u ndeshën në shtresat sipërme janë analizuar bashkë me gjetjet e vogla në Kapitullin 10 me gëzhojat dhe plumbat në Shojcën 3.

KAPITULLI 2.
GËRMIMI I TUMËS

John K. Papadopoulos, Lorenc Bejko, dhe Sarah P. Morris, me kontribut nga Alket Islami dhe Richard MacDonald

Gërmimi i tumës

Gërmimet në Lofkënd filluan në 20 Qershor, 2004 pas përgatitjes së hartës së kontureve të tumës dhe rrethinave të saj të afërta. Pozicionimi i tumës në hartat ushtarake ishte i pamundur (**Fig. 2.1**). Sidoqoftë një hartë e Ushtrisë Shqiptare e vitit 1986 na siguroi të dhëna të dobishme për studimin e dherave të rajonit (**Fig. 16.18**), vëzhgimin të zgjeruar rreth tumës (**Fig. 18.1**), dhe studimin e Lofkëndit si vend i kultivuar (**Fig. 20.1**).

Tetë piketa të fiksuara në beton u vendosën në pikën (ST01) në JL të tumës brenda një zone të rrethuar, dhe ST02 i përdorur për matjet e përditshme me stacionin total. Nga këto pika Max Farrar arriti të rindërtonte një plan konturi 2 dhe 3 Dimensional të zonës rreth tumës në një rrjet 100 × 100 m (**Fig. 2.2**), i mbivendosur nga një tjetër rrjet i ndarë 1 × 1 m (**Fig. 2.3**). Qendra Eksperimentale e Teknologjisë në UCLA mori koordinatat e GPS në çdo piketë për të projektuar modelin dixhital në një sistem gjeografik koordinatash. Tuma e Lofkëndit është vendosur në UTM Zona 34T, drejt lindjes 391927, dhe drejt veriut 4500348.

Tuma u nda në katër sektorë secili me trape/ndarës të gjerë 0.50 m; në 2006 sektorët 2 dhe 3 u bashkuan në një (Sektori 2/3). Gërmimet në çdo sektor u bashkë-mbikqyrën nga studentë pasuniversitarë Amerikanë dhe Shqiptarë me shënime të marra në Anglisht dhe të dhëna të regjistruara në arkivin dixhital të GSHASH. Për gërmimin e tumës u përdorën vegla metali, ndërsa për varret u përdorën vegla prej druri (**Fig. 2.4**). Të gjitha materialet u kaluan në sitë të thatë me kampione për çdo grup dheu. Përmbajtja e enëve të varreve u vendos në sitë të ujshme për fluotacion dhe analiza paleo-etnobotanike (afërsisht 8–10 litra vëllim) (Kapitulli 16.1). Kampione u morën nga çdo varr, shtresë, njësi, njolla të zeza me copëza druri të karbonizuar për analizat e datimit C^{14} (Kapitulli 4, 16). Mbetjet e djegura bimore ishin të pakta (një farë e karbonizuar në sezonin e parë), edhe pas sitjes në ujë (Kapitulli 16.1). Të gjitha gjetjet e vogla të mbushjes së tumës u ekzaminuan, dokumentuan dhe ruajtën; gjetjet e metalit, qeramika e dekoruar, dhe fragmentet u trajtuan nga konservimi (Kapitulli 5). Çdo objekt tjetër u pastrua, klasifikua, dhe inventarizua. Të gjitha materialet e pa-inventarizuara u numëruan dhe peshuan sipas njësive përkatëse. Materialet u regjistruan në arkivin dixhital të projektit me kopje në UCLA dhe Tiranë. Kockat njerëzore hiqeshin në terren shpesh nga antropologët fizikë për t'u pastruar dhe klasifikuar (Kapitulli 6). Materiali skeletor u ekzaminua për shenja të trajtimit postmodern (në kafkë) por pa ndonjë të dhënë të qartë në asnjë sezon. Në 2004 një projekt pilot duke përdorur ADN mitokrondriale dhe nukleare nuk mundi të gjente ndonjë marrëdhënie familjare (Shtojca 1, Kapitulli 6). Gjithashtu u mblodh një grup kockash kafshësh (Kapitulli 16.1) dhe shumë guaska kërmijsh toke (Kapitulli 16.2).

Një grup prej 28 varresh u zbulua në 2004 (**Fig. 2.5a–b**), në 2005 u rrit me 66 dhe në 2007 një total prej 100 varresh u mblodhën shumë prej të cilëve varrime të shumëfishta (150 individë). Forma ovale e tumës në 2004 dhe 2005 (**Fig. 2.3**) ndryshoi në sezonin e fundit (**Figs. 2.8–2.11**). Mbetjet njerëzore paraqiten në Kapitullin 6, analizat e izotopeve të qëndrueshme të kockave njerëzore në Kapitullin 7, dhe praktikat mortore prehistorike në Kapitullin 8 (praktikat mortore moderne janë trajtuar në Kapitullin 3.2).

Varret e gërmuara i përkasin dy fazave me 15 varrime moderne të rriturish dhe fëmijësh me vendosje trupi, afërsisht 1800 e.s. (Kapitulli 4), në zonën verilindore (**Fig. 3.287**, Kapitulli 19) orientuar nga lindja në perëndim me prani kockash kafshësh (Kapitulli 3.2). Afër me një varrezë Myslimane të shekullit të 20-të e lidhur me Ngrançijen (**Fig. 2.6**), këto varre në tumë mbanin një komunitet Kristian. Tetëdhjetë e pesë varre prehistorike datojnë nga rreth 1400 deri në 800 p.e.s. sipas edhe të dhënave të radiokarbonit (Kapitulli 4). Varret individualë u numëruan sipas rradhës së zbulimit në numra Arab dhe u rinumëruan në rend kronologjik me numra Romak (Kapitulli 3: konkordanca, Kapitulli 2.2). Trapet u lëvizën në 2007 (**Figs. 22.3, 22.4**) dhe u etiketuan: Sektori 5 mes 1 dhe 4, Sektori 6 mes 1 dhe 2, Sektori 7 mes 2 dhe 3 dhe Sektori 8 mes 3 dhe 4.

Forma përfundimtare e tumës në shkëmbin natyror (**Fig. 2.8**) kishte përmasat rreth 20 m e gjerë lindje-perëndim dhe 25 m e gjatë (**Figs. 1.9, 2.9, 22.3a–b, 22.4**), dhe formën e një gungëze që binte drejt jugut, lindjes dhe më tepër nga perëndimi, ndërsa ana veriore mbeti e pandryshuar (**Figs. 2.10, 22.4**). Gjatë gërmimeve, shkëmbi natyror u arrit me shpejtësi në anët veriore, verilindore dhe veriperëndimore, madje edhe në jugperëndim niveli i varreve më të hershme ishte relativisht i lartë. Vetëm në juglindje kishte mbushje dhe varre në thellësi dhe këtu gërmuam deri në shkëmbin natyror për t'i gjetur të gjithë (**Figs. 2.9, 2.10**). Gjithashtu gërmuam një kuadrat 10 x 4 m në juglindje të tumës në zonë të sheshtë (Sektori 9: **Fig. 2.11a–b**) por nuk gjetëm ndonjë varr.

E vetmja strukturë e ndërtuar ishte një linjë prej rreth 25 gurësh përgjatë tumës, nga PJP në LVL në sektorët 4 dhe 1 më pak gurë në lindje (**Figs. 2.8, 2.12, 2.21, 2.22**). Gurët me përmasa mesatare dhe të vogla krijojnë një lakore me vetëm një rradhë gurësh (**Fig. 2.12**). Muri 1 shtrihet goxha lart (poshtë varreve të Fazës Va dhe b dhe një varri modern) mund të shënonte ndonjë moment themelimi ose ishte mur mbajtës, apo ndoshta ndarës i tipeve ose fazave të varreve të veriut dhe jugut. Pak gurë të tjerë vendoseshin mbi grupin e varreve prehistorikë (Kapitulli 8), dhe njëri më i dukshëm rrethonte disa varre moderne (**Figs. 3.288, 3.289, 3.290a, 3.305, 3.306a–b, 3.320a, 3.321a–b, 3.322a, 3.323, 3.324, 3.326a, 3.327a–b**). Gurët individualë të mbushjes ishin të mëdhenj ose "mengjir", disa u gjetën në trapin e sipërm të Sektorit 8 (**Figure 2.13**), ca në një skutë në perëndim afër Varrit XXVIII (77) (**Fig. 3.89a**), dhe një përqëndrim tjetër gurësh më i dukshëm në juglindje ("Muri" 2, **Fig. 3.14**); asnjë syresh nuk mund të lidhej me ndonjë varr apo ndonjë njësi tjetër. Disa gurë në juglindje ishin ose të rastësishëm ose të shkëputur nga shkëmbi natyror (**Fig. 2.11a–b.**). Ata u përdorën më vonë për rimbushjen e tumës (**Fig. 2.14**, Kapitulli 22). Gurët që shërbenin si shënjues varresh ishin të rrallë; tre rastet e ndeshura diskutohen në Kapitullin 3 (**Figs. 3.146, 3.147**). Një tjetër veçori me gurë u ndesh në kanalin e shkëmbit natyror me drejtimin nga lindja në perëndim (**Figs. 2.8, 2.15, 2.16, 2.17, 3.4**), afër me Murin 1 por pa ndonjë lidhje me të, ndoshta një element natyror.

Stratigrafia dhe formimi i tumës

Tuma e Lofkëndit përbëhej nga një kodër me dhé me 100 varre, shumë prej të cilëve të ndërlidhur (Kapitulli 4) e të formuar në mbi 600 vjet (rreth 1400–800 p.e.s.) të rihapur dhe sërish të ripërdorur. Në mbushjet e dallueshme me dhé (Kapitulli 16.4, **Figs. 2.16– 2.18**) u gërmuarn 595 njësi stratigrafike (Kapitulli 2.3); Harris matrix paraqet varret dhe mbushjet kryesore (**Fig. 2.19a–c**). Ndonëse formimi i tumës fillon në fazën I, të gjitha varret e hershme (përveç Varrit 1 [64]) zgjaten drejt juglindjes poshtë kulmit të mëvonshëm (**Figs. 2.18**, **4.2**, dhe **19.2, 19.3**), kodërzimi i vërtetë filloi me Fazën II (Kapitujt 4 dhe 19). Në disa zona dhe periudha vendi ishte sheshuar për një varrim me dheun e hedhur përsipër pa ndonjë gropë të qartë (Kapitulli 2.3, Kapitulli 3). Ndërsa varret më të vjetër ishin rihapur me vendosjen e varrimeve të mëvonshëm (Kapitujt 3 dhe 8), gropat për varret prehistorikë ishin të rralla dhe nuk para ndërhynin në ato të mëhershmet (për një përjashtim shih **Figs. 3.151–3.153**).

Prania e skutave të veçanta me dhé të dukshme gjatë formimit të tumës (**Figs. 2.16, 2.17**) jo gjithnjë mund të ndiqej gjatë gërmimit. Kjo ngatërresë në stratigrafi u sqarua nga rimbushjet tona vjetore të tumës, seksionet e së cilës treguan diferenca mes dheut të tumës dhe hedhjeve të veçanta të grumbujve të dheut. Gërmimi i vazhdueshëm i varreve të rinj, rihapja e të vjetërve dhe mbushja me dhé nga sipër e çdo varri me kalimin e kohës i dha tumës profil karateristik.

Pjesa më e madhe e kulturës materiale u gjet në mbushjen e tumës, një pjesë e saj e zhvendosur nga varret e vjetra gjatë rihapjes për varre të reja. Njëzet e tetë gjetje prej bronzi dhe hekuri vijnë nga mbushja e tumës dhe shtresa e sipërme (**Tabela 2.1**), përfshi edhe objekte ndoshta nga varret (p.sh., **10/18, 10/51, 10/75–10/78, 10/61**). Dhjetë objekte prej terrakote që u gjetën në mbushjen e tumës dhe shtresën sipërfaqësore (**Tabela 2.2**), jo të të njëjtit tip me ato të ndeshura në varre. Tre unaza për bosht tjerrjeje, rruaza ose kopsa erdhën nga varret. Pjesa më e madhe e materialit të ndeshur në mbushjen e tumës dhe shtresën sipërfaqësore përbëhej nga fragmentet e qeramikës, veglat e gurit (strallet, përfshi debitazhet) dhe mbetje qerpiçi me shumë material prehistorik gjetur në shtresën e sipërme edhe sipërfaqe (Kapitulli 1.2) dhe rastësisht ndonjë objekt modern (Shtojca 3, Kapitulli 10).

Kategoria më e madhe e materialit nga mbushja e tumës i përkiste qeramikës së Epokës së Bronzit dhe Hekurit bashkëkohore me atë të gjetur në varre. E gjithë qeramika e kataloguar është paraqitur në Kapitullin 9 dhe të gjitha fragmentet me numrin e një Gjetje Speciale (GS) në formë Tabele (Kapitulli 2.4). Në total, 433 fragmentesh qeramike janë inventarizuar dhe 341 janë kataloguar në Kapitullin 9; mbi 5.000 fragmente, ose 38.6 kg u numëruan dhe peshuan pa u inventarizuar apo kataloguar (**Tabela 2.3**). Qeramika e fragmentuar ishte standarte, në sasi të madhe dhe në njësi të mëdha (shtresa sipërfaqësore, nivelet e sipërme dhe mbushja e tumës). Disa grupe u emërtuan si "njësi qeramike" sepse i përkisnin enëve të mëdha (P283; **10/259**, **14/12**; **Figs. 3.214**, **15.4**—të gjetura afër Varrit LXIII [35], me përmbajtje bitumi të lëngshëm që ishte tharë: Kapitulli 5). E gjithë qeramika e pa-inventarizuar u klasifikua në gjashtë kategori (e ashpër e punuar me dorë; gjysëm e ashpër e punuar me dorë; e padekoruar e punuar me çark; e dekoruar e punuar me dorë; e pa diagnostikuar, ose copëza të vogla qeramike). Pak qeramikë ishte e punuar me çark, kryesisht fragmente Korintase (Shtojca 2) dhe fragmente moderne nga shtresat e sipërme dhe njësitë sipërfaqësore. Fragmentet e qeramikës së ashpër të punuar me dorë u ndeshën thuajse në të njëjtin numër me kategorinë gjysmë të ashpër, por ishin dyfish më të rënda. Copëzat e qerpiçit (Kapitulli 14) nga tuma ishin po aq të papritura sa edhe veglat prehistorike të gurit (Kapitulli 13); fragmentet e zgjedhura janë kataloguar dhe shtjelluar në Kapitullin 14, me një shtesë prej 15.875 fragmentesh nga shtresa e sipërme dhe mbushja e tumës. Kallama, thupra ose impresione me shkop ruheshin në fragmentet e plitharëve. Të njëjtat copëza më herët janë gjetur në tumën e Patosit por nuk janë publikuar, ndoshta mund të jenë ndeshur edhe në të tjera varreza tumulare në Shqipëri por kanë mbetur pa u vërejtur.

Ashklat e veglave prej guri në tumë ishin domethënëse (598 gjetje: 94 vegla dhe bërthama, 495 copëza debitazhi: Kapitulli 13) dhe datojnë në periudhat e Paleolitit dhe Mezolitit të mesëm, dhe Epokat e Neolitit dhe Bronzit. Straje kaq të hershëm janë karakteristike në rajonin e Apollonisë dhe Fierit dhe të shpeshtë në të tjera tuma Ilire. Nga materiali organik u dokumentuan 358 mostra me kocka kafshësh në mbushje brenda disa varreve (Kapitulli 16.1) dhe shumë pak materiale të florës. Një tjetër komponent domethënës i ndeshur në tumë ishte mullusku, të gjitha thuajse guaska toke (Kapitulli 16.2) të lidhura me procese natyrore. Në total, 7.835 kg (2.308 nga shtresat sipërme ose sipërfaqa, 3.552 nga mbushja e tumës dhe pjesa tjetër nga 70 varre: **Tabela 16.5** dhe Kapitulli 16) përfaqësojnë koleksionin e parë të guaskave të mbledhura, dokumentuara dhe studiuara nga një tumë në Ballkan.

Ndërthurja e qeramikës, me vegla guri, qerpiç dhe kocka kafshësh ose gjendej në sasi të mëdha në afërsi ose mund të jetë sjellë nga vendndodhje më të largëta (Papadopoulos 2006), ende nuk ka prani të mjaftueshme të faunës dhe florës për një vendbanim "nomal" prehistorik, kështu që mund të jenë zgjedhur ose veçuar si materiale të veçanta (si në disa vendndodhje tumulare në Shtetet e Bashkuara). Pra, strajet e Epokave të Neolitit dhe Bronzit mund të kenë arritur në tumë përmes aktivitetit të varrimit ose paketave të dheut të përgatitura që në vendbanim, ndërsa ato të Paleolitit dhe Mezolitit janë mbledhur bashkë me dherat e pasur me argjilë (Kapitulli 16.4) të zgjedhur për të stabilizuar tumën (Kapitulli 13, Kapitulli 18) kundër erozionit prej gurëve të shumtë me përmbajtje gëlqerore (Kapitulli 16.4). Balta e kasolleve në tumë mund të ketë patur origjinë artkitekturore dhe ndoshta e kuruar ose e thekur nga zjarri fillimisht për të ruajtur një strukturë të humbur e jo të përhershme në "banesën e përjetshme," shtëpia e të vdekurit (Kapitulli 20). Të gjitha materialet e mbushjes së tumës–me përjashtim të veglave të gurit të Paleolitit dhe Mezolitit–shpërndaheshin përreth tumës dhe popullsia e luginës së Lumit të Gjanicës që varrosi të vdekurit në Lofkënd mes 1400 dhe 800 p.e.s. mund të ketë sjellë me vete nga vendbanimi(et) materialet e saj, ose më të vjetër në këtë zonë. Rritja e elementeve përkujtimorë në tumë, të mbijetuarit e të vdekurve me qëllim përzgjodhën të gërshetonin komponentë të mjedisit të tyre jetësor për të ngritur lart kujtimin e të vdekurit të tyre.

Kapitujt gjithashtu paraqesin një Konkordancë të Varreve dhe numrat e Varreve, një listë përmbledhëse dhe spjegime të të gjitha njësive stratigrafike gjithashtu edhe regjistrimin e gjetjeve të inventarizuara sipas kontekstit.

Fotografimi Ajror

Në vitin 2005 ne u përpoqëm të realizonim fotografimin ajror për të zgjeruar dokumentimin e tumës me të gjitha veçoritë e saj të ekspozuara, si varret individuale dhe Murin 1. Fotografimi i ulët vertikal me një balonë me litar (si ato të përdoura në Greqi), aeroplan të lehtë ose helikopter nuk ishte i disponueshëm në Shqipëri. Në vend të tyre ne kërkuam Alket Islamin dhe parashutën e tij me fuqi motorike për 60 fotografi me nivel të ulët të tumës (**Fig. 2.20**), i pari mbulim ajror për një projekt arkeologjik në Shqipëri. Të vetmet veçori të ekspozuara ishin Muri 1 dhe Varri XLVIII (52 (**Figs. 2.21, 2.22**); dy pamje e kapën tumën me mjedisin përreth (**Fig. 2.23** nga juglindja, **Fig. 2.24** nga veriu). Përpjekja qëlloi të ishtë një sukses edhe pse në të njëjtën kohë kishte edhe gërmime të tjera në Shqipëri që nuk ishin në gjendje të pajtonin të njëjtin shërbim.

Dy Minuta në jetën e Lofkëndit: Panorama e vendvarrimit nga Richard MacDonald, 4, Korrik 2006

Gjatë katër vitesh gërmim vërejtëm efekte të ndryshme ndriçmi. Ndarëset shfaqnin hije sfiduese dhe drita e drejtëpërdrejtë e diellit ishte e preferuar për detajet. Efektet e ndryshme të dritës mund të japin katër pamje të ndryshme për një varr në të njëjtin mëngjes (**Figs. 3.168a–d**, Varri LII [69]). Përveç dokumentimit me video (*http://www.sscnet.ucla.edu/ioa/staff/papadopoulos/lofkend/index.html*) dhe fotografi të gërmimit, përfshi edhe rindërtimin e tumës (Kapitulli 22) ne lidhëm bashkë dhjetra fotografi për një pamje panoramike (**Fig. 2.25**) që përfaqësonte dy minuta në jetën e tumës në 4 Korrik, 2006 e marrë nga pika më e lartë e katër sektorëve.

KAPITULLI 3.
KATALOGU I VARREVE DHE PËRMBAJTJA E TYRE

John K. Papadopoulos, Lorenc Bejko, dhe Sarah P. Morris, me kontribut nga Shpresa Gjongecaj

Ky kapitull paraqet katalogim të plotë të të gjithë varreve bashkë me përmbajtjen e tyre dhe ndahet në dy pjesë: e para merret me 85 varret prehistorikë, dhe e dyta me 15 varret modern. Çdo varr tregohet i vendosur brenda planit të përgjithshëm të tumës, i përshkruar në detaje me përmbledhje për përmbajtjen e cila trajtohet më e plotë në kapitujt e mëvonshëm bashkë me një tjetër përmbledhje për bioarkeologjinë e invididëve të varrosur. Varret janë paraqitur sipas një rendi paraprak kronologjik (për këtë shih Kapitullin 4).

KAPITULLI 4.
KRONOLOGJIA RELATIVE DHE ABSOLUTE E TUMËS

Brian N. Damiata dhe John Southon, me kontribut nga John K. Papadopoulos

Ky kapitull fillon me një pasqyrim të përgjithshëm të kronologjisë relative të varreve bazuar në ndërlidhjet vertikale mes varreve dhe stratigrafisë së tumës. Varret janë organizuar sistematikisht në 6 faza të emërtuara Fazat I-IV dhe gjithashtu si Fazat Va dhe Vb.

Shtylla kryesore e kapitullit është analiza e AMS (aceleratori mas spektometër) e kryer në kampinonet e kolagjenit njerëzor e marrë nga mbetjet skeletore të ruajtura më mirë dhe e mbështetur nga disa datime AMS të kampioneve të drurit të karbonizuar. Një total prej 37 datimesh C^{14} radiokarboni AMS u siguruan duke përfaqësuar kështu grupin më të madh të të analizave të një materiali funerar në Shqipëri. Duke qëne se kampionet bazoheshin në kolagjenin njerëzor, fenomenet dëmtues që patën ndikuar studime më të hershme, veçanërisht efektet e drurit të vjetër këtu u minimizuan. Menjëherë pasi datimet AMS u përfunduan ato u vendosën përkrah listës kronologjike të varreve me përcaktim stratigrafik dhe kombinimi mes datimit relativ dhe atij absolut ishte i jashtëzakonshëm.

Seksioni përfundimtar i kapitullit merret me ndikimin e datimit të ri absolut dhe përfshin datimin AMS të marrë nga varrezat tumulare në Apolloni nga Epoka e Bronzit të hershëm në varret e Tumës 10, bashkë me varrimet nga periudha Arkaike në atë Helenistike. Datimet AMS u morën edhe nga varret moderne në Lofkënd dhe Apolloni. Fillimisht janë konsideruar ndikimet e këtyre datimeve të reja për zhvillimin e brëndshëm të tumës së Lofkëndit, më pas ndiqet një diskutim i gjerë rreth kronologjisë absolute në Shqipëri. Deri tani konteksti i brendshëm i Lofkëndit përqëndrohet tek varret më të hershme të datuar në shekullin e 14-të p.e.s., ndoshta diku nga fundi i shekullit të pesëmbëdhjetë, ndërsa varret e vona janë qartësisht të grupuara rreth 800 p.e.s. Datimet e reja AMS të marra nga Lofkëndi dhe Apollonia e ngrejnë datimin e Bronzit të hershëm me disa shekuj duke e sjellë kronologjinë absolute të kësaj periudhe në Shqipëri në të njëjtën linjë me atë të Egjeut. Po aq shtjelluese qe edhe kronologjia absolute e Bronzit të vonë dhe Hekurit të hershëm e cila u ngrit me 300 vjet më herët sesa kronologjia e hamendësuar relative.

KAPITULLI 5.
KONSERVIMI NË PROJEKTIN ARKEOLOGJIK TË LOFKËNDIT

Vanessa Muros

Pas një hyrje të shkurtër, ky kapitull ofron një pasqyrim të pajisjeve të laboratorit dhe materialeve të konservimit të përdorura në Projektin Arkeologjik të Lofkëndit bashkë me një trajnim për konservimin. Kjo ndiqet nga një përmbledhje e metodologjisë së ekzaminimit dhe identifikimit e materialeve dhe të dhëna të plota për trajtimin e objekteve sipas përbërjes (qeramika, metalet, kocka, gurë gjysmë-të-çmuar, guaska, gjithashtu xham dhe materiale të tjera prej qelqi). Trajtimet konservuese në vend-gjetjen e objekteve janë përshkruar, përfshi heqjen në formë blloku të objekteve të veçanta delikate dhe të dhëna të plota të teknikave të paketimit dhe ruajtjes dhe protokolli të dokumentimit (fotografim, raporte të shkruara, arkivi dixhital i konservimit dhe emërtesat e gjetjeve). Kapitulli përfundon me një përmbledhje të studimeve teknike të realizuara me materialin e Lofkëndit dhe një listë me pajisjet e konservimit të përdorura në laborator.

KAPITULLI 6.
BIOARKEOLOGJIA E TUMËS SË LOFKËNDIT

Lynne A. Schepartz

Ky kapitull ka një raportim të detajuar të popullsisë njerëzore të tumës. Fillon me qëllimet e studimit dhe problemet e ngritura për shqyrtim. Kjo ndiqet nga përshkrimi i metodave të përdorura në studim, veçanërisht përgatitjen dhe kurimin e materialit si edhe etapat e analizave skeletore. Demografia e popullatës adresohet e plotë me përfshirjen e shpërndarjes së madhësisë së kampionit, moshën dhe seksin. Diskutimi mbi veçoritë e përgjithshme të popullatës përqëndrohet tek forma e kafkës, shtati dhe bimorfizmin seksual.

Disa prej patologjive të përgjithshme të ndeshura në popullatë janë smalti hypoplasia, cribra orbitale, porotic hyperostosis, sëmundjet degjenerativetë nyjeve, trauma, dhe sëmundjet infektuese. Gjithashtu në diskutim përfshihen patologjitë dentare si karies, humbja e dhëmbëve pas vdekjes, mbushja dhe qarkullimi.

Nga aspektet më interesante të studimit janë përcaktimet e lidhjeve familjare të popullatës sipas analizave të tipareve jo-metrike. Për këtë qëllim, treguesit më të rëndësishëm përfshinin nofullën dhe tiparet e dhëmbëve prerës, foramen caecum molare dhe veçoritë e Carabellit. Në kapitull përfshihet një krahasim interesant i popullatës së Lofkëndit me kampionin e Apollonisë. Pjesa e mbetur e kapitujve konsiston në përshkrime të detajuara të të gjitha mbetjeve njerëzore në varret individualë.

SHTOJCA 1.
PËRMBLEDHJA E REZULTATEVE TË ANALIZAVE TË ADN SË KOCKAVE TË VJETRA NJERËZORE NGA TUMA E LOFKËNDIT

Laura Menez, me kontribut nga John K. Papadopoulos

Kjo shtojcë e shkurtër ofron të dhënat e një projekti pilot të filluar në nisje të gërmimeve me Qendrën e Vjetër Biomolekurare Henry Hellcome në Universitetin e Oxford. Pavarësisht vendosjes së protokollit të rreptë dhe nivelet e ndryshme të sterilitetit të ndjekura në marrjen e kampioneve, kockat e vjetra u kontaminuan me ADN moderne. Për disa arsye rezultatet dolën të ngatërruara dhe jo mjaftueshmërisht të sakta për të piketuar ndonjë model të kënaqshëm dhe nuk mund të riprodhoheshin.

KAPITULLI 7.
REZULTATET E ANALIZAVE TË IZOTOPEVE TË QËNDRUESHME TË KAMPIONËVE TË KOCKAVE NJERËZORE NGA LOFKËNDI

Brian N. Damiata dhe John Southon

Kapitulli 7 paraqet rezultatet e analizave të izotopeve të qëndrueshme të kockave njerëzore nga Lofkëndi. Qëllimi i analizave ishte marrja e informacionit për paleodietën e kësaj pjese të Shqipërisë. Kjo u perceptua si pjesë e një studimi më të madh i cili në vazhdim do të krahasonte dietën me banorët e kolonisë Greke të Apollonisë (Kapitulli 6) me një popullatë në prapatokën e kolonisë (jo vetëm Lofkëndi por edhe të tjera popullsi Ilire) dhe mundësisht dietën e metropolit ose qytetit-mëmë të kolonisë, në këtë rast atë të Korintit antik. Ky kapitull përmbledh rezultatet e karbonit (C) dhe Nitrogjenit (N) analizat e izotopeve të qëndrueshme të marra nga totali i 25 matjeve. Analizat u kryen bashkë me ato të datimit absolut AMS.

KAPITULLI 8.
PRAKTIKAT MORTORE PREHISTORIKE

Lyssa C. Stapleton

Ky kapitull merret me aspektet e varreve para-moderne nga tuma e Lofkëndit përfshi ndërtimin e tyre, orientimin, vendosjen e varreve dhe skeleteve njerëzore. Trajtimet normative të varrimeve janë përcaktuar dhe përfundimet lidhen me rëndësinë e larmisë ndërmjet varreve në përgjithësi, veçanërisht varreve normativë dhe ata të pasur.

Gjithashtu është konsideruar një varg i gjerë spjegimesh të mundshme lidhur me demografinë e pazakontë të popullatës së tumës dhe diskutimin e disa interpretimeve për modele në depozitimin e takëmit mortor jo-normativ. Njëri prej aspekteve më intriguese të kësaj varreze, ishte kontrasti mes takëmit mortor në varret e fëmijëve dhe grave adoleshente dhe pjesës tjetër të popullatës, kjo e konsoliduar me vazhdimin e gërmimit. Grupi i varreve relativisht të pasur prehistorikë vendoset në qëndër të vëmendjes në kapitull. Gjashtë varre përjashtohen nga të tjerët për shkak të tipit, vendndodhjes, vendosjes së takëmit mortor, moshës dhe seksit të individëve. Në pesë prej këtyre varreve, disa objekte të vogla të shoqëruara me një objekt kryesor ose diçka e tillë u gjetën mbi ose rreth kafkës së kufomës. Ato mbanin një femër adoleshente, një fëmijë ose foshnjë e zbukuruar me një diadem bronzi ose të tjera stoli koke. Varri i gjashtë kishte më pak takëme por tipet e objekteve dhe pozicioni i skeletit lejonin që edhe ky të përfshihej në grupin e varreve më të pasura. Gjithashtu përmenden varrime në pjesë të tjera të Mesdheut, përfshi Egjeun, Italinë dhe Qipron.

Studimet mbi mortin përpiqen të arrijnë të kuptuarin e strukturës sociale nëpërmjet interpretimit të ritualit funerar. Sidoqoftë varrimet në vetvete përfaqsojnë më shumë sesa thjesht projeksionin e ndërtimit të hirearkisë sociale; ato shprehin një varg të gjatë mesazhesh sociale që përfshijnë sërën, statusin, moshën, gjininë, zanatin ose rolin social, anomalitë fizike ose dobësitë, shkaqet e vdekjes, marrëdhëniet me kufomën kryesore (në rastin e varrimeve të shumëfishta) dhe tabiatet përpara vdekjes.

Seksi dhe mosha janë karakteristika integrale që formojnë pjesën më të madhe të riteve të rrugëtimit, ngjarjet kritike të jetës, si fillimi i pubertetit, martesa, shtatëzania lindja e fëmijës, dhe vdekja. Vetë vdekja imponon transformimin e rolit social të individit brenda komunitetitit, dhe kjo e shdërron aktin funerar në një zakon boshtor të shprehur nga ritet e rrugëtimit dhe karakteristikat e statusit të të vdekurit, sërës së tij hierarkike dhe namit në shoqëri të perceptuara nga familja dhe komuniteti në jetë. Funerali mbetet një ceremoni e krijuar nga personat në jetë dhe zgjedhjet lidhur me ku, si dhe me çfarë të të vdekurit në varrim bien pikërisht mbi ta.

Në pjesën cilësore të këtij studimi janë përdorur metoda standarte të analizave mortore përfshi ekzaminimin e strukturave fizike të varreve dhe individëve të varrosur në to dhe analizat krahasuese të takëmit mortor. Përdorimi i analizave të komponentëve dhe marrëdhënia midis karakteristikave të varreve përcaktojnë cilat praktika janë normative dhe cilat, në rast se ka, jo-normative. Përmasat e varrit, bashkë me punën e bërë për ndërtimin e tij, përpunimi i shënjuesit të varrit, përgatitja dhe pozicionimi i trupit dhe numri dhe tipet e takëmit mortor janë të gjitha atribute sasiore. Gjithashtu paralelet etnografike dhe ndër-kulturore paraqiten dhe diskutohen.

KAPITULLI 9.
QERAMIKA NGA VARRET DHE MBUSHJA E TUMËS

Seth Pevnick dhe Esmeralda Agolli

Ky kapitull paraqet të gjithë qeramikën e inventarizuar nga gërmimet e tumës së Lofkëndit, përfshi enët e plota dhe afërsisht të plota të gjetura si dhurime në varre bashkë me shumë copëza dhe pjesë të fragmentuara të gjetura brenda shtresës së sipërme, mbushjes së tumës dhe varreve. Lexuesit e familjarizuar me publikimet e qeramikës prehistorike dhe protohistorike të punuar me dorë nga varrezat bashkëkohore dhe tumat në rajonet fqinje të Shqipërisë si Greqia veriore dhe Italia jugore do të njohin shumë ngjajshmëri të rëndësishme me fragmentet e Lofkëndit, veçanërisht tek enët e plota. Për shkak të mungesës së vendbanimit bashkëkohor me varrezën në rajon dhe sepse materiali i fragmentuar qeramik mbetet disi i kufizuar, njohuritë shkencore për një pjesë të konsiderueshme të qeramikës së

Bronzit të vonë dhe Hekurit të hershëm mbeten të paplota. Me qëllim, korrigjimin e kësaj mungese-dhe ofrimin e një ideje më të mirë të të gjithë larmisë së brumit dhe tipeve të enëve të përdorura në Epokat e Bronzit të vonë dhe Hekurit të hershëm në Shqipëri–vendosëm të publikonim një sasi më të plotë se zakonisht të copëzave të fragmentuara. Në këtë mënyrë, afërsisht të gjitha pjesët, të gjetura si në pozicion fillestar ose të fragmentuara që mund të klasifikoheshin nga brumi ose përgjithësisht duke iu refereuar tipit dhe formës së enëve të plota janë përfshirë në katalog. Pikërisht kjo kategori përbën edhe korpusin më të madh të kapitullit. Katalogu gjithashtu përmban shumë fragmente që mund të klasifikohen vetëm prej brumit, ndoshta përmasave të enës dhe pjesës së ruajtur të enës (p.sh. vegje, buzë, ose fund); të tjera copëza mund t'i përkasin tipeve simotra me enët e plota, shumë të tjera rrjedhin nga tipe të panjohura tek enët e plota të Lofkëndit apo gjetiu. Përveç një seksioni të shkurtër në fund kushtuar disa fragmenteve jo-prehistorike të gjetura në Lofkënd, dhe një shtojce (Shtojca 2) për ato pak fragmente të qeramikës Korintase, ky kapitull merret tërësisht në qeramikën prehistorike.

Pas seksioneve të klasifikimit dhe terminologjisë vjen trajtimi i shkurtër i tipologjisë së repertorit të Epokës së Bronzit të vonë dhe Hekurit të hershëm në Lofkënd. Për shkak se tipologjia bazohet kryesisht në enë të plota ose thuajse të plota të ndeshura si takëm mortor nga varret me datim të sigurt në Lofkënd, ne arritëm të klasifikonim në rend kronologjik tipet brenda secilit grup brumi. Fragmentet nga tipet e pasigurta në çdo grup brumi gjithashtu janë diskutuar, por për shkak se të gjitha këto fragmente janë gjetur në shtresën e sipërme ose mbushjen e tumës datimi për to kryesisht bazohet në krahasimet rajonale me tumat e tjera. Pas tipologjisë radhiten seksionet e tipeve të dekorimit, motiveve dhe në fund shpërndarja e brendshme e qeramikës në tumë.

SHTOJCA 2.
ILIRIA CAPTA: QERAMIKA ME ÇARK NGA TUMA E LOFKËNDIT

Sarah P. Morris

Ndonëse mbushja e tumës dhe shtresat e sipërme dominoheshin nga qeramika lokale Bronzit të vonë dhe Hekurit (Ilire) e punuar me dorë, tek gjetjet më të vona të qeramikës u ndeshën pak fragmente të punuara me çark, padyshim baltë Korintase ndoshta të një kotile ose të një filxhani me anë të drejta, të zakonshme në Greqinë Arkaike dhe Klasike. Fragmentet ishin shumë të vogla dhe delikate për t'u pastruar, bashkuar apo skicuar plotësisht: gjurmët e një rozete të inçizuar dhe rrezet paralele në fund të trupit ndihmuan datimin e tyre në periudhën Arkaiko-Klasike, diku rreth shekullit të gjashtë p.e.s. Ky datim i vendos fragmentet dy ose më shumë shekuj më vonë nga koha e varrimit të fundit në tumë. Gjithsesi ato ofrojnë një lidhje të rëndësishme të historisë afatgjatë të rajonit kur qyteti i Korintit i themeluar si koloni, i dha emrin Apollonisë diku nga fundi i shekullit të shtatë ose atij të gjashtë p.e.s. në grykëderdhjen e lumit të Vjosës/Aoös dhe afër me lumin e Semanit. Importe të tjera Korintase të shekullit të shtatë p.e.s. në formën e Tipit A të amforave të transportit janë dokumentuar në Margëlliç. Qeramika Korintase nga Lofkëndi është diskutuar dhe krahasuar edhe me atë të Apollonisë. Identiteti i atyre që e kanë lënë qeramikën Korintase në Lofkënd mund vetëm të spekulohet si rënie ose rastësore ose e qëllimshme e ofruar ndaj të vdekurve. Kushdo të kenë qënë këta vizitorë, rënia rastësore ose depozitimi i enës Korintase në Lofkënd ka lënë shenjat e kontaktit të një kulture ndërhyrëse e cila ndryshoi njëherë e përgjithmonë faqen e historisë Ilire dhe kulturën materiale.

KAPITULLI 10.
OBJEKTET PREJ TERRAKOTE, METALI (AR/ELEKTRUM, BRONZI, HEKURI DHE DYMETALIK) GURËT GJYSMË TË ÇMUAR, FAJANCA, XHAMI DHE KOCKAT E PËRPUNUARA

John K. Papadopoulos dhe Rovena Kurti, me kontribut nga Vanessa Muros

Të gjitha gjetjet e vogla me përjashtim të qeramikës paraqiten në këtë kapitull. Materiali përfshin kategoritë e depozituara në varre dhe ato në tumë dhe në shtresën e sipërme. Kapitulli fillon me objektet që janë antike duke iu referuar kontekstit të tyre dhe ato shtresës së sipërme që mendohet të jenë antike bazuar në stil, brum dhe prekje. Vëzhgimi i objekteve antike ndiqet nga ato moderne (materiali antik dhe modern i mbledhur në sipërfaqe përpara fillimit të gërmimeve paraqitet i veçantë në Kapitullin 1). Klasat individuale të objekteve shfaqen në një tipologji me detaje të plota të kronologjisë dhe afërsive kulturore të përfshira në hyrjen e çdo kategorie.

Kapitulli fillon me objektet prej terrakote dhe veçanërisht gurët e tezgjahut, rruazat, ose kopsat të cilat vijnë pas qeramikës së trajtuar në Kapitullin 9. Pjesa më e madhe e terrakotave janë gurët e tezgjahut, rruazat ose kopsat. Përcaktimi nëse objektet individuale i përkisnin zbukurimeve personale ose veglave dhe implementeve ishte i pamundur veçanërisht për ato copëza që nuk u ndeshën brenda kontekstit të varreve. Pjesa tjetër e materialit antik paraqitet në dy kategori të mëdha: zbukurime dhe objekte të tjera zbukurimi dhe vegla/armë. Brenda secilës prej këtyre kategorive, materiali paraqitet sipas llojit të objektit (p.sh. fibul, gjilpërë, varëse, pafta në formë disku, mbështjellëse spiralike, unaza, rrathë koke [diadema], rruaza, maja shtize, maja shigjete, thika dhe kështu me rradhë) dhe kurdo që ishte e mundur edhe sipas tipit.

Krahasime me materialet nga vendndodhje të tjera në Shqipëri, Egje, Europën qëndrore dhe juglindore përfshi edhe gadishullin Italian dhe Lindjen e Afërme antike paraqiten sistematikisht dhe diskutohen për çdo tip objekti.

KAPITULLI 11.
STUDIMET ANALITIKE TË OBJEKTEVE METALIKE NGA LOFKËNDI

Vanessa Muros dhe David A. Scott

Studimi i gjetjeve metalike të gërmuara në tumën e Lofkëndit ofron mundësinë unike për të ekzaminuar numrin e objekteve nga i njëjti kontekst dhe periudhën kohore duke i shtuar të dhënat e publikuara për gjetjet metalike nga Shqipëria dhe Europa juglindore. Ky studim, përdor mikroskopinë optike, rrezet-X të fluorencës spektroskopike (pXRF), rrezet-X të diffraksionit (XRD) dhe ekzaminimet metalografike dhe ofron rezultate paraprake nga puna e realizuar deri tani me metalet e Lofkëndit. Studime të tjera janë në vazhdim, por rezultatet fillestare ndihmojnë për të kuptuar më mirë teknikat metalurgjike të përdorura në këtë zonë gjatë Bronzit të vonë dhe Hekurit të hershëm.

KAPITULLI 12.
TEKSTILET DHE MBETJET E TJERA TË MINERALIZUARA NË LOFKËND

Vanessa Muros

Gjetjet prej materiali organik nuk para mbijetojnë në kontekstet arkeologjike për shkak se ruajtja e tyre kërkon kushte të veçanta depozitimi. Shpeshherë informacioni për këto objekte merret jo vetëm nga vetë objektet por edhe nga gjurmët lëna ose *pseudomorfet* (zëvendësimi i materialeve organike me korrozion të produkteve minerale ose metalike) e ndeshura në gjetje të tjera. Në varret e Lofkëndit nuk janë gjetur materiale prej cohe, druri, ose litari por të dhënat për këto materiale janë gjetur në formë pseudomorfe në disa gjetje metalike duke ofruar informacion lidhur me teknologjinë e objekteve dhe njerëzve që përdorën tumën. Ky raport dokumenton tekstilet pseudomorfe të gjetura në objektet e Lofkëndit dhe përpiqet të ofrojë interpretime paraprake për të identifikuar materialet organike dhe ndonjë informacion mbi teknologjinë aq as mund të arrihej nga kaq pak mbetje.

KAPITULLI 13.
STRALLET

Jamie D. Aprile

Në tumën e Lofkëndit objektet prej ciflash guri u gjetën si në varre edhe në dheun e mbushjes apo shtresën e sipërme. Datimi tyre ka kufinj të gjerë e shtrihet në epokat e Paleolitit (i mesëm dhe i lartë), Mezolitit, Neolitit dhe Bronzit. Ndeshja e këtij tipi materiali nuk ishte e papritur përderisa gjetje të tilla janë raportuar nga shumë tuma në Shqipëri përfshi ato të Vajzës, Pazhokut, Shkrelit dhe Apollonisë. Sidoqoftë në studimet e mëparshme vetëm një kampion i vogël (zakonisht nga dy në dhjetë) të objekteve të strajeve janë përshkruar dhe ilustruar në tekst. Prendi, dhe më vonë Jubani pohojnë se ashklat e gurëve përfaqësojnë ndërhyrje në mbushjen e tumave sepse tipologjikisht objektet përshtaten me periudha më të hershme se ato të varrimeve në tuma. E njëjta situatë duket të ndeshet edhe në tumën e Lofkëndit. Sidoqoftë proceset e shumëfishta depozituese për praninë e tyre pa dyshim duhet të merren parasysh.

Nëntëdhjetë e katër vegla (përfshi bërthamat) u identifikuan dhe kataloguan ndërsa 495 copëza debitazhi u mblodhën duke formuar një total prej 589 objektesh ciflash prej guri. Dyzet e një vegla dhe 158 fragmente debitazhi, rreth 34% e koleksionit të përgjithshëm u mblodhën nga sipërfaqja dhe konteksti i shtresës së sipërme. Dyzet e katër vegla dhe 244 fragmente debitazhi, afërsisht 49% e korpusit u gjetën në tumë dhe mbushjen e gropave të varreve. Një sasi tjetër prej nëntë copësh debitazhi u tërhoqën nga gërmimi në dheun e butë dhe ranor në shkëmbin natyror poshtë tumës kjo gjë la të hamendësohet burimin e disave prej tyre në dherat lokale. Tetë nga varret e tumës dhanë një total prej nëntë veglash stralli, ndërsa 84 fragmente depitazhi u tërhoqën nga 42 varre; pra afërsisht vetëm 16% e të gjithë sasisë së objekteve të strallit vijnë nga konteksti i varreve. Duke u referuar tek analizat e këtyre veglave dhe të dherave ku ato u ndeshën, këto gjetje me shumë mundësi kanë mbërritur në varre nga dherat e përdorur për ndërtimin e tumës gjatë aksionit të ndërtimit ose veprimtarisë së vazhdueshme të varrimeve.

Pavarësisht numrit të madh të objekteve prej ciflash guri, prania e tyre nuk ofron informacion të përcaktuar kronologjik lidhur me ndërtimin e tumës. Më shumë se gjysma (57%) e veglave të kataloguara nuk mund t'u caktohet një datim i besueshëm dhe shumë janë diagnostikuar nga një varg periudhash ndonjëherë në shtrirje prej mijëra vjetësh. Për shkak të këtij fakti, këto objekte nuk përfaqësojnë copëza që mund të përdoren për një datim të besueshëm të tumës apo të ndonjë individi brenda saj. Me shumë mundësi ato përfaqësojnë një varg datimesh që fillojnë me depozitimin e hershëm të dherave të përdorur për ndërtimin e tumës dhe gjithë hapësirën kohore kur tuma është përdor për varrime. Objektet që datojnë në Epokat e Neolitit dhe/ose Bronzit si prerëset dhe elementet majë draperi janë të një standarti, tërësisht të formës funksionale që mund të jetë prodhuar dhe përdorur ose përpara ose pas hyrjes së veglave prej metali në përdorim. Këto objekte mund të jenë bashkëkohore me ndërtimin e tumës gjatë Epokës së Bronzit të vonë. Duke marrë parasysh shfaqjen e qerpiçit (Kapitulli 14) dhe mungesën relative të kocakve të faunës dhe atë të mbetjeve florale të karbonizuara (Kapitulli 16.1) në mbushjen e tumës është e vështirë të përcaktohet nëse dherat e përdorur për ndërtimin e tumës janë sjellë me qëllim nga konteksti i vendbanimit. Karakteri i vendodhjeve të ndeshura gjatë vëzhgimit sipërfaqësor në afërsi dhe shpërndarja e objekteve prej stralli në tumë sugjeron se veglat e Epokave të Neolitit/Bronzit kanë arritur në tumë prej veprimtarisë së varrosjes ose si pjesë e paketave të dheut të përgatitura që në vendbanim. Ndërsa ato të datuara në Paleolit dhe Mezolit janë mbledhur nga dherat lokalë të pasur me argjilë e të përdorur për stabilizimin e tumës.

KAPITULLI 14.
BALTA E KASOLLEVE (BYKU)

John K. Papadopoulos

Një prej gjetjeve më të spikatura në mbushjen e tumës dhe me siguri më e papritura ishte qerpiçi. Copëzat e shumta të këtij materiali fillimisht u regjistruan në arkivin dixhital të të dhënave të Lofkëndit me emërtimin "baltë e pjekur, jo qeramikë" dhe u numëruan e peshuan bashkë me qeramikën, strajet, dhe të tjera gjetje. Pak prej tyre u përzgjodhën për t'u inventarizuar duke iu referuar kontekstit ose ndonjë gjurme ende të ruajtur. Në shumë nga këto copa koklat e baltës janë amorfe, tek shumë të tjera janë ruajtur fije kallamash, thuprash ose gjurmë shkopinjsh që tregojnë se balta është përdorur si element lidhës për arkitekturën e thurjes-me-qerpiç/ plitharë. Pjesë të tilla qerpiçi janë të zakonshme për vendndodhjet me vendbanime prehistorike. Gjithsesi prania e kaq shumë qerpiçi në një tumë (shih **Tabela 14.1**) ishte e papritur dhe kërkonte spjegim.

Gjetiu në Shqipëri, qerpiçi është vërejtur në larmi kontekstesh vendbanimi veçanërisht për periudhën e Neolitit dhe gjithashtu nga Epokat e Bronzit në Hekurin e hershëm tek shtresat e Sovjanit në fushëgropën e Korçës. Vlen të përmendet që qerpiçi është i zakonshëm dhe vazhdon të përdoret si material ndërtimi kudo në Ballkanin prehistorik. Në Shqipëri dhe gjetkë qerpiçi nuk përbën vetëm një gjetje arkeologjike por është përdorur gjatë kohëve të ndryshme të erës historike duke mbijetuar deri në ditët tona. Në vitin 1904, Carl Patsch dokumentoi një shtëpi moderne të thurjes-me-qerpiç në Muzaki. Disa dekada më pas të njëjtën gjë e bëri edhe Luigi Ugolini. Shtëpi të njëjta ende ekzistojnë në Shqipëri dhe veçanërisht në rajonin e Lofkëndit (**Figs. 14.1–14.3**) ndonëse me kalimin e kohës janë rralluar. I njëjti tip qerpiçi është përdorur në Shqipëri për arsye të tjera, përfshi mbulesat për ruajtjen nga uji të mullarëve (**Fig. 14.4a–b**).

Sa për kronologjinë, datimi i tij është i pamundur të hamendësohet, vetëm se qerpiçi duhet të jetë më i hershëm sesa konteksti i gjetjes në Lofkënd përfshi ndeshjen rastësore në mbushjet e varreve dhe kryesisht në mbushjen e tumës. Se sa herët nuk mund të vlerësohet meqënëse arkitektura e thurjes-me-qerpiç ka qënë në përdorim qysh nga periudha e Neolitit deri në Epokën së Hekuri të hershëm e përtej saj.

Në kontekstin e stratigrafisë dhe etapave formuese të tumës, copat e qerpiçit nga Lofkëndi duhet të merren së bashku me një tjetër kategori të spikatur materiali të ndeshur si në sipërfaqe edhe në mbushjen e tumës së Lofkëndit: ciflat e veglave prej guri (Kapitulli 13). Kombinimi i ciflave të veglave prej guri me mbetjet e arkitekturës së thurjes-me-qerpiç sugjeron se ky material ose ka qënë mjaftueshmërisht i bollshëm në mjedisin afër për t'u përdorur për mbushjet e varreve ose të tumës, ose ngre mundësinë intriguese që ata që kanë varrosur të vdekurin e kanë sjellë me qëllim materialin nga vendndodhje të lashta jo domosdoshmërisht të afërta me tumën. Më gjerësisht këto dy kategori materialesh të ndeshura mbi dhe në tumë diskutohen të plota në Kapitujt 2 dhe 20. Në këtë kapitull ofrohet një pasqyrim i prehistorisë së qerpiçit arkeologjik e ndjekur nga një shqyrtim i detajuar i sasisë për çdo kontekst të qerpiçit në tumën e Lofkëndit përfshi numërimin dhe peshën e të gjithë fragmenteve të ndeshura bashkë me katalogimin e disa prej copëzave më përfaqësuese të këtij materiali.

KAPITULLI 15.
BITUMI NË LOFKËND: DEPOZITA, FRAGMENTET, DHE ENËT

Sarah P. Morris

Mjedisi që rrethon tumën e Lofkëndit dhe veçanërisht luginën e Gjanicës është i pasur me depozita prej fosili hidrokarboni dhe përbërës petroli që variojnë nga fraksionet e rëndë (asfalt) të lëngshëm (petrol) dhe të lehtë (gaz metan, naftë, etj.). Qyteza e afërt e Ballshit, është banuar qysh në periudhën e antikitetit të vonë si pasardhëse e Bylisit, ky i fundit vendi i Bazilikës së madhe të Kristianizmit të hershëm, sot është qendra moderne e industrisë Shqiptare të petrolit dhe i jep emrin naftës së saj autoktone ("nafta e Ballshit"). Jugperëndimi i luginës së Gjanicës, përgjatë bregut të Mallakastrës e ndan atë nga rrjedha e Vjosës së poshtme dhe më poshtë përgjatë Bylisit antik, Lumi i Shushicës bashkohet me Vjosën afër Selenicës moderne. Ky lokalitet në tërësi e bën Shqipërinë të njohur për përmbajtjen e një forme autoktone asfalti (alizh natyror me 80–85% karbon dhe 9–10% hidrogjen bashkuar me gjurmë sulfuri, nitrogjeni dhe oksigjeni të cilët shkrihen në 90–120° C) që është veçanërisht i pastër në depozita të vlefshme nënsipërfaqësore me 86% bitum natyror. Përmbajtja dhe përpunimi dëshmohen nga udhëtarët e hershëm modern, shumë prej të cilëve kishin dijeni për namin e tij në antikitet. Në kohët post-antike ky mineral i rëndësishëm nxirrej për përdorim industrial, kozmetik, dhe dekorativ jashtë vendit (për interesat e Venedikasve, Otomanëve, Britanikëve, Francezëve dhe Italianëve). Kërkesat për të u rritën sidomos pas hyrjes së prodhimit të rrugëve të asfaltuara (**Fig. 15.1**) dhe u venit gjatë shekullit të njëzetë me favorizimin e produkteve të naftës të së njëjtës zonë.

Në kohën post-komuniste, operacionet në minierat e pasura nëntokësore janë rijetëzuar nga një sipërmarrje Franceze (*www.selenicebitumi.com*), e cila tregton produktet kryesore si Selenizza SLN120®. Për të gjitha këto arsye një problem madhor për ekipin e Lofkëndit ishte të kuptonim rolin potencial të kësaj forme hidrokarburi në ekonominë antike.

Kjo zonë gjatë antikitetit është e njohur për minierat e saj të bitumit, veçanërisht tek romakët të cilët shfrytëzuan kombinimin lokal të burimeve minerale dhe drurit për ndërtimin dhe riparimin e anijeve. Madje hamendësime të kohëve moderne e atribuojnë tërheqjen fillestare të Korintasve në bregdetin Ilir bashkë-pranisë së asfaltit natyror, afër lumenjve gjë që çoi edhe në pasurimin me lëndë drusore të prapatokës, kjo mund të ketë mbështetur kërkimet e tyre nismëtare në industrinë e ndërtimit të anijeve përpara se sa Romakët ta bënin atë të famshme. Cilësia e lartë e *Illyrica pix* lavdërohej në kohët Romake dhe Apollonia (që ndoshta mund të nënkuptojë burimet e Selenicës) dhe Dyrrhachium (Durrësi modern) përmenden si burimet që e prodhonin këtë material. Kjo substancë dhe shfaqja e saj në sipërfaqe (në copëza, pellgje, burime dhe depozita të lëngshme ose të flakëta) tërhoqi vëmendjen e veçantë të kërkuesve antikë dhe hartuesve të kureshjtes shkencore. Burimet antike gjithashtu përmendin një aspekt të veçantë që e shdërroi këtë material në mrekulli natyrore, orakulli i famshëm bashkuar me burimin e zjarrtë në Nymfe ndoshta lëshonte metanin njëherazi me ziftin dhe vajin. Nga shekulli i tretë e.s. kjo vendndodhje është nderuar në anën e pasme të monedhës vendase që tregonte tre Nimfa eponime si figura të femrave kërcimtare pranë një kazani me zjarr (**Fig. 15.2**).

Bitumi ishte tejet i njohur në prehistorinë e Botës së Vjetër edhe si ngjitës (llaç) midis radhëve të tullave të baltës, agjent lidhës i shtuar bashkë me kashtën dhe baltën në prodhimin e tullave ose në aplikime të përzgjedhura për suvatimin e sipërfaqeve për papërshkueshmërinë e ujit. Gjithashtu ai ka shërbyer si agjent lidhës dhe izolator për qeramikën dhe për dorezat e veglave të gurit që në Epokën e Gurit të vjetër, kryesisht në Lindjen e Afërme. Në rastet kur me të përziheshin dhe nxeheshin edhe materiale tjera si argjila, mineralet, materialet e shtypyra dhe përbërësit organik formohej një mase elastike që më pas përdorej në prodhimin rruazave, pllakave, figurinave dhe enëve. Diku afër me Shqipërinë përdorej për dekorimin e qeramikës si shtresë për zbukurimet me shkop në sipërfaqe ose motivet e incizuara, në fushëgropën e Karpateve (Hungari) në kulturën Tisza ose në Epokën e hershme të Bakrit. Në prehistorinë e Shqipërisë i vetmi përdorim i tij deri tani ka qënë për ringjitjen e fragmeteve të qeramikës në vendbaniminn e Cakranit, jo larg nga Lofkëndi. Veglat prej guri nuk tregojnë ndonjë shenjë të vendosjes së dorezave prej druri me bitum.

Në kohët Klasike, bitumi fosil përdorej gjerësisht për të niveluar brëndësinë e enëve lokale Greke dhe Romake të transportit. Disa pitosa të ripërdorur në varret e nekropolit të Apollonisë, të identifikuar si tipe lokale Korintase të shekujve të gjashtë dhe të pestë p.e.s. janë lyer me bitum, në disa raste në sipërfaqe dhe në brendësinë e enëve. Studimet më të fundit të ekipit Gjermano-Shqiptar në teatrin e Apollonisë nxorrën rreth 1.000 fragmente qeramike Helenistike dhe Romake me gjurmë bitumi gjithashtu të gjetura në "trupin e amforave". Ndoshta kjo substancë përmirësonte përshkueshmërinë e ujit në enët e përdorura për mbajtjen e verës, më shumë se sa të lidhet me mbushjen enëve me zift të lëngshëm (e përdorur në mure dhe varre). Një studim gjithëpërfshirës për këto enë në Shqipëri dhe industritë e mundshme lokale në nxjerrjen dhe shpërndarjet e asfaltit mbetet për t'u kryer. Duke marë parasysh këtë tërësi interpretimesh rreth bitumit në mjedisin e Lofkëndit, shfaqja e tij në formë asfalti nga materiali i tumës dhe vëzhgimi sipërfaqësor në terren nuk është e papritur.

Njëzetë shembuj qeramike të regjistruara në këtë kapitull përfaqësojnë fragmente të gjetura në gërmimin e tumës (dhe një nga heqja e përfundimtare e trapit), të gjitha enët e punuara me dorë me gjurmët e bitumit (asfalt) gjenden në sipërfaqen e brëndshme të vendosura si element mbrojtës nga uji (**Figs. 15.3–15.5**). Vetëm pak syresh ruajnë njolla në sipërfaqen e jashtme gjë që mund të dëshmojë një formë dekorimi ose thjesht për papërshkueshmëri nga uji. Disa mund të datojnë në Epokën e Bronzit dhe ndoshta janë mbuluar nga një shtresë e përzierë me bitum për të shtuar llustrën. Fragmentet janë të shumë vogla dhe të gërryera për të përcaktuar përmasat e sakta, formën dhe dukjen e enëve por në të shumtën e rasteve format dhe brumi tregojnë të jenë më të vjetra se sa varret. Të tjera janë gjetur me njolla ose spërkatje të bitumit që mbulonin thyerjet si të ishin në kontakt me një formë të lëngshme bitumi pas thyerjes (por jo qëllimisht të përdorura për riparim). Pjesa më e madhe u gjet në shtresat e mbushjes së tumës ose shtresës sipërfaqësore, pak kishin hyrë në mbushjet e varreve të Hekurit të hershëm, por asnjëra nuk u gjet në shtresa më të thella dhe varret më të hershme të Epokës së Bronzit të vonë të ndeshura gjatë sezonit të fundit (Fazat I-II). Fakti se bitumi ishte gjetje e zakonshme e fragmenteve të ndeshura në zonën që mbuloi vëzhgimi sipërfaqësor (Kapitulli 18) dhe mbledhja informale e sipërfaqësore sugjerojnë se mjedisi natyror përreth tumës mund të ketë sjellë shumë gjetje në kontakt natyror me këtë mineral pa ndonjë ndërhyrje njerëzore.

Përveç gjetjeve rastësore të ridepozituara dhe me sipërfaqe zifti, gërmimi në varret e Lokëndit nxori në dritë një zbulim tërësisht të pazakontë. Një enë e madhe (**9/259**) u gjet e thyer në vend në rrethinën e Varrit LXIII mjaft afër për t'u konisderuar si pjesë e takëmit mortor të varrit: shtrihej në një depozitë (NJS: 280) 20 cm poshtë dhe në veri të varrit, e identifikuar si pjesë e mbushjes së tumës (**Fig. 15.4a**; shih gjithashtu **Fig. 3.214**). Se për çfarë qëllimesh kjo enë ka shërbyer në afërsi të varreve protohistorike mbetet e vështirë të përcaktohet, ndoshta pjesërisht mund t'i jepet përgjigje duke krahasuar zbulimet e tjera domethënëse nga tuma e Lofkëndit. Të paktën një varr në zonën jug-qëndrore (**Figs. 3.61, 3.62**), Varri XX (50) nga Faza e II-të i datuar në Bronzin e vonë dhe Hekurin e hershëm (shekujt dymbëdhjetë-njëmbëdhjetë p.e.s.) kishin skeletin e një femre të rritur (20–25 vjeçe) me një depozitë bitumi që ishte tharë përgjatë brinjëve të poshtme në anën e majtë, krahun e majtë të përthyer në zonën e legenit dhe anën e majtë të legenit (**Fig. 15.6a–b**). Dikush mund vetëm të hamendësojë që kjo njollë e parregullt përfaqëson aplikimin e një substance ngjitëse ose dekorative në lëkurë ose veshje të trupit të depozituar në varr, ndoshta për të mbyllur ose lidhur qefinin (asnjë gjurmë tekstili nuk ishte e dukshme në fragmentet e marra, edhe nën zmadhim).

KAPITULLI 16.
ARKEOLOGJIA MJEDISORE NË LOFKËND

16.1. ARKEOLOGJIA MJEDISORE NË TUMËN E LOFKËNDIT: REZULTATET E ZOOARKEOLOGJISË, FLUOTACIONI, DHE ANALIZAT E QYMYRIT TË DRURIT

John M. Marston

Tuma e Lofkëndit nuk ishte një kantidat i mirë për shqyrtimin e gjithanshëm të arkeologjisë mjedisore: tërësisht e përbërë nga depozita dytësore sedimentesh nga zona të tjera të mjedisit kulturor fqinj, kishte pak arsye të besoje se mbetjet e bimëve dhe kafshëve e nxjerra nga tuma do të ishin tregues të ndërveprimit të mjedisit njërëzor bashkëkohor me ndërtimin e tumës. Pavarësisht kësaj përshtypje, mbledhja gjithëpërfshirëse e kockave të kafshëve, kampionëve të fluotacionit, dhe qymyrit të drurit u bë pjesë e planit të gërmimit për tre arsye: për t'iu mësuar studentëve Amerikanë dhe Shqiptarë metodat dhe strategjitë e kampionimit arkeologjik, për t'u siguruar që kategori të rëndësishme informacioni nuk humbnin gjatë gërmimit dhe për të zgjeruar të dhënat e pakta të kampioneve dhe analizat e mbetjeve bimore dhe shtazore arkeologjike nga gërmimet në Shqipëri.

Tre tipe të kampionimit të arkeologjisë së mjedisit u kryen në tumë: analizat zooarkeologjike të kockave të kafshëve, analizat anatomike të fragmenteve të të drurit të karbonizuar, dhe kampionimi i grupeve të vogla të dheut për fluotim për të mbledhur mbetjet e mëdha bimore të karbonizuara. Kockat e kafshëve ishin më të zakonshmet në të gjitha mbetjet dhe analizat zooarkeologjike për ato ishin më informueset veçanërisht të lidhura me zbulimin e dy varrimeve të kafshëve. Analizat e drurit të karbonizuar ishin të kufizuara në qëllim por sugjerojnë ngjajshmëri midis strukturës pyjore të sotme me atë të prehistorisë. Disi më ndryshe, analizat e fluotacionit, dhanë pak mbetje bimore. Mungesa e sasive të cilësuara të farave të djegura dhe mbetjeve të drurit të karbonizuar nga tuma është udhëzuese në vetvete duke treguar se gjetjet e djegura në vend ose flijimet funerare nuk kanë qënë komponent i riteve të varrimit në Lofkënd. Ky kapitull paraqet sistematikisht rezultatet e këtyre tre analizave përfshi listat e plota me të gjithë speciet e ndeshura në formë tabele brenda kapitullit. Inventari i plotë i kockave të identifikuara të kafshëve është regjistruar veçmas në **Tabelën 16.5**.

16.2. MBJETJE MULLUSKE GUASKASH NGA LOFKËNDI

Evi G. Vardala-Theodorou

Në këtë kapitull paraqitet materiali mullusk nga gërmimet arkeologjike në Lofkënd. Gërmimet e ushtruara mes viteve 2004 dhe 2007 nxorrën në dritë 100 varre shumë prej të cilave i përkisnin sasive të ndryshme të mulluskëve: gjithashtu nga mbushja e tumës u mblodhën edhe guaska të tjera. Mulluskët u grumbulluan me kujdes sipas stratigrafisë dhe përfaqësojnë, thuajse të vetmet guaska toke që i përkasin specieve të mëposhtme: *Pomatias elegans*, *Helicella itala*, *Xeromunda vulgarissima*, *Lindholmiola girva corcyrensis*, *Cernuella virgata*, *Monacha cartusiana*, *Trochoidea pyramidata*, *Siciliaria stigmatica*, *Strigilodelima conspersa*, *Mastus pupa*, *Poiretia delesserti*, *Aegopis cf. verticillus*, dhe *Cerastoderma edule*. Me përjashtim të një mullusku të vetëm prej uji të njelmët shfaqja tyre i atribuohet proceseve natyrore pa ndonjë përkatësi të veprimtarisë njerëzore. Mesa jemi në dijeni, kjo është hera e parë që një material i tillë mblidhet në të gjithë sasinë e tij dhe dokumentohet e studiohet prej një tumë në Shqipëri.

Në përpjekjen për të siguruar të dhëna lidhur me paleomjedisin mulluskët me guaskat e tyre i kanë shërbyer njerëzve në disa mënyra gjatë antikitetit. Përveç sigurimit të informacionit rreth preferencave të dietës, mulluskët mund të jenë përdorur nga njerëzit në pjesë të ndryshme të botës si sende me vlerë këmbimi ose thjesht ruajtje; gjithashtu ato mund të jenë përdorur si vegla ose zbukurime për stoli personale. Për më tepër, mulluskët sigurojnë informacion për kushtet e bregdetit, lidhur me tipet e vijës bregdetare, luhatjet e nivelit të detit në periudhën e depozitimit të materialit arkeologjik duke lejuar rindërtimin e mjediseve të vjetër arkeologjikë. Guaskat gjithashtu ofrojnë informacion për kronologjinë absolute me metoda të ndryshme si mostrat me guacka të trasha që mund të datohen nga termoluminishenca (TL). Ndryshe nga gjithë këto përdorime të ndryshme të mullusqeve në Lofkënd ato vetëm shërbejnë për të informuar karakteristikat e mjedisit në tumë.

Zakonisht qëllimi për të studiuar materialin maiakologjik bëhet për të sqaruar aspektet lidhur me mënyrën e jetesës në antikitet dhe përshtatshmërinë e njerëzve në paleomjedis. I të njëjtës rëndësi është edhe interpretimi i shfaqjes së mulluskëve të grumbulluar pas depozitimit të materialit arkeologjik. Këto mbetje duhet të trajtohen veçmas prej atyre që janë pjesë e koleksionit arkeologjik dhe janë drejtëpërdrejtë të lidhura me aktivitetet njerëzore siç janë mulluskët e detit. Ndonjëherë shfaqja e specieve të tokës në varre kushtëzohet edhe prej kushteve klimatike (temperaturat, lagështia), kimia e dherave dhe kështu me radhë. Në këto raste ato nuk lidhen me veprimtarinë njerëzore. Është mëse e mundur që në kontekstin e varreve të gjenden fosile detare, njelmët, ose mostra nga ujrat e freskëta. Ndonjëherë këto mund të jenë pjesë e sedimentit në të cilin hapet varri.

16.3. VËREJTJE MBI SHFAQJEN E FOSILEVE TË DRURIT NË LOFKËND

George Theodorou dhe Evangelos Velitzelos

Gjatë vëzhgimit sipërfaqësor, disa mostra fosilesh prej fragmentesh druri u mblodhën në zonën e Lofkëndit (**Fig. 16.16a–b**). Mostrat e mëdha të fosileve të drurit, në rast se do të ishin të disponueshme në zonë lehtësisht mund të jenë konsideruar si copëza guri veçanërisht gjatë procesit të mbledhjes së mostrave mundësisht të lidhura me veprimtarinë njerëzore. Materiali i mbledhur gjithashtu përfshinte pak mostra detare me fosile të Miocenit që i përkisnin faunës detare Neogjene të zonës (*Pleurotoma* sp. dhe *Ostrea gryphoides*) (Koskeridou et al. 2005) (**Fig. 16.17**).

Shfaqja e fragmenteve të vogla të fosilit të drurit është qartësisht më interesante sesa fosilet e zakonshme detare. Ky material u analizua në seksionet e imta shlifet (**Fig. 16.16a–b**) i përgatitur në Departamentin e Gjeologjisë Historike dhe Paleontologjisë në Universitetin e Athinës. Për shkak të përmasave të vogla vetëm seksionet që mbulonin zonat e brendashkruara mund të realizoheshin. Kjo gjë nuk lejoi saktësi në përcaktimin e specieve. Fragmentet e disponueshme mund t'i përkisnin të njëjtës njësi gjeologjike me fosile pemësh të ndeshur në pjesën perëndimore të Greqisë veriore. Vëzhgimi gjeologjik në zonë duhet të kryhet në të ardhmen për të lokalizuar vendin e tyre të origjinës. Kjo do të lejonte edhe zbulimin e fosilit në pyjet e petrifikuar dhe do të ofronte rastin e një studimi të plotë dhe efektiv me seksione më të mëdhenj të trungjeve të pemëve.

16.4. SHQYRTIMET E DHERAVE NË VENDVARRIMIN ARKEOLOGJIK TË LOFKËNDIT DHE MJEDISI PËRRETH

J. E. Foss

Shqyrtimi i dherave dhe hapësirave në tumën e Lofkëndit filloi në Qershor 2005 dhe vazhdoi gjatë sezoneve të 2006 dhe 2008. Studimi u përqëndrua në dy faza: (1) karakterizimi i dherave të Lofkëndit dhe (2) vlerësimi shpërhapjes rajonale të peisazheve dhe dherave në zonën e studimit. Ky hulumtim përfshinte identifikimin e sedimeteve gjeologjikë të materialit mëmë të dherave dhe karakteristikat morfologjike të dherave të ndeshur në gërmim. Pjesa më e madhe e dherave të shfaqura në tumë u emërtuan sipas metodave standarte të nomeklaturës. Përshkrimi i dherave përfshin ngjyrën, teksturën, strukturën, kompaktësinë, kufinjtë e horizonteve, dhe përmbajtjen relative të karboneve. Duke u bazuar në këto karakteristika, shtresave të ndryshme iu caktuan emërtimet e horizontit të cilat tregojnë zhvillimin dhe historinë e dherave.

Shpërndarja rajonale e dherave u përcaktua nga përshkrimi në këmbë i zonës vëzhgimet e peisazhit, dherave dhe prerjeve anës rrugëve, gërmimeve të ndryshme dhe përdorimit të trapanit kudo ku ishte e mundur për analizimin e dherave në thellësi 1.5 m. Ky vëzhgim i përgjithshëm i dherave dhe studimet dhe analizat më të detajuara janë të nevojshme për karakterizimin e vendndodhjeve të veçanta.

Kampionet e dheut u morën në dherat tipikë në gërmimet e Lofkëndit. Tre profile u përzgjodhën për analiza laboratorike; këto analiza përfshinin agjentët kimikë (përbërjen e elementëve) dhe karakteristikat fizike (tekstura). Disa kampionë nga trapet u morën për shqyrtimin e mikromorfologjisë dhe disa materialeve fibroze të panjohura.

KAPITULLI 17.
HULUMTIMI MBI TUMAT NË ARKEOLOGJINË SHQIPTARE

Lorenc Bejko

Më shumë se 65 vjet më parë, Shtjefën Konstantin Gjeçovi (shih **Fig. 21.2**), një prift Katolik i shkolluar, ishte i pari Shqiptar që gërmoi një varrezë të Epokës së Hekurit me qëllim akademik. Ai gjithashtu për herë të parë në historinë e arkeologjisë Shqiptare ndërrmori hapa koherente për një gërmim sistematik. Gërmimet e Gjeçovit në katër tumat afër Kuvendit të Toroshanit u kryen me kujdes, të shoqëruara me përshkrim të detajuar të inventareve përfshi përmasat dhe pozicionin e varre-ve, pozicionin e skeleteve dhe tipet e varreve. Gjeçovi me siguri kishte dijeni për tumat e gërmuara të Glasinacit në Ballkanin qëndror në fillim të shekullit dhe e përdori këtë informacion për të bërë një studim krahasues për zakonet mortore të përdorura nga Ilirët në pjesë të ndryshme të rajonit. Klasifikimi dhe datimi i materialit arkeologjik bazohej në ngjajshmëritë tipologjike me çka ishte gjetur në Glasinac, pjesa tjetër e aktivitetit të Gjeçovit në gërmime, vëzhgime, koleksione e kështu me radhë e shdërron atë në të parin arkeolog Shqiptar me ndikim të madh në studimet e mëpasme të tumave në vend.

Më shumë se 30 vjet pas punës së Gjeçovit jo shumë iu shtua kësaj fushe. Megjithatë gjatë viteve 1950, gërmime sistematike të një shkalle të gjerë filluan në disa tuma rreth vendit. Mes më të rëndësishmeve ishin gërmimet në një sërë tumash në rajonin e Matit (Rrethe, Bajzë, Kokërdhok, Bushkash, Bruç, Urakë, Shtogj, Perlat, Klos), gjatë periudhës 1952–1960, tumat në Vajzë në 1953 dhe 1954, tumat në rajonin e Dropullit të Sipërm në 1955 dhe 1956 dhe tumat e Mjedës (Shkodër) në 1958.

Gjatë viteve 1950-të u gërmua një total prej 44 tumash prej të cilave rreth 35 i përkisin vetëm rajonit të Matit. Këto gërmime ofruan të dhëna të bollshme lidhur me ndërtimin e tumave, zakoneve mortore dhe ritualet, tipet e varreve, takëmet mortore, dhe kronologjinë e pjesës më të mirë të zhvillimit të tyre (për vendndodhjen e tumave shih **Fig. 17.1**).

Gjatë viteve 1960 dhe në fillim të 1970 hapësira gjeografike e varrezave të gërmuara u zgjerua me projekte të reja të ndërrmara në Pazhok, Barç (**Fig. 17.2**) dhe Kuçi i Zi nga 1969 në 1971, Çepunë në 1969, Bajkaj në 1970 dhe në Krumë dhe Çinamak (**Fig. 17.3**) dhe në rajonin e Kukësit ndërmjet 1969 dhe 1971. Me këto gërmime të reja u përftuan shumë të dhëna në fushën e studimeve të tumave në vend. Ky informacion shërbeu si bazë për studime më të përgjithshme të shkallës rajonale kjo e reflektuar në punën e Prendit. Gjithashtu është e rëndësishme të përmendet gjatë kësaj kohe, puna e arkeologëve Shqiptarë dhe publikimet e tyre fituan vëmendjen e studiuesve të huaj si N.G.L. Hammond dhe Kenneth Wardle. Gjë që pati interest të veçantë për të ardhmen e zhvillimeve në kërkimet mbi tumat në vend sepse për herë të parë rezultatet e hulumtimeve të shumë viteve u vendosën në kontekst të gjerë rajonal. Çështje si origjina e varrezave Shqiptare tumulare dhe lidhja e tyre me Mikenën dhe ishujt Jonianë u adresuan.

Gjatë viteve 1970 dhe 1980 të tjera varreza u gërmuan jo vetëm në rajonet tashmë të njohura si të pasur në varrezat tumulare por edhe në zona të reja. Të tjera tuma u gërmuan në Pazhok në 1973, dhe në rajonet e Matit dhe Kukësit si vazhdim i gërmimeve të mëparshme. Megjithatë të dhënat më të rëndësishme të kësaj periudhe erdhën nga varrezat të panjohura më parë si ato të zonës së Kolonjës me tumat e saj në Prodan, Rehovë, Psar, Shtikë, Luaras, dhe varrezën e sheshtë në Borovë, ato në luginën e Lumit Vjosë me tumat e Piskovës, Rapckës dhe Grabovës dhe në zonën e Shkodrës me tumat e Shtojit dhe Shkrelit. Varreza të tjera të izoluara u zbuluan dhe gërmuan gjatë fundit të viteve 1970 dhe fillimit të viteve 1980 në Patos (**Figs. 17.4, 17.5**), Gërmenj, Katundas, Cerrujë, Hamallaj, Bujan dhe Mujaj.

Një total prej 156 tumash u gërmuan në të gjithë vendin nga viti 1952 deri në 1987. Vendndodhja e tyre dhe numri janë dhënë në **Tabelën 17.1**. Ky sigurisht është një numër i madh vendndodhjesh të gërmuara që nënvizon interesimin e dukshëm të kushtuar nga Arkeologjia shqiptare hulumtimit të tumave. Megjithatë gjatë viteve 1990 thuajse asgjë nuk iu bashkua listës së vendndodhjeve të gërmuara me përjashtim të një tume të vetme të gërmuar në nekropolin e Apollonisë në 1996 nga bashkëpunimi i një ekipi Shqiptaro-Fancez.

Në dekadën e parë të shekullit të njëzet e një, një rijetëzim relativ i kërkimeve në tuma u vërejt me tre projekte të rëndësishme dhe komplekse të organizuara në Kamenicë (**Fig. 17.6**), Apolloni (**Figs. 17.7, 17.8**), dhe në Lofkënd (paraqitur në këtë volum).

Këto projekte të reja qartësisht u bazuan në problematikë krejt të re shkencore dhe metodologji solide. Thellësia e informacionit që ka rezultuar nga qasja multidisiplinore në këto projekte përfaqësojnë zhvillimin e një tradite të rëndësishme në hulumtimet mbi tumat në vend.

Tradita relativisht e gjatë e studimeve të tumave në Shqipëri përgjithësisht është karakterizuar nga një zhvillim përparues i teknikave të gërmimit dhe gjithashtu strukturës së të dhënave të vëzhguara që nga varret individuale deri tek varrezat në tërësi. Megjithatë fokusi kryesor i kërkimeve ka qënë kultura materiale, kronologjia relative, dhe zakonet mortore si shprehi të rëndësishme të identitetit etnik për komunitetet prehistorike. Diskutimet mbi organizimin social të vërejtura nga të dhënat e varrezave kanë qënë jo vetëm të pakta në numër por gjithashtu të bazuara në themele të dobta teorike. Marrëdhëniet mes karakteristikave sociale dhe formës së varreve të diskutuara gjerësisht në vitet 1960 dhe 1970 gjetiu nuk para u prekën ose u vunë në dyshim në kontekstin Shqiptar. Përgjithësimet lidhur me sërën dhe organizimin shoqëror të individëve të varrosur shfaqen rastësisht të bazuara në pretendime dhe jo parime të mirë-mbështetura. Megjithatë larmia e zakoneve mortore në rajone të ndryshme të vendit është studiuar pjesërisht dhe ka qënë e përqëndruar më shumë tek dallimet e ndërtimit të tumave dhe varret ndërmjet rajoneve të ndryshme duke injoruar ndryshimet e programeve mortore dhe modelet e strukturës shoqërore të varrezave.

Një tjetër mangësi e punës në arkeologjinë e shekullit të fundit ka qënë mungesa e mbledhjes sistematike dhe studimi i mbetjeve skeletore nga varrezat. Vetëm në 1990 studimi i hollësishëm dhe sistematik i mbetjeve skeletore fitoi interes të spikatur.

KAPITULLI 18.
NJË VËZHGIM INTENSIV, SISTEMATIK ARKEOLOGJIK I MJEDISIT RRETH TUMËS SË LOFKËNDIT

Jamie D. Aprile

Gërmimi në tumën e Lofkëndit nxorri një numër të rëndësishëm problematikash rreth lidhjes mes varrezës dhe mjedisit përreth që vetëm mund adresoheshin nga mbledhja e të dhënave arkeologjike të detajuara nga një vëzhgim intensiv sipërfaqësor. Një sërë tumash të Hekurit të hershëm janë gërmuar në Shqipëri si ato të Pazhokut, Barçit dhe Kamenicës (Kapitulli 17), por pa vendbanime të lidhura me to. Teknikat e zgjeruara të vëzhgimit zakonisht kanë identifikuar fortifikime, qyteza dhe varreza. Projekti i Lofkëndit ofronte një rast të jashtëzakonshëm për bashkimin e gërmimit me vëzhgimin për të kuptuar rrethinat e tumës dhe me qëllim të veçantë të identifikonte natyrën e vendbanimit të Bronzit të vonë dhe Hekurit të hershëm në afërsi. Për më tepër, veç të mësuarit më shumë rreth vetë tumës, ndërrmarja e një vëzhgimi në luginën e Lumit të Gjanicës ofronte të dhëna të rëndësishme lidhur me zhvillimin e vendbanimit në kohë në brendësi të një lugine me orientim lindje-perëndim larg nga vendndodhjet e tjera të rëndësishme. Ky raport është një përmbledhje e rezultateve të vëzhgimit të ndërrmarrë kryesisht në 2008; një vlerësim më i plotë me punë plotësuese në terren synohet në të ardhmen.

Në Shqipëri vëzhgimi intensiv sipërfaqësor është kryer veçanërisht në rajonet e rëndësishme historike për të kuptuar natyrën e venbanimit afër vendndodhjeve të trajtuara më shumë. Ndryshe nga këto, vëzhgimi i Lofkëndit kampionoi një zonë që kishte marrë pak vëmendje. Tre vëzhgimet më të rëndësishme të dhjetë viteve të fundit të realizuar në Shqipëri kryesisht janë përqëndruar në të kuptuarin e prapatokës së qyteteve Greke dhe Romake të Butrintit, Apollonisë, dhe Dyrrachium (Durrësi modern)–të treja përgjatë bregdetit–dhe mënyrës sesi ato ndërvepruan me popullsinë lokale. Një vëzhgim i rëndësishëm në rajonin e Luginës së Shalës në Shqipëri ka kontribuar në mënyrë domethënëse të kuptuarin e periudhave më të vona të historisë Shqiptare si edhe në natyrën e vendbanimit në Alpet veriore. Një tjetër vëzhgim i ndërrmarrë në 2005–2007 në luginën më të madhe bujqësore të juglindjes, në Fushëgropën e Korçës ka nxjerrë të dhëna të rëndësishme lidhur me natyrën e vendbanimit në këtë rajon që ka shërbyer si zonë e prodhimit të ushqimit dhe si prapatokë e rëndësishme për lidhjet mes Shqipërisë dhe fqinjëve Ballkanik në jug dhe lindje. Lugina e Lumit të Gjanicës ofron damarë të rëndësishëm komunikimi mes fushave bregdetare dhe qafave të maleve. Asnjë vëzhgim intensiv arkeolgojik nuk është ndërrmarrë në këtë zonë të kufizuar. Në këtë mënyrë të dhënat e përftuara nga ky projekt do të ndihmojnë për të ofruar modele fillestare të historisë afatgjatë të vendbanimit në Shqipëri.

Siç edhe vërehet në Kapitullin 20, tuma ka luajtur rolin e rëndësishëm të një pike referimi në historinë afatgjatë të peisazhit lokal. Të dhënat e mbledhura nga vëzhgimi si një pjesë e integruar e analizave sigurojnë një dritare për ndërtimin e mjedisit dhe të së shkuarës në një vend ku banorët modernë bashkëveprojnë në strukturën e perceptimeve të tyre dhe përdorimit të peisazhit të sotëm.

KAPITULLI 19.
Modeli Tre-Dimensional: Gërmimi në Projektin e Informimit

Christopher Johanson

Pamjet dixhitale shumë dimensionale në vendndodhjet arkeologjike mbeten një praktikë fillestare. Përshtatja e fuqishme që herët dhe tashmë gjithëpranimi virtual në vendvarrim dhe laborator ka ndikuar në lulëzimin e jashtëzakonshëm të pamjeve që për nga qëllimi variojnë nga plane të thjeshta dhe në lartësi deri në aplikime ndërvepruese të bazuara në rrjet përfshi, aplikimet e informacionit të sistemit gjeografik (GIS), kompleksiteti tri dimensional, dhe modelet në kohë reale. Mënyrat retorike dhe prioritet e projektimit ende merren si të mirëqëna nga studiuesit të cilët krijojnë argumente të bazuara në tekst duke injoruar krijimin e këtyre imazheve. Qëllimi, synimi dhe projektimi, teza, struktura dhe argumenti shpesh varen nga disponimi i drejtëpërdrejtë i aftësive teknologjike të vendosura afër rezultateve të parakonfigurara dhe parapërcaktuara. Kjo është funksionalisht e barazvlefshme me efektin e PowerPoint i artikuluar me elokuencë nga Tufte (2006) në të cilën shablloni i software prodhon pamje që dështojnë të komunikojnë efektivisht. Modelet e pasqyruara nga defektet janë norma dhe kjo traditë kërkimore e vendosur pengon përparimin produktiv kur modelet e disponueshme për matje realizohen dobët. Ka pamje që ose krijohen me pak reflektim kritik në projektim ose formohen nga defektet në radhitje të një kompjuteri pamendje duke i dhënë në epërsi zhurmës dhe "tabelës së vjetër" për shembull, dendësi të jashtëzakonshme të vijave të palexueshme topografike, teksturat dhe hijëzimet defektive të padobishme ose gjetje të thjeshta të dukshme si linjat e vazhdueshme të dhëmbëzuara të krijuara nga pakujdesia në shumë paketa software të GIS. Këto përfaqësime pamore përhapen dhe ndërtohen pa ndonjë kujdes dhe ndonjëherë pa qëllim vetëm për shkak të lidhjeve të shkurtra mes një projekti arkeologjik dhe ekipit teknik.

Problemi rëndohet nga edhe mungesa e teorive me aplikim rigoroz në raportet e një vendndodhje arkeologjike. Një studiues si Burrell e nënvizon këtë problem me një shqyrtim kritik për gjendjen e raportimeve moderne për një vendndodhje. Pasi vëren peshën e madhe të një ekipi për të prodhuar të dhëna solide dhe të përhershme format e larmishme që këto raporte marrin dhe pamundësia e kryerjes së punës kur autorët përballen me kufizime buxhetore, përqëndrohet në sinteza që mund të jenë më të shkurtra, më pak të kushtueshme, edhe në këtë rast–nëse është e mundur?–apo interesante. Ende ky raport i dobët kur bashkohet me arkiva shoqërues të "të dhënave" duhet të përmbledhë gërmimin, "ri-krijojë" stratigrafinë e shkatërruar, spjegojë zhvillimin e vendndodhjes në fazat e tij dhe të rinumërojë gjetjet. Edhe pas retorikës kritike, të raporteve të mëdha, një qëllim i padyshimtë i kësaj ndërrmarrje është të ofrojë "ri-krijimin" e asaj çka ishte. Ky qëllim gjithnjë do të mbetet i paarritshëm, por përparimet në teknologji ofrojnë iluzionin për arritje, duke siguruar mjetet e mundshme që dokumentojnë dhe rizhvillojnë çdo gjë.

Një model tri-dimensional i një vendndodhje arkeologjike bazohet në ndërprerjen e prurjes së të dhënave dhe nevojës për përpunim të përfundimeve. Rezultati do të jetë po aq kokëfortë sa edhe të dhënat, por virtualisht është i palexueshëm në rast se ka përpjekje për të përfaqësuar prekjen totale të materialit. Ai mund të shërbejë si burim për aksonometri të thjeshtë, kryerje dixhitale skicash, në një arkiv dixhital kompleks të stratigrafisë dhe gjetjeve të vogla deri te një pasqyrim tejet-realist të peisazhit dhe banorëve. Kur qëllimi për ruajtjen dhe shpërndarjen e sa më shumë të dhënave të mundshme përballet me këtë kapacitet të pakufizuar dixhital, kufinjtë duhet të përcaktohen.

Tuma e Lofkëndit paraqet një subjekt veçanërisht të vështirë për rindërtimin dixhital dhe pamor. Thuajse e tëra është organike me përmbajtje komplekse dhe mbivendosje stratigrafike dhe strehon një prej gjetjeve më të vështira për modelimin tredimensional, mbetjet skeletore. Teknika më e zakonshme përfaqësuese për zbulimin e varreve të brëndshme të tumës përdor vizatime me dorë, skica tre-dimensionale ose të ndërtuara me dorë, modele fizike tre-dimensionale me seksione tërthore ose prerje çerek seksionesh. Megjithatë kur ndarjet e seksioneve tërthore të tumave, kodrave të varreve, dhe karrocat ekzistojnë ato pashmangshmërisht tregojnë një formë të qartë të brendshme arkitekturore. Si shembull që provon këtë rregull është prerja seksionit tërthor e Tumës së Madhe të Verginës, i ashtuquajturi varri i Filipit të II-të janë qëndërzuar në një formë arkitekturore varrimi, varri mbretëror, i ndërtuar nga mur guri. Këto përfaqësime shmangin vështirësitë për të shquar gropat e varreve si ato të ndeshura në Lofkënd. Rindërtime të tilla më shumë ndeshen në bashkimin mes akademisë dhe kulturës popullore, ndoshta përpjekja më e suksesshme për pamjen komplekse, në varret e mbivendosura në varg imazhesh u prodhuan nga Ned Seidler për *National Geographic* për ilustrimin e shtresave tre-dimensionale të varreve të Lordit të Sipanit në bregun verior të Perusë. Edhe në këtë rast, mbetjet skeletore ishin paketuar në tekstile dhe vendosur në varr në rregull të qëllimshëm.

Tuma e Lofkëndit përbëhet nga njëqind varre, gjashtë faza varrimi dhe asnjë formë monumentale ose arkitekturore ndonëse në vetvete ajo përbën një monument. Dikush nuk mundet vetëm të ndërtojnë një përfaqësim tre-dimensional të kësaj vendndodhje pa një artikulim të qartë të përfundimeve të kërkuara. Prirja për të ruajtur dhe treguar çdo gjë e bën këtë sipërmarrje edhe më të vështirë. Në krahasim, rindërtimet dixhitale të vendndodhjeve arkeologjike që përmbajnë arkitekturë monumentale ofrojnë disa nivele që tregojnë dobi të menjëhershme edhe nëse konceptualizimi i projektit mund të ketë qënë disi i paqartë. Krijimi i mureve hipotetikë dhe mbulimi lejon analizat e vijave të shikueshme, akustikën dhe përafërimet e paqarta të ekperiencës urbane. Më në fund, disa modele edhe kur janë ndërtuar pa themele të qarta teorike i japin studiuesve mjete të qarta për të kuptuarin e marrëdhënieve hapësinore në vendndodhje. Detyra për përfaqësimin e një tume nuk lidhet me rindërtim por me shikueshmërinë dhe paraqitjen e informacionit.

KAPITULLI 20.
LOFKËNDI SI VEND I KULTIVUAR

Samantha L. Martin-MacAuliffe

Pjesa e parë e këtij kapitulli konsideron në detaje pozicionin e tumës: fizionominë e saj, mjedisin dhe marrëdhënien me topografinë përreth. Një lexim i ngushtë i kodër-varrezës brenda terrenit është kritik për një sërë arsyesh. Nga njëra anë qartëson kushtet aktuale të tumës duke krijuar një hartë përnjohëse të mirë-përcaktuar, jo vetëm të tumës në vetvete por edhe të vendit të saj. Në të njëjtën kohë, analiza të kujdesshme hapën shtigje të reja çështjesh dhe diskutimesh duke na lejuar të dallonim marrëdhëniet delikate fizike dhe gjeografike. Kjo linjë hulumtimi shpresuam të zhvillonte çështje rreth rolit të tumës përtej funksionit të saj parësor si vendvarrim. Ndonëse tuma ndoshta integrohet me topografinë e saj, njëherazi ajo është produkt i veprimtarisë njëzore. Për këtë arsye studimi i paraqitur do të sqarojë si kjo kodër-varrezë dhe rrethinat e saj mund të përcaktohen si peisazh. Pjesa përfundimtare e seksionit sugjeron se Lofkëndi mbart para-kuptimin e një vendi, e për më tepër në vetvete është parapërgatitur për ndërtimin e tumës.

Duke ndjekur studimin e pozicionit të tumës, ky kapitull do të këqyr edhe kultivimin e vendit. Seksioni i dytë fillon me rishqyrtimin e etimologjisë së fjalës kultivim, duke parë me vëmendje të veçantë se si ajo mbart konotacion të drejtëpërdrejtë dhe metaforik. Diskutimi sugjeron që marrëdhënia mes tumës së Lofkëndit dhe topografisë së saj është e shtresëzuar. Në këtë vend në veçanti ne shohim shprehi pragmatike dhe themelore të kultivimit si ushqim dhe ende ka kushte që i japin jetë ndërlidhjeve më formale që kanë të bëjnë me venbanimin. Përtej kësaj, varreza tumulare ka kapacitetin të mbart kuptime figurative që shoqërojnë nocionet e kujtesës dhe nderit. Përfundimisht seksioni i mesit propozon që varreza tumulare u kultivua me qëllim, dhe që njerëzit–banorët, përdoruesit, kalimtarët e rastit–katalizuan disa kushte që i dhanë formë ndërtimit të saj lokal dhe fizik në kuptim të gjerë.

Seksioni i tretë dhe i fundit i këtij kapitulli hulumton se si ne mund të kuptojmë dhe përshkruajmë një varrezë tumulare si vend-mbartës. Pjesa më e madhe e vendvarrimeve mund të referohen si mbajtësit e të vdekurve, ende situata në Lofkënd kërkon konsiderim të veçantë. Për shembull, ndryshe nga një mazoleum me prerje në gur, forma e tumës së Lofkëndit është drejtëpërdejt e varur dhe e formuar nga përmbajtja. Në mënyrë kategorike, kjo varrezë tumulare nuk mund të boshatiset nga përbërësit pa humbur integritetin e saj fizik. Kjo situatë ngre shumë pyetje të natyrës praktike dhe filozofike rreth natyrës së kufinjve në përgjithësi. Duke filluar nga vetë tuma, ky seksion shkon përtej monumentit për të shqyrtuar mënyrën e influencës së tumës në peisazhin e saj përfshi marrëdhëniet me tuma të tjera dhe piketimet natyrore. Pjesa e fundit e këtij kapitulli zgjeron më tej diskutimin e kufinjve. Ndërsa termi kufi shpesh ngjall cilësi konkrete në pikëpamje të parë, gjithashtu mund të mbart kuptime abstrakte dhe metaforike. Varreza tumulare e Lofkëndit qartësisht lidhet me jetën e përtejme dhe kështu është e detyrueshme që ne të studiojmë kufinjtë që ajo ndjek në kuadrin e kohës.

KAPITULLI 21.
FILLIMI I FUNDIT TË TUMËS SË LOFKËNDIT DHE PREHISTORIA E *KANUNIT*

John K. Papadopoulos

Kur këqyrnim të gjithë materialin nga tuma e Lofkëndit të varrosur me të vdekurit, një veçori dalluese është mungesa e armëve (shih Kapitullin 10). Sasia e armëve është fare e pakët: një majë shtize e vetme nga Varri XLV (**10/118**); dy maja shigjete nga Varri XXXII (**10/119**, **10/120**); dhe pak thika hekuri (njëra nga Varri LXXXIV [**10/121**] dhe maja e një thikë nga Varri XXXVIII [**10/122**]; një thikë e tretë u ndesh në shtresën e sipërme dhe ndoshta është moderne [**10/123**]). E vetmja majë shtize dhe thikat e hekurit përsërisin shembuj të ndeshur në shumë varreza në Epirin fqinj dhe Greqinë veriore ku pothuaj çdo mashkull varrosej me dy ose më shumë shtiza. Po ashtu, thika prej hekuri thuajse e plotë në Varrin LXXXIV përsërit një shembull të vërejtur gjetkë në Ballkan, por afër Epirit thikat e hekurit–ndryshe nga shtizat të cilat lidhen gjithnjë me meshkujt e rritur–janë të zakonshme në varret e burrave dhe grave.

Të dy varret janë përshkruar plotësisht në detaje në Kapitullin 3. Në rastin e Varrit XXXII, të dyja majat e heshtave u gjetën të lidhura me një prej dy meshkujt e rritur në varr, një skelet plotësisht i artikuluar i identifikuar si mashkull i rritur në moshë 20–30 vjeç. Të dyja majat e heshtave u gjetën mbi torso, një në lindje dhe tjetra në perëndim; më e vona në zonën e barkut, e para në mes/djathtas torsos. Këto qenë të vetmet maja shigjete në të gjithë tumën dhe pozicioni i tyre mbi torson e kufomës ishte i tillë që ato ose mund të interpretohen si dy shigjeta të vendosura-apo të mbajtura nga kufoma ose si maja që nga dhuna ndërpersonale kanë çuar në vdekjen e individit. Kockat ishin tejet të fragmentuara për të treguar ndonjë të dhënë për vdekje të shkaktuar nga maja e shigjetave. Nga këto dy interpretime, vendosja e majave të shigjetës në varr si "dhurim" duket goxha e çuditshme për sa kohë që pjesën më të madhe të varreve të Bronzit të vonë dhe në Hekurit të hershëm ku majat e shigjetave janë ndeshur, ato zakonisht gjenden të lidhura me armë të tjera veçanërisht shpata dhe shpesh në një këllëf me shumë shigjeta. Mundësia se ky mashkull është vrarë nga shigjetat nuk mund të anashkalohet lehtë.

Në rastin e majës së thikës prej hekuri të gjetur në Varrin XXXVIII, rrethanat ishin tejet të pazakonta. Varri ishte varrim i dyfishtë i një mashkulli të rritur 30 deri 40 vjeç dhe një individ i djegur, ndoshta femër. Mashkulli me vendosje trupi ruhej mirë dhe është i përshkruar si i madh dhe i shëndoshë (shih Kapitullin 6). Më i pazakontë ishte fakti se këmbët ishin shtrënguar dhe përthyer kryq njëra pas tjetrës. Këndi i mbledhjes ishte jo vetëm i qëllimtë por sugjeron se trupi ishte shtrënguar si një dëng për të mbajtur kufomën në drejtim të tumës. Ky interpretim u vërtetua nga të dhënat e një litari të zi të gjetur në vend që me shumë mundësi lidhej me mbetje tekstili ose lëkure, ndoshta qefin që mesa duket mbulonte trupin dhe gjithashtu kockat e djegura. Siç edhe detajohet në Kapitullin 3, djegia ka qënë sekondare dhe mbetjet djegura janë mbledhur dhe vendosur menjëherë në juglindje të vendosjes së trupit dhe mund të jenë fshtjellë me tekstil ose lëkurë. Kjo ishte njëra prej dy djegieve në të gjithë tumën. Por varri u bë edhe më interesant gjatë heqjes së mbetjeve skeletore: fragmenti i tehut të një thike hekuri, sigurisht, vetëm maja e saj u gjetën menjëherë poshtë vertebrës së kraharorit në mashkullin me vendosje trupi. Gjendja e fragmentuar e tehut bashkë e pozicionin e saj në vend sugjerojë se mashkulli i rritur mund të ketë vdekur nga një goditje në bark, anë apo kurriz. Ky kombinim veçorishmbledhja e shtrënguar dhe ndoshta në dëng e trupit brenda asaj çfarë dukej të ishte pjesë tekstili ose lëkure, djegia sekondare e vendosur bashkë me trupin e varrosur dhe ndoshta një vdekje me dhunë sugjerojnë se të dy individët mund të kenë vdekur disi larg nga tuma dhe janë mbledhur në dëng dhe transportuar për t'u prehur në banesën e fundit.

Të dy meshkujt në Varret XXXII dhe XXXVIII përfaqësojnë të vetmit shembuj ku ka të dhëna që nga armët të sugjerohet tek e fundit mundësia e vdekjes me anë të dhunës. Në të dy rastet, një vdekje e dhunshme nuk mund të vërtetohet nga studimi i mbetjeve të tyre skeletore por nëse kocka nuk do të ishte goditur nga thika apo shigjetat e ndeshura në secilin varr. Në të dy rastet skenari i një vdekje të dhunshme mbetet bindës. Sidoqoftë këto dy varre nuk janë të vetmit të pazakontë dhe jo-normativë në Tumën e Lofkëndit. Dy të tjerë, Varret I (Varri 64) dhe LXXIV (Varri 29) janë jo vetëm të veçantë në kontekstin e Lofkëndit por edhe në rast se krahasohen me çfarë njihen si zakone mortore Ilire dhe Ballkanike në përgjithësi. Për më tepër njëri prej këtyre varreve është ndoshta varrimi më i hershëm në Lofkënd–ngjarja e themelimit–dhe tjetri prej më të vonëve, ndonëse qartësisht jo më i voni. Ndoshta nuk do ta kuptojmë kurrë se si të tre burrat në varrin LXXIV e panë fundin e tyre, dhe në mungesë të analizave të ADN, ndoshta nuk do ta dimë ndonjëherë nëse ata kishin lidhje familjare. Gjithëçka dimë është gjinia e tyre, mosha e përafërt dhe varrimi i përbashkët në një varr, si pjesë e të njëjtës ngjarje rituale. Pas tyre komuniteti që varrosi të vdekurit në këtë tumë në kodrat e Mallakastrës, la pas atë çka mund të përshkruhet si trashëgimni disi e shpejtë.

Për të qënë të qartë nga këto të dhëna nuk ka ndonjë gjë bindëse, por modeli i përgjithshëm është intrigues sepse sjell në vëmendje mundësinë e dhunës ndërpersonale në disa prej varreve të Fazave Va and Vb dhe të dhëna më bindëse për vdekje me anë të dhunës në rastet e varreve XXXII dhe XXXVIII në Fazat II dhe III të përdorimit të tumës.

Për një shoqëri fisnore, bazuar në klan, një dhunë e tillë bëhej sipas rregullash ose ligjesh të cilat si edhe këngët e folkut në Shqipëri nuk shkruheshin–sigurisht, deri në një kohë më të vonë–por këndoheshin të kaluara gojë më gojë dhe brez pas brezi. Siç edhe Camaj vëren, ligjet zakonore shqiptare kanë lënë gjurmë në karakterin e njerëzve veçanërisht në "kuptimin mbi nderin, hakmarrjen, kurajon dhe vendosmërinë në situata kritike, dhe ndjenjën e afërsisë brenda familjes, vllazërinë dhe klanin." Ky kapitull sugjeron se ligji tribal Shqiptar si i përmbledhur në *Kanun* mund të ketë një të shkuar më të largët, ndoshta të përçuar deri në prehistori.

KAPITULLI 22. MENAXHIMI I TRASHËGIMNISË DHE E ARDHMJA E TUMËS

John K. Papadopoulos, Lorenc Bejko, dhe Sarah P. Morris

Pamja e dukshme e tumës së Lofkëndit ka qënë një çështje vazhdimisht e theksuar në këtë volum (Kapitujt 1, 20, Epilogu), çka e shndërroi atë në një objekt hulumtimi ishin dukshmëria dhe materialiteti i tumës që nxitën vendimarrjen nga zanafilla e projektit deri në rindërtimin e tumës. Ne nuk mund ta zhduknim këtë monument nga mjedisi. Prandaj, financimi për rindërtimin ose rivendosjen përfundimtare të tumës sa më afër pamjes së saj origjinale u caktua që në fillim të projektit.

Duke konsideruar vendndodhjen e saj dominuese në hapësirën përreth, mbi një breg dhe në një vend imponues është me vlerë të përmendet se tuma ndenji veçanërisht e palëvizur për rreth dy mijë vjet e gjysëm pas fillimit të përdorimit të saj. Pavarësisht erozionit të varreve të vërejtur përpara fillimit të gërmimit në anët e tumës veçanërisht ato përgjatë rrëpirës së anës jugore, tuma mbijetoi thuajse e paprekur.

Shkaku që tuma nuk u shemb ose gërrye më shumë, ose zvogëlua nga stuhi periodike të cilat për këtë pjesë të Shqipërisë mund të jenë shkatërruese na bëri të dyshonim që në fillim të gërmimeve që diçka në përbërjen e saj kish ndikuar për të mbajtur bashkë dhe për këtë arsye u zotuam të hulumtonim në tërësi dherat e tumës dhe ato rreth mjedisit të saj (Kapitulli 16.4).

Ishte e qartë se për të qënë të suksesshëm çdo zgjidhje duhet të përqëndrohej në materialet tanimë të gatshëm afër tumës dhe ne me qëllim i shmangëm zgjedhjet e "teknologjisë së lartë." Duke konsideruar njohurinë e dherave të rajonit të Lofkëndit dhe eksperiencën tonë në mbulimin e tumës për tre vjet të njëpasnjëshëm ne vendosëm të aplikonim një tekonologji të thjeshtë, e pakushtueshme dhe e shpejtë për rindërtimin e trapeve dhe përdorimin e tyre si kornizë për mbajtjen e dheut të gërmuar. Ne shqyrtuam një sërë mundësish lidhur me natyrën e materialit që do të përdorej për trapet. Në fund ne zgjodhëm metodën për bërjen e tullave të baltës me dheun tumës në rindërtimin e trapeve (ndonëse tullat e baltës nuk ishin përdorur asnjëherë në monumentin original). Ideja themelore për projektin tonë të rindërtimit ishte se me kalimin e kohës tullat e baltës të përdorura do të shpërbëheshin duke u bërë pjesë efektive e tumës, si tipi i një skeleti të kalbur që e mbante bashkë strukturën e saj. Ky kapitull ofron kronikën e bërjes së 2.000 tullave prej balte të përdorura për rindërtimin e trapeve dhe përdorimin e këtyre trapeve për ta kthyer edhe një herë tumën e Lofkëndit në formën e mëparshme.

EPILOG

Sarah P. Morris dhe John K. Papadopoulos

EPILOG 1. MËNYRAT E JETESËS SË NDËRTUESVE TË TUMËS

John K. Papadopoulos

Në Kapitullin 1 ne sugjeruam se për të kuptuar peisazhin e Lofkëndit, dikush duhet të kuptonte natyrën e tumës, duke qënë se përfaqësonte të vetmen të dhënë të qartë të përdorimit e mjedisit gjatë

Bronzit të vonë dhe Hekurit të hershëm në këtë vend të veçantë.

Ne gjithashtu e kontekstualizuam këtë çështje kundrejt sfondit të vendosjes së kolonive Greke, veçanërisht në Apolloni diku rreth 600 p.e.s. dhe lindjes së aglomeratit urban i cili nga shekulli i katërt p.e.s. ishte tërësisht i përforcuar në mjedisin Ilir me vendet fqinje të Klos-Nikajës dhe Bylisit por edhe pak më larg në jug dhe perëndim si vendndodhja e Amantias. Lofkëndi gjithashtu qarkohej nga një sërë vendndodhjesh "protourbane" si Mashkjeza, Gurëzeza, dhe Dimali–kronologjia dhe karakteri i të cilave janë ende pak të kuptuara–përfshi edhe kodrën e Margëlliçit e cila të paktën mund të shquhet për atë pak material Mikenas. Ne gjithashtu vërejtëm informacionin e mbledhur nga venbanimet e njohura në burimet Klasike si *komai* (p.sh. fshatrat e parrethuar) më në jug, në Epirin Grek. Sigurisht që bashkëveprimi midis fshatrave të parrethuara, metropoleve koloniale dhe qendrave urbane krijonte një situatë dinamike që lejonte çdo lloj eksperimentimi politik dhe shoqëror. Për pasojë, ne arsyetuam se hulumtime lidhur me natyrën e vendbanimit në rajon do të kishin ndikim më të gjerë nga sa kishim parashikuar në fillim. Tuma ndalon të përdoret si vendvarrim prehistorik diku rreth 800 p.e.s. (për fundin e tumës shih Kapitujt 4 dhe 21), afërsisht dy shekuj para themelimit të kolonisë Greke në Apolloni. Ndonëse vëzhgimi sistematik rreth zonës së tumës ka plotësuar shumë boshllëqe natyra e vendbanimit të Epokës së Bronzit të vonë dhe Hekurit të hershëm vazhdon të jetë e pakapshme (shih Kapitullin 18). Çfarë mbetet të përcaktohet është mënyra e përshtatjes dhe mjetet e jetesës për banorët e Bronzit të vonë dhe Hekurit të hershëm në zonën rreth tumës, çka i lejoi ata të popullonin këtë mjedis për 600 vjet.

Siç edhe Samantha Martin-McAuliffe thekson në përfundimet e saj në Kapitullin 20, ndërtimi tumës së Lofkëndit në lashtësi fizikisht formalizoi një zotim njerëzor të mirë-formuar në luginën e Lumit të Gjanicës. Ajo gjithashtu vëren se fakti që shumë çështje që lidhen me ndërtuesit e tumës ne mund t'i perceptojmë si banues të luginës dhe si të tillë ata u identifikuan me këtë zonë dhe topografinë e saj më gjerë. Më tepër se kaq, këta "individë ishin kultivuesit e vendit dhe zgjedhja e vendit të tumës reflektoi lidhjen e tyre me luginën" (Martin-McAuliffe, Kapitulli 20).

Gjithashtu edhe informacioni i mbledhur nga studimi fizik i popullatës së Lofkëndit (Kapitulli 6), bashkë me informacionin për paleo dietën të sugjeruara nga izotopet qendrore (Kapitulli 7). Për shembull, afërsitë biofizike të Lofkëndit dhe marrëdhëniet e tyre me popullatën prehistorike nga Apollonia tregojnë se ndonëse edhe Lofkëndasit dhe Apolloniatët ishin vazhdimisht subjekt i niveleve të ulta të stresit gjatë fëmijërisë, siç edhe sugjerohet nga dalja e dhëmbëve, popullsia e Lofkëndit duhet të jetë ekspozuar me stres të një intensiteti më të ulët (gjithashtu u ndikuan më pak nga malaria sesa popullsia e Apollonisë) (Kapitulli 6). Për më tepër ndryshimet mes popullsive në Lofkënd dhe Apolloni shënjuan mes të tjerash, larmia në dietë lidhur me mbështetjen tek drithërat mund të kenë sjellë lindjen e shkallës së karies për popullsitë moderne. Vlerat e izotopeve (Kapitulli 7) për popullsinë prehistorike të Lofkëndit tregojnë se proteinat dietike nga terreni lëshohen kryesisht nga burimet e C_3 (p.sh. gruri, elbi, bimët rrënjore, bishtajoret, perimet, arrorët, mjalti dhe frutat). Ushqimi mund të ketë përfshirë produkte të bulmetit dhe mishit por pak ose fare burime të detit. Krahasuar me dietat e Egjeut Neolitik dhe Minoan edhe ato të Lofkëndasve tregojnë pak proteina nga bulmeti dhe mishi ose ndoshta më shumë konsum të bishtajoreve apo mishgrënës që konsumonin bishtajore. Në kontrast, popullsia moderne konsumon kryesisht bimët C_4 dhe ndoshta misër.

Duke konsideruar përmasat e vogla të kampionit, pak mund të përmendet për shëndetin e përgjithshëm dhe dietën e Lofkëndasve. Më shumë mund të thuhet rreth mënyrës së jetesës së banorëve që varrosën të vdekurit në Lofkënd dhe kontekstualizimi i këtyre mënyrave me çka njihet më larg drejt jugut. Sigurisht studimi i *komait* Epirot, i paraqitur në Kapitullin 1, veçanërisht në zonat e larta krijuan përshtypjen se banorët e pjesës më të madhe të Greqisë veriperëndimore dhe Ilirisë jugore ishin mbarështrues të kafshëve nomade. Shumë studiues kanë sugjeruar shfrytëzim sezonal të mjedisit fizik, atë çka imponohej nga gjeomorfologjia e rajonit, që favorizonte zhvillimin dhe rritjen e një ekonomie të bazuar gjerësisht në blegtori. Në këtë mënyrë prodhimi bujqësor i drithërave, fasuleve dhe bishtajave duhet gjithnjë të ketë qënë i kufizuar në qëllim dhe i dorës së dytë pas blegtorisë. Kjo pikëpamje u përkrah nga studiues si Hammond dhe Ioulia Vokotopoulou, mes të tjerësh pavarësisht faktit se një numër studiuesish me ndikim që në shekullin nëntëmbëdhjetë e kishin kundërshtuar në tërësi këtë pikëpamje. Analizat e izotopeve të qëndrueshme të paraqitura në Kapitullin 7 gjithashtu nënvlerësojnë mendimet rreth rëndësisë së nomadizmit.

Një numër studimesh të rëndësishme nga Paul Halstead dhe John Cherry, në mënyrë krejt të pavarur kanë sfiduar mendimin e përhapur se mjedisi malor dhe afrimiteti pan-Ballkanas i kulturës materiale të Pindit sugjeron baritorë nomadë ose transhumancë. I njëjti fakt mbetet i vërtetë edhe për Ilirinë e Jugut. Halstead në mënyrë efektive tregoi se dy ose tre modelet e ndryshme të jetesës së sotme njohur si baritorë nomad të specializuar të praktikuara nga Sarakatsani, dhe pastoralizmi i larmishëm transhumant sipas modelit Vllah thjesht nuk ekzistojnë në prehistorinë e vonë. Sigurisht shfaqja e Vllehëve në Greqinë veriore dhe zonat fqinje diku rreth fundit të mijëvjeçarit të parë e.s. mund të reflektojë kolonizimin e një forme të re ekonomike. Për më tepër, të dhënat arkeologjike për përhapjen e vendbanimeve në Pind rreth fundit të mijëvjeçarit të dytë p.e.s. janë interpretuar si bujqësore sendentare të kombinuara, duke përsëritur të njëjtat strategji jetese që dominuan në ultësirat greke gjatë Epokave të Neolitit dhe Bronzit. Halstead ishte i kujdesshëm të vërente se një ekonomi e kombinuar bujqësore nuk parandalon as përdorimin sezonal të kullotave të largëta as ndryshimet brenda dhe mes komunititeteve dhe rëndësinë relative të punimit të tokës dhe mbajtjes së kafshëve. Ai vazhdon të theksojë që të dhënat palinologjike për ndikimin e kultivimit të hershëm edhe në bimësinë e ultësirës janë tejet të vështira për t'u gjetur dhe nga maja e vargut të Pindit këto të dhëna thjesht nuk ndeshen. Megjithatë bërthamat e polenit nga vendndodhjet në gjerësi të mesme gjeografike sugjerojnë që shpyllëzimi i shkallës së sotme nuk ka ndodhur deri pak shekuj më parë.

Në të njëjtën linjë, Cherry argumenton se pastoralizmi i specializuar është mjet i pamundur jetese në kontekstin e tipeve të kuadrit të sistemit mjedisor dhe bujqësor në Greqinë e hershme prehistorke dhe sigurisht, është një përshtatje që historikisht ka lindur vetëm në rrethana të veçanta shoqërore ekonomike dhe politike. Në këtë mënyrë, Cherry rrëzoi dy supozime që kishin pllakosur në këtë fushë, e para që pastoralizmi në Greqi ishte tërësisht i përcaktuar nga mjedisi dhe e dyta që mjedisi ka mbetur në thelb i pandryshuar që në antikitet. Në një fokus interesant, Cherry, me referim nga puna e hershme e Eric Higgs dhe kolegët e tij solli në vëmendje faktin se studentët e Paleolitit në Epir janë ndikuar nga etnografia në mënyrën që blegtorët modernë në Greqinë veriperëndimore janë përdorur si model i drejtëpërdrejtë për modelin e transhumancën sezonale mes gjuetarëve të Paleolitit të Lartë në vendndodhje të tilla si Asprochaliko dhe Kastritsa. Gjithashtu, Claudia Chang ka shprehur rezerva lidhur me projektimin direkt të modelit të Sarakatsanit në Epokën e Bronzit dhe Hekurit të hershëm.

Në rishikimin e tij mbi blegtorët C.R. Whittaker nënvizon dy pika të rëndësishme: e para është se blegtoria është një tip ideal që nuk ka ekzistuar asnjëherë në formë të pastër qoftë edhe mes Masai ose Berberëve; dhe e dyta që blegtoria duhet që gjithnjë të fillojë nga bujqësia. Për pikën e dytë, Whittaker ndjek argumentin e Ferdinand Braudel sipas të cilit ishte pikërisht vendosmëria e kultivuesve për të pluguar ajo që u la rrugë të lirë barinjve. Siç edhe Peregrine Horden dhe Nicholas Purcell theksojnë, është e gabuar që blegtoria të lidhet me aktivitetin karakteristik të maleve jomikpritëse. Avantazhet për fermerët sendentarë për mbajtjen e kafshëve janë të shumta. E para, kafshët mund të jenë "ushqim i ruajtur" kundër mungesës së zarzavateve dhe e dyta, në në veçanti në mjedisin Mesdhetar kafshët nuk nevojitet të konkurojnë me njerëzit për tokë. Siç edhe Horden dhe Purcell vërejnë, kafshët e organizuara janë një burim fleksibël i pakrahasueshëm. Përveç mishit (dhe të mirave post-moderne si lëkura, kocka dhe brirët), speciet e zbutura në Mesdhe kanë ofruar qumësht, veshje, pleh dhe energji për tërheqje, transport në tokë dhe ujë. Cherry i bën jehonë këtij fakti duke vërejtur se kopetë e dhenve pastrojnë dhe fekondojnë fushat e kultivuara afër vendbanimeve dhe në këtë proces shndërrojnë materialet e papërdorshme bimore në mish dhe qumësht që shërbejnë si plotësime dietike në kohët e varfra dhe mungesës së të korrurrave.

Në rastin e veçantë të Epirit (një rajon, ndryshe nga kodrat e Mallakastrës që kanë mjedis të ndryshëm dhe larmi vertikale), Konstantinos Zachos ka sugjeruar se në rastin e fushës ose fushëgropës së Janinës (Helopia antike) ndoshta nuk ka qënë nevoja për planifikim dimrin-në-fushë dhe verën-në-mal pasi vetë ekosistemi i siguronte të gjitha burimet e nevojshme për banim sedentar gjatë gjithë-vitit. I njëjti përfundim është argumentuar edhe nga Thomas Tartaron për luginën e poshtme të

Acheronit në Epirin jugor. Gjithashtu Angelika Douzougli dhe Zachos argumentojnë se, në shumë pjesë të Epirit kombinimi i zonave të ulta dhe të larta siguron bollëk dhe larmi burimesh që lejon një ekonomi të përzierë me bujqësi dhe blegtori lokale.

Shembujt arkeologjikë për Bronzin e vonë dhe Hekurin e hershëm që janë shfaqur kohët e fundit nga vendndodhje Epirote si Vitsa Zagoriou, Pogoni, dhe Liatovouni të marra së bashku me materialin e nxjerrë nga Shqipëria duket të jenë qarta. Ato sugjerojnë se që nga periudha e Neolitit deri gjatë viteve të shekujve të pestë dhe fillimit të të katërtit p.e.s., një ekonomi e bazuar në bujqësi sedentare e lidhur me kullotat lokale mbeti mënyra kryesore e jetesës.

Banuesit e luginës siç edhe Samantha Martin-McAuliffe i quan këta të Lumit të Gjanicës në Lofkënd nuk bëjnë përjashtim. Demografia e tumës me pak më shumë se 150 individë të shpërndarë në rreth gjashtë shekuj sugjeron për popullsi të vogël, e shumta një familje e zgjeruar. Vendbanimi i tyre duhet të ketë qënë ngjajshmërisht i vogël, duke reflektuar fshatrat e vogla të parrethuar si atë të *komait* në Vitsa dhe Liatovouni në Greqinë verilindore ose vendndodhjen e Teqesë së Melanit në Shqipërinë jugore. Ne mund të imagjinojmë mirë një rrjet të tillë vendbanimesh si pjesë e një klani ose fisi që bashkarisht ndërtuan streha të fortifikuara dhe vend takime që janë njohur si qendra "protourbane" pavarëshisht mungesës së ndonjë veçorie ose karakteristike "urbane". Sigurisht riorganizimi i parë themelor ekonomik, shoqëror, dhe politik i kësaj mënyre jetese ndodh në shekullin e katërt p.e.s kohë në të cilën grupet e shpërndara janë përqëndruar brenda dhe rreth qyteteve të reja të fortifikuara si Klos-Nikaja dhe Bylis të cilat përfaqësojnë të dhënat më të hershme të zhvillimit të plotë urban në rajon.

Në rast se popullsia që varrosi të vdekurit në tumën e Lofkëndit ishin banorë të luginës, ka shumë gjasa që afërsia me Lumit e Gjanicës të ketë qënë kritike për furnizimin me ujë gjatë vitit për njerëzit, kafshët edhe bujqësinë. Natyra dinamike e mjedisit lumor mund të ketë fshirë gjurmën e ndonjë vendbanimi të vogël ose katundi të vendosur afër bregut të Gjanicës. I njëjti mjedis lumor në formën e dy lumenjve afërsisht paralel, Vjosë/Aoös dhe Seman e kanë spostuar sot portin e qytetit të Apollonisë rreth 26 km larg detit. Si edhe Epiri në jug, ky mjedis dominohej prej lumenjve dhe banorët që jetonin në këtë hapësirë i shfrytëzuan ata në dobi të tyre nga periudha e Paleolitit deri në kohët moderne.

Epilog 2 Nga Epoka e Gurit në të Shkuarën e Afërt: Biografia Kulturore e Mjedisit dhe një Tumë Ilire

Sarah Morris dhe John K. Papadopoulos

Historia afatgjatë e një tume modeste në Shqipërinë jugqëndrore nuk është e kufizuar vetëm në hiatusin kohor të varrimeve (Kapitulli 4) nga Epoka e Bronzit të vonë deri në shekullin e tetë p.e.s. Për më tepër, forma dhe përmbajtja e saj mbledhin në vetvete shumë eksperienca që nga konteksti mjedisor i Paleolitit tek një ngrehinë moderne për ushtarët e cila përqëndroi dhe zgjeroi para dhe pas-vdekjen e jetës së banorëve të saj. Siç edhe jemi përpjekur të kuptojmë themelimin dhe pasjetën e një tume si një vend i kultivuar me qëllim e që vendos rregull në hapësirën përreth (Kapitulli 20). Në këtë seksion ne synojmë të rrefëjmë gërmimin e tumës në kohë edhe më gjerë kuadrin saj në historinë e Ilirisë.

Epokat e Gurit dhe Bronzit

Natyra ka luajtur rol themelor në këtë luginë të ngushtë, lumenjtë e të cilës presin shpatet e mprehta në të dyja krahët me depozitime pjellore në secilin prej tyre siç edhe depërtojnë për të gjetur rrugën të bashkohen me Lumin e Vjosës/Aoös e më tej të ecin drejt detit. Me dobësimin e Epokës së fundit akullnajore, këto shpate u mbuluan nga një klimë më e freskët dhe e lagësht me pyje të dendur lisi dhe panje (Kapitulli 16.1). Gjatë Paelolitit të vonë (rreth. 200,000–50,000 PT) të paktën një bandë e vogël gjuetarësh frekuentonte lumin bashkë me burimet e tij ku kafshët vinin për t'u ushqyer dhe pirë duke lënë gjurmët e një vendndodhje të vogël në hapësirë në ose mbi bregun lindor të Lumit të Gjanicës, poshtë fshatit modern të Belishovës (Kapitulli 18, Vendndodhja 003). Lloji i veglave të gurit që ata krijuan e përdorën dhe copëzat e dala nga etapat e prodhimit dhe ripërdorimi i tyre u shpërndanë përgjatë zonave të ndryshme në shpatet lindore të lumit përfshi afërsinë direkte me tumën dhe sipërfaqen e saj (Kapitulli 1). Një sasi domethënëse prej 589 ashklash prej guri nga Epokat e Paleolitit drejt Bronzit gjetën rrugën drejt mbushjen e tumës (Kapitulli 13). Kjo mund të duket e parëndësishme, por për një varrezë tumulare një sasi e tillë gjetjesh të stralleve është e konsiderueshme. Veglat e strallit ndoshta janë mbledhur me qëllim ose me shumë mundësi kanë lëvizur me llojet e ndryshme të dherave (veçanërisht argjila) të përzgjedhura t'i jepnin zhvillim dhe masë të qëndrueshme kodrës (Kapitujt 13, 16.4 dhe 20). Në këtë mënyrë kjo duhet të ketë qënë e lidhur me materialet (dherat kompakt dhe të ngjeshur) të cilët vendosën pasjetën e këtyre veglave duke i tërhequr ato në formimin e një tume të Bronzit të vonë dhe Hekurit të hershëm. Paskëtaj, vendi fillimisht i zgjedhur për varrim rreth 1400 p.e.s. (Kapitulli 4) mbante në të prejardhjen e habitatit më të vjetër njerëzor dhe aktiviteti në mjedisin përreth duke shkuar pas në kohë rreth 50.000 vjet dhe ndoshta edhe më herët, përpara se një grup familjar solli të varroste të vdekurit e parë tek ky breg i spikatur. Duke marrë parasysh rezultatet e Lofkëndit vështirë të përmëndësh më shumë për vendndodhjet dhe jetën gjatë Epokës së Gurit dhe hershëm në atë të Bronzit në këtë rajon. Periudha Neolitike është e vështirë të gjurmohet në këtë zonë, ndoshta sepse vendbanimet e reja tërhoqën bujqit dhe blegtorët larg drejt venbanimeve të përhershme në brendësi të luginës në vendndodhje të tilla si ajo e Cakranit. Nga Epoka e Bronzit të mesëm, fortifikimi i Krapsit (emri modern) në një breg në skajin verior të luginës së Gjanicës tregon popullim, qartësisht të lidhur me një "katund Ilir" gjatë Bronzit të vonë deri në Hekur. Edhe më e madhe dhe jetëgjatë ishte vendndodhja e Margëlliçit që dominonte zonën nga Bronzi i vonë, e shquar si nga Krapsi edhe tuma në Lofkënd. Sigurisht sipas të dhënave që vendbanimi dhe vëzhgimet tregojnë, ai mbetet më i dukshmi mes vendndodhjeve të tjera në zonë duke ofruar popullim nga Bronzi i vonë në Hekurin e hershëm ndonëse me nekropolin e tij. Kjo e lë luginën e Gjanicës ende të varfër në vendbanime, veçanërisht për ata që varrosën të vdekurit e tyre në Lofkënd.

Periudha e Përdorimit të Tumës
(1400–800 p.e.s)

Në një moment në Epokën e Bronzit të vonë (gjatë shekullit, diku rreth dhe pas 1400 p.e.s.), një grup i vogël familjar ose klan i zgjeruar zunë bregun sipër lumit dhe sollën aty një koleksion mbetjesh të të vdekurve të tyre. Nëpërmjet një riti themelor për caktimin e një banese të re për të pajetët e të shkuarës dhe të ardhmes duke hapur një gropë afërsisht katërkëndëshe poshtë shkëmbit natyror në zonën me të lartë të bregut dhe vendosën aty mbetjet e pjesshme të skeleteve të tre burrave (një i madh, dhe dy të rritur me lidhje të ngushtë mes tyre; shih **Tabela 6.14**), bashkë me kocka derrash, delesh dhe dhish (**Fig. 3.8**, Kapitulli 16.1). Duke bërë këtë, ata pasqyruan të njëjtat veprime të qëllimshme të grupeve të tjera gjatë kësaj periudhe ku zanafilla e një vendvarrimi të ri lidhej me kockat e "paraardhësve" të tyre, kjo e ndeshur të paktën në dy vendndodhje në Epirin e Jugut (Greqia veriore, Ephyra dhe Pogoni). Gjetjet e reja në atë çfarë ne quajtëm "Varri" I (Varri 64) na lejuan t'a piketonim kohën e këtij momenti: dy gjilpërat prej kocke (**10/52**, **10/56**) mund ose nuk i përkasin të varrosurve, dhe një copëz e vetme prej druri të karbonizuar (datuar në 1373 ± 57 p.e.s.) ndoshta mund të ishte një cope dru e vjetër–ose më–e vonë në kohën kur u përzie me mbetjet tjera kockore prej njerëzish dhe kafshësh. Çka mbetet kritike është sa thelbësore ishte kjo veprimtari për shndërrimin e një hapësire natyrore në një vend kulturor siç edhe mbetet sot e kësaj dite.

Për të gjitha synimet dhe qëllimet, ata që rivarrosën të vdekurit këtu krijuan një lokacion të ri dhe të përhershëm për lidhjet familjare përmes së shkuarës në të ardhmen. Pak kohë më vonë, grupe individësh–burra, gra, dhe fëmijë (ndonjëherë nënë dhe fëmijë bashkë si në Varrin XII)–u varrosën në pjerrësinë e shpatit në jug dhe lindje të kësaj grope qëndrore, mbi majën e shkëmbit natyror (**Tabela 4.1**). Varret II dhe III datojnë brenda së njëjtës gjeneratë (Kapitulli 3.1, Kapitulli 4), grupi i parë, në një total prej 12 varresh gjatë shekullit të katërmbëdhjetë p.e.s., ose ndoshta pak më vonë. Duke konsideruar gjetjet e përzgjedhura për banesën e fundit, të rriturit dhe fëmijët ishin mbledhur me veshje të mbërthyera me gjilpëra kocke (**10/53–10/55**) dhe bronzi (**10/27, 10/28**), disa prej këtyre gjilpërave mund të kenë shërbyer edhe si kapëse flokësh. Tekstilet tashmë janë humbur (pseudomorfet janë të rralla jashtë kategorisë së objekteve të hekurit, vetëm disa ishin ruajtur tek objektet e bronzit: Kapitulli 12) dhe gjilpërat mund të kenë qënë kufiri i zbukurimit funerar. Një fëmijë ishte varrosur ndoshta duke mbajtur një rruazë të vogël bronzi (**10/65**, Varri VIII), pa ndonjë enë qeramike, stoli ose zbukurime të cilat nisin të përdoren në fazën e mëpasshme të përdorimit të tumës. As qeramika nuk përdoret në Fazën I të tumës. Enët ose përmbajtja e tyre nuk janë konsideruar të rëndësishme ose e kundërta shumë me vlerë për t'iu dhënë të vdekurit, ndërsa pjesa më e madhe e zbukurimeve në përdorim është funksionale. Përshtypja që të krijon faza e parë e varrimeve është ajo e një komuniteti modest me lidhje të ngushta. Të saporriturit dhe adoleshentët mungojnë dukshëm në grupin e hershëm të popullatës, që kryesisht konsistojnë në të rritur dhe foshnja/fëmijë; jeta e tyre ka qënë e vështirë, kjo e vërejtur nga shenjat e lëna në kocka dhe dhëmbë (Kapitulli 6). Varret dhe përmbajtja formojnë një kontrast të habitshëm me varret spektakolarë të "luftëtarëve" të Epokës së Bronzit të vonë me prani shpatash, mburojash dhe enësh.

Gjatë fazës së dytë të përdorimit (Varret XIV-XXXIII), tuma u rrit me varret të shpërndarë larg nga depozitat origjinale, veçanërisht e zgjeruar drejt zonës së sheshtë në veri dhe perëndim të Varrit 1 (**Fig. 4.1** dhe Kapitulli 19). Gjatë kësaj periudhe u varros një numër i lartë fëmijësh, të rinjsh, dhe femrash, fillimi i një modeli që zgjati për disa faza varrimi dhe është veçanërisht i pazakontë për standartet demografike të popullsisë prehistorike ku pritet një numër i madh foshnjesh dhe të rriturish të maturuar (Kapitulli 6). Mbetet e vështirë të përcaktohet në rast se kjo reflekton nivelet e larta të vdekshmërisë tek të rinjtë bashkë me zëvendësimin e ulët të popullsisë si pasojë e humbjeve të pazakonta tek moshat e reja ose është përzgjedhje e qëllimtë e fëmijëve dhe të rinjve për varrim (Kapitulli 8). Më shumë varrime të shumëfishta me vendosje trupi janë realizuar në Fazën II (p.sh. Varret XXI, XXVII, XXXII), dhe dy varrime sekondare me djegie u futën në tumë (Varri XXX, një foshnjë). Për të parën herë disa individë (të rinj) ishin pajisur me enë qeramike (**9/59, 9/60**, ndoshta plot me ushqim dhe pije) dhe grupi më i hershëm dhe tërheqës i dhurimeve shoqëruar me dy fëmijë (të moshës 7 dhe 4 vjeç) të varrosur bashkë në Varrin XVII. Dy enë dhe disa objekte zbukurimesh prej bronzi përfshi një rreth koke, vath dhe tela spiralikë shënojnë modelin e parë të dhurimeve luksoze për një moshë kaq të re (**Figs. 3.50–3.54**).

Varri XVIII njëtrajtësisht bashkonte zbukurimet e bronzit me foshnjat e vogla dhe kur një femër e re (15 vjeçe) iu bashkua një mashkulli të vjetër (prind, vëlla, bashkëshort?) në Varrin XXI, ajo ose ata morën objekte elegante bronzi dhe për herën e parë zbukurime prej hekuri. Siç diskutohet në Kapitullin 8 kjo fazë shënon periudhën e parë kur të rriturit dhe meshkujt marrin më pak takëme mortore (ndonjëherë asgjë) krahasuar me të vdekurit në moshë të re (p.sh., Varri XXVIII, një fëmijë 3 vjeç i kurorëzuar me rruaza xhami dhe bronzi) dhe gratë e reja, ndonëse jo çdo fëmijë trajtohej me kaq pajisje (p.sh., Varret XXIII-XXV). Njëri prej dy meshkujve të rritur në Varrin XXXII ishte varrosur me një grup majash heshte prej hekuri. Depozitimi i majave ishte i tillë që ato mund të interpretoheshin si dy maja të vendosura sipër ose ndoshta të mbajtura nga i vdekuri ose si mjete vrasëse (kocka ishte shumë e fragmentuar dhe nuk ruante ndonjë të dhënë të qartë vdekje nga maja e shigjetës). Këto ishin të paktat armë që shënjuan këtë periudhë kalimtare nga Epoka e Bronzi të vonë në Hekur, datuar nga një kampion i vetëm druri të karbonizuar nga Varri XXIII në rreth 1070 p.e.s. Armët janë gjetje të rralla për çdo fazë të tumës dhe përbëhen nga një majë heshte prej hekuri e gjetur me një mashkull të vjetër në Varrin XLV (Faza III); të paktat thika hekuri të ndeshura në varre shihen më shumë si vegla shumëqëllimëshe sesa armë në kuptimin e plotë të fjalës. Pjesa më e madhe e këtyre modeleve vazhdojnë në Fazën III (Varret XXXIV-LIV, Hekuri i hershëm, sërish të datuara nga një kampion i vetëm druri i karbonizuar në varrin XLV në 953 ± 53 p.e.s.), pa shenja të qarta të ndonjë vlere të bashkangjitur për luftën dhe armët, veglat ose sende të tregtuara: sigurisht gurët e peshës, rruazat ose kockat ishin të vlera për varrimet, bashkë me ndonjë thikë (Varri XXXV) dhe në përgjithësi dhurimet e brëndshme formojnë përshtypjen e një komuniteti që u përkujdes për zbukurimet e metaleve kryesisht për individët që humbën jetën përpara se të fitonin një status të plotë në jetë (Kapitulli 8), ose një rit kritik i kalimit në jetën e përtejme. Një shembull i dytë i djegies në Lofkënd, në Varrin XXXVIII shënon një shtesë të pazakontë të djegies së një individi (ndoshta femër) në varrin e një mashkulli të rritur që mund të ketë vdekur si pasojë e plagës nga një goditje dhe ka qënë "mbështjellë" (mbajtur në pozicion të kruspullt), ndoshta të dy të transportuar nga vendi i vdekjes së dhunshme ose të papritur dhe djegies në vatrën funerare të një varreze të paqtë dhe të vetmuar (familja ose klani) në Lofkënd. Si varrim i trefishtë (Varri XLV, kampioni i kolagjenit njerëzor dha një datë absolute 953 ± 53 p.e.s.) që përmbante vetëm një majë shtize (një prej të paktave armë të vërteta) në tumën e Lofkëndit rrethana të pazakonta mund të kenë sjellë tre anëtarë të një familje bashkë në vdekje në të dy këto varre. Pavarësisht numrit më të lartë të meshkujve të rritur (krahasuar me femrat) në Fazat III, Va, dhe Vb (**Tabela 6.4**), thuajse se asnjëri nuk varrosej me elementë të një karakteri ushtarak apo ndonjë arritje tjetër në jetë: kjo jo domosdoshmërisht nënkupton se klani i Lofkëndit gëzonte jetë të paqtë, por sepse kjo shenjë dalluese thjesht nuk ka qënë e rëndësishme për farefisin e tyre. Të paktën në vdekje, ka fare pak përpjekje për të shënjuar sërën e meshkujve të rritur: "burrat e mëdhenj" nuk janë pjesë e riteve funerare në Lofkënd. Këto modele përshtaten me zakonet e "Ilirëve" modern (Rascianët e zonës së Moesisë) të vëzhguar nga udhëtari kurioz i shekullit të nëntëmbëdhjetë, Shoberl i cili në 1827 vërente:

> Kur një burrë vdes, ata bëjnë gati me rrobat, i veshin çizmet, dhe vendosin afër tij torbën e duhanit, llullën, thikën, pirunin dhe të tjerë sende të cilat ai i përdorte gjatë jetës. [Më pas e veja e pyet atë nëse i duhet diçka tjetër!]

Çfarë mbetet e panjohur është vendi se ku popullsia e Lofkëndit jetonte. Mbetjet e materialeve të ndërtimit nga mbushja e varreve dhe tumës (qerpiçi i prekur nga zjarri: Kapitulli 14) nuk janë të lidhura me ndonjë periudhë të veçantë të së shkuarës: sigurisht, njësoj si veglat e gurit, ato mund të datohen përpara ose pas kohëve të varrimeve (Kapitulli 14).

Nga statura fizike individët prehistorikë nga tuma kanë qënë më të gjatë sesa shokët e tyre modern, si krahasuar me popullatën e tanishme por edhe me atë të varrosur në Varret LXXXVI-C (Kapitulli 6). Së bashku ata tregojnë shenja të bimorfizmit seksual konvencional (**Tabela 6.5**) me larmi më të madhe të bimorfizmit mes meshkujve, ndoshta sepse detyra dhe nivele të ndryshme të ushqimit formuan jetët dhe fizikun e individual të burrave. Lidhur me dietën dhe shëndetin popullata e Lofkëndit ka vuajtur nivele të larta smaltit linear të hipoplasisë tek meshkujt dhe femrat (**Tabela 6.6**), kjo ndoshta nga ekspozimi i përbashkët ndaj stresit (sëmundjet dhe/ose periudhat me ushqim të varfër) gjatë fëmijërisë. Pak individë (një fëmijë dhe pesë të rritur: Kapitulli 6) tregojnë shenja të mundshme poroziteti të kafkës në zonën e cibrias orbitale sidoqoftë nuk mund të konfirmohet nëse ata që vuanin nga anemia kequshqyese, infeksionet ose parazitët ose nuk kanë mbijetuar për t'u varrosur në këtë vend ose pjesa më e madhe e komunitetit iu largua këtyre sfidave. Disa individë vuanin nga osteoartriti dhe osteofizi i vertebrës që kryesisht ndikoi në plakje (p.sh., individi 495 në Varrin XLV, burri më i vjetër i Lofkëndit). Kariet e dhëmbëve janë relativisht më të larta në krahasim me popullsi të tjera të Epokës së Hekurit, duke iu afruar më shumë atyre karakteristike për Neolitin, me ndodhshmëri më të lartë për burrat sesa gratë (ndoshta për shkak të numrit më të lartë të meshkujve në këtë popullatë: **Tabela 6.7**), dhe ka mbushje frekuente të dhëmbëve të një lloji të që i atribuohet dietës së butë. Veçoritë jo-metrike të dhëmbëve sugjerojnë bindshëm marrëdhënie të ngushta farefisnore mes anëtarëve të komunitetit matur si në varrin e parë dhe atë të fundit (**Fig. 6.14**). Sipas historisë së jetës, përtej disa plagosjeve të këmbëve dhe krahëve, një individ (139, në Varrin LXXVII) ka qëndruar duke i shpëtuar një plage të rëndë në kokë mbi syrin e djathtë (**Fig. 6.14**), të paktën një burrë mund të ketë vdekur nga një plagë shigjete (Individi 508, në Varrin XXXII, i afërm i një tjetër mashkulli në të njëjtin varr: **Tabela 6.14**) por pak skelete tregojnë shenja kacafytjesh të vazhdueshme ose plagosje të shumëfishta.

Si krahasohet ky komunitet njerëzor me zonat e tjera dhe të varrosurit më vonë? Çka mbetet kritike rreth popullsisë së Lofkëndit është kontrasti i shëndetit të tyre fizik me individët e hershëm modern të varrosur në të njëjtën tumë (Kapitujt 3 dhe 6–7). Të pesë të rriturit modern (varrosur rreth 1700–1900 e.s.) kishin sëmundje degjenerative të nyjeve të vertebrës së rruazës së kraharorit dhe pjesa më e madhe e dhëmbëve ishin me karie ose të humbur ndoshta për shkak edhe të moshës së tyre (**Tabelat 6.7, 6.9**). Mes të varrosurve në Apolloni gjatë periudhës prehistorike (60 individë, Epoka e Bronzit të hershëm përgjatë Hekurit të vonë në Tumat 9, 10 dhe 11) meshkujt e vjetër ishin më dominues në numër me raport të lartë të të rriturve të rinj krahasuar me popullsitë standarte prehistorike sikur e njëjta përzgjedhje varrimesh të ishte praktikuar në Lofkëndin para dhe proto-historik. Në të tjera konvergjenca individët Klasikë dhe Helenistikë në Apolloni ndajnë veçori (të tilla si kafka e zgjatur) dhe disa faktorë biodistantë me të dyja popullsitë prehistorike, ndërsa popullsia moderne në tumat e Lofkëndit dhe Apollonisë ndryshon prej tyre e shfaqin më shumë ngjajshmëri me individët modernë Ballkanas. Në të njëjtën kohë, individët e varrosur në Apolloni në kohët prehistorike dhe koloniale (Klasike dhe Helenistike) tregojnë nivele të larta të stresit, sëmundje ose kequshqyerje (në kocka dhe dhëmbë): ndoshta në deltën dhe moçalishtet në grykëderdhjen e lumenjve të Vjosës/Aoös dhe Semanit u shfaqën popullsinë lokale me malarie dhe të tjera sëmundje. Këto shenja të stresit apo sëmundjeve janë edhe më të larta mes individëve Klasikë dhe Helenistikë në Apolloni, kjo edhe me ndryshimet e stilit të jetës nën kolonizimin Grek, densitetit të lartë të popullsisë ose një kombinimi faktorësh shoqëror dhe mjedisor.

Në përgjithësi, ata që jetuan afër Lofkëndit kanë qënë disi më të shëndetshëm sesa të ngjajshmit e tyre në Apolloni në kohën vdekjes, por kanë vdekur në moshë më të re. Më kritike është rënia e shëndetit në popullsitë moderne të Lofkëndit dhe Apollonisë, kjo e përforcuar nga analizat qëndrore të izotopeve të kampionëve të kockave të kolagjenit njerëzor (Kapitulli 7) gjë që vë në dukje mbështetje të lartë në bimë me sasi të madhe të C_4 por të varfra në proteina dhe hekur (ndoshta misër) karakteristike e kohëve moderne. Lofkëndi prehistorik nga ana tjetër me shumë mundësi konsumonte më tepër bishtajore (bimë C_3) sesa produktet e bulmetit dhe mishit. Një gjë e tillë e nënvlerëson nocionin mbi ekonominë antike blegtorale e bazuar në mish dhe qumësht duke sugjeruar në të kundërt bimët e kultivuara si burimi kryesor i ushqimit në Epokat e Bronzit të vonë dhe Hekurit të hershëm.

Kultura materiale e mbyllur me të vdekurit, sidomos zbukurimet prej metali dhe enët e qeramikës e lidhin këtë komunitet ngushtë me ata që përdorën të njëjtat objekte në skajin verior të Ballkanit, me ato të Epirit jugor, Maqedonisë, dhe Thesalisë dhe përtej detit në Itali pa ndonjë saktësi lidhur me identitetin e tyre ndër-Ilir. Lidhur me modelet shoqërore, ndonjë shenjë pozite ose hierarkie vështirë se mund të dallohet në varre, as në gjetje ose në faktorët e shëndetit dhe duket të kenë munguar në jetën e përditshme: dikush mund ta cilësonte si një komunitet jo më i madh sesa një familje e zgjeruar në kufi me shoqëritë më të prosperuara dhe më në dukje, por mundësisht, ndoshta viktimë e një konflikti të largët dhe në rritje (Kapitulli 21).

Pikërisht gjatë fazës përfundimtare të përdorimit të Epokës së Hekurit (900–800 p.e.s.) tuma duket se akumulon numër të lartë varrimesh gjithashtu edhe ngritjen me varrime me vendosje trupi dyfishe dhe shumëfishe: a ishte kjo për shkak të rritjes së popullsisë, ngritja e niveleve të vdekshmërisë, ose rritja të vdekjeve në grup nga dhuna? Një prej arsyeve mund të jetë rritja e nivelit të stresit ose konflikti, i llojit me ndikim të mundshëm në fundin e jetëve të dy burrave të rinj bashkë me një burrë të vjetër habitshëm, dhe në mënyrë jonormale në pozicion të shtrirë, paralel dhe ngushtë me njëri-tjetrin pa takëme mortore (Varri LXXIV: Kapituli 21). E vështirë të mos peshohen këto statistika përkundër mbylljes së papritur të fundit të këtij vendvarrimi, rreth ose pak më pas se 800 p.e.s., menjëherë gjatë kohës kur vendndodhje në zona të ngritura fillojnë të fortifikohen përreth (Gurëzeza, Mashkjeza) dhe në veriun e largët (Grunas, Shalë). Ndoshta të zhvendosur nga tribu të reja intrusive aq sa për të imponuar migrime përtej Adriatikut në Italinë e jugut (Stocker 2009: përmbledhja). Komuniteti që varrosi të vdekurit në Lofkënd pushoi ta bëjë këtë gjë në shekullin e tetë p.e.s. Nëse u absorbuan me vendndodhjet e fortifikuara (më të sigurta?) si Margëlliçi, ose u rivendosën përtej bregut në Italinë e jugut–për të cilën nuk ka ndonjë të dhënë bindëse–fati i tyre i përket një historie të fshehtë që mbetet fare pak e kuptuar, por sigurisht ende autoktonë në origjinën e zhvillimit të tyre, dhe e pavarur nga ambiciet Greke në këtë zonë.

ILLIRIA CAPTA: PËRPARIMI I KORINTASVE

Deri në kohën kur një vizitor i panjohur la pak fragmente të një kotile Korintase mbi sipërfaqen e tumës (Shtojca 2, dhe shih **9/332–9/336**, dhe **Tabela 9.3**), askush nuk ishte varrosur atje prej mëse 200 vjetësh dhe në këtë kohë tuma kishte fituar rolin e saj si monument. Me këtë kapacitet, dhe prej formës dalluese tuma ndoshta ka tërhequr vëmendjen e këtij vizitori dhe frymëzoi një prej tyre ose të tjerë që këto tuma në hapësirën Ilire i përdorën për të varrosur të vdekurit të formuara njësoj siç edhe bënë në Apolloni (Amore 2010). Duke ngritur shumë varreza tumulare në luginën e Kryegjatës në Apolloni. Sharon Stocker dhe Jack Davis (2006:91) pyesin:

> A mundet që ardhësit Grekë në Apolloni të kenë ndërrmarrë një vendim të ndërgjegjshëm për të imituar një formë varrimi të gjetur në hapësirë–formë që në atë kohë nuk ishte zakon për zonat e Greqisë jugore nga e cila ata emigruan? Me praktikimin e varrimeve në tuma a mundën ata me qëllim të dallonin zakonet e tyre mortore nga ato të vendbanimeve të tjera Greke në Shqipëri, veçanërisht në Epidamnos-Dyrrachium? A e vërejtën ata këtë formë varrimi si reflektimin heroik të praktikave familjare me to nga Homeri epik?

Në këtë ndërrmarrje, Stocker dhe Davis citojnë Ian Morris që nënvizon se eposi dhe të dhënat materiale janë të dyja produkt i njerëzve real, që veprojnë sipas qëllimeve të tyre në një periudhë ndryshimesh dramatike....[N]ë ritualet që siguruan të dhëna arkeologjike, Grekët e Epokës së Hekurit manipuluan kulturën materiale si një gjuhë jo-verbale nëpërmjet të cilës ata debatonin se cilët ishin dhe ku qëndronin në raport me të tjerët, më gjerë në botën e Mesdheut lindor dhe të shkuarës së humbur heroike.

Korintasit që u vendosën në Apolloni e futën në një erë të re dhe nivel të pranisë njerëzore në këtë pjesë të Shqipërisë bregdetare, gjë që ndryshoi rrethinat dhe hapësirën pakthyeshmërisht. Ndonëse disa prej tumave të luginës së Kryegjatës në Apolloni qartësisht datohet deri në Epokën e Bronzit të hershëm, pjesa më e madhe e tyre datojnë nga periudhat Arkaike drejt asaj Helenistike dhe Romake. Sipas cilësdo mënyrë numërimi; Vangjel Dimo vërejti rreth 450 tuma në Apolloni rreth 1995–1996, ndërsa Stocker dhe Davis gjatë punës në terren nxorrën një numër më të vogël me rreth 100 tuma. Numri i tyre vetëm në luginën e Kryegjatës është mbresëlënës dhe nëse merret parasysh çfarë ka ekzistuar më parë ai është një tregues i rëndësishëm për njerëzit që populluan rajonin. Sado që fragmentet e pakta Korintase në Lofkënd mund të duken një reflektim i dobët i asaj çfarë po ndodhte më afër me deltën e Lumenjve të Vjosës dhe Semanit, gjithsesi ato mbeten simptoma të erës së kolonizimit Grek.

Ndërsa ne kemi propozuar disa skenarë për fundin e tumës tonë Ilire dhe lindjen e formave të reja fillimisht si fortresa protourbane dhe më pas si qytete Helenistike, hulumtimi i tumës së Lofkëndit nuk e ofron hapur kontributin e fundit për të kuptuar këtë zhvillim. Në fakt, ato pak fragmete qeramike vetëm ofrojnë një lidhje të mundimshme nga fundi i tumave në fillimin e jetës në qytetet e stilit Grek. Ndërsa është fare e pak e mundur që varrime të vona (Arkaike) të jenë futur në tumë duke trazuar e ndërhyrë mes shekullit gjashtë p.e.s. dhe erës moderne të varreve, për këtë rast ne do të prisnim të dhëna më thelbësore mbetjesh nga varret e trazuar bashkë me takëmet e tyre në vitet e ndërmjetme.

Zona e Lofkëndit në Periudhat Greke dhe Romake

Pak dihet rreth kësaj zone gjatë periudhave Klasike në Romake, dhe njohuritë tona kufizohen në gjetje sipërfaqësore por dy vendndodhje në afërsi dokumentojnë banim të përhershëm me pritshmëri për varrime. Nga veriu i tumës, në spostim lindor dhe vetëm pak qindra metra larg, një dalje e shkëmbit natyror është zgjedhur një zgavër natyrore për mbajtje varresh disa edhe të mbuluara me tjegulla gjatë periudhave Helenistike dhe Romake (Kapitulli 18, Vendndodhja 001, Mezhdat e Kuqe: **Figs. 18.1–18.3**). Afër Gjinoqarës, të paktën një vendndodhje dha qeramikë Romake dhe Mesjetare dhe më vonë u shdërrua në vendndodhjen e një kishe që rrëzua në 1924. Sidoqoftë për pjesën më të madhe të periudhave Klasike dhe Helenistike duket se Margëlliçi mbeti vendbanimi më i madh në zonë dhe aty nga periudha Romake jo vetëm Bylis por një komunitet i ri Kristian në Ballsh e dominoi këtë rajon.

Për vetë tumën materiali që daton midis pas Korintasve dhe para periudhës moderne ishte tejet i kufizuar (shih **Tabelae 9.3**). Mes enëve të qeramikës të paraqitura në Kapitullin 9 ka një fragment fundi me shumë mundësi Athinas, ndoshta të një skifosi (**9/330**), mbledhur në sipërfaqen e tumës; një tjetër fragment nga shtresa e sipërme mund të jetë Grek (**9/331**). Pjesa tjetër është tërësisht e pa diagnostikuar me tre fragmente gri brumi të reduktuar (**9/337–9/339**), dhe një fragment i djegur i dëmtuar (**9/340**), rreth të cilit shumë pak mund të thuhet. Prej këtyre, njëri (**9/338**) që vjen nga shtresa sipërfaqësore dhe pjesa tjetër nga mbushja e tumës shumë mirë mund të jenë të Epokës së Bronzit të vonë dhe Hekurit të hershëm. Përmes qeramikës së painventarizuar padekorim e të punuar me çark e përmbledhur në Kapitullin 2 (shih veçanërisht **Tabela 2.3** dhe diskutimin për të), një fragment i vërejtur mundësisht Helenistik por shumë pak i ruajtur për t'i dhënë ndonjë datim të sigurt.

Nga Lofkënd në Likofone: Kristianët dhe Muslimanët e Luginës së Gjanicës

Për shekuj tuma qëndroi e patrazuar nga ndërhyrje serioze veçanërisht pas rënies së Bylisit dhe Ballshit fqinj (vendi i pasardhësve të peshkopëve Kristian gjatë Kohëve Mesjetare), dhe shndërrimit të pjesës më të madhe të popullsisë Ortodokse në Islam (Kapitujt 3.2). Sidoqoftë pak Kristianë mbetën në zonë ose emigruan në të, si Vllahu i vetëm i regjistruar nga censusi Austriak i shekullit të njëzetë si Ortodoks Lindor (Kapitulli 3.2). Një shekull më vonë banorët lokalë, ndoshta Kristian të cilën nuk ndanin varrezat Muslimane me fqinjët e tyre (në Ngraçije, juglindje të tumës ose në Mashkullorë: **Fig. 18.4**) zgjodhën fundin verior të tumës për të varrosur disa prej të vdekurve të tyre. Ende mbetet e panjohur nëse ata tërhiqeshin nga dukshmëria e tumës, largësia prej praktikave Muslimane ose kishin ndonjë njohuri për strehimin e të vdekurve aty në të shkuarën e largët. Ata që varrosën aty i përshtaten një profil më të parashikueshëm demografik: foshnja ose të sapolindur, të rritur të moshuar, pak syresh të varrosur me takëm mortor, një mbërtheckë prej tunxhi në veshjen e një fëmije (**TXCVIII-1: Figs. 3.325a–b**) ose një monedhë për të paguar çuesin e të vdekurve në varret e dy të rriturve dhe një foshnje (**TXCII-1, Figs. 3.306c–d, 3.308a–b; TXCIV-1, Figs. 3.314c, 3.315a–b; TC-1, Figs. 3.329a, 3.330**). Nga pozicioni (nga lindja drejt Jerusalemit) ata e identifikojnë veten si Kristian por asnjë dëshmi e mbijetuar nuk mund t'u japë identitet më të përafërt ose spjegojë rrethanat sepse aty u varrosën më të lëndueshmit nga vdekja (të rriturit e vjetër dhe foshnjat e sapolindura) vetëm për pak vjet. Çfarë i bën varrimet e vona në Lofkënd të duken më pak anomale janë rastet e shumta të ndërhyrjeve mesjetare dhe të hershme moderne të vërejtura edhe në tumat të tjera prehistorike në Europën juglindore dhe Anatoli (shih diskutimin në Kapitullin 3.2)

Prej materialit sipërfaqësor ose shtresës së sipërme të tumës, një fragment (**9/341**) është qartësisht modern dhe vjen nga një enë e hershme moderne e pikturuar nga Grottaglie në ishullin e afërt Grek të Korfuzit. Të paktat fragmente me çark të padekoruara nga sipërfaqa ose njësitë e sipërme të tumës janë përmbledhur në Kapitullin 2, thuajse të gjitha moderne, por pa ndonjë datim të saktë.

Kontinenti i Errët: Shqipëria dhe Europa në një Shekull Dhune

Në periudhën e saj përfundimtare të përdorimit, tuma e Lofkëndit u shdërrua në një pikë avantazhuese për luftimet e shekullit të njëzetë. Gëzhojat bosh të nxjerra nga pothuaj 30 gjuajtje ishin lënë mbi tumë, shumë syresh kishin depërtuar edhe në shtresat e sipërme të tumës prehistorike (Shojca 3, pas Kapitullit 10), dhe një ose dy edhe më poshtë se sa kaq. Siç fati e solli, një Grek i lindur në Shqipëri që kishte lëvizur përtej kufirit në moshë të re pikërisht gjatë nisjes së këtyre ngjarjeve, u shndërrua në këshilluesin tonë për këto plumba, falë edhe pozicionit të tij në Muzeun e Luftës në Athinë dhe bujarisë për kohën dhe ekspertizën e tij.

Shekulli i njëzetë shënjoi pavarësinë e kombeve si Shqipëria nga perandoria Otomane, lindjen e shtetit pas konfliktit të parë botëror, eksperiencat e tij nën monarkinë, komunizmin dhe demokracinë dhe Luftën e Dytë Botërore që e solli Shqipërinë në rrugën e një fqinji ekspansionist, Italisë Fashiste. Komponentët arkeologjikë të vizionit Musolian dërguan Luigi Ugolinin dhe kolegët e tij të gërmonin në Butrint dhe hulumtonin Romën "Trojane" dhe pasardhësit e Augustit, dhe më qartësisht nisi një ekspeditë strategjike për të ri-kolonizuar çfarë Romakët pushtuan si Iliri në të dyja luftrat. Lufta e Parë Botërore e përhumbi tumën tonë përmes legjendave vendase të ushtarëve të rënë të portretizuar në romanin të pavdekshëm të Kadaresë (*Gjenerali i Ushtrisë së Vdekur*), po gjithashtu solli ushtarët Italian, ose rregullat e tyre në tumë. Lufta e Dytë Botërore solli Italinë në kufinjtë e Greqisë ku disa prej betejave më të vështira u kryen në dëborë të thellë dhe në malet e larta (të tilla si në Malin e Gramozit), por gjithashtu tërhoqën Greqinë thellë në territorin Shqiptar deri në Vlorë dhe mesa duket lart në luginën e Gjanicës siç edhe dëshmohet nga gëzhojave standarte të pushkëve greke.

Shqipëria Post-Komuniste: Teqeja dhe Pema e Dardhës

Në një vizitë në tumë përpara gërmimeve në 2003, Bejko dhe Papadopoulos rastisën me pronarin lokal të tokës ku qëndronte tuma, një farë Rrapi Malaj, shtëpia e të cilit ndodhej poshtë shpatit të tumës në lindje. Pas diskutimit me të, përgjigja pohuese e tij për projektin sipas sugjerimit tonë na gëzoi. Aprovimi i zotëruesit të tokës është kritik për suksesin e çdo projekti arkeologjik, dhe në këtë rast ne e morëm me mend se kjo do të sigurohej përmes një tarife "qeraje" (e njohur më mirë nga termi Persian *bakshish*), dhe sigurisht kështu rezultoi. Kur u kthyem me ekipin e plotë të gërmuesve dhe specialistëve në vitin pasues në ditën e parë të gërmimit takuam jo më Rrapi Malajn por Baki Ymerin i cili po punonte tokën bashkë me familjen e tij. Pasi iu afruam për të kuptuar marrëdhënien e tij me tokën, përgjigja e tij na surprizoi duke qënë se ai pohoi të ishte pronari i tokës. Pas diskutimesh të tjera me Rrapi Malaj qe e qartë se të dy zotërinjtë kishin kërkesa dhe pretendime naive pronësie. Ndonëse u dorëzuam drejt një tarife dyfishe "qeraje" kërkuam t'i shkonim deri në fund kësaj situate për të kuptuar natyrën e këtyre dy pretendimeve, dhe pikërisht këtu diplomacia dhe takti i atij që gjatë asaj kohe ishte

Drejtor i Institutit të Arkeologjisë së Akademisë së Shkencave të Shqipërisë tregoi të ishte kritike duke qënë se ai kishte përjetuar në vendin e tij kohët Komuniste dhe post-Komuniste. Siç edhe rezultoi, njëri pronar përfaqësonte një organizatë lokale fetare, *teqe* (shih Kapitulli 18, Vendndodhja V008, **Fig. 18.1**) dhe kishte qënë kujdestari i tokës që i përkiste këtij objekti pavarësisht faktit se *teqeja* nuk ishte më në përdorim ndërsa një tjetër fermer pretendonte pronësinë për shkak të kultivimit të tokës. Pretendimet e tyre të dyfishta ndoshta mundësisht mund të zgjidhen nga vendimet e gjyqeve në Shqipëri, të nevojshme në kohën post-Komuniste gjatë procesit të vendosjes në regjistrin kombëtar. Ndërkohë që projekti ynë respektoi pretendimet e dyfishta të pronësisë dhe si pasojë kompesimi në frymën e bashkëpunimit me vendasit dhe ndërrmarrjes intelektuale me nocionin që hapësirat si antike dhe moderne gjenerojnë grupe interesi të shumëfishta. Rastësia e fundit e projektit tonë qe pema e dardhës e shartuar prej një lloji të egër dhe të kultivuar që kacavjerrej në kulmin e tumës përpara se të fillonim gërmimin (**Fig. 2.5a**: pamja e tumës në 2004). Pema u desh që të lëvizej gjatë sezonit tonë të dytë dhe për këtë arsye u pagua edhe kompesim. Sidoqoftë kjo vetëm provokoi një fushatë të re mbjellje pemësh nga pronari (Rrapi Malaj) duke trazuar të paktën një prej varreve prehistorike. Në këtë pikë pronarët dhe arkeologët mbërritën në një marrëveshje për të ndërprerë ndonjë mbjellje të mundshme dhe çrrënjosje të pemëve frutore për hir të të vdekurve në tumë.

Pas Arkeologjisë: Rindërtimi dhe Kthimi i një Monumenti në Hapësirën e Tij

Përkushtimi dhe detyrimi ynë përfundimtar përfshiu restaurimin e një fenomeni natyror dhe njërëzor në vendin e saj kritik në hapësirë. Siç edhe detajohet në Kapitullin 22 dhe në publikime të hershme, i njëjti ekip që shkatërroi tumën iu kthye rindërtimit të saj në sezonin e fundit në një proces që qarkulloi hedhjet e dheut të një tume të ndërtuar nga njerëzit por pa banorët e saj ose zotërimet e tyre në një monument të riformuar (**Figs. 22.12, 22.13**). Ndërsa nuk mundeshim të kthenim në banesën e fundit popullsinë e lashtë që zgjodhi të varroste këtu, një përkujtim për jetët e tyre dhe jetën e përtejme do të lulëzojë në hapsirën lokale.

Çfarë ne rindërtuam në hapësirën që nuk ishte kodër varreza por dukshmëria e tumës. Në kohëzgjatjen e ekzistencës së saj tuma shërbeu si vendvarrim vetëm për pak kohë-rreth 1400–800 p.e.s. dhe sërish për një periudhë të shkurtër kohe rreth 1800 e.s. Brenda këtij grumbulli dheu ishin mbyllur gjetje Paleolitike bashkë me plumba dhe gëzhoja bosh nga Lufta e Dytë Botërore dhe të tjera nëpër të. Në sipërfaqen e saj ne ndeshëm edhe më shumë hedhje moderne (paraqitur në Kapitullin 1.2). Në kohë të ndryshme të ekzistencës së saj ky grumbull dheu shërbeu si fokusi i përkujtimit, kujtesës shoqërore edhe vetë identitetit. Tuma e Lofkëndit në Epokën e Bronzit të vonë dhe Hekurit të hershëm u dha banorëve të luginës së Gjanicës jo vetëm një imazh por ankorimin e së shkuarës së tyre duke i ofruar një pikë konceptuale tek e cila ata i qaseshin të ardhmes. Për banorët përreth tumës sot, grumbulli i gërmuar dhe rindërtuar i dheut përfaqëson një fokus të ndryshëm identiteti që është bërë pjesë e trashëgimnisë kulturore të tyre dhe tonës.

Mendime Përfundimtare: Arkeologjia në Vetvete

Arkeologjia moderne i ktheu banorët e një varreze të vetme të një familje të zgjeruar të kohës prehistorike në një luginë të largët të Shqipërisë në qytetarë të epokës globale. Në këtë proces shkatërrues ata u zbuluan nga një ekip, disa prej një kontinenti të panjohur në Europë në jetën e tyre dhe analizuan me instrumenta dhe metoda të kohës nukleare, prezantuan tek paraardhësit e Epokës së Gurit veglat e të cilëve ishin palosur në banesën e fundit, për t'u rivarrosur në kuti në muzeun e të afërme të tyre në Apolloni. I fundit prej tyre u varros përpara se Grekët të adoptonin alfabetin dhe ta sillnin atë në Shqipëri (i vërejtur hershëm në Apolloni, jo më pak në një sarkofag Korintas ose të stilit-Korintas) dhe të gjithë janë tashmë të dukshëm në formë dixhitale për sa kohë interneti të zgjasë.

Për ekipin modern që hulumtoi tumën, shumë pasazhe të jetës shënuan 5 vitet e gërmimeve për anëtarët: gjashtë arkeologë të rinj nga Shtetet e Bashkuara dhe Shqipëria u bënë prindër të paracaktuar për jetë më të shëndetshme se ajo e fëmijëve të lindur në Ilirinë e Epokës së Bronzit dhe Hekurit; disa studentë fituan diploma pasuniversitare dhe shumë filluan karriera; njëri prej pronarëve celebroi martesën e vajzës, e cila ishte anëtare e ekipit të terrenit të Lofkëndit. Pra ritet e lindjes dhe martesës mbajtën biografi njerëzore përpara dhe pas ndeshjeve të shumëfishta me fundit e jetës. Drejtuesit e gërmimit erdhën në tumë me të tjera pritshmëri nga të tjera vendndodhje (varr qëndror luftëtari, enë dyshe ose treshe, shpata bronzi ishin paramenduar) dhe ne parashikuam e shumta një sezon të vetëm terreni. Më vonë, pas pesë sezonesh pune të rëndë, ne jemi mirënjohës për ndërhyrjen tonë në historinë e hershme të Ilirisë.

Concordance of Inventory and Catalogue Numbers

Compiled by John K. Papadopoulos and Seth Pevnick

Inv. No.	Cat. No.	Inv. No.	Cat. No.	Inv. No.	Cat. No.
P 001	**9/88**	P 033	**9/149**	P 064	**9/30**
P 002	**9/330**	P 034	**9/241**	P 065	**9/300**
P 003	**9/279**	P 035	**9/182**	P 066	**9/36**
P 004	**9/11**	P 036	**9/261**	P 067	**9/146**
P 005	**9/28**	P 037	**9/177**	P 068	**9/134**
P 006	**9/29**	P 038	**9/179**	P 069	**9/214**
P 007	**9/139**	P 039	**9/178**	P 070	**9/239**
P 008	**9/163**	P 040	**9/224**	P 071	**9/61**
P 010	**9/308**	P 041	**9/111**	P 072	**9/212**
P 011	**9/305**	P 042	**9/309**	P 073	**9/6**
P 012	**9/65**	P 043	**9/108**	P 074	**9/22**
P 013	**9/44**	P 044	**9/140**	P 075	**9/226**
P 014	**9/296**	P 045	**9/152**	P 076	**9/188**
P 015	**15/1**	P 046	**9/152**	P 077	**9/4**
P 016	**9/234**	P 047	**9/16**	P 078	**9/17**
P 017	**9/222**	P 048	**9/48**	P 079	**9/162** and **15/4**
P 018	**15/2**	P 049	**9/100**	P 080	**9/33**
P 019	**9/62**	P 050	**9/99**	P 081	**9/283**
P 020	**9/9**	P 051	**9/102**	P 082	**9/15**
P 021	**9/120**	P 052	**9/262**	P 083	**9/219**
P 022	**9/293**	P 053	**9/225**	P 084	**9/235**
P 023	**9/131**	P 054	**9/201**	P 085	**9/295**
P 024	**9/260**	P 055	**9/117**	P 086	**9/203**
P 025a, b	**15/3**	P 056	**9/118**	P 087	**9/312**
P 026	**9/294**	P 057	**9/221**	P 088	**9/189**
P 027	**9/114**	P 058	**9/72**	P 089	**9/127**
P 028	**9/238**	P 059	**9/236**	P 090	**9/320**
P 029	**9/164**	P 060	**9/209**	P 091	**9/199**
P 030	**9/84**	P 061	**9/264**	P 092	**9/60**
P 031	**9/32**	P 062	**9/183**	P 093	**9/55**
P 032	**9/98**	P 063	**9/100**	P 094a	**9/206**

Inv. No.	Cat. No.	Inv. No.	Cat. No.	Inv. No.	Cat. No.
P 094b	**9/167**	P 144	**15/8**	P 202	**9/13**
P 095	**9/223**	P 145	**9/299**	P 203	**9/171**
P 096	**9/55**	P 146	**9/323**	P 204	**9/128**
P 097	**9/87**	P 147	**9/324**	P 205	**9/174**
P 098	**9/190**	P 149	**9/147**	P 206	**9/141**
P 099	**9/184** and **15/5**	P 150	**9/132**	P 207	**9/240**
P 100	**9/317**	P 151	**9/153**	P 208	**9/210**
P 101	**15/6**	P 153	**9/327**	P 209	**9/237**
P 102	**15/6**	P 154	**9/285**	P 210	**9/157**
P 103	**9/110**	P 155	**9/202**	P 211	**9/113**
P 104	**9/78**	P 156	**9/218**	P 212	**9/2**
P 105	**9/251**	P 157	**9/192**	P 213	**9/10**
P 106	**9/69**	P 158	**9/180**	P 214	**9/257**
P 107	**9/26**	P 159	**9/42**	P 215	**9/137**
P 108	**9/86**	P 160	**9/129**	P 216	**9/196**
P 109	**9/143**	P 161	**9/289**	P 217	**9/280**
P 110	**9/194**	P 163	**9/51**	P 218	**9/200**
P 111	**9/181**	P 165	**9/68**	P 219	**9/268**
P 112	**9/83**	P 166	**9/7**	P 220	**9/43**
P 113	**9/21**	P 167	**9/79**	P 221	**9/20**
P 114	**9/31**	P 168	**9/220**	P 222	**9/242**
P 115	**9/79**	P 169	**9/154**	P 223	**9/93**
P 116	**9/168**	P 170	**9/249**	P 224	**9/290**
P 117	**9/256**	P 171	**9/133**	P 226	**9/325**
P 118	**9/269**	P 172	**9/39**	P 227	**9/284**
P 119	**9/169**	P 173	**9/45**	P 228	**9/95**
P 120	**9/166**	P 174	**9/252** and **15/9**	P 229	**9/119**
P 121	**9/216**	P 175	**9/332**	P 230	**9/286**
P 122	**9/229**	P 176	**9/288**	P 231	**9/287**
P 123	**9/97**	P 178	**9/208**	P 232	**9/89**
P 124	**9/136**	P 179	**9/270**	P 233	**9/104**
P 125	**9/34**	P 180	**9/321**	P 234	**9/248**
P 126	**9/125**	P 181	**9/109**	P 235	**9/105**
P 127	**9/71**	P 182	**9/193**	P 236	**9/38**
P 128	**9/266**	P 185	**9/301**	P 237	**9/87**
P 131	**9/12**	P 186	**9/213**	P 238	**9/275**
P 132	**9/75**	P 189	**9/304**	P 239	**9/281**
P 133	**9/54**	P 191	**9/126**	P 240	**9/278**
P 134	**9/103**	P 192	**9/148**	P 242	**9/106**
P 135	**9/274**	P 193	**9/79**	P 243	**9/243**
P 136	**9/265**	P 194	**9/150**	P 244	**9/291**
P 137	**9/101**	P 195	**9/77**	P 245	**9/185**
P 138	**15/7**	P 196	**9/247**	P 246	**9/52**
P 139	**9/340**	P 197	**9/151**	P 247	**9/55**
P 140	**9/170**	P 198	**9/18**	P 248	**9/142**
P 141	**9/306**	P 199	**9/138**	P 249	**9/246**
P 142	**9/107**	P 201	**9/161** and **15/10**	P 250	**9/66**
P 143	**9/197**			P 251	**9/278**

Inv. No.	Cat. No.	Inv. No.	Cat. No.	Inv. No.	Cat. No.
P 252	**9/74**	P 310	**9/211**	P 367	**9/81**
P 253	**9/91**	P 311	**15/17**	P 368	**9/195**
P 254	**15/11**	P 313	**9/302**	P 369	**9/329**
P 255	**9/83**	P 314	**9/173**	P 370	**9/55**
P 256	**9/14**	P 316	**9/156**	P 371	**9/56**
P 257	**9/333**	P 317	**9/292**	P 372	**9/335**
P 258	**9/47**	P 318	**9/217**	P 373	**9/25**
P 259	**9/64**	P 319	**15/18**	P 375	**9/57**
P 260	**9/322**	P 320	**9/198**	P 376	**9/278**
P 261	**9/310**	P 321	**9/207**	P 377	**9/232**
P 262	**9/59**	P 322	**9/94**	P 380	**9/233**
P 267	**9/228**	P 323	**9/231**	P 383	**9/326**
P 268	**9/46**	P 324	**9/75**	P 384	**9/336**
P 269	**9/258**	P 325	**9/316**	P 386	**9/19**
P 272	**9/70**	P 326	**9/5**	P 387	**9/314**
P 273	**9/341**	P 327	**9/328**	P 388	**9/86**
P 275	**9/41**	P 328	**9/334**	P 389	**9/49**
P 276	**9/96**	P 332	**9/273**	P 391	**9/72**
P 277	**9/8**	P 333	**9/313**	P 392	**9/271**
P 278	**9/23**	P 334	**9/112**	P 394	**9/277**
P 279	**9/40**	P 335	**9/253**	P 395	**9/50** and **15/21**
P 281	**9/315**	P 336	**9/303**	P 397	**9/282**
P 282	**9/290**	P 337	**9/215**	P 398	**9/115**
P 283	**9/259** and **15/12**	P 338	**9/339**	P 399	**9/191**
		P 339	**9/123**	P 400	**9/272**
P 284	**9/80**	P 340	**9/337**	P 401	**9/172**
P 285	**9/87**	P 342	**9/37**	P 404	**9/145**
P 286	**15/13**	P 343	**9/144**	P 405	**9/205**
P 287	**9/35**	P 344	**9/186**	P 406	**9/135**
P 288	**9/244**	P 345	**9/187**	P 408	**9/314**
P 289	**9/176**	P 346	**9/85**	P 411	**9/290**
P 290	**9/20**	P 347	**9/165** and **15/19**	P 413	**9/230**
P 291	**9/298**			P 414	**9/53**
P 292	**15/14**	P 349	**9/158**	P 416	**9/63**
P 293	**9/175**	P 350	**9/1**	P 417	**9/204**
P 294	**15/15**	P 351	**9/58**	P 418	**9/24**
P 295	**9/130**	P 353	**9/79**	P 421	**9/155**
P 296	**9/67**	P 354	**9/318**	P 422	**9/90**
P 297	**15/16**	P 355	**15/20**	P 423	**9/245**
P 299	**9/124**	P 356	**9/73**	P 424	**9/331**
P 301	**9/338**	P 357a	**9/116**	P 428	**9/297**
P 302	**9/255**	P 360	**9/311**	P 429	**9/250**
P 303	**9/159**	P 361	**9/122**	P 430	**9/254**
P 304	**9/3**	P 362	**9/160**	P 431	**9/76**
P 305	**9/82**	P 363	**9/267**	P 432	**9/263**
P 306	**9/307**	P 364	**9/92**		
P 307	**9/121**	P 365	**9/276**	SF 003	**A3/12**
P 309	**9/319**	P 366	**9/227**	SF 005	**13/3**

Inv. No.	Cat. No.	Inv. No.	Cat. No.	Inv. No.	Cat. No.
SF 006	**13/37**	SF 086	**10/140**	SF 163e	**10/75**
SF 007	**13/89**	SF 087	**13/86**	SF 163f	**10/76**
SF 008	**13/82**	SF 089	**10/87**	SF 163g	**10/77**
SF 009	**13/59**	SF 091	**10/87**	SF 163h	**10/78**
SF 010	**10/137**	SF 091d	**10/73**	SF 164a	**10/101**
SF 013	**10/100**	SF 091e	**10/74**	SF 164b	**10/89**
SF 014	**A3/3**	SF 092	**10/64**	SF 168	**10/4**
SF 015	**A3/20**	SF 098	**10/136**	SF 169	**10/3**
SF 018	**10/40**	SF 103	**10/139**	SF 170	**10/14**
SF 019	**10/121**	SF 104	**10/129**	SF 171	**10/33**
SF 020	**13/4**	SF 105	**10/30**	SF 172	**10/71**
SF 022	**13/89**	SF 106	**10/15**	SF 173	**14/17**
SF 023	**13/82**	SF 107	**10/13**	SF 180	**13/42**
SF 026	**13/38**	SF 108	**A3/25**	SF 181	**13/42**
SF 028	**13/5**	SF 110	**10/22**	SF 182	**13/2**
SF 029	**13/30**	SF 111	**10/50**	SF 190	**13/44**
SF 032	**13/27**	SF 113	**10/104**	SF 192	**13/57**
SF 036	**13/18**	SF 114	**10/142**	SF 193	**13/75**
SF 037	**14/2**	SF 116	**10/38**	SF 194	**13/43**
SF 038	**14/3**	SF 118	**13/80**	SF 195	**13/58**
SF 039	**14/11**	SF 119	**13/19**	SF 197	**13/77**
SF 040	**13/60**	SF 121	**13/88**	SF 199	**13/68**
SF 041	**13/36**	SF 123	**13/20**	SF 206	**13/83**
SF 042	**A3/4**	SF 124	**10/99**	SF 209	**13/21**
SF 044	**10/125**	SF 125	**10/110**	SF 210	**13/8**
SF 045	**10/10**	SF 126	**10/93**	SF 211	**A3/29**
SF 046	**10/37**	SF 127	**14/13**	SF 212	**10/126**
SF 050	**13/61**	SF 128	**14/6**	SF 213	**13/22**
SF 051	**13/62**	SF 129	**13/6**	SF 214	**13/53**
SF 052	**13/49**	SF 130	**10/6**	SF 215	**13/72**
SF 054	**13/50**	SF 132	**13/56**	SF 217	**10/134**
SF 057	**A3/5**	SF 134	**10/131**	SF 218	**14/12**
SF 059	**13/51**	SF 135	**14/10**	SF 220	**13/9**
SF 064	**13/33**	SF 138	**10/8**	SF 225	**10/48**
SF 065	**10/5**	SF 139	**13/7**	SF 226	**10/45**
SF 066	**10/39**	SF 144	**13/31**	SF 227	**10/31**
SF 067	**10/51**	SF 145	**13/63**	SF 228	**10/143**
SF 069	**13/28**	SF 147	**13/41**	SF 229	**A3/27**
SF 070	**A3/11**	SF 149	**13/92**	SF 230	**10/123**
SF 072	**A3/2**	SF 150	**13/32**	SF 231	**10/26**
SF 073	**A3/17**	SF 151	**13/26**	SF 232	**13/46**
SF 077	**14/1**	SF 152	**13/1**	SF 236	**13/69**
SF 078	**14/4**	SF 155	**10/47**	SF 240	**13/47**
SF 079	**14/5**	SF 157	**10/18**	SF 241	**13/16**
SF 080	**14/7**	SF 160	**A3/21**	SF 242	**13/71**
SF 083	**13/74**	SF 162	**10/24**	SF 244	**13/10**
SF 085	**10/49**	SF 163a-d	**10/61**	SF 247	**13/11**

Inv. No.	Cat. No.	Inv. No.	Cat. No.	Inv. No.	Cat. No.
SF 249	**10/32**	SF 305	**10/43**	SF 372	**A3/18**
SF 250	**A3/22**	SF 306	**10/141**	SF 373	**13/15**
SF 251	**10/21**	SF 307	**10/70**	SF 377	**13/81**
SF 253	**10/44**	SF 308	**A3/23**	SF 378	**10/130**
SF 254	**10/34**	SF 310	**14/15**	SF 380	**A3/7**
SF 255	**10/86**	SF 311	**A3/6**	SF 381	**A3/8**
SF 256	**10/132**	SF 313a	**10/80**	SF 384	**10/133**
SF 257a	**10/92**	SF 313b	**10/66**	SF 385	**13/65**
SF 257b	**10/91**	SF 313c	**10/81**	SF 386	**A3/19**
SF 258	**10/114**	SF 313d	**10/67**	SF 387	**14/14**
SF 259	**10/63**	SF 315	**10/16**	SF 388	**10/2**
SF 260	**10/42**	SF 316	**10/88**	SF 389	**10/1**
SF 261	**10/23**	SF 317	**10/85**	SF 390	**10/35**
SF 262	**10/25**	SF 318	**10/60**	SF 391	**A3/24**
SF 263	**10/118**	SF 320	**13/23**	SF 393	**10/128**
SF 264	**10/72**	SF 321	**13/40**	SF 394	**A3/9**
SF 265	**10/69**	SF 322	**13/29**	SF 395	**13/79**
SF 266a	**10/98**	SF 323	**13/34**	SF 396	**13/14**
SF 266b	**10/97**	SF 324	**13/24**	SF 402	**A3/13**
SF 267	**10/68**	SF 326	**13/73**	SF 404	**10/55**
SF 269	**13/13**	SF 333	**10/135**	SF 405	**A3/1**
SF 274	**13/70**	SF 335	**10/58**	SF 408	**10/127**
SF 278	**13/64**	SF 336	**10/59**	SF 409	**10/120**
SF 280	**13/39**	SF 337	**10/20**	SF 410	**10/119**
SF 282	**13/55**	SF 338	**10/112**	SF 411	**10/52**
SF 283	**10/113**	SF 339	**10/115**	SF 412	**10/56**
SF 284	**A3/14**	SF 340	**10/102**	SF 413	**10/9**
SF 287a	**10/95**	SF 341	**10/108**	SF 414	**10/41**
SF 287b	**10/96**	SF 342	**10/106**	SF 422	**A3/26**
SF 288	**10/94**	SF 343	**10/59**	SF 423	**10/36**
SF 289	**10/138**	SF 344	**10/116**	SF 424	**10/90**
SF 290	**10/11**	SF 347	**10/124**	SF 425	**13/66**
SF 291	**10/12**	SF 348	**10/66**	SF 428	**A3/31**
SF 292a	**10/79**	SF 349	**10/84**	SF 430	**A3/10**
SF 292b	**13/48**	SF 350	**A3/30**	SF 431a	**10/19**
SF 293	**10/7**	SF 351	**10/117**	SF 431b	**10/83**
SF 294	**10/109**	SF 353	**13/17**	SF 434	**10/17**
SF 295	**10/107**	SF 354	**10/29**	SF 436	**13/76**
SF 296	**10/103**	SF 356	**14/16**	SF 437	**10/53**
SF 297	**10/62**	SF 359	**13/67**	SF 438	**10/27**
SF 298	**10/105**	SF 360	**13/52**	SF 439	**A3/28**
SF 299	**10/82**	SF 361	**13/25**	SF 440	**10/28**
SF 300	**10/57**	SF 362	**13/85**	SF 442	**A3/32**
SF 301	**10/144**	SF 364	**13/84**	SF 443	**10/54**
SF 302	**A3/16**	SF 366	**13/45**	SF 444	**10/65**
SF 303	**A3/15**	SF 370	**10/122**	SF 445	**14/9**
SF 304a-c	**10/46**	SF 371	**13/54**	SF 446	**14/8**

Inv. No.	Cat. No.	Inv. No.	Cat. No.	Inv. No.	Cat. No.
SF 447	**13/94**	SF 450	**10/111**	SF 453	**13/12**
SF 448	**13/78**	SF 451	**13/35**	SF 454	**13/93**
SF 449	**13/90**	SF 452	**13/91**	Bulk collection	**13/87**

Ancient Authors and Texts Cited

Ancient Sources

Aelian *Varia historia*
Ampelius
Aristotle *Physica*
Poetica
Politica
Aristotle (auctor ignotus)
De mirabilius auscultationibus
Arrian *Anabais*
Caesar *Bellum Civile*
Cicero *Tusculanae disputations*
Dio Cassius
Diodorus Siculus
Dioscorides *Materia medica*
Hekataios (of Abdera)
Herodian Grammaticus
Herodiani technici reliquiae
Herodotos
Hieronymus of Cardia
Homer *Iliad*
Odyssey
Ovid *Ars amatoria*
Epsitulae ex Ponto
Pausanias
Pliny *Naturalis historia*
Plutarch *Pyrrhus*
Sulla
Pseudo-Scymnus
Stephanos Byzantinos
Strabo
Teucer (of Kyzikos)
Theopompus Historicus
Thucydides
Vitruvius *De architectura*
Xenophon *Cyropedia*

References Cited

Åberg, N.
1930 *Bronzezeitliche und früheisenzeitliche Chronologie, I: Italien*. Stockholm: Verlag der Akademie.
1931 *Bronzezeitliche und früheisenzeitliche Chronologie, II: Hallstattzeit*. Stockholm: Verlag der Akademie.
1932 *Bronzezeitliche und früheisenzeitliche Chronologie, III: Kupfer- und Frühbronzezeit*. Stockholm: Verlag der Akademie.
1933 *Bronzezeitliche und früheisenzeitliche Chronologie, IV: Griechenland*. Stockholm: Verlag der Akademie.
1935 *Bronzezeitliche und früheisenzeitliche Chronologie, V: Mitteleuropäische Hochbronzezeit*. Stockholm: Verlag der Akademie.

Aberle, D.F.
1961 Navaho. In *Matrilineal kinship*, edited by D.M. Schneider and K. Gough, 96–201. Berkeley and Los Angeles: University of California Press.

Acsádi, G., and J. Nemeskéri
1970 *History of human life span and mortality*. Budapest: Akadémiai Kiadó.

Agolli, E.
2009 The evolution of household pottery production in the Iron Age pottery of the Kamenicë Tumulus in southeast Albania: toward new principles, theories and methodologies in pottery studies. M.A. thesis, Cotsen Institute of Archaeology, University of California, Los Angeles.
2014 The shaping of social complexity, networks and cultural transmissions: pottery from the Bronze and Iron Age communities of southern Illyria and northern Epirus (2500–500 BC). Ph.D. dissertation, University of California, Los Angeles.

Alcock, S.
2002 *Archaeologies of the Greek past: landscape, monuments and memories.* Cambridge: Cambridge University Press.

Alexander, J.
1964 The pins of the Jugoslav Early Iron Age. *Proceedings of the Prehistoric Society* 30:159–185.
1965 The spectacle fibulae of southern Europe. *American Journal of Archaeology* 69:7–23.
1972 *Jugoslavia before the Roman conquest.* London: Thames and Hudson.
1973a The history of the fibula. In *Archaeological theory and practice*, edited by D.E. Strong, 217–230. London: Seminar Press.
1973b The study of fibulae (safety-pins). In *The explanation of culture change: models in prehistory*, edited by C. Renfrew, 185–194. London: Duckworth.

Alexander, J., and S. Hopkins
1982 The origins and early development of European fibulae. *Proceedings of the Prehistoric Society* 48: 401–416.

Alexandri, O.
1968 Νυκτερινή ανασκαφή σκάμματος κατά μήκος της οδού Κριεζή. *Athens Annals of Archaeology* 1:20–30.

Aliu, S.
1984 Tuma e Prodanit. *Iliria* 14(1):27–67.
1985 Aspekte të kulturës ilire në krahinat e Kolonjës në shek. VII–V p.e. sonë. *Iliria* 15(2):271–280.
1988 Gjetje të rastit nga Plasa. *Iliria* 18(1):204–208.
1994 Varreza e Borovës. *Iliria* 24(1–2):5–86.
1995 Tuma e Psarit. *Iliria* 25(1–2):119–148.
1996 Tuma e Shtikës. *Iliria* 26(1–2):57–78.
2004 *Tuma e Luarasit.* Tiranë: Akademia e Shkencave e Shqipërisë, Instituti i Arkeologjisë.
2006 Recent prehistoric research in southeast Albania: a review. In *New directions in Albanian archaeology: studies presented to Muzafer Korkuti*, edited by L. Bejko and R. Hodges, 43–55. Tirana: International Centre for Albanian Archaeology.
2007 Les bijoux de la période du Bronze Récent et de la période du Fer Ancien dans la region de Kolonja. In *Between the Aegean and Baltic seas: prehistory across borders. Proceedings of the International conference Bronze and Early Iron Age interconnections and contemporary developments between the Aegean and the regions of the Balkan Peninsula, central and northern Europe, University of Zagreb, 11–14 April 2005* (Aegaeum 27), edited by I. Galanaki, H. Tomas, Y. Galanakis, and R. Laffineur, 239–247. Liège: Université de Liège.
2012 *Tuma e Rehovës.* Akademi e Shkencave e Shqipërisë. Instituti i Arkeologjisë. Korçë: Libraria Kotti.

Allen, S.E.
2002 The paleoethnobotany of Sovjan, Albania: preliminary results. In *L'Albanie dans l'Europe préhistorique. Actes du Colloque international organisé par l'École Française d'Athènes et l'Université de Bretagne-Sud, Lorient, 8–10 juin 2000 (Bulletin de Correspondance Hellénique*, Supplément 42), edited by G. Touchais and J. Renard, 61–72. Paris: de Boccard.
2005 A living landscape: the paleoethnobotany of Sovjan, Albania. Ph.D. dissertation, Boston University.

Alva, W.
1988 Discovering the New World's richest unlooted tomb. *National Geographic* 174(4):510–548.

Ambrose, S.H.
1990 Preparation and characterization of bone and tooth collagen for isotopic analysis. *Journal of Archaeological Science* 17:431–451.
1993 Isotopic analysis of paleodiets: methodological and interpretive considerations. In *Investigations of ancient human tissue: chemical analysis in anthropology*, edited by M.K. Sandford, 59–130. New York: Gordon and Breach.

Ambrose, S.H., and L. Norr
1992 On stable isotope data and prehistoric subsistence in the Soconusco region. *Current Anthropology* 33:401–404.

Amore, M.G.
2003–2004 Necropolis of Apollonia: new results from Tumulus 9. *Iliria* 31(1–2):267–280.
2004 Necropolis of Apollonia: Tumulus 9. In *L'Illyrie méridionale et l'Epirus dans l'antiquité*, IV. *Actes du IVe colloque internationale de Grenoble (10–12 octobre 2002)*, edited by P. Cabanes and J.-L. Lamboley, 307–310. Paris: de Boccard.
2011 The complex of tumuli 9, 10, amd 11 in the necropolis of Apollonia: new light on the history of the colony. In *L'Illyrie méridionale et l'Épire dans l'Antiquité*, V. *Actes du Ve colloque international de Grenoble (8–11 octobre 2008)*, edited by J.-L. Lamboley and M. Castiglioni, 181–188. Paris: de Boccard.

Amore, M.G. (editor)
2010 *The complex of tumuli 9, 10, and 11 in the necropolis of Apollonia (Albania)* (British Archaeological Reports International Series 2059). Oxford: Archaeopress.

Amore, M.G., C. Balandier, and P. Cabanes
1995 Apollonia d'Illyrie (Albanie). *Bulletin de Correspondance Hellénique* 119:761–781.

Amore, M.G., L. Bejko, and V. Dimo
2006 Apollonia Tumulus 10. The Albanian Rescue Archaeology Unit. *http://icaa.org.al/Anglisht/tumulus%2010.html* (2006).

Amore, M.G., and V. Dimo
2010 Pottery catalogue. In *The complex of tumuli 9, 10, and 11 in the necropolis of Apollonia (Albania)* (British Archaeological Reports Interna-

tional Series 2059), edited by M.G. Amore, 352–596. Oxford: Archaeopress.

Amyx, D.A., and P. Lawrence
1975 *Corinth* VII.ii. *Archaic Corinthian pottery and the Anaploga well.* Princeton, New Jersey: American School of Classical Studies.

Anamali, S.
1949 Vjosa dhe qendrat antike në luginën e saj. *Buletin i Institutit të Shkencave* 3(4):53–65.
1972 Amantia. *Iliria* 2:67–120.
1977–1978 Bazilika e Ballshit. *Iliria* 7–8:301–306.
1981 Rreth skulpturës dekorative arkitektonike të bazilikës së Ballshit. *Iliria* 11(2):237–242.
1987 Katër mbishkrime ndërtimi nga Bylisi. *Monumentet* 33(1):63–72.

Andersen, H.D., H.W. Horsnæs, S. Houby-Nielsen, and A. Rathje (editors)
1997 *Urbanization in the Mediterranean in the 9th to the 6th centuries BC* (Acta Hyperborea 7). Copenhagen: Museum Tusculanum Press.

Anderson, R.
2005 An annotated list of the non-marine mollusca of Britain and Ireland. *Journal of Conchology* 38(6): 607–637.

Andrea, Zh.
1976a La civilisation tumulaire du Bassin de Korçë et sa place dans les Balkans du sud-est. *Iliria* 4:133–155.
1976b Tumat e Kuçit të Zi. *Iliria* 6:166–233.
1977–1978 Tumat e Kuçit të Zi. *Iliria* 7–8:127–155.
1981a Një varrezë e dystë në fshatin Gërmenj (Rrethi i Lushnjës). *Iliria* 11(1):219–226.
1981b Gërmime në Cërrujë (Gramsh). *Iliria* 11(1): 270.
1984 Bujan (Tropojë). *Iliria* 14(2):261.
1985 *Kultura ilire e tumave në Pellgun e Korçës.* Tirana: Akademia e Shkencave e RPS të Shqipërise, Qendra e Kërkimeve Arkeologjike.
1986 Bujan (Tropojë). *Iliria* 16(2):254.
1995 Varreza tumulare e Bujanit. *Iliria* 25(1–2):91–118.
1996 Vendbanimi i Zagorës. *Iliria* 26(1–2):21–55.
2005 Varre të periudhës antike të vonë dhe asaj mesjetare në varrezat tumulare të pellgut të Korçës dhe në varrezën e sheshtë të Gërmenjit. *Candavia* 2:349–360.
2005–2006 Shtresa me qeramikë të pikturuar të stilit devollit në sektorin C të vendbanimit prehistorik të Maliqit (gërmime të viteve 1988–90). *Iliria* 32:5–38.
2009–2010 Tuma I e Shuecit (Rrethi i Korçës). Rezultatet e fushatës së parë të gërmimeve. *Iliria* 34: 223–279.

Andrefsky, W., Jr.
2005 *Lithics: macroscopic approaches to analysis*, 2nd ed. Cambridge: Cambridge University Press.

Andreou, E.
1981 Πωγώνι· Κάτω Μερόπη. *Archaiologikon Deltion* 36, B'2 (Chronika): 271–273.
1982a Μερόπη-Παλιόπυργος Πωγωνίου. *Archaiologikon Deltion* 37, B'2 (Chronika): 259.
1982b Οι τύμβοι Πωγωνίου Μερόπης. *Archaiologia* 3: 54–60.
1983 Τύμβος Κάτω Μερόπης. *Archaiologikon Deltion* 38, B'2 (Chronika): 229–230.
1985 Ανασκαφή τύμβων Πωγωνίου. *Archaiologia* 16: 86–87.
1997 Μερόπη Πωγωνίου. *Archaiologikon Deltion* 52, B'2 (Chronika): 553–557.

Andreou, E., and I. Andreou
1987 Une nécropole tumulaire à Pogoni de l'Épire. In *L'Illyrie méridionale et l'Épire dans l'Antiquité. Actes du Colloque international de Clermont-Ferrand (22–25 Octobre 1984)*, edited by P. Cabanes, 47–49. Clermont-Ferrand: Adosa.
1999a Η κοιλάδα του Γορμού στο Πωγώνι της Ηπείρου, κέντρο ζωής και ανάπτυξης κατά την πρώϊμη εποχή του σιδήρου. In *Η περιφέρια του Μυκηναϊκού κόσμου. Α΄ διεθνές διεπιστημονικό συμπόσιο, Λαμία, 25–29 Σεπτεμβρίου, 1994*, 77–90. Lamia: Greek Ministry of Culture.
1999b Les villages préhistoriques fortifiés de la vallée de Gormos à Pogoni d'Épire. In *L'Illyrie méridionale et l'Épire dans l'Antiquité*, III. *Actes du IIIe colloque international de Chantilly (16–19 octobre, 1996)*, edited by P. Cabanes, 51–56. Paris: de Boccard.

Andronikos, M.
1969 *Βεργίνα Ι: το νεκροταφείον των τύμβων*. Athens: Athens Archaeological Society.
1984 *Vergina: The royal tombs and the ancient city.* Athens: Ekdotike Athenon.

Andronikos, M. (editor)
1988 *Ancient Macedonia: Australian exhibition catalogue.* Athens: Ministry of Culture.

Angel, J.L.
1944 Greek teeth: ancient and modern. *Human Biology* 16:283–297.
1945 Skeletal material from Attica. *Hesperia* 14(4): 279–363.
1966 Porotic hyperostosis, anemias, malarias, and marshes in the prehistoric eastern Mediterranean. *Science* n.s. 153(3737):760–763.
1972 Ecology and population in the eastern Mediterranean. *World Archaeology* 4(1):88–105.

Anhegger, R.
1943 *Beiträge zur Geschichte des Bergbaus im Osmanischen Reich. I: Europäische Turkei.* Istanbul: Istanbuler Schriften nr. 2.

Anheuser, K., and M. Roumeliotou
2003 Characterisation of mineralized archaeological textile fibers through chemical staining. *The Conservator* 27:23–33.

Arbuckle, B.S.
2012 Pastoralism, provisioning, and power at Bronze Age Acemhöyük, Turkey. *American Anthropologist* 114:462–476.

Arensburg, B., and I. Hershkovitz
1989 Artificial "skull" treatment in the PPNB period: Nahal Hemar. In *People and culture in change: Proceedings of the Second symposium of Upper Paleolithic, Mesolithic, and Neolithic populations of Europe and the Mediterranean Basin* (BAR International Series 508), edited by I. Hershkovitz, 115–131. Oxford: British Archaeological Reports.

Arnold, B.
2001 The limits of agency in the analysis of elite Iron Age Celtic burials. *Journal of Social Archaeology* 1(2):210–224.
2002 "Sein und Werden": gender as process in mortuary ritual. In *In pursuit of gender: worldwide archaeological approaches*, edited by S.M. Nelson and M. Rosen-Ayalon, 239–256. Walnut Creek, California: AltaMira Press.

Arnold, B., and N.L. Wicker (editors)
2001 *Gender and the archaeology of death*. Walnut Creek, California: AltaMira Press.

Anund, J., W. Duczko, H. Göthberg, N. Price, and J. Wikborg
1998 *Gamla Uppsala: centralplats och omland. Arkeologisk förundersökning på ostkustbanan 1996–97. Gamla Uppsalaprojektet* (UV Uppsala, rapport 1997:26). Uppsala: Riksantikvarieämbetet.

Ashmore, W., and A.B. Knapp (editors)
1999 *Archaeologies of landscape: contemporary perspectives*. Malden, Massachusetts: Blackwell.

Asouti, E., and P. Austin
2005 Reconstructing woodland vegetation and its exploitation by past societies, based on the analysis and interpretation of archaeological wood charcoal macro-fossils. *Environmental Archaeology* 10:1–18.

Atchity, K.J., and E.J.W. Barber
1987 Greek princes and Aegean princesses: the role of women in Homeric poems. In *Critical essays on Homer*, edited by K.J. Atchity, R. Hogart, and D. Price, 15–36. Boston: K.J. Hall.

Athanasiadis, N.
1975 Zur postglazialen Vegetationsentwicklung von Litochoro Katerinis und Pertouli Trikalon (Griechenland). *Flora* 164:99–132.

Avila, R.A.J.
1983a *Bronzene Lanzen- und Pfeilspitzen der griechischen Spätbronzezeit* (*PBF* V:1). Munich: Beck.
1983b Das Kuppelgrab von Volos-Kapakli (Kapakli 1). *Prähistorische Zeitschrift* 58:15–60.

Avramea, A.P.
1974 *Η βυζαντινή Θεσσαλία μέχρι το 1204*. Athens: University of Athens.

Bader, T.
1983 *Die Fibeln in Rumänien* (*PBF* XIV:6). Munich: Beck.

Bailey, D.W.
2000 *Balkan prehistory: exclusion, incorporation and identity*. London: Routledge.

Bailey-Schmidt, S.E.
1995 Population distribution of the tuberculum dentale complex and anomalies of the maxillary anterior teeth. M.A. thesis, Arizona State University, Tempe.

Bakalakis, G., and A. Sakellariou
1981 *Paradimi* (Heidelberger Akademie der Wissenschaften. Internationale Interakademische Kommission für die Erforschung der Vorgeschichte des Balkans, Bd. II). Mainz: Philipp von Zabern.

Balasubramaniam, R., M.N. Mungole, V.N. Prabhakar, D.V. Sharma, and D. Banerjee
2002 Studies on ancient Indian OCP period copper. *Indian Journal of History of Science* 37(1):1–15.

Balandier, C., L. Koço, and P. Leinhardt
2007 Les fortifications d'Apollonia. In *Apollonie d'Illyrie. 1. Atlas archéologique et historique* (Collection de l'École Française de Rome 391), edited by V. Dimo, P. Lenhardt, and F. Quantin, 159–186. Recherches archéologiques franco-albanaises 1. Athens: École Française d'Athènes; Paris: Ministère des affaires étrangères; Rome: École Française de Rome.

Bank, R., P. Bouchet, G. Falkner, E. Gittenberger, B. Hausdorf, T. Proschwitz, and T. Ripken
2001 CLECOM-PROJECT: supraspecific classification of European non-marine mollusca (CLECOM Sections I and II). *Heldia* 4:77–128.

Barber, E.J.W.
1991 *Prehistoric textiles: the development of cloth in the Neolithic and Bronze Ages, with special reference to the Aegean*. Princeton, New Jersey: Princeton University Press.
1995 *Women's work: the first 20,000 years. Women, cloth, and society in early times*. New York: Norton.

Barrett, J.C.
1988 The living, the dead and the ancestors: Neolithic and Early Bronze Age mortuary practices. In *The archaeology of context in the Neolithic and Early Bronze Age: recent trends*, edited by J.C. Barrett and I.A. Kinnes, 30–41. Sheffield: Department of Archaeology and Prehistory, University of Sheffield.
1990 The monumentality of death: the character of Early Bronze Age mortuary mounds in southern Britain. *World Archaeology* 22(2):179–189.
1994 *Fragments from antiquity: an archaeology of social life in Britain, 2900–1200 BC*. Oxford: B. Blackwell.
1999 The mythical landscapes of the British Iron Age. In *Archaeologies of landscape: contemporary per-*

spectives, edited by W. Ashmore and A.B. Knapp, 253–265. Malden, Massachusetts: Blackwell.

Bartoloni, G., A.M. Bietti Sestieri, M.A. Fugazzola Delpino, C. Morigi Govi, and F. Parise Badoni (editors)

1980 *Dizionari terminologici: materiali dell'età del Bronzo finale e della prima età del Ferro*. Firenze: Centro Di.

Bartoloni, G., and F. Delpino (editors)

2005 *Oriente e occidente. Metodi e disciplini a confronto. Riflessioni sulla cronologia dell'età del ferro in Italia. Atti dell'incontro di studi, Roma, 30–31 ottobre 2003* (*Mediterranea* 1). Pisa: Istituti Editoriali e Poligrafici Internazionali.

Bar-Yosef Mayer D.E. (editor)

2005 *Archaeomalacology: molluscs in former environments of human behavior. Proceedings of the 9th conference of the International Council of Archaeozoology, Durham, August 2002*. Oxford: Oxbow Books.

Basso, K.H.

1996 Wisdom sits in places: notes on a western Apache landscape. In *Senses of place*, edited by S. Feld and K.H. Basso, 53–90. Santa Fe, New Mexico: School of American Research Press.

Bather, A.G.

1892–1893 The bronze fragments of the Acropolis, II. Ornamented bands and small objects. *Journal of Hellenic Studies* 13:232–271.

Batziou-Eustathiou, A.

1984 Πρωτογεωμετρικά από τη δυτική Θεσσαλία. *Athens Annals of Archaeology* 17:74–87.

1999 Το νεκροταφείο της Νέας Ιωνίας Βόλου κατά τη μετάβαση από την ΥΕΙΙΙΓ στην ΠΓ εποχή. In *Η περιφέρεια του Μυκηναϊκού κόσμου. Α΄ Διεθνές Διεπιστημονικό Συμπόσιο, Λαμία, 25–29 Σεπτεμβρίου 1994*, 117–130. Lamia: Greek Ministry of Culture.

Baxter, J.E.

2004 *The archaeology of childhood: children, gender, and material culture*. Walnut Creek, California: AltaMira Press.

Beaumont, L.

2000 The social status and artistic presentation of "adolescence" in fifth-century Athens. In *Children and material culture*, edited by J.S. Derevenski, 39–55. London: Routledge.

Beaumont, R.L.

1952 Corinth, Ambracia, Apollonia. *Journal of Hellenic Studies* 72:62–73.

Beck, A.

1980 *Beiträge zur frühen und älteren Urnenfelderkultur im nordwestlichen Alpenvorland* (*PBF* XX:2). Munich: Beck.

Beck, H.C.

1928 Classification and nomenclature of beads and pendants. *Archaeologia* 77:1–76.

Becker, C.

1986 *Kastanas: die Tierknochenfunde* (Prähistorische Archäologie in Südosteuropa, Bd. 5). Berlin: V. Spiess.

Bedrick E.J., J. Lapidus, and J.F. Powell

2000 Estimating the Mahalanobis distance from mixed continuous and discrete data. *Biometrics* 56:394–401.

Beecher, R.M., and R.S. Corruccini

1981 Effects of dietary consistency on craniofacial and occlusal development in the rat. *The Angle Orthodontist* 51(1):61–69.

Bejko, L.

1993 Mbi praninë e qeramikës mikene në Shqipërinë Jugore dhe probleme lidhur me të. *Iliria* 23(1–2): 101–122.

1998 Vështrim mbi mendimin arkeologjik Shqiptar dhe kontekstin e tij social. *Iliria* 28(1–2):194–208.

1999–2000 Mortuary customs in the Late Bronze Age of south-eastern Albania. *Iliria* 29(1–2):155–181.

2002a Mortuary customs in the Late Bronze Age of southeastern Albania. In *L'Albanie dans l'Europe préhistorique. Actes du Colloque international organisé par l'École Française d'Athènes et l'Université de Bretagne-Sud, Lorient, 8–10 juin 2000* (*Bulletin de Correspondance Hellénique*, Supplément 42), edited by G. Touchais and J. Renard, 171–198. Paris: de Boccard.

2002b Mycenaean presence and influence in Albania. In *Greek influence along the east Adriatic coast: Proceedings of the conference held in Split from September 24th to 26th 1998*, edited by N. Cambi, S. Čače, and B. Kirigin, 9–24. Književni Krug, Split.

2004 The Iron Age tumulus at Kamenica (Albania). In *L'Illyrie méridionale et l'Épire dans l'Antiquité, IV. Actes du IVe colloque international de Grenoble (10–12 octobre 2002)*, edited by P. Cabanes and J.-L. Lamboley, 39–44. Paris: de Boccard.

2007 Expressions of identities in the Late Bronze and Early Iron Age southeastern Albania. In *Between the Aegean and Baltic seas: prehistory across borders. Proceedings of the International conference Bronze and Early Iron Age interconnections and contemporary developments between the Aegean and the regions of the Balkan Peninsula, central and northern Europe, University of Zagreb, 11–14 April 2005* (Aegaeum 27), edited by I. Galanaki, H. Tomas, Y. Galanakis, and R. Laffineur, 203–209. Liège: Université de Liège.

Forthcoming *The tumulus of Kamenicë*.

Bejko, L., S. Aliu, and S. Lela

2004 Varreza mesjetare në Rëmbec të Korçës. *Candavia* 1:211–243.

Bejko, L., M.G. Amore, and S. Aliu

2006 Korça Basin Archaeological Survey Project: preliminary report for the season of 2005. The

Albanian Rescue Archaeology Unit. *http://www.icaa.org. al/Anglisht/kobas_2005.html* (2006).

Bejko, L., M.G. Amore, S. Aliu, E. Agolli, T. Fenton, S. Lela, and J. Vroom
2005 Varreza mesjetare në Rëmbec të Korçës (gërmime të vitit 2004). *Candavia* 2:327–348.

Bejko, L., T. Fenton, and D. Foran
2006 Recent advances in Albanian mortuary archaeology, human osteology, and ancient DNA. In *New directions in Albanian archaeology: studies presented to Muzafer Korkuti*, edited by L. Bejko and R. Hodges, 309–322. Tirana: International Centre for Albanian Archaeology.

Bela, M.
1987 Tumat e Mujajve (Kukës). *Iliria* 17(2):243–244.
1990 Varreza tumulare e Myç-Hasit. *Iliria* 20(2):97–136.

Bell, D., and S. Song
1994 Explaining the level of bridewealth. *Current Anthropology* 35(3):311–316.

Bell, L.S., G. Cox, and J. Sealy
2001 Determining isotopic life history trajectories using bone density fractionation and stable isotope measurements: a new approach. *American Journal of Physical Anthropology* 116:66–79.

Benac, A., and B. Čović
1956 *Glasinac I: Bronzezeit*. Sarajevo: Verlag des Landesmuseum.
1957 *Glasinac II: Eisenzeit*. Sarajevo: Verlag des Landesmuseum.

Bender, B. (editor)
1993 *Landscape: politics and perspectives*. Providence, Rhode Island: Berg.

Bender, B.
2002 Time and landscape. *Current Anthropology* 43 (supplement):103–112.

Bentley, R.A., N. Tayles, C. Higham, C. Macpherson, and T.C. Atkinson
2007 Shifting gender relations at Khok Phanom Di, Thailand: isotopic evidence from the skeletons. *Current Anthropology* 48(2):301–314.

Benton, S.
1934–1935 Excavations in Ithaca, III: the cave at Polis, I. *Annual of the British School at Athens* 35:45–73.
1939–1940 Bronzes from Palaikastro and Praisos. *Annual of the British School at Athens* 40:51–59.
1950 The dating of horses on stands and spectacle fibulae in Greece. *Journal of Hellenic Studies* 70:16–22.
1952 Note on spectacle fibulae and horses. *Journal of Hellenic Studies* 72:119.
1953 Further excavations at Aetos. *Annual of the British School at Athens* 48:255–361.

Béquignon, Y.
1937 *Recherches archéologiques à Phères de Thessalie* (Publications de la Faculté des Lettres de l'Université de Strasbourg, 78). Paris: Les belles lettres.

Bérard, C.
1970 *Eretria, fouilles et recherches, III: l'hérôon à la porte de l'ouest*. Bern: A. Francke.

Bereti, V., V. Dimo, J.-L. Lamboley, and B. Vrekaj
2007 La céramique d'Apollonia. In *Apollonie d'Illyrie. 1. Atlas archéologique et historique* (Collection de l'École Française de Rome 391), edited by V. Dimo, P. Lenhardt, and F. Quantin, 129–146. Recherches archéologiques franco-albanaises 1. Athens: École Française d'Athènes; Paris: Ministère des affaires étrangères; Rome: École Française de Rome.

Bereti, V., G. Gonsagra, J.-P. Descoeudres, S. Shruza, and C. Zindel
2011 Orikos: première colonie grecque en Adriatique? La première campagne de fouille Albano-Suisse. In *L'Illyrie méridionale et l'Épire dans l'Antiquité, V. Actes du Ve colloque international de Grenoble (8–11 octobre 2008)*, edited by J.-L. Lamboley and M. Castiglioni, 419–430. Paris: de Boccard.

Berger, J.
1972 *Ways of seeing*. London: Harmondsworth.

Berranger-Auserve, D. (editor)
2007 *Épire, Illyrie, Macédoine . . . Mélanges offerts au Professor Pierre Cabanes* (Collection ERGA Recherches sur l'antiquité). Clermont-Ferrand: Presses Universitaires Blaise Pascal.

Bertin, J.
2011 [1967] *Semiology of graphics? Diagrams, networks, maps*. Redlands, California: ESRI Press.

Betzler, P.
1974 *Die Fibeln in Süddeutschland, Österreich und der Schweiz I (Urnenfeldzeitliche Typen)* (*PBF* XIV: 3). Munich: Beck.

Bielefeld, E.
1968 *Schmuck* (*Archaeologia Homerica*, Bd. I, Kap. C). Göttingen: Vandenhoeck and Ruprecht.

Bietti Sestieri, A.M.
1992 *The Iron Age community of Osteria dell'Osa: a study of socio-political developments in central Tyrrhenian Italy*. Cambridge: Cambridge University Press.

Binford, L.R.
1971 Mortuary practices: their study and their potential. In *Approaches to the social dimensions of mortuary practices*, edited by J.A. Brown, 6–29. Washington, D.C.: Society for American Archaeology.
1972 *An archaeological perspective: studies in archaeology*. New York: Seminar Press.

Bintliff, J.L.
1977 *Natural environment and human settlement in prehistoric Greece* (BAR International Series 28). Oxford: British Archaeological Reports.

Bird, D.W., J.L. Richardson, P.M. Veth, and A.J. Barham
2002 Explaining shellfish variability in middens on the Meriam Islands, Torres Strait, Australia. *Journal of Archaeological Science* 29(5):457–469.

Blegen, C.W.
1928 *Zygouries: a prehistoric settlement in the valley of Cleonae*. Cambridge, Massachusetts: Harvard University Press.
1937 *Prosymna: the Helladic settlement preceding the Argive Heraeum*. Cambridge: Cambridge University Press.
1952 Two Athenian grave groups of about 900 B.C. *Hesperia* 21:279–294.

Blegen, C.W., J.L. Caskey, M. Rawson, and J. Sperling
1950 *Troy I: General introduction, the first and second settlements*. Princeton, New Jersey: Princeton University Press.

Blegen, C.W., H. Palmer, and R.S. Young
1964 *Corinth: results of the excavations conducted by the American School of Classical Studies at Athens. XIII: the north cemetery*. Cambridge, Massachusetts: Harvard University; and Princeton, New Jersey: Princeton University.

Blinkenberg, C.
1926 *Lindiaka V: fibules grecques et orientales* (Det Kgl. Danske Videnskabernes Selskab. Historik-filologiske Meddelelser. XIII, 1). Copenhagen: Bianco Lunos Bogtrykkeri.

Boardman, J.
1960 Protogeometric graves at Agios Ioannis near Knossos. *Annual of the British School at Athens* 55:128–148.
1961 *The Cretan collection in Oxford: the Dictaean Cave and Iron Age Crete*. Oxford: Clarendon Press.
1967 *Excavations in Chios, 1952–1955: Greek Emporio* (BSA Supplement 6). Oxford: Thames and Hudson.

Bodinaku, N.
1977–1978 Të dhëna arkeologjike nga rrethi i Përmetit. *Iliria* 7–8:354–357.
1981 Kërkime arkeologjike në rrethin e Përmetit. *Iliria* 11(2):243–262.
1982 Varreza tumulare e Pazhokut (Gërmime të vitit 1973). *Iliria* 12(1): 49–101.
1984 Fibula harkore prej bronzi të Hekurit të hershëm. *Iliria* 14(2):47–58.
1988 Dy armë bronzi nga Këlcyra. *Iliria* 18(1):33–52.
1989 Rreth kronologjisë së qeramikës së pikturuar mat të kohës së vonë të Bronzit dhe të Hekurit. *Iliria* 19 (1):53–63.
1990a Rreth origjinës dhe bartësve të qeramikës së pikturuar mat të kohës së vonë të bronzit dhe asaj të Hekurit. *Iliria* 21(2):65–95.
1990b Kama bronzi nga Shqipëria jugore. *Iliria* 20(1): 85–97.
2001–2002 Varreza tumulare e Dukatit në rrethin e Vlorës (gërmime të viteve 1973–1974). *Iliria* 30(1–2):9–100.

Boeda, E., J. Connan, and S. Mehesen
1998 Bitumen as hafting material on Middle Paleolithic artifacts from the El Kowm Basin, Syria. In *Neanderthals and modern humans in western Asia*, edited by T. Akazawa, K. Aoki, and O. Bar-Yosef, 181–204. New York: Plenum Press.

Bökönyi, S.
2005 Appendix B: the animal remains found in tombs. In J.K. Papadopoulos, *The Early Iron Age cemetery at Torone* (Monumenta Archaeologica 24), 317–320. Los Angeles: Cotsen Institute of Archaeology.

Bonogofsky, M.
2001a Cranial sanding, not defleshing of two plastered skulls from 'Ain Ghazal. *Paléorient* 27:141–146.
2001b An osteo-archaeological examination of the ancestor cult during the pre-pottery Neolithic B period in the Levant. Ph.D. dissertation, University of California, Berkeley.
2003 Neolithic plastered skulls and railroading epistemologies. *Bulletin of the American Schools of Oriental Research* 331:1–10.
2005 A bioarchaeological analysis of plastered skulls from Anatolia: new discoveries and interpretations. *International Journal of Osteoarchaeology* 15:124–135.
2011 *The bioarchaeology of the human head: decapitation, decoration, deformation*. Gainesville: University Press of Florida.

Borga, E., and P. Càssola Guida
2007 At the fringe of the tumulus culture: Bronze Age tumuli of north-eastern Italy between Europe and the Aegean. In *Between the Aegean and Baltic Seas: prehistory across borders. Proceedings of the International conference Bronze and Early Iron Age interconnections and contemporary developments between the Aegean and the regions of the Balkan peninsula, central and northern Europe, University of Zagreb, 11–14 April 2005* (Aegaeum 27), edited by I. Galanaki, H. Tomas, Y. Galanakis, and R. Laffineur, 47–56. Liège: Université de Liège.

Boskey, A.L.
1999 Mineralization, structure and function of bone. In *Dynamics of bone and cartilage metabolism*, edited by M.J. Seibel, S.P. Robins and J.P. Bilezikian, 153–164. San Diego: Academic Press.

Bottema, S.
1974 *Late Quaternary vegetation history of northwestern Greece*. Groningen: V.R.B. Offsetdrukkerij.
1979 Pollen analytical investigations in Thessaly (Greece). *Palaeohistoria* 21:19–40.
1982 Palynological investigations in Greece with special reference to pollen as an indicator of human activity. *Palaeohistoria* 24:257–289.

Bouzek, J.
1982 The simple repoussé decoration on early Greek bronzes. *Eirene* 19:99–109.
1985 *The Aegean, Anatolia and Europe: cultural interrelations in the second millennium B.C.* (Studies

in Mediterranean Archaeology 29). Göteborg: Paul Åströms Forlag.

Boyd, M.C.
1993 Burial tumuli of northern Greece. *Parnassos* 35: 561–568.

Bradley, R.
1985 *Consumption, change and the archaeological record: the archaeology of monuments and the archaeology of deliberate deposits* (University of Edinburgh, Department of Archaeology Occasional Papers 13). Edinburgh: Edinburgh University Press.
1995 Introduction. In *Regional approaches to mortuary analysis*, edited by L.A. Beck, i–xix. New York: Plenum Press.
1998 *The significance of monuments*. London: Routledge.

Braka, K.
1987 Katundas–Varreza. *Iliria* 16(2):252–253.
1987 Një varrezë e periudhës së hekurit në fshatin Katundas të rrethit të Beratit. *Iliria* 17(2):37–47.

Braudel, F.
1995 *The Mediterranean and the Mediterranean world in the age of Philip II*, Vol. 1. Berkeley and Los Angeles: University of California Press.

Bräuning, A., and I. Kilian-Dirlmeier
2013 *Die eisenzeitlichen Grabhügel von Vergina. Die Ausgrabungen von Photis Petsas 1960–1961. Monographien des Römisch-Germanische Zentralmuseums, Bd. 119.* Mainz: Verlag des Römisch-Germanischen Zentralmuseums.

Broad, W.
2006 *The oracle: the lost secrets and hidden message of ancient Delphi*. New York: Penguin.

Brock, J.K.
1957 *Fortetsa: early Greek tombs near Knossos* (BSA Supplement 2). Cambridge: Cambridge University Press.

Broholm, H.C., and M. Hald
1940 *Costumes of the Bronze Age in Denmark: Contributions to the archaeology and textile-history of the Bronze Age*. Copenhagen: NYT Nordisk Forlag.

Broneer, O.
1958 Excavations at Isthmia, third campaign, 1955–1956. *Hesperia* 27:1–37.

Bronk Ramsey, C.
2009 Bayesian analysis of radiocarbon dates. *Radiocarbon* 51:337–360.

Brown, T.A., D.E. Nelson, J.S. Vogel, and J.R. Southon
1988 Improved collagen extraction by modified Longin method. *Radiocarbon* 34:279–291.

Brück, J.
2006 Death, exchange and reproduction in the British Bronze Age. *European Journal of Archaeology* 9 (1):73–101.

Brückner, A., and E. Pernice
1893 Ein attischer Friedhof. *Mitteilungen des Deutschen Archäologischen Instituts, Athenische Abteilung* 18:73–191.

Buccellati, F.
2005 How to vacuum-pack cleaned metals for storage or museum. Unpublished report. Mozan/Urkesh Archaeological Project.

Buchholz, H.-G.
1962 Der Pfeilglätter aus dem VI. Schachtgrab von Mykene und die helladischen Pfeilspitzen. *Jahrbuch des deutschen archäologischen Instituts* 77:1–58.

Budina, Dh.
1969 Tuma e Çepunës (Gjirokastër). *Buletini Arkeologjik* 1:49–55.
1971 Gërmime arkeologjke në varrezën tumulare të Bajkajt. *Buletini Arkeologjik* 1:57–66.
1975 La carte archéologique de la vallée de Drino. *Iliria* 3:355–392.

Buikstra, J.E., and D.K. Charles
1999 Centering the ancestors: cemeteries, mounds and sacred landscapes of the ancient North American midcontinent. In *Archaeologies of landscape: contemporary perspectives*, edited by W. Ashmore and A.B. Knapp, 201–228. Malden, Massachusetts: Blackwell.

Buikstra, J.E., and D.H. Ubelaker (editors)
1994 *Standards for data collection from human skeletal remains. Proceedings of a seminar at the Field Museum of Natural History* (Arkansas Archaeological Survey Research Series No. 44). Fayetteville: Arkansas Archaeological Survey.

Bulanda, E.
1913 *Bogen und Pfeil bei den Völkern des Altertums*. Vienna and Leipzig: Alfred Hölder.

Bunguri, A.
1989 Armë ilire në krahinën e Dibrës. *Iliria* 19(1):67–75.
1993 Vendbanimi neolitik i Topojanit (faza Topojan II). *Iliria* 23(1–2):33–94.
1994 Vendbanimi ilir i Pesjakës (Dibër). *Iliria* 24 (1–2): 87–104.

Burnham, H.B.
1965 Çatal Hüyük: the textiles and twined fabrics. *Anatolian Studies* 15:169–174.

Burke, J.
1996 Anoxic microenvironments: a simple guide. *The Society for the Preservation of Natural History Collections Leaflets* 1(1):1–4.

Burrell, B.
2011 Thoughts on writing the final report, from the example at Paneas. *Journal of Roman Archaeology* 24:791–796.

Buschor, E., and W. von Massow
1927 Vom Amyklaion. *Mitteilungen des Deutschen Archäologischen Instituts, Athenische Abteilung* 52:1–64.

Buys, S., and V. Oakley
1996 *Conservation and restoration of ceramics.* Oxford: Butterworth Heinemann.
Byock, J.L.
1982 *Feud in the Icelandic saga.* Berkeley and Los Angeles: University of California Press.
Cabanes, P.
1983 Vendi i gruas në Epirin antik. *Iliria* 13(2):193–209.
2002 La présence grecque sur la côte orientale de l'Adriatique en Illyrie du sud. In *Greek influence along the east Adriatic coast. Proceedings of the international conference held in Split from September 24th to 26th 1998,* edited by N. Cambi, S. Cace, and B. Kirigin, 51–62. Split: Knijzevni Krug.
2007 Les sources littéraires grecques: le bitume et le Nymphaion. In *Apollonia d'Illyrie: atlas archéologique et historique* (Mission épigraphique et historique en Albanie, Recherches archéologiques franco-albanaises 1), 42–44. Athens: École Française d'Athènes.
2011 Les confines illyro-épirotes du Ve au IIe siècle av. J.-C. In *L'Illyrie méridionale et l'Épire dans l'Antiquité, V. Actes du Ve colloque international de Grenoble (8–11 octobre 2008),* edited by J.-L. Lamboley and M. Castiglioni, 81–92. Paris: de Boccard.
Cabanes, P., N. Ceka, and O. Deslondes
1994 Apollonia d'Illyrie (Albanie). *Bulletin de Correspondance Hellénique* 118:521–529.
Cabanes, P., F. Drini, and J.-L. Lamboley
1997 Apollonia d'Illyrie (Albanie). *Bulletin de Correspondance Hellénique* 121:848–860.
Cabanes, P., M. Korkuti, A. Baçe, and N. Ceka
2008 *Carte archéologique de l'Albanie.* Tirana: Pegi.
Cabanes, P., and J.-L. Lamboley (editors)
2004 *L'Illyrie méridionale et l'Épire dans l'Antiquité, IV. Actes du IVe colloque international de Grenoble (10–12 octobre 2002).* Paris: de Boccard.
Cabanes, P., J.-L. Lamboley, and B. Vrekaj
1999 Apollonia d'Illyrie (Albanie). *Bulletin de Correspondance Hellénique* 123:569–580.
2000 Apollonia d'Illyrie (Albanie). *Bulletin de Correspondance Hellénique* 124:620–630.
Camaj, M.
1989 Foreword. In *Kanuni i Lekë Dukagjinit: the code of Lekë Dukagjini.* New York: Gjonlekaj Publishing Company.
Cambitoglou, A., and J.K. Papadopoulos
2001 The excavations of 1975, 1976, and 1978. In *Torone I: the excavations of 1975, 1976, and 1978,* edited by A. Cambitoglou, J.K. Papadopoulos, and O. Tudor Jones, 89–272. Athens: Athens Archaeological Society.
Caneva, G., M.P. Nugari, and O. Salvadori
2008 *Plant biology for cultural heritage* (translated by Helen Glanville). Los Angeles: Getty Conservation Institute.
Capasso, L., and G. Di Tota
1996 Foramen caecum molare as a developmental defect of the enamel. In *Recent contributions to the study of enamel defects,* edited by L. Capasso and A. Goodman, 91–106. Teramo (Italy): Edigrafital.
Carington Smith, J., and I. Vokotopoulou
1988 Ανασκαφή στον Κούκο Συκιάς, Νόμος Χαλκιδικής. *To Archaiologikon Ergo ste Makedonia kai Thrake* 2:357–370.
Carr, C.
1995 Mortuary practices: their social, philosophical-religious, circumstantial, and physical determinants. *Journal of Archaeological Method and Theory* 2(2):105–200.
Carrió, V., and S. Stevenson
2003 Assessment of materials used for anoxic microenvironments. In *Conservation science 2002: papers from the conference held in Edinburgh, Scotland, 22–24 May 2002,* edited by J.H. Townsend, K. Eremin, and A. Adriaens, 32–38. London: Archetype Publications.
Carroll, J.
1987 *Wallace Stevens' supreme fiction: a new romanticism.* Baton Rouge and London: Louisiana State University Press.
Carver, M.
1998 *Sutton Hoo: burial ground of kings?* London: British Museum Press.
2005 *Sutton Hoo: a seventh-century princely burial ground and its context* (Reports of the Research Committee of the Society of Antiquaries of London 69). London: British Museum Press.
Casey, E.S.
1993 *Getting back into place: toward a renewed understanding of the place-world.* Bloomington: Indiana University Press.
1997 *The fate of place: a philosophical history.* Berkeley and Los Angeles: University of California Press.
Casson, S.
1923–1925 Excavations in Macedonia, 2. *Annual of the British School at Athens* 26:1–29.
Catling, H.W.
1964 *Cypriot bronzework in the Mycenaean world.* Oxford: Clarendon Press.
Cavanagh, W., and C. Mee
1984 Mycenaean tombs as evidence for social and political organization. *Oxford Journal of Archaeology* 3(3):45–64.
1998 *A private place: death in prehistoric Greece* (Studies in Mediterranean Archaeology). Jonsered: Paul Åströms Förlag.
Ceka, N.
1975 Un tumulus à Dukat. *Iliria* 3:139–161.
1976 Gërmimet arkeologjike të viteve 1974–75. *Iliria* 6:334–335, 371–372.

1977–1978 Vendbanime protourbane në Ilirinë e jugut. *Iliria* 7–8:249–262.
1983a Lindja e jetës qytetare tek Ilirët e jugut. *Iliria* 13 (2):135–192.
1983b Gërmime të vitit 1983: Mashkjezë (Cakran). *Iliria* 13(2):251–254.
1983c Processi di transformazioni nell'Illiria del sud durante il periodo archaico. In *Modes de contact et processus de transformation dans les société anciennes. Actes du Colloque de Cortone* (24–30 mai 1981), 203–218. Pisa: Scuola Normale Superiore.
1985a Kultura protoqytetate Ilire (La civilization protourbaine illyrienne). *Iliria* 15(1):111–150.
1985b Fortifikimet parahistorike Ilire. *Monumentet* 29: 27–58.
1986a Amfora antike nga Margëlliçit. *Iliria* 16(2): 71–98.
1986b Fortifikimet parahistorike Ilire. *Monumentet* 31: 49–84.
1987 Arkitektura e qytezës së Margëlliçi. *Monumentet* 33:5–25.
1990a Fortifikimet antike të bashkësisë byline. *Iliria* 20 (1):99–146.
1990b Periudha paraqytertare në trevën byline. *Iliria* 21 (2):137–160.
2009–2010 Atintanët: një rivështrim mbi territorin dhe historinë e tyre. *Iliria* 34:5–23.
2011 Les fortifications dans les villes d'Illyrie méridionale et d'Épire. In *L'Illyrie méridionale et l'Épire dans l'Antiquité,* V. *Actes du Ve colloque international de Grenoble (8–11 octobre 2008),* edited by J.-L. Lamboley and M. Castiglioni, 649–661. Paris: de Boccard.

Ceka, N., and S. Muçaj
2005 *Byllis: its history and monuments.* Tirana: Migjeni.

Chang, C.
1994 Sheep for the ancestors: ethnoarchaeology and the study of ancient pastoralism. In *Beyond the site: regional studies in the Aegean area,* edited by P.N. Kardulias, 353–371. Lanham, Maryland: University Press of America.

Chapman, R.W.
1977 Burial practices: an area of mutual interest. In *Archaeology and Anthropology: areas of mutual interest,* edited by M. Spriggs, 19–33. Oxford: British Archaeological Reports.

Chenal-Velarde, I., and J. Studer
2003 Archaeozoology in a ritual context: the case of a sacrificial altar in Geometric Eretria. In *Zooarchaeology in Greece: recent advances,* edited by E. Kotjabopoulou, Y. Hamilakis, P. Halstead, C. Gamble, and P. Elefanti, 215–220. London: British School at Athens.

Cherry, J.F.
1988 Pastoralism and the role of animals in the pre- and protohistoric economies of the Aegean. In *Pastoral economies in Classical antiquity* (The Cambridge Philological Society Supplement 14), edited by C.R. Whittaker 6–34. Cambridge: Cambridge University Press.
2003 Archaeology beyond the site: regional survey and its future. In *Theory and practice in Mediterranean archaeology: Old World and New World perspectives* (Cotsen Advanced Seminars 1), edited by J.K. Papadopoulos and R.M. Leventhal, 137–159. Los Angeles: Cotsen Institute of Archaeology.

Cherry, J.F., and J.L. Davis
2007 The other finds. In *Excavations at Phylakopi in Melos 1974–77* (BSA Supplement 42), edited by C. Renfrew, N. Brodie, C. Morris, and C. Scarre, 401–464. London: British School at Athens.

Cherry, J. F., J. L. Davis, and E. Mantzourani
1991 *Landscape archaeology as long-term history: northern Keos in the Cycladic Islands from earliest settlement until modern times.* Los Angeles: University of California, Institute of Archaeology.

Childe, V.G.
1929 *The Danube in prehistory.* Oxford: Clarendon Press.

Choyke, A.M., and L. Bartosiewicz (editors)
2001 *Crafting bone: skeletal technologies through time and space.* Oxford: Archaeopress.

Chrysostomou, A.
2008a *Αρχαία Έδεσσα.* Edessa: University Studio Press.
2008b Agricultural installations on the route of the Edessa southern by-pass. In *National road Thessaloniki-Florina: rescue excavations in 2006–2008 at Nea Halkidona-Edessa segment,* edited by A. Chrysostomou, 89–102. Thessalonike: Region of Central Macedonia, Department of Public Works.

Claassen, C.
1998 *Shells* (Cambridge Manuals in Archaeology). Cambridge: Cambridge University Press.

Clemmer, V.
2005 Maternal relatedness within double burials of an ancient Albanian tumulus. M.S. thesis, Michigan State University, East Lansing.

Coldstream, J.N.
1973 *Knossos: the sanctuary of Demeter* (BSA Supplement 8). Oxford: Thames and Hudson.
1977 *Geometric Greeece.* London: Methuen.

Coldstream, J.N., and H.W. Catling (editors)
1996 *Knossos north cemetery: early Greek tombs* (BSA Supplement 28). London: British School at Athens.

Colvin, H.
1992 *Architecture and afterlife.* New Haven, New Jersey and London: Yale University Press.

Connan, J., and O. Deschesne
1996 *Le bitumen à Suse.* Paris: Collection du Louvre.

Coon, C.
1950 *The mountains of giants: a racial and cultural study of the north Albanian mountain Ghegs*

(Paper of the Peabody Museum of American Archaeology and Ethnology, Harvard University, Vol. XXIII, No. 3). Cambridge, Massachusetts: Harvard University.

Cooney, G.
2006 Newgrange: a view from the platform. *Antiquity* 80:697–708.

Corbett, P.E.
1949 Attic pottery of the later fifth century from the Athenian Agora. *Hesperia* 18:298–351.

Cosgrove, D.
1984 *Social formation and symbolic landscape*. London: Croom Helm.
1985 Prospect, perspective and the evolution of the landscape idea. *Transactions of the Institute of British Geographers* n.s. 10:45–62.
1989 Geography is everywhere: culture and symbolism in human landscapes. In *Horizons in human geography*, edited by D. Gregory and E. Walford, 118–135. London: Macmillan.

Cosmopoulos, M.B., H.J. Greenfield, and D. Ruscillo
2003 Animal and marine remains from the new excavations at Eleusis: an interim report. In *Zooarchaeology in Greece: recent advances*, edited by E. Kotjabopoulou, Y. Hamilakis, P. Halstead, C. Gamble, and P. Elefanti, 145–152. London: British School at Athens.

Cossignani, V., and T. Cossignani
1995 *Atlante delle conchiglie terrestri e dulciacquicole italiane*. Ancona: L'Informatore Piceno.

Costello, J.A.
1969 The corrosion of metals by micro-organisms: a literature survey. *International Biodeterioration Bulletin* 5(3):101–118.

Covey, R.A.
2008 The Inca empire. In *The handbook of South American archaeology*, edited by H. Silverman and W.H. Isbell, 1047–1060. New York: Springer.

Craddock, P.T.
1976 The composition of the copper alloys used by the Greek, Etruscan and Roman civilizations, 1. The Greeks before the Archaic period. *Journal of Archaeological Science* 3:93–113.
1977 The composition of the copper alloys used by the Greek, Etruscan and Roman civilizations, 2. The Archaic, Classical and Hellenistic Greeks. *Journal of Archaeological Science* 4:103–123.
1978 The composition of the copper alloys used by the Greek, Etruscan and Roman civilizations, 3. The origins and early use of brass. *Journal of Archaeological Science* 5:1–16.
1988 Problems and possibilities for provenancing bronzes by chemical composition. In *Bronzeworking centres of western Asia c. 1000–539 B.C.*, edited by J. Curtis, 317–327. London: Routledge.

Craddock, P.T., and N.D. Meeks
1987 Iron in ancient copper. *Archaeometry* 29(2):187–204.

Craig, H.
1957 Isotopic standards for carbon and oxygen and correction factors for mass-spectrometric analysis of carbon dioxide. *Geochimica et Cosmochimica Acta* 12:133–149.

Crawford, S.
2000 Children, grave goods and social status in early Anglo-Saxon England. In *Children and material culture*, edited by J. Moore and E. Scott, 160–179. London: Routledge.

Cronyn, J.M.
1990 *The elements of archaeological conservation*. London: Routledge.

Cross, G.N.
1932 *Epirus: a study in Greek constitutional development*. Cambridge: Cambridge University Press.

Crumley, C.
1994 Cultural ecology: a multidimensional ecological orientation. In *Historical ecology: cultural knowledge and changing landscapes*, edited by C. Crumley, 1–16. Santa Fe, New Mexico: School for American Research Press.

Dakaris, S.
1964 *Οι γενεαλογικοί μύθοι των Μολοσσών*. Athens: Athens Archaeological Society.
1976 *Η κτηνοτροφία στην αρχαία Ήπειρο* (lecture, University of Ioannina, November 28, 1976). Ioannina: University of Ioannina.

Dakoronia, Ph.
2003 Ερευνώντας δύο 'περιφέρειες': ανατολική Λοκρίδα και ΒΑ Φωκίδα. Ομοιότητες και διαφορές. In *The periphery of the Mycenaean world. Proceedings of the 2nd international interdisciplinary colloquium, 26–30 September, 1999, Lamia*, edited by N. Kyparissi-Apostolika and M. Papakonstantinou, 339–349. Athens: Greek Ministry of Culture.
2006 Early Iron Age elite burials in east Lokris. In *Ancient Greece: from the Mycenaean palaces to the age of Homer* (Edinburgh Leventis Studies 3), edited by S. Deger-Jalkotzy and I.S. Lemos, 483–504. Edinburgh: Edinburgh University Press.

Damiata, B., J.K. Papadopoulos, M.G. Amore, S.P. Morris, L. Bejko, J.M. Marston, and J. Southon
2007–2008 Towards an absolute chronology of Albanian archaeology: AMS radiocarbon dates from Apollonia and Lofkënd. *Iliria* 33 (2009):135–185.

Damiata, B.N., and J. Southon
2010 Results of AMS dating and stable-isotope analyses of skeletal remains: Tumuli 9 and 10, Apollonia (Albania). In M.G. Amore, *The complex of Tumuli 9, 10, and 11 in the necropolis of Apollonia (Albania)*, Vol. I (BAR International Series 2059 [I]), 330–335. Oxford: Archaeopress.

Dandoy, J.R., P. Selinsky, and M.M. Voigt
2002 Celtic sacrifice. *Archaeology* 55(1):44–49.

Danielidou, D.
2008 *Ανασκαφές Μυκηνών, II. Το 'εργαστήριο' των Μυκηνών*. Athens: Athens Archaeological Society.

Daniels, S., and D. Cosgrove
1988 Introduction: iconography and landscape. In *The iconography of landscape*, edited by D. Cosgrove and S. Daniels, 1–10. Cambridge: Cambridge University Press.

Dausse, M.-R.
2007 Les villes molosses: bilan et hypothèses sur les quatre centres mentionnés par Tite-Live. In *Épire, Illyrie, Macédoine...Mélanges offerts à Professeur Pierre Cabanes* (Collections ERGA Recherches sur l'Antiquité 10), edited by D. Berranger-Auserve, 197–233. Clermont-Ferrand: Presses Universitaires Blaise Pascal.

Dautaj, B.
1965 La découverte de la cite illyrienne de Dimale. *Studia Albanica* 1:65–71.
1971 Rezultatet e gërmimeve arkeologjike në Dimal: raport për gërmimet në Sektorin A (1971). *Buletini Arkeologjik* 1971:73–84.
1973 Dimali në dritën e të dhënave arkeologjike. *Studime Historike* 2:87–103.
1974 Dimali në dritën e të dhënave arkeologjike. In *Kuvendi I i Studimeve Ilire* 1, 475–505. Tirana: Akademia e Shkencave.
1976 Dimale à la lumière des données archéologiques. *Iliria* 4:385–409.

David, W.
2002 *Studien zu Ornamentik und Datierung der bronzezeitlichen Depotfundgruppe Hajdúsámson-Apa-Ighiel-Zajta, I–II* (Bibliotheca Musei Apulensis, XVIII). Karlsberg: Verlag Altip.

Davidson, G.R.
1952 *Corinth XII: the minor objects*. Princeton, New Jersey: American School of Classical Studies.

Davies, N.
1981 *Human sacrifice: in history and today*. New York: Morrow.

Davies, P.
1972 *Snails: archaeology and landscape change*. Oxford: Oxbow Books.

Davis, E.N.
1986 Youth and age in the Thera frescoes. *American Journal of Archaeology* 90(4):399–406.

Davis, J., S. Alcock, J. Bennet, Y. Lolos, and C. Shelmerdine
1997 The Pylos regional archaeological project, 1. Overview and the archaeological survey. *Hesperia* 66(3):391–494.

Davis, J.L., V. Dimo, I. Pojani, S.R. Stocker, K.M. Lynch, T. Gerke, and E. Gorogianni
2006 Bonjakët excavations, Apollonia: 2004–2005. In *New directions in Albanian archaeology: studies presented to Muzafer Korkuti*, edited by L. Bejko and R. Hodges, 118–127. Tirana: International Centre for Albanian Archaeology.

Davis, J.L, A. Hoti, I. Pojani, S. Stocker, A. Wolpert, P. Acheson, and J. Hayes
2003 The Durrës regional archaeological project: archaeological survey in the territory of Epidamnus/Dyrrachium in Albania. *Hesperia* 72(1): 41–119.

Davis, J.L., M. Korkuti, M. Galaty, S. Muçaj, and S.R. Stocker
2007 The hinterlands of Apollonia. In *Apollonia d'Illyrie. 1. Atlas archéologique et historique* (Collection de l'École Française de Rome 391), edited by V. Dimo, P. Lenhardt, and F. Quantin, 13–23. Athens: École Française d'Athènes.

Davis, J.L., M. Korkuti, L. Bejko, M.L. Galaty, S. Muçaj, and S.R. Stocker
2003–2004 The hinterlands of Apollonia. *Iliria* 31(1–2): 305–322.

Davis, J.L., S.R. Stocker, I. Pojani, V. Dimo, K.M. Lynch, and T. Gerke
2011 Archaic Apollonia: new light from the Bonjakët site. In *L'Illyrie méridionale et l'Épire dans l'Antiquité, V. Actes du Ve colloque international de Grenoble (8–11 octobre 2008)*, edited by J.-L. Lamboley and M.P. Castiglioni, 209–214. Paris: de Boccard.

Davis, S.J.M.
2005 Why domesticate food animals? Some zoo-archaeological evidence from the Levant. *Journal of Archaeological Science* 32(9):1408–1416.

Dawkins, R.M. (editor)
1929 *The sanctuary of Artemis Orthia at Sparta*. London: Macmillan.

Dawkins, R.M., A.J.B. Wace, J.P. Droop, G. Dickens, M.N. Tod, H.J.W. Tillyard, and A.M. Woodward
1906–1907 Laconia, I: excavations at Sparta, 1907. *Annual of the British School at Athens* 13:1–218.

Day, J.
2006 Practical application of the Revolutionary Preservation (RP) System® for marcasite. *Preprints of the 14th triennial meeting of the ICOM Committee for Conservation in The Hague, 12–16 September 2005*, Vol. 1:435–442.

Day, L.P.
1984 Dog burials in the Greek world. *American Journal of Archaeology* 88(1):21–32.

Debénath, A., and H.L. Dibble
1994 *Handbook of Palaeolithic typology, Vol. 1: Lower and Middle Palaeolithic of Europe*. Philadelphia: University Museum of the University of Pennsylvania.

de Boer, J.Z., J.R. Hale, and J.P. Chanton
2001 New evidence for the geological origin of the ancient Delphic oracle. *Geology* 29(8):707–711.

de la Genière, J.
1968 *Recherches sur l'Âge du Fer en Italie méridionale: Sala Consilina.* Naples: Institut Français de Naples, Centre Jean Bérard.

Delamotte, M., and E. Vardala-Theodorou
1994 A preliminary check list of marine gastropods from Greece. *Epeteris Mouseiou Goulandre (Annalles Musei Goulandris)* 9:375–427.

de la Torre, M.
1997 *The conservation of archaeological sites in the Mediterranean region: an international conference organized by the Getty Conservation Institute and the J. Paul Getty Museum, 6-12 May 1995.* Los Angeles: Getty Conservation Institute.

Delougaz, P.
1933 *Plano-convex bricks and the methods of their employment* (Studies in Ancient Oriental Civilization 7). Chicago: Oriental Institute, University of Chicago.

Delouis, O., J.-L. Lamboley, P. Lenhardt, F. Quantin, A. Skenderaj, S. Verger, and B. Vrekaj
2007 La ville haute d'Apollonia d'Illyrie: étapes d'une recherche en cours. In *Épire, Illyrie, Macédoine . . . Mélanges offerts à Professeur Pierre Cabanes* (Collections ERGA Recherches sur l'Antiquité 10), edited by D. Berranger-Auserve, 37–53. Clermont-Ferrand: Presses Universitaires Blaise Pascal.

Demand, N.
1994 *Birth, death, and motherhood in Classical Greece.* Baltimore: The Johns Hopkins University Press.

Denèfle, M., A.M. Lézine, E. Fouache, and J.J. Dufaure
2000 A 12,000-year pollen record from Lake Maliq, Albania. *Quaternary Research* 54(3):423–432.

DeNiro, M.J.
1985 Postmortem preservation and alteration of *in-vivo* bone collagen isotope ratios in relation to paleodietary reconstruction. *Nature* 317:806–809.
1987 Stable isotopes and archaeology. *American Scientist* 75:182–191.

Dentler, R.A., and K.T. Erikson
1959 The function of deviance in groups. *Social Problems* 1:98–107.

Deonna, W.
1938 *Exploration archéologique de Délos, XVIII: le mobilier délien.* Paris: de Boccard.

Derevenski, J.S.
1997 Engendering children, engendering archaeology. In *Invisible people and processes: writing gender and childhood into European archaeology*, edited by J. Moore and E. Scott, 192–202. London: Routledge.
2000 Material culture shock: confronting expectations in the material culture of children. In *Children and material culture*, edited by Joanna Sofaer Derevenski, 3–16. London and New York: Routledge.

Desborough, V.R.d'A.
1952 *Protogeometric pottery.* Oxford: Clarendon Press.
1954 Mycenae 1939–1953, part V: four tombs. *Annual of the British School at Athens* 49:258–266.
1955 Mycenae 1939–1954, part VI: three Geometric tombs. *Annual of the British School at Athens* 50: 239–247.
1964 *The Last Mycenaeans and their successors: an archaeological survey c. 1200–c. 1000 B.C.* Oxford: Oxford University Press.
1972 *The Greek dark ages.* London: Ernest Benn.

Descoeudres, J.-P. (editor)
1990 *Greek colonists and native populations: Proceedings of the First Australian Congress of Classical archaeology held in honour of Emeritus Professor A.D. Trandall, Sydney, 9–14 July 1985.* Oxford: Clarendon Press.

Dhima, A.
1979–1980 Të dhëna mbi tipin antropologjik të shqiptarëve gjatë mesjetës. *Iliria* 9–10:301–351.
1983 Vëzhgime antropologjike mbi formimin e popullit shqiptar. *Iliria* 13(1):251–260.
1984 Vlerësimi i lidhjeve antropologjike në disa seri mesjetare me metodën e distancës përafruese. *Iliria* 14(2):227–235.
1985a Materiale skeletore nga nekropoli i Dyrrahut. *Iliria* 15(1):247–270.
1985b Vështrim mbi tiparet antropologjike të ilirëve. *Iliria* 15(2):293–301.
1992 Analizë paleodemografike e varrezës së Prosekut. *Iliria* 22(1–2):263–287.
1993 Evolutionary trends in Albanian anthropological material. *International Journal of Anthropology* 8:221–231.

Dhora, D., and F. Welter-Schultes
1996 List of species and atlas of the non-marine mollusks of Albania. *Schriften zur Malakozoologie* 9: 90–107.

Dida, M.
2003 *State of forest tree genetic resources in Albania* (Forest Genetic Resources Working Paper No. FGR/62E. Food and Agriculture Organization of the United Nations, Forest Resources Development Service). Rome: United Nations.

Dikaios, P.
1969 *Enkomi, Excavations 1948–1958.* Mainz: Philipp von Zabern.

Dillon, M.
2002 *Girls and women in Classical Greek religion.* London and New York: Routledge.

Dimacopoulos, J.E.
1997 *A shelter in the style of a tumulus: Vergina, an underground archaeological site and museum in the type of a crypt.* 2nd ed. Athens: Greek Ministry of Culture.

Dimo, V.
2004 *Apollonia e Ilirisë: guidë e ilustruar historiko-arkeologjike.* Tirana: Institute of Archaeology.

Dimo, V., and A. Fenet
1996 Nouvelles recherches sur la nécropole tumulaire d'Apollonia. In *L'Illyrie méridionale et l'Epire dans l'Antiquité*. Vol. III. *Actes du IIIe colloque international de Chantilly*, edited by P. Cabanes, 217–223. Paris: de Boccard.

Dimo, V., P. Lenhardt, and F. Quantin (editors)
2007 *Apollonia d'Illyrie: atlas archéologique et historique* (Mission épigraphique et historique en Albanie). Athens: École Française d'Athènes.

Donnan, C.B.
1995 Moche funerary practice. In *Tombs for the living: Andean mortuary practices*, edited by T.D. Dillehay, 111–158. Washington, D.C.: Dumbarton Oaks Research Library and Collection.

D'Onofrio, A.M. (editor)
2007 *Tallies, tokens and counters: from the Mediterranean to India. Proceedings of the meeting held at the Università degli Studi di Napoli 'L'Orientale,' Naples 31st May 2004*. Naples: Università degli Studi di Napoli 'L'Orientale,' Dipartimento di Studi del Mondo Classico e del Mediterraneo Antico.

Dothan, M., and D. Ben-Shlomo
2005 *Ashdod VI: the excavations of area H and K (1968–1969)*. Jerusalem: Israel Antiquities Authority.

Doumas, C., and L. Marangou
1978 *Catalogue of the exhibition of ancient Greek art from the N.P. Goulandris collection, Benaki Museum*. Athens: Benaki Museum.

Douzougli, A.
1996 Η κοιλάδα του Αώου. Αρχαιολογικές μαρτυρίες για την ανθρώπινη δραστηριότητα από την προϊστορική εποχή ως την ύστερη αρχαιότητα. In *Η επαρχία Κόνιτσας στο χώρο και χρόνο*, 11–61. Konitsa: Deme of Konitsa.

Douzougli, A., and J.K. Papadopoulos
2010 Liatovouni: a Molossian cemetery and settlement in Epirus. *Jahrbuch des Deutschen Archäologischen Instituts* 125, 2010 [2011]:1–87.

Douzougli, A., and K.L. Zachos
2002 L'archéologie des zones montagneuses: modèles et interconnexions dans le Néolithique de l'Épire et de l'Albanie méridionale. In *L'Albanie dans l'Europe préhistorique. Actes du Colloque international organisé par l'École Française d'Athènes et l'Université de Bretagne-Sud, Lorient, 8–10 juin 2000* (*Bulletin de Correspondance Hellénique*, Supplément 42), edited by G. Touchais and J. Renard, 111–143. Paris: de Boccard.

Dowden, K.
1989 *Death and the maiden: girls' initiation rites in Greek mythology*. London and New York: Routledge.

Dragendorff, H., W. Dörpfeld, F.F. Hiller von Gaertringen, A. Schiff, C. Watzinger, P. Wilski, and R. Zahn
1903 *Theräische Gräber*. Berlin: Georg Reimer.

Droop, J.P.
1905–1906 Dipylon vases from the Kynosarges site. *Annual of the British School at Athens* 12:80–92.

Dufraisse, A. (editor)
2006 *Charcoal analysis: new analytical tools and methods for archaeology. Papers from the table-ronde held in Basel 2004*. Oxford: Archaeopress.

Dugas, C.
1921 Le sanctuaire d'Aléa Athéna à Tégée avant le IVe siècle. *Bulletin de Correspondance Hellénique* 45: 335–435.

Dunbabin, T.J. (editor)
1962 *Perachora, the sanctuaries of Hera Akraia and Limenia, II: pottery, ivories, scarabs, and other objects from the votive deposit of Hera Limenia*. Oxford: Clarendon Press.

Duncan, S.J., and H. Ganiaris
1987 Some sulphide corrosion products on copper alloys and lead alloys from London waterfront sites. In *Recent advances in the conservation and analysis of artifacts*, edited by J. Black, 109–118. London: Summer Schools Press, University of London.

Durić, N., J. Glišić, and J. Todorović
1975 *Romaja Parahistorike*. Beograd: Association des sociétés archéologiques de Yugoslavie.

Duru, R.
2008 *From 8000 B.C. to 2000 B.C.: six thousand years of the Burbur-Antalya region*. Antalya: Suna and İnan Kıraç Research Institute on Mediterranean Civilizations.

Dusenbery, E.B.
1998 *Samothrace 11: the nekropoleis. Catalogues of objects by categories*. Princeton, New Jersey: Princeton University Press.

Earle, T.
1990 Style and iconography as legitimation in complex chiefdoms. In *The uses of style in archaeology*, edited by M.W. Conkey and C. Hastorf, 73–81. Cambridge: Cambridge University Press.

Eggebrecht, A.
1988 *Albanien: Schätze aus dem Land der Skipetaren* (exhibition catalogue, Hildesheim, July 18–November 20, 1988). Mainz am Rhein: P. von Zabern.

Eisman, M.M., and L. Turnbull
1978 Robinson's kiln-skyphos. *American Journal of Archaeology* 82:394–399.

Ekroth, G.
2002 *The sacrificial remains of Greek hero-cults in the Archaic to the early Hellenistic period* (*Kernos* Supplement 12). Liège: Centre international d'étude de la religion grecque antique.

Eliade, M.
1954 *The myth of the eternal return*. Princeton, New Jersey: Princeton University Press.
1957 *The sacred and the profane*. New York: Harcourt.

Eliakis, C., C.E. Eliakis, and P. Iordanidis
1966 Sur la determination de la taille d'après la mensuration des os longs. *Annales de Médicine Légale* 46:403–421.
Elsie, R.
2001 *A dictionary of Albanian religion, mythology and folk culture*. London: Hurst and Co.
Elster, E.S., and C. Renfrew (editors)
2003 *Prehistoric Sitagroi: excavations in northeast Greece, 1968–1970, Vol. 2: the final report*. Los Angeles: Cotsen Institute of Archaeology.
Emery, I.
1994 [1966] *The primary structure of fabrics*. London: Thames and Hudson.
Emre, K.
1978 *Yanarlar: Afyon yöresinde bir hitit mezarliği (Yanarlar: a Hittite cemetery near Afyon)*. Ankara: Türk Tarih Kurumu Basimevi.
Eriksen, P.
2006 The rolling stones of Newgrange. *Antiquity* 80: 709–710.
Erikson, K.T.
1966 *Wayward Puritans: a study in the sociology of deviance*. New York: Wiley.
1975 On the sociology of deviance. In *The collective definition of deviance*, edited by F.J. Davis and R. Stivers, 11–21. New York: Free Press.
Eslick, C.
1992 *Elmalı-Karataş I: the Neolithic and Chalcolithic periods: Bağaşı and other sites*. Bryn Mawr, Pennsylvania: Bryn Mawr College.
Evangelides, D.
1935 Η ανασκαφή της Δωδώνης 1935. *Epeirotika Chronika* 10:192–260.
1947 *Οι αρχαίοι κάτοικοι της Ηπείρου*. Athens: Epeirotike Zoe.
Evans, A.J.
1892–1893 A Mykênæan treasure from Ægina. *Journal of Hellenic Studies* 13:195–226.
Evans, J.G.
1972 *Land snails in archaeology, with special reference to the British Isles*. London: Seminar Press.
Evely, D. (editor)
2006 *Lefkandi IV. The Bronze Age: the Late Helladic IIIC settlement at Xeropolis* (BSA Supplement Vol. 39). Oxford: British School at Athens.
Fabrizi, M., H. Ganiaris, S. Tarling, and D. Scott
1989 The occurrence of sampleite, a complex copper phosphate, as a corrosion product on copper alloy objects from Memphis, Egypt. *Studies in Conservation* 39(1):45–51.
Falkner, G., R. Bank, and T. von Proschwitz
2001 CLECOM-PROJECT: check-list of the non-marine molluscan species-group taxa of the states of northern, Atlantic and central Europe (CLECOM I). *Heldia* 4:1–76.
Faustman, A., and R. Palmer
2005 Wings over Armenia: use of a paramotor for archaeological aerial survey. *Antiquity* 79:402–410.
Favro, D.
2006 In the eye of the beholder: virtual reality re-creations and academia. In *Imaging ancient Rome: documentation, visualization, imagination. Proceedings of the Third Williams Symposium on Classical architecture, held at the American Academy in Rome, the British School at Rome, and the Deutsches Archäologisches Institut, Rome, on May 20–23, 2004*, edited by L. Haselberger and J. Humphrey (*Journal of Roman Archaeology* Supplementary Series 61), 321–334. Portsmouth, Rhode Island: Journal of Roman Archaeology.
Favro, D., and C. Johanson
2010 Death in motion: funeral processions in the Roman forum. *Journal of the Society of Architectural Historians* 69:12–37.
Felsch, R.C.S.
1979 Boiotische Ziegelwerkstätten archaischer Zeit. *Mitteilungen des Deutschen Archäologischen Instituts, Athenische Abteilung* 94:1–40.
1990 Further stamped roof tiles from central Greece, Attica, and the Peloponnese. *Hesperia* 59:301–323.
2007 Die Bronzefunde. In *Kalapodi. Ergebnisse der Ausgrabungen im Heiligtum der Artemis und Apollon von Hyampolis in der antiken Phokis, Bd. II*, edited by R.C.S. Felsch, 28–422. Mainz: Philipp von Zabern.
Ferguson, A.R.
1971 *The collected works of Ralph Waldo Emerson*, Vol. I: *nature, addresses and lectures*. Cambridge, Massachusetts: Harvard University Press.
Ferrari, G.
2003 What kind of rite of passage was the ancient Greek wedding? In *Initiation in ancient Greek rituals and narratives*, edited by D.B. Dodd and C.A. Faraone, 27–42. London and New York: Routledge.
Fiedler, M., and G. Döhner.
2013 Bitumenrückstände an Keramik aus der griechisch-illyrischen Stadt Apollonia. In *Networks in the Hellenistic world according to the pottery in the eastern Mediterranean and beyond*, edited by N. Finn and C. Römer-Strehl, 131–140. Oxford: Archaeopress.
Filipovic, V., and A. Starovic
2005 Excavations at a Bronze Age mound necropolis site at Illyrian Graveyard, western Serbia. *Antiquity* 79. Available at *http://www.antiquity.ac.uk/projgall/filipovic304/*.
Finkelberg, M.
1991 Royal succession in heroic Greece. *The Classical Quarterly* n.s. 41(2):303–316.
Fistani, A.
1989 Gjetje të paleolitit në afërsi të shpellës së Gajtanit (Shkodër). *Iliria* 19(1):5–23.

Fitter, C.
1995 *Poetry, space, landscape*. Cambridge: Cambridge University Press.
Fleming, K.E.
1999 *The Muslim Bonaparte: diplomacy and orientalism in Ali Pasha's Greece*. Princeton, New Jersey: Princeton University Press.
Flint-Hamilton, K.B.
1999 Legumes in ancient Greece and Rome: food, medicine, or poison? *Hesperia* 68:371–385.
Foltiny, S.
1961 Athens and the east Hallstatt region: cultural interrelations at the dawn of the Iron Age. *American Journal of Archaeology* 65:283–297.
Forbes, R. J.
1936 *Bitumen and petroleum in antiquity*. Leiden: E.J. Brill.
Forsén, J.
2009 The "Dark Age" in the Kokytos valley—not so dark after all. In *Thesprotia expedition I: towards a regional history* (Papers and Monographs of the Finnish Institute at Athens, Vol. 15), edited by B. Forsén, 55–72. Helsinki: Foundation of the Finnish Institute at Athens.
Forstenpointner, G.
2003 Promethean legacy: investigations into the ritual procedure of "Olympian" sacrifice. In *Zooarchaeology in Greece: recent advances*, edited by E. Kotjabopoulou, Y. Hamilakis, P. Halstead, C. Gamble, and P. Elefanti, 203–213. London: British School at Athens.
Foster, K.P.
1979 *Aegean faience of the Bronze Age*. New Haven, Connecticut, and London: Yale University Press.
Foss, J.E., and M.E. Timpson
2007 Appendix 2: preliminary report on the soils of the Lofkënd archaeological site. *American Journal of Archaeology* 111:140–144.
Fotiadis, M.
1995 Modernity and the past-still-present: politics of time in the birth of regional archaeological projects in Greece. *American Journal of Archaeology* 99:59–78.
Fouache, E.
2007 Contexte physique de la plaine de Museqë et du site d'Apollonia. In *Apollonie d'Illyrie, 1. Atlas Archéologique et historique* (Collection de l'École Française de Rome 391), edited by V. Dimo, P. Lenhardt, and F. Quantin, 3–13. Recherches archéologiques franco-albanaises 1. Athens: École Française d'Athènes; Paris: Ministère des affaires étrangères; Rome: École Française de Rome.
Fouache, E., J.J. Dufaure, M. Denèfle, A.M. Lézine, P. Léra, F. Prendi, and G. Touchais
2001 Man and environment around Lake Maliq (southern Albania) during the Late Holocene. *Vegetation History and Archaeobotany* 10(2):79–86.
Fox, L.
1989 Introduction. In *Kanuni i Lekë Dukagjinit: the code of Lekë Dukagjini*, xvi–xix. New York: Gjonlekaj Publishing Company.
Francis, K.
2005 *Explorations in Albania, 1930–1939. The notebooks of Luigi Cardini, prehistorian with the Italian Archaeological Mission* (BSA Supplement Vol. 73). Oxford: British School at Athens.
Franke, P.R.
1955 *Alt-Epirus und das Königtum der Molosser*. Kallmunz: M. Lassleben.
Frayer, D.W., and M.H. Wolpoff
1985 Sexual dimorphism. *Annual Review of Anthropology* 14:429–473.
Frederiksen, R.
2011 *Greek city walls of the Archaic period, 900–480 BC*. Oxford: Oxford University Press.
Furmánek, V.
1980 *Die Anhänger in der Slowakei* (*PBF* XI:3). Munich: Beck.
Furmánek, V., L. Veliačik, and J. Vladár
1999 *Die Bronzezeit im slowakischen Raum* (Prähistorische Archäologie in Südosteuropa, Bd. 15). Rahden: Verlag Marie Leidorf.
Furtwängler, A.
1890 *Olympia IV. Die Bronzen und die übrigen kleineren Funde*. Berlin: A. Asher and Co.
Furtwängler, A., E.R. Fiechter, and H. Thiersch
1906 *Aegina: das Heiligtum der Aphaia*. Munich: Verlag der Akademie der Wissenschaften.
Gadamer, H.G.
1989 *Truth and method*. 2nd rev. ed. Translation revised by J. Weinsheimer and D.G. Marshall. London: Sheed and Ward.
Galaty, M.L.
2002 Modeling the formation and evolution of an Illyrian tribal system: ethnographic and archaeological analogs. In *The archaeology of tribal societies*, edited by W.A. Parkinson, 109–122. Ann Arbor, Michigan: International Monographs in Prehistory.
2006 From points A to B: the Shala Valley Project and the Albanian Middle Palaeolithic. In *New directions in Albanian archaeology: studies presented to Muzafer Korkuti*, edited by L. Bejko and R. Hodges, 18–30. Tirana: International Centre for Albanian Archaeology.
2007 "There are prehistoric cities up there": the Bronze and Iron Ages in northern Albania. In *Between the Aegean and Baltic seas: prehistory across borders. Proceedings of the International conference Bronze and Early Iron Age interconnections and contemporary developments between the Aegean and the regions of the Balkan Peninsula, central and northern Europe, University of Zagreb, 11–14 April 2005* (Aegaeum 27), edited by I. Galanaki,

H. Tomas, Y. Galanakis, and R. Laffineur, 133–141. Liège: Université de Liège.
Galaty, M.L., O. Lafe, W.E. Lee, Z. Tafilica (editors)
2013 *Light and shadow: isolation and interaction in the Shala Valley of northern Albania* (Monumenta Archaeologica 28). Los Angeles: Cotsen Institute of Archaeology.
Galaty, M.L., O. Lafe, and Z. Tafilica
2005 Shala Valley Project: final report of the 2005 field season. Department of Sociology and Anthropology, Millsaps College, Jackson, Mississippi; Albanian Ministry of Tourism, Culture, Youth and Sports; Shkodra Historical Museum.
2011 Grunas, Shala: a new fortified prehistoric site in the northern Albanian mountains. In *L'Illyrie méridionale et l'Épire dans l'Antiquité, V. Actes du Ve colloque international de Grenoble (8–11 octobre 2008)*, edited by J.-L. Lamboley and M.P. Castiglioni, 13–24. Paris: de Boccard.
Garašanin, D.
1954 *Katalog der vorgeschichtlichen Metalle (Vorgeschichte I)*. Belgrade: Nationalmuseum.
1973 Les tombes tumulaires préhistoriques de la peninsule Balkanique et leur attribution ethnique et chronologique. *Studia Albanica* 1:178–184.
Gardeisen, A., L. Garcia Petit, and G. Piques
2002 La recherché archéozoologique en Albanie: un état de la question à Sovjan (basin de Korçë). In *L'Albanie dans l'Europe préhistorique. Actes du Colloque international organisé par l'École Française d'Athènes et l'Université de Bretagne-Sud, Lorient, 8–10 juin 2000* (*Bulletin de Correspondance Hellénique*, Supplément 42), edited by G. Touchais and J. Renard, 3–42. Paris: de Boccard.
Garland, R.
1985 *The Greek way of death*. Ithaca: Cornell University Press.
Garoufalias, P.
1979 *Pyrrhus, King of Epirus*. London: Stacey International.
Garside, P., and P. Wyeth
2003 Identification of cellulosic fibers by FTIR spectroscopy: thread and single fibre analysis by attenuated total reflectance. *Studies in Conservation* 48:269–275.
Garson, J.G.
1886 The cephalic index. *Journal of the Royal Anthropological Institute* 16:11–17.
Gedl, M.
2004 *Die Fibeln in Polen* (*PBF* XIV:10). Munich: Beck.
Gergova, D.
1987 *Früh- und ältereisenzeitliche Fibeln in Bulgarien* (*PBF* XIV:7). Munich: Beck.
Gettens, R.J.
1970 Patina: noble and vile. In *Art and technology: a symposium on Classical bronzes*, edited by S. Doeringer, D.G. Mitten, and A. Steinberg, 57–69. Cambridge, Massachusetts: MIT Press.
Gilkes, O.
2003 *The theatre at Butrint: Luigi Maria Ugolini's excavations at Butrint, 1928–1932* (Albania Antica IV), edited by O. Gilkes. London: British School at Athens.
2013 *Albania: an archaeological guide*. London and New York: I.B. Tauris.
Gimatzidis, S.
2010 *Die Stadt Sindos: Eine Siedlung von der späten Bronze- bis zur klassischen Zeit am thermaischen Golf in Makedonien* (Prähistorische Archäologie in Südosteuropa, Bd. 26). Rahden: Verlag Marie Leidorf GmbH.
Gimbutas, M.
1965 *Bronze Age cultures in central and eastern Europe*. The Hague: Mouton and Co.
Giusti, F., G. Manganelli, and P.J. Schembri
1995 *The non-marine mollusks of the Maltese Islands* (Monografie 15, Museo Regionale di Scienze Naturali). Turin: Museo Regionale di Scienze Naturali.
Gjeçovi, Sh. K.
1920 Një argëtim arkeologjik. *Zani i Shna Ndout* 6: 109–112.
Gjerstad, E., J. Lindros, E. Sjöqvist, and A. Westholm
1934 *The Swedish Cyprus Expedition, I: finds and results of excavations in Cyprus, 1927–1931*. Stockholm: Swedish Cyprus Expedition.
1935 *The Swedish Cyprus Expedition, II: finds and results of excavations in Cyprus, 1927–1931*. Stockholm: Swedish Cyprus Expedition.
Gjipali, I.
1995 Vendbanimi neolitik i Rashtanit (Librazhd). *Iliria* 25(1–2):17–52.
1999–2000 Përmbajtja dhe përkatësia kulturore kronologjike e vendbanimi neolitik të Rajcës (Rajcë II). *Iliria* 29(1–2):29–77.
2006 Recent research on the Palaeolithic and Mesolithic archaeology of Albania. In *New directions in Albanian archaeology: studies presented to Muzafer Korkuti*, edited by L. Bejko and R. Hodges, 31–42. Tirana: International Centre for Albanian Archaeology.
2011 Prehistoric settlements in the territory of western Chaonia. In *l'Illyrie méridionale et l'Épire dans l'Antiquité, V. Proceedings of the Ve Colloque International, Grenoble, France, octobre 8–12, 2008*, edited by J.-L. Lamboley and M. Castiglioni, 25–40. Paris: de Boccard.
Gjongecaj, S., and O. Picard
2003–2004 Monnaies d'Apollonia. *Iliria* 31(1–2):259–262.
2007 Les monnaies d'Apollonie. In *Apollonia d'Illyrie: atlas archéologique et historique* (Mission épigraphique et historique en Albanie, Recherches

archéologiques franco-albanaises 1), 81–106. Athens: École Française d'Athènes.

Glogović, D.
2003 *Fibeln im kroatischen Küstengebiet (Istrien, Dalmatien)* (*PBF* XIV:13). Munich: Beck.

Goldman, H.
1931 *Excavations at Eutresis in Boeotia*. Cambridge, Massachusetts: Harvard University Press.
1940 The acropolis of Halae. *Hesperia* 9:381–514.
1956 *Excavations at Gözlü Kule, Tarsus,* Vol. II: *from the Neolithic through the Bronze Age*. Princeton, New Jersey: Princeton University Press.
1963 *Excavations at Gözlü Kule, Tarsus,* Vol. III: *the Iron Age*. Princeton, New Jersey: Princeton University Press.

Goldstein, L.
1981 One-dimensional archaeology and multidimensional people: spatial organization and mortuary analysis. In *The archaeology of death*, edited by R. Chapman, I. Kinnes, and K. Randsborg, 53–69. Cambridge: Cambridge University Press.

Goodwin, D., B. Schöne, and D. Dettman
2003 Resolution and fidelity of oxygen isotopes as paleotemperature proxies in bivalve mollusk shells: models and observations. *Palaios* 18(2):110–125.

Goody, J.R.
1973 Bridewealth and dowry in Africa and Eurasia. In *Bridewealth and dowry*, edited by M. Fortes, J.R. Goody, E.R. Leach, and S.J. Tambiah, 1–58. Cambridge: Cambridge University Press.

Gosden C., and L. Head
1994 Landscape—a usefully ambiguous concept. *Archaeology in Oceania* 29:113–116.

Graedel, T.E.
1987 Copper patinas formed in the atmosphere, III: a semi-quantitative assessment of the rates and constraints in the greater New York metropolitan area. *Corrosion Science* 27(7):741–769.

Graedel, T.E., K. Nassau, and J.P. Franey
1987 Copper patinas formed in the atmosphere, I: introduction. *Corrosion Science* 27(7):639–657.

Graham, A.J.
1964 *Colony and mother city in ancient Greece*. Manchester: Manchester University Press.

Grant, A.
1984 Survival or sacrifice? A critical appraisal of animal burials in Britain in the Iron Age. In *Animals and archaeology*, edited by C. Grigson and J. Clutton-Brock, 221–227. Oxford: British Archaeological Reports.

Gräslund, B.
2000 Gamla Uppsala during the Migration period. In *Myth, might, and man: ten essays on Gamla Uppsala*. Stockholm: Riksantikvarieämbetet.

Grayson, D.K.
2001 The archaeological record of human impacts on animal populations. *Journal of World Prehistory* 15(1):1–68.

Grbić, M.
1960 *Porodin: Eine spätneolithische Ansiedlung auf der Tumba bei Bitolj*. Bitolj: National Museum and the Archaeological Institute of the Serbian Academy of Science in Belgrade.

Greifenhagen, A.
1970 *Schmuckarbeiten in Edelmetall, Bd. I: Fundgruppen*. Berlin: Mann.
1975 *Schmuckarbeiten in Edelmetall, Bd. II: Einzelstücke*. Berlin: Mann.

Greig, J.R.A., and J. Turner
1974 Some pollen diagrams from Greece and their archaeological significance. *Journal of Archaeological Science* 1:177–194.

Guzzo, P.G., and M. Carrara
1981 Roggiano Gravina (Cosenza), località Prunetta: scavo di una necropoli dell'età del ferro. *Notizie degli Scavi* 1981:443–490.

Haber, W.
1995 Concept, origin and meaning of "landscape." In *Cultural landscapes of universal value: components of a global strategy*, edited by B. von Droste, H. Plachter, and M. Rossler, 38–42. Stuttgart and New York: Gustav Fisher Verlag Jena and UNESCO.

Haevernick, T.E.
1981 Urnenfelderzeitliche Glasperlen. In *Beiträge zur Glasforschung: die wichtigsten Aufsätze von 1938 bis 1981*, Mainz, 375–383 (originally published in *Zeitschrift für Schweizerische Archäologie und Kunstgeschichte* 35, 1978:145–157).
1983 *Glasperlen der vorrömischen Eisenzeit I* (*Marburger Studien zur Vor- und Frühgeschichte*, Bd. 5. Mainz: Philipp von Zabern.
1987 *Glasperlen der vorrömischen Eisenzeit, II: Ringperlen und verwandte Perlengruppen* (*Marburger Studien zur Vor- und Frühgeschichte*, Bd. 9), Marburg/Lahn: Dr. Wolfram Hitzeroth Verlag.

Hage, P., and J. Marck
2003 Matrilineality and the Melanesian origin of Polynesian Y chromosome. *Current Anthropology* 44: 121–127.

Hahn, E.
1891 Waren die Menschen der Urzeit zwischen der Jägerstufe und der Stufe des Ackerbus Nomaden? *Das Ausland* 64:54–59.

Hale, J.R, J. de Boer, J. Chanton, and H.A. Spiller
2003 Questioning the Delphic oracle. *Scientific American* 298(2):57–63.

Hall, J.M.
1997 *Ethnic identity in Greek antiquity*. Cambridge: Cambridge University Press.

Halstead, P.

1981 Counting sheep in Neolithic and Bronze Age Greece. In *Patterns of the past: studies in honour of David Clarke*, edited by I. Hodder, G. Isaac, and N. Hammond, 307–339. Cambridge: Cambridge University Press.

1987a Traditional and ancient rural economy in Mediterranean Europe: plus ça change? *Journal of Hellenic Studies* 107:77–87.

1987b Man and other animals in later Greek prehistory. *Annual of the British School at Athens* 82:71–83.

1988 On redistribution and the Minoan-Mycenaean palatial economies. In *Problems in Greek prehistory: papers presented at the centenary conference of the British School of Archaeology at Athens, Manchester, April 1986*, edited by E.B. French and K.A. Wardle, 519–530. Bristol: Bristol Classical Press.

1989 The economy has a normal surplus: economic stability and social change among early farming communities of Thessaly, Greece. In *Bad year economics: cultural responses to risk and uncertainty*, edited by P. Halstead and J. O'Shea, 68–80. Cambridge: Cambridge University Press.

1990 Present to past in the Pindhos: diversification and specialization in mountain economies. *Rivista di Studi Liguri* 1:61–80.

2000 Land use in postglacial Greece: cultural causes and environmental effects. In *Landscape and land use in postglacial Greece*, edited by P. Halstead and C. Frederick, 110–128. Sheffield: Sheffield Academic Press.

2005 Resettling the Neolithic: faunal evidence for seasons of consumption and residence at Neolithic sites in Greece. In *(Un)Settling the Neolithic*, edited by D.W. Bailey, A.W.R. Whittle, and V. Cummings, 38–50. Oxford: Oxbow Books.

Ham, G.

1999 The Choes and Anthesteria reconsidered: male maturation rites and the Peloponnesian wars. In *Rites of passage in ancient Greece: literature, religion, society*, edited by M.W. Padilla, 201–220. Lewisburg, Pennsylvania: Bucknell University Press.

Hamilakis, Y.

2003 The sacred geography of hunting: wild animals, social power and gender in early farming societies. In *Zooarchaeology in Greece: recent advances*, edited by E. Kotjabopoulou, Y. Hamilakis, P. Halstead, C. Gamble, and P. Elefanti, 239–247. London: British School at Athens.

Hamilakis, Y., and E. Konsolaki

2004 Pigs for the gods: burnt animal sacrifices as embodied rituals at a Mycenaean sanctuary. *Oxford Journal of Archaeology* 23(2):135–151.

Hamilton, R.

1992 *Choes and Anthesteria*. Ann Arbor: University of Michigan Press.

Hammond, N.G.L.

1931–1932 Prehistoric Epirus and the Dorian invasion. *Annual of the British School at Athens* 32:131–179.

1966 The opening campaigns and the battle of the Aoi Stena in the Second Macedonian war. *Journal of Roman Studies* 56:39–54.

1967a *Epirus: the geography, the ancient remains, the history, and the topography of Epirus and adjacent areas*. Oxford: Clarendon Press.

1967b Tumulus-burial in Albania, the grave circles at Mycenae, and the Indo-Europeans. *Annual of the British School at Athens* 62:77–105.

1971 The dating of some burials in tumuli in south Albania. *Annual of the British School at Athens* 66:229–241.

1972 *A history of Macedonia*. Oxford: Clarendon Press.

1974 The tumulus-burials of Leucas and their connections in the Balkans. *Annual of the British School at Athens* 69:129–144.

1976 Tumulus-burial in Albania and problems of ethnogenesis. *Iliria* 4:127–132.

1982 Illyris, Epirus and Macedonia in the Early Iron Age. In *The Cambridge ancient history*, 2nd ed. 3.1:619–656.

1992 The relations of Illyrian Albania with the Greeks and Romans. In *Perspectives on Albania*, edited by T. Winnifrith, 29–39. New York: St. Martin's.

2000 The ethne in Epirus and upper Macedonia. *Annual of the British School at Athens* 95: 345–352.

Hänsel, B.

1985 Wanderungen in Südosteuropa während der späten Bronzezeit und Verhältnis zum Territorium Albaniens. *Iliria* 15(1):224–239.

1987 Der mittelalterliche Grabfund im Nordwestteil der Grabungsfläche. In A. Hochstetter, *Kastanas. Ausgrabungen in einem Siedlungshügel der Bronze- und Eisenzeit Makedoniens, 1975–1979: Die Kleinfunde*, 113–120. Berlin: Wissenschaftsverlag Volker Spiess.

1989 *Kastanas. Ausgrabungen in einem Siedlungshügel der Bronze- und Eisenzeit Makedoniens, 1975–1979: Die Grabung und der Baubefund* (Prähistorische Archäologie in Südosteuropa, Bd. 7, 1). Berlin: Volker Spiess.

Hänsel, B., and I. Aslanis

2010 *Das prähistorische Olynth. Ausgrabungen in der Toumba Agios Mamas 1994–1996. Die Grabung und der Baubefund* (Prähistorische Archäologie in Südosteuropa Bd. 23). Rahden/Westfalen: Verlag Marie Leidorf GmbH.

Hansen, J.

1999 Konispol Cave plant remains. In *Meletemata: studies in Aegean archaeology presented to Malcolm H. Wiener as he enters his 65th Year* (Aegaeum 20), edited by P. P. Betancourt, V. Karageorghis,

R. Laffineur, and W.-D. Niemeier, 333–337. Liège: Université de Liège and UT-PASP.
2001 Macroscopic plant remains from Mediterranean caves and rockshelters: avenues of interpretation. *Geoarchaeology: An International Journal* 16(4): 401–432.

Harding, A.F.
1984 *The Mycenaeans and Europe*. London and Orlando: Academic Press.
1992 The prehistoric background of Illyrian Albania. In *Perspectives on Albania*, edited by T. Winnifrith, 14–28. New York: St. Martin's Press.
2007 Interconnections between the Aegean and continental Europe in the Bronze and Iron Ages: moving beyond skepticism. In *Between the Aegean and Baltic Seas: prehistory across borders. Proceedings of the international conference Bronze and Early Iron Age interconnections and contemporary developments between the Aegean and the regions of the Balkan Peninsula, central and northern Europe, University of Zagreb, 11–14 April 2005* (Aegaeum 27), edited by I. Galanaki, H. Tomas, Y. Galanakis, and R. Laffineur, 47–56. Liège: Université de Liège.

Harries, K.
1998 *The ethical function of architecture*. Cambridge, Massachusetts: MIT Press.

Harris, E.F.
2007 Carabelli's trait and tooth size of human maxillary first molars. *American Journal of Physical Anthropology* 132(2):238–246.

Harris, E.F., and T. Sjovøld
2004 Calculation of Smith's mean measure of divergence for intergroup comparison using nonmetric data. *Dental Anthropology* 17(3):83–93.

Harrison, R.P.
2001 Hic Jacet. *Critical Inquiry* 2:393–407.
2003 *The dominion of the dead*. Chicago: University of Chicago Press.

Hasluck, M.
1954 *The unwritten law in Albania*. Cambridge: Cambridge University Press.

Hasluck, M.M., and G.M. Morant
1929 Measurements of Macedonian men. *Biometrika* 21:322–336.

Hatipoglu, G.H., H.N. Ozcan, U.M. Hatipoglu, and E. Yuksei
2008 Age, sex and body mass index in relation to calvarial diploe thickness and craniometric data on MRI. *Forensic Science International* 182(1):46–51.

Hausdorf, B.
1988 The *Xeromunda* species from the Greek mainland (Gastropoda, Hygromyidae). *Archiv für Molluskenkunde der Seckenbergisher Naturforschenden Gesellschaft* 119(4–6):107–132.

Head, B., G. H. Hill, G. MacDonald, and W. Wrote
1911 *Historia numorum: a manual of Greek numismatics*. Oxford: Clarendon Press.

Hedges, R.E.M.
2000 Appraisal of radiocarbon dating of kiore bones (Pacific rat *Rattus exulans*). *Journal of the Royal Society of New Zealand* 30:385–398.

Heidegger, M.
1971 Building dwelling thinking. In *Poetry, language, thought*. Translated by A. Hofstadter. New York: Harper and Row.

Heiden, J.
2000 Korinthische Keramik. In *Archaische Keramik aus Olympia*. E. Kunze-Götte, J. Heiden, and J. Burow, 167–197. *Olympische Forschungen XXVIII*. Berlin: de Gruyter.

Heilmeyer, W.-D.
1994 Frühe olympische Bronzefiguren: Die Wagenvotive. In E. Kunze, et al., *IX Bericht über die Ausgrabungen in Olympia*, 172–208. Berlin: de Gruyter.

Helbig, W.
1887 *Das homerische Epos aus den Denkmälern erläutert*. 2nd ed. Leipzig: B.G. Teubner.

Hellström, P. (editor)
1987 *Paradeisos: a late Neolithic settlement in Aegean Thrace*. Stockholm: Medelhavsmuseet.

Henderson, J.
1988 Glass production and Bronze Age Europe. *Antiquity* 62:435–451.

Hertz, R.
1907 [1960] *Death and the right hand*. New York: Free Press. (Originally published as *Contribution à une étude sur la représentation collective de la mort*. Translated by R. and C. Needham. Paris: L'Année sociologique, 1907.)

Heurtley, W.A.
1923–1925 Pottery from Macedonian mounds. *Annual of the British School at Athens* 26:30–37.
1925 Report of an excavation at the Toumba of Vardino, Macedonia. *Liverpool Annals of Anthropology and Archaeology* 12:15–36.
1926–1927 A prehistoric site in western Macedonia and the Dorian invasion. *Annual of the British School at Athens* 28:158–194.
1927 Early Iron Age pottery from Macedonia. *Antiquaries Journal* 7:44–59.
1929 Prehistoric Macedonia. *Antiquity* 3:318–323.
1939 *Prehistoric Macedonia: an archaeological reconnaissance of Greek Macedonia (west of the Struma) in the Neolithic, Bronze, and Early Iron Ages*. Cambridge: Cambridge University Press.

Heurtley, W.A., and R.W. Hutchinson
1925–1926 Report on excavations at the toumba and tables of Vardaróftsa, Macedonia, 1925, 1926. Part 1: the toumba. *Annual of the British School at Athens* 27:1–66.

Heurtley, W.A., and C.A.R. Radford
1927–1928 Two prehistoric sites in Chalcidice. *Annual of the British School at Athens* 29:117–186.
1929–1930 Report on excavations at the toumba of Saratse in Macedonia, 1929. *Annual of the British School at Athens* 30:113–150.

Heurtley, W.A., and T.C. Skeat
1930–1931 The tholos tombs of Marmariane. *Annual of the British School at Athens* 31:1–55.

Hewlett, B.S.
1991 Demography and childcare in preindustrialized societies. *Journal of Anthropological Research* 47 (1):1–37.

Higgins, R.
1980 *Greek and Roman jewellery*. 2nd ed. Berkeley and Los Angeles: University of California Press.

Higgs, E.S., C. Vita-Finzi, D.R. Harris, and A.E. Fagg
1967 The climate, environment and industries of Stone Age Greece: part III. *Proceedings of the Prehistoric Society* 33:1–29.

Hiller, S., and V. Nikolov (editors)
1997 *Karanovo Bd. I: Die Ausgrabungen im Südsektor 1984–1992*. Salzburg and Sofia: Verlag Ferdinand Berger and Söhne.

Hillson, S.
1996 *Dental anthropology*. Cambridge: Cambridge University Press.

Hirsch, E., and M. O'Hanlon
1995 *The anthropology of landscape: perspectives on place and space*. Oxford: Clarendon Press.

Hochstetter, A.
1982 Die mattbemalte Keramik in Nordgriechenland, ihre Herkunft und lokale Ausprägung. *Praehistorische Zeitschrift* 57:201–219.
1984 *Kastanas. Ausgrabungen in einem Siedlungshügel der Bronze- und Eisenzeit Makedoniens, 1975–1979: Die handgemachte Keramik, Schichten 19 bis 1*. Berlin: Verlang Volker Spiess.
1987 *Kastanas. Ausgrabungen in einem Siedlungshügel der Bronze- und Eisenzeit Makedoniens, 1975–1979: Die Kleinfunde*. Berlin: Wissenschaftsverlag Volker Spiess.

Hodder, I. (editor)
2000 *Towards reflexive method in archaeology: the example at Çatalhöyük by members of the Çatalhöyük teams* (British Institute of Archaeology at Ankara Monograph 28). Cambridge: McDonald Institute for Archaeological Research.

Hoffman, M.
1964 *The warp-weighted loom: studies in the history and technology of an ancient implement* (Studia Norvegica 14). Oslo: Universitetsforlaget.

Hogarth, D.G.
1908 *Excavations at Ephesos: the Archaic Artemisia*. London: British Museum.

Holden, C., and R. Mace
1999 Sexual dimorphism in stature and women's work: a phylogenetic cross-cultural analysis. *American Journal of Physical Anthropology* 110 (1):27–45.

Holland, H.
1815 *Travels in the Ionian Isles, Albania, Thessaly, Macedonia, &c. during the years 1812 and 1813*. London: Longman, Hurst, Rees, Orme and Brown.

Holste, F.
1953 *Die Bronzezeit in Süd- und Westdeutschland* (Handbuch der Urgeschichte Deutschlands, Bd. 1). Berlin: Walter de Gruyter.

Hood, S.
1960 Tholos tombs of the Aegean. *Antiquity* 34:166–176.

Hood, M.S.F., and J.N. Coldstream
1968 A Late Minoan tomb at Ayios Ioannis near Knossos. *Annual of the British School at Athens* 63:205–218.

Hood, S., G. Huxley, and N. Sandars
1958–1959 A Minoan cemetery on upper Gypsades (Knossos Survey 156). *Annual of the British School at Athens* 53–54:194–262.

Hoppa, R.D.
2002 Paleodemography: looking back and thinking ahead. In *Paleodemography: age distributions from skeletal samples*, edited by R.D. Hoppa and J.W. Vaupal, 9–28. Cambridge: Cambridge University Press.

Horden, P., and N. Purcell
2000 *The corrupting sea: a study of Mediterranean history*. Oxford: Blackwell.

Horejs, B.
2007a The phenomenon of mattpainted pottery in the northern Aegean: introduction, overview and theories. *http://www.aegeobalkanprehistory.net/article.php?id_art=8*.
2007b *Das prähistorische Olynth: Ausgrabungen in der Toumba Agios Mamas, 1994–1996. Die spätbronzezeitliche handgemachte Keramik der Schichten* 31 bis 1 (Prähistorische Archäologie in Südosteuropa, Bd. 21). Rahden/Westfalen: Verlag Marie Leidorf.

Hoti, A.
1981 Tumat e fshatit Kënetë. *Iliria* 11(1):211–217.
1982a Varreza tumulare e Bardhocit në rrethin e Kukësit. *Iliria* 12(1):15–48.
1982b Gërmime në Hamallaj (Durrës). *Iliria* 12(1): 255–257.
1986 Tumat V e VI të Kënetës. *Iliria* 16(2):41–70.
1993 Kërkime arkeologjike në fshatin Hamallaj (Rrethi i Durrësit). *Iliria* 23:123–131.

Houby-Nielsen, S.
2000 Child burials in ancient Athens. In *Children and material culture*, edited by J. Moore and E. Scott, 151–166. London: Routledge.

Huber, S.
2003 *Eretria XIV. L'Aire sacrificielle au nord du Sanctuaire d'Apollon Daphnéphoros: un ritual des époques géométriques et archaïque.* Montreux: École Suisse d'archéologie en Grèce.
Hubert, J.
2000 Introduction. In *Madness, disability, and social exclusion: the archaeology and anthropology of "difference,"* edited by J. Hubert, 1–8. New York: Routledge.
Hughes, D.D.
1991 *Human sacrifice in ancient Greece.* London and New York: Routledge.
2005 Sacrifice and the cult of the dead in ancient Argos. In *Greek sacrificial ritual, Olympian and Chthonian: proceedings of the sixth international seminar on ancient Greek cult, organized by the Department of Classical Archaeology and Ancient History, Göteborg University, 25–27 April 1997,* edited by R. Hägg and B. Alroth, 75–83. Stockholm: Paul Åströms Forlag.
Hundt, H.-J.
1959 Vorgeschichtliche Gewebe aus dem Hallstätter Salzberg. *Jahrbuch des Römisch-germanischen Zentralmuseums Mainz* 6:66–100.
1960 Vorgeschichtliche Gewebe aus dem Hallstätter Salzberg. *Jahrbuch des Römisch-germanischen Zentralmuseums Mainz* 7:126–150.
1961 Neunzehn Textilreste aus dem Dürrnberg in Hallein. *Jahrbuch des Römisch-germanischen Zentralmuseums Mainz* 8:7–25.
1962 Textilreste aus dem Hohmichele. In *Der Hohmichele, ein Fürstengrabhügel der späten Hallstattzeit bei der Heuneburg,* edited by G. Riek, 119–214. Berlin: W. de Gruyter.
1967 Vorgeschichtliche Gewebe aus dem Hallstätter Salzberg. *Jahrbuch des Römisch-germanischen Zentralmuseums Mainz* 14:38–67.
1969 Über vorgeschichtliche Seidenfunde. *Jahrbuch des Römisch-germanischen Zentralmuseums Mainz* 16:59–71.
1974 Die Textilreste aus den Gräbern vom Dürrnberg. In *Der Dürrnberg bei Hallein,* edited by F. Moosleitner, L. Pauli, and E. Penninger, 135–142. Munich: C.H. Beck.
Hunt, J.D.
1987 *William Kent: landscape garden designer. An assessment and catalogue of his designs.* London: Zwemmer.
Huntington, R., and P. Metcalf
1979 *Celebrations of death: the anthropology of mortuary ritual.* Cambridge: Cambridge University Press.
Iakovidis, S.
1969–1970 *Περατή. Το νεκροταφείον.* Vols. 1–3. Athens: Athens Archaeological Society.
1977 On the use of Mycenaean "buttons." *Annual of the British School at Athens* 72:113–119.
Immerwahr, S.A.
1971 *The Athenian Agora XIII: the Neolithic and Bronze Ages.* Princeton, New Jersey: American School of Classical Studies at Athens.
İnalcık, H.
1954 *Hicrî 835 tarihli Sûret-I defter-I sancak-I Arvanid.* Ankara: Türk Tarh Kurumu Basımevi.
Ingo, G.M., T. de Caro, C. Riccucci, E. Angelini, S. Grassini, S. Balbi, P. Bernardini, D. Salvi, L. Bousselmi, A. Çilingiroğlu, M. Gener, V.K. Gouda, O. al Jarrah, S. Khorsroff, Z. Mahdjoub, Z. al Saad, W. el-Saddik, and P. Vassiliou
2006 Large scale investigation of chemical composition, structure and corrosion mechanism of bronze archaeological artefacts from the Mediterranean basin. *Applied Physics A* 83:513–520.
Ingram, R.S.
2005 Faience and glass beads from the Late Bronze Age shipwreck at Uluburun. M.A. thesis, Texas A&M University, College Station.
Ioannidou, E.
2003 Taphonomy of animal bones: species, sex, age and breed variability of sheep, cattle and pig bone density. *Journal of Archaeological Science* 30 (3):355–365.
Irish, J., and L. Konigsberg
2007 The ancient inhabitants of Jebel Moya redux: measures of population affinity based on dental morphology. *International Journal of Osteoarchaeology* 7:138–156.
Islami, A.
2006 *Albania from the air,* Vol. 1. Tirana: Ministry of Tourism, Culture, Youth and Sports.
Islami, S., and H. Ceka
1965 Të dhëna të reja mbi lashtësinë ilire në territorin e Shqipërisë. *KISI (Première conférence des études des albanologiques)* 1:446–456.
Islami, S., H. Ceka, F. Prendi, and S. Anamali
1955 Zbulime të kulturës Ilire në luginën e Matit. *Buletin për Shkencat Shoqërore* 1:110–135.
Ito, G., S. Mitani, and J.H. Kim
1988 Effect of soft diets on craniofacial growth in mice. *Anatomischer Anzeiger* 165:151–166.
Jackson, J.B.
1984 The word itself. In J.B. Jackson, *Discovering the vernacular landscape.* New Haven, Connecticut, and London: Yale University Press.
Jacobsthal, P.
1956 *Greek pins and their connexions with Europe and Asia.* Oxford: Clarendon Press.
Jiang, Y.
2004 Ritual practice, status, and gender identity: Western Zhou tombs at Baoji. In *Gender and Chinese archaeology,* edited by K.M. Linduff and

Y. Sun, 17–36. Walnut Creek, California: AltaMira Press.

Jiao, T.
2001 Gender studies in Chinese Neolithic archaeology. In *Gender and the archaeology of death*, edited by B. Arnold and N.L. Wicker, 51–64. Walnut Creek, California: AltaMira Press.

Johanson, C.
2009 Visualizing history: modeling in the eternal city. *Visual Resources: An International Journal of Documentation* 25:403–418.

Johnson, J.K.
1989 The utility of production trajectory modeling as a framework for regional analysis. (Special issue: Alternative approaches to lithic analysis, edited by D.O. Henry and G.H. Odell.) *Archeological Papers of the American Anthropological Association* 1:119–138.

Johnston, R.
1998 Approaches to the perception of landscape: philosophy, theory, methodology. *Archaeological Dialogues* 5:54–68.

Jones-Bley, K.
2000 Shintashta burials and their western European counterparts. In *Kurgans, ritual sites, and settlements: Eurasian Bronze and Iron Age*, edited by J. Davis-Kimball, E.M. Murphy, L. Koryakova, and L.T. Yablonsky, 126–143. Oxford: Archaeopress.

Jørgensen, L.
1987 Family burial practices and inheritance systems: the development of an Iron Age society from 500 B.C. to A.D. 1000 on Bornholm, Denmark. *Acta Archaeologica* 58:17–53.

Jørgensen, L.B.
2003 Industries of early historic Europe and the Mediterranean: Europe. In *The Cambridge history of Western textiles*, Vol. I, edited by David Jenkins, 52–70. Cambridge: Cambridge University Press.

Jubani, B.
1969 Tipare të përbashkëta në ritin e varrimit tek ilirët e trevës së Shqipërisë. In *Ilirët dhe gjeneza e shqiptarëve*, 91–101. Tirana: Universiteti Shtetëror i Tiranës.
1971 Varreza tumulare e Çinamakut (Kukës). *Buletini Arkeologjik* 1:37–51.
1974 Të dhëna për kulturën tumulare të Shqipërisë verilindore. In *Kuvendi I i studimeve ilire*, 189–209. Tirana: Akademia e Shkencave.
1982 Tumat e Krumës (rrethi i Kukësit). *Iliria* 12(2): 147–195.
1983 Tumat ilire të Kënetës. *Iliria* 13(2):77–133.
1985 Tipare të kulturës ilire në trevën e Dardanisë. *Iliria* 15(2):211–222.
1992 Gërmime të tumat nr. 2 dhe 3 të Fushës së Shtojit. *Iliria* 22(1–2):23–58.
1995 Kultura e bronzit të hershëm në tumat e Shkrelit. *Iliria* 25(1–2):53–90.

Jung, R.
2006 *Χρονολογία comparata. Vergleichende Chronologie von Südgriechenland und Süditalien von ca. 1700/1600 bis 1000 v.u.Z* (Veröffentlichungen der Mykenischen Kommission 26). Vienna: Österreichische Akademie der Wissenschaften.

Kadare, I.
1970 *The general of the dead army*. London: W.H. Allen.
2003 *Broken April*. London: Vintage.

Kaniuth, K.
2006 *Metallobjeckte der Bronzezeit aus Nordbaktrien* (Archäologie in Iran und Turan, Bd. 16). Mainz: Philipp von Zabern.

Karageorghis, V.
1983 *Palaepaphos-Skales: an Iron Age cemetery in Cyprus* (*Ausgrabungen in Alt-Paphos auf Cypern*, Bd. 3). Konstanz: Universitätsverlag Konstanz GMBH.

Karaiskaj, G.
1976 Les fortifications illyriennes du premier Âge du Fer dans les environs de Korçë. *Iliria* 4:197–221.

Karamitrou-Mentessidi, G.
2008 *Aiani: a guide to the archaeological site and museum*. Athens: Ministry of Culture, 30th Ephorate of Prehistoric and Classical Antiquities.

Karkanas, P., M.K. Dabney, R.A.K. Smith, and J.C. Wright
2012 The geoarchaeology of Mycenaean chamber tombs. *Journal of Archaeological Science* 39: 2722–2732.

Karo, G.
1930–1933 *Die Schachtgräber von Mykenai*. Munich: F. Bruckmann.

Karydas, A.G.
2007 Application of a portable XRF spectrometer for the non-invasive analysis of museum metal artefacts. *Annali di Chimica* 97:419–432.

Katzenberg, M.A.
2000 Stable isotope analysis: a tool for studying past diet, demography, and life history. In *The biological anthropology of the human skeleton*, edited by M.A. Katzenberg and S.R. Saunders, 305–327. New York: John Wiley and Sons.

Kavvadias, P.
1912 Περί των εν Κεφαλληνία ανασκαφών. *Praktika tis Archaiologikis Hetairias* 1912:247–268.

Keene, S., and C. Orton
1985 Stability of treated archaeological iron: an assessment. *Studies in Conservation*, 30(3):136–142.

Kemp, B.
2000 Soil (including mud-brick architecture). In *Ancient Egyptian materials and technology*, edited by P.T. Nicholson and I. Shaw, 78–103. Cambridge: Cambridge University Press.

Kennedy, B.V.E.
1988 Variation in $\delta^{13}C$ values of post-medieval Europeans. Ph.D. dissertation, University of Calgary.
Kerr, J., P. Ratiu, and M. Sellberg
1996 Volume rendering of visible human data for an anatomical virtual environment. *Studies in Health Technology and Informatics* 29:352–370.
Kilian, K.
1970 *Früheisenzeitliche Funde aus der Südostnekropole von Sala Consilina (Provinz Salerno).* (Archäologische Forschungen in Lukanien III, Mitteilungen des Deutschen Archäologischen Instituts, Römische Abteilung, Ergänzungsheft 15). Heidelberg: F.H. Kerle.
1972 Zur mattbemalten Keramik der ausgehenden Bronzezeit und der Früheisenzeit aus Albanien. *Archäologisches Korrespondenzblatt* 2:115–123.
1973a Zum italischen und griechischen Fibelhandwerk das 8. und 7. Jahrhunderts. *Hamburger Beiträge zu Archäologie* 3(1):1–39.
1973b Zur eisenzeitliche Transhumanz in Nordgriechenland. *Archäologisches Korrespondenzblatt* 3: 431–435.
1974 Koha e Hershme e Hekurit në Shqipëri. In *Kuvendi I i Studimeve Ilire* 1:255–261.
1975a *Fibeln in Thessalien von der mykenischen bis zur archaischen Zeit* (*PBF* XIV:2). Munich: Beck.
1975b Trachtzubehör der Eisenzeit zwischen Ägäis und Adria. *Prähistorische Zeitschrift* 50:9–140.
1981 Ausgrabungen in Tiryns 1978, 1979. *Archäologischer Anzeiger* 1981:17–194.
1982 Ausgrabungen in Tiryns 1980: Mykenische Keramik und Kleinfunde. *Archäologischer Anzeiger* 1982:412–418.
1983 Weihungen aus Eisen und Eisenverarbeitung im Heiligtum zu Philia (Thessalien). In *The Greek renaissance of the eighth century B.C.: tradition and innovation. Proceedings of the Second international symposium at the Swedish Institute in Athens, 1–5 June, 1981*, edited by R. Hägg, 131–147. Stockholm: Swedish Institute at Athens.
Kilian-Dirlmeier, I.
1972 *Die hallstattzeitlichen Gürtelbleche und Bleckgürtel Mitteleuropas* (*PBF* XII:1). Munich: Beck.
1979 *Anhänger in Griechenland von der mykenischen bis zur spätgeometrischen Zeit (Griechisches Festland, Ionische Inseln, dazu Albanien und Jugoslawisch Mazedonien)* (*PBF* XI:3). Munich: Beck.
1984a *Nadeln der frühhelladischen bis archaischen Zeit von der Peloponnes* (*PBF* XIII:8). Munich: Beck.
1984b Gjilpërat e kohës së hekurit në Shqipëri (Die Nadeln der Eisenzeit in Albanien). *Iliria* 14(1): 69–109.
1984c Der dorische Peplos: Ein archäologisches Zeugnis der Dorischen Wanderung? *Archäologisches Korrespondenzblatt* 14:281–291.
1997 *Das mittelbronzezeitliche Schachtgrab von Ägina.* Mainz: Philipp von Zabern.
2002 *Kleinfunde aus dem Athena Itonia-Heiligtum bei Philia (Thessalien).* Mainz: Römisch-Germanischen Zentralmuseum.
Klingmark, E.H.
2003 *Gamla Uppsala* (Svenska Kulturminnen 59). Stockholm: Riksantikvarieämbetet.
Klotzsch, C.
1911 *Epirotische Geschichte bis zum Jahre 280 v. Chr.* Berlin: Weidmannsche Buchhandlung.
Knapp, A.B.
1999 Ideational and industrial landscape on prehistoric Cyprus. In *Archaeologies of landscape: contemporary perspectives*, edited by W. Ashmore and A.B. Knapp, 229–252. Malden, Massachusetts: Blackwell.
Knudson, K.J., and C.M. Stojanowski
2008 New directions in bioarchaeology: recent contributions to the study of human social identities. *Journal of Archaeological Research* 16:397–432.
Koçi, J.
1991 Fortifikimet parahistorike në bregdetin Kaon. *Iliria* 21(1–2):39–61.
Koka, A.
1985 Kultura e varrezës tumulare të Shtojit. *Iliria* 15 (2):241–250.
1990 Tuma 6 e Shtojit. *Iliria* 20(1):27–73.
2012 *Kultura Ilire e tumave të Shtojit Shkodër*. Tirana: Shtëpia Botuese Botart.
Konigsberg, L.W., and J.E. Buikstra
1995 Regional approaches to the investigation of past human biocultural structure. In *Regional approaches to mortuary analysis*, edited by L.A. Beck, 191–216. New York: Plenum Press.
Konsolaki-Yiannopoulou, E.
2010 The Middle Helladic Establishment at Megali Magoula, Galatas (Troizinia). In *Mesohelladika: la Grèce continental au Bronze Moyen. Actes du Colloque international organisé par l'École Française d'Athènes, en collaboration avec l'American School of Classical Studies at Athens et le Netherlands Institute in Athens, Athènes, 8–12 mars 2006 (Bulletin de Correspondance Hellénique*, Supplément 52), edited by A. Philippa-Touchais, G. Touchais, S. Voutsaki, and J. Wright, 67–76. Paris: de Boccard.
Korkuti, M.
1969 Qeramika e pikturuar e kohës së vonë të Bronzit dhe e kohës së Hershme të Hekurit dhe karakteri Ilir i bartësve të saj. *Studime Historike* 3:159–171.
1970 Marrëdhëniet kulturore iliro-egjeane në epokën e Bronzit të vonë dhe mbijetesa e disa objekteve të tipit miken në epokën e Hekurit. *Studia Albanica* 2:43–59.

1971 Vendbanimi prehistorik i Trenit. *Iliria* 1:31–64.
1974 Stacioni i neolitit të mesëm në Cakran të Fierit. *Iliria* 3:45–101.
1977–1978 Vendbanimi prehistorik i Kolshit. *Iliria* 7–8: 239–248.
1981 Tuma e Patosit. *Iliria* 11(1):7–55.
1982a Vashtëmia: një vendbanimi i neolitit të hershëm. *Iliria* 12(2):91–146.
1982b Die Siedlungen der späten Bronze- und der frühen Eisenzeit in Südwestalbanien. In *Südosteuropa zwischen 1600 und 100 v. Chr.*, edited by B. Hänsel, 234–253. Berlin: Moreland Editions.
1983a Vendbanimi neolitik i Kolshit. *Iliria* 13(2): 11–75.
1983b Gjetje të reja të paleolitit të mesëm nga stacioni i Xarës (Sarandë). *Iliria* 13(1):39–49.
1984 Piktura Shkëmbore e Lepenicës. *Iliria* 14(1):5– 14.
1985 Marrëdhëniet midis dy brigjeve të Adriatikut në epokën e bronzit dhe të hekurit. *Iliria* 15(1):93–109.
1995a *Neolithikum und Chalkolithikum in Albanien.* Mainz: Philipp von Zabern.
1995b Paleoliti në Shqipëri. *Iliria* 25(1–2):5–15.
1998 Fifty years of Albanian archaeology. *Iliria* 28 (1–2):39–47.

Korkuti, M., and Zh. Andrea
1974 La station du néolithique moyen à Cakran de Fieri. *Iliria* 3:49–107.

Korkuti, M., and A. Bunguri
1996 Kultura e epokës së bronzit në vendbanimin e Cetushit. *Iliria* 26(1–2):5–19.

Korkuti, M., and K. Petruso
1993 Archaeology in Albania. *American Journal of Archaeology* 97(4):703–743.

Korkuti, M., K.M. Petruso, L. Bejko, B.B. Ellwood, J.M. Hansen, F.B. Harrold, N. Russell, and S. Bottema
1996 Shpella e Konispolit (raport paraprak për gërmimet e viteve 1992–1994). *Iliria* 26(1–2):183–224.

Koskeridou E., E. Vardala-Theodorou, E. Tsiolakis, I. Panayides, G. Theodorou, and A. Antonarakou
2005 The marine faunistic assemblages of the Pliocene of Paleometocho and Athienou (Cyprus, east Mediterranean): some palaeoecological aspects. *Abstracts of the 12th international congress, Regional Committee on Mediterranean Neogene Stratigraphy*, 134–136. Vienna: Wechselnde Verlagsorte.

Koukouli-Chrysanthaki, Ch.
1992 *Πρωτοϊστορική Θάσος. Τα νεκροταφεία του οικισμού Καστρί.* Athens: Greek Ministry of Culture.

Kourouniotis, K.
1910 Το εν Βάσσαις αρχαιότερον ιερόν του Απόλλωνος. *Archaiologike Ephemeris* 1910:273–332.
1913 Goldschmuck aus Eretria. *Mitteilungen des Deutschen Archäologischen Instituts, Athenische Abteilung* 38:289–328.

Kovács, T.
1977 *The Bronze Age in Hungary.* Budapest: Corvina Press.

Kraiker, W., and K. Kübler
1939 *Kerameikos I: Die Nekropolen des 12. bis 10. Jahrhunderts.* Berlin: Walter de Gruyter.

Kritsas, Ch.B.
1972 Αρχαιότητες και μνημεία Αργολιδοκορινθίας: Ἄργος. *Archaiologikon Deltion* 27, Chr. B'1: 192–219.

Kroll, H.J.
1983 *Kastanas: Ausgrabungen in einem Siedlungshügel der Bronze- und Eisenzeit Makedoniens, 1975–1979. Die Pflanzenfunde.* Berlin: Volker Spiess.

Kron, U.
1998 Sickles in Greek sanctuaries: votives and cultic instruments. In *Ancient Greek cult practice from the archaeological evidence: proceedings of the fourth international seminar on ancient Greek cults, organized by the Swedish Institute at Athens, 22–24 October 1993*, edited by R. Hägg, 187–215. Stockholm: Swedish Institute at Athens.

Krzyszkowska, O.H.
1996 Asine Chamber Tomb I:2: the ivories. In *Asine III: Supplementary Studies on the Swedish Excavations, 1922–1930*, edited by R. Hägg, G.C. Nordquist, and B. Wells, 81–90. Stockholm: Swedish Institute at Athens.

Kübler, K.
1943 *Kerameikos IV: Neufunde aus der Nekropole des 11. und 10. Jahrhunderts.* Berlin: Walter de Gruyter.
1954 *Kerameikos V,1: Die Nekropole des 10. bis 8. Jahrhunderts.* Berlin: Walter de Gruyter.

Kuo, C.H.
1983 The influence of the tooth to denture base discrepancy on cervical dental caries. *Journal of Dental Health* 32:439–457.

Kurti, D.
1976 Nouveaux elements sur la civilization illyrienne des tumuli de Mati. *Iliria* 4:237–248.
1977–1978 Tumat e Burrelit (Tuma I). *Iliria* 7–8: 157–187.
1983 Tumat ilire të Burrelit, Tuma III. *Iliria* 13(1): 85–108.
1987 Tumat IV, V, VI, VII të Burrelit. *Iliria* 17(1):85–115.

Kurti, R.
2006 Appearance and modes of dress: theory and practice in the case of Barç tumulus. In *New directions in Albanian archaeology: studies presented to Muzafer Korkuti*, edited by L. Bejko and R. Hodges, 65–74. Tirana: International Centre for Albanian Archaeology,.

Kurtz, D.C., and J. Boardman
1971 *Greek burial customs.* Ithaca, New York: Cornell University Press.

Kyparissi, N.
1919 "Κεφαλληνιακά," *Archaiologikon Deltion* 5:83–122.

Lafe, O.
2003 The earliest urbanized settlement in the hinterland of Apollonia (Albania): 7th–mid 5th century BC. M.A. thesis, University of Cincinnati.
2006 An overview of extensive and intensive survey in Albania (1945–2005). In *New directions in Albanian archaeology: studies presented to Muzafer Korkuti*, edited by L. Bejko and R. Hodges, 331–338. Tirana: International Centre for Albanian Archaeology,.

Lahi, B., and M. Fiedler
2010 Ausgrabungen im Zentrum von Apollonia (Albanien). Vorläufige Ergebnisse zu der sog. Amphorenmauer und ihrer Umgebung. *Mitteilungen des Deutschen Archäologischen Instituts, Römische Abteilung* 116:213–255.

Lamb, W.
1936 *Excavations at Thermi in Lesbos*. Cambridge: Cambridge University Press.

Lamboley, J.-L.
1987 Les fouilles actuelles. In *Apollonie d'Illyrie. 1. Atlas archéologique et historique*, Collection de l'École Française de Rome 391, edited by V. Dimo, P. Lenhardt, and F. Quantin, 163–187. Recherches archéologiques franco-albanaises 1. Athens: École Française d'Athènes; Paris: Ministère des affaires étrangères; Rome: École Française de Rome.

Lamboley, J.-L., and B. Vrekaj
2003–2004 Les travaux de la mission épigraphique et archéologique Française. Les fouilles Franco-Albanaises d'Apollonia: 1994–2004. *Iliria* 31(1–2): 163–187.

Langdon, S.
2008 *Art and identity in Dark Age Greece, 1100–700 B.C.E.* Cambridge: Cambridge University Press.

Larsen, C.S.
1997 *Bioarchaeology: interpreting behaviour from the human skeleton*. Cambridge: Cambridge University Press.

Larsen, C.S., R. Shavit, and M.C. Griffin
1991 Dental caries evidence for dietary change: an archaeological context. In *Advances in Dental Anthropology*, edited by M.A. Kelley and C.S. Larsen, 179–202. New York: Wiley-Liss.

Laux, F.
1973 *Die Fibeln in Niedersachsen* (*PBF* XIV:1). Munich: Beck.

Leggett, B.J.
1992 *Early Stevens: the Nietzschean intertext*. Durham, North Carolina: Duke University Press.

Lehoux, D.
2007 Drugs and the Delphic oracle. *Classical World* 101(1):41–56.

Lemmer, G.M.
1972 The cleaning and protective coating of ferrous metals. *Bulletin of the American Group. International Institute for Conservation of Historic and Artistic Works* 12(2):97–108.

Lepore, E.
1962 *Richerche sull'antico Epiro: le origine storiche e gli interesse greci*. Naples: Libreria scientifica editrice.

Lera, P.
1987a Vendbanimi i neolitit të vonë në Barç (Faza Barç II). *Iliria* 17(1):25–69.
1987b Gërmimet arkeologjike të vitit 1987—Drenovë (Korçë). *Iliria* 17(2):242–243.
1988 Vendbanimi i neolitit të vonë në Dërsnik (Résumé: L'Habitat du Neolithique Récent à Dërsnik). *Iliria* 2:21–68.
1992 Vendbanimi ilir në Gradishtën e Symizës. *Iliria* 22(1–2):177–208.

Lera, P., F. Prendi, and G. Touchais
1996 Sovjan (Albanie). *Bulletin de Correspondance Hellénique* 120:995–1026.

Lera, P., C. Oberweiler, and G. Touchais
2011 Le passage du Bronze Récent au Fer Ancien sur le site de Sovjan (Bassin de Korçë, Albanie): nouvelles données chronologiques. In *L'Illyrie méridionale et l'Épire dans l'Antiquité*, V. *Actes du Ve colloque international de Grenoble (8–11 octobre 2008)*, edited by J.-L. Lamboley and M.P. Castiglioni, 41–52. Paris: de Boccard.

Leshnik, L.S.
1972 Pastoral nomadism in the archaeology of India and Pakistan. *World Archaeology* 4(2): 150–166.

Lévêque, P.
1957 *Pyrrhos*. Paris: de Boccard.

Levi, D.
1931 Arkades: una città cretesi all'alba della civiltà ellenica. *Annuario della Scuola Archeologica di Atene* 10–12:1–723.

Lewartowski, K.
2000 *Late Helladic simple graves: a study of Mycenaean burial customs* (BAR International Series 878). Oxford: Archaeopress.

Lewthwaite, J.
1981 Plain tails from the hills: transhumance in Mediterranean archaeology. In *Economic archaeology: towards an integration of ecological and social approaches* (BAR International Series 96), edited by A. Sheridan and G. Bailey, 57–66. Oxford: British Archaeological Reports.

Lightfoot, K.G., A. Martinez, and A.M. Schiff
1998 Daily practice and material culture in pluralistic social settings: an archaeological study of culture change and persistence from Fort Ross, California. *American Antiquity* 63(2):199–222.

Lima, S.
2013 Butrint and the Pavllas River Valley in the Late Bronze and Early Iron Age. In *Butrint 4: The archaeology and histories of an Ionian town*, edited by I.L. Hansen, R. Hodges, and S. Leppard, 31–46. Butrint Archaeological Monographs. Oxford: Oxbow Books for the Butrint Foundation.

Lincoln, B.
1991 *Death, war, and sacrifice: studies in ideology and practice*. Chicago: University of Chicago Press.

Liston, M.A.
2007 Secondary cremation burials at Kavousi Vronda, Crete: symbolic representation in mortuary practice. *Hesperia* 76:57–71.

Liston, M.A., and J.K. Papadopoulos
2004 The "rich Athenian lady" was pregnant: the anthropology of a Geometric tomb reconsidered. *Hesperia* 73:7–38.

Little, L.M., and J.K. Papadopoulos
1998 A social outcast in Early Iron Age Athens. *Hesperia* 67:375–404.

Livingston, R.A.
1991 Influence of the environment on the patina of the statue of liberty. *Environmental Science and Technology* 25(8):1400–1408.

Lofland, J.
1969 *Deviance and identity*. Englewood Cliffs, New Jersey: Prentice-Hall.

Lohmann, H.
2007 Forschungen und Ausgrabungen in der Mykale 2001–2006. *Istanbuler Mitteilungen* 57:59–178.

Lomas, K.
2000 Cities, states and ethnic identity in southeast Italy. In *The emergence of state identities in Italy in the first millennium BC*, edited by E.B. Herring and K. Lomas, 79–90. London: Accordia Research Institute, University of London.

Longin, R.
1971 New method for collagen extraction for radiocarbon dating. *Nature* 230:241–242.

Lo Schiavo, F.
1970 *Il gruppo Liburnico-Japodico per una definizione nell'ambito della protostoria balcanica*. Rome: Accademia Nazionale dei Lincei.
1983–1984 Fibule da Francavilla nel Museo Civico di Cosenza dono del Dott. Agostino de Santis. *Atti e Memori Magna Grecia* 24–25:135–138.

Loud, G., and C.B. Altman
1938 *Khorsabad II: the citadel and the town*. Chicago: Oriental Institute, University of Chicago.

Lowenthal, D.
1979 Age and artifact: dilemmas of appreciation. In *The interpretations of ordinary landscapes: geographical essays*, edited by D.W. Meinig, 103–128. Oxford: Oxford University Press.

Lucas, A., and J.R. Harris
1962 [1934] *Ancient Egyptian materials and industries*. 4th rev. ed. London: E. Arnold.

Luedtke, B.E.
1992 *An archaeologist's guide to chert and flint*. Los Angeles: Institute of Archaeology, University of California.

Luppino, S.
1982 Il versante nord-occidentale del fiume Savuto. In *Temesa e il suo territorio. Atti del colloguio di Perugia e Trevi (30–31 maggio 1981)*, edited by G. Maddoli, 75–78. Taranto: Istituto per la storia e l'archeologia della Magna Grecia.

Lyons, C.L.
2003 Archaeology, conservation, and the ethics of sustainability. In *Theory and practice in Mediterranean archaeology: Old World and New World perspectives*, edited by J.K. Papadopoulos and R. Leventhal, 299–308. Los Angeles: Cotsen Institute of Archaeology.

Macnamara, E.
2006 Pithecusan gleanings II: other bronze objects. In *Across frontiers: Etruscans, Greeks, Phoenicians and Cypriots. Studies in honour of David Ridgway and Francescsa Romana Serra Ridgway*, edited by E. Herring, I. Lemos, F. Lo Schiavo, L. Vagnetti, R. Whitehouse, and J. Wilkes, 267–279. London: Accordia Research Institute, University of London.

Mahalanobis, P.C.
1936 On the generalized distance in statistics. *Proceedings of the National Institute of Science, India* 2(1):49–55.

Malkin, I.
1998 *The returns of Odysseus: colonization and ethnicity*. Berkeley and Los Angeles: University of California Press.
2011 *A small Greek world: networks in the ancient Mediterranean*. Oxford: Oxford University Press.

Mano, A.
1971 Nekropoli i Apollonisë—Tuma 1 (Gërmime të viteve 1958–1959). *Iliria* 1:103–207.
1974 Nekropoli i dystë i Apolonisë (Gërmime të vitit 1962). *Iliria* 3:153–256.
1977–1978 Disa mendime rreth nekropolit të Apollonisë. *Iliria* 7–8:59–70.
2003–2004 Apollonia dans les annales de l'histoire. *Iliria* 31(1–2):157–162.

Marinatos, S.N.
1932 Αι ανασκαφαί Goekoop εν Κεφαλληνία. *Archaiologike Ephemeris* 1932:1–47.

Mariotti, A.
1983 Atmospheric nitrogen is a reliable standard for natural ^{15}N abundance measurements. *Nature* 303:685–687.

Markgraf, F.
1932 *Pflanzengeographie von Albanien, ihre Bedeutung für Vegetation und Flora der Mittelmeerländer*. Stuttgart: Schweizerbart.

Marston, J.M.
2009 Modeling wood acquisition strategies from archaeological charcoal remains. *Journal of Archaeological Science* 36(10):2192–2200.

Martin, S.
2006 Archaeology at the crossroads: the future of Albania's past. In *New directions in Albanian archaeology: studies presented to Muzafer Korkuti*, edited by L. Bejko and R. Hodges, 367–378. Tirana: International Centre for Albanian Archaeology.

Marston, J.M., J. d'Alpoim Guedes, and C. Warinner (editors)
Forthcoming *Method and theory in paleoethnobotany*. Boulder: University Press of Colorado.

Martin-McAuliffe, S.
Forthcoming Defining landscape: the prehistoric tumulus at Lofkënd, Albania. In *Tumulus as sema: space, politics, culture and religion in the first millennium BC* (Topoi: Berlin Studies of the Ancient World), edited by O. Henry and U. Kelp . Berlin: Walter de Gruyter.

Mason, P.
1996 *The Early Iron Age of Slovenia* (BAR International Series 643). Oxford: British Archaeological Reports.

Matthews, R.
1996 Surface scraping and planning. In *On the surface: Çatalhöyük 1993–1995* (British Institute of Archaeology at Ankara Monograph 22), edited by I. Hodder, 77–99. Cambridge: MacDonald Institute of Archaeology.

Maxwell-Hyslop, K.R.
1956 Notes on some distinctive bronzes from Populonia, Etruria. *Proceedings of the Prehistoric Society* 22:126–142.

McDonald, W.A.
1972 Excavations at Nichoria in Messenia: 1969–71. *Hesperia* 41:218–273.

McDonald, W.A., S. Aschenbrenner, H. Blitzer, S.R.B. Cooke, W.D.E. Coulson, O.T.P.K. Dickinson, W.P. Donovan, F.V. Grady, R. Hope Simpson, R.J. Howell, H. Hughes-Brock, G. Rapp, Jr., J. Rosser, C.T. Shay, J. Shay, R.E. Sloan, W.D. Wade, N. Wilkie, and D.L. Wolberg
1975 Excavations at Nichoria in Messenia: 1972–1973. *Hesperia* 44:69–141.

McDonald, W.A., W.D.E. Coulson, and J. Rosser (editors)
1983 *Excavations at Nichoria in southwest Greece*, Vol. 3. *Dark Age and Byzantine occupation*. Minneapolis: University of Minnesota Press.

McDonald, W.A., and R.H. Simpson
1969 Further explorations in southwestern Peloponnese: 1964–1968. *American Journal of Archaeology* 73(2):123–177.

McGeorge, P.J.P.
1992 Part II. The burials. In *Late Minoan III burials at Khania: the tombs, finds and deceased in Odos Palama* (Studies in Mediterranean Archaeology, Vol. 93), edited by B.P. Hallager and P.J.P. McGeorge, 29–44. Göteborg: Paul Åströms Verlag.

McGovern, P.
2003 *Ancient wine: the search for the origins of viticulture*. Princeton, New Jersey: Princeton University Press.

Metallinou, G. (editor)
2008 *Atlas historiko-arkeologjik i zones kufitare Greko-Shqiptare*. Athens: Ministry of Culture.

Metcalf, P., and R. Huntington
1991 *Celebrations of death: the anthropology of mortuary ritual*. 2nd ed. Cambridge: Cambridge University Press.

Michael, C., N. Zacharias, D. Dimotikali, E. Vardala-Theodorou, and Y. Bassiakos
1998 Συνδυασμένες απόλυτες χρονολογήσεις TL και EPR απολιθωμένων κελυφών μαλακίων. In *Πρακτικά XIV Πανελληνίου συνεδρίου φυσικής στερεάς κατάστασης, Ιωάννινα*, 363–366. Ioannina: Hellenic Society for Solid State Physics.

Mielke, J.H., L.W. Konigsberg, and J.H. Relethford
2006 *Human biological variation*. Oxford and New York: Oxford University Press.

Mijksenaar, P.
1997 *Visual function: an introduction to information design*. New York: Princeton Architectural Press.

Mikulčič, I.
1966 *Pelagonija u Svetlosti arheoloških Nalaza*. Skopje: Arheološko Društvo Jugoslavije Arheološki Muzej.

Miller, N.F., M.A. Zeder, and S.R. Arter
2009 From food and fuel to farms and flocks: the integration of plant and animal remains in the study of the agropastoral economy at Gordion, Turkey. *Current Anthropology* 50:915–924.

Milojcić, V.
1961 *Samos I: Die prähistorische Siedlung unter dem Heraion, Grabung 1953 und 1955*. Bonn: Rudolf Habelt.

Mockers, O., M. Aubry, and B. Mafart
2004 Dental crowding in a prehistoric population. *European Journal of Orthodontics* 26(2):151–156.

Montelius, O.
1895 *La civilisation primitive en Italie, I*. Stockholm and Berlin: A. Asher and Co.
1904 *La civilization primitive en Italie, II*, Planche 2^1 et 2^2. Stockholm and Berlin: A. Asher and Co.
1910 *La civilization primitive en Italie, II, Texte*. Stockholm and Berlin: Behrend and Co.

Mooder, K.P., T.G. Schurr, F.J. Bamforth, V.I. Bazaliiski, and N.A. Savel'ev
2006 Population affinities of Neolithic Siberians: a snapshot from prehistoric Lake Baikal. *American Journal of Physical Anthropology* 129(3):349–361.

Mooder, K.P., A.W. Weber, F.J. Bamforth, A.R. Lieverse, T.G. Schurr, V.I. Bazaliiski, and N.A. Savel'ev
2005 Matrilineal affinities and prehistoric Siberian mortuary practices: a case study from Neolithic Lake Baikal. *Journal of Archaeological Science* 32 (4):619–634.

Moorey, P.R.S.
1994 *Ancient Mesopotamian materials and industries: the archaeological evidence.* Oxford: Clarendon Press.
1999 *Ancient Mesopotamian materials and industries: the archaeological evidence.* Winona Lake, Indiana: Eisenbrauns.

Morricone, L.
1965–1966 Eleona e Langada: sepolcreti della tarda Età del Bronzo a Coo. *Annuario della Scuola Archeologica di Atene* 43–44 (n.s. 27–28):5–311.

Morris, I.
1987 *Burial and ancient society: the rise of the Greek city-state.* Cambridge: Cambridge University Press.
1989 Attitudes toward death in Archaic Greece. *Classical Antiquity* 8(2):296–320.
1992 *Death-ritual and social structure in Classical antiquity.* Cambridge: Cambridge University Press.
1997 Homer and the Iron Age. In *A new companion to Homer*, edited by I. Morris and B. Powell, 535–559. Leiden: Brill.

Morris, S.P.
2006 *Illyrica pix*: the exploitation of bitumen in ancient Albania. In *New directions in Albanian archaeology: studies presented to Muzafer Korkuti*, edited by L. Bejko and R. Hodges, 94–106. Tirana: International Centre for Albanian Archaeology.
2008 Wine and water in the Bronze Age: fermenting, mixing, and serving vessels. In *DAIS: the Aegean feast. Proceedings of the 12th International Aegean Conference, University of Melbourne, Centre for Classics and Archaeology, March 2008* (Aegaeum 29), edited by L. Hitchcock, R. Laffineur, and J. Crowley, 113–123. Liège: Université de Liège.
2009–2010 Prehistoric Torone: a Bronze Age emporion in the northern Aegean. Preliminary report on the Lekythos excavations 1986 and 1988–1990. *Mediterranean Archaeology* 22–23:1–67.

Morris, S.P., and J.K. Papadopoulos
2007 Excavations at the prehistoric burial tumulus of Lofkënd in Albania. *Backdirt: Annual Review* 48–49.

Mossey, P.A.
1999a The heritability of malocclusion: Part 1—genetics, principles and terminology. *British Journal of Orthodontics* 26:103–113.
1999b The heritability of malocclusion: Part 2—The influence of genetics in malocclusion. *British Journal of Orthodontics* 26:195–203.

Mountjoy, P.A.
1993 *Mycenaean pottery: an introduction.* Oxford: Oxford University Press.

Muçaj, S.
1979–1980 Vendbanime të antikitetit të vonë e Mallakastrës. *Iliria* 9–10:279–299.
1986 Dy monumente të antikitetit të vonë në Bylis. *Iliria* 16(1):321–327.
1987 Bazilika A e Bylisit. *Iliria* 17(1):167–202.
1990 Sistemi fortifikues i qytetit të Bylisit në antikitetin e vonë. *Iliria* 20(1):169–200.

Muçaj, S., and E. Hobdari
2004 Manastiri i Shën Mërisë, Ballsh (Glavenicë) gërmime të vitit 2003. *Candavia* 1:189–210.
2005 Teqeja e Melanit. *Candavia* 2:29–92.

Müller-Karpe, H.
1959 *Beiträge zur Chronologie der Urnenfelderzeit nördlich und südlich der Alpen* (Römisch-Germanische Forschungen, Bd. 22). Berlin: Walter de Gruyter.
1962 Die Metallbeigaben der früheisenzeitlichen Kerameikos-Gräber. *Jahrbuch des Deutschen archäologischen Instituts* 77:59–129.

Mulville, J., and A.K. Outram (editors)
2005 *The zooarchaeology of fats, oils, milk and dairying.* Oxford: Oxbow Books.

Munsell Color
2000 *Munsell soil color charts.* Rev. ed. New Windsor, New York: Munsell Color.

Muros, V., V. Parry, and A. Whyte
2002 Conservation laboratory research projects. *The Oriental Institute News and Notes* 175 (Fall): 1–8.

Muros, V., J. Lehner, and A. Bardho
2009 The sampling of archaeological metals for lead isotope analysis using EDTA—a "minimally destructive" alternative. Poster presented at the annual meeting of the American Institute for Conservation of Historic and Artistic Works, Los Angeles. *http://jwlehner.bol.ucla.edu/EDTA-icpms-poster.pdf.*

Muros, V., A. Lewis, and A. Bardho
2006 Lofkënd 2006: final conservation report, June 19–July 29, 2006. Unpublished report. Lofkënd Archaeological Project.

Murray, L.
2006 Genetic analysis of the great circle from the tumulus at Kamenica, Albania. M.S. thesis, Michigan State University, East Lansing.

Musgrave, J.H.
2005 Appendix A: an anthropological assessment of the inhumations and cremations from the Early Iron Age cemetery at Torone. In J.K. Papadopoulos, *The Early Iron Age cemetery at Torone* (Monumenta Archaeologica 24), 243–315. Los Angeles: Cotsen Institute of Archaeology.

Myers, J.W.
1978 Balloon survey field season, 1977. *Journal of Field Archaeology* 5:145–159.

Myers, J.W., and E.E. Myers
1980 The art of flying: balloon archaeology. *Archaeology* 33:33–40.
1990 Low-altitude aerial photography in Crete. *Expedition* 32(3):31–33.

Myers, J.W., E.E. Myers, and G. Cadogan (editors)
1992 *The aerial atlas of ancient Crete.* Berkeley and Los Angeles: University of California Press.

Mylona, D.
2003 Archaeological fish remains in Greece: general trends of the research and a gazetteer of sites. In *Zooarchaeology in Greece: recent advances*, edited by E. Kotjabopoulou, Y. Hamilakis, P. Halstead, C. Gamble, and P. Elefanti, 193–200. London: British School at Athens.

Mylonas, G.E.
1948 Homeric and Mycenaean burial customs. *American Journal of Archaeology* 52:56–81.
1975 *Το δυτικόν νεκροταφείον της Ελευσίνος*, I–III. Athens: Athens Archaeological Society.

Myrto, H.
1998 *Albania archeologica: bibliografia sistematica dei Centri Antichi* I. Bari: Edipuglia.

Nelson, S.M.
2004 *Gender in archaeology: analyzing power and prestige.* 2nd ed. Walnut Creek, California: AltaMira Press.

Nemeskéri, J., and A. Dhima
1988 Essai de reconstitution paléobiologique de la population Apolloniate des Ier–IIIe siècles de n. ère. *Iliria* 18(1):119–155.

Nenci, G.
1953 *Pirro: aspirazioni egemoniche ed equilibro Mediterraneo.* Turin: Università di Torino.

Netterstrøm, J.B., and B. Poulsen (editors)
2007 *Feud in medieval and early modern Europe.* Aarhus: Aarhus University Press.

Newton, M., K.A. Wardle, and P.I. Kuniholm
2003 Dendrochronology and radiocarbon determinations from Assiros and the beginning of the Greek Iron Age. *Το Αρχαιολογικό Έργο στη Μακεδονία και Θράκη* 17:173–190.

Nicholson, P., and I. Shaw
2000 *Ancient Egyptian materials and technology.* Cambridge: Cambridge University Press.

Nikita, K., and J. Henderson
2006 Glass analyses from Mycenaean Thebes and Elateia: compositional evidence for a Mycenaean glass industry. *Journal of Glass Studies* 48: 71–120.

Nikitović, L., M. Stojić, and R. Vasić
2002 *Mojsinje: a Bronze and Iron Age mound necropolis.* Čačak: National Museum.

Nilsson, M.P.
1909 *Studien zur Geschichte des alten Epeiros.* Lund: H. Ohlssons Buchdruckerei.

Nobis, G.
1993 Archäologische Untersuchungen von Tierresten aus dem "Palast des Nestors" bei Pylos in Messenien/SW-Peloponnes. *Zeitschrift für Assyriologie und vorderasiatische Archäologie* 27:151–173.

Nopsca, F.B.
1925 *Albanien: Bauten, Trachten und Geräte Nordalbaniens.* Berlin: Walter de Gruyter.

Nord, A.G., E. Mattsson, and K. Tronner
2005 Factors influencing the long-term corrosion of bronze artefacts in soil. *Protection of Metals* 41 (4):309–316.

Nordquist, G.C.
1987 *A Middle Helladic village: Asine in the Argolid* (Boreas, Uppsala Studies in Ancient Mediterranean and Near Eastern Civilizations 16). Uppsala: Universitatis Upsaliensis.

Novotná, M.
2001 *Die Fibeln in der Slowakei* (*PBF* XIV:11). Munich: Beck.

Nuttonson, M.Y.
1947 *Ecological plant geography of Albania, its agricultural crops and some North American climatic analogues.* Washington, D.C.: American Institute of Crop Ecology.

Oakley, J.H.
2003 Death and the child. In *Coming of age in ancient Greece: images of childhood from the classical past*, edited by J. Neils and J.H. Oakley, 164–194. New Haven, Connecticut: Yale University Press.

Oakley, J.H., and R.H. Sinos
1993 *The wedding in ancient Athens.* Madison: University of Wisconsin Press.

Oates, D.
1990 Innovations in mud-brick: decorative and structural techniques in ancient Mesopotamia. *World Archaeology* 21:388–406.

Ochsenschlager, E.
1992 Ethnographic evidence for wood, boats, bitumen, reeds in southern Iraq: ethnoarchaeology at al-Hiba. *Bulletin on Sumerian Agriculture* 6: 47–78.
2004 *Iraq's marsh Arabs in the garden of Eden.* Philadelphia: University of Pennsylvania Museum of Anthropology and Archaeology.

Odegaard, N., S. Carroll, and W.S. Zimmt
2005 *Material characterization tests for objects of art and archaeology.* 2nd ed. London: Archetype Books.

Oeftiger, C.
1984 *Mehrfachbestattungen im Westhallstattkreis: zum Problem der Totenfolge* (*Antiquitas* 3, Bd. 26). Bonn: Rudolf Habelt.

O'Gorman, J.A.
2001 Life, death, and the longhouse: a gendered view of Oneonta social organization. In *Gender and the archaeology of death*, edited by B. Arnold and N.L. Wicker, 23–50. Walnut Creek, California: AltaMira Press.

O'Grady, C., and A. Bardho
2005 Lofkënd 2005: end of season conservation report 20 June–29 July, 2005. Unpublished report. Lofkënd Archaeological Project.

Ohly, D.
1953 *Griechische Goldbleche des 8. Jahrhunderts v. Chr.* Berlin: Mann.

O'Kelly, M.J.
1979 The restoration of Newgrange. *Antiquity* 53:205–210.

Olson, B.R., R.A. Placchetti, J. Quartermaine, and A.E. Killebrew
2013 The Tel Akko Total Archaeology Project (Akko, Israel): assessing the suitability of multi-scale 3D field recording in archaeology. *Journal of Field Archaeology* 38(3):244–262.

Olsson, I.
1986 Radiometric methods. In *Handbook of Holocene palaeoecology and palaeohydrology*, edited by B. Berglund, 273–312. Chichester: John Wiley and Sons.

Olujić, B.
2011 Two Bronze Age drystone burial mounds at the base of Biokovo Mountain. In *L'Illyrie méridionale et l'Épire dans l'Antiquité, V. Actes du Ve colloque international de Grenoble (8–11 octobre 2008)*, edited by J.-L. Lamboley and M. Castiglioni, 53–64. Paris: de Boccard.

Onassoglou, A.
1981 Οι γεωμετρικοί τάφοι της Τραγάνας στην ανατολική Λοκρίδα. *Archaiologikon Deltion* 36, A' (Mel.):1–57.

Orsi, P.
1899 Pantalica e Cassibile. *Monumenti Antichi* 9:33–146.
1926 Le necropolis preelleniche calabresi di Torre Galli e di Canale Janchina, Patariti. *Monumenti Antichi* 31:1–376.
1933 *Templum Apollinis Alaei ad Crimisa promontorium.* Rome: Società Magna Grecia.

O'Shea, J.M.
1981 Coping with scarcity: exchange and social storage. In *Economic archaeology: towards an integration of ecological and social approaches* (BAR International Series 96), edited by A. Sheridan and G. Bailey, 167–183. Oxford: British Archaeological Reports.
1984 *Mortuary variability: an archaeological investigation.* New York: Academic Press.
1996 *Villagers of the Maros: a protrait of an Early Bronze Age society.* New York: Plenum Press.

Pader, E.-J.
1982 *Symbolism, social relations and the interpretation of mortuary remains.* Oxford: British Archaeological Reports.

Paine, R.R., and J.L. Boldsen
2002 Linking age-at-death distributions and ancient population dynamics: a case study. In *Paleodemography: age distributions from skeletal samples*, edited by R.D. Hoppa and J.W. Vaupal, 169–180. Cambridge: Cambridge University Press.

Panagiotaki, M.
1997 A study of vitreous materials from Minoan Crete. In *Αρχαία ελληνική τεχνολογία. Πρακτικά 1º διεθνές συνεδρίο*, 303–311. Thessalonike: Society for the Study of Ancient Greek Technology and the Technical Museum of Thessalonike.
2000 Crete and Egypt: contacts and relationships seen through vitreous materials. In *Κρήτη-Αίγυπτος· Πολιτισμικοί δεσμοί τριών χιλιετιών*, edited by A. Karetsou, 154–161. Athens: Greek Ministry of Culture.
2001 Faience, jewellery and luxury objects. In *Crete-Egypt: three thousand years of cultural links, catalogue*, edited by A. Karetsou and M. Andreadaki-Vlazaki, 93–131. Herakleion and Cairo: Greek Ministry of Culture.

Papadopoulos, J.K.
2000 Skeletons in wells: towards an archaeology of social exclusion in the ancient Greek world. In *Madness, disability, and social exclusion: the archaeology and anthropology of 'difference,'* edited by J. Hubert, 96–118. London and New York: Routledge.
2001 Magna Achaea: Akhaian Late Geometric and Archaic pottery in south Italy and Sicily. *Hesperia* 70:373–460.
2002 Παίζω η χέζω? A contextual approach to *pessoi* (gaming pieces, counters, or convenient wipes?). *Hesperia* 71:423–427.
2003 *La Dea di Sibari e il santuario ritrovato. Studi sui rinvenimenti dal Timpone Motta di Francavilla Marittima, II.1: the Archaic votive metal objects* (*Bollettino d'Arte*, volume speciale). Rome: Ministero per i Beni e le Attività Culturali.
2005 *The Early Iron Age cemetery at Torone*, 2 vols. (Monumenta Archaeologica 24). Los Angeles: Cotsen Institute of Archaeology.
2006 Mounds of memory: burial tumuli in the Illyrian landscape. In *New directions in Albanian archaeology: studies presented to Muzafer Korkuti*, edited by L. Bejko and R. Hodges, 75–84. Tirana: International Centre for Albanian Archaeology.
2010a The bronze headbands of prehistoric Lofkënd and their Aegean and Balkan connections. *Opuscula* 3:33–54.

2010b A new type of Early Iron Age fibula from Albania and northwest Greece. *Hesperia* 79:233–252.

Papadopoulos, J.K., L. Bejko, and S.P. Morris
2007 Excavations at the prehistoric burial tumulus of Lofkënd in Albania: a preliminary report for the 2004-2005 seasons. *American Journal of Archaeology* 111:105–147.
2008 Reconstructing the prehistoric burial tumulus of Lofkënd in Albania. *Antiquity* 82:686–701.
2009 Rindërtimi i tumës prehistorike në Lofkënd. *Monumentet* 2007 (2009):19–33.

Papadopoulos, J.K., B.N. Damiata, and J.M. Marston
2011 Once more with feeling: Jeremy Rutter's plea for the abandonment of the term Submycenaean revisited. In *Our cups are full: pottery and society in the Aegean Bronze Age. Papers presented to Jeremy B. Rutter on the occasion of his 65th birthday*, edited by W. Gauss, M. Lindblom, R.A.K. Smith, and J.C. Wright, 187–202. Oxford: Archaeopress.

Papadopoulos, J.K., and E.L. Smithson
Forthcoming *The Athenian Agora. The Early Iron Age, Part 1: four cemeteries.* Princeton, New Jersey: American School of Classical Studies at Athens.

Papadopoulos, J.K., and G. Urton (editors)
2012 *The construction of value in the ancient world.* Los Angeles: Cotsen Institute of Archaeology.

Papajani, L.
1974 Gërmimet arkeologjike të vitit 1973 në qytetin ilir në fshatin Klos të Mallakastrës. *Iliria* 3:455–460.
1976 La cite illyrienne de Klos. *Iliria* 4:411–422.

Papathanasiou, A.
2003 Stable isotope analysis in Neolithic Greece and possible implications on human health. *International Journal of Osteoarchaeology* 13:314–324.

Papazovska, A.
2005 Iron Age layers on Vardarski Rid. In *Vardarski Rid*, Vol. 1, edited by D. Mitrevski 115–157. Skopje: Foundation Vardarski Rid.

Pappa, M.
1998 Η Χαλκιδική κατά τους προϊστορικούς χρόνους. In *Η ιστορία της Χαλκιδικής*, M. Pappa, K. Sismanides, I.A. Papangelos, and E. Zelliou-Mastrokosta, 11–29. Thessalonike: Nomarchiarki Autodioikisi Chalkidikis.
2010 Die Nekropolen der frühen Bronzezeit bei der Toumba Agios Mamas. In *Das prähistorische Olynth: Ausgrabungen in der Toumba Agios Mamas 1994–1996. Die Grabung und der Baubefund* (Prähistorische Archäologie in Südosteuropa 23), edited by B. Hänsel and I. Aslanis, 382–440. Rahden/Westfalen: Marie Leidorf.

Parker Pearson, M.
1982 Mortuary practices, society and ideology: an ethnoarchaeological study. In *Symbolic and structural archaeology*, edited by I. Hodder, 99–113. Cambridge: Cambridge University Press.
1999 *The archaeology of death and burial.* College Station: Texas A&M University Press.

Parkinson, W.A. (editor)
2002 *The archaeology of tribal societies.* Ann Arbor, Michigan: International Monographs in Prehistory.

Parlama, L., M. Theochari, S. Bonatsos, Ch. Romaiou, and Y. Manos
2010 Παλαμάρι Σκύρου: η πόλη της Μέσης Χαλκοκρατίας. In *Mesohelladika: la Grèce continental au Bronze Moyen. Actes du colloque international organié par l'École Française d'Athènes, en collaboration avec l'American School of Classical Studies at Athens et le Netherlands Institute in Athens, Athènes, 8–12 mars 2006* (*Bulletin de Correspondance Hellénique*, Supplément 52), edited by A. Philippa-Touchais, G. Touchais, S. Voutsaki, and J. Wright, 281–289. Paris: de Boccard.

Parlama, L., and N. Ch. Stampolidis (editors)
2000 *The city beneath the city: finds from excavations for the Metropolitan Railway of Athens.* Athens: Greek Ministry of Culture/N.P. Goulandris Foundation.

Pasqui, A.
1888 Territorio di Sibari. Scavi della necropoli di Torre Mordillo nel commune di Spezzano Albanese. *Notizie degli Scavi* 1888:239–268, 462–480, 565–592, 648–671.

Patsch, C.
1904 *Das Sandschak Berat in Albanien* (Schriften der Balkankommission, antiquarische Abteilung III). Vienna: Alfred Hölder.

Patterson, C.C.
1971 Native copper, silver and gold accessible to early metallurgists. *American Antiquity* 36(3):286–321.

Pauli, L.
1975 *Keltischer Volksglaube. Amulette und Sonderbestattungen am Dürnberg bei Hallein und im eisenzeit Mitteleuropa.* Munich: Beck.

Payne, A.
2011 Laser scanning for archaeology: a guide to good practice. In *Archaeology data service/digital antiquity guides to good practice.* Archaeology Data Service, University of York, UK. *http://guides.archaeologydataservice.ac.uk/g2gp/LaserScan_Toc.*

Payne, H.
1940 *Perachora, the sanctuaries of Hera Akraia and Limenia, I: architecture, bronzes, terracottas.* Oxford: Clarendon Press.

Pearsall, D.M.
2000 *Paleoethnobotany: a handbook of procedures*. 2nd ed. San Diego: Academic Press.

Pelon, O.
1976 *Tholoi, tumuli et cercles funéraires: recherches sur les monuments funéraires de plan circulaire dans l'Égée de l'âge du bronze (IIIe et IIe millénaires av. J.-C.)* (Bibliothèque des Écoles Françaises d'Athènes et de Rome 229). Paris: de Boccard.

Pendlebury, H.W., J.D.S. Pendlebury, and M.B. Money-Coutts
1937–1938 Excavations in the plain of Lasithi, III. Karphi: a city of refuge of the Early Iron Age in Crete. *Annual of the British School at Athens* 38: 57–145.

Penninger, E.
1972–1978 *Der Dürrnberg bei Hallein: Katalog der Grabfunde der Hallstatt- und Latènenzeit* (Münchner Beiträge zur Vor- und Frühgeschichte, Bd. 16–18). Munich: C.H. Beck.

Pennington, R.
2001 Hunter-gatherer demography. In *Hunter-gatherers: an interdisciplinary perspective*, edited by C. Panter-Brick, R.H. Layton, and P. Rowley-Conwy, 170–204. Cambridge: Cambridge University Press.

Perdrizet P.
1908 *Fouilles de Delphes, V,1: monuments figurés, petit bronzes, terres-cuites, antiquités diverses*. Paris: Albert Fontemoing and Écoles Françaises d'Athènes et de Rome.

Peregrine, P.N.
2001 Cross-cultural comparative approaches in archaeology. *Annual Review of Anthropology* 30:1–18.

Peristeri, K.
2004 Η πρώιμη εποχή του Σιδήρου στο Νομό Δράμας. In *Το Αιγαίο στην πρώιμη εποχή του Σιδήρου. Πρακτικά του διεθνούς συμποσίου, Ρόδος, 1-4 Νοεμβρίου 2002*, edited by N.Ch. Stampolides and A. Giannikouri, 259–264. Athens: University of Crete, Archaeological Institute of Aegean Studies.

Perlès, C.
1990 *Excavations at Franchthi Cave, Greece, Fasc. 5 : les industries lithiques taillées de Franchthi (Argolide, Grèce)*. Bloomington: Indiana University Press.

Peroni, R., and F. Trucco (editors)
1994a *Enotri e Micenei nella Sibaritide, I: Broglio di Trebisacce*. Taranto: Istituto per la storia e l'archeologia della Magna Grecia.
1994b *Enotri e Micenie nella Sibaritide, II: altri siti della Sibaritide*. Taranto: Istituto per la storia e l'archeologia della Magna Grecia.

Peters, K.E., C.C. Walters, and J.M. Moldowan
2005 *The biomarker guide*. Cambridge: Cambridge University Press.

Petropoulos, M.
1987–1988 Τρίτη ανασκαφική περίοδος στο Άνω Μαζαράκι (Ρακίτα) Αχαΐας. In *Acts of the Third international congress of Peloponnesian studies, Kalamata, 8–15 September, 1985* (*Peloponnesiaka* Supplement 13), 81–96. Athens: Peloponnesiaka.

Petroutsa, E.I., and S.K. Manolis
2010 Reconstructing Late Bronze Age diet in mainland Greece using stable isotope analysis. *Archaeological Science* 37:614–620.

Petruso, K.M., M. Korkuti, L. Bejko, S. Bottema, B.B. Ellwood, J.M. Hansen, F.B. Harrold, and N. Russell
1996 Konispol Cave, Albania: a preliminary report on excavations and related studies, 1992–1994. *Iliria* 26(1–2):183–224.

Petsas, Ph.
1961–1962 Ανασκαφή αρχαίου νεκροταφείου Βεργίνης. *Archaiologikon Deltion* 17, A':218–288.

Philipp, H.
1981 *Olympische Forschungen*, Bd. XIII: *Bronzeschmuck aus Olympia*. Berlin: Walter de Gruyter.

Pietrusewsky, M., and M.T. Douglas
2002 *Ban Chiang, a prehistoric site in northeast Thailand. I: the human skeletal remains*. Philadelphia: University of Pennsylvania Museum of Archaeology and Anthropology.

Pliakou, G.
2007 Το λεκανοπέδιο των Ιωαννίνων και η ευρύτερη περιοχή της Μολοσσίας στην κεντρική Ήπειρο. Αρχαιολογικά κατάλοιπα, οικιστική οργάνωση και οικονομία. Ph.D. dissertation, University of Thessalonike.
2011 *Cômai* et *ethne*: l'organisation spatiale du Bassin d'Ioannina à la lumière du matériel archéologique. In *L'Illyrie méridionale et l'Épire dans l'Antiquité*, V. *Actes du Ve colloque international de Grenoble (8–11 octobre 2008)*, edited by J.-L. Lamboley and M. Castiglioni, 631–647. Paris: de Boccard.

Pluciennik, M.
2004 The environs of Butrint 2: the 1995–1996 field survey. In *Byzantine Butrint: excavations and surveys 1994–1999*, edited by R. Hodges, W. Bowden and K. Lako, 47–63. Cambridge: Cambridge University Press.

Pollard, A.M., R.G. Thomas, and P.A. Williams
1990 Connellite: stability relationships with other secondary copper minerals. *Mineralogical Magazine* 54:425–430.

Pomadère, M.
2007 Des enfants nourris au biberon à l'Âge du Bronze? In *Cooking up the past: food and culinary practices in the Neolithic and Bronze Age Aegean*, edited by C. Mee and J. Renard, 270–289. Oxford: Oxbow Books.

Pomeroy, S.B.

1975 *Goddesses, whores, wives, and slaves: women in Classical antiquity.* New York: Schocken Books.

Popham, M.R., and I.S. Lemos

1996 *Lefkandi III: the Toumba cemetery. The excavations of 1981, 1986, and 1992-4* (BSA Supplement 29). Oxford: British School at Athens.

Popham, M.R., L.H. Sackett, and P.G. Themelis (editors)

1979–1980 *Lefkandi I: the Iron Age settlement and the cemeteries* (BSA Supplement 11). Oxford: Thames and Hudson.

Popescu, D.

1944 *Die frühe und mittlere Bronzezeit in Siebenbürgen.* Bucharest: Biblioteca Museului National de Antichitati din Bucureşti.

Popović, P., and M. Vukmanović

1998 *Vajuga-Pesak: Early Iron Age cemetery.* Belgrade: Institut Archéologique.

Porter, A.

2002 The dynamics of death: ancestors, pastoralism, and the origins of a third-millennium city in Syria. *Bulletin of the American Schools of Oriental Research* 325:1–36.

Pottenger, F.M., Jr.

1942 Nutritional aspects of the orthodontic problem. *The Angle Orthodontist* 12(4):184–189.

Powell, A.

2010 Animal bones from "Grave" 62, Tumulus 9 (Necropolis of Apollonia). In M.G. Amore, *The complex of tumuli 9, 10, and 11 in the necropolis of Apollonia (Albania)* (British Archaeological Reports International Series 2059), 853–856. Oxford: Archeopress.

Praschniker, C.

1922–1924 Muzakhia und Mallakastra. *Jahreshefte des Österreichischen Archäologischen Institutes* 21–22:1–224.

Prendi, F.

1956 Mbi rezultatet e gërmimeve në fshatin Vodhinë të rrethit të Gjirokastrës. *Buletin për Shkencat Shoqërore* 1:180–187.

1957 Tumat në fushën e fshatit Vajzë (Vlorë). *Buletin për Shkencat Shoqërore* 2:76–110.

1966 La civilisation préhistorique de Maliq. *Studia Albanica* 1:255–280.

1974 Vështrim mbi kulturën e periudhës së parë të Hekurit në Shqipëri. *Iliria* 3:103–130.

1975 Un aperçu sur la civilization de la première période du Fer en Albanie. *Iliria* 3:109–138.

1977–1978 Epoka e Bronzit në Shqipëri. *Iliria* 7–8: 5–58.

1982 The prehistory of Albania. In *Cambridge ancient history*, 2nd ed., Vol. III, Part 1, 187–237. Cambridge: Cambridge University Press.

1985a Alcune considerazioni sulla ceramica dipinta "Devolliana" della Tarda Eta del Bronzo e del Ferro Antico in Albania. *Magna Greca, Epiro e Macedonia. Atti del ventiquattresimo convengo di studi sulla Magna Grecia, Taranto 5–10 octobre, 1984*, 181–197. Taranto: Istituto per la storia e l'archeologia della Magna Grecia.

1985b Unitet dhe veçori në kulturën Ilire të epokës së Hekurit në territorin e Shqipërisë. *Iliria* 15(1): 63–88.

1988 Kërkimet arkeologjike në fushën e kulturës pre dhe protohistorike Ilire në Shqipëri. *Iliria* 18(1): 5–33.

1993 La Chaonie préhistorique et ses rapports avec les regions de l'Illyrie du Sud. In *L'Illyrie méridionale et l'Épire dans l'Antiquité*, II. *Actes du IIe colloque international de Clermont-Ferrand (25–27 octobre 1990)*, edited by P. Cabanes, 17–28. Paris: de Boccard.

1998 Epoka e bronzit dhe e hekurit në kërkimet Shqiptare. *Iliria* 28(1–2):73–103.

Prendi, F., and S. Aliu

1971 Vendbanimi neolitik në fshatin Kamnik të rrethit Kolonjës. *Iliria* 1:13–30.

Prendi, F., P. Lera, and G. Touchais

1996 Gërmime arkeologjike të vitit 1995: Sovjan. *Iliria* 26(1–2):225–272.

Prendi, F., and G. Touchais

1999 Sovjan: un habitat de l'Âge du Bronze et du début de l'Âge du Fer dans la plaine de Korça. In *L'Illyrie méridionale et l'Épire dans l'Antiquité*, III. *Actes du IIIe colloque international de Chantilly (16–19 octobre 1996)*, edited by P. Cabanes, 19–27. Paris: de Boccard.

Price, T.D. (editor)

2000 *Europe's first farmers.* Cambridge: Cambridge University Press.

Protonotariou-Deïlaki, E.

2009 *Η τύμβοι του Άργος (Άργος I).* Athens: Ephoreia Palaioanthropologias–Spilaiologias Notias Elladas.

Psaraki, K., and S. Andreou

2010 Regional processes and interregional interactions in northern Greece during the early 2nd millennium BC. In *Mesohelladika: the Greek mainland in the Middle Bronze Age. Actes du Colloque international, Athens 8–12 March, 2006* (*Bulletin de Correspondance Hellénique*, Supplément 52), edited by A. Philippa-Touchais, G. Touchais, S. Voutsaki, and J. Wright, 995–1003. Paris: de Boccard.

Pulak, Ç.

2001 The cargo of the Uluburun ship and evidence for trade with the Aegean and beyond. In *Italy and Cyprus in Antiquity 1500–450 BC*, edited by L. Bonfante and V. Karageorghis, 13–60. Nicosia: Costakis and Leto Severis Foundation.

Raczky, P.

1987 Öcsöd-Kováshalom: a settlement of the Tisza culture. In *The Late Neolithic of the Tisza region:*

a survey of recent excavations and their findings, edited by N. Kalicz and P. Racsky, 61–83. Budapest: Szolnok.

Radt, W.
1974 Die früheisenzeitliche Hügelnekropole bei Vergina in Makedonien. In *Beiträge zu italienischen und griechischen Bronzefunden*, edited by H. Müller-Karpe (*PBF* XX:1), 98–147. Munich: Beck.

Raftopoulou, S.
1998 New finds from Sparta. In *Sparta in Laconia: the archaeology of a city and its countryside. Proceedings of the 19th British Museum Classical colloquium held with the British School at Athens and King's and University Colleges, London, 6–8, December 1995* (BSA Studies 4), edited by W.G. Cavanagh and S.E.C. Walker, 125–140. London: British School at Athens.

Rahmstorf, L.
2003 Clay spools from Tiryns and other contemporary sites: an indication of foreign influence in LH IIIC? In *The periphery of the Mycenaean world. Proceedings of the 2nd international interdisciplinary colloquium, 26–30 September, Lamia 1999*, edited by N. Kyparissi-Apostolika and M. Papakonstantinou, 397–415. Athens: Greek Ministry of Culture.

Randall-MacIver, D.
1924 *Villanovans and early Etruscans: a study of the Early Iron Age in Italy as it is seen near Bologna, in Etruria and in Latium*. Oxford: Clarendon Press.
1927 *The Iron Age of Italy: a study of those aspects of the early civilization which are neither Villanovan nor Etruscan*. Oxford: Clarendon Press.

Randsborg, K.
1980 *The Viking age in Denmark: the formation of a state*. London: Duckworth.

Randsborg, K., and K. Christensen
2006 *Bronze Age oak-coffin graves: archaeology and dendro-dating* (Acta Archaeologica Supplementa VII, Centre of World Archaeology 3). Copenhagen: Blackwell and Munksgaard.

Raubitschek, I.K.
1998 *Isthmia VII: The Metal Objects (1952–1989)*. Princeton, New Jersey: American School of Classical Studies.

Rega, E.
1997 Biological reality in the Early Bronze Age cemetery at Mokrin. In *Invisible people and processes: writing gender and childhood into European archaeology*, edited by J. Moore and E. Scott, 229–247. London and New York: Leicester University Press.

Rehak, P.
1996 Aegean breechcloths, kilts, and the Keftiu paintings. *American Journal of Archaeology* 100:35–51.

Rehm, R.
1994 *Marriage to death: the conflation of wedding and funeral rituals in Greek tragedy*. Princeton, New Jersey: Princeton University Press.

Reichel, W.
1942 *Griechisches Goldrelief*. Berlin: Archäologisches Institut des Deutschen Reiches.

Reichel, W., and A. Wilhelm
1904 Das Heiligtum der Artemis zu Lousoi. *Jahreshefte des Österreichischen Archäologischen Instituts* 4:1–89.

Reichstein, H.
1979 Erste Ergebnisse von Untersuchungen an Tierknochen aus bronzezeitlichen Siedlungsschichten im nördlichen Griechenland. *Jahrbuch des Römisch-Germanischen Zentralmuseums* 1979:239–270.

Reimer, P.J., M.G.L. Baillie, E. Bard, A. Bayliss, J.W. Beck, C.J.H. Bertrand, P.G. Blackwell, C.E. Buck, G.S. Burr, K.B. Cutler, P.E. Damon, R.L. Edwards, R.G. Fairbanks, M. Friedrich, T.P. Guilderson, A.G. Hogg, K.A. Hughen, B. Kromer, G. McCormac, S. Manning, C.B. Ramsey, R.W. Reimer, S. Remmele, J.R. Southon, M. Stuiver, S. Talamo, F.W. Taylor, J. van der Plicht, and C.E. Weyhenmeyer
2004 IntCal04 terrestrial radiocarbon age calibration, 0–26 cal kyr BP. *Radiocarbon* 46:1029–1058.

Reineck H.-E., and I.B. Singh
1975 *Depositional sedimentary environments, with reference to terrigenous clastics*. Berlin: Springer Verlag.

Reinholdt, C.
2008 *Der frühbronzezeitliche Schmuckhortfund von Kap Kolonna: Ägina und die Ägäis im Goldzeitalter des 3. Jahrtausends v. Chr.* (Ägina-Kolonna Forschungen und Ergebnisse, Bd. II). Vienna: Österreichische Akademie der Wissenschaften.

Reischütz, P., and H. Sattmann
1990 Beiträge zur Molluskenfauna des Epirus, II. *Annales des naturhistorishes Museum Wien* 91 (B): 253–272.

Reitz, E.J., and E.S. Wing
1999 *Zooarchaeology*. Cambridge: Cambridge University Press.

Renfrew, C., M. Gimbutas, and E.S. Elster (editors)
1986 *Excavations at Sitagroi: a prehistoric village in northeast Greece*, Vol. I. Los Angeles: Institute of Archaeology, University of California.

Rennick, S.
2005 Genetic analysis of a monumental structure with the Kamenica tumulus, Albania. M.S. thesis, Michigan State University, East Lansing.

Rey, L.
1932 Fouilles de la Mission Française à Apollonie d'Illyrie (1930–1931): exploration de la nécropole. *Albania* 4:7–27.

Reyer-Völlenklee, S.
2005 Antike Flickungen aus bituminöser Masse an apulish-rotfiguriger Keramik. In *Vis Imaginum:*

Festschrift für Elisabeth Walde zum 65. Geburtstag, edited by G. Grabherr, B. Kainrath, A. Larcher, and B. Welte, 399–404. Innsbruck: Institut für Klassische und Provinzialrömische Archäologie der Universitat Innsbruck.

Rhomaios, K.A.
1929 Η κάθαρσις της Δήλου και το εύρημα του Σταυροπούλου. *Archaiologikon Deltion* 12:181–223.

Rhomiopoulou, K.
1971 Ταφαί πρωΐμου εποχής του Σιδήρου εις ανατολικήν Πίνδου. *Athens Annals of Archaeology* 4:37–42.

Rhomiopoulou, K., and I. Kilian-Dirlmeier
1989 Neue Funde aus der eisenzeitlichen Hügelnekropole von Vergina, Griechisch Makedonien. *Prähistorische Zeitschrift* 64:86–145.

Rhomiopoulou, K., and I. Touratsoglou
2002 *Μίεζα: Νεκροταφείο υστεροαρχαϊκών χρόνων*. Athens: TAPA.

Rice, M.P.
1987 *Pottery analysis: a sourcebook*. Chicago: University of Chicago Press.

Richards, M.P., and R.E.M. Hedges
2008 Stable isotope evidence of past human diet at the sites of the Neolithic cave of Gerani; the Late Minoan III cemetery of Armenoi; Grave Circles A and B at the palace site of Mycenae; and Late Helladic chamber tombs. In *Archaeology meets science: biomolecular investigations in Bronze Age Greece: the primary scientific evidence 1997–2003*, edited by Y. Tzedakis, H. Martlew and M.K. Jones, 220–230. Oxford: Oxbow Books.

Ridgway, D.
1992 *The first western Greeks*. Cambridge: Cambridge University Press.

Ridley, C., K.A. Wardle, and C.A. Mould
2000 *Servia I: Anglo-Hellenic rescue excavations, 1971–1973* (BSA Supplement 32). Oxford: British School at Athens.

Říhovský, J.
1993 *Die Fibeln in Mähren* (*PBF* XIV:9). Munich: Beck.

Robert, L.
1928 Notes d'épigraphie Hellénistique. *Bulletin de Correspondance Hellénique* 52:426–443.

Robinson, D.M.
1941 *Excavations at Olynthus X: metal and minor miscellaneous finds, an original contribution to Greek life*. Baltimore: The Johns Hopkins Press.

Ross, A.H., and L.W. Konigsberg
2002 New formulae for estimating stature in the Balkans. Technical report. *Journal of Forensic Sciences* 47:165–167.

Rothschild, B.M.
1985 Diffuse idiopathic skeletal hyperostosis (DISH): misconceptions and reality. *Clinical Rheumatology* 4:207–212.

Rotroff, S.
2013 *Industrial religion: the saucer pyres of the Athenian Agora* (*Hesperia* Supplement 47). Princeton, New Jersey: American School of Classical Studies.

Rotroff, S.I., L.M. Little, and L.M. Snyder
1999 The reanalysis of a well deposit from the second century B.C. in the Athenian Agora: animal sacrifice and infanticide in late Hellenistic Athens? *American Journal of Archaeology* 103:284–285.

Rowlands, M.
1993 The role of memory in the transmission of culture. *World Archaeology* 25:141–151.

Ruffo, M.
1994–1995 Anagni (Frosinone), località S. Cecilia. Indagini nel santuario ernico: Il deposito votivo arcaico. *Notizie degli Scavi* 1994–1995:5–164.

Rullkötter, J., and A. Nissenbaum
1988 Dead Sea asphalt in Egyptian mummies: molecular evidence. *Naturwissenschaften* 75(12):618–621.

Runnels, C.
2009 Mesolithic sites and surveys in Greece: a case study from the southern Argolid. *Journal of Mediterranean Archaeology* 22:57–73.

Runnels, C., M. Korkuti, M.L. Galaty, M.E. Timpson, J.C. Whittaker, S.P. Stocker, J.L. Davis, L. Bejko, and S. Muçaj
2004 The Paleolithic and Mesolithic of Albania: survey and excavation at the site of Kryegjata B (Fier district). *Journal of Mediterranean Archaeology* 17:3–29.

Ruppenstein, F.
2007 *Kerameikos XVIII. Die submykenische Nekropole: Neufunde und Neubewertung*. Munich: Hirmer.

Ruscillo, D.
2005 Appendix C: marine remains and land mollusks from Terrace V. In J.K. Papadopoulos, *The Early Iron Age cemetery at Torone*, 2 vols. (Monumenta Archaeologica 24), 321–338. Los Angeles: Cotsen Institute of Archaeology.

Russell, N.
1998a Cattle as wealth in Neolithic Europe: where's the beef? In *The archaeology of value: essays on prestige and the processes of valuation* (BAR International Series 730), edited by D.W. Bailey and S. Mills, 42–54. Oxford: British Archaeological Reports.
1998b The Mesolithic-Neolithic transition in the faunal assemblage from Konispol Cave, Albania. In *Archaeozoology of the Near East III: Proceedings of the Third international symposium on the archaeozoology of southwestern Asia and adjacent areas* (ARC-Publications 18), edited by H. Buitenhuis, L. Bartosiewicz, and A.M. Choyke, 145–159. Groningen: Centre for Archeological Research and Consultancy.

Ryan, K., and P.J. Crabtree (editors)
1995 *The symbolic role of animals in archaeology.* Philadelphia: MASCA, University of Pennsylvania Museum of Archaeology and Anthropology.

Safran, Y., and W. Wang (editors)
1985 *The architecture of Adolf Loos.* London: Arts Council of Great Britain.

Sakashita, R., M. Inoue, N. Inoue, Q. Pan, and H. Zhu
1997 Diet and discrepancy between tooth and jaw size in the Yin-Shang period of China. *American Journal of Physical Anthropology* 103(3):497–505.

Sandars, N.K.
1955 The antiquity of the one-edged bronze knife in the Aegean. *Proceedings of the Prehistoric Society* 21: 174–197.

Šašel Kos, M.
2011 Peoples on the northern fringes of the Greek world: Illyria as seen by Strabo. In *L'Illyrie méridionale et l'Épire dans l'Antiquité,* V. *Actes du Ve colloque international de Grenoble (8–11 octobre 2008),* edited by J.-L. Lamboley and M. Castiglioni, 617–629. Paris: de Boccard.

Sattmann, H., and P. Reischütz
1994 Notes on land snails of Epirus (Greece) (Gastropoda: Pulmonata, Prosobranchia), II. *Annales des naturhistorishes Museum Wien* 96(B):39–48.

Saunders, J.W., R.D. Mandel, C.G. Sampson, C.M. Allen, E.T. Allen, D.A. Bush, J.K. Feathers, K.J. Gremillion, C.T. Hallmark, H.E. Jackson, J.K. Johnson, R. Jones, et al.
2006 Watson Brake, a middle archaic mound complex in northeast Louisiana. *American Antiquity* 70: 631–668.

Savvopoulou, Th.
1988 Νέα στοιχεία από το νεκροταφείο του Παλαιού Γυναικόκαστρο. *To Archaiologiko Ergo stin Makedonia kai Thrake* 2:219–229.
2004 Η περιοχή του Αξιού στην πρώιμη εποχή του Σιδήρου. In *Το Αιγαίο στην πρώιμη εποχή του Σιδήρου. Πρακτικά του διεθνούς συμποσίου, Ρόδος, 1–4 Νοεμβρίου 2002,* edited by N.Ch. Stampolides and A. Giannikouri, 307–316. Athens: University of Crete, Archaeological Institute of Aegean Studies.

Saxe, A.A.
1970 Social dimensions of mortuary practices. Ph.D. dissertation, University of Michigan, Ann Arbor.

Scarpitti, F.R., and P.T. McFarlane (editors)
1975 *Deviance: action, reaction, interaction.* Reading, Massachusetts: Addison-Wesley.

Schalk, E.
2008 Die Bronzenadeln in der Berliner Sammlung. In *Schliemanns Sammlung Trojanischer Altertümer—Neuvorlage. Bd. 1: Forschungsgeschichte, keramische Funde der Schichten VII bis IX, Nadeln, Gewichte und durchlochte Tongeräte* (Berliner Beiträge zur Vor- und Frühgeschichte, Neue Folge, Bd. 14), edited by M. Wemhoff, D. Hertel, and A. Hänsel, 183–226. Berlin: Staatliche Museen zu Berlin.

Schaumberg, A.
1910 *Bogen und Bogenschütze der Griechen mit besonderer Rücksicht auf die Denkmäler bis zum Ausgang des archaischen Stils.* Nürnberg: Benedikt Hilz.

Scheibler, I.
1995 *Griechische Töpferkunst: Herstellung, Handel und Gebrauch der antiken Tongefässe,* 2nd ed. Munich: C.H. Beck.

Schepartz, L.A.
1989 Modeling subsistence pattern change. In *People and culture in change* (BAR International Series 508[i]), edited by I. Hershkowitz, 199–217. Oxford: British Archaeological Reports.
2010 The skeletal biology of tumulus 9, 10, 11 and the appendices. In M.G. Amore, *The complex of tumuli 9, 10, and 11 in the necropolis of Apollonia* (BAR International Series 2059[i–ii]), 48–77. Oxford: Archaeopress.

Schepartz, L.A., S. Miller-Antonio, and J.M.A. Murphy
2009 Differential health and status among Mycenaeans of Messenia: the status of females at Pylos. In *New directions in the skeletal biology of Greece* (*Hesperia* Supplement 43, Occasional Wiener Laboratory Series), edited by L.A. Schepartz, S. Fox and C. Bourbou, 155–174. Princeton, New Jersey: American School of Classical Studies at Athens.

Schliemann, H.
1880 *Ilios: the city and country of the Trojans.* London: John Murray.

Schlörb-Vierneisel, B.
1966 Eridanos-Nekropole, I: Gräber und Opferstellen hS 1–204. *Mitteilungen des Deutschen Archäologischen Instituts, Athenische Abteilung* 81:4–111.

Schmidt, H.
1902 *Heinrich Schliemann's Sammlung trojanischer Altertümer.* Berlin: Georg Reimer.

Schmidt, H.
1997 Reconstruction of ancient buildings. In *The conservation of archaeological sites in the Mediterranean region: an international conference organized by the Getty Conservation Institute and the J. Paul Getty Museum, 6–12 May, 1995,* edited by M. de la Torre, 41–50. Los Angeles: Getty Conservation Institute.

Schmidt, R.R.
1945 *Die Burg Vučedol* (Ausgabe des kroatischen archäologischen Staatsmuseums in Zagreb). Zagreb: Kroatische Staatsdruckerei.

Schmitt, H.-O.
2007 Die Angriffswaffen. In *Kalapodi. Ergebnisse der Ausgrabungen in Heiligtum der Artemis und*

Apollon von Hyampolis in der antiken Phokis, Bd. II, edited by R.C.S. Felsch, 423–551. Mainz: Philipp von Zabern.

Schoch, W., I. Heller, F.H. Schweingruber, and F. Kienast
2004 Wood anatomy of central European species. *http://www.woodanatomy.ch/key.pdf.*

Schuldenrein, J.
1998 Konispol cave, southern Albania, and correlations with other Aegean caves occupied in the Late Quaternary. *Geoarchaeology: An International Journal* 13(5):501–526.

Schwandner-Sievers, S.
2004 Times past: references for the construction of local order in present-day Albania. In *Balkan identities: nation and memory*, edited by M. Todorova, 103–128. New York: New York University Press.

Schweingruber, F.H.
1990 *Anatomy of European woods.* Stuttgart: Haupt.

Schweingruber, F.H., A. Börner, and E.-D. Schulze
2006 *Atlas of woody plant stems: evolution, structure, and environmental modifications.* Berlin: Springer.

Schweizer, F.
1995 Bronze objects from lake sites: from patina to "biography." In *Ancient and historic metals: conservation and scientific research*, edited by D. Scott, J. Podany and B. Considine, 33–50. Los Angeles: Getty Publications.

Scott, D.A.
1985 Periodic corrosion phenomena in bronze antiquities. *Studies in Conservation* 3(2):49–57.
1991 *Metallography and microstructure of ancient and historic metals.* Los Angeles: Getty Conservation Institute.
1994 An examination of the patina and corrosion morphology of some Roman bronzes. *Studies in Conservation* 33(1):1–23.
2000 A review of copper chlorides and related salts in bronze corrosion and as painting pigments. *Studies in Conservation* 45(1):39–53.
2002 *Copper and bronze in art: corrosion, colorants, conservation.* Los Angeles: Getty Trust Publications.
2005 Fractal forms and the deterioration of artefacts. *Studies in Conservation* 50(3):179–189.

Scott, D.A., and J. Maish
2010 A Lydian bed of iron, bronze and copper: technical examination of a metallurgical masterpiece. *Studies in Conservation* 55(1):1–17.

Scott, D.A., R. Finnerty, Y. Taniguchi, E. Koseto, R. Schmidtling, and Z.A. Stos
2003 Analytical studies of the Francavilla Archaic votive metal objects. In J.K. Papadopoulos, *La Dea di Sibari e il santuario ritrovato. Studi sui rinvenimenti dal Timpone Motta di Francavilla Marittima. II.1 The Archaic votive metal objects*, 167–234. Rome: Ministero per I Beni e le Attivita Culturali.

Scott, E.
1999 *The archaeology of infancy and infant death.* Oxford: British Archaeological Reports.

Scott, G.R.
1980 Population variance of Carabelli's trait. *Human Biology* 53:63–78.

Sease, C.
1978 Benzotriazole: a review for conservators. *Studies in Conservation* 23(2):76–85.

Seeher, J.
2005 Bohren wie die Hethiter: Rekonstruktion von Bohrmaschinen der Spätbronzezeit und Beispiele ihrer Verwendung. *Istanbuler Mitteilungen* 55:17–36.
2007 Sägen wie die Hethiter: Rekonstruktion einer Steinschneidetecknik im bronzezeitlichen Bauhandwerk. *Istanbuler Mitteilungen* 57:27–43.
2008 Bauen wie die Hethiter. *Antike Welt* 5:55–62.

Segal, T.
1998 Collections management: marking. In *The new museum registration methods*, edited by R.A. Buck and J.A. Gilmore, 65–77. Washington, D.C.: American Association of Museums.

Selmeier, A., and E. Velitzelos
2000 Neue Aufsammlungen von verkieselten Holzresten aus tertiären Schichten Griechenlands (Lesbos, Kastoria, Thrakien). *Mitteilungen Bayerische Staatssammlung Paläontologie und historische Geolologie* 40:213–227.

Selwyn, L.S., P.J. Sirois, and V. Argyropoulos
1999 The corrosion of excavated archaeological iron with details on weeping and akaganéite. *Studies in Conservation* 44(4):217–232.

Serpico, M. (with a contribution by R. White)
2000 Resins, amber and bitumen. In *Ancient Egyptian materials and technology*, edited by P.T. Nicholson and I. Shaw, 430–474. Cambridge: Cambridge University Press.

Shaffer, G.D.
1993 An archaeomagnetic study of a wattle and daub building collapse. *Journal of Field Archaeology* 20:59–75.

Shashoua, Y
1999 Ageless® oxygen absorber: from theory to practice. In *Preprints of the 12th triennial meeting, Lyon, of the ICOM Committee for Conservation*, edited by J. Bridgland, 881–887. London: James and James.

Shaw, B.D.
1982–1983 "Eaters of flesh, drinkers of milk": the ancient Mediterranean ideology of the pastoral nomad. *Ancient Society* 13–14:5–31.

Shay, T.
1985 Differentiated treatment of deviancy at death as revealed in anthropological and archaeological

material. *Journal of Anthropological Archaeology* 4(3):221–241.

Shear, T.L.
1930 Excavations in the north cemetery at Corinth in 1930. *American Journal of Archaeology* 34:403–431.

Sherratt, A.G.
1981 Plough and pastoralism: aspects of the secondary products revolution. In *Pattern of the past: studies in honour of David Clark*, edited by I. Hodder, G.L. Isaac, and N. Hammond, 261–305. Cambridge: Cambridge University Press.

Shoberl, F.
1827 *Illyria and Dalmatia, containing a description of the manners, customs, habits, dress and other peculiarities of the inhabitants and those of the adjacent countries.* London: R. Ackermann.

Sinopoli, C.
1991 *Approaches to archaeological ceramics.* New York: Plenum Press.

Sivignon, M.
2007 Histoire de la frontière gréco-albanaise. In *Épire, Illyrie, Macédoine... Mélanges offerts à Professeur Pierre Cabanes* (Collections ERGA Recherches sur l'Antiquité 10), edited by D. Berranger-Auserve, 375–385. Clermont-Ferrand: Presses Universitaires Blaise Pascal.

Sloan, R.E., and M.A. Duncan
1978 Zooarchaeology of Nichoria. In *Excavations at Nichoria in southwest Greece,* Vol. I: *site, environs and techniques*, edited by G. Rapp, Jr. and S.E. Aschenbrenner, 60–77. Minneapolis: University of Minnesota Press.

Smiris, G.
2001 *Το δίκτυο των οχυρώσεων στο πασαλίκι των Ιωαννίνων (1788–1822). Ιστορική, πολιτική, οικονομική, και χωριταξυκή θεώρηση.* Athens: self-published.

Smith, R.J., and H.L. Bailit
1977 Problems and methods in research on the genetics of dental occlusion. *The Angle Orthodontist* 47(1):65–77.

Smolar, P.
2011 Where revenge is in the blood: a brutal code of retribution is still tearing families apart in Albania. *The Guardian Weekly*, July 8, 2011.

Snodgrass, A.M.
1962 Iron Age Greece and central Europe. *American Journal of Archaeology* 66:408–410.
1964 *Early Greek armour and weapons from the end of the Bronze Age to 600 B.C.* Edinburgh: Edinburgh University Press.
1971 *The dark age of Greece: an archaeological survey of the eleventh to the eighth centuries B.C.* Edinburgh: Edinburgh University Press.
1980 *Archaic Greece: the age of experiment.* 2nd ed. Berkeley: University of California Press.

Snyder, L.M., and W.E. Klippel
1996 The Cretan badger (*Meles meles*) as a food resource at Late Bronze/Early Iron Age Kavousi-Kastro. In *Pleistocene and Holocene fauna of Crete and its first settlers*, edited by D.S. Reese, 283–294. Madison: Prehistory Press.
2003 Dog burials in the Greek world: the animals from an Early Iron Age Grave at Kavousi. *American Journal of Archaeology* 101(2):368.
2003 From Lerna to Kastro: further thoughts on dogs as food in ancient Greece: perceptions, prejudices and reinvestigations. In *Zooarchaeology in Greece: recent advances*, edited by E. Kotjabopoulou, Y. Hamilakis, P. Halstead, C. Gamble, and P. Elefanti, 221–231. London: British School at Athens.

Sordinas, A.
1970 *Stone implements from northwestern Corfu, Greece* (Memphis State University Anthropological Research Center, Occasional Papers, No. 4). Memphis, Tennessee: Memphis State University.

Sørensen, A.C.
2001 *Ladby: a Danish ship-grave from the Viking Age.* Roskilde: Viking Ship Museum.

Soueref, K.
1989 Presenza micenea in Albania e in Epiro: problemi ed osservazioni. *Iliria* 19(2):65–86.

Souyoudzoglou-Haywood, C.
1999 *The Ionian islands in the Bronze Age and Early Iron Age, 3000–800 B.C.: an area study.* Liverpool: Liverpool University Press.

Srejović, D., and L. Babović
1981 *Lepenski Vir: Menschenbilder einer frühen europäischen Kultur.* Köln and Munich: Römisch-Germanisches Museum and Prähistorische Staatssammlung.

Stahl, A.
1976 Animal sacrifices in the Balkans. In *The realm of the extra-human: ideas and actions*, edited by A. Bharati, 443–451. The Hague: Mouton Publishers.

Stallo, J.R., L.A. Schepartz, V. Grimes, and M.P. Richards
2010 Strontium isotope ratios and mobility reconstruction. In M.G. Amore, *The complex of tumuli 9, 10, and 11 in the necropolis of Apollonia (Albania)* (British Archaeological Reports International Series 2059), 78–84. Oxford: Archeopress.

Stalsberg, A.
2001 Visible women made invisible: interpreting Varangian women in Old Russia. In *Gender and the archaeology of death*, edited by B. Arnold and N.L. Wicker, 65–80. Walnut Creek, California: AltaMira Press.

Stampolides, N.Ch. (editor)
2012 *"Princesses" of the Mediterranean at the dawn of history.* Athens: Museum of Cycladic Art.

Štamol, V.
2004 Terrestial snails (Mollusca: Gastropoda terrestrial) of the Telaščica Nature Park (Dugi otok, Croatia). *Natura Croatica* 13(2):95–113.

Stanley Price, N.
2003 Site preservation and archaeology in the Mediterranean region. In *Theory and practice in Mediterranean archaeology: Old World and New World perspectives*, edited by J.K. Papadopoulos and R. Leventhal, 269–283. Los Angeles: Cotsen Institute of Archaeology.

Stanzel, M.
1991 Die Tierreste aus dem Artemis/Apollon Heiligtum bei Kalapodi in Boötien, Griechenland. Ph.D. dissertation, University of Munich.

Stark, F.
1949 *Ionia: a quest*. London: John Murray.

Stavropoullos, Ph.D.
1965 Αθήνα-Αττική. 19. Οδού Καβαλόττι. *Archaiologikon Deltion* 20, Chr. B'1 :75–80.

Steadman, D.W., and P.S. Martin
2003 The Late Quaternary extinction and future resurrection of birds on Pacific islands. *Earth-Science Reviews* 61(1–2):133–147.

Stears, K.
1998 Death becomes her: gender and Athenian death ritual. In *The sacred and the feminine in ancient Greece*, edited by S. Blundell and M. Williamson, 113–127. London: Routledge.

Steckel, R.H.
1995 Stature and the standard of living. *Journal of Economic Literature* 33:1903–1940.

Steckel, R.H., and J.C. Rose (editors)
2002 *The backbone of history: health and nutrition in the Western Hemisphere*. Cambridge and New York: Cambridge University Press.

Stephani, E.
2010 Αγγελοχώρι Ημαθίας· οικισμός της Ύστερης Εποχής Χαλκού, Vol. 1. Thessalonike: Adelphon Kyriakidi.

Stevens, W.
2006 *Collected poems*. London: Faber and Faber.

Stiner, M.C., N.D. Munro, and T.A. Surovell
2000 The tortoise and the hare: small-game use, the broad-spectrum revolution, and Paleolithic demography. *Current Anthropology* 41(1)39–73.

Stocker, S.R.
2009 Illyrian Apollonia: toward a new ktisis and developmental history of the colony. Ph.D. dissertation, University of Cincinnati.

Stocker, S.R., and J.L. Davis
2006 The earliest history of Apollonia: heroic reflections from beyond the acropolis. In *New directions in Albanian archaeology: studies presented to Muzafer Korkuti*, edited by L. Bejko and R. Hodges, 85–93. Tirana: International Centre for Albanian Archaeology.

Stojanowski, C.M., C.S. Larsen, T.A. Tung, and B.G. McEwan
2007 Biological structure and health implications from tooth size at Mission San Luis de Apalachee. *American Journal of Physical Anthropology* 132(2):207–222.

Stol, M.
2012 Bitumen in ancient Mesopotamia: the textual evidence. *Bibliotheca Orientalis* 69(1–2):49–59.

Stone, L.
2000 *Kinship and gender: an introduction*. 2nd ed. Westview Press.

Stoop, M.W.
1980 Note sugli scavi nel santuario di Atena sul Timpone della Motta (Francavilla Marittima-Calabria), 3. *Bulletin Antieke Beschaving* 55(2): 163–189.
1987 Note sugli scavi nel santuario di Atena sul Timpone della Motta (Francavilla Marittima-Calabria), 7: oggetti di bronzo vari (animali, ornamenti personali, armi, varia). *Bulletin Antieke Beschaving* 62:21–31.

Stravopodi, E., S.K. Manolis, S. Kousoulakos, V. Aleporou, and M.P. Schultz
2009 Porotic hyperostosis in Neolithic Greece: new evidence and further implications. In *New Directions in the skeletal biology of Greece* (Hesperia Supplement 43, Occasional Wiener Laboratory Series), edited by L.A. Schepartz, S. Fox, and C. Bourbou, 257–270. Princeton, New Jersey: American School of Classical Studies at Athens.

Stuart, B.
2007 Infrared spectroscopy. In *Analytical techniques in materials conservation*, 110–136. London: John Wiley and Sons, Ltd.

Stuart-Macadam, P.
1985 Porotic hyperostosis: representative of a childhood condition. *American Journal of Physical Anthropology* 66(4):391–398.
1992 Porotic hyperostosis: a new perspective. *American Journal of Physical Anthropology* 87(1):39–47.

Stuiver, M., and H.A. Polach
1977 Discussion: reporting of ^{14}C data. *Radiocarbon* 19:355–363.

Subai, P.
1980 Revision der lebenden Arten der Gattung Poiretia (Gastropoda: Oleacinidae). *Archiv für Molluskenkunde* 110(4/6):151–172.

Sullivan, S.
1997 A planning model for the management of archaeological sites. In *The conservation of archaeological sites in the Mediterranean region: an international conference organized by the Getty*

Conservation Institute and the J. Paul Getty Museum, 6–12 May, 1995, edited by M. de la Torre, 15–26. Los Angeles: Getty Conservation Institute.

Sundwall, J.
1943 *Die älteren italischen Fibeln*. Berlin: Walter de Gruyter.

Sutter, R.C., and R.J. Cortez
2005 The nature of Moche human sacrifice: a bio-archaeological perspective. *Current Anthropology* 46(4):521–549.

Szilgágyi, J.G.
1957 Some problems of Greek gold diadems. *Acta Antiqua* 5:45–93.

Tainter, J.A.
1975 Social inference and mortuary practice: an experiment in numerical classification. *World Archaeology* 7:1–15.
1977 Modeling change in prehistoric social systems. In *For theory building in archaeology*, edited by L. Binford, 327–351. New York: Academic Press.
1978 Mortuary practices and the study of prehistoric social systems. In *Advances in archaeological method and theory*, edited by M.B. Schiffer, 105–141. New York: Academic Press.

Tam, M.D.
2010 Building virtual models by postprocessing radiology images: a guide for anatomy faculty. *Anatomical Sciences Education* 3:261–266.

Tartaron, T.F.
2004 *Bronze Age landscape and society in southern Epirus, Greece* (BAR International Series 1290). Oxford: Archaeopress.

Taylour, W.D.
1983 *The Mycenaeans*. 2nd ed. London: Thames and Hudson.

Taylour, W.D., and R. Janko
2008 *Ayios Stephanos: excavations at a Bronze Age and medieval settlement in southern Laconia* (BSA Supplement 44). London: British School at Athens.

Team Tanah Datar
2012 Tanah Datar Archaeological Project: up, up and away! Tanah Datar Archaeological Project. *http://tanahdatar-archaeology.blogspot.com/2012/03/blog-post.html*.

Temkin, B., E. Acosta, A. Malvankar, and S. Vaidyanath
2006 An interactive three-dimensional virtual body structures system for anatomical training over the Internet. *Clinical Anatomy* 19(3):267–274.

Teržan, B.
1984 Nosnja kao obiljezje drustvenih skupina izmedu Drima i Devolla u starije zeljezno doba (Romaja-Burrel-Kuç i Zi). In *Duhovna Kultura Ilira*, 197–214. Sarajevo.
1985 Poskus rekonstrukcije halstatske druzbene strukture v dolenjskem kulturnem krogu. *Arheoloski Vestnik* 36:77–106.
1995 *Hoards and individual metal finds from the Eneolithic and Bronze Ages in Slovenia*, Vol. I. Ljubljana: Narodni Muzej.

Themelis, P.G.
1969 Ερετριακά. *Archaiologike Ephemeris* 1969:143–178.
1983 An 8th century goldsmith's workshop at Eretria. In *The Greek renaissance of the eighth century B.C.: tradition and innovation. Proceedings of the second international symposium at the Swedish Institute in Athens, 1–5 June, 1981*, edited by R. Hägg, 157–165. Stockholm: Swedish Institute at Athens.

Theochari, M.D.
1966 Πρωτογεωμετρικά Θεσσαλίας. *Thessalika* 5:37–53.

Thiébault, S. (editor)
2002 *Charcoal analysis: methodological approaches, palaeoecological results and wood uses. Proceedings of the second international meeting of anthracology, Paris, September 2000*. Oxford: Archaeopress.

Thomas, E.
2007 *Monumentality and the Roman Empire: architecture in the Antonine age*. Oxford: Oxford University Press.

Thomas, J.
1991 *Rethinking the Neolithic*. Cambridge: Cambridge University Press.

Thompson, H.A.
1947 The excavation of the Athenian Agora, 1940–1946. *Hesperia* 16:193–213.

Thoreau, H.D.
1960 *Walden and civil disobedience*. New York: Signet Classics.

Tierney, P.
1989 *The highest altar*. New York: Penguin.

Tildesley, M.L.
1933 The Albanians of the north and south: (1) introductory account of measurements and photographs taken in 1929. *Biometrika* 25(1/2):21–51.

Tilley, C.
1994 *A phenomenology of landscape: places, paths and monuments*. Oxford: Berg.
1996 The power of rocks: topography and monument construction on Bodmin Moor. *World Archaeology* 28:161–176.

Toffolo, M.B., A. Fantalkin, I.S. Lemos, R.C.S. Felsch, W.-D. Niemeier, G.D.R. Sanders, I. Finkelstein, and E. Boaretto
2013 Towards an absolute chronology of the Aegean Iron Age: new radiocarbon dates from Lefkandi,

Kalapodi and Corinth. *PLoS ONE* 8(12): e83117. doi:10.1371/journal.pone.0083117.

Touchais, G., and P. Lera
2007 L'Albanie méridionale et le monde égéen à l'Âge du Bronze: problemes chronologiques et rapports culturels. In *Between the Aegean and Baltic seas: prehistory across borders. Proceedings of the international conference Bronze and Early Iron Age interconnections and contemporary developments between the Aegean and the regions of the Balkan Peninsula, central and northern Europe, University of Zagreb, 11–14 April 2005* (Aegaeum 27), edited by I. Galanaki, H. Tomas, Y. Galanakis, and R. Laffineur, 141–148. Liège: Université de Liège.

Trani, S.
2007 *L'unione fra l'Albania e l'Italia. Censimento delle fonti (1939–1945) conservate negli archivi pubblici e privati di Roma.* Ministero per i Beni e le Attività Culturali. Direzione Generale per gli Archivi. Pubblicazioni degli Archivi di Stato Strumenti CLXXIII.

Trotter, M.H.
1970 Estimation of stature from intact long bones. In *Personal identification in mass disasters*, edited by T.D. Stewart, 71–83. Washington, D.C.: Smithsonian Institution Press.

Truhelka, Ć.
1904 Der vorgeschichtliche Pfahlbau im Savebette bei Donja Dolina (Bezirk Bosnisch-Gradiška): Bericht über die Ausgrabungen bis 1904. *Wissenschaftliche Mitteilungen aus Bosnien und der Herzegowina* 9:3–170.

Tsonos, A.
2009 *Σκάβοντας στην Αλβανία. Ιστορία και ιδεολογία των αρχαιολογικών ερευνών κατά το 19ο και 20ο αιώνα.* Ioannina: Ekdoseis Isnaphi.

Tsountas, Ch.D.
1888 Ανασκαφαί τάφων εν Μυκήναις. *Archaiologike Ephemeris* 1888:119–180.
1898 Κυκλαδικά. *Archaiologike Ephemeris* 1898:137–212.
1899 Κυκλαδικά, II. *Archaiologike Ephemeris* 1899: 73–134.
1908 *Αι προϊστορικαί ακρόπολεις Διμινίου και Σέσκλου.* Athens: Athens Archaeological Society.

Tuan, Y.-F.
1971 Geography, phenomenology and the study of human nature. *Canadian Geographer* 15:181–192.
1974 *Topophilia: a study of environmental perception, attitudes, and values.* Englewood Cliffs, New Jersey: Prentice-Hall.
1977 *Space and place: the perspective of experience.* Minneapolis: University of Minnesota Press.

Tufte, E.R.
1983 *The visual display of quantitative information.* Cheshire, Connecticut: Graphics Press.
2006 *The cognitive style of PowerPoint? Pitching out corrupts within.* Cheshire, Connecticut: Graphics Press.

Turner, J.
1978 The vegetation of Greece during prehistoric times: the palynological evidence. In *Thera and the Aegean world 1. Papers presented at the second international scientific congress, Santorini, Greece, August 1978*, edited by C. Doumas and H.C. Puchelt, 765–773. London: Thera and the Aegean World.

Turner, J., and J. Greig
1975 Some Holocene pollen diagrams from Greece. *Review of Palaeobotany and Palynology* 20:171–204.

Turrill, W.B.
1929 *The plant-life of the Balkan Peninsula: a phytogeographical study.* Oxford: Clarendon Press.

Tuten, B.S., and T.L. Billado (editors)
2010 *Feud, violence and practice: essays in medieval studies in honor of Stephen D. White.* Farnham (U.K.): Ashgate.

Tutin, T.G.
1968–1980 *Flora Europaea.* 5 vols. Cambridge: Cambridge University Press.

Tutin, T.G., N.A. Burges, A.O. Chater, J.R. Edmonson, V.H. Heywood, D.M. Moore, D.H. Valentine, S.M. Walters, and D.A. Webb (editors)
1993 *Flora Europaea.* 2nd ed., Vol. 1. Cambridge: Cambridge University Press.

Tylecote, R.F.
1976 *A history of metallurgy.* London: The Metals Society.

Tzortzatou, A., and L. Fatsiou
2009 New Early Iron Age and Archaic sites in Thesprotia. In *Thesprotia expedition I: towards a regional history* (Papers and Monographs of the Finnish Institute at Athens, Vol. 15), edited by B. Forsén, 39–53. Helsinki: Foundation of the Finnish Institute at Athens.

Ubelaker, D.H.
1989 *Human skeletal remains: excavation, analysis, interpretation*, Vol. 2 (Manuals on Archeology). Washington, D.C.: Smithsonian Institution.

Ubelaker, D.H., and I. Pap
1998 Skeletal evidence for health and disease in the Iron Age of northeastern Hungary. *International Journal of Osteoarchaeology* 8(4):231–251.

Ucko, P.
1969 Ethnography and archaeological interpretation of funerary remains. *World Archaeology* 1:262–280.

Ugolini, L.M.
1927 *Albania antica,* Vol. I: *recherche archeologiche.* Rome and Milan: Società Editrice d'Arte Illustrata.
1942 *Albania antica,* Vol. III: *l'acropoli di Butrinto.* Rome: Scalia Editore.

Undet, J.
1889 Zu den ältesten Fibeltypen. *Zeitschrift für Ethnologie* 21:205–234.

Unruh, J.
2001 A revised endpoint for ceramics desalination at the archaeological site of Gordion, Turkey. *Studies in Conservation* 46(2):81–92.

Valentini, G.
1945 La famiglia nel diritto tradizionale Albanese. *Annali Lateranensi* 9:9–212.
1956 *Il diritto della comunità nella traditione giuridica Albanese: generalità*. Firenze: Vallecchi.

Van Dyke, R.M.
2003a *Lived landscapes: memory, phenomenology, and Chacoan society*. Santa Fe, New Mexico: School of American Research Press.
2003b Memory and the construction of Chacoan society. In *Archaeologies of memory*, edited by R.M. Van Dyke and S.E. Alcock, 180–200. Oxford: Blackwell.
2008 *The Chaco experience: landscape and ideology at the center place*. Santa Fe, New Mexico: School of American Research Press.

van Gennep, A.
1908 [1960] *The rites of passage*. Translated by M.B. Vizedom and G.L. Caffee. Chicago: University of Chicago Press. (Originally published as *Rites du Passage*. Paris: A. and J. Picard, 1908.)

van Klinken, G.J.
1999 Bone collagen quality indicators for palaeodietary and radiocarbon measurements. *Journal of Archaeological Science* 26:687–695.

van Straten, F.T.
1988 The god's portion in Greek sacrificial representations: is the tail doing nicely? In *Early Greek cult practice. Proceedings of the fifth international symposium at the Swedish Institute at Athens, 26–29 June 1986*, edited by R. Hägg, N. Marinatos, and G.C. Nordquist, 51–68. Stockholm: Swedish Institute at Athens.
1995 *Hierà kalá: images of animal sacrifice in Archaic and Classical Greece*. Leiden: E.J. Brill.

Vardala-Theodorou, E.
1998 Study of recent and fossil (Quaternary) benthic mollusca of Vouliagmeni Lake at Perachora Loutraki, Corinthian Gulf. Ph.D. dissertation, University of Athens.

Vasić, R.
1999 *Die Fibeln im Zentralbalkan (Vojvodina, Serbien, Kosovo und Makedonien)* (*PBF* XIV:12). Stuttgart: Franz Steiner Verlag.

Velo, E., A. Bino, A. Shehi, and P. Vasili
2002 Albania. In *Epidemiological surveillance of malaria in countries of central and eastern Europe and newly selected independent states*, 5–6. Copenhagen: World Health Organization Regional Office for Europe.

Verano, J.W.
2008 Trophy head-taking and human sacrifice in Andean South America. In *The handbook of South American archaeology*, edited by H. Silverman and W.H. Isbell, 1047–1062. New York: Springer.

Verdelis, N.M.
1963 Neue geometrische Gräber in Tiryns. *Mitteilungen des Deutschen Archäologischen Instituts, Athenische Abteilung* 78:1–62.

Vérilhac, A.-M., and C. Vial
1998 *Le marriage grec du VIe siècle av. J.-C. à l'époque d'Auguste* (*Bulletin de Correspondance Hellénique*, Supplément 32). Athens and Paris: École Française d'Athènes.

Veseli, S.
2006 Archaeology, nationalism and the construction of national identity in Albania. In *New directions in Albanian archaeology: studies presented to Muzafer Korkuti*, edited by L. Bejko and R. Hodges, 323–330. Tirana: International Centre for Albanian Archaeology.

Videski, Z.
2005 The Bronze Age at Vardarski Rid. In *Vardarski Rid*, Vol.1, edited by D. Mitrevski, 91–112. Skopje: Foundation Varda.
2007 Mycenaean influences in the FYRO Macedonia identified in the Late Bronze Age cemeteries. In *Between the Aegean and Baltic seas: prehistory across borders. Proceedings of the international conference Bronze and Early Iron Age interconnections and contemporary developments between the Aegean and the regions of the Balkan Peninsula, central and northern Europe, University of Zagreb, 11–14 April 2005* (Aegaeum 27), edited by I. Galanaki, H. Tomas, Y. Galanakis, and R. Laffineur, 211–214. Liège: Université de Liège.

Vogeikoff-Brogan, N.
2007 *Triumph over time (1947): the American School in post-war Greece*. Princeton, New Jersey: American School of Classical Studies at Athens.

Vogel, J.S., J.R. Southon, D.E. Nelson, and T.A. Brown
1984 Performance of catalytically condensed carbon for use in accelerator mass spectrometry. *Nuclear Instruments and Methods in Physics Research* B5:289–293.

Vokotopoulou, I.
1969a Αρχαιότητες και μνημεία Ήπειρου. *Archaiologikon Deltion* 24, B':246–257.
1969b Νέοι κιβωτιόσχημοι τάφοι της ΥΕ ΙΙΙΒ-Γ περιόδου εξ Ηπείρου. *Archaiologike Ephemeris* 1969: 179–207.
1973 *Οδηγός Μουσείου Ιωαννίνων*. Athens: Tameio Archaiologikon Poron kai Apallotrioseon.
1985 Les armes en fer de Vitsa: rapports et divergences avec la region illyrienne. *Iliria* 15(2):195–204.

1986 *Βίτσα. Τα νεκροταφεία μίας μολοσσικής κώμης.* Athens: Tameio Archaiologikon Poron kai Apallotrioseon.

von den Driesch, A., and J. Boessneck
1990 Die Tierreste von der mykenischen Burg Tiryns bei Nauplion/Peloponnes. In *Tiryns: Forschungen und Berichte,* Bd. 11, 87–164. Mainz: Philipp von Zabern.

Von der Osten, H.H.
1937 *The Alishar Hüyük, seasons of 1930–32, Part I.* Chicago: University of Chicago Press.

von Eles Masi, P.
1986 *Le fibule dell'Italia settentrionale* (*PBF* XIV:5). Munich: Beck.

von Freytag gen. Löringhoff, B.
1974 Ein spätgeometrisches Frauengrab vom Kerameikos. *Mitteilungen des Deutschen Archäologischen Instituts, Athenische Abteilung* 89:1–25.

von Hassell, U.
1947 *Pyrrhus.* Munich: Münchner Verlag und graphische Kunstanstalten.

Von Sacken, F.
1868 *Das Grabfeld von Hallstatt in Oberösterreich und dessen Alterthümer.* Vienna: Wilhelm Braumüller.

Voutsaki, S.
1995 Social and political processes in the Mycenaean Argolid: the evidence from the mortuary practices. In *Politeia: society and state in the Aegean Bronze Age. Proceedings of the 5th international symposium at the University of Heidelberg, Archäologisches Institut, 10–13 April, 1994* (Aegaeum 12), edited by R. Laffineur and W.-D. Niemeier, 55–66. Liège: Université de Liège.

Vroom, J.
1998 Early modern archaeology in central Greece: the contrast of artefact-rich and sherdless sites. *Journal of Mediterranean Archaeology* 11:131–164.
2003 *After antiquity: ceramics and society in the Aegean from the 7th to 20th century A.C. A case study from Boeotia, Greece.* Leiden: Leiden University Press.
2005 *Byzantine to modern pottery in the Aegean: 7th to 20th century. An introduction and field guide.* Utrecht: Parnassus Press.

Vukmanović, M., and N. Radojčić
1995 *Catalogue of metalwork II (prehistory III).* Belgrade: National Museum.

Vulić, N.
1933 Neue Gräber bei Trebenischte. *Archäologischer Anzeiger* 48:459–482.

Wace, A.J.B.
1932 *Chamber tombs at Mycenae* (Archaeologia 82). Oxford: Society of Antiquaries.

Wace, A.J.B., and J.P. Droop
1906–1907 Excavations at Theotokou, Thessaly. *Annual of the British School at Athens* 13:309–327.

Wace, A.J.B., and M.S. Thompson
1911–1912 Excavations at Halos. *Annual of the British School at Athens* 18:1–29.
1912 *Prehistoric Thessaly: being some account of recent excavations and explorations in north-eastern Greece from Lake Kopais to the borders of Macedonia.* Cambridge: Cambridge University Press.

Waksman, S.A., and R.L. Starkey
1931 *The soil and the microbe: an introduction to the study of the microscopic population of the soil and its role in soil processes and plant growth.* New York: John Wiley and Sons

Waldron, T.
1994 *Counting the dead: the epidemiology of skeletal populations.* New York: John Wiley & Sons.

Waldstein, C.
1905 *The Argive Heraeum,* II: *terra-cotta figurines, terra-cotta relief, vase fragments, bronzes, engraved stones, gems, and ivories, coins, Egyptian, or Graeco-Egyptian, objects.* Boston and New York: Houghton, Mifflin and Co.

Walsh, V.A.
1980 A computer simulation of the house construction activity system at Nichoria in southwest Greece. Ph.D. dissertation, University of Minnesota.

Walter, H., and H.-J. Weisshaar
1993 Alt-Ägina: die prähistorische Innenstadt westlich des Apollontempels. *Archäologisches Anzeiger* 1993:293–297.

Walton, M.S., A.J. Shortland, S. Kirk, and P. Degryse
2009 Evidence for the trade of Mesopotamian and Egyptian glass to Mycenaean Greece. *Journal of Archaeological Science* 36(7):1496–1503.

Wapler, U., E. Crubézy, and M. Schultz
2004 Is cribra orbitalia synonymous with anemia? Analysis and interpretation of cranial pathology in Sudan. *American Journal of Physical Anthropology* 123(4):333–339.

Wardle, K.A.
1972 The Greek Bronze Age of the Pindus. Ph.D. dissertation, University of London.
1977 Cultural groups of the Late Bronze and Early Iron Age in north-west Greece. *Godišnjak* 13:153–199.
1980 Excavations at Assiros, 1975–9: a settlement site in central Macedonia and its significance for the prehistory of south-east Europe. *Annual of the British School at Athens* 75:229–267.

Wardle. K.A. (editor)
1996 *Nea Nikomedeia I: the excavation of an Early Neolithic village in northern Greece 1961–1964* (BSA Supplement 25). Oxford: British School at Athens.

Wardle, K.A., M. Newton, and P.I. Kuniholm
2007 Troy VIIb2 revisited: the date of the transition from Bronze Age to Iron Age in the northern Aegean. In *The Stryma/Strymon River valley in prehistory. Proceedings of the international sym-*

posium Strymon Praehistoricus, Kjustendil-Blagoevgrad (Bulgaria) and Serres-Amphipolis (Greece), 27.09–1.20.2004 (in the steps of James Harvey Gaul 2), edited by H. Todorova, M. Stefanovich, and G. Ivanov, 451–479. Sofia: Gerda Henkel Stiftung.

Wardle, K.A., and D. Wardle
2000 Assiros Toumba: remains of the later Iron Age. In *Μύρτος. Μνήμη Ιουλίας Βοκοτοπούλου*, 653–673. Thessalonike: Thanasis Altintzis.

Warneke, T.F.
1999 *Hallstatt- und frühlatènezeitlicher Anhängerschmuck. Studien zu Metallanhängern des 8.-5. Jahrhunderts v. Chr. zwischen Main und Po* (Internationale Archäologie, Bd. 50). Rahden: Verlag Marie Leidorf.

Warner, J.L.
1994 *Elmalı-Karataş II: the Early Bronze Age village of Karataş*. Bryn Mawr, Pennsylvania: Bryn Mawr College.

Wason, D.
1994 *The archaeology of rank*. Cambridge: Cambridge University Press.

Watkinson, D., and M.T. Lewis
2005 Desiccated storage of chloride-contaminated archaeological iron objects. *Studies in Conservation* 50(4):241–252.

Weber, M.
1894 Der Streit um den Charakter der altgermanischen Sozialverfassung in der deutschen Literatur des letzten Jahrzehnts. *Jahrbücher für Nationalökonomie und Statistik* 28:433–470.

Weglian, E.
2001 Grave goods do not a gender make: a case study from Singen am Hohentwiel, Germany. In *Gender and the archaeology of death*, edited by B. Arnold and N.L. Wicker, 137–155. Walnut Creek, California: AltaMira Press.

Weinberg, S.S.
1943 *Corinth* VII.1. *The Geometric and Orientalizing pottery*. Cambridge, Massachusetts: American School of Classical Studies.
1948 A cross-section of Corinthian antiquities (excavations of 1940). *Hesperia* 17:197–241.

Weiss, K.M.
1973 *Demographic models for anthropology* (Memoirs 27, Society for American Anthropology). Washington, D.C.: Society for American Anthropology.

Wells, B.
1983a *Asine II: Results of the excavations east of the acropolis, 1970–1974*. Fasc. 4: *the Protogeometric period*, Part 2: *an analysis of the settlement*. Stockholm: Svenska Institutet i Athen.
1983b *Asine II: Results of the excavations east of the acropolis, 1970–1974*. Fasc. 4: *the Protogeometric period*, Part 3: *catalogue of pottery and other artefacts*. Stockholm: Svenska Institutet i Athen.

Welter-Schultes, F.
2005 AnimalBase Project Group. Early zoological literature online. *http://www.animalbase.org*.

Wendt, C.J.
2009 The scale and structure of bitumen processing in Early Formative Olmec households. In *Housework: craft production and domestic economy in ancient Mesoamerica* (Archeological Papers of the American Anthropological Association No. 19), edited by K.G. Hirth, 33–44. Washington, D.C.: American Anthropological Association.

Weninger, B., and R. Jung
2009 Absolute chronology of the end of the Aegean Bronze Age. In *LH IIIC chronology and synchronisms* III. *LH IIIC Late and the tranisition to the Early Iron Age. Proceedings of the international workshop held at the Austrian Academy of Sciences at Vienna, February 23rd and 24th, 2007* (Veröffentlichungen der mykenischen Kommission 30), edited by S. Deger-Jalkotzy and A.E. Bächle, 373–416. Vienna: Verlag der Österreichischen Akademie der Wissenschaften.

West, C.F.
2009 Kodiak Island's prehistoric fisheries: human dietary response to climate change and resource availability. *Journal of Island and Coastal Archaeology* 4:223–239.

Westphal, T., W. Tegel. K.-U. Heussner, P. Lera, and K.F. Rittershofer
2010 Erste dendrochronologische Datierungen historischer Hölzer in Albanien. *Archäologischer Anzeiger* 2010(2):75–95.

Whitaker, I.
1968 Tribal structure and national politics in Albania, 1910–1950. In *History and social anthropology*, edited by I.M. Lewis, 253–293. London: Tavistock Publications.

White, T.D., and P.A. Folkens
2000 *Human osteology*. 2nd ed. San Diego: Elsevier.

White, T.E.
1952 Observations on the butchering techniques of some Aboriginal peoples: 1. *American Antiquity* 17(4):337–338.
1953 A method for calculating the dietary percentages of various food animals utilized by Aboriginal peoples. *American Antiquity* 18(4):396–398.

Whitehouse, R.
1973 The earliest towns in peninsular Italy. In *The exploration of culture change: models in prehistory*, edited by C. Renfrew, 617–624. London: Duckworth.

Whitley, J.
1991 *Style and society in Dark Age Greece*. Cambridge: Cambridge University Press.

Whittaker, C.R.
1988 Introduction. In *Pastoral economies in Classical antiquity* (CPS Supplement 14), edited by C.R. Whittaker, 1–5. Cambridge: Cambridge Philological Society.
Wide, S.
1910 Gräberfunde aus Salamis. *Mitteilungen des Deutschen Archäologischen Instituts, Athenische Abteilung* 35:17–36.
Wiencke, M.H.
2000 *Lerna IV: the architecture, stratification, and pottery of Lerna III.* Princeton, New Jersey: American School of Classical Studies at Athens.
Wild, J.P., and P. Walton Rogers
2003 Introduction. *The Cambridge history of western textiles,* Vol. I, edited by D. Jenkins, 9–29. Cambridge: Cambridge University Press.
Wilkes, J.J.
1992 *The Illyrians.* Oxford: Blackwell.
Wilkins, L.T.
1964 *Social deviance: social policy, action, and research.* London: Tavistock Publications.
Williams, J. (editor)
1997 *Money: a history.* London: British Museum.
Willis, K.J.
1989 Late Quaternary vegetation history of Epirus, northwest Greece. Ph.D. dissertation, University of Cambridge.
1992 The late Quaternary vegetational history of northwest Greece (Part I: Lake Grammousti, Part II: Rezina marsh, Part III: a comparative study of two contrasting sites). *New Phytologist* 121:101–155.
Willkomm, H.
1989 Radiokohlenstoffdatierungen des Siedlungshügel Kastanas. In B. Hänsel, *Kastanas. Ausgrabungen in einem Siedlungshügel der Bronze- und Eisenzeit Makedoniens, 1975–1979: Die Grabung und der Baubefund* (Prähistorische Archäologie in Südosteuropa, Bd. 7, 1), 395–411. Berlin: Volker Spiess.
Willvonseder, K.
1937 *Die mittlere Bronzezeit in Österreich.* Vienna and Leipzig: Anton Schroll and Heinrich Keller.
Wilson, A.S., T. Taylor, M.C. Ceruti, J.A. Chavez, J. Reinhard, V. Grimes, W. Meier-Augenstein, L. Cartmell, B. Stern, M.P. Richards, M. Worobey, I. Barnes, and M.T.P. Gilbert
2007 Stable isotope and DNA evidence for ritual sequences in Inca child sacrifice. *Proceedings of the National Academy of Sciences* 104(42):16456–16461.
Wilson, B.
1999 Displayed or concealed? Cross cultural evidence for symbolic and ritual activity depositing Iron Age animal bones. *Oxford Journal of Archaeology* 18(3):297–305.
Winnifrith, T.
2002 *Badlands, borderlands: a history of northern Epirus/southern Albania.* London: Duckworth.
Witthoft, J.
1967 Glazed polish on flint tools. *American Antiquity* 32(3):383–388.
Wobst, H.M.
1977 Stylistic behavior and information exchange. In *For the director: research essays in honor of James B. Griffin,* edited by C.E. Cleland, 317–344. Ann Arbor: Museum of Anthropology, University of Michigan.
Wood, J.W., D.J. Holman, K.A. O'Connor, and R.J. Ferrell
2002 Mortality models for paleodemography. In *Paleodemography: age distributions from skeletal samples,* edited by R.D. Hoppa, and J.W. Vaupal, 129–168. Cambridge: Cambridge University Press.
Wood, J.W., G.R. Milner, H.C.H. Weiss, and K.M. Weiss
1992 The osteological paradox: problems of inferring prehistoric health from skeletal samples. *Current Anthropology* 33(4):343–370.
Woolf, G.
1993 Rethinking the oppida. *Oxford Journal of Archaeology* 12(2):223–234.
Wrede, W.
1916 Kriegers Ausfahrt in der archaisch-griechischen Kunst. *Mitteilungen des Deutschen Archäologischen Instituts, Athenische Abteilung* 41:221–374.
Wylie, A.
1985 The reaction against analogy. In *Advances in archaeological method and theory,* edited by M.B. Schiffer, 63–107. New York: Academic Press.
1992 The interplay of evidential constraints and political interests: recent archaeological research on gender. *American Antiquity* 57(1):15–35.
Xenaki-Sakellariou, A.
1985 *Οι θαλαμωτί τάφοι των Μυκηνών. Ανασκαφής Χρ. Τσούντα (1887–1898).* Paris: de Boccard.
Xhuveli, L., and J. Schultze-Motel
1995 Neolithic cultivated plants from Albania. *Vegetation History and Archaeobotany* 4(4):245–248.
Yannouli, E.
1997 Dimitra, a Neolithic and Late Bronze Age village in northern Greece: the faunal remains. In *Neolithiki Makedonia,* edited by D. Grammenos. Athens: Greek Ministry of Culture.
Yildirim, B., and M.-H. Gates
2007 Archaeology in Turkey, 2004–2005. *American Journal of Archaeology* 111:275–356.
Ylli, L.
1988 Vendbanimi prehistorik i Liqethit. *Iliria* 18(2): 93–103.
Yntema, D.
1990 *The matt-painted pottery of southern Italy: a general survey of the matt-painted pottery styles of southern Italy during the Final Bronze Age and*

the Iron Age (Universita di Lecce Dipartimento di Scienze della Antichita Settore Storico-Archeologico. Collana del Dipartimento, 4). Lecce: Congedo Editore.

Young, R.S.

1939 *Late Geometric graves and a seventh century well in the Agora* (*Hesperia* Supplement 2). Athens and Vienna: American School of Classical Studies at Athens.

1942 Graves from the Phaleron cemetery. *American Journal of Archaeology* 46(1):23–57.

1949 An early Geometric grave near the Athenian Agora. *Hesperia* 18(4):275–297.

1951 Sepulturae intra urbem. *Hesperia* 20(2):67–134.

Yphantidis, Ph.

2006 Τα κοσμήματα του Νεολιθικού οικισμού Δισπηλιού Καστορίας· παραγωγή και χρήση μίας "αισθητική εργαλειοθήκη." Ph.D. dissertation, Aristotelian University, Thessalonike.

Zacharias, N., C. Michael, D. Dimotikali, E. Vardala-Theodorou, and Y. Bassiakos

1999 Application of TL and EPR advances in palaeoenvironmental studies. *"LED" September 1999, Rome*, 115 (abstract only). Rome: LED.

2002 A modified TL technique (foil technique) for dating calcite shells from marine Upper Pleistocene. In *Archaeometry 98: Proceedings of the 31st symposium, Budapest, April 26–May 3, 1998* (BAR International Series 1043), edited by E. Jerem, K.T. Biró, and E. Rudner, 151. Oxford: Archaeopress.

Zachos, K.L.

1997 Τοπογραφικά Ελοπίας. Το λεκανοπέδιο των Ιωαννίνων κατά την ύστερη Χαλκοκρατία και την Πρώιμη Εποχή του Σιδήρου. In *Αφιέρωμα στον N.G.L. Hammond* (Παράρτημα Μακεδονικών 7), 153–167. Thessalonike: Society for Macedonian Studies.

Zancani Montuoro, P.

1974–1976 Francavilla Marittima. Necropoli I. Tre notabili enotri dell'VIII sec. a.C. *Atti e Memorie della Società Magna Grecia* 15–17:9–82.

1977–1979 Francavilla Marittima. Necropoli di Macchiabate: Saggi e scoperte in zone varie. *Atti e Memorie della Società Magna Grecia* 18–20:7–91.

1980–1982 Francavilla Marittima. Necropoli e ceramico a Macchiabate zona T. (Temparella). *Atti e Memorie della Società Magna Grecia* 21–23:7–129.

Zarinebaf, F., J. Bennet, and J.L. Davis

2005 *A historical and economic geography of Ottoman Greece: the southwest Morea in the 18th century* (*Hesperia* Supplement 34). Princeton, New Jersey: American School of Classical Studies.

Zeder, M.A.

1991 *Feeding cities: specialized animal economy in the ancient Near East.* Washington, D.C.: Smithsonian Institution Press.

Zeder, M.A., E. Emshwiller, B.D. Smith, and D.G. Bradley

2006 Documenting domestication: the intersection of genetics and archaeology. *Trends in Genetics* 22 (3):139–55.

Zenetos, A., E. Vardala-Theodorou, and C. Alexandrakis

2005 Update of the marine bivalvia mollusca checklist in Greek waters. *Journal of the Marine Biological Association of the United Kingdom* 85:993–998.

Index

Note: Page numbers in ***bold italics*** indicate illustrations or tables.